Genome Stability

Translational Epigenetics Series

Trygve O. Tollefsbol, Series Editor

Transgenerational Epigenetics
Edited by Trygve O. Tollefsbol, 2014

Personalized Epigenetics
Edited by Trygve O. Tollefsbol, 2015

Epigenetic Technological Applications
Edited by Y. George Zheng, 2015

Epigenetic Cancer Therapy
Edited by Steven G. Gray, 2015

DNA Methylation and Complex Human Disease
By Michel Neidhart, 2015

Epigenomics in Health and Disease
Edited by Mario F. Fraga and Agustin F. Fernández, 2015

DNA Biomarkers and Diagnostics
Edited by José Luis García-Giménez, 2015

Drug Discovery in Cancer Epigenetics
Edited by Gerda Egger and Paola Arimondo, 2015

Medical Epigenetics
Edited by Trygve O. Tollefsbol, 2016

Chromatin Signaling
Edited by Olivier Binda and Martin Fernandez-Zapico, 2016

Genome Stability

From Virus to Human Application

Edited by

Igor Kovalchuk
Olga Kovalchuk
University of Lethbridge, Lethbridge, AB, Canada

ELSEVIER

AMSTERDAM • BOSTON • HEIDELBERG • LONDON • NEW YORK • OXFORD • PARIS
SAN DIEGO • SAN FRANCISCO • SINGAPORE • SYDNEY • TOKYO

Academic Press is an imprint of Elsevier

Academic Press is an imprint of Elsevier
125 London Wall, London EC2Y 5AS, United Kingdom
525 B Street, Suite 1800, San Diego, CA 92101-4495, United States
50 Hampshire Street, 5th Floor, Cambridge, MA 02139, United States
The Boulevard, Langford Lane, Kidlington, Oxford OX5 1GB, United Kingdom

Library of Congress Cataloging-in-Publication Data
A catalog record for this book is available from the Library of Congress

British Library Cataloguing-in-Publication Data
A catalogue record for this book is available from the British Library

ISBN: 978-0-12-803309-8

For information on all Academic Press publications
visit our website at https://www.elsevier.com/

Publisher: Mica Haley
Acquisition Editor: Peter Linsley
Editorial Project Manager: Lisa Eppich
Production Project Manager: Edward Taylor
Designer: Mark Rogers

Typeset by TNQ Books and Journals

Transferred to Digital Printing in 2017

Contents

Section II
Genome Instability in Bacteria and Archaea

Section III
Genome Stability of Unicellular Eukaryotes

32. Chemical Carcinogens and Their Effect on Genome and Epigenome Stability

O. Kovalchuk

33. Environmental Sources of Ionizing Radiation and Their Health Consequences

A.A. Goodarzi, A. Anikin and D.D. Pearson

Section VIII
Bystander and Transgenerational Effects – Epigenetic Perspective

34. Sins of Fathers Through a Scientific Lens: Transgenerational Effects

M. Merrifield and O. Kovalchuk

35. Genomic Instability and the Spectrum of Response to Low Radiation Doses

C. Mothersill and C. Seymour

36. Transgenerational Genome Instability in Plants

I. Kovalchuk

List of Contributors

A. Anikin University of Calgary, Calgary, AB, Canada

J.N. Barr University of Leeds, Leeds, United Kingdom

A. Bilichak Agriculture and Agri-Food Canada, Lethbridge, AB, Canada

D. Bonatto Federal University of Rio Grande do Sul, Porto Alegre, Brazil

L. Boteva The University of Edinburgh, Edinburgh, United Kingdom

Y. Cheng University of British Columbia, Vancouver, BC, Canada; Xiamen University, Xiamen, P.R. China

J. Cobb University of Calgary, Calgary, AB, Canada

J. de Faria Poloni Federal University of Rio Grande do Sul, Porto Alegre, Brazil

F. Eudes Agriculture and Agri-Food Canada, Lethbridge, AB, Canada

R. Fearns Boston University School of Medicine, Boston, MA, United States

B.C. Feltes Federal University of Rio Grande do Sul, Porto Alegre, Brazil

L.R. Ferguson University of Auckland, Auckland, New Zealand

D.V. Firsanov Institute of Cytology of the Russian Academy of Sciences, Saint-Petersburg, Russia; Nikiforov Russian Center of Emergency and Radiation Medicine of the EMERCOM of Russia, Saint-Petersburg, Russia

Y. Gao University of North Carolina at Chapel Hill, Chapel Hill, NC, United States

C. Gelot Université Paris-Saclay, Villejuif, France

N. Gilbert The University of Edinburgh, Edinburgh, United Kingdom

R.M. Golsteyn University of Lethbridge, Lethbridge, AB, Canada

A. Golubov University of Lethbridge, Lethbridge, AB, Canada

V. Gomez University College London, London, United Kingdom

A.A. Goodarzi University of Calgary, Calgary, AB, Canada

R. Gundogdu University College London, London, United Kingdom

T. Hatkevich University of North Carolina, Chapel Hill, NC, United States

A. Hergovich University College London, London, United Kingdom

W. Hernandez-Sanchez Case Western Reserve University, Cleveland, OH, United States

M. Hetman University of Louisville, Louisville, KY, United States

J.K. Holsclaw University of North Carolina, Chapel Hill, NC, United States

P. Hsieh National Institutes of Health, Bethesda, MD, United States

T.C. Humphrey University of Oxford, Oxford, United Kingdom

T. Izumi University of Kentucky, Lexington, KY, United States

R.E. Jones University of Oxford, Oxford, United Kingdom

F. Jönsson University of Witten/Herdecke, Witten, Germany

I. Kovalchuk University of Lethbridge, Lethbridge, AB, Canada

O. Kovalchuk University of Lethbridge, Lethbridge, AB, Canada

Y. Kulaberoglu University College London, London, United Kingdom

T. Le-Guen Université Paris-Saclay, Villejuif, France

A.F.C. Lopes University of Cologne, Cologne, Germany

B.S. Lopez Université Paris-Saclay, Villejuif, France

I. Mellon University of Kentucky, Lexington, KY, United States

M. Merrifield University of Lethbridge, Lethbridge, AB, Canada

J.-E. Messling University of Cologne, Cologne, Germany

V.M. Mikhailov Institute of Cytology of the Russian Academy of Sciences, Saint-Petersburg, Russia

K.N. Miyamoto Federal University of Rio Grande do Sul, Porto Alegre, Brazil

P. Moskwa Medical University of Greifswald, Greifswald, Mecklenburg-Vorpommern, Germany

C. Mothersill McMaster University, Hamilton, ON, Canada

E. Mutter-Rottmayer University of North Carolina at Chapel Hill, Chapel Hill, NC, United States

T. Negishi Okayama University, Okayama, Japan

D.D. Pearson University of Calgary, Calgary, AB, Canada

M. Pereira-Gómez Universitat de València, Paterna, Spain

S. Ragu Université Paris-Saclay, Villejuif, France

M. Renaud-Young University of Calgary, Calgary, AB, Canada

K. Riabowol University of Calgary, Calgary, AB, Canada

M. Rieckher University of Cologne, Cologne, Germany

J. Risso Universitat de València, Paterna, Spain

K.D. Robertson Mayo Clinic, Rochester, MN, United States

R. Sanjuán Universitat de València, Paterna, Spain

B. Schumacher University of Cologne, Cologne, Germany

J. Sekelsky University of North Carolina, Chapel Hill, NC, United States

C. Seymour McMaster University, Hamilton, ON, Canada

C. Sidler ETH Zürich, Zürich, Switzerland

L.V. Solovjeva Institute of Cytology of the Russian Academy of Sciences, Saint-Petersburg, Russia

M.P. Svetlova Institute of Cytology of the Russian Academy of Sciences, Saint-Petersburg, Russia

L.H. Swift University of Lethbridge, Lethbridge, AB, Canada

S. Tateishi Kumamoto University, Kumamoto, Japan

D.J. Taylor Case Western Reserve University, Cleveland, OH, United States

C. Vaziri University of North Carolina at Chapel Hill, Chapel Hill, NC, United States

B. Wang University of Lethbridge, Lethbridge, AB, Canada

W. Wei Massachusetts General Hospital Cancer Center, Boston, MA, United States

A.B. Williams University of Cologne, Cologne, Germany

M. Xu Case Western Reserve University, Cleveland, OH, United States

M. Yang National Institutes of Health, Bethesda, MD, United States

D. Zhou Mayo Clinic, Rochester, MN, United States

Introduction

We have made our best efforts to collect chapters covering research on DNA repair and genome stability/instability in various species belonging to all domains of life. Moreover, we thought it would be very beneficial to give an epigenetics perspective to understanding of the regulation of DNA repair and genome stability. Therefore, several chapters discuss the role of the environment in DNA repair and genome stability. We also attempted to summarize genome changes during the evolution of new species, with an emphasis on randomness and nonrandomness of changes in various species. Finally, we have attempted to compare DNA-repair pathways in different organisms and to summarize what can be potentially done in the future to know more about DNA repair and genome stability.

The opening chapter (Chapter 1) of this book is "Genome stability: an evolutionary perspective" by Igor Kovalchuk. In this chapter, he discusses various aspects of genome evolution, by presenting the case of both Lamarckian and Darwinian models of evolution, describing the role of the environment in evolution and in particular in genome evolution, and explaining how the genetic and epigenetic components of DNA repair and genome stability have evolved. The chapter also touches upon some interesting controversies over the randomness or nonrandomness of changes in the genome, as well as the rate of DNA repair in various genomic regions in the same or different species.

In the next several chapters, we discuss the aspects of DNA repair and genome stability across all domains of life. Specifically, the chapters by John Barr and Rachel Fearns, Rafael Sanjuán, Marianoel Pereira-Gómez, and Jennifer Risso describe DNA-repair mechanisms and the aspects of genome instability in RNA and DNA viruses, respectively. In the chapter covering genome stability in RNA viruses (Chapter 2), Barr and Fearns describe how RNA viruses are able to withstand challenge with antiviral drugs and cause epidemics in previously exposed human populations. They also describe an extraordinary degree of genetic heterogeneity in RNA viruses; RNA viruses exist not as single genotypes but as swarms of related variants, and this genomic diversity is an essential feature of their biology. It appears that RNA viruses have a variety of mechanisms that act in combination to determine their genetic heterogeneity, such as polymerase fidelity, error-mitigation mechanisms, genomic recombination, and different modes of genome replication. Finally, Barr and Fearns present the evidence that some RNA viruses operate close to a threshold where the polymerase error rate has evolved to maximize the availability of the possible sequence space, while avoiding the accumulation of lethal deleterious mutations.

Describing genome stability of DNA viruses (Chapter 3), Sanjuán and colleagues show that DNA viruses can exhibit rapid sequence changes that are often found in loci involved in dynamic host–virus interactions. In particular, they describe that DNA viruses are capable of promoting genomic instability of certain specific genes, thus boosting the diversity where it is needed. Among specific mechanisms maintaining the viral diversity are diversity-generating retro-elements, a recombination-driven gene amplification, and DNA editing by host-encoded deaminases.

The second section of the book describes DNA repair and genome instability in prokaryotes. This section includes chapters by Ashley Williams and colleagues in which they compare genome stability of bacteria and archaea (Chapter 4), and show the role of the environment in bacterial mutagenesis (Chapter 5). Jan-Erik Messling and Ashley B. Williams demonstrate that through a careful regulation, prokaryotic cells employ the complex and sophisticated pathways that maintain a balance between genome stability and instability to allow not only self-preservation in future generations but also a genetic innovation for continued adaptation and evolution. They discuss the basic mechanisms that bacteria and archaea use to limit genome instability and also compare different approaches used by these two domains of life to face assaults by both endogenous and exogenous forces. In Chapter 5, Ashley Williams discusses various sources of genome instability in bacteria, including stress responses, mutator genotypes, recombination, specialized genetic sequences, and mobile genetic elements. They also show the consequences of these different pathways, with a specific focus on those ones that impact pathogens and their interactions with their hosts. In Chapter 6, Adnrey Golubov describes an ancient bacterial immune system—CRISPR/Cas9. He discusses the history of research on CRISPR/Cas9, the genetic structure of this system and its molecular mechanisms of action and regulation. The discovery of CRISPRs and a path to its application in genetic engineering is an example of a rare breakthrough that revolutionizes science and technology. It is not surprising that in 2015, the CRISPRs technology has been named one of the breakthrough technologies of the year by *Science* magazine.

The third section of this book covers genome stability of unicellular eukaryotes by presenting two chapters which address genome stability in ciliates by Franziska Jönsson (Chapter 7) and genome stability in yeasts by Rebecca Jones and Timothy Humphrey (Chapter 8). Ciliates are unique organisms as far as genome stability is concerned because they involve massive developmentally regulated changes in the genome structure driven by noncoding RNAs (ncRNAs). Franziska Jönsson compares the macronuclear development of two distantly related ciliate classes to show how the nuclear organization reflects different types of adaptation and regulation mechanisms exhibiting the power of ncRNAs in genome evolution.

In Chapter 8, Rebecca Jones and Timothy Humphrey describe the details of double-strand break (DSB) repair pathways in yeast.

Section 4 of the book focuses on DNA repair and genome stability in multicellular eukaryotes, excluding mammals. In Chapter 9, Julie Korda Holsclaw, Talia Hatkevich, and Jeff Sekelsky describe the major mitotic DSB-repair pathways, homologous recombination (HR) repair, and end joining in *Drosophila*. Specifically, they discuss the key regulatory mechanisms of meiotic recombination that promote repair of meiotic DSBs in the form of crossovers and present a novel model recently introduced for meiotic recombination in *Drosophila*. Research in *Drosophila* has been complemented by Chapter 10 presented by Tomoe Negishi who describes the role of mismatch repair (MMR) activities in genome stability of *Drosophila*, reviewing the role of MMR in meiotic recombination, mitotic DNA replication, and the induction of mutations.

In Chapter 11, Matthias Rieckher, Amanda Franqueira Lopes, and Björn Schumacher provide an overview of DNA repair and DNA-damage response mechanisms in *Caenorhabditis elegans* and various methodologies that have been employed to gain an insight into metazoan genome stability.

Chapters 12 and 13 cover research in plants. In Chapter 12, Andriy Bilichak provides a review of the most important DNA-repair pathways in plants that work together during plant development for an efficient maintenance of genome stability and plant survival, whereas in Chapter 13, Andriy Bilichak and Francois Eudes describe the use of three modern genome-editing technologies, zinc-finger nucleases, transcription activator-like effector nucleases, and clustered regularly interspaced short palindromic repeats/Cas9, that are used for permanent modifications of plants. The authors present an overview of the discovery of each of the technologies, their benefits, limitations, and their potential effectiveness for the future development of genetic engineering in plants.

Section 5 of the book is the largest one covering details of various DNA-repair pathways utilized by mammals to repair DNA. The section begins with Chapter 14 by Valenti Gomez and Alexander Hergovich focusing on the regulation of the mammalian cell cycle in the context of DNA-damage signaling and DNA damage–repair mechanisms. The authors also discuss some molecular aspects in the context of diseases. In Chapter 15, Yavuz Kulaberoglu, Ramazan Gundogdu, and Alexander Hergovich provide an overview of an important role of p53, p21, and p16 in cell-intrinsic checkpoints as well as signaling and repair responses to prevent genome instability. They also discuss the significance of these tumor suppressors in the context of cancer and DNA-damaging therapies designed to specifically target tumor cells.

Further details of DNA repair and genome stability in mammals are presented by Cyrus Vaziri, Satoshi Tateishi, Elizabeth Mutter-Rottmayer, and Yanzhe Gao in Chapter 16 that describes the role of RAD18 in DNA replication and post-replication repair (PRR). The authors review the mechanisms of RAD18 activation in response to DNA damage, its participation in translesion DNA synthesis and template switching, and the basis for the integration of RAD18 activity with other elements of the DNA-damage response and the cell cycle. The authors also discuss the potential impact of RAD18 on genome stability and tumorigenesis.

In Chapter 17, Tadahide Izumi and Isabel Mellon introduce the history and concepts of DNA-excision repair, discuss types of DNA damage processed by base-excision repair (BER) and nucleotide-excision repair (NER), and describe the repair mechanisms. The authors present a recent evidence indicating that DNA-damage and excision-repair proteins are key signal transducers and play critical roles in DNA-damage responses and communications between nuclei and mitochondria.

In Chapter 18, Mingzhang Yang and Peggy Hsieh summarize current knowledge of the mechanism of MMR in mammalian cells, its role in DNA-damage signaling, and its contribution to the maintenance of genome stability and tumor suppression.

In Chapter 19, Patryk Moskwa introduces nonhomologous end joining (NHEJ) repair pathways of DSBs and their components in mammals. He covers the details of their function, structure, binding partners, and possible regulations.

In Chapter 20, Camille Gelot, Tangui Le-Guen, Sandrine Ragu, and S. Bernard summarize different molecular mechanisms of HR in mammals. They discuss the role of HR in DNA DSB repair, the reactivation of arrested replication forks, and their consequences (radiation sensitivity, meiosis, and the application in genome manipulation). The authors propose a model for the choice of the DSB-repair pathway that functions in two steps: ssDNA resection versus the canonical NHEJ pathway and alternative end-joining pathway versus HR on resected DNA. Finally, the authors present the role of HR in the molecular evolution of multigene families in the process of concerted evolution.

In Chapter 21, the role of telomere maintenance in genome stability is introduced by Wilnelly Hernandez-Sanchez, Mengyuan Xu, and Derek J. Taylor. The authors describe recent advances that suggest the involvement of dysfunctional telomeres or the improper regulation of telomerase activity in the development of a host of human diseases that include premature aging syndromes, bone marrow failures, and cancer.

In Chapter 22, Lucy H. Swift and Roy M. Golsteyn describe the relationship between checkpoint adaptation and mitotic catastrophe in cancer cells. They review factors that contribute to checkpoint adaptation, such as the cell cycle, checkpoints, mitotic catastrophe, and cell death, and describe their relationship to genome instability. The authors suggest that distinguishing checkpoint adaptation from apoptosis will aid in the study of how genome instability is generated and amplified in cancer cells.

The following three chapters of this section introduce the role of epigenetic components in the regulation of genome stability. Chapter 23 by Lora Boteva and Nick Gilbert describe an indispensable role of the chromatin structure and nuclear organization in DNA repair and the maintenance of genome stability. The authors present a modern view in which genome stability aims to integrate the influence of fundamental cellular processes such as transcription and replication with the chromatin context in order to get a better understanding of the processes that shaped the human genome. In Chapter 24, Dan Zhou and Keith Robertson describe the role of DNA methylation in the maintenance of genome stability. The authors describe how DNA methylation and DNA methyltransferases assist in recognizing mismatches, while also contributing to genome stability by regulating MMR gene transcription. The authors show how intensive DNA methylation at heterochromatin repeats stabilizes such domains from their translocation and undesired spreading, ensuring the appropriate functions of centromeres and telomeres as well as the genetic integrity. Finally, in Chapter 25, Igor Kovalchuk explains how ncRNAs play direct and indirect roles in the regulation of genome stability.

Section 6 covers human genetic diseases and conditions associated with genome instability. In Chapter 26, Bruno César Feltes, Joice de FariaPoloni, Kendi Nishino Miyamoto, and DiegoBonatto describe human diseases associated with genome instability. The authors review the repair machineries underlying rare genetic diseases and different types of cancer. Chapters 27 and 28 present genome instability in cancer. Specifically, in Chapter 27, Wei Wei, Yabin Cheng, and Bo Wang provide a comprehensive coverage of various DNA-repair pathways and their contribution to almost all types of human malignancies. In this chapter, they describe the current understanding of defects in DNA-repair pathways in the development of cancer as well as DNA-repair pathway targeting as a potential strategy for cancer therapy. In Chapter 28, Margaret Renaud-Young, Karl Riabowol, and Jennifer Cobb describe the epigenetic regulation of the cell cycle and DNA repair in cancer. Since the ability of proteins in different repair pathways to find and repair lesions is critical in the maintenance of genome stability and prevention of cancer, work in the chromatin field is particularly important to understand the chromatin dynamics during DNA-damage recognition and repair. The authors summarize the factors regulating chromatin changes that are critical in mediating the repair of DNA damage.

Chapter 29 by Corinne Sidler introduces us to the world of aging and the role that genome instability plays in this process. The author presents several lines of evidence showing that the accumulation of unrepaired DNA damage and the resulting genomic instability play a central role in promoting aging.

In Chapter 30, Michal Hetman describes the function of the nucleolus in the DNA-damage response and the genomic (in)stability in the nervous system. In this chapter, the mechanisms and potential consequences of nucleolar involvement in the genomic instability are discussed in detail.

Section 7 focuses on the effects of the environment on genome stability. In Chapter 31, Lynnette Ferguson describes the association between variations in the diet, changes in the level of free radicals, and DNA damage. The author discusses how diets enriched with various nutrients and phytochemicals may provide DNA protection. Furthermore, the author discusses how polymorphisms in genes for nutrient uptake, metabolism, and excretion can determine the optimal dietary intake for an individual. In Chapter 32, Olga Kovalchuk describes the role of chemicals and chemical mutagenesis in DNA repair and genome instability. In Chapter 33, Aaron Goodarzi, Alexander Anikin, and Dustin Pearson describe how the ionizing radiation (IR) from environmental sources affects human health. They review the primary sources of human IR exposure, including a nuclear attack, civilian nuclear disasters, aerospace travel, medical radiation (radiotherapy and computed tomography imaging), and radon gas inhalation, and their health consequences.

Section 8 covers epigenetic perspective of bystander and transgenerational effects. In Chapter 34, Matt Merrifield and Olga Kovalchuk introduce the concept of transgenerational response in animals and describe how this phenomenon is controlled by epigenetic components. They specifically focus on the effect of IR, describing direct and indirect effects and introducing the concept of bystander and nontargeted effects of IR. In Chapter 35, Carmel Mothersill and Colin Seymour discuss whether low doses of radiation are harmful or good. The authors present some of the reasons for the controversy and mistrust. They propose that we as humans need to accept uncertainty; they also suggest some approaches to developing protection frameworks which might be more fruitful.

In Chapter 36, Igor Kovalchuk introduces the concept of transgenerational response in plants. The author describes changes that occur in response to stress at the level of genome stability and epigenome regulation and discusses the role of transgenerational response to stress in plants for the regulation of genome stability.

Chapter 37 by Firsanov, Solovjeva, Mikhailov, and Svetlova presents various methods that can be used to analyze DNA damage and changes in genome stability. They pay a special attention to immunological methods of monitoring DNA damage and repair in single cells.

Finally, Chapter 38 by Igor Kovalchuk presents a logical summary of the book "Conserved and divergent features of DNA repair. Future perspectives in genome-instability research." In this chapter, the author describes differences in DNA repair among bacteria, archaea, and eukarya and suggests potential future directions in DNA-damage repair and genome-stability research.

Igor Kovalchuk
Olga Kovalchuk

Chapter 1

Genome Stability: An Evolutionary Perspective

I. Kovalchuk
University of Lethbridge, Lethbridge, AB, Canada

Chapter Outline

1. INTRODUCTION

Genome stability is a feature of every organism to preserve and faithfully transmit the genetic material from generation to generation or from one somatic cell to another. This includes an error-free replication of genetic material (DNA or RNA) and the repair of replication mistakes or damaged DNA/RNA (see the corresponding chapters in this book). In contrast, genome instability covers a broad range of topics referring to an increased rate of DNA damage and the associated mutations, the role of potential direct and indirect mutagens, the role of external and internal factors contributing or mitigating such instability, and so on.

In this chapter, we discuss genome stability/instability from the perspective of evolution, and specifically genome evolution. Genome evolution refers to changes in the genome structure or genome size over time. Usually, such changes are changes in the DNA (RNA) sequence that are passed on to the progeny and the accumulation/fixation of such genetic variants or their rejection in a population. Genome evolution is normally discussed in the context of evolution of new species, and thus it is closely associated with the appearance of new traits, new phenotypical and morphological features, and so on.

Considering that new traits and phenotypes may develop due to epigenetic changes, it is plausible to think that genome evolution in part also involves changes in genome structure due to epigenetic modifications. In contrast to genetic changes that include mutations in the DNA sequence, epigenetic changes involve heritable but potentially reversible changes in gene expression due to changes in DNA methylation and histone modifications, among others.

Hence, how does the genome actually evolve? What are driving forces in genome evolution? Does evolution have the spontaneous and random nature? Is it actually directed and driven by changes in the external and internal environment? In the following, we contrast Darwin's and Lamarck's views on evolution and discuss the mechanisms that impact rate of evolution and the crucial role of the environment, especially as far as evolution of genome is concerned.

Genome Stability. http://dx.doi.org/10.1016/B978-0-12-803309-8.00001-X

2. EVOLUTION THEORIES AND MY REFLECTION ON THEM

The first theory of evolution was proposed about 200 years ago by a prominent French biologist Jean-Baptiste Lamarck (1744–1829). Although he was the first who built a somewhat coherent theory of evolution, he initially referred to the process as "transformation," explaining that organisms transform as a result of "a new need that continues to make itself felt." Evolution, according to Lamarck, was also the process in which less complex species evolve into more complex ones. Lamarck's understanding of evolution was summarized by him in four major laws. He used these laws to explain the two forces that he believed comprise evolution; a force driving animals from simple to complex forms, and a force adapting animals to their local environments. He formulates:

> Law 1*: Life, by its own forces, continually tends to increase the volume of every body which possesses it and to enlarge the size of its parts up to a limit which it brings about itself.* Law 2*: The production of a new organ in an animal body results from the appearance of a new want or need, which continues to make itself felt, and from a new movement which this want gives birth to and maintains.* Law 3*: The development of the organs and their strength of action are constantly in proportion to the use of these organs.* Law 4*: All that has been acquired, impressed upon, or changed in the organization of individuals during the course of their life is preserved by generation and transmitted to the new individuals that come from those which have undergone those changes.*

Let's briefly discuss all four laws. *Law 1* suggests that in the process of evolution organisms become larger; by this Lamarck also meant—more complex. Although there is some evidence that in the process of evolution organisms increased in size, primarily on transition to multicellularity, there is no evidence that larger organisms are more complex in general. Also, there is no real correlation between the size and complexity of an organism; although we must admit that it is difficult to define how complex certain species are. However, there appears to be evidence that during the history of life, the complexity of organisms increased. Hence, the law may stand with limitations. *Law 2* postulates that an organism can change when there is a need or want. The need to change when the environment changes is understandable, but what Lamarck meant here was that organisms could drive their changes in the direction they want. This is definitely an overstatement, as there is no evidence that "desire" to change plays any role in evolution. Thus, it is difficult to defend this law, but we come back to it later on in the review. *Law 3* is a famous law of "use and disuse." There are a lot of interpretations of this law and a lot of debates around it. To me, it suggests that if a certain trait is constantly required (used) in a given environment, it becomes prominent and fixed in the progeny, whereas a trait that is not used (it may have been important in the previous environment), it slowly disappears upon transition from progeny to progeny. Although it seems that Lamarck links the law of "use and disuse" to the law of "want and need," the law in my opinion stands, as use or disuse of certain traits due to the presence or absence of environment that calls upon them may result in appearance or disappearance of this trait in evolution. *Law 4* is known as "inheritance of acquired characteristics." Lamarck suggested that every change acquired by an organism is passed on to the progeny. We have plenty of evidence that this is not true, but many articles, especially in the past 10–15 years, clearly demonstrate that the acquired characteristics can be inherited. We discuss this later on in the review. The law stands in part. Therefore, as it appears, no matter how simplistic and naive the postulates may have been, they have a lot of merit. Lamarck definitely deserves a credit for recognizing the environment as a shaping force of evolution.

What about Charles Darwin's theory of evolution? There is no doubt that Darwin's work was influenced by Lamarck's work. Darwin summarized his voyage on the Beagle into the book *"On the Origin of Species by Means of Natural Selection, or the Preservation of Favoured Races in the Struggle for Life"* [1]. Darwin's theory of evolution suggested that organisms in a given population evolve through a process of natural selection. His main idea was that among individuals within a population there are always those ones who are fitter than others, and fitter organisms have higher chances to reproduce and leave more progeny, passing their heritable traits on to the next generation. As the environment changes due to the migration of a population of a given species or due to some unpredictable changes in the environment such as large temperature shifts, changes in water availability, or the appearance of competitors, a certain part of population that is more fit for the new environment adapts passing its genes/alleles on to the progeny, and eventually even may become a separate species. Seeing undisputable role of environment in the process of evolution, Darwin accepted a version of the inheritance of acquired characteristics proposed earlier by Lamarck. This is perhaps due to the fact that genetics and genes per se were not known to Darwin, and he had a difficult time coming up with a detailed mechanism for passing on traits from one generation to the next.

Evolution by Natural Selection proposed by Darwin did not transform much from the time he published his theory, but it was completed by additional knowledge of genes and genetics, thus transforming into the currently accepted *modern evolutionary synthesis* (*MES*) (also referred to as modern synthesis). One of the descriptions of MES was done by Futuyma. He writes:

> *The major tenets of the evolutionary synthesis, then, were that populations contain genetic variation that arises by random (i.e., not adaptively directed) mutation and recombination; that populations evolve by changes in gene frequency brought about by random genetic drift, gene flow, and especially natural selection; that most adaptive genetic variants have individually slight phenotypic*

effects so that phenotypic changes are gradual (although some alleles with discrete effects may be advantageous, as in certain color polymorphisms); that diversification comes about by speciation, which normally entails the gradual evolution of reproductive isolation among populations; and that these processes, continued for sufficiently long, give rise to changes of such great magnitude as to warrant the designation of higher taxonomic levels (genera, families, and so forth) [2].

First, whereas Darwin's theory of evolution suggested natural selection as the main (the only?) driving force, MES proposed several other possibilities, including genetic drifts and gene flow, although still giving the main emphasis to natural selection. *Second*, MES recognizes that traits are passed on from one generation to another in a discrete form of inheritance—genes. It further postulates that potential variants among individuals within the same species are due to the presence of multiple alleles of the same gene. *Third*, MES postulates that evolution is a gradual process, where small changes accumulate at the gene level becoming big changes leading to a speciation event. Therefore, macroevolution is, strictly speaking, multiple events of microevolution.

Understanding the process of evolution and what it involves still triggers many debates and disagreements that arise primarily from different perspectives, depending on whether you are a geneticist, a naturalist, a population biologist, an epigeneticist, or a paleontologist. It is still not clear whether evolution has gradual nature—whether macroevolution is a result of multiple steps of microevolution. Most of the paleontological findings suggest that speciation events occur rapidly. In 1972, the theory of punctuated equilibrium was put forward by Gould and Niles Eldredge; they proposed that evolutionary changes occur in relatively rapid spurts coincident with an increase in speciation rates [3]. They argued that such disruptions of equilibrium occur when a selective pressure is increased and when organisms adapted to a particular environment are no longer able to cope with the changing environment.

The appearance of new species may require several steps of microevolution. Since this is a continuous process, we hardly can trace the beginning and the end of the evolution process between, say, species A and B. Heterogeneity of individuals in a large population of certain species is already an initial step of microevolution, as a large pool of variations allows the population to acquire certain beneficial mutations that is favored in a certain environment much faster. *Species* are defined in many different ways. One definition is that they are a large group of individuals that carry similar phenotypic characteristics capable of interbreeding and giving rise to the fertile progeny. The ability to interbreed is the most important part of this definition. Therefore, until different populations of the same species are unable to interbreed, they are the same species, no matter how large the differences in their genetic material are. Is this actually true?

The analysis of taxonomic trees and the actual ability of certain species to interbreed and give rise to a viable progeny gave surprising results. In animals, less than 40% of animal taxa and less than 70% plant taxa represent reproductively independent lineages [4]. These data suggest that the definition of species as those that are incapable of interbreeding is at least outdated.

3. THE ROLE OF SYMBIOSIS IN GENOME EVOLUTION

As species do not live in isolation and constantly interact with each other, it is not surprising that they have learned to coexist in *parasitic*, *commensalistic*, and *mutualistic* ways. Altogether, such coexistence is referred to as a *symbiotic interaction*, although some scientists still consider symbiosis as a mutualistic relationship. Symbiosis had a great influence on evolution, including genome evolution. The very appearance of eukaryotic organisms is believed to be the result of symbiotic relationships between aerobic and anaerobic bacteria. Chloroplasts and mitochondria as well as possibly peroxisomes, flagella, cilia, centrioles, and maybe even the nucleus itself may have been independent organisms at one point, coexisting inside of a larger bacterial ancestor. The endosymbiotic theory of the appearance of eukaryotic organisms is supported by several lines of evidence [5].

First of all, the mitochondrial and chloroplast genomes resemble the genomes of some of currently existing bacteria; sequences in mitochondria have a similarity to alphaproteobacteria, whereas sequences of chloroplasts resemble those of cyanobacteria [6]. *Second*, as expected, there are organisms alive today that are in symbiotic relationships and may be in transition to becoming a new form of multicellular organisms. Such organisms are called the living intermediates. For example, the giant multinucleated protist amoeba *Pelomyxa* uses aerobic bacteria instead of mitochondria for aerobic respiration [7]. Also, the organisms such as corals, clams, and some *Paramecium* species host algae cells. Each coral polyp has a zooxanthellae algae cell within itself that carries out photosynthesis. Some clam species have a special type of cells known as iridocytes; microscopic towers of algae cells grow under iridocytes, resembling the stack of grana in chloroplasts [8]. The role of iridocytes is to filter the wavelengths that the algae prefer. In both described cases, algae cells gain protection by sharing the product of photosynthesis with host cells.

3.1 Changes in the Structure of the Organellar Genome Over Time

The mitochondrial and chloroplast genomes have undergone substantial changes since the time they have been bacterial organisms, but they have retained a circular structure characteristic of prokaryotic organisms, and genes located in them in most cases lack introns, unlike the nuclear genomes. Most of the chloroplast genomes are about 150kb in size represented as a single circular molecule, although in some rare cases, for example, in dinophytic algae, the chloroplast genome is distributed over as many as 40 small plasmids, each carrying only several genes [9]. Through the process of evolution, the chloroplast genomes lost many genes either due to gene elimination or gene transfer to the nuclear genome, the process known as an endosymbiotic gene transfer. As a result, on average, chloroplasts contain fewer than 100 genes, much less than the original number of genes (likely between 1000 and 5000) in endosymbiotic bacteria such as cyanobacteria. Plant nuclei contain many genes originating from chloroplasts, perhaps as many as 10–15%, with *Arabidopsis* having as much as 18% of all protein-coding genes stemming from the chloroplast [10].

Was the reduction of gene number in chloroplasts a gradual process, or was it a result of several massive rapid changes? Although there is no clear answer, it is believed that most of the genes were transferred early on [11]. However, one of the reports suggests that at least some endosymbiotic gene transfers may have occurred in land plants fairly recently. For example, the gene rpl22 encoding the chloroplast ribosomal protein CL22 is present in the chloroplast genome of all plants examined except legumes, while a functional copy of rpl22 is located in the nucleus of the legume pea [12], suggesting that the transfer occurred after speciation event in legumes.

Although it is commonly accepted that chloroplasts lost genes to the nucleus, there is very little evidence that they gained genes from the nucleus or from the environment. One such example includes the horizontal gene transfer (horizontal transfer, HT) of four genes between *Bacteroides* species and minicircles of plastid genomes of dinoflagellate species *Ceratium horridum* and *Pyrocystis lunula* [13].

Mitochondrial genomes are much more complex and diverse as compared to plastid genomes. They can be either circular as in most multicellular animals, linear as in fungi, protozoa, and algae, or a combination of circular and linear chromosomes as in many plants. While animal mitochondrial genomes are fairly small, less than 20kb in size, plant genomes are much larger, between 200kb and 2000kb. In animals, most of the genes are intronless, with coding regions representing over 90%, whereas in plants, many genes contain introns and only 10% of DNA represent coding regions. Cucumber has one of the largest mitochondrial genomes (~1700kb), and unlike most plant mitochondrial genomes, it has three circular chromosomes instead of one [14]. The smaller size of mitochondrial genomes in animals compared to plants can be due to the higher rate of endosymbiotic gene transfer in animals versus plants [15].

One of the theories explaining the relative compactness of organelle genomes is an advantage in the replication process. Smaller genomes replicate faster [15], allowing the organelle to divide faster and therefore to be overrepresented in the cytoplasm. Thus, organelles with smaller genomes may have had an advantage in a natural selection process.

3.2 Mutation Rates in Organellar Genomes and Adaptive Evolution

Do the organellar genomes still evolve? Do they evolve at the same rate in different species? In part this can be answered by the analysis of mutation rate in different organellar genomes. Plastid-bearing eukaryotes typically have lower rates of silent substitutions in plastid genomes versus mitochondrial genomes [16]; however, it is generally accepted that for land plants, the plastid and nuclear genomes have up to a 10-fold greater mutation rate than the mitochondrial genome. At the same time, several plants have higher point mutation rates in mitochondria than in plastid genomes [17,18].

In animals, the mitochondrial genome mutation rate is much higher than that in the nuclear genome [19]. In addition, the mutation rate in mtDNA varies greatly among different animals. Two major theories of such variations are the generation time and metabolic rate hypotheses. The generation time hypothesis suggests that short-lived species have a higher number of replication errors per a certain time unit (eg, a year) due to a higher number of DNA replication rounds. Obviously, this model only explains the difference in replication-dependent mutations, assuming that the replication error is constant across species [20]. The metabolic rate hypothesis expresses that the mitochondrial mutation rate is a reflection of metabolic rates and the level of produced free radicals, which are different in animals of different body mass [21]. This model may be applicable to all type of mutations, regardless of the replication rate.

The detailed analysis of nucleotide substitution rates at the gene encoding cytochrome *b* across 1696 mammalian species revealed a two orders of magnitude variation between the tested lineages [20]. It was found that cytochrome *b* third codon positions are renewed every 1–2Myr in the fastest evolving mammals and over 100Myr in slow-evolving ones. The authors further demonstrated that the generation time and metabolic rate hypothesis could not fully explain the data. It was suggested that mitochondrial mutation rates decrease in long-lived species, which is in agreement with the mitochondrial theory of aging, which suggests that organisms age in part due to mutations accumulating in mitochondrial [20]. This finding also suggests that mitochondrial mutation rates may have the partial adaptive nature. A more substantial proof of the adaptive nature of mutations in mtDNA comes from the work of James et al. [22]. The authors used a pairwise comparison and computer algorithms for the mitochondrial genome sequence data from 500 animals and found the evidence that mitochondria generally experience a substantial level of adaptive evolution. In addition, they found some weak evidence that the level of adaptive evolution in mitochondria correlates with the *effective population size* (Ne).

3.3 Symbiotic Interactions Between Viruses, Prokaryotes, and Eukaryotes: The Role of Transposable Elements

Symbiotic interactions between viruses and prokaryotic or eukaryotic cells as well as between prokaryotes and eukaryotes continue to shape the genomes of modern species.

Horizontal transfer (HT)—the transfer of genetic material between asexually reproducing organisms—is one of the mechanisms shaping the genomes of various species. This is a common mechanism among prokaryotes that contains passive (transduction, transformation) and active (conjugation, transformation) ways of HT. In contrast, HT is relatively rare in eukaryotes, partially due to the presence of barriers such as the presence of a nuclear envelope and the separation of gametic and somatic cells in multicellular eukaryotes. Despite this fact, many HT events have been recorded, most of them giving a selective adaptive advantage [11]. HT events can include the transfer of nongenic regions, genes, and transposable elements (TEs). HT of TE has been documented to occur across multiple phylogenetic levels both within prokaryotes and eukaryotes (reviewed in Ref. [23]). In contrast to HT of genes or TEs within prokaryotic or eukaryotic species, HT between the domains of life is less common for both genes and TEs. Moreover, interdomain HT of genes is substantially more common than interdomain HT of TEs. Hundreds of cases of prokaryote-to-eukaryote gene HTs have been characterized (reviewed in Ref. [23]). Gilbert and Cordaux carried out a comprehensive search for various groups of prokaryotic insertion sequences in 430 eukaryotic genomes [23]. They have identified 80 sequences integrated in the genomes of 14 different eukaryotes, all belonging to four distinct phyla (Amoebozoa, Ascomycetes, Basidiomycetes, and Stramenopiles). Further analysis revealed that these insertions were relatively recent events.

TEs have been initially discovered in 1948 by Barbara McClintock, a Nobel Prize winner in 1983. From the time of their discovery in maize to current days, they continue to be a puzzle. It is still unclear whether they are parasitic elements or symbionts of prokaryotic and eukaryotic species (perhaps at least some of them). Regardless of a clear definition, TEs have a powerful effect on genomes and evolution. TEs are capable of shaping host genomes through insertions, deletions, and recombination.

They can be broadly divided into *transposons*, elements capable of excision and reinsertion (conservative transposition) or elements capable of making a DNA copy first (replicative transposition) and *retrotransposons*, elements that amplify in the genome through RNA intermediates and reverse transcription followed by reinsertion. TEs are abundant among many species in various domains of life. Eubacteria, for example, may contain up to 20% of transposons [24], although many species have a very small number of transposons, and most of them are believed to be recent insertions [25]. Archaea are also believed to be similar in this matter [26]. In eukaryotes, especially in multicellular eukaryotes, TEs occupy a much larger genomic area: 45% in humans [27] and up to 85% in maize [28].

If TEs behave as any other genetic element, they have to evolve as any other genomic region under a certain selection pressure. If they are deleterious to the genome (species), they should gradually disappear, whereas if they are beneficial, they should be fixed. The fact that TEs exist in abundance in nearly all lineages suggests that transposons are likely to be under a positive selection pressure. Lineages that have a large number of TEs activated in response to stress may have a great advantage over those that either do not have them or have mobile elements that are irresponsive to environmental pressures. The mobilization of TEs may result in beneficial mutations, no matter how rare they are. These rare advantageous mutations may balance the fitness cost associated with maintaining and propagating TEs. Despite certain differences in the TE structure and the mode of activation in prokaryotes and eukaryotes, TEs have the same properties, and nearly in all cases, they are regulated by environmental factors.

Does the rate of occurrence of TEs depend on the genome size or reproduction mode? It was shown that TEs accumulate much faster in species with a low population size such as multicellular eukaryotes as compared to prokaryotes that have a large population size [29,30]. A larger population size may allow a more efficient elimination of slightly deleterious insertions by natural selection (see later), whereas a smaller population size may allow to retain them longer. The work by Startek et al. suggests that TE accumulation is also more likely in asexual eukaryotic populations that are under the constant environmental pressure than in populations living in the normal stable environment [31]. Similarly, it was suggested that in bacteria, the genomic TE content might also be influenced by stress [24].

The jury is still out whether TEs are simple cellular parasites (like viruses) or they are symbionts. The fact that cells have developed various strategies to suppress the TE activity and that during the process of evolution the majority of transposons have lost the ability to transpose strongly suggests that TEs are DNA parasites. On the other hand, TEs may very likely be symbionts of cellular organisms because they give them the advantage of fitness in the time of adverse environmental pressures, while being propagated themselves.

4. FIXATION OF A MUTANT ALLELE IN A POPULATION

The importance of the maintenance of genome integrity is reflected by the fact that mutations (changes in DNA/RNA sequence) are very rare, and they are often in the range of 10^{-9}–10^{-10} per single nucleotide per cell division. Many occurring mutations are silent, that is they do not result in the change of the phenotype (trait) because they are either *synonymous* resulting in no changes in amino acid composition of the encoded protein (due to the degenerate nature of the genetic code) or *nonsynonymous* resulting in changes to amino acid with similar properties (charge, binding capacity, and so on). Considering a low mutation rate and the fact that only a fraction of mutations changes certain inherited characteristics, 1 in 10,000,000 individuals (10^{-7}–10^{-8}) in the population would have a phenotype change due to a mutation in a defined gene of an average size of 1000 coding nucleotides. Any such mutation has a chance to be lost or fixed in the population. According to population genetics, the fixation of a mutation (or an allele) results from an increase of the frequency of such mutation from being present only in a few individuals to occurring in all individuals in the population carrying such mutation (an occurrence of 100% or near 100% in the population). Mutations in the population that spread through random genetic drift and without a positive selection pressure (the neutral theory of evolution) may have a very low chance of fixation in the population.

Let us assume that we have a population of 1000 individuals. Then, a mutation occurring in one of two alleles of these 1000 individuals will arise with a frequency of ½×1000 (1/2000). In a genetic drift, the probability that this allele will be fixed by chance is equal to its frequency in the population, ½×1000 or 5×10^{-4}. If the population is larger, let's say 1,000,000 individuals, the probability of fixing a single nucleotide mutation is ½×1,000,000 or 5×10^{-7}. With the neutral selection pressure, the rate of mutation fixation equals the rate of the introduction of such mutations. Since the fixation depends on the size of the population, the fixation of alleles in the population can only have a realistic chance either in a very small population or at a high initial frequency of occurrence of such alleles in the population. Genome position may have a substantial influence on the frequency of fixation; when mutation is positioned in the region with the high initial frequency of crossing over, the chance of fixation increases due to the higher frequency of homozygotization of de novo mutations [32]. According to Monte Carlo simulation performed by Kimura, a rare mutant gene in the effective population Ne can be fixed in 4Ne generations while being under the neutral selection pressure [33,34]. Since the population size also has a tendency to change, the decreasing population would lead to a decrease in the time required for the fixation, and the increasing population would increase the fixation time.

Selection is also an extremely important factor contributing to the fixation time of a mutation. In a deterministic model, an initially rare beneficial mutation occurring in a relatively small population will increase in frequency in each generation reaching fixation [35]. The more beneficial is the mutation, the higher selection coefficient for fixation it will have. Natural fluctuations in the frequency of the occurrence of this allele, known as *genetic drift*, especially in a larger population, may however lead to the extinction of such allele, regardless of how beneficial it is. An increase of the frequency of a beneficial allele to a certain meaningful number (10–20%) typically allows it to get fixed, and the time to fixation can be estimated using a deterministic model. In the nature, populations are in a constant flux, and the growth or decline in the population size may have a dramatic effect on beneficial mutations. Selection coefficients may be more effective for the growing population and less effective for the declining one. In the case of deleterious mutations, the effect may be opposite, selection coefficients may be more effective for the declining population and less effective for the growing population. This means that when populations grow, beneficial mutations have more chances to get fixed, while deleterious mutations will be lost. When populations decline in size, beneficial mutations are lost more frequently, while deleterious mutations are fixed more frequently.

A complete fixation of neutral or deleterious alleles is difficult to demonstrate in a research experiment due to the length of time required to achieve it when a de novo mutation with a low frequency is considered. Most of the examples of fixation of mutations come from studies in bacterial populations propagated under the selective pressure; however, advances in sequencing technology during 2010s allowed performing the so-called "evolve-and-resequence" experiments where the fixation of certain mutations is demonstrated through sequencing and mainly for point mutations. Several experimental evolution studies were performed in *Drosophila*, demonstrating the near fixation of mutations under the selective pressure [36–38].

Experiments since 2000, using partial outcrossing in worms for 50 generations also demonstrated the near fixation (near 90%) of several de novo mutations developed in a population with known fixed deleterious mutations [39]. The deleterious mutations were introduced in genes rendering sex determination temperature sensitive. Ten founder populations were evolved in intermediate temperatures for 50 generations. In these conditions, the temperature-sensitive mutants had up to a 75% reduction in fecundity and a 50% reduction in fertility. After 50 generations of experimental evolution, fertility rates have recovered in some populations almost to the wild-type levels, likely due to the selection of alleles that allowed the suppression of deleterious mutant phenotypes. Simulations performed by the author demonstrated that the fixation of de novo mutations in several populations could be due to low rates of outcrossing and a high selection pressure existing in laboratory conditions. It is interesting that a deleterious mutation can survive the selection pressure and get fixed through the co-selection with compensatory alleles (which are also favored and will likely get fixed in such population). This process resembles genetic hitchhiking, a mechanism of selection and fixation of alleles that are in a close proximity from other alleles that are under a high selection pressure.

In contrast, another experiment performed by Chelo and Teotonio using 100 generations of *Caenorhabditis elegans* failed to demonstrate the fixation of mutations, although it showed the substantially increased allelic variations [40]. The differences between this study and the study of Chandler [39] could be explained by the fact that Chelo and Teotonio [40] used a substantially larger number of genetic isolates as the initial founding population and at a much higher frequency of genetic outcrossings (over 20% versus under 5%).

In another example of fixation, Schwartz has performed a simple experiment allowing him to select wild-type maize plants over mutant plants impaired in the alcohol dehydrogenase activity encoded by the Adh1 gene [41]. Since the germination of maize seeds in flooding conditions resulted in the inability of the *adh1* mutant to germinate, a repetitive exposure of the progeny to flooding resulted in the elimination of a mutant allele from the experimental population. Similar experiments can be easily performed with plant mutants that are impaired in the germination under normal or induced conditions, also with mutants that are impaired in self-fertilization or outcrossing, and with partially or completely sterile mutants, and so on. The propagation of such plants for multiple generations may quickly eliminate homozygous mutants and severely reduce the presence of heterozygous ones, likely leading to a complete fixation.

All these examples support the notion that whereas selection has a great influence on fixation of traits, the effective population size (Ne) affects the rate of fixation to a large degree.

5. EVOLUTION OF MUTATION RATES

As stated earlier, the rate of evolution and fixation of traits largely depends on the rate of the introduction of new mutations. Mutations are an absolute prerequisite for the evolution of species and speciation (with a rare exception of epimutations that also can result in speciation). Were the mutation rate and the evolution rate constant across 3.5 billion years of life on our planet? Two major theories were proposed to describe the rate of evolution. *First*, *uniformitarianism* coined by William Whewell and originally applied to geology suggests that all laws of the universe operate at a steady constant rate. This theory contrasted *catastrophism*, a theory that suggests that the geological features of our planet have been influenced by rapid violent catastrophic events. Catastrophism as a theory can also be applied to the understanding of evolution rates.

It is still not very clear whether the rate of evolution remained constant. If it did, then the rate of mutations and the appearance of new species would have a constant rate throughout the history of life. However, several examples of rapid speciation events are known, one of them representing the Cambrian period. This period is characterized by the explosion of life diversity on our planet—the appearance of most modern animal phyla. Lee et al. employed Bayesian and maximum likelihood phylogenetic clock methods on an extensive anatomical and genomic data set for arthropods and found that the phenotypic evolution was about 4 times faster and the molecular evolution about 5.5 times faster during the Cambrian period compared to all subsequent periods [42]. The authors state that although the evolution of arthropods was much faster than the normal one, it is still within the range that can be considered acceptable and consistent with the evolution by natural selection.

The question remains whether mutation rates were constant throughout all periods and in all domains of life. An extensive research in the area of the patterning of mutation rates across various organisms was done by John Drake. In his work in 1991, John Drake analyzed the mutation rate per nucleotide per generation (u) in seven species, four DNA bacteriophage species, a bacterium, an yeast, and a filamentous fungus [43]. He found that while the average mutation rates per base pair varied by about 16,000-fold, mutation rates per genome (G) varied only by about 2.5-fold. Drake concluded that the average mutation rates per nucleotide inversely correlate with the genome size, and that the mutation rate per genome is nearly constant across all microbes [43].

Further analysis performed by Lynch shows a strong support for the initial analysis provided by John Drake [44]. A more detailed analysis that included RNA viruses, DNA viruses, archaea, and eubacteria showed a negative correlation between u and G (Fig. 1.1A). The same analysis performed only for species that have a defined cell structure, including bacteria, archaea, and various eukaryotes, showed a completely opposite picture—an increase in the genome size resulted in an increase in mutations per nucleotide (nt) per genome per generation; a positive correlation between u and G was found (Fig. 1.1B). It is interesting to note that the mutation rate of dsDNA viruses per replication was comparable to the average mutation rate per generation in mammals.

How would one explain higher mutation rates in high eukaryotes compared to single cell eukaryotes or prokaryotes? One of the theories proposed by Drake suggests that organisms under strong selection for rapid replication cannot maximize the fidelity of DNA replacement without limiting the rate of DNA synthesis necessary for daughter cell production [43]. If this cost-of-repair hypothesis is correct, then the replication in multicellular species occurs at higher mutation costs as compared to prokaryotes [44]. This is likely true, however, if one considers that mutations in somatic cells of multicellular organisms do not have the same impact as mutations in unicellular organisms, there should be an alternative explanation for higher mutation rates in multicellular eukaryotes.

An intriguing hypothesis was proposed by Lynch [44]. He suggested that there is a certain lower limit to the mutation rate per generation in any given species, and this rate is defined by molecular or biochemical properties of a cell or by the physiology of organisms. Lynch rather suggests that the lower bound on mutation rates is set by the intrinsic inability of natural selection to push the rates any lower [44]. In neutral evolution, a spread of mutations in the population occurs through genetic drift or a random chance (see earlier). Genetic drift restricts the influence of natural selection on any mutation, and when the mutation rate is reduced to the level when any further incremental improvement conveys a fitness advantage that is smaller than the power of drift, selection will not reduce the mutation rate any lower.

Lynch further used the data on the equilibrium level of heterozygosity at silent nucleotide substitutions from major phylogenetic groupings and the data on the average mutation rates (Fig. 1.1) to calculate the average effective population size (Ne) [44]. Plotting the average base substitution rate per generation (u) versus Ne revealed a significant negative correlation between u and Ne (Fig. 1.2A). A similar negative correlation was found by Lynch [44] for mammalian mitochondrial genomes using the data from Piganeau and Eyre-Walker [45] (Fig. 1.2B).

A comparison of the nuclear and mitochondrial genome mutation rates per generation shows a much larger u in mitochondrial genomes. A higher mutation rate in mitochondria is typically explained by a high metabolic rate and the presence of free radicals as well as a lower efficiency of DNA repair. In addition, mitochondria have a very low level of homologous

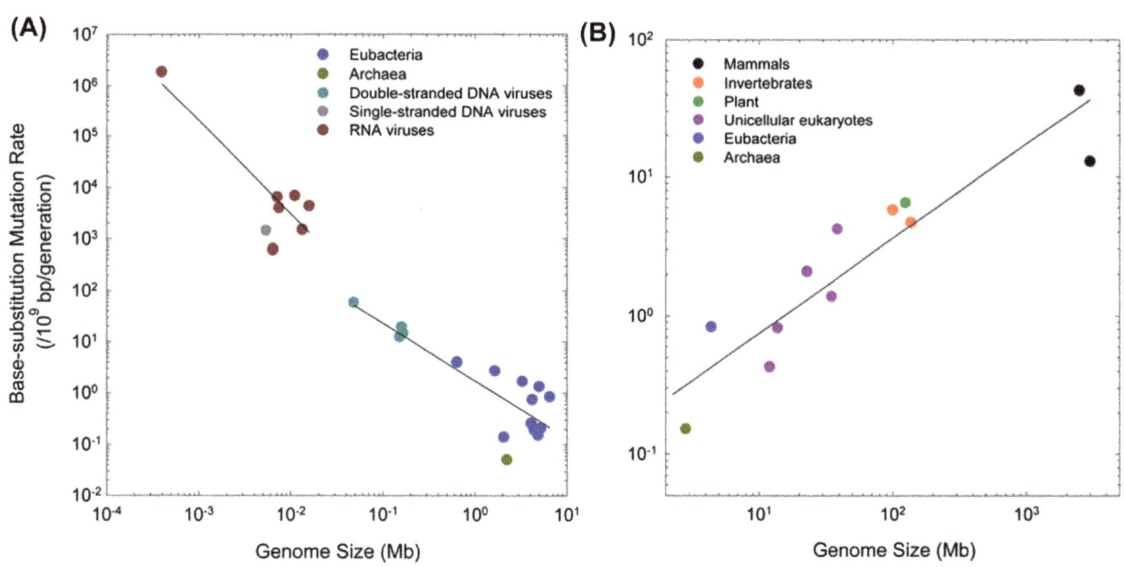

FIGURE 1.1 **The scaling of base substitution rate/nucleotide site/generation with genome size.** Each data point represents the average estimate for a separate taxon, although the results for most microbes are based on just one or a few reporter constructs (and hence, have a high sampling error), whereas those for most multicellular taxa are based on very large data sets (in several cases, whole genome sequences). (A) For noneukaryotes, two separate regressions are provided, one for RNA viruses alone, and the other for the pooled data from double-stranded DNA viruses, eubacteria, and archaea. The respective regressions of the $\log_{10}$ plotted mutation rates are $-0.17-1.83\log_{10}(G)$ and $0.24-1.12\log_{10}(G)$, with G denoting the genome size in megabases, and $r^2=0.78$ and 0.72, respectively. (B) The regression for cellular organisms is $-0.81+0.68\log_{10}(G)$, with $r^2=0.80$. Here, the results for various eubacteria (excluding *Buchnera* which has an unusually small genome) are averaged into a single point. The pattern is quite similar if prokaryotes are excluded (the slope $=0.59$ and $r^2=0.83$). *Reproduced with permission from Lynch M. The rate, molecular spectrum, and consequences of human mutation. Proc Natl Acad Sci USA 2010;107:961–68.*

recombination and therefore do not have an efficient mechanism of elimination of deleterious mutations. Mitochondria are uniparentally inherited, and mtDNA is distributed to several primary oocytes where mtDNA is replicated during maturation of oocytes. This segregation of mtDNA variants represents a genetic bottleneck that allows a rapid propagation of at least some mutations [46]. At the same time, the mutation rate in plant mitochondria is significantly smaller than in animal mitochondria, likely due to the fact that plant mitochondria are larger in size than animal mitochondria. However, the mutation rate in plant mitochondria is much lower than in the nucleus, despite the fact that the nuclear genome is much larger in size. Hence, it is still unclear whether a low mutation rate in plant mitochondria is either due to a more efficient repair rate, or a higher level of recombination, or perhaps due to a different mechanism of inheritance of plant mitochondria by the progeny.

5.1 Evolution of Somatic Mutation Rates

Somatic mutations in multicellular organisms may not alter the germline mutation rate; however, the accumulation of somatic mutations may contribute to a reproductive success of an organism due to the development of cancer or other severe diseases. Since plants do not have a predetermined germline, but rather develop it from somatic cells, somatic mutations occurring early in plant development may actually be inherited. For other higher eukaryotes, it is not clear whether somatic mutations can substantially affect germline mutations or their heritability. It is possible that the somatic mutation rate influences the evolution of the germline mutation rate in animals. Also, it is possible that the germline mutation rate actually influences the somatic mutation rate [44].

In order to compare the somatic mutation rate with the germline mutation rate, Lynch has prorated the germline mutation rates to a single cell division [44]. This comparison was very revealing: in humans, the rate of nucleotide substitution per cell division in the germline is very low, although not as low as in microbes (Table 1.1). The somatic mutation rates in metazoans are much higher than the germline mutation rates. For example, in humans, the average somatic mutation rate calculated from four tissue types was 17-fold higher than in the germline, 1.02×10^{-9}/nt/cell division versus 0.6×10^{-10}. This somatic mutation rate is also higher (3.5-fold) than the average one in yeast and *Escherichia coli* [44]. The analysis of mutation rates in mice and rats using LacZ or LacI transgenes also showed that somatic cells have up to a sixfold higher mutation rate compared to the mutation rate in testes (Table 1.1). Moreover, it was clearly demonstrated that somatic mutations accumulated with age, whereas germline mutations remain relatively constant (Fig. 1.3).

Considering that mammalian genomes are large (often several Gbp) and adult organisms consist of billion cells, the genetic load of new mutations in somatic cells of adult organisms is enormous. With the diploid genome size of a human being about 6.5×10^{9} nt and the average cell number being about 10^{13}, an adult human might carry as many as 10^{16} point

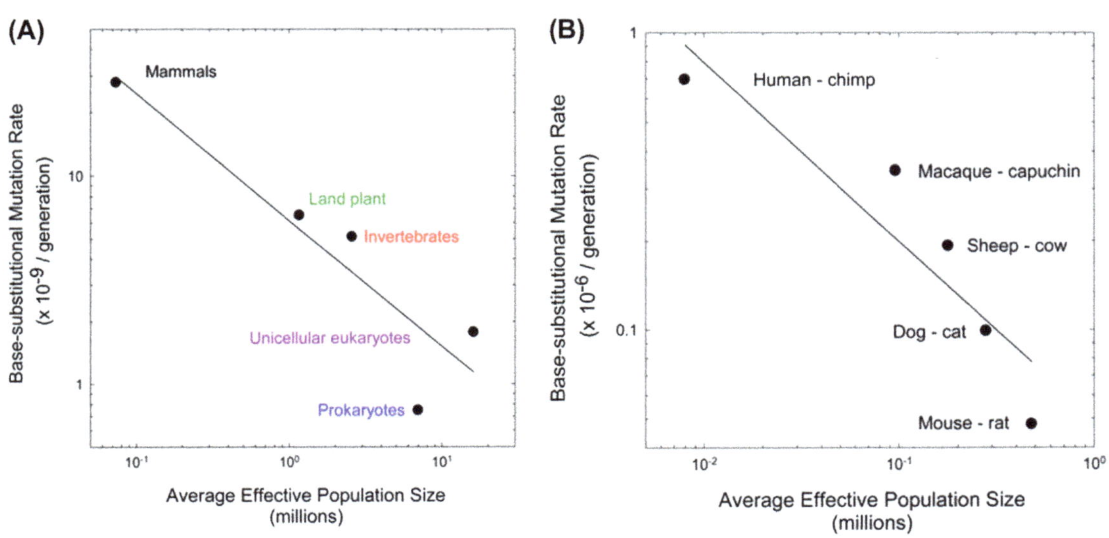

FIGURE 1.2 **The scaling of the base substitution mutation rate per generation (*u*) and the effective number of genes per locus (2Ne for diploids, and Ne for haploids).** (A) The slope of the log–log regression for the nuclear genome of major phylogenetic groupings is –0.60 (0.16), where the number in parentheses denotes the standard error, with $r^2=0.84$, although if the estimated Ne for prokaryotes is assumed to be 10 times too low [29], the slope is modified to –0.52 (0.02) with $r^2=0.99$. (B) The slope of the log–log regression for the mitochondrial genome of mammalian lineages is –0.60 (0.15), with $r^2=0.84$. The data are the average estimates from analyses assuming fixed and variable substitution rates in Piganeau and Eyre-Walker [45]. *Reproduced with permission from Lynch M. The rate, molecular spectrum, and consequences of human mutation. Proc Natl Acad Sci USA 2010;107:961–68.*

TABLE 1.1 Mutation Rates per Nucleotide Site (×10^{-9}) in a Variety of Tissues

Species	Tissue	Cell Divisions per Generation[a]	Mutation Rates[b]	
			Per Generation	Per Cell Division
Homo sapiens	Germline	216	12.85	0.06
	Retina	55	54.45	0.99
	Intestinal epithelium	600	162	0.27
	Fibroblast (culture)			1.34
	Lymphocytes (culture)			1.47
Mus musculus	Male germline	39	38	0.97
	Brain		76.94	
	Colon		83.35	
	Epidermis		90.38	
	Intestine		117.69	
	Liver		237.88	
	Lung		166.83	
	Spleen		130	
Rattus norvegicus	Colon		178.38	
	Kidney		167.45	
	Liver		179.92	
	Lung		223.22	
	Mammary gland		57.7	
	Prostate		448.9	
	Spleen		101.62	
Drosophila melanogaster	Germline	36	4.65	0.13
	Whole body		380.92	
Caenorhabditis elegans	Germline	9	5.6	0.62
Arabidopsis thaliana	Germline	40	6.5	0.16
Saccharomyces cerevisiae		1	0.33	0.33
Escherichia coli		1	0.26	0.26

[a]*References to data on the number of germline cell divisions: human [81]; D. melanogaster and mouse [82]; C. elegans [83]; and A. thaliana [84]. The number of cell divisions is unknown for the mouse and rat rates.*
[b]*Mammalian tissue-specific rates are given only for tissues in which at least two independent estimates have been acquired. All data on human mutation rates are taken from Lynch (2010) [85]. Data for somatic mutation rates in mice and rats are derived from references contained within the Supplementary Material, see Lynch (2010) [85]. References to data on germline mutation rates: D. melanogaster [86]; C. elegans [87]; A. thaliana [59]; S. cerevisiae [53]; and E. coli [19].*
Reproduced with permission from Lynch M. The rate, molecular spectrum, and consequences of human mutation. Proc Natl Acad Sci USA 2010;107:961–68.

mutations [44]. Although only a fraction of these mutations occur in coding regions, and even smaller fractions may be deleterious, adult organisms easily accumulate over 10^{10} mutations.

This analysis also showed that tissues accumulate more mutations with age. Similar data are reported for medaka fish [47] and *Drosophila melanogaster* [48]. Lynch concluded that if not for the separation of generations via the germline, the heritable per generation mutation rates for animals would be much higher than reported now [44]. This

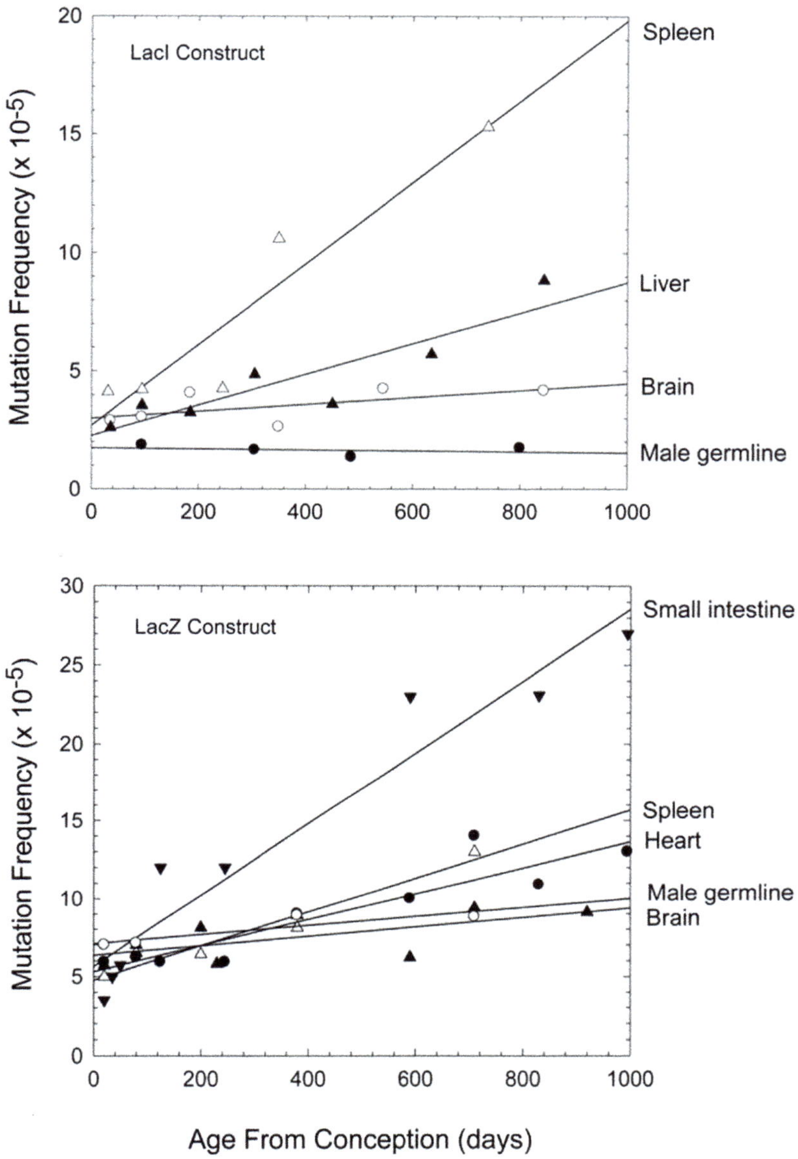

FIGURE 1.3 **Tissue-specific frequencies of mutations as a function of age in mouse lines carrying Lac reporter constructs.** Results are averaged over multiple studies. *Reproduced with permission from Lynch M. The rate, molecular spectrum, and consequences of human mutation. Proc Natl Acad Sci USA 2010;107:961–68.*

likely would have dramatic and likely negative effect on animals, due to substantial increase in heterogeneity and deleterious mutations in relatively stable environment. It would be interesting to run simulations to confirm or disprove this hypothesis.

6. GENOME INSTABILITY: IS IT RANDOM?

As we discussed earlier, mutations (whether they are genetic or epigenetic in nature) are an absolute prerequisite of evolution. Assuming that all mutations have a random nature, it is only the selection process that decides which mutation is beneficial or deleterious, and thus it is ultimately up to the environment to fix the mutation or make it disappear. If the environment has an impact on fixation of certain mutations, does this mean that evolution as a process is directed toward the fixation of beneficial traits that give advantage in a specific environment? Does it mean that the mutation process might be random, whereas the evolution process is directed—directed to the survival of the fittest? The answers to these questions are not necessarily obvious.

But are mutations actually truly random? If this was the case, then a mutation of a nucleotide in any genomic position would always have the same frequency. Nowadays, we know that this is not the case as certain genomic regions evolve faster than others. However, this is the result of mutations and forces of evolution acting upon them (genetic drift or/and natural selection). This is where the problem lies; it is difficult to separate the two events—mutations and an evolutionary force. For example, it is well known that cancers demonstrate higher mutation rates in genes and genomic regions, clustering at recurrent mutation hot spots that actually aid cancer cells in their immortality [49,50]. It is also possible, however, that nonrandomness of appearance of mutations in different genomic regions is unrelated to randomness of accumulation of mutations in the population under selective pressure.

6.1 A Bias in Mutations in Different Genomic Regions

Mutations may occur in certain regions more frequently due to the intrinsic properties of endogenous and exogenous DNA-damaging agents and chemical modifications of nucleotides (oxidation, spontaneous deamination, and so on), and the associated DNA repair processes [51]. Purines are damaged by alkylation more often than pyrimidines; as a result, depurination is a frequent process. A bias in the type of DNA damage results in a bias in the type of mutations that occurs in the genome. For example, for every two possible transversions (changes from purine to pyrimidine and vice versa), there is one possible transition (a change from purine to purine or from pyrimidine to pyrimidine), giving the rate of transitions (ti) to transversions (tv) as 0.5. Despite this fact, the ti/tv rate in the human genome is about 2.0, and even higher at the exons—about 3.0 [52]. For other organisms, it has been reported that the ti/tv ratio is 0.62 for yeast [53] and 1.5 for maize [54]. Curiously, mitochondrial DNA (mtDNA) has been reported to have a much stronger bias toward transitions over transversions compared with nuclear genes [55,56]. In the human genome, A:T→G:C (A→G) and G:C→A:T (G→A) changes are predominant among other types of point mutations, including T→C transitions [44]; this difference is even more dramatic in the transcribed regions, likely influenced by transcription-coupled repair [57].

In addition to a bias in the mutation rate at a single nucleotide site, there is a substantial bias in point mutations at dinucleotides. For example, in mammals, and specifically in primates, C→T transitions that arise at CpG dinucleotides are 15 times more frequent than at other dinucleotides [58]; this is likely explained by the common spontaneous oxidative deamination of methylated cytosines at CpGs. A high frequency of C→T mutations at CpGs may explain an extremely rare occurrence of CpGs in the human genome. Considering that the human genome is about 42% GC rich, the occurrence of G or C nucleotide should be 21%, and thus, the frequency of occurrence of CpG should be $0.21 \times 0.21 = 4.41\%$. Instead, there is less than 1% of CpGs in the human genome. Similar data were also reported for *Arabidopsis* plants; the mutation rate at methylated cytosines at CpGs is substantially higher than at nonmethylated ones. There is also a higher rate of G→A substitutions at CpG sites, which cannot be easily explained [59].

Also, the rate of nucleotide substitutions varies at the base pair scale. In primates, substitution rates at the G:C base pairs (excluding the CpG sites) are up to 85% higher than at the A:T base pairs [60,61], it is possibly because cytosine is intrinsically more mutable than other bases [62].

Finally, it is well known that some short sequence motifs (such as minisatellite and microsatellite repeats, 1–100 nt repeated multiple times) are highly mutable. The mutation rate at some microsatellite loci is 10^3–10^5 higher than in the coding regions [63]. Among the possible mechanisms of such high mutability are replication slippage, a gain or loss of one or more repeat units [64], an unequal crossing over, nucleotide substitutions, and duplication events [65].

The existence of cryptic mutation hot spots was demonstrated by Hodgkinson et al. [66]. The authors compared the pattern of single nucleotide polymorphism (SNP) in the human and chimpanzee genomes. They hypothesized that if some genomic regions are more mutable than others, there should be more sites that are similarly polymorphic in both species (the so-called coincident SNPs). Indeed, they have found that the number of coincident SNPs was three times higher than the number of SNPs that are randomly distributed in the two genomes.

One of the most common and perhaps the easiest ways to test the randomness of mutations was to subject cells to nonlethal selection and allow them to mutate in a right direction, so that they are able to grow and multiply. The most commonly known experiment demonstrated that the *E. coli* strain with a nonsense mutation in the *lacZ* gene rendering it unable to use the lactose mutated back to the wild-type allele only when the lactose was present in the medium; the absence of the lactose in the medium did not induce this mutation [67]. Another experiment demonstrated that a strain in which *lacZ* was deleted gained two mutations that allowed to utilize the lactose; the first mutation occurred in a cryptic gene *ebgA* in such a way that it was able to hydrolyze the lactose, and the second mutation inactivated the *ebgR* gene, a negative regulator of *ebgA*. Considering that either of these mutations requires the frequency of 10^{-8}, these two mutations can coincide in the same cell with the frequency of 10^{-16} [68]. Both of these examples demonstrate the fact that mutations can likely occur in a nonrandom manner, and that such mutations are triggered by the environmental factor.

A very interesting work (that still continues) was done using *E. coli* that has demonstrated how the environment drives the appearance of certain type of mutations. *E. coli* cells cannot grow on citrate. In the experiment that started in 1988, Blount et al. propagated 12 identical populations of *E. coli* in the abundance of citrate and a limited amount of glucose [69]. They have frozen bacterial cultures every 500 generations. This gave them the possibility to analyze the potential genetic changes. The propagation of all 12 populations for 30,000 generations did not result in the appearance of the capacity to metabolize citrate, although such number of generations should have been sufficient to mutate every possible nucleotide more than once. After 31,500 generations, one of the populations has developed the ability to use citrate. It took another 2000 generations for this ability in the population to become very common. Although Cit+ cells continued to use glucose, they did not drive Cit− cells to extinction because the Cit− cells were superior competitors for glucose (even though it was present in low amounts). The authors performed replay evolution starting from different generations (at 500 increments). If they had started regrowing bacteria from stocks frozen before generation 20,000, they would not be able to obtain Cit+ cells. Between 20,000 and 27,000 generations, Cit+ cells were obtained, but they were extremely rare, after the 27,000th generation, their appearance was just rare [69].

Therefore, the authors concluded that the appearance of mutations leading to Cit+ cells was contingent upon the appearance of one or several previous enabling mutations. A population in which Cit+ cells appeared was no hypermutable, suggesting that the potentiation of the appearance of Cit+ cells was the event that was specific to the Cit function. The authors proposed two possible mechanisms [69]. The first mechanism suggests the epistatic interaction whereby the locus where a mutation leading to the Cit+ phenotype has occurred likely interacts with another locus where an earlier mutation has occurred. The second mechanism suggests that the first mutation(s) create a new sequence that allows the Cit+ mutation to occur; this can be due to the insertion/deletion of a sequence or the rearrangement or the insertion/excision of a TE, and so on. The exact sequence of the enabling mutation and a mutation leading to Cit+ are not known; the authors are currently sequencing bacteria from before and after generation 20,000.

Our research also demonstrated some directionality in the appearance of mutations. We used two types of tobacco plants, one was resistant to infection with tobacco mosaic virus (TMV) due to the presence of the resistance gene *N*, and the other one that was sensitive to TMV because plants lacked the *N* gene [70]. We collected seeds from the infected plants and analyzed genomic rearrangements in genes that carried homology to the *N* gene, the gene that conferred resistance to TMV. In order to do this, we performed the Southern blot analysis using the genomic DNA of progeny plants with DNA of fourth exon (the conservative region encoding leucine-rich repeats required for the recognition of pathogens) of the *N* gene as a probe. The experiment showed that tobacco plants carried as many as 30 loci with a certain degree of sequence homology (about 50–70%) to the fourth exon. The analysis showed great variations in the pattern of bands in the progeny of infected plants that did not contain the *N* gene; in fact, an eightfold higher rearrangement frequency was found in the progeny of infected plants as compared to the progeny of control plants [70]. The same analysis using probes against actin loci did not show any difference between the progeny of infected plants and the progeny of control plants, suggesting that an increase in the rearrangement frequency was locus specific. This analysis allowed us to conclude that the environmental pressure can induce potentially beneficial mutations (as it was not confirmed whether rearrangements in these loci resulted in the appearance of a viable resistance gene). Unfortunately, at the time of experiment, no comparison was done between mutation rates in the progeny of plants that had the *N* gene versus plants that did not have the *N* gene. Therefore, we do not know whether or not the absence of the active resistance gene (*N*) that triggers an increase in rearrangements at the loci mentioned earlier in response to TMV infection. More details on transgenerational changes in response to the environmental pressure, including changes in genome stability can be found in Chapter 35 and Chapter 36.

7. GENOME EVOLUTION MAY START FROM CHANGES AT THE LEVEL OF DNA METHYLATION OR CHROMATIN MODIFICATION

So far, we have only mentioned the possibility that epigenetic modifications can be critical components (or perhaps even a driving force) of evolution. If genetic variations in a population can be fixed either through random genetic drift or through selection, the same can likely happen with epigenetic variations. Epigenetic modifications such as DNA methylation and histone modifications (see Chapter 36) regulate gene expression and thus contribute to the phenotype appearance. Epigenetic modifications are highly sensitive to environmental changes allowing organisms to respond to the environmental cues more efficiently. Epigenetic modifications are inherited from somatic cell to somatic cell and from generation to generation. However, epigenetic modifications are reversible; therefore, changes in phenotypes that occur due to epigenetic modifications may also be reversible. Also, epigenetic changes typically occur in a large part of a population (perhaps in an entire population if it was exposed to certain environmental changes). Hence, epigenetic modifications may bring many critical advantages as far as the mechanisms of evolution are concerned (Table 1.2).

TABLE 1.2 Comparison of Genetic and Epigenetic Mechanisms of Evolution

	Genetic	Epigenetic
The frequency of occurrence	Extremely low	Very high
Response to the environment	Slow and likely random	Immediate
The directionality of changes (in terms of immediate benefits)	Likely neutral	Directed
The specificity of a response	Unspecific	Highly specific
The reversibility of changes	Rarely reversible	Mostly reversible
The directionality of changes (in terms of immediate benefits)	Likely neutral	Directed
Requirements of a constant environmental pressure	Required for selection	Required for the maintenance of changes
Costs to an organism	Very low (unless mutation is deleterious)	Very high
Long-term solution	Typical	Problematic

Since epigenetic changes occur immediately in response to the environment, they may represent the first level of defense, they may also be the first step in the process of evolution [71]. However, since epigenetic changes are reversible and need to be maintained, they may not represent an ideal long-term solution for an organism or a population. Yet, because the presence of certain epigenetic modifications may increase the chance for certain genetic mutations to occur, epigenetic modifications may aid genetic changes at certain loci, directing changes toward them. For example, methylated cytosines are prone to deamination leading to C→T mutations; when such mutations occur, they eliminate cytosines; therefore, they may, for example, change the sequence of certain genetic elements in the promoter region or other regions with regulatory elements, thus changing the ability of regulatory proteins to bind and affect gene expression [72]. In addition, DNA methylation influences the chromatin structure—higher methylation levels lead to a more closed chromatin structure, and vice versa [73]. Similarly, histone modifications also alter the chromatin structure; some of them make it more open, whereas the others make it more closed (see Chapter 36). Since genomic rearrangements occur more frequently in open chromatin, epigenetic changes that alter the chromatin structure may alter the frequency of rearrangements [74].

Spontaneous epigenetic variations are far more common in *Arabidopsis* than genetic variations. The propagation of 10 *Arabidopsis* lines for 30 generations resulted in a substantial genetic variation, but a far larger epigenetic variation was observed; specific epimutations were detected in all lines tested [59,75]. Johannes et al. used two parental *Arabidopsis* lines that had DNA sequence differences, but substantial DNA methylation differences demonstrate the stability of epigenetic variations [76]. They propagated these lines for eight generations by selfing, thus obtaining epigenetic Recombinant Inbred Lines (epiRILs). The epiRILs showed a variation and a high heritability of flowering time and plant height that lasted at least eight generations. This work demonstrated the fact that numerous epialleles can be stable over many generations in the absence of selection or an extensive DNA sequence variation [76].

Both of the abovementioned examples derive from highly inbred lines that were selfed and propagated in the laboratory environment. Nevertheless, even these lines grown in similar environments quickly gain a substantial epigenetic diversity. The effect of the environment on epigenetic variations should be far greater. The analysis of two nearby habitats of *Laguncularia racemosa* grown at a riverside or near a salt marsh revealed great dissimilarities, with individual plants showing little genetic but abundant DNA methylation differentiation. Moreover, plants grown near a salt marsh had the hypomethylated genome as compared to plants grown in a riverside [77]. This study showed that phenotypic variations in plants grown in the contrasting environments can be primarily triggered by epigenetic changes. It is possible that these two populations will eventually undergo genetic differentiation, and new species may arise as a consequence of both epigenetic and genetic changes, with epigenetic changes being a primary cause.

Differences in methylation patterns in plants grown in the contrasting environments can be propagated for a number of generations. In oil seed crop, *Brassica napus*, variations in methylation patterns analyzed by methylation-sensitive amplification polymorphism (me-AFLP) were demonstrated to be extremely stable. Only 5% of 627 me-AFLP markers identified in several independent parental lines were variable at different developmental stages following the growth in the contrasting environments (11 different natural environments with winter and spring planting). Moreover, for two distinct parental lines, 97% of epialleles were transmitted through five meioses and segregated in a mapping population [78].

In the study by Yi et al., five populations of *Jatropha curcas* plants collected from China (CN), Indonesia (MD), Suriname (SU), Tanzania (AF), and India (TN) were planted in one farm under the same agronomic practices (and the environmental conditions) [79]. The analysis of polymorphism showed a very low level of genetic diversity (a polymorphic band <0.1%) within and among populations, despite the fact that plants were grown in the contrasting environments. In contrast, intermediate but significant epigenetic diversity was detected (25.3% of bands were polymorphic) within and among populations. The authors identified 39 different polymorphic epimarkers, with 30 of them being heritable and following Mendelian segregation [79].

8. CONCLUSION

DNA damage caused by external and internal stresses must be efficiently resolved through the activity of various DNA repair pathways. DNA damage repair is the primary mechanism that preserves the intactness of the plant genome and insures its stability. Noteworthy, many of DNA repair pathways lack high fidelity, and their widespread activity may actually destabilize the genome, compromise its integrity, and may even be lethal for a cell [80].

At the same time, a certain level of genome flexibility is absolutely required for the successful evolution of plant species. In fact, mistakes made during DNA repair may serve as the raw material of evolution. Similarly, the rearrangement and duplication of existing DNA sequences may lead to the generation of new traits conferring a selective advantage under new growth conditions. In view of this, a very delicate balance must exist between different DNA repair pathways to ensure the continuous generation of new DNA sequence variants without affecting genome functions and cell vitality. In fact, different groups of organisms seem to preferentially use specific DNA repair pathways, depending on their genome content. Some organisms with a small genome size and a low content of repetitive sequences preferentially use HR that ensures the minimum number of mistakes being made. An increase in genome size leads to an increase in the probability that a wrong template may be chosen for the HR-mediated repair of DNA damage. The presence of a high number of repetitive DNA elements further complicates this problem as an improper recombination event may result in large deletions and duplications. Hence, a shift occurs toward using a more error-prone pathway which is less dangerous to the genome. For example, plants usually prefer using nonhomologous end joining (NHEJ) to HR. At the same time, HR is mainly used for the reshaping of the plant genome. Indeed, in the genome of *Arabidopsis thaliana*, a substantial part of genetic material was created by the duplication of existing DNA sequences which was possibly mediated by the HR pathway.

The emerging evidences suggest that epigenetic marks and stress-induced sequence-specific signals such as small RNAs (sRNAs) may play an important role in the maintenance of genome stability and chromatin. Decreased DNA methylation and an open chromatin structure attract the activity of the HR pathway and lead to the activation of transposons, thus decreasing genome stability. At the same time, frequent rearrangements at unmethylated loci may help accelerate the evolution of targeted sequences. On the contrary, increased genome methylation permits to preserve genome stability at the time of stress and prevents undesirable rearrangements and transpositions.

The epigenetic landscape of eukaryotic genomes is rather complex, and it is subject to continuous modifications in response to various stimuli. The forgoing leads to an interesting speculation that epigenetic marks are not only used to control the transcriptional activity of the chromatin helping to adjust transcriptional profiles to new growth conditions, but they also contribute to maintain genome stability and direct genome evolution. The fact that stress can cause sequence-specific epigenetic changes permits further speculation about the involvement of epigenetic modifications in stress-directed genome evolution. Unfortunately, there is still no direct evidence that links together stress, epigenetic modifications, and mechanisms that control genome stability in a multifaceted system that regulates and directs genome evolution. It is the challenge for future studies to unravel these links and provide a better understanding of how stress-directed genome evolution and adaptation occur in nature.

GLOSSARY

Catastrophism A theory that suggests that geologic features of our planet have been influenced by rapid violent catastrophic events.

Coincident SNPs Polymorphisms found at similar genomic positions upon comparison between closely related species.

Commensalistic interaction Represents a symbiotic interaction between two organisms where one organism benefits from it and another one is not affected by it.

Endosymbiotic gene transfer A transfer of genes from the chloroplast to the nuclear genome; refers to the evolutionary process.

Fecundity Fertility of an organism.

Fixation Reaching 100% or near 100% occurrence of a certain mutation/allele in a population.

Genetic drift Refers to variations in the frequency of certain alleles in a population due to randomness of sampling. For haploid and diploid species, the variance in allele frequency resulting from drift is proportional to 1/Ne and 1/(2Ne), respectively [40].

Genetic hitchhiking A mechanism of selection and fixation of alleles that are in close proximity from alleles that are under a high selection pressure.

Macroevolution Typically reflects larger changes, across the boundaries of species; often referred to major changes, such as *the appearance of new substantially different species*, the appearance of flowering plants, placental animals, and so on.

Microevolution Reflects changes in traits/phenotypes within a population or a species.

Modern evolutionary synthesis The modern theory of evolution that was synthetized on the basis of Darwin's theory of evolution and completed by additional knowledge of genes and genetics.

Mutualistic interaction Represents a symbiotic interaction between two organisms where both organisms benefit.

Ne An effective population size which reflects the size of a population with respect to a random distribution of alleles. Ne is typically smaller than the actual size of a population due to variations in sex-ratio bias, variations in family size, the nonrandomness of mating, to name a few factors. The smaller is the population, the closer is Ne to the actual population size [44].

Nonsynonymous A mutation that changes the amino acid composition of the encoded protein.

Silent mutation (silent site) A nucleotide substitution that does not result in changes in the protein sequence.

Single nucleotide polymorphism Variations at specific genomic positions existing in a population of certain species.

Somatic mutation A mutation in somatic (nongermline) cells of a multicellular organism. The somatic mutation rate may correlate and influence the germline mutation rate and may influence the fitness of an organism leading to changes in inheritance.

Synonymous mutation A mutation that does not change the amino acid composition of the encoded protein.

Transposable element A mobile element of a transposon, a genetic element capable of moving/duplicating itself (either through excision or through copying) from one genomic position to another.

u The mutation rate per nucleotide per generation.

Uniformitarianism A theory that suggests that all laws of the universe operate at a steady constant rate.

LIST OF ABBREVIATIONS

HT Horizontal transfer
MES Modern evolutionary synthesis
TE Transposable element

REFERENCES

[1] Darwin CR. On the origin of species by means of natural selection, or the preservation of favoured races in the struggle for life. 1st ed. London: John Murray; 1859.

[2] Futuyma DJ. Evolutionary biology. Sinauer Associates; 1986. p. 12.

[3] Eldredge N, Gould SJ. Punctuated equilibria: an alternative to phyletic gradualism. In: Schopf TJM, editor. Models in paleobiology. San Francisco: Freeman Cooper; 1972. p. 82–115. Reprinted in N. Eldredge Time frames. Princeton: Princeton Univ. Press, 1985, pp. 193–223.

[4] Rieseberg LH, Wood TE, Baack EJ. The nature of plant species. Nature 2006;440(7083):524–7.

[5] Martin WF, Garg S, Zimorski V. Endosymbiotic theories for eukaryote origin. Philos Trans R Soc Lond B Biol Sci 2015;370(1678):20140330.

[6] Timmis JN, Ayliffe MA, Huang CY, Martin W. Endosymbiotic gene transfer: organelle genomes forge eukaryotic chromosomes. Nat Rev Genet February 2004;5(2):123–35.

[7] Martin W, Hoffmeister M, Rotte C, Henze K. An overview of endosymbiotic models for the origins of eukaryotes, their ATP-producing organelles (mitochondria and hydrogenosomes), and their heterotrophic lifestyle. Biol Chem November 2001;382(11):1521–39.

[8] Holt AL, Vahidinia S, Gagnon YL, Morse DE, Sweeney AM. Photosymbiotic giant clams are transformers of solar flux. J R Soc Interface December 6, 2014;11(101):20140678.

[9] Sandelius AS. The chloroplast: interactions with the environment. Springer; 2009. ISBN: 978-3-540-68696-5. p. 18.

[10] Martin W, Rujan T, Richly E, Hansen A, Cornelsen S, Lins T, et al. Evolutionary analysis of arabidopsis, cyanobacterial, and chloroplast genomes reveals plastid phylogeny and thousands of cyanobacterial genes in the nucleus. Proc Natl Acad Sci USA September 17, 2002;99(19):12246–51.

[11] Keeling PJ, Palmer JD. Horizontal gene transfer in eukaryotic evolution. Nat Rev Genet 2008;9:605–18.

[12] Gantt JS, Baldauf SL, Calie PJ, Weeden NF, Palmer JD. Transfer of rpl22 to the nucleus greatly preceded its loss from the chloroplast and involved the gain of an intron. EMBO J October 1991;10(10):3073–8.

[13] Mackiewicz P, Bodył A, Moszczyński K. The case of horizontal gene transfer from bacteria to the peculiar dinoflagellate plastid genome. Mob Genet Elem 2013;3(4):e25845.

[14] Alverson AJ, Rice DW, Dickinson S, Barry K, Palmer JD. Origins and recombination of the bacterial-sized multichromosomal mitochondrial genome of cucumber. Plant Cell 2011;23(7):2499–513.

[15] Selosse M, Albert B, Godelle B. Reducing the genome size of organelles favours gene transfer to the nucleus. Trends Ecol Evol 2001;16(3):135–41.

[16] Smith DR. Mutation rates in plastid genomes: they are lower than you might think. Genome Biol Evol 2015;7(5):1227–34.

[17] Mower JP, Touzet P, Gummow JS, Delph LF, Palmer JD. Extensive variation in synonymous substitution rates in mitochondrial genes of seed plants. BMC Evol Biol 2007;7:135.

[18] Sloan DB, Alverson AJ, Chuckalovcak JP, et al. Rapid evolution of enormous, multichromosomal genomes in flowering plant mitochondria with exceptionally high mutation rates. PLoS Biol 2012;10:e1001241.
[19] Lynch M. The origins of eukaryotic gene structure. Mol Biol Evol 2006;23:450–68.
[20] Nabholz B, Glémin S, Galtier N. Strong variations of mitochondrial mutation rate across mammals–the longevity hypothesis. Mol Biol Evol January 2008;25(1):120–30.
[21] Gillooly JF, Allen AP, West GB, Brown JH. The rate of DNA evolution: effects of body size and temperature on the molecular clock. Proc Natl Acad Sci USA 2005;102:140–5.
[22] James JE, Piganeau G, Eyre-Walker A. The rate of adaptive evolution in animal mitochondria. Mol Ecol November 17, 2015. http://dx.doi.org/10.1111/mec.13475.
[23] Gilbert C, Cordaux R. Horizontal transfer and evolution of prokaryote transposable elements in eukaryotes. Genome Biol Evol 2013;5(5):822–32.
[24] Newton ILG, Bordenstein SR. Correlations between bacterial ecology and mobile DNA. Curr Microbiol 2011;62:198–208.
[25] Wagner A. Cooperation is fleeting in the world of transposable elements. PLoS Comput Biol 2006;2:1522–9.
[26] Filée J, Siguier P, Chandler M. Insertion sequence diversity in archaea. Microbiol Mol Biol Rev March 2007;71(1):121–57.
[27] Lander ES, Linton LM, Birren B, Nusbaum C, Zody MC, Baldwin J, et al. Initial sequencing and analysis of the human genome. Nature 2001;409:860–921.
[28] Schnable PS, Ware D, Fulton RS, et al. The B73 maize genome: complexity, diversity, and dynamics. Science 2009;326:1112–5.
[29] Lynch M. The origins of genome architecture. Sunderland (MA, USA): Sinauer Associates Inc.; 2007.
[30] Whitney KD, Garland Jr T. Did genetic drift drive increases in genome complexity? PLoS Genet 2010;6:e1001080.
[31] Startek M, Le Rouzic A, Capy P, Grzebelus D, Gambin A. Genomic parasites or symbionts? Modeling the effects of environmental pressure on transposition activity in asexual populations. Theor Popul Biol December 2013;90:145–51.
[32] Haddrill PR, Halligan DL, Tomaras D, Charlesworth B. Reduced efficacy of selection in regions of the *Drosophila* genome that lack crossing over. Genome Biol 2007;8(2):R18.
[33] Kimura M. Evolutionary rate at the molecular level. Nature February 17, 1968;217(5129):624–6.
[34] Kimura M, Ohta T. The average number of generations until fixation of a mutant gene in a finite population. Genetics 1968;61:763–71.
[35] Patwa Z, Wahl LM. The fixation probability of beneficial mutations. J R Soc Interface 2008;5(28):1279–89.
[36] Burke MK, Dunham JP, Shahrestani P, Thornton KR, Rose MR, et al. Genome-wide analysis of a long-term evolution experiment with *Drosophila*. Nature 2010;467:587–90.
[37] Tobler R, Franssen SU, Kofler R, Orozco-terWengel P, Nolte V, et al. Massive habitat-specific genomic response in *D. melanogaster* populations during experimental evolution in hot and cold environments. Mol Biol Evol 2014;31:364–75.
[38] Turner TL, Miller PM. Investigating natural variation in *Drosophila* courtship song by the evolve and resequence approach. Genetics 2012;191: 633–642.
[39] Chandler CH. Parallel genome-wide fixation of ancestral alleles in partially outcrossing experimental populations of *Caenorhabditis elegans*. G3 2014;4(9):1657–65.
[40] Chelo IM, Teotonio H. The opportunity for balancing selection in experimental populations of *Caenorhabditis elegans*. Evolution 2013;67:142–56.
[41] Schwartz D. An example of gene fixation resulting from selective advantage in suboptimal conditions. Am Nat 1969;103(933):479–81.
[42] Lee MS, Soubrier J, Edgecombe GD. Rates of phenotypic and genomic evolution during the Cambrian explosion. Curr Biol October 7, 2013;23(19):1889–95.
[43] Drake JW. A constant rate of spontaneous mutation in DNA-based microbes. Proc Natl Acad Sci USA 1991;88:7160–4.
[44] Lynch M. Evolution of the mutation rate. Trends Genet August 2010;26(8):345–52. http://dx.doi.org/10.1016/j.tig.2010.05.003.
[45] Piganeau G, Eyre-Walker A. Evidence for variation in the effective population size of animal mitochondrial DNA. PLoS One 2009;4:e4396.
[46] Wai T, Teoli D, Shoubridge EA. The mitochondrial DNA genetic bottleneck results from replication of a subpopulation of genomes. Nat Genet 2008;40:1484–8.
[47] Winn RN, Norris MB, Brayer KJ, Torres C, Muller SL. Detection of mutations in transgenic fish carrying a bacteriophage lambda cII transgene target. Proc Natl Acad Sci USA 2000;97:12655–60.
[48] Edman U, Garcia AM, Busuttil RA, Sorensen D, Lundell M, Kapahi P, et al. Lifespan extension by dietary restriction is not linked to protection against somatic DNA damage in *Drosophila melanogaster*. Aging Cell 2009;8:331–8.
[49] Wang K, Yuen ST, Xu J, Lee SP, Yan HH, Shi ST, et al. Whole-genome sequencing and comprehensive molecular profiling identify new driver mutations in gastric cancer. Nat Genet June 2014;46(6):573–82.
[50] Weinhold N, Jacobsen A, Schultz N, Sander C, Lee W. Genome-wide analysis of noncoding regulatory mutations in cancer. Nat Genet November 2014;46(11):1160–5.
[51] Helleday T, Eshtad S, Nik-Zainal S. Mechanisms underlying mutational signatures in human cancers. Nat Rev Genet September 2014;15(9):585–98.
[52] Guo Y, Ye F, Sheng Q, Clark T, Samuels DC. Three-stage quality control strategies for DNA re-sequencing data. Brief Bioinform November 2014;15(6):879–89.
[53] Lynch M, Sung W, Morris K, et al. A genome-wide view of the spectrum of spontaneous mutations in yeast. Proc Natl Acad Sci USA 2008;105:9272–7.
[54] Morton BR, Bi IV, McMullen MD, et al. Variation in mutation dynamics across the maize genome as a function of regional and flanking base composition. Genetics 2006;172:569–77.
[55] Lanave C, Tommasi S, Preparata G, et al. Transition and transversion rate in the evolution of animal mitochondrial DNA. Biosystems 1986;19:273–83.
[56] Guo Y, Cai Q, Samuels DC, et al. The use of next generation sequencing technology to study the effect of radiation therapy on mitochondrial DNA mutation. Mutat Res 2012;744:154–60.

[57] Green P, Ewing B, Miller W, Thomas PJ, Green ED. Transcription-associated mutational asymmetry in mammalian evolution. Nat Genet 2003;33:514–7.
[58] Kim SH, Elango N, Warden C, Vigoda E, Yi SV. Heterogeneous genomic molecular clocks in primates. PLoS Genet 2006;2:e163.
[59] Ossowski S, Schneeberger K, Lucas-Lledó JI, Warthmann N, Clark RM, Shaw RG, et al. The rate and molecular spectrum of spontaneous mutations in *Arabidopsis thaliana*. Science January 1, 2010;327(5961):92–4.
[60] Arndt PF, Hwa T, Petrov DA. Substantial regional variation in substitution rates in the human genome: importance of GC content, gene density, and telomere-specific effects. J Mol Evol 2005;60:748–63.
[61] Duret L, Arndt PF. The impact of recombination on nucleotide substitutions in the human genome. PLoS Genet 2008;4(5):e1000071.
[62] Birdsell JA. Integrating genomics, bioinformatics, and classical genetics to study the effects of recombination on genome evolution. Mol Biol Evol 2002;19:1181–97.
[63] Li YC, Korol AB, Fahima T, Beiles A, Nevo E. Microsatellites: genomic distribution, putative functions and mutational mechanisms: a review. Mol Ecol 2002;11(12):2453–65.
[64] Schlötterer C, Tautz D. Slippage synthesis of simple sequence DNA. Nucleic Acids Res 1992;20(2):211–5.
[65] Hancock JM. Microsatellites and other simple sequences: genomic context and mutational mechanisms. In: Goldstein D, Schlötterer C, editors. Microsatellites: evolution and applications. New York: Oxford University Press; 1999. p. 1–9.
[66] Hodgkinson A, Ladoukakis E, Eyre-Walker A. Cryptic variation in the human mutation rate. PLoS Biol February 3, 2009;7(2):e1000027.
[67] Cairns J, Overbaugh J, Miller S. The origin of mutants. Nature 1988;335:142–5.
[68] Hall BG, Hartl DL. Regulation of newly evolved enzymes 1. Selection of a novel lactase regulated by lactose in *Escherichia coli*. Genetics 1975;76:391–400.
[69] Blount ZD, Borland CZ, Lenski RE. Inaugural article: Historical contingency and the evolution of a key innovation in an experimental population of *Escherichia coli*. Proc Natl Acad Sci USA 2008;105(23):7899–906.
[70] Boyko A, Kathiria P, Zemp FJ, Yao Y, Pogribny I, Kovalchuk I. Transgenerational changes in the genome stability and methylation in pathogen-infected plants: (virus-induced plant genome instability). Nucleic Acids Res 2007;35(5):1714–25.
[71] Boyko A, Kovalchuk I. Epigenetic control of plant stress response. Environ Mol Mutagen January 2008;49(1):61–72.
[72] Boyko A, Kovalchuk I. Genome instability and epigenetic modification–heritable responses to environmental stress?. Curr Opin Plant Biol June 2011;14(3):260–6. http://dx.doi.org/10.1016/j.pbi.2011.03.003. Epub 2011 Mar 26.
[73] Jin B, Li Y, Robertson KD. DNA methylation: superior or subordinate in the epigenetic hierarchy? Genes Cancer June 2011;2(6):607–17.
[74] Déjardin J. How chromatin prevents genomic rearrangements: locus colocalization induced by transcription factor binding. Bioessays February 2012;34(2):90–3.
[75] Becker C, Hagmann J, Müller J, Koenig D, Stegle O, Borgwardt K, et al. Spontaneous epigenetic variation in the *Arabidopsis thaliana* methylome. Nature September 20, 2011;480(7376):245–9.
[76] Johannes F, Porcher E, Teixeira FK, Saliba-Colombani V, Simon M, Agier N, et al. Assessing the impact of transgenerational epigenetic variation on complex traits. PLoS Genet 2009;5(6):e1000530.
[77] Lira-Medeiros CF, Parisod C, Fernandes RA, Mata CS, Cardoso MA, Ferreira PC. Epigenetic variation in mangrove plants occurring in contrasting natural environment. PLoS One April 26, 2010;5(4):e10326.
[78] Long Y, Xia W, Li R, Wang J, Shao M, Feng J, et al. Epigenetic QTL mapping in *Brassica napus*. Genetics 2011;189:1093–102.
[79] Yi C, Zhang S, Liu X, Bui HT, Hong Y. Does epigenetic polymorphism contribute to phenotypic variances in *Jatropha curcas* L.? BMC Plant Biol November 23, 2010;10:259.
[80] Wu J, Starr S. Low-fidelity compensatory backup alternative DNA repair pathways may unify current carcinogenesis theories. Future Oncol 2014;10(7):1239–53.
[81] Crow JF. The origins, patterns and implications of human spontaneous mutation. Nat Rev Genet October 2000;1(1):40–7.
[82] Drost JB, Lee WR. Biological basis of germline mutation: comparisons of spontaneous germline mutation rates among *Drosophila*, mouse, and human. Environ Mol Mut 1995;25(Suppl. 26):48–64.
[83] Wilkins AS. Genetic analysis of animal development. 2nd ed. NewYork: Wiley-Liss; 1992.
[84] Hoffman PD, Leonard JM, Lindberg GE, Bollmann SR, Hays JB. Rapid accumulation of mutations during seed-to-seed propagation of mismatch-repair-defective *Arabidopsis*. Genes Dev 2004;18:2676–85.
[85] Lynch M. The rate, molecular spectrum, and consequences of human mutation. Proc Natl Acad Sci USA 2010;107:961–8.
[86] Keightley PD, Trivedi U, Thomson M, Oliver F, Kumar S, Blaxter ML. Analysis of the genome sequences of three *Drosophila melanogaster* spontaneous mutation accumulation lines. Genome Res 2009;19:1195–201.
[87] Denver DR, Dolan PC, Wilhelm LJ, Sung W, Lucas-Lledó JI, Howe DK, et al. A genome-wide view of *Caenorhabditis elegans* base substitution mutation processes. Proc Natl Acad Sci USA 2009;106:16310–4.

Section I

Genome Instability of Viruses

Chapter 2

Genetic Instability of RNA Viruses

J.N. Barr[1], R. Fearns[2]
[1]*University of Leeds, Leeds, United Kingdom;* [2]*Boston University School of Medicine, Boston, MA, United States*

Chapter Outline

1. INTRODUCTION

Viruses are enormously successful. They have been identified in organisms within all domains of life. Despite decades of scientific effort to combat viruses that cause disease in humans and economically important crops and animals, there are relatively few cases in which we have succeeded. Viruses have shown they are able to adapt and multiply to overcome almost any obstacle that is imposed on them. This remarkable adaptability can be attributed to their extremely high replication rate and their propensity for mutation. This is particularly true of the viruses that have RNA genomes: the riboviruses and retroviruses. This chapter will focus on these RNA viruses and on the exciting research that has provided valuable insight into how RNA viruses benefit from their genetic variability. In the first two sections of the chapter, two fundamental concepts are introduced: the intimate relationship between RNA viruses and their hosts, and the idea that viruses behave as quasispecies. Having introduced these concepts, the remainder of the chapter discusses the viral and host mechanisms that govern RNA virus genetic variability and the ability of viruses to withstand mutation. We then discuss evidence that at least some RNA viruses have a replication fidelity that is poised to maximize genome sequence space without incurring catastrophic lethal mutations and describe how this can be exploited to control viral infections. Throughout the chapter, we attempt to convey the diversity of RNA virus biology and mutation frequency and we conclude by speculating that each RNA virus has evolved an error rate that complements its genome replication strategy and mode of transmission.

2. OVERVIEW OF RNA VIRUS MULTIPLICATION

RNA viruses are very simple entities with small genomes that vary in length from about 2 to 32 kb, depending on the virus. Thus, they have very limited coding capacity and so, similarly to DNA viruses, they are obligate intracellular parasites, depending on a host cell to provide energy generating systems, ribo- and deoxyribonucleotides, cellular translation machinery, tRNAs and amino acids to translate their mRNAs, cellular enzymes to posttranslationally modify their proteins, and

Genome Stability. http://dx.doi.org/10.1016/B978-0-12-803309-8.00002-1

cellular structures such as membranes, vesicular compartments, and/or cytoskeleton networks to act as a scaffold for assembling and transporting components required to make virus particles. There are many RNA viruses and they vary enormously in their genome structures and mechanisms of replication. However, in its most distilled and generic form, the RNA virus infection cycle consists of the steps shown in Fig. 2.1.

First a protein on the surface of the virus particle attaches to a receptor molecule on a host cell enabling the viral genome to be delivered into the cell. The genome is expressed to produce viral proteins and replicated multiple times to produce progeny genomes. The progeny genomes are packaged with the proteins that make up the virus particle and are released to infect new cells. Thus, viruses multiply by a process of genome replication, expression, and assembly, rather than division, and a cell infected by a single infectious virus particle could release thousands of progeny virions in a matter of hours. This enables viruses to multiply very rapidly and to achieve large population sizes. Because viruses depend on a host cell to be able to replicate, their ability to multiply is heavily influenced by the biology of each cell that they encounter, such as the nature and density of surface molecules that can act as viral receptors, the cell's metabolic rate and availability of macromolecules, as well as the cell's innate antiviral defenses that have evolved to suppress viral replication. In addition to being able to replicate within a single cell type, most viruses require the capacity to replicate and spread within a multicellular host organism, which has tissues with varied cellular characteristics, physiological and anatomical constraints, and an adaptive immune response. While some viruses might only require the ability to infect one tissue to be successfully maintained in the environment, some viruses need to infect and multiply in different tissues to be spread to a new host and complete their transmission cycles. For example, measles virus initially infects alveolar macrophages and dendritic cells in the lung. It is then transferred to T- and B-lymphocytes and is amplified and spread systemically throughout the body. Infected lymphocytes can then transfer the virus to the basolateral surface of lung epithelial cells by attaching to an epithelial cell receptor. The virus multiplies further in the lung epithelium and is spread to new hosts by coughing and sneezing [1,2]. Thus, measles virus requires the ability to infect multiple cell types to complete its transmission cycle. Viruses must also be capable of replicating within populations of hosts whose immune responses are shaped by different histories of virus exposure and some viruses even require the ability to replicate in different host species. For example, West Nile virus transmission is dependent on the virus being able to replicate efficiently in both mosquitos and birds. In mosquitos the virus multiples in the salivary glands and is transmitted to birds when the mosquito takes a blood meal. The virus is amplified in

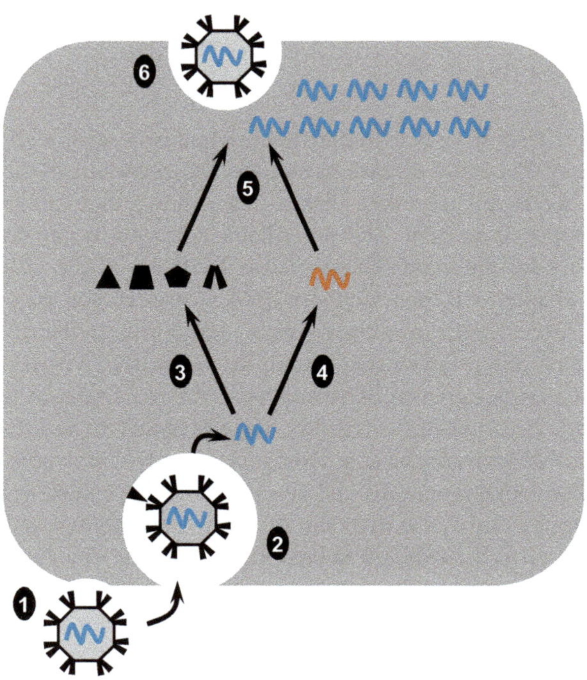

FIGURE 2.1 **A schematic diagram showing the infection cycle of an RNA virus.** An RNA virus particle, or virion, consists of an RNA genome (blue) surrounded by a protein coat or capsid (black). Some viruses also have a lipid envelope studded with viral proteins surrounding the capsid (not shown). The virus particle attaches to a receptor on the surface of a susceptible host cell (1), and becomes internalized (2). The viral genome codes for viral proteins (black shapes) (3) and is replicated via a replication intermediate (red) (4). Newly synthesized genomes and proteins assemble together (5) and newly made virus particles are released (6).

birds and can be transferred to further mosquitos [3]. Because RNA viruses need to replicate in these highly variable and dynamic environments, they need to be highly adaptable to maintain their existence. This adaptability is conferred by their genetic heterogeneity.

3. VIRUSES AS QUASISPECIES

In the late 1970s it was discovered that the nucleotide sequences of RNA bacteriophage, Qβ, are highly heterogeneous [4], and this observation has since been extended to all RNA viruses. Accurate quantification of RNA virus mutation rates is challenging, but they have been estimated at 10^{-4} to 10^{-6} per nucleotide per round of copying [5,6]. This equates to approximately one mutation per genome replication event, which is a considerably higher rate than that of bacteria, estimated at one mutation per 300 genome replication events [5]. In addition to point mutations, recombination between viral genomes can occur in high frequency in some RNA viruses, resulting in replacement of different regions of genome sequence. Any particular RNA virus population is always in flux, with new mutations arising and deleterious mutations being lost through selection. The high mutation rate of RNA viruses, coupled with their very high levels of replication and the large population sizes that they can achieve means that RNA viruses exist as a swarm of variants rather than as a single genotype entity. Thus, RNA viruses are a genetic paradox: they are in one sense very simple entities, having very limited genetic information, but on the other hand, they are genetically complex, having the capability to access millions of sequence combinations. Adding to this complexity, there is evidence for some RNA viruses that they can exist as quasispecies in which the related genome sequences can complement each other and function cooperatively [7,8]. Thus, when a virus spreads from cell to cell and host to host, it is the properties of a swarm of genetically related but distinct viruses that enables this to occur, not the properties of a single, isogenic virus. As described in detail later in this chapter, RNA viruses require a high mutation rate to enable them to survive the varied environments that they encounter in the course of their transmission cycle. Interestingly, they also have evolved genome sequences that have a bias that allows them to rapidly adapt. However, there is also evidence that at least some have a mutation rate that is so high that they are poised at the edge of a threshold of viability, with small increases in mutation rate causing them to accumulate so many lethal mutations that they are extinguished. Together, these findings suggest that RNA viruses have evolved to have a specific mutation rate and mutation bias to enable them to survive in the particular environments in which they need to exist.

4. OVERVIEW OF RNA VIRUS REPLICATION MECHANISMS

There are several sources of genetic variability in RNA viruses, some are inherent to the biology of the virus and others are consequences of the cellular environment. The viral mutation rate is the rate at which a viral genome acquires mutations per genome replication event and is determined by the viral polymerase and any proofreading activities that the virus encodes. The mutation frequency of a virus is the frequency with which mutations accumulate over a virus infection cycle and can be impacted by the mode of virus replication, and cellular factors. Thus, to understand how viral genetic heterogeneity arises, it is helpful to have an appreciation of the mechanisms by which RNA viruses replicate their genomes. RNA viruses can be divided into different classes by virtue of their distinct genome structures and strategies of genome replication [9] (Fig. 2.2).

The riboviruses replicate their genomes via an RNA intermediate synthesized by a viral RNA-dependent RNA polymerase (RdRp). Riboviruses can have single or double-stranded RNA genomes; those with single-stranded genomes can be further characterized by being either positive or negative stranded (ie, having a genome that is of the same sense, or the opposite sense to mRNA, respectively). Riboviruses can also have genomes contained within a single piece of RNA, or a genome that is divided into multiple segments. Another class of RNA virus is the retroviruses. These viruses have an RNA genome, which is reverse transcribed by the viral reverse transcriptase enzyme into double-stranded DNA. The virus-specific double-stranded DNA then integrates into the host genome and becomes a template for cellular RNA polymerase II, which synthesizes multiple copies of RNA to generate the progeny viral genomes. It is important to appreciate that this classification system does not relate in any way to the tissues or hosts that a virus can infect, or the way in which it is transmitted to new hosts. For example, hepatitis C virus (HCV) and West Nile virus are both positive-strand RNA viruses, but they cause very different diseases and are spread in different ways.

5. THE VIRAL POLYMERASE AS A SOURCE OF ERROR

Both RdRps and reverse transcriptases have the potential to introduce deletions, insertions, and nucleotide mismatches into the nucleic acid product [10–12]. Unlike DNA-based life forms, most RNA viruses have no mechanisms to identify and repair mismatches [11,13] and so polymerase error is not corrected. The error-prone nature of polymerase activity, coupled

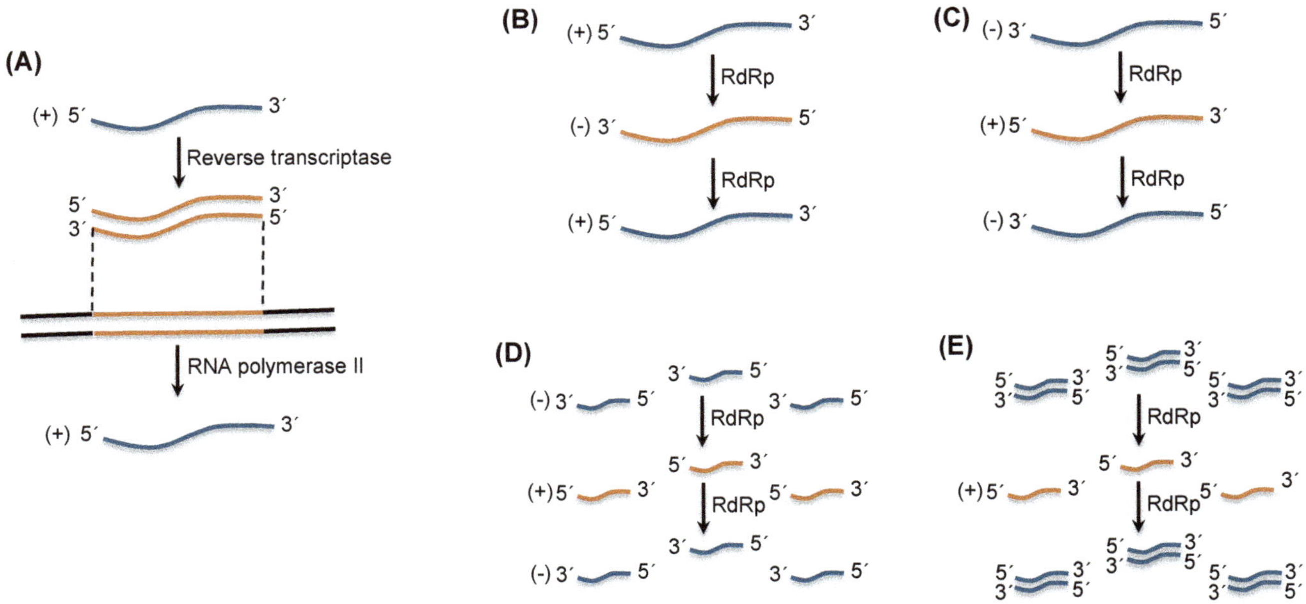

FIGURE 2.2 **Examples of genome structures and genome replication strategies of RNA viruses.** The examples shown are a retrovirus (A), a nonsegmented, positive-strand RNA virus (B), a nonsegmented, negative-strand RNA virus (C), a segmented, negative-strand RNA virus (D), and double-stranded RNA virus (E). The virus genomes are shown in blue and the replicative intermediates in orange. The (+) and (−) symbols indicate if the RNA is positive or negative sense (this is also indicated by the polarity of the RNA). The enzyme that performs each step of nucleic acid synthesis is shown. The replicative intermediate for the retrovirus (panel A) is DNA, whereas for the other viruses shown (panels B–E) it is RNA. Note that not all RNA virus genome structures are illustrated.

with the absence of a proofreading mechanism, is the key reason why RNA virus genomes acquire mutations and exist as a swarm of genetic variants. Although all RdRps and reverse transcriptases are capable of introducing mutations, they are not equally error prone. For example, the viral mutation rate inversely correlates with genome size, such that viruses with larger genomes have a lower per nucleotide mutation rate than those with small genomes [14]. This is intuitively logical as a high mutation rate in a virus with a large genome would increase the chance of genomes acquiring a lethal mutation and so viruses with low fidelity polymerases could not be sustained. This suggests that viruses with larger genomes have evolved to limit their mutation rate and some RNA viruses encode proteins that function to mitigate polymerase error, as described in the following. However, even when related viruses with similar genome lengths are compared, there are differences in polymerase fidelity [11,15]. For example, in a side-by-side comparison, using in vitro biochemical assays, the RdRp of coxsackievirus B is of higher fidelity than that of poliovirus, even though these are highly related viruses [16]. In sum, these facts suggest that polymerase error rate is determined by selection pressures related to viral genome size and other facets of virus biology.

The molecular mechanisms that govern polymerase fidelity have been elucidated by detailed enzyme kinetics studies of wild-type polymerases and by studying mutant versions of polymerase with altered fidelity [14,17–24]. These studies have shown that the error rate of the polymerase can be modulated by single amino acid substitutions in the enzyme, and that substitutions outside the active site can have an effect. Thus, the structure of the polymerase is tuned to enable it to manifest a particular fidelity. In addition to controlling the rate of replication error, polymerase determinants can also influence what substitution mutations are introduced. In a landmark study, a novel sequencing approach was employed to identify low-frequency mutations that accrued in the poliovirus genome under relatively constant conditions [25]. Viral populations present at different times were analyzed to determine what mutations accumulated in this stable environment, where selection pressure was minimized. This analysis showed that transitions occurred more frequently than transversions, and within these categories there was variation: C-to-U and G-to-A transitions accumulated more frequently than U-to-C or A-to-G. Thus, these studies indicate that there is directionality to the mutation pattern of the viral swarm. Similar findings had been made with HIV [10] and studies with West Nile virus have shown that different polymerase variants have different mutational biases [26]. Thus, RNA viruses do not incur substitutions randomly, but have a mutation bias that is likely governed by the molecular determinants of fidelity in the polymerase. This bias might play an important role in allowing the virus to generate a favorable spectrum of sequences following a genetic bottleneck.

6. OTHER VIRAL DETERMINANTS OF MUTATION RATE

Although the viral polymerase is typically the key viral factor that determines how faithfully the viral genome will be copied, it is not the only factor and there are examples of viruses in which other proteins can come into play to reduce polymerase error. As noted earlier, the genomes of all RNA viruses are relatively small compared to those of the largest DNA viruses, and it is thought that the high mutation rate of RNA virus polymerases imposes an upper limit on genome size. However, there is a wide range of genome sizes within the RNA viruses, with the largest being those of the coronaviruses, at up to 32 kb. This is more than twofold longer than most other RNA viruses. It has now become apparent that the reason why the coronaviruses can sustain this relatively large RNA genome is that they have an RNA proofreading activity [27,28] facilitated by an exonuclease that probably functions by removing incorrect insertions at the 3′- end of the RNA product during RNA synthesis [29]. Interestingly, the activity of the exonuclease is significantly enhanced by an additional coronavirus protein [30,31]. The fact that a multipartite complex performs the polymerization and RNA proofreading activities raises the intriguing possibility that coronavirus fidelity could be regulated.

Some of the nonsegmented, negative-strand RNA viruses also have an additional protein that might function to limit RdRp error. The pneumovirus subfamily has genomes of approximately 15 kb, and so they are in the midrange of RNA virus genome sizes. These viruses encode a small protein called M2-2. It has been found that deletion of M2-2 of human meta-pneumovirus results in increased accumulation of transitions, transversions, deletions, and insertions in the viral RNA, suggesting that M2-2 serves to increase the fidelity of the viral polymerase [32]. The mechanism by which M2-2 functions is not known, but it has no known enzyme activity and so it is unlikely to function as an exonuclease, but instead might serve to increase fidelity by altering RdRp structure.

A deoxyuridine-triphosphatase (dUTPase) enzyme is expressed by some, but not all, retroviruses [33,34]. This enzyme hydrolyzes dUTP and maintains low dUTP:TTP ratios, thus limiting misincorporation of deoxyuridine into viral DNA. The viral dUTPase has been shown to limit the mutation rate of feline immunodeficiency virus and caprine arthritis-encephalitis virus [35,36]. Interestingly, the primate lentiviruses, including HIV, do not encode a dUTPase, but might package a cellular DNA repair enzyme, uracil DNA glycosylate, into their virions to help limit the mutation rate [37,38].

7. RECOMBINATION

In addition to the mutations that can be introduced when the polymerase selects an incorrect nucleotide during RNA synthesis, genetic variation can also arise by recombination. Recombination can occur when two or more viral genomes enter the same cell and a part of one genome is incorporated into the other. This can result in significant changes in genome composition with dramatic impact on virus biology. For example, there is evidence that recombination might have been a factor that enabled the emergence of SARS coronavirus [39] and it is a key factor in emergence of pandemic influenza viruses [40]. However, while recombination can impact diversity, there is debate as to whether it has evolved as a means to generate variability, or is merely a consequence of viral genome replication [41]. In this respect, it is interesting that even viruses with similar genome structures can undergo different rates of recombination, perhaps suggesting that recombination is also finely tuned by evolutionary pressure. There are three mechanisms by which RNA viruses can recombine: template-switching recombination, nonreplicative recombination, and re-assortment (Fig. 2.3).

Template-switching, otherwise known as copy-choice recombination, can occur during the process of RNA synthesis if the viral polymerase transfers from one template to another, while remaining attached to the nascent nucleic acid chain [42]. This results in production of a mosaic genome. Template switching tends to occur between sequences of close similarity to give rise to a homologous recombination event. Nonhomologous recombination can also occur, but this typically results in defective genomes and is observed less frequently. Viruses differ significantly in the rate with which they can recombine by template-switching [41]. It can be highly frequent in retroviruses, particularly HIV, and also in coronaviruses. The high frequency of recombination in these viruses may be due to the replication strategies that they have. In the case of retroviruses, the reverse transcription process is highly complex and the reverse transcriptase must switch from one template to another during DNA synthesis [43]. Likewise, in coronaviruses, transcription of the genome, to allow gene expression, requires the RdRp to transfer from one site to another on the genomic template [44]. Thus, the fact that the polymerases of these viruses have evolved to transfer to a different template sequence probably means they are more likely to do so during other aspects of RNA synthesis. Recombination can occur in other positive-strand RNA viruses besides coronaviruses, and in double-stranded RNA viruses, although the recombination frequency apparently varies between viruses. For example, it occurs frequently in the positive-stranded enteroviruses, such as poliovirus, but less so in the flaviviruses, such as HCV [45]. Template switch recombination is much less frequent in the negative-strand RNA viruses [46], probably because their genomes are not naked RNA, but rather are encapsidated, or buried, in protein called nucleoprotein [47]. The polymerase

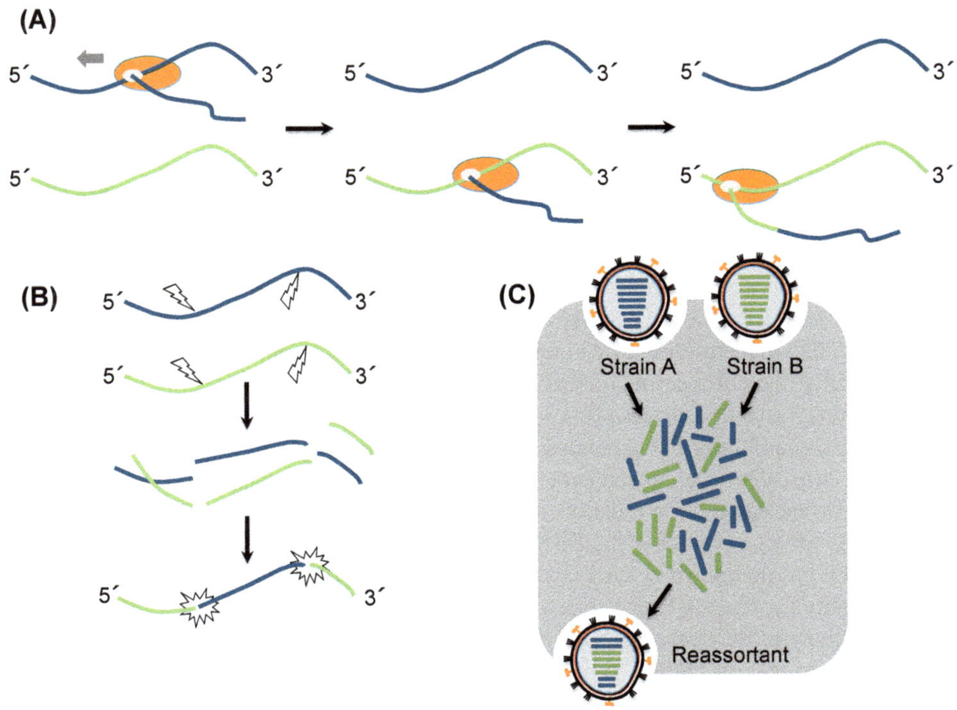

FIGURE 2.3 **Mechanisms of RNA virus recombination.** (A) The template-switching mechanism of recombination. The polymerase (orange) begins replicating the genome of one virus (blue), but then transfers to a genome of another related virus (green), resulting in a mosaic genome (blue and green). The *gray arrow* shows the direction of polymerase movement. (B) The nonreplicative mechanism of recombination. The genomes of two viruses (blue and green) are cleaved, either by nucleases, physical shearing, or cryptic ribozyme activity, resulting in partial genome fragments. The fragments are joined together to form a mosaic genome (blue and green). (C) A cell is infected with two viruses with different segmented genomes (blue and green). During the infection cycle the genome segments are amplified and a re-assortant virus containing a mixture of blue and green genome segments is released.

transiently displaces nucleoprotein as it moves along the template and only recognizes RNA sequences as the nucleoprotein is displaced. This probably prevents the RdRp from transferring from one genome to a similar sequence in another genome to yield functional recombinants. However, negative-strand RNA viruses containing gene duplications have arisen naturally [48] and defective interfering genomes, which contain promoters and partial genome sequences, but not complete genomes, are often detected. These findings suggest that the RdRps of the negative-strand RNA viruses are capable of jumping from one sequence to another, or that nonreplicative recombination (described later) can come into play, but perhaps most products of these events are nonviable and so are not detected.

In contrast to template switch recombination, nonreplicative recombination seems to be a relatively rare event that to date has only been described for a few positive-strand RNA viruses. This mechanism was documented by recovery of viable viruses (or replicative templates) following cotransfection of cells with two viral RNA fragments, each of which was unable to function in replication independently. The fragments recombined to form functional RNAs [49–52]. This mechanism of nonreplicative recombination might not involve a viral enzyme activity. Instead, it seems that the two RNA strands are joined together either by a transesterification reaction [49], or by ligation, presumably by cellular ligases [51–53]. Thus, RNA genome fragments created by physical shearing, nuclease cleavage, or cryptic ribozyme activity have the potential to be joined to form a novel viral genome, which can be further refined by homologous recombination to remove duplicated sequence [54].

Re-assortment is a process that can occur during coinfection of a cell with viruses with segmented genomes [55]. During re-assortment, a virus can exchange one of its own segments for that of another related virus. This process is well studied in influenza virus, in which it occurs frequently. Influenza A virus has eight genome segments that all need to be packaged into a virion for that virus particle to be infectious. This process is not completely random; there are packaging signals in the RNA segments and epistatic interactions enable the correct complement of segments to be incorporated into virions. There are many subtypes of influenza virus, but if the packaging signals of two viral subtypes coinfecting a cell are sufficiently similar, this enables a segment from one virus to be incorporated into another, resulting in release of virions with a new genome composition. Because different viral subtypes have different antigenic properties, this process has significant impact on influenza virus epidemiology [40].

8. THE EFFECT OF REPLICATION MODE ON MUTATION FREQUENCY

As described earlier, the mutation frequency of a virus differs from the mutation rate, in that it refers to the accumulation of mutations over a virus infection cycle, for example, from the point of entry of a virus into a cell until release of infectious progeny. In addition to having different genome structures and nucleic acid intermediates, different RNA viruses have different numbers of replication events per infection cycle and so this can impact on mutation frequency. Retrovirus reverse transcriptase only copies the genome twice during an infection cycle: once to generate cDNA from the RNA template and a second time to synthesize the complementary DNA strand to generate double-stranded DNA. Thus, these are the only two occasions in the retrovirus infection cycle where the viral polymerase can introduce mutations. The cellular RNA polymerase II enzyme is responsible for generating multiple copies of genome RNA that become packaged into viral particles, and while there is the potential for error to be introduced by RNA polymerase II, cellular proofreading mechanisms come into play at this step and so the major source of mutation during retrovirus genome replication is the reverse transcriptase [56]. In the riboviruses, the viral RdRp is responsible for all genome replication events and it copies the genome multiple times. Thus, in this case, there are many more opportunities for mutations to be introduced by polymerase error. Within the riboviruses, there are different modes of genome replication, referred to as a stamping machine or geometric modes, and the degree to which a virus employs one mode versus the other will affect mutation frequency [57]. In stamping machine mode, the infecting genome template is used to make multiple progeny genomes, but these genomes are not used as templates until they have been delivered into another cell. It is thought that double-stranded RNA viruses use this mode primarily. In contrast, in geometric replication, an incoming genome template acts as a template to make multiple complementary strands (or antigenomes), which in turn act as templates to make multiple genome sense strands, within the same infection cycle. In this case, there are many more opportunities for mismatch errors to be introduced than in the stamping machine mode. Positive and negative sense RNA viruses probably use a combination of both modes, but the exact contribution of each to the output virus is not well characterized, except in a few cases [57]. The mutation rate of the viral polymerase, coupled with the replication mode that the virus employs (and extrinsic factors, described in the following text) will determine the extent of genetic variability of viruses released from an infected cell.

9. THE EFFECT OF CELLULAR FACTORS ON VIRUS MUTATION RATE

The cellular environment can impact virus mutation rates and frequency. For example, dNTP pool imbalances can affect retrovirus mutation rates [58], and it has been suggested that differences in substitution rates between RNA viruses is a consequence of differences in virus RNA synthesis rates in different cell types [59]. In addition to these effects, there are also cellular factors that can result in increased mutation in RNA viruses. Adenosine-to-inosine modification by enzymes called adenosine deaminase acting on RNA (ADAR) is the most common form of RNA base modification that occurs in mammals. A-to-I conversion has important consequences in the coding potential of substrate RNAs, as inosine is decoded as a G by polymerases during template copying. The A-to-I conversion in a dsRNA duplex also has consequences to stability of RNA secondary structures, as the A:I pairing is less stable than a canonical A:U pair. This can have important consequences for RNAs that depend on their structure rather than sequence for their function [60]. ADAR modification of cellular double-stranded RNA was shown to prevent its recognition by the cytoplasmic sensor of nonself RNAs that would otherwise lead to chronic activation of innate immune pathways [61]. There is also evidence that ADAR can modify viral RNAs. Sequence analysis of RNA virus genomes has revealed that they preferentially accumulate A-to-G transitions, which are characteristic hallmarks of ADAR activity. Measles virus is a negative-stranded RNA virus, responsible for an acute disease predominantly in infants, but in rare instances associated with a fatal latent infection of the CNS known as subacute sclerosing panencephalitis (SSPE). Analysis of measles virus genomes from SSPE victims has revealed abundant A-to-G transitions, suggesting a role for ADAR in establishment of SSPE [62]. Consistent with an antiviral role for ADAR, measles virus infection of ADAR knock-out cell lines displayed increased cellular pathology, and similar findings were reported for other RNA viruses, implicating ADAR as a cellular restriction factor for a wide range of negative-stranded RNA viruses [63]. Direct evidence of ADAR modification of a viral RNA genome comes from studies of hepatitis delta virus (HDV). HDV is the smallest of the RNA viruses and encodes just two proteins, HDAg-L and HDAg-S, both of which are essential for virus viability. HDAg-L and HDAg-S share the same amino terminal open reading frame, but HDAg-L possesses a carboxyl terminal extension that is accessed when the stop codon at the end of the HDAg-S ORF is bypassed. Early during infection only the truncated HDAg-S is expressed, but then at later times expression of HDAg-L increases due to the site-specific modification of the stop codon by ADAR [64]. This editing event is highly specific and is promoted by the highly secondary structured HDV RNA genome. This action by ADAR is clearly proviral, in that without the activity of ADAR, no infectious HDV particles would form.

Another family of cellular factors that can modify the sequence of viral genomes is the APOBEC family of enzymes. These comprise an extensive arm of the innate immune system [65]. They are responsible for the modification by deamination of cytosine residues to uracil, which is an activity largely performed on single-stranded DNA substrates, leading to the phenomenon of hypermutation. APOBEC activity can affect the retroviruses. HIV infection is blocked by APOBEC, unless it expresses the viral infectivity factor (Vif). The mechanism for this blockade relies on the packaging of multiple APOBEC family members within HIV virions, which can act on the HIV genome once it has been copied by reverse transcriptase into a complementary DNA. The effect of APOBEC activity can be the modification of up to 10% of susceptible cytosine residues, resulting in a drop in infectivity of up to 100-fold.

10. MECHANISMS UNDERLYING GENETIC ROBUSTNESS IN RNA VIRUSES

Together, the studies described earlier show that there is a range of viral and host factors that combine to alter mutation frequency. The question that arises is: How do RNA viruses withstand mutation? The ability of a genome to withstand genetic or environmental perturbations without a change in phenotype is referred to as genetic robustness [66]. The high mutation rate that RNA viruses incur comes at a cost. It has been estimated that 30–40% of virus genomes generated during infection are defective [6] and so at an individual level, most viral genomes are not robust. This is not surprising: the small size of RNA virus genomes limits their coding potential and so they have limited genetic redundancy. Moreover, RNA virus are highly compact, often containing overlapping reading frames, and nucleotide sequences that function at the RNA level, for example, as *cis*-acting elements that enable genome replication, as well as at the protein level. However, robustness is influenced by the genetic background in which it operates and so in the case of RNA viruses, genetic robustness is considered in the context of the viral swarm, rather than individual genotypes. RNA viruses are not all equally robust, and even closely related viruses can exhibit different degrees of robustness [15]. There are several factors that contribute toward this [67,68] which are described in the following paragraphs.

Robustness is conferred if a virus has an ability to more readily arrive at a new optimal or adapted genotype in the face of a changed environment, and the genetic composition of the viral swarm can facilitate this. Because the majority of nucleotide changes in RNA virus genomes are either strongly deleterious or lethal, the population is perpetually refined as deleterious genomes become purged through selection, leaving only mutations with phenotypically neutral or advantageous consequences to persist [69,70]. The neutral mutations can impact robustness. An explanation for this is that if the virus encounters a new environment, multiple nucleotide changes might need to occur for it to arrive at an optimal genotype. If some of these changes are already in place, then the jump to the new genotype is more likely to occur. This means that a population that includes a high proportion of neutral mutations will be more adaptable in the face of environmental change, as genomes with neutral mutations can act as stepping-stones toward reaching the new adapted genotype [71] (Fig. 2.4).

Thus one viral determinant of robustness is their high mutation frequency, which results in a more extensive neutral network [66,72–74]. Consequently, factors that affect mutation frequency, such as polymerase fidelity and replication mode, will also impact robustness. Interestingly, there is evidence that some RNA virus genomes have evolved to enable rapid adaptation. Experiments in which synonymous mutations were introduced into RNA virus genomes and fitness was assessed showed that the RNA nucleotide sequence has an effect on fitness, independently of its effect on protein sequence [75]. This could be due to effects on RNA structures and *cis*-acting elements. However, experiments with poliovirus showed that this might not be the only explanation. In this case, a region of the poliovirus genome that does not contain *cis*-acting RNA structures was recoded with synonymous mutations. The virus variant containing the synonymous mutations had reduced robustness and was attenuated in an animal model [76]. This finding suggests that wild-type poliovirus occupies a sequence space that enables it to rapidly adapt to environmental pressure.

Another viral determinant of robustness relates to the ability of RNA viruses to generate large numbers of genomes within individual infected cells. A consequence of the resulting polyploidy is that a genome containing a detrimental change can be complemented by the properties of another genome that is unaltered. This mechanism also has a downside in that it reduces the ability to purge poorly adapted genotypes, and thus their persistence in a population may lead to a reduction in its overall fitness. Interestingly, the huge range in the extent of polyploidy that occurs throughout the infection cycle may allow different levels of robustness at different times of the virus life cycle, with more opportunity for complementation at later stages of infection when the copy number of viral genomes is at its highest. Such a scenario may have important consequences for viruses that stimulate innate immunity early on in the infection cycle. The innate immune response poses a high adaptive requirement at a time when viral genome numbers are at their lowest. Conversely, persistent viruses that maintain high copy numbers for extended periods of time without inducing cell death, such as HCV, may be particularly robust due to the wide range of genotypes contained with the massive population of persisting RNAs. The presence of multiple genomes within the same cell can also enable recombination. Recombination is another factor that influences robustness, as it can result in purging of multiple mutations from a genome in a single recombination or re-assortment event [73].

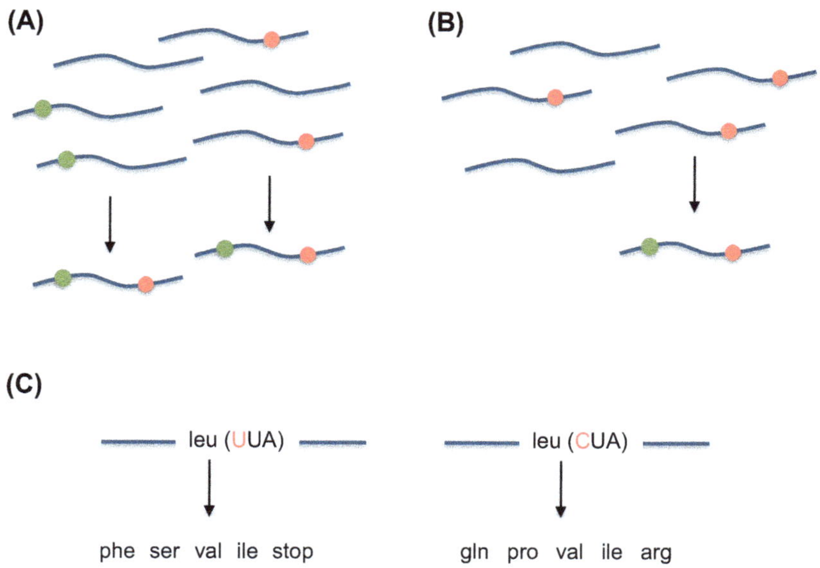

FIGURE 2.4 **Mechanisms by which neutral networks can enable new genotypes to arise.** (A) A neutral network containing a mixture of genomes containing green or red mutations provides a stepping-stone to a double mutant containing both mutations. (B) A neutral network can provide a stepping-stone to a new genotype by epistatic interactions. In this example, the green mutation alone is deleterious, but is neutral or beneficial in combination with the red mutation. If the neutral network contains genomes with red mutations, it provides a stepping-stone to enable introduction of the green mutation. (C) A neutral network containing genomes that have different codons for the same amino acid can provide a stepping-stone to genomes containing different spectra of amino acids following a single nucleotide substitution. In this example, a neutral network contains genomes coding for leucine at a given position, but the genomes differ by coding for leucine with either UUA or CUA. This expands the range of amino acids that could arise following a single nucleotide change.

RNA virus robustness can also be impacted by host cell factors. The ability of chaperones to buffer mutations was first proposed for the GroEL molecular chaperone [77]. Subsequently it has been experimentally observed that chaperones, such as members of the heat shock protein 70 and 90 families, play important roles in the infection cycles of many RNA viruses. It has been proposed that viruses have evolved the ability to interact with chaperones in order to buffer the effects of deleterious coding mutations that would otherwise prevent their correct folding [67,68]. This provision is particularly important as viruses depend on assembly of high-order multimers to build their capsids, a major component of the virions that are released. In these cases, a single misfolded protein has the potential to disrupt the function of the entire complex and so mechanisms that facilitate appropriate protein folding can have a significant impact.

Although there are a number of properties of RNA viruses that contribute to genetic robustness, the role of robustness in the natural history of RNA viruses is a controversial topic. A virus population with increased neutral genotypic diversity and thus high robustness can readily adapt to new environments due to its inherent diversity, and increased availability of adaptive pathways. This has important implications for viral pathogenesis and robustness has been shown to increase virulence in host organisms [76]. However, it appears that the converse can also be true and under certain conditions the neutral network can be composed of genotypes that are unable to reach a high level of fitness in the new environment [78]. This suggests that it may be difficult to make generalizations over how robustness shapes virus adaptability.

11. RNA VIRUSES ON THE EDGE

As mentioned at the beginning of this chapter, genetic heterogeneity of RNA viruses is such a key facet of their biology that it brings up the question of whether their high mutation rates have been selected for and are of evolutionary benefit. Fidelity comes at the price of elongation efficiency [16]. Thus, it is possible that the high mutation rates of RNA viruses are simply a consequence of polymerases that are under selective pressure to replicate genomes very rapidly to ensure efficient viral infection [79–81]. According to this view, RNA viruses have evolved a balance between rapid genome synthesis and error, such that the mutations that they incur are tolerable and on occasion advantageous, but are not necessary for virus survival. However, while genome synthesis rate is certainly an important factor in virus fitness [82]; for some viruses there is also evidence that the high mutation rate is beneficial and that RNA virus polymerase fidelity is tuned, enabling the virus to maximize sequence space while avoiding the accumulation of so many deleterious mutations that the genomes become nonviable. This is the concept that RNA viruses are “on the edge.”

As described earlier, most mutations that arise are deleterious and so there is a significant cost to having an error-prone polymerase. Furthermore, while complementation between defective genomes can occur, enabling genetic robustness, it is also possible for defective genomes to have an antagonistic effect, for example, by expressing mutant proteins that function as dominant negatives. Nonetheless, despite these disadvantages, it is possible to generate mutant viruses that have changes in the polymerase that result in its increased accuracy; these are known as high-fidelity mutants. Elegant studies performed with a poliovirus high-fidelity mutant showed that efficient spread within a host requires a quasispecies, and an error-prone RdRp to generate it [83,84]. Naturally, poliovirus replicates in the gut, but it can replicate in other tissues and spread to the spinal cord and brain. The ability to infect this variety of tissues requires poliovirus to overcome significant barriers to replication [85]. Experiments comparing the growth characteristics of wild type and a variant of poliovirus with a high-fidelity RdRp showed that the high-fidelity variant could replicate relatively efficiently compared to the wild-type virus in a single multiplication cycle in cell culture [83,84], and if introduced into mice intravenously, it could replicate efficiently in the spleen, kidney, and small intestine [84]. Thus, in this case, genome replication was not significantly delayed by the increased accuracy in RNA synthesis. However, in contrast to wild-type poliovirus, this high-fidelity mutant virus could not efficiently spread to the central nervous system (CNS), hence the 50% lethal dose (LD_{50}) was increased 300-fold [83,84]. To examine if the defect in virus spread was due specifically to the mutation (perhaps this variant of the RdRp could not function in a neuronal environment), or to the lack of genome diversity within its population, Vignuzzi and coworkers increased the diversity of the high-fidelity virus by treating it with mutagens. This had the dramatic effect of increasing the ability of the high-fidelity virus to replicate in the spinal cord and brain, and the LD_{50} was restored to the same level as wild-type poliovirus. This result showed that poliovirus spread to the CNS is dependent on the virus being able to establish a highly diverse population. In addition, it was shown that coinfection of mice with wild-type and high-fidelity mutant virus enabled the high-fidelity virus to reach the brain [84]. This indicates that different viral genotypes in the quasispecies can complement each other to facilitate infection spread. It is not known exactly how complementation functions in this case, but it is easy to imagine that one variant of a virus might be more efficient at subverting innate immune defenses (which could impact virus genomes within the same cell and neighboring cells), whereas another variant might express a capsid protein better adapted to bind to a new cell receptor. In its natural context, poliovirus is spread through ingestion of contaminated water and so there is no necessity for poliovirus to be able to spread to the CNS to be able to complete its transmission cycle. However, these studies are important because they show that viruses can benefit from polymerase infidelity and a high mutation rate, particularly under conditions where they encounter a change in environment [86,87]. Studies with a number of viruses indicate that these findings are widely applicable in RNA virology and so it seems likely that RNA viruses have evolved a high mutation rate that enables them to rapidly adapt to the dynamic and varied environments in which they exist.

The studies described earlier show that RNA viruses benefit by having an error-prone polymerase to enable them to adapt to new conditions. However, there is also a cost if the polymerase has mutations that decrease its fidelity, so that the error rate is increased. Experiments performed with coxsackievirus B3 and poliovirus showed that low-fidelity mutants were able to replicate efficiently in cell culture when propagated at high multiplicity of infection (ie, when the population size was large), but were extinguished when the viruses were propagated under low multiplicity conditions, which mimics conditions when a virus first establishes infection in a host or when it has overcome a barrier, such as adaptation to a new host cell type. Consistent with these findings, both the coxsackievirus B3 and poliovirus low-fidelity mutants were attenuated in vivo. In the case of the coxsackievirus B3, low-fidelity mutants were unable to establish productive infection in the heart, the usual site for coxsackievirus B3 virus replication, and in the case of poliovirus they were unable to reach the CNS [82,88]. Comparison of the high- and low-fidelity poliovirus variants indicates how much latitude there is in the mutation rate for this virus. The high-fidelity RdRp had an approximately twofold decrease in nucleotide misincorporation rate, and the low-fidelity RdRp had a twofold increase [82]. Thus, the range in misincorporation rate that can lead to virus extinction in an animal host is not that substantial, even in a virus that is relatively genetically robust. This indicates that the fidelity of the polymerase, coupled with the impact that accuracy has on the rate of RNA synthesis, is optimized to enable viruses to adapt to the many environments in which they need to exist while avoiding extinction [82].

12. VIRUS GENETIC VARIABILITY AND THE VIRUS–HOST "ARMS RACE"

The propensity that RNA viruses have for mutation seems to have opened this up as an avenue for host cell defense. Pathogenic viruses and their hosts are engaged an epochal "arms race" in which the host evolves immune defenses to suppress virus infection and the virus in turn evolves countermeasures to disable host defenses. The existence of APOBEC and ADAR, cellular proteins that can increase virus mutation frequency, suggests that mammalian hosts have taken advantage of the high mutation rate of viruses and evolved mechanisms to induce further mutations in the viral genomes and push viruses toward extinction [89]. Conversely, primate lentiviruses have evolved vif, a protein that can target APOBEC for proteosomal degradation, indicating that these retroviruses have evolved a mechanism to counter this cellular defense [90,91]. Likewise, the nonsegmented, negative-strand RNA viruses, which are susceptible to ADAR, maintain their genomes encased in a ribonucleoprotein complex throughout the infection cycle, reducing the opportunity for them to adopt double-stranded RNA structures, the substrate for ADAR. This perhaps prevents ADAR causing as much damage as it otherwise might.

13. TAKING ADVANTAGE OF THE MUTABILITY OF RNA VIRUSES

The high mutation rate of RNA viruses has often been an impediment to drug and vaccine development as viruses can rapidly gain resistance to antiviral drugs and to the immune response elicited by vaccines. However, our increasing understanding of function and consequences of genetic variability has opened new avenues for controlling viral infection. As described earlier, small decreases in polymerase fidelity can have dramatic effects on viral infectivity. Similarly, studies have shown that small increases in viral mutation rate caused by treatment with mutagenic compounds can result in significant decreases in viral fecundity [92,93]. Thus, treatment with mutagens that increase the accumulation of mutations in the viral genome can lead directly to virus extinction, or can reduce virus infection to enable effective clearance with other inhibitors, given in combination, or by host immune responses [94,95]. The identification of high-fidelity mutant viruses that can infect animals has also suggested a means to exploit these mutants as vaccine candidates. Live-attenuated virus vaccines can be highly effective, but have the disadvantage that they can potentially revert to a wild-type pathogenic phenotype. By engineering recombinant viruses with increased fidelity, it is possible to generate viruses that are attenuated, as described earlier, and that elicit protective immune responses, with reduced risk of reversion [96].

14. CONCLUSION

The RNA viruses are hugely diverse, not only in their genome structures and replication strategies, but also in their "lifestyles," which can differ significantly, even between closely related viruses. What has emerged from studies of virus genetics is that RNA viruses are also highly divergent in terms of their polymerase fidelity, recombination rates, replication modes, and genetic robustness. We speculate that RNA viruses have evolved such that there is an intricate balance between these factors that is tuned to match the "lifestyle" of each virus, enabling it to occupy the niche in which it exists. There is some evidence to support this idea. For example, a side-by-side comparison of influenza virus and HIV polymerase fidelity showed that influenza virus RdRp is much less error prone than HIV reverse transcriptase. This may be a reflection of the fact that the influenza virus RdRp performs many more genome replication events during an infection cycle than the HIV reverse transcriptase and needs to be less error prone to avoid having a mutation frequency that is too high [97]. Another example comes from studies of West Nile virus. While the fidelity of the West Nile virus RdRp has not been directly compared to that of other viruses, there is a greater difference in fidelity between the wild-type West Nile virus RdRp and a high-fidelity mutant than has been found for most RdRps [26]. This could suggest that West Nile virus RdRp is naturally more error prone than most. This could be a necessary feature of West Nile virus to enable it to cycle back and forth between mosquito and avian hosts. Understandably, much of the work that has been performed so far has focused on viruses that are "model" viruses—those that are relatively easy to culture in vitro and replicate rapidly. However, a fuller understanding of how the factors that influence genetic diversity intertwine with virus biology will come from extending the work that has been performed so far and applying it to other viruses that have similar genome structures and replication strategies, but diverse lifestyles, such as West Nile virus and HCV, or vesicular stomatitis virus and measles virus. Research in this area will potentially be transformed by new sequencing techniques, such as CirSeq, which can detect low-level genetic variants above the background of errors introduced during RNA sequencing [25], and BAsE-Seq, a method for obtaining long stretches of sequence that can be used to identify haplotypes [98]. Ultimately, application of cutting-edge sequencing technologies, mathematical analyses, and virology studies to a range of viruses will enhance our understanding of the genesis and functional consequences of RNA virus genetic instability.

GLOSSARY

Antigenome The RNA replicative intermediate formed during viral genome replication.

Capsid A protein coat that surrounds the viral genome in the packaged virus particle. The capsid is important for protecting the viral nucleic acid and for delivering it to cells and/or to the appropriate location within cells.

***cis*-Acting RNA signal** An element in the RNA genome that has a functional role in the replication (or transcription or translation) of the genome.

Complementation The ability of the products of one viral genome to provide a function that cannot be performed by the products of another viral genome.

Copy-choice recombination A recombination event that occurs when the viral polymerase switches to another template while remaining attached to the nascent RNA. Also known as template switch recombination.

Epistatic mutation A phenomenon in which mutations have different effects in combination than individually.

Fidelity The accuracy with which the polymerase copies the template. A high-fidelity polymerase will make fewer errors than a low-fidelity polymerase.

Genetic robustness The degree to which a genome can withstand environmental or genetic perturbation.

Geometric mode A mode of genome replication in which the newly synthesized genomes become templates for further rounds of genome replication during the infection cycle.

Infection cycle The cycle of events by which an infectious virus particle infecting a cell results in release of virus progeny. In the virology field, this is often referred to as the virus replication cycle, but infection cycle was used here to avoid confusion with genome replication.

Lethal dose 50 (LD_{50}) The quantity of infectious virus that is required to cause death in 50% of inoculated hosts.

Live-attenuated virus vaccine A vaccine that consists of a live (infection-competent) virus that contains mutations that reduce the disease symptoms, usually by impairing its ability to efficiently complete its infection cycle.

Mutation rate The rate at which a viral genome acquires mutations per genome replication event.

Mutation frequency The frequency at which a viral genome acquires mutations per viral infectious cycle. This frequency could be affected by cellular factors and the mode of viral replication, as well as by polymerase fidelity.

Nonreplicative recombination A recombination event in which two RNA fragments are joined together by either a *trans*-esterification reaction, or ligation by cellular ligases.

Persistent virus A virus that can infect a host and maintain the infection for extended periods of time. HIV and HCV are examples of persistent viruses.

Polyploidy The presence of multiple viral genomes within the same cell.

Quasispecies A collection of closely related viral genomes, genetically linked through mutation, that compete within a highly mutagenic environment, interact cooperatively, and collectively contribute to the population phenotype.

Re-assortment A recombination event that can occur with viruses with segmented genomes, in which a genome segment from one virus is packaged into a virus particle in place of a genome segment from another virus, thus producing a virus with a novel complement of genome segments.

Retrovirus A class of RNA viruses that replicate their genomes via a double-stranded DNA intermediate.

Reverse transcriptase A viral enzyme encoded by retroviruses that is responsible for generating a double-stranded DNA copy of the viral RNA genome.

Ribovirus RNA viruses that replicate their genomes via an RNA intermediate.

RNA-dependent RNA polymerase A viral enzyme encoded by riboviruses that is responsible for generating the viral genome RNA and the RNA replication intermediates.

RNA virus A virus that carries a genome composed of RNA in the virus particle.

Stamping machine mode A mode of genome replication in which the incoming genome is reiteratively used as a template to produce multiple copies of replication product.

Swarm A population of closely related viruses, connected through mutation, similarly to a quasispecies. We have used the term swarm in many instances here because a population of virus variants might not always fully fulfill the definition of quasispecies.

Synonymous mutation A nucleotide substitution that does not result in an amino acid change.

Template switch recombination A recombination event that occurs when the viral polymerase switches to another template while remaining attached to the nascent RNA, also known as copy-choice recombination.

Transmission cycle The cycle of events by which a virus is transmitted from one host to another host in the same species.

LIST OF ABBREVIATIONS

ADAR RNA-specific adenosine deaminase
APOBEC Apolipoprotein B mRNA editing enzyme, catalytic polypeptide-like
CNS Central nervous system
dUTPase Deoxyuridine-triphosphatase
HCV Hepatitis C virus
HDV Hepatitis delta virus
HIV Human immunodeficiency virus
kb Kilobase
LD_{50} 50% lethal dose
RdRp RNA-dependent RNA polymerase
SARS coronavirus Severe acute respiratory syndrome coronavirus
SSPE Subacute sclerosing panencephalitis
Vif Viral infectivity factor

REFERENCES

[1] de Vries RD, Mesman AW, Geijtenbeek TB, Duprex WP, de Swart RL. The pathogenesis of measles. Curr Opin Virol 2012;2(3):248–55.
[2] Noyce RS, Richardson CD. Nectin 4 is the epithelial cell receptor for measles virus. Trends Microbiol 2012;20(9):429–39.
[3] Chancey C, Grinev A, Volkova E, Rios M. The global ecology and epidemiology of West Nile virus. Biomed Res Int 2015;2015:376230.
[4] Domingo E, Sabo D, Taniguchi T, Weissmann C. Nucleotide sequence heterogeneity of an RNA phage population. Cell 1978;13(4):735–44.
[5] Drake JW, Charlesworth B, Charlesworth D, Crow JF. Rates of spontaneous mutation. Genetics 1998;148(4):1667–86.
[6] Sanjuan R, Nebot MR, Chirico N, Mansky LM, Belshaw R. Viral mutation rates. J Virol 2010;84(19):9733–48.
[7] Andino R, Domingo E. Viral quasispecies. Virology 2015;479–480:46–51.
[8] Holland JJ, De La Torre JC, Steinhauer DA. RNA virus populations as quasispecies. Curr Top Microbiol Immunol 1992;176:1–20.
[9] Baltimore D. Expression of animal virus genomes. Bacteriol Rev 1971;35(3):235–41.
[10] Preston BD, Poiesz BJ, Loeb LA. Fidelity of HIV-1 reverse transcriptase. Science 1988;242(4882):1168–71.
[11] Roberts JD, Bebenek K, Kunkel TA. The accuracy of reverse transcriptase from HIV-1. Science 1988;242(4882):1171–3.
[12] Castro C, Arnold JJ, Cameron CE. Incorporation fidelity of the viral RNA-dependent RNA polymerase: a kinetic, thermodynamic and structural perspective. Virus Res 2005;107(2):141–9.
[13] Steinhauer DA, Domingo E, Holland JJ. Lack of evidence for proofreading mechanisms associated with an RNA virus polymerase. Gene 1992;122(2):281–8.
[14] Bradwell K, Combe M, Domingo-Calap P, Sanjuan R. Correlation between mutation rate and genome size in riboviruses: mutation rate of bacteriophage Qβ. Genetics 2013;195(1):243–51.
[15] Graci JD, Gnadig NF, Galarraga JE, Castro C, Vignuzzi M, Cameron CE. Mutational robustness of an RNA virus influences sensitivity to lethal mutagenesis. J Virol 2012;86(5):2869–73.
[16] Campagnola G, McDonald S, Beaucourt S, Vignuzzi M, Peersen OB. Structure-function relationships underlying the replication fidelity of viral RNA-dependent RNA polymerases. J Virol 2015;89(1):275–86.
[17] Arias A, Arnold JJ, Sierra M, Smidansky ED, Domingo E, Cameron CE. Determinants of RNA-dependent RNA polymerase (in)fidelity revealed by kinetic analysis of the polymerase encoded by a foot-and-mouth disease virus mutant with reduced sensitivity to ribavirin. J Virol 2008;82(24):12346–55.
[18] Arnold JJ, Cameron CE. Poliovirus RNA-dependent RNA polymerase ($3D^{pol}$): pre-steady-state kinetic analysis of ribonucleotide incorporation in the presence of Mg^{2+}. Biochemistry 2004;43(18):5126–37.
[19] Arnold JJ, Vignuzzi M, Stone JK, Andino R, Cameron CE. Remote site control of an active site fidelity checkpoint in a viral RNA-dependent RNA polymerase. J Biol Chem 2005;280(27):25706–16.
[20] Garforth SJ, Domaoal RA, Lwatula C, Landau MJ, Meyer AJ, Anderson KS, et al. K65R and K65A substitutions in HIV-1 reverse transcriptase enhance polymerase fidelity by decreasing both dNTP misinsertion and mispaired primer extension efficiencies. J Mol Biol 2010;401(1):33–44.
[21] Gohara DW, Arnold JJ, Cameron CE. Poliovirus RNA-dependent RNA polymerase ($3D^{pol}$): kinetic, thermodynamic, and structural analysis of ribonucleotide selection. Biochemistry 2004;43(18):5149–58.
[22] Moustafa IM, Korboukh VK, Arnold JJ, Smidansky ED, Marcotte LL, Gohara DW, et al. Structural dynamics as a contributor to error-prone replication by an RNA-dependent RNA polymerase. J Biol Chem 2014;289(52):36229–48.
[23] Skasko M, Weiss KK, Reynolds HM, Jamburuthugoda V, Lee K, Kim B. Mechanistic differences in RNA-dependent DNA polymerization and fidelity between murine leukemia virus and HIV-1 reverse transcriptases. J Biol Chem 2005;280(13):12190–200.
[24] Weiss KK, Chen R, Skasko M, Reynolds HM, Lee K, Bambara RA, et al. A role for dNTP binding of human immunodeficiency virus type 1 reverse transcriptase in viral mutagenesis. Biochemistry 2004;43(15):4490–500.
[25] Acevedo A, Brodsky L, Andino R. Mutational and fitness landscapes of an RNA virus revealed through population sequencing. Nature 2014;505(7485):686–90.
[26] Van Slyke GA, Arnold JJ, Lugo AJ, Griesemer SB, Moustafa IM, Kramer LD, et al. Sequence-specific fidelity alterations associated with West Nile virus attenuation in mosquitoes. PLoS Pathog 2015;11(6):e1005009.
[27] Denison MR, Graham RL, Donaldson EF, Eckerle LD, Baric RS. Coronaviruses: an RNA proofreading machine regulates replication fidelity and diversity. RNA Biol 2011;8(2):270–9.
[28] Eckerle LD, Lu X, Sperry SM, Choi L, Denison MR. High fidelity of murine hepatitis virus replication is decreased in nsp14 exoribonuclease mutants. J Virol 2007;81(22):12135–44.
[29] Sevajol M, Subissi L, Decroly E, Canard B, Imbert I. Insights into RNA synthesis, capping, and proofreading mechanisms of SARS-coronavirus. Virus Res 2014;194:90–9.
[30] Bouvet M, Imbert I, Subissi L, Gluais L, Canard B, Decroly E. RNA 3'-end mismatch excision by the severe acute respiratory syndrome coronavirus nonstructural protein nsp10/nsp14 exoribonuclease complex. Proc Natl Acad Sci USA 2012;109(24):9372–7.
[31] Smith EC, Case JB, Blanc H, Isakov O, Shomron N, Vignuzzi M, et al. Mutations in coronavirus nonstructural protein 10 decrease virus replication fidelity. J Virol 2015;89(12):6418–26.
[32] Schickli JH, Kaur J, Macphail M, Guzzetta JM, Spaete RR, Tang RS. Deletion of human metapneumovirus M2-2 increases mutation frequency and attenuates growth in hamsters. Virol J 2008;5:69.
[33] Elder JH, Lerner DL, Hasselkus-Light CS, Fontenot DJ, Hunter E, Luciw PA, et al. Distinct subsets of retroviruses encode dUTPase. J Virol 1992;66(3):1791–4.
[34] Threadgill DS, Steagall WK, Flaherty MT, Fuller FJ, Perry ST, Rushlow KE, et al. Characterization of equine infectious anemia virus dUTPase: growth properties of a dUTPase-deficient mutant. J Virol 1993;67(5):2592–600.

[35] Lerner DL, Wagaman PC, Phillips TR, Prospero-Garcia O, Henriksen SJ, Fox HS, et al. Increased mutation frequency of feline immunodeficiency virus lacking functional deoxyuridine-triphosphatase. Proc Natl Acad Sci USA 1995;92(16):7480–4.
[36] Turelli P, Guiguen F, Mornex JF, Vigne R, Querat G. dUTPase-minus caprine arthritis-encephalitis virus is attenuated for pathogenesis and accumulates G-to-A substitutions. J Virol 1997;71(6):4522–30.
[37] Chen R, Wang H, Mansky LM. Roles of uracil-DNA glycosylase and dUTPase in virus replication. J Gen Virol 2002;83(Pt 10):2339–45.
[38] Selig L, Benichou S, Rogel ME, Wu LI, Vodicka MA, Sire J, et al. Uracil DNA glycosylase specifically interacts with Vpr of both human immunodeficiency virus type 1 and simian immunodeficiency virus of sooty mangabeys, but binding does not correlate with cell cycle arrest. J Virol 1997;71(6):4842–6.
[39] Graham RL, Baric RS. Recombination, reservoirs, and the modular spike: mechanisms of coronavirus cross-species transmission. J Virol 2010;84(7):3134–46.
[40] Yoon SW, Webby RJ, Webster RG. Evolution and ecology of influenza A viruses. Curr Top Microbiol Immunol 2014;385:359–75.
[41] Simon-Loriere E, Holmes EC. Why do RNA viruses recombine? Nat Rev Microbiol 2011;9(8):617–26.
[42] Kirkegaard K, Baltimore D. The mechanism of RNA recombination in poliovirus. Cell 1986;47(3):433–43.
[43] Hu WS, Hughes SH. HIV-1 reverse transcription. Cold Spring Harb Perspect Med 2012;2(10).
[44] Sawicki SG, Sawicki DL, Siddell SG. A contemporary view of coronavirus transcription. J Virol 2007;81(1):20–9.
[45] Gonzalez-Candelas F, Lopez-Labrador FX, Bracho MA. Recombination in hepatitis C virus. Viruses 2011;3(10):2006–24.
[46] Spann KM, Collins PL, Teng MN. Genetic recombination during coinfection of two mutants of human respiratory syncytial virus. J Virol 2003;77(20):11201–11.
[47] Ruigrok RW, Crepin T, Kolakofsky D. Nucleoproteins and nucleocapsids of negative-strand RNA viruses. Curr Opin Microbiol 2011;14(4):504–10.
[48] Trento A, Galiano M, Videla C, Carballal G, Garcia-Barreno B, Melero JA, et al. Major changes in the G protein of human respiratory syncytial virus isolates introduced by a duplication of 60 nucleotides. J Gen Virol 2003;84(Pt 11):3115–20.
[49] Chetverin AB, Chetverina HV, Demidenko AA, Ugarov VI. Nonhomologous RNA recombination in a cell-free system: evidence for a transesterification mechanism guided by secondary structure. Cell 1997;88(4):503–13.
[50] Gallei A, Orlich M, Thiel HJ, Becher P. Noncytopathogenic pestivirus strains generated by nonhomologous RNA recombination: alterations in the NS4A/NS4B coding region. J Virol 2005;79(22):14261–70.
[51] Gmyl AP, Belousov EV, Maslova SV, Khitrina EV, Chetverin AB, Agol VI. Nonreplicative RNA recombination in poliovirus. J Virol 1999;73(11):8958–65.
[52] Gmyl AP, Korshenko SA, Belousov EV, Khitrina EV, Agol VI. Nonreplicative homologous RNA recombination: promiscuous joining of RNA pieces? RNA 2003;9(10):1221–31.
[53] Austermann-Busch S, Becher P. RNA structural elements determine frequency and sites of nonhomologous recombination in an animal plus-strand RNA virus. J Virol 2012;86(13):7393–402.
[54] Holmblat B, Jegouic S, Muslin C, Blondel B, Joffret ML, Delpeyroux F. Nonhomologous recombination between defective poliovirus and coxsackievirus genomes suggests a new model of genetic plasticity for picornaviruses. MBio 2014;5(4):e01114–9.
[55] Vijaykrishna D, Mukerji R, Smith GJ. RNA virus reassortment: an evolutionary mechanism for host jumps and immune evasion. PLoS Pathog 2015;11(7):e1004902.
[56] O'Neil PK, Sun G, Yu H, Ron Y, Dougherty JP, Preston BD. Mutational analysis of HIV-1 long terminal repeats to explore the relative contribution of reverse transcriptase and RNA polymerase II to viral mutagenesis. J Biol Chem 2002;277(41):38053–61.
[57] Safari M, Roossinck MJ. How does the genome structure and lifestyle of a virus affect its population variation? Curr Opin Virol 2014;9:39–44.
[58] Julias JG, Pathak VK. Deoxyribonucleoside triphosphate pool imbalances in vivo are associated with an increased retroviral mutation rate. J Virol 1998;72(10):7941–9.
[59] Hicks AL, Duffy S. Cell tropism predicts long-term nucleotide substitution rates of mammalian RNA viruses. PLoS Pathog 2014;10(1):e1003838.
[60] Samuel CE. ADARs: viruses and innate immunity. Curr Top Microbiol Immunol 2012;353:163–95.
[61] Liddicoat BJ, Piskol R, Chalk AM, Ramaswami G, Higuchi M, Hartner JC, et al. RNA editing by ADAR1 prevents MDA5 sensing of endogenous dsRNA as nonself. Science 2015;349(6252):1115–20.
[62] Bass BL, Weintraub H, Cattaneo R, Billeter MA. Biased hypermutation of viral RNA genomes could be due to unwinding/modification of double-stranded RNA. Cell 1989;56(3):331.
[63] Ward SV, George CX, Welch MJ, Liou LY, Hahm B, Lewicki H, et al. RNA editing enzyme adenosine deaminase is a restriction factor for controlling measles virus replication that also is required for embryogenesis. Proc Natl Acad Sci USA 2011;108(1):331–6.
[64] Casey JL. Control of ADAR1 editing of hepatitis delta virus RNAs. Curr Top Microbiol Immunol 2012;353:123–43.
[65] Harris RS, Dudley JP. APOBECs and virus restriction. Virology 2015;479–480:131–45.
[66] de Visser JA, Hermisson J, Wagner GP, Ancel Meyers L, Bagheri-Chaichian H, Blanchard JL, et al. Perspective: evolution and detection of genetic robustness. Evolution 2003;57(9):1959–72.
[67] Elena SF. RNA virus genetic robustness: possible causes and some consequences. Curr Opin Virol 2012;2(5):525–30.
[68] Lauring AS, Frydman J, Andino R. The role of mutational robustness in RNA virus evolution. Nat Rev Microbiol 2013;11(5):327–36.
[69] Martinez F, Lafforgue G, Morelli MJ, Gonzalez-Candelas F, Chua NH, Daros JA, et al. Ultradeep sequencing analysis of population dynamics of virus escape mutants in RNAi-mediated resistant plants. Mol Biol Evol 2012;29(11):3297–307.
[70] Wright CF, Morelli MJ, Thebaud G, Knowles NJ, Herzyk P, Paton DJ, et al. Beyond the consensus: dissecting within-host viral population diversity of foot-and-mouth disease virus by using next-generation genome sequencing. J Virol 2011;85(5):2266–75.
[71] Draghi JA, Parsons TL, Wagner GP, Plotkin JB. Mutational robustness can facilitate adaptation. Nature 2010;463(7279):353–5.

[72] Codoner FM, Daros JA, Sole RV, Elena SF. The fittest versus the flattest: experimental confirmation of the quasispecies effect with subviral pathogens. PLoS Pathog 2006;2(12):e136.
[73] Montville R, Froissart R, Remold SK, Tenaillon O, Turner PE. Evolution of mutational robustness in an RNA virus. PLoS Biol 2005;3(11):e381.
[74] Sanjuan R, Cuevas JM, Furio V, Holmes EC, Moya A. Selection for robustness in mutagenized RNA viruses. PLoS Genet 2007;3(6):e93.
[75] Cuevas JM, Domingo-Calap P, Sanjuan R. The fitness effects of synonymous mutations in DNA and RNA viruses. Mol Biol Evol 2012;29(1):17–20.
[76] Lauring AS, Acevedo A, Cooper SB, Andino R. Codon usage determines the mutational robustness, evolutionary capacity, and virulence of an RNA virus. Cell Host Microbe 2012;12(5):623–32.
[77] Fares MA, Ruiz-Gonzalez MX, Moya A, Elena SF, Barrio E. Endosymbiotic bacteria: groEL buffers against deleterious mutations. Nature 2002;417(6887):398.
[78] Stern A, Bianco S, Yeh MT, Wright C, Butcher K, Tang C, et al. Costs and benefits of mutational robustness in RNA viruses. Cell Rep 2014;8(4):1026–36.
[79] Furio V, Moya A, Sanjuan R. The cost of replication fidelity in an RNA virus. Proc Natl Acad Sci USA 2005;102(29):10233–7.
[80] Furio V, Moya A, Sanjuan R. The cost of replication fidelity in human immunodeficiency virus type 1. Proc Biol Sci 2007;274(1607):225–30.
[81] Regoes RR, Hamblin S, Tanaka MM. Viral mutation rates: modelling the roles of within-host viral dynamics and the trade-off between replication fidelity and speed. Proc Biol Sci 2013;280(1750):20122047.
[82] Korboukh VK, Lee CA, Acevedo A, Vignuzzi M, Xiao Y, Arnold JJ, et al. RNA virus population diversity, an optimum for maximal fitness and virulence. J Biol Chem 2014;289(43):29531–44.
[83] Pfeiffer JK, Kirkegaard K. Increased fidelity reduces poliovirus fitness and virulence under selective pressure in mice. PLoS Pathog 2005;1(2):e11.
[84] Vignuzzi M, Stone JK, Arnold JJ, Cameron CE, Andino R. Quasispecies diversity determines pathogenesis through cooperative interactions in a viral population. Nature 2006;439(7074):344–8.
[85] Kuss SK, Etheredge CA, Pfeiffer JK. Multiple host barriers restrict poliovirus trafficking in mice. PLoS Pathog 2008;4(6):e1000082.
[86] Borderia AV, Stapleford KA, Vignuzzi M. RNA virus population diversity: implications for inter-species transmission. Curr Opin Virol 2011;1(6):643–8.
[87] Lancaster KZ, Pfeiffer JK. Viral population dynamics and virulence thresholds. Curr Opin Microbiol 2012;15(4):525–30.
[88] Gnadig NF, Beaucourt S, Campagnola G, Borderia AV, Sanz-Ramos M, Gong P, et al. Coxsackievirus B3 mutator strains are attenuated in vivo. Proc Natl Acad Sci USA 2012;109(34):E2294–303.
[89] Dapp MJ, Patterson SE, Mansky LM. Back to the future: revisiting HIV-1 lethal mutagenesis. Trends Microbiol 2013;21(2):56–62.
[90] Bonnac LF, Mansky LM, Patterson SE. Structure-activity relationships and design of viral mutagens and application to lethal mutagenesis. J Med Chem 2013;56(23):9403–14.
[91] Strebel K. HIV accessory proteins versus host restriction factors. Curr Opin Virol 2013;3(6):692–9.
[92] Crotty S, Cameron CE, Andino R. RNA virus error catastrophe: direct molecular test by using ribavirin. Proc Natl Acad Sci USA 2001;98(12):6895–900.
[93] Loeb LA, Essigmann JM, Kazazi F, Zhang J, Rose KD, Mullins JI. Lethal mutagenesis of HIV with mutagenic nucleoside analogs. Proc Natl Acad Sci USA 1999;96(4):1492–7.
[94] Anderson JP, Daifuku R, Loeb LA. Viral error catastrophe by mutagenic nucleosides. Annu Rev Microbiol 2004;58:183–205.
[95] Graci JD, Cameron CE. Therapeutically targeting RNA viruses via lethal mutagenesis. Future Virol 2008;3(6):553–66.
[96] Vignuzzi M, Wendt E, Andino R. Engineering attenuated virus vaccines by controlling replication fidelity. Nat Med 2008;14(2):154–61.
[97] Aggarwal S, Bradel-Tretheway B, Takimoto T, Dewhurst S, Kim B. Biochemical characterization of enzyme fidelity of influenza A virus RNA polymerase complex. PLoS One 2010;5(4):e10372.
[98] Hong LZ, Hong S, Wong HT, Aw PP, Cheng Y, Wilm A, et al. BAsE-Seq: a method for obtaining long viral haplotypes from short sequence reads. Genome Biol 2014;15(11):517.

Chapter 3

Genome Instability in DNA Viruses

R. Sanjuán, M. Pereira-Gómez, J. Risso
Universitat de València, Paterna, Spain

Chapter Outline

1. OVERVIEW

DNA viruses comprise important pathogens such as herpesviruses, smallpox viruses, adenoviruses, and papillomaviruses, among many others. DNA viruses are divided into three major categories: double-stranded DNA viruses (eg, poxviruses), single-stranded DNA viruses (eg, parvoviruses), and pararetroviruses (eg, hepadnaviruses) which replicate their genome through an RNA intermediate. Large DNA viruses (>10 kb) have double-stranded DNA, whereas small DNA viruses have circular single- or double-stranded DNA. These broad viral groups differ in their rates of spontaneous mutation, defined as the probability that an unrepaired genetic change is passed on to the viral progeny in each cell infection cycle [1,2]. For instance, single-stranded DNA microviruses such as bacteriophage ϕX174 and innoviruses produce about 10^{-6} spontaneous mutations per nucleotide per cell infection cycle (m/n/c), a mutation rate which is close to those of some RNA viruses. In contrast, the double-stranded DNA herpes simplex virus (HSV) and bacteriophage T4, both of which have genome sizes exceeding 150 kb, show clearly lower mutation rates (10^{-8}–10^{-7} m/n/c). As a result, there is an inverse correlation between genome size and per-base mutation rate in DNA viruses, while the per-genome mutation rate stays approximately constant. This correlation extends to unicellular organisms and is known as Drake's rule [3,4] (Fig. 3.1). RNA viruses also exhibit an inverse relationship between genome size and mutation rate, albeit with a different slope [5]. The main feature that distinguishes RNA viruses from DNA viruses in terms of genome stability is probably the absence of 3′-exonuclease proofreading activity from most RNA virus-encoded polymerases, which makes them particularly error prone [6]. The 3′-exonuclease activity leads to roughly 10-fold to 100-fold increase in replication fidelity [7,8]. In turn, differences in mutation rate among DNA viruses should be determined by their ability to access postreplicative repair. For instance, bacteriophage ϕX174 lacks sequence motifs required for methyl-directed mismatch repair (MMR) in *Escherichia coli*, therefore excluding the phage DNA from this major repair system [9]. In contrast, the interplay between viral replication and host postreplicative repair pathways is far more complex and less well understood in eukaryotic DNA viruses. Molecular evolution studies indicate that the classical dichotomy between fast-evolving RNA and slow-evolving DNA viruses becomes blurred when full-genome datasets are considered [10], suggesting that DNA viruses probably have other mechanisms for promoting genetic diversity. Some of these mechanisms have already been characterized and include gene amplification [11] and diversity-generating retro-elements (DGRs) [12], both of which act on specific genome regions. Additionally, retroviruses

Genome Stability. http://dx.doi.org/10.1016/B978-0-12-803309-8.00003-3

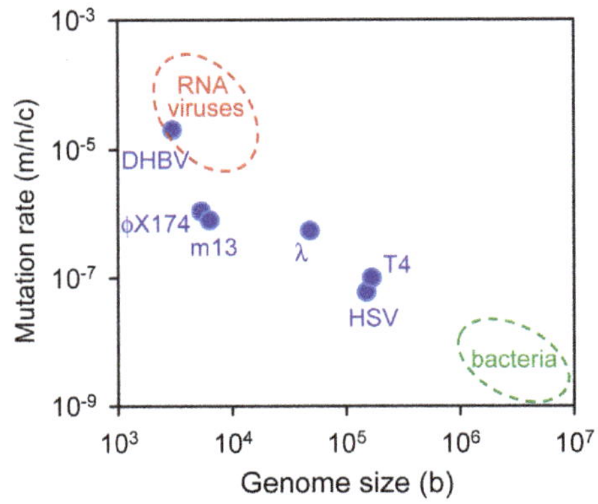

FIGURE 3.1 The relationship between genome size and the rate of spontaneous mutation in DNA viruses. *Dots* correspond to bacteriophages ϕX174, m13, λ, and T4, duck hepatitis B virus (DHBV), and herpes simplex virus (HSV). DHBV is a pararetrovirus, ϕX174 and m13 are single-stranded DNA viruses, and λ T4 and HSV are double-stranded DNA viruses. The approximate location of RNA viruses and bacteria is shown. See text for references from which these estimates are taken.

and several DNA viruses are subject to host-encoded DNA editing by enzymes of the APOBEC3 family, which can produce hypermutated viral genomes. Like all other biological systems, DNA viruses have to keep a balance between the avoidance of deleterious mutations and the production of diversity, and genome instability mechanisms probably play a central role in the maintenance of this balance.

2. RATES OF SPONTANEOUS MUTATION AND GENETIC DIVERSITY OF DNA VIRUSES

Although DNA viruses were traditionally believed to evolve slowly, analysis of sequences from field isolates with known sampling dates spanning years to decades suggested fast molecular evolution rates for several single-stranded DNA viruses, including emerging canine parvovirus strains [13], human parvovirus B19 [14], tomato yellow leaf curl geminivirus [15], and beak-and-feather disease circovirus [16]. During 2010s, it was further suggested that large double-stranded DNA viruses can also evolve fast. For instance, analysis of samples of the African swine fever virus (ASFV) spanning 70 years yielded estimated evolution rates in the order of 10^{-4} substitutions per nucleotide per year (s/n/y) [17], a value that falls within the typical range exhibited by many RNA viruses. In HSV, frameshift mutations, insertion/deletions, and large complex rearrangements are major sources of genetic diversity as well [18]. Next-generation sequencing (NGS) of HSV laboratory samples has detected the appearance of new mutations after few transfers, further questioning the long-believed genetic stability of DNA viruses [19]. It has also been found that serial plaque-to-plaque transfers can rapidly reduce the fitness of this virus by promoting the accumulation of deleterious mutations, thus echoing the results obtained with RNA viruses in 1990s [20,21]. Plaque-to-plaque transfers reduce viral effective population sizes dramatically, thus favoring the action of random genetic drift, but to set these processes into motion, spontaneous mutations need to occur at a relatively high rate. However, our current knowledge of DNA virus mutation rates is far more limited than for RNA viruses. These rates have been measured directly for a handful of DNA viruses, including HSV and bacteriophages ϕX174, m13, λ, and T4 [2,3,22–25]. A better understanding of DNA virus mutation rates is thus needed to evaluate baseline levels of genome instability. Probably, the main reason for this scarcity of data is of a technical nature. While, in many RNA viruses, mutation rates have been estimated based on sequence analysis, this has not been possible so far for DNA viruses in which mutations are less frequent. Classical Sanger sequencing does not provide deep-enough information to sample low-frequency, new spontaneous mutations. In turn, NGS platforms have the capacity to yield hundreds of billions of nucleotides of DNA sequences in a single experiment, but they are limited by their high per-read error rates which can be orders of magnitude higher than the mutation rate to be measured [26]. However, high-fidelity NGS techniques, developed during early 2010s, such as duplex sequencing [27,28] or circular sequencing [29,30] offer a promising solution for these limitations and should enable a much deeper understanding of spontaneous mutation rates and genome instability in DNA viruses (Fig. 3.2).

3. MUTATOR PHENOTYPES PRODUCED BY LOW-FIDELITY DNA VIRUS POLYMERASES

The most extreme form of genomic instability is achieved by mutators in which genome-wide rates of spontaneous mutation are elevated by orders of magnitude. Mutator strains and their evolutionary and clinical implications have been extensively

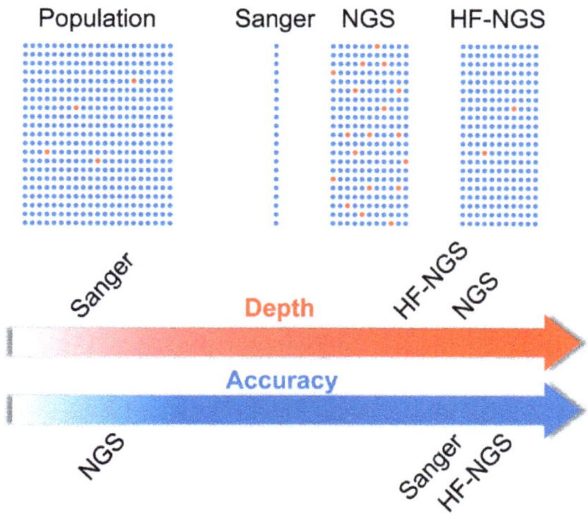

FIGURE 3.2 **Benefits of high-fidelity NGS.** Sequencing has been used for estimating the genetic diversity of viral populations and for characterizing hypermutation and other genome instability processes. Classical Sanger sequencing of PCR molecular clones is a reliable approach, but its coverage is typically limited to 10–100 reads per site, thus preventing sampling of low-frequency mutations. Using NGS platforms, sequencing coverage can be easily increased to >1000 reads per site, but the per-read per-base technical error rate is relatively high (0.1–1.0%). In contrast, the recently developed high-fidelity (HF) NGS technologies, such as duplex sequencing or circular sequencing, can achieve high coverage with an extremely low error rate (<0.001%). See text for references.

studied in bacteria [31–34] and constitute a highly active topic in cancer research [35,36]. DNA viruses can also adopt a mutator phenotype, as shown decades ago using the model bacteriophage T4. Although T4 has a low per-base spontaneous mutation rate compared to other DNA viruses, this rate is still about 30-fold higher than that of its host and is mainly determined by the fidelity of the viral polymerase [37,38]. Most T4 mutators are produced by replacements in the N-terminal domain of the polymerase where the 3′-exonuclease activity resides and can reduce replication fidelity by up to 400-fold [39]. This contrasts with RNA viruses including HIV-1, hepatitis C virus, and influenza virus in which natural mutators have not been described, probably because the wild-type mutation rate is already in the order of 0.1–1.0 new mutations per genome copying, a value that is presumably very close to the theoretical maximum level compatible with virus survival, also termed error threshold [40–42]. Changes in other genes involved in replication, including single-stranded DNA-binding proteins and helicase and clamp proteins, can also produce a mutator phenotype in T4, albeit typically more modestly than low-fidelity polymerases [43]. T4 antimutators showing 100-fold increase in replication fidelity have also been described and often map to the central exonuclease and palm subdomains and the carboxyl-terminal thumb subdomain of the viral polymerase [39]. It has also been noted that T4 antimutator polymerases tend to replicate DNA more slowly than wild-type polymerases, therefore negatively impacting viral fitness. This cost suggests that there is an upper limit for replication fidelity which is determined by the need to replicate fast. On the other hand, mutator phenotypes are also costly because mutations falling at essential genes inflate the genetic load of the population [44]. However, in theory, mutators may still be favored in populations that are maladapted or subject to rapidly changing environments because they boost the production of genetic diversity. However, their rise should be transient, particularly in recombining populations where the mutator locus rapidly unlinks from loci where positively selected mutations are found [45,46]. While in bacteria, these predictions have been largely confirmed, less is known about the evolutionary dynamics of mutators in DNA viruses. In addition to the well-studied T4 system, low-fidelity polymerases may play a central role in the production of diversity in other large DNA viruses such as, for instance, ASFV. Besides the replicative DNA polymerase, ASFV encodes a simple DNA repair system consisting of an endonuclease, a repair polymerase termed pol X and an ATP-dependent DNA ligase. Pol X, which belongs to the same family as the mammalian base-excision repair pol β, exhibits a high error rate, which is determined by the lack of 3′-exonuclease activity and a poor base discrimination capacity [47]. It has been suggested that the relatively high diversity found among ASFV isolates may in part have originated during mutagenic repair involving the highly error-prone DNA pol X [48].

4. DNA COLIPHAGES AND THE MMR SYSTEM

Compared with proofreading, relatively little attention has been paid to the role played by postreplicative repair in determining the genomic stability of DNA viruses. Inasmuch as the lack of 3′-exonuclease proofreading is believed to be a major determinant of RNA virus error-prone replication, access to postreplicative repair may dictate to a large extent the rate of spontaneous mutation of DNA viruses. An excellent model for addressing this question is the *E. coli* MMR system which affords up to 1000-fold reduction in the rate of spontaneous mutations [49]. MMR is carried out by the MutHLS proteins which perform a strand-specific bidirectional repair [50]. Base mismatches or small insertion/deletion loops are recognized by MutS which interacts with MutL and leads to the activation of the MutH endonuclease which excises the daughter strand. DNA resynthesis from the parental strand is then carried out by DNA pol III followed by ligation of the nicked DNA. For this process to operate, the daughter and parental strands need to be distinguished. This is made possible because the parental strand has a methyl group in the adenosine of GATC sequence motifs which is added by *Dam* methylase. Therefore, MMR requires the presence of GATC motifs in the genome, which are normally found at a high frequency in the bacterial chromosome (about 1 in every 250 bases are expected by chance). However, GATC motifs are strikingly absent from the 5.4 kb genome of bacteriophage ϕX174. This strong GATC avoidance necessarily impairs MMR in the phage and should produce a major effect on the ϕX174 mutation rate. Supporting this, the mutation rate of bacteriophage ϕX174 (c. 10^{-6} m/n/c) is three orders of magnitude higher than that of *E. coli* [22]. However, the introduction of 20 GATC motifs in the ϕX174 genome using site-directed mutagenesis reduced the phage mutation rate only by eight fold, with varying effects of these motifs depending on their genome location, the lower-than-expected effect of GATC motifs on phage mutation rate being probably due to an inefficient methylation of the phage DNA [9]. Fast replication or the transient nature of double-stranded replicative intermediates may offer fewer chances for *Dam* methylation in the phage DNA compared to the bacterial chromosome. Although less marked, GATC depletion extends to other coliphages and plasmids, but it is still unclear whether this is a consequence of selection acting on either mutation rates or unrelated traits [51].

5. THE INTERACTION BETWEEN DNA VIRUSES AND THE EUKARYOTIC DNA DAMAGE RESPONSE

The DNA damage response (DDR) comprises a set of signaling pathways for the detection and repair of DNA damage and includes the MMR system for mispaired bases, the base excision repair system for small base modifications, the nucleotide excision repair for intrastrand crosslinks and pyrimidine dimers, the single-strand break repair and double-strand break (DSB) repair pathways involving homologous recombination and nonhomologous end joining [52]. The DDR is primarily controlled by two protein kinases, ataxia-telangiectasia mutated (ATM) and ATM/Rad3-related (ATR) protein kinases. ATM is mainly implicated in the repair of DSBs sensed by the protein complex MRN [53], whereas ATR responds to various types of DNA damage that have in common the presence of single-stranded DNA [54]. Numerous studies have demonstrated that viruses interact with DDR pathways and that, whereas many viruses evade DDR, others appear to benefit from it [55,56]. DNA viruses have developed different strategies to modulate DDR by altering the localization or promoting the degradation of DDR components. For instance, the adenovirus E4orf6 protein recruits a ubiquitin ligase and promotes the proteasomal degradation of TOPBP1, an activator of ATR [57]. Defects in the adenoviral E4 gene lead to the formation of genome concatemers constituted by ligated viral DNA with heterogeneous junctions [58], underscoring the importance of DDR evasion for adenoviruses (Fig. 3.3).

Similarly, HSV proteins such as the regulatory factor ICP0, antagonize DDR by promoting the mislocalization of ATR-interacting protein (ATRIP) [59]. As a result, mutants with ICP0 defects show very poor growth. DDR induction can produce undesirable effects for the virus, such as premature entry into apoptosis. Hence, the inhibition of downstream DDR pathways that stimulate apoptosis is also a common feature of DNA viruses. For instance, the HSV latency-associated protein M2 induces ATM activation, which results in p53 phosphorylation, the inhibition of DNA repair, the blockade of DNA damage-induced apoptosis, and the induction of G1 cell-cycle arrest [60]. Although DNA viruses tend to produce genomic instability in the infected cell, it is still poorly understood how DDR impairment affects DNA virus genomic stability and mutation rates. Highlighting the complexity of virus–DDR interactions, some DNA viruses also use DDR for their own benefit. For example, polyomaviruses induce and exploit the ATM signaling pathway [61]. The T antigen protein expressed by the SV40 polyomavirus activates ATM kinase and downstream targets that are required to obtain unit length viral replication products [62]. Other small DNA viruses such as papillomaviruses [63] and parvoviruses [64] also need to activate the DDR pathways for an efficient replication. These viruses share the common property of having small circular DNA genomes which do not encode their own polymerases and, therefore, they depend strictly on cellular polymerases for replication, as opposed to larger DNA viruses such as adenoviruses, herpesviruses, and poxviruses which encode autonomous replication complexes. Therefore, a possible explanation for why some small viruses promote DDR is that they need to prolong the S cell-cycle phase to create a more favorable environment for replication. By adopting circular genomes, these viruses would also avoid the formation of DDR-associated concatemers as those found in adenoviruses. Given the effects of repair avoidance in the mutation rates of prokaryotic viruses, changes in the expression, and localization of DDR repair-associated proteins might also have major effects on the genomic stability of eukaryotic DNA viruses.

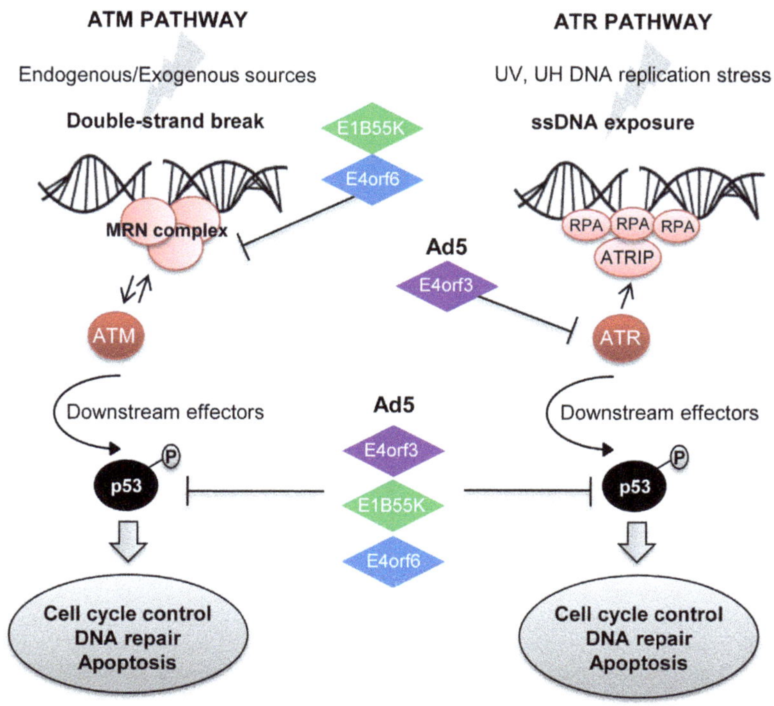

FIGURE 3.3 **Inhibition of ATM and ATR pathways by adenoviruses.** E1b55K–E4orf6 complexes promote the degradation of the MRN complex by recruiting cellular ubiquitin ligases, which prevents ATM signaling. These complexes also recruit ubiquitin ligases to promote the degradation of p53, avoiding apoptosis. The Ad5 E4orf3 protein abolishes the MRN-dependent activation of ATR, resulting in the inhibition of the ATR pathway, and it also inhibits p53. The Ad12 E1B55K protein recruits an E3 ubiquitin ligase to promote TOPB1 degradation, leading to ATR pathway suppression.

6. DIVERSITY-GENERATING RETRO-ELEMENTS IN BACTERIOPHAGES

Some DNA viruses have evolved the ability to target mutations to specific genome regions, thus avoiding the cost of genome-wide hypermutation. A unique and fascinating mechanism of mutation targeting is provided by DGRs. These elements are located in genome regions involved in host attachment and tropism, a trait which is frequently subject to rapidly changing selective pressures dictated by host availability. The first and best-studied DGR was found in the *Bordetella* BPP-1 bacteriophage [65]. This DGR consists of two sequence repeats of about 150 bp each and two ORFs (Fig. 3.4). The first repeat is called the variable repeat (VR) and is located in the 3′-end of the *mtd* gene (major tropism determinant) which encodes a tail fiber protein. Downstream of the VR is located the template repeat (TR) which, contrarily to VR, has a highly conserved sequence. A second ORF (*brt*) encodes a reverse transcriptase which synthesizes cDNA from the VR transcript. During this process, extensive mutagenesis occurs whereby adenines are systematically substituted for random bases by an as yet unknown mechanism. VR cDNA is then transferred to TR, thus producing a large number of variants of the *mtd* gene potentially capable of interacting with new ligands [12]. For this transfer to occur, several *cis*-acting elements are required, including an IMH (initiation of mutagenic homing) region which contains a 15-bp GC-only sequence identical to a portion of TR and a 21-bp sequence similar but not identical to another TR fragment followed by inverted repeats that can adopt a cruciform secondary structure [66]. Using a metagenomics approach, DNA viruses present in the human lower gastrointestinal tract were found to harbor hot spots of hypervariation in genes showing homology to BPP-1 DGR, along with other loci encoding the Ig-superfamily proteins, most of which were linked to genes encoding reverse transcriptases [67]. DGRs have also been found in plasmids, bacterial chromosomes, archaea, and archaeal viruses [68–70]. Although their absolute abundance is low, their powerful mutagenic effect may have a significant impact on the adaptability of prokaryotic viruses.

7. RECOMBINATION-DRIVEN GENOME INSTABILITY IN DNA VIRUSES

DGRs have not been described in eukaryotes or their viruses. The latter may thus use different mechanisms of targeted hypermutation. One such possible mechanism has been demonstrated in poxviruses and is based on recombination-mediated gene amplification. For instance, the inverted terminal repeats of the vaccinia virus genome are known to experience rapid changes in size [71]. This region contains abundant repeats of 10–100 bp sequence motifs that undergo frequent unequal crossover events [72]. While other regions of poxvirus genomes are believed to exhibit greater genome stability, diversity

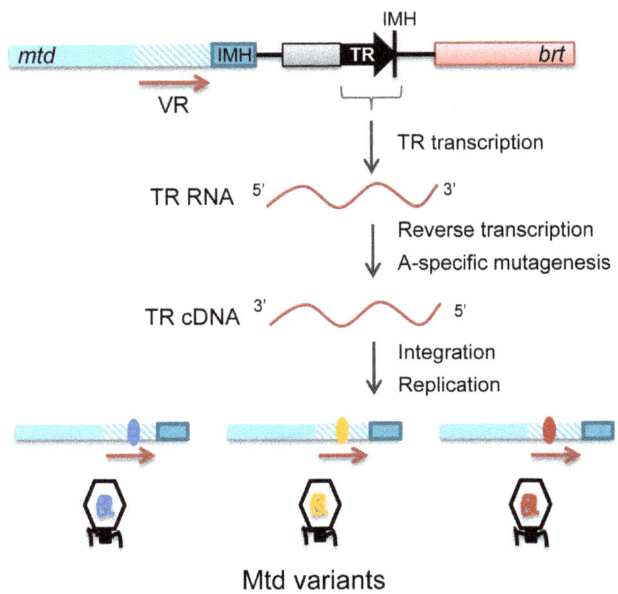

FIGURE 3.4 **Organization and function of DGRs.** The prototypic DGR of phage BPP-1 is linked to the major tropism determinant (*mtd*) gene which encodes a tail fiber protein. The TR transcript is converted to cDNA by the *brt*-encoded reverse transcriptase, and in this process, extensive mutagenesis of adenosines takes place. The cDNA then displaces the VR of the *mtd* gene, a process that is dependent on the IMH motif. As a result, a large number of Mtd variants are produced, allowing for rapid changes in host tropism at the viral population level.

is nevertheless required in these regions for immune escape and for the colonization of novel hosts. This requirement originates from the species-specific selective pressure exerted by host immunity. A central component of innate immunity is protein kinase R (PKR), which induces translational shutoff by phosphorylating the eukaryotic translation initiation factor 2 and mediates additional antiviral responses through its effects on protein phosphorylation status, mRNA stability, and apoptosis [73]. The host–pathogen evolutionary arms race has led to the coevolution of PKR and poxvirus proteins such as K3L and E3L which counteract PKR and contribute to virus-host specificity [74,75]. To investigate the plasticity of these genes, experimental evolution of a vaccinia virus deleted for E3L was carried out to impose a strong selection pressure favoring gain-of-function mutations in the other PKR suppressor, K3L [11]. The virus became adapted to this deletion by increasing the copy number of the K3L gene, inflating its total genome size by up to 10%. Low-frequency variants in the viral population carrying recombination breakpoints were identified as the most likely founders of these genomic expansions. The beneficial effect of gene amplification was two pronged. First, it increased K3L levels, thus providing a direct fitness advantage. Second, it also increased the number of targets available for the appearance of spontaneous gain-of-function mutations. Once these mutations were positively selected and became fixed in the viral population, K3L copy numbers were again reduced, a process which was probably driven by the cost of increased genome size. This thus led to accordion-like evolutionary dynamics whereby copy numbers expand and contract through time. Genomic accordions may also be relevant to the evolution of other poxviruses, such as for instance adaptive gene duplications found in myxomavirus [76]. More broadly, recombination plays a central role in DNA virus biology, including replication, the production of genetic diversity, and the preservation of genome integrity, and it has been associated with host range expansion, the emergence of new viruses, modifications of transmission vector specificity, pathogenesis, and host immunity evasion [77–79]. Early work suggested a nonhomologous recombination hot spot in the replication origin of phage m13 [80]. In phage λ, recombination can occur independently of DNA replication and is active even in cells deficient for the RecA protein (the main protein involved in *E. coli* recombination), which allowed for the identification of a phage-encoded homologous recombination system termed Red [81]. Herpesviruses also have their own recombination machinery used both for replication and DNA repair [81]. Sources of genome instability including DSBs and single-strand DNA breaks are sensed by DNA virus proteins and repaired using different recombination pathways depending of the type of DNA damage. Since these repair pathways are generally error prone, recombination hot spots may drive targeted genomic instability.

8. APOBEC3 PROTEINS AND DNA VIRUS GENOME INSTABILITY

The induction of viral genome instability by host-encoded factors is best illustrated by the action of the apolipoprotein-B mRNA-editing catalytic polypeptide-like 3 (APOBEC3) family of cytidine deaminases which constitutes an innate cellular defense mechanism against retroviruses, endogenous retro-elements, and some DNA viruses [82]. APOBEC3 proteins produce mutations on viral DNA by deaminating cytidine to uracil [83,84]. The first studies showing the antiviral effect of APOBEC3 proteins were carried out in HIV-1 more than 10 years ago [85,86]. However, several subsequent studies have shown that APOBEC3 members can also edit the genomes of hepatitis B virus (HBV) and other DNA viruses that do not undergo a reverse transcription step. HBV has a partially double-stranded DNA genome of 3.2 kb which contains four highly overlapped open reading frames. The HBV genomic DNA is synthesized by the reverse transcription of a pregenomic RNA inside the nucleocapsid, and since reverse transcriptases are highly error prone, this is believed to be the main source of diversity in HBV. APOBEC3 proteins were first shown to inhibit HBV in a mutagenesis-independent manner when the amount of viral RNA associated with core particles was found to be reduced in the presence of APOBEC3 due to the inhibition of the pregenomic RNA encapsidation [87]. However, transfection experiments have shown that different APOBEC3 forms edit both plus and minus DNA strands [88] and have been found to produce hypermutated viral genomes in vivo [89], particularly in cirrhotic patients [90]. As opposed to HIV-1, HBV hypermutated genomes cannot usually be detected by molecular clone sequencing of conventional PCR products and require an ad hoc modified PCR protocol in which a lower denaturation temperature is used to favor the selective amplification of APOBEC3-edited A/T-rich sequences [88]. Interestingly, APOBEC3 footprints have also been detected in non-reverse transcribing DNA viruses. Human papillomavirus, a circular double-stranded DNA virus, has been found to be subject to APOBEC3 editing of both DNA strands in cotransfection experiments and in vivo, producing hypermutated viruses in benign and precancerous lesions [91]. Single-stranded DNA parvoviruses have also been found to be inhibited by APOBEC3, although in this case, this was not accompanied by hypermutation [92]. In contrast, transfusion-transmitted virus, another single-stranded DNA virus with no known homology to previously described viral families, has been found to be susceptible to hypermutation caused by APOBEC3 proteins [93]. Finally, APOBEC3-mediated editing has also been described in large double-stranded DNA viruses such as HSV and Epstein–Barr virus [94]. The primary effect of hypermutation is antiviral because a large number of deleterious missense or nonsense mutations are produced. For instance, one of the preferred APOBEC3G targets is the TGG trinucleotide which, after editing, can lead to TAG premature stop codons, most of which are lethal to the virus. In HIV-1, APOBEC3 expression levels can determine disease progression, with higher APOBEC3 activity associated with higher CD4 counts and slower progression [95,96]. However, a fraction of the edited genomes might be viable and could contribute to immune escape or drug resistance. In HIV-1, it has been shown that many drug-resistance mutations [97] and CTL-escape mutants [98] are located within typical APOBEC3G targets. Similarly, the 3TC-resistance M184I replacement in the HIV-1 reverse transcriptase arose faster in APOBEC3-expressing cells [99]. However, the role played by APOBEC3 proteins in the genetic diversity and virulence of DNA viruses still remains poorly characterized.

9. CONCLUSIONS AND FUTURE DIRECTIONS

The long-accepted genetic stability and slow evolution of DNA viruses have been challenged by multiple reports showing that DNA viruses can exhibit levels of genetic diversity approaching those of some RNA viruses. While for some small DNA viruses, this could be explained by a relatively high rate of spontaneous mutation, the few available estimates for large DNA viruses support a lower average mutation rate. High-fidelity NGS techniques should provide a powerful tool for the study of DNA virus mutation rates and genomic instability in the near future. Despite current uncertainties, DNA virus mutation rates appear to be higher than those of their hosts, probably because the former undergo less efficient DNA repair. The depletion of GATC motifs found in some coliphages provides evidence for repair avoidance, but further work is needed to clarify the evolutionary forces driving such avoidance. The relationship between DNA viruses and cellular repair pathways is much more complex in eukaryotes. It is well established that DNA virus infections modify DDR pathways, but the cause–effect relationships of these changes remain poorly understood. From the virus perspective, it appears that DDR is sometimes an undesired yet unavoidable cellular response to infection, whereas in other cases, DDR is a beneficial or even necessary cellular resource for the virus. Furthermore, the effects of DDR activation/inhibition on viral genomic stability as well as the implications for DNA virus genetic diversity remain obscure. Similar dualities apply to APOBEC3-mediated hypermutation of DNA virus genomes which, despite being primarily an antiviral response, can promote the appearance of immune escape or drug-resistance mutations. Another important realization is that although large DNA viruses show a higher average genomic stability than small DNA viruses and RNA viruses, mutational hot spots can be found at specific genome regions involved in dynamic virus–host interactions, and transient boosts of diversity may also be afforded by short-lived genome-wide mutators in DNA viruses. While the selective pressures acting on bacterial mutators have been well studied, much less is known about the fate of DNA virus mutators, particularly for eukaryotic viruses. DGRs provide a clear mechanistic basis for the ability of some DNA bacteriophages to target mutations to specific genome regions, and their in-depth characterization has both basic and practical implications for directed evolution purposes. Different mechanisms appear to be used for targeted hypermutation in large eukaryotic DNA viruses, in which recombination-driven genomic instability appears to play a central role.

GLOSSARY

Error threshold The theoretical maximal mutation rate tolerated by a given population. Above this threshold, natural selection fails to preserve the sequence of the fittest variants and other, less fit variants reach high population frequencies and may become fixed. This is expected to favor population extinction, although extinction is not a necessary consequence of error threshold crossing.

Genomic accordion An evolutionary expansion/contraction of a gene's copy number, typically associated with strong selection acting on this specific gene.

Hypermutation Strong elevation of the rate of spontaneous mutation which, in viruses, is typically associated with host-mediated edition of the viral genome and tends to be specific to some bases or sequence contexts.

Mutational hot spot Elevation of the spontaneous mutation rate at a specific genome region.

Mutator phenotype A highly increased rate of spontaneous mutation affecting the entire genome and caused by loss of fidelity mechanisms, including proofreading activity and/or postreplicative repair.

Plaque-to-plaque transfer A virus culture technique whereby a single viral plaque is picked and used to seed a fresh culture, in which new plaques develop, and so on. By passaging a virus in this manner, the effective population size is strongly reduced, thereby allowing for the accumulation of mutations under random genetic drift.

Rate of spontaneous mutation The probability that new genetic changes appear and are passed to the next generation. In viruses, a generation is typically defined as one cell infection cycle.

LIST OF ACRONYMS AND ABBREVIATIONS

APOBEC3 Apolipoprotein-B mRNA-editing catalytic polypeptide-like 3 (protein)
ASFV African swine fever virus
ATM Ataxia–telangiectasia mutated (protein)
ATR ATM/Rad3-related (protein)
DDR DNA damage response
DGR Diversity-generating retro-element
DSB Double-strand break
HBV Hepatitis B virus
HSV Herpes simplex virus
IMH Initiation of mutagenic homing (DGR element)
MMR Methyl-directed mismatch repair
NGS Next-generation sequencing
PKR Protein kinase R
TR Template repeat (DGR element)
VR Variable repeat

REFERENCES

[1] Duffy S, Shackelton LA, Holmes EC. Rates of evolutionary change in viruses: patterns and determinants. Nat Rev Genet 2008;9:267–76.
[2] Sanjuán R, Nebot MR, Chirico N, Mansky LM, Belshaw R. Viral mutation rates. J Virol 2010;84:9733–48.
[3] Drake JW. A constant rate of spontaneous mutation in DNA-based microbes. Proc Natl Acad Sci USA 1991;88:7160–4.
[4] Drake JW, Charlesworth B, Charlesworth D, Crow JF. Rates of spontaneous mutation. Genetics 1998;148:1667–86.
[5] Bradwell K, Combe M, Domingo-Calap P, Sanjuán R. Correlation between mutation rate and genome size in riboviruses: mutation rate of bacteriophage Qβ. Genetics 2013;195:243–51.
[6] Steinhauer DA, Domingo E, Holland JJ. Lack of evidence for proofreading mechanisms associated with an RNA virus polymerase. Gene 1992;122:281–8.
[7] Kunkel TA. DNA replication fidelity. J Biol Chem 2004;279:16895–8.
[8] Schaaper RM. Antimutator mutants in bacteriophage T4 and *Escherichia coli*. Genetics 1998;148:1579–85.
[9] Pereira-Gómez M, Sanjuán R. Effect of mismatch repair on the mutation rate of bacteriophage ϕX174. Virus Evol 2015;1:vev010.
[10] Biek R, Pybus OG, Lloyd-Smith JO, Didelot X. Measurably evolving pathogens in the genomic era. Trends Ecol Evol 2015;30:306–13.
[11] Elde NC, Child SJ, Eickbush MT, Kitzman JO, Rogers KS, Shendure J, et al. Poxviruses deploy genomic accordions to adapt rapidly against host antiviral defenses. Cell 2012;150:831–41.
[12] Medhekar B, Miller JF. Diversity-generating retroelements. Curr Opin Microbiol 2007;10:388–95.
[13] Shackelton LA, Parrish CR, Truyen U, Holmes EC. High rate of viral evolution associated with the emergence of carnivore parvovirus. Proc Natl Acad Sci USA 2005;102:379–84.
[14] Shackelton LA, Holmes EC. Phylogenetic evidence for the rapid evolution of human B19 erythrovirus. J Virol 2006;80:3666–9.
[15] Duffy S, Holmes EC. Phylogenetic evidence for rapid rates of molecular evolution in the single-stranded DNA begomovirus tomato yellow leaf curl virus. J Virol 2008;82:957–65.

[16] Sarker S, Patterson EI, Peters A, Baker GB, Forwood JK, Ghorashi SA, et al. Mutability dynamics of an emergent single stranded DNA virus in a naive host. PLoS One 2014;9:e85370.
[17] Michaud V, Randriamparany T, Albina E. Comprehensive phylogenetic reconstructions of African swine fever virus: proposal for a new classification and molecular dating of the virus. PLoS One 2013;8:e69662.
[18] Szpara ML, Gatherer D, Ochoa A, Greenbaum B, Dolan A, Bowden RJ, et al. Evolution and diversity in human herpes simplex virus genomes. J Virol 2014;88:1209–27.
[19] Szpara ML, Tafuri YR, Parsons L, Shamim SR, Verstrepen KJ, Legendre M, et al. A wide extent of inter-strain diversity in virulent and vaccine strains of alphaherpesviruses. PLoS Pathog 2011;7:e1002282.
[20] Chao L. Fitness of RNA virus decreased by Muller's ratchet. Nature 1990;348:454–5.
[21] Duarte E, Clarke D, Moya A, Domingo E, Holland J. Rapid fitness losses in mammalian RNA virus clones due to Muller's ratchet. Proc Natl Acad Sci USA 1992;89:6015–9.
[22] Cuevas JM, Duffy S, Sanjuán R. Point mutation rate of bacteriophage FX174. Genetics 2009;183:747–9.
[23] Drake JW, Hwang CB. On the mutation rate of herpes simplex virus type 1. Genetics 2005;170:969–70.
[24] Kunkel TA. The mutational specificity of DNA polymerase-beta during in vitro DNA synthesis. Production of frameshift, base substitution, and deletion mutations. J Biol Chem 1985;260:5787–96.
[25] Luria SE. The frequency distribution of spontaneous bacteriophage mutants as evidence for the exponential rate of phage reproduction. Cold Spring Harb Symp Quant Biol 1951;16:463–70.
[26] Quinones-Mateu ME, Avila S, Reyes-Teran G, Martinez MA. Deep sequencing: becoming a critical tool in clinical virology. J Clin Virol 2014;61:9–19.
[27] Kennedy SR, Schmitt MW, Fox EJ, Kohrn BF, Salk JJ, Ahn EH, et al. Detecting ultralow-frequency mutations by Duplex Sequencing. Nat Protoc 2014;9:2586–606.
[28] Schmitt MW, Kennedy SR, Salk JJ, Fox EJ, Hiatt JB, Loeb LA. Detection of ultra-rare mutations by next-generation sequencing. Proc Natl Acad Sci USA 2012;109:14508–13.
[29] Acevedo A, Brodsky L, Andino R. Mutational and fitness landscapes of an RNA virus revealed through population sequencing. Nature 2014;505:686–90.
[30] Acevedo A, Andino R. Library preparation for highly accurate population sequencing of RNA viruses. Nat Protoc 2014;9:1760–9.
[31] Gonzalez C, Hadany L, Ponder RG, Price M, Hastings PJ, Rosenberg SM. Mutability and importance of a hypermutable cell subpopulation that produces stress-induced mutants in *Escherichia coli*. PLoS Genet 2008;4:e1000208.
[32] Jolivet-Gougeon A, Kovacs B, Le Gall-David S, Le BH, Bousarghin L, Bonnaure-Mallet M, et al. Bacterial hypermutation: clinical implications. J Med Microbiol 2011;60:563–73.
[33] LeClerc JE, Li B, Payne WL, Cebula TA. High mutation frequencies among *Escherichia coli* and *Salmonella* pathogens. Science 1996;274:1208–11.
[34] Sniegowski PD, Gerrish PJ, Lenski RE. Evolution of high mutation rates in experimental populations of *E. coli*. Nature 1997;387:703–5.
[35] Negrini S, Gorgoulis VG, Halazonetis TD. Genomic instability–an evolving hallmark of cancer. Nat Rev Mol Cell Biol 2010;11:220–8.
[36] Roberts SA, Gordenin DA. Hypermutation in human cancer genomes: footprints and mechanisms. Nat Rev Cancer 2014;14:786–800.
[37] Drake JW, Allen EF, Forsberg SA, Preparata RM, Greening EO. Genetic control of mutation rates in bacteriophageT4. Nature 1969;221:1128–32.
[38] Santos ME, Drake JW. Rates of spontaneous mutation in bacteriophage T4 are independent of host fidelity determinants. Genetics 1994;138:553–64.
[39] Mansky LM, Cunningham KS. Virus mutators and antimutators: roles in evolution, pathogenesis and emergence. Trends Genet 2000;16:512–7.
[40] Bull JJ, Sanjuán R, Wilke CO. Theory of lethal mutagenesis for viruses. J Virol 2007;81:2930–9.
[41] Crotty S, Cameron CE, Andino R. RNA virus error catastrophe: direct molecular test by using ribavirin. Proc Natl Acad Sci USA 2001;98:6895–900.
[42] Andino R, Domingo E. Viral quasispecies. Virology 2015;479–480C:46–51.
[43] Mufti S. Mutator effects of alleles of phage T4 genes 32, 41, 44, and 45 in the presence of an antimutator polymerase. Virology 1979;94:1–9.
[44] Sniegowski PD, Gerrish PJ, Johnson T, Shaver A. The evolution of mutation rates: separating causes from consequences. BioEssays 2000;22:1057–66.
[45] Taddei F, Radman M, Maynard-Smith J, Toupance B, Gouyon PH, Godelle B. Role of mutator alleles in adaptive evolution. Nature 1997;387:700–2.
[46] Raynes Y, Sniegowski PD. Experimental evolution and the dynamics of genomic mutation rate modifiers. Heredity 2014;113:375–80.
[47] Oliveros M, Yanez RJ, Salas ML, Salas J, Vinuela E, Blanco L. Characterization of an African swine fever virus 20-kDa DNA polymerase involved in DNA repair. J Biol Chem 1997;272:30899–910.
[48] Lamarche BJ, Kumar S, Tsai MD. ASFV DNA polymerase X is extremely error-prone under diverse assay conditions and within multiple DNA sequence contexts. Biochemistry 2006;45:14826–33.
[49] Fijalkowska IJ, Schaaper RM, Jonczyk P. DNA replication fidelity in *Escherichia coli*: a multi-DNA polymerase affair. FEMS Microbiol Rev 2012;36:1105–21.
[50] Jiricny J. Postreplicative mismatch repair. Cold Spring Harb Perspect Biol 2013;5:a012633.
[51] McClelland M. Selection against dam methylation sites in the genomes of DNA of enterobacteriophages. J Mol Evol 1984;21:317–22.
[52] Ciccia A, Elledge SJ. The DNA damage response: making it safe to play with knives. Mol Cell 2010;40:179–204.
[53] Lee JH, Paull TT. ATM activation by DNA double-strand breaks through the Mre11-Rad50-Nbs1 complex. Science 2005;308:551–4.
[54] Cimprich KA, Cortez D. ATR: an essential regulator of genome integrity. Nat Rev Mol Cell Biol 2008;9:616–27.

[55] Luftig MA. Viruses and the DNA damage response: activation and antagonism. Annu Rev Virol 2014;1:605–25.
[56] Weitzman MD, Lilley CE, Chaurushiya MS. Genomes in conflict: maintaining genome integrity during virus infection. Annu Rev Microbiol 2010;64:61–81.
[57] Blackford AN, Patel RN, Forrester NA, Theil K, Groitl P, Stewart GS, et al. Adenovirus 12 E4orf6 inhibits ATR activation by promoting TOPBP1 degradation. Proc Natl Acad Sci USA 2010;107:12251–6.
[58] Weiden MD, Ginsberg HS. Deletion of the E4 region of the genome produces adenovirus DNA concatemers. Proc Natl Acad Sci USA 1994;91: 153–7.
[59] Wilkinson DE, Weller SK. Herpes simplex virus type I disrupts the ATR-dependent DNA-damage response during lytic infection. J Cell Sci 2006;119:2695–703.
[60] Liang X, Pickering MT, Cho NH, Chang H, Volkert MR, Kowalik TF, et al. Deregulation of DNA damage signal transduction by herpesvirus latency-associated M2. J Virol 2006;80:5862–74.
[61] Dahl J, You J, Benjamin TL. Induction and utilization of an ATM signaling pathway by polyomavirus. J Virol 2005;79:13007–17.
[62] Sowd GA, Li NY, Fanning E. ATM and ATR activities maintain replication fork integrity during SV40 chromatin replication. PLoS Pathog 2013;9:e1003283.
[63] Gillespie KA, Mehta KP, Laimins LA, Moody CA. Human papillomaviruses recruit cellular DNA repair and homologous recombination factors to viral replication centers. J Virol 2012;86:9520–6.
[64] Luo Y, Qiu J. Parvovirus infection-induced DNA damage response. Future Virol 2013;8:245–57.
[65] Liu M, Deora R, Doulatov SR, Gingery M, Eiserling FA, Preston A, et al. Reverse transcriptase-mediated tropism switching in *Bordetella* bacteriophage. Science 2002;295:2091–4.
[66] Guo H, Tse LV, Nieh AW, Czornyj E, Williams S, Oukil S, et al. Target site recognition by a diversity-generating retroelement. PLoS Genet 2011;7:e1002414.
[67] Minot S, Grunberg S, Wu GD, Lewis JD, Bushman FD. Hypervariable loci in the human gut virome. Proc Natl Acad Sci USA 2012;109:3962–6.
[68] Arambula D, Wong W, Medhekar BA, Guo H, Gingery M, Czornyj E, et al. Surface display of a massively variable lipoprotein by a *Legionella* diversity-generating retroelement. Proc Natl Acad Sci USA 2013;110:8212–7.
[69] Doulatov S, Hodes A, Dai L, Mandhana N, Liu M, Deora R, et al. Tropism switching in *Bordetella* bacteriophage defines a family of diversity-generating retroelements. Nature 2004;431:476–81.
[70] Paul BG, Bagby SC, Czornyj E, Arambula D, Handa S, Sczyrba A, et al. Targeted diversity generation by intraterrestrial archaea and archaeal viruses. Nat Commun 2015;6:6585.
[71] Moss B, Winters E, Cooper N. Instability and reiteration of DNA sequences within the vaccinia virus genome. Proc Natl Acad Sci USA 1981;78:1614–8.
[72] Baroudy BM, Moss B. Sequence homologies of diverse length tandem repetitions near ends of vaccinia virus genome suggest unequal crossing over. Nucleic Acids Res 1982;10:5673–9.
[73] Munir M, Berg M. The multiple faces of proteinkinase R in antiviral defense. Virulence 2013;4:85–9.
[74] Elde NC, Child SJ, Geballe AP, Malik HS. Protein kinase R reveals an evolutionary model for defeating viral mimicry. Nature 2009;457:485–9.
[75] Rothenburg S, Seo EJ, Gibbs JS, Dever TE, Dittmar K. Rapid evolution of protein kinase PKR alters sensitivity to viral inhibitors. Nat Struct Mol Biol 2009;16:63–70.
[76] Kerr PJ, Hone J, Perrin L, French N, Williams CK. Molecular and serological analysis of the epidemiology of myxoma virus in rabbits. Vet Microbiol 2010;143:167–78.
[77] Martin DP, Biagini P, Lefeuvre P, Golden M, Roumagnac P, Varsani A. Recombination in eukaryotic single stranded DNA viruses. Viruses 2011;3:1699–738.
[78] Simon-Loriere E, Holmes EC. Why do RNA viruses recombine? Nat Rev Microbiol 2011;9:617–26.
[79] Thiry E, Meurens F, Muylkens B, McVoy M, Gogev S, Thiry J, et al. Recombination in alphaherpesviruses. Rev Med Virol 2005;15:89–103.
[80] Ehrlich SD, Bierne H, d'Alencon E, Vilette D, Petranovic M, Noirot P, et al. Mechanisms of illegitimate recombination. Gene 1993;135:161–6.
[81] Weller SK, Sawitzke JA. Recombination promoted by DNA viruses: phage lambda to herpes simplex virus. Annu Rev Microbiol 2014;68:237–58.
[82] Harris RS, Dudley JP. APOBECs and virus restriction. Virology 2015;479–480C:131–45.
[83] Armitage AE, Katzourakis A, de OT, Welch JJ, Belshaw R, Bishop KN, et al. Conserved footprints of APOBEC3G on Hypermutated human immunodeficiency virus type 1 and human endogenous retrovirus HERV-K(HML2) sequences. J Virol 2008;82:8743–61.
[84] Vieira VC, Soares MA. The role of cytidine deaminases on innate immune responses against human viral infections. Biomed Res Int 2013;2013:683095.
[85] Harris RS, Bishop KN, Sheehy AM, Craig HM, Petersen-Mahrt SK, Watt IN, et al. DNA deamination mediates innate immunity to retroviral infection. Cell 2003;113:803–9.
[86] Mangeat B, Turelli P, Caron G, Friedli M, Perrin L, Trono D. Broad antiretroviral defence by human APOBEC3G through lethal editing of nascent reverse transcripts. Nature 2003;424:99–103.
[87] Turelli P, Mangeat B, Jost S, Vianin S, Trono D. Inhibition of hepatitis B virus replication by APOBEC3G. Science 2004;303:1829.
[88] Suspène R, Guetard D, Henry M, Sommer P, Wain-Hobson S, Vartanian JP. Extensive editing of both hepatitis B virus DNA strands by APOBEC3 cytidine deaminases in vitro and in vivo. Proc Natl Acad Sci USA 2005;102:8321–6.
[89] Baumert TF, Rosler C, Malim MH, von WF. Hepatitis B virus DNA is subject to extensive editing by the human deaminase APOBEC3C. Hepatology 2007;46:682–9.
[90] Vartanian JP, Henry M, Marchio A, Suspène R, Aynaud MM, Guetard D, et al. Massive APOBEC3 editing of hepatitis B viral DNA in cirrhosis. PLoS Pathog 2010;6:e1000928.

[91] Vartanian JP, Guetard D, Henry M, Wain-Hobson S. Evidence for editing of human papillomavirus DNA by APOBEC3 in benign and precancerous lesions. Science 2008;320:230–3.
[92] Chen H, Lilley CE, Yu Q, Lee DV, Chou J, Narvaiza I, et al. APOBEC3A is a potent inhibitor of adeno-associated virus and retrotransposons. Curr Biol 2006;16:480–5.
[93] Tsuge M, Noguchi C, Akiyama R, Matsushita M, Kunihiro K, Tanaka S, et al. G to A hypermutation of TT virus. Virus Res 2010;149:211–6.
[94] Suspène R, Aynaud MM, Koch S, Pasdeloup D, Labetoulle M, Gaertner B, et al. Genetic editing of herpes simplex virus 1 and Epstein-Barr herpesvirus genomes by human APOBEC3 cytidine deaminases in culture and in vivo. J Virol 2011;85:7594–602.
[95] Land AM, Ball TB, Luo M, Pilon R, Sandstrom P, Embree JE, et al. Human immunodeficiency virus (HIV) type 1 proviral hypermutation correlates with CD4 count in HIV-infected women from Kenya. J Virol 2008;82:8172–82.
[96] Cuevas JM, Geller R, Garijo R, López-Aldeguer J, Sanjuán R. Extremely high mutation rate of HIV-1 in vivo. PLoS Biol 2015;13:e1002251.
[97] Berkhout B, de Ronde A. APOBEC3G versus reverse transcriptase in the generation of HIV-1 drug-resistance mutations. AIDS 2004;18:1861–3.
[98] Wood N, Bhattacharya T, Keele BF, Giorgi E, Liu M, Gaschen B, et al. HIV evolution in early infection: selection pressures, patterns of insertion and deletion, and the impact of APOBEC. PLoS Pathog 2009;5:e1000414.
[99] Kim EY, Bhattacharya T, Kunstman K, Swantek P, Koning FA, Malim MH, et al. Human APOBEC3G-mediated editing can promote HIV-1 sequence diversification and accelerate adaptation to selective pressure. J Virol 2010;84:10402–5.

Section II

Genome Instability in Bacteria and Archaea

Chapter 4

Genome Instability in Bacteria and Archaea: Strategies for Maintaining Genome Stability

J.-E. Messling, A.B. Williams
University of Cologne, Cologne, Germany

Chapter Outline

1. INTRODUCTION

The genomes of all organisms are constantly challenged by DNA damaging forces, from their own internal metabolic byproducts to various outside forces. DNA, as the carrier of the genetic details of an organism, is undeniably the most important macromolecule, both for individual cells and for the species. Unlike RNA and protein, which can be replaced using the information stored in DNA, any loss of DNA sequence due to DNA damage is essentially irreplaceable without extraordinary measures. Besides this catastrophic outcome, DNA damage can also interfere with important ongoing cellular processes in individual cells, such as DNA replication and transcription, potentially leading to cell death. Because of the importance of maintaining both the fidelity of the heritable genetic information and the integrity of the DNA for ongoing cellular processes, a complex network of DNA-repair systems has evolved. In a testament to the primal need to maintain genetic integrity, these pathways are conserved in all known organisms. In some cases, the components of different DNA pathways are clearly orthologs derived from a common evolutionary ancestor; in other cases, the different players may have evolved independently to serve the same function. For comparison and reference, Table 4.1 compares orthologous and functionally homologous genes in bacteria, archaea, and the simple eukaryotic nematode *Caenorhabditis elegans*.

Like many biological processes, DNA repair proteins have been systematically organized into mostly linear pathways for the purpose of understanding how cells repair particular types of lesions; however, one must bear in mind that, in fact, DNA repair pathways represent a dazzlingly complex network in which proteins function in multiple DNA repair pathways, or even in entirely different biochemical processes. The following discussion addresses the fundamental processes that preserve genome integrity in bacteria and archaea, including how cells respond to DNA damage to support DNA repair,

Genome Stability. http://dx.doi.org/10.1016/B978-0-12-803309-8.00004-5

TABLE 4.1 DNA Repair Homologs in Selected Model Organisms

Function/Enzyme	*Escherichia coli*	*Archea*	*Caenorhabditis elegans*
Mismatch Repair (MMR)			
Mismatch recognition	MutS	MutS (TTHA1892) (*Thermus thermophilus*)	MutSα (MSH-1 + MSH-6)
			MutSβ (MSH-2 + MSH-3)
Match-making ATPase	MutL		
Endonuclease ATPase		MutL (TTHA1892) (*T. thermophilus*)	MutLα (MLH-1 + PMS-2)
			MutLβ (MLH-1 + PMS-1)
			MutLγ (MLH-1 + MLH-3)
Activation of MutL homolog		DNA Pol III β subunit (TTHA0001) (*T. thermophilus*)	PCN-1 (f.h.)[a]
		DNA Pol III δ,δ,γ,τ subunits (TTHA0001) (*T. thermophilus*)	RFC (RFC-1, -2, -3, -4, -5) (f.h.)
Crossover promotion during meiosis			Not identified
Mismatch repair during mitosis and meiosis			Not identified
Endonuclease	MutH	?	Not identified
DNA helicase	UvrD	UvrD (TTHA0244) (*T. thermophilus*)	?
Strand excision (5′–3′)	RecJ	RecJ (TTHA1892) (*T. thermophilus*)	
Strand excision (3′–5′)	ExoI	ExoI (TTHB187) (*T. thermophilus*)	
Strand excision (5′–3′)	ExoVII		
Strand excision (5′ –3′)	ExoX		
Strand excision (5′ –3′)			EXO-1
Single-strand binding	SSB	SSB (TTHA0244) (*T. thermophilus*)	?
DNA polymerase	Pol III holoenzyme (subunits: α,γ,δ,δ′,Ψ,λ,τ,ε,θ)	Pol III?	Pol δ (**F10C2.4, F12F6.7?**)
Ligation	LigA	?	LIG-1
Nucleotide Excision Repair (NER)			
Binding of damaged DNA			XPC-1
Endonuclease			XPG-1
Removal of transcription blocking lesions	Mfd	Mfd (TTHA0889) (*T. thermophilus*)	CSB-1 (f.h.)
			CSA-1
Replication factor C			RFC (consists of RFC-1 to -5)
			RPA-1 and RPA-2
Binding of DNA and proteins in preincision complex			XPA-1
3′-incision nuclease	Cho	Not identified	
3′-incision nuclease			ERCC-5
5′-incision nuclease			ERCC-1
5′-incision nuclease			ERCC-4

TABLE 4.1 DNA Repair Homologs in Selected Model Organisms—cont'd

Function/Enzyme	*Escherichia coli*	*Archea*	*Caenorhabditis elegans*
Nucleotide excision and basal transcription	No prokaryotic homolog	No prokaryotic homolog	
			XBP-1
			XPD-1
			R02D3.3
			T16H12.4
			Zk1128.4
			Y73F8A.24
			Y55B1AL.2
			CDK-7
			CYH-3
			MNAT-1
Involved in single strand break repair			Not identified
E3 ligase interaction			DDB-1
Recruitment of TLS polymerase in TC NER (proliferating cell nuclear antigen)			PCN-1
DNA synthesis	Pol III holoenzyme (subunits: α,γ,δ,δ′,Ψ,λ,τ,ε,θ)		**F10C2.4, F12F6.7?**
Binding of damaged DNA	UvrA	UvrA (*Halobacterium* sp.)	Not identified
DNA unwinding	UvrB	UvrB (*Halobacterium* sp.)	Not identified
3′- and 5′-incision nuclease	UvrC	UvrC (*Halobacterium sp.*)	Not identified
Base Excision Repair (BER)			
3MeA DNA glycosylase	AlkA, TagA	AlkA (TTHA0392) (*T. thermophilus*)	?
Uracil DNA glycosylase	Ung	UDGA (TTHA0718) (*T. thermophilus*)	UNG-1
Removal of Hydroxymethyl U			?
8 oxoguanine DNA glycosylase/AP lyase	MutY	MutY (TTHA1898) (*T. thermophilus*)	?
8 oxoguanine DNA glycosylase/AP lyase	Fpg (MutM)	MutM (TTHA1806) (*T. thermophilus*)	
Thymine glycol DNA glycosylase/AP lyase	Nth		?
TDG T:G mismatch DNA glycosylase			?
Exonuclease III	XthA		
Endonuclease IV	Nfo		?
Incision 3′ of hypoxanthine and uracil	Nfi (EndoV)	EndoV (TTHA1374) (*T. thermophilus*)	?
Removal of thymine glycol	Nei (EndoVIII)		?

Continued

TABLE 4.1 DNA Repair Homologs in Selected Model Organisms—cont'd

Function/Enzyme	*Escherichia coli*	*Archea*	*Caenorhabditis elegans*
Removal of oxidative products of C and U			?
Endonuclease IIIV like glycosylase 3			?
Flap endonuclease	PolI	Pol (TTHA1054) (*T. thermophilus*)	
Proliferating cell nuclear antigen			PCN-1
Replication factor C			RFC (consists of RFC-1 to -5)
DNA polymerases			
			Pol δ (**F10C2.4, F12F6.7 ?**)
			?
DNA ligase	LigA	LigA (TTHA1097) (*T. thermophilus*)	LIG-1
ATP-dependent DNA ligase			?
Poly (ADP ribose) Polymerase			PARP-1
ADPRT-like enzyme			PARP-2
Homologous Recombination (HR)			
Recombinase	RecA	RecA (TTHA1818) (*T. thermophilus*)	RAD-51
Branch migration complex	RuvA	RuvA (TTHA0291) (*T. thermophilus*)	
	RuvB	RuvB (TTHA0406) (*T. thermophilus*)	
DNA helicase	RecG	RecG (TTHA1266) (*T. thermophilus*)	
RecQ-like DNA helicase	RecQ		?
RecQ family DNA helicase			WRN-1
SMC-like ATPase		SbcC (TTHA1288) (*T. thermophilus*)	RAD-50
3′–5′-exonuclease (in complex with SbcC)	SbcD	SbcD (TTHA1289) (*T. thermophilus*)	MRE-11
Accessory protein for MR complex			
RecA-like ATPase	RadA/Sms	RadA/Sms (TTHA0541) (*T. thermophilus*)	
RadA paralog		RadC1 (*S. islandicus*)	
Rad51 like recombinase mediator			?
			?
			?
			?
RAD54 family DNA translocase, recombinase mediator			?
			RAD-54
Strand excision (5′–3′)	RecJ	RecJ (TTHA1892) (*T. thermophilus*)	EXO-1 (f.h.)
Strand excision (5′–3′)			

TABLE 4.1 DNA Repair Homologs in Selected Model Organisms—cont'd

Function/Enzyme	*Escherichia coli*	Archea	*Caenorhabditis elegans*
Helicase/nuclease complex	RecB		
	RecC		
	RecD		
5′–3′ exonuclease	RecE		
ssDNA annealing	RecT		
Single-strand binding	SSB	SSB (TTHA0244) (*T. thermophilus*)	?
DNA-binding complex	RecF	RecF (TTHA0264) (*T. thermophilus*)	
	RecO	RecO (TTHA0623) (*T. thermophilus*)	
	RecR	RecR (TTHA1600) (*T. thermophilus*)	
HJ resolvase	RuvC	RuvC (TTHA1090) (*T. thermophilus*)	
	RusA		
			SLX-1
			GEN-1
Recombinase inhibitor	RecX	RecX (TTHA0848) (*T. thermophilus*)	
DNA helicase	UvrD	UvrD (TTHA1427) (*T. thermophilus*)	
Accessory protein for structure-specific nucleases			HIM-18
Structure-specific endonuclease			MUS-81
		MutS2 (TTHA1645) (*T. thermophilus*)	
			Not identified
Complex with ERCC4 (Rad1)			ERCC-1

[a]*f.h. indicates functional homolog (ie, no obvious sequence homology).*

the actual molecular transactions at the DNA that lead to damage repair, and ways that cells preserve their genomes when challenged with parasitic foreign DNA.

2. REPONSES TO DNA DAMAGE

2.1 The SOS Response: A Primitive Cell-Cycle Checkpoint

Since its discovery and early characterization by Evelyn Witkin and Miroslav Radman in the early to mid-1970s, the SOS response has become a paradigm for the bacterial DNA damage response (for a detailed review, see Ref. [1]). While the SOS pathway has proven to be extremely complex, at its core are only two proteins: the LexA repressor and the RecA activator. Under normal conditions, LexA binds to a special sequence, the SOS box, in the promoters of SOS-regulated genes and blocks their expression. One of the most common outcomes of DNA damage is the formation of single-strand DNA (ssDNA) through any number of possible processes. The ssDNA is rapidly coated with RecA protein to form the RecA filament; while this RecA filament may go on to participate in homologous recombination, it has another function: to induce the autocleavage of LexA, relieving its repressive activity, and allowing expression of the SOS regulon. As an aside, despite over 30 years of work, how the RecA filament forms is still an active area of research and discussion (eg, see Ref. [2]). The SOS response is rapid and within just a few minutes the amount of LexA decreases by nearly 10-fold. During this time, cell division is blocked by the cytokinesis inhibitor SulA (SfiA) (whose gene is expressed robustly after SOS induction) and cells form distinctive filaments as they grow without division. The outcome of SOS induction is a massive transcriptional reprogramming. Quite remarkably, evolution has endowed the SOS response with a logical order. The more rapidly induced genes are from the nucleotide excision repair (NER) pathway (discussed in the following text), one of the most versatile DNA repair mechanisms. Subsequently, *recA* and other genes for homologous recombination are induced, supporting high-fidelity repair and amplification of the response. The *lexA* gene itself is also induced, preparing the cell to turn off the SOS response when the activating signal wanes, probably because the SOS response is demanding on cellular resources and that it blocks the formation of progeny cells. Finally, as late as 40 min into the SOS response, the error-prone DNA polymerase Pol V (encoded by the *umuDC* genes) is expressed and activated (the latter process requires a RecA filament-dependent autocleavage event of the UmuD subunit). A second error-prone polymerase, Pol IV (*dinB*), is also upregulated. These polymerases replicate damaged DNA to allow continued cell growth with the hope that other DNA repair pathways will catch up; however, these polymerases can also introduce mistakes that can be preserved as mutations that can have either deleterious, or sometimes advantageous, outcomes (see Chapter X for a discussion of the latter).

The importance of the SOS response for bacteria is highlighted by its vast conservation among divergent species; however, some notable deviations from the *Escherichia coli* model have been observed. There is a disconnection between RecA-ssDNA binding and activation of the SOS response in *Baccilus subtilus* [3]. In *Caulobacter crescentus*, following a similar cell-cycle arrest [4] (albeit by a different molecular mechanism), the SOS response also triggers a more sophisticated programmed cell death pathway, akin to eukaryotic apoptosis [5]. Mycobacteria have evolved a more complex regulatory system for DNA damage-responsive genes. In this species, most of the DNA damage genes are regulated by a second factor, ClpR [6–8]; however, conserved regulation does exist, as *recA* is controlled by LexA (in addition to ClpR). Large-scale transcriptional reprogramming occurs in the extremophilic species *Deinococcus radiodurans*, although it depends mostly on the PprI protein instead of the two LexA homologs [9]. PprI binds to damage-responsive promoters after exposure to ionizing radiation to induce expression, including the *recA* promoter. Some archaea cope quite differently with DNA damage. The hyperthermophile *Sulfolobus* does not induce a large number of genes following UV exposure; however, it has a sophisticated alternative coping mechanism (see later).

Growing evidence suggests that the SOS response has many other functions in addition to this basic checkpoint control, including functions in horizontal gene transfer, the development of antibiotic resistance, and pathogenesis (see Ref. [1]). Nevertheless, it is clear that the SOS response is a first line of defense in the preservation of genome stability for bacteria.

2.2 An Archaeal UV Response Based on DNA Sharing

Invocation of a large-scale transcriptional reprogramming after UV exposure does not appear to be universal in archaea, but a UV-induced stress response has been characterized in *Sulfolobus* (Fig. 4.1).

After UV exposure, the cells induce expression of the *ups* genes, which encode a specialized type IV pilus system that enables efficient DNA transfer between cells [10–12]. The relocated DNA can then be used as a template for homologous recombination (HR)-dependent DNA repair—a generally, but not exclusively, accurate DNA repair pathway. That this

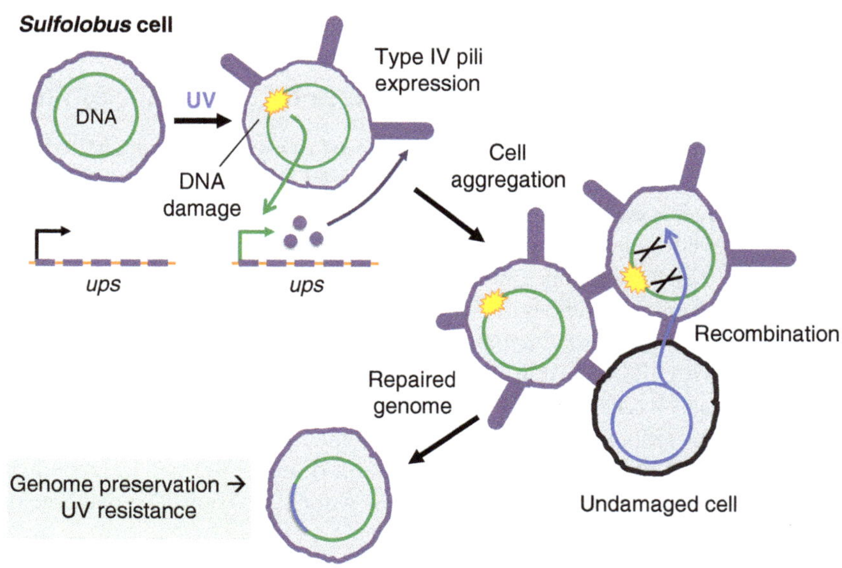

FIGURE 4.1 ***Sulfolobus* copes with UV-induced DNA damage via a type IV pilis-dependent DNA exchange pathway.** After DNA damage, *Sulfolobus* induces the *ups* operon to express type IV pili. After pili form, the cells aggregate and exchange DNA (both undamaged and damaged cells can act as DNA donors). Undamaged homologous DNA can be recombined to replace damaged segments, thus rescuing cells from the deleterious effects of DNA damage.

DNA sharing somehow protects the cells, presumably by dampening UV-induced genome instability, is supported by the observation that strains capable of expressing the type IV pili have higher survival rates after UV exposure. A 2015 work had further characterized this response by demonstrating additional involvement of four genes adjacent to the *ups* locus: an endonuclease III, a ParB-like protein, a glycosyltransferase, and a RecQ-like helicase [13]. With the exception of the ParB-like protein (which likely participates in the DNA transfer), these proteins are proposed to function in a homologous recombination-dependent DNA repair process downstream of the DNA transfer. While a bona fide SOS response is clearly absent in *Sulfolobus*, this system illustrates a novel genetic innovation for dealing with UV-induced DNA damage.

3. DNA REPAIR PATHWAYS

While the SOS response provides a genome-stabilizing function, it has no inherent DNA repair capacity. Instead, cells have evolved several intertwined molecular pathways comprised by the actual molecular transactions leading to damage repair: direct reversal (the only DNA synthesis–independent pathway); base excision repair (BER); NER; mismatch repair (MMR); and HR-dependent repair. The fundamentals of DNA repair have been most intensely studied in *E. coli*, thus its molecular biology forms the foundation of the discussion; however, important deviations in other species are also highlighted.

3.1 Direct Reversal of DNA Damage

One way to repair DNA damage is to simply undo the particular molecular changes, that is, to directly reverse the damage. Evolution has endowed life with (at least) three direct reversal pathways: photolyases, which repair UV-induced damage, and two mechanisms that repair alkylated bases, O^6-alkylguanine alkyl transferases (AGTs) and AlkB-family dioxygenases. While the molecular mechanisms vary drastically, the end result of all of these pathways is the restoration of the original molecular structure without the need for new DNA synthesis.

UV irradiation leads to two main types of DNA lesions that can disrupt many DNA-related processes, most importantly, replication and transcription: pyrimidine (6-4) photoproducts (6-4PPs) and cyclobutane pyrimidine dimers (CPDs). Due to the different structures of these lesions, different photolyase enzymes are required for their repair; however, a common feature of photolyases is that they obtain energy from light to fuel the reaction (hence the classical name "light reactions") and use flavin adenine dinucleotide (FAD) for catalysis. In general, it is thought that direct reversal occurs via a multistep process in which light energy is harnessed to drive the full reduction of FAD to $FADH^-$, followed by an electron transfer to the lesion leading to the breakage of covalent bonds, and finally a retransfer of the free electron back to the FADH radical to produce $FADH^-$ (reviewed in [14] and [15]). In the end, the photolyase reaction is simply a stepwise transfer of energy that reconfigures the covalent bonds in the original bases to restore the original structure.

The *E. coli* K-12 photolyase is encoded by the *phr* gene (also known as *phrB*) [16] and the protein is maintained at low levels in stationary phase cells [17]. It is interesting to note that, despite its involvement in repairing UV-induced CPD dimers, *phr* is not regulated by the SOS response [18]. While CPD photolyases were one of the earliest characterized DNA repair mechanisms and have been found in all three domains of life, (6-4) PP photolyases remained elusive until only recently. The first bacterial (6-4) PP photolyase was reported in 2013 in *Agrobacterium tumefaciens* and is encoded by the *phrB* gene [19]. Photolyases have also been studied in various archaea, including halophiles, methanogenic species, and thermophiles [20–22]. Besides some structural differences and utilization of different chromophores for light collection, the functions of archaeal photolyases are conserved from their bacterial counterparts.

Alkylating agents interact with atoms in DNA bases leading to the formation of a variety of cytotoxic and potentially mutagenic adducts. These adducts can be as simple as methyl groups or larger bulky adducts. Without proper repair, these lesions represent a significant threat to genome stability. Lesions caused by alkylation are efficiently repaired by the BER pathway (discussed in the following section); however, they can also be directly repaired by alkyl transferases and AlkB-family dioxygenases (reviewed in Refs [23] and [24]).

Direct reversal of alkylation damage in *E. coli* is mediated by either the general housekeeping alkyltransferase Ogt, or an adaptive response controlled by the Ada protein that is mediated by its targets *alkA* and *alkB* [25]. O^6-methylguanine (6-meG) is mutagenic due to its ability to induce G:C to A:T transitions during DNA replication due to faulty base pairing. The Ada protein is a bifunctional alkyltransferase: the N-terminal domain (N-Ada) repairs methylphosphotriester lesions (damage to the DNA backbone that is generally innocuous to cells), while the C-terminal domain (C-Ada) repairs the much more potent alkyl lesions at the O^6 position of guanine. The direct reversal reaction occurs via the transfer of the alkyl group from the damaged base onto a reactive cysteine residue via an S_N2 reaction, thus permanently inactivating the protein.

The AlkB dioxygenase is similar to Ada in that it catalyzes the direct reversal of base alkylation damage. While the exact function of AlkB was difficult to determine (discussed in detail by Mishina et al. [24]), it was finally shown to catalyze the direct reversal of 1-methyladenine (1-meA) and 3-methylcytosine (3-meC) to adenine and cytosine. The mechanisms of these two direct reversal pathways are different: while alkyltransferases use an S_N2 reaction, AlkB uses an oxidative dealkylation reaction that depends on an active site iron(II) atom.

At high temperatures, alkylated bases can be converted to abasic sites or DNA breaks that can cause irreversible chromosome fragmentation. From this perspective, it seems that alkylation repair should be highly developed and efficient in thermophilic archaea. In fact, however, the literature on alkylation repair in archaea is quite limited. Alkyltransferases from two thermophiles (*Methanococcus jannaschii* and *Sulfolobus taokadaii*) have been purified and crystalized and they show distinct structural similarities to the human homologs (PDB entry 1WRJ) [26]. Overexpression of a *Pyrococcus* methyltransferase in an *E. coli ogt* mutant can rescue its sensitivity to alkylating agents [27], confirming a functional conservation between bacterial and this archaeal methyltransferase. Other in vivo information on archaeal alkytransferase function has demonstrated that this functional conservation is probably a general feature (eg, see Refs [28] and [29]).

3.2 Base Excision Repair and Removal of Uracil from DNA

In addition to direct reversal, many organisms have another highly conserved (Table 4.1) pathway to repair damaged bases: BER (reviewed in detail in Ref. [30]). BER was first discovered in *E. coli* by Tomas Lindahl when he attempted to elucidate the pathway for the repair of genomic uracil, a byproduct of cytosine deamination [31]. Subsequent research over many years revealed that the cognate lesions for BER are base damage that does not cause major distortions in the DNA double helix, including oxidized bases (eg, 8-oxoguanine), alkylated bases (eg, 3-meA), deaminated bases (eg, hypoxanthine and xanthine), and uracil.

BER is initiated when a damage-specific DNA glycosylase recognizes a damaged base. *E. coli* has at least six glycosylases, while higher organisms tend to have more [32,33] (Table 4.1). The glycosylase activity hydrolyzes the *N*-glycosydic bond that connects the altered base to its sugar ring, leaving behind an abasic (AP) site. There are two types of glycosylases: monofunctional (such as *E. coli* AlkB) and bifunctional (such as *E. coli* Nei). Removal of bases by monofunctional glycosylases forms AP sites identical to those from spontaneous depurinations or depyrimidinations. These sites require further processing by an AP endonuclease. A bifunctional glycosylase also excises its cognate base (albeit by a different molecular mechanism), but can also incise the phosphodiester backbone at the abasic site to generate a single-strand break (SSB), excluding the need for a separate AP endonuclease. The incision in the phosphodiester backbone provides a 3′ hydroxyl group that is ultimately a substrate for DNA Pol I. The exonuclease function of Pol I removes the damaged strand and the polymerase activity synthesizes replacement DNA. Finally, the nick is sealed by DNA ligase.

An important consequence of BER is the suppression of mutagenesis due to biochemical properties of the damaged bases or uracil in the DNA. Occasional misincorporation of uracil-adjacent adenine during DNA replication is not inherently mutagenic; however, uracil formed via hydrolytic deamination of cytosine is unequivocally mutagenic. In *E. coli*, this misplaced uracil is removed mostly by the monofunctional uracil-DNA-glycosylase UNG (UDG) to avoid transition mutations.

BER enzymes have been found in archaea and the fundamental pathway is conserved [34], although archaeal BER components and their molecular biology are more similar to eukaryotes than to bacteria (for a detailed discussion of this topic, see Ref. [34]). The BER pathways of some archaea have novel features, while others use an additional mechanism to prevent mutation due to genomic uracil. *Ferroplasma acidarmanus* encodes a novel AGT protein (AGTendoV) that has an O^6-methyltransferase domain fused to an endonuclease V domain [35]. This bifunctional enzyme has been found in other archaeal genomes suggesting that it may be a general adaptation to their harsh environments. A more extreme deviation from the canonical BER pathway is the use of uracil-scanning DNA polymerases. In most cases, as noted earlier, genomic uracil is removed by uracil-DNA-glycosylases. Bacterial polymerases, in general, replicate past uracil by inserting an adenine (preserving the sequence); cases where uracil forms via cytosine deamination lead to CG-to-TA mutations [36]. In contrast, some archaeal replicative polymerases stall before misplaced uracils, representing a "read-ahead" proofreading function not found in bacteria or eukaryotes [37,38]. As uracil nucleotides are efficiently removed from these strains, the polymerase must somehow pass off the uracil to another protein for repair before continuing DNA replication. This idea was supported experimentally by Dionne and Bell when they demonstrated that in *Sulfolobus solfataricus,* a uracil glycosylase (UDG1) interacts with the DNA replication processivity factor PCNA (*E. coli* beta clamp), potentially recruiting it to replication forks [39]. One hypothesis to explain this unusual phenomenon is that it provides a "time stamp" to distinguish newly synthesized DNA strands from their templates, analogous to the GATC methylation used by *E. coli* for strand discernment during MMR (see later); however, this idea remains to be explored [40]. While in most respects archaeal BER seem unremarkable, these unique features hint that other interesting discoveries remain to be made.

3.3 Nucleotide Excision Repair: A Versatile DNA-Repair Pathway

NER is a tremendously versatile DNA-repair system that is highly conserved from bacteria to humans (Table 4.1). It consists of two subpathways: global-genome NER (GG-NER), which monitors the entire genome for damage; and transcription-coupled NER (TC-NER), which repairs damage that specifically interferes with transcription (reviewed in detail in [41]). As mentioned in the previous section, expression of NER components (with the exception of UvrC) is regulated by the SOS response and the genes are some of the earliest to be expressed after the detection of DNA damage. The versatility of NER is largely due to its mode of damage recognition. Unlike BER, which recognizes and repairs specific lesions that have little to no effect on the structure of the DNA double helix, the NER pathway monitors the DNA for even small structural distortions. From one perspective, it could be said that the NER pathway repairs distortions, and as a consequence, removes the causal damage (for a list of damage repaired by NER, see Table 4.2).

Both branches of NER consist of four distinct stages: damage detection, damage verification, excision, and ligation.

In *E. coli*, damage is detected via collaboration between UvrA and UvrB in the GG-NER pathway. Alternatively, if the damage is first encountered by RNA polymerase (RNAP), leading to transcription stalling, the Mfd protein (also known as the transcriptional-repair coupling factor, or TRCF) displaces the stalled RNAP and recruits UvrAB to the damage site (TC-NER) [42]. How exactly UvrA and UvrB bind to the damaged DNA remains a challenging experimental problem that is discussed in detail in [41].

After this initial detection step, UvrB takes over and separates the two strands to verify the position of the lesion, simultaneously, leading to the release of UvrA. UvrB remains tightly associated with the DNA and acts as scaffold for UvrC, an enzyme with two nuclease domains. UvrC makes two cuts in the damaged strand, one eight nucleotides 5′ to the lesion and the other four to five nucleotides 3′ to the lesion. *E. coli, Mycobacterium, Salmonella,* and *Clostridium* (at least) have a relatively recently discovered alternative endonuclease called Cho [43], which is also SOS regulated. Cho differs from UvrC, in that it has a single nuclease domain and makes the 3′ incision four nucleotides further from the lesion than UvrC. The exact biological basis for this redundancy is not clear; however, it has been speculated that Cho may be required for especially large lesions that can't be accommodated by UvrC [43]. After incision, UvrC is ejected by the helicase II UvrD and DNA polymerase I (Pol I). Together, UvrD and Pol I displace the damaged strand and Pol I synthesizes the replacement strand using the undamaged sequence as a template, leading to highly accurate repair. Finally, DNA ligase seals the nick at the end of the newly synthesized strand.

TABLE 4.2 Selected Lesions Recognized and Repaired by NER

Damaging Agent	Lesion
4-Nitroquinolone oxide	N^2-deoxyguanosine adducts, and others
Aflatoxin-B1	Purine adducts, N^7-guanine, formamidopyrimidine
Anthramycin	N^2-guanine
N-acetoxy-2-acetylaminofluorene (AAF), *N*-hydroxyaminofluorene (AF)	C^8-guanine
N′-methyl-*N*-nitronitrosguanidine (MNNG)	O^6-methylguanine
Polycyclic aromatic hydrocarbons	N^2-guanine, benzo(a)pyrene diol epoxide, and others
Psoralin	Monoadducts (8-methoxypsoralen, 8-MOP)
Cisplatin	N^7-guanine, interstrand crosslinks
Chemically induced	DNA-protein/DNA-peptide linkages
Mitomycin C	N^7-guanine, O^6-methylguanine, N^2-guanine
Nitrogen mustard	Alkylation
UV radiation	Pyrimidine dimers (6-4) photoproducts

Adapted from Truglio JJ, Croteau DL, Van Houten B, Kisker C. Prokaryotic nucleotide excision repair: the UvrABC system. Chem Rev 2006;106:233–52.

The NER pathways in bacteria and archaea are functionally similar; however, despite the ancient nature of archaea, some aspects of their NER pathways are more similar to eukaryotic versions than to bacterial versions and they may or may not have *uvr* homologs [40,44,45]. The presence of clear *uvr* homologs seems to coincide with lifestyle: mesophilic archaea tend to have *uvr* genes, while hyperthermophilic archaea (HA) do not. A universal feature, however, seems to be the presence of homologs of eukaryotic factors. In two mesophilic species, which have both eukaryotic-like proteins and *uvr* homologs, *Methanobacterium thermoautotrophicum*, and *Halobacterium salinarum*, experimental evidence suggests that they use mostly or entirely the prokaryotic NER proteins [46,47]. Beyond the DNA damage detection stage, most eukaryotic NER factors have additional nonrepair functions; thus, the conservation of these proteins in archaea may simply reflect other cellular functions [48]. The genomes of HA do not encode *uvr* homologs [40]; furthermore, deletion of any of their eukaryotic-like NER genes has little to no effect on UV resistance [40]. These observations force the question: How do archaea that thrive in such harsh environments cope with the absence of such a versatile DNA-repair pathway? One hypothesis proposed by Dennis Grogan is that HA do not attempt to remove lesions before DNA replication and, instead, rely on interactions between replication forks and lesions for repair [40]. Grogan suggests that upon the collision between the replication fork and a blocking lesion, a cut is made in the ssDNA liberated by the unimpeded helicase. This cut would result in the formation of a double-strand end that would be a substrate for end processing; degradation of the double-strand end would remove the lesion. Homologous recombination would then be used to restore the fork for continued replication. This model remains to be fully tested; however, if it is proven true, it would establish a novel paradigm for the repair of many types of lesions.

3.4 Correcting Mismatched Bases: Cleanup After DNA Replication

The primary function of MMR is to remove bases incorrectly inserted by DNA polymerase during DNA replication and its importance is emphasized by its cross-domain functional and homologous conservation (Table 4.1). In *E. coli,* MMR can improve the accuracy of DNA replication up to 400-fold [49]. The *E. coli* MMR pathway has been reconstituted in vitro with only three MMR-specific proteins: MutS, MutL, and MutH [50]. The initiating step of the MMR pathway is the recognition and binding of a mismatched base in the dsDNA helix by a MutS dimer. A MutL dimer subsequently binds to the MutS–DNA complex, thereby stabilizing it and activating the MutH restriction endonuclease. MutH then nicks the strand containing the incorrectly incorporated base. The errant strand is then removed via helicase (UvrD) and exonuclease activities (ExoI/ExoVII/RecJ) and a new strand is synthesized by DNA polymerase III using the undamaged strand as a template. Finally, the nick is sealed by DNA ligase.

An obvious challenge for MMR is to identify which DNA strand has the misincorporated base. Given that the bases themselves are not informative in this respect, *E. coli* meets this challenge by monitoring the methylation status of the two DNA strands (although there is ongoing discussion on the absolute necessity of this activity [51]). As the fork proceeds during DNA replication, the daughter strand is methylated at GATC sites by the DNA adenine methyltransferase Dam. During a transient period, the newly synthesized dsDNA is hemimethylated, that is, only one strand is methylated. Different values for how long this hemimethylated state persists have been obtained using different experimental systems (see Refs [52–54] and [55] and references therein); however, it seems clear that hemimethylation can exist for just minutes after the replication fork passes and the period of hemimethylation limits the window of opportunity for MMR to discern the daughter strand. While GATC sites are overrepresented in the *E. coli* genome, one may not be in the direct proximity of the mismatched base. How the MMR complex can discern the strands in this situation continues to be discussed and debated [51,56,57]; however, a consensus seems to be that the reading of distant GATC sites may occur by two mechanisms: a *cis*-model, in which MutS translocates along the DNA, or a *trans*-model, in which a loop forms between the sites. These details remain an open question and further work is required to fully understand this aspect of MMR.

In the preceding discussions, the *E. coli* repair pathways have generally been used as basic models; however, *E. coli* MMR may be the exception, rather than the rule. Homologs of MutS and MutL are widely distributed, but MutH seems to be rare in other bacteria and archaea. In bacteria that lack MutH, the MutH nuclease activity seems to be replaced by a nuclease activity in MutL [58]. In this way, the MMR of *mutH*-less bacteria are more reminiscent of eukaryotic MMR, where MutLα is required for the incision step. For detailed information on eukaryotic MMR and some comparisons with *E. coli* MMR, see Ref. [59].

Mesothermophilic archaea tend to have MMR pathways that mirror the canonical bacterial pathways [40,60], although they likely originated from horizontal gene transfer [61]. In contrast, the HA lack MutS and MutL homologs (the same group that lacks canonical *uvr* homologs); however, despite the lack of MutS and MutL, genome replication is accurate in these organisms [40]. This lack of increased mutagenesis suggests that the hyperthermophiles have some mechanism that accomplishes the same net outcome as MMR. Solutions to this problem have been hypothesized, but not experimentally validated [40]. One idea is that reconfiguration of progressing replication forks into "chicken foot" structures might expose mismatched bases in the newly synthesized strands for removal by end-processing enzymes—in effect giving the replication fork a "do over" (analogous to the model proposed earlier to replace NER).

Clearly, more work remains in order to understand how HA ensure the stability of their genome sequences under the harsh environmental conditions in which they live. It is difficult to understand why these organisms, which thrive in conditions that may pose the greatest threats to genome stability, lack some of the most universally conserved DNA-repair components (NER and MMR in particular), some of which have even evolved entirely independently in the three domains of life. Understanding the biological implications of this paradox may represent one of the greatest challenges in the DNA repair field and, while it is being addressed by only a relatively small number of groups, persistence may yield some of the most novel future breakthroughs in understanding the sources of genome stability.

A brief statement is necessary regarding the interaction between MMR and homologous recombination [62]. As both of these pathways function in tight association with the replication fork, they share both space and time. It is well established that MMR suppresses illegitimate recombination, especially highlighted by the observation that loss of MMR increases the frequency of interspecies DNA exchange between *E. coli* and *Salmonella* during conjugation [63,64]. Similar observations were also noted for transduction and transformation [65–67]. In this way, MMR can limit the impact of foreign DNA on genome stability, similar to restriction-modification systems (discussed in the following section). Despite the time since these observations, the molecular mechanisms underlying them were fully worked out in 2013 [68].

3.5 Recombination Repair: Dealing With Double-Strand Breaks

It is generally agreed that double-strand breaks (DSBs) in DNA represent the greatest threat to genome stability. Many exogenous and endogenous agents, including cosmic radiation and ionizing radiation; reactive oxygen species; replication fork malfunctions; and chemicals, can cause DSBs. In humans, defects in DSB repair can be potent precursors to cancer development. Because of the extreme importance of this pathway, it has been dissected in remarkable detail and reviewed extensively (for exhaustive reviews, see Refs [69,70]). This chapter presents a discussion focused on variations in the classical recombination pathways and highlights some interesting and important recent discoveries.

HR-dependent repair of DSBs can be distilled into discrete steps that are conserved from bacteria and archaea to eukaryotes (although the players in each step vary):

1. End resection. The broken ends of the DNA must be prepared for the subsequent molecular transactions.
2. Strand invasion. A single-stranded stretch of DNA terminating in a 3′-OH is guided into the duplex of a homologous molecule. This process is mediated by recombinases, including RecA (*E. coli*), Rad51 (many eukaryotes), and RadA (archaea).
3. Branch migration. Strand invasion leads to a four-strand branched intermediate. This intermediate is remodeled to facilitate new DNA synthesis and other molecular processes.
4. Holliday junction resolution. This step leads to the restoration of two DNA duplexes via strand cutting.

One of the most puzzling and unresolved aspects of homologous recombination is how homologous loci are located and brought together before the strand invasion can occur, and in particular, how far apart can the homologous molecules be before the homology search fails. This problem has been addressed in a study in *E. coli* that also highlights the power of superresolution microscopy to understand events in single bacterial cells. Lesterlin et al. [71] demonstrated that DSB-induced pairing of homologous sequences can occur even between distantly separated sister loci (already positioned for segregation into daughter cells). One interesting implication of this discovery is that HR-dependent repair of DSBs is not limited to the short period of time when two newly produced chromosomes are linked after DNA replication. A similar study in *C. crescentus* reinforced that this distant pairing ability is likely a general phenomenon on bacteria [72].

Resistance to radiation, which likely corresponds to an organism's ability to manage the deleterious outcomes of exposure, is not universal. The extremophilic bacteria *D. radiodurans* grows in environments with high levels of ionizing radiation (IR) supported by an exceptional DNA-repair faculty [73]. It was unclear for some time whether *D. radiodurans* possessed additional DNA-repair pathways, or whether it had more potent versions of known pathways. Michael Cox's lab set out to further understand the molecular basis of its radiation resistance by examining several evolved lines of *E. coli* obtained by repeatedly subjecting cultures to selection by IR exposure [74]. Quite remarkably, they were able to recover *E. coli* lines with three to four orders of magnitude higher resistance to 3000 Gy (a high dose) than the parental strain, on par with *D. radiodurans*. While these strains carried multiple mutations, it was ultimately determined that the increased resistance was conferred by mutations in only three genes: *recA*, *dnaB*, and *yfjK* [75]. The first two genes have well-understood functions in DNA repair demonstrating that extreme radiation resistance can arise via genetic innovations in existing pathways, and that additional protein machinery is not necessarily needed. Logically, increased DNA-repair activity would be advantageous; however, this example shows that an organism may not realize its full potential in the absence of the proper selective pressures. In this case, *E. coli* has evolved a DNA-repair system that is just good enough to ensure the stability of its genome within its natural environments.

In most bacteria, mutations that completely abolish HR (eg, *recA* nulls in *E. coli*) are tolerated to varying degrees. Similarly, HR seems to be dispensable in some archaea; however, HR is an essential function in hyperthermophiles, as *radA* deletions are lethal [40]. Furthermore, archaeal homologs responsible for other central steps in HR are also essential (ie, Mre11, Rad50, HerA, and NurA). As discussed earlier, hyperthermophiles lack a conserved NER pathway; therefore, they should accumulate more DNA lesions that could inhibit DNA replication, transcription, or other DNA-related processes. It has been hypothesized that these functions are replaced by novel HR-dependent pathways [40] and their necessity in this pathway may underlie their essentiality.

Nonhomologous end joining (NHEJ) is another pathway for DSB repair. This pathway is error prone as it mediates the direct attachment of two DNA double-strand ends independent of extensive homology; thus, it is a last-resort effort as it almost certainly leads to heritable loss of significant amounts of genetic information. This pathway is perhaps best studied in eukaryotic models, but some bacteria have simplified versions (reviewed in Refs [76,77]). Not due to a lack of effort, a pathway for NHEJ remained elusive in *E. coli* and it was generally accepted for many years that no pathway exists. A 2010 work, however, has demonstrated that *E. coli* strains do possess an end-joining mechanism, now called alternative end joining (A-EJ) [78]. This novel pathway, which does not share conserved factors with canonical NHEJ pathways, depends on bidirectional strand resection, frequent use of microhomology, and nontemplated DNA synthesis. Although conserved components of NHEJ were readily identified in archaea [79–81], a functional repair pathway was identified in 2013 in a mesophilic archaeon [82]. Certainly, further study of end-joining in bacteria and archaea will yield further insight, and perhaps some additional surprises, into this complex DNA-repair pathway.

4. RESTRICTION-MODIFICATION SYSTEMS: PROTECTING THE GENOME FROM INVADERS

DNA damage at the atomic level can be catastrophic, potentially leading to mutations and loss of genetic information; however, genome stability is also threatened on a larger scale by various genetic parasites, including bacteriophage, plasmids, and other specialized genetic elements. In 1978, Werner Arber, Daniel Nathans, and Hamilton Smith won the Nobel Prize for Physiology or Medicine "for the discovery of restriction enzymes and their application to problems of molecular genetics." The first observations of the phenotypic readouts of restriction-modification systems were quite early in the 1950s when it was noticed that some *E. coli* strains were more resistant to bacteriophages (bacterial viruses) than others, leading to the use of the term "restriction" [83,84]. It was also noted that some bacteriophage escaped restriction and were able to infect their host. These modified strains could then be propagated on the original bacterial host, but the resistance was absent when the bacteriophages were transferred to new hosts. Thus, it was clear that the ability of bacteriophage to productively infect their host was controlled by a two-part process in which a pathway restricting infection competed with some type of modification that alleviated the restriction (Fig. 4.2).

Along with later work by Hamilton Smith, in which he purified the first restriction enzyme [85], these discoveries formed the foundation of our current understanding of restriction-modification systems: that cells can distinguish self and foreign DNA, and that the latter is destroyed. In this way, cells could protect their genomes from alteration due to the introduction of nonself-DNA. We now know that restriction-modification systems are widespread in prokaryotes and they have been found in both Bacteria and Archaea. Wild isolates of *E. coli* K-12 express the EcoKI enzyme, encoded by the *hsdRMS* genes (missing in most laboratory strains), as well as three other systems encoded by the *mcrA, mcrBC,* and *mrr* genes (one or more of which exist in many laboratory strains). The first restriction enzyme purified by Smith came from *Haemophilus influenzae* and the extreme thermophilic archaeon *Pyrococcus* encodes a thermostable restriction-modification system [86].

The precise molecular components of restriction-modification systems are diverse and they have been divided into four major groups (I–IV) based on several properties: structure, energy requirement, and cleavage mechanism. In general, all restriction-modification systems function on the same basic molecular principle to distinguish self and foreign DNA. One enzyme encodes a methyltransferase that modifies self-DNA via the addition of methyl groups to specific sequences. Another complementary enzyme recognizes the same sequences and, when they are unmodified, cuts the DNA by hydrolyzing the phosphodiester backbones of both strands. Depending on the group, some restriction-modification systems include additional factors for more complex biochemical activities. The genetic loci that encode restriction-modification systems can be quite complex and often encode variable but coregulated genes [87]—some of which may include additional protective functions (eg, the *E. coli* anticodon nuclease, a defense against T4 phage infection [88]).

The stability of prokaryotic genomes is challenged by three processes that allow the intercellular transfer of genetic material: transformation, transduction, and conjugation. It is clear from sequence analysis of bacterial genomes that genetic

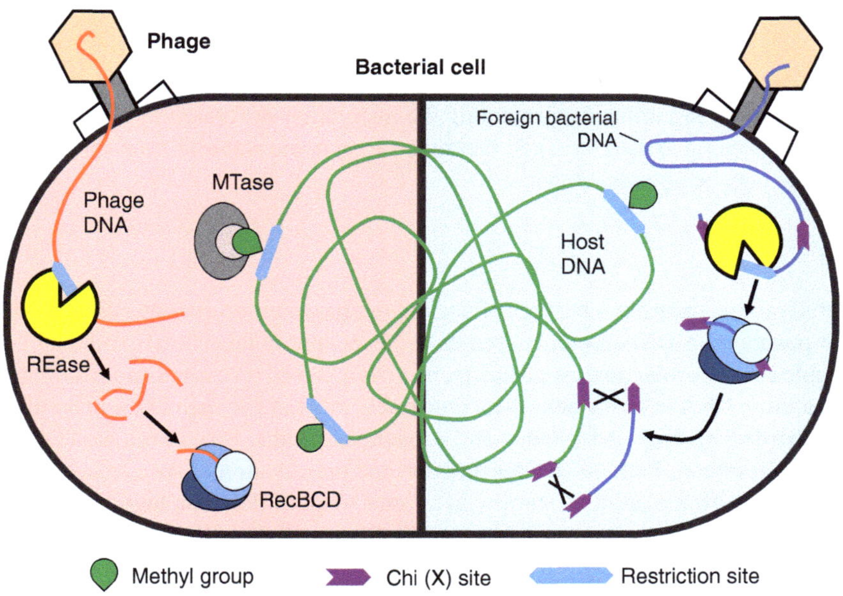

FIGURE 4.2 **Restriction-modification systems control the flux of foreign DNA.** *Red:* Unmethylated bacteriophage DNA is recognized and cleaved by restriction endonuclease (REase). The host DNA is not cut because the restriction enzyme recognition sites are methylated by a sequence-specific DNA methyltransferase (MTase). After cleavage by the REase, the DNA is degraded by the RecBCD complex. *Blue:* A bacteriophage injects nonphage DNA derived from other bacteria (transduction). Unless the DNA donor has a compatible restriction-endonuclease system, the DNA is cleaved by the donor REase. The cleaved DNA becomes a substrate for RecBCD, which degrades the DNA until it reaches a Chi site, a sequence-specific attenuator. DNA transferred from closely related species may recombine due to sequence homology. Since Chi sites are somewhat conserved between different bacterial species, homologous recombination between Chi sites in the foreign DNA and the host DNA can lead to integration of the foreign DNA into the genome.

exchange by these mechanisms has been extensive [89]. Invasions by foreign DNA can induce genome instability via interactions (ie, recombination) with the host chromosome. For example, upon infection, bacteriophages inject their genome into the host cell and, in many cases, the bacteriophage genome inserts into the host chromosome either at specific loci or nonspecifically. These insertions can represent powerful threats to genome integrity since they can disrupt coding or regulatory sequences, potentially disrupting genes or inducing potentially harmful gene expression changes (see Chapter X for further discussion).

Restriction-modification systems can protect cells from plasmids [90–92] and from DNA taken up by transformation (for an example, see Ref. [93]), but the best characterized example is the containment of incoming bacteriophage DNA. Upon entry into a host cell, bacteriophage DNA is, under normal conditions, unmethylated. At this junction, two outcomes are possible: the first option is that the DNA is rapidly cleaved by the restriction enzyme to prevent infection; the second option is that the DNA is methylated by the host methylase, preventing cleavage and supporting infection. As restriction enzymes tend to be more active than methylases, the balance is generally shifted in favor of protection. Support for the idea that restriction-modification systems protect host DNA against invading bacteriophage DNA is further provided by the presence of extensive countermeasures in bacteriophage to circumvent the protective functions [94].

DNA fragmentation by restriction enzymes can also stimulate recombination (eg, see Refs [95,96]), suggesting an alternative way that restriction-modification systems can influence genome stability. In this case, instead of limiting the effects of foreign genetic material on the genome, a restriction-modification system could support the incorporation of novel DNA via recombination [97]. McKane and Milkman [98] demonstrated that when chromosomal DNA from divergent *E. coli* strains was transduced in the laboratory strain K-12, the recombinational replacements were smaller (8–14 kb) than the fragment of DNA injected by the phage (about 100 kb) and that the foreign DNA was inserted in discrete units. These results suggested a model in which the incoming DNA was cleaved by the host restriction-modification system into smaller fragments that were subsequently integrated into the host genome via recombination. As restriction-modification-stimulated recombination in *E. coli* seems to be primarily mediated by the RecBCD end-processing complex [96], it is likely that the incorporation of foreign DNA via this mechanism may be limited to exchanges between closely-related species. One idea is that RecBCD degrades the fragmented DNA until it reaches its control element Chi. Since Chi sequences are well conserved among bacteria, they may serve as substrates for recombination. Ongoing research has revealed even greater complexity in the bacterial response to invading DNA, most notable the CRISPR/Cas system. This pathway represents a primitive type of adaptive immunity and has led to rapid advancements in genome-editing capabilities in several model organisms.

These examples illustrate how restriction-modification systems could simultaneously protect the genome against foreign DNA and introduce genome instability by promoting the integration of foreign DNA. Since restriction-modification systems are present in most bacteria and archaea and the threats to genome stability from foreign parasitic DNA are ubiquitous, it is likely that these may be universal functions.

5. CONCLUSION

DNA has been successfully extracted from a number of ancient organic samples (in one study, as old as 13,000 years [99]) and, despite the lack of postmortem DNA-repair processes and exposure to harsh environmental conditions, some of this DNA has still been suitable for molecular analysis. That ancient DNA can be recovered and manipulated is a testament to its remarkable molecular stability. DNA in living cells is incomparably more stable than DNA in nonliving tissue, even though it is also constantly bombarded by both endogenous and exogenous insults. Since even small defects in DNA sequence can have profound deleterious effects, both on cellular function and preservation of species, natural selection has bestowed highly efficient DNA-repair pathways upon all known life forms. Within domains and across the three domains of life (bacteria, archaea, and Eukarya), a basic core of DNA-repair pathways exists. Remarkably, however, some functionally equivalent pathways appear to have evolved entirely independently. That organisms have converged on this common set of pathways affirms the concept that challenges to genome stability are universal. The examples discussed in this chapter specifically illustrate some variations on these common themes, suggesting that the evolution of DNA repair pathways in different species was influenced by specific challenges experienced in their environments. It is clear that there are exciting opportunities for continued research in the fields of DNA repair and genome stability, even the most humble organisms—bacteria and archaea. Their cells offer many riddles to solve and many possibilities for new, exciting, and beneficial discoveries.

GLOSSARY

Extremophilic bacteria Bacteria that thrive in unusually extreme environments.
Hyperthermophilic archaea Archaea that thrive in high-heat environments.
Mesothermophilic archaea Archaea that live in moderate temperatures.
Processivity The ability of an enzyme to catalyze sequential reactions without disassociating from its substrate.

LIST OF ABBREVIATIONS

1-meA 1-Methyladenine
3-meC 3-Methylcytosine
6-4PP Pyrimidine (6-4) photoproduct
6-meG O^6-methylguanine
A-EJ Alternative end joining
AGT Alkylguanine alkyl transferase
BER Base excision repair
CPD Cyclobutane pyrimidine dimer
DSB Double-strand break
FAD Flavin adenine dinucleotide
GG-NER Global-genome NER
HR Homologous recombination
IR Ionizing radiation
MMR (Methyl-directed) mismatch repair
NER Nucleotide excision repair
Pol I DNA polymerase I
Pol IV DNA polymerase IV
Pol V DNA polymerase V
RNAP RNA polymerase
TC-NER Transcription-coupled NER
TRCF Transcriptional-repair coupling factor
UNG/UDG Uracil-DNA-glycosylase

REFERENCES

[1] Kreuzer KN. DNA damage responses in prokaryotes: regulating gene expression, modulating growth patterns, and manipulating replication forks. Cold Spring Harb Perspect Biol 2013;5:a012674.

[2] Indiani C, O'Donnell M. A proposal: source of single strand DNA that elicits the SOS response. Front Biosci (Landmark Ed) 2013;18:312–23.

[3] Simmons LA, et al. Comparison of responses to double-strand breaks between *Escherichia coli* and *Bacillus subtilis* reveals different requirements for SOS induction. J Bacteriol 2009;191:1152–61.

[4] Modell JW, Hopkins AC, Laub MT. A DNA damage checkpoint in *Caulobacter crescentus* inhibits cell division through a direct interaction with FtsW. Genes Dev 2011;25:1328–43.

[5] Bos J, Yakhnina AA, Gitai Z. BapE DNA endonuclease induces an apoptotic-like response to DNA damage in *Caulobacter*. Proc Natl Acad Sci USA 2012;109:18096–101.

[6] Rand L, et al. The majority of inducible DNA repair genes in *Mycobacterium tuberculosis* are induced independently of RecA. Mol Microbiol 2003;50:1031–42.

[7] Gamulin V, Cetkovic H, Ahel I. Identification of a promoter motif regulating the major DNA damage response mechanism of *Mycobacterium tuberculosis*. FEMS Microbiol Lett 2004;238:57–63.

[8] Wang Y, Huang Y, Xue C, He Y, He Z-G. ClpR protein-like regulator specifically recognizes RecA protein-independent promoter motif and broadly regulates expression of DNA damage-inducible genes in mycobacteria. J Biol Chem 2011;286:31159–67.

[9] Lu H, Chen H, Xu G, Shah AM-U-H, Hua Y. DNA binding is essential for PprI function in response to radiation damage in *Deinococcus radiodurans*. DNA Repair 2012;11:139–45.

[10] Fröls S, et al. Response of the hyperthermophilic archaeon *Sulfolobus solfataricus* to UV damage. J Bacteriol 2007;189:8708–18.

[11] Ajon M, et al. UV-inducible DNA exchange in hyperthermophilic archaea mediated by type IV pili. Mol Microbiol 2011;82:807–17.

[12] Grogan DW. Exchange of genetic markers at extremely high temperatures in the archaeon *Sulfolobus acidocaldarius*. J Bacteriol 1996;178: 3207–11.

[13] van Wolferen M, Ma X, Albers S-V. DNA processing proteins involved in the UV-induced stress response of *Sulfolobales*. J Bacteriol 2015. http://dx.doi.org/10.1128/JB.00344-15.

[14] Müller M, Carell T. Structural biology of DNA photolyases and cryptochromes. Curr Opin Struct Biol 2009;19:277–85.

[15] Brettel K, Byrdin M. Reaction mechanisms of DNA photolyase. Curr Opin Struct Biol 2010;20:693–701.

[16] Husain I, Sancar A. Photoreactivation in *phr* mutants of *Escherichia coli* K-12. J Bacteriol 1987;169:2367–72.

[17] Harm W, Harm H, Rupert CS. Analysis of photoenzymatic repair of UV lesions in DNA by single light flashes. II. In vivo studies with *Escherichia coli* cells and bacteriophage. Mutat Res 1968;6:371–85.

[18] Payne NS, Sancar A. The LexA protein does not bind specifically to the two SOS box-like sequences immediately 5' to the *phr* gene. Mutat Res 1989;218:207–10.

[19] Zhang F, Scheerer P, Oberpichler I, Lamparter T, Krauß N. Crystal structure of a prokaryotic (6-4) photolyase with an Fe-S cluster and a 6,7-dimethyl-8-ribityllumazine antenna chromophore. Proc Natl Acad Sci USA 2013;110:7217–22.

[20] Fitt PS, Sharma N, Castellanos G. A comparison of liquid-holding recovery and photoreactivation in halophilic and non-halophilic bacteria. Biochim Biophys Acta 1983;739:73–8.

[21] Kiener A, Gall R, Rechsteiner T, Leisinger T. Photoreactivation in *Methanobacterium thermoautotrophicum*. Arch Microbiol 1985;143:147–50.

[22] Wood ER, Ghané F, Grogan DW. Genetic responses of the thermophilic archaeon *Sulfolobus acidocaldarius* to short-wavelength UV light. J Bacteriol 1997;179:5693–8.

[23] Yi C, He C. DNA repair by reversal of DNA damage. Cold Spring Harb Perspect Biol 2013;5:a012575.

[24] Mishina Y, Duguid EM, He C. Direct reversal of DNA alkylation damage. Chem Rev 2006;106:215–32.

[25] Kleibl K. Molecular mechanisms of adaptive response to alkylating agents in *Escherichia coli* and some remarks on O(6)-methylguanine DNA-methyltransferase in other organisms. Mutat Res 2002;512:67–84.

[26] Roberts A, Pelton JG, Wemmer DE. Structural studies of MJ1529, an O^6-methylguanine-DNA methyltransferase. Magn Reson Chem 2006;44. Spec No, S71–S82.

[27] Leclere MM, et al. The O6-methylguanine-DNA methyltransferase from the hyperthermophilic archaeon *Pyrococcus* sp. KOD1: a thermostable repair enzyme. Mol Gen Genet 1998;258:69–77.

[28] Kanugula S, Pegg AE. Alkylation damage repair protein O^6-alkylguanine-DNA alkyltransferase from the hyperthermophiles *Aquifex aeolicus* and *Archaeoglobus fulgidus*. Biochem J 2003;375:449–55.

[29] Perugino G, et al. Activity and regulation of archaeal DNA alkyltransferase: conserved protein involved in repair of DNA alkylation damage. J Biol Chem 2012;287:4222–31.

[30] Krokan HE, Bjørås M. Base excision repair. Cold Spring Harb Perspect Biol 2013;5:a012583.

[31] Lindahl T. An N-glycosidase from *Escherichia coli* that releases free uracil from DNA containing deaminated cytosine residues. Proc Natl Acad Sci USA 1974;71:3649–53.

[32] Fromme JC, Banerjee A, Verdine GL. DNA glycosylase recognition and catalysis. Curr Opin Struct Biol 2004;14:43–9.

[33] Jacobs AL, Schär P. DNA glycosylases: in DNA repair and beyond. Chromosoma 2012;121:1–20.

[34] Grasso S, Tell G. Base excision repair in Archaea: back to the future in DNA repair. DNA Repair 2014;21:148–57.

[35] Kanugula S, Pauly GT, Moschel RC, Pegg AE. A bifunctional DNA repair protein from *Ferroplasma acidarmanus* exhibits O^6-alkylguanine-DNA alkyltransferase and endonuclease V activities. Proc Natl Acad Sci USA 2005;102:3617–22.

[36] Lindahl T. Instability and decay of the primary structure of DNA. Nature 1993;362:709–15.
[37] Greagg MA, et al. A read-ahead function in archaeal DNA polymerases detects promutagenic template-strand uracil. Proc Natl Acad Sci USA 1999;96:9045–50.
[38] Wardle J, et al. Uracil recognition by replicative DNA polymerases is limited to the archaea, not occurring with bacteria and eukarya. Nucleic Acids Res 2008;36:705–11.
[39] Dionne I, Bell SD. Characterization of an archaeal family 4 uracil DNA glycosylase and its interaction with PCNA and chromatin proteins. Biochem J 2005;387:859–63.
[40] Grogan DW. Understanding DNA repair in hyperthermophilic archaea: persistent gaps and other reasons to focus on the fork. Archaea 2015:1–12. http://dx.doi.org/10.1155/2015/942605.
[41] Kisker C, Kuper J, Van Houten B. Prokaryotic nucleotide excision repair. Cold Spring Harb Perspect Biol 2013;5:a012591.
[42] Selby CP, Sancar A. Molecular mechanism of transcription-repair coupling. Science 1993;260:53–8.
[43] Moolenaar GF, van Rossum-Fikkert S, van Kesteren M, Goosen N. Cho, a second endonuclease involved in *Escherichia coli* nucleotide excision repair. Proc Natl Acad Sci USA 2002;99:1467–72.
[44] White MF. Archaeal DNA repair: paradigms and puzzles. Biochem Soc Trans 2003;31:690–3.
[45] Grogan DW. The question of DNA repair in hyperthermophilic archaea. Trends Microbiol 2000;8:180–5.
[46] Ogrünç M, Becker DF, Ragsdale SW, Sancar A. Nucleotide excision repair in the third kingdom. J Bacteriol 1998;180:5796–8.
[47] Crowley DJ, et al. The *uvrA, uvrB* and *uvrC* genes are required for repair of ultraviolet light induced DNA photoproducts in *Halobacterium* sp. NRC-1. Saline Syst 2006;2:11.
[48] Rouillon C, White MF. The evolution and mechanisms of nucleotide excision repair proteins. Res Microbiol 2011;162:19–26.
[49] Schaaper RM. Base selection, proofreading, and mismatch repair during DNA replication in *Escherichia coli*. J Biol Chem 1993;268:23762–5.
[50] Lahue RS, Au KG, Modrich P. DNA mismatch correction in a defined system. Science 1989;245:160–4.
[51] Sneppen K, Semsey S. Mismatch repair at stop codons is directed independent of GATC methylation on the *Escherichia coli* chromosome. Sci Rep 2014;4:7346.
[52] Razin A, Szyf M. DNA methylation patterns. Formation and function. Biochim Biophys Acta 1984;782:331–42.
[53] Szyf M, et al. DNA methylation pattern is determined by the intracellular level of the methylase. Proc Natl Acad Sci USA 1984;81:3278–82.
[54] Szyf M, Gruenbaum Y, Urieli-Shoval S, Razin A. Studies on the biological role of DNA methylation: V. The pattern of *E.coli* DNA methylation. Nucleic Acids Res 1982;10:7247–59.
[55] Barras F, Marinus MG. The great GATC: DNA methylation in *E. coli*. Trends Genet 1989;5:139–43.
[56] Wang H, Hays JB. Signaling from DNA mispairs to mismatch-repair excision sites despite intervening blockades. EMBO J 2004;23:2126–33.
[57] Kolodner RD, Mendillo ML, Putnam CD. Coupling distant sites in DNA during DNA mismatch repair. Proc Natl Acad Sci USA 2007;104:12953–4.
[58] Fukui K, Nishida M, Nakagawa N, Masui R, Kuramitsu S. Bound nucleotide controls the endonuclease activity of mismatch repair enzyme MutL. J Biol Chem 2008;283:12136–45.
[59] Jiricny J. Postreplicative mismatch repair. Cold Spring Harb Perspect Biol 2013;5:a012633.
[60] Grogan DW. Stability and repair of DNA in hyperthermophilic Archaea. Curr Issues Mol Biol 2004;6:137–44.
[61] Lin Z, Nei M, Ma H. The origins and early evolution of DNA mismatch repair genes–multiple horizontal gene transfers and co-evolution. Nucleic Acids Res 2007;35:7591–603.
[62] Spies M, Fishel R. Mismatch repair during homologous and homeologous recombination. Cold Spring Harb Perspect Biol 2015;7:a022657.
[63] Rayssiguier C, Dohet C, Radman M. Interspecific recombination between *Escherichia coli* and *Salmonella typhimurium* occurs by the RecABCD pathway. Biochimie 1991;73:371–4.
[64] Rayssiguier C, Thaler DS, Radman M. The barrier to recombination between *Escherichia coli* and *Salmonella typhimurium* is disrupted in mismatch-repair mutants. Nature 1989;342:396–401.
[65] Zahrt TC, Maloy S. Barriers to recombination between closely related bacteria: MutS and RecBCD inhibit recombination between *Salmonella typhimurium* and *Salmonella typhi*. Proc Natl Acad Sci USA 1997;94:9786–91.
[66] Zahrt TC, Mora GC, Maloy S. Inactivation of mismatch repair overcomes the barrier to transduction between *Salmonella typhimurium* and *Salmonella typhi*. J Bacteriol 1994;176:1527–9.
[67] Majewski J, Zawadzki P, Pickerill P, Cohan FM, Dowson CG. Barriers to genetic exchange between bacterial species: *Streptococcus pneumoniae* transformation. J Bacteriol 2000;182:1016–23.
[68] Tham K-C, et al. Mismatch repair inhibits homeologous recombination via coordinated directional unwinding of trapped DNA structures. Mol Cell 2013;51:326–37.
[69] Kowalczykowski SC, Dixon DA, Eggleston AK, Lauder SD, Rehrauer WM. Biochemistry of homologous recombination in *Escherichia coli*. Microbiol Rev 1994;58:401–65.
[70] Persky NS, Lovett ST. Mechanisms of recombination: lessons from *E. coli*. Crit Rev Biochem Mol Biol 2008;43:347–70.
[71] Lesterlin C, Ball G, Schermelleh L, Sherratt DJ. RecA bundles mediate homology pairing between distant sisters during DNA break repair. Nature 2014;506:249–53.
[72] Badrinarayanan A, Le TBK, Laub MT. Rapid pairing and resegregation of distant homologous loci enables double-strand break repair in bacteria. J Cell Biol 2015;210:385–400.
[73] Cox MM, Battista JR. *Deinococcus radiodurans* – the consummate survivor. Nat Rev Microbiol 2005;3:882–92.

[74] Harris DR, et al. Directed evolution of ionizing radiation resistance in *Escherichia coli*. J Bacteriol 2009;191:5240–52.
[75] Byrne RT, et al. Evolution of extreme resistance to ionizing radiation via genetic adaptation of DNA repair. eLife 2014;3:e01322.
[76] Shuman S, Glickman MS. Bacterial DNA repair by non-homologous end joining. Nat Rev Microbiol 2007;5:852–61.
[77] Pitcher RS, Brissett NC, Doherty AJ. Nonhomologous end-joining in bacteria: a microbial perspective. Annu Rev Microbiol 2007;61:259–82.
[78] Chayot R, Montagne B, Mazel D, Ricchetti M. An end-joining repair mechanism in *Escherichia coli*. Proc Natl Acad Sci USA 2010;107:2141–6.
[79] Weller GR, et al. Identification of a DNA nonhomologous end-joining complex in bacteria. Science 2002;297:1686–9.
[80] Nair PA, Smith P, Shuman S. Structure of bacterial LigD 3'-phosphoesterase unveils a DNA repair superfamily. Proc Natl Acad Sci USA 2010;107:12822–7.
[81] Aravind L, Koonin EV. Prokaryotic homologs of the eukaryotic DNA-end-binding protein Ku, novel domains in the Ku protein and prediction of a prokaryotic double-strand break repair system. Genome Res 2001;11:1365–74.
[82] Bartlett EJ, Brissett NC, Doherty AJ. Ribonucleolytic resection is required for repair of strand displaced nonhomologous end-joining intermediates. Proc Natl Acad Sci USA 2013;110:E1984–91.
[83] Luria SE, Human ML. A nonhereditary, host-induced variation of bacterial viruses. J Bacteriol 1952;64:557–69.
[84] Bertani G, Weigle JJ. Host controlled variation in bacterial viruses. J Bacteriol 1953;65:113–21.
[85] Smith HO, Wilcox KW. A restriction enzyme from *Hemophilus influenzae*. I. Purification and general properties. J Mol Biol 1970;51:379–91.
[86] Morgan R, Xiao JP, Xu SY. Characterization of an extremely thermostable restriction enzyme, PspGI, from a *Pyrococcus* strain and cloning of the PspGI restriction-modification system in *Escherichia coli*. Appl Environ Microbiol 1998;64:3669–73.
[87] Makarova KS, Wolf YI, Snir S, Koonin EV. Defense islands in bacterial and archaeal genomes and prediction of novel defense systems. J Bacteriol 2011;193:6039–56.
[88] Kaufmann G. Anticodon nucleases. Trends Biochem Sci 2000;25:70–4.
[89] Ochman H, Lawrence JG, Groisman EA. Lateral gene transfer and the nature of bacterial innovation. Nature 2000;405:299–304.
[90] Pinedo CA, Smets BF. Conjugal TOL transfer from *Pseudomonas putida* to *Pseudomonas aeruginosa*: effects of restriction proficiency, toxicant exposure, cell density ratios, and conjugation detection method on observed transfer efficiencies. Appl Environ Microbiol 2005;71:51–7.
[91] Purdy D, et al. Conjugative transfer of clostridial shuttle vectors from *Escherichia coli* to *Clostridium difficile* through circumvention of the restriction barrier. Mol Microbiol 2002;46:439–52.
[92] Moser DP, Zarka D, Kallas T. Characterization of a restriction barrier and electrotransformation of the cyanobacterium *Nostoc* PCC 7121. Arch Microbiol 1993;160:229–37.
[93] Stein DC, Gregoire S, Piekarowicz A. Restriction of plasmid DNA during transformation but not conjugation in *Neisseria gonorrhoeae*. Infect Immun 1988;56:112–6.
[94] Labrie SJ, Samson JE, Moineau S. Bacteriophage resistance mechanisms. Nat Rev Microbiol 2010;8:317–27.
[95] Thaler DS, Stahl MM, Stahl FW. Tests of the double-strand-break repair model for red-mediated recombination of phage lambda and plasmid lambda dv. Genetics 1987;116:501–11.
[96] Stahl MM, Kobayashi I, Stahl FW, Huntington SK. Activation of Chi, a recombinator, by the action of an endonuclease at a distant site. Proc Natl Acad Sci USA 1983;80:2310–3.
[97] Price C, Bickle TA. A possible role for DNA restriction in bacterial evolution. Microbiol Sci 1986;3:296–9.
[98] McKane M, Milkman R. Transduction, restriction and recombination patterns in *Escherichia coli*. Genetics 1995;139:35–43.
[99] Pääbo S. Ancient DNA: extraction, characterization, molecular cloning, and enzymatic amplification. Proc Natl Acad Sci USA 1989;86:1939–43.
[100] Truglio JJ, Croteau DL, Van Houten B, Kisker C. Prokaryotic nucleotide excision repair: the UvrABC system. Chem Rev 2006;106:233–52.

Chapter 5

Genome Instability in Bacteria: Causes and Consequences

A.B. Williams

University of Cologne, Cologne, Germany

Chapter Outline

1. INTRODUCTION

DNA repair pathways in bacteria are extremely efficient, as the typical mutation rate in *Escherichia coli* cultured under normal conditions is about 2×10^{-10} mutations per base pair (bp) per generation [1]. Nevertheless, various forces can perturb the efficiency or accuracy of DNA repair leading to increased genome instability. This chapter discusses the causes and consequences of some of the best characterized genome instability pathways in bacteria, focusing mostly on *E. coli* and incorporating important findings from other models when possible. From a human health perspective, genome instability is often associated with negative outcomes, such as tumor formation and congenital birth defects; however, from the perspective of a bacterial cell, genome instability can have important positive outcomes, including improved adaptation to host environments, nutrient utilization, and the generation of genetic diversity. Genome instability due to endogenous and exogenous DNA-damaging agents is not discussed here; for more information on this topic, see Chapter 17 and references therein.

2. EFFECTS OF STRESS RESPONSES ON GENOME INSTABILITY

Whether or not prokaryotes possess programmed pathways to modulate mutation rates under stressful conditions remains a controversial topic [2–9]; nevertheless, it is certainly clear (and generally accepted) that several stress responses induce changes in cellular metabolism that can lead to increased genome instability, including the SOS response, the RpoS-regulated general stress response, the stringent response to amino acid starvation, the heat and cold shock responses, and

Genome Stability. http://dx.doi.org/10.1016/B978-0-12-803309-8.00005-7

the polyphosphate-mediated starvation response. While these responses are discussed separately here, attempts were made to make interconnections since it is clear that these distinctions are largely artificial constructions and that bacterial stress responses are actually an extremely complex, interconnected system that can fine-tune stress responses to a diverse array of different types of stress factors (Fig. 5.1).

In this way, bacteria can monitor threats to genome stability from a similar array of challenges and react accordingly by inducing coordinated and concerted responses to promote genome maintenance.

2.1 The SOS Response

Maintenance of the genome during exposure to DNA-damaging agents is critical for the preservation of species. Unlike higher organisms, bacteria lack sophisticated signaling cascades to regulate cell cycle progression in response to DNA damage; nevertheless, many species have evolved a tightly controlled pathway to minimize the deleterious effects of DNA damage known as the SOS response (reviewed in detail in Ref. [10]). While the specifics of the SOS response vary from organism to organism, a conserved feature is that the primary agonist is single-stranded DNA (ssDNA) coated with the RecA recombinase, a complex that is generally absent under normal conditions. That ssDNA plays this important role should not come as a surprise, as it is a byproduct of many types of DNA damage caused by exogenous sources and is an intermediate in several important DNA repair pathways. It can also be formed through defects in endogenous processes, including chromosome segregation, DNA replication and recombination, maintenance of nucleotide pools, and as a byproduct of harmful metabolic intermediates. The SOS response can also be induced by exposure to environmental conditions that do not directly damage DNA. For example, when starved cells grow on glycerol as the primary carbon source, a cyclic-AMP-dependent pathway can lead to SOS induction [11]. Importantly, exposure to certain antibiotics can also lead to SOS induction via the DpiBA two-component signaling pathway (discussed later) [12] and the SOS response can be involved in the development of antibiotic resistance [13]. Furthermore, the activity of the primary regulator of the SOS response, the LexA repressor, is sensitive to changes in the environment. For example, it becomes unstable in alkaline conditions [14] and it is inactivated in ageing colonies [15] and as cells enter stationary phase [16].

The outcome of activation of the SOS response is a massive transcriptional reprogramming enriched in genes involved in DNA repair, chromosome maintenance, and cell division regulators. During the *E. coli* SOS response, these transcriptional changes include the induction of the genes encoding its two Y-family error-prone DNA polymerases: *dinB*, DNA polymerase IV (Pol IV) and *umuDC*, DNA polymerase V (Pol V) (reviewed in Ref. [17]). The mutagenic potential of these polymerases, particularly Pol IV, is a common feature in several of the pathways discussed in more detail later. The primary function of these polymerases is to insert nucleotides opposite bulky lesions in the DNA, an activity that requires relaxed DNA replication fidelity. Nucleotide incorporation by Pol IV and Pol V is largely error free at their cognate lesions; however, their replication of undamaged DNA can be mutagenic due to improper nucleotide discernment and a lack of proofreading. Mutagenesis by Pol IV and Pol V has been evaluated genetically using a *lac* reporter for mutagenesis in cells with constitutive SOS induction [18]. In this system, loss of

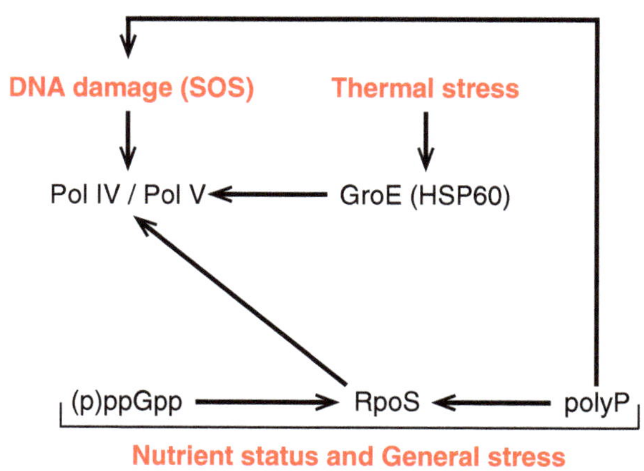

FIGURE 5.1 **Interplay between bacteria stress responses that influence genome stability.** Bacterial stress responses are not isolated in the cell. As cells respond to different stress stimuli, cross-talk between the pathways leads to tight integration of the input signals and allows the cells to cope with the stress as efficiently as possible via concerted response outputs.

Pol IV reduced the frequency of base substitution mutations by 50–70% and loss of Pol V completely eliminated SOS mutagenesis. Together, these results suggest that while the SOS response functions to stabilize the genome, it can also have potentially mutagenic consequences via the induction of mutator genes.

While the mutagenic outcomes of the SOS-dependent induction of *dinB* and *umuDC* are mostly point mutations [19,20] (although Pol IV can also extend misaligned primers leading to insertions and deletions—see later), SOS induction can also promote higher order genome instability. Integrons are mobile genetic elements (MGEs) that can capture and rearrange open reading frames leading to the formation of novel transcriptional units (operons) controlled by an integron-encoded promoter (reviewed in Ref. [21]). The formation of these new sequences is mediated by an integron-encoded site-specific recombinase, IntI, and through their recombination activities, integrons can be important factors in promoting genome instability. Since the *intI* gene is a target of the SOS response [22], SOS induction can enhance integron activity. Because integrons can be easily transmitted between hosts via mobilization in transmissible elements, such as insertion sequences, transposons, or conjugative plasmids (see later), they can serve as important vectors for the dissemination of genetic information, including antibiotic resistance genes [23]. The SOS response also influences genome stability via effects on other MGEs [24]. For example, SOS induction stimulates the activity of the *Vibrio cholerae* SXT integrative conjugative element (ICE), which can transfer and integrate into recipient genomes carrying with it resistance to several antibiotics [24]. Further discussion of SOS-independent cases of horizontal gene transfer (HGT) follows.

Genome instability leading to the development of antibiotic resistance has also been shown to be a possible consequence of SOS activation by antibiotics, both via DNA damage and independent of DNA damage. In some cases, increased SOS-dependent mutagenesis may accelerate the development of antibiotic resistance [13]. Quinolone antibiotics target two type II topoisomerases that have central functions in DNA replication: DNA gyrase and topoisomerase IV [25]. Disruption of these enzymes by quinolones leads to DNA double-strand breaks (DSBs) that can induce the SOS response [26], thereby increasing the chance that mutations occur in these same genes that lead to antibiotic resistance. In one case, ciprofloxacin treatment promoted the formation of antibiotic-resistant strains of pathogenic *E. coli* via a pathway that required several DNA repair and SOS-regulated proteins: the RecA recombinase, the RecBCD end-processing machinery, and the SOS-induced DNA polymerases II, IV, and V (genetically similar to Cairns–Foster mutagenesis, discussed later). Similarly, deletion of the SOS-regulated DNA polymerase DnaE2 in *Mycobacterium tuberculosis* reduced the virulence of the pathogen and decreased the frequency of mutations conferring rifampicin resistance [27]. While the molecular basis of the reduction of virulence is not entirely clear, one hypothesis is that under normal conditions, the genome destabilizing activity of DnaE2 can promote the accumulation of mutations supporting adaptation to the host immune response and increase the frequency of mutations leading to antibiotic resistance.

β-lactam antibiotics bind to and inactivate penicillin-binding proteins, which are involved in cell wall biosynthesis; thus, these antibiotics are generally only effective in dividing cells. Defects in cell wall synthesis can induce the SOS-response via the two-component signaling system encoded by DpiBA [12]. The SOS-induced gene *sulA (sfiA)* blocks cell division conferring a temporary antibiotic resistance phenotype [12]. The simultaneous increase in the levels of Pol IV and Pol V can then increase the chance of mutations conferring antibiotic resistance. The plausibility of this scenario has been demonstrated in a system in which Pol IV-dependent mutations are significantly increased after exposure to β-lactams [28].

2.2 The RpoS-Mediated General Stress Response

The general stress response, regulated by the alternative RNA polymerase sigma factor RpoS (σ^{38}) in *E. coli* (reviewed in Ref. [29]), can also lead to genome destabilizing outcomes as cells experience nutrient deprivation or other stress. For example, induction of the RpoS regulon increases the expression of the error-prone DNA Pol IV independently from the SOS response [30,31]. Higher Pol IV levels play an important role in the development of mutations in the Cairns–Foster system (discussed later), in which mutations are detected in cells under nutritional deprivation. Entry into stationary phase also modulates the methyl-directed mismatch repair (MMR) pathway (see Chapter X). While MMR is active in stationary phase cells, the levels of two components of the pathway, MutS (the mismatch binding protein) and MutH (the initiating endonuclease), decrease via an RpoS-dependent pathway [32]. Overexpression of MMR proteins in stationary phase cells can suppress mutations [32], suggesting that the decline in MMR after starvation results in a decreased genome stability. While decreased MMR activity cannot cause mutations, it can support the preservation of errors made by other processes as mutations.

Finally, RpoS also influences larger scale genomic rearrangements. For example, under starvation conditions, RpoS is involved in the formation of gene amplifications in the Cairns–Foster system [32]. It is also required for genetic rearrangements that can lead to *araB–lacZ* fusions that occur under carbon-limiting conditions [32]. While other examples exist in the literature, these exemplary cases clearly demonstrate that RpoS can be a potent regulator of genome stability, especially during starvation.

2.3 The Stringent Response

During amino acid deprivation, *E. coli* cells induce a well-characterized adaptive mechanism called the stringent response, which is regulated by the alarmone (p)ppGpp, a multi-phosphorylated guanine derivative [33–35]. During the stringent response, transcription of genes involved in coping with the nutrient limitation is enhanced. Demonstrating the complex intermingling of various stress responses possible in bacteria, the levels of the RpoS sigma factor also increase during the stringent response [32]. In this way, the signals leading to (p)ppGpp production and the general stress response may be coordinated to promote cell survival. As a consequence, the genome maintenance functions of the general stress response may be recruited as part of the stringent response.

More direct connections between (p)ppGpp and genome maintenance have also been reported. In *E. coli* and *Bacillus subtilis*, (p)ppGpp can provide a link between DNA replication and nutrient availability [36]. In *E. coli* (p)ppGpp can block the initiation of DNA replication [37], and in *B. subtilis* it can inhibit replication progression [38].

2.4 Heat and Cold Shock Responses

As bacteria experience shifts in their environment, a central adaptive response is the utilization of alternative sigma factors [56]. As already discussed, the RpoS sigma factor is important for survival of starvation conditions. Two additional sigma factors, RpoH ($\sigma^{32/H}$) and RpoE ($\sigma^{24/E}$), regulate the heat shock response [39]. During exposure to high temperatures, the expression of GroE protein, a highly conserved HSP60 homolog is up-regulated, promoting protein folding and stability during temperature stress and, in some cases, during normal growth [41]. Interestingly, GroE is important for the maintenance of the normal levels of both SOS-induced DNA polymerases, Pol IV and Pol V [32]. Under conditions that reduce the levels of GroE, UV mutagenesis is similarly reduced [32]. This requirement for GroE in stabilizing error-prone polymerases suggests that the heat shock response has some function during heat stress to protect the genome against heat-associated insults. Alternatively, progeny cells may benefit from mutations that could arise as side effects from their activities, suggesting that genome fluidity under such conditions may be advantageous. A less-understood connection also exists between cold shock and genome maintenance. The *E. coli* small histone-like protein HU, which consists of two homologous subunits HUα and HUβ, exists in three dimeric forms: HUα_2, HUβ_2, and Hu$\alpha\beta$ [42]. The cellular composition of these three dimers varies under different stress conditions [43]. While HUα_2 and Hu$\alpha\beta$ seem to be most important under normal conditions, the expression of *hupB*, but not *hupA*, increases during shifts to cold temperatures [32], thus increasing the relative levels of HUβ_2 and HU$\alpha\beta$. A shift to low temperature also robustly induces the expression of the transcription factor NusA [32]. Interestingly, both HU$\alpha\beta$ and NusA are required for mutagenesis during starvation in the Cairns–Foster system [32]. Taken together, these observations suggest that both the heat and cold shock responses may modulate genome stability under specific conditions, although the exact biological implications of these potential functions remain unclear.

2.5 Polyphosphate-Mediated Starvation Response

Inorganic polyphosphate (polyP) in *E. coli* is a polymer of orthophosphates that can be tens to hundreds of residues long and are synthesized by polyphosphate kinase (Ppk). While polyP appears to be an energy storage molecule in some organisms, its abundance in *E. coli* seems to be too low to serve such a purpose [44]. Instead, it may function as a gauge for the nutritional status of the bacterial cells. When *E. coli* cells are starved for amino acids, nitrogen, or experience osmotic stress, polyP levels increase due to inhibition of exopolyphosphatase (Ppx), the polyP degradative enzyme and this inhibition depends on increased levels of (p)ppGpp (discussed earlier) [44]. PolyP levels also increase during nutritional downshifts [44] and upon entry into stationary phase and polyP is required for the expression of the *rpoS* gene during stationary phase [45]. In this capacity, polyP may act as an indirect regulator of the repertoire of genes co-regulated in the RpoS regulon and it could serve as rheostat to fine-tune the genome protective capacity of the general stress response under more specific stress conditions.

Quite remarkably, polyP levels also seem to regulate SOS-responsive genes independent of DNA damage [46]. Decreased levels of polyP due to overexpression of *ppx* cause a decrease in DNA damage resistance and block the induction of *recA* and *umuDC* (DNA Pol V)—typical functional markers for SOS activation—after exposure to DNA-damaging agents [32]. Overexpression of *ppk*, which increases the concentration of polyP, induces the expression of *recA* independently of DNA damage or canonical SOS activation by ssDNA-RecA [32]. These observations suggest that the genome protective effects of the SOS response are also recruited in response to stresses unconnected with DNA damage.

3. GENOME INSTABILITY DUE TO STABLE MUTATOR GENOTYPES

Constitutive mutator bacteria have an increased spontaneous mutation rate caused by defects in genes encoding DNA repair factors or other components of other genome monitoring and protective pathways that leads to destabilization of the genome. Of the typical DNA repair pathways (see Chapter 4), the majority of strong mutator phenotypes are due to mutations in the MMR pathway. The MMR pathway is extremely important for maintaining genome stability in *E. coli*, as inactivation of the pathway by mutations in any of the central genes can increase mutation rates between 100- and 200-fold [1]. This large increase in the spontaneous mutation rate in MMR-defective strains reflects the diversity in the types of damage recognized and repaired by this pathway, including incorrectly paired bases (especially due to misincorporation and proofreading failure during DNA replication) and small insertions and deletions. In a 2012 mutation accumulation experiment [1], an MMR-deficient *E. coli* strain had a 138-fold increase in the number of base-pair substitutions compared to the isogenic wild-type strain. The MMR-defective strain also had a 288-fold increase in the formation of insertions and deletions, typically ≤4 nucleotides.

An interesting, and sometimes underappreciated, function of the MMR pathway is to suppress improper recombination [47]. *E. coli* homologous recombination (HR)-mediated by RecA and the RecBCD complex requires perfect or near-perfect homology in the recombining DNA sequences; however, the level of homology required for productive recombination is relaxed in MMR-defective strains. For example, transduction between *Salmonella enterica* serovar Typhimurium and *E. coli* is limited by the recipient's MMR system, which detects and disrupts the formation of heteroduplexes by recognizing sequence divergence [48]. As this example shows, the regulation of recombination by MMR may act as a barrier to HGT between closely related species, thus serving an additional role in preserving genome integrity.

Mutator phenotypes are not due exclusively to defects in the MMR pathway. A key ancillary factor to MMR is the DNA adenine methylase Dam [49]. This protein is required for the DNA methylation that facilitates strand discernment during MMR and has other important roles in DNA replication and gene regulation. Inactivation of *dam* or *drg* (*dam*-replacing genes) in *Pasteurella multocida* leads to robust mutator phenotypes [50]. Mutations in *dnaQ*, which encodes the proofreading (epsilon) subunit of the replicative DNA polymerase III, cause remarkable increases in mutation rate due to defective removal of misincorporated nucleotides during DNA replication. Mutations in genes encoding the GO system (*mutM*, *mutY*, and *mutT*), which repairs oxidized guanines (8-oxodG), lead to medium to high mutator phenotypes (eg, see Ref. [51]). Other mutator genes encode proteins that prevent DNA damage via detoxification, rather than repair damages, such as *oxyR* [52] and *sodA* [53].

Mutator strains are estimated to make up to about 1% of the natural *E. coli* population, and mutator phenotypes occur in both commensal and pathogenic strains [54]. Constitutive mutators represent a powerful challenge for the medical field as they tend to be common in infectious diseases, such as cystic fibrosis (discussed here), urinary tract infections [55], and food-related diseases [56]. This section focuses on one of the best-characterized examples of the impact of mutator strains on clinical practice: *Pseudomonas aeruginosa* colonization of cystic fibrosis (CF) patients.

CF is caused by mutations in the gene encoding the CF transmembrane conductance regulator (CFTR) and is the most common autosomal-recessive genetic diseases in Caucasians [57]. The physical effects of CF lead to a strong predisposition for chronic respiratory infections (CRIs), which are the main cause of high morbidity and premature mortality in CF patients [58]. While infections with a range of bacteria can occur in CF patients, including *Haemophilus influenzae* and *Staphylococcus aureus*, *P. aeruginosa* infection is the most common [59]. Quite remarkably, an early study revealed that up to 20% of the isolates from 37% of the patients examined were chronically colonized by mutator strains of *P. aeruginosa* [60]. In contrast, patients presenting with acute infections did not show such enrichment for mutators [61]. These observations suggest that mutators were especially associated with CRIs. This concept was confirmed by a later study that showed that the proportion of mutator isolates increased from 0% to 65% after 20 years of chronic infection [62]. The genetic basis for these mutator phenotypes was, as might be expected, largely due to defects in the MMR pathway, as between 60% and 90% of the mutator isolates had mutations in MMR genes, most commonly *mutS* [63–65].

One intensely studied consequence of CRI by mutator strains of *P. aeruginosa* is the development of antibiotic resistance. After many years of aggressive antibiotic treatment, antibiotic resistance is more common in *P. aeruginosa* strains isolated from chronically infected CF patients compared to isolates from acute infections [66]. Oliver et al. (2000) first characterized the prevalence of mutator strains among antibiotic-resistant *P. aeruginosa* strains isolated from CF patients. They demonstrated that mutator strains were more frequently resistant to eight commonly used therapeutics against *P. aeruginosa* infection: up to 40% of mutators were resistant compared to only 5% of nonmutators. This correlation between mutator phenotypes and antibiotic resistance were subsequently corroborated by follow-up studies [62,67]. In one particularly interesting study, Ferroni et al. (2009) demonstrated that mutator strains acquired additional antibiotic resistance more rapidly than nonmutator strains. The role of mutator genotypes in promoting antibiotic resistance is not unique to *P. aeruginosa*, as similar relationships have been found in *S. aureus* and *H. influenzae* strains isolated from CF patients [68,69].

A CF patient is normally colonized by a single strain of *P. aeruginosa* that persists through their lifetime [70]. The host respiratory tract represents a microenvironment in which different selective pressures can act on the strain leading to the divergence and fixation of phenotypic variants. In many cases, these novel phenotypes are due to loss-of-function mutations that contribute to host adaptation and support chronic infection. For example, mutations that attenuate virulence genes may shift the pathological outcomes of the infection away from acute damage and toward chronic effects that support persistence. The genetic underpinning of such adaptation has been temporally characterized using whole-genome sequencing [71]. Comparison of the genomes from early and late isolates revealed the accumulation of up to 68 mutations that, in many cases, resulted in the loss of function of virulence genes, representing a virulence adaptation that favored long-term colonization (as mentioned earlier). Not surprisingly, subsequent work demonstrated that this rapid genetic adaptation was driven by a mutator phenotype [63].

This example demonstrates that constitutively higher mutation rates and the corresponding increase in genome instability can be beneficial to an organism as it faces the challenges of its environment; however, such genomic fluidity can also have detrimental outcomes for the organism. For example, adapted strains of *P. aeruginosa* isolated from CF patients have reduced transmissibility [72]. Furthermore, while nonmutator strains can spread between CF patients, the spread of mutator strains has not been observed [65]. Finally, highly adapted mutator isolates have decreased fitness and virulence in secondary environments [73]. Thus, it is clear that, in fact, the genome instability induced by mutator genotypes results in a tug-of-war between potentially beneficial and detrimental outcomes, which are sorted out based on the selective pressures of the environment. While this consequence of genome instability is specifically demonstrated in this example, it likely represents a universal attribute of organisms with elevated mutation rates.

4. GENOME INSTABILITY DUE TO HOMOLOGOUS AND ILLEGITIMATE RECOMBINATION

The genomes of bacteria carry sequences that are especially prone to participate in recombination transactions. Depending on the context, such recombination can lead to intrinsic genomic instability. This section focuses on just a few well-studied examples with important biological outcomes, especially for adaptation and virulence.

4.1 Microsatellite Instability

Microsatellites, or simple sequence repeats (SSRs), consist of repetitions of sequences less than 5 or 6 bps. In bacteria, these sequences are unstable and can expand and contract and have higher spontaneous mutation rates and can thus be potent drivers of adaptation [74]. While the causes of this instability are varied, illegitimate recombination due to strand slippage is a driving factor. Depending on the location in the genome, expansion and contraction of microsatellites can have varying phenotypic outcomes. Contingency loci in bacteria are unstable genetic sequences that can lead to heritable genotypic switching [75]. In some cases, this genetic instability is due to the presence of microsatellite DNA, whose length can vary leading to changes in gene expression and sometimes phase variation. Phase variation is a programmed alteration of the genome that leads to heritable and reversible phenotypic outcomes at frequencies higher than the background spontaneous mutation rate [76]. These phenotypic changes are often important for adaptation to specific niches, including colonization of hosts by pathogens, and survival in new environments. Fimbrial expression in the opportunistic pathogen *H. influenzae* presents one of the best-characterized examples of microsatellite-mediated phase variations (Fig. 5.2A) [77]. *Haemophilus influenzae* cells exhibit different levels of fimbriae expression depending on their biological niche: nasopharyngeal isolates tend to be fimbriated, while isolates from systemic infections are not. Fimbrial phase transitions depend on the differential regulation of the *hifA* and *hifB* genes, which encode the fimbrial structural protein and chaperone, respectively. This switch in transcriptional activity depends on the number of dinucleotide TA repeats between the −10 and −35 sigma factor recognition sequences in the overlapping promoters of *hifA* and *hifB*. Expansions and contractions in the number of TA repeats due to slipped strand mispairing determine whether the spacing of the −10 and −35 sequences in each gene's promoter are optimal for binding by RNA polymerase.

4.2 Gene Conversion

Gene conversion occurs when two related but divergent sequences exist in the same cell and can be substrates for recombination [78]. The outcome of gene conversion is a unidirectional transfer of genetic sequence information from a donor sequence into a highly similar recipient sequence. One of the most studied outcomes of gene conversion in bacteria is the type IV pilus antigenic variation in *Neisseria gonorrhoeae* [79]. Antigenic variation leads to phenotypic heterogeneity within a genetically clonal bacterial population, as different cells can express one of a several possible antigenic forms

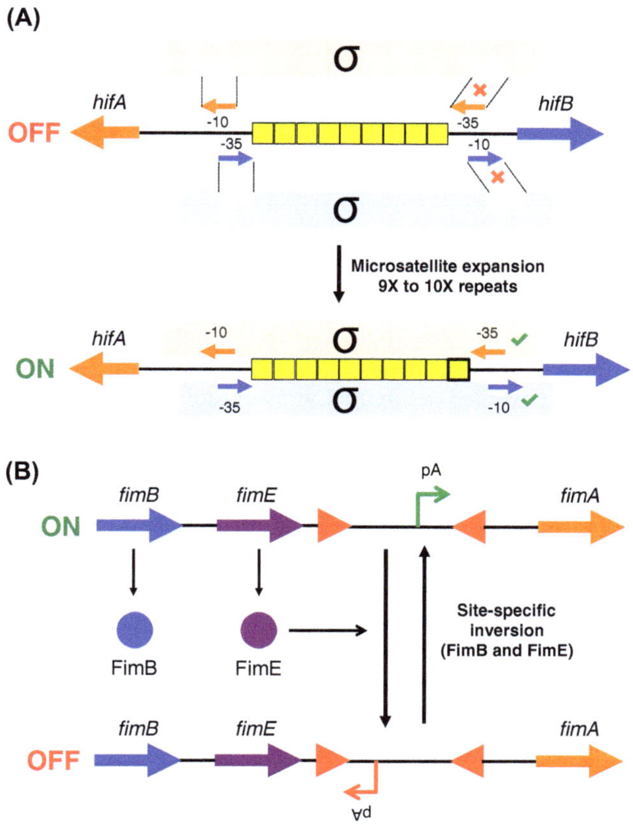

FIGURE 5.2 **Recombination-mediated genome instability leads to changes in gene expression.** Bacteria have harnessed genome instability mediated by specialized DNA sequences to tightly control gene expression. (A) Microsatellite expansions and contractions in *Haemophilus influenzae* regulate the expression of the pilus by altering the spacing between the −10 and −35 promoter elements. Changes in this spacing control the binding ability of the RNA polymerase sigma factors. Different numbers of TA repeats (*yellow boxes*) either block expression (9X repeats), lead to optimal expression (10X repeats), or promote limited expression (11X repeats). (B) Site-specific inversion controls the expression of fimbriae in *Escherichia coli* by switching the orientation of the promoter for the structural gene for type-1 fimbrial protein, *fimA*. This switching, mediated by the FimB and FimE recombinases (encoded by adjacent genes), leads to an off–on switch in gene expression.

of a protein. In *N. gonorrheae*, the antigenic pilus variants are due to the differences in the structure of the pilin protein. Each pilin variant shares a conserved N-terminal region but they differ in the C terminus. The conserved part of pilin is encoded by the *pilE* locus and the variable segment of the protein is encoded by up to six different nontranscribed or weakly transcribed *pilS* loci [80,81]. The expression of a full length, functional pilin protein requires a gene conversion event in which one of the silent *pilS* loci is transferred to the *pilE* locus via a recombination reaction mediated by RecA and RecOR. While the *pilE* locus is genetically unstable, the gene conversion event has no effect on the *pilS* loci. The biological forces driving pilin antigenic variation in *N. gonorrhoeae* are not fully understood; however, iron levels may influence the frequency of antigenic variation and it may be important during transfer into new hosts [82].

4.3 Site-Specific Inversion Systems

A common feature in bacterial genomes are site-specific inversion systems [83–85]. These genetic features range in size from about 100 bps to 35 kilobase pairs (kbps) and are flanked by two terminal inverted repeats. At a frequency of between 10^{-3} and 10^{-5} per cell per generation, these repeats are recognized by an invertase (a specialized recombinase), encoded either in the fragment that is inverted, or elsewhere in the genome, and undergo an inversion event. In general, the inversions result in an on-off toggle and can have a variety of biological consequences, especially in regulating the production of proteins that form structures on the surfaces of cells (eg, flagella and pili). Site-specific inversion at the *fim* locus is the underlying mechanism for fimbrial phase variation in *E. coli* [86–88], which can be an important determinant of virulence in uropathogenic *E. coli* strains (Fig. 5.2B) [89–93]. In this system, the *fimA* gene, which encodes the fimbrial structural protein, is under the control of a σ^{70} promoter that lies on the inverting fragment. The FimE and FimB invertases can mediate the inversion of this fragment, placing the promoter in either the on or the off position, depending on the orientation. This inversion system depends on several *E. coli* small histone-like proteins, such as H-NS and IHF [94,95] and can be influenced by environmental conditions, including changes in nutrient availability [96] and exposure to human urine [97]. The stress factor (p)ppGpp also influences the frequency of inversion [98], demonstrating yet another interaction between stress signaling and genome instability. Site-specific inversion represents an example in which the bacterial cell has exploited genomic instability to integrate information from complex environmental signals to finely regulate gene expression.

4.4 Error-Prone Double-Strand Break Repair

In most normal circumstances, the repair of DNA DSBs is considered an error-free repair process, since an undamaged homologous molecule is used as a template for the repair; however, a collection of work has revealed that under stressful conditions bacteria can switch from a high-fidelity DSB repair pathway to a lower fidelity pathway. This switch to low-fidelity repair has been most extensively worked out in the Cairns–Foster adaptive mutation Lac reversion assay [99,100]. This system was mostly used in *E. coli*, but it was also adapted for use in *S. enterica*, and while the features are generally similar in both species, certain divergent features have been reported [9]. This discussion focuses on the *E. coli* model, as it has been more extensively analyzed.

In the Cairns–Foster system, a Lac$^-$ *E. coli* strain carrying a +1 bp frameshift in a *lac* allele on an F′ conjugative plasmid is grown in minimal medium with glycerol (or another nonlactose carbon source) to stationary phase. During this growth period, typical Luria-Delbrück spontaneous, selection-independent mutations occur. These stationary phase cells are then plated on minimal medium containing lactose as the sole carbon source so that only cells that revert to Lac$^+$ can form colonies. Over the course of several days, Lac$^+$ colonies form continuously with more or less linear kinetics. Extensive analysis of the genetic requirements for this reversion under selection has revealed that several stress responses are required: the RpoS general stress response, the SOS response, and the RpoE envelope stress response (reviewed in Ref. [32]). During the first 5 days of incubation, most colonies have a compensatory −1 bp frameshift in a homopolymeric sequence of guanine residues [32]. In subsequent days, colonies containing amplifications of the *lac* allele become the predominant type [32]. While the point mutations and the amplifications happen independently [32], it is clear that both are stimulated by stress since the RpoS response is required in both cases. Analysis of the molecular requirements for the point mutations has revealed that proteins involved in DSB repair via HR are central to the process [32]. While dissenting opinions do exist [9], it is generally accepted that the stress-induced point mutations are due to an RpoS-mediated switch from high-fidelity HR-dependent DSB repair to a low-fidelity mode of HR-dependent DSB repair at a persistently generated DSB [101,102]. When cells are growing normally without stress, DSB repair by HR is a multistep process that ends with the synthesis of homologous DNA by the high-fidelity replicative DNA polymerase III (Pol III) (see Chapter X for a review of this pathway). During starvation, and upon activation of the RpoS response, it is thought that the SOS and RpoS-regulated error-prone DNA Pol IV (see the preceding paragraphs) replaces Pol III leading to a switch to mutagenic DSB repair [30,32]. Via its ability to extend misaligned primers, DNA synthesis by Pol IV is the main source of the −1 frameshifts in the mutant *lac* allele. It is important to bear in mind that while certain models for the formation of these point mutations require the *lac* allele to be carried on the F′ plasmid and invoke selection as the driving force (rather than a stress-induced increase in the mutation rate) [9], ongoing research has largely ruled out that the F′ plasmid is an essential component of stress-induced mutations. It is now clear that Cairns–Foster mutagenesis can also occur in starved cells with no plasmid and in vastly different, *lac*-independent experimental systems (eg, a tetracycline resistance reversion assay) [101,102]; thus, it is likely that this mutational pathway is not a phenomenon specific to the original Lac reversion system, and that it likely has broad implications for genome stability under stressful conditions.

Since its discovery, research on Cairns–Foster mutagenesis has largely focused on unraveling the complex basic genetic requirements of the pathway in *E. coli* and *S. enterica*, without much emphasis on the implications of the pathway. Interestingly, some findings hint that Cairns–Foster mutagenesis may have important clinical consequences. As discussed earlier, exposure to ciprofloxacin causes resistance to the drug via a mutagenic pathway that shares most genetic requirements with the low-fidelity DSB repair underlying Cairns–Foster mutagenesis [13]. In another example, exposure of pathogenic *Salmonella* to bile can lead to accumulation of mutations that confer bile resistance [103]. Quite remarkably, this mutagenic pathway also requires DSB repair proteins, Pol IV, and SOS proteins. These examples suggest that stress-induced Cairns–Foster mutagenesis may have important implications in understanding the evolutionary forces driving the development of antibiotic resistance and may provide new targets for drugs to slow the formation of novel antibiotic-resistant pathogenic strains. As our ability to dissect complex molecular mechanisms continues to improve, it is critical that we are willing to sever ties to or to refine classical models to avoid missing such potentially important biological processes.

5. GENOME INSTABILITY DUE TO SPECIALIZED GENETIC ELEMENTS

Despite the relative simplicity of bacteria, their genomes are extraordinarily complex and consist of a number of functional elements in addition to the core coding genes and their respective basic regulatory sequences. Many of these ancillary elements have properties that allow them to move around in the genome and these special sequences are called mobile elements. In many cases, the movement of mobile elements within the genome can have mutagenic outcomes and, thus, they represent an important threat to genome stability. Mobile elements are very diverse and, for the sake of brevity, only four types of intrinsic elements are considered here: insertion sequences, transposons, integrons (discussed with the SOS response earlier), and miniature inverted-repeat transposable elements (MITEs) (Fig. 5.3).

The current discussion focuses on intrinsic genome instability caused by mobile elements. Genome alterations via transmissible mobile elements between cells are discussed in the following section.

5.1 Insertion Sequences

Insertion sequences are small (<2.5 kb) DNA segments delimited by short terminal inverted repeats that contain one (or sometimes two) open reading frames that encode proteins specifically required for the mobility of the insertion sequence, that is, a transposase [104]. Insertion sequences can represent a severe threat to genome instability as their insertion always changes the bacterial chromosome; however, excision can either restore the original sequence of a chromosome, or generate a mutation. Insertion results in the introduction of foreign DNA sequence (the transposase gene) and often the molecular exchanges leading to insertion result in the formation of direct repeats. Insertion can interrupt a gene or can alter genetic regulatory sequences leading to changes in gene expression or even inactivation of genes. Improper excision of insertion sequences can either leave some insertion sequences behind, resulting in an insertion, or remove some host DNA, resulting in a deletion. Recombination between insertion sequences and homologous DNA in the host can also lead to genomic rearrangements. Despite their small size and relative simplicity, the biology of insertion sequences is complex.

The genomic instability resulting from insertion and excision of insertion sequences can have important consequences for bacterial cells, including some pathogens. One biological outcome of insertion sequence mobility is phase variation (introduced earlier). Insertion of IS*492* in the *eps* locus of *Pseudoalteromonas atlantica*, a pathogen of crabs, prevents expression of extracellular proteins involved in biofilm formation [105,106]. Precise excision,

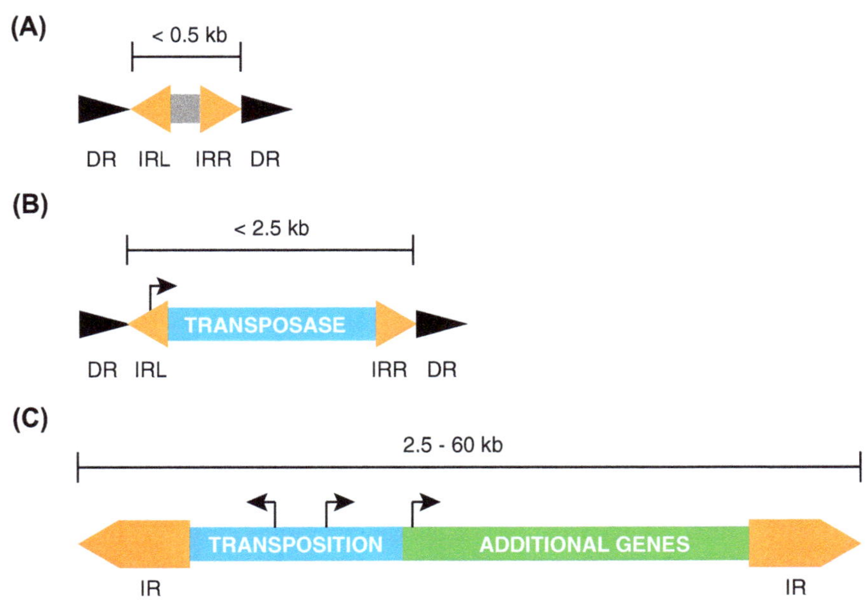

FIGURE 5.3 **Typical structures of mobile genetic elements.** (A) MITEs are small elements delineated by inverted repeats (IRL and IRR) and bordered by direct repeats that are formed during insertion (DR). MITEs do not encode additional genes. (B) Insertion sequences are larger elements that, like MITEs, are also flanked by direct repeats and inverted repeats. Unlike MITEs, insertion sequences encode at least one gene for a transposase. (C) Transposons have more complex structures and are much larger than MITEs and insertion sequences. They are flanked by long inverted repeats and typically encode several genes. One gene cassette encodes several genes required for transposition (blue) and the other encodes additional genes with various functions, especially common are genes that confer drug resistance (green).

mediated by the transposase MooV, produces a circular insertion sequence-derived molecule and allows expression of the *eps* locus [107,108]. In this case, genomic instability introduced by the insertion sequence offers the bacterial cells phenotypic variability to adjust to different environmental conditions.

5.2 Transposons (Nonconjugative)

Transposons are similar to insertion sequences except that they encode additional factors independent of their mobility functions, including genes for antibiotic resistance, virulence, and fitness (eg, heavy metal resistance and enhanced metabolic capabilities). As a consequence, transposons tend to be much larger, ranging in size from 2.5 to 60 kb, and are flanked by terminal repeats. For the most part, the genomic instability induced by transposons is similar to that caused by insertion sequences, that is, gene disruption or deregulation and larger scale genome alterations, such as deletions, duplications, and inversions within one cell. These alterations depend on the mobilization of transposons; thus, the root of transposon-induced genome instability is how the transposon activity is regulated. Any stimulus that induces transposon activity, even if it does not cause DNA damage itself, can induce genome instability secondarily in cells that carry certain transposons. Such stimuli are diverse and range from intrinsically regulated factors, to regulation by host elements [109–111].

Because transposons can carry large amount of genetic information, their movement can have remarkable biological consequences. Two well-known complex transposons from *E. coli* are particularly well-known vectors for antibiotic resistance: Tn*10* [112–114] encodes tetracycline resistance and Tn*5* [115,116] encodes resistance to kanamycin, bleomycin, and streptomycin. These transposons are more complex as they are flanked by insertion sequences (IS*10* and IS*50*, respectively) and can integrate into chromosomes from divergent bacterial species. Some even more complex transposon-like sequences, often called ICEs, encode conjugation functions and are discussed in the following section.

5.3 Miniature Inverted-Repeat Transposable Elements

MITEs are short AT-rich sequences (<0.5 kb) that contain terminal inverted-repeat sequences and, in many cases, lie within a stretch of target site duplications [104]. Some of the first MITEs were the Correia elements of *N. gonorrhoeae* and *Neisseria meningitides* [117,118] and they are widespread in eukaryotic genomes, but in 2011 only, their distribution in bacterial genomes was fully appreciated. Depending on the model system, MITEs are also known by alternative terms, including RU elements (enterobacteria) and RUP, BOX, or SPRITE elements (*Streptococcus*) (for details, see Ref. [119]). While similar to transposons, a distinctive feature of MITEs is that they do not encode a transposase; however, they can commandeer transposases from other mobile elements for mobilization [120,121].

MITEs can induce several types of genome instability, including the introduction of genetic material, gene inactivation, changes in gene regulation, or even deletions and chromosomal rearrangements [120]. Studies on Correia elements have provided some particularly clear and interesting outcomes of their activity. Correia insertion points are hotspots for recombination and rearrangement [122,123] and they can alter the stability of the mRNA from neighboring genes [124,125] or act as transcriptional terminators [126]. Quite remarkably, Correia elements also have a −35 sequence compatible with the vegetative σ^{70} RNA polymerase appropriately positioned with a TATA sequence [127]; thus, they can form ectopic promoters at their insertion sites. In fact, Correia elements have been shown to control the transcription of several genes in *Neisseria* species. As work continues on bacterial MITEs, more consequences of their genome destabilizing properties are sure to emerge.

6. GENOME INSTABILITY DUE TO GENETIC EXCHANGE

Bacterial genomes have a remarkable ability to accommodate foreign DNA, either from related strains or even highly divergent species, which can confer a selective advantage to the organism. This process of sharing of relatively large pieces of genetic information is called HGT and it is mediated by MGEs. HGT commonly occurs by three mechanisms: transduction, conjugation, and transformation. As discussed earlier, the nature of MGEs is diverse and they differ in their specific molecular properties [109]. Common MGEs that are subject to transmission include plasmids, bacteriophages, pathogenicity islands (PAIs), insertion sequences and transposons (discussed earlier), and the broad class of ICEs, which includes some conjugative transposons. While bacteria have mechanisms to protect their genomes against invasion and modification by MGEs, the potential advantages of acquiring foreign DNA has certainly led to the evolution of some flexibility in these systems to take advantage of the potential benefits of the genome destabilizing effects of HGT. Diverse species of bacteria share a core set of mechanisms for the dissemination and sharing of genetic information: transduction, conjugation, and transformation. Through these different processes, DNA can be directly shared between donor and recipient cells (conjugation) or can be transmitted via cell-independent mechanisms (transduction and transformation). The outcome of each process for the recipient cells is a genomic expansion via the acquisition of often nonessential genetic material.

6.1 Transduction

Transduction is a process mediated by bacterial viruses called bacteriophages (or phages) in which they transfer DNA by an infectious process. Transduction comes in two basic flavors: generalized transduction and specialized transduction. While the precise processes of generalized and specialized transduction differ, the end outcome is the same: host DNA is inadvertently packaged into phage particles. Upon subsequent rounds of infection, this DNA is then injected into the recipient cell leading not to a productive infection, but to the exchange of nonphage genetic material. Phage can carry virulence factors as part of their core genomes that can be expressed under specific conditions, such as the transition from a *lysogenic phase*, when they are integrated into the host chromosome, to the *lytic stage*, when they actively replicate for reinfection [128]. HGT by phages can lead to the conversion of a nonvirulent bacterial strain into a virulent strain when phage-borne genes are delivered into potential proto-pathogens. The potential complexity of phage-mediated virulence acquisition is exemplified by the relationship between *V. cholerae* and its phage CTXϕ (Fig. 5.4) [129].

The primary virulence factors of *V. cholerae* are encoded by the cholera toxin genes that are carried in the genome of the CTXϕ filamentous phage; thus, *V. cholerae* virulence depends on the delivery of these genes into bacterial cells via phage infection. Phages target their hosts via specific interactions with receptor proteins on the exterior of the bacteria and the receptor for CTXϕ is the intestinal colonization factor TcpA. Quite remarkably, TcpA is encoded in the VP1 PAI that is, in fact, encoded by a lysogenic phage (VPIϕ). Thus, acquisition of the cholera toxin genes by *V. cholerae* requires prior lysogenization of the bacterial cells by the VP1-encoding phage. Other examples of phage-mediated transfer of virulence factors can be found, among others, in *S. enterica*, *E. coli*, and *Streptococcus pyogenes*.

6.2 Conjugation

Conjugation, often called bacterial mating, generally requires direct contact between the donor and recipient cells and, in many cases, depends on the formation of specialized mating structures for the transfer of DNA—the *sex pilus* in Gram-negative bacteria, and an *adhesion-mediated cell–cell attachment mechanism* along with construction of a transfer apparatus in Gram-positive bacteria [130]. Conjugation is normally associated with the transfer of plasmids, generally circular, extra-chromosomal DNA molecules that self-replicate and are partitioned to daughter cells during cell division. Some plasmids, typified by the *E. coli* F plasmid, encode the complex molecular machinery required for conjugation, while others can transfer via piggy-backing

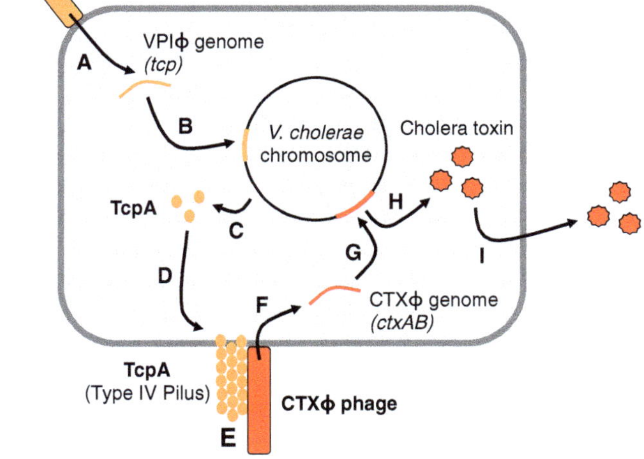

FIGURE 5.4 **Virulence acquisition of *Vibrio cholerae* via phage infection.** Virulence in *V. cholerae* is conferred by the secreted cholera toxin protein, encoded by the phage CTXϕ. Infection by CTXϕ requires previous infection and lysogeny by the VPIϕ phage and occurs via a sequential process. (A) VPIϕ injects its genome into the host cell. (B) The phage DNA integrates into the host genome leading to the stable expression of the TcpA protein from the *tcp* gene cluster (C), which forms the type IV pilus on the cell surface (D). The type IV pilus acts as the receptor for CTXϕ, which binds (E) and injects its genome into the host cell (F). The CTXϕ is integrated into the host genome (G) where it expresses the cholera toxin encoded by the *ctxA* and *ctxB* genes (H). Finally, the CtxAB dimer is secreted leading to the virulent effects on the host (I).

with other conjugative plasmids. In some cases, plasmids can integrate into the host genome resulting in stable heritability and providing an additional mechanism for the transmission of host DNA during conjugation (exemplified by the Hfr lifestyle of the F plasmid). Plasmids can encode a range of nonessential factors that can confer phenotypic variation including changes in virulence, antibiotic resistance, and adaptability to different niches. Conjugative plasmids that carry antibiotic resistance genes are an important agent of antibiotic resistance dissemination in clinical settings. This phenomenon is well illustrated by the RP1 plasmid, first identified in a clinical isolate of *P. aeruginosa* [131]. Quite remarkably, RP1 appears to be transmissible to most, if not all, Gram-negative bacteria and is a potent disseminator of antibiotic resistance, as it encodes resistance to carbenicillin (ampicillin), neomycin, kanamycin, cephaloridine, and tetracycline [132].

In contrast to plasmids, which can self-replicate, ICEs, including certain complex transposons, can also be transmitted by conjugation. These elements integrate into host chromosomes for replication, but then excise and transfer themselves from one cell to another. Insertion and excision of ICEs can cause similar chromosomal changes as insertion sequence elements (discussed earlier). The conjugative transposon Tn*5397*, originally identified in *Clostridium difficile* [133], confers tetracycline resistance and can transfer between *C. difficile* (where it has a strong insertion site preference) and *B. subtilis* (where its integration sites appear to be nonspecific) [134,135]. Tn*5397* has also been found in *Enterococcus faecalis* [136] and oral *Streptococcus* [137], suggesting that this ICE likely transfers in natural environments and could mediate horizontal transfer of antibiotic resistance.

6.3 Transformation

Transformation is a process by which naturally competent bacteria take up naked DNA from the surrounding environment [138]. The foreign DNA then typically integrates into the host chromosome either by HR, or via nonHR encoded by the foreign DNA. While the ability to take up DNA by transformation varies between different bacterial species (*competence*), transformation is a potent mechanism for the transfer of DNA between vastly divergent species. In some cases, transformation is mediated by specific recognition sequences (eg, *H. influenzae* and *N. gonorrhoeae*), but it can also be sequence independent (eg, *B. subtilis* and *Streptococcus pneumoniae*).

Natural transformation was first discovered by Frederick Griffith when he studied how *S. pneumoniae* could switch between virulent and nonvirulent strains [139]. This transformation occurred when a capsule-free nonvirulent strain took up free DNA from a heat-killed virulent strain that encoded a protective capsule. A more common function for transformation in virulence dissemination comes from the oral pathogen *Porphyromonas gingivalis* [140]. When the bacteria form biofilms on the teeth and gums, free DNA can be transferred between virulent and nonvirulent cells.

7. CONCLUSION

Bacteria live in complex environments that require rapid responses to changing conditions, as well as continued adaptation to new niches and hosts. Genome damage and resulting instability due to exposure to DNA-damaging agents, such as chemical mutagens and endogenous metabolic byproducts (eg, reactive oxygen species), is generally considered a threat to bacteria survival; however, the examples discussed here present a different side of genome instability. Bacteria have harnessed genome instability as an important element in regulating their dynamic lifestyle, by controlling gene expression in response to various stresses and stimuli. Furthermore, genome instability is an important part of the generation of genetic diversity, especially via HGT, and represents an important component of bacterial adaptation and evolution. These examples have further demonstrated the importance of genome instability in pathogenic organisms and highlighted the potential impacts on human health. Ongoing research is sure to reveal new facets of these systems, expand our basic knowledge of emerging systems (eg, bacterial MITEs) and long-studied models (eg, stress-induced mutagenesis). Almost certainly new examples of bacterial genome instability will surface as science delves deeper and deeper into the often-underappreciated complexities of bacterial genomes.

GLOSSARY

Contingency loci Simple sequence repeats often located within genes or regulatory regions that are involved in the production of surface proteins. In many cases, contingency loci facilitate responses to environmental cues through genetic rearrangements.

Integrons Mobile genetic elements that capture gene cassettes via recombination.

Phase variation Changes in protein expression, often via on–off systems controlled by genetic rearrangement. They are often involved in responding to rapidly changing environments.

SOS response A bacterial response, first characterized in *E. coli*, to DNA damage. Functionally, it is somewhat analogous to a eukaryotic cell cycle checkpoint.

LIST OF ABBREVIATIONS

bp Base pair
CF Cystic fibrosis
CRI Chronic respiratory infection
DSB Double-strand break
HGT Horizontal gene transfer
ICE Integrative conjugative element
kb Kilobase
MGE Mobile genetic element
MITE Miniature inverted-repeat transposable elements
MMR Methyl-directed mismatch repair
PAI Pathogenicity island
Pol III DNA polymerase III
Pol IV DNA polymerase IV
Pol V DNA polymerase V
polyP Inorganic polyphosphate
Ppk Polyphosphate kinase
Ppx Exopolyphosphatase
ssDNA single-stranded DNA
SSR Simple sequence repeat
$\sigma^{24/E}$ RpoE sigma factor
$\sigma^{32/H}$ RpoH sigma factor
σ^{38} RpoS sigma factor

REFERENCES

[1] Lee H, Popodi E, Tang H, Foster PL. Rate and molecular spectrum of spontaneous mutations in the bacterium *Escherichia coli* as determined by whole-genome sequencing. Proc Natl Acad Sci USA October 9, 2012;109(41):E2774–83.

[2] Rosenberg SM, Hastings PJ. Adaptive point mutation and adaptive amplification pathways in the *Escherichia coli* Lac system: stress responses producing genetic change. J Bacteriol August 1, 2004;186(15):4838–43.

[3] Roth JR, Andersson DI. Adaptive mutation: how growth under selection stimulates Lac(+) reversion by increasing target copy number. J Bacteriol August 2004;186(15):4855–60.

[4] Foster PL. Adaptive mutation in *Escherichia coli*. J Bacteriol August 2004;186(15):4846–52.

[5] Rosenberg SM, Hastings PJ. Rebuttal: adaptive mutation in *Escherichia coli* (Foster). J Bacteriol August 1, 2004;186(15):4853.

[6] Roth JR, Andersson DI. Rebuttal: adaptive mutation in *Escherichia coli* (Foster). J Bacteriol August 2004;186(15):4854.

[7] Foster PL. Rebuttal: adaptive point mutation (Rosenberg and Hastings). J Bacteriol August 2004;186(15):4845.

[8] Stumpf JD, Poteete AR, Foster PL. Amplification of *lac* cannot account for adaptive mutation to Lac^{+} in *Escherichia coli*. J Bacteriol March 2007;189(6):2291–9.

[9] Maisnier-Patin S, Roth JR. The origin of mutants under selection: how natural selection mimics mutagenesis (adaptive mutation). Cold Spring Harb Perspect Biol 2015;7(7):a018176.

[10] Baharoglu Z, Mazel D. SOS, the formidable strategy of bacteria against aggressions. FEMS Microbiol Rev November 1, 2014;38(6):1126–45.

[11] Janion C, Sikora A, Nowosielska A, Grzesiuk E. Induction of the SOS response in starved *Escherichia coli*. Environ Mol Mutagen 2002;40(2):129–33.

[12] Miller C, Thomsen LE, Gaggero C, Mosseri R, Ingmer H, Cohen SN. SOS response induction by beta-lactams and bacterial defense against antibiotic lethality. Science September 10, 2004;305(5690):1629–31.

[13] Cirz RT, Chin JK, Andes DR, de Crécy-Lagard V, Craig WA, Romesberg FE. Inhibition of mutation and combating the evolution of antibiotic resistance. PLoS Biol May 10, 2005;3(6):e176.

[14] Little JW. Mechanism of specific LexA cleavage: autodigestion and the role of RecA coprotease. Biochimie April 1, 1991;73(4):411–21.

[15] Taddei F, Matic I, Radman M. cAMP-dependent SOS induction and mutagenesis in resting bacterial populations. Proc Natl Acad Sci USA December 5, 1995;92(25):11736–40.

[16] Dri AM, Moreau PL. Control of the LexA regulon by pH: evidence for a reversible inactivation of the LexA repressor during the growth cycle of *Escherichia coli*. Mol Microbiol May 1, 1994;12(4):621–9.

[17] Fuchs RP, Fujii S. Translesion DNA synthesis and mutagenesis in prokaryotes. Cold Spring Harb Perspect Biol December 2, 2013;5(12):a012682.

[18] Kuban W, Banach-Orlowska M, Schaaper RM, Jonczyk P, Fijalkowska IJ. Role of DNA polymerase IV in *Escherichia coli* SOS mutator activity. J Bacteriol November 1, 2006;188(22):7977–80.

[19] Wagner J, Nohmi T. *Escherichia coli* DNA polymerase IV mutator activity: genetic requirements and mutational specificity. J Bacteriol August 1, 2000;182(16):4587–95.

[20] Tang M, Shen X, Frank EG, O'Donnell M, Woodgate R, Goodman MF. UmuD'(2)C is an error-prone DNA polymerase, *Escherichia coli* pol V. Proc Natl Acad Sci USA August 3, 1999;96(16):8919–24.

[21] Gillings MR. Integrons: past, present, and future. Microbiol Mol Biol Rev June 2014;78(2):257–77.

[22] Guerin E, Cambray G, Sanchez-Alberola N, Campoy S, Erill I, Da Re S, et al. The SOS response controls integron recombination. Science May 22, 2009;324(5930):1034.

[23] Carattoli A. Importance of integrons in the diffusion of resistance. Vet Res (EDP Sciences) May 2001;32(3–4):243–59.

[24] Beaber JW, Hochhut B, Waldor MK. SOS response promotes horizontal dissemination of antibiotic resistance genes. Nature 2004;427(6969):72–4.

[25] Drlica K, Zhao X. DNA gyrase, topoisomerase IV, and the 4-quinolones. Microbiol Mol Biol Rev (American Society for Microbiology) September 1997;61(3):377–92.

[26] Lewin CS, Howard BM, Ratcliffe NT, Smith JT. 4-quinolones and the SOS response. J Med Microbiol June 1989;29(2):139–44.

[27] Boshoff HIM, Reed MB, Barry CE, Mizrahi V. DnaE2 polymerase contributes to in vivo survival and the emergence of drug resistance in *Mycobacterium tuberculosis*. Cell April 18, 2003;113(2):183–93.

[28] Pérez-Capilla T, Baquero M-R, Gómez-Gómez J-M, Ionel A, Martín S, Blázquez J. SOS-independent induction of *dinB* transcription by beta-lactam-mediated inhibition of cell wall synthesis in *Escherichia coli*. J Bacteriol February 1, 2005;187(4):1515–8.

[29] Battesti A, Majdalani N, Gottesman S. The RpoS-mediated general stress response in *Escherichia coli*. Annu Rev September 22, 2011;65(1):189–213.

[30] Layton JC, Foster PL. Error-prone DNA polymerase IV is controlled by the stress-response sigma factor, RpoS, in *Escherichia coli*. Mol Microbiol October 2003;50(2):549–61.

[31] Storvik KAM, Foster PL. RpoS, the stress response sigma factor, plays a dual role in the regulation of *Escherichia coli's* error-prone DNA polymerase IV. J Bacteriol July 2010;192(14):3639–44.

[32] Williams A, Foster P. Stress-induced mutagenesis. EcoSal Plus 2012. http://dx.doi.org/10.1128/ecosalplus.7.2.3.

[33] Wu J, Xie J. Magic spot: (p) ppGpp. J Cell Physiol August 2009;220(2):297–302.

[34] Boutte CC, Crosson S. Bacterial lifestyle shapes stringent response activation. Trends Microbiol April 2013;21(4):174–80.

[35] Starosta AL, Lassak J, Jung K, Wilson DN. The bacterial translation stress response. FEMS Microbiol Rev November 2014;38(6):1172–201.

[36] Gourse RL, Keck JL. Magic spots cast a spell on DNA primase. Cell March 9, 2007;128(5):823–4.

[37] Schreiber G, Ron EZ, Glaser G. ppGpp-mediated regulation of DNA replication and cell division in *Escherichia coli*. Curr Microbiol 1995;30(1):27–32.

[38] Wang JD, Sanders GM, Grossman AD. Nutritional control of elongation of DNA replication by (p)ppGpp. Cell March 9, 2007;128(5):865–75.

[39] Feklístov A, Sharon BD, Darst SA, Gross CA. Bacterial sigma factors: a historical, structural, and genomic perspective. Annu Rev Microbiol 2014;68(1):357–76.

[40] Deleted in review.

[41] Fayet O, Ziegelhoffer T, Georgopoulos C. The *groES* and *groEL* heat shock gene products of *Escherichia coli* are essential for bacterial growth at all temperatures. J Bacteriol March 1989;171(3):1379–85.

[42] Rouvière-Yaniv J, Kjeldgaard NO. Native *Escherichia coli* HU protein is a heterotypic dimer. FEBS Lett October 15, 1979;106(2):297–300.

[43] Claret L, Rouvière-Yaniv J. Variation in HU composition during growth of *Escherichia coli*: the heterodimer is required for long-term survival. J Mol Biol October 17, 1997;273(1):93–104.

[44] Rao NN, Gómez-García MR, Kornberg A. Inorganic polyphosphate: essential for growth and survival. Annu Rev Biochem 2009;78(1):605–47.

[45] Shiba T, Tsutsumi K, Yano H, Ihara Y, Kameda A, Tanaka K, et al. Inorganic polyphosphate and the induction of *rpoS* expression. Proc Natl Acad Sci USA October 14, 1997;94(21):11210–5.

[46] Tsutsumi K, Munekata M, Shiba T. Involvement of inorganic polyphosphate in expression of SOS genes. Biochim Biophys Acta September 7, 2000;1493(1–2):73–81.

[47] Spies M, Fishel R. Mismatch repair during homologous and homeologous recombination. Cold Spring Harb Perspect Biol March 2015;7(3):a022657.

[48] Zahrt TC, Mora GC, Maloy S. Inactivation of mismatch repair overcomes the barrier to transduction between *Salmonella typhimurium* and *Salmonella typhi*. J Bacteriol March 1994;176(5):1527–9.

[49] Marinus MG. DNA methylation and mutator genes in *Escherichia coli* K-12. Mutat Res October 2010;705(2):71–6.

[50] Chen L, Paulsen DB, Scruggs DW, Banes MM, Reeks BY, Lawrence ML. Alteration of DNA adenine methylase (Dam) activity in *Pasteurella multocida* causes increased spontaneous mutation frequency and attenuation in mice. Microbiology (Reading, England) August 2003;149(Pt 8):2283–90.

[51] Michaels ML, Cruz C, Grollman AP, Miller JH. Evidence that MutY and MutM combine to prevent mutations by an oxidatively damaged form of guanine in DNA. Proc Natl Acad Sci USA August 1, 1992;89(15):7022–5.

[52] Storz G, Tartaglia LA. OxyR: a regulator of antioxidant genes. J Nutr March 1992;122(3 Suppl.):627–30.

[53] Miller A-F. Superoxide dismutases: ancient enzymes and new insights. FEBS Lett March 9, 2012;586(5):585–95.

[54] Matic I, Radman M, Taddei F, Picard B, Doit C, Bingen E, et al. Highly variable mutation rates in commensal and pathogenic *Escherichia coli*. Science September 19, 1997;277(5333):1833–4.

[55] Denamur E, Bonacorsi S, Giraud A, Duriez P, Hilali F, Amorin C, et al. High frequency of mutator strains among human uropathogenic *Escherichia coli* isolates. J Bacteriol January 2002;184(2):605–9.

[56] Le Gall S, Desbordes L, Gracieux P, Saffroy S, Bousarghin L, Bonnaure-Mallet M, et al. Distribution of mutation frequencies among *Salmonella enterica* isolates from animal and human sources and genetic characterization of a *Salmonella* Heidelberg hypermutator. Vet Microbiol June 12, 2009;137(3–4):306–12.

[57] Cutting GR. Cystic fibrosis genetics: from molecular understanding to clinical application. Nat Rev Genet January 2015;16(1):45–56.

[58] Gibson RL, Burns JL, Ramsey BW. Pathophysiology and management of pulmonary infections in cystic fibrosis. Am J Respir Crit Care Med October 15, 2003;168(8):918–51.

[59] Lyczak JB, Cannon CL, Pier GB. Lung infections associated with cystic fibrosis. Clin Microbiol Rev April 2002;15(2):194–222.

[60] Oliver A, Cantón R, Campo P, Baquero F, Blázquez J. High frequency of hypermutable *Pseudomonas aeruginosa* in cystic fibrosis lung infection. Science May 19, 2000;288(5469):1251–4.

[61] Gutiérrez O, Juan C, Pérez JL, Oliver A. Lack of association between hypermutation and antibiotic resistance development in *Pseudomonas aeruginosa* isolates from intensive care unit patients. Antimicrob Agents Chemother September 2004;48(9):3573–5.

[62] Ciofu O, Riis B, Pressler T, Poulsen HE, Høiby N. Occurrence of hypermutable *Pseudomonas aeruginosa* in cystic fibrosis patients is associated with the oxidative stress caused by chronic lung inflammation. Antimicrob Agents Chemother June 2005;49(6):2276–82.

[63] Mena A, Smith EE, Burns JL, Speert DP, Moskowitz SM, Perez JL, et al. Genetic adaptation of *Pseudomonas aeruginosa* to the airways of cystic fibrosis patients is catalyzed by hypermutation. J Bacteriol December 2008;190(24):7910–7.

[64] Oliver A, Baquero F, Blázquez J. The mismatch repair system (*mutS*, *mutL* and *uvrD* genes) in *Pseudomonas aeruginosa*: molecular characterization of naturally occurring mutants. Mol Microbiol March 2002;43(6):1641–50.

[65] Ciofu O, Mandsberg LF, Bjarnsholt T, Wassermann T, Høiby N. Genetic adaptation of *Pseudomonas aeruginosa* during chronic lung infection of patients with cystic fibrosis: strong and weak mutators with heterogeneous genetic backgrounds emerge in *mucA* and/or *lasR* mutants. Microbiology (Reading, England) April 2010;156(Pt 4):1108–19.

[66] Henwood CJ, Livermore DM, James D, Warner M. Pseudomonas Study Group. Antimicrobial susceptibility of *Pseudomonas aeruginosa*: results of a UK survey and evaluation of the British Society for Antimicrobial Chemotherapy disc susceptibility test. J Antimicrob Chemother June 2001;47(6):789–99.

[67] Ferroni A, Guillemot D, Moumile K, Bernede C, Le Bourgeois M, Waernessyckle S, et al. Effect of mutator *P. aeruginosa* on antibiotic resistance acquisition and respiratory function in cystic fibrosis. Pediatr Pulmonol August 2009;44(8):820–5.

[68] Prunier A-L, Malbruny B, Laurans M, Brouard J, Duhamel J-F, Leclercq R. High rate of macrolide resistance in *Staphylococcus aureus* strains from patients with cystic fibrosis reveals high proportions of hypermutable strains. J Infect Dis June 1, 2003;187(11):1709–16.

[69] Román F, Cantón R, Pérez-Vázquez M, Baquero F, Campos J. Dynamics of long-term colonization of respiratory tract by *Haemophilus influenzae* in cystic fibrosis patients shows a marked increase in hypermutable strains. J Clin Microbiol April 2004;42(4):1450–9.

[70] Römling U, Fiedler B, Bosshammer J, Grothues D, Greipel J, von der Hardt H, et al. Epidemiology of chronic *Pseudomonas aeruginosa* infections in cystic fibrosis. J Infect Dis December 1994;170(6):1616–21.

[71] Smith EE, Buckley DG, Wu Z, Saenphimmachak C, Hoffman LR, D'Argenio DA, et al. Genetic adaptation by *Pseudomonas aeruginosa* to the airways of cystic fibrosis patients. Proc Natl Acad Sci USA May 30, 2006;103(22):8487–92.

[72] Mena A, Maciá MD, Borrell N, Moya B, de Francisco T, Pérez JL, et al. Inactivation of the mismatch repair system in *Pseudomonas aeruginosa* attenuates virulence but favors persistence of oropharyngeal colonization in cystic fibrosis mice. J Bacteriol May 2007;189(9):3665–8.

[73] Hogardt M, Hoboth C, Schmoldt S, Henke C, Bader L, Heesemann J. Stage-specific adaptation of hypermutable *Pseudomonas aeruginosa* isolates during chronic pulmonary infection in patients with cystic fibrosis. J Infect Dis January 1, 2007;195(1):70–80.

[74] Zhou K, Aertsen A, Michiels CW. The role of variable DNA tandem repeats in bacterial adaptation. FEMS Microbiol Rev January 2014;38(1):119–41.

[75] Moxon R, Bayliss C, Hood D. Bacterial contingency loci: the role of simple sequence DNA repeats in bacterial adaptation. Annu Rev Genet 2006;40(1):307–33.

[76] van der Woude MW. Phase variation: how to create and coordinate population diversity. Curr Opin Microbiol April 2011;14(2):205–11.

[77] van Ham SM, van Alphen L, Mooi FR, van Putten JP. Phase variation of *H. influenzae* fimbriae: transcriptional control of two divergent genes through a variable combined promoter region. Cell June 18, 1993;73(6):1187–96.

[78] Santoyo G, Romero D. Gene conversion and concerted evolution in bacterial genomes. FEMS Microbiol Rev April 2005;29(2):169–83.

[79] Hill SA, Davies JK. Pilin gene variation in *Neisseria gonorrhoeae*: reassessing the old paradigms. FEMS Microbiol Rev May 2009;33(3):521–30.

[80] Haas R, Meyer TF. The repertoire of silent pilus genes in *Neisseria gonorrhoeae*: evidence for gene conversion. Cell January 17, 1986;44(1):107–15.

[81] Wachter J, Masters TL, Wachter S, Mason J, Hill SA. *pilS* loci in *Neisseria gonorrhoeae* are transcriptionally active. Microbiology (Reading, England) May 2015;161(Pt 5):1124–35.

[82] Cahoon LA, Seifert HS. Focusing homologous recombination: pilin antigenic variation in the pathogenic *Neisseria*. Mol Microbiol September 2011;81(5):1136–43.

[83] Dybvig K. DNA rearrangements and phenotypic switching in prokaryotes. Mol Microbiol November 1993;10(3):465–71.

[84] van der Woude MW, Bäumler AJ. Phase and antigenic variation in bacteria. Clin Microbiol Rev July 2004;17(3):581–611.

[85] Grindley NDF, Whiteson KL, Rice PA. Mechanisms of site-specific recombination. Annu Rev Biochem 2006;75(1):567–605.

[86] Klemm P. Two regulatory *fim* genes, *fimB* and *fimE*, control the phase variation of type 1 fimbriae in *Escherichia coli*. EMBO J June 1986;5(6):1389–93.

[87] Kulasekara HD, Blomfield IC. The molecular basis for the specificity of *fimE* in the phase variation of type 1 fimbriae of *Escherichia coli* K-12. Mol Microbiol February 1999;31(4):1171–81.

[88] Sohanpal BK, Kulasekara HD, Bonnen A, Blomfield IC. Orientational control of *fimE* expression in *Escherichia coli*. Mol Microbiol October 2001;42(2):483–94.

[89] Hultgren SJ, Schwan WR, Schaeffer AJ, Duncan JL. Regulation of production of type 1 pili among urinary tract isolates of *Escherichia coli*. Infect Immun December 1986;54(3):613–20.

[90] Schaeffer AJ, Schwan WR, Hultgren SJ, Duncan JL. Relationship of type 1 pilus expression in *Escherichia coli* to ascending urinary tract infections in mice. Infect Immun February 1987;55(2):373–80.

[91] Connell I, Agace W, Klemm P, Schembri M, Mărild S, Svanborg C. Type 1 fimbrial expression enhances *Escherichia coli* virulence for the urinary tract. Proc Natl Acad Sci USA September 3, 1996;93(18):9827–32.

[92] Langermann S, Palaszynski S, Barnhart M, Auguste G, Pinkner JS, Burlein J, et al. Prevention of mucosal *Escherichia coli* infection by FimH-adhesin-based systemic vaccination. Science April 25, 1997;276(5312):607–11.
[93] Holden NJ, Totsika M, Mahler E, Roe AJ, Catherwood K, Lindner K, et al. Demonstration of regulatory cross-talk between P fimbriae and type 1 fimbriae in uropathogenic *Escherichia coli*. Microbiology (Reading, England) April 2006;152(Pt 4):1143–53.
[94] Corcoran CP, Dorman CJ. DNA relaxation-dependent phase biasing of the *fim* genetic switch in *Escherichia coli* depends on the interplay of H-NS, IHF and LRP. Mol Microbiol December 2009;74(5):1071–82.
[95] O'Gara JP, Dorman CJ. Effects of local transcription and H-NS on inversion of the *fim* switch of *Escherichia coli*. Mol Microbiol April 2000;36(2):457–66.
[96] Lahooti M, Roesch PL, Blomfield IC. Modulation of the sensitivity of FimB recombination to branched-chain amino acids and alanine in *Escherichia coli* K-12. J Bacteriol September 2005;187(18):6273–80.
[97] Greene SE, Hibbing ME, Janetka J, Chen SL, Hultgren SJ. Human urine decreases function and expression of type 1 pili in uropathogenic *Escherichia coli*. mBio 2015. 6(4):e00820–15.
[98] Åberg A, Shingler V, Balsalobre C. (p)ppGpp regulates type 1 fimbriation of *Escherichia coli* by modulating the expression of the site-specific recombinase FimB. Mol Microbiol June 2006;60(6):1520–33.
[99] Cairns J, Foster PL. Adaptive reversion of a frameshift mutation in *Escherichia coli*. Genetics August 1, 1991;128(4):695–701.
[100] Foster PL. Adaptive mutation: implications for evolution. Bioessays December 2000;22(12):1067–74.
[101] Ponder RG, Fonville NC, Rosenberg SM. A switch from high-fidelity to error-prone DNA double-strand break repair underlies stress-induced mutation. Mol Cell September 16, 2005;19(6):791–804.
[102] Shee C, Gibson JL, Darrow MC, Gonzalez C, Rosenberg SM. Impact of a stress-inducible switch to mutagenic repair of DNA breaks on mutation in *Escherichia coli*. Proc Natl Acad Sci August 16, 2011;108(33):13659–64.
[103] Prieto AI, Ramos-Morales F, Casadesús J. Repair of DNA damage induced by bile salts in *Salmonella enterica*. Genetics October 2006;174(2):575–84.
[104] Siguier P, Gourbeyre E, Varani A, Ton-Hoang B, Chandler M. Everyman's guide to bacterial insertion sequences. Microbiol Spectr April 2015;3(2):555–90.
[105] Bartlett DH, Silverman M. Nucleotide sequence of IS*492*, a novel insertion sequence causing variation in extracellular polysaccharide production in the marine bacterium *Pseudomonas atlantica*. J Bacteriol March 1989;171(3):1763–6.
[106] Bartlett DH, Wright ME, Silverman M. Variable expression of extracellular polysaccharide in the marine bacterium *Pseudomonas atlantica* is controlled by genome rearrangement. Proc Natl Acad Sci USA June 1988;85(11):3923–7.
[107] Higgins BP, Carpenter CD, Karls AC. Chromosomal context directs high-frequency precise excision of IS*492* in *Pseudoalteromonas atlantica*. Proc Natl Acad Sci USA February 6, 2007;104(6):1901–6.
[108] Perkins-Balding D, Duval-Valentin G, Glasgow AC. Excision of IS*492* requires flanking target sequences and results in circle formation in *Pseudoalteromonas atlantica*. J Bacteriol August 1999;181(16):4937–48.
[109] Craig N, Gragie R, Gellert M, Lambowitz AM, editors. Mobile DNA II. 2nd ed. American Society for Microbiology; 2002.
[110] Labrador M, Corces VG. Transposable element-host interactions: regulation of insertion and excision. Annu Rev Genet 1997;31(1):381–404.
[111] Nagy Z, Chandler M. Regulation of transposition in bacteria. Res Microbiol June 2004;155(5):387–98.
[112] Kleckner N. Regulating Tn*10* and IS*10* transposition. Genetics March 1990;124(3):449–54.
[113] Kleckner N, Bender J, Gottesman S. Uses of transposons with emphasis on Tn*10*. Methods Enzymol 1991;204:139–80.
[114] Schlör S, Kemmer G, Reidl J. Transposon Tn*10*. Methods Mol Med 2003;71:211–24.
[115] Reznikoff WS. The Tn*5* transposon. Annu Rev Microbiol 1993;47(1):945–63.
[116] Reznikoff WS. Tn*5* as a model for understanding DNA transposition. Mol Microbiol March 2003;47(5):1199–206.
[117] Correia FF, Inouye S, Inouye M. A family of small repeated elements with some transposon-like properties in the genome of *Neisseria gonorrhoeae*. J Biol Chem September 5, 1988;263(25):12194–8.
[118] Correia FF, Inouye S, Inouye M. A 26-base-pair repetitive sequence specific for *Neisseria gonorrhoeae* and *Neisseria meningitidis* genomic DNA. J Bacteriol September 1986;167(3):1009–15.
[119] Delihas N. Impact of small repeat sequences on bacterial genome evolution. Genome Biol Evol 2011;3(0):959–73.
[120] Delihas N. Small mobile sequences in bacteria display diverse structure/function motifs. Mol Microbiol February 2008;67(3):475–81.
[121] Bardaji L, Añorga M, Jackson RW, Martínez-Bilbao A, Yanguas-Casás N, Murillo J. Cordaux R, editor. Miniature Transposable Sequences Are Frequently Mobilized in the Bacterial Plant Pathogen *Pseudomonas syringae* pv. phaseolicola. PLoS ONE 2011;6(10):e25773.
[122] Buisine N, Tang CM, Chalmers R. Transposon-like Correia elements: structure, distribution and genetic exchange between pathogenic *Neisseria* sp.. FEBS Lett July 3, 2002;522(1–3):52–8.
[123] Liu SV, Saunders NJ, Jeffries A, Rest RF. Genome analysis and strain comparison of correia repeats and correia repeat-enclosed elements in pathogenic *Neisseria*. J Bacteriol November 2002;184(22):6163–73.
[124] De Gregorio E, Abrescia C, Carlomagno MS, Di Nocera PP. The abundant class of nemis repeats provides RNA substrates for ribonuclease III in *Neisseriae*. Biochim Biophys Acta June 7, 2002;1576(1–2):39–44.
[125] De Gregorio E, Abrescia C, Carlomagno MS, Di Nocera PP. Ribonuclease III-mediated processing of specific *Neisseria meningitidis* mRNAs. Biochem J September 15, 2003;374(Pt 3):799–805.
[126] Francis F, Ramirez-Arcos S, Salimnia H, Victor C, Dillon JR. Organization and transcription of the division cell wall (*dcw*) cluster in *Neisseria gonorrhoeae*. Gene June 27, 2000;251(2):141–51.
[127] Black CG, Fyfe JA, Davies JK. A promoter associated with the neisserial repeat can be used to transcribe the *uvrB* gene from *Neisseria gonorrhoeae*. J Bacteriol April 1995;177(8):1952–8.

[128] Penadés JR, Chen J, Quiles-Puchalt N, Carpena N, Novick RP. Bacteriophage-mediated spread of bacterial virulence genes. Curr Opin Microbiol February 2015;23:171–8.

[129] Faruque SM, Mekalanos JJ. Phage-bacterial interactions in the evolution of toxigenic *Vibrio cholerae*. Virulence (Taylor & Francis) November 15, 2012;3(7):556–65.

[130] Smillie C, Garcillán-Barcia MP, Francia MV, Rocha EPC, de la Cruz F. Mobility of plasmids. Microbiol Mol Biol Rev September 2010;74(3): 434–52.

[131] Grinsted J, Saunders JR, Ingram LC, Sykes RB, Richmond MH. Properties of a R factor which originated in *Pseudomonas aeruginosa 1822*. J Bacteriol May 1972;110(2):529–37.

[132] Olsen RH, Shipley P. Host range and properties of the *Pseudomonas aeruginosa* R factor R1822. J Bacteriol February 1973;113(2):772–80.

[133] Hächler H, Kayser FH, Berger-Bächi B. Homology of a transferable tetracycline resistance determinant of *Clostridium difficile* with *Streptococcus* (Enterococcus) *faecalis* transposon Tn*916*. Antimicrob Agents Chemother July 1987;31(7):1033–8.

[134] Mullany P, Wilks M, Lamb I, Clayton C, Wren B, Tabaqchali S. Genetic analysis of a tetracycline resistance element from *Clostridium difficile* and its conjugal transfer to and from *Bacillus subtilis*. J Gen Microbiol July 1990;136(7):1343–9.

[135] Wang H, Smith MCM, Mullany P. The conjugative transposon Tn*5397* has a strong preference for integration into its *Clostridium difficile* target site. J Bacteriol July 2006;188(13):4871–8.

[136] Roberts AP, Braun V, von Eichel-Streiber C, Mullany P. Demonstration that the group II intron from the clostridial conjugative transposon Tn*5397* undergoes splicing in vivo. J Bacteriol February 2001;183(4):1296–9.

[137] Roberts AP, Pratten J, Wilson M, Mullany P. Transfer of a conjugative transposon, Tn*5397* in a model oral biofilm. FEMS Microbiol Lett August 1, 1999;177(1):63–6.

[138] Johnston C, Martin B, Fichant G, Polard P, Claverys J-P. Bacterial transformation: distribution, shared mechanisms and divergent control. Nat Rev Microbiol March 2014;12(3):181–96.

[139] Griffith F. The significance of pneumococcal types. J Hyg (Lond) January 1928;27(2):113–59.

[140] Kerr JE, Abramian JR, Dao D-HV, Rigney TW, Fritz J, Pham T, et al. Hensel M, editor. Genetic Exchange of Fimbrial Alleles Exemplifies the Adaptive Virulence Strategy of *Porphyromonas gingivalis*. PLoS ONE 2014;9(3):e91696.

Chapter 6

CRISPR: Bacteria Immune System

A. Golubov
University of Lethbridge, Lethbridge, AB, Canada

Chapter Outline

1. INTRODUCTION

All living forms from archaea and bacteria domains have various highly ingenious systems to defend themselves from foreign genetic material, which can invade them in the form of a virus particle (bacteriophage) or plasmid DNA. There is an antagonistic coevolution between bacteriophage and host cell that promotes rapid evolution of the diverse types of prokaryotic defense systems. It can be described as a combination of an arms race of defense and counter-defense between a host and an invader, and fluctuating selection on their rare genotypes [1]. To date, there are few known defensive strategies in archaea and bacteria (reviewed in Ref. [2]) that can be referred to as the multilayer prokaryotic immune system: the restriction-modification system (RMS), the adsorption inhibition, abortive infection, blocking DNA injection, and clustered regularly interspaced short palindromic repeats (CRISPRs) [3]. The initial discovery of a CRISPR structure was made accidentally in *Escherichia coli* by Ishino and colleagues in 1987 [4], but the acronym CRISPR was born in 2002 after Jansen and colleagues observed similar structures in archaeal and bacterial genomes [3]. Since 2002, our understanding of this defense system as an adaptive and heritable archaeal/bacterial immunity has made a major step forward. The highly diverse CRISPRs defense system has been found in half of the bacterial and almost all the archaeal genomes sequenced to date [5]. It is comprised of the CRISPRs and the CRISPR-associated (cas) genes (CRISPR/Cas system) that protect cells against selfish invading DNA [3].

In this chapter, I review the history of the discovery, organization in different species, and the mode of work of CRISPR, and its significance for bacterial immunity against foreign genetic material. I focus on the findings that have been made since 2000, and speculate on future perspectives of this fascinating discovery and the CRISPRs-based genome-editing technology.

2. HISTORY OF THE CRISPR/CAS DISCOVERY

The study of phage biology has a long history: bacteriophages were discovered in 1917 by the French-Canadian microbiologist Félix d'Hérelle, working at the Pasteur Institute in Paris. Since then there have been many exciting findings in that field, and it may look weird that the CRISPR/Cas immune system was not discovered before 2002. However, there are two good reasons for that: the tight regulation of the CRISPR/Cas systems in model organisms that hid it from researches, and the absence of high throughput genomic tools (such as cheap and reliable sequencing methods). Indeed,

Genome Stability. http://dx.doi.org/10.1016/B978-0-12-803309-8.00006-9

since the development of the next-generation sequencing methods, CRISPR/Cas research has been flourishing, resulting in the discovery of this highly diverse defense system in many organisms and rapid gain of knowledge about the CRISPR/Cas molecular mechanisms.

In 1987, Ishino et al. [4] were studying the *iap* gene, responsible for alkaline phosphatase isozyme conversion in *E. coli*. Analysis of the *E. coli* chromosomal DNA fragment containing the *iap* gene revealed an unusual structure in its 3′-end flanking region: it contains "five highly homologous sequences of 29 nucleotides that were arranged as direct repeats with 32 nucleotides as spacing." At that time, due to the lack of available genome sequences, no sequence homologs have been found elsewhere in prokaryotes.

During the next 15 years, similarly organized direct repeats were found in nine archaeal genomes (eg, *Haloferax. volcanii*, *Methanocaldococcus jannaschii*, *Methanocaldococcus thermoautotrophicum*), 10 bacterial genomes (eg, *Thermotoga maritima*, *Aquifex aeolicus*, *Yersinia pestis*), and in one mitochondrial genome (*Vicia faba*) [6]. Based on the available data, Mojica et al. proposed new family of repeated short elements that are organized in clusters and named them short regularly spaced repeats (SRSRs), but their function remained unknown [6].

In 2002, Jansen et al. performed in silico analysis of a novel family of repetitive DNA sequences (SRSRs) and found that they are present in more than 40 archaeal and bacterial genomes, but absent from eukaryotes or viruses [3]. They also found that these loci were flanked on one side by a common leader sequence of 300–500 bp, and identified four genes that were invariably located adjacent to the repetitive locus, which might be an indication of their functional relationship. In order to avoid confusing nomenclature, Mojica et al. and Jansen et al. have agreed to use the acronym CRISPR, which reflects the characteristic features of this family of clustered regularly interspaced short palindromic repeats. Four adjacent genes were named cas ("CRISPR-associated"), and Jansen et al. suggested that Cas proteins play important role in the genesis of CRISPR loci. Also, they provided evidence that the putative CRISPR in *V. faba* mitochondrial genome should be considered an imperfect direct repeat and not a CRISPR.

In 2005, Bolotin et al., Mojica et al., and Pourcel et al. published their discovery that spacer sequences match viruses and plasmids, and suggested a defense function for CRISPR/Cas systems in the prevention of phage infection and plasmid conjugation [7–9].

The year of 2006 was the year of the major breakthrough: Makarova et al. incorporated all these data into a model for CRISPR/Cas immunity that is similar to the RNA interference system (RNAi) in eukaryotes and provided the first complete model for the molecular mechanism of CRISPR/Cas defense system [10]. In this model, the small CRISPR RNAs fragments had an antisense function against bacteriophage or plasmid transcripts as prokaryotic siRNAs (psiRNA), by base pairing with the target mRNAs and promoting their degradation or translation shutdown. Also, they classified numerous cas gene products into ~25 distinct protein families with several new functional and structural predictions. Cas proteins in this model have a function of the effectors of the immunity mechanism.

Two further fundamental studies uncovered two more features of CRISPR/Cas system that were not foreseen by Makarova et al.: adaptation and the targeting of the invading plasmids. In 2007, Barrangou et al. published the first evidence that the CRISPR/Cas system provides acquired resistance against viruses in prokaryotes and, thus, is adaptive [11]. To test their hypothesis about an involvement of the CRISPR/Cas system in anti-phage defense, authors infected a phage-sensitive wild-type *Streptococcus thermophilus* strain widely used in the dairy industry, DGCC7710, with two distinct but closely related virulent bacteriophages isolated from industrial yogurt samples, phage 858 and phage 2972. They have made several interesting observations about the acquired resistance: the addition of new spacers in response to phage infection seemed to be polarized toward one end of the CRISPR/Cas locus; there is similarity between the additional spacers inserted in the CRISPR/Cas locus to sequences found within the genomes of the phages used in the challenge; no particular sequence, gene, or functional group of the phages seemed to be targeted specifically. In 2008, Marraffini and Sontheimer demonstrated that the CRISPR/Cas system limits horizontal gene transfer (plasmid conjugation) in staphylococci by targeting DNA. They rigorously tested their model that consisted of wild types and deletion mutants of *Staphylococcus epidermidis* and *S. aureus* strains, and two variants of the conjugative plasmid pG0400 (wild type and mutant) [12]. The study revealed that even host cell transformation by plasmid electroporation is subject to CRISPR/Cas interference. The other important finding was that the CRISPR/Cas system provides immunity by targeting DNA, rather than RNA. Also, they were first who speculated about possibility to use the CRISPR/Cas system in genetic engineering and a clinical setting.

3. STRUCTURE OF THE CRISPR LOCI

CRISPR loci have several common features that can be found in studied organisms (Fig. 6.1): (1) multiple direct repeats with identical or nearly identical, often palindromic sequences; (2) nonrepetitive similar-sized spacer sequences; (3) a leader sequence flanking the repeats at one end; (4) the absence of functional open reading frames within the repeat arrays

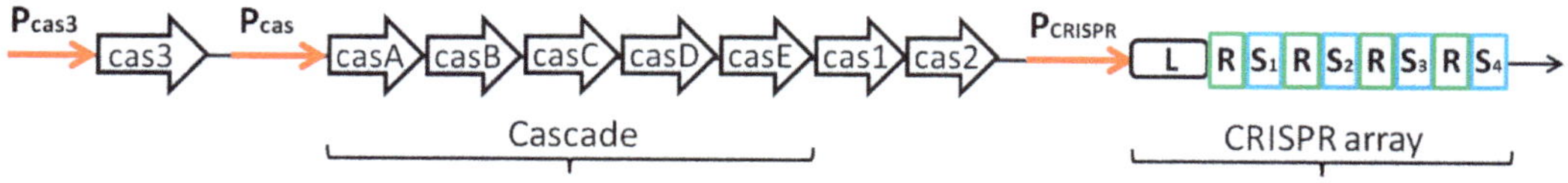

FIGURE 6.1 **cas operon and CRISPR array organization in *E. coli* K12.** The Type I system of *E. coli* K12 consists of eight cas genes and an immediately adjacent CRISPR locus. Promoters driving expression of the cas genes (P_{cas3} and P_{cas}) and the CRISPR locus (P_{CRISPR}) are indicated by *red arrows*. casABCDE genes form Cascade complex after transcription/translation. There are three distinctive elements in the CRISPR array: leader sequence (L), repeats (R), and spacers (S). One repeat and one spacer form one CRISPR unit.

and leader sequences; and (5) the genetic association of the direct repeats with cas genes. The number of CRISPR arrays varies in different organisms as well, even in different strains of the same species. It was shown that, on average, an archaeal genome contains about five CRISPR arrays, whereas three CRISPR arrays are found per bacterial genome [5]. The highest number of uninterrupted CRISPR arrays per prokaryotic genome is 20 loci in *M. jannaschii* [13].

1. Multiple direct repeats are 24–37 bp in length, they typically show weak dyad symmetry and their number per CRISPR locus can vary considerably from a few in one species to hundreds in another [14]. The largest CRISPR locus was found in *Verminephrobacter eiseniae* consisting of 245 repeats on one side and 45 repeats on the other side of an insertion sequence (IS) element [5]. Interestingly, in both archaea and bacteria, three well-separated size classes are observed: small direct repeats (24–25 bp), medium size (28–30 bp), and large (36–37 bp). The smaller direct repeats group is more represented in archaea (42% versus less than 2% for this size class in bacteria) and it is also where the differences between direct repeat and spacer size are the largest [5]. The repeat sequences have been classified in 12 different clusters, some of which are predicted to form an RNA hairpin secondary structure, whereas others are predicted to be unstructured. Kunin et al. have shown that the stems of these RNAs are well conserved and different Cas protein subtypes appeared to prefer one or more repeat types [15].
2. Nonrepetitive similar-sized spacer sequences are highly diverse elements of the CRISPR loci ranging from 26 bp to 72 bp in length. The spacers have similar lengths within a single CRISPR locus. There are no two identical spacers in the same CRISPR array, with the exception of spacer duplications in larger CRISPR arrays in some species [13]. There is a growing body of evidences suggesting that chromosomal spacers are taken up directly and, probably, randomly and nondirectionally from plasmid and bacteriophage DNA [9]. For example, the chromosomal spacers show a high level of matches (~30%) with bacteriophage or plasmid genomes in the crenarchaeal thermoacidophiles [16]. At first sight, 30% similarity to the bacteriophage or plasmid genomes looks relatively low. But the number of currently sequenced bacteriophages and plasmids is extremely small compared to the huge number of unknown mobile genetic elements that occur in nature [17].
3. A leader sequence flanking the repeats at 5′-end is an adenine/thymine (AT)-rich DNA sequence consisting of hundreds of nucleotides. It has been confirmed that it contains active promoter elements where transcription of the CRISPR array is initiated, and binding sites for regulatory proteins, such as Cas proteins involved in spacer integration, and, thus, CRISPRs adaptation [18,19].
4. So far, there is no single evidence about the presence of the functional open reading frames within the repeat arrays and leader sequences. It is not surprising, taken in mind the repetitive nature of the CRISPRs arrays and the presence of the active promoter elements in the leader sequences.
5. The possible genetic association of the direct repeats with cas genes was shown in 2002 by Jansen et al., who found that the cas genes are exclusively present in genomes that contain CRISPRs and are often located in close proximity to CRISPR arrays [3]. The cas1–4 core genes were originally marked out by Jansen et al. and are characterized by their close proximity to the CRISPR loci and their broad distribution among studied prokaryotic species. These genes are always located near a repeat locus, usually oriented head-to-tail as if cotranscribed, with the most common arrangement cas3–cas4–cas1–cas2 [3]. Haft et al. have also defined two additional core cas genes (cas5 and cas6) [14]. They have shown that the cas1–6 core gene families are not restricted to certain Cas subtypes and only two of them (cas1 and cas2) are universal. Makarova et al. have found that Cas1 (COG1518) might be the best marker of the CRISPR/Cas systems. This gene encodes a highly conserved protein and is represented in all cas neighborhoods, with the single exception of *Pyrococcus abyssii* [10]. Several studies that are based on sequence analysis predicted that Cas proteins can function as nucleases, helicases, integrases, and polymerases [20–22].

4. CRISPR/CAS CLASSIFICATION

Based on the previous observations, Haft et al. identified 45 cas gene families and eight different CRISPR/Cas subtypes that were named based on the name of organism in which the particular subtype has first been characterized: Ecoli, Ypest,

Nmeni, Dvulg, Tneap, Hmari, Apern, and Mtube [14]. There is also the RAMP module (repair-associated mysterious protein) that includes cmr1–6 genes, which is the most diverse class of cas genes, and that has been named after the RAMP superfamily.

In 2006, Makarova et al. have refined Haft's classification of 45 cas gene families by unifying them into 23 superfamilies, have tried to simplify confusing CRISPR/Cas classification, and also expanded the list of CRISPR-linked genes to include those that are found in cas operons less. In addition to the previously identified five distinct families of RAMPs, Makarova et al. detected five new ones, namely BH0337-like, y1726-like, YgcH-like, y1727-like, and MJ0978-like families. They have shown that despite the evident sequence difference, all these proteins contain the RAMP signature, the G-rich loop at the C-terminus [10].

In 2011, Makarova et al. proposed a new, polythetic classification of CRISPR/Cas systems in which the cas1 and cas2 genes form the core of three distinctive types of system (Type I, II, and III) with 10 subtypes. They have also shown that Cas1 and Cas2 are present in all CRISPR–Cas systems that are predicted to be active, and suggested to consider the Cas1/Cas2 as the information-processing subsystem that is involved in spacer integration during the adaptation stage [23]. The three types of CRISPR/Cas systems show a noticeably nonuniform distribution among the studied lineages of prokaryotes. To date, the Type II systems have been found in the bacteria only, whereas Type III systems are more common in the archaea. Also, there is a trend of overrepresentation of CRISPR in the archaea compared to the bacteria. Interestingly, the majority of sequenced archaeal genomes have more than one CRISPR/Cas system with unrelated modules within the same genome [23].

5. COMPOSITION OF THE CRISPR/CAS SYSTEMS

The *E. coli* K12 CRISPR/Cas system consists of eight cas genes: cas3, five genes designated casABCDE, cas1 (predicted integrase), and the endoribonuclease gene cas2 [21]. Five Cas proteins (Cse1, Cse2, Cas7, Cas5, and Cas6e), translated from casABCDE transcript, form a complex called Cascade. The hallmark of the Type I CRISPR/Cas systems is the cas3 gene, which encodes a protein that works as a helicase/DNase comprising of a histidine/aspartate (HD)-nuclease domain and a DExH helicase domain, in addition to genes encoding proteins that could form Cascade-like complexes [23]. These complexes contain various proteins from the RAMP superfamily (eg, Cas5, Cas6, and Cas7 families), based on the extensive sequence and structure analysis. Also, the Cascade-like complexes involved in the CRISPR/Cas/DNA interaction can contain other proteins from Cse1, Cse2, and BH0338-like families, or other, less conserved subunits [23].

Typical Type II CRISPR/Cas systems contain very large Cas9 protein, in addition to the ubiquitous Cas1 and Cas2. Cas9 has two nuclease domains and seems to be sufficient for generating crRNA and cleaving the target. It has been shown by Makarova et al. that the Type II systems include the "HNH"-type system (*Streptococcus*-like; also known as the Nmeni subtype, for *Neisseria meningitidis* serogroup A str. Z2491, or CASS4) [23]. Targeting of plasmid and phage DNA has been demonstrated in vivo using the *S. thermophilus* Type II CRISPR/Cas system, and inactivation of Cas9 has been shown to abolish interference [11,24].

The Type III CRISPR/Cas systems are the most complex and poorly understood to date, they include analogous to the Cascade complex RNA polymerase and RAMP modules, in which RAMPs seem to be involved in the processing of transcripts originated from the spacer/repeat arrays [23]. Type III systems contain, apart from the universal Cas2 protein and in addition to Cas6, at least two RAMPs that are involved in spacer/repeat transcript processing. In many studied specie, Type III CRISPR/*cas* operons lack the cas1/cas2 gene pair. Interestingly, in all these organisms, an additional either Type I or Type II CRISPR/Cas system is also present in the corresponding genome, indicating that Cas1 and Cas2 could act in trans. In all other organisms, the cas1/cas2 gene pair is combined into a single operon with the polymerase/RAMP modules, forming either a locus with the typical architecture in *S. epidermidis* and *Mycobacterium tuberculosis* (a Type IIIA module) or a distinct variant in *Halorhodospira halophila* (a Type IIIB module). The Type IIIA system in *S. epidermidis* RP62a harbors nine cas/csm genes [25]. The Type IIIB module consists of six different Cas proteins (Cmr1, Cas10, Cmr3, Cmr4, Cmr5, and Cmr6) and mature crRNA of either 39 nucleotide (nt) or 45 nt [26]. It was shown that *Sulfolobus solfataricus* contains similar crRNA-loaded Cmr complex, comprising of seven Cmr proteins (Cmr1–7), with manganese-dependent endoribonuclease activity on complementary RNA [27]. There are no other CRISPR/Cas systems found in these organisms, suggesting that the Type III system is fully functional and autonomous when the polymerase/RAMP module is present with Cas1 and Cas2 [23].

Despite the fact that the most of the CRISPR/*cas* loci can be classified into the three CRISPR/Cas types and their corresponding subtypes, there are loci that cannot fall into any specific system, even at the type level. Makarova et al. proposed to name such loci "Type U," for example, the CRISPR/Cas system in *Acidithiobacillus ferrooxidans* str. ATCC 23270 [23].

6. MOLECULAR MACHINES OF CRISPR/CAS SYSTEMS

Molecular mechanisms of the Type I CRISPR/Cas system are mediated by the multiunit Cascade complex and the Cas3 nuclease (Fig. 6.2). The Cascade complex includes Cas6e and Cas8 (also known as CasA or Cse1) subunits. The precursor crRNA (CRISPR RNAs), which is generated by transcription of the full CRISPR array, should be cut by the repeat-specific endoribonuclease Cas3 in the "crRNA biogenesis" phase. This reaction produces short crRNAs that stay associated with the Cascade complex and that are used by the complex to find a protospacer, which is a complementary sequence in the target DNA. Cas8 recognizes a short sequence motif located upstream of the target sequence recognized by the crRNA. Sequence motifs flanking the targets specified by CRISPR spacers were named as "protospacer adjacent motif," or PAM [28]. It has been shown by Semenova et al. that the PAM motif recognition is required for prevention of an autoimmune reaction for the Type I CRISPR/Cas immunity, because the absence of a PAM in the spacer/repeat array prevents the targeting of the host chromosomal DNA by their complementary crRNAs [29]. The Cascade complex binds to its target in the presence of the PAM motif, which promotes the binding and the formation of an R-loop between the crRNA spacer sequence and the target double-stranded DNA (dsDNA) [30,31]. At the final stage, the formation of the Cascade complex/target DNA structure triggers activation of the Cas3 nuclease, which introduces single-stranded DNA (ssDNA) breaks into the target DNA of bacteriophage or plasmid, thus initiating their degradation [32,33]. Semenova et al. defined a "seed" sequence within the target and showed that the first 8 bp at the 5′-end of the crRNA/DNA duplex are critical for immunity. It was shown that mutations in this region protect bacteriophages from Type I CRISPR immunity in *E. coli*. There is exclusion though: mutations in the 6th nt of the seed sequence have no effect on CRISPR immunity [29].

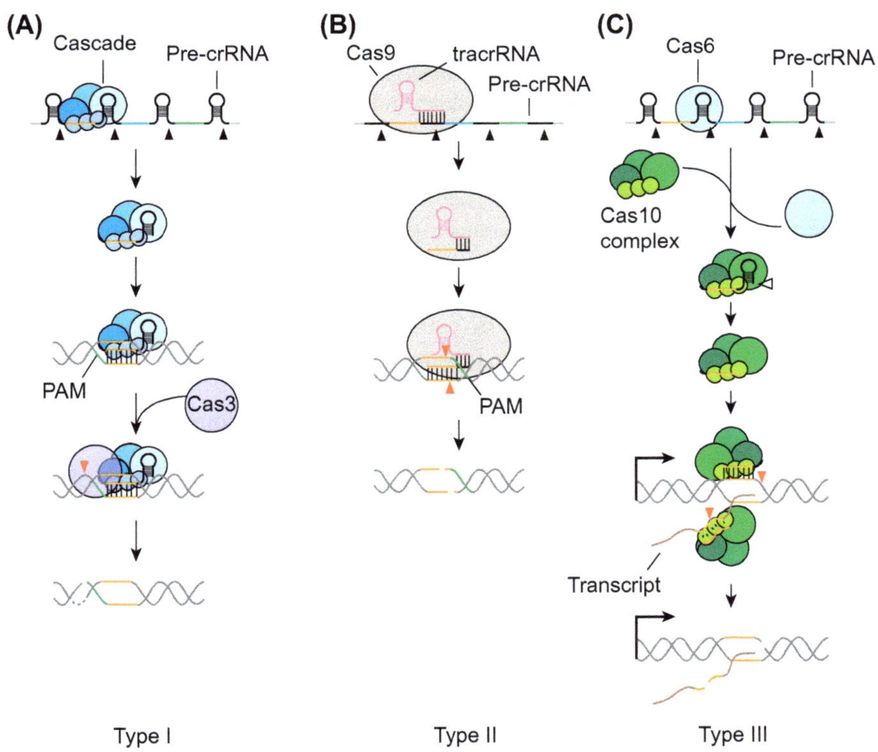

FIGURE 6.2 **Immunity mechanisms of the different CRISPR-Cas types.** (A) Type I systems. A Cas protein complex known as Cascade cleaves at the base of the stem–loop structure of each repeat in the long precursor crRNA (pre-crRNA, *black arrowheads*), which generates short crRNA guides. The Cascade– crRNA complex scans the target DNA for a matching sequence (known as protospacer), which is flanked by a protospacer-adjacent motif (PAM, in green). Annealing of the crRNA to the target strand forms an R-loop; the Cas3 nuclease is recruited and cleaves the target downstream of the PAM (*red arrowhead*) and also degrades the opposite strand. (B) Type II systems. These systems encode another small RNA known as trans-encoded crRNA (tracrRNA) which is bound by Cas9 and has regions of complementarity to the repeat sequences in the pre-crRNA. The repeat/tracrRNA/dsRNA is cleaved by RNase III to generate crRNA guides for the Cas9 nuclease (*black arrowheads*). This nuclease cleaves both strands of the protospacer/crRNA R-loop (*red arrowhead*). A PAM (in green) is located downstream of the target sequence. (C) Type III systems. Cas6 is a repeat-specific endoribonuclease that cleaves the pre-crRNA at the base of the stem–loop structure of each repeat (*black arrowhead*). The crRNA is loaded into the Cas10 complex where it is further trimmed at the 3*black arrowhead′* end to generate a mature crRNA (*white arrowhead*). The Cas10 complex requires target transcription to cleave the nontemplate strand of the protospacer DNA and it is also capable of crRNA-guided transcript cleavage (*red arrowheads*). *Reproduced from Marraffini LA. CRISPR-Cas immunity in prokaryotes. Nature 2015;526:55–61. http://dx.doi.org/10.1038/nature15386.*

In contrast to the Type I, to perform its immunity functions, the Type II CRISPR/Cas system requires only cas9 gene, either Csn2 (Type IIA) or Cas4 (Type IIB) proteins, the presence of a targeting spacer sequence, a PAM, and two small RNAs: the crRNA and the trans-encoded crRNA (tracrRNA) [34,35]. Compared to the Type I CRISPR/Cas system, the PAM is recognized by a Cas9 PAM-binding domain and is located downstream of the target sequence [36]. The tracrRNA has two regions: one region forms a secondary structure that mediates its association with Cas9 and the other one is complementary to the repeat sequence of the CRISPR array [36]. The tracrRNA and the precursor crRNA form dsRNA, which should be digested by RNase III. Resulted cleavage products of the long CRISPR transcript are small crRNA guides [35]. Sapranauskas et al. demonstrated that there are two nuclease domains in Cas9: RuvC/RNaseH and McrA/HNH [34]. These domains are required for the Type II CRISPR/Cas immunity, which results in the introduction of crRNA-specific dsDNA breaks in the invading DNA. The tracrRNA is absolutely required for cleavage by nuclease domains of Cas9; each of these domains cleaves one DNA strand of the protospacer sequence [24,37]. It was shown that the McrA/HNH nuclease domain of Cas9 cleaves the base-pairing strand, and the RuvC/RNaseH nuclease domain cleaves the displaced strand, resulting in a blunt-end cleavage product [37]. DNA target recognition starts with the transient binding of Cas9 to PAM motifs within the target DNA, which results in the separating of the two DNA strands immediately upstream of the PAM sequences [38]. This process, in turn, triggers the creation of an R-loop and cleavage of the DNA target between 6–8 bases of the spacer sequence of the crRNA guide and the melted DNA (the "seed" region) [31].

In Type III CRISPR/Cas systems, a repeat-specific endoribonuclease, Cas6, digests the precursor crRNA at the base of a putative stem-loop structure in the pre-crRNA repeat. The result of this processing step is a sequence known as the crRNA tag, which is eight nucleotides of the repeat sequence remaining at the 5′-end of the spacer sequence in the crRNA [39]. Depending on a system subtype (IIIA or IIIB), the small crRNAs generated after Cas6 digestion are transferred to a larger complex, the Cas10/Csm or Cas10/Cmr, respectively [40]. Further, crRNAs go through a process of maturation within these complexes, by which the 3′- end is trimmed at 6-nt intervals. In the Type IIIA systems, for example, *S. epidermidis*, pre-crRNA maturation resulted in mature crRNA of two defined lengths (37 nt and 43 nt) [41]. Interestingly, The Type IIIB system of *Pyrococcus furiosus* is the first example of a prokaryotic immune system targeting RNA. It was shown that the components of the Type IIIB system can cleave RNA both in vitro and in vivo [26,42]. Recent studies of the Type IIIB system revealed that the endoribonuclease Cas6 binds crRNA at the 5′-end of the unstructured repeat sequence and cleaves it through metal-independent endoribonuclease activity, yielding 67-nt fragments. These fragments are further trimmed at the 3′-end by unknown mechanisms to 39-nt and 45-nt mature crRNA fragments that are bound to Cas proteins [22,42,43]. In the Type III CRISPR/Cas systems, the Cas10 complex contains endoribonuclease and endodeoxyribonuclease activities: subunits Csm3 and Cmr4 are responsible for cleavage of the RNA transcripts (Type IIIB system) and the palm domain of Cas10 is responsible for digestion of the nontemplate DNA strand [44–46]. To date, nobody could show that the Type III CRISPR/Cas system requires PAM for targeting activity. It was reported by Marraffini and Sontheimer that Type III system in *S. epidermidis* depends on extended base pairing between crRNA and CRISPR DNA repeats flanking the protospacer, in order to avoid autoimmunity [47]. Despite the progress in the RNA-directed immunity, the biological significance and underlying mechanisms for the RNA-targeting phenomenon has to be still elucidated.

7. CRISPR/CAS SYSTEMS AT WORK

In 2009, van der Oost et al. proposed three main stages of the development and execution of CRISPR/Cas system (Fig. 6.3): (1) adaptation (immunization), where the alien nucleic acids are encountered, and resistance is acquired by integration of an invader-derived new spacer sequence in a CRISPR array; (2) expression, where cas genes and the CRISPR array are transcribed; cas transcripts are translated, CRISPR transcripts are processed into pre-crRNAs that are subsequently digested either by a Cas6 (Type I and Type III CRISPR/Cas systems) or by an RNase III (Type II CRISPR/Cas systems); (3) interference, where the invader's nucleic acid is recognized and eliminated by the Cas/crRNA complex (crRNA recognizes complementary sequence and guides one or more Cas proteins that cleave alien DNA or RNA) [48]. We have to note that our understanding of the CRISPR/Cas systems is still partial, schemes of work are hypothetical and far from complete at the moment.

7.1 The CRISPR Adaptation

The CRISPR adaptation process was first observed in the Type II system of *S. thermophilus*, but the spacer acquisition mechanism has been studied in detail in the *E. coli* Type I CRISPR/Cas system. This is because in a report by Yosef et al., a very simple and a robust assay in *E. coli* was described that would allow to explore the process of adaptation [49]. The adaptation process can be arbitrarily divided into two stages: (1) the selection of protospacer sequences from the invader DNA and (2) integration of the protospacer sequences into the CRISPR array.

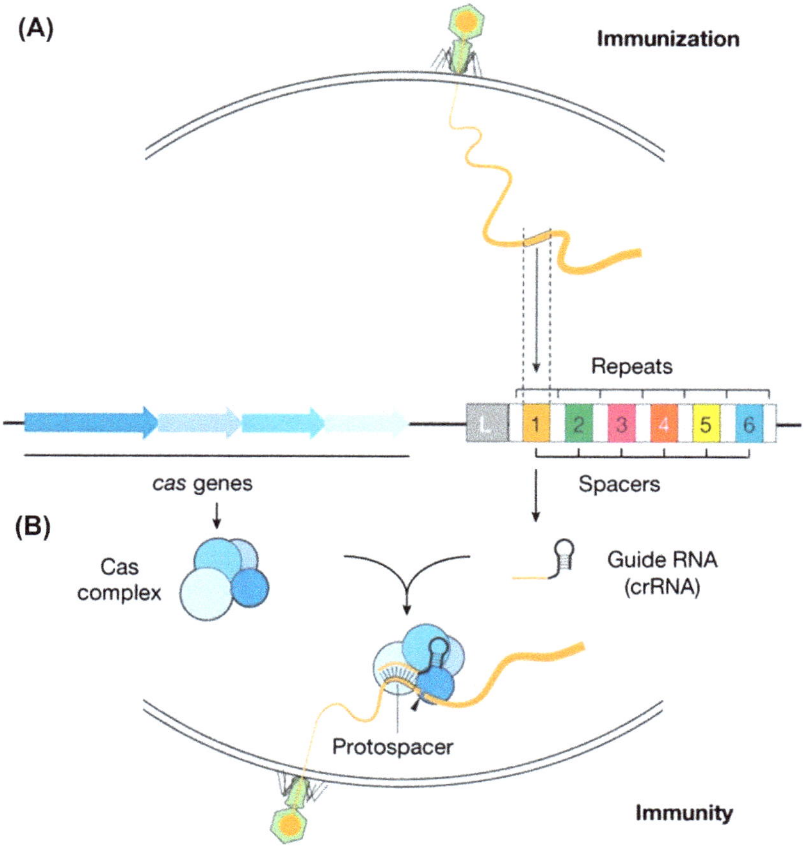

FIGURE 6.3 **Stages of CRISPR-Cas immunity.** CRISPR loci are a cluster of short DNA repeats (*white boxes*) separated by equally short spacer sequences of phage and plasmid origin (*colored, numbered boxes*). This repeat/spacer array is flanked by an operon of CRISPR-associated (cas) genes (*blue-tone arrows*) that encode the machinery for the immunization and immunity stages of the system. The CRISPR array is preceded by a leader sequence (*gray box*) containing the promoter for its expression. (A) In the immunization (adaptation) stage, spacer sequences are captured upon entry of the foreign DNA into the cell and integrated into the first position of the CRISPR array. (B) In the immunity stage (expression + interference), the spacer is used to target invading DNA that carries a cognate sequence for destruction. Spacers are transcribed and processed into small CRISPR RNAs (crRNAs) in the "crRNA biogenesis" phase. These small RNAs act as antisense guides for Cas RNA-guided nucleases (which usually form a complex) that locate and cleave the target sequence (*black arrowhead*) in the invader's genome during the "targeting" phase. *Reproduced from Marraffini LA. CRISPR-Cas immunity in prokaryotes. Nature 2015;526:55–61. http://dx.doi.org/10.1038/nature15386.*

The main aspect of the new spacer sequences selection is the ability of the prey acquisition machinery to distinguish self from nonself-DNA in order to prevent autoimmunity. Recently, it has been shown by Levy et al. in their elegant work that spacer acquisition in *E. coli* is replication-dependent, and that spacer acquisition is promoted by DNA breaks formed at stalled replication forks [50]. Numerous Chi sites (8-bp sequences) in *E. coli* restrict chromosomal hotspots of spacer acquisition, suggesting that these sites prevent spacer acquisition from self DNA. They have also shown that the autoimmunity is mediated by the RecBCD dsDNA break repair complex. They hypothesized that this model explains the strong preference of the CRISPR/Cas machinery to acquire spacers both from high-copy plasmids and from phages [50]. The second question of spacer selection relates to the PAM requirement for targeting. In *Streptococcus pyogenes*, Cas9 cleavage of bacteriophage genome requires the presence of a 5′-NGG-3′ PAM sequence immediately downstream of the viral target [35]. Heler et al. have demonstrated that Cas9 recognizes PAMs of functional spacers and, thus, provides their selection and new spacer acquisition [51]. In 2012, Datsenko et al. discovered a phenomenon that they referred to as "priming" [52]. Using phage M13 as a model system, they have shown that the presence of preexisting spacers with partial homology to the ssDNA increases the rate of spacer acquisition by several times. It has been shown that primed adaptation requires the Cascade complex, which plays dual role in immunity allowing for efficient degradation of bona fide targets and priming of mutated DNA targets [53,54].

During the second stage of the CRISPR adaptation, spacers are integrated into the CRISPR array in a reaction that is similar to retroviral integration. In 2015, it was shown by Nuñez et al. that the Cas1/Cas2 complex integrates DNA substrates into acceptor DNA to yield products similar to those generated by retroviral integrases and transposases [55]. In this complex, Cas1 is the catalytic subunit and Cas2 substantially increases integration activity. The integration events occur preferentially at the ends of CRISPR repeats and at sequences adjacent to cruciform structures adjacent to AT-rich regions, similar to the CRISPR leader sequence [55].

7.2 The Expression Stage

The expression stage consists of three parts: regulation of the *cas* gene and CRISPR expression, Cas complex formation, and pre-crRNA maturation [48]. It has been shown by Pul et al. that transcription of pre-crRNA is directed by a promoter (P_{CRISPR}) located within the CRISPR leader sequence [18]. Transcription of the genes encoding proteins of the Cascade complex is driven by a second promoter (P_{cas}), which is located upstream of the cas genes and the activity of which resulted in a polycistronic transcript encoding Cascade (Cse1, Cse2, Cas7, Cas5, and Cas6e proteins), Cas1, and Cas2 (Fig. 6.1). Also, they have demonstrated the role of the DNA-binding protein H-NS (a global transcriptional repressor) in silencing of the CRISPR/cas promoters [18]. In contrast to other cas genes, the cas3 gene is transcribed from its own constitutive promoter (Fig. 6.1). The CRISPR/Cas system in *E. coli* K12 is also regulated by LeuO protein (a LysR-type transcription factor), which works as a transcriptional activator [56]. LeuO binds the cse1 upstream region in two sites, which flank the cse1 promoter and the H-NS nucleation site, resulting in derepression of transcription of cas genes [56]. It has been proposed by Westra et al. that regulation of transcription depends on the H–NS feature to bind AT-rich DNA, which may lead to H–NS titration from its own genome when cells encounter arrival of a mobile genetic element with AT-rich DNA [57]. This event releases repression of both the LeuO promoter (which is positively regulated by LeuO itself) and the P_{cas} promoter, initiating a CRISPR-based immune response [57].

In *E. coli* K12, the Cas6e is an endoribonuclease that cleaves the pre-crRNA in a metal-independent way, yielding mature 61-nt crRNA with an 8-nt repeat-derived 5′ handle (psi-tag) and a 21-nt 3′ handle [58]. The crRNA remains Cascade bound, forming a ribonucleoprotein complex with an attention-grabbing stoichiometry of $Cse1_1Cse2_2Cas7_6Cas5_1Cas6e_1crRNA_1$ and an asymmetrical seahorse-like shape [59]. Interestingly, in studies on different species, Cas6 homologous proteins yield similar mature crRNA with an 8-nt 5′ handle, despite the differences in protein structure and crRNA-binding ability [12,21,39,60–62].

7.3 The CRISPR Interference

CRISPR interference is a multistep process, which starts with crRNAs that forms the surveillance complex with the Cas proteins and guides it to PAM and protospacer seed regions in invading nucleic acids. The protospacer is usually dsDNA, with the exception of the Type IIIB system where the Cascade complex targets complementary ssRNA [27,63]. In the Type II and Type IIIB systems, target cleavage is carried out by the Cas/crRNA ribonucleoprotein complex, in contrast to the Type I and Type IIIA systems which require a Cas nuclease [64]. In Type I systems, the surveillance complex binds to dsDNA, recruits the Cas3 nuclease, and degrades the target [63]. In Type II systems, the Cas9/crRNA/tracrRNA complex binds to and cleaves target dsDNA [37]. The Type IIIA Csm/crRNA complex and Csm6 protein bind and most likely degrade invader dsDNA, whereas Type IIIB Cmr/crRNA complex cleaves complementary RNA [65]. The Csm complex in the Type IIIA system includes six different proteins but the nuclease is not yet identified [66]. The Cmr complex in the Type IIIB system includes six or seven different proteins and the Cmr4 protein works as a nuclease [67]. It has been shown that in *Thermus thermophilus* and *S. thermophiles*, the Csm complex targets RNA, and in *T. thermophilus*, which harbors both Type IIIA and IIIB systems, the Csm and Cmr complexes share crRNA [45,46]. In *S. islandicus*, Cmr complex targets both RNA and DNA [68,69]. Another interesting CRISPR-Cas interference mechanism was recently found in *S. epidermidis*: it can prevent lytic infection but tolerate lysogenization by temperate phages. It was shown that conditional tolerance is achieved through transcription-dependent DNA targeting, and ensures that targeting is resumed upon induction of the prophage lytic cycle [70].

8. OTHER ROLES OF THE CRISPR/CAS SYSTEMS

There is a growing body of evidence suggesting that the CRISPR/Cas system, besides its immunity function, can be a part of other cellular processes such as DNA repair and regulation of virulence.

Taking in mind the ability of the CRISPR/Cas systems to shape a bacterial genome landscape by acquisition of new spacers, it was quite obvious to hypothesize that these systems might have an impact on the stability and evolution of bacterial genomes. Indeed, recent study of *S. thermophilus* revealed that the CRISPR/Cas systems target mobile genetic elements (bacteriophages, transposons, and plasmids), which likely contributed to gene acquisition and loss during evolutionary adaptation to milk, thus limiting genetic diversity and stabilizing of the *S. thermophilus* genome [71]. On the contrary, the genome analysis of *T. maritima* MSB8 and *Thermotoga neapolitana* NS-E provided evidence that the CRISPR/Cas systems might be a cause of numerous CRISPR-associated large-scale DNA rearrangements that destabilize and reshape genomes [72].

It was shown that the Cas1 protein of *E. coli* interacts with RecB, RecC, and RuvB, it can process single-stranded and branched DNA species, replication forks and 5′ flaps [73]. In *Francisella novicida*, Cas9 protein uses a unique, small, CRISPR/Cas-associated RNA (scaRNA) to repress transcription of a bacterial lipoprotein (FTN_1103). As bacterial lipoproteins trigger a proinflammatory innate immune response in a host, CRISPR/Cas-mediated transcriptional repression of FTN_1103 is important for *F. novicida* to reduce this host response and promote virulence [74]. It has been demonstrated in *Campylobacter jejuni* that inactivation of the Type II CRISPR/Cas marker gene csn1 effectively reduced virulence in primarily cst-II-positive *C. jejuni* isolates [75]. cas2 mutants in *Legionella pneumophila*, although they grew typically in macrophages, were significantly impaired for infection of both *Hartmannella* and *Acanthamoeba* species. Given that infection of amebae is critical for *L. pneumophila* persistence in water systems, these data indicate that cas2 might play a role in the transmission of Legionnaires' disease [76].

To date, there is not enough data to draw a conclusion on whether the CRISPR/Cas systems are mainly involved in the bacterial immunity. Above-mentioned examples raise interesting questions about the evolution of CRISPR/Cas function, which require more in-depth research.

9. CONCLUSION

CRISPR/Cas research experienced tremendous boost during the last decade. It should be appreciated that the CRISPR/Cas discovery has had a huge impact on bacteriology and genetic engineering, which can be compared to, for example, discovery of the polymerase chain reaction or development of the next-generation sequencing technology. However, many molecular details and mechanisms of action CRISPR/Cas systems remain to be determined. It would be very interesting to see whether cross-talk exists in species with multiple CRISPR/Cas systems, their regulation, effect on virulence in the case of animal pathogens, and many others. Another baffling problem of CRISPR/Cas systems is their high diversity: there are many subtypes of the "main" three CRISPR types with different sets of Cas proteins, structural organization, and regulation. The biological significance of this diversity remains to be determined.

GLOSSARY

Dyad symmetry Dyad symmetry refers to two areas of a DNA strand whose base-pair sequences are inverted repeats of each other. They are often described as palindromes. For example, the following shows dyad symmetry between sequences GAATAC and GTATTC which are reverse complements of each other.

Innate immune response Innate immune response is the first line of defense against invading microbial pathogens and relies on a large family of pattern recognition receptors (PRRs), which detect distinct evolutionarily conserved structures on pathogens, termed pathogen-associated molecular patterns (PAMPs). Among the PRRs, the Toll-like receptors have been studied most extensively.

IS element IS element (also known as an IS, an insertion sequence element, or an insertion sequence) is a short DNA sequence that acts as a simple transposable element. IS have two major characteristics: they are small relative to other transposable elements (generally about 700–2500 bp in length) and only code for proteins implicated in the transposition activity (they are thus different from other transposons, which also carry accessory genes such as antibiotic-resistance genes).

Next-generation sequencing (NGS) Next-generation sequencing also known as high-throughput sequencing, is the catch-all term used to describe a number of different modern sequencing technologies, including Illumina (Solexa) sequencing, Roche 454 sequencing, Ion torrent (Proton/PGM) sequencing, and SOLiD sequencing. These technologies allow us to sequence DNA and RNA much more quickly and cheaply than the previously used Sanger sequencing, and as such have revolutionized the study of genomics and molecular biology.

Polythetic classification Polythetic classification is defined in terms of a broad set of criteria that are neither necessary nor sufficient. Each member of the category must possess a certain minimal number of defining characteristics, but none of the features has to be found in each member of the category. This way of defining classes is associated with Wittgenstein's concept of "family resemblances."

Protospacer Protospacer is the sequence complementary to the crRNA (CRISPR RNA) spacer.

RNA interference (RNAi) RNA interference is an important pathway that is used in many different organisms to regulate gene expression. This is a biological process in which small RNA molecules inhibit gene expression, typically by causing the destruction of specific mRNA molecules. Historically, it was known by other names, including *cosuppression*, *posttranscriptional gene silencing* (PTGS), and *quelling*. Two types of small RNA molecules—microRNA (miRNA) and small interfering RNA (siRNA)—are central to RNA interference.

Seed region Seed region is a noncontiguous 7-nt region of a protospacer (positions 1–5, 7, and 8 of the spacer) near the 5′-end of the crRNA.

Virulence Virulence is the degree of pathogenicity within a group or species of pathogens as indicated by case fatality rates and/or the ability of the organism to invade the tissues of the host.

LIST OF ACRONYMS AND ABBREVIATIONS

bp Base pairs
cas genes CRISPR-associated genes
CRISPRs Clustered regularly interspaced short palindromic repeats
crRNA CRISPR RNA
dsDNA Double-stranded DNA
dsRNA Double-stranded RNA
nt Nucleotides
PAM Protospacer adjacent motif
psiRNA Prokaryotic small interfering RNA
RAMP Repair-associated mysterious protein
RMS Restriction-modification system
RNAi RNA interference
SRSRs Short regularly spaced repeats
ssDNA Single-stranded DNA
tracrRNA Transencoded crRNA

REFERENCES

[1] Hall AR, Scanlan PD, Morgan AD, Buckling A. Host-parasite coevolutionary arms races give way to fluctuating selection. Ecol Lett July 2011;14(7):635–42.
[2] Labrie SJ, Samson JE, Moineau S. Bacteriophage resistance mechanisms. Nat Rev Microbiol May 2010;8(5):317–27.
[3] Jansen R, Embden JD, Gaastra W, Schouls LM. Identification of genes that are associated with DNA repeats in prokaryotes. Mol Microbiol March 2002;43(6):1565–75.
[4] Ishino Y, Shinagawa H, Makino K, Amemura M, Nakata A. Nucleotide sequence of the iap gene, responsible for alkaline phosphatase isozyme conversion in *Escherichia coli*, and identification of the gene product. J Bacteriol December 1987;169(12):5429–33.
[5] Grissa I, Vergnaud G, Pourcel C. The CRISPRdb database and tools to display CRISPRs and to generate dictionaries of spacers and repeats. BMC Bioinf 2007;8:172.
[6] Mojica FJ, Diez-Villasenor C, Soria E, Juez G. Biological significance of a family of regularly spaced repeats in the genomes of archaea, bacteria and mitochondria. Mol Microbiol April 2000;36(1):244–6.
[7] Bolotin A, Quinquis B, Sorokin A, Ehrlich SD. Clustered regularly interspaced short palindrome repeats (CRISPRs) have spacers of extrachromosomal origin. Microbiology August 2005;151(Pt 8):2551–61.
[8] Mojica FJ, Diez-Villasenor C, Garcia-Martinez J, Soria E. Intervening sequences of regularly spaced prokaryotic repeats derive from foreign genetic elements. J Mol Evol February 2005;60(2):174–82.
[9] Pourcel C, Salvignol G, Vergnaud G. CRISPR elements in *Yersinia pestis* acquire new repeats by preferential uptake of bacteriophage DNA, and provide additional tools for evolutionary studies. Microbiology March 2005;151(Pt 3):653–63.
[10] Makarova KS, Grishin NV, Shabalina SA, Wolf YI, Koonin EV. A putative RNA-interference-based immune system in prokaryotes: computational analysis of the predicted enzymatic machinery, functional analogies with eukaryotic RNAi, and hypothetical mechanisms of action. Biol Direct 2006;1:7.
[11] Barrangou R, Fremaux C, Deveau H, Richards M, Boyaval P, Moineau S, et al. CRISPR provides acquired resistance against viruses in prokaryotes. Science March 23, 2007;315(5819):1709–12.
[12] Marraffini LA, Sontheimer EJ. CRISPR interference limits horizontal gene transfer in staphylococci by targeting DNA. Science December 19, 2008;322(5909):1843–5.
[13] Lillestol RK, Redder P, Garrett RA, Brugger K. A putative viral defence mechanism in archaeal cells. Archaea August 2006;2(1):59–72.
[14] Haft DH, Selengut J, Mongodin EF, Nelson KE. A guild of 45 CRISPR-associated (Cas) protein families and multiple CRISPR/Cas subtypes exist in prokaryotic genomes. PLoS Comput Biol November 2005;1(6):e60.
[15] Kunin V, Sorek R, Hugenholtz P. Evolutionary conservation of sequence and secondary structures in CRISPR repeats. Genome Biol 2007;8(4):R61.
[16] Shah SA, Hansen NR, Garrett RA. Distribution of CRISPR spacer matches in viruses and plasmids of crenarchaeal acidothermophiles and implications for their inhibitory mechanism. Biochem Soc Trans February 2009;37(Pt 1):23–8.
[17] Snyder JC, Bateson MM, Lavin M, Young MJ. Use of cellular CRISPR (clusters of regularly interspaced short palindromic repeats) spacer-based microarrays for detection of viruses in environmental samples. Appl Environ Microbiol November 2010;76(21):7251–8.
[18] Pul U, Wurm R, Arslan Z, Geissen R, Hofmann N, Wagner R. Identification and characterization of *E. coli* CRISPR-cas promoters and their silencing by H-NS. Mol Microbiol March 2010;75(6):1495–512.
[19] Al-Attar S, Westra ER, van der Oost J, Brouns SJ. Clustered regularly interspaced short palindromic repeats (CRISPRs): the hallmark of an ingenious antiviral defense mechanism in prokaryotes. Biol Chem April 2011;392(4):277–89.
[20] Wiedenheft B, Zhou K, Jinek M, Coyle SM, Ma W, Doudna JA. Structural basis for DNase activity of a conserved protein implicated in CRISPR-mediated genome defense. Structure June 10, 2009;17(6):904–12.
[21] Brouns SJ, Jore MM, Lundgren M, Westra ER, Slijkhuis RJ, Snijders AP, et al. Small CRISPR RNAs guide antiviral defense in prokaryotes. Science August 15, 2008;321(5891):960–4.

[22] Carte J, Pfister NT, Compton MM, Terns RM, Terns MP. Binding and cleavage of CRISPR RNA by Cas6. RNA November 2010;16(11):2181–8.

[23] Makarova KS, Haft DH, Barrangou R, Brouns SJ, Charpentier E, Horvath P, et al. Evolution and classification of the CRISPR-Cas systems. Nat Rev Microbiol June 2011;9(6):467–77.

[24] Garneau JE, Dupuis ME, Villion M, Romero DA, Barrangou R, Boyaval P, et al. The CRISPR/Cas bacterial immune system cleaves bacteriophage and plasmid DNA. Nature November 4, 2010;468(7320):67–71.

[25] Hatoum-Aslan A, Maniv I, Samai P, Marraffini LA. Genetic characterization of antiplasmid immunity through a type III-A CRISPR-Cas system. J Bacteriol January 2014;196(2):310–7.

[26] Hale CR, Zhao P, Olson S, Duff MO, Graveley BR, Wells L, et al. RNA-guided RNA cleavage by a CRISPR RNA-Cas protein complex. Cell November 25, 2009;139(5):945–56.

[27] Zhang J, Rouillon C, Kerou M, Reeks J, Brugger K, Graham S, et al. Structure and mechanism of the CMR complex for CRISPR-mediated antiviral immunity. Mol Cell February 10, 2012;45(3):303–13.

[28] Mojica FJ, Diez-Villasenor C, Garcia-Martinez J, Almendros C. Short motif sequences determine the targets of the prokaryotic CRISPR defence system. Microbiology March 2009;155(Pt 3):733–40.

[29] Semenova E, Jore MM, Datsenko KA, Semenova A, Westra ER, Wanner B, et al. Interference by clustered regularly interspaced short palindromic repeat (CRISPR) RNA is governed by a seed sequence. Proc Natl Acad Sci USA June 21, 2011;108(25):10098–103.

[30] Rutkauskas M, Sinkunas T, Songailiene I, Tikhomirova MS, Siksnys V, Seidel R. Directional R-loop formation by the CRISPR-Cas surveillance complex Cascade provides efficient off-target site rejection. Cell Rep March 3, 2015. http://dx.doi.org/10.1016/j.celrep.2015.01.067.

[31] Szczelkun MD, Tikhomirova MS, Sinkunas T, Gasiunas G, Karvelis T, Pschera P, et al. Direct observation of R-loop formation by single RNA-guided Cas9 and Cascade effector complexes. Proc Natl Acad Sci USA July 8, 2014;111(27):9798–803.

[32] Hochstrasser ML, Taylor DW, Bhat P, Guegler CK, Sternberg SH, Nogales E, et al. CasA mediates Cas3-catalyzed target degradation during CRISPR RNA-guided interference. Proc Natl Acad Sci USA May 6, 2014;111(18):6618–23.

[33] Mulepati S, Bailey S. In vitro reconstitution of an *Escherichia coli* RNA-guided immune system reveals unidirectional, ATP-dependent degradation of DNA target. J Biol Chem August 2, 2013;288(31):22184–92.

[34] Sapranauskas R, Gasiunas G, Fremaux C, Barrangou R, Horvath P, Siksnys V. The *Streptococcus thermophilus* CRISPR/Cas system provides immunity in *Escherichia coli*. Nucleic Acids Res November 2011;39(21):9275–82.

[35] Deltcheva E, Chylinski K, Sharma CM, Gonzales K, Chao Y, Pirzada ZA, et al. CRISPR RNA maturation by trans-encoded small RNA and host factor RNase III. Nature March 31, 2011;471(7340):602–7.

[36] Anders C, Niewoehner O, Duerst A, Jinek M. Structural basis of PAM-dependent target DNA recognition by the Cas9 endonuclease. Nature September 25, 2014;513(7519):569–73.

[37] Jinek M, Chylinski K, Fonfara I, Hauer M, Doudna JA, Charpentier E. A programmable dual-RNA-guided DNA endonuclease in adaptive bacterial immunity. Science August 17, 2012;337(6096):816–21.

[38] Sternberg SH, Redding S, Jinek M, Greene EC, Doudna JA. DNA interrogation by the CRISPR RNA-guided endonuclease Cas9. Nature March 6, 2014;507(7490):62–7.

[39] Carte J, Wang R, Li H, Terns RM, Terns MP. Cas6 is an endoribonuclease that generates guide RNAs for invader defense in prokaryotes. Genes Dev December 15, 2008;22(24):3489–96.

[40] Sokolowski RD, Graham S, White MF. Cas6 specificity and CRISPR RNA loading in a complex CRISPR-Cas system. Nucleic Acids Res June 2014;42(10):6532–41.

[41] Hatoum-Aslan A, Samai P, Maniv I, Jiang W, Marraffini LA. A ruler protein in a complex for antiviral defense determines the length of small interfering CRISPR RNAs. J Biol Chem September 27, 2013;288(39):27888–97.

[42] Hale CR, Majumdar S, Elmore J, Pfister N, Compton M, Olson S, et al. Essential features and rational design of CRISPR RNAs that function with the Cas RAMP module complex to cleave RNAs. Mol Cell February 10, 2012;45(3):292–302.

[43] Wang R, Preamplume G, Terns MP, Terns RM, Li H. Interaction of the Cas6 riboendonuclease with CRISPR RNAs: recognition and cleavage. Structure February 9, 2011;19(2):257–64.

[44] Samai P, Pyenson N, Jiang W, Goldberg GW, Hatoum-Aslan A, Marraffini LA. Co-transcriptional DNA and RNA Cleavage during type III CRISPR-Cas immunity. Cell May 21, 2015;161(5):1164–74.

[45] Staals RH, Zhu Y, Taylor DW, Kornfeld JE, Sharma K, Barendregt A, et al. RNA targeting by the type III-A CRISPR-Cas Csm complex of *Thermus thermophilus*. Mol Cell November 20, 2014;56(4):518–30.

[46] Tamulaitis G, Kazlauskiene M, Manakova E, Venclovas C, Nwokeoji AO, Dickman MJ, et al. Programmable RNA shredding by the type III-A CRISPR-Cas system of *Streptococcus thermophilus*. Mol Cell November 20, 2014;56(4):506–17.

[47] Marraffini LA, Sontheimer EJ. Self versus non-self discrimination during CRISPR RNA-directed immunity. Nature January 28, 2010;463(7280):568–71.

[48] van der Oost J, Jore MM, Westra ER, Lundgren M, Brouns SJ. CRISPR-based adaptive and heritable immunity in prokaryotes. Trends Biochem Sci August 2009;34(8):401–7.

[49] Yosef I, Goren MG, Qimron U. Proteins and DNA elements essential for the CRISPR adaptation process in *Escherichia coli*. Nucleic Acids Res July 2012;40(12):5569–76.

[50] Levy A, Goren MG, Yosef I, Auster O, Manor M, Amitai G, et al. CRISPR adaptation biases explain preference for acquisition of foreign DNA. Nature April 23, 2015;520(7548):505–10.

[51] Heler R, Samai P, Modell JW, Weiner C, Goldberg GW, Bikard D, et al. Cas9 specifies functional viral targets during CRISPR-Cas adaptation. Nature March 12, 2015;519(7542):199–202.

[52] Datsenko KA, Pougach K, Tikhonov A, Wanner BL, Severinov K, Semenova E. Molecular memory of prior infections activates the CRISPR/Cas adaptive bacterial immunity system. Nat Commun 2012;3:945.

[53] Blosser TR, Loeff L, Westra ER, Vlot M, Kunne T, Sobota M, et al. Two distinct DNA binding modes guide dual roles of a CRISPR-Cas protein complex. Mol Cell April 2, 2015;58(1):60–70.

[54] Fineran PC, Gerritzen MJ, Suarez-Diez M, Kunne T, Boekhorst J, van Hijum SA, et al. Degenerate target sites mediate rapid primed CRISPR adaptation. Proc Natl Acad Sci USA April 22, 2014;111(16):E1629–38.

[55] Nunez JK, Lee AS, Engelman A, Doudna JA. Integrase-mediated spacer acquisition during CRISPR-Cas adaptive immunity. Nature March 12, 2015;519(7542):193–8.

[56] Westra ER, Pul U, Heidrich N, Jore MM, Lundgren M, Stratmann T, et al. H-NS-mediated repression of CRISPR-based immunity in *Escherichia coli* K12 can be relieved by the transcription activator LeuO. Mol Microbiol September 2010;77(6):1380–93.

[57] Westra ER, Swarts DC, Staals RH, Jore MM, Brouns SJ, van der Oost J. The CRISPRs, they are a-changin': how prokaryotes generate adaptive immunity. Annu Rev Genet 2012;46:311–39.

[58] Jore MM, Brouns SJ, van der Oost J. RNA in defense: CRISPRs protect prokaryotes against mobile genetic elements. Cold Spring Harb Perspect Biol June 2012;4(6).

[59] Jore MM, Lundgren M, van Duijn E, Bultema JB, Westra ER, Waghmare SP, et al. Structural basis for CRISPR RNA-guided DNA recognition by Cascade. Nat Struct Mol Biol May 2011;18(5):529–36.

[60] Haurwitz RE, Jinek M, Wiedenheft B, Zhou K, Doudna JA. Sequence- and structure-specific RNA processing by a CRISPR endonuclease. Science September 10, 2010;329(5997):1355–8.

[61] Lintner NG, Kerou M, Brumfield SK, Graham S, Liu H, Naismith JH, et al. Structural and functional characterization of an archaeal clustered regularly interspaced short palindromic repeat (CRISPR)-associated complex for antiviral defense (CASCADE). J Biol Chem June 17, 2011;286(24):21643–56.

[62] Terns RM, Terns MP. The RNA- and DNA-targeting CRISPR-Cas immune systems of *Pyrococcus furiosus*. Biochem Soc Trans December 2013;41(6):1416–21.

[63] Westra ER, van Erp PB, Kunne T, Wong SP, Staals RH, Seegers CL, et al. CRISPR immunity relies on the consecutive binding and degradation of negatively supercoiled invader DNA by Cascade and Cas3. Mol Cell June 8, 2012;46(5):595–605.

[64] Reeks J, Naismith JH, White MF. CRISPR interference: a structural perspective. Biochem J July 15, 2013;453(2):155–66.

[65] Westra ER, Buckling A, Fineran PC. CRISPR-Cas systems: beyond adaptive immunity. Nat Rev Microbiol May 2014;12(5):317–26.

[66] Rouillon C, Zhou M, Zhang J, Politis A, Beilsten-Edmands V, Cannone G, et al. Structure of the CRISPR interference complex CSM reveals key similarities with cascade. Mol Cell October 10, 2013;52(1):124–34.

[67] Benda C, Ebert J, Scheltema RA, Schiller HB, Baumgartner M, Bonneau F, et al. Structural model of a CRISPR RNA-silencing complex reveals the RNA-target cleavage activity in Cmr4. Mol Cell October 2, 2014;56(1):43–54.

[68] Deng L, Garrett RA, Shah SA, Peng X, She Q. A novel interference mechanism by a type IIIB CRISPR-Cmr module in *Sulfolobus*. Mol Microbiol March 2013;87(5):1088–99.

[69] Peng W, Feng M, Feng X, Liang YX, She Q. An archaeal CRISPR type III-B system exhibiting distinctive RNA targeting features and mediating dual RNA and DNA interference. Nucleic Acids Res January 2015;43(1):406–17.

[70] Goldberg GW, Jiang W, Bikard D, Marraffini LA. Conditional tolerance of temperate phages via transcription-dependent CRISPR-Cas targeting. Nature October 30, 2014;514(7524):633–7.

[71] Selle K, Klaenhammer TR, Barrangou R. CRISPR-based screening of genomic island excision events in bacteria. Proc Natl Acad Sci USA June 30, 2015;112(26):8076–81.

[72] DeBoy RT, Mongodin EF, Emerson JB, Nelson KE. Chromosome evolution in the Thermotogales: large-scale inversions and strain diversification of CRISPR sequences. J Bacteriol April 2006;188(7):2364–74.

[73] Babu M, Beloglazova N, Flick R, Graham C, Skarina T, Nocek B, et al. A dual function of the CRISPR-Cas system in bacterial antivirus immunity and DNA repair. Mol Microbiol January 2011;79(2):484–502.

[74] Sampson TR, Saroj SD, Llewellyn AC, Tzeng YL, Weiss DS. A CRISPR/Cas system mediates bacterial innate immune evasion and virulence. Nature May 9, 2013;497(7448):254–7.

[75] Louwen R, Horst-Kreft D, de Boer AG, van der Graaf L, de Knegt G, Hamersma M, et al. A novel link between *Campylobacter jejuni* bacteriophage defence, virulence and Guillain-Barre syndrome. Eur J Clin Microbiol Infect Dis February 2013;32(2):207–26.

[76] Gunderson FF, Cianciotto NP. The CRISPR-associated gene cas2 of *Legionella pneumophila* is required for intracellular infection of amoebae. MBio 2013;4(2):e00074–13.

[77] Marraffini LA. CRISPR-Cas immunity in prokaryotes. Nature October 1, 2015;526(7571):55–61.

Section III

Genome Stability of Unicellular Eukaryotes

Chapter 7

From Micronucleus to Macronucleus: Programmed DNA Rearrangement Processes in Ciliates Are Regulated Epigenetically by Small and Long Noncoding RNA Molecules

F. Jönsson
University of Witten/Herdecke, Witten, Germany

Chapter Outline

1. INTRODUCTION

Ciliates, unicellular eukaryotes, have developed into powerful model systems for studying pathways involved in genome remodeling processes. Very uniquely, ciliates have evolved a separation of germline and soma by differentiating two morphologically and functionally different types of nuclei, the micronucleus (mic) and the macronucleus (mac), in the same cell [1]. In the course of developing a new mac, extensive DNA rearrangement processes have to take place including DNA amplification, DNA fragmentation and elimination, in some species reordering of sequences, and, finally, de novo telomere addition. More details on the regulation of these processes have been elucidated over the years showing a strong connection to processes regulated by noncoding RNA (ncRNA). Variation in nuclear organization between different ciliate classes, such as the Oligohymenophoreans (eg, *Tetrahymena*) and the only distantly related Spirotrichs (eg, *Oxytricha* and *Stylonychia*) is reflected in different adaptations of these regulation processes and shows the huge power of ncRNA-regulated mechanisms in genome dynamics.

In ciliates, diploid, generative mics are required for sexual reproduction and therefore are often considered as analogs of "germline" nuclei. During vegetative growth (asexual reproduction by binary fission), mics are transcriptionally almost inactive with their DNA being organized in heterochromatic structures (Fig. 7.1A and B).

The macs, on the other hand, are DNA rich and transcriptionally highly active during vegetative growth, supplying the cell with all transcripts required for its maintenance and vegetative growth [1]. The macs therefore often are referred to as somatic nuclei. In Fig. 7.1A, a vegetative *Stylonychia* cell was stained with antibodies against acetylated histone H3. These histone H3 acetylations are connected with permissive chromatin, showing the transcriptionally active state of macs in the

Genome Stability. http://dx.doi.org/10.1016/B978-0-12-803309-8.00007-0

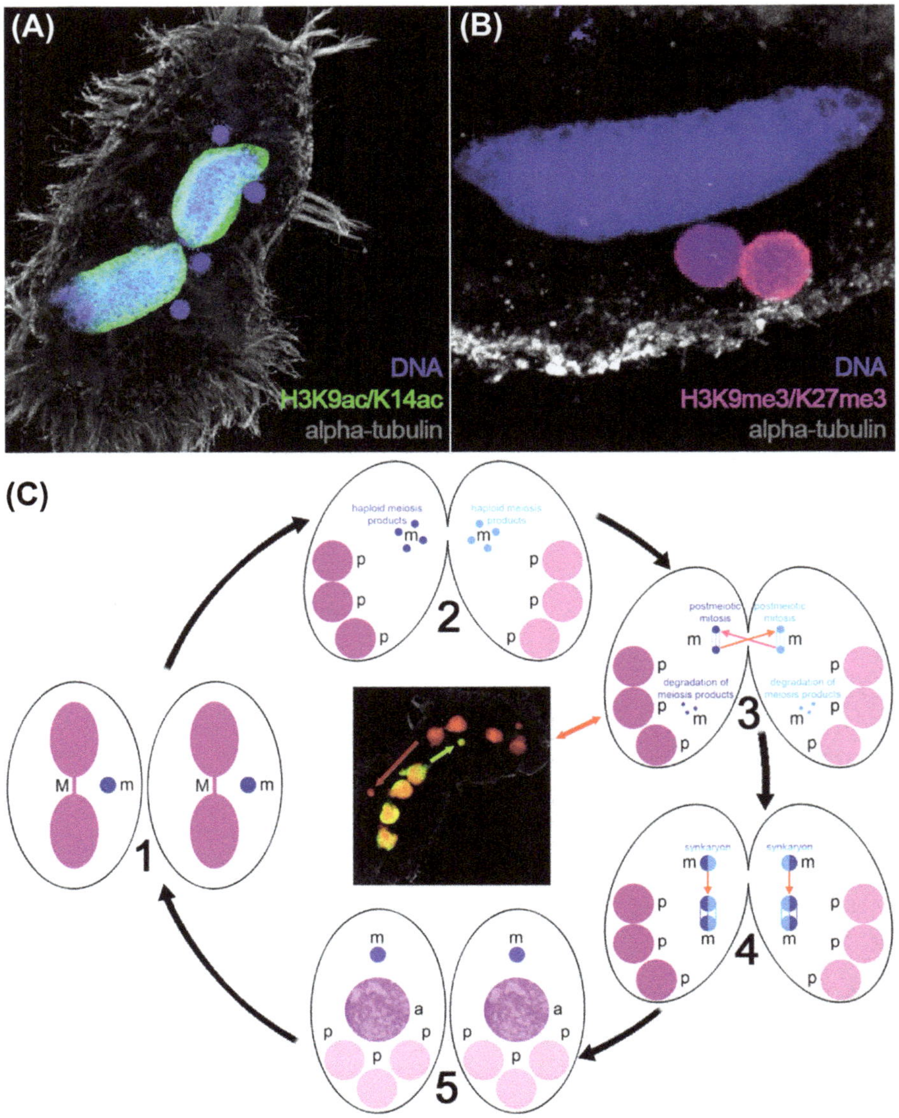

FIGURE 7.1 **Nuclear dimorphism in the stichotrichous ciliate *Stylonychia lemnae* and its sexual life cycle.** (A,B) Nuclear dimorphism in vegetative *Stylonychia* cells, in both panels DNA was counterstained in *blue* showing several micronuclei as well as the macronucleus which consists of two parts connected by a thin nucleoplasmic bridge (A), cellular shapes are visualized in *gray* by using an α-tubulin antibody. In (A) permissive chromatin of the macronucleus (*green*) was detected by staining with H3K9ac/K14ac antibodies. In (B) transcriptionally inert, heterochromatic micronuclei are stained with H3K9me3/K27me3 (*pink*). (C1–5) Schematic diagram of conjugation in ciliates *(Modified after Grell KG. Protozoology. Berlin, Heidelberg, New York: Springer Verlag; 1973.)*. For explanation see the text. The microscopic panel shows two conjugating *Stylonychia* cells. DNA was stained in *red*. In one conjugation partner nuclei were furthermore marked by incorporation of bromodeoxyuridine in *green*.

vegetative cell [2]. The macs differentiate from derivatives of mics in an elaborated developmental process, resulting in the removal of specifically selected sequences from the developing mac while all sequences encoding genes and regulatory sequences required for their transcription and replication are retained. This elimination of specific sequences can be seen as most extreme form of gene silencing and its regulation shares features with processes involved in RNAi silencing [3,4].

The number of mics and macs in one cell varies between different ciliate species and during the stages in their life cycle. While mics divide by conventional mitosis during vegetative growth, macs undergo a process called amitosis during which they divide without spindle formation or apparent chromosome condensation [5] resulting in daughter nuclei which obtain roughly but not necessarily the same amount of DNA.

2. THE SEXUAL LIFE CIRCLE OF CILIATES

During sexual reproduction (conjugation), which is induced by mixing ciliates of two different mating types, a new mac is generated from a micronuclear derivative. Fig. 7.1C shows the events taking place during conjugation of stichotrichous ciliates as, for example, *Oxytricha* or *Stylonychia*: first, the mic (m) of each conjugating cell undergoes a meiotic, followed by a mitotic division resulting in four haploid mics (Fig. 7.1C2). Then, in each conjugation partner one of these haploid mics divides mitotically into a stationary and a migratory nucleus, while the other three haploid mics disintegrate. Via a cytoplasmic bridge, connecting the two conjugation partners during conjugation, both migratory nuclei move into the other partner (Fig. 7.1C3) and fuse with the stationary nucleus forming a diploid zygotic nucleus (synkaryon) (Fig. 7.1C4). After mitotic division of the zygote nucleus, one of the daughter nuclei becomes the new mic, while the other differentiates into a new mac (M). During this differentiation process the developing mac is also called macronuclear anlage (a) (Fig. 7.1C5).

In order to differentiate a new mac from the zygotic nucleus the genome has to undergo extreme DNA processing processes, including DNA fragmentation, DNA elimination, DNA reordering (in some ciliate species), and DNA amplification processes [1]. As a first step of macronuclear differentiation the genome of the macronuclear anlage is endoreplicated to a copy number specific for each ciliate species. These endoreplicated chromosomes then become fragmented and a large part of sequences becomes eliminated. This sequence elimination is a very specific process ensuring that all sequences encoding genes and their regulatory sequences remain in the new mac. Depending on the ciliate species, the amount of sequences eliminated during macronuclear development varies. Remaining sequences in the developing macs are thereby processed into short DNA fragments, with the average size and copy numbers of these fragments varying between different ciliate species. To each of these newly formed macronuclear fragments, telomeric sequences are added de novo and (at least in stichotrichous ciliates) in a second amplification process, these macronuclear fragments are amplified, each to its specific copy number [6,7]. Early in conjugation, the old mac (parental mac (p)) is first fragmented and later starts to disintegrate until it disappears from the cell (Fig. 7.1C). Therefore, during a large part of conjugation, the parental macs as well as the anlage coexist in the same cell. This, as discussed later, enables the conjugating cell to exchange factors between these two nuclei and thereby epigenetically transmit information from parent to its offspring [8].

In the following, similarities and differences in the genome structures of the two different ciliate classes, Oligohymenophorea (eg, *Tetrahymena*) and Stichotrichia (eg, *Oxytricha* and *Stylonychia*) are described and discussed in the context of different models proposed for the epigenetic regulation of genome rearrangements in these ciliates.

3. ORGANIZATION OF THE MICRO- AND MACRONUCLEAR GENOMES

In mics, as in conventional eukaryotic nuclei, the genome is organized in long chromosomes with the genes being scattered along the DNA molecules and separated by long stretches of intergenic DNA. A 2014 draft assembly of micronuclear sequences of the stichotrichous ciliate *Oxytricha trifallax* estimated a size of the micronuclear genome of approximately 500 Mbp [9] organized in about 120 chromosomes, while the micronuclear genome of the only distantly related oligohymenophorean ciliate *Tetrahymena thermophila* contains about 157 Mbp in five chromosomes [1,10].

The genome of macs, on the other hand, is organized in shorter linear molecules. Each of these short macronuclear molecules carries all sequences necessary for its replication and all are terminated with telomeric sequences of homogeneous length at both ends, but lack centromeric DNA sequences [1]. The DNA complexity of macronuclear genomes is much lower than in their corresponding mics [11]. Most extremely, in stichotrichous ciliates as, for example, *Oxytricha* or *Stylonychia,* up to over 95% of the micronuclear DNA sequences are eliminated during development of a new mac, leaving behind only sequences coding for genes and sequences necessary for their transcription and replication. In stichotrichous ciliates these remaining sequences are organized in 15,000–20,000 different short molecules, with sizes between several hundred base pairs up to more than 20 kbp with an average length of about 2–4 kbp [12,13]. Because of their small size, macronuclear molecules are often called nanochromosomes in these ciliates. Usually they only encode one gene; in *Oxytricha*, only 10% of the nanochromosomes contain sequences coding for more than one gene [13].

In the oligohymenophorean ciliate *Tetrahymena*, a much lower percentage of micronuclear DNA sequences are eliminated during macronuclear development: only about 34% of the micronuclear genome is removed from the macronuclear genome, the remaining sequences are organized in 187 macronuclear molecules with an average length of about 500 kbp [14], always encoding many genes separated by spacer DNA. In these ciliates, a well-conserved 15 nt chromosome breakage site (CBS) has been identified which is necessary and sufficient for fragmentation of micronuclear sequences into macronuclear DNA molecules to occur. After breakage, the CBS and about 20 additional base pairs on both sides are lost [15,16] and telomeric sequences are added de novo to both ends of each fragment. Until now, proteins which recognize this CBS have not been identified, but it has been speculated that a homing endonuclease could have been domesticated to perform fragmentation [17]. In *Tetrahymena*, loss of DNA caused by fragmentation only accounts for a minor amount of sequences eliminated during development of a new mac. In fact, most DNA is eliminated as internal DNA deletion, by removal of specific sequences from internal regions of chromosomes without generating new stable ends (Fig. 7.2A).

Ciliate micronuclear genomes contain numerous internal eliminated sequences (IESs), interrupting macronuclear precursor sequences in the micronuclear genome [18–20]. A 2010 genome sequence study identified more than 10,000 IESs,

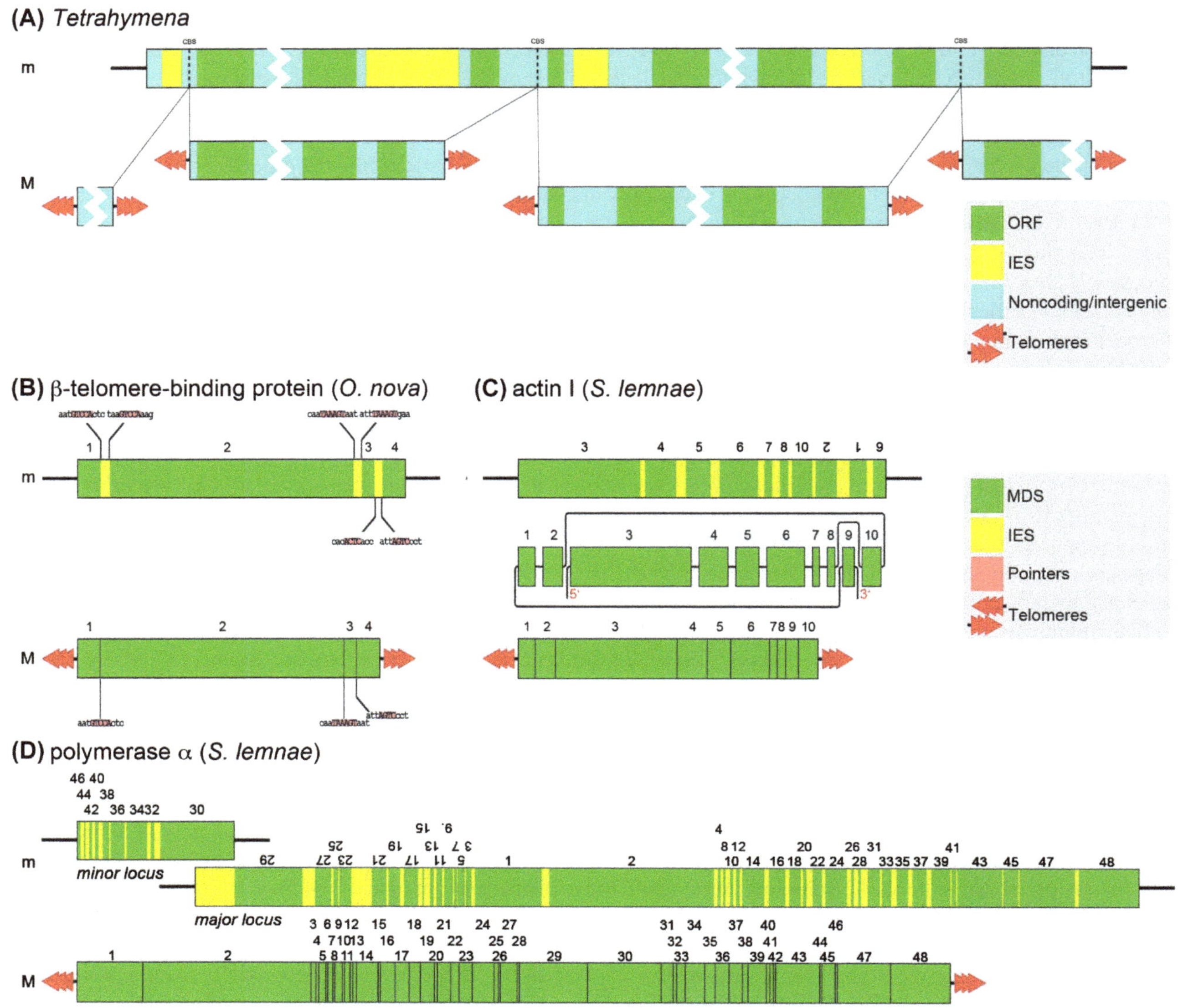

FIGURE 7.2 **Examples of the differing micro- and macronuclear genome organization in oligohymenophorean and stichotrichous ciliates.** (A) Scheme of the different organization of a micronuclear locus (*m*) and its corresponding macronuclear molecules (*M*) in *Tetrahymena*. Open reading frames (ORF) are shown in green, internal eliminated sequences (IESs) in *yellow*, noncoding, intergenic sequences in *blue*, telomeres as *red arrowheads* and nonspecified flanking sequences as *black lines*. (B–D) Examples of the different organization of genes in the micronuclear (*m*) and the macronuclear (*M*) genome of stichotrichous ciliates. MDSs are shown in green, IESs in *yellow*, pointers in *light red*, telomeres as *red arrowheads* and flanking micronuclear sequences as *black lines*. Numbers show the order of MDSs in the macronuclear nanochromosome, upside down numbers indicate an inverted orientation of the MDS in the micronuclear genome. (B) The linear *β-telomere-binding* gene in *O. nova* [47]. At the borders between MDSs and IESs both copies of individual direct repeats (pointers) are indicated, of which only one copy remains in the macronucleus. (C) The scrambled *actin I* gene in *Stylonychia lemnae* [48]. The rearrangement processes necessary to descramble the micronuclear *actin I* gene into a functional nanochromosome [51] are indicated in the middle scheme. (D) The extensively scrambled *polymerase* α gene is encoded in two different loci in the micronuclear genome of *S. lemnae* [49].

with sizes ranging from several hundred base pairs up to 10 kbp (on average 2–3 kbp), residing in the micronuclear genome of *Tetrahymena*, most, if not all, in noncoding regions [10]; many of these IESs resemble transposable elements and most contain short terminal direct repeats. In the early stages of macronuclear development, before becoming eliminated, these micronuclear-specific sequences adopt heterochromatic chromatin organization: chromatin to become removed has been shown to be under-acetylated [21]. Moreover, di- and tri-methylation at lysine 9 of histone H3 (H3K9me2,3), which are known to be consistent histone modifications assigned to heterochromatin organization in many eukaryotes, as well as tri-methylation at lysine 27 (H3K27me3) appear to be associated to IESs in the developing mac and disappear after elimination has occurred [22]. One of three methyltransferase genes identified in the genome of *Tetrahymena*, *EZL1*, is expressed only during conjugation and seems to be involved in introducing the H3K27me3 modification, which in turn seems to facilitate methylation of H3K9 [23]. Moreover, a chromodomain protein, Pdd1p, related to the conserved heterochromatin protein 1 (HP1), could be identified to be expressed only during conjugation. It localizes to the developing anlage, where in later stages of conjugation it colocalizes with aggregates of IESs [24]. Gene knockout studies showed that Pdd1p is required for DNA deletion and, when tethered to sequences normally remaining in the mac, caused these sequences to become eliminated [22]. Pdd1p contains two chromodomains and one chromoshadow domain, with the first chromodomain presumably being responsible for DNA targeting, while the other two domains seem to be involved in histone modifications and Pdd1p aggregation [25].

During 2010, two genes, *TPB1* and *TPB2*, were identified which show high similarity to piggyback transposases and lack other transposon features. Differently to other transposons and transposon-like elements residing in the micronuclear genome of *Tetrahymena*, they are not removed from the developing mac, but instead are expressed specifically during conjugation [26]. Namely, Tpb2p seems to be involved in DNA deletion: it colocalizes with Pdd1p forming aggregates, and knockdown strains are deficient in DNA deletion and chromosome fragmentation. When expressed in bacteria, Tpb2p exhibits a weak endonuclease activity. This endonucleolytic activity generates DNA ends resembling the termini of DNA molecules which occur as intermediate products during removal of the M-element, an IES often used as model for deletion studies in *Tetrahymena* [27]. This suggests that Tpb2p could be required for DNA cutting in the process of DNA elimination. During the process of domestication, *TPB2* seems to have acquired further exons in its C-terminus in addition to piggyback domains, leaving it nearly twice as long as most other piggyback transposases. Presumably through these additional C-terminal sequences, including a zinc-finger domain, Tpb2p is now targeted to heterochromatic sequences [28].

Even though IESs in *Tetrahymena* are located in noncoding regions, therefore not necessarily requiring exact excision of sequences, and are marked only imprecisely by adopting heterochromatic chromatin structure, most IESs are nevertheless removed rather precisely from the genome of the developing mac, with only minor occasional junction variations of less than 10 bp [29,30]. Only some IESs are eliminated with junctions varying by several hundred nucleotides [31]. IESs eliminated precisely seem to be flanked by pairs of *cis*-acting sequences, for example, the M-element is flanked on both sides by a 10 bp sequence motive, 45 bp away from each end. This sequence motive seems to be responsible for setting distinct deletion boundaries as without it deletion of the M-element becomes variable. Moreover, new boundaries are induced if the motif is inserted within the M-element [32]. Similar flanking sequence motives have also been identified near other IESs. This led to the speculation, that in *Tetrahymena* precisely excised IESs are subdivided into many families, with each family sharing a specific sequence motive for determining IES elimination boundaries [33]. It has been suggested that these sequence motives flanking the heterochromatic sequences destined to be eliminated could help to target Tpb2p and presumably also other similar nucleases to these heterochromatin boundaries, where they would cut the DNA according to their specific sequence requirements [34].

So far, no distinct CBSs necessary for fragmentation could be identified in stichotrichous ciliates. In *Stylonychia*, a 6 nt sequence localized in the 5′- and 3′-subtelomeric region of the nanochromosome coding for a 1.3 kbp gene of unknown function seems to be necessary for its fragmentation [35]. This putative CBS is very similar to a conserved CBS (E-CBS) identified in the hypotrichous ciliate *Euplotes* [36]. In contrast to *Tetrahymena*, in stichotrichous ciliates micronuclear-specific sequences are eliminated largely as long intergenic DNA stretches, separating the macronuclear precursor sequences from each other. These sequences eliminated as bulk DNA consist in large parts of satellite repeats, transposons, and, of some micronuclear-specific coding sequences [9].

Earlier studies in *Stylonychia* already showed by electron microscopy that in the developing mac at the end of polytenization, a large part of the genome adopts heterochromatic features and is excised as ring-like chromatin structures [37]. This appearance of heterochromatic chromatin in the macronuclear anlage was confirmed by studies on histone modification patterns and chromatin plasticity during macronuclear differentiation [2,38]: very early in macronuclear development, repressive histone modifications typical for the germline mic (eg, H3K27me3) are removed, and instead, the anlagen genome becomes associated with histone modifications, such as histone H3 acetylations, typical for "open" chromatin. Sequences which will be retained in the developing mac stay associated with these histone modifications, while repressive markers (such as H3K9me3 or H3K27me3) are introduced to sequences to become removed. Furthermore, similar to *Tetrahymena*, chromatin-modifying proteins, the heterochromatin-specific *Stylonychia* HP1 homolog Spdd1p and a putative E(z)kmt6-like histone methyl-transferase could be localized in the developing anlage simultaneously to the appearance of repressive markers. During mid-2010s it was proposed that in *Stylonychia*, the fate of specific DNA sequences targeted to become organized into either permissive or repressive chromatin, leading to retention and then respectively elimination, could be determined not only by introducing specific posttranslational histone modifications, but also by a preceding deposition of specific histone H3 variants [39]. In *Stylonychia*, eight different histone H3 variants could be identified; some of these variants were shown to be differentially expressed during macronuclear development and to be specifically targeted by posttranslational modifications.

Besides chromatin modifications acting on histones, a further epigenetic modification, methylation of DNA, could be involved in marking sequences for elimination from the anlagen genome [40]. In stichotrichous ciliates, extensive methylation of cytosines as well as hydroxymethylation has been observed to occur in micronuclear-specific transposons and satellite repeat sequences [41,42].

In addition to sequences removed as bulk DNA in the process of fragmentation, similar to *Tetrahymena*, stichotrichous ciliates contain more than 200,000 IESs in their micronuclear genome [9]. In contrast to *Tetrahymena*, IESs in stichotrichous ciliates are smaller, with sizes generally between only 0 and 100bp [9,43,44] residing in noncoding as well as coding regions. According to the number of interrupting IESs, macronuclear precursor sequences in the micronuclear genome are separated into several blocks of so-called macronuclear destined sequences (MDSs) (Fig. 7.2B–D). IESs in stichotrichous ciliates always are flanked by "pointer" sequences, direct repeats of 2–20bp in length, with one of the two pointer copies at the 3′-end of MDS n and the second copy of the pointer at the 5′-end of MDS $n+1$ according to their order in the macronuclear nanochromosome (Fig. 7.2B). IESs with a size of 0bp therefore just consist of tandem repeats of two pointers in the micronuclear genome. After excision of IESs, only one copy of each pointer can be found at the junction between neighboring MDSs, suggesting a homology directed repair mechanism to be involved in IES elimination [43,45]. In *Oxytricha*, thousands of active transposase genes were identified to reside in the micronuclear genome-encoding proteins which are thought to be involved in the excision of IESs. Each of these transposases would be targeted to its specific pointer sequence and then be responsible for the excision of all IESs flanked by this specific pointer sequence [46]. In general, MDSs in the micronuclear genome of ciliates occur in the same order as in the corresponding macronuclear chromosome of the adult mac, they are linearly arranged (eg, Fig. 7.2B, micronuclear *β-telomere binding protein* locus [47]) but in stichotrichous ciliates, more than 30% of the MDSs are arranged in permuted disorder or inverted orientation in the micronuclear chromosomes [9,45] (eg, Fig. 7.2C, micronuclear *actin I* locus [48]) with MDSs sometimes being even located on different chromosomes (eg, Fig. 7.2D, *polymerase α* locus [49]). In order to form functional nanochromosomes, in these cases not only the IESs have to be excised precisely during macronuclear development, as in linearly arranged MDSs, but furthermore scrambled MDSs have to be reordered into their correct macronuclear order and orientation (Fig. 7.2C) (for review: [45]).

In stichotrichous ciliates, each nanochromosome is amplified to its specific copy number in the course of macronuclear development. These copy numbers vary between a few 100 to up to 10^6 copies [1,6,7]. Copy numbers are generated in two amplification steps. First, as in *Tetrahymena*, early in conjugation, before fragmentation into nanochromosomes takes place, anlagen chromosomes are endoreplicated into polytene chromosomes. Already during this first amplification stage, transposon-like elements as well as the IESs become excised from the anlagen genome [50–52]. After this first amplification stage, fragmentation follows and the intervening micronuclear-specific bulk DNA sequences are eliminated and telomere sequences are added de novo to both ends of each nanochromosome. In a second amplification step, all nanochromosomes become amplified, each to its specific copy number. How this specific amplification could be regulated is discussed later.

In the first part of this chapter the organization of the micro- and macronuclear genome of two different ciliate classes, the oligohymenophorean ciliate *Tetrahymena* and the stichotrichous ciliates *Oxytricha* and *Stylonychia* was described, as well as some of the processes and the proteins identified so far which in the course of macronuclear differentiation are necessary in order to develop a new mac from a micronuclear derivative. But how are sequences selected to become organized into heterochromatin and finally deleted?

In the second part, the actual knowledge on the regulation of these selection processes is discussed.

4. EPIGENETIC REGULATION OF MACRONUCLEAR DEVELOPMENT IN *TETRAHYMENA*

One of several early hints that genomic rearrangement during macronuclear development could be regulated epigenetically by information from the parental mac came from studies of two differing cell lines of *Paramecium*, like *Tetrahymena* belonging to the oligohymenophorean ciliates. These two cell lines contained identical mics but differed in their macronuclear genomes. While one cell line kept a specific IES in its macronuclear genome, the second cell line reproducibly excised this IES during macronuclear development. In progenies of crosses of these two cell lines, it could be shown that the genomic alternative, that is, retaining the IES or eliminating it from the macronuclear chromosome, was maternally inherited. Furthermore, when parental macs in these two cell lines were transformed with fragments of either version of the respective macronuclear chromosome, presence of the IES in the parental mac resulted in retention, while absence resulted in elimination of this IES in the newly developed macronuclear genome of its sexual progeny. This suggested that information from the parental mac is involved in selecting sequences for deletion [53].

In 2002, Mochizuki et al [54]. presented a model, the scanRNA model, explaining regulation of macronuclear development in *Tetrahymena*. This model was based mainly on the identification of a population of small RNA (sRNA) molecules, about

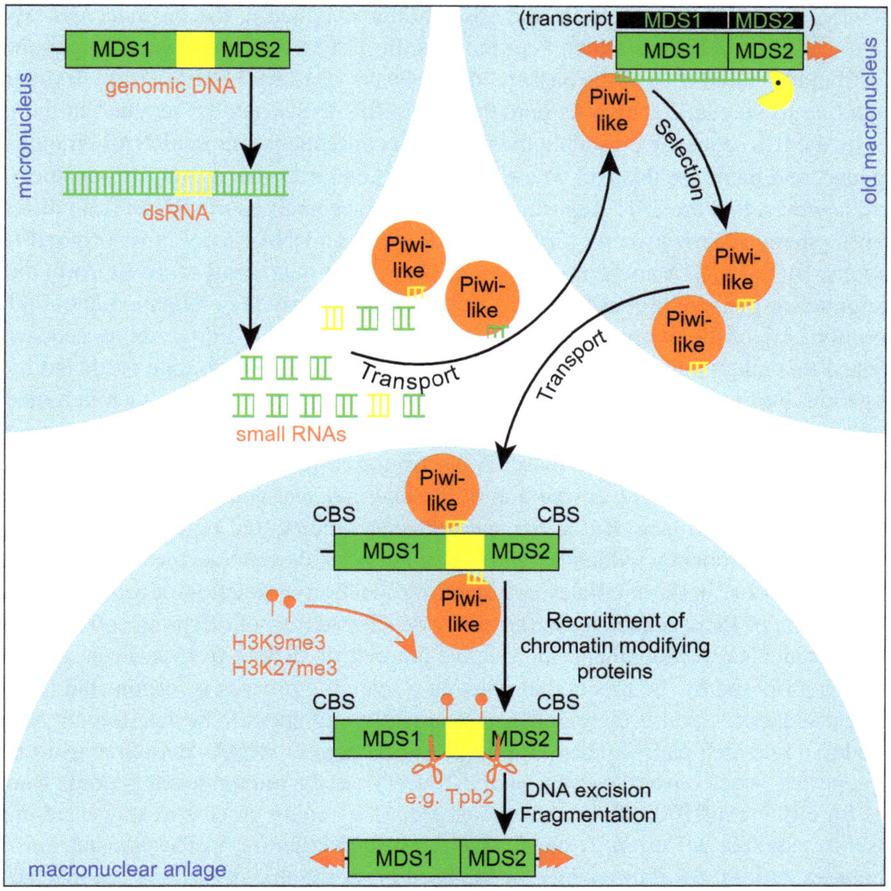

FIGURE 7.3 The scanRNA model. sRNAs are processed from long dsRNA molecules deriving from bidirectional transcription of the micronuclear genome. After association to piwi-like proteins, sRNAs invade the parental macronucleus. By interacting with the macronuclear genome [or maybe transcripts of the macronuclear genome (*black, in brackets*)], complementary sequences (*green*) are retained in the parental macronucleus and become eventually degraded. sRNAs homologous to micronuclear sequences (*yellow*) remain free to invade the macronuclear anlage. By recruiting chromatin-modifying enzymes to their corresponding sequences in the anlagen genome, sRNAs mark these sequences for deletion by inducing heterochromatin formation. Heterochromatic sequences are targeted by endonucleases, such as Tpb2, a domesticated piggyback transposase, excised and finally degraded. The anlagen chromosomes become fragmented at sites defined by the conserved CBSs and telomeres (*red arrowheads*) are added de novo to each end of the new macronuclear chromosomes. *Modified after Mochizuki K, Fine NA, Fujisawa T, Gorovsky MA. Analysis of a piwi-related gene implicates small RNAs in genome rearrangement in Tetrahymena. Cell 2002;110(6):689–99.*

27 nt in size, appearing in the early stages of macronuclear development and linked regulation of macronuclear development to the mechanism of RNA interference (RNAi) which only shortly before had been described in nematodes [55]. Very early in the course of macronuclear development, appearing sRNAs were shown to be homologous to both kinds of sequences present in the micronuclear genome, to the MDSs as well as the micronuclear-specific sequences. In later stages of macronuclear development, sRNAs homologous to micronuclear-specific sequences become enriched. Furthermore, they identified proteins expressed during these stages of macronuclear development and required for elimination, which were homologous to proteins known to play key roles in RNAi-related pathways in other organisms. In the scanRNA model (Fig. 7.3), very early in macronuclear development, the entire micronuclear genome is transcribed bidirectionally into double-stranded (ds) RNA molecules. These dsRNA molecules are then processed by Dcc1, a dicer-like protein [56,57], into sRNAs, called scan RNAs (scnRNAs), with sizes of about 27 nt. scnRNAs are associated with Twi1, a member of the Piwi family [54], which during conjugation was shown first to appear in the parental mac and in later stages to localize to the developing anlage. According to the model, scnRNAs/Twi1 complexes assemble in the cytoplasm and are then shifted into the parental mac. scnRNAs homologous to macronuclear sequences are retained in the parental mac by their complementary macronuclear DNA sequences (hence the term scnRNA) and become degraded. In contrast, scnRNAs homologous to micronuclear-specific sequences stay free to leave the parental mac and to invade the developing anlage. There they recruit chromatin-modifying enzymes to their complementary DNA sequences, that is, the micronuclear-specific sequences, marking them to be excised by recruiting chromatin-modifying enzymes, organizing these sequences into heterochromatin.

The scnRNA model very conclusively explains the observations obtained in the *Paramecium* experiments mentioned earlier: offspring of ciliates from cell lines which kept the specific IES in their macronuclear genome or of ciliates, into which this IES was artificially introduced in the parental mac, always retained the IES during development of a new mac. According to the model, in these cases, scnRNAs homologous to this IES would be retained in the paternal mac as they would be scanned for by the IES sequence present in the mac. As a consequence, no scnRNA homologous to the IES would reach the developing mac, and therefore this IES would not be marked for excision from its genome.

It should be noted, however, that the 2012 high-throughput sequence analysis of sRNAs from different time points during macronuclear development of *Tetrahymena* [58] showed that the scnRNA model in its original form seem at least to need some modifications. By this sRNA analysis, it could be shown that in contrast to the uniform bidirectional transcription of the entire micronuclear genome proposed in the scnRNA model, early in conjugation about 80% of all 27 nt sRNA could be assigned to only 25% of the micronuclear sequences coding for IESs. Furthermore, sequences to be retained in the mac were transcribed to a much lower extent (only 15% of all sRNAs analyzed) than predicted by the model for these early stages of conjugation. Furthermore, such bias of sRNA transcription could also be seen in mutants of *TWI1*. In these mutants, according to the model, the scanning pathway should be impeded, and therefore no enrichment should be obtained. This argues for the presence of an already highly biased transcription resulting in an enrichment of sRNAs homologous to some IESs sequences, instead of a uniform transcription of the micronuclear genome, followed by a subsequent scanning process of the sRNAs in the paternal mac. But so far mechanisms which could regulate such a biased transcription are unknown. Until now, direct experimental evidences arguing in favor for a scanning mechanism in the parental mac only could be obtained in *Paramecium*. In these ciliates, artificial introduction or deletion of sequences from the parental mac led to retention or elimination of these sequences in the newly developed mac of offspring cells [53,59].

In later stages of macronuclear development, the relative amount of sRNAs from macronuclear precursor sequences decreases significantly as proposed by the model, but whether a scanning process is responsible for this decrease or transcription of micronuclear sequences is already regulated dynamically, still needs to be determined. As one alternative to the original scanRNA model, a two-step pathway resulting in an enrichment of sRNAs homologous to micronuclear-specific sequences has been suggested. A selective transcription of dsRNAs from the micronuclear genome would be the first step to determine a sequence for elimination [60,61]. In a secondary step, a scanning process as suggested in the scanRNA model would than augment this sequence selection. A mechanism how such scanning in the parental mac could be performed still needs to be elucidated. At least for *Paramecium*, it has been proposed that in the parental mac an interaction between invading sRNAs and transcripts of all macronuclear chromosomes instead of the macronuclear genome itself could be responsible for the retention of sRNAs homologous to macronuclear sequences [59,62]. This would then leave sRNAs homologous to micronuclear-specific sequences free to invade the developing anlage and to recruit chromatin-modifying enzymes to their corresponding sequences marking them for elimination, similar as in RNA-induced transcriptional gene silencing in *Schizosaccharomyces pombe* and other organisms [4,63]. By evolving pathways to excise these sequences, elimination can thereby be seen as most extreme form of gene silencing.

5. EPIGENETIC REGULATION OF MACRONUCLEAR DEVELOPMENT IN STICHOTRICHOUS CILIATES

sRNAs with sizes of about 27 nt have also been detected during early stages of macronuclear development in stichotrichous ciliates [38,64,65]. In contrast to the findings in *Tetrahymena*, 2012, in high-throughput sequence analyses of conjugation-specific sRNAs in *Oxytricha*, no micronuclear-specific sRNAs were observed accumulating during macronuclear development, but instead a high level of a macronuclear-specific class of sRNAs were identified appearing solely during early macronuclear development [64,65]. These 27 nt long sRNAs originate from bidirectional transcription of the parental mac. They preferentially cover the open reading frames of all nanochromosomes including macronuclear-specific pointer recombination junctions which are only present in the rearranged macronuclear genome. As in oligohymenophorean ciliates, they are associated with Piwi-like proteins (Mdp1 in *Stylonychia* [66] and Otiwi1, one of 13 Twi1 homologs identified in *Oxytricha* [64]), which as in *Tetrahymena* were shown to first appear in the parental mac and subsequently to localize to the developing macronuclear anlage [2,64,67]. In contrast to the oligohymenophorean scanRNA model, in which micronuclear-specific scnRNAs target sequences for elimination, the macronuclear-specific sRNAs of stichotrichous ciliates seem to mark sequences for retention in the developing mac. Microinjection of synthetic sRNAs complementary to IESs into conjugating cells leading to retention of these IESs and therefore reprogramming their developmental fate [64] confirmed this suggestion. Moreover, their retention was shown to be stably inherited over asexual as well as sexual generations. In stichotrichous ciliates, the macronuclear-specific sRNAs could protect their corresponding sequences in the anlagen genome against introduction of epigenetic marks as DNA or histone methylation. Thereby all macronuclear destined sequences would stay in an open chromatin state, while all other sequences would be marked for deletion by packaging them into heterochromatin.

Marking sequences by chromatin modifications or protecting sequences by a pool of 27 nt sRNA molecules can only be imprecise. Such imprecise marking may be sufficient for sequences excised as bulk DNA and residing in noncoding sequences; however, IESs, most of them interrupting sequences destined to contribute to coding sequences in the macronuclear genome, need to be excised precisely in order to generate nanochromosomes encoding functional open reading frames. Indeed, so far in mature mac nanochromosomes, junctions between consecutive MDS were always observed to be correct accounting for a precise elimination of IESs in stichotrichous ciliates. But a different picture emerges when looking at intermediate molecules, generated as transient products during the process of genome rearrangement. A surprisingly high percentage of such intermediates show traces of IESs which had been incorrectly excised from their macronuclear precursor DNA [51]. In these cases, either too much or not enough sequences were excised leading to nanochromosomes with either additional or missing sequences. The fact that this does not lead to a high percentage of nonfunctional nanochromosomes in the mature macs suggests that either incorrectly rearranged nanochromosomes are selectively eliminated or that they are corrected by a proofreading mechanism. All cases of imprecise excision observed nevertheless always took place between pairs of direct repeats, called cryptic pointers, in contrast to the original pointers which should have been used for precise excision. Imprecise marking of specific sequences by modifying chromatin structure could explain the observed frequent occurrence of imprecise IES excision in intermediates. Excision could take place by homologous recombination between any direct repeats localized in the vicinity of the endogenous correct pointer repeats using them as cryptic pointers.

A previous experiment had already suggested before that in *Stylonychia* a proofreading mechanism requiring template sequences from the parental macronuclear genome could be involved in macronuclear development of this ciliate. In this experiment, various constructs of a micronuclear locus containing precursor sequences corresponding to an entire macronuclear nanochromosome as well as its IESs and additional flanking micronuclear-specific sequences were injected into anlagen before fragmentation took place [35,68]. Injection of constructs containing deletion of the micronuclear locus, missing up to more than 200 bp of one end of the nanochromosomal precursor sequences, resulted in fragmentation of these sequences from their flanking micronuclear-specific sequences. Surprisingly, after finishing macronuclear development, sequences missing in the injected deletion constructs had been filled up to correct full-length nanochromosomes. This suggested that a proofreading mechanism could exist ensuring that nanochromosomes, which had been truncated after fragmentation from their neighboring micronuclear-specific sequences, are filled up to their correct length. In order to supplement missing sequences correctly, such a proofreading mechanism would require full-length templates of all macronuclear nanochromosomes.

Moreover, as described earlier, in stichotrichous ciliates up to 30% of the genes occur in a scrambled disorder in the micronuclear genome. With pointer sequences between 2 and 20 bp in length, pointers are not specific enough to provide sufficient information to assure correct alignment of consecutive MDSs, especially if MDSs are scrambled in the mic, and therefore can be located far apart or even as in the case of the *polymerase α* in *Stylonychia* on different chromosomes (Fig. 7.2C and D).

In 2003, a theoretical model, the template-guided model of recombination, was proposed by Prescott et al. [69] explaining how processes necessary to rearrange the micronuclear genome into the genome of the mature mac could be regulated in stichotrichous ciliates (Fig. 7.4). They suggested that templates containing all sequences of the macronuclear genome are produced from the parental mac and transported into the early macronuclear anlage. In the anlage, templates align to their homologous DNA sequences of the early anlagen genome, thereby bringing corresponding direct pointer repeats into very close vicinity, while intervening micronuclear-specific sequences extrude as loops from the DNA–template complexes. Homologous recombination between the aligned pointer sequences then would allow excision of the loops containing the micronuclear-specific sequences, leaving behind only one copy of the pointers in the mature macronuclear genome.

In 2008, strong experimental support was provided that indeed in stichotrichous ciliates, DNA processing during macronuclear development could be guided by templates, presumably RNA molecules, originating from transcription of the parental mac [70]: degrading putative RNA templates of specific macronuclear nanochromosomes by applying RNAi techniques during macronuclear development resulted in aberrant gene unscrambling of the corresponding micronuclear loci. Furthermore, when early in macronuclear development, artificial RNA template molecules were injected, in which the template sequences corresponding to MDSs of a specific macronuclear nanochromosome were lined up in a different order to the MDS order of the endogenous template molecules, these artificial templates led to reprogrammed DNA rearrangements resulting in nanochromosomes with MDSs arranged according to the order of the artificial template.

The template-guided model conclusively can explain how rearrangement processes, necessary for reordering of scrambled genes into their corresponding macronuclear nanochromosomes as well as correcting imprecise excision of IESs and filling up truncated nanochromosomes after fragmentation, could be regulated. But why such imprecise IES excision

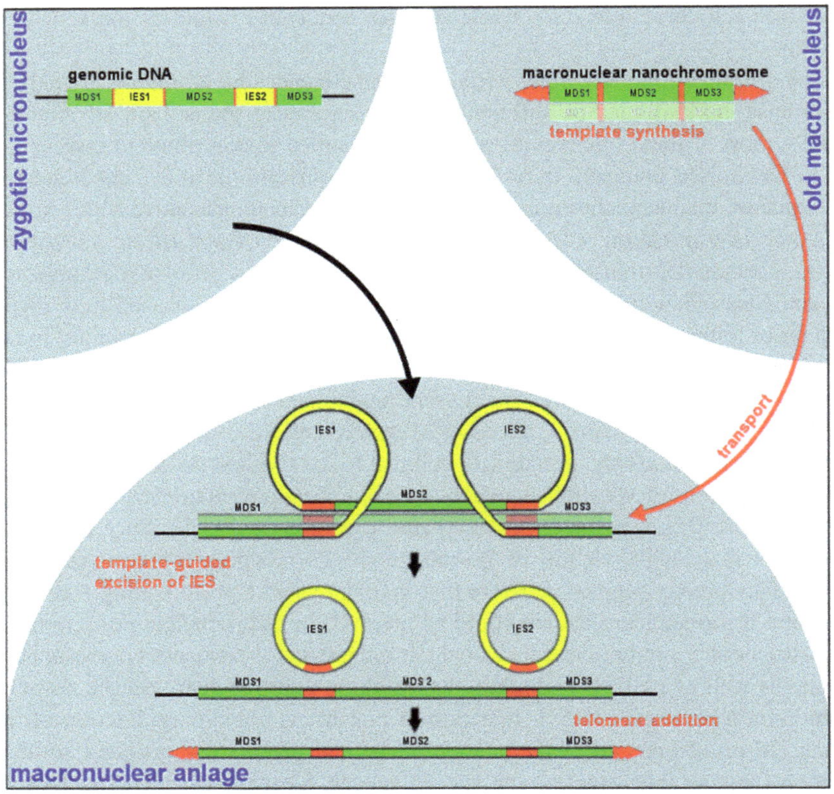

FIGURE 7.4 **The template guided model.** Genome rearrangement processes are guided by templates (*faint green* and *red*) of all nanochromosomal sequences. These templates derive from transcription of all nanochromosomes in the old (parental) macronucleus. Templates are transported into the macronuclear anlage and align to their corresponding sequences of the anlagen genome, thereby aligning both pointer repeats (*red*) at the junction of consecutive blocs of macronuclear destined sequences (MDSs, *green*). IESs (in *yellow*) looping out as pointers align, are excised by homologous recombination between the two pointer repeats. Finally, fragmentation into nanochromosomes occurs at the ends of template covered regions and telomeres are added de novo to each end of the newly arranged nanochromosome *(Modified after Prescott DM, Ehrenfeucht A, Rozenberg G. Template-guided recombination for IES elimination and unscrambling of genes in stichotrichous ciliates. J Theor Biol 2003;222(3):323–30.).* In this scheme, descrambling, arranging scrambled MDS according to their order in the correct nanochromosomal order is not shown. Fig. 7.2C gives a hint on how sequences of a scrambled micronuclear gene would need to bend and twist in order to unscramble by aligning to its nanochromosomal template. (More information on the gymnastics of DNA processing in ciliates can be found in Ref. [45].)

should occur is difficult to explain. In the template-guided model, the macronuclear precursor sequences should align precisely according to the template sequences. If imprecise IES excision should take place, correction by a proofreading mechanism should occur concomitantly with the excision process, while the template molecule is still aligned, not leading to imprecise intermediate molecules detectable by PCR methods.

Therefore, for stichotrichous ciliates, a third model was suggested combining aspects of both the scanRNA model as well as the template-guided model of recombination [71]. This model was modified according to the finding that in these ciliates, sRNAs originate from the parental mac targeting sequences for retention in the developing mac [64], as discussed earlier. In this model (Fig. 7.5), MDSs in the early anlagen genome are marked by homologous sRNAs which originate from bidirectional transcription of the parental macronuclear genome early during conjugation. After processing of the dsRNA molecules into sRNAs, they are associated to Piwi-like homologs and subsequently invade the developing mac. In the developing mac, these sRNAs protect their corresponding sequences of the anlagen genome from being marked for excision by chromatin-modifying enzymes [64,65]. Excision of IESs takes place by homologous recombination between direct repeats located between MDSs protected by sRNAs. As marking of sequences by chromatin modification is very imprecise, not only the correct pointer repeats, but also cryptic pointers, random direct repeats located near the correct pointer sequences, can be used as sites for excision by homologous recombination, thereby leading to imprecise excision of IESs. Later during conjugation all nanochromosomes of the parental mac are transcribed into long RNA molecules, presumably also covering the telomeric ends, which then migrate into the developing anlage too. There, these RNA molecules guide the alignment of MDSs according to their order in the parental nanochromosomes. Furthermore, they serve as templates for a proofreading mechanism repairing imprecise excision and filling up

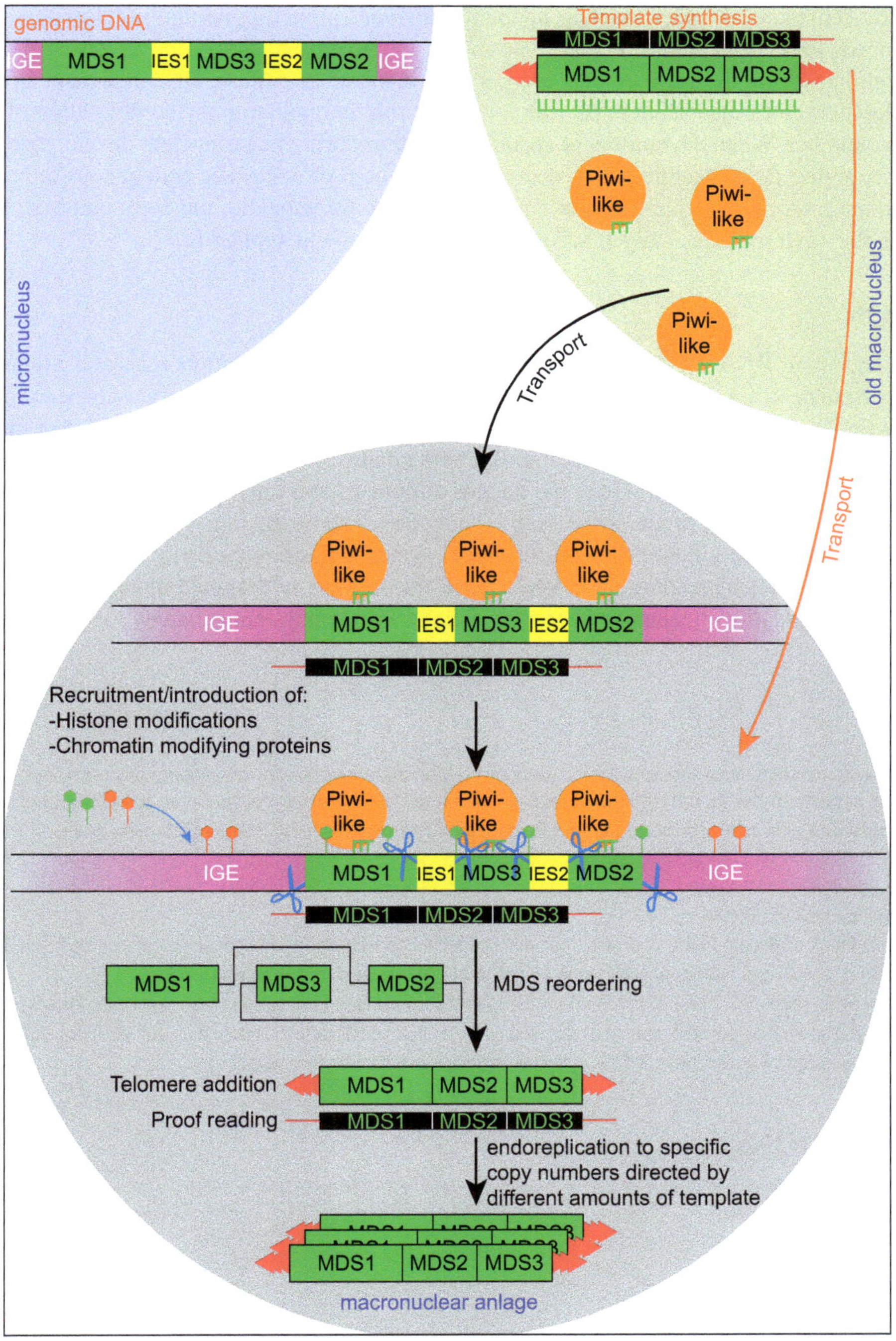

FIGURE 7.5 Model explaining macronuclear development in stichotrichous ciliates. Early in macronuclear development, bidirectional transcripts of the old (parental) macronucleus are processed into sRNAs (green) and associated to piwi-like proteins. These complexes invade the early macronuclear anlage and mark corresponding sequences of the anlagen genome as macronucleus destined sequences (*green*). Chromatin of all other sequences [intergenic sequences (IGE, *purple*) and IESs (*yellow*)] is modified by chromatin-modifying enzymes and thereby marked for deletion. Excision of IESs takes place (possibly by not yet characterized different tranposases) between the two pointer repeats of consecutive MDSs or any direct repeats in their vicinity. In addition to the early transcription of the old macronucleus, generating the sRNA molecules, the old macronucleus is further transcribed into the long probably full length transcripts (*black*) including the telomeres. They are transported into the anlage and align to corresponding sequences of the anlagen genome thereby guiding the reordering of scrambled MDSs. Furthermore, they act as templates for correcting imprecise excision of IESs or for complementing nanochromosomes which became truncated by fragmentation. Telomeres are added de novo to each nanochromosomal end. Finally, by a still unknown mechanism, the amount of template molecules, which reflects the copy number of each nanochromosome in the old macronucleus, then determines the level of amplification of each nanochromosome, thereby defining the specific copy number of each nanochromosome in the mature new macronucleus. *Modified after Fang W, Wang X, Bracht JR, Nowacki M, Landweber LF. Piwi-interacting RNAs protect DNA against loss during Oxytricha genome rearrangement. Cell 2012;151(6):1243–55; Juranek SA, Lipps HJ. New insights into the macronuclear development in ciliates. Int Rev Cytol 2007;262:219–51.*

missing subtelomeric sequences of macronuclear nanochromosomes after fragmentation. How this proofreading could be performed still needs to be determined.

Finally, RNA templates from the parental mac are not only involved in guiding DNA rearrangement processes during macronuclear development, they also seem to play an important role in regulating the amplification of nanochromosomes to their specific copy number. When the number of specific templates during macronuclear development is changed experimentally, either by applying RNAi techniques to decrease the amount of a specific endogenous template or by injecting artificial RNA template molecules to increase the amount of a specific template, the copy number of the corresponding nanochromosome in the adult macs is changed according to the experiment applied [6,7].

6. CONCLUSION

At first sight, the contrast in RNA regulation mechanisms between Oligohymenophorean and Stichotrichs seems to be surprising: sRNAs originate from different nuclei (mic versus parental mac), they target sequences inducing opposing developmental fates (deletion versus retention) and, in the case of Stichotrichs, an additional RNA-regulated mechanism is required to guide genome reordering and proofreading. But both ciliate classes are only very distantly related, their lineages separating over a billion years ago comparable to the linages of humans and fungi [72,73]. Moreover, this variability in the regulation of genome rearrangement processes in both ciliate classes show the high adaptability of RNA-induced regulation in genome dynamics making ciliates very useful model systems to study pathway regulated by ncRNA molecules. Regulation of genome dynamics depending on ncRNAs have shown to be widespread in eukaryotes. Ciliates, extensively using variations of these regulation mechanisms, provide excellent model systems to study the pathways involved in these processes.

GLOSSARY

Anlage During development of a new macronucleus from a derivative of the micronucleus, the developing macronucleus is called anlage.

Cryptic pointers Direct repeats in the vicinity of the correct pointer repeats. Use of cryptic pointers for IES elimination leads to imprecise elimination: These mistakes have to be corrected in order to generate functional macronuclear nanochromosomes.

Endoreplication DNA replication without nuclear division.

Macronucleus Somatic nucleus of ciliates.

Micronucleus Germline nucleus of ciliates.

Nanochromosome Short DNA molecule in the stichotrichous macronucleus, terminated by telomeric sequences and containing all sequences necessary for its replication, mostly encoding only one gene and all sequences necessary for its transcription.

Pointer repeats Direct repeats (between 2 and 20bp) flanking IESs in the micronuclear genome. One copy of the direct pointer repeat is localized at the 3′ end of MDS n, the second pointer repeat is localized at the 5′ end of MDS $n+1$. After excision of IESs, only one copy of the pointer repeats remains at the junction of consecutive MDSs in the macronuclear nanochromosome.

LIST OF ABBREVIATIONS

CBS Chromosome breakage site
IES Internal eliminated sequence
Mac Macronucleus
MDS Macronucleus destined sequence
Mic Micronucleus
scnRNA Scan RNA

REFERENCES

[1] Prescott DM. The DNA of ciliated protozoa. Microbiol Rev 1994;58(2):233–67.

[2] Postberg J, Heyse K, Cremer M, Cremer T, Lipps HJ. Spatial and temporal plasticity of chromatin during programmed DNA-reorganization in *Stylonychia* macronuclear development. Epigenetics Chromatin 2008;1:3.

[3] Verdel A, Jia S, Gerber S, Sugiyama T, Gygi S, Grewal SI, et al. RNAi-mediated targeting of heterochromatin by the RITS complex. Science 2004;303(5658):672–6.

[4] Volpe TA, Kidner C, Hall IM, Teng G, Grewal SI, Martienssen RA. Regulation of heterochromatic silencing and histone H3 lysine-9 methylation by RNAi. Science 2002;297(5588):1833–7.

[5] Cervantes MD, Coyne RS, Xi X, Yao MC. The condensin complex is essential for amitotic segregation of bulk chromosomes, but not nucleoli, in the ciliate *Tetrahymena thermophila*. Mol Cell Biol 2006;26(12):4690–700.

[6] Heyse G, Jonsson F, Chang WJ, Lipps HJ. RNA-dependent control of gene amplification. Proc Natl Acad Sci USA 2010;107(51):22134–9.
[7] Nowacki M, Haye JE, Fang W, Vijayan V, Landweber LF. RNA-mediated epigenetic regulation of DNA copy number. Proc Natl Acad Sci USA 2010;107(51):22140–4.
[8] Nowacki M, Landweber LF. Epigenetic inheritance in ciliates. Curr Opin Microbiol 2009;12(6):638–43.
[9] Chen X, Bracht JR, Goldman AD, Dolzhenko E, Clay DM, Swart EC, et al. The architecture of a scrambled genome reveals massive levels of genomic rearrangement during development. Cell 2014;158(5):1187–98.
[10] Fass JN, Joshi NA, Couvillion MT, Bowen J, Gorovsky MA, Hamilton EP, et al. Genome-scale analysis of programmed DNA elimination sites in *Tetrahymena thermophila*. G3 2011;1(6):515–22.
[11] Ammermann D, Steinbruck G, von Berger L, Hennig W. The development of the macronucleus in the ciliated protozoan *Stylonychia mytilus*. Chromosoma 1974;45(4):401–29.
[12] Aeschlimann SH, Jonsson F, Postberg J, Stover NA, Petera RL, Lipps HJ, et al. The draft assembly of the radically organized *Stylonychia lemnae* macronuclear genome. Genome Biol Evol 2014;6(7):1707–23.
[13] Swart EC, Bracht JR, Magrini V, Minx P, Chen X, Zhou Y, et al. The *Oxytricha trifallax* macronuclear genome: a complex eukaryotic genome with 16,000 tiny chromosomes. PLoS Biol 2013;11(1):e1001473.
[14] Eisen JA, Coyne RS, Wu M, Wu D, Thiagarajan M, Wortman JR, et al. Macronuclear genome sequence of the ciliate *Tetrahymena thermophila*, a model eukaryote. PLoS Biol 2006;4(9):e286.
[15] Fan Q, Yao MC. A long stringent sequence signal for programmed chromosome breakage in *Tetrahymena thermophila*. Nucleic Acids Res 2000;28(4):895–900.
[16] Yao MC, Zheng K, Yao CH. A conserved nucleotide sequence at the sites of developmentally regulated chromosomal breakage in *Tetrahymena*. Cell 1987;48(5):779–88.
[17] Fan Q, Yao M. New telomere formation coupled with site-specific chromosome breakage in *Tetrahymena thermophila*. Mol Cell Biol 1996;16(3):1267–74.
[18] Callahan RC, Shalke G, Gorovsky MA. Developmental rearrangements associated with a single type of expressed alpha-tubulin gene in *Tetrahymena*. Cell 1984;36(2):441–5.
[19] Yao MC, Choi J, Yokoyama S, Austerberry CF, Yao CH. DNA elimination in *Tetrahymena*: a developmental process involving extensive breakage and rejoining of DNA at defined sites. Cell 1984;36(2):433–40.
[20] Yao MC, Gorovsky MA. Comparison of the sequences of macro- and micronuclear DNA of *Tetrahymena pyriformis*. Chromosoma 1974;48(1):1–18.
[21] Duharcourt S, Yao MC. Role of histone deacetylation in developmentally programmed DNA rearrangements in *Tetrahymena thermophila*. Eukaryot Cell 2002;1(2):293–303.
[22] Taverna SD, Coyne RS, Allis CD. Methylation of histone h3 at lysine 9 targets programmed DNA elimination in *Tetrahymena*. Cell 2002;110(6):701–11.
[23] Liu Y, Taverna SD, Muratore TL, Shabanowitz J, Hunt DF, Allis CD. RNAi-dependent H3K27 methylation is required for heterochromatin formation and DNA elimination in *Tetrahymena*. Genes Dev 2007;21(12):1530–45.
[24] Madireddi MT, Coyne RS, Smothers JF, Mickey KM, Yao MC, Allis CD. Pdd1p, a novel chromodomain-containing protein, links heterochromatin assembly and DNA elimination in *Tetrahymena*. Cell 1996;87(1):75–84.
[25] Schwope RM, Chalker DL. Mutations in Pdd1 reveal distinct requirements for its chromodomain and chromoshadow domain in directing histone methylation and heterochromatin elimination. Eukaryot Cell 2014;13(2):190–201.
[26] Cheng CY, Vogt A, Mochizuki K, Yao MC. A domesticated piggyBac transposase plays key roles in heterochromatin dynamics and DNA cleavage during programmed DNA deletion in *Tetrahymena thermophila*. Mol Biol Cell 2010;21(10):1753–62.
[27] Saveliev SV, Cox MM. Developmentally programmed DNA deletion in *Tetrahymena thermophila* by a transposition-like reaction pathway. EMBO J 1996;15(11):2858–69.
[28] Vogt A, Mochizuki K. A domesticated PiggyBac transposase interacts with heterochromatin and catalyzes reproducible DNA elimination in *Tetrahymena*. PLoS Genet 2013;9(12):e1004032.
[29] Austerberry CF, Snyder RO, Yao MC. Sequence microheterogeneity is generated at junctions of programmed DNA deletions in *Tetrahymena thermophila*. Nucleic Acids Res 1989;17(18):7263–72.
[30] Austerberry CF, Yao MC. Nucleotide sequence structure and consistency of a developmentally regulated DNA deletion in *Tetrahymena thermophila*. Mol Cell Biol 1987;7(1):435–43.
[31] Chau MF, Orias E. Developmentally programmed DNA rearrangement in *Tetrahymena thermophila*: isolation and sequence characterization of three new alternative deletion systems. Biol Cell 1996;86(2–3):111–20.
[32] Godiska R, James C, Yao MC. A distant 10-bp sequence specifies the boundaries of a programmed DNA deletion in *Tetrahymena*. Genes Dev 1993;7(12A):2357–65.
[33] Chalker DL, La Terza A, Wilson A, Kroenke CD, Yao MC. Flanking regulatory sequences of the *Tetrahymena* R deletion element determine the boundaries of DNA rearrangement. Mol Cell Biol 1999;19(8):5631–41.
[34] Yao MC, Chao JL, Cheng CY. Programmed genome rearrangements in *Tetrahymena*. Microbiol Spectr 2014;2(6).
[35] Jonsson F, Steinbruck G, Lipps HJ. Both subtelomeric regions are required and sufficient for specific DNA fragmentation during macronuclear development in *Stylonychia lemnae*. Genome Biol 2001;2(2).
[36] Klobutcher LA. Characterization of in vivo developmental chromosome fragmentation intermediates in *E. crassus*. Mol Cell 1999;4(5):695–704.
[37] Meyer GF, Lipps HJ. Chromatin elimination in the hypotrichous ciliate *Stylonychia mytilus*. Chromosoma 1980;77(3):285–97.

[38] Juranek SA, Rupprecht S, Postberg J, Lipps HJ. snRNA and heterochromatin formation are involved in DNA excision during macronuclear development in stichotrichous ciliates. Eukaryot Cell 2005;4(11):1934–41.

[39] Forcob S, Bulic A, Jonsson F, Lipps HJ, Postberg J. Differential expression of histone H3 genes and selective association of the variant H3.7 with a specific sequence class in *Stylonychia* macronuclear development. Epigenetics Chromatin 2014;7(1):4.

[40] Bracht JR. Beyond transcriptional silencing: is methylcytosine a widely conserved eukaryotic DNA elimination mechanism? Bioessays 2014;36(4):346–52.

[41] Bracht JR, Perlman DH, Landweber LF. Cytosine methylation and hydroxymethylation mark DNA for elimination in *Oxytricha trifallax*. Genome Biol 2012;13(10):R99.

[42] Juranek S, Wieden HJ, Lipps HJ. De novo cytosine methylation in the differentiating macronucleus of the stichotrichous ciliate *Stylonychia lemnae*. Nucleic Acids Res 2003;31(5):1387–91.

[43] Prescott DM, DuBois ML. Internal eliminated segments (IESs) of Oxytrichidae. J Eukaryot Microbiol 1996;43(6):432–41.

[44] Prescott JD, DuBois ML, Prescott DM. Evolution of the scrambled germline gene encoding alpha-telomere binding protein in three hypotrichous ciliates. Chromosoma 1998;107(5):293–303.

[45] Prescott DM. Genome gymnastics: unique modes of DNA evolution and processing in ciliates. Nat Rev Genet 2000;1(3):191–8.

[46] Nowacki M, Higgins BP, Maquilan GM, Swart EC, Doak TG, Landweber LF. A functional role for transposases in a large eukaryotic genome. Science 2009;324(5929):935–8.

[47] DuBois ML, Prescott DM. Volatility of internal eliminated segments in germ line genes of hypotrichous ciliates. Mol Cell Biol 1997;17(1):326–37.

[48] Mollenbeck M, Cavalcanti AR, Jonsson F, Lipps HJ, Landweber LF. Interconversion of germline-limited and somatic DNA in a scrambled gene. J Mol Evol 2006;63(1):69–73.

[49] Landweber LF, Kuo TC, Curtis EA. Evolution and assembly of an extremely scrambled gene. Proc Natl Acad Sci USA 2000;97(7):3298–303.

[50] Klobutcher LA, Herrick G. Developmental genome reorganization in ciliated protozoa: the transposon link. Prog Nucleic Acid Res Mol Biol 1997;56:1–62.

[51] Mollenbeck M, Zhou Y, Cavalcanti AR, Jonsson F, Higgins BP, Chang WJ, et al. The pathway to detangle a scrambled gene. PLoS One 2008;3(6):e2330.

[52] Wen J, Maercker C, Lipps HJ. Sequential excision of internal eliminated DNA sequences in the differentiating macronucleus of the hypotrichous ciliate *Stylonychia lemnae*. Nucleic Acids Res 1996;24(22):4415–9.

[53] Duharcourt S, Butler A, Meyer E. Epigenetic self-regulation of developmental excision of an internal eliminated sequence on *Paramecium tetraurelia*. Genes Dev 1995;9(16):2065–77.

[54] Mochizuki K, Fine NA, Fujisawa T, Gorovsky MA. Analysis of a piwi-related gene implicates small RNAs in genome rearrangement in *Tetrahymena*. Cell 2002;110(6):689–99.

[55] Fire A, Xu S, Montgomery MK, Kostas SA, Driver SE, Mello CC. Potent and specific genetic interference by double-stranded RNA in *Caenorhabditis elegans*. Nature 1998;391(6669):806–11.

[56] Malone CD, Anderson AM, Motl JA, Rexer CH, Chalker DL. Germ line transcripts are processed by a Dicer-like protein that is essential for developmentally programmed genome rearrangements of *Tetrahymena thermophila*. Mol Cell Biol 2005;25(20):9151–64.

[57] Mochizuki K, Gorovsky MA. A Dicer-like protein in *Tetrahymena* has distinct functions in genome rearrangement, chromosome segregation, and meiotic prophase. Genes Dev 2005;19(1):77–89.

[58] Gao S, Liu Y. Intercepting noncoding messages between germline and soma. Genes Dev 2012;26(16):1774–9.

[59] Garnier O, Serrano V, Duharcourt S, Meyer E. RNA-mediated programming of developmental genome rearrangements in *Paramecium tetraurelia*. Mol Cell Biol 2004;24(17):7370–9.

[60] Schoeberl UE, Kurth HM, Noto T, Mochizuki K. Biased transcription and selective degradation of small RNAs shape the pattern of DNA elimination in *Tetrahymena*. Genes Dev 2012;26(15):1729–42.

[61] Yao MC, Chao JL. RNA-guided DNA deletion in *Tetrahymena*: an RNAi-based mechanism for programmed genome rearrangements. Annu Rev Genet 2005;39:537–59.

[62] Lepere G, Betermier M, Meyer E, Duharcourt S. Maternal noncoding transcripts antagonize the targeting of DNA elimination by scanRNAs in *Paramecium tetraurelia*. Genes Dev 2008;22(11):1501–12.

[63] Grewal SI, Moazed D. Heterochromatin and epigenetic control of gene expression. Science 2003;301(5634):798–802.

[64] Fang W, Wang X, Bracht JR, Nowacki M, Landweber LF. Piwi-interacting RNAs protect DNA against loss during *Oxytricha* genome rearrangement. Cell 2012;151(6):1243–55.

[65] Zahler AM, Neeb ZT, Lin A, Katzman S. Mating of the stichotrichous ciliate *Oxytricha trifallax* induces production of a class of 27 nt small RNAs derived from the parental macronucleus. PLoS One 2012;7(8):e42371.

[66] Fetzer CP, Hogan DJ, Lipps HJ. A PIWI homolog is one of the proteins expressed exclusively during macronuclear development in the ciliate *Stylonychia lemnae*. Nucleic Acids Res 2002;30(20):4380–6.

[67] Postberg J, Juranek SA, Feiler S, Kortwig H, Jönsson F, Lipps HJ. Association of the telomere-telomere binding protein-complex of hypotrichous ciliates with the nuclear matrix and dissociation during replication. J Cell Sci 2001;114(10):1861–6.

[68] Jonsson F, Wen JP, Fetzer CP, Lipps HJ. A subtelomeric DNA sequence is required for correct processing of the macronuclear DNA sequences during macronuclear development in the hypotrichous ciliate *Stylonychia lemnae*. Nucleic Acids Res 1999;27(14):2832–41.

[69] Prescott DM, Ehrenfeucht A, Rozenberg G. Template-guided recombination for IES elimination and unscrambling of genes in stichotrichous ciliates. J Theor Biol 2003;222(3):323–30.

[70] Nowacki M, Vijayan V, Zhou Y, Schotanus K, Doak TG, Landweber LF. RNA-mediated epigenetic programming of a genome-rearrangement pathway. Nature 2008;451(7175):153–8.

[71] Juranek SA, Lipps HJ. New insights into the macronuclear development in ciliates. Int Rev Cytol 2007;262:219–51.

[72] Bracht JR, Fang W, Goldman AD, Dolzhenko E, Stein EM, Landweber LF. Genomes on the edge: programmed genome instability in ciliates. Cell 2013;152(3):406–16.

[73] Parfrey LW, Lahr DJ, Knoll AH, Katz LA. Estimating the timing of early eukaryotic diversification with multigene molecular clocks. Proc Natl Acad Sci USA 2011;108(33):13624–9.

[74] Grell KG. Protozoology. Berlin, Heidelberg, New York: Springer Verlag; 1973.

Chapter 8

Homologous Recombination and Nonhomologous End-Joining Repair in Yeast

R.E. Jones, T.C. Humphrey
University of Oxford, Oxford, United Kingdom

Chapter Outline

1. INTRODUCTION

DNA double-strand breaks (DSBs) are the most cytotoxic of DNA lesions a cell can incur, with a single unrepaired break being potentially lethal [1]. Further, misrepaired DSBs can result in various genetic alterations such as deletions, translocations, loss of heterozygosity (LOH), and chromosome loss, all of which can cause genetic diseases such as cancer [2]. DSBs occur when breaks are formed proximally on both strands of a DNA duplex simultaneously and the two ends of a DSB become physically separated. These lesions can arise through programmed biological events such as mating-type switching, class-switch recombination, restriction enzyme digestion, V(D) J recombination, and meiotic recombination, which are responsible for genetic variation [3]. DSBs can also arise from endogenous metabolic errors, such as stalled replication forks or reactive oxygen species (a byproduct of respiration), or from exposure to exogenous agents such as ionizing radiation (IR) and IR mimetics (eg, bleomycin), ultraviolet radiation (UV), alkylating agents, or other clastogens [4]. Indeed, generating irreparable DSBs through exposure to such exogenous agents is the aim of both radiotherapy and many forms of chemotherapy. Moreover, introducing targeted DSBs through the use of zinc finger nucleases or more promisingly CRISPR-CAS9 technology is the basis of genome editing, and holds considerable biomedical promise [5]. For these reasons, there is considerable interest in understanding the mechanisms of DSB repair (DSBR) and recombination.

Here we describe DSBR in budding and fission yeasts, the study of which has provided unparalleled insights into these processes. The yeasts have provided excellent models for studying repair pathways due to their simplicity, the ease with which

Genome Stability. http://dx.doi.org/10.1016/B978-0-12-803309-8.00008-2

they can be cultured, and because the functions of genes can be elucidated by screening for various phenotypes [6]. As such, findings made in yeast, and in particular the budding yeast *Saccharomyces cerevisiae*, have formed the basis for much of our understanding of DSBR pathways. Models for homologous recombination (HR) repair were first proposed based on observations in yeast and when the proteins involved were subsequently identified, many were found to be evolutionarily conserved in higher eukaryotes [7]. In this chapter, we describe the two DSBR pathways, HR and nonhomologous end-joining (NHEJ), together with their sub-pathways. Furthermore, we discuss the proteins that underpin the various steps in these repair pathways. While we cite original papers, we also highlight helpful reviews throughout the chapter where further details can be obtained.

1.1 A Brief History

In 1949, Latarjet and Ephrussi found a correlation between the level of ploidy of yeast and their sensitivity to radiation [8]. This suggested that when a sister chromosome was available for repair cells were less susceptible to damage by radiation. However, it was not until the 1960s that radiation-sensitive mutants in yeast were isolated, and the 1970s when studies in yeast began to provide an extensive genetic analysis of DNA repair [9]. In 1967, Nakai and Matsumoto discovered yeast mutants that were sensitive to ultraviolet (UV) radiation [10], and Snow [11] and Resnick [12] isolated further UV-sensitive mutants. Resnick also identified X-ray-sensitive yeast mutants [12], some of these caused by mutations in the same genes identified by Rodarte et al. as conferring recombination deficiency when mutated [10,12–14]. At the International Conference on Yeast Genetics in 1970, it was decided that all of the radiation-sensitive mutants in yeast should be labeled as "*rad*" and that mutants that are UV sensitive or UV and IR sensitive should be numbered 1–49. It was also agreed that Rad50 and onwards would be used to describe mutants that primarily affect IR sensitivity [9]. In fact, many of the genes involved in recombination were initially identified as causing sensitivity to IR when mutated [15].

2. HOMOLOGOUS RECOMBINATION MODELS

HR provides a high-fidelity form of DSBR that uses homologous DNA sequences as a template. It is ubiquitous to all organisms and can be initiated by various DNA lesions, including DSBs and interstrand crosslinks (ICLs) [16,17]. It is used to increase genetic diversity during meiosis, but in mitotic cells the main purpose of HR is to repair DSBs [18,19]. In yeast, HR is also required for mating-type switching, a process by which yeast can switch their mating type in order for haploid cells of opposite mating type to mate with each other, forming diploid cells that undergo meiosis [20].

2.1 Holliday Model

The current models for HR were developed from models of meiotic recombination in yeast from over half-a-century ago. In 1964, Robin Holliday proposed a molecular model of HR that was based on observations during meiosis in the fungus *Ustilago maydis* [21]. Based on the structure of DNA, he proposed that unraveling DNA strands anneal with complementary bases within the homologous chromosome. A key feature of this model was the prediction that the homologous chromosomes would become covalently bound through a DNA structure at this exchange point that would form a symmetrical four-way junction. Importantly, this model accounted for the genetic phenotype of crossovers as these eponymously termed "Holliday" junctions (HJs) could be resolved in either of two orientations resulting in DNA molecules with either parental (noncrossover, NCO), or recombinant (crossover, CO) configurations with respect to distal genetic markers (Fig. 8.1, double Holliday junction (dHJ)). Moreover, if crossovers occurred at sites where parental molecules differed, this could result in gene conversion, following correction of the mismatched region.

Evidence supporting the Holliday model materialized in the early 1970s from studies involving bacteriophages where HJs were visualized using electron microscopy [22–24]. In 1979, Bell and Byers provided evidence of HJs forming in *S. cerevisiae* during recombination [25,26]. Significantly, this model formed the basis of the prevailing HR models proposed to accommodate new genetic data generated over the next half-century.

2.2 Double-Strand Break Repair Model

A mechanism of DSBR using HR was first proposed by Resnick [14] and was based on direct studies of repair of chromosomal DSBs [1] that were induced by X-rays. The model extended that of Holliday in two significant ways: the initiating event was a DSB rather than a single-strand break, and it proposed that when a DSB occurs in DNA the broken DNA ends are resected to produce 3′ single-stranded tails which could then engage a homologous chromosome or sister chromatid. Another DSB-based HR model, the DSBR model proposed by Szostak and Stahl in 1983 [27], arose from studies where

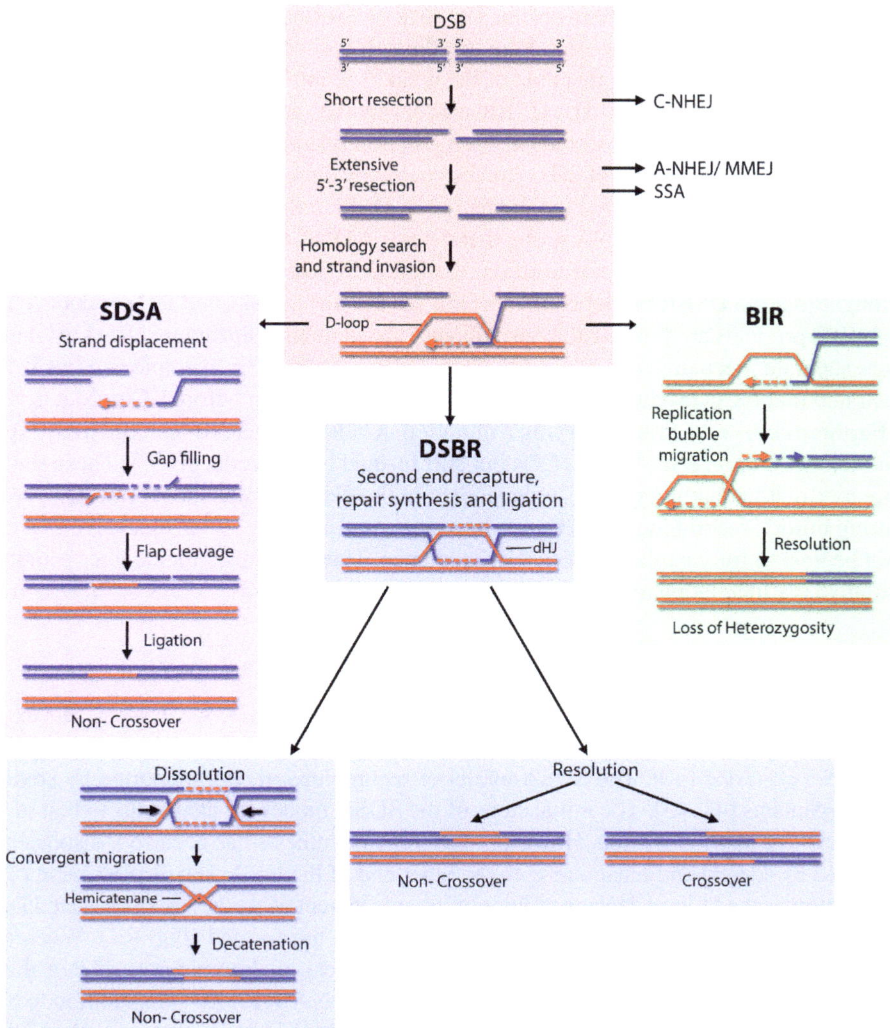

FIGURE 8.1 **The various models of homologous recombination (HR) and nonhomologous end-joining (NHEJ)-mediated double-strand break (DSB) repair in yeast.** After a DSB has formed in DNA, ends can be repaired via classical NHEJ (C-NHEJ) if no resection occurs. Alternatively, short resection can occur which enables repair via the alternative NHEJ and single-strand annealing (SSA) pathways. If extensive 5′–3′ resection is carried out, repair is committed to HR pathways. Homology search and strand invasion lead to D-loop formation. The main models of HR repair in yeast are double-strand break repair (DSBR), synthesis-dependent strand annealing (SDSA), and break-induced replication (BIR). These homologous recombination pathways have various outcomes. If repair is carried out via SDSA, after repair synthesis, the invading strand is displaced from the D-loop. The ends of the DSB are then annealed, possibly with flaps of excess DNA copied from the homologous template. Flap cleavage occurs to remove these, followed by ligation of the two DNA ends. This HR repair pathway results in noncrossover (NCO). In the DSBR model, once the D-loop has formed and repair synthesis carried out, second end recapture occurs forming a double Holliday junction (dHJ). The dHJ can be resolved to produce either crossover (CO) or NCO products. Alternatively, dissolution of the dHJ can take place via convergent migration of the HJs resulting in a hemicatenane, leading to noncrossover. BIR repair occurs when only one end of the DSB is available, such as at stalled replication forks. BIR involves migration of the D-loop replication "bubble" along the chromosome until another replication fork or the end of the chromosome is reached. This form of repair causes both strands of the homologous donor sequence to be copied resulting in loss of heterozygosity (LOH).

it was observed that, when plasmids containing yeast genes were linearized to simulate a DSB, increased integration of the genetic material into the yeast genome was observed compared to nonlinearized plasmids [27–29]. The DSBR model included elements of both the Holliday and Resnick models to explain these observations, and suggested that when a DSB occurs in DNA, the broken DNA ends are resected to produce 3′ single-stranded tails [30] (Fig. 8.1). Through subsequent findings, the specific role of the 3′ tail was elaborated as well as downstream events leading to HJs and resolution of recombinant structures. RPA (replication protein A), a heterotrimeric single-stranded DNA (ssDNA)-binding protein, binds to the 3′ ssDNA tails. RPA is then dissociated from the single-stranded tail by Rad51 [18]. Rad51 binds to ssDNA and forms a helical nuclear protein filament that seeks out a homologous sequence and binds to it [31]. The ssDNA nucleoprotein filament invades the homologous sequence and base pairs (bps) with the template strand, displacing the complementary homologous strand in a process known as strand invasion. The structure formed is called a displacement loop (D-loop) [27]. DNA synthesis occurs via DNA polymerase to extend the invading strand, copying a portion of the homologous sequence, and allowing the invading strand to rejoin the other side of the break (second end recapture). Further DNA synthesis occurs and the broken ends are ligated to form a dHJ. The DSBR model revolves around the generation of dHJs [32]. A dHJ can be resolved by either cutting the same pair of strands at each junction resulting in NCO, or by cutting different strands at each junction resulting in CO [33]. Resolution of dHJs can also occur by dissolution resulting in NCO (Fig. 8.1). The DSBR model is supported by studies demonstrating that the formation of DSBs initiates HR during yeast meiosis [34], as well as experiments using Southern blotting and an RNA probe that detects ssDNA that show end resection resulting in 3′ ssDNA tails [30]. Additionally, evidence for the existence of dHJs was found in yeast using 2D gel electrophoresis [35].

It was initially thought that the DSBR model could account for the formation of all HR products based on the observation that both CO and NCO products are formed at approximately the same time during meiosis [36]. However, later studies indicated that there are separate mechanisms for forming NCO DSB products. For example, studies in *S. cerevisiae* during meiosis have demonstrated that NCO products can be detected more than 30 min before CO products, and at the same time as dHJs are present. Furthermore, yeast strains carrying a mutation in Ndt80, a meiosis-specific transcription factor required for dHJ resolution in meiotic *S. cerevisiae*, had no COs but still formed NCO products [37]. These findings suggested that NCOs could also arise by a distinct pathway that is independent of the dHJ intermediates formed as predicted by the DSBR model. Moreover, during mitotic recombination, COs are infrequent and can give rise to extensive LOH, suggesting that the DSBR model may not be able to fully explain mitotic HR outcomes. These studies indicate that recombination resulting in NCOs may arise through a distinct pathway [38].

2.3 Synthesis-Dependent Strand Annealing Model

An alternative model for HR, known as synthesis-dependent strand annealing (SDSA), explains the experimental evidence indicating that NCOs can arise before and independently of COs [37,39] which comprise only approximately 5% of mitotic recombination events in *S. cerevisiae* [40]. The SDSA model of recombination was supported by studies in the T4 phage, *S. cerevisiae*, and other organisms [41–43]. The initial steps of the SDSA model are analogous to that of the DSBR model in the processing of DSB ends and strand invasion. However, once repair synthesis has occurred, displacement of the invading strand takes place followed by second end recapture with the other end of the DSB. The displacement of the invading strand can occur either by migration of the D-loop causing continuous displacement of the newly synthesized DNA [44], or through the activity of DNA helicases [45]. Unlike in the DSBR model, no HJs are formed (Fig. 8.1). When the displaced strand reanneals with the other end of the DSB, this may result in a flap of excess nonhomologous ssDNA that is removed by the Rad1–Rad10 endonuclease [46] and the nick that is formed is sealed by DNA ligase [47]. In addition to being responsible for meiotic NCOs, it is thought that yeast mating-type switching occurs via SDSA [17]. Furthermore, the SDSA model is thought to be the predominant mechanism of repair during mitotic recombination, which has a fundamentally different purpose [48].

2.4 Break-Induced Replication Model

Break-induced replication (BIR) (also known as recombination-dependent DNA replication) is similar in the initial steps to the DSBR and SDSA models but occurs when only one end of the DSB is available for recombination or the two ends of a DSB are not coordinated. It is primarily used to reestablish stalled replication forks and repair degraded telomeres, as in both of these instances the DSBs are one-ended and so the other repair pathways cannot be used. The recombination-dependent DNA replication model was initially described in the bacteriophage T4 [49], but was later described in yeast as the BIR model using a chromosome fragmentation vector [50]. The BIR model predicts that after strand invasion, DNA synthesis extends the invading strand, copying the template via a unidirectional replication fork or replication "bubble" that migrates along the template sequence [50] (Fig. 8.1). Repair synthesis and bubble migration continue until it encounters a converging replication fork or reaches the end of the chromosome [51]. This results in an extensive nonreciprocal transfer of genetic information from the template chromosome to the invading chromosome [45], which can cause extensive LOH and is highly mutagenic [51–53]. Unlike SDSA, which uses leading strand DNA synthesis, BIR requires both leading and lagging strand synthesis. BIR restoration of collapsed replication forks is Rad51 dependent [54]; however, Rad51-independent BIR has been reported in both budding and fission yeast [52,55,56].

2.5 Single-Strand Annealing Model

Single-strand annealing (SSA) is a form of HR repair that occurs when repetitive DNA sequences lay either side of the break, and differs from most other HR pathways in that it is independent of Rad51 in yeast [57]. As in HR, 5′–3′ resection

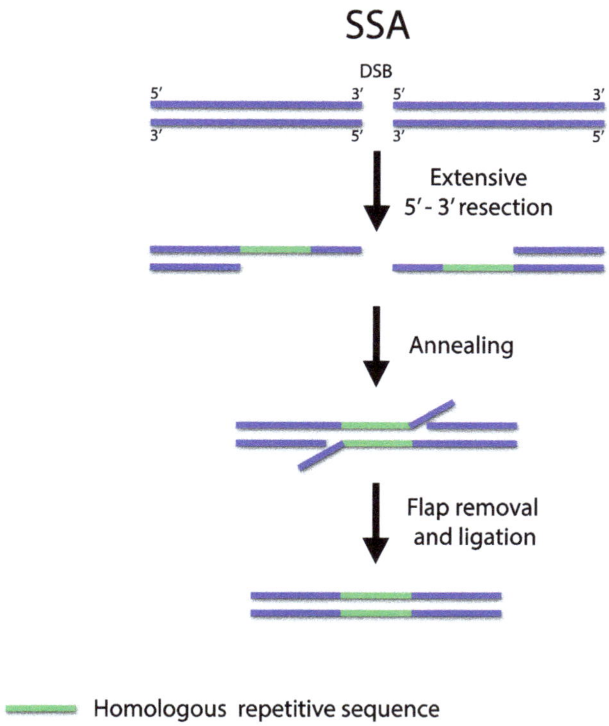

FIGURE 8.2 **The single-strand annealing (SSA) pathway of homologous recombination (HR) repair.** Extensive 5′–3′ resection of double-strand break (DSB) ends reveals homologous repetitive sequences on either side of the DSB. These sequences anneal and flaps of nonhomologous DNA can form. The flaps are cleaved and the annealed ends ligated. This model of repair differs from other HR pathways, as it does not involve a homologous donor sequence as a template.

occurs but SSA takes place when resection exposes repetitive homologous sequences on either side of the break site, leaving 3′ ssDNA tails. The two complementary ends of the DSB anneal and any unique sequence present between the repeats forms 3′ overhanging flaps [45], which are cleaved by the Rad1–Rad10 endonuclease. Ligation occurs, sealing the break, resulting in deletion of the DNA that existed between the repeats (Fig. 8.2). Thus, SSA is a nonconservative HR pathway, but it is an efficient form of DSBR when the repeats are greater than 200bp in length [58]. Indeed, SSA remains efficient even if up to 15kb of unique sequence lies between the repeats [17]. However, SSA is likely to occur at far higher rates in higher eukaryotes than in yeast, because the yeast genome contains comparatively low levels of repetitive sequences, and the other HR mechanisms in *S. cerevisiae* are extremely efficient [59].

3. COMMON HOMOLOGOUS RECOMBINATION STEPS

HR can be divided into several functionally distinct steps, which occur before or after synapsis (ie, the formation of joint molecule intermediates as a result of homology search and strand invasion [60]). Presynaptic steps include end resection, nucleofilament formation, and homology search. Postsynaptic steps include DNA synthesis, branch migration, HJ resolution/dissolution, and strand annealing [61]. Here we describe these key steps, together with the relevant repair factors that facilitate them (Table 8.1).

3.1 End Resection

End resection, or end processing, refers to the removal of bases from the 5′ end to reveal a 3′ overhang. This is a two-step process, the first of which requires the MRX complex [comprised of the proteins Mre11, Rad50, and Xrs2 (Nbs1 in *Schizosaccharomyces pombe* and *Homo sapiens*) [62]] in conjunction with Sae2 to initiate end resection of approximately 100 nucleotides, resulting in a short 3′ ssDNA overhang [63]. Accordingly, null mutants of Mre11, Rad50, or Xrs2 in yeast are highly sensitive to IR and exhibit slowed DSB processing [7]. Mre11 is an 83kDa protein with phosphoesterase motifs that are involved in its 3′–5′ double-stranded DNA (dsDNA) exonuclease activity and require manganese as a cofactor [64,65]. Sae2 and the nuclease activity of Mre11 are required to remove covalent adducts from DSB ends or "dirty ends," but are

TABLE 8.1 Proteins Involved in the Homologous Recombination (HR) Repair Pathways in *Saccharomyces cerevisiae* and *Schizosaccharomyces pombe*

Saccharomyces cerevisiae	*Schizosaccharomyces pombe*[b]	Function
(a) Presynaptic[a]		
Rad52	Rad52/Rad22	Facilitates nucleoprotein filament formation.
Mre11	Mre11/Rad32	Part of the MRX complex and involved in end resection. This is known as the MRN complex in *S. pombe*.
Rad50	Rad50	Rad50 is required for formation of the MRX/MRN complex and is vital for end resection.
Xrs2	Nbs1	A subunit of the MRX complex.
Sae2	Ctp1	Works with MRX to resect 5′ ends of a DSB.
Exo1	Exo1	5′–3′ Exonuclease required for extensive end resection.
Dna2	Dna2	DNA-dependent ATPase required for extensive resection.
RPA (Replication protein A)	Rpa1	Single-stranded DNA-binding protein. RPA coats ssDNA preventing secondary structures forming. It is required for Rad51 filament formation.
Rad51	Rhp51	Forms helical filaments with single-stranded DNA. Involved in strand exchange and homology seeking.
Rad55	Rhp55	Paralogue of Rad51 and forms a heterodimer with Rad57. This complex stabilizes Rad51 filaments.
Rad57	Rad57	Paralogue of Rad51 and forms a heterodimer with Rad55. This complex stabilizes Rad51 filaments.
Rad54	Rhp54	Facilitates strand exchange and D-loop formation.
Rdh54	Rdh54	Facilitates strand exchange and D-loop formation.
Srs2	Srs2	DNA helicase and DNA-dependent ATPase. Srs2 is anti-recombinogenic by interrupting Rad51 filaments and inhibiting DNA strand exchange.
Shu1	No known homolog	Shu1 is a component of the Shu complex, consisting of Psy3, Csm2, Shu2 and Shu1. It is involved in inhibiting Srs2 and stabilizing Rad51 filaments.
Shu2	Sws1	Shu2 is a component of the Shu complex, consisting of Psy3, Csm2, Shu2 and Shu1. It is involved in inhibiting Srs2 and stabilizing Rad51 filaments.
Psy3	Rld1	Psy3 is a component of the Shu complex, consisting of Psy3, Csm2, Shu2, and Shu1. It is involved in inhibiting Srs2 and stabilizing Rad51 filaments.
Csm2	Swi3	Csm2 is a component of the Shu complex, consisting of Psy3, Csm2, Shu2, and Shu1. It is involved in inhibiting Srs2 and stabilizing Rad51 filaments.
Rad59	No known homolog	Part of the Rad52 epistasis group.
(b) Postsynaptic		
Pol3/DNA polymerase δ	Cdc6	DNA polymerase δ is used during leading strand repair synthesis.
Pol2/DNA polymerase ε	Cdc20	DNA polymerase ε is used during leading strand repair synthesis.
Pol1/DNA polymerase α	Pol1	DNA polymerase α is used in BIR as both leading and lagging strand synthesis is required.

TABLE 8.1 Proteins Involved in the Homologous Recombination (HR) Repair Pathways in *Saccharomyces cerevisiae* and *Schizosaccharomyces pombe*—cont'd

Saccharomyces cerevisiae	*Schizosaccharomyces pombe*[b]	Function
PCNA complex	PCNA complex	Acts as a sliding clamp for DNA polymerase δ.
Sgs1	Rqh1	DNA helicase that works with Dna2 in extensive resection. It is also required for dHJ dissolution and forms a complex with Top3 and Rmi1 in *Saccharomyces cerevisiae*.
Top3	Top3	Topoisomerase III is a type IA topoisomerase that unwinds single stranded negatively supercoiled DNA. Required for dHJ dissolution.
Rmi1	Rmi1	Subunit of the Sgs1–Top3–Rmi1 complex. It stimulates decatenation.
Yen1	No known homolog	Holliday junction resolvase.
Ndt80	No known homolog	Meiosis-specific transcription factor implicated in dHJ resolution to form COs.
Rad1	Rad16	Forms a structure-specific endonuclease with Rad10 (Rad1–Rad10) that is involved in removing excess nonhomologous ssDNA after recombination resolution.
Rad10	Swi10	Forms a structure-specific endonuclease with Rad1 (Rad10–Rad10).
Mms4	Eme1	Forms a structure-specific nuclease with Mus81. Involved in resolution of HR-intermediates.
Mus81	Mus81	Forms a structure-specific nuclease with Mms4. Involved in resolution of HR-intermediates.
Mph1	Fml1 and Fml2 are orthologs	A helicase involved in strand displacement during SDSA.

CO, Crossover; *dHJ*, double Holliday junction; *DSB*, double-strand break; *SDSA*, synthesis-dependent strand annealing; *ssDNA*, single-stranded DNA.
[a]*The proteins have been categorized into presynaptic (a) and postsynaptic (b).*
[b]*The* S. pombe *homologs were found by searching for the systematic name.*
Taken from http://www.yeastgenome.org/ on http://www.pombase.org/.

not required to resect "clean ends." Similarly, the Sae2 ortholog, Ctp1 in fission yeast is required for removal of covalently bound Topoisomerase I and II from DNA [66]. The nuclease activities of Mre11 are enhanced by the binding of Rad50 [67], which is part of the structural maintenance of chromosomes (SMC) family and has a long coiled-coil domain that is necessary for Mre11 binding and MRX complex function. Mre11 binds to Rad50 at the coiled-coil in a zinc-dependent manner forming an MR subcomplex [67]. A Rad50 mutation (Rad50S) causes 3′ ssDNA tails to be abolished, indicating that it is also required for end resection [30]. The Xrs2 subunit of MRX binds Mre11 but not Rad50 and translocates Mre11 to the nucleus, which is necessary for mitotic DNA repair [62]. Xrs2 also enhances the exonuclease activity of Mre11 and the MR subcomplex [67]. The MRX complex is also able to stimulate resection in an indirect manner by recruiting Dna2 and/or Exo1 nucleases [68].

Dna2 together with Sgs1, or Exo1 alone, can catalyze extensive resection. They can directly process and resect DSB ends, but not those that contain covalent adducts. Additionally, Top3 and Rmi1, which interact with Sgs1, are necessary for end resection; however, their functions are not catalytic but structural. A model has been proposed based on studies in meiosis that MRX and Sae2 create a nick in the 5′ strand at a point distant from the DSB end that is resected in both directions by Mre11, which has 3′–5′ exonuclease activity, and Exo1, which is a 5′–3′ exonuclease. This theory is based on the observation that the Mre11 3′–5′ exonuclease has a role in end resection, which would not be otherwise possible as it is unable to resect in the 5′–3′ direction [69]. Experiments have demonstrated that the short 3′ ssDNA tails that result from MRX/Sae2-dependent resection are adequate for HR to occur. Because this occurs without Exo1 or Sgs1–Dna2, it suggests that extensive resection may not be required for HR. Additionally, a lack of MRX, Mre11, or Sae2 causes complete inhibition of end resection and cell death. However, despite extensive end resection not being necessary for HR to occur, it is required for the DNA damage checkpoint to be activated [45].

3.2 Nucleofilament Formation

Once nucleases begin to resect DSB ends, RPA binds to the resulting ssDNA. RPA is needed to prevent ssDNA tail degradation by MRX-Sae2 and the formation of hairpin-capped ends that would hinder HR repair [70]. RPA is also required to stimulate Rad51 nucleoprotein filament formation by preventing secondary structures from forming in the 3′ ssDNA tails [71]. Although RPA can inhibit Rad51 filament formation if both bind to ssDNA simultaneously, mediator proteins (eg, Rad52) can overcome this inhibition [72–74].

Rad51 nucleoprotein filaments are important for homology searching and strand invasion [3]. Rad51 is part of the Rad52 epistasis group of proteins and polymerizes to form helical filaments on ssDNA in an ATP-dependent manner [45,75,76]. Rad51 has been shown to catalyze strand exchange alone in vitro, but other proteins are required for this to occur efficiently in vivo. These mediator proteins have various roles such as facilitating Rad51 filament formation, filament stabilization, and strand exchange [77,78]. The Rad51 paralogues Rad55 and Rad57 form a heterodimer (Rad55–Rad57) that has been shown to form a complex with Rad51 and Srs2 simultaneously in vitro and may be incorporated into Rad51 filaments which acts to stabilize the Rad51 filament and inhibit displacement of Rad51 from ssDNA by Srs2 [79]. Srs2 is a translocase/helicase that interacts with Rad51 and activates Rad51's ATPase activity, causing Rad51 filament disassembly and acting as a negative regulator of HR [76,79]. The Shu proteins Shu1, Shu2, Psy3, and Csm2 form a complex that is structurally similar to the Rad51 dimer and it has been suggested that the Shu complex is integrated into Rad51 filaments in a manner similar to the Rad55–Rad57 heterodimer. Furthermore, the Shu complex has been shown to stabilize Rad51 filaments on ssDNA and to inhibit Srs2 anti-recombination activity [80].

3.3 Homology Search and Strand Invasion

Homology seeking is the process by which the nucleoprotein filament searches for a homologous sequence within the genome to use as a template for repair. Sister chromatids are preferred over homologs as substrates for recombinational repair in *S. cerevisiae* [81]. This is due to cohesins providing physical links between sister chromatids [82]. Despite chromatid cohesion, chromatin immunoprecipitation (ChIP) experiments demonstrated that Rad51 signals can be found at great distances from the site of a DSB and it was suggested that these signals are caused by homology probing by the Rad51 nucleoprotein filament transiently interacting with distant and nonhomologous sequences. However, these signals were lower in regions of the genome more distant to where the DSB occurred, suggesting that homology search/genome probing is not equally efficient throughout the genome [83]. In addition, physical proximity of homologous regions within the nucleus and the DSB site is an important factor in efficient and fast repair. For example, if both a DSB and a homologous sequence lie close to a centromere, homology search and repair occur more rapidly because in yeast the centromeres of chromosomes cluster in the nucleus and are tethered to the spindle pole bodies. The three-dimensional (3D) organization of chromosomes within the nucleus is not random and this organization greatly affects the efficiency of recombination events [83,84].

Homology search is also stimulated by the dsDNA-dependent ATPases Rad54 and Rdh54 (a Rad54 homolog) [85], and when both are absent genome probing is abolished. These proteins stimulate D-loop formation by interacting with Rad51 [83]. When bound to dsDNA, Rad54 can trigger conformational changes in the DNA, resulting in positive and negative supercoils. The negative supercoiling (or unwinding) of DNA may allow transient separation of homologous dsDNA strands, facilitating strand invasion and D-loop formation [85,86]. Rad54 is also important for displacement of Rad51 once strand invasion takes place, thereby facilitating DNA synthesis by allowing DNA polymerases to access and extend the invading strand [87].

3.4 DNA Repair Synthesis

DNA repair synthesis is important for creating a homologous sequence to which the second end can anneal. Using a homologous chromatid or chromosome as its replication template facilitates the high fidelity of HR repair [88]. DNA repair synthesis involves extension of the 3′ end of the invading strand within the D-loop by DNA polymerase δ and/or ε. DNA pol δ and ε are redundant with each other as studies have demonstrated that repair synthesis can occur when either is deleted but not when both are absent [47]. However, experiments studying *MAT* locus recombination have demonstrated that repair synthesis does not require a number of the proteins necessary for "standard" DNA replication. For example, some of the proteins required for initiating replication, such as ORC (origin recognition complex) and Cdc7–Dbf4 kinase, are not needed. Furthermore, DNA pol α and Okazaki fragment processing proteins are mostly not essential for HR repair synthesis, presumably because many HR pathways use only leading strand synthesis [45,51,89]. However, BIR requires DNA pol δ, ε, and α (the latter being the primary polymerase responsible for replication) due to both leading and lagging strand synthesis taking place in this repair pathway [90,91].

Additionally, the trimeric PCNA complex is required and forms a ring around DNA, acting as a sliding clamp for DNA pol δ. It has been demonstrated that posttranslational modification of PCNA affects its function. For example, polyubiquitination of PCNA is necessary for its role in error-free HR replication synthesis, whereas SUMOylation (attachment of a small ubiquitin-related modifier to the protein) inhibits its repair replication functions. Both SUMO and ubiquitin competitively bind to PCNA, thereby acting as a molecular switch to either activate or antagonize its repair synthesis activity [47,88].

In addition to repair synthesis within the D-loop, DNA synthesis also occurs to fill any gaps after recombination intermediates have been resolved. This requires deoxyribonucleotide triphosphate (dNTP) synthesis, which is induced by the DNA damage checkpoint in response to DNA damage to promote HR [92].

3.5 Strand Annealing

Rad52 has been shown to stimulate ssDNA annealing in vitro [93] which may be important for second end recapture in DSBR and SDSA and for recombination via SSA. It has been demonstrated that SUMOylation of Rad52 promotes repair via HR but suppress SSA and BIR in yeast [94]. RPA ensures that SSA is not able to occur in the absence of Rad52 by preventing annealing between short microhomologies that cannot be annealed in a Rad52-dependent manner [45,95]. Once annealing takes place, there may be flaps of excess DNA where the strands have annealed, which are cleaved by the Rad1–Rad10 endonuclease [46].

3.6 Resolution and Dissolution of Recombination Intermediates

The SDSA model predicts that only NCO products are created via this pathway, by resolving D-loop intermediates via displacement of the extended invading strand. This is now understood to be facilitated through the activities of RecQ, Srs2, Fbh1, Mph1, and possibly other helicases.

In *S. cerevisiae*, the Mph1 helicase has been shown in vitro to dissociate D-loop intermediates generated by Rad51. Furthermore, mutations in *MPH1* lead to an increase in CO products formed in response to DSBR, indicating that D-loop dissociation is impaired in these mutants, likely causing dHJ formation. Furthermore, 2D gel electrophoresis experiments have demonstrated that when Mph1 is not present, HJs transiently accumulate. Overexpressing Mph1 suppresses BIR and mutation of Mph1 causes an increase in the incidence of BIR. This is likely due to Mph1-dependent D-loop intermediate dissociation. However, BIR does require strand displacement via Pif1 during repair synthesis, which facilitates D-loop migration rather than dissociation [45].

The helicase Srs2 has been shown to facilitate NCO formation, more specifically by promoting SDSA [96]. Srs2 has also been shown to be capable of resolving D-loop intermediates in vitro; however, it is less efficient than Mph1 [97]. Additionally, the *S. pombe* Srs2 ortholog Fbh1 has been shown to suppress CO formation [98]. The budding yeast RecQ helicase Sgs1 is an ortholog to human BLM and, as well as having a role in end resection, has been shown to promote repair via the SDSA pathway. Human BLM has been shown to dissociate D-loops but the primary role of Sgs1 in resolving HR intermediates is in dissolution [96].

dHJs can be resolved via dissolution, resulting in NCO. Dissolution involves convergent migration of the two HJs, forming a hemicatenane (the joining of two DNA duplexes where one strand of a DNA duplex is wound around the strand of another duplex), followed by unlinking (decatenation) of the two duplexes without CO (Fig. 8.1). The Sgs1–Top3–Rmi1 complex has been shown to dissolve dHJs in vitro [99]. Sgs1 is a helicase that causes branch migration of the HJs leading to a hemicatenane and creating ssDNA as a substrate for Top3. RPA binds the ssDNA generated by Sgs1 and stabilizes it, facilitating unlinking. Top3 is a type IA topoisomerase that is responsible for decatenation. Rmi1 acts to stimulate Sgs1–Top3-mediated dissolution, specifically the decatenation stage [100].

Resolution of HR intermediates can also occur via endonucleolytic cleavage.

Mus81 forms a structure-specific nuclease with Mms4, becoming Mus81–Mms4 (Mus81–Eme1 in *S. pombe*), and has been implicated in the production of CO products [38]. However, Mus81–Mms4 uses nicked HJs or 3′ flaps and replication fork substrates preferentially and it has therefore been suggested that it does not act on dHJs, but instead cleaves D-loops. This type of HR intermediate resolution differs from both the SDSA and DSBR models and it has been proposed that dHJs are not formed [101,102]. Another resolvase, Yen1, has been implicated in resolution of dHJs, which can result in both CO and NCO products [103]. This type of resolution is consistent with the DSBR model of recombination. However, *yen1Δ* mutants do not exhibit severe deficiencies in recombination in both mitotic and meiotic cells compared to when both Yen1 and Mms4 are deleted. If dHJs were the primary form of HR intermediates in mitotic cells, it would be expected that deletion of Yen1 would have a greater effect on recombination efficiency. Experimental data has demonstrated that Mms4 is able to compensate for the activity of Yen1 in *yen1Δ* strains, but Yen1 cannot replace the activity of Mms4. This further supports the existence of alternative recombination intermediates to dHJs [101].

4. NONHOMOLOGOUS END-JOINING

The term NHEJ was coined by Moore and Haber in 1996 when they used it to describe DSBR in yeast that occurs when a homologous donor sequence is not present [104]. It is the process by which the two ends of a DSB are re-ligated together and is considered to be a more error-prone method of repair compared to HR [105]. In contrast to HR, most of the research relating to NHEJ has come from mammalian studies, which reflects the finding that NHEJ is the predominant repair pathway in higher eukaryotes. In yeast, however, HR is the predominant method used for DSBR. This has made identifying the proteins involved in NHEJ in yeast more difficult, because if NHEJ genes are simply deleted in a background where HR is possible, it is harder to detect repair defects [106]. NHEJ is the predominant form of repair during G1 phase of the cell cycle in yeast [107]. This further complicates the identification of yeast NHEJ repair genes as yeast are usually grown under exponential growth conditions where G1 is short, and because DSBs incurred in G1 are fixed via HR pathways in S phase. Furthermore, fission yeast lack a G1/S checkpoint, which would be expected to delay cell cycle progression in response to DSBs in other organisms. Despite these complexities, many NHEJ genes have been identified in yeast.

The basic process of NHEJ involves the processing and alignment of the two ends of a DSB, followed by ligation. In brief, NHEJ is initiated by the binding of Yku to DSB ends, which then facilitates the recruitment of nucleases to remove damaged DNA, polymerases to facilitate repair, and ligase to ligate the ends together [108].

The initial steps of C-NHEJ (classical-NHEJ) are evolutionarily conserved and involve binding of Yku (the yeast equivalent of Ku) and MRX to the DSB ends in yeast. Yku is a heterodimeric protein, and its binding to DNA ends prevents resection of the DSB ends by Exo1, which would otherwise result in a loss of genetic material [45]. Additionally, extensive 5′ resection of DSB ends inhibits NHEJ by irreversibly committing that DSB to be repaired via HR [109]. In order for the NHEJ reaction to occur, the two DSB ends need to be in close proximity to one another. It has been suggested that end-bridging occurs, whereby protein–protein interactions between each DSB end physically connect the two DNA molecules. The MRX complex is involved in NHEJ as well as HR and has been shown to have end-bridging activity. While such end-bridging is thought to result from the formation of Rad50 dimers via its zinc hooks, experimental data indicate that all the subunits of the MRX complex are required for end-bridging [109–111]. Perhaps surprisingly, *S. pombe* does not require MRN (the MRX homolog) for C-NHEJ [112].

4.1 Core Nonhomologous End-Joining Machinery

The core machinery of NHEJ are the factors considered essential for C-NHEJ (Table 8.2) [109]. The Yku heterodimer, made up of Yku70 and Yku80, is vital for NHEJ. It binds and forms a ring around DNA at DSB ends with the Yku80 subunit oriented toward the DSB end to enable it to interact with Dnl4. Dnl4 also forms a ring structure around DNA and creates a complex with Lif1, making up DNA ligase IV; interestingly, *S. pombe* lacks Lif1. DNA ligase IV joins together the two broken DNA ends in a DSB by creating a phosphodiester bond between the 3′ end of one end of the DSB and the 5′ end of the other via adenylation [112–114]. Dnl4 has two tandem BRCT (BRCA1 C-terminal) domains that are connected by a linker that allows Dnl4 to bind to Lif1. Interaction between the linker and Lif1 is required for NHEJ and without Lif1, Dnl4 is not recruited to DSB ends. It was thought that this interaction between the BRCT linker of Dnl4 and Lif1 is what made Lif1 and the BRCT domains of Dnl4 necessary for NHEJ but, in reality, interaction between Lif1 and the BRCT domains themselves (not just their linker) is required [114]. Lif1 is predicted to have a coiled-coil domain that allows it to bind DNA and interact with the BRCT linker of Dnl4. Xrs2 of the MRX complex recruits and interacts with Lif1 via the forkhead-associated (FHA) domain of Xrs2. FHA domains bind phosphorylated threonine residues, indicating that Lif1 may be phosphorylated; similarly, the human homolog of Lif1, XRCC4, is phosphorylated during NHEJ. The FHA domain of Xrs2 is specific to its NHEJ activity and is not required for its role in HR [109,111,115].

Nej1 interacts with the Dnl4–Lif1 complex and is essential for efficient NHEJ in yeast. It has been shown to facilitate nuclear localization of Lif1 and to promote efficient adenylation of Dnl4–Lif1. Furthermore, it has a greater role in the deadenylation of Dnl4–Lif1, which is required for Dnl4–Lif1 molecules to be able to catalyze more than one ligation reaction. Without this deadenylation activity, the efficiency of NHEJ would be limited by the abundance of Dnl4–Lif1 molecules. It has also been suggested that Nej1 may have a role earlier on in NHEJ by stabilizing the binding of Yku70–Yku80 to DSB ends [109,116,117].

4.1.1 End Processing During Nonhomologous End-Joining

Although many of the end-processing components in mammalian NHEJ have been identified, only some homologs have been discovered in yeast. For example, the main nucleases in mammalian C-NHEJ (eg, Artemis/DNA-PKc complex) have

TABLE 8.2 Core Proteins Involved in the Classical NHEJ (C-NHEJ) Pathway in *Saccharomyces cerevisiae* and *Saccharomyces pombe* [109]

Saccharomyces cerevisiae	*Schizosaccharomyces pombe*[a]	Function
Yku70	Pku70	Subunit of the Ku complex, binds to double-strand break (DSB) ends.
Yku80	Pku80	Subunit of the Ku complex, binds to DSB ends.
Lif1	Xrc4	Forms a heterodimer with Dnl4.
Dnl4	Lig4	Forms a heterodimer with Lif1 and is required for ligation.
Nej1	Xlf1	Interacts with Dnl4–Lif1 complex to localize Lif1 to the nucleus.
Mre11	-	Forms MRX complex which functions similarly to DNA-PKc in mammalian cells. N/A in *S. pombe*.
Rad50	–	Forms MRX complex which functions similarly to DNA-PKc in mammalian cells. N/A in *S. pombe*.
Xrs2	–	Forms MRX complex which functions similarly to DNA-PKc in mammalian cells. N/A in *S. pombe*.
Pol4	Pol4	Required for filling gaps in the DNA where DSB ends have been misaligned.
Rad27	Rad2/FEN1	5′ Flap endonuclease recruited by Nej1 and Dnl4–Lif1 to remove 5′ flaps of noncomplementary DNA after DSB end annealing.

[a]*The* S. pombe *homologs were found by searching for the systematic name.*
Taken from http://www.yeastgenome.org/ on http://www.pombase.org/.

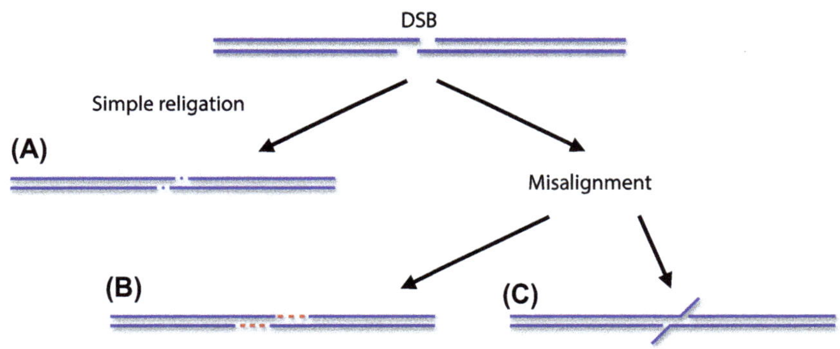

FIGURE 8.3 **The classical NHEJ (C-NHEJ) repair pathway.** Double-strand break (DSB) ends are bound by Yku70–80 heterodimer that protects the ends from resection, thus facilitating C-NHEJ. (A) Simple re-ligation of DSB ends requires Yku, MRX, and Dnl4/Lif1/Nej1. (B) Misalignment of DSB ends requires Yku, MRX, Dnl4/Lif1/Nej1, and Pol4 to fill gaps. This gap filling is error prone. (C) Misalignment of DSB ends resulting in flaps of excess DNA. This requires Yku, MRX, Dnl4/Lif1/Nej1, and Rad27 to cleave flaps, resulting in deletions.

no homologs in yeast. However, it has been suggested that MRX may not only have a role in early NHEJ, but may also be involved in the removal of adducts from DSB ends in NHEJ [118]. Pol4 is a DNA polymerase that has been implicated in NHEJ in yeast and is required to fill gaps in the DNA where DSB ends have been misaligned. This acts to stabilize the annealed DNA ends before ligation takes place [118]. Pol4 also reads through mismatched bases, even when the priming 3′ base is mismatched, as it lacks proofreading exonuclease activity [119]. This contributes to the error-prone nature of NHEJ. Rad27 is a 5′ flap endonuclease that interacts with and is recruited by Nej1 and Dnl4–Lif1. It is involved in cleaving 5′ flaps of noncomplementary DNA that can form when DSB ends anneal [120]. The combined activity of Rad27 and Pol4 facilitates efficient gap filling and end processing in NHEJ [109]. However, whenever Rad27 is required to cleave 5′ flaps of excess DNA or Pol4 is required to fill in gaps, it is possible that the misalignment of the two DSB ends has occurred, resulting in deletions (former) or insertions (latter) (see Fig. 8.3).

4.1.2 Ligation

If the ends of a DSB are "clean" with complementary overhangs (such as the DSBs caused by nucleases), NHEJ can occur via simple re-ligation. However, DSBs in DNA often result in "dirty" DNA ends, meaning that the ends lack 5′ phosphates and 3′ hydroxyls and thus cannot simply be re-ligated [109]. Furthermore, DSBs that occur as a result of DNA-damaging agents such as IR or bleomycin (a radiomimetic) may have further damage such as adducts or missing, damaged, or altered bases, making re-ligation more difficult. If simple re-ligation is not possible, microhomologies (short region of complementary bases at either end of a DSB) at each of the DSB ends can be aligned, but this can lead to deletions and occasionally small insertions [109,118,121,122].

4.2 Alternative End-Joining

Alternative NHEJ (A-EJ) pathways have been proposed based on observations that end joining can occur independently of Yku, and were first discovered in yeast [123]. Curiously, in *yku70Δ* cells, joining of complementary ends (such as those generated by HO endonuclease) is inefficient while noncomplementary ends are joined with great efficiency. However, these joins require greater microhomology, have more extensive deletions [124], and are prone to base substitutions [112]. This type of repair is termed microhomology-mediated end-joining (MMEJ) (see Fig. 8.4), and falls under the category of A-EJ. MMEJ differs from SSA, in that it is Rad52 independent. However, if the microhomologies are greater than 8 bp in length, Rad52 is required [45,122]. Additionally, evidence suggests that the Rad1–Rad10 endonuclease is required for MMEJ, similar to SSA [109]. Studies have demonstrated that while MMEJ can occur independently of MRX, the majority of MMEJ events are MRX dependent [122]. In *S. cerevisiae*, while DNA ligase IV is required for all C-NHEJ events, it is only partially required for A-EJ, where ligase I can also be employed, albeit less efficiently [105]. However, in *S. pombe* (as in mammals), A-EJ can occur entirely independently of ligase IV [112].

Evidence suggests that blunt ends (DSBs that do not result in overhangs) may be repaired by a different mechanism from C-NHEJ. Unlike in mammalian cells, blunt end joining via NHEJ in yeast is very inefficient and occurs independently of Yku70 [105]. This form of repair is inaccurate, requires Rad50 and Rad52, and is partially dependent on Srs2. However, when Yku70 is present, blunt end repair is not dependent on Srs2 or Rad52 and occurs with greater accuracy [125].

It has been argued that these A-EJ pathways are not separate pathways to C-NHEJ, but are simply the result of cells utilizing other analogous proteins when the core NHEJ machinery is not available, resulting in less-efficient and more error-prone repair [118]. In wild-type *S. pombe*, only 1 in 48 NHEJ events occurred via MMEJ in an extrachromosomal DSBR assay where HO endonuclease was used to induce DSBs. This suggests that repair by MMEJ is repressed in the presence of core C-NHEJ factors (Yku and Dnl4) [112]. These findings support the argument that A-EJ pathways are redundant mechanisms used only when the core NHEJ machinery is not available. Additionally, the dependence on Rad1–Rad10 in some MMEJ events raises the possibility that MMEJ is a class of SSA and not a separate pathway. However, Yku-dependent end joining (C-NHEJ) occurs when microhomologies are less than 5 bp, whereas Rad52-dependent end joining occurs only

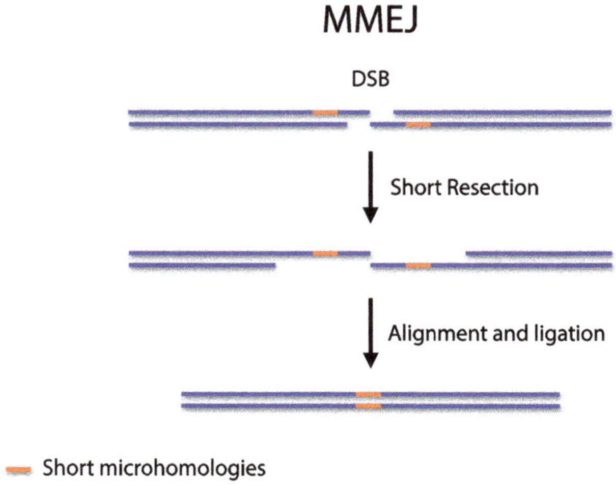

FIGURE 8.4 **The microhomology-mediated end-joining (MMEJ) repair pathway.** Unlike classical NHEJ, this alternative NHEJ pathway involves short resection of double-strand break (DSB) ends to reveal short homologous sequences of 5–25 bp [122]. These microhomologies are used to align the DSB ends before ligation.

when microhomologies are greater than 8 bp. When homologies of 6–8 bp exist between DSB ends, end joining is Yku and Rad52 independent. These findings support the notion that MMEJ is a separate pathway from SSA and C-NHEJ [122].

5. CELL CYCLE REGULATION OF HOMOLOGOUS RECOMBINATION AND NONHOMOLOGOUS END-JOINING

The cell cycle plays an important role in regulating DSBR pathway choice. In yeast, NHEJ repair pathways are restricted to G1 phase, while HR repair takes place in the S and G2 phases of the cell cycle. This is logical since, due to the haploid nature of yeast, sister chromatids are only available as a homologous template during S and G2.

A key mechanism by which DSBR pathways are regulated through the cell cycle is through CDK1-dependent phosphorylation and activation of Sae2, which facilitates DSB end-resection. Moreover, CDK1 activation is required for RPA recruitment and Rad51 nucleoprotein filament formation, and is thus required for HR to occur in yeast [126]. As CDK activity is low during G1 but increases in S-phase and G2, this helps ensure DSBR pathway choice.

Another mechanism of pathway choice is employed during meiosis, where Nej1, a negative regulator of DNA ligase IV, is transcriptionally repressed by *MAT*a1/α2. This repressor is induced in meiotic diploid cells, where both alleles of the *MAT* locus are expressed, thus, HR repair is favored as a result of DNA ligase IV inhibition [116].

Chromatin remodeling also influences DSBR pathway choice. Histones that form chromatin with DNA can be methylated or acetylated, affecting the structure of chromatin [127]. Posttranslational methylation of histone H3 on lysine 36 (H3K36) by Set2 has been shown to reduce the accessibility of chromatin, diminish resection of DSB ends, and promote the recruitment of Ku to DSBs, thereby promoting NHEJ. In contrast, Gcn5-dependent H3K36 acetylation increases accessibility of DNA within chromatin, enhances end resection, and encourages repair via HR. Accordingly, trimethylated H3K36 is increased in G1, while acetylated H3K36 is increased in S phase [128]. Thus, cell cycle regulation of DSBR pathway choice is regulated by multiple factors.

The importance of regulating DSBR pathway choice during the cell cycle is likely to reflect the consequences of inappropriate choice on genome stability. HR in G1 in haploid yeast may result in duplication of inappropriate genetic material or chromosomal rearrangements if HR repair is associated with a crossover. Furthermore, it has been suggested that HR occurring during G1 phase results in LOH events [129]. Moreover, NHEJ repair of a one-ended break resulting from replication fork collapse will result in deletions or chromosomal rearrangements.

6. CONCLUSION

DSBs can be caused by a multitude of events and the pathways that repair them, which are outlined in this chapter, are vital for maintaining both cell viability and genomic stability [130]. In yeast, HR pathways are of particular importance as they are the primary mechanism by which DSBs are repaired and are thus vital for maintaining the integrity of the genome during replication, where DSBs can arise following replication fork collapse [131]. Furthermore, HR plays an important role in meiosis, where it not only promotes genetic diversity by creating COs, but also provides physical connections between sister chromatids which are vital for proper chromosome segregation and the prevention of aneuploidy [132]. The advances in our understanding of DSBR using yeast model systems have contributed to our understanding of DSBR in all organisms. This is particularly important in the field of cancer research where, in mammalian cells, chromosomal rearrangements contribute to the transformation of normal cells to cancerous cells via tumor-suppressor gene loss or oncogene activation [133]. Indeed, many orthologs of both the C-NHEJ and HR pathways function as tumor suppressors in mammals.

While considerable advances have been made in elucidating the factors required for C-NHEJ and HR repair mechanisms, there are a number of issues that remain to be resolved. There is still a lack of clarity regarding the mechanisms leading to A-EJ, and the functional relationship between C-NHEJ and A-EJ pathways. Moreover, while the impact of chromatin on DSBR is becoming more apparent, a detailed mechanistic understanding of the role of chromatin remodeling and histone variants, or their modifications on DSBR, and how these events influence pathway choice, requires further elucidation. A comprehensive understanding of DSBR in yeast and higher eukaryotes will be expected to impact on a range of related disciplines, including cancer therapy and genome editing, and thus remains an important goal.

GLOSSARY

Aneuploidy An abnormal number of chromosomes within a cell.

Bubble migration Movement of the D-loop and accompanying replication machinery along a chromosome.

Clastogen A mutagenic agent that causes chromosome breaks.

Clean ends The two DNA ends of a double-strand break where each end has a complementary overhang and no lesions.
Covalent adduct A type of DNA lesion where a chemical is covalently bound to DNA.
Crossover Reciprocal exchange of a region of DNA between homologous chromosomes as a result of homologous recombination.
Decatenation Unlinking of two DNA duplexes where one strand of a DNA duplex is wound around a strand of another duplex (unlinking of a hemicatenane).
Dirty ends DNA ends of a double-strand break that lack 5′ phosphates and/or 3′ hydroxyls or exhibit other forms of DNA damage.
End bridging Physical connection of two DNA molecules (each end of a double-strand break) via protein–protein interactions between each molecule.
Endonuclease Enzyme that cleaves a polynucleotide molecule by cleaving nucleotides within the polynucleotide chain.
Endonucleolytic cleavage The process by which an endonuclease cuts a polynucleotide sequence.
Exonuclease Enzyme that cleaves nucleotides from the end of a polynucleotide molecule.
Helicase Enzyme that unwinds the double-stranded helical structure of polynucleotide molecules.
Hemicatenane Joining of two DNA duplexes where one strand of a DNA duplex is wound around a strand of another duplex.
Holliday junctions Homologous recombination intermediate containing four double-stranded arms joined together to form a joint molecule.
Homologous recombination A high-fidelity form of DNA damage repair that makes use of a homologous DNA sequence as a template.
Interstrand crosslinks A type of DNA lesion that prevents separation of the strands within a DNA duplex.
Loss of heterozygosity An event where either one copy of a gene is lost or is replaced with the allele from the sister chromosome.
Microhomology-mediated end-joining A subset of alternative end-joining repair that relies on regions of microhomology on either side of the break, which anneal following limited resection.
Noncrossover The resulting structure when recombination intermediates are resolved with no reciprocal exchange of regions of DNA between homologous chromosomes.
Nonhomologous end-joining A form of double-strand break repair whereby two broken DNA ends are ligated together without the need for a homologous template sequence.
Nucleoprotein filament A complex comprised of helical chains of protein(s) bound to DNA.
Okazaki fragment Short segments of DNA newly synthesized on the lagging template strand as part of DNA replication.
Ploidy The number of pairs of chromosomes within a cell.
Postsynapsis The steps of homologous recombination following joint molecule formation.
Presynapsis The steps of homologous recombination preceding strand invasion and joint molecule formation.
Replication fork A point in a DNA duplex where the strands are separated to allow replication to occur.
Resolvase In respect to homologous recombination, an enzyme that acts to resolve Holliday junctions and recombination intermediates.
SUMOylation A type of posttranslational modification by which a small ubiquitin-related modifier is attached to a protein.
Synapsis The process of strand invasion and joint molecule formation as part of homologous recombination.
Translocase Protein that assists in the movement of another molecule.
Translocation Transfer of part of a chromosome to a different position such as a nonhomologous chromosome.

LIST OF ACRONYMS AND ABBREVIATIONS

2D Two-dimensional
3D Three-dimensional
A-EJ Alternative end-joining
ATP Adenosine triphosphate
BIR Break-induced replication
BLM Bloom syndrome protein
CDK Cyclin-dependent kinase
ChIP Chromatin immunoprecipitation
C-NHEJ Classical nonhomologous end-joining
CO Crossover
CRISPR Clustered regularly interspaced short palindromic repeat
dHJ Double Holliday junction
D-loop Displacement loop
DNA pol DNA polymerase
dNTP Deoxyribonucleotide triphosphate
DSB Double-strand break
DSBR Double-strand break repair
dsDNA Double-stranded DNA
FHA Forkhead associated
G1 phase Growth 1 phase of the cell cycle
G2 phase Growth 2 phase of the cell cycle
H3K36 Histone 3 lysine 36

HJ Holliday junction
HR Homologous recombination
ICL Interstrand crosslinks
IR Ionizing radiation
LOH Loss of heterozygosity
MAT Mating locus
MMEJ Microhomology-mediated end-joining
MR Subcomplex comprised of Mre11 and Rad50
MRX Complex comprised of Mre11, Rad50, and Xrs2
NCO Noncrossover
NHEJ Nonhomologous end-joining
ORC Origin recognition complex
PCNA Proliferation cell nuclear antigen
RPA Replication protein A
S phase Synthesis phase of the cell cycle
SDSA Synthesis-dependent strand-annealing
SMC Structural maintenance of chromosomes
ssDNA Single-stranded DNA
Top3 Topoisomerase III
UV Ultraviolet

REFERENCES

[1] Resnick MA, Martin P. The repair of double-strand breaks in the nuclear DNA of *Saccharomyces cerevisiae* and its genetic control. MGG 1976;143(2):119–29.
[2] Kasparek TR, Humphrey TC. DNA double-strand break repair pathways, chromosomal rearrangements and cancer. Sem Cell Dev Biol 2011;22(8):886–97.
[3] Shrivastav M, De Haro LP, Nickoloff JA. Regulation of DNA double-strand break repair pathway choice. Cell Res 2008;18:134–47.
[4] Pfeiffer P, Goedecke W, Obe G. Mechanisms of DNA double-strand break repair and their potential to induce chromosomal aberrations. Mutagenesis 2000;15(4):289–302.
[5] Doudna JA, Charpentier E. Genome editing. The new frontier of genome engineering with CRISPR-Cas9. Science 2014;346(6213):1258096.
[6] Aylon Y, Kupiec M. DSB repair: the yeast paradigm. DNA Repair 2004;3:797–815.
[7] Symington LS. Role of RAD52 epistasis group genes in homologous recombination and double-strand break repair. Microbiol Mol Biol Rev 2002;66:630–70.
[8] Latarjet R, Ephrussi B. Courbes de survie de levures haploides et diploides soumises aux rayons-X. C R Acad Sci 1949;229(4):306–8.
[9] Friedberg EC. Deoxyribonucleic acid repair in the yeast *Saccharomyces cerevisiae*. Microbiol Rev 1988;52:70–102.
[10] Nakai S, Matsumoto S. Two types of radiation-sensitive mutant in yeast. Mutat Res 1967;4(2):129–36.
[11] Snow R. Mutants of yeast sensitive to ultraviolet light. J Bacteriol 1967;94(3):571–5.
[12] Resnick MA. Genetic control of radiation sensitivity in *Saccharomyces cerevisiae*. Genetics 1969;62(3):519–31.
[13] Rodarte U, Fogel S, Mortimer R, editors. Detection of Recombination Defective Mutants in *Saccharomyces cerevisiae*. Genetics 1968. 428 EAST Preston St, Baltimore, MD 21202.
[14] Resnick MA. The repair of double-strand breaks in DNA; a model involving recombination. J Theor Biol 1976;59(1):97–106.
[15] Moore CW. Responses of radiation-sensitive mutants of *Saccharomyces cerevisiae* to lethal effects of bleomycin. Mutat Res 1978;51:165–80.
[16] Li X, Heyer W-D. Homologous recombination in DNA repair and DNA damage tolerance. Cell Res 2008;18:99–113.
[17] Pâques F, Haber JE. Multiple pathways of recombination induced by double-strand breaks in *Saccharomyces cerevisiae*. Microbiol Mol Biol Rev 1999;63:349–404.
[18] Dupaigne P, Le Breton C, Fabre F, Gangloff S, Le Cam E, Veaute X. The Srs2 helicase activity is stimulated by Rad51 filaments on dsDNA: implications for crossover incidence during mitotic recombination. Mol Cell 2008;29:243–54.
[19] Yeeles JTP, Poli J, Marians KJ, Pasero P. Rescuing stalled or damaged replication forks. Cold Spring Harb Perspect Biol 2013;5:a012815.
[20] Haber JE. Mating-type genes and MAT switching in *Saccharomyces cerevisiae*. Genetics 2012;191:33–64.
[21] Holliday R. A mechanism for gene conversion in fungi. Genet Res 1964;5:282–304.
[22] Doniger J, Warner RC, Tessma I. Role of circular dimer DNA in the primary recombination mechanism of bacteriophage S13. Nat New Biol 1973;242(114):9–12.
[23] Thompson BJ, Escarmis C, Parker B, Slater WC, Doniger J, Tessman I, et al. Figure-8 configuration of dimers of S13 and phiX174 replicative form DNA. J Mol Biol 1975;91(4):409–19.
[24] Benbow RM, Zuccarelli AJ, Sinsheimer RL. Recombinant DNA molecules of bacteriophage phi chi174. Proc Natl Acad Sci USA 1975;72(1):235–9.
[25] Bell L, Byers B. Occurrence of crossed strand-exchange forms in yeast DNA during meiosis. Proc Natl Acad Sci USA 1979;76:3445–9.
[26] Liu Y, West SC. Happy Hollidays: 40th anniversary of the Holliday junction. Nat Rev Mol Cell Biol 2004;5:937–44.

[27] Szostak JW, Orr-Weaver TL, Rothstein RJ, Stahl FW. The double-strand-break repair model for recombination. Cell 1983;33:25–35.
[28] Orr-Weaver TL, Szostak JW, Rothstein RJ. Yeast transformation: a model system for the study of recombination. Proc Natl Acad Sci USA 1981;78:6354–8.
[29] Orr-Weaver TL, Szostak JW. Yeast recombination: the association between double-strand gap repair and crossing-over. Proc Natl Acad Sci USA 1983;80:4417–21.
[30] Sun H, Treco D, Szostak JW. Extensive 3′-overhanging, single-stranded DNA associated with the meiosis-specific double-strand breaks at the ARG4 recombination initiation site. Cell 1991;64:1155–61.
[31] Mazin AV, Bornarth CJ, Solinger JA, Heyer W-D, Kowalczykowski SC. Rad54 protein is targeted to pairing loci by the Rad51 nucleoprotein filament. Mol Cell 2000;6:583–92.
[32] Sung P, Klein H. Mechanism of homologous recombination: mediators and helicases take on regulatory functions. Nat Rev Mol Cell Biol 2006;7:739–50.
[33] Whitby MC. Making crossovers during meiosis. Biochem Soc Trans 2005;33:1451.
[34] Cao L, Alani E, Kleckner N. A pathway for generation and processing of double-strand breaks during meiotic recombination in *S. cerevisiae*. Cell 1990;61:1089–101.
[35] Schwacha A, Kleckner N. Identification of double Holliday junctions as intermediates in meiotic recombination. Cell 1995;83:783–91.
[36] Storlazzi A, Xu L, Cao L, Kleckner N. Crossover and noncrossover recombination during meiosis: timing and pathway relationships. Proc Natl Acad Sci USA 1995;92:8512–6.
[37] Allers T, Lichten M. Differential timing and control of noncrossover and crossover recombination during meiosis. Cell 2001;106:47–57.
[38] Mitchel K, Zhang H, Welz-Voegele C, Jinks-Robertson S. Molecular structures of crossover and noncrossover intermediates during gap repair in yeast: implications for recombination. Mol Cell 2010;38(2):211–22.
[39] Nassif N, Penney J, Pal S, Engels WR, Gloor GB. Efficient copying of nonhomologous sequences from ectopic sites via P-element-induced gap repair. Mol Cell Biol 1994;14:1613–25.
[40] Ira G, Malkova A, Liberi G, Foiani M, Haber JE. Srs2 and Sgs1-Top3 suppress crossovers during double-strand break repair in yeast. Cell 2003;115:401–11.
[41] Mueller JE, Clyman J, Huang YJ, Parker MM, Belfort M. Intron mobility in phage T4 occurs in the context of recombination-dependent DNA replication by way of multiple pathways. Genes Dev 1996;10:351–64.
[42] McGill C, Shafer B, Strathern J. Coconversion of flanking sequences with homothallic switching. Cell 1989;57:459–67.
[43] Nelson HH, Sweetser DB, Nickoloff JA. Effects of terminal nonhomology and homeology on double-strand-break-induced gene conversion tract directionality. Mol Cell Biol 1996;16:2951–7.
[44] Formosa T, Alberts BM. DNA synthesis dependent on genetic recombination: characterization of a reaction catalyzed by purified bacteriophage T4 proteins. Cell 1986;47:793–806.
[45] Symington LS, Rothstein R, Lisby M. Mechanisms and regulation of mitotic recombination in *Saccharomyces cerevisiae*. Genetics 2014;198: 795–835.
[46] Lyndaker AM, Goldfarb T, Alani E. Mutants defective in Rad1-Rad10-Slx4 exhibit a unique pattern of viability during mating-type switching in *Saccharomyces cerevisiae*. Genetics 2008;179:1807–21.
[47] Wang X, Ira G, Tercero JA, Holmes AM, Diffley JFX, Haber JE. Role of DNA replication proteins in double-strand break-induced recombination in *Saccharomyces cerevisiae*. Mol Cell Biol 2004;24:6891–9.
[48] Andersen SL, Sekelsky J. Meiotic versus mitotic recombination: two different routes for double-strand break repair. BioEssays 2010;32:1058–66.
[49] Luder A, Mosig G. Two alternative mechanisms for initiation of DNA replication forks in bacteriophage T4: priming by RNA polymerase and by recombination. Proc Natl Acad Sci USA 1982;79(4):1101–5.
[50] Morrow DM, Connelly C, Hieter P. "Break copy" duplication: a model for chromosome fragment formation in *Saccharomyces cerevisiae*. Genetics 1997;147:371–82.
[51] Llorente B, Smith CE, Symington LS. Break-induced replication: what is it and what is it for? Cell Cycle 2008;7:859–64.
[52] Cullen JK, Hussey SP, Walker C, Prudden J, Wee BY, Dave A, et al. Break-induced loss of heterozygosity in fission yeast: dual roles for homologous recombination in promoting translocations and preventing de novo telomere addition. Mol Cell Biol 2007;27(21):7745–57.
[53] Sakofsky CJ, Roberts SA, Malc E, Mieczkowski PA, Resnick MA, Gordenin DA, et al. Break-induced replication is a source of mutation clusters underlying kataegis. Cell Rep 2014;7(5):1640–8.
[54] Davis AP, Symington LS. RAD51-dependent break-induced replication in yeast. Mol Cell Biol 2004;24:2344–51.
[55] Anand RP, Lovett ST, Haber JE. Break-induced DNA replication. Cold Spring Harb Perspect Biol 2013;5(12):a010397.
[56] Tinline-Purvis H, Savory AP, Cullen JK, Dave A, Moss J, Bridge WL, et al. Failed gene conversion leads to extensive end processing and chromosomal rearrangements in fission yeast. EMBO J 2009;28(21):3400–12.
[57] Ivanov EL, Sugawara N, Fishman-Lobell J, Haber JE. Genetic requirements for the single-strand annealing pathway of double-strand break repair in *Saccharomyces cerevisiae*. Genetics 1996;142(3):693–704.
[58] Sugawara N, Ira G, Haber JE. DNA length dependence of the single-strand annealing pathway and the role of *Saccharomyces cerevisiae* RAD59 in double-strand break repair. Mol Cell Biol 2000;20(14):5300–9.
[59] Perez C, Guyot V, Cabaniols J-P, Gouble A, Micheaux B, Smith J, et al. Factors affecting double-strand break-induced homologous recombination in mammalian cells. Biotechniques 2005;39:109–15.
[60] Liu J, Ehmsen KT, Heyer WD, Morrical SW. Presynaptic filament dynamics in homologous recombination and DNA repair. Crit Rev Biochem Mol Biol 2011;46(3):240–70.

[61] Schipler A, Iliakis G. DNA double-strand-break complexity levels and their possible contributions to the probability for error-prone processing and repair pathway choice. Nucleic Acids Res 2013;41(16):7589–605.

[62] Tsukamoto Y, Mitsuoka C, Terasawa M, Ogawa H, Ogawa T. Xrs2p regulates Mre11p translocation to the nucleus and plays a role in telomere elongation and meiotic recombination. Mol Biol Cell 2005;16:597–608.

[63] Mimitou EP, Symington LS. DNA end resection: many nucleases make light work. DNA Repair 2009;8:983–95.

[64] Furuse M, Nagase Y, Tsubouchi H, Murakami-Murofushi K, Shibata T, Ohta K. Distinct roles of two separable in vitro activities of yeast Mre11 in mitotic and meiotic recombination. EMBO J 1998;17:6412–25.

[65] Sharples GJ, Leach DR. Structural and functional similarities between the SbcCD proteins of *Escherichia coli* and the RAD50 and MRE11 (RAD32) recombination and repair proteins of yeast. Mol Microbiol 1995;17:1215–7.

[66] Hartsuiker E, Neale MJ, Carr AM. Distinct requirements for the Rad32(Mre11) nuclease and Ctp1(CtIP) in the removal of covalently bound topoisomerase I and II from DNA. Mol Cell 2009;33(1):117–23.

[67] Trujillo KM, Roh DH, Chen L, Komen SV, Tomkinson A, Sung P. Yeast Xrs2 binds DNA and helps target Rad50 and Mre11 to DNA ends. J Biol Chem 2003;278:48957–64.

[68] Shim EY, Chung WH, Nicolette ML, Zhang Y, Davis M, Zhu Z, et al. *Saccharomyces cerevisiae* Mre11/Rad50/Xrs2 and Ku proteins regulate association of Exo1 and Dna2 with DNA breaks. EMBO J 2010;29(19):3370–80.

[69] Paull TT, Gellert M. The 3′ to 5′ exonuclease activity of Mre11 facilitates repair of DNA double-strand breaks. Mol Cell 1998;1(7):969–79.

[70] Chen H, Lisby M, Symington LS. RPA coordinates DNA end resection and prevents formation of DNA hairpins. Mol Cell 2013;50(4):589–600.

[71] Sugiyama T, Zaitseva EM, Kowalczykowski SC. A single-stranded DNA-binding protein is needed for efficient presynaptic complex formation by the *Saccharomyces cerevisiae* Rad51 protein. J Biol Chem 1997;272(12):7940–5.

[72] New JH, Sugiyama T, Zaitseva E, Kowalczykowski SC. Rad52 protein stimulates DNA strand exchange by Rad51 and replication protein A. Nature 1998;391(6665):407–10.

[73] Shinohara A, Ogawa T. Stimulation by Rad52 of yeast Rad51-mediated recombination. Nature 1998;391(6665):404–7.

[74] Sung P. Function of yeast Rad52 protein as a mediator between replication protein A and the Rad51 recombinase. J Biol Chem 1997;272(45):28194–7.

[75] Sung P, Robberson DL. DNA strand exchange mediated by a RAD51-ssDNA nucleoprotein filament with polarity opposite to that of RecA. Cell 1995;82:453–61.

[76] Antony E, Tomko EJ, Xiao Q, Krejci L, Lohman TM, Ellenberger T. Srs2 disassembles Rad51 filaments by a protein-protein interaction triggering ATP turnover and dissociation of Rad51 from DNA. Mol Cell 2009;35:105–15.

[77] Sung P. Catalysis of ATP-dependent homologous DNA pairing and strand exchange by yeast RAD51 protein. Science 1994;265(5176):1241–3.

[78] Sheridan SD, Yu X, Roth R, Heuser JE, Sehorn MG, Sung P, et al. A comparative analysis of Dmc1 and Rad51 nucleoprotein filaments. Nucleic Acids Res 2008;36(12):4057–66.

[79] Liu J, Renault L, Veaute X, Fabre F, Stahlberg H, Heyer W-D. Rad51 paralogues Rad55-Rad57 balance the antirecombinase Srs2 in Rad51 filament formation. Nature 2011;479:245–8.

[80] Bernstein KA, Reid RJ, Sunjevaric I, Demuth K, Burgess RC, Rothstein R. The Shu complex, which contains Rad51 paralogues, promotes DNA repair through inhibition of the Srs2 anti-recombinase. Mol Biol Cell 2011;22(9):1599–607.

[81] Kadyk LC, Hartwell LH. Sister chromatids are preferred over homologs as substrates for recombinational repair in *Saccharomyces cerevisiae*. Genetics 1992;132(2):387–402.

[82] Sjogren C, Nasmyth K. Sister chromatid cohesion is required for postreplicative double-strand break repair in *Saccharomyces cerevisiae*. Curr Biol 2001;11(12):991–5.

[83] Renkawitz J, Lademann Claudio A, Kalocsay M, Jentsch S. Monitoring homology search during DNA double-strand break repair in vivo. Mol Cell 2013;50:261–72.

[84] Renkawitz J, Lademann CA, Jentsch S. Mechanisms and principles of homology search during recombination. Nat Rev Mol Cell Biol 2014;15:369–83.

[85] Petukhova G, Komen SV, Vergano S, Klein H, Sung P. Yeast Rad54 promotes Rad51-dependent homologous DNA pairing via ATP hydrolysis-driven change in DNA double helix conformation. J Biol Chem 1999;274:29453–62.

[86] Petukhova G, Sung P, Klein H. Promotion of Rad51-dependent D-loop formation by yeast recombination factor Rdh54/Tid1. Genes Dev 2000;14:2206–15.

[87] Li X, Heyer WD. RAD54 controls access to the invading 3′-OH end after RAD51-mediated DNA strand invasion in homologous recombination in *Saccharomyces cerevisiae*. Nucleic Acids Res 2009;37(2):638–46.

[88] Matunis MJ. On the road to repair: PCNA encounters SUMO and ubiquitin modifications. Mol Cell 2002;10:441–2.

[89] Owens JC, Detweiler CS, Li JJ. CDC45 is required in conjunction with CDC7/DBF4 to trigger the initiation of DNA replication. Proc Natl Acad Sci USA 1997;94(23):12521–6.

[90] Bhaumik D, Wang TS-F. Mutational effect of fission yeast Polα on cell cycle events. Mol Biol Cell 1998;9:2107–23.

[91] Lydeard JR, Jain S, Yamaguchi M, Haber JE. Break-induced replication and telomerase-independent telomere maintenance require Pol32. Nature 2007;448:820–3.

[92] Moss J, Tinline-Purvis H, Walker CA, Folkes LK, Stratford MR, Hayles J, et al. Break-induced ATR and Ddb1-Cul4(Cdt)(2) ubiquitin ligase-dependent nucleotide synthesis promotes homologous recombination repair in fission yeast. Genes Dev 2010;24(23):2705–16.

[93] Mortensen UH, Bendixen C, Sunjevaric I, Rothstein R. DNA strand annealing is promoted by the yeast Rad52 protein. Proc Natl Acad Sci USA 1996;93(20):10729–34.

[94] Krejci L, Altmannova V, Spirek M, Zhao X. Homologous recombination and its regulation. Nucleic Acids Res 2012;40(13):5795–818. http://dx.doi.org/10.1093/nar/gks270.

[95] Deng SK, Gibb B, de Almeida MJ, Greene EC, Symington LS. RPA antagonizes microhomology-mediated repair of DNA double-strand breaks. Nat Struct Mol Biol 2014;21(4):405–12.

[96] Mitchel K, Lehner K, Jinks-Robertson S. Heteroduplex DNA position defines the roles of the Sgs1, Srs2, and Mph1 helicases in promoting distinct recombination outcomes. PLoS Genet 2013;9(3):e1003340.

[97] Sebesta M, Burkovics P, Haracska L, Krejci L. Reconstitution of DNA repair synthesis in vitro and the role of polymerase and helicase activities. DNA Repair (Amst) 2011;10(6):567–76.

[98] Tsutsui Y, Kurokawa Y, Ito K, Siddique MS, Kawano Y, Yamao F, et al. Multiple regulation of Rad51-mediated homologous recombination by fission yeast Fbh1. PLoS Genet 2014;10(8):e1004542.

[99] Wu L, Hickson ID. The Bloom's syndrome helicase suppresses crossing over during homologous recombination. Nature 2003;426(6968):870–4.

[100] Cejka P, Plank JL, Bachrati CZ, Hickson ID, Kowalczykowski SC. Rmi1 stimulates decatenation of double Holliday junctions during dissolution by Sgs1–Top3. Nat Struct Mol Biol 2010;17:1377–82.

[101] Osman F, Dixon J, Doe CL, Whitby MC. Generating crossovers by resolution of nicked Holliday junctions: a role for Mus81-Eme1 in meiosis. Mol Cell 2003;12(3):761–74.

[102] Agmon N, Yovel M, Harari Y, Liefshitz B, Kupiec M. The role of Holliday junction resolvases in the repair of spontaneous and induced DNA damage. Nucleic Acids Res 2011;39(16):7009–19.

[103] Ip SC, Rass U, Blanco MG, Flynn HR, Skehel JM, West SC. Identification of Holliday junction resolvases from humans and yeast. Nature 2008;456(7220):357–61.

[104] Moore JK, Haber JE. Cell cycle and genetic requirements of two pathways of nonhomologous end-joining repair of double-strand breaks in *Saccharomyces cerevisiae*. Mol Cell Biol 1996;16:2164–73.

[105] Lieber MR. The mechanism of double-strand DNA break repair by the nonhomologous DNA end-joining pathway. Annu Rev Biochem 2010;79:181–211.

[106] Jazayeri A, Jackson SP. Screening the yeast genome for new DNA-repair genes. Genome Biol 2002;3. reviews1009.

[107] Ferreira MG, Cooper JP. Two modes of DNA double-strand break repair are reciprocally regulated through the fission yeast cell cycle. Genes Dev 2004;18(18):2249–54.

[108] Lieber MR. The mechanism of human nonhomologous DNA end joining. J Biol Chem 2008;283(1):1–5.

[109] Daley JM, Palmbos PL, Wu D, Wilson TE. Nonhomologous end joining in yeast. Annu Rev Genet 2005;39:431–51.

[110] Hefferin ML, Tomkinson AE. Mechanism of DNA double-strand break repair by non-homologous end joining. DNA Repair 2005;4:639–48.

[111] Chen L, Trujillo K, Ramos W, Sung P, Tomkinson AE. Promotion of Dnl4-catalyzed DNA end-joining by the Rad50/Mre11/Xrs2 and Hdf1/Hdf2 complexes. Mol Cell 2001;8:1105–15.

[112] Li P, Li J, Li M, Dou K, Zhang MJ, Suo F, et al. Multiple end joining mechanisms repair a chromosomal DNA break in fission yeast. DNA Repair (Amst) 2012;11(2):120–30.

[113] Lehnman IR. DNA ligase: structure, mechanism, and function. Science 1974;186:790–7.

[114] Chiruvella KK, Renard BM, Birkeland SR, Sunder S, Liang Z, Wilson TE. Yeast DNA ligase IV mutations reveal a nonhomologous end joining function of BRCT1 distinct from XRCC4/Lif1 binding. DNA Repair 2014;24:37–45.

[115] Durocher D, Jackson SP. The FHA domain. FEBS Lett 2002;513:58–66.

[116] Valencia M, Bentele M, Vaze MB, Herrmann G, Kraus E, Lee SE, et al. NEJ1 controls non-homologous end joining in *Saccharomyces cerevisiae*. Nature 2001;414:666–9.

[117] Chen X, Tomkinson AE. Yeast Nej1 is a key participant in the initial end binding and final ligation steps of nonhomologous end joining. J Biol Chem 2011;286:4931–40.

[118] Pannunzio NR, Li S, Watanabe G, Lieber MR. NHEJ often uses microhomology: implications for alternative end joining. DNA Repair 2014;17:74–80.

[119] Pardo B, Ma E, Marcand S. Mismatch tolerance by DNA polymerase Pol4 in the course of nonhomologous end joining in *Saccharomyces cerevisiae*. Genetics 2006;172(4):2689–94.

[120] Yang H, Matsumoto Y, Trujillo KM, Lees-Miller SP, Osley MA, Tomkinson AE. Role of the yeast DNA repair protein Nej1 in end processing during the repair of DNA double strand breaks by non-homologous end joining. DNA Repair 2015;31:1–10.

[121] McKinney JS, Sethi S, Tripp JD, Nguyen TN, Sanderson BA, Westmoreland JW, et al. A multistep genomic screen identifies new genes required for repair of DNA double-strand breaks in *Saccharomyces cerevisiae*. BMC Genomics 2013;14:251.

[122] McVey M, Lee SE. MMEJ repair of double-strand breaks (director's cut): deleted sequences and alternative endings. Trends Genet 2008;24(11):529–38.

[123] Boulton SJ, Jackson SP. *Saccharomyces cerevisiae* Ku70 potentiates illegitimate DNA double-strand break repair and serves as a barrier to error-prone DNA repair pathways. EMBO J 1996;15(18):5093–103.

[124] Ma JL, Kim EM, Haber JE, Lee SE. Yeast Mre11 and Rad1 proteins define a Ku-independent mechanism to repair double-strand breaks lacking overlapping end sequences. Mol Cell Biol 2003;23(23):8820–8.

[125] Hegde V, Klein H. Requirement for the SRS2 DNA helicase gene in non-homologous end joining in yeast. Nucleic Acids Res 2000;28(14):2779–83.

[126] Ira G, Pellicioli A, Balijja A, Wang X, Fiorani S, Carotenuto W, et al. DNA end resection, homologous recombination and DNA damage checkpoint activation require CDK1. Nature 2004;431(7011):1011–7.

[127] Wurtele H, Verreault A. Histone post-translational modifications and the response to DNA double-strand breaks. Curr Opin Cell Biol 2006;18(2): 137–44.

[128] Pai CC, Deegan RS, Subramanian L, Gal C, Sarkar S, Blaikley EJ, et al. A histone H3K36 chromatin switch coordinates DNA double-strand break repair pathway choice. Nat Commun 2014;5:4091.

[129] Choi YE, Pan Y, Park E, Konstantinopoulos P, De S, D'Andrea A, et al. MicroRNAs down-regulate homologous recombination in the G1 phase of cycling cells to maintain genomic stability. Elife 2014;3:e02445.

[130] Burma S, Chen BP, Chen DJ. Role of non-homologous end joining (NHEJ) in maintaining genomic integrity. DNA Repair (Amst) 2006; 5(9–10):1042–8.

[131] Haber JE. DNA recombination: the replication connection. Trends Biochem Sci 1999;24(7):271–5.

[132] Roeder GS. Meiotic chromosomes: it takes two to tango. Genes Dev 1997;11:2600–21.

[133] Hanahan D, Weinberg RA. Hallmarks of cancer: the next generation. Cell 2011;144(5):646–74.

Section IV

Genome Stability in Multicellular Eukaryotes

Chapter 9

Meiotic and Mitotic Recombination: First in Flies

J.K. Holsclaw[a], T. Hatkevich[a], J. Sekelsky
University of North Carolina, Chapel Hill, NC, United States

Chapter Outline

1. INTRODUCTION

1.1 Recombination in *Drosophila*: The First 100 Years

Many species of *Drosophilids* have been adapted for the laboratory since their induction as a model organism over a century ago, but none are as commonly used and widely known as the species we discuss here, *Drosophila melanogaster*. The rediscovery of Gregor Mendel's work around 1900 sparked a sudden and intense interest in the field of genetics and with that came the need for animal models. *Drosophila* proved well suited to the task, requiring little space and simple husbandry. Heredity could be studied at a much faster pace than in plants or mammals due to the short generation time and the vast array of phenotypic markers that obeyed Mendelian rules of inheritance. As a consequence, flies boast an impressive list of firsts in the areas of genome structure and recombination, including the first evidence of meiotic recombination (1911); the first meiotic map (1913); the first use of ionizing radiation to make chromosome breaks (1927); the first physical map of chromosomes (1929); the first evidence of mitotic recombination (1936). Given this list, it is no surprise that important contributions to our understanding of how double-strand breaks (DSBs) are repaired have been made in flies, including, for example, the first model of synthesis-dependent strand annealing (SDSA) (1994).

DSB repair has a dichotic nature in complex organisms: in mitotic cells, recombination can be detrimental, causing loss of heterozygosity and chromosome rearrangements that affect viability; yet in meiosis, recombination is important for accurate chromosome segregation. In both mitosis and meiosis, unrepaired DSBs are deleterious, causing chromosome fragmentation and cell death. Here we discuss the major pathways for DSB repair in mitotic cells: homologous recombination, with an emphasis on SDSA, and end joining; we also explore the regulation of these pathways to promote the

a. These authors contributed equally to this work.

Genome Stability. http://dx.doi.org/10.1016/B978-0-12-803309-8.00009-4

formation of noncrossover (NCO) products. This chapter also examines the meiosis-specific modifications to DSB-repair pathways that facilitate crossover (CO) formation, including homolog preference and regulation of recombination intermediates, and we describe a novel model for meiotic recombination in *Drosophila*.

1.2 *Drosophila* as a Model Organism: The Basics

D. melanogaster is well suited to a variety of laboratory and experimental conditions. Flies can survive in temperatures ranging from 15°C to 34°C, with the optimal temperature at 25°C (roughly room temperature). Their diet is simple, consisting mainly of sugar and yeast. Flies develop from zygote to sexual maturity in 7–10 days and a single female can lay as many as 3000 eggs in her 45–60-day lifespan [1].

D. melanogaster has four chromosomes that comprise its ~180 Mb genome: a sex chromosome and three autosomes (Fig. 9.1). The sex of a fly is determined by the ratio of *X* to autosomes, not the presence of a *Y* [1]. The *Y* chromosome is predominantly repetitive in sequence, entirely heterochromatic in content, and mostly genetically inert [2–4]. The *X*, *2*, and *3* chromosomes make up the majority of the euchromatin, while chromosome *4* is a mere 4.3 Mb in size and contains only ~100 genes interspersed between regions of heterochromatin [5]. The metacentric chromosomes *2* and *3* are subdivided into left and right arms, designated *2L*, *2R*, *3L*, and *3R*, respectively.

The *Drosophila* nucleus is highly ordered, with centromeres clustering at one pole of the nucleus and most telomeres clustered at the opposing pole [6]. Additionally, homologous chromosomes pair not just during meiosis, but in somatic tissues and the premeiotic germline. Centromere clustering and pairing of homologous chromosomes are independent of one another [7,8].

The ovaries of the female fruit fly are uniquely ordered as well. Each of the two ovaries consists of 12–16 ovarioles and each ovariole has a germarium-containing germline and somatic stem cells at the anterior end. Egg chambers increase in maturity as they migrate toward the posterior end of the ovariole, with a mature egg making up the final chamber [9]. Very early in the development of *Drosophila* as a model organism, researchers observed that meiotic recombination occurred only in female flies. Male flies have an alternate system for the proper segregation of their chromosomes that does not rely on crossing over of homologous chromosomes [10]. Thus, in one model system we have the means to study both meiotic recombination events (female germline) *and* exclusively mitotic recombination events (male germline).

2. MITOTIC RECOMBINATION

DSBs can arise from a variety of exogenous sources such as gamma radiation and chemical mutagens, as well as endogenous sources like collapsed replication forks, making this type of damage a common threat to genome integrity in both

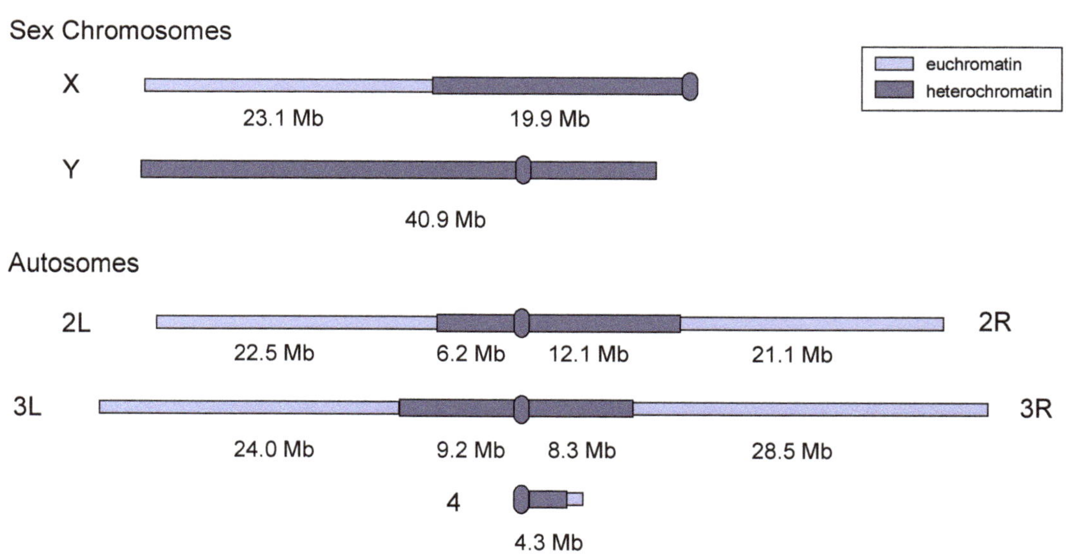

FIGURE 9.1 **Basic structure of *D. melanogaster* chromosomes.** Schematic representation of the four *Drosophila* chromosomes with approximate length in megabases. Size for *X*, *2*, and *3* chromosomes is separated into euchromatin and heterochromatin content. Size for *Y* and *4* chromosomes represents the entire chromosome. Light gray: euchromatin; dark gray: heterochromatin; oval: centromere. Heterochromatin sizes based on cytological evidence as reported in [104]. Euchromatin size from *Drosophila* genome assembly release 6.

mitotically cycling and quiescent cells. Improperly repaired DSBs can lead to mutations and genetic rearrangements; left unrepaired, DSBs cause chromosome fragmentation and cell death. It is critical that mitotic DSBs are properly identified and repaired in such a way that the integrity of the genome is restored and recombination is avoided.

2.1 Mitotic Recombination: A Historical Perspective

Mitotic recombination was first introduced by Curt Stern in 1939 using *Drosophila* as a model organism [11]. While J.T. Patterson had been inducing genomic rearrangements in somatic tissues via X-ray since 1930 [12], it was Stern's elegant and encompassing work that first proved reciprocal genetic exchange between chromosomes occurred outside the germline. From Patterson's work came the knowledge that recombination could be induced in the male germline through X-ray treatment (later determined to cause DSBs), a process still utilized to study DSB repair today. By using the male germline, it is possible to recover not only fully repaired, single mitotic recombination events, but also reciprocal products of a single event in the recombinant progeny of males [13]. Once transmitted to progeny, the event becomes fixed, allowing for molecular analysis from whole flies.

The male germline was effectively used for several decades to study spontaneous and induced mitotic recombination; however, flies lacked a system for generating site-specific DSBs until a powerful tool utilizing transposable elements became available in the 1980s. Evidence of what appeared to be male meiotic recombination in crosses between laboratory females and wild males, but not in the reciprocal cross between wild females and laboratory males, was reported in the 1970s and dubbed "hybrid dysgenesis" [14,15]. In 1982, Rubin, Kidwell, and Bingham showed that hybrid dysgenesis was the result of a transposable element, the P-element, which had been introduced into the wild population after laboratory strains were isolated. Wild populations had developed repression mechanisms to prevent transposition that were passed to the zygote from the mother through silencing the gene on P-elements responsible for transposition: *P-transposase* [16–18]. P-elements were quickly engineered to remove *P-transposase* and contain any sequence of interest. Once integrated into the genome, they are fixed until exposed to an alternate source of P-transposase [19–21]. This discovery revolutionized genome engineering in *Drosophila*, providing fly researchers with a site-specific DSB induction system similar to those in yeast, but with novel attributes. In addition to the genome-editing capabilities of P-elements, DSBs induced by the activity of "tame" P-transposase usually occur in only one sister chromatid, allowing for a more biologically relevant system than previous work with *I-SceI* or HO in yeast, which cut both sisters and required ectopic repair templates.

2.2 Mechanisms of Mitotic Recombination

In *Drosophila*, as in other eukaryotes, DSBs are repaired through either a template-mediated pathway or an end-joining pathway. Template-mediated repair, known as homologous recombination repair (HRR), necessitates access to an undamaged copy of DNA—either a sister chromatid or a homologous chromosome. HRR can have multiple outcomes including both NCO and CO products, but mitotic regulation in *Drosophila* favors NCO formation. End joining (EJ) involves direct ligation of the broken ends, often after processing that can result in small insertions or deletions. *Drosophila* actively uses at least three variations of EJ, depending on the context of the break.

2.3 Initial Response and Pathway Choice

Extensive work in yeast and mammalian cells has established the MRN complex (Mre11–Rad50–Nbs in *Drosophila*) as the DSB sensor for mitotic cells. The MRN complex activates the DNA damage response protein kinase ATM (*Drosophila* tefu), which then phosphorylates many downstream factors to initiate repair, one of which is the histone variant H2AX (*Drosophila* H2AV) [22,23]. The γH2AV signal peaks within 5 minutes of gamma irradiation in flies and provides a scaffold to recruit additional proteins to amplify the repair signal [24].

In yeast and mammalian cells, pathway choice is cell cycle dependent. In G0-G1, phosphorylated 53BP1 binds the broken ends of a DSB to block 5′–3′ resection of the ends preventing HRR. During S–G2, when the genome has been replicated and a sister chromatid is available as a template, BRCA1 is phosphorylated by ATM leading to degradation of 53BP1, freeing the ends for resection [22,23]. Thus, it appears that the choice of HRR or EJ is decided by whether or not resection occurs. Additionally, the role of 53BP1 suggests the default repair mechanism for DSBs is HRR and only by blocking HRR can EJ occur.

It is less clear how pathway choice is made in *Drosophila*. While the core components of the early response are conserved, flies lack many of the regulatory controls such as 53BP1 and BRCA1 (Table 9.1). According to limited studies, pathway choice is somewhat age dependent, with HRR strongly favored in older flies (>2 weeks), while EJ is utilized more

TABLE 9.1 Orthologous Repair Genes in Fly, Human, and Yeast

D. melanogaster	*H. sapiens*	*S. cerevisiae*
Resection		
tosca	*EXO1*	*EXO1*
nbs	*NBS1*	*XRS2*
mre11	*MRE11*	*MRE11*
rad50	*RAD50*	*RAD50*
CG5872	*CtIP*	*SAE2*
CG2990	*DNA2*	*DNA2*
MCMs		
rec	*MCM8*	–
mei-217	*C8ORF45/MCMDC2*	–
mei-218		–
Recombinases		
spn-A	*RAD51*	*RAD51*
–	*RAD52*	*RAD52*
–	*BRCA1*	–
Brca2	*BRCA2*	–
–	*DMC1*	*DMC1*
Helicases and Associated Proteins		
Blm	*BLM*	*SGS1*
Top3α	*TOPO3α*	*TOP3*
–	*RMI1*	*RMI1*
–	*RMI2*	*RMI2*
Fancm	*FANCM*	*MPH1*
Checkpoints		
mei-41	*ATR*	*MEC1*
mus304	*ATRIP*	*DDC2*
tefu	*ATM*	*TEL1*
grp	*CHK1*	*CHK1*
Lok	*CHK2*	*RAD53*
p53	*P53*	–
Nucleases		
mei-9	*XPF*	*RAD1*
Ercc1	*Ercc1*	*RAD10*
Gen	*GEN1*	*YEN1*
mus81	*MUS81*	*MUS81*
mus312	*SLX4/BTBD12*	*SLX4*
hdm	*MEIOB*	–
mms4		

TABLE 9.1 Orthologous Repair Genes in Fly, Human, and Yeast—cont'd

D. melanogaster	*H. sapiens*	*S. cerevisiae*
End Joining		
Irbp	*KU70*	*YKU70*
Ku80	*KU80*	*YKU80*
Lig4	*DNA LIG4*	*DNL4*
–	*XRCC4*	–
–	*XLF*	–
mus308	*DNA pol θ*	–

Comparison of major repair genes in multiple pathways. Although not all inclusive, this table highlights major areas of conservation or divergence. Dashes indicate gene is not present.

in young flies (<2 weeks) [25]. There are two caveats worth noting in this study: (1) the dominant repair pathway was single strand annealing (SSA), a pathway strongly favored by the 157-bp repeats flanking the cut site in the reporter construct; (2) age-dependent pathway choice has only been studied in the male germline and these studies may reveal cell type–specific pathway choice (mature sperm-EJ vs. stem cells-HRR) rather than a true age correlation. Interestingly, tumorigenesis in epithelial cells of older flies correlates with errors in HRR, but not EJ, suggesting adverse effects on fitness with utilization of HRR as flies age [26].

Pathway choice does not seem to be affected by chromatin environment in *Drosophila*. It has been proposed that heterochromatin is naturally more resistant to DSBs due to compaction; and when breaks occur, EJ is the preferred repair pathway to avoid illegitimate recombination due to the highly repetitive nature of heterochromatic DNA. Chiolo et al. showed in 2011 that neither of these hypotheses is supported in *Drosophila*: heterochromatin is as susceptible to DSB formation via ionizing radiation as euchromatin, and HRR is still the dominant pathway for repair. *Drosophila* heterochromatin forms a distinct region within the nucleus and using high-resolution microscopy, Chiolo and colleagues were able to show γH2AV within the heterochromatin domain in response to gamma radiation. It was further shown that resection occurred within the heterochromatin domain but the remaining steps of HRR were suspended until the break physically moved to the outer periphery of the heterochromatin and was stripped of the heterochromatin marker HP1a, presumably to reduce compaction and enable repair factors access to the lesion [27]. These data indicate that HRR is the dominant pathway in *Drosophila* regardless of chromatin environment; that EJ and HRR have a dynamic and contextual relationship; and that flies may utilize spatial positioning as a means to regulate repair outcomes.

2.4 Synthesis-Dependent Strand Annealing: A Model Consummated in Flies

SDSA is now the predominant model for DSB repair by HRR [28]. SDSA was first proposed in *Drosophila* to explain repair products arising from P-element-induced mitotic DSBs. Throughout the 1980s, P-elements had been used to generate mutations through a method commonly called "imprecise excision," but probably actually arising from rare imprecise repair events. Sved, Eggleston, and Engels showed that recombination could be induced via P-element in the male germline, that it clustered around the site of a P-element, that the events were premeiotic (and therefore mitotic) in nature, and they could recover reciprocal clusters [19]. Over the next 4 years, Engels' group showed that P-element excision resulted in the formation of a DSB that was repaired via HRR, predominately using the sister chromatid as a template [29,30]. This type of repair requires extensive synthesis to accurately "replace" the missing P-element, and thus the lesion could more accurately be described as a double-strand gap rather than a break. Repair events were dependent on homology between the resected ends and the template as well as highly sensitive to single-base mismatches within the homology, suggesting that a mechanism existed that was capable of finding the precise and correct template for repair anywhere in the genome. Most importantly, they found that these events were rarely associated with COs [31].

Work in yeast suggested that DSBs repaired via HRR formed a joint molecule called a double Holliday junction (dHJ) that can form COs when resolved [32]. This molecule consists of two chromosomes concatenated into a four-stranded structure at two locations and requires endonucleolytic cleavage, either by a type I topoisomerase (coupled with a helicase) or a dsDNA nuclease, to separate the strands. In this model, often called the DSBR model but here referred to as the dHJ model, CO events would be predicted at a higher rate than those observed by Engels. The popularity of the dHJ model can be attributed, in part, to a difference in experimental method. The assays that definitively show dHJ formation are studies of meiosis—a process biased *toward* CO outcomes (discussed in detail further in this chapter) and gene replacement, which also requires COs. Assays used in other model systems to specifically study mitotic events cut the genome at a variety of locations and relied on a template that was either on the same chromosome separated by few kilobases, or on an ectopic circular plasmid (as discussed in [30,31]). In contrast, Engels' system used P-element-induced DSBs that could repair off of the endogenous sister or homolog.

It was also possible to use tailored templates located at ectopic sites to recover and molecularly analyze repair products. Nassif et al. observed the same fidelity of repair and lack of COs when the template was inserted on a nonhomologous chromosome [33]. Using a variety of template cassette sizes, they were able to recover insertions of up to 8 kb at the repaired locus. Most striking, however, was the preponderance of "conversion-duplication" events that contained sequence from the template followed by sequence from the original P-element. These complex events could only be explained if both the ectopic site and the sister chromatid were used as templates for repair and then annealed at sequence that was complementary, which did not fit the dHJ model. From these conclusions, Nassif et al. combined models from a diverse collection of experiments in bacteria, fungi, and mouse cells to build a model they called SDSA [33]. SDSA was the most parsimonious model to fit the emerging data of the time, though it did not gain wide acceptance in the field until Haber's work in yeast was published 4 years later (his previous work considered homology annealing and strand invasion as two distinct and separate pathways) [34,35].

SDSA is sometimes considered to be a truncated form of the dHJ model because both pathways have the same early steps, which begin with 5′–3′ end resection (Fig. 9.2). The resulting 3′ tail is coated with Spn-A (Rad51 in yeast and humans) to form a stable and flexible filament proficient at finding homology in the dsDNA template. The filament invades

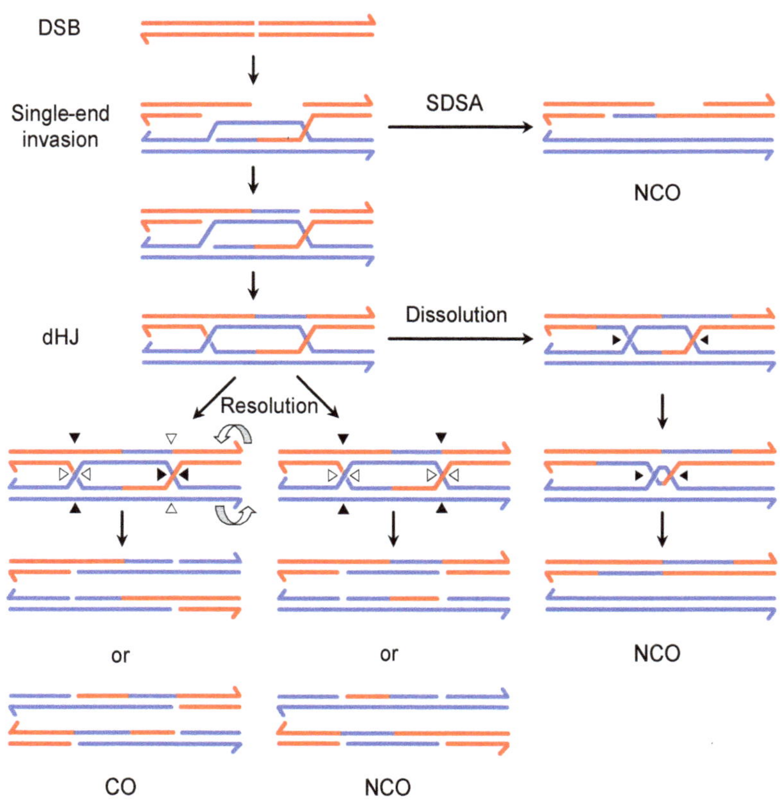

FIGURE 9.2 **Homologous recombination repair (HHR) model.** HRR begins with a DSB that is resected to form 3′ tails that invade a dsDNA template to form a D-loop (single end invasion). SDSA occurs when the D-loop is dismantled and the complementary ends anneal, followed by gap filling to yield an NCO. If the D-loop is not dismantled, second-end capture occurs and primes synthesis to yield a ligated dHJ product that can be disentangled through migration and decantenation (dissolution) to yield an NCO. Alternatively, the ligated dHJ can be cleaved through unbiased endonucleolytic cleavage of the HJs to form CO products in either orientation (*open arrowheads* versus *black arrowheads*) or NCO products in either orientation (*open arrowheads* versus *black arrowheads*). *Figure adapted from Crown KN, McMahan S, Sekelsky J. Eliminating both canonical and short-patch mismatch repair in* Drosophila melanogaster *suggests a new meiotic recombination model. PLoS Genet September 2014;10(9):e1004583.*

the duplex template, displacing the nontemplate strand (strand exchange) and facilitating extension of the invading strand via synthesis; this forms a structure called a D-loop. It is at this point that the two HRR pathways diverge: in SDSA, the D-loop is dismantled, freeing the newly synthesized end to find its complement from the opposite side of the break. In *Drosophila* P-element assays, Blm helicase is necessary to dismantle the D-loop [20]. If complementarity is not found between the nascent strand and the processed end at the other side of the break, reinvasion of the template occurs. This process of invasion, dismantling, complementarity search, and reinvasion occurs until annealing is achieved or the cycle is terminated and the ends are joined [36–38].

These findings are corroborated using other reporter systems as well. The activity of *I-Sce*I produces a DSB with 4 nt complementary overhangs, in contrast to the P-element system which generates 17 nt overhangs that are not complementary. Preston et al. found that a variation of SDSA that does not require synthesis, SSA, was strongly preferred over the dHJ model when *I-Sce*I was used [25]. In fact, EJ and SSA worked in a compensatory fashion to facilitate repair while dHJ events were exceedingly rare, suggesting that SDSA is the dominant form of HRR and not simply an aborted version of the dHJ model [25].

2.5 End Joining in *Drosophila*

Drosophila utilize multiple forms of end joining to repair DSBs. Canonical end joining, called nonhomologous end joining (NHEJ), involves ligation of DSB ends without synthesis. In vertebrates, the Ku70/Ku80 heterodimer binds the ends of a DSB and recruits DNA-PKcs. Autophosphorylation of DNA-PKcs activates the complex, recruiting accessory factors to process damaged nucleotides and single-base overhangs. Lastly, the ligation complex consisting of Lig4 and XRCC4 (LIG4, XRCC4, and XLF) ligates the ends of the break [39]. Because it is untemplated and does not rely on complementary overhangs or resection, it is thought to be more error prone (though much faster) than other pathways.

In P-element systems, end joining is observed when strand exchange is prevented via *spn-A* mutations. These events are independent of Lig4 and rely on microhomology, and are thus categorized as microhomology-mediated end joining (MMEJ) [40,41]. Assays using zinc finger nucleases also provide evidence for end joining in both *wild-type* and *lig4* mutants. Lig4-independent events were not microhomology mediated, suggesting that a third type of end joining, alternative end joining (alt-EJ), is also possible [42]. *Drosophila*, like other invertebrates, lack the key regulator of canonical NHEJ, DNA-PKcs, yet they retain orthologs of Ku70/Ku80, Lig4, and XRCC4, suggesting that canonical NHEJ is still utilized, though how it is regulated or how ends are processed remains unknown.

Work with *mus308* (PolQ in humans) suggests MMEJ or alt-EJ is used regularly by *Drosophila* to repair DSBs. Chan, Yu, and McVey showed that MMEJ was *mus308* dependent [43]. They also showed that *mus308* mutations are synergistic with mutations in *spn-A*, suggesting that mus308-mediated MMEJ is a compensatory response to inactivated HRR. In contrast, *lig4 spn-A* double mutants had no viability, fertility, or morphological defects, indicating that NHEJ is dispensable in the absence of HRR. They further showed that MMEJ occurred in wild-type backgrounds and increased in frequency in *lig4*-deficient backgrounds [43,44]. Collectively, these data indicate that *Drosophila* actively utilize multiple forms of EJ to repair DSBs and that MMEJ can compensate for both HRR and NHEJ, perhaps providing an alternative to DNA-PKcs-mediated end processing. This role for PolQ in *lig4*-independent EJ was corroborated in mammals [45–47].

2.6 Mitotic COs and the dHJ Model

The data presented thus far points to SDSA being strongly favored in DSB repair events, with some form of EJ providing a back-up mechanism; however, mitotic COs are observed in certain genetic backgrounds, such as *Blm* mutants, suggesting that the dHJ model is still a valid and utilized pathway for repair in *Drosophila* [13]. dHJ formation occurs when the second resected end of a DSB anneals to the D-loop and begins synthesis (Fig. 9.2). This process is thought to occur sequentially, with strand invasion occurring first to open the D-loop, followed by synthesis and ligation of the nascent end to the opposing 5′ strand without dismantling of the D-loop. Once ligated, the second resected end is "captured" by the single-stranded D-loop. As synthesis continues, the nascent strand eventually meets the opposing side and ligates to the remaining 5′ end to form a concatenated joint molecule—the dHJ. dHJs are toxic structures that prevent proper segregation during mitosis and block transcription; it is imperative that the chromosomes are separated accurately, preferably without exchange of genetic information in the form of COs. There are two possible mechanisms for disentanglement: dissolution via migration and decatenation or resolution via endonucleolytic cleavage.

Migration and decatenation is carried out by the BTR complex in humans (BLM, TOPO3α, RMI1/2) and the STR complex in yeast (Sgs1, Top3α, Rmi1) (Table 9.1) [48,49]. BLM helicase migrates the junctions toward each other and TOPO3α (a type I topoisomerase) decatenates the strands through nicking and religating one strand of the dsDNA. The RMI proteins are thought to provide stability to the complex as well as facilitate decatenation through coordination with TOPO3α. Mitotic COs are elevated in *Blm* mutant flies, suggesting that the function of the complex is conserved in *Drosophila* [13]. Interestingly, flies do not have orthologs to the RMI proteins; the C-terminal region of Top3α has a large insertion that may play a similar role but this hypothesis has not been tested [50].

The presence of mitotic COs in *Blm* mutants, rather than an increase in lethality, suggests unbiased resolution of dHJs by structure-specific endonucleases called resolvases. Andersen et al. showed that *Blm* mutations are lethal when combined with mutations in the genes *mus81*, *mus312*, or *Gen* (MUS81, SLX4, GEN1, respectively, in humans), all of which encode subunits of putative HJ resolvases [51]. The synthetic lethality of the double mutants could be partially rescued by mutating *spn-A*, (in the case of *mus81 Blm* double mutant, fully rescued) suggesting that the phenotype was strand invasion dependent, and therefore related to a toxic HRR product [51]. The absence of mitotic COs in flies with wild-type *Blm*, combined with the viability of single endonuclease mutants, indicate that the primary pathway for disentangling dHJs is Blm-mediated dissolution with endonuclease cleavage serving as a back-up mechanism.

3. MEIOTIC RECOMBINATION

It is clear that somatic cells have a complex system with multiple interacting pathways to prevent dHJ formation and COs during DSB repair. Yet in germ cells undergoing meiosis, crossing over of genetic material between homologous chromosomes is required for proper segregation of chromosomes, suggesting that a completely separate regulatory network exists to promote dHJ formation and crossing over during meiotic recombination. Much of the research investigating the genetic basis of meiotic recombination began using *Drosophila* as a model organism.

3.1 Meiotic Recombination: A Historical Perspective

Drosophila researchers have been making pioneering discoveries in the field of meiotic recombination for over a century. In 1910, Thomas Hunt Morgan was the first to report meiotic recombination when he observed progeny that could arise only from maternal crossing over between the homologous sex chromosomes [52]. Following the discovery of meiotic recombination, Morgan hypothesized that genes are arranged linearly along chromosomes [53]. In 1913, Morgan's student Alfred Sturtevant reasoned that if Morgan's linear arrangement hypothesis is correct, he could determine the relative location of genes by measuring CO frequency [54]. By mapping six genes in a linear arrangement, Sturtevant did in fact prove Morgan's hypothesis to be true, and as a consequence, Sturtevant was the first to build a meiotic map. In this landmark study, Sturtevant also observed that the occurrence of one CO reduces the formation of a nearby CO, a phenomenon referred to as CO interference. Although CO interference was first observed over a century ago, the mechanism in which interference acts is still largely unknown.

In 1930, Theodosius Dobzhansky used chromosomal translocations induced by X-rays to construct a cytological map of *D. melanogaster* chromosome *2*. During this study, he noticed that there was a discrepancy between cytological distance (ie, physical distance) and genetic distance through the observation that genes in the middle of the chromosome arm undergo more recombination than the genes at the ends of the arm [55]. George Beadle performed a similar experiment using CO rates from translocations of chromosome *3* in 1932. Beadle's data indicated that the spindle fiber attachment region (now referred to as the centromere) impedes crossing over in adjacent regions on the chromosome. This reduction in COs occurred even when genomic regions located in the middle of the arm were experimentally placed adjacent to the centromere via translocations [56]. This phenomenon is now referred to as the centromere effect and has been observed in fungi, plants, and vertebrates.

By the early 1960s, much about the process of meiosis had been described through studies from *Drosophila*, maize, and fungi; however, surprisingly little was known regarding meiotic regulation. It was understood that recombination during meiosis is important for proper separation of chromosomes (meiotic disjunction), so Larry Sandler and colleagues screened natural *Drosophila* populations to find mutations that increased meiotic nondisjunction (improper separation of homologous chromosomes) [57]. Fifteen naturally occurring mutations that affected disjunction in one or both sexes were recovered. Baker and Carpenter performed a second screen, this time inducing mutations of the *X* chromosome via ethylmethane sulfonate (EMS) and uncovered additional novel meiotic mutants [58]. Together, these screens provided the scientific community with valuable resources still being used today; most importantly, the subsequent analysis of these mutants revealed new principles surrounding the mechanisms of meiotic recombination.

3.2 Mechanisms of Meiotic Recombination

Meiotic recombination is initiated by the formation of programmed DSBs, which are resected to yield 3′ DNA overhangs that invade the homologous chromosome, giving rise to a D-loop structure (Fig. 9.2). Similar to mitotic recombination, after synthesis the D-loop can either be unwound through SDSA to generate a NCO or can be stabilized so it can mature into a dHJ or other joint molecules. Unlike in mitotic recombination, joint molecules are preferentially resolved to form COs, which are vital for proper meiotic disjunction [59,60]. Because the meiotic recombination pathway utilizes many of the same repair proteins used during mitotic DSB repair, meiotic recombination has long been thought to have evolved from mitotic pathways [60–63]. Nonetheless, the fundamental purpose of these two processes are distinct: the outcome of mitotic recombination is complete and error-free repair of DSBs, while the primary goal of meiotic recombination is to carefully form stable COs between two homologs to ensure proper bipolar orientation of homologous chromosomes at the meiotic spindle [64]. For CO formation between homologs to occur during meiotic recombination, several meiosis-specific modifications to the somatic DSB-repair program have to transpire, as discussed later [60,62,63].

3.3 Initiation of Recombination

DSBs occur at a much higher frequency during meiosis when compared to the somatic cell cycle [65]. This increase is required to ensure that sufficient amounts of meiotic COs are formed to achieve proper segregation of homologous chromosomes. Accordingly, an important feature of meiotic recombination is deliberate and controlled DSB formation to initiate the repair process. In most, if not all, sexually reproducing organisms, Spo11, a type II-like topoisomerase conserved throughout eukaryotes, is the nuclease responsible for creating these meiosis-specific DSBs [66,67]. In most organisms, including yeast and mouse, Spo11 is not only responsible for creating meiotic DSBs, but it also initiates recombination by promoting repair through interactions with the MRN complex [68,69]. The MRN complex, along with Exo1, is responsible for resection of the break, marking the beginning of the repair process.

3.4 Preference of Homolog as Repair Template

In contrast to mitotically dividing cells that use the sister chromatid, cells undergoing meiotic recombination use an intact homologous chromatid as a repair template. This preference ensures CO formation between homologs, which prevents nondisjunction and promotes genetic diversity [70]. Invasion of a homologous duplex is promoted by DNA strand-exchange proteins of the RecA family. In most eukaryotes, there are two RecA homologs that aid in strand exchange during meiotic recombination, Rad51 and Dmc1 [71]. *RAD51* and *DMC1* diverged during the separation of the prokaryotic and eukaryotic kingdoms. While Rad51 participates in both mitotic and meiotic recombination, Dmc1 is meiosis specific, suggesting its function is to promote recombination preferentially between homologs [72]. The *DMC1* gene is found in most eukaryotes, including *Saccharomyces cerevisiae*, plants, mice, and humans. Interestingly, all Dipteran insects, including *Drosophila*, are missing DMC1, and it appears to have been lost independently in other clades, including fission yeast and some nematodes [73]. One explanation for this loss of *DMC1* in Dipteran insects may be the timing of formation of the synaptonemal complex (SC) in *Drosophila*, as discussed below.

The SC is a tripartite proteinaceous structure that connects paired homologous chromosomes along the length of their axes to provide an environment suitable for successful recombination during meiosis. Although the true function of the SC is unknown, it was initially thought to aid in the pairing of homologs before recombination could begin. However, this initial hypothesis was refuted when Spo11-dependent DSBs were shown to appear before formation of the SC during recombination in yeast, plants, and mammals, indicating that SC formation is not a prerequirement for recombination in these organisms [74].

Surprisingly, it was later found that in *Drosophila*, the SC is formed before the occurrence of DSBs, and in fact, normal levels of Spo-11 DSBs are dependent on the proper formation of the SC [75]. The only other organism known to exhibit this reversal of SC formation and DSB appearance in meiosis is the nematode *Caenorhabditis elegans* [76], which also lacks the *DMC1*. This being said, the structure of the SC may provide enough restraint on the chromosomes to ensure invasion of the homolog rather than the sister, negating the need for Dmc1 in *Drosophila* and *C. elegans*, whereas later formation of the SC in yeast, plants, and mammals necessitates a specialized strand invasion protein to facilitate homolog preference [73]. This hypothesis is supported by the finding that Ord, a *Drosophila* sister chromatid cohesin protein that promotes proper assembly of the SC, also promotes homolog bias during meiotic recombination [77]. The contrast between the *Drosophila* and the yeast/mammal recombination initiation suggests that mechanisms for homolog preference are not necessarily equivalent across model organisms; however, the fact that each species has a mechanism for it reinforces the importance of recombination between homologs in meiosis.

3.5 Promoting CO Formation: Pro-CO Complexes

Formation of COs is necessary to achieve proper chromosomal disjunction in meiosis I, but there are more DSBs than COs; surplus DSBs are repaired into NCOs. In *S. cerevisiae*, most NCOs are formed earlier than COs via SDSA, and are dependent on Sgs1 [78]. To promote COs, specialized proteins antagonize the activity of Sgs1 [79], and in most organisms, these specialized pro-CO proteins are MSH4 and MSH5, the subunits of MutSγ [80–83]. Interestingly, neither MSH4 nor MSH5 have roles in gene conversion or mismatch repair (MMR), but without either, CO formation is severely reduced, implicating them in the maturation of CO products [81,82]. Through biochemical studies, it has been shown that MSH4 and MSH5 form a heterodimer that preferentially binds to dHJs to form a sliding clamp, presumably to stabilize and protect recombination intermediates from disassembly by helicases, thereby promoting the dHJ pathway and CO formation [84]. The use of MSH4–5 as a pro-CO complex in meiotic recombination is highly conserved, yet it is absent in *Drosophila* [85]. In fact, a meiosis-specific pro-CO complex in *Drosophila* was not identified until 2012 by Kohl et al. in a landmark study [85].

Kohl studied three *Drosophila* genes, *mei-218*, *mei-217*, and *rec*, whose functions at the time were unknown. The gene *mei-218* was first discovered in the Baker and Carpenter screen in 1972 [58], while *mei-217* was discovered by Liu and McKim decades later [86]. *mei-217* and *mei-218* are transcribed as a dicistronic message and mutations in these genes result in 80–90% reduction of COs [85,86]. Studies suggest that female mutants for *mei-218* may fail to produce recombination intermediates, yet these mutants do not show a significant change in NCOs. Together, these observations suggest that the formation of CO-fated recombination intermediates is impeded when *mei-217* and *mei-218* are disrupted [87,88].

The *rec* gene was discovered in 1984 by Rhoda Grell through an EMS screen for temperature-sensitive meiotic mutants [89]. Interestingly, *rec* mutants display the exact phenotype of *mei-217* and *mei-218* mutants, such that REC is required for a majority of COs yet does not affect NCO formation. REC was shown to be the *Drosophila* ortholog of MCM8 and has no apparent role outside of meiosis [90]. Although MEI-217 and MEI-218 have no obvious sequence similarities, Kohl showed through structural analysis that these two proteins are predicted to fold like MCM proteins and have apparently evolved from an ancestral MCM-like protein [85]. Further, Kohl showed that MEI-217 interacts with both REC and MEI-218, together forming a complex referred to as the mei-MCM complex. In budding yeast, the CO defect in *msh4* mutants in *S. cerevisiae* is suppressed by eliminating Sgs1, suggesting that Msh4–5 promotes CO formation by antagonizing Sgs1 [79]. Paralleling this result, the removal of Blm in *Drosophila* suppresses the CO defects seen in *mei-MCM* mutants [85]. This observation indicates that the mei-MCM complex functionally replaces Msh4–5 in *Drosophila*, and more importantly, suggests that the general strategy of promoting CO formation in meiotic recombination may be universal to all sexually reproducing organisms.

3.6 Promoting CO Formation: Meiotic Resolvases

Somatic cells utilize resolvases as a last resort for dHJ resolution; this can still result in NCO formation through unbiased cleavage. In meiosis, recombination intermediates need to be resolved with a bias toward CO products, requiring a specialized set of resolvases. In *S. cerevisiae*, the primary meiotic resolvase is MLH1–3, the MutLγ heterodimer [91]. In *mlh1* or *mlh3* mutants, joint molecules are formed normally, but COs are severely reduced [92]. In humans, as well as in mice, MLH1–3 has also been implicated as the major meiotic resolvase [93]. In contrast, the primary meiotic resolvase in *Drosophila* is a complex containing MEI-9, MUS312, ERCC1, and HDM.

The gene *mei-9* was also discovered by Baker and Carpenter [58]. Females mutant for *mei-9* show a 90% reduction in COs but NCOs are not reduced [58,87]. The protein encoded from *mei-9* is an ortholog of the *S. cerevisiae* nucleotide excision repair protein Rad1 and the human structure-specific endonuclease XPF [94,95]. Females mutant for the gene *mus312* show meiotic phenotypes similar to that of *mei-9* mutants, with the formation of COs being reduced by 90% of *wild type* [95,96]. Through a yeast two-hybrid screen, it was shown that the proteins MUS312 and MEI-9 physically interact, and this interaction is required for formation of meiotic COs. Interestingly, MUS312 is shown to participate in interstrand crosslink repair, but not in nucleotide excision repair, while its meiotic binding partner, MEI-9, participates in both processes [95,97].

The product of the *Ercc1* gene physically interacts with MEI-9, as initially shown via yeast two hybrid, and is required for the role of MEI-9 in nucleotide excision repair [98]. Its role was implicated in the generation of meiotic COs with MEI-9 and MUS312 when Radford et al. demonstrated that all three proteins physically interact, and that ERCC1 is required for a subset of meiotic COs [99]. Lastly, the gene *hdm* encodes HDM, a protein that contains three OB fold domains, which are often associated with single-stranded DNA-binding capabilities. HDM physically interacts with MEI-9 and ERCC1 and is also required for a subset of meiotic COs [100].

In *Drosophila* meiotic recombination, as in other organisms, meiosis-specific features enable sufficient number of COs between homologs; yet the details of meiotic recombination seem to be vastly different in *Drosophila* as compared to other models: DSBs are primarily dependent on the formation of the SC, presumably negating the requirement for DMC1; the mei-MCM complex functionally replaces MSH4–5; and the meiotic nuclease complex is MEI-9, MUS312, ERCC1, and HDM. Together, these differences in proteins raise the question: *Is the dHJ model, elucidated primarily in yeast, applicable to meiotic recombination in Drosophila?*

3.7 Meiotic Recombination in *Drosophila*: Double-End Engagement Model

The dHJ model of meiotic recombination was largely elucidated in *S. cerevisiae* using recombination hotspots, which are loci with a high frequency of recombination. By molecularly manipulating hotspots, yeast geneticists could recover recombination intermediates for molecular analysis [101,102]. High conservation of meiotic proteins has led to an assumption that the dHJ model is also conserved across species; however, the model had never been directly tested in a metazoan because of an inability to reproduce a system for physical analysis of recombination intermediates like that in yeast. In 2014, Crown et al. used molecular analysis of *Drosophila* heteroduplex DNA (hDNA) to provide the first evidence suggesting that some features of the dHJ model differ *Drosophila*; instead, they proposed that unligated dHJs give rise to both COs and a substantial fraction of NCOs [59].

During recombination, strand invasion and subsequent synthesis create recombination intermediates that contain hDNA, in which each strand of the duplex is derived from a different parental chromosome (Fig. 9.3). hDNA is repaired by MMR machinery to yield NCOs and COs without mismatches. By inactivating MMR machinery, hDNA generated during recombination can be preserved in these products. The hDNA tracts can then be molecularly analyzed to determine the orientation of the hDNA tracts, and through this, the structure of the recombination intermediate can be inferred.

In an attempt to recover and analyze hDNA in CO and NCO products in *Drosophila*, Radford et al. eliminated the canonical MMR machinery by mutating Msh6 [103]. Surprisingly, the hDNA recovered was not continuous, meaning that some patches of hDNA were repaired and some were not, even in the same recombination event. From these data, Radford proposed the noncontinuous hDNA tracts resulted from a short-patch MMR system that was able to repair some mismatches in hDNA in concert with the canonical MMR machinery, and this short-patch MMR may include NER proteins, as shown in fission yeast. In 2014, Crown tested this hypothesis by inactivating both MMR and NER pathways through mutations in *Msh6* and *Xpc*, respectively, and found that all hDNA were preserved [59].

According to the dHJ model, NCOs are formed primarily through SDSA. hDNA tracts by SDSA are predicted to be in *cis*-orientation, meaning all of the markers from the donor are on one strand of the product (Fig. 9.2); however, Crown et al. found that only half of the NCO synthesis tracts were associated with *cis*-hDNA. Surprisingly, the other half of NCO tracts

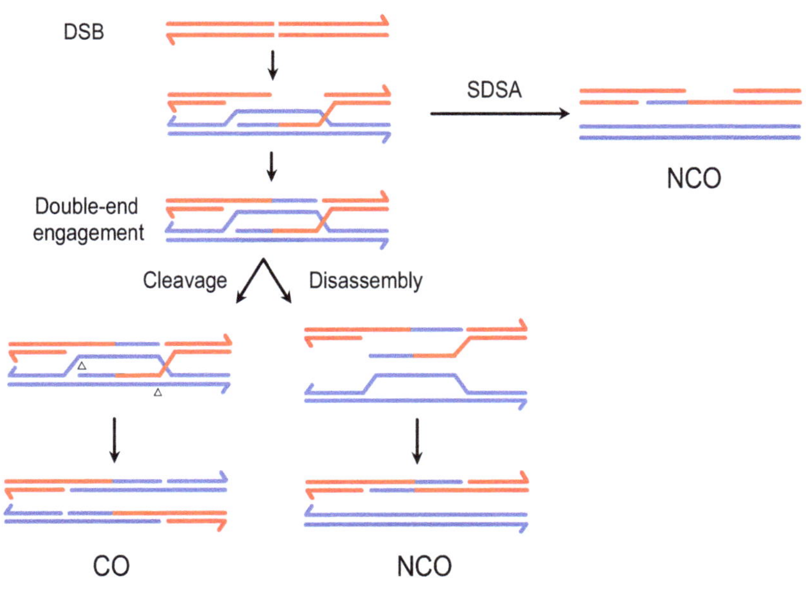

FIGURE 9.3 **Double-end engagement (DEE) model.** In *Drosophila* meiosis, a Spo11-generated DSB is resected and one 3′ tail invades the homologous chromosome to form a D-loop. If the D-loop is dismantled, an NCO product is formed through SDSA. If the D-loop is protected, both 3′ tails anneal to the same template and prime synthesis to form an unligated dHJ, termed DEE. The DEE can be processed by disassembly through migration of the single ligated HJ to yield an NCO or through biased cleavage (at open *arrowheads*) by meiotic resolvases to generate a CO. *Figure adapted from Crown KN, McMahan S, Sekelsky J. Eliminating both canonical and short-patch mismatch repair in* Drosophila melanogaster *suggests a new meiotic recombination model. PLoS Genet September 2014;10(9):e1004583.*

had two adjacent tracts of hDNA in *trans*-orientation, meaning the markers from the donor are on both strands of the product, which is not predicted by the SDSA model. Additionally, the dHJ model predicts that COs are formed by the resolution of dHJ in either of two orientations, both of which are equally likely (Fig. 9.2). However, the COs that were recovered with hDNA only appeared in one orientation. Based on these data, Crown proposed a new model with an unligated dHJ as an intermediate as opposed to the fully ligated dHJ model.

Together, these results suggest a novel and unified model for CO and NCO formation in *Drosophila*, referred to as the double-end engagement (DEE) model. In this model, up to half of all NCOs may arise from SDSA, giving rise to NCO products associated with hDNA in *cis*-orientation. The intermediates that do not undergo SDSA are processed into an unligated dHJ, referred to as a DEE intermediate, which can either be cleaved by MEI-9 to form a CO or can be disassembled by a helicase, such as Blm, to form an NCO. It is possible that the SDSA-mediated NCOs are early events similar to NCO formation in yeast, while nicked dHJ-mediated NCOs represent later recombination events, which may suggest a fine-tuning mechanism to ensure that proper number of COs per meiosis exists in *Drosophila*. Additionally, the DEE intermediate represents a joint molecule that is both simple to resolve and regulate.

Regardless, COs and NCOs arising from the same intermediate sharply contrast the dHJ model. It remains unclear if the structural difference in joint molecules explains the difference in meiotic resolvases between *Drosophila* and other eukaryotes, or if unligated dHJs are more common in metazoans than previously suspected. Pathways responsible for short-patch MMR have not been determined in other organisms, so it is not yet possible to do the type of analysis that Crown et al. did [59]. Likewise, there is not yet biochemical data on substrate preferences for the MEI-9 complex used in flies or for the MutLγ complex from fungi, plants, and mammals.

4. *DROSOPHILA*: THE NEXT 100 YEARS

A wealth of insight into both mitotic and meiotic recombination has been found using the unique traits of *Drosophila* as a model organism, yet much remains unknown. The field of *Drosophila* EJ is still in its infancy and the interplay of NHEJ, MMEJ, and alt-EJ—both the individual pathway regulation and the mechanisms by which they compensate for each other--is a field ripe for discovery. Likewise, mechanisms of intermediate steps in HRR are imperfectly understood. While SDSA is the dominant mechanism of DSB repair in mitotically dividing cells, it is still unclear how complementarity is found and annealing facilitated during the final stages. In the absence of complementarity, how is the choice made to reinvade versus capture the second end to form a dHJ? Are there mitotic dHJ agonists that prevent D-loop dissociation or is second end capture a stochastic event? Is the choice dependent on physical restraints within the highly ordered nucleus?

With regard to meiotic recombination, *Drosophila* is both intriguingly different and astonishingly similar to other model systems. Certain events must occur for successful meiosis in all sexually reproducing organisms: CO formation, homolog bias, and controlled joint molecule resolution; nonetheless, the mechanisms by which those events occur can vary widely between organisms, both through temporal alterations and divergence or outright replacement of meiosis-specific proteins. A benefit to this is that each system can enhance our overall understanding of the universal mechanisms governing meiotic recombination; still, many uncertainties remain. One major unanswered question is: *How are COs regulated?*

The development of sophisticated tools such as the CRISPR/Cas9 system and ultra-resolution microscopy, combined with further engineering of established assays and the versatile fly genome, may provide fine-tuned tools with which to ask these nuanced questions. Through continued study in *Drosophila*, we have the opportunity to examine universal properties of mitotic and meiotic recombination that affect all complex organisms.

GLOSSARY

Alternative end joining End joining that is independent of Lig4 and does not rely on microhomology.
Autophosphorylation The ability of a kinase to phosphorylate a residue within itself.
Autosome A chromosome that is not a sex chromosome.
***cis*-hDNA** A region of hDNA that contains all of the markers from the donor on one strand of the recombination product.
Concatenate DNA that is linked together and must be nicked or cut to disentangle.
Crossover Reciprocal exchange of genetic material between chromosomes.
Decatenate To disentangle concatenated or linked DNA.
Disjunction The proper segregation of homologous chromosomes.
D-loop A three-stranded DNA structure formed when a 3′ single strand of DNA invades a duplex template, displacing one strand.
Donor strand A single-strand of DNA involved in recombination that is used as a template during synthesis.
Double Holliday junction A recombination intermediate containing two Holliday junctions.
Double-strand break A break in both strands of a duplex DNA molecule.

End joining Generic term for ligation of the ends of a double-strand break.
Endogenous Originating from within.
Endonuclease An enzyme that cuts DNA between two bases.
Euchromatin Decompacted chromatin that often contains actively transcribed genes.
Exogenous Originating from the exterior environment.
Gene conversion Change of DNA sequence on one chromosome to the donor sequence (typically the same locus on the homologous chromosome).
Germ cells Egg and sperm cells.
Heterochromatin Densely compacted chromatin that contains silenced genes and repetitive sequences.
Heteroduplex DNA A region of double-strand DNA where each strand of the duplex is derived from a different parental chromosome and originates from recombination.
Holliday junction A four-stranded DNA structure in which the strands swap pairing partners.
Homologous chromosomes A pair of chromosomes that contains a maternal chromosome and a paternal chromosome.
Homologous recombination repair The process of DSB repair that uses an intact duplex DNA template to restore genetic information to the broken chromosome.
Hotspot A locus with a high frequency of meiotic recombination.
Hybrid dysgenesis Mating between strains that produces unidirectional lethality; in the context of P-elements, progeny from wild males and laboratory females are inviable or sterile, whereas the reciprocal cross yields viable progeny.
Meiosis Specialized type of reductive cell division.
Microhomology-mediated end joining End joining of a DSB that is Lig4 independent and relies on small homologies (4–8 nt) of the broken ends for ligation and repair.
Mitosis Nonreductive cell division.
Noncrossover Nonreciprocal exchange of genetic material between chromosomes; most noncrossovers are detected as gene conversion.
Nondisjunction Missegregation of homologous chromosomes.
Nonhomologous end joining Canonical end joining that is Lig4 dependent.
P-element A DNA (cut-and-paste) transposable element in *Drosophila* that requires a source of transposase for excision.
Progeny Descendants from a mating.
Recombination The rearrangement of genetic material due to DNA repair.
Resection Enzymatic activity that removes bases from one strand of duplex DNA in a 5′ to 3′ direction to yield 3′ ssDNA tails.
Resolvase An enzyme that cuts Holliday junctions or similar recombination intermediates.
Single-strand annealing A form of DSB repair that utilizes direct annealing of complementary resected ends without template invasion or synthesis.
Somatic cells Nonreproductive cells.
Synthesis-dependent strand annealing A type of DSB repair that utilizes a template and synthesis but does not utilize a double Holliday junction intermediate.
Transposable element A piece of DNA with the capacity to excise and/or integrate into the genome of its host; can be autonomous or require enzymatic activity from a different locus for mobility.
***trans*-hDNA** A region of hDNA that contains the markers from the donor on both strands of the recombination product.

LIST OF ACRONYMS AND ABBREVIATIONS

alt-EJ Alternative end joining
CO Crossover
dHJ Double Holliday junction
DSB Double-strand break
dsDNA Double-stranded DNA
EJ End joining
hDNA Heteroduplex DNA
HRR Homologous recombination repair
kb Kilobase
Mb Megabase
MMEJ Microhomology-mediated end joining
MMR Mismatch repair
NCO Non-crossover
NHEJ Nonhomologous end joining
nt Nucleotide
SC Synaptonemal complex
SDSA Synthesis-dependent strand annealing
SSA Single-strand annealing
ssDNA Single-stranded DNA

REFERENCES

[1] Ashburner M, Golic KG, Hawley RS. Drosophila: a laboratory handbook. 2nd ed. NY: Cold Spring Harbor Laboratory; 2005.

[2] Gilbert SF. Chromosomal sex determination in *Drosophila*. In: Developmental biology. 6th ed. Sunderland, MA: Sinauer Associates; 2000.

[3] Aldrich JC, Maggert KA. Simple quantitative PCR approach to reveal naturally occurring and mutation-induced repetitive sequence variation on the *Drosophila* Y chromosome. PLoS One 2014;9(10):e109906.

[4] Smith CD, Shu S, Mungall CJ, Karpen GH. The Release 5.1 annotation of *Drosophila melanogaster* heterochromatin. Science June 15, 2007;316(5831):1586–91.

[5] dos Santos G, Schroeder AJ, Goodman JL, Strelets VB, Crosby MA, Thurmond J, et al. FlyBase: introduction of the *Drosophila melanogaster* Release 6 reference genome assembly and large-scale migration of genome annotations. Nucleic Acids Res January 2015;43:D690–7.

[6] Wesolowska N, Amariei FL, Rong YS. Clustering and protein dynamics of *Drosophila melanogaster* telomeres. Genetics October 2013;195(2):381–91.

[7] Henikoff S. Nuclear organization and gene expression: homologous pairing and long-range interactions. Curr Opin Cell Biol June 1997;9(3):388–95.

[8] Fung JC, Marshall WF, Dernburg A, Agard DA, Sedat JW. Homologous chromosome pairing in *Drosophila melanogaster* proceeds through multiple independent initiations. J Cell Biol April 6, 1998;141(1):5–20.

[9] Kirilly D, Xie T. The *Drosophila* ovary: an active stem cell community. Cell Res January 2007;17(1):15–25.

[10] Hawley RS. Meiosis: how male flies do meiosis. Curr Biol October 1, 2002;12(19):R660–2.

[11] Stern C. Somatic crossing over and segregation in *Drosophila melanogaster*. Genetics November 1936;21(6):625–730.

[12] Patterson JT. Proof that the entire chromosome is not eliminated in the production of somatic variations by X-Rays in *Drosophila*. Genetics March 1930;15(2):141–9.

[13] LaFave MC, Andersen SL, Stoffregen EP, Holsclaw JK, Kohl KP, Overton LJ, et al. Sources and structures of mitotic crossovers that arise when BLM helicase is absent in *Drosophila*. Genetics January 2014;196(1):107–18.

[14] Hiraizumi Y. Spontaneous recombination in *Drosophila melanogaster* males. Proc Natl Acad Sci USA February 1971;68(2):268–70.

[15] Kidwell MG, Kidwell JF. Selection for male recombination in *Drosophila melanogaster*. Genetics October 1976;84(2):333–51.

[16] Rio DC, Laski FA, Rubin GM. Identification and immunochemical analysis of biologically active *Drosophila* P element transposase. Cell January 17, 1986;44(1):21–32.

[17] Misra S, Rio DC. Cytotype control of *Drosophila* P element transposition: the 66 kd protein is a repressor of transposase activity. Cell July 27, 1990;62(2):269–84.

[18] Bingham PM, Kidwell MG, Rubin GM. The molecular basis of P-M hybrid dysgenesis: the role of the P element, a P-strain-specific transposon family. Cell July 1982;29(3):995–1004.

[19] Sved JA, Eggleston WB, Engels WR. Germ-line and somatic recombination induced by in vitro modified P elements in *Drosophila melanogaster*. Genetics February 1990;124(2):331–7.

[20] Engels WR, Benz WK, Preston CR, Graham PL, Phillis RW, Robertson HM. Somatic effects of P element activity in *Drosophila melanogaster*: pupal lethality. Genetics December 1987;117(4):745–57.

[21] Gloor GB, Nassif NA, Johnson-Schlitz DM, Preston CR, Engels WR. Targeted gene replacement in *Drosophila* via P element- induced gap repair. Science September 6, 1991;253(5024):1110–7.

[22] Mimitou EP, Symington LS. DNA end resection: many nucleases make light work. DNA Repair September 2, 2009;8(9):983–95.

[23] Daley JM, Niu H, Miller AS, Sung P. Biochemical mechanism of DSB end resection and its regulation. DNA Repair August 2015;32:66–74.

[24] Madigan JP, Chotkowski HL, Glaser RL. DNA double-strand break-induced phosphorylation of *Drosophila* histone variant H2Av helps prevent radiation-induced apoptosis. Nucleic Acids Res September 1, 2002;30(17):3698–705.

[25] Preston CR, Flores C, Engels WR. Age-dependent usage of double-strand-break repair pathways. Curr Biol October 24, 2006;16(20):2009–15.

[26] Dekanty A, Barrio L, Milan M. Contributions of DNA repair, cell cycle checkpoints and cell death to suppressing the DNA damage-induced tumorigenic behavior of *Drosophila* epithelial cells. Oncogene February 19, 2015;34(8):978–85.

[27] Chiolo I, Minoda A, Colmenares SU, Polyzos A, Costes SV, Karpen GH. Double-strand breaks in heterochromatin move outside of a dynamic HP1a domain to complete recombinational repair. Cell March 4, 2011;144(5):732–44.

[28] Heyer WD, Ehmsen KT, Liu J. Regulation of homologous recombination in eukaryotes. Annu Rev Genet 2010;44:113–39.

[29] Engels WR, Johnson-Schlitz DM, Eggleston WB, Sved J. High-frequency P element loss in *Drosophila* is homolog dependent. Cell August 10, 1990;62(3):515–25.

[30] Johnson-Schlitz DM, Engels WR. P-element-induced interallelic gene conversion of insertions and deletions in *Drosophila melanogaster*. Mol Cell Biol November 1993;13(11):7006–18.

[31] Nassif N, Engels W. DNA homology requirements for mitotic gap repair in *Drosophila*. Proc Natl Acad Sci USA February 15, 1993;90(4):1262–6.

[32] Bzymek M, Thayer NH, Oh SD, Kleckner N, Hunter N. Double Holliday junctions are intermediates of DNA break repair. Nature April 8, 2010;464(7290):937–41.

[33] Nassif N, Penney J, Pal S, Engels WR, Gloor GB. Efficient copying of nonhomologous sequences from ectopic sites via P-element-induced gap repair. Mol Cell Biol March 1994;14(3):1613–25.

[34] Sugawara N, Haber JE. Characterization of double-strand break-induced recombination: homology requirements and single-stranded DNA formation. Mol Cell Biol February 1992;12(2):563–75.

[35] Pâques F, Leung WY, Haber JE. Expansions and contractions in a tandem repeat induced by double-strand break repair. Mol Cell Biol April 1998;18(4):2045–54.

[36] McVey M, Andersen SL, Broze Y, Sekelsky J. Multiple functions of *Drosophila* BLM helicase in maintenance of genome stability. Genetics 2007;176:1979–92.

[37] McVey M, Adams MD, Staeva-Vieira E, Sekelsky JJ. Evidence for multiple cycles of strand invasion during repair of double-strand gaps in *Drosophila*. Genetics June 2004;167(2):699–705.

[38] McVey M, Larocque JR, Adams MD, Sekelsky JJ. Formation of deletions during double-strand break repair in *Drosophila* DmBlm mutants occurs after strand invasion. Proc Natl Acad Sci USA November 2, 2004;101(44):15694–9.

[39] Williams GJ, Hammel M, Radhakrishnan SK, Ramsden D, Lees-Miller SP, Tainer JA. Structural insights into NHEJ: building up an integrated picture of the dynamic DSB repair super complex, one component and interaction at a time. DNA Repair May 2014;17:110–20.

[40] McVey M, Radut D, Sekelsky J. End-joining repair of double-strand breaks in *Drosophila* is largely DNA ligase IV-independent. Genetics 2004;168:2067–76.

[41] Adams MD, McVey M, Sekelsky J. *Drosophila* BLM in double-strand break repair by synthesis-dependent strand annealing. Science January 10, 2003;299(5604):265–7.

[42] Bozas A, Beumer KJ, Trautman JK, Carroll D. Genetic analysis of zinc-finger nuclease-induced gene targeting in *Drosophila*. Genetics July 2009;182(3):641–51.

[43] Chan SH, Yu AM, McVey M. Dual roles for DNA polymerase theta in alternative end-joining repair of double-strand breaks in *Drosophila*. PLoS Genet July 2010;6(7):e1001005.

[44] Yu AM, McVey M. Synthesis-dependent microhomology-mediated end joining accounts for multiple types of repair junctions. Nucleic Acids Res September 2010;38(17):5706–17.

[45] Ceccaldi R, Liu JC, Amunugama R, Hajdu I, Primack B, Petalcorin MI, et al. Homologous-recombination-deficient tumours are dependent on Polθ-mediated repair. Nature February 12, 2015;518(7538):258–62.

[46] van Schendel R, Roerink SF, Portegijs V, van den Heuvel S, Tijsterman M. Polymerase θ is a key driver of genome evolution and of CRISPR/Cas9-mediated mutagenesis. Nat Commun 2015;6:7394.

[47] Mateos-Gomez PA, Gong F, Nair N, Miller KM, Lazzerini-Denchi E, Sfeir A. Mammalian polymerase theta promotes alternative NHEJ and suppresses recombination. Nature February 12, 2015;518(7538):254–7.

[48] Wu L, Bachrati CZ, Ou J, Xu C, Yin J, Chang M, et al. BLAP75/RMI1 promotes the BLM-dependent dissolution of homologous recombination intermediates. Proc Natl Acad Sci USA March 14, 2006;103(11):4068–73.

[49] Plank JL, Wu J, Hsieh TS. Topoisomerase IIIα and Bloom's helicase can resolve a mobile double Holliday junction substrate through convergent branch migration. Proc Natl Acad Sci USA July 25, 2006;103(30):11118–23.

[50] Chen SH, Wu CH, Plank JL, Hsieh TS. Essential functions of C terminus of *Drosophila* Topoisomerase IIIα in double holliday junction dissolution. J Biol Chem June 1, 2012;287(23):19346–53.

[51] Andersen SL, Kuo HK, Savukoski D, Brodsky MH, Sekelsky J. Three structure-selective endonucleases are essential in the absence of BLM helicase in *Drosophila*. PLoS Genet October 2011;7(10):e1002315.

[52] Morgan TH. The method of inheritance of two sex-limited characters in the same animal. Proc Soc Exp Biol Med 1910;8:17–9.

[53] Morgan TH. Random segregation versus coupling in Mendelian inheritance. Science September 22, 1911;34(873):384.

[54] Sturtevant AH. The linear arrangement of six sex-linked factors in *Drosophila*, as shown by their mode of association. J Exp Zool January 1913;14(1):43–59.

[55] Dobzhansky T. Translocations involving the third and the fourth chromosomes of *Drosophila melanogaster*. Genetics July 1930;15(4):347–99.

[56] Beadle GW. A possible influence of the spindle fibre on crossing-over in *Drosophila*. Proc Natl Acad Sci USA February 1932;18(2):160–5.

[57] Sandler L, Lindsley DL, Nicoletti B, Trippa G. Mutants affecting meiosis in natural populations of *Drosophila melanogaster*. Genetics 1968;60:525–58.

[58] Baker BS, Carpenter ATC. Genetic analysis of sex chromosomal meiotic mutants in *Drosophila melanogaster*. Genetics 1972;71(2):255–86.

[59] Crown KN, McMahan S, Sekelsky J. Eliminating both canonical and short-patch mismatch repair in *Drosophila melanogaster* suggests a new meiotic recombination model. PLoS Genet September 2014;10(9):e1004583.

[60] Kohl KP, Sekelsky J. Meiotic and mitotic recombination in meiosis. Genetics June 2013;194(2):327–34.

[61] Marcon E, Moens PB. The evolution of meiosis: recruitment and modification of somatic DNA-repair proteins. Bioessays August 2005;27(8):795–808.

[62] Villeneuve AM, Hillers KJ. Whence meiosis? Cell September 21, 2001;106(6):647–50.

[63] Wilkins AS, Holliday R. The evolution of meiosis from mitosis. Genetics January 2009;181(1):3–12.

[64] Nicklas RB. Chromosome segregation mechanisms. Genetics September 1974;78(1):205–13.

[65] Pâques F, Haber JE. Multiple pathways of recombination induced by double-strand breaks in *Saccharomyces cerevisiae*. Microbiol Mol Biol Rev June 1999;63(2):349–404.

[66] Keeney S, Giroux CN, Kleckner N. Meiosis-specific DNA double-strand breaks are catalyzed by Spo11, a member of a widely conserved protein family. Cell 1997;88(3):375–84.

[67] Keeney S. Spo11 and the formation of double-strand breaks in meiosis. Mol Biol 2007;2:81–123.

[68] Roeder GS. Meiotic chromosomes: it takes two to tango. Genes Dev October 15, 1997;11(20):2600–21.

[69] Romanienko PJ, Camerini-Otero RD. The mouse Spo11 gene is required for meiotic chromosome synapsis. Mol Cell November 2000;6(5):975–87.

[70] Koehler KE, Boulton CL, Collins HE, French RL, Herman KC, Lacefield SM, et al. Spontaneous X chromosome MI and MII nondisjunction events in *Drosophila melanogaster* oocytes have different recombinational histories. Nat Genet 1996;14(4):406–14.

[71] Brown MS, Bishop DK. DNA strand exchange and RecA homologs in meiosis. Cold Spring Harb Perspect Biol January 2015;7(1):a016659.

[72] Schwacha A, Kleckner N. Interhomolog bias during meiotic recombination: meiotic functions promote a highly differentiated interhomolog-only pathway. Cell September 19, 1997;90(6):1123–35.

[73] Blanton H, Sekelsky J. Unique invasions and resolutions: DNA repair proteins in meiotic recombination in *Drosophila melanogaster*. Cytogenet Genome Res 2004;107(3–4):172–9.

[74] Page SL, Hawley RS. The genetics and molecular biology of the synaptonemal complex. Annu Rev Cell Dev Biol 2004;20:525–58.

[75] Jang JK, Sherizen DE, Bhagat R, Manheim EA, McKim KS. Relationship of DNA double-strand breaks to synapsis in *Drosophila*. J Cell Sci August 1, 2003;116(Pt 15):3069–77.

[76] Dernburg AF, McDonald K, Moulder G, Barstead R, Dresser M, Villeneuve AM. Meiotic recombination in *C. elegans* initiates by a conserved mechanism and is dispensable for homologous chromosome synapsis. Cell 1998;94:387–98.

[77] Webber HA, Howard L, Bickel SE. The cohesion protein ORD is required for homologue bias during meiotic recombination. J Cell Biol March 15, 2004;164(6):819–29.

[78] Allers T, Lichten M. Intermediates of yeast meiotic recombination contain heteroduplex DNA. Mol Cell July 2001;8(1):225–31.

[79] Jessop L, Rockmill B, Roeder GS, Lichten M. Meiotic chromosome synapsis-promoting proteins antagonize the anti-crossover activity of Sgs1. PLoS Genet September 22, 2006;2(9):e155.

[80] Zalevsky J, MacQueen AJ, Duffy JB, Kemphues KJ, Villeneuve AM. Crossing over during *Caenorhabditis elegans* meiosis requires a conserved MutS-based pathway that is partially dispensable in budding yeast. Genetics 1999;153(3):1271–83.

[81] Hollingsworth NM, Ponte L, Halsey C. *MSH5*, a novel MutS homolog, facilitates meiotic reciprocal recombination between homologs in *Saccharomyces cerevisiae* but not mismatch repair. Genes Dev 1995;9:1728–39.

[82] Ross-Macdonald P, Roeder GS. Mutation of a meiosis-specific MutS homolog decreases crossing over but not mismatch correction. Cell December 16, 1994;79(6):1069–80.

[83] Bocker T, Barusevicius A, Snowden T, Rasio D, Guerrette S, Robbins D, et al. hMSH5: a human MutS homologue that forms a novel heterodimer with hMSH4 and is expressed during spermatogenesis. Cancer Res 1999;59(4):816–22.

[84] Snowden T, Acharya S, Butz C, Berardini M, Fishel R. hMSH4-hMSH5 recognizes Holliday Junctions and forms a meiosis-specific sliding clamp that embraces homologous chromosomes. Mol Cell August 13, 2004;15(3):437–51.

[85] Kohl KP, Jones CD, Sekelsky J. Evolution of an MCM complex in flies that promotes meiotic crossovers by blocking BLM helicase. Science December 7, 2012;338(6112):1363–5.

[86] Liu H, Jang JK, Graham J, Nycz K, McKim KS. Two genes required for meiotic recombination in *Drosophila* are expressed from a dicistronic message. Genetics 2000;154(4):1735–46.

[87] Carpenter ATC. Mismatch repair, gene conversion, and crossing-over in two recombination-defective mutants of *Drosophila melanogaster*. Proc Natl Acad Sci USA 1982;79:5961–5.

[88] Bhagat R, Manheim EA, Sherizen DE, McKim KS. Studies on crossover-specific mutants and the distribution of crossing over in *Drosophila* females. Cytogenet Genome Res 2004;107(3–4):160–71.

[89] Grell RF. Time of recombination in the *Drosophila melanogaster* oocyte. III. Selection and characterization of temperature-sensitive and -insensitive, recombination-defecient alleles in *Drosophila*. Genetics October 1, 1984;108(2):425–43.

[90] Blanton HL, Radford SJ, McMahan S, Kearney HM, Ibrahim JG, Sekelsky J. REC, *Drosophila* MCM8, drives formation of meiotic crossovers. PLoS Genet September 2005;1(3):e40.

[91] Kolas NK, Cohen PE. Novel and diverse functions of the DNA mismatch repair family in mammalian meiosis and recombination. Cytogenet Genome Res 2004;107(3–4):216–31.

[92] Zakharyevich K, Tang S, Ma Y, Hunter N. Delineation of joint molecule resolution pathways in meiosis identifies a crossover-specific resolvase. Cell April 13, 2012;149(2):334–47.

[93] Baudat F, Imai Y, de Massy B. Meiotic recombination in mammals: localization and regulation. Nat Rev Genet November 2013;14(11):794–806.

[94] Sekelsky JJ, McKim KS, Chin GM, Hawley RS. The *Drosophila* meiotic recombination gene *mei-9* encodes a homologue of the yeast excision repair protein Rad1. Genetics October 1995;141(2):619–27.

[95] Yıldız Ö, Majumder S, Kramer BC, Sekelsky J. Drosophila MUS312 interacts with the nucleotide excision repair endonuclease MEI-9 to generate meiotic crossovers. Mol Cell 2002;10:1503–9.

[96] Green MM. *mus(3)312*^{D1}, a mutagen sensitive mutant with profound effects on female meiosis in *Drosophila melanogaster*. Chromosoma 1981;82:259–66.

[97] Boyd JB, Golino MD, Setlow RB. The *mei-9*^{a} mutant of *Drosophila melanogaster* increases mutagen sensitivity and decreases excision repair. Genetics 1976;84:527–44.

[98] Sekelsky JJ, Hollis KJ, Eimerl AI, Burtis KC, Hawley RS. Nucleotide excision repair endonuclease genes in *Drosophila melanogaster*. Mutat Res April 28, 2000;459(3):219–28.

[99] Radford SJ, Goley E, Baxter K, McMahan S, Sekelsky J. Drosophila ERCC1 is required for a subset of MEI-9-dependent meiotic crossovers. Genetics 2005;170:1737–45.

[100] Joyce EF, Tanneti SN, McKim KS. *Drosophila* hold'em is required for a subset of meiotic crossovers and interacts with the DNA repair endonuclease complex subunits MEI-9 and ERCC1. Genetics January 2009;181(1):335–40.

[101] Hunter N, Kleckner N. The single-end invasion: an asymmetric intermediate at the double-strand break to double-holliday junction transition of meiotic recombination. Cell July 13, 2001;106(1):59–70.

[102] Schwacha A, Kleckner N. Identification of double Holliday junctions as intermediates in meiotic recombination. Cell 1995;83:783–91.

[103] Radford SJ, Sabourin MM, McMahan S, Sekelsky J. Meiotic recombination in *Drosophila Msh6* mutants yields discontinuous gene conversion tracts. Genetics May 2007;176(1):53–62.

[104] Hoskins RA, Smith CD, Carlson JW, Carvalho AB, Halpern A, Kaminker JS, et al. Heterochromatic sequences in a *Drosophila* whole-genome shotgun assembly. Genome Biol 2002;3(12):RESEARCH0085.

Chapter 10

Genome Stability in *Drosophila*: Mismatch Repair and Genome Stability

T. Negishi
Okayama University, Okayama, Japan

Chapter Outline

1. INTRODUCTION

Living organisms are subjected to a variety of endogenous and exogenous damages. Since DNA alterations caused by such factors are highly risky not only for individuals, but also for the continuity of species, many protective systems such as DNA repair are present to counter the effects of each damaging factor. Among these systems, mismatch repair (MMR) has been recognized to play a very important role in the preservation of genome stability. Historically, MMR was proposed to play a role in the accurate processing of genetic recombination during meiosis [1]. The importance of MMR was highlighted by the observation that MMR systems are well conserved from prokaryotes to higher eukaryotes [2,3]. To affect the repair of damaged DNA, MMR is thought to engage in crosstalk with other repair systems [4,5], and it has also been proposed that signaling cascades leading to cell cycle arrest and the induction of apoptosis might be regulated by MMR pathways depending on the damaging factors [6]. *Drosophila* also possesses a similar MMR system to *Escherichia coli* and mammals. MMR activity that is responsible for the repair of heteroduplex DNA containing mismatched base pairs has been demonstrated in extracts from cultured cells, embryos and adult flies [7]. The proteins and genes involved in MMR have been identified, and evidence that MMR systems play an important role in maintaining genome stability during both mitotic replication and meiotic recombination has accumulated.

2. MMR ACTIVITY IN *DROSOPHILA*

Holmes et al. demonstrated the strand-specific mismatch correction activity in *Drosophila Kc* cells [8]. They performed an in vitro MMR assay using nuclear extracts from cultured *Drosophila Kc* cells and human fibroblast HeLa cells. A similar MMR activity was detected in both *Kc* and HeLa cell nuclei, and the repair was almost limited to nicked strands of heteroduplex DNA containing mismatched base pairs. Mispairs were repaired with efficiencies in the order G·T>G·G≃A·C>C·C. On the other hand, Bhuki-Kaur et al. observed that MMR activity was higher in *Drosophila* tissue extracts than in HeLa cells, and that MMR activity was expressed continuously throughout the *Drosophila* life span, from the embryo to the adult fly [7]. They prepared cell extracts from wild-type Oregon-R embryos (after 0–18 h oviposition), young adults (4–5 days after eclosion), and aged senescent adults (35 days after eclosion) and measured the repair activity of each extract according to the methods described by Thomas et al. [9]. Heteroduplex DNA containing mismatched base pairs, and 1 and 5 bp loops were prepared from a replicative form of bacteriophage M13mp2. Following the incubation of the heteroduplex phage DNA with *Drosophila* tissue extracts, heteroduplex phage DNA was transfected into competent *E. coli* in an effort

Genome Stability. http://dx.doi.org/10.1016/B978-0-12-803309-8.00010-0

to identify which strand was repaired as determined by the observation of plaque phenotypes. It was discovered that MMR activity was constantly present, from the embryo to the adult fly, and it was at higher levels than in Hela cells. T·G and G·G mispairs were efficiently repaired in a nick-dependent manner consistent with the findings of a previous report [7], whereas the repair of A·A, C·C, C·T, T·T, C·A, G·A, and A·G mispairs and both loops was not nick dependent. The A·A mismatch was the most efficiently repaired, and the efficiency of repair was in the order as described earlier. *Drosophila* appears to require MMR activity throughout its life span, although the reasons are unclear. Bhui-Kaur et al. also observed that the nick-dependent repair was reduced in the extract of the *Drosophila mei-9* mutant which is defective in nucleotide excision repair (NER) in somatic cells [10] and defective in crossover during meiotic recombination [11]. These results suggested that MMR collaborates with other repair systems to complete an accurate repair in an effort to maintain genome stability and introduce the possibility that the Mei-9 protein acts as an endonuclease to incise the strand possessing misincorporated bases or loops, whereas the *mei-9* gene has been shown to encode the human XPF homolog-protein [10].

3. MMR GENES IN *DROSOPHILA*

The genes encoding proteins involved in MMR have been identified in *Drosophila*. In many eukaryotes, two sets of MMR initiation complexes, MSH2–MSH6 and MSH2–MSH3, bind to DNA lesions whose binding property corresponds to the type of the mismatch; for example, the human MSH2–MSH6 complex can bind to the mismatch region, and the MSH2–MSH3 complex binds to loops, but not to mismatched base pairs [2]. In *Drosophila*, the MSH2 ortholog is encoded by the *spellchecker1* gene referred to as *spel1* [12], and the MSH6 ortholog referred to as the *Msh6* gene was identified from the complete *Drosophila* genome sequence (reviewed in 13). The *spel1* gene is positioned at 35A4–35B1 on the left hand of the second chromosome [13], and the *Msh6* gene is at 71B6 on the left hand of the third chromosome [14]. However, in *Drosophila*, a homologous sequence of the gene encoding the MSH3 protein is absent [15]. Therefore, only the Spel1–MSH6 complex might be engaged in the recognition of mismatched heteroduplex DNA, including base–base mismatches, small loops, and possibly large loops. The presence of *E. coli* MutL orthologs is inferred from sequence homology where *Mlh1* and *Pms2* genes are positioned at 44B8 and 51F11, respectively, on the right hand of the second chromosome [16,17]. In *Drosophila*, a gene encoding the *E. coli* MutH homologous protein has not been identified as it has been found in other eukaryotes. However, it is estimated that two nucleases, *tos* and *mei-9* gene encoding products, might play important roles in MMR. The sequence analysis has revealed that the *tos* gene encodes a protein referred to as TOSCA which is highly related to the Exo1 protein and is a double-stranded DNA 5′–3′ exodeoxyribonuclease specifically induced in meiotic prophase I in *Schizosaccharomyces pombe*, and it is a member of the RAD2 protein family that plays a role in NER [18]. TOSCA is selectively expressed in *Drosophila* developing oocytes. Therefore, the *tos* gene may play an important role in the maintenance of genome stability by repairing mismatches that may occur during replication or recombination in oogenesis [19]. Mei-9, a product of the *mei-9* gene and an ortholog of mammalian XP-F, might act as a substitute for MutH and engage its incision activity during NER as mentioned earlier. It is considered that similar to TOSCA, Mei-9 plays a role in the repair of damaged DNA during both meiosis and mitosis.

4. MMR AND MICROSATELLITE INSTABILITY

MMR systems play important roles in maintaining the high fidelity of genomic DNA by the recognition and repair of mismatched base pairs during DNA replication [20]. It is well documented that a lack of MMR increases genomic instability and the risk of certain types of cancer such as hereditary non-polyposis colorectal cancer (HNPCC) [21,22]. Microsatellite instability (MSI), a typical genomic instability, caused, for example, by frameshift mutation leads to the mutation of various target genes and can lead to the development of cancer by the inactivation of responsible genes [23]. A deficiency in MMR leads to MSI manifested by the alteration of repeat lengths not only in mammals, but also in *Drosophila*. In the MutS-deficient mutant of *Drosophila* (*spel1−/−*) constructed by Flores and Engels, the rate of MSI in long runs of dinucleotide repeats increased [12]. They observed alterations in the repeat number of microsatellites after 10–12 fly generations in *spel1*-null offspring. From the results of seven loci of microsatellites, the length of dinuleotide microsatellite loci altered with a variation of 3.1–26.5%, but it was not scored in microsatellite loci comprising trinucleotide repeats. In wild-type *Drosophila*, the mutation rate of microsatellites is averaged as 6.3×10^{-6} with 24 loci, and is lower than in several mammalians in which the rates are estimated to be in the order of 10^{-3} to 10^{-5} [24–27]. This discrepancy is considered to be due to the shorter length of microsatellites in *Drosophila* compared to mammals. The highest mutation rate was shown in the longest microsatellite region (28 repeats of the CA dinucleotide), at a similar level to the mutation rate found in mammalians (3×10^{-4}) [25]. The frequency of microsatellite alteration depends on the repeat sequence and the expression of MMR [26]. The G·T repeat sequence was subjected to the highest alteration rate in the presence of MMR, while the alteration rate of the A·T repeat sequence was higher in the absence of MMR.

In conclusion, the lack of the MutS ortholog frequently increases the mutation rates of microsatellite loci even in *Drosophila*, which is consistent with previous reports pertaining to bacteria, yeast, and mammalians. Significant changes in microsatellite length also occurred during the repair of double-strand DNA breaks in the *spel1*-null mutant, where a greater than fivefold increase in the rate of repeat length changes was observed [12].

Numerous proteins other than MMR proteins involved in DNA repair should contribute to genome stability during somatic replication and trans-generation events. Velázquez and collaborators reported that in the *Drosophila* PCNA mutant (*mus209*) germline, genomic instability is induced through MSI at a lesser extent than that in the *spel1* mutant. The rate of MSI in *mus209* was higher in heterozygotes than in homozygotes with PCNA mutation [27,28]. On the other hand, the product of the *mus201* gene, a mammalian XPG ortholog essential for the excision repair of the global genome, is not associated with the MMR process [29].

5. THE ROLE OF MMR IN MEIOTIC RECOMBINATION

The functions of genes other than the *spel1* gene involved in MMR have yet to be delineated. Since 2000, it has been revealed that MMR proteins play an important role in an accurate crossover generated through meiotic recombination in yeast [30] and mice [31]. In *Drosophila*, Radford et al. reported the involvement of *Msh6* in meiotic recombination [32]. Crossovers between homologous chromosomes are indispensable for the accurate chromosome segregation during meiosis [33] and also in *Drosophila* [34]. In an effort to understand the processes involved in meiotic recombination including DNA double-strand breaks (DSBs), crossover and chromosome segregation during meiotic cell division, many studies have been performed using crossover-defective mutants. The undesirable postmeiotic segregation occurs when heteroduplex DNA formed during meiotic recombination is not repaired correctly at the first meiotic division. In the *Drosophila Msh6* mutant [32], the frequency of postmeiotic segregation is higher compared to the wild-type and *mei-9* mutant in which the frequency of crossovers is reduced due to the lack of nicking activity of Holliday junction formed during meiotic recombination [35]. Crown et al. proposed a new meiotic recombination model for *Drosophila*. When *Drosophila* is defective in canonical and short-patch MMR (the Msh6 mutant), the XPC homolog encoded by the *mus210* gene, a damage recognition factor in NER, is involved in the repair of mismatched heteroduplex DNA together with the Mei-9 protein [36]. MMR is speculated to repair not only mismatches during replication, but also heterogeneous DNA duplexes that result during meiotic recombination.

6. MMR AND SOMATIC CELL MUTATION

DNA repair by the MMR system is best investigated in *E. coli*, and the lack of MMR increases genomic instability by generating a mutator phenotype with the increased spontaneous and induced mutation rates, as previously mentioned and also observed in our investigations [37]. In mammals, MMR deficiency is responsible for an increased cancer risk and causes HNPCC that accompanies genome instability in humans [22,23]. Although it is widely recognized that cancers can develop by the gradual accumulation of somatic cell mutations, it remains to be revealed whether MMR deficiency can affect the frequency of somatic cell mutations, including chromosomal alterations. Flores and Engels cloned a *mutS* ortholog gene from *Drosophila* referred to as *spellchecker1* (*spel1*) and constructed two lines that possess a deletion of DNA tract including the *spel1* gene at different regions [12]. When they examined the sensitivity of *spel1*-null mutant flies to methyl methanesulfonate (a methylating agent) or γ-irradiation, the *spel1*-null mutant was insensitive to such genotoxic factors, although the mutant exhibited a significant MSI without treatment with damaging factors [12]. Williams et al. in 2011 revealed that the *spel1*-null mutant was hypermutable to diepoxybutane (a crosslinking agent) [38]; in this assay, the number of tumors caused by mutation of the tumor-suppressor gene (*lats*) served as the mutation frequency. They also demonstrated that a product of the Fanconi anemia (FA)-related gene in *Drosophila* plays an important role in the repair of DNA crosslinks, although the function of the FA-related gene and Spel1 gene products appeared to be epistatic [38].

There have been no systems presently at hand to investigate somatic cell mutations directly in MMR-deficient *Drosophila*. To examine the involvement of MMR in somatic cell mutations, we have generated a new *Drosophila* strain in which the *spel1* gene is heterozygotically deleted and *mwh* genes are homozygotically mutated. We have developed a mutation assay referred to as the somatic mutation and recombination test (SMART) [39], using the newly generated flies and wild-type flies. In SMART, the recessive *mwh* gene imparts a multiple wing hair phenotype on wings when chromosomal recombination, chromosomal non-disjunction, and gene mutations are induced during somatic cell division. We examined whether genomic instability was induced in the MMR-deficient *spel1*-null flies (*spel1*−/−) generated from a cross between a newly generated strain and another existing heterozygotic *spel1* mutant according to Flores and Engels [12]. Several microsatellite sequences were analyzed by PCR using each specific primer. The results showed that even after the fifth generation, microsatellite sequences were more frequently altered in MMR-deficient flies (*spel1*−/−) than in MMR-proficient flies (*spel1*+/−), as shown in Fig. 10.1 (Miyamoto: unpublished data).

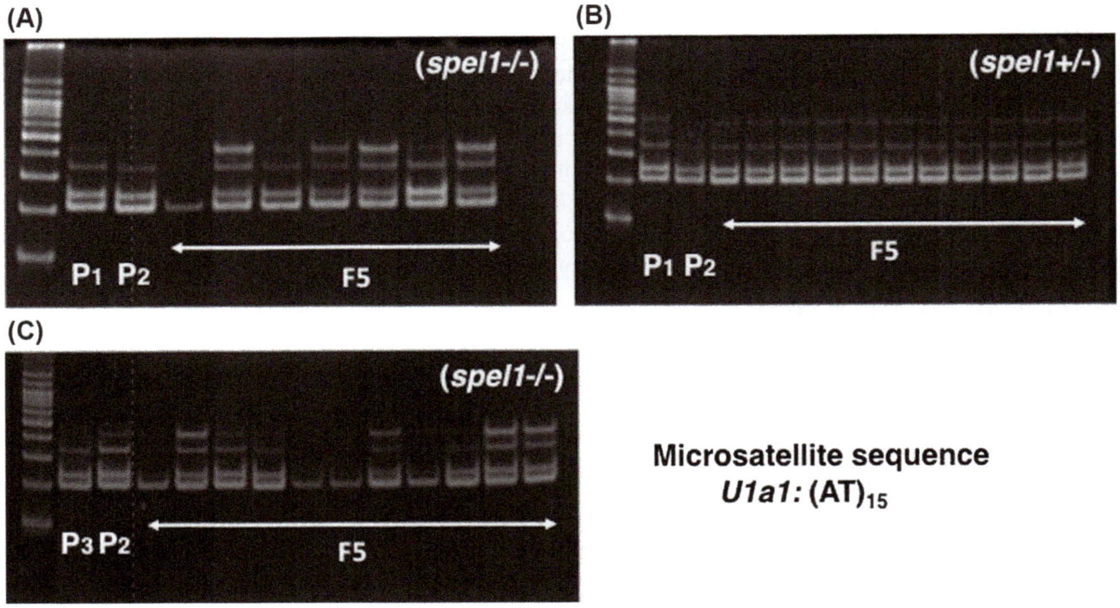

FIGURE 10.1 **Alterations in microsatellite repeats detected by PCR using primers for the *U1a1* microsatellite sequence.** (A) The fifth-generation flies (spel1−/−) from a cross between each existing heterozygote (*spel1*+/−) (P1 and P2). (B) Flies (*spel1*+/−) from a cross between each existing heterozygote (*spel1*+/−) (P1 and P2). (C) The fifth-generation flies (*spel1*−/−) from a cross between the existing heterozygote (*spel1*+/−) (P2) and the newly generated heterozygote (P3).

X-ray irradiation induces DNA DSBs and oxidative damage resulting in somatic cell mutations. We observed that chromosomal recombination was accounted for over 70% of mutations induced by X-ray irradiation in *Drosophila* (Toyoshima-Sasatani, unpublished data) in agreement with the previous report in which chromosomal recombination was mainly accounted for the mutation detected in SMART [39]. When we examined mutations induced by X-ray irradiation, the mutagenicity of X-rays was unexpectedly found to be lower in MMR-deficient flies than in MMR-proficient flies, as shown in Table 10.1 (Miyamoto, unpublished data).

A lack of MMR is presently thought to induce a mutator phenotype. However, *Drosophila* appears to lose its mutator phenotype since the spontaneous mutation rate remains unchanged in *spel1*-null and *spel1*+/− flies. During somatic cell division of *Drosophila*, the pairing of each homologous chromosome occurs, and daughter cells obtain the assortment of paternal and maternal chromosomes. When DNA is subjected to the damaging factors such as X-rays, chromosomal recombination occurs between maternal and paternal chromosomes paired at damaged sites or in the neighborhood through DNA strand breaks [39]. The mutagenicity in this assay is assessed according to the extent of chromosomal recombination, and the mutagenicity decreases if recombination is blocked. Our results suggested that MMR is also required for homologous recombination through strand breaks induced by DNA damage.

Alkylated DNA bases are well-known lesions that can induce mutations followed by carcinogenesis. MMR is involved in the repair of base pairs consisting of alkylated and normal bases, following the recognition of the mismatched base pair by MutS or MutS homolog proteins. The efficiency of recognition appears to be dependent on the alkyl group [40]. When we examined the mutation rate of *N*-nitrosodimethylamine (NDMA) and *N*-nitrosodiethylamine (NDEA) using the *Drosophila* wing spot test (SMART), NDMA was found to be more recombinogenic than NDEA [41]. When we performed mutation assays using the newly generated flies, the mutagenicity of NDMA was found to be significantly lower in the MMR-deficient flies (*spel1*−/−) than in the MMR-proficient flies (*spel1*+/−). The converse was observed in the case of NDEA. These results suggested that the MutS protein recognizes DNA methylated lesions more frequently than ethylated lesions even in *Drosophila*, and that the MutS homolog protein functions to induce chromosomal recombination following DNA strand breaks and gene mutations in *Drosophila*. These findings were unexpected and contrast the results of the *E. coli* mutation assays where the mutation rate of alkylating agents was markedly higher compared to wild type, as shown by many investigations including those performed in our laboratory [37]. Zhang et al. examined human fibroblast mutants and suggested that the MNNG-induced homologous recombination requires functional MMR [42]. In their experiments, the MNNG-induced recombination decreased, although the MNNG-induced gene mutations at the *hprt* gene were elevated. The elevation in gene mutations can be accounted for by considering the canonical function of MMR in which the methylated guanine or thymine residues are targets of repair, and MMR deficiency results in the absence of mismatched base-pair

TABLE 10.1 Mutagenicity of X-ray Irradiation in MMR-deficient and -Proficient *Drosophila* as Determined by the Wing Spot Test

	spel1−/−		*spel1+/−*	
X-ray Dose (Gy)	Survival (%)[a]	Mutagenicity[b]	Survival (%)[a]	Mutagenicity[b]
0	100	0.25	100	0.24
5	107	1.96[c]	96	2.67
10	130	3.58[c]	89	5.76
15	117	3.31[c]	92	8.82
20	92	5.31[c]	95	9.16

[a]*Survival (%) = the number of flies from nontreated larvae/the number of flies from irradiated larvae × 100.*
[b]*Mutagenicity is represented by the number of mwh mutant cell colonies per wing.*
[c]*P < .01, a significant difference from the corresponding spel1+/− flies.*

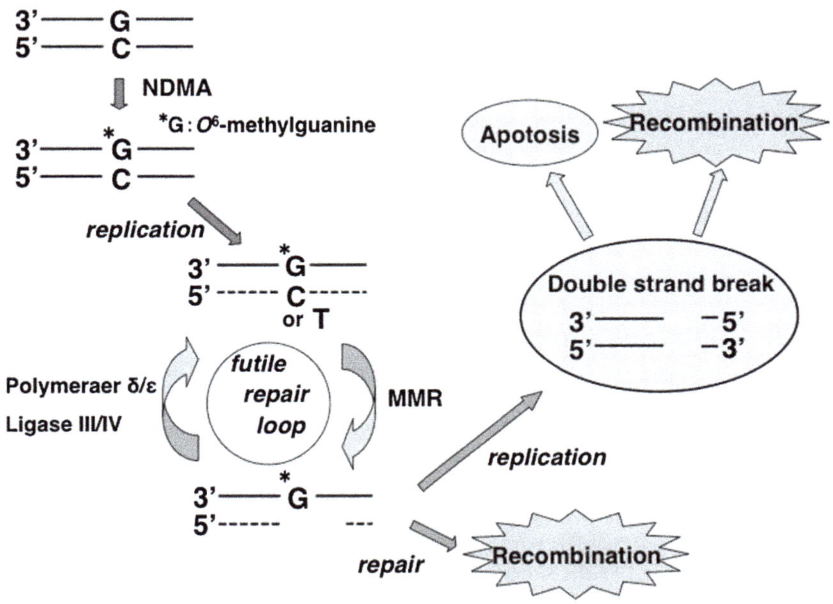

FIGURE 10.2 **Scheme outlining the involvement of MMR in chromosomal recombination induced by the methylating agent.** C or T is incorporated at the opposite site of methylated G during the first replication. Both normal bases are recognized as mismatches and removed from the newly synthesized strand. If the same event occurs next time, a so-called futile repair cycle is induced, and the strand possessing the methylated G is sustained in the single strand. This unstable state of DNA leads to double-strand breaks, and the recombination or apoptosis is then induced.

repairs, thereby leading to mutations. On the other hand, MMR can induce strand breaks during the repair process, and MMR might be required to facilitate chromosomal recombination through strand breaks. As shown in Fig. 10.2, the methyl-G·T pair is recognized as a base-pair mismatch by MutS or its homolog (Spel1·Msh6 heterodimer in *Drosophila*), and the methyl-G·C pair is also recognized as a mismatch, although at a lower frequency [40]. This repair step is known as a futile repair loop that occurs in the presence of functional MMR; its activity results in continuous strand breaks leading to recombination or apoptosis. Although further investigations are required, we speculate that the role of MMR in the somatic mutation recombination is as follows: if MMR is deficient, futile repair loops may be absent and thus are not activating chromosomal recombination. As a result, the recombination rate in somatic cells might decrease.

7. CONCLUSION

Living organisms continue efforts to maintain their genome by employing various devices. The induced and spontaneous DNA damage represents severe risks to genome stability. Therefore, organisms possess many countermeasures such as repair systems to guard against DNA damage. Among these systems, MMR plays an indispensable role in both somatic and germ cells, and *Drosophila* is no exception. In this section, the characteristics of MMR in *Drosophila* for maintaining genome stability have been addressed. Genes involved in MMR and diverse functions are conserved in *Drosophila*, and while only one recognition complex is present in *Drosophila*, almost all other eukaryotes possess two sets of complexes. It has been revealed that MMR does not work alone but collaborates with other repair systems such as NER during meiotic and mitotic recombination. A lack of MMR induces genome instability and is generally represented by changes in microsatellite repeats. However, there is the possibility that the requirement of MMR in chromosomal recombination might facilitate mutation and chromosomal recombination in the damaged DNA.

GLOSSARY

Eclosion The emergence of an adult from the pupa.
Futile repair loop The MMR-dependent repair loop involved in the induction of strand breaks and the activation of recombination.
Oviposition Laying eggs.

LIST OF ABBREVIATIONS

FA Fanconi anemia
HNPCC Hereditary non-polyposis colorectal cancer
MMR Mismatch repair
MSI Microsatellite instability
NDEA *N*-nitrosodiethylamine
NDMA *N*-nitrosodimethylamine
NER Nucleotide excision repair
SMART Somatic mutation and recombination test

REFERENCES

[1] Friedberg EC. DNA repair. New York: W.H. Freeman and Company; 1985.
[2] Marti TM, et al. DNA mismatch repair and mutation avoidance pathway. J Cell Phys 2002;191(1):28–41.
[3] Kunkel TA, Erie DA. DNA mismatch repair. Annu Rev Biochem 2005;74:681–710.
[4] Kirkpatrick DT, Petes TD. Repair of DNA loops involves DNA-mismatch and nucleotide-excision repair proteins. Nature 1997;387(6636):929–31.
[5] Meira LB, et al. Mice defective in the mismatch repair gene Msh2 show increased predisposition to UVB radiation-induced skin cancer. DNA Repair (Amst) 2002;1(11):929–34.
[6] Stojic L, et al. Mismatch repair and DNA damage signalling. DNA Repair (Amst) 2004;3(8–9):1091–101.
[7] Bhui-Kaur A, et al. DNA mismatch repair catalyzed by extracts of mitotic, postmitotic, and senescent *Drosophila* tissues and involvement of *mei-9* gene function for full activity. Mol Cell Biol 1998;18(3):1436–43.
[8] Holmes J, et al. Strand-specific mismatch correction in nuclear extracts of human and *Drosophila melanogaster* cell lines. Proc Natl Acad Sci USA 1990;87(15):5837–41.
[9] Thomas DC, et al. Heteroduplex repair in extracts of human HeLa cells. J Biol Chem 1991;266(6):3744–51.
[10] Boyd JB, et al. The *mei-9*^{a} mutant of *Drosophila melanogaster* increases mutagen sensitivity and decreases excision repair. Genetics 1976;84(3):527–44.
[11] Carpenter ATC. Mismatch repair, gene conversion, and crossing-over in two recombination-defective mutants of *Drosophila melanogaster*. Proc Natl Acad Sci USA 1982;79(19):5961–5.
[12] Flores C, Engels W. Microsatellite instability in *Drosophila spelchecker1* (MutS homolog) mutants. Proc Natl Acad Sci USA 1999;96(6):2964–9.
[13] http://flybase.org/reports/FBgn0015546.html.
[14] http://flybase.org/reports/FBgn0036486.html.
[15] Sekelsky JJ, et al. DNA repair in *Drosophila*: insights from the *Drosophila* genome sequence. J Cell Biol 2000;150(2):F31–6.
[16] http://flybase.org/reports/FBgn0011659.html.
[17] http://flybase.org/reports/FBgn0011660.html.
[18] http://flybase.org/reports/FBgn0015553.html.
[19] Digilio FA, et al. *Tosca*: a *Drosophila* gene encoding a nuclease specifically expressed in the female germline. Dev Biol 1996;178(1):90–100.
[20] Hsieh P, Yamane K. DNA mismatch repair: molecular mechanism, cancer, and ageing. Mech Ageing Dev 2008;129(7–8):391–407.
[21] Eshleman JR, Markowitz SD. Mismatch repair defects in human carcinogenesis. Hum Mol Genet 1996;5:1489–94.
[22] Müller A, Fishel R. Mismatch repair and the hereditary non-polyposis colorectal cancer syndrome. Cancer Invest 2002;20(1):102–9.
[23] Imai K, Yamamoto H. Carcinogenesis and microsatellite instability: the interrelationship between genetics and epigenetics. Carcinogenesis 2008;29(4):673–80.
[24] Schug MD, et al. Low mutation rates of microsatellite loci in *Drosophila melanogaster*. Nat Genet 1997;15(1):99–102.

[25] Schöllerer C, et al. High mutation rate of a long microsatellite allele in *Drosophila melanogaster* provides evidence for allele-specific mutation rate. Mol Biol Evol 1998;15(10):1269–74.

[26] Harr B, et al. Mismatch repair-driven mutational bias in *D. melanogaster*. Mol Cell 2002;10(1):199–205.

[27] Baida A, et al. Germline mutations at microsatellite loci in homozygous and heterozygous mutants for mismatch repair and *PCNA* genes in *Drosophila*. DNA Repair (Amst) 2003;2(7):827–33.

[28] López A, et al. Germline genomic instability in *PCNA* mutants of *Drosophila*: DNA fingerprinting and microsatellite analysis. Mutat Res 2005;570(2):253–65.

[29] López A, et al. Analysis of genomic damage in the mutagen-sensitive mus-201 mutant of *Drosophila melanogaster* by arbitrarily primed PCR (AP-PCR) fingerprinting. Mutat Res 1999;435(1):63–75.

[30] Borts RH, et al. The many faces of mismatch repair in meiosis. Mutat Res 2000;451(1–2):129–50.

[31] Kolas NK, Cohen PE. Novel and diverse functions of the DNA mismatch repair family in mammalian meiosis and recombination. Cytogenet Genome Res 2004;107(3–4):216–31.

[32] Radford SJ, et al. Meiotic recombination in *Drosophila Msh6* mutants yields discontinuous gene conversion tracts. Genetics 2007;176(1):53–62.

[33] Villeneuve AM, Hillers KJ. Whence meiosis? Cell 2001;106(6):647–50.

[34] McKim KS, et al. Meiotic recombination and chromosome segregation in *Drosophila* females. Annu Rev Genet 2002;36:205–32.

[35] Radford SJ, et al. Heteroduplex DNA in meiotic recombination in *Drosophila mei-9* mutants. Genetics 2007;176(1):63–72.

[36] Crown KN, et al. Eliminating both canonical and short-patch mismatch repair in *Drosophila melanogaster* suggests a new meiotic recombination model. PLoS Genet 2014;10(9):e1004583.

[37] Taira K, et al. Distinct pathways for repairing mutagenic lesions induced by methylating and ethylating agents. Mutagenesis 2013;28(3):341–50.

[38] Williams SA, et al. Functional and physical interaction between the mismatch repair and FA-BRCA pathways. Hum Mol Genet 2011;20(22):4395–410.

[39] Graf U, et al. Somatic mutation and recombination test in *Drosophila melanogaster*. Environ Mutagen 1984;6(2):153–88.

[40] Taira K, et al. Binding of MutS protein to oligonucleotides containing a methylated or an ethylated guanine residue, and correlation with mutation frequency. Mutat Res 2008;640(1–2):107–12.

[41] Goto Y, et al. Mutagenicities of *N*-nitrosodimethylamine and *N*-nitrosodiethyl-amine in *Drosophila* and their relationship to the levels of *O*-alkyl adducts in DNA. Mutat Res 1999;425(1):125–34.

[42] Zhang H, et al. Mismatch repair is required for O^6-methylguanine-induced homologous recombination in human fibroblasts. Carcinogenesis 2000;21(9):1639–45.

Chapter 11

Genome Stability in *Caenorhabditis elegans*

M. Rieckher[a], A.F.C. Lopes[a], B. Schumacher
University of Cologne, Cologne, Germany

Chapter Outline

1. INTRODUCTION

The maintenance of genome integrity is essential for a healthy life of an individual and assures reproductive success of a species. Genomic DNA is constantly threatened by a plethora of environmental and cell-intrinsic genotoxic agents that inflict a variety of DNA lesions, with thousands of events estimated to occur in each individual cell per day [1]. Failure in DNA repair can interrupt or alter gene functions, resulting in cell death, senescence, or cancer; conversely, error-prone repair in the germline is the driving force of genome evolution and intraspecies genome diversity. All organisms have evolved lesion-specific DNA-repair mechanisms to keep the genome in check: helix-distorting lesions, such as ultraviolet (UV) light irradiation-induced 6–4 photoproducts (6–4 PPs) or cyclobutane pyrimidine dimers (CPDs), are repaired by nucleotide excision repair (NER), which either operates genome-wide by instrumenting global genome NER (GG-NER), or upon RNA polymerase II stalling during transcription by employing the transcription-coupled NER (TC-NER). Intrinsic or extrinsic reactive oxygen species (ROS) induce a diverse number of oxidative lesions that are mostly targeted by base excision repair (BER) or mismatch repair (MMR). While oxidative and helix-distorting lesions and single-strand breaks (SSBs) are the most abundant type of damage, double-strand breaks (DSBs) are the most toxic form and are rejoined by the error-prone nonhomologous end joining (NHEJ) or by the high-fidelity homologous recombination (HR) repair. Beyond DNA-repair pathways, specialized mechanisms for DNA-damage signaling have been addressed in great detail, commonly referred to as DNA-damage response (DDR). Upon DNA injury, DNA-damage checkpoint signaling can pause cell cycle progression at various phases, allowing the cell time for repair or, alternatively, induce signaling events that drive the cell into apoptosis. The various DDR mechanisms are well conserved across species and a large body of knowledge rears from studies in bacteria, yeast, and mammalian cell lines.

a. These authors contributed equally.

Genome Stability. http://dx.doi.org/10.1016/B978-0-12-803309-8.00011-2

2. THE *CAENORHABDITIS ELEGANS* MODEL

The transparent nematode *Caenorhabditis elegans* has proven instrumental in providing insights into the mechanisms underlying numerous cellular and developmental processes, including cell differentiation and apoptosis, organismal aging, host–pathogen interactions, and even molecular aspects of neurodegenerative diseases and tumorigenesis [2–7]. Since the description of *C. elegans* genetics in the 1970s by Sydney Brenner, the nematode was rapidly adopted as a powerful model organism, resulting in a fully sequenced genome, in which 60–80% of the genes have a human counterpart, the affluence of data made available at Wormbase (http://www.wormbase.org/) about gene structure, mutant and RNAi phenotypes, microarray data, transcription factor binding sites, protein–protein interaction networks, and the availability of a vast collection of mutants, for example, at the *Caenorhabditis* Genetics Center (CGC) or the National Bioresource Project for the nematode (NBRP-*C. elegans*) [8]. In 2009, the nematode was also employed as model organism in space biology to study the molecular mechanisms underlying muscle adaptation, space radiation response, and gene expression patterns at zero gravity at the International Space Station (ISS) [9]. In the laboratory, *C. elegans* is easily handled: animals feed on *E. coli* bacteria have a rapid reproductive life cycle of 2.5 days at 22°C, during which they progress through four larval stages to develop into hermaphroditic adults, which have a life span of about 2 weeks and lay 300 eggs (compare Fig. 11.1A). Upon completion of development, adults have 959 postmitotic somatic cells that comprise tissues, such as muscles, intestine, epidermis, and 302 neurons, which form a neuronal network with a fully deciphered wiring plan and largely completed connectome [10]. The dominant sexual form of *C. elegans* is hermaphrodite (XX), but males (XO) can be isolated and used for genetic crosses to produce strains carrying multiple mutations. The adult hermaphrodite reproductive system consists of two U-shaped gonad arms that contain both male and female germ cells, which undergo mitotic and meiotic cell divisions and comprise an immortal and totipotent cell lineage. Hence, *C. elegans* reproduces by self-fertilization, and populations are genetically identical and do not suffer from inbreeding depression. The advantages of a simple body plan, transparency of eggs and cuticle, and the invariance of cell divisions and developmental stages have expedited a highly detailed developmental and anatomical description of the animal, which is well documented in open access resources (eg, http://www.wormatlas.org/ and http://wormbook.org/).

As DDR is highly conserved from worms to man, *C. elegans* serves as a relevant model to study the consequences of DNA-repair deficiency [11,12]. In the nematode, DNA damage induces a vigorous response of DNA repair and signaling pathways in dividing germ cells, which is distinct in mitotic and meiotic compartments of the germline [13]. Conversely, somatic cells are entirely postmitotic and display remarkably high resistance to ionizing radiation (IR)-induced DNA damage [14]. The DDR in germ cells highlights the importance for ensuring the stable passage of genomic material through an immortal germline across generations, while somatic tissues are not important for species-survival after successful reproduction (discussed in Ref. [15]). The disposable soma theory poses that the resources of the organism need to be allocated between the soma and germline to maximize fitness before the soma can be disposed of upon reproduction (summarized in Ref. [16]). In *C. elegans*, the "germline DNA damage–induced systemic stress resistance" (GDISR) illustrates particularly well how somatic maintenance adapts to the requirements of germ cells: GDISR elevates stress resistance in somatic tissues to allow delay of progeny production until DNA damage in germ cells is repaired [17]. Hence, DDR in the nematode can be highly instructive for understanding systemic response mechanisms (discussed in Ref. [18–20]). This chapter presents a comprehensive collection of methodologies that are currently employed to study DNA repair and DDR in the nematode. Further, we provide an overview of the various well-conserved DNA repair mechanisms that are activated in *C. elegans* to counteract genomic instability. Further, we expand on recent reports that exemplify the relevance and advantages of the nematode to the field of systemic DDR during development and aging.

3. POWERFUL GENETIC TOOLS TO EXPLORE DDR DYNAMICS

C. elegans is genetically malleable by various applied methodologies (compiled in WormMethods on http://www.wormbook.org). Most commonly, forward genetic screens are implemented by creating mutant libraries through the use of mutagenic chemicals, such as ethyl methanesulfonate (EMS), or combined treatment with trimethylpsoralen (TMP) and UV irradiation that create random deletion, point mutation, and insertion events throughout the genome [21]. Pioneering studies by Hartman and colleagues in the 1980s led to the genetic identification and characterization of *C. elegans* mutants that show hypersensitivity to various genotoxic agents [22]. The advent of whole genome sequencing allows rapid analysis of single mutants or large mutant libraries and has been successfully employed to define mutation accumulation in DNA-repair-deficient *C. elegans* [12,23]. In 2014, research in the nematode profits from the successful adaptation of the CRISPR/Cas9-based genome engineering system for targeted genomic alterations (summarized in Ref. [24]).

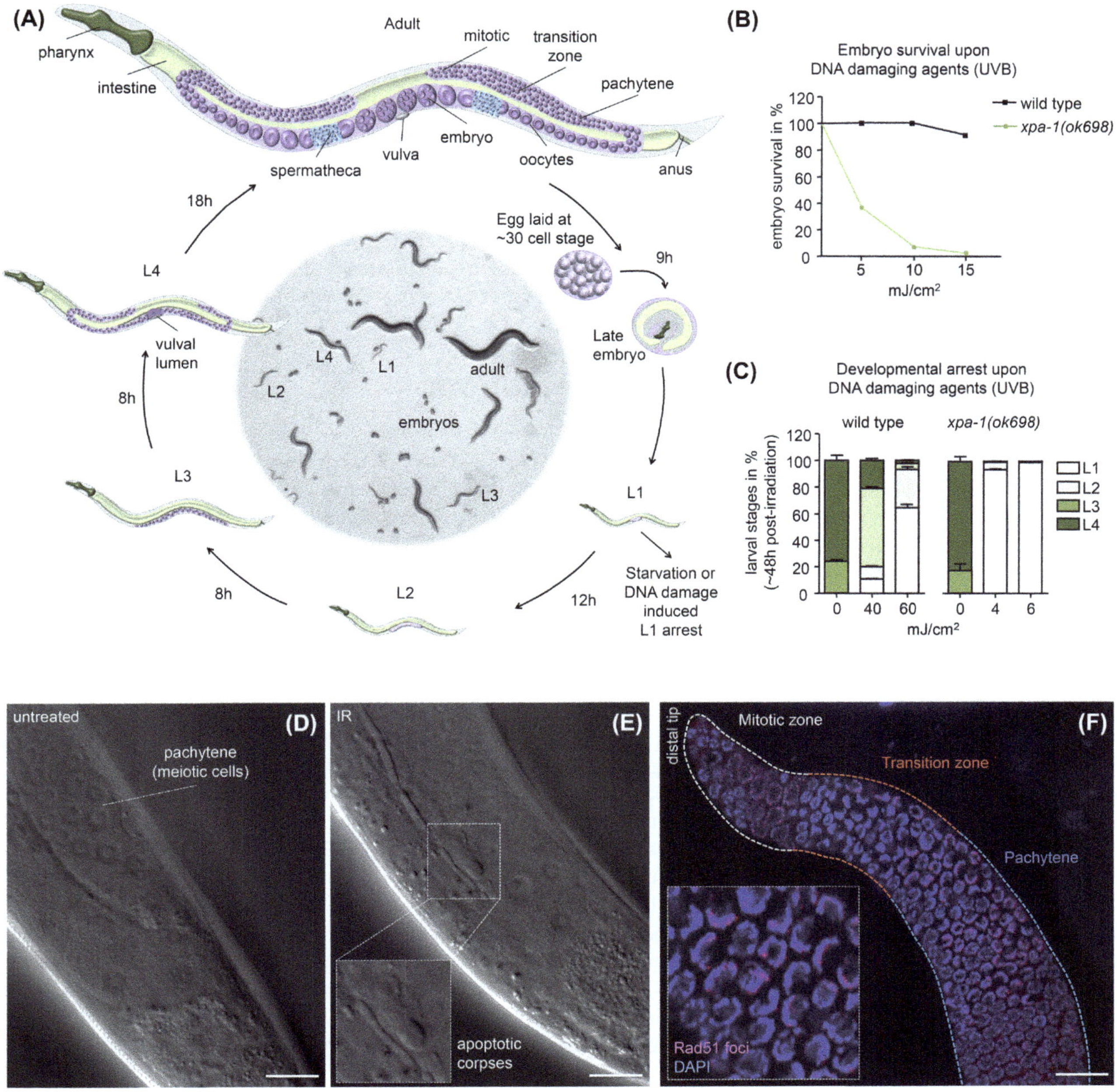

FIGURE 11.1 ***C. elegans* as a model organism to study DNA repair and DDR mechanisms.** All size bars correspond to 25 µm. (A) The *C. elegans* life cycle from fertilized embryo, through the larval stages L1–L4 until fertile adult, is completed in about 2.5 days when grown on bacteria seeded agar plates (22°C). An adult produces about 300 eggs, which require about 9 h of ex utero development before hatching of L1 larvae. Stereoscopic bright field imaging allows clear distinction of mixed *C. elegans* developmental stages on an agar plate (inlay). Detailed descriptions on *C. elegans* development from germ cell to fertile adult are available on http://www.wormbook.org and http://www.wormatlas.org. (B) Representative graph displaying embryo survival of wild type and NER mutant *xpa-1(ok698)* upon different UVB irradiation intensities. Typically, adult animals are exposed to DNA damaging insults and egg-laying and hatching rate are determined. (C) Representative result of larval development 48-h post-irradiation (UVB). Animals are first staged as L1 larvae, and then exposed to DNA damaging insults and developmental stages can be determined 48–72 h later. Note that wild type requires 10-fold UVB intensities as compared to the *xpa-1* mutant to achieve a partial developmental delay or arrest. (D) DIC image of the pachytene region of the meiotic germline in the posterior gonad arm of adult *C. elegans*. Nuclei and outline of germ cells can clearly be distinguished. (E) DIC image of the pachytene region of an animal irradiated with IR resulting in apoptosis induction of meiotic germ cells. Apoptotic corpses are clearly visible as button-like structures (inlay) (D, E, *courtesy of Najmeh Soltanmohammadi*). (F) Fluorescent microscopy image of a dissected germline stained with DAPI (*blue*) and Rad51 antibody (*red*) to visualize DSB events (compare inlay). The germline contains a clearly distinguishable mitotic zone, followed by a transition zone and the meiotic pachytene. The spatiotemporal distribution of mitotic and meiotic germ cells allows the study of distinct DNA repair or checkpoint mechanisms in response to DSB formation (F, *courtesy of Ashley B. Williams*).

Research on the nematode has pioneered the application of RNA interference (RNAi)-mediated gene downregulation, which is achieved by double-stranded DNA (dsDNA) delivery via feeding, soaking, or injection, and resulted in the compilation of genome-wide RNAi libraries [25,26]. Systematic RNAi screens have revealed whole DDR networks that govern genome stability and maintenance [11,27,28]. RNAi is highly efficient in the germline and most somatic tissues with the exception of the neuronal system, which can be overcome by the application of various RNAi hypersensitive transgenic lines or mutants [29]. *C. elegans* transgenes can be rapidly and cost-effectively obtained by genetic transformation through DNA microinjection or DNA-coated microparticle bombardment, allowing the introduction and stable inheritance of exogenous DNA into the genome [30]. A collection of DDR-specific genes can be obtained at the "*C. elegans* TransGeneome" project, a genome-scale transgenic project for fluorescent- and affinity-tagged proteins [31]. Transcriptome and proteome profiling, of whole animals or specific tissues, is standardized in *C. elegans* and has helped to unravel the dynamics of DDR during development and aging in DNA repair deficient mutants or upon DNA-damage induction (eg, see Refs. [32–34]).

4. GENOTOXIC AGENTS FOR DNA DAMAGE INDUCTION

A number of procedures for *C. elegans* are available to evoke genome instability, which in turn mounts a lesion-specific response [35]. UV irradiation is applied to study NER in the soma and the germline, since it results in the formation of bulky photolesions 6–4 PP and CPDs. UV-B (320–290 nm) displays a higher penetrance and thus induces DNA damage throughout the animal as opposed to UV-C (290–100 nm), whose shorter wavelengths are absorbed by water and cellular biopolymers [36]. In addition, UV-induced lesions indirectly give rise to DSB formation, for example, when replication forks break down at unrepaired CPDs and can lead to apoptotic demise of germ cells [37]. UV-A (320–400 nm) mostly damages DNA indirectly through the formation of ROS, which results in the mild induction of germ cell apoptosis and DNA-damage checkpoint activation [38]. Most frequently, IR (X-rays or γ irradiation) is used to cause DSB formation and chromosome rearrangements, which triggers a highly reproducible response of DSB repair, cell cycle arrest, and apoptosis in germ cells (see Figs. 11.1D–F) [39].

Except electromagnetic waves, various chemicals can be used to induce DNA damage in *C. elegans*. Photosensitizers, such as ethidium bromide or bromodeoxyuridine, or the light-sensitive cross-linking agent TMP, enhance the genotoxic havoc caused by UV irradiation [40]. Alkylating agents, such as EMS or MMS, are highly potent mutagens but have been less frequently employed in *C. elegans* to study specific DDR [41–43]. Hydroxyurea (hydroxycarbonate) is known to destabilize the replication fork, thus resulting in DSBs in mitotic germ cells [44]. Illudins are chemical compounds that induce DNA lesions that hamper with transcription and also in the nematode require TC-NER for their removal [45].

Metals, such as silver nanoparticles and cadmium, cause measurable oxidative damage to DNA and result in mitogen-activated protein kinase (MAPK)-dependent germline apoptosis, respectively, though the exact nature of DDR remains to be determined [46,47]. Other prooxidant compounds, including paraquat, sodium azide, or menadione bisulfite, can cause 8-oxo-G formation, which are predominantly cleared by BER [48]. In addition, several studies also show decreased tolerance of NER-deficient animals to oxidative stress [49,50]. In addition, some mitochondrial mutants display hypersensitivity to oxidative stress and accumulate oxidative lesions in their genome [51]. Generally, *C. elegans* exposure to most chemical compounds produces phenotypic readouts that are easily scored, qualifying the nematode as a high-throughput platform for environmental genotoxins [52].

5. METHODS FOR DNA DAMAGE DETECTION

Genome instability in *C. elegans* manifests in various morphological, developmental, and behavioral phenotypes that can be scored in vivo, additionally several in vitro methods are available to directly visualize and quantify various DNA lesions (see Fig. 11.1). The significant differences between repair and response to DNA damage in mitotically and meiotically dividing germ cells as compared to the postmitotic somatic cell types allow the study of tissue-specific as well as systemic DDR.

The germline is easily discerned from somatic tissues by differential interference contrast microscopy (DIC; or Nomarski microscopy) and clearly separated in a distal mitotic zone, where germ stem cells proliferate, and the transition zone, in which germ cells enter meiosis prophase I (Fig. 11.1). Meiotic recombination is completed by late pachytene, before germ cells progress through the diplotene until oocytes arrest in diakenesis, and resume meiosis only upon fertilization in the spermatheca upon which embryogenesis commences in the uterus (see sketch in Fig. 11.1A). Physiological germ cell death occurs in the late pachytene and is enhanced in response to IR or UV, which can be scored via DIC, since apoptotic corpses display a distinct cellular morphology (see Fig. 11.1D–F) [13]. Similarly, cell cycle arrest is easily visualized and quantified in the mitotic region, since size and number of mitotic germ cells is significantly altered upon DNA damage [53]. DNA-damage checkpoint activity can also be quantified during early embryogenesis by monitoring the asynchronous cell divisions, which are timed during checkpoint activation [42]. Genetic screens based on those phenotypes have contributed to the identification of comprehensive genetic pathways regulating DNA damage–induced apoptosis and checkpoints (summarized in Ref. [54]).

C. elegans lays a defined number of 300 eggs and develops through four well-timed larval stages before reaching adulthood. DNA-damaging insults can significantly reduce offspring number and viability, and delay developmental timing, allowing the distinction between germline and somatic DDR (see Fig. 11.1B,C) [55]. Further, germline development can be genetically suppressed to distinguish between somatic and germline-specific repair [32]. The accumulating effect of DNA damage can be quantified by monitoring animal survival or tissue decline, for example, gross morphology, muscle function in the pharynx or locomotion, in individuals of a defined population from birth to death [56–58].

DNA repair can be directly assessed by immunological methods: antibodies are available for specific lesions (eg, 6-4 PP or CPDs), and can be applied to genomic DNA extracts via slot blot or whole-animal or tissue-specific immunostaining [59]. However, both cuticle and egg shell of the nematode represent barriers for staining that need to be disrupted chemically and/or mechanically. DAPI and BrdU staining to visualize DNA or newly synthesized DNA, respectively, are highly efficient in fixed or alive *C. elegans* [60]. The dissected germline is easily accessible for immunostaining, and DSB-repair processes can be visualized with anti-RAD-51 or anti-CDK-1 phophotyrosine antibodies (see Fig. 11.1F) [61]. Complexes of DNA-repair factors and DNA molecules can be analyzed by electrophoretic mobility shift assays (EMSA) and further resolved by immuno-gold electron microscopy [62]. Fluorescence in situ hybridization (FISH) is readily applied to study chromosome-pairing events upon DNA damage [63].

Several protocols implement quantitative PCR (qPCR) to quantify DNA damage across the genome by exploiting the capacity of many lesions to block or inhibit the progression of DNA polymerases [32,64]. In addition, this method allows the distinction between nuclear and mitochondrial DNA damage, but is not lesion specific [58,65]. The mutation accumulation across many generations is promoted in DNA-repair-deficient animals, which is easily assessed in *C. elegans* due to its rapid reproductive cycle and the availability of transgenes carrying balancer chromosomes that allow for morphological read-outs of mutations [66].

6. EXCISION REPAIR

Three major excision repair pathways are highly conserved between worms and man: NER, BER, and MMR [67]. The importance of NER, BER, or MMR activity for genome stability maintenance has been demonstrated by experiments that follow mutation accumulation over many generations and has revealed that particularly MMR protects the genomes from mutations, followed by NER and eventually by BER [68,69]. However, the most comprehensive studies on single-stranded DNA repair have been performed on *C. elegans* NER and are discussed further in more detail.

6.1 Nucleotide Excision Repair

NER removes bulky nucleotide lesions 6–4 PPs, CPDs, and their Dewar valence isomers, which can be induced upon irradiation with UVB and UVC [70]. Already in the 1980s, the analysis of radiation-sensitive (*rad*) mutants revealed a significantly reduced repair capacity for 6–4 PPs and CPDs in *rad-3* mutants [71]. Indeed, almost two decades later, genome sequence analysis for NER homologs and RNAi-mediated gene knock-down revealed that *rad-3* encodes the *C. elegans* homolog of the mammalian xeroderma pigmentosum complementation group A (XPA) [72,73].

Based on studies in *Saccharomyces cerevisiae* and mammalian systems, NER can be divided into four consecutive steps, in which the lesion is first detected, which triggers the recruitment of factors necessary for unwinding the DNA strand, followed by excision of the area containing the damage and finalized by filling the gap through DNA synthesis and ligation. NER is initiated by two distinct mechanisms of DNA-damage detection, which activate the same downstream core machinery to repair the damage: (I) transcription-coupled NER (TC-NER) is activated by stalling of RNA polymerase II during transcription and requires the recruitment of the chromatin remodeling protein Cockayne syndrome protein B (CSB) and Cockayne syndrome protein A (CSA); (II) global genome NER (GG-NER) is initiated upon lesion detection by the UV-damaged DNA-binding protein (UV-DDB) complex and xeroderma pigmentosum group C (XPC), which subsequently recruits several other NER proteins and removes lesions throughout the genome [74]. In humans, inherited mutations in GG-NER genes result in xeroderma pigmentosum (XP), which includes severe UV sensitivity and an increased risk for skin cancer; TC-NER deficiency causes Cockayne syndrome (CS), which is characterized by a variety of neurodevelopmental symptoms and premature aging [75].

Experimental efforts during the last decade have enlarged the group of mammalian NER orthologs in *C. elegans*: a number of mutants for homologs of *CSB, CSA, XPC, XPF, XPG, RAD23,* and *ERCC1* have been isolated, all of them displaying increased sensitivity to UVB [45,55,73,76]. These findings paint the convincing image of a well-conserved NER pathway in *C. elegans*, making the nematode an increasingly important model for studying in vivo NER activity in a time-dependent fashion and in the context of a whole organism [15].

Particularly important is the discovery that the GG-NER and TC-NER sub-pathways perform differential tissue-specific roles in response to UV irradiation: animals carrying mutations in genes of the core NER machinery (*xpa-1*, *xpf-1*, and *xpg-1*) and specifically in the GG-NER sensor *xpc-1* show severely decreased germ cell and embryo survival, and a diminished CPD repair capacity upon UV irradiation [55]. Similar effects are apparent upon loss of RAD-23 that is homologous to HR23A and HR23B, which are responsible for stabilizing and enhancing the binding of XPC in mammalian cells [77]. Conversely, loss of CSB-1 or CSA-1 does not result in germline-specific defects upon UVB irradiation but significantly reduce larval development timing and survival [45,55,56]. However, some studies report partial redundancy for the tissue-specific NER repair function: depending on the UV dose and source applied, CSB-1-deficient animals can display increased levels of germ cell apoptosis, morphological abnormalities, and decreased hatching rates [78]. In summary, GG-NER is the major pathway mediating UV response in early development, germ cells and embryos, while TC-NER mediates somatic repair in juvenile and adult animals [55].

6.1.1 The Role of NER in Development and Aging

In human cell lines, moderate UVB and UVC irradiation (<100 mJ/cm^{2}UVB or <1 mJ/cm^{2}UVC) result in a transient decrease in cell division and DNA replication activity, whereas higher doses of UV irradiation lead to a permanent arrest of DNA replication [79–81]. UV irradiation of repair-proficient *C. elegans* at L1 larval stage (at which worms can be synchronized to allow assessment of developmental growth on a large scale) causes delayed larval development, while NER deficiency exacerbates UV sensitivity during development (see Fig. 11.1C) [56]. Interestingly, TC-NER is particularly important for withstanding UV-induced lesions during development [55]. Most cell divisions in the development of *C. elegans* occur already during embryogenesis after which differentiated cells mostly grow in size. During active transcription, UV lesions lead to the stalling of RNA pol II and the subsequent degradation of its subunit AMA-1, which is mediated by the putative E3 ubiquitin ligase WWP-1, ortholog of the yeast Rsp5 [73].

TC-NER deficiency in humans leads to the devastating disease of CS that is characterized by developmental growth retardation and a variety of premature ageing symptoms [75]. Mammalian cells respond to DNA lesions that stall RNA polymerase II-mediated transcription by downregulating the growth hormone receptor (GHR) and the insulin-like growth factor-1 receptor (IGF-1R), which in turn promotes IGF-1 resistance and defense to cellular oxidative stress [82]. Both GHR and IGF-1R are regulators not only of postnatal growth, but also of the aging process as mice with reduced GH/IGF-1 signaling show dwarfism and extended life span [83]. Similarly, defects in the IGF-1R homolog *daf-2* lead to greatly extended longevity in *C. elegans* [84]. Ever since the discovery of the *daf-2* mutant longevity, *C. elegans* has served as an important model for the genetics of aging [3].

In 2014, a crosstalk between NER and the insulin/insulin-like growth factor signaling (IIS) in *C. elegans* had become apparent. In the nematode, the conserved IIS pathway is a major regulator of starvation-induced L1 arrest, development, stress resistance, and lifespan [85,86]. Key players of the IIS in the nematode are the insulin/IGF receptor DAF-2 that, upon self-phosphorylation, recruits the phosphatidylinositol 3-kinase (PI3K) subunit AGE-1, which generates phosphatidylinositol-3,4,5-triphosphate (PIP3) molecules that activate kinases of the AKT family. The latter phosphorylates DAF-16, member of the FOXO transcription factor family [87]. DAF-16 remains inactive in its phosphorylated form in the cytoplasm. When the insulin signaling cascade is inactive, DAF-16 is hypophosphorylated and localizes from the cytoplasm to the nucleus, where it governs the transcriptional regulation of a plethora of genes regulating aging, stress response, metabolism, thermotolerance, and pathogen resistance [88]. Thus, knockout of *daf-2* or *age-1* results in the constitutive DAF-16 nuclearization and increased stress resistance and life span (summarized in Ref. [89]).

Transcriptome analysis of UV-treated wild-type and arrested *xpa-1* mutant L1 larvae has revealed that, similar to mammals, IIS is attenuated in response to UV-induced DNA damage. DAF-16 is efficiently activated upon DNA damage during development while its responsiveness, specifically to DNA lesions, declines with aging. Functionally, DAF-16 activity alleviates developmental arrest and enhances somatic tissue functionality in response to UV-induced DNA damage [56]. It was suggested that the longevity assurance factor DAF-16 might thus antagonize DNA damage–driven aging by enhancing tolerance of genotoxic stress [20].

DAF-16 acts in specific tissues to execute differential outputs, which is governed by a number of cofactors and coregulators, including the heat-shock factor HSF-1 and the Nrf-like transcription factor SKN-1 (summarized in Ref. [90]). The

GATA transcription factor EGL-27 genetically interacts with DAF-16 to promote both longevity and stress response [91]. Importantly, upon UV damage induction, DAF-16 functions together with EGL-27 to mount the DDR in *C. elegans*, which does not require HSF-1 or SKN-1 activity [56].

Intriguingly, transcriptomic and proteomic profiling of *xpa-1* mutants shows increased induction of antioxidant defenses and higher ROS levels, in comparison to wild type, which might indicate elevated levels of oxidative DNA damage [34,92]. Surprisingly, loss of *ercc-1* or *xpf-1* prolongs the lifespan of long-lived *daf-2* mutants, which was suggested to rear from an active signaling by DNA damage–detection proteins to implement a hormetic response that promotes survival [57]. An alternative explanation is that meiotic defects in *ercc-1* and *xpf-1* mutants might influence the lifespan of *daf-2* mutants that display compromised egg-laying activity [55,84,93–95].

6.1.2 NER Deficiency in Mitochondrial Diseases

Mitochondria are the powerhouse of the cell and contain their own 16.5 kb genome (mtDNA) that cooperates with the nuclear genome (ncDNA) to encode the proteins of the OXPHOS system [96]. mtDNA is subjected to environmental toxins, and exogenous or endogenous ROS, which is typically repaired by BER (summarized in Ref. [97]). To date there is no conclusive evidence for NER activity in mitochondria leaving them incapable of repairing CDPs or 6-4 PPs in mtDNA and highly susceptible to UV irradiation. Thus, UV-induced lesions potentially persist and stall DNA replication and transcription, and lead to the depletion of mtDNA and mitochondrial proteins, resulting in mitochondrial breakdown [98]. In primary human fibroblasts and in *C. elegans*, UVC-induced lesions in mtDNA are removed by clearing mitochondria via autophagy, mitophagy, and mitochondrial fission and fusion events [99,100]. Findings in 2012 in mice indicate a presence of CSB in mitochondria, where it might act as DNA-damage sensor, signaling the clearance of mitochondria, with damaged genome, by autophagy [101]. Mitochondrial dysfunction is associated with a large number of human neurodegenerative disorders, including the major DNA-repair disorders CS, ataxia–telangiectasia (AT), and XP, which can be qualitatively and quantitatively predicted in silico by specifically designed databases for mitochondrial pathologies [101–103]. Importantly, XPA deficiency results in a well-conserved mitochondrial decline, which is induced by the activation of poly-ADP-ribose polymerase-1 (PARP-1) [104]. PARPs and poly(ADP-ribose)glycohydrolases (PARGs) perform the posttranslational modification poly(ADP-ribosyl)ation upon NAD^+ consumption, which regulates cellular stress responses by mediating DNA repair, chromatin structure, DNA synthesis, apoptosis, and mitochondrial homeostasis [105]. *C. elegans* carries three homologs for PARPs and two homologs for PARGs that maintain a conserved function in DNA repair [106,107]. In addition, studies in cells and the nematode reveal that PARP activity reduces NAD^+ availability, which in turn suppresses the sirtuin-signaling (SIRT1) pathway, a known regulator of proper mitochondrial homeostasis under stress conditions. This intricate mechanism can be attenuated by supplementing PARP inhibitors or NAD^+ precursors, which leads to an activation of the mitochondrial unfolded protein response (UPR^{mt}) and consequently a boost of mitochondrial function resulting in increased health and life span [108]. XPA deficiency in cells, mice, and *C. elegans* trigger a similar pathway: overactivation of the DNA-damage sensor PARP drains cells of NAD^+, suppresses SIRT1, and leads to defective mitophagy, which might explain the neurodegenerative phenotypes in CS, AT, and XP patients [104]. The pathway is analogous in CSB mutant cells, nematodes, and mice: PARP inhibition or treatment with NAD^+ precursors increases SIRT1 expression, rescues the shortened life span of *csb-1* mutant *C. elegans* and restores metabolic, mitochondrial, and transcriptional alterations in *Csb*-deficient mice. In a similar fashion, high fat diet induced by supplementing the ketone β-hydroxybutarate (β-OHB) rescues the reduced life span of *csb-1* mutant nematodes [109].

In summary, TC-NER deficiency leads to neurodegenerative phenotypes that are linked to mitochondrial dysfunction and involve conserved mechanisms. *C. elegans* is a highly versatile system and suitable for high-throughput drug screenings, which will be a relevant feature for future therapeutic target identification in NER-deficiency disorders.

6.2 Base Excision Repair

The recognition and excision of oxidized nucleotides in BER is executed by DNA glycosylases that cleave the *N*-glycosolic bond between the DNA base and the sugar phosphate backbone. Subsequently, the baseless site (apurinic/apyrimidinic site or AP site) is transformed into an SSB by the activity of an AP endonuclease, leaving a 3′-hydroxyl and 5′-phosphate gap that is filled by the DNA polymerase β (Pol β) (summarized in Ref. [110]). To date, 11 human damage-specific DNA glycosylases are known and a branchy BER system in mammals is resolved in great detail [111].

In *C. elegans*, only two DNA glycosylases have been identified, the uracil-DNA glycosylase homolog UNG-1 and the DNA N-glycosylase homolog NTH-1 [112,113]. Transcriptome analysis revealed that NTH-1 deficiency activates oxidative stress response and lowers IIS activity, which does not result in a clear phenotype related to oxidative DNA damage, such as resistance to oxidative stress or altered life span [92]. Activity measurements of UNG-1 in *C. elegans* embryonic protein extract demonstrate its capability to specifically cleave U:G mispairing, which requires the Mg^{2+}-dependent hydrolytic AP endonuclease EXO-3 [114,115]. There are two AP endonucleases homologs in the nematode, EXO-3 and APN-1, and cross-species complementation studies show the rescue of DNA-repairdeficiency in yeast [115,116]. Both AP endonucleases are known to differentially function in the MMR pathway to induce toxicity in response to the cancer therapeutic 5-Fluorouracil (5-FU): EXO-3 is required for RPA-1 filament formation, indicating its requirement for MMR activation, while APN-1 acts in checkpoint activation [117]. Furthermore, APN-1, but not EXO-3, is specifically required for resistance to a variety of DNA-damaging agents, including UVC, oxidative stressors *tert*-butylhydroperoxide (*tert*-BH) and H_2O_2, and the methylating agent MMS [118].

In the *C. elegans* genome, no homolog for Pol β could be identified but BER is finalized by the gap-filling activity of a Pol β polymerase homolog, which acts as a lesion bypass polymerase and a backup BER polymerase in vertebrate systems [119]. Thus far, the *C. elegans* BER mechanism appears to be of much lower complexity as compared to mammalian systems, and BER deficiency generally results in relatively mild stress response or age-related phenotypes [120].

7. MISMATCH REPAIR

Major advances in defining MMR activity come from studies in *E. coli*, but the mechanisms of this repair machinery are conserved between bacteria and eukaryotes. MMR removes mismatched base pairs, mainly resulting from replication errors, and is therefore critical for DNA replication accuracy and genome stability maintenance across generations, which is highlighted by increased cancer occurrence in MMR-deficient humans [121].

C. elegans expresses orthologs of the central human MMR factors MSH2, MSH6, MLH1, and PMS2. MMR deficiency results in elevated somatic DNA instability and germline mutagenesis, when monitored across 40 generations in transgenes carrying a heat shock-promoter-driven frame-shifting DNA repeat that interrupts *LacZ* expression, which upon mutation events, can shift back in frame, resulting in a *LacZ*-positive readout [27,66]. Similar in-depth approaches relying on PCR-based genome analysis confirmed this observation [68,122].

Thus far, with the exception of the above-mentioned crosstalk between BER and MMR, the nematode has been less intensely utilized to elucidate mechanistic information on MMR activity [117]. However, 2015 results demonstrate that MLH-1-deficient animals display decreased meiotic cell death in the germline upon DNA damage, and an elevated resistance to alkylating and oxidizing agents. Further, MMR induces autophagy-mediated cell death of nondividing adult somatic cells independently of the checkpoint signaling factor ATL-1 [123].

8. DOUBLE-STRAND BREAK REPAIR IN *C. ELEGANS*

DNA DSBs are considered to be the most toxic form of DNA damage: in yeast, a single DSB can lead to cell cycle arrest or cell death, if left unrepaired [124]. The impact of DSB-repair deficiency in humans is highlighted by the appearance of disorders including cancer predisposition and infertility [125]. DSB repair is of high importance for the production of germ cells which must carry the correct genetic material to the next generation. Uncontrolled DSB repair can lead to tumorigenesis, previously reported in mice and humans, or serious inborn diseases due to chromosomal aberrations [54,126].

Cells employ different DNA damage–response (DDR) mechanisms depending on the nature of DSBs: breaks in the DNA can be caused upon exposure to IR or stalling of DNA replication forks [127]. DSBs most commonly occur during the production of gametes in the process of meiotic recombination (compare Fig. 11.1F). Depending on the cause leading to DSBs, different intermediates and substrates are formed, which are target of a delicately balanced array of repair proteins [128].

The two major mechanisms for DSB repair are homologous recombination (HR) and nonhomologous end joining (NHEJ). HR is characterized as an error-free method, which uses the sister chromatid or homologous chromosome as an undamaged template for repair [126,129,130]. In *C. elegans*, HR functions in DSB repair in proliferating somatic cells during early embryogenesis and remains active in germ cells during adulthood [129]. NHEJ is known for its efficient, yet error-prone repair that joins damaged DNA ends regardless of homology, which can result in the addition or removal of nucleotides [126,129–131]. NHEJ functions predominantly in somatic cells in *C. elegans* starting from late embryogenesis [15,129]. Various studies report that NHEJ and HR can cooperate, compete or act in parallel to repair DSBs, as evidenced by experiments where HR is impaired in germ cells or somatic cells, but not after IR [11,132–137].

8.1 Homologous Recombination

HR repair in *C. elegans* has been extensively studied during meiotic recombination, which takes place in the transition zone of the gonad [15]. Fig. 11.2 summarizes the HR-repair mechanism and functions of known homologs in *C. elegans*.

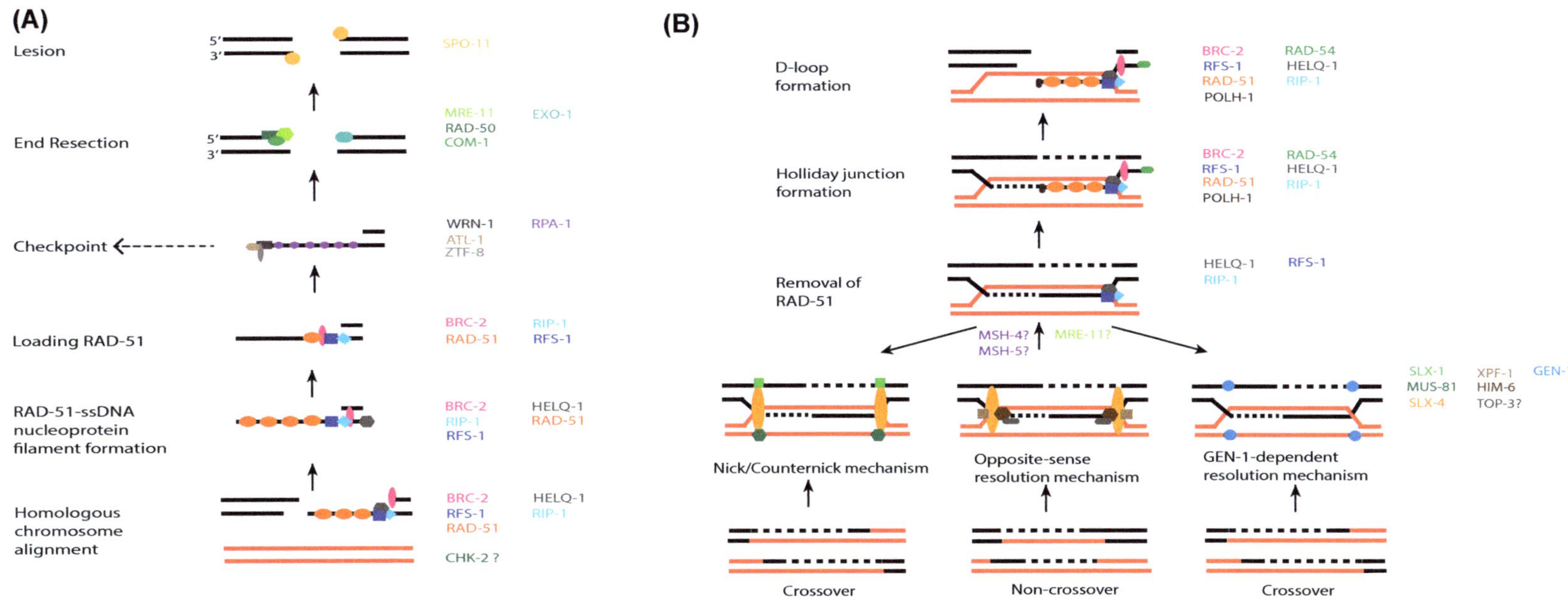

FIGURE 11.2 **Schematic diagram of the *C. elegans* homologous recombination pathway.** All colors of the proteins match the colors of the names of the proteins (A) Homologous recombination starts when the protein SPO-11 introduces breaks around the double-stranded DNA damage. The MRN complex composed of MRE-11, RAD-50, and COM-1, and the nuclease EXO-1 resect the DNA producing an area of ssDNA. The ssDNA is bound by the replication protein A (RPA), which is later replaced by RAD-51 with the help of BRC-2. Proteins, such as WRN-1, ATL-1, and ZTF-8 also act in the HR pathway; however, they have been found to function at the divergence point between checkpoint repair and DSB repair. Once the RAD-51-ssDNA nucleoprotein filament is formed, CHK-2 is speculated to mediate the alignment of the damaged chromosome with a template chromosome. (B) Strand invasion is mediated by a number of proteins: the helicase HELQ-1, the complex formed by RFS-1 and RIP-1, the translesion synthesis polymerase POLH-1, and the recombination protein RAD-54. After strand invasion, a D-loop is formed, followed by a Holliday junction structure. Once new DNA has been synthesized, the DNA is marked by MSH-4 and MSH-5 for resolution by nucleases. The Holliday junction is then resolved to form CO and NCO products. Three different mechanisms have been found at this step: the nick/counternick mechanism involves the scaffold protein SLX-4 acting as a platform for the interactions of SLX-1 and MUS-81, and produces CO in the chromosomes. The opposite-sense resolution mechanism involves, again, SLX-4, as a platform for the interactions of XPF-1 and HIM-6, and the topoisomerase TOP-3 also plays a role, although its functions have still to be elucidated. This mechanism results in chromosomes showing no CO events. The GEN-1-dependent resolution mechanism is known to produce CO events independent of SLX-4. MRE-11 is also involved in the next step of chromosomal CO, although its function is still unclear.

During meiosis I, before recombination, homologous chromosomes must first align with their pairs and associate via the synaptonemal complex (SC) [138]. Subsequently, the topoisomerase II-like enzyme SPO-11 introduces around 11–12 meiotic DSBs per nucleus in a programmed fashion [138,139]. The area around the DSBs is then resected by an endonuclease activity of the MRN complex, which causes the release of SPO-11 and the formation of a 3′-single-stranded DNA (ssDNA) tail [140,141]. This ssDNA tail is bound by replication protein A (RPA) that is later replaced with RAD-51 proteins, which are recombinases with individually weak activity [142,143].

The protein complex consisting of MRE11, RAD50, and NSB1 (MRN complex) in mammals, or Mre11, Rad50, and Xrs2 (MRX complex) in yeast, derives its name from the proteins required for meiotic cell division. Presently, *C. elegans* homologs for MRE-11, RAD-50, and a putative candidate that shares significant homology to the human NBS1 protein have been identified [129,138,140]. Apart from initiating DSB repair by resecting the damaged DNA strand, MRE-11 is involved in downstream processes of repair, which are crucial for chromosomal crossover. In addition, MRE-11 plays a role in NHEJ [144]. In 2013, Lemmens and colleagues identified the 5′ to 3′ acting nuclease EXO-1 to be involved in DSB repair [126]. The proposed model predicts that bidirectional DNA end resection takes place with MRE-11 starting resection in a 3′ to 5′ direction and allowing the efficient recruitment of EXO-1, which facilitates resection from 5′ to 3′. The MRN complex only aids in the recruitment of EXO-1 during the early prophase I of meiosis, and not in later stages of cell division [130].

Further, the model suggests that MRE-11 causes the release of SPO-11 and, together with the meiotic recombination factor COM-1, triggers the release or blockage of the protein complex Ku, which consists of CKU-70, an ortholog of human XRCC6, and CKU-80, an ortholog of human XRCC5. This mechanism allows HR to repair up to 97% of all the meiotic DSBs and suppresses compensatory NHEJ activation [126,130]. Ku proteins have been suggested to be toxic when unbound, as they can cause chromosomal aggregates and cause a decrease of crossing-over events and thus of meiotic recombination. Despite Ku protein toxicity, germline cells in *C. elegans* have a bias of at least one hundred-fold toward HR compared to NHEJ repair [126].

The *C. elegans* gene rdh-1/rad-51 codes for RAD-51 protein monomers, which bind the damaged DNA and form a nucleoprotein filament that is responsible for two functions: finding a homologous template and invading the homologous DNA strand [138,140,141,145]. RAD-51 activity requires BRC-2, the *C. elegans* homolog of the human breast cancer type 2 susceptibility protein BRCA2 [137]. BRC-2 transports RAD-51 from the cytoplasm to the nucleus, where it is recruited to the sites of DSBs, and loaded onto the ssDNA. RAD-51 function depends on BRC-2 activity but not vice versa. BRC-2 has been shown to interact with the RAD-51 paralog RFS-1, which mediates binding the DSB and displacing RPA, before recruiting RAD-51 onto the DNA [146].

Upon binding, RAD-51 proteins convey the ssDNA to invade the homologous double-strand DNA (dsDNA), which will be used as a template for synthesis of new DNA on the resected end of the invading ssDNA. This invasion causes the formation of a D-loop structure, which is mediated by BRC-2 and the translesion synthesis polymerase POLH-1 [147]. RAD-51 is also known to catalyze ATP hydrolysis, which leads to the depolymerization of nucleoprotein filaments. This process is suppressed by BRC-2 activity [137,140,141]. After completion of DSB repair, the RAD-51-like protein RFS-1 acts together with the helicase HELQ-1 to release RAD-51 from the DNA by directly interacting with the DNA via distinct mechanisms [146].

RFS-1 and HELQ-1 display additional functions: RFS-1 has the prorecombinogenic role of loading RAD-51 onto ssDNA, while HELQ-1 stabilizes the ssDNA-RAD-51 filaments prior to strand invasion [146]. In addition, RFS-1 stabilizes the HR mechanism in replication fork barriers during S-phase by mediating the loading of RAD-51 to one-ended DSBs, that occur upon replication fork regression, and to ssDNA stalled replication forks [128]. RFS-1 activity requires complex formation with RIP-1 (RFS-1 Interacting Protein), which is crucial for optimal HR function in *C. elegans*. In addition, this complex stimulates the recombinase activity of RAD-51, which in turn remodels and stabilizes RAD-51-ssDNA filaments to take a more flexible conformation. Thereby, RFS-1/RIP-1 facilitates the search for a homologous DNA template and strand exchange with the template via displacement loop (D-loop) formation [62]. In yeast, RAD-51-mediated strand exchange, cross-bridging double-stranded DNA, and remodeling the chromatin to facilitate for HR repair, is aided by RAD-54, which is proposed to be conserved in *C. elegans* [146]. Subsequently, the D-loop structure induces a so-called "double Holliday junction" (dHJ or HJ), and later a crossover (CO) that is visible as a chiasmata and needed for the continuation of meiosis. Studies in *C. elegans* show that interhomolog crossovers are tightly regulated, limiting the occurrence to one crossover event between homologous chromosomes, while the other DSBs are repaired as non-crossovers [93,148].

Both HJ and CO formation are dependent on meiosis-specific members of the MutS homolog family, namely HIM-14/MSH-4 and MSH-5 proteins in the nematode [140]. Some evidence shows that MSH-4 and MSH-5 mark DSBs for resolution by nucleases but their exact functions have not been elucidated [148]. Another important element in DSB repair is the checkpoint kinase CHK-2, which is speculated to mediate alignment of homologous chromosomes and act epistatic to RAD-51, thus influencing the formation of the chiasmata [63].

COs can form via the opposite-sense resolution of double Holliday junctions or via a "nick/counternick" mechanism [149]. COs at collapsed replication forks in mitotic and meiotic nuclei are resolved by a protein complex consisting of the scaffold protein SLX-4 (other name: HIM-18), which acts as a platform for the coordination of multiple enzymes involved in processing recombinant intermediates [95,149]. SLX-4 either interacts with the endonucleases SLX-1 and MUS-81 involved in the "nick/counternick" mechanism, or the ATP-dependent helicase HIM-6 and the endonuclease XPF-1, the *C. elegans* ortholog of ERCC4, involved in the "opposite-sense resolution" of HJs [93,95]. SLX-1 and MUS-81 have a different substrate preference as compared to XPF-1 [148]. According to the "nick/counternick" model, SLX-4 coordinates a symmetrical cleavage, in which SLX-1 nicks HJs that are subsequently processed by MUS-81 endonuclease [93]. Conversely, the "opposite-sense resolution" pathway employs HIM-6 to unwind thermodynamically unstable HJs, while XPF-1 acts as an HJ-nicking enzyme [150]. In addition, HIM-6 is important for recombination initiation, for which it requires interaction with the nuclease DNA-2 [95].

Consistently, MUS-81 acts redundantly to HIM-6 in limiting the accumulation of double HJs during early meiosis, and to XPF-1 in the production of interhomolog crossovers. MUS-81 also has a nonredundant but overlapping role with SLX-4 in processing recombinant intermediates, such as dHJs. Several studies show that SLX-1 regulates CO distribution along the chromosomes by locally suppressing the formation of COs at the center of the chromosomes via "same sense resolution" of HJs to produce non-crossover products, or via synthesis-dependent strand annealing [93,148–150].

In a parallel pathway, the *C. elegans* homolog of the human HJ resolving enzyme GEN1 is involved in the repair of DSBs upon IR-induced DNA damage by producing same sense HJ nicks at later steps of the repair [127,149]. GEN1-mediated DSB repair is the major pathway to resolve HJs and form COs upon DNA damage in mammals and flies [151–153].

Consequently, several DSB repair factors are involved in DNA-damage checkpoint signaling in response to IR, including GEN-1 and HIM-6 [127]. Here, HIM-6 has a crucial role in ensuring normal mitotic function and processes 80–90% of all meiotic recombination intermediates, generating non-crossover products [95]. To this end, HIM-6 acts partially redundant when compared with the topoisomerase TOP-3 downstream of SPO-11 and RAD-51, preventing a toxic accumulation of recombination intermediates [154]. TOP-3 is required for meiotic recombination and interacts both genetically and physically to process DNA damage during normal mitotic germ cell divisions and to form non-crossovers [93,154].

HIM-6 physically interacts with the *C. elegans* ATR homolog ATL-1 during DNA damage–checkpoint responses, while their exact role in DSB repair is not known [154]. However, several studies indicate that ATL-1 functions in DSB repair requires the simultaneous activation of CHK-1 by the helicase WRN-1, ortholog of human Werner's syndrome factor WRN. Studies in *C. elegans* reveal that in response to IR-induced DSBs, WRN-1 functions upstream of ATL-1 and ATM-1 to trigger cell cycle arrest, by regulating proper RPA-1 stabilization [140,155]. Further, WRN-1 causes nuclear accumulation of ATM-1 and takes the role of the human regulatory partner of ATR, ATRIP, in recruiting both ATL-1 and ATM-1 to the replication fork [156,157].

The checkpoint kinases of the phosphatidylinositol-3 kinase-related kinases (PIKK) family, ATM-1 and ATL-1, regulate CO interference and are involved in the initial steps of DSB recognition in mitotic and meiotic germ cells, triggering checkpoint response and cell cycle arrest, or apoptosis [54]. ATM-1 is involved in IR-induced DSB response independent of HR and NHEJ [140]. Upon activation, via autophosphorylation, ATM-1 gets recruited to the DNA-damage site by interacting with the MRN complex factor MRE-11 and RPA-1. UV/IR induces replication fork stalling and DSB formation, causing the recruitment of ATL-1, by RPA-1-bound-ssDNA, to the damaged site and activation by WRN-1 activity. ATL-1 requires MRE-11 to bind to the resected ssDNA–RPA-1 complex. This preprocessing step, before ATL-1 recruitment, is apparently required upon IR-induced DSBs but not observed during replication stress [140].

The presence of ATL-1 and ATM-1 in mitotic and meiotic germline cells with DSBs is of importance for the factor ZTF-8, a functional homolog of the mammalian RHINO, which is involved in DDR and cell cycle regulation [158]. ZTF-8 is a player of meiotic recombination as it is involved in processing stalled replication forks and has a supportive function in intersister repair when a homologous chromosome is not available. ZTF-8 also aids in recruiting the 9-1-1 complex and other proteins to the site of DSB. 9-1-1 is formed by the proteins HPR-9, HUS-1, and MRT-2 (RAD9, HUS1, and RAD1 in mammals) and is commonly known as the cell cycle checkpoint complex. Both MRT-2 and HUS-1 has been found to directly interact with ZTF-8, highlighting its importance in the checkpoint response, specifically in DNA-damage checkpoint-induced apoptosis. ZTF-8 activation and localization, as well as interaction with ATL-1 and ATM-1, require SUMOylation. ZTF-8 is proposed to act upstream of the divergence point of the DSB repair and the checkpoint repair pathways, suggesting that the 9-1-1 complex is required to interact with ZTF-8 for either of the repair pathways to take place [159].

8.2 Nonhomologous End Joining

During meiosis I and II, the spindles pull apart the DNA strands, which can tear apart HJs, by force, leading to DSBs that are commonly repaired by NHEJ [93]. NHEJ depends on the conserved Ku proteins CKU-70 and CKU-80, which form heterodimers at the damaged site to protect the DNA from HR-mediated resection. NHEJ is a highly error-prone repair mechanism that is suppressed in germ cells by the protein COM-1, the *C. elegans* homolog for the human tumor-suppressor CtIP, to ensure maintenance of the genetic material for the next generation. During meiosis, COM-1 blocks the toxic effects of CKU-70 and CKU-80 by misplacing them from the damaged DNA. COM-1-deficient animals are able to repair SPO-11-induced DSBs via HR and NHEJ, indicating that COM-1 is not required for meiotic recombination per se, but for DNA end resection and CO formation [54,126].

In addition, DNA repair of induced meiotic DSBs, in the absence of the MRN complex, takes place in the middle to late-prophase phases of meiosis I and is not carried out via HR but via NHEJ. This indicates that the MRN complex is essential for HR activity but can be bypassed by EXO-1-dependent resection [130].

8.3 Other Conserved DSB-Repair Mechanisms

Alternatively to HR and NHEJ, DSBs can be repaired by single-strand annealing (SSA) and alternative end joining (alt-EJ), which is also termed microhomology-mediated end joining (MMEJ). Both act in somatic cells and can make use of a large stretch of homology (30–400 bp) or a very small stretch of homology (5–15 bp) between damaged and template strands [54]. SSA employs a genetic network overlapping with HR, including XPF-1 activity and RPA binding to the resected ends of the damaged ssDNA. Subsequently, BRC-2 binds to the ssDNA to displace RPA from the DSB [160]. In *C. elegans*, no clear mechanism for alt-EJ has been identified yet [131].

Meiotic recombination can also make use of a sister chromatid instead of a homologous chromosome as a template, and not lead to CO. This mechanism consists of a sub-repair route of HR that is called synthesis-dependent strand annealing (SDSA) or intersister HR. SDSA occurs early in meiosis, and is independent of recombinant intermediates. It takes place after the disassembly of the D-loop when the ssDNA has been resected, and interacts with the other broken DNA end [93].

During *C. elegans* meiosis, only one CO event per chromosome can be observed, while at least two DSBs occur, indicating a second level of DSB repair mediated by non-crossover events (NCO) [161]. During SDSA, RTEL-1, the homolog of the human regulator of telomere elongation helicase 1, is employed to limit CO formation by dissociating strand-invasion events [162]. The elongated invading strand is then annealed to the complementary ssDNA tail on the other side of the DSB, the single-stranded gaps are filled and the nicks ligated, which bypasses CO product formation [54].

Intersister HR is crucial to repair a number of meiotic DSBs to ensure genomic integrity. This sub-pathway is of major importance at the arm regions of the chromosomes where most DSBs are converted into NCOs in an RTEL-1-dependent manner [148]. Intersister HR involves the *C. elegans* homolog of the breast cancer tumor suppressor BRCA1, which interacts with SMC-5 and SMC-6, the homologs of the structural maintenance of chromosome proteins in humans. Single mutants of *smc-5* and *smc-6* show phenotypes similar to *brc-1* mutant, the proteins localize to condensed nuclear chromosomes and also appear to be involved in meiotic intersister HR [163].

Alt-EJ (or MMEJ) acts independently of other DSB-repair pathways, highlighting its importance in avoiding developmental arrest of the animal [131]. Both SSA and alt-EJ require DNA replication and/or cell cycle progression occurrence for their function and they produce similar end products.

Interstrand crosslink (ICL) is another type of DNA damage characterized by interlinking the two strands of the DNA double helix, which blocks replication, and interrupts the translocation of crucial proteins along the DNA required for transcription. Nucleolytic processing of ICL lesions has been shown to lead to DSBs, which creates a substrate for HR. However, ICL repair involves factors from various repair pathways including the Fanconi anemia (FA) proteins and members of the NER pathway [164]. The ubiquitin ligase RNF-113 is important for the repair of ICL-induced DSBs: it acts epistatically to RFS-1 and mediates RAD-51 binding to ssDNA. RFS-1 binds to the ssDNA together with RPA-1. RNF-113 ubiquitinates an unknown factor, most likely RFS-1 or RPA-1, causing the release of RPA-1 from the DNA and subsequent RAD-51 binding to the ssDNA, which initiates strand invasion and the HR pathway [165]. Once RAD-51 is loaded onto the ssDNA, the putative histone demethylase JMJD-1.1 modulates the chromatin structure and influences RAD-51 dissociation from the single strand. The exact mode of action of JMJD-1.1 is not known, although some synergism with RAD-54 in the HR regulation has been demonstrated [166]. The endonucleases MUS-81 and XFP-1 generate ICL-induced DSBs by causing dual incisions around the damage site in order to separate the covalently bound DNA strands [149].

The ICL-specific repair pathway, known as Fanconi anemia (FA), employs the DNA helicase HEL-308 that is known to act in HR [164]. Further, FA involves FCD-2, the homolog of human FANCD, which is known to guide DSBs to HR repair, instead of NHEJ [165]. FCD-2 binds to the DNA together with BRC-2 and RAD-51. HEL-308 partly contributes to the loading and stabilization of RAD-51 on the ssDNA [167].

The NER machinery cooperates with HR damage repair of UV-C lesions in the germ line: the current hypothesis proposes that damage is either directly repaired via HR activity and, in parallel, the NER pathway produces repair intermediates which are subsequently processed and repaired by GG-NER. Alternatively, the NER machinery activates the well-conserved core HR factors RPA-1, MRE-11, and RAD-54, and employs the 9-1-1 complex to induce p53/CEP-1-mediated germ cell apoptosis [37].

9. DNA-DAMAGE CHECKPOINTS

DNA-damage checkpoint is a signal transduction pathway that halts cell cycle progression upon detection of different DNA lesions. Various sensors recognize DNA damage, including the ring structure complex Rad9, Hus1, and Rad1 (9-1-1 complex) or the ataxia telangiectasia (AT) mutated (ATM) protein, and the ATM and Rad3-related (ATR) response complex. Depending on the quality and extent of the DNA damage, specific signal transducer proteins mediate the activation of effector protein networks that respond by arresting the cell cycle, triggering DNA repair, or leading to apoptotic cell death. Genetic networks in eukaryotes that define DNA-damage checkpoint and apoptotic response to DNA damage are evolutionarily conserved and well defined in nematodes, flies, and mammals [13,168–171].

DNA damage–checkpoint responses during mitosis and meiosis are genetically distinct: somatic cells arrest proliferation to allow time for DNA repair, while meiotic germ cells that carry DNA-damage or -display asynapsis are removed by apoptosis to ensure genomic stability across generations [61]. The *C. elegans* germline is specifically suitable to study both mitotic and meiotic checkpoint mechanisms, since cell cycle arrest and apoptosis are spatially defined, easily quantified, and controlled by a well-known developmental gene network [13]. Cell cycle arrest in the mitotic germline can occur during G1/S, S, and G2/M checkpoint phases of cell cycle progression in order to allow repair before DNA replication or cell division [171–174]. The importance of apoptotic events in maintaining genomic stability across generations is emphasized in two waves of cell death during *C. elegans* development. The first wave occurs during embryogenesis, where a sequence of well-defined apoptosis events in somatic cells determines tissue development and shapes the organism. The second wave occurs during oogenesis in the adult germline to eliminate cells that could compromise the transfer of genetic material to the offspring. Several reviews summarize the genetic pathway of programmed cell death in *C. elegans* [175,176]. Apoptosis events in the germline occur to maintain tissue homeostasis and can be triggered in response to bacterial pathogens or genotoxic stresses, in each case employing genetically distinct pathways.

9.1 Sensors of the DNA Damage Response

The major checkpoint sensor complex 9-1-1 is well conserved in *C. elegans*: homologs of the yeast Rad9, Hus1, and Rad1 proteins are HPR-9, HUS-1, and MRT-2 in the nematode [44,177]. The 9-1-1 scans the chromatin and senses DNA damage– or unrepaired-- recombination intermediates, for example, produced during the DSB-repair procedure [44]. The complex acts as a recruitment platform for the translesion synthesis (TLS) machinery to act on stalled replication forks. TLS repair is able to replicate the DNA amid unrepaired lesions [158,178,179].

During checkpoint activation, HPR-9 and MRT-2 interact to achieve proper nuclear localization of the HUS-1 and the 9-1-1 complex [44,172]. The 9-1-1 complex interacts with the transducer factor ZTF-8, homolog of the mammalian protein RHINO, to resolve DNA damage that leads to replication fork stalling, meiotic checkpoint activation, as well as the repair of meiotic and mitotic DSBs. To that end, transducer activity of the checkpoint kinase members of the PI3K superfamily, ATM-1 and ATL-1, homologous to the mammalian ATM and ATR proteins, phosphorylate ZTF-8 at the chromatin or in the nucleolus. ZTF-8 acts as a TLS platform during S-phase and is required for the optimal regulation of the DNA damage–induced apoptotic pathway [158].

RAD-5 (also called CLK-2), ortholog of the *S. cerevisiae* telomere length–regulating protein Tel2p, acts independently of 9-1-1 as a checkpoint-sensing factor in response to endogenous and exogenous DNA damage [171]. This pathway is absent in yeast, suggesting that it developed as a novel pathway during evolution in metazoans [172]. Studies in the early 2000s localized CLK-2/RAD-5 at DNA-damage sites, either as a primary sensor for damage or as a repair protein, affecting both the DNA-damage checkpoint as well as the S-phase replication checkpoint downstream of ATL-1 [168,169].

9.2 Checkpoint Sensor Proteins in Telomere Length Maintenance

The 9-1-1 factors HUS-1 and MRT-2 are implicated in telomere length maintenance, by facilitating telomerase-mediated telomere replication and acting to prevent telomere shortening. This role of 9-1-1 appears to be an evolutionary adaptation as it is presently undiscovered in yeast but highly conserved in multicellular organisms. Several studies show that HUS-1 and MRT-2 recognize telomeres either during replication fork stalling at telomeres, or during S phase when unfolding of chromatin or telomere-binding proteins at the T-loop can lead to a structure similar to a recombination intermediate. Subsequent to recognition, checkpoint proteins are recruited [180,181]. Upon telomere binding, the 9-1-1 complex and the *C. elegans* Rad17 RFC clamp loader homolog, HPR-17, stimulate the recruitment of telomerase. To mediate repair, telomeric chromatin is kept in an open conformation by the activity of PME-5 (also called TANK-1), which contains a C-terminal PARP regulatory and catalytic domain, which is upregulated by HUS-1 [182].

9.3 Effectors of DNA-Damage Checkpoints

After DNA-damage detection the signal has to be passed from the transducers to the effectors. To this end, transducers often amplify and diversify the signal via phosphorylation of multiple effectors. Two main transducers in *C. elegans* are ATM-1 and ATL-1. One effector that is being phosphorylated is CHK-2, which is a member of the Cds1/Chk2 checkpoint kinase family that acts at the checkpoints G1/S, G2/M, and S phase by transforming information detected by the sensors and translated by the transducers, into specific biological responses. CHK-2 is required for pairing and spatial reorganization of homologous chromosomes during early meiotic prophase. To this end, CHK-2 induces the co-localization of homolog sister chromatids and organizes the chromatin into a form more receptive to pairing. Further, it regulates the length of the premeiotic S phase and mediates completion of replication and pairing of chromosomes [63].

Another effector phosphorylated by ATL-1 is CHK-1, a CHK1-like serine threonine protein kinase, which has a conserved role from basal metazoans to humans mediating cell cycle arrest at the S/M checkpoint during early embryogenesis and in the postembryonic germline cell cycles [183,184].

ATM-1 and ATL-1 are also known to connect the checkpoint pathway to the apoptosis stimulation in the germline: as discussed earlier, DNA damage induced by gamma irradiation leads to the recruitment of HUS-1 and ZTF-8.

DNA damage–checkpoint activation in meiotic pachytene cells leads to activation of the *C. elegans* p53 homolog CEP-1. Prior to late pachytene, CEP-1/p53 is kept at bay by translational repression through GLD-1, thus preventing the DSBs from triggering apoptosis aberrantly during normal meiotic recombination [185]. Once activated, CEP-1/p53 induces the expression of the proapoptotic gene *egl-1* and *ced-13* [44,186]. Both of those BH3-only domain proteins trigger the apoptotic demise of meiotic pachytene cells by removing the Bcl2 homolog CED-9 from the Apaf-1 homolog CED-4, which in turn activates the CED-3 caspase [187–192].

Germ cell survival and suppression of apoptosis upon minor DNA-damage events caused by environmental factors or meiotic recombination is mediated by the anti-apoptotic factor ABL-1 that negatively regulates CEP-1, thereby allowing the system to distinguish between different types of DNA damage, such as DSBs and DNA adducts [193].

9.4 Cytokinesis Checkpoint

APC/C, the anaphase-promoting complex/cyclosome, is the major regulator of chromosome segregation in eukaryotes. It is an E3 ubiquitin ligase, which facilitates polyubiquitination of its substrates for degradation via the ubiquitin–proteasome system [194]. During the metaphase-to-anaphase transition APC/C acts to degrade the protein IFY-1 (interactor with FZY-1), a *C. elegans* securin, which inhibits the activity of the separase SEP-1 [195]. Separases are enzymes important for cleaving cohesin, which is the complex holding the sister chromatids together [196,197]. Cohesin produces a tension in the spindle microtubles, opposite to the pull produced by the spindle pole on microtubules attached to the kinetochore of each sister chromatid. Upon faulty attachment of kinetochores or the absence of tension in the spindle, APC activity is inhibited via sequestration of FZY-1, the ortholog of the *S. cerevisiae* Cdc20p [195]. This inhibition causes a delay in the onset of anaphase and is called the spindle assembly checkpoint (SAC). Only when the spindle is correctly assembled, the cell continues through its cycle [198–200].

FZY-1 is inhibited via the involvement of the kinetochore-bound MDF-1–MDF-2 complex interfering with free MDF-2 proteins and changing their conformation, from an open to a closed MDF-2 form, which in turn inhibits FZY-1 [201]. This complex, however, is not the sole inhibitor of FZY-1. The kinase BUB-1 has also been suggested to either bind and phosphorylate FZY-1 or function via the MDF-1–MDF-2 complex to inhibit FZY-1, as well as regulate kinetochore function and chromatin cohesion [199–201]. Besides BUB-1, the kinetochore scaffold protein KNL-1 regulates two other components of checkpoint activation: the NDC-80 complex and the RZZ (Rod/Zwilch/Zw10) complex, of which only the kinetochore components ROD-1 and CZW-1, homologs of Rod and Zwilch, respectively, have been described in *C. elegans*. BUB-1 and KNL-1 interact with HCP-3, homolog of the centromere CENP-A protein, and with HCP-4, homolog of the centromere CENP-C protein, to mediate the localization of HCP-1, a centromere CENP-F protein homolog, and of HCP-2, an ortholog of the human CAGE1 protein. The functions of HCP-1 and HCP-2 proteins are not fully elucidated but, similarly to the spindle checkpoint component BUB-1, they overlap in spindle checkpoint regulation by interacting with the SAC proteins SAN-1 and MDF-2 to facilitate the correct pairing of chromosomes and their segregation [198,201].

HCP-1 and HCP-2 regulate the levels of free MDF-2, which is rate limiting for the folding of MDF-2 in its closed conformation, thus inhibiting checkpoint. The rate-limiting levels allow for the integration of yet another branch of SAC, in which SAN-1 and BUB-3 interact in the cytoplasm to inhibit APC/C. Both checkpoint branches involving MDF-1/MDF-2 and SAN-1/BUB-3 are not sufficient to cause cell cycle arrest on their own [201].

The APC/C pathway for chromosomal alignment and segregation only takes place in meiosis. Alignment and segregation of the chromosomes generally can differ during meiosis and mitosis. However, the aurora kinases, also known as chromosomal passengers, act in both mitosis and meiosis [202]. *C. elegans* expresses two aurora kinases AIR-1 and AIR-2, but only AIR-2 acts in SAC. AIR-2 is a kinase involved in chromosome alignment during metaphase I in mitosis, where it localizes to the point of contact between sister chromatids [203]. During metaphase II of meiosis, AIR-2 is involved in chromosomal separation, where it localizes to chromosome arms distal to the chiasmata [204]. AIR-2 moves along the microtubules from the chromosomes to the midzone microtubules during division and is required for mechanistically similar processes, such as polar body extrusion and stabilization and completion of cytokinesis. During this process, AIR-2 promotes proper localization of other midbody microtubule components, such as ZEN-4, an MKLP-1-related kinesin. AIR-2 acts upstream of ZEN-4 and physically interacts with ZEN-4, allowing association with the spindle midzone and aiding in polar body extrusion and cytokinesis [205].

AIR-2 also regulates proper localization of BMK-1, a BimC kinesin. The same study shows that AIR-2 kinase activity and movement depends on ICP-1, a chromosomal passenger protein, that mediates physical interaction between AIR-2 and BMK-1 via phosphorylation events at three residues, causing BMK-1 relocalization from the kinetochore microtubules to the midzone microtubules [205].

For chromosomal segregation to take place during meiosis AIR-2 must phosphorylate the meiosis cohesin REC-8. Phosphorylation of this cohesin by AIR-2 is negatively regulated by CeGLC-7 α/β phosphatases. These phosphatases antagonize AIR-2 by blocking its localization, thereby inhibiting AIR-REC-8 dephosphorylation [202]. Although AIR-2 phosphorylates cohesin during meiosis, it remains unclear whether it also executes this role during mitosis. Instead, during mitosis in embryonic and postembryonic development, AIR-2 promotes the bi-orientation of sister kinetochores, the association of condensin to chromosomes, and the function of some condensins, such as chromosomal organization [204,206].

10. CONCLUDING REMARKS

The nematode model has been firmly established as an important model system for studying DNA repair. *C. elegans* has played a major role as a model organism for a large variety of biological processes, including programmed cell death, neurobiology, RNA interference, development, and aging. The traceable genetics and host of methodologies have made the worm also a crucial system for investigating DDR ranging from mechanistic discoveries of DSB repair, also in the context of meiosis, to the systemic responses, on the organismal level. In many ways, *C. elegans* has closed the gap between the traditional "work horses" of genome stability research, the powerful genetic yeast system and the mouse as a disease model. The vibrant and ever-expanding community of *C. elegans* research will continue to gain new and unexpected insights into cellular and organismal mechanisms of genome stability.

GLOSSARY

6-4 photoproducts The consequence of a covalent bond formation of carbons at position six and four between adjacent thymine bases upon exposure to UV. The resulting distortion of the DNA helix can be removed by NER mechanism.

9-1-1 complex A ring structure complex formed by Rad9, Hus1, and Rad1 (gene names taken from yeast) which act as a sensor complex for recognizing DNA damage.

Alternative end joining Also known as microhomology-mediated end joining, is a type of double-strand break repair which is found only in somatic cells and can use as many as 30–400 bp or as little as 5–15 bp of homology between the damaged and the template strand for repair. No clear mechanism for this type of repair has been identified in *C. elegans* yet.

Anaphase-promoting complex/cyclosome An E3 ubiquitin ligase, which facilitates polyubiquitination of its substrates for degradation via the ubiquitin–proteosome system. It is the major regulator of chromosome segregation in eukaryotes, and acts by tagging specific proteins for degradation, such as the protein IFY-1, in *C. elegans*.

Ataxia telangiectasia A rare inherited neurodegenerative disease that is defined by impaired coordination and small, dilated blood vessels. Patients display uncoordinated movements, a weakened immune system, and DNA-repair deficiency of double-strand breaks based on defects in the ATM gene.

Cockayne syndrome A rare autosomal recessive neurodegenerative disorder underlying DNA-repair defect that includes devastating characteristics, such as growth failure, misdevelopment of the nervous system, high sensitivity to sunlight, and premature aging, but no cancer predisposition.

Crossover An event that occurs after Holliday junction formation. While the newly synthesized DNA strand and the template DNA strand cross over, genetic information can be exchanged, producing recombinant chromosomes. This exchange of genetic material is called crossover.

Cyclobutane pyrimidine dimers Arise upon UV irradiation that causes the coupling of C=C double bonds of pyrimidines in thymine or cytosine. The resulting four-membered ring structure leads to distortion of the DNA helix structure that is primarily repaired by NER in eukaryotic cells.

Displacement loop The structure formed by the damaged strand of DNA and the template DNA being used for repair. It forms after strand invasion has taken place, and refers to the shape the template DNA strand forms.

DNA double-strand breaks A type of DNA damage that cause both DNA strands to break. It is considered to be the most toxic form of DNA damage and can be repaired by a variety of different ways, all of which involve the simultaneous repair of both strands at the same time.

DNA-damage response The mechanisms in which each organism detects the DNA damage and initiates its repair.

DNA-damage checkpoint Is a type of DNA-damage response which causes the cell to halt its cell cycle progression, and trigger either DNA repair, or apoptotic cell death.

DNA glycosylase Mediates the repair of single damaged bases in DNA in base excision repair by flipping the damaged base out of the double helix and subsequently cleave the N-glycosidic bond. This creates an apurinic/apyrimidinic site and leaves the sugar-phosphate backbone intact.

Fanconi anemia A genetic disorder that causes bone marrow failure.

Holliday junction The structure formed consecutive to the displacement loop, after new DNA has been synthesized using the template DNA strand and two points are formed where the newly synthesized DNA strand and the template DNA strand cross over.

Homologous recombination A major type of double-strand break repair that is characterized as an error-free method, and uses either the sister chromatid or homologous chromosomes as undamaged templates for repair.

Ionizing radiation Majorly gamma rays, X-rays, and to some extent UV radiation that carry enough energy to free electrons from atoms or molecules, thereby resulting in their ionization. Causes a broad range of damage to DNA, including double-strand breaks that are repaired by a variety of DNA-repair mechanisms.

Intersister homologous recombination A type of double-strand break repair which uses a sister chromatid instead of a homologous chromosome as a template. It does not lead to crossover events. It is also called synthesis-dependent strand annealing.

Interstrand crosslink Another type of DNA damage characterized by interlinking the two strands of the DNA double helix, thus blocking replication and interrupting the translocation of crucial proteins along the DNA required for transcription.

Microhomology-mediated end joining Also known as alternative end joining, is a type of double-strand break repair which is found only in somatic cells and can use as many as 30–400 bp or as little as 5–15 bp of homology between the damaged and the template strand for repair. No clear mechanism for this type of repair has been identified in *C. elegans* yet.

MRN complex Named after the proteins that for this complex are found in mammals, subsequently MRE11, RAD50, and NSB1. It has also been found in yeast, this time named as the MRX complex, after the proteins Mre11, Rad50, and Xrs2. This complex is important during the initiation of the repair of double-strand breaks.

Non-crossover The situation in which the Holliday junction is resolved and does not result in a crossover event. Since crossover events only occur once per chromosome, non-crossover events are common.

Nonhomologous end joining A major type of double-strand break repair that is known for its efficient, yet error-prone repair which joins damaged DNA ends regardless of their homology, leading to the addition or removal of nucleotides.

Nick/counternick mechanism A mechanism used by cells to resolve the Holliday junction structure and can lead to the occurrence of crossovers. The mechanism involves the proteins SLX-1 and MUS-81 that act by nicking the Holliday junction twice, one after the other, in a symmetrical manner.

Opposite-sense resolution A mechanism used by cells to resolve the Holliday junction structure and can lead to the occurrence of crossovers. The mechanism involves the proteins HIM-6 and XPF-1, which unwind thermodynamically unstable Holliday junctions, and nick the Holliday junction, respectively.

Poly-ADP-ribose polymerase An enzyme that mediates single-strand DNA break repair and programmed cell death and requires NAD^+. Upon single-strand break detection, PARP binds to the DNA and synthesizes a poly ADP-ribose (PAR) chain to signal a DNA repair mechanism involving XRCC1. Upon repair, PAR chains are degraded by Poly(ADP)-ribose) glycohydrolase.

Synaptonemal complex The protein structure that forms between homologous chromosomes during meiosis. This complex is important for chromosome alignment and pairing, synapsis and recombination.

Single-strand annealing A type of double-strand break repair which is found only in somatic cells and can use as many as 30–400 bp or as little as 5–15 bp of homology between the damaged and the template strand for repair.

Synthesis-dependent strand annealing A type of double-strand break repair which uses a sister chromatid instead of a homologous chromosome as a template. It does not lead to crossover events. It is also called intersister homologous recombination.

Spindle assembly checkpoint A type of cell cycle halt which causes a delay in the onset of anaphase. It occurs in mitosis and ensures the fidelity of chromosome segregation, since only when the spindle is correctly assembled is the cell allowed to continue through its cycle. Upon faulty attachment of kinetochores or the absence of tension in the spindles, the spindle assembly checkpoint is activated.

Sister chromatids Identical chromatids that were produced by replication of only one copy, and can be found together bound by a centromere.

Sumoylation A posttranslational modification, similar to ubiquitylation, however with the addition of SUMOs (small ubiquitin-like modifiers) instead of ubiquitin. This modification can affect both protein structure and its subcellular localization.

Translesion synthesis A type of DNA damage–repair mechanism that is able to replicate the DNA amid unrepaired lesions. This type of repair uses more specialized translesion polymerases which are able to insert new bases next to damaged nucleotides.

Xeroderma pigmentosum An autosomal recessive genetic disorder in which the ability to repair DNA damage, including pyrimidine dimers and 6-4 photoproducts, caused by UV light is deficient. Patients early on develop basal cell carcinomas and most commonly die upon the occurrence of metastatic malignant melanomas and squamous cell carcinoma.

LIST OF ABBREVIATIONS

53BP1 p53 binding protein 1
6-4PP 6-4 photoproducts
alt-EJ Alternative end joining
AP site Apurinic/apyrimidinic site
AT Ataxia-telangiectasia
BER Base excision repair
C. elegans *Caenorhabditis elegans*
CO Crossover
CPDs Cyclobutane pyrimidine dimers
CS Cockayne syndrome
CSA Cockayne syndrome protein A
CSB Cockayne syndrome protein B
DDR DNA-damage response
D-loop Displacement loop
DNA Deoxyribonucleic acid
DSBs Double-strand breaks
dsDNA Double-stranded DNA
ERCC1 Excision repair cross complementation group 1e
FA Fanconi anemia
GG-NER Global genome NER
HJ Holliday junction
HR Homologous recombination
ICL Interstrand crosslink
IFY-1 Interactor with FZY-1
IIS Insulin/insulin-like growth factor signaling
IR Ionizing radiation
MMEJ Microhomology-mediated end joining
MMR Mismatch repair
MMS Methyl methanesulfonate
mtDNA Mitochondrial DNA
ncDNA Nuclear DNA
NCO Non-crossover
NER Nucleotide excision repair
NHEJ Nonhomologous end joining
OXPHOS Oxidative phosphorylation
PARGs Poly(ADP-ribose)glycohydrolases
PARP Poly(ADP-ribose) polymerase
PARP-1 Poly-ADP-ribose polymerase-1
PI3K Phosphatidylinositol 3-kinase
PIKK Phosphatidylinositol-3 kinase-related kinases
PIP3 Phosphatidylinositol--3,4,5-triphosphate
Pol β DNA polymerase β
Pol θ DNA polymerase θ
RZZ Rod/Zwilch/Zw10 complex
SAC Spindle assembly checkpoint
SC Synaptonemal complex
SDSA Synthesis-dependent strand annealing
SSA Single-strand annealing

SSB Single-strand break
ssDNA Single-stranded DNA
TC-NER Transcription-coupled NER
TLS Translesion synthesis
TTD Trichothiodystrophy
UV Ultraviolet light
UVB Ultraviolet light type B
UVC Ultraviolet light type C
UV-DDB UV-damaged DNA-binding protein
XP Xeroderma pigmentosum
XPA Xeroderma pigmentosum complementation group A
XPC Xeroderma pigmentosum complementation group A
XPF Xeroderma pigmentosum complementation group F
XPG Xeroderma pigmentosum complementation group F

ACKNOWLEDGMENTS

We thank Najmeh Soltanmohammadi and Ashley B. Williams for providing images. MR is supported by FP7 ITN CodeAge 316354, AFCL by FP7 ITN MARRIAGE 316964, BS acknowledges funding from the Deutsche Forschungsgemeinschaft (CECAD, SFB 829, SFB 670, and KFO 286), the European Research Council (ERC Starting grant 260383), Marie Curie (FP7 ITN CodeAge 316354, aDDRess 316390, MARRIAGE 316964), the German-Israeli Foundation (GIF 1104–68.11/2010), the Deutsche Krebshilfe (109453), and the Bundesministerium fu¨r Forschung und Bildung (Sybacol FKZ0315893A-B).

REFERENCES

[1] De Bont R, van Larebeke N. Endogenous DNA damage in humans: a review of quantitative data. Mutagenesis May 2004;19(3):169–85.
[2] Kirienko NV, Mani K, Fay DS. Cancer models in *Caenorhabditis elegans*. Dev Dyn 2010;239(5):1413–48.
[3] Antebi A. Genetics of aging in *Caenorhabditis elegans*. PLoS Genet 2007;3(9):e129.
[4] Brenner S. The genetics of *Caenorhabditis elegans*. Genetics May 1974;77(1):71–94.
[5] Sulston JE, Schierenberg E, White JG, Thomson JN. The embryonic cell lineage of the nematode *Caenorhabditis elegans*. Dev Biol Nov 1983;100(1):64–119.
[6] Ewbank JJ, Zugasti O. *C. elegans*: model host and tool for antimicrobial drug discovery. Dis Models Mech May 9, 2011;4(3):300–4.
[7] Alexander AG, Marfil V, Li C. Use of *Caenorhabditis elegans* as a model to study Alzheimer's disease and other neurodegenerative diseases. Front Genet 2014;5:279.
[8] Harris TW, Baran J, Bieri T, Cabunoc A, Chan J, Chen WJ, et al. WormBase 2014: new views of curated biology. Nucleic Acids Res Jan 2014;42:D789 (Database issue).
[9] Adenle AA, Johnsen B, Szewczyk NJ. Review of the results from the International *C. elegans* first experiment (ICE-FIRST). Adv Space Res Jul 15, 2009;44(2):216.
[10] White JG, Southgate E, Thomson JN, Brenner S. The structure of the nervous system of the nematode *Caenorhabditis elegans*. Philos Trans R Soc Lond, B, Biol Sci Nov 12, 1986;314(1165):1–340.
[11] Boulton SJ, Gartner A, Reboul J, Vaglio P, Dyson N, Hill DE, et al. Combined functional genomic maps of the *C. elegans* DNA damage response. Science Jan 4, 2002;295(5552):127–31.
[12] Meier B, Cooke SL, Weiss J, Bailly AP, Alexandrov LB, Marshall J, et al. *C. elegans* whole genome sequencing reveals mutational signatures related to carcinogens and DNA repair deficiency. Genome Res Jul 16, 2014:1624–36.
[13] Gartner A, Milstein S, Ahmed S, Hodgkin J, Hengartner MO. A conserved checkpoint pathway mediates DNA damage–induced apoptosis and cell cycle arrest in *C. elegans*. Mol Cell Mar 2000;5(3):435–43.
[14] Hartman PS, Simpson VJ, Johnson T, Mitchell D. Radiation sensitivity and DNA repair in *Caenorhabditis elegans* strains with different mean life spans. Mutat Res Jun 1988;208(2):77–82.
[15] Lans H, Vermeulen W. Accepted Manuscript. Elsevier B.V; Apr 30, 2015. p. 1–22.
[16] Kirkwood TB. Understanding the odd science of aging. Cell Feb 25, 2005;120(4):437–47.
[17] Ermolaeva MA, Segref A, Dakhovnik A, Ou H-L, Schneider JI, Utermöhlen O, et al. DNA damage in germ cells induces an innate immune response that triggers systemic stress resistance. Nature Sep 19, 2013;501(7467):416–20.
[18] Ermolaeva MA, Schumacher B. Systemic DNA damage responses: organismal adaptations to genome instability. Elsevier Ltd Trends Genet Mar 1, 2014;30(3):95–102.
[19] Schumacher B, Garinis GA, Hoeijmakers JHJ. Age to survive: DNA damage and aging. Trends Genet Feb 2008;24(2):77–85.
[20] Castells-Roca L, Mueller MM, Schumacher B. Longevity through DNA damage tolerance. Cell Cycle Feb 16, 2015;14(4):467–8.
[21] Barstead RJ, Moerman DG. *C. elegans* deletion mutant screening. In: C. elegans. New Jersey: Humana Press; 2006. p. 51–8.
[22] Hartman PS, Herman RK. Radiation-sensitive mutants of *Caenorhabditis elegans*. Genetics Oct 1982;102(2):159–78.
[23] Hobert O. The impact of whole genome sequencing on model system genetics: get ready for the ride. Genetics Feb 2010;184(2):317–9.

[24] Waaijers S, Boxem M. Engineering the *Caenorhabditis elegans* genome with CRISPR/Cas9. Methods Aug 1, 2014;68(3):381–8.

[25] Tabara H, Grishok A, Mello CC. RNAi in *C. elegans*: soaking in the genome sequence. Science Oct 16, 1998;282(5388):430–1.

[26] Kamath RS, Fraser AG, Dong Y, Poulin G, Durbin R, Gotta M, et al. Systematic functional analysis of the *Caenorhabditis elegans* genome using RNAi. Nature Jan 16, 2003;421(6920):231–7.

[27] Identification of genes that protect the *C. elegans* genome against mutations by genome-wide RNAi. Genes Dev Feb 15, 2003;17(4):443–8.

[28] Identification of conserved pathways of DNA-damage response and radiation protection by genome-wide RNAi. Jul 2006;16(13):1344–1350. Available from: http://linkinghub.elsevier.com/retrieve/pii/S0960982206016307.

[29] Enhanced neuronal RNAi in *C. elegans* using SID-1. Nat Methods May 30, 2010;7(7):554–9.

[30] Rieckher M, Kourtis N, Pasparaki A, Tavernarakis N. Transgenesis in *Caenorhabditis elegans*. Totowa, NJ: Humana Press; 2009. p. 21–39.

[31] Sarov M, Murray JI, Schanze K, Pozniakovski A, Niu W, Angermann K, et al. A genome-scale resource for in vivo tag-based protein function exploration in *C. elegans*. Cell Aug 17, 2012;150(4):855–66.

[32] Meyer JN, Boyd WA, Azzam GA, Haugen AC, Freedman JH, Van Houten B. Decline of nucleotide excision repair capacity in aging *Caenorhabditis elegans*. Genome Biol 2007;8(5):R70.

[33] Greiss S, Schumacher B, Grandien K, Rothblatt J, Gartner A. Transcriptional profiling in *C. elegans* suggests DNA damage dependent apoptosis as an ancient function of the p53 family. BMC Genomics 2008;9:334.

[34] Arczewska KD, Tomazella GG, Lindvall JM, Kassahun H, Maglioni S, Torgovnick A, et al. Active transcriptomic and proteomic reprogramming in the *C. elegans* nucleotide excision repair mutant xpa-1. Nucleic Acids Res Apr 10, 2013;41(10):5368–68.

[35] Kessler Z, Yanowitz J. Methodological considerations for mutagen exposure in *C. elegans*. Methods Aug 1, 2014;68(3):441–9.

[36] Lans H, Vermeulen W. Nucleotide excision repair in *Caenorhabditis elegans*. Mol Biol Int 2011;2011:542795.

[37] Stergiou L, Eberhard R, Doukoumetzidis K, Hengartner MO. NER and HR pathways act sequentially to promote UV-C-induced germ cell apoptosis in *Caenorhabditis elegans*. Cell Death Differ May 2011;18(5):897–906.

[38] Guo X, Bian P, Liang J, Wang Y, Li L, Wang J, et al. Synergistic effects induced by a low dose of diesel particulate extract and ultraviolet-a in *Caenorhabditis elegans*: DNA damage-triggered germ cell apoptosis. Chem Res Toxicol Jun 16, 2014;27(6):990–1001.

[39] Sakashita T, Takanami T, Yanase S, Hamada N, Suzuki M, Kimura T, et al. Radiation biology of *Caenorhabditis elegans*: germ cell response, aging and behavior. JRR 2010;51(2):107–21.

[40] Mills DK, Hartman PS. Lethal consequences of simulated solar radiation on the nematode *Caenorhabditis elegans* in the presence and absence of photosensitizers. Photochem Photobiol Dec 1998;68(6):816–23.

[41] Flibotte S, Edgley ML, Chaudhry I, Taylor J, Neil SE, Rogula A, et al. Whole-genome profiling of mutagenesis in *Caenorhabditis elegans*. Genetics Jun 2010;185(2):431–41.

[42] Holway AH, Kim S-H, La Volpe A, Michael WM. Checkpoint silencing during the DNA damage response in *Caenorhabditis elegans* embryos. J Cell Biol Mar 27, 2006;172(7):999–1008.

[43] Gengyo-Ando K, Mitani S. Characterization of mutations induced by ethyl methanesulfonate, UV, and trimethylpsoralen in the nematode *Caenorhabditis elegans*. Biochem Biophys Res Commun Mar 5, 2000;269(1):64–9.

[44] Hofmann ER, Milstein S, Boulton SJ, Ye M, Hofmann JJ, Stergiou L, et al. *Caenorhabditis elegans* HUS-1 is a DNA damage checkpoint protein required for genome stability and EGL-1-mediated apoptosis. Curr Biol Nov 19, 2002;12(22):1908–18.

[45] Babu V, Hofmann K, Schumacher B. A *C. elegans* homolog of the Cockayne syndrome complementation group A gene. Elsevier B.V DNA Repair (Amst) Nov 7, 2014;24:1–6.

[46] Wang S, Tang M, Pei B, Xiao X, Wang J, Hang H, et al. Cadmium-induced germline apoptosis in *Caenorhabditis elegans*: the roles of HUS1, p53, and MAPK signaling pathways. Toxicol Sci Apr 2008;102(2):345–51.

[47] Hunt PR, Marquis BJ, Tyner KM, Conklin S, Olejnik N, Nelson BC, et al. Nanosilver suppresses growth and induces oxidative damage to DNA in *Caenorhabditis elegans*. J Appl Toxicol May 2, 2013;33(10):1131–42.

[48] Oxidative DNA damage and nucleotide excision repair. Jun 20, 2013;18(18):2409–2419. Available from: http://eutils.ncbi.nlm.nih.gov/entrez/eutils/elink.fcgi?dbfrom=pubmed&id=23216312&retmode=ref&cmd=prlinks.

[49] Hyun M, Lee J, Lee K, May A, Bohr VA, Ahn B. Longevity and resistance to stress correlate with DNA repair capacity in *Caenorhabditis elegans*. Nucleic Acids Res Mar 2008;36(4):1380–9.

[50] Sanada Y, Zhang-Akiyama QM. An increase of oxidised nucleotides activates DNA damage checkpoint pathway that regulates post-embryonic development in *Caenorhabditis elegans*. Mutagenesis Mar 2014;29(2):107–14.

[51] Hartman P, Ponder R, Lo H-H, Ishii N. Mitochondrial oxidative stress can lead to nuclear hypermutability. Mech Ageing Dev Jun 2004;125(6):417–20.

[52] Leung MCK, Williams PL, Benedetto A, Au C, Helmcke KJ, Aschner M, et al. *Caenorhabditis elegans*: an emerging model in biomedical and environmental toxicology. Toxicol Sci Aug 18, 2008;106(1):5–28.

[53] Craig AL, Moser SC, Bailly AP, Gartner A. Methods for studying the DNA damage response in the *Caenorhabdatis elegans* germ line. Methods Cell Biol 2012;107:321–52.

[54] DNA double-strand break repair in *Caenorhabditis elegans*. Chromosoma Feb 2011;120(1):1–21.

[55] Involvement of global genome repair, transcription coupled repair, and chromatin remodeling in UV DNA damage response changes during development. May 2010;6(5):e1000941. Available from: http://dx.plos.org/10.1371/journal.pgen.1000941.s004.

[56] Mueller MM, Castells-Roca L, Babu V, Ermolaeva MA, Müller R-U, Frommolt P, et al. DAF-16/FOXO and EGL-27/GATA promote developmental growth in response to persistent somatic DNA damage. Nat Cell Biol Nov 24, 2014;16(12):1168–79.

[57] Lans H, Lindvall JM, Thijssen K, Karambelas AE, Cupac D, Fensgård OFA, et al. DNA damage leads to progressive replicative decline but extends the life span of long-lived mutant animals. Nature Publishing Group Cell Death Differ Aug 30, 2013;20:1–10.

[58] Boyd WA, Crocker TL, Rodriguez AM, Leung MCK, Wade Lehmann D, Freedman JH, et al. Nucleotide excision repair genes are expressed at low levels and are not detectably inducible in *Caenorhabditis elegans* somatic tissues, but their function is required for normal adult life after UVC exposure. In: Mutat Research/Fundamental Mol Mech Mutagen [Internet], ;683(1–2). Jan 2010. p. 57–67. Available from: http://www.ncbi.nlm.nih.gov/entrez/query.fcgi?cmd=Retrieve&db=PubMed&dopt=Citation&list_uids=19879883.

[59] Wolters S, Ermolaeva MA, Bickel JS, Fingerhut JM, Khanikar J, Chan RC, et al. Loss of *Caenorhabditis elegans* BRCA1 promotes genome stability during replication in smc-5 mutants. Genetics Apr 2014;196(4):985–99.

[60] Duerr JS. Antibody staining in *C. elegans* using "freeze-cracking". J Vis Exp 2013;(80). http://www.jove.com/video/50664/antibody-staining-in-c-elegans-using-freeze-cracking.

[61] Germ cell apoptosis and DNA damage responses. In: Advances in Experimental Medicine and Biology [Internet]. New York, NY: Springer; 2012. p. 249–76. Available from: http://link.springer.com/10.1007/978-1-4614-4015-4_9.

[62] Taylor MRG, Špírek M, Chaurasiya KR, Ward JD, Carzaniga R, Yu X, et al. Rad51 paralogs remodel pre-synaptic Rad51 filaments to stimulate homologous recombination. Cell Jul 16, 2015;162(2):271–86.

[63] MacQueen AJ, Villeneuve AM. Nuclear reorganization and homologous chromosome pairing during meiotic prophase require *C. elegans chk-2*. Genes Dev Jul 1, 2001;15(13):1674–87.

[64] Neher DA, Stürzenbaum SR. Extra-long PCR, an identifier of DNA adducts in single nematodes (*Caenorhabditis elegans*). Comp Biochem Physiol C Toxicol Pharmacol Nov 2006;144(3):279–85. Available from: http://eutils.ncbi.nlm.nih.gov/entrez/eutils/elink.fcgi?dbfrom=pubmed&id=17112787&retmode=ref&cmd=prlinks.

[65] Hunter SE, Jung D, Di Giulio RT, Meyer JN. The qPCR assay for analysis of mitochondrial DNA damage, repair, and relative copy number. Methods Aug 2010;51(4):444–51.

[66] Tijsterman M, Pothof J, Plasterk RHA. Frequent germline mutations and somatic repeat instability in DNA mismatch-repair-deficient *Caenorhabditis elegans*. Genetics Jun 2002;161(2):651–60.

[67] O'neil N. DNA Repair. WormBook. 2006. p. 1–12.

[68] Denver DR, Feinberg S, Estes S, Thomas WK, Lynch M. Mutation rates, spectra and hotspots in mismatch repair-deficient *Caenorhabditis elegans*. Genetics May 2005;170(1):113.

[69] Denver DR, Feinberg S, Steding C, Durbin M, Lynch M. The relative roles of three DNA repair pathways in preventing *Caenorhabditis elegans* mutation accumulation. Genetics Sep 2006;174(1):57–65.

[70] Molecular mechanisms of ultraviolet radiation-induced DNA damage and repair. Hindawi Publishing Corporation; 2010;2010(6551):1–32. Available from: http://www.hindawi.com/journals/jna/2010/592980/.

[71] Hartman PS, Hevelone J, Dwarakanath V, Mitchell DL. Excision repair of UV radiation-induced DNA damage in *Caenorhabditis elegans*. Genetics Jun 1989;122(2):379–85.

[72] Park HK, Yook J-S, Koo H-S, Choi IS, Ahn B. The *Caenorhabditis elegans* XPA homolog of human XPA. Mol Cells Aug 31, 2002;14(1):50–5.

[73] Astin JW, O'Neil NJ, Kuwabara PE. Nucleotide excision repair and the degradation of RNA pol II by the *Caenorhabditis elegans* XPA and Rsp5 orthologues, RAD-3 and WWP-1. DNA Repair (Amst) Feb 1, 2008;7(2):267–80.

[74] Kamileri I, Karakasilioti I, Garinis GA. Nucleotide excision repair: new tricks with old bricks. Elsevier Ltd Trends Genet Nov 1, 2012;28(11): 566–73.

[75] Wolters S, Schumacher B. Genome maintenance and transcription integrity in aging and disease. Front Genet 2013;4:19.

[76] Park HK, Suh D, Hyun M, Koo H-S, Ahn B. A DNA repair gene of *Caenorhabditis elegans*: a homolog of human XPF. DNA Repair (Amst) Oct 5, 2004;3(10):1375–83.

[77] Bergink S, Toussaint W, Luijsterburg MS, Dinant C, Alekseev S, Hoeijmakers JHJ, et al. Recognition of DNA damage by XPC coincides with disruption of the XPC-RAD23 complex. J Cell Biol Mar 19, 2012;196(6):681–8.

[78] Lee MH, Ahn B, Choi IS, Koo HS. The gene expression and deficiency phenotypes of Cockayne syndrome B protein in *Caenorhabditis elegans*. FEBS Lett Jul 3, 2002;522(1–3):47–51.

[79] Latonen L, Taya Y, Laiho M. UV-radiation induces dose-dependent regulation of p53 response and modulates p53-HDM2 interaction in human fibroblasts. Oncogene 2001;20(46):6784–93.

[80] The effect of UV irradiation on proliferation and life span of human diploid fibroblast-like cells. Aug 1982;18(8):703–707. Available from: http://eutils.ncbi.nlm.nih.gov/entrez/eutils/elink.fcgi?dbfrom=pubmed&id=7129484&retmode=ref&cmd=prlinks.

[81] Gentile M. Cell cycle arrest and apoptosis provoked by UV radiation-induced DNA damage are transcriptionally highly divergent responses. Nucleic Acids Res Aug 15, 2003;31(16):4779–90.

[82] Garinis GA, Uittenboogaard LM, Stachelscheid H, Fousteri M, van Ijcken W, Breit TM, et al. Persistent transcription-blocking DNA lesions trigger somatic growth attenuation associated with longevity. Nat Cell Biol May 2009;11(5):604–15.

[83] Bartke A. The somatotropic axis and aging: mechanisms and persistent questions about practical implications. Exp Gerontol Jun–Jul 2009;44(6–7):372–4.

[84] Kenyon C, Chang J, Gensch E, Rudner A, Tabtiang R. A *C. elegans* mutant that lives twice as long as wild type. Nature Dec 2, 1993;366(6454): 461–4.

[85] Kenyon CJ. The genetics of ageing. Nature Publishing Group Nature 2010;464(7288):504–12.

[86] Baugh LR, Sternberg PW. DAF-16/FOXO regulates transcription of cki-1/Cip/Kip and repression of lin-4 during *C. elegans* L1 arrest. Curr Biol Apr 18, 2006;16(8):780–5.

[87] Ogg S, Paradis S, Gottlieb S, Patterson GI, Lee L, Tissenbaum HA, et al. The fork head transcription factor DAF-16 transduces insulin-like metabolic and longevity signals in *C. elegans*. Nature Oct 30, 1997;389(6654):994–9.

[88] Murphy CT, McCarroll SA, Bargmann CI, Fraser A, Kamath RS, Ahringer J, et al. Genes that act downstream of DAF-16 to influence the lifespan of *Caenorhabditis elegans*. Nature Jul 17, 2003;424(6946):277–83.

[89] Tullet JMA. DAF-16 target identification in *C. elegans*: past, present and future. Biogerontology Aug 26, 2014:221–34.

[90] Integration of diverse inputs in the regulation of *Caenorhabditis elegans* DAF-16/FOXO. 2010;NA–NA. Available from: http://doi.wiley.com/10.1002/dvdy.22244.

[91] Xu X, Kim SK. The GATA transcription factor *egl-27* delays aging by promoting stress resistance in *Caenorhabditis elegans*. Public Library of Science PLoS Genet 2012;8(12):e1003108.

[92] Fensgård Ø KH, Bombik I, Rognes T, Lindvall JM, Nilsen H. A two-tiered compensatory response to loss of DNA repair modulates aging and stress response pathways. Aging (Albany NY) Mar 2010;2(3):133–59.

[93] Agostinho A, Meier B, Sonneville R, Jagut M, Woglar A, Blow J, et al. Combinatorial regulation of meiotic holliday junction resolution in *C. elegans* by HIM-6 (BLM) helicase, SLX-4, and the SLX-1, MUS-81 and XPF-1 nucleases. PLoS Genet 2013;9(7):e1003591.

[94] Saito TT, Youds JL, Boulton SJ, Colaiacovo MP. *Caenorhabditis elegans* HIM-18/SLX-4 interacts with SLX-1 and XPF-1 and maintains genomic integrity in the germline by processing recombination intermediates. PLoS Genet Nov 2009;5(11):e1000735.

[95] O'Neil NJ, Martin JS, Youds JL, Ward JD, Petalcorin MIR, Rose AM, et al. Joint molecule resolution requires the redundant activities of MUS-81 and XPF-1 during *Caenorhabditis elegans* meiosis. PLoS Genet 2013;9(7):e1003582.

[96] Scheffler IE. Mitochondria. 978 ed. Wiley; 2007.

[97] Bohr VA, Stevnsner T, de Souza-Pinto NC. Mitochondrial DNA repair of oxidative damage in mammalian cells. Gene Mar 2002;286(1):127–34.

[98] Cline SD. Mitochondrial DNA damage and its consequences for mitochondrial gene expression. Biochim Biophys Acta (BBA) – Gene Regul Mech Sep 2012;1819(9–10):979–91.

[99] Bess AS, Crocker TL, Ryde IT, Meyer JN. Mitochondrial dynamics and autophagy aid in removal of persistent mitochondrial DNA damage in *Caenorhabditis elegans*. Nucleic Acids Res Sep 12, 2012;40(16):7916–31.

[100] Bess AS, Leung MCK, Ryde IT, Rooney JP, Hinton DE, Meyer JN. Effects of mutations in mitochondrial dynamics-related genes on the mitochondrial response to ultraviolet C radiation in developing *Caenorhabditis elegans*. Worm Jan 1, 2013;2(1):e23763.

[101] Scheibye-Knudsen M, Ramamoorthy M, Sykora P, Maynard S, Lin PC, Minor RK, et al. Cockayne syndrome group B protein prevents the accumulation of damaged mitochondria by promoting mitochondrial autophagy. J Exp Med Apr 9, 2012;209(4):855–69.

[102] Mitochondrial dysfunction in ataxia-telangiectasia. Feb 9, 2012;119(6):1490–1500. Available from: http://eutils.ncbi.nlm.nih.gov/entrez/eutils/elink.fcgi?dbfrom=pubmed&id=22144182&retmode=ref&cmd=prlinks.

[103] A novel diagnostic tool reveals mitochondrial pathology in human diseases and aging. Mar 2013;5(3):192–208. Available from: http://eutils.ncbi.nlm.nih.gov/entrez/eutils/elink.fcgi?dbfrom=pubmed&id=23524341&retmode=ref&cmd=prlinks.

[104] Fang EF, Scheibye-Knudsen M, Brace LE, Kassahun H, SenGupta T, Nilsen H, et al. Defective mitophagy in xpa via PARP-1 hyperactivation and NAD. Elsevier Inc Cell May 8, 2014;157(4):882–96.

[105] Biology of poly(ADP-Ribose) polymerases: the factotums of cell maintenance. Jun 18, 2015;58(6):947–958. Available from: http://eutils.ncbi.nlm.nih.gov/entrez/eutils/elink.fcgi?dbfrom=pubmed&id=26091343&retmode=ref&cmd=prlinks.

[106] Gagnon SN, Hengartner MO, Desnoyers S. The genes *pme-1* and *pme-2* encode two poly(ADP-ribose) polymerases in *Caenorhabditis elegans*. Biochem J Nov 15, 2002;368(Pt 1):271.

[107] St-Laurent J-F, Gagnon SN, Dequen F, Hardy I, Desnoyers S. Altered DNA damage response in *Caenorhabditis elegans* with impaired poly(ADP-ribose) glycohydrolases genes expression. DNA Repair (Amst) Mar 1, 2007;6(3):329–43.

[108] Mouchiroud L, Houtkooper RH, Moullan N, Katsyuba E, Ryu D, Cantó C, et al. The NAD(+)/sirtuin pathway modulates longevity through activation of mitochondrial UPR and FOXO signaling. Cell Jul 18, 2013;154(2):430–41.

[109] Scheibye-Knudsen M, Mitchell SJ, Fang EF, Iyama T, Ward T, Wang J, et al. A high-fat diet and NAD. Elsevier Inc Cell Metab Nov 4, 2014;20(5):840–55.

[110] David SS, O'Shea VL, Kundu S. Base-excision repair of oxidative DNA damage. Nature Jun 21, 2007;447(7147):941–50.

[111] Dianov GL, Hübscher U. Mammalian base excision repair: the forgotten archangel. Nucleic Acids Res Apr 1, 2013;41(6):3483–90.

[112] Morinaga H, Yonekura S-I, Nakamura N, Sugiyama H, Yonei S, Zhang-Akiyama QM. Purification and characterization of *Caenorhabditis elegans* NTH, a homolog of human endonuclease III: essential role of N-terminal region. DNA Repair (Amst) Jul 4, 2009;8(7):844–51.

[113] Nakamura N, Morinaga H, Kikuchi M, Yonekura S-I, Ishii N, Yamamoto K, et al. Cloning and characterization of uracil-DNA glycosylase and the biological consequences of the loss of its function in the nematode *Caenorhabditis elegans*. Mutagenesis Sep 2008;23(5):407–13.

[114] Shatilla A, Ramotar D. Embryonic extracts derived from the nematode *Caenorhabditis elegans* remove uracil from DNA by the sequential action of uracil-DNA glycosylase and AP (apurinic/apyrimidinic) endonuclease. Biochem J Jul 15, 2002;365(Pt 2):547–53.

[115] Shatilla A, Ishchenko AA, Saparbaev M, Ramotar D. Characterization of *Caenorhabditis elegans* exonuclease-3 and evidence that a Mg2+-dependent variant exhibits a distinct mode of action on damaged DNA. Biochemistry Sep 27, 2005;44(38):12835.

[116] Masson JY, Tremblay S, Ramotar D. The *Caenorhabditis elegans* gene CeAPN1 encodes a homolog of *Escherichia coli* and yeast apurinic/apyrimidinic endonuclease. Gene Nov 14, 1996;179(2):291–3.

[117] SenGupta T, Torgersen ML, Kassahun H, Vellai T, Simonsen A, Nilsen H. Base excision repair AP endonucleases and mismatch repair act together to induce checkpoint-mediated autophagy. Nat Commun 2013;4:2674.

[118] Zakaria C, Kassahun H, Yang X, Labbé J-C, Nilsen H, Ramotar D. *Caenorhabditis elegans* APN-1 plays a vital role in maintaining genome stability. DNA Repair (Amst) Feb 4, 2010;9(2):169–76.

[119] Asagoshi K, Lehmann W, Braithwaite EK, Santana-Santos L, Prasad R, Freedman JH, et al. Single-nucleotide base excision repair DNA polymerase activity in *C. elegans* in the absence of DNA polymerase. Nucleic Acids Res Jan 13, 2012;40(2):670–81.

[120] Kassahun H, Nilsen H. Active transcriptomic and proteomic reprogramming in the *C. elegans* nucleotide excision repair mutant *xpa-1*. Worm Oct 1, 2013;2(4):e27337.

[121] Fukui K. DNA mismatch repair in eukaryotes and bacteria. J Nucleic Acids 2010;2010(32):1–16.

[122] Degtyareva NP, Greenwell P, Hofmann ER, Hengartner MO, Zhang L, Culotti JG, et al. *Caenorhabditis elegans* DNA mismatch repair gene *msh-2* is required for microsatellite stability and maintenance of genome integrity. Proc Natl Acad Sci USA Feb 19, 2002;99(4):2158–63.

[123] Moriwaki T, Kato Y, Nakamura C, Ishikawa S, Zhang-Akiyama QM. A novel DNA damage response mediated by DNA mismatch repair in *Caenorhabditis elegans*: induction of programmed autophagic cell death in non-dividing cells. Genes Cancer Jul 1, 2015:1–15.

[124] Bennett CB, Lewis AL, Baldwin KK, Resnick MA. Lethality induced by a single site-specific double-strand break in a dispensable yeast plasmid. Proc Natl Acad Sci USA Jun 15, 1993;90(12):5613–7.

[125] Hoeijmakers JHJ. DNA damage, aging, and cancer. N Engl J Med [Internet] Oct 8, 2009;361(15):1475–85. Available from: http://eutils.ncbi.nlm.nih.gov/entrez/eutils/elink.fcgi?dbfrom=pubmed&id=19812404&retmode=ref&cmd=prlinks.

[126] Lemmens BBLG, Johnson NM, Tijsterman M. COM-1 promotes homologous recombination during *Caenorhabditis elegans* meiosis by antagonizing Ku-mediated non-homologous end joining. PLoS Genet 2013;9(2):e1003276.

[127] Bailly AP, Freeman A, Hall J, Déclais A-C, Alpi A, Lilley DMJ, et al. The *Caenorhabditis elegans* homolog of Gen1/Yen1 resolvases links DNA damage signaling to DNA double-strand break repair. Public Library of Science PLoS Genet Jul 2010;6(7):e1001025.

[128] Ward JD, Barber LJ, Petalcorin MI, Yanowitz J, Boulton SJ. Replication blocking lesions present a unique substrate for homologous recombination. EMBO J Jul 25, 2007;26(14):3384–96.

[129] Clejan I, Boerckel J, Ahmed S. Developmental modulation of nonhomologous end joining in *Caenorhabditis elegans*. Genetics Jul 2006;173(3):1301–17.

[130] Yin Y, Smolikove S. Impaired resection of meiotic double-strand breaks channels repair to nonhomologous end joining in *Caenorhabditis elegans*. Mol Cell Biol Jul 2013;33(14):2732–47.

[131] Pontier DB, Tijsterman M. A robust network of double-strand break repair pathways governs genome integrity during *C. elegans* development. Curr Biol Aug 25, 2009;19(16):1384–8.

[132] Allen C, Halbrook J, Nickoloff JA. Interactive competition between homologous recombination and non-homologous end joining. Mol Cancer Res Oct 2003;1(12):913–20.

[133] Adachi N, Ishino T, Ishii Y, Takeda S, Koyama H. DNA ligase IV-deficient cells are more resistant to ionizing radiation in the absence of Ku70: implications for DNA double-strand break repair. Proc Natl Acad Sci USA Oct 9, 2001;98(21):12109–13.

[134] Schär P, Fäsi M, Jessberger R. SMC1 coordinates DNA double-strand break repair pathways. Nucleic Acids Res 2004;32(13):3921–9.

[135] Siede W, Friedl AA, Dianova I, Eckardt-Schupp F, Friedberg EC. The *Saccharomyces cerevisiae* Ku autoantigen homologue affects radiosensitivity only in the absence of homologous recombination. Genetics Jan 1996;142(1):91–102.

[136] Smolikov S, Eizinger A, Hurlburt A, Rogers E, Villeneuve AM, Colaiácovo MP. Synapsis-defective mutants reveal a correlation between chromosome conformation and the mode of double-strand break repair during *Caenorhabditis elegans* meiosis. Genetics Aug 2007;176(4):2027–33.

[137] Martin JS, Winkelmann N, Petalcorin MI, McIlwraith MJ, Boulton SJ. RAD-51-dependent and -independent roles of a *Caenorhabditis elegans* BRCA2-related protein during DNA double-strand break repair. Mol Cell Biol Apr 2005;25(8):3127–39.

[138] Alpi A, Pasierbek P, Gartner A, Loidl J. Genetic and cytological characterization of the recombination protein RAD-51 in *Caenorhabditis elegans*. Chromosoma Jul 2003;112(1):6–16.

[139] Dernburg AF, McDonald K, Moulder G, Barstead R, Dresser M, Villeneuve AM. Meiotic recombination in *C. elegans* initiates by a conserved mechanism and is dispensable for homologous chromosome synapsis. Cell Aug 7, 1998;94(3):387–98.

[140] Garcia-Muse T, Garcia-Muse T, Boulton SJ, Boulton SJ. Distinct modes of ATR activation after replication stress and DNA double-strand breaks in *Caenorhabditis elegans*. EMBO J Dec 21, 2005;24(24):4345–55. EMBO Press.

[141] Takanami T, Mori A, Takahashi H, Horiuchi S, Higashitani A. *Caenorhabditis elegans Ce-rdh-1/rad-51* functions after double-strand break formation of meiotic recombination. Chromosome Res 2003;11(2):125–35.

[142] Ogawa T, Yu X, Shinohara A, Egelman EH. Similarity of the yeast RAD51 filament to the bacterial RecA filament. Science Mar 26, 1993;259(5103):1896–9.

[143] Petalcorin MI, Sandall J, Wigley DB, Boulton SJ. CeBRC-2 stimulates D-loop formation by RAD-51 and promotes DNA single-strand annealing. J Mol Biol Aug 11, 2006;361(2):231–42.

[144] Chin GM, Villeneuve AM. *C. elegans mre-11* is required for meiotic recombination and DNA repair but is dispensable for the meiotic G(2) DNA damage checkpoint. Genes Dev Mar 1, 2001;15(5):522–34.

[145] Takanami T, Sato S, Ishihara T, Katsura I, Takahashi H, Higashitani A. Characterization of a *Caenorhabditis elegans* recA-like gene Ce-rdh-1 involved in meiotic recombination. DNA Res Dec 31, 1998;5(6):373–7.

[146] Ward JD, Muzzini DM, Petalcorin MI, Martinez-Perez E, Martin JS, Plevani P, et al. Overlapping mechanisms promote postsynaptic RAD-51 filament disassembly during meiotic double-strand break repair. Mol Cell 2010;37(2):259–72.

[147] Youds JL, Barber LJ, Ward JD, Collis SJ, O'Neil NJ, Boulton SJ, et al. DOG-1 is the *Caenorhabditis elegans* BRIP1/FANCJ homologue and functions in interstrand cross-link repair. Mol Cell Biol Mar 2008;28(5):1470–9.

[148] Saito TT, Lui DY, Kim H-M, Meyer K, Colaiácovo MP. Interplay between structure-specific endonucleases for crossover control during *Caenorhabditis elegans* meiosis. PLoS Genet 2013;9(7):e1003586.

[149] SLX-1 is required for maintaining genomic integrity and promoting meiotic noncrossovers in the *Caenorhabditis elegans* germline. Aug 2012;8(8):e1002888. Available from: http://eutils.ncbi.nlm.nih.gov/entrez/eutils/elink.fcgi?dbfrom=pubmed&id=22927825&retmode=ref&cmd=prlinks.

[150] Bellendir SP, Sekelsky J. An elegans solution for crossover formation. PLoS Genet 2013;9(7):e1003658.
[151] Svendsen JM, Smogorzewska A, Sowa ME, O'Connell BC, Gygi SP, Elledge SJ, et al. Mammalian BTBD12/SLX4 assembles a Holliday junction resolvase and is required for DNA repair. Cell Jul 10, 2009;138(1):63–77.
[152] Fekairi S, Scaglione S, Chahwan C, Taylor ER, Tissier A, Coulon S, et al. Human SLX4 is a Holliday junction resolvase subunit that binds multiple DNA repair/recombination endonucleases. Cell Jul 10, 2009;138(1):89.
[153] Andersen SL, Bergstralh DT, Kohl KP, LaRocque JR, Moore CB, Sekelsky J. *Drosophila* MUS312 and the vertebrate ortholog BTBD12 interact with DNA structure-specific endonucleases in DNA repair and recombination. Mol Cell Jul 10, 2009;35(1):135.
[154] Wicky C, Alpi A, Passannante M, Rose A, Gartner A, Müller F. Multiple genetic pathways involving the *Caenorhabditis elegans* Bloom's syndrome genes *him-6, rad-51*, and *top-3* are needed to maintain genome stability in the germ line. Mol Cell Biol Jun 2004;24(11):5016–27.
[155] Lee S-J, Gartner A, Hyun M, Ahn B, Koo H-S. The *Caenorhabditis elegans* Werner syndrome protein functions upstream of ATR and ATM in response to DNA replication inhibition and double-strand DNA breaks. PLoS Genet Jan 2010;6(1):e1000801.
[156] Hyun M, Bohr VA, Ahn B. Biochemical characterization of the WRN-1 RecQ helicase of *Caenorhabditis elegans*. Biochemistry Jul 15, 2008;47(28):7583–93.
[157] Doherty KM, Sommers JA, Gray MD, Lee JW, Kobbe von C, Thoma NH, et al. Physical and functional mapping of the replication protein a interaction domain of the werner and bloom syndrome helicases. J Biol Chem Aug 19, 2005;280(33):29494–505.
[158] Kim H-M, Colaiácovo MP. ZTF-8 interacts with the 9-1-1 complex and is required for DNA damage response and double-strand break repair in the *C. elegans* germline. PLoS Genet Oct 2014;10(10):e1004723.
[159] Kim HM, Colaiacovo MP. New insights into the post-translational regulation of DNA damage response and double-strand break repair in *Caenorhabditis elegans*. Genetics Mar 26, 2015:495–504.
[160] Garcia-Muse T, Boulton SJ. Meiotic recombination in *Caenorhabditis elegans*. Chromosome Res 2007;15(5):607–21.
[161] Condensins regulate meiotic DNA break distribution, thus crossover frequency, by controlling chromosome structure. Oct 2, 2009;139(1):73–86. Available from: http://eutils.ncbi.nlm.nih.gov/entrez/eutils/elink.fcgi?dbfrom=pubmed&id=19781752&retmode=ref&cmd=prlinks.
[162] Youds JL, Mets DG, McIlwraith MJ, Martin JS, Ward JD, ONeil NJ, et al. RTEL-1 enforces meiotic crossover interference and homeostasis. Science 2010;327(5970):1254–8.
[163] Bickel JS, Chen L, Hayward J, Yeap SL, Alkers AE, Chan RC. Structural maintenance of chromosomes (SMC) proteins promote homolog-independent recombination repair in meiosis crucial for germ cell genomic stability. Public Library of Science PLoS Genet 2010;6(7):e1001028.
[164] McVey M. Strategies for DNA interstrand crosslink repair: insights from worms, flies, frogs, and slime molds. Environ Mol Mutagen Jul 2010;51(6):646–58.
[165] Lee H, Alpi AF, Park MS, Rose A, Koo H-S. *C. elegans* ring finger protein RNF-113 is involved in interstrand DNA crosslink repair and interacts with a RAD51C homolog. PLoS One 2013;8(3):e60071.
[166] Lee C, Hong S, Lee MH, Koo H-S. A PHF8 homolog in *C. elegans* promotes DNA repair via homologous recombination. PLoS One 2015;10(4):e0123865.
[167] Muzzini DM, Plevani P, Boulton SJ, Cassata G, Marini F. *Caenorhabditis elegans* POLQ-1 and HEL-308 function in two distinct DNA interstrand cross-link repair pathways. DNA Repair (Amst) Jun 1, 2008;7(6):941–50.
[168] Ahmed S, Alpi A, Hengartner MO, Gartner A. *C. Elegans* RAD-5/CLK-2 defines a new DNA damage checkpoint protein. Curr Biol Dec 11, 2001;11(24):1934–44.
[169] Harris J, Lowden M, Clejan I, Tzoneva M, Thomas JH, Hodgkin J, et al. Mutator phenotype of *Caenorhabditis elegans* DNA damage checkpoint mutants. Genetics Oct 2006;174(2):601–16.
[170] Jaramillo-Lambert A, Harigaya Y, Vitt J, Villeneuve A, Engebrecht J. Meiotic errors activate checkpoints that improve gamete quality without triggering apoptosis in male germ cells. Curr Biol Dec 7, 2010;20(23):2078–89.
[171] Meier B, Ahmed S. Checkpoints: chromosome pairing takes an unexpected twist. Curr Biol Oct 30, 2001;11(21):R865–8.
[172] Stergiou L, Hengartner MO. Death and more: DNA damage response pathways in the nematode *C. elegans*. Cell Death Differ Jan 2004;11(1):21–8.
[173] Rhind N, Russell P. Chk1 and Cds1: linchpins of the DNA damage and replication checkpoint pathways. J Cell Sci Nov 2000;113(Pt 22):3889–96.
[174] Murakami H, Nurse P. DNA replication and damage checkpoints and meiotic cell cycle controls in the fission and budding yeasts. Biochem J Jul 1, 2000;349(Pt 1):1–12.
[175] Liu QA, Hengartner MO. The molecular mechanism of programmed cell death in *C. elegans*. Ann NY Acad Sci 1999;887:92–104.
[176] Conradt B, Xue D. Programmed cell death. In: WormBook, editor. The *C. elegans* Research Community. WormBook; 2005. Available from: http://www.wormbook.org.
[177] Dengg M, Garcia-Muse T, Gill SG, Ashcroft N, Boulton SJ, Nilsen H. Abrogation of the CLK-2 checkpoint leads to tolerance to base-excision repair intermediates. EMBO Rep Oct 2006;7(10):1046–51.
[178] Kemp M, Sancar A. DNA distress: just ring 9-1-1. Curr Biol Sep 15, 2009;19(17):R733–4.
[179] Evidence that DNA damage detection machinery participates in DNA repair. Apr 2005;4(4):529–532. Available from: http://eutils.ncbi.nlm.nih.gov/entrez/eutils/elink.fcgi?dbfrom=pubmed&id=15876866&retmode=ref&cmd=prlinks.
[180] Boerckel J, Walker D, Ahmed S. The *Caenorhabditis elegans* Rad17 homolog HPR-17 is required for telomere replication. Genetics May 2007;176(1):703–9.
[181] Ahmed S, Hodgkin J. MRT-2 checkpoint protein is required for germline immortality and telomere replication in *C. elegans*. Nature Jan 13, 2000;403(6766):159–64.
[182] Gravel C. The *C. elegans* gene *pme-5*: molecular cloning and role in the DNA-damage response of a tankyrase orthologue. DNA Repair (Amst) Feb 3, 2004;3(2):171–82.

[183] Brauchle M, Baumer K, Gönczy P. Differential activation of the DNA replication checkpoint contributes to asynchrony of cell division in *C. elegans* embryos. Curr Biol May 13, 2003;13(10):819–27.

[184] chk-1 is an essential gene and is required for an S-M checkpoint during early embryogenesis. Taylor & Francis; Oct 28, 2014;3(9):1194–1198. Available from: http://www.tandfonline.com/doi/abs/10.4161/cc.3.9.1116.

[185] Schumacher B, Hanazawa M, Lee M-H, Nayak S, Volkmann K, Hofmann ER, et al. Translational repression of *C. elegans* p53 by GLD-1 regulates DNA damage-induced apoptosis. Cell Feb 11, 2005;120(3):357–68.

[186] Schumacher B, Schertel C, Wittenburg N, Tuck S, Mitani S, Gartner A, et al. *C. elegans* ced-13 can promote apoptosis and is induced in response to DNA damage. Cell Death Differ Feb 2005;12(2):153–61.

[187] Silva N, Adamo A, Santonicola P, Martinez-Perez E, La Volpe A. Pro-crossover factors regulate damage-dependent apoptosis in the *Caenorhabditis elegans* germ line. Cell Death Differ Sep 2013;20(9):1209–18.

[188] Chen F, Hersh BM, Conradt B, Zhou Z, Riemer D, Gruenbaum Y, et al. Translocation of *C. elegans* CED-4 to nuclear membranes during programmed cell death. Science Feb 25, 2000;287(5457):1485–9.

[189] Structural, biochemical, and functional analyses of CED-9 recognition by the proapoptotic proteins EGL-1 and CED-4. Sep 24, 2004;15(6):999–1006. Available from: http://eutils.ncbi.nlm.nih.gov/entrez/eutils/elink.fcgi?dbfrom=pubmed&id=15383288&retmode=ref&cmd=prlinks.

[190] Conradt B, Horvitz HR. The *C. elegans* protein EGL-1 is required for programmed cell death and interacts with the Bcl-2-like protein CED-9. Cell May 15, 1998;93(4):519–29.

[191] Jiang H-S, Wu Y-C. LIN-3/EGF promotes the programmed cell death of specific cells in *Caenorhabditis elegans* by transcriptional activation of the pro-apoptotic gene *egl-1*. PLoS Genet Aug 2014;10(8):e1004513.

[192] Yang X, Chang HY, Baltimore D. Essential role of CED-4 oligomerization in CED-3 activation and apoptosis. Science Aug 28, 1998;281(5381): 1355–7.

[193] Deng X, Hofmann ER, Villanueva A, Hobert O, Capodieci P, Veach DR, et al. *Caenorhabditis elegans* ABL-1 antagonizes p53-mediated germline apoptosis after ionizing irradiation. Nat Genet Aug 2004;36(8):906–12.

[194] Stein KK, Davis ES, Hays T, Golden A. Components of the spindle assembly checkpoint regulate the anaphase-promoting complex during meiosis in *Caenorhabditis elegans*. Genetics Soc America Genet 2007;175(1):107–23.

[195] Kitagawa R, Law E, Tang L, Rose AM. The Cdc20 homolog, FZY-1, and its interacting protein, IFY-1, are required for proper chromosome segregation in *Caenorhabditis elegans*. Curr Biol Dec 23, 2002;12(24):2118–23.

[196] Musacchio A, Salmon ED. The spindle-assembly checkpoint in space and time. Nat Rev Mol Cell Biol May 2007;8(5):379–93.

[197] Tarailo-Graovac M, Wang J, Chu JS, Tu D, Baillie DL, Chen N. Spindle assembly checkpoint genes reveal distinct as well as overlapping expression that implicates MDF-2/Mad2 in postembryonic seam cell proliferation in *Caenorhabditis elegans*. BMC Cell Biol 2010;11(1):71.

[198] Hajeri VA, Stewart AM, Moore LL, Padilla PA. Genetic analysis of the spindle checkpoint genes *san-1, mdf-2, bub-3* and the CENP-F homologues *hcp-1* and *hcp-2* in *Caenorhabditis elegans*. Cell Div 2008;3:6.

[199] Wang X, Liu M, Li W, Suh CD, Zhu Z, Jin Y, et al. The function of a spindle checkpoint gene *bub-1* in *C. elegans* development. PLoS One 2009;4(6):e5912.

[200] Tarailo M, Kitagawa R, Rose AM. Suppressors of spindle checkpoint defect (such) mutants identify new *mdf-1*/MAD1 interactors in *Caenorhabditis elegans*. Genetics Feb 4, 2007;175(4):1665–79.

[201] Essex A, Dammermann A, Lewellyn L, Oegema K, Desai A. Systematic analysis in *Caenorhabditis elegans* reveals that the spindle checkpoint is composed of two largely independent branches. Mol Biol Cell Feb 2009;20(4):1252–67.

[202] Rogers E, Bishop JD, Waddle JA, Schumacher JM, Lin R. The aurora kinase AIR-2 functions in the release of chromosome cohesion in *Caenorhabditis elegans* meiosis. JCB Apr 15, 2002;157(2):219–29.

[203] Schumacher JM, Golden A, Donovan PJ. AIR-2: an Aurora/Ipl1-related protein kinase associated with chromosomes and midbody microtubules is required for polar body extrusion and cytokinesis in *Caenorhabditis elegans* embryos. JCB Dec 14, 1998;143(6):1635–46.

[204] Kaitna S, Pasierbek P, Jantsch M, Loidl J, Glotzer M. The aurora B kinase AIR-2 regulates kinetochores during mitosis and is required for separation of homologous chromosomes during meiosis. Curr Biol May 14, 2002;12(10):798–812.

[205] Bishop JD, Han Z, Schumacher JM. The *Caenorhabditis elegans* Aurora B kinase AIR-2 phosphorylates and is required for the localization of a BimC kinesin to meiotic and mitotic spindles. Mol Biol cell Feb 2005;16(2):742–56.

[206] Woollard A, Hodgkin J. Stu-7/*air-2* is a *C. elegans* aurora homologue essential for chromosome segregation during embryonic and post-embryonic development. Mech Dev Apr 1999;82(1–2):95–108.

Chapter 12

Genetic Engineering of Plants Using Zn Fingers, TALENs, and CRISPRs

A. Bilichak, F. Eudes

Agriculture and Agri-Food Canada, Lethbridge, AB, Canada

Chapter Outline

1. INTRODUCTION

The improvement of crop traits has been an ongoing practice since plant domestication. A conventional breeding technique utilizes an available pool of natural genetic variation combined with extensive backcrossing to introduce traits into an elite background. The existence of valuable alleles in nature and the inability to introduce traits selectively limit the outcome of this approach.

The discovery of X-ray mutagenesis in the 1920s [1,2] has initiated the development of mutation breeding for the artificial introduction of new traits in crops of interest. In the process, the mutagenized population of plants was screened using forward genetic approaches for the direct identification of specific phenotypes. This method has a significant drawback because most of the random mutations are recessive, and the polyploidy of crop species can mask any phenotypic effects resulted from a given sequence mutation [3]. The lack of technology to target mutations to predefined positions in the genome also impeded the utilization of reverse genetic screening for a fast and efficient linkage between a gene and a phenotype. Hence, further progress of plant biotechnology and breeding necessitates the discovery of new tools for targeted genetic engineering.

Genetic engineering is an umbrella term that covers a precise modification of the genome by means of the targeted insertion, replacement, or editing of the selected locus/loci (Fig. 12.1). Historically, homologous recombination (HR) was the method of choice to achieve gene targeting (GT) in model organisms [4]. Efficient HR in eukaryotes has been overall limited to yeast, chicken DT40 cells, mouse embryonic stem (ES) cells, and moss *Physcomitrella patens* (reviewed in Refs. [5,6]). At the dawn of plant biotechnology, a major impediment to genetic engineering in vascular plants was the limited frequencies of HR ranging from 10^{-4} to 10^{-6} (reviewed in Ref. [7]). The implementation of positive/negative selection markers [8,9] and the labor-intensive screening of the generated transgenic plants had to be performed to identify putative GT events in the plant population [7]. The pioneer studies on the utilization of a rare cutting yeast enzyme I-*Sce*-I in plants and animals have revealed that the cleavage of DNA at the artificially created endogenous position increases the rate of HR and GT events by 1000 folds or more at this locus [10–13]. Most of the DNA double-strand breaks (DSBs) in plants are repaired through the error-prone nonhomologous end joining (NHEJ) pathway that results in the introduction of insertion/deletions at the cut site [14]. This natural effect is utilized to produce lines with the desired gene knockouts in a time- and cost-efficient manner.

Genome Stability. http://dx.doi.org/10.1016/B978-0-12-803309-8.00012-4

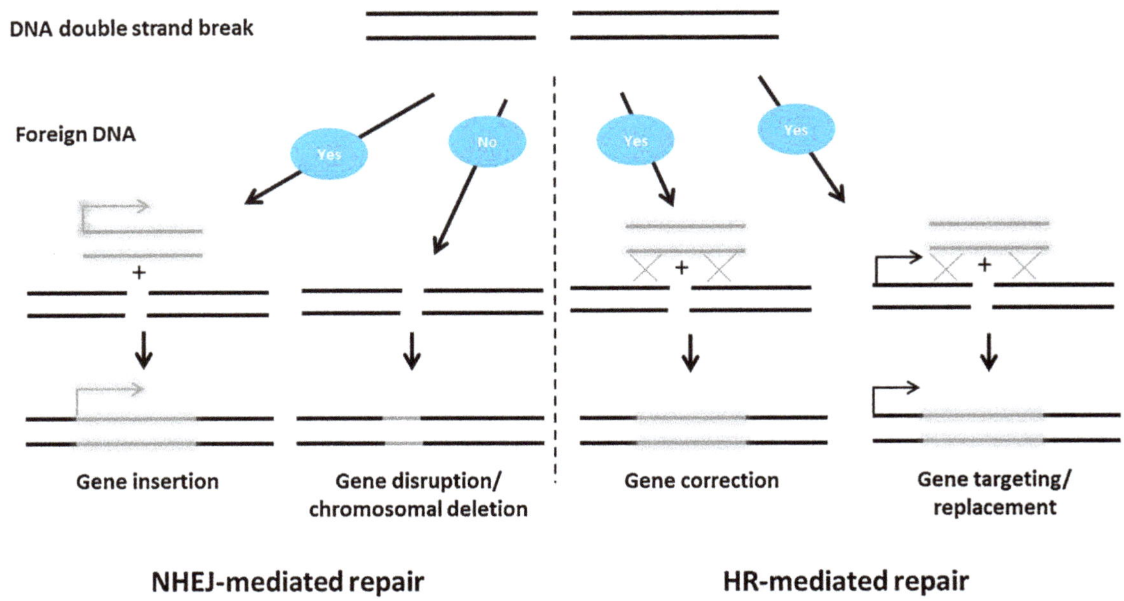

FIGURE 12.1 **Possible strategies for genome editing in plants using designed endonucleases.** The induction of double-strand breaks by endonucleases typically increases the frequency of genome editing by hundred times as compared to those resulted from spontaneous homologous recombination. Supplementing the donor DNA (shown in gray) either with or without homology to the endogenous region can lead to different outcomes depending on the DNA repair pathway involved.

The generation of targeted DSBs requires a protein or nucleoprotein complex that can be designed to bind to any sequence of interest [15]. A fusion of a programmable DNA-binding motif to the nonspecific endonuclease domain allows for a precise introduction of DSBs at the preselected positions [16]. In the late 1990s, the first artificial endonucleases appeared on the horizon that set a stage for the rapid development of novel enzymes with a specific cleavage activity called the designed or engineered endonucleases. Currently, four types of engineered nucleases are used for genome editing: engineered homing endonucleases/meganucleases (EMNs), zinc-finger nucleases (ZFNs), transcription activator-like effector nucleases (TALENs), and clustered regularly interspaced short palindromic repeats (CRISPR)/CRISPR-associated 9 (Cas9). All of them have been successfully used in plants to introduce modifications at the predefined positions in the genome. Nevertheless, challenges with the design, verification, and prohibitive licensing fees associated with some of the engineered endonucleases made their utilization less frequent as compared to other endonucleases. In particular, as of 2016, TALENs and CRISPR/Cas9 are the most widely used technologies in plants [17]. In this chapter, we provide a brief overview of the current status of the genome-editing technology in plants using the designed endonucleases and the future perspective of the possible technology application in plant genetic engineering.

2. ZINC FINGER NUCLEASES FOR GENOME ENGINEERING OF PLANTS

The Cys2–His2 zinc-finger motif is one of the most common types of DNA-binding domains present in eukaryotes. In addition, it is observed in almost half of transcription factors in humans [18–20]. The invention of ZFNs was a gradual process that included deciphering the interaction of zinc-finger motifs with DNA and the examination of the most efficient strategy of fusion of a DNA-binding domain to the nonspecific endonuclease *Fok*I (Fig. 12.2). The *Fok*I protein is a type II restriction enzyme produced by *Flavobacterium okeanokoites* [21]. The N-terminal end of the protein is a DNA-binding motif, and the C-terminal end acts as a nonspecific cleavage domain. Upon binding to its target sequence and in the presence of divalent metal ions, the *Fok*I enzyme dimerizes [22]. The *Fok*I nonspecific cleavage domain is also used in chimeric endonucleases, such as ZFNs, TALENs, and in the specialized CRISPR/dCas9 enzymes.

A single zinc-finger unit consists of three or four binding modules, and each module recognizes a nucleotide (nt) triplet. Two ZFN monomers can bind to the unique 18–24 bp-long sequences spaced by a 5–6 bp gap between them. Upon *Fok*I dimerization, DSB is created with 4–5 bp 5′-overhangs [23]. Since the first demonstration of the yellow gene disruption in a fruit fly in 2002 [24], various ZFNs have been applied for genome editing in a number of plants, including *Arabidopsis*, tobacco, maize, and soybean [25–29].

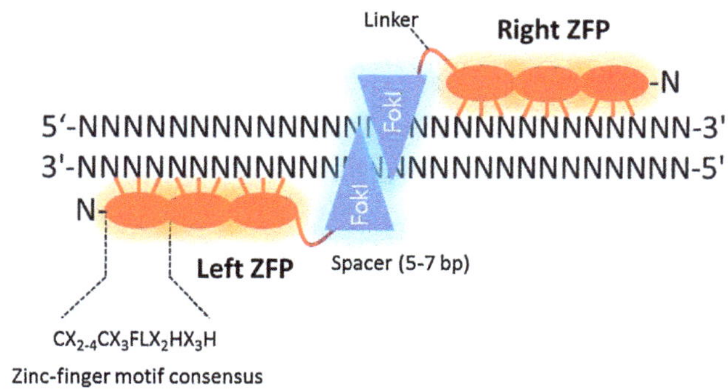

FIGURE 12.2 **Schematic representation of a zinc-finger nuclease (ZFN) protein.** Each ZFN consists of a zinc-finger protein (ZFP) at the N-terminus and a *Fok*I nuclease domain at the C-terminus. ZFN typically can target 18–36 bp long sequences. In the zinc-finger-motif consensus, X represents any amino acid.

2.1 Zinc Finger Nucleases Application in Model Plant Species

The first pioneering studies on ZFN-induced targeted mutagenesis in plants involved the modification of previously integrated transgenes carrying the ZFN cleavage sites in *Arabidopsis* [30–32]. Mutation frequencies ranging from 2% [31] to as high as 19.6% were observed in *Arabidopsis* depending on the ZFN-induction system used (constitutive versus inducible expression). A nontransgenic virus-based ZFN delivery in tobacco and petunia plants resulted in a targeted modification of reporter construct in somatic cells that was stably inherited in the following generation [33]. In most of the cases, the authors observed simple deletions at the target sites of 1–80 bp, thus reinforcing the hypothesis that imprecise NHEJ repair prevails in higher plants [30].

Some examples of endogenous genomic loci mutations mediated by ZFNs in *Arabidopsis* include the disruption of *ABAINSENSITIVE-4* (*ABI4*) [27], *ALCOHOL DEHYDROGENASE-1* (*ADH1*), and *TRANSPARENT TESTA-4* (*TT4*) genes [28]. In both of the studies, the inducible promoters were used for the activation of the previously integrated ZFN constructs. Upon induction, the observed mutation frequencies in somatic cells were 3%, 7%, and 16% for *ABI4*, *ADH1*, and *TT4*, respectively. Mutations were stably transmitted to the progeny, and the associated phenotype was observed for all genes. Curiously, for both of the genes, *ADH1* and *TT4*, homozygous mutants in the T1 generation were recovered, suggesting simultaneous biallelic mutations. At the same time, no potential off-target effects were observed in the edited plants [28].

A broader application of NHEJ-mediated gene disruption includes the replacement of an endogenous locus preceded by its cleavage at the 5′- and 3′-termini using ZFNs. Successful deletions of 2.8 and 4.3 kb fragments at the transgene loci were reported in tobacco plants [34,35]. The expression of ZFN in transgenic tobacco plants containing a GREEN FLUORESCENT PROTEIN (GFP) recombination construct with a 2.8 kb interrupting DNA sequence and a ZFN cleavage site resulted in the targeted DSB formation, recombination between GFP gene fragments, and deletion of the intervening 2.8 kb sequence. Moreover, the successful targeted deletions, inversions, and duplications of multiple gene clusters mediated by ZFNs have also been reported in *Arabidopsis* [36]. The simultaneous deletion of eight resistance (*R*) genes that compose a *RPP4* gene cluster was achieved by the inducible expression of ZFNs that targeted the regions of 55 kb apart [36]. The frequency of deletions in somatic cells was about 1%. Furthermore, the authors have even achieved deletions larger than 9 Mb on the chromosome 1 with the frequency of less than 1%. The feasibility of targeted deletions of large chromosomal regions offers an opportunity of a precise removal of a particular trait when it is regulated by a few clustered genes in crops. In addition, a gene cluster can be replaced with the genes of interest when the HR-mediated integration is engaged.

An additional precision of genome engineering can be achieved when ZFN cassettes are co-delivered into plant cells with the donor DNA that has homology to the endogenous sequence flanking the ZFN cut site. In this case, the HR pathway can be involved in DSB repair, and the donor DNA can be used as a template in the process of synthesis-dependent strand annealing [37]. Sequence modifications from the donor DNA can be copied into the targeted cut site with modifications that can vary from single- to few base-pair modifications (ie, gene editing) to the integration of complete transgene expression cassettes (ie, site-specific integration) [3]. Since the frequency of HR repair in somatic plant cells appears to be extremely low, the identification and isolation of such modifications is usually achieved by applying a selection pressure [3]. In one of the examples, specific mutations of the *SULFONYLUREA RECEPTOR* genes *SuRA* and *SuRB* in tobacco render cells insensitive to imidazolinone and sulfonylurea herbicides [26]. The co-delivery of ZFNs and the donor DNA template for the correction of *SuRA* and *SuRB* genes into tobacco protoplasts resulted in the recovery of herbicide-resistant calli at the frequency of 2%. Moreover, mutations as far as 1.3 kb from the ZFN cleavage site were obtained, suggesting that plant genes can be edited even when the DNA sequence excludes the ZFN recognition sites near the desired locus of modifications [26]. Similarly, two specific point mutations in a *PROTOPORPHYRINOGEN OXIDASE* (*PPOX*) gene of *Arabidopsis* confer the plant's resistance to the herbicide butafenacil [38]. Floral dip transformation of wild-type plants and plants that constitutively express ZFN was performed using *Agrobacterium* carrying a binary vector. The plasmid contained a donor template with the *PPO* gene missing the 5′ coding region but having two necessary mutations to confer resistance to the herbicide butafenacil in the edited plants. Selection of T1 plants on butafenacil yielded GT frequencies of 0.8×10^{-3} and 3.1×10^{-3} per transformation event, in wild-type and ZFN lines, respectively.

The future advancement of gene-editing technology through HR will require the development of tools for high-throughput screening of the generated plant populations. This will allow a selection-free elimination of wild-type plants in order to find the desired modification [3].

2.2 Zinc Finger Nucleases Application in Crops

Genome editing in crops presents a particular challenge because it relies on the availability of a highly efficient transformation method and the ability to design a unique engineered endonuclease for targeting a distinct locus, or loci, in the complex polyploid genome. The first successful report of genome editing in crops has been published in 2009. It involved the ZFN-mediated disruption of an *IPK1* gene that encodes an inositol-1,3,4,5,6-penta-kisphosphate 2-kinase, an enzyme that catalyzes the final step in phytate biosynthesis in maize seeds [25]. Phytate accounts for 75% of the total seed phosphorus and is an antinutritional component of feed grains that contributes to the environmental pollution through the waste stream. Reducing the level of phytate is agriculturally important for both increasing the bioavailability of phosphorus in corn grains and decreasing the negative environmental impact. Four ZFN pairs designed to cut *IPK1* at two positions in exon 2 were transformed into embryogenic callus of maize using a whisker-mediated DNA delivery [39]. Along with the ZFN cassettes, two HR repair templates were transformed that contained short homology arms to the *IPK1* gene and either an autonomous herbicide-tolerance gene expression sequence (*PAT*) or a nonautonomous donor that relied on a precise integration under the endogenous *IPK1* promoter for the expression of the marker gene. The frequencies of successful GT events ranged from 3.4% to 100%, depending on the ZFN pair used and the donor template. No off-target mutations were observed at the noncognate homologous sites in T0 plants carrying GT events at the *IPK1* gene. The effect of gene disruption on *IPK1* expression was transmitted through two rounds of meiosis, and the edited plants had a significant number of seeds with reduced phytate levels and a concomitant increase in inorganic phosphate as compared to plants with random integration of the donor template [25].

In another example, a targeted mutagenesis of a transgene and nine endogenous soybean (*Glycine max*) genes was performed using ZFNs [29]. Soybean has a highly duplicated paleopolyploid genome that jeopardized the development of ZFNs which recognize distinctive sequences in the genome. A number of ZFN constructs were constructed to target either unique or duplicated paralogs of epigenetic-related genes. Following *Agrobacterium rhizogenes*-mediated hairy-root transformation, somatic mutations were detected for the following genes: *DICER-LIKE1a* (*DCL1a*), *DCL1b*, *DCL4a*, *DCL4b*, *RNA-DEPENDENT RNA POLYMERASE 6a* (*RDR6a*), *RDR6b*, and *HUA ENHANCER 1a* (*HEN1a*). The whole-plant transformation of soybean using a cassette under the control of an estrogen-inducible promoter and encoding ZFNs targeting two paralogous genes, *DCL4a* and *DCL4b*, resulted in the recovery of three T0 plants from the hormone-treated explants. Sequence analysis of PCR-amplified products revealed that one of the plants had an adenine base insertion at the *DCL4a* locus, and another one had a two-base thymine and adenine insertion at the *DCL4b* locus. Both plants appeared to be heterozygous for the mutation. The plant with the *dcl4a* mutation exhibited phenotypic abnormalities, including aborted seed development. The *dcl4b* plant appeared to be normal and produced T1 progeny in which the *dcl4b* mutation segregated in a Mendelian fashion as1:2:1 [29]. These results provide the clear evidence that the designed endonucleases can be successfully implemented in the paleopolyploid crop species for a precise genome editing.

The development of a single crop variety with disease resistance, abiotic stress tolerance, yield enhancement, and quality traits requires the involvement of a labor- and resource-intensive introgression via conventional breeding. The appearance of tools for targeting DNA sequence to a selected locus may apparently eliminate problems of unpredicted cassette expression due to chromatin composition and segregation problems following meiosis. Combining two or more traits in one variety can now be achieved by molecular trait stacking in a single transgene locus. This can be done by using a transformation vector carrying the trait genes with homology sequences to the target region and a ZFN expression cassette designed to target the desired integration locus upon expression of an active protein. The proof-of-concept study has provided an example of successful on-demand transgene integration and trait stacking in the maize genome [40]. The authors used modular "trait landing pads" (TLPs) that flanked the herbicide-resistance gene, *pat*, and had ZFN target sites with sequences homologous to an incoming DNA. Separate cotransformation of transgenic plants with a donor DNA containing a second herbicide-resistance gene, *aad1*, flanked by sequences homologous to the integrated TLP along with the corresponding ZFN expression construct allowed for the *aad1* transgene to be precisely integrated at TLP, directly adjacent to the *pat* transgene. The frequency of up to 5% in the embryo-derived transgenic events was achieved, and both herbicide-resistance genes co-segregated in the subsequent generations.

2.3 Potential Limitations of the Zinc Finger Nucleases Technology

Although the ZFN technology has proven itself as an efficient tool for the genome editing in a number of model and crop species, the design of multi–zinc finger modules is challenging due to complex interactions between amino acid residues and base pairs of the target sequence [41]. In addition, the assembly and testing of ZFNs is usually expensive. Moreover, the availability of endogenous targets is restricted due to a limited number of modules for the context-dependent assembly platform [42].

A major drawback of the broad usage of ZFNs in plants is a prohibitive licensing fee that restricts an access to the required design tools developed by the company Sangamo Bioscience [43].

3. TALENs FOR THE GENOME ENGINEERING OF PLANTS

Following pioneering studies on ZFNs, the genome engineering using TALENs in plants has progressed rapidly [44]. The transcription activator-like effector (TALE) DNA-binding domains were "borrowed" from plant pathogens in the genus *Xanthomonas* which deliver the proteins to plant cells during infection through the type III secretion pathway [45,46]. The TALE proteins can bind to the effector-specific DNA sequences and transcriptionally activate gene expression of host genes. This makes plants more susceptible to the pathogen attack in most of the cases. Binding of the TALE protein to DNA sequence is mediated by a middle region that contains 30 tandem repeats of a 33–35 amino acid–sequence motif. Each repeat has a mostly consistent amino acid sequence, with the exception of two adjacent amino acids (the repeat variable diresidue or RVD) at positions 12 and 13. Distinct RVDs within the repeats dictate the specificity of the repeat to recognize nts in the target sequence. In 2009, the cipher was decoded by two research groups who showed a clear relation between RVDs in the repeat domain and the nts in the target DNA sequence [45,47,48]. Using the current ZFN architecture, TALEs were fused to the catalytic domain of the *Fok*I restriction enzyme, and the resulting chimeric endonucleases also demonstrated a specific cleavage activity in the yeast LacZ assay [45] (Fig. 12.3). The off-target effects of TALENs seem to be fewer than those of ZFNs due to the longer target recognition site [7]. The assembly of TALENs has been simplified by the Golden Gate-based cloning method that allows directional and seamless assembly of multiple DNA fragments [49]. The availability of the tool kit along with the freely distributed module assembly plasmids allowed a number of groups to design and construct TALENs for the specific genome-editing objectives.

3.1 The Application of TALENs in the Model Plant Species

The in planta testing of designed TALENs was first performed in tobacco by transient cotransformation of a *uidA* reporter construct carrying the recognition sequence and the corresponding TALEN using *Agrobacterium* [50]. The cleavage at the

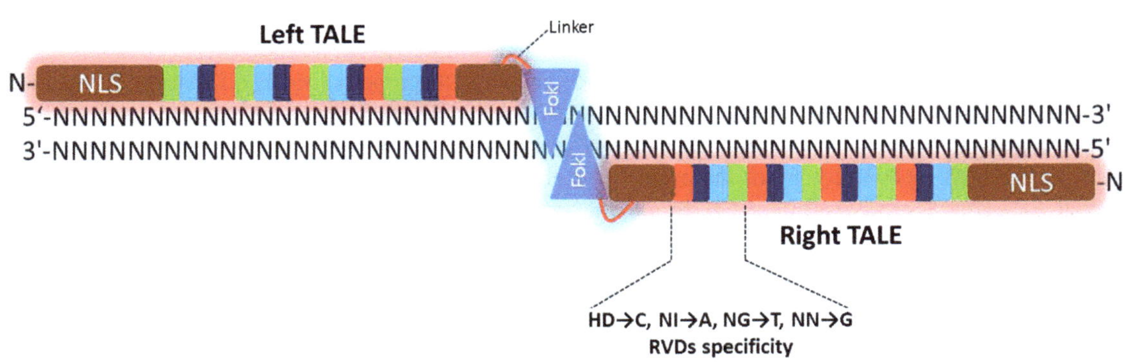

FIGURE 12.3 Schematic representation of a transcription activator-like effector nuclease (TALEN). Each monomer contains a DNA-binding domain at the amino terminus and a *Fok*I nuclease domain at the carboxyl terminus. Each TALE module (shown as small colored boxes) can recognize only one nucleotide through its 13th amino acid. Each TALE module typically contains 34 amino acids with the 12th and 13th residues being responsible for the specificity (repeated variable diresidues). A recognition pattern of modules is shown in the figure. *NLS*, a nuclear localization signal.

recognition sequence followed by a subsequent repair mediated by the cellular repair machinery would remove the stop codon in the reporter coding sequence and allowed for the expression of the *uidA* reporter. Following the co-delivery of both constructs into tobacco leaves, the authors observed blue sectors in the infiltration regions, and the resulting products of DSB repair were confirmed by sequencing [50]. Similarly, a transient expression of custom-designed TALEN targeting an *ALCOHOL DEHYDROGENASE1* (*ADH1*) gene in *Arabidopsis* resulted in the recovery of six independent mutations consisting of deletions ranging from 4 to 15 bp [51]. To assess the TALEN activityin planta, a yellow fluorescent protein (YFP)-based single-strand annealing (SSA) reporter construct has been developed [7]. The reporter has a TALEN recognition sequence flanked by a 255 bp direct repeat of the YFP-coding sequence. A successful cleavage of the construct by TALEN results in the recombination of homologous sequences and the reconstitution of a functional YFP gene. The co-delivery of both TALEN and the reporter construct into tobacco protoplasts allows for a fast screening of the TALEN activity using flow cytometry. The TALEN activities observed in the protoplast SSA assay demonstrated a high correlation with mutagenesis frequencies detected at the endogenous loci for the same TALENs. The mutagenesis efficiencies after TALEN delivery were in the range of 30% for an *ACETOLACTATE SYNTHASE* (*ALS*) gene that allowed the recovery of calli with targeted mutations without applying a selection pressure. In addition, 4% of calli showed an evidence of targeted gene replacement when a 322 bp donor molecule with 6 bp difference from the *ALS*-coding sequence was co-delivered with TALEN.

A stable integration of TALEN constructs designed to target separately five different genes in the *Arabidopsis* genome resulted in somatic mutagenesis frequencies ranging from 2% to 15% at the selected loci for all tested TALENs [52]. Furthermore, mutations were transmitted to the next generation at the rate of 1.5–12%. A stable germline transmission of somatic mutations in *Arabidopsis* caused by TALEN activity was also confirmed in a separate study [53]. The expression of TALENs under control of a shoot apical meristem–specific promoter resulted in targeting a *CLAVATA3* (*CLV3*) gene at the rate that allowed a recovery of biallelic mutants already in the T1 generation.

The successful application of TALENs in two monocot model species, rice and *Brachypodium*, demonstrated the utility of the tool for gene disruption in cereal crops [54]. When a stable *Agrobacterium*-mediated transformation of embryonic cells was performed with TALENs, the mutation frequencies of resistant calli were recovered at the rate from 3.8% to 100%, depending on TALEN and the species. Most of the mutations were small deletions ranging from 1 to 20 bp, and biallelic modifications were recovered as a result of action 5 of 13 TALENs tested. Moreover, a large deletion was detected when two TALENs with recognition sequences of more than 1.3 kb apart were co-delivered into rice protoplasts.

An alternative approach to the stable integration of the TALEN cassette was proposed by the Daniel Voytas Lab in 2014 [55]. Transient expression of sequence-specific nucleases in tobacco leaves using a geminivirus resulted in the recovery of NHEJ events at the target regions of the three nucleases tested (ZFN, TALEN, and CRIPSPR/Cas9). Moreover, the co-delivery of DNA repair templates using the *bean yellow dwarf* virus resulted in GT events at the rates from one to two orders of magnitude over the conventional *Agrobacterium tumefaciens* T-DNA delivery. Interestingly, the authors observed a low level of NHEJ events and a high frequency of GT in the cells. Based on the experiments, they speculated that the effect was caused by a combination of targeted DSBs, a high replication of a repair template and a pleiotropic effect of the *trans*-acting replication-initiation protein (Rep) and RepA. With this technique, it was possible to regenerate plants with a desired change in the DNA sequence in less than 6 weeks. The proposed approach holds a big promise for the genome editing in monocots because some of the geminiviruses belonging to a genus *Mastrevirus* (eg, wheat dwarf virus and maize streak virus) have been successfully used for protein expression in monocots [55].

3.2 The Application of TALENs in Crops

The utilization of TALENs for crop improvement were clearly shown in few reports [56–59]. The most prominent improvements were the disruption of two fatty desaturase genes (*FAD2-1A* and *FAD2-1B*) in soybean [56], targeted mutations of three homoeoalleles that encode the MILDEW-RESISTANCE LOCUS (MLO) proteins in wheat [57], the mutation at the promoter site of the barley phytase gene of the purple acid phosphatase group named *HvPAPhy_a* [58], and the disruption of a *PROCERA* (*PRO*) gene in tomato [59]. Simultaneous mutations in the *FAD2-1A* and *FAD2-1B* genes resulted in the generation of lines low in polyunsaturated fats that have an economic value for increasing oil shelf life and improving oxidative stability. After segregation, mutant plants were isolated that lacked the TALEN transgene and carried only the targeted mutations. Furthermore, a new trait not found in nature was developed using TALENs after simultaneous targeting of three homoalleles in wheat [57]. TALEN-induced disruption of all three *TaMLO* homologs in the same plant conferred heritable broad-spectrum resistance to powdery mildew.

An increase in the cold storage and processing of potato tubers was achieved by targeted disruption of a *VACUOLAR INVERTASE* gene (*VInv*) that encodes an enzyme involved in hydrolysis of sucrose to glucose and fructose [60]. Full *VInv*-knockout plants had undetectable levels of reducing sugars that can form a potential carcinogen when reacted with free amino acids upon high-temperature processing. As in the case of the soybean study, the authors managed to select plants that did not contain TALEN transgenes in the genome but only mutations in *VInv* alleles. The edited potato is void of the regulation covering GMO crops in the USA and may soon enter a market as the first crop edited with designed endonucleases (http://www.aphis.usda.gov/biotechnology/downloads/reg_loi/aphis_response_cellectis_potato.pdf).

3.3 Potential Limitations of the TALEN Technology

The number of endogenous sequences that can be targeted by TALENs are limited by the need of a thymidine nt at the 5′ position [61]. Each TALEN must be experimentally validated since not all de novo assembled TALEN pairs work efficiently in vivo [62]. In addition, conventional TALENs are not able to cleave DNA containing 5-methylcytosine. Since methylated cytosine is indistinguishable from thymidine in the major groove, a repeat that recognizes cytosine can be substituted for one that binds to thymine. This approach, however, can reduce the target specificity [63,64].

The construction of multiple repeat sequences to assemble the DNA-binding domains remains a challenging task. The repetitive nature of TALE arrays makes it difficult to amplify them with PCR, and the assembled TALENs can be mutated by recombination in vivo [65]. Different methods have been developed to simplify the cloning of repeat arrays [49,51,66], and various computer programs are available for efficient design of TALEs and target prediction [67]. The most popular assembly method is a Golden Gate platform which offers a rapid, inexpensive and user-friendly protocol for TALEN assembly.

4. THE CRISPR/CAS9 SYSTEM FOR THE GENOME ENGINEERING OF PLANTS

An RNA-based and very efficient genome-editing tool was developed using the bacterial CRISPR and Cas9 protein. The CRISPR arrays were first identified in the *Escherichia coli* genome in 1987 [68], but their biological relevance was not known. In 2005, it was shown that some of the regions of the CRISPR sequence were homologous to viral and plasmid DNA, suggesting a role in adaptive immunity [69–71]. Later on, the CRISPR arrays were confirmed to provide protection against invading viruses when combined with the Cas genes, and the mechanism of this RNA-mediated DNA-targeting immune system was demonstrated [72–75].

Although the CRISPR/Cas system is present in most of the archaeal and many bacterial genomes [76], the most used CRISPR/Cas genome-editing tool originates from *Streptococcus pyogenes*. It contains the minimal CRISPR machinery composed of a single Cas9 protein, CRISPR RNA (crRNA) with a complementary sequence to the target site, and a *trans*-activating RNA (tracrRNA) that forms a hairpin with crRNA [41]. The CRISPR/Cas system is a part of an adaptive immune system that protects bacteria and archaea from viruses by digesting their DNA in a sequence-specific manner. The immunity is attained by the incorporation of short fragments of the viral DNA known as spacers at the proximal end of the CRISPR locus between two repeat arrays [77]. The CRISPR sequence is transcribed during subsequent infections with the virus and is sliced into 40 nt-long crRNAs. Eventually, crRNAs are combined with the tracrRNA to activate and guide the Cas9 nuclease to the invading DNA. The Cas9 enzyme cleaves the homologous DNA sequences into fragments called protospacers [72]. Binding specificity is provided by the so-called "seed sequence" of about 12 bases and a short DNA sequence termed a protospacer adjacent motif (PAM). The PAM usually contains a sequence of 5′-NGG-3′ (less frequently 5′-NAG-3′ [78]) and is situated downstream of the target DNA [79] (Fig. 12.4).

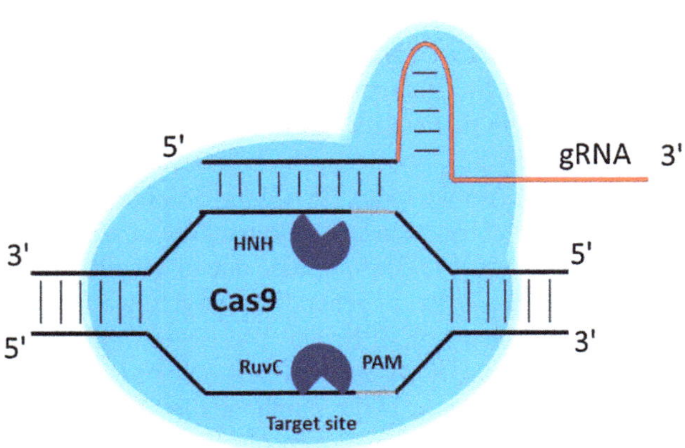

FIGURE 12.4 **The CRISPR/Cas system.** The system consists of a guided RNA (gRNA) and a Cas9 endonuclease. While gRNA is responsible for the specificity, the Cas9 protein mediates the cleavage of a complementary transcript. Cas9 requires the presence of a correct protospacer adjacent motif (PAM) sequence at the 3′ end of the target transcript, and upon binding to DNA, the nuclease unwinds the duplex and cleaves strands using the catalytic domains HNH and RuvC.

The progress for establishing the CRISPR/Cas system as a genome-editing tool was achieved when it was demonstrated that the target DNA sequence could be reprogrammed simply by replacing 20 nt in crRNA. In addition, crRNA could be combined with tracrRNA in a chimeric single-guide RNA (gRNA), thus reducing the system from three to two components and making it more efficient [80,81]. In comparison to the ZFN and TALEN technology, the CRISPR/Cas system relies on a simple Watson–Crick base pairing between gRNA and the target DNA sequence; therefore, the sophisticated protein engineering of each target is omitted [77]. The digestion of the target DNA sequence is performed by two cleavage domains (RuvC and HNH) of Cas9. The cleavage domains produce DSB at a position that is 3 nt upstream of PAM leaving in most of the cases blunt ends [80].

Another unique feature of the CRISPR/Cas system that sets it apart from other designed nucleases is the ability to selectively target either DNA or RNA. For instance, the Type III-B CRISPR/Cas system from *Pyrococcus furiosus* mediates the homology-dependent degradation of complementary RNA guided by an engineered crRNA [82]. The posttranscriptional control of gene expression would possibly be a more powerful alternative to RNA interference when the binding of the designed endonuclease to the target DNA is inhibited either by chromatin structure or by the presence of other bound proteins. In addition, the target elimination of only one of several splice variants from a single transcript could be possibly achieved. This is the gene expression regulation that is currently impossible to obtain by targeted DNA mutagenesis [77].

Everything that can be achieved with ZFNs and TALENs can in general be achieved with the CRISPR/Cas technology. The first publications on the utilization of the CRIPSR/Cas system in eukaryotes (human, mouse, and zebrafish) demonstrated that it is a simple, inexpensive, and versatile tool for genome editing [62,83–85].The target mutation efficiency of the CRISPR/Cas9 system in zebrafish embryos was found to be similar to that of ZFNs and TALENs [62]. The design and assembly of the CRISPR/Cas9 cassettes is relatively straightforward, currently devoid of intellectual property barriers, and thus can be preferred over other designed nucleases for genome-editing applications both in basic and applied studies [41].

4.1 The Application of the CRISPR/Cas System in Model Plant Species

In 2013, five reports demonstrated the use of the CRISPR/Cas9 system for gene disruption/targeting in *Arabidopsis*, tobacco, and rice [86–89]. All studies used a range of transformation platforms, including protoplast transformation, transient and stable *Agrobacterium*-mediated DNA delivery into leaves as rapid methods for the CRISPR/Cas9 system screening. A stable integration of the CRISPR/Cas cassette into the *Arabidopsis* and rice genomes resulted in the recovery of mutants with an expected phenotype already in the T1 generation of multiple genes that were targeted [86]. The mutation frequency was high in both *Arabidopsis* and rice, ranging in most of the cases from 26% to 84%. Similar to other designed endonucleases, multiple mutated alleles with different indels were recovered from transgenic plants, indicating DNA repair through the NHEJ pathway. The successful application of the CRISPR/Cas system for the targeted mutagenesis in monocots (rice and sorghum) and dicots (*Arabidopsis* and tobacco) was shown in a separate study [90]. Overall, when stably integrated, the CRIPSR/Cas technique can generate detectable mutations at a frequency of 50–89% for a single locus and 68–74% for double loci in plants [91]. In line with other designed endonucleases, it was possible to isolate transgene-free *Arabidopsis* plants with specific and heritable genome-editing events. In addition, the main practical advantage of CRISPR/Cas9 over ZFNs and TALENs is the ease of multiplexing. It simply requires the monomeric Cas9 protein and selected sequence-specific gRNAs [91]. On the other hand, multiplex editing with either ZFNs or TALENs demands separate dimeric proteins assembled for each target site [77]. The simultaneous introduction of targeted mutations at multiple sites can be used either to knock out redundant genes, parallel pathways or to create large genomic deletions/inversions [81,86,92]. More importantly, it has been noted that a high mutation frequency observed in rice (up to 91.6%) is apparently due to the unique feature of the CRISPR/Cas system (unlike ZFN and TALEN) to tolerate DNA methylation at cleavage sites [78,93]. This makes the CRISPR/Cas technology more favorable over other designed endonucleases because about 70% of the CG/CNG sites are methylated in plants [94]. The CRISPR/Cas9 system is therefore more useful for genome editing in plants, such as monocots with high genomic GC content [57,93].

In plants, gRNAs can be expressed under the control of different promoters that are recognized by RNA polymerase II and III, such as U6-26, *AtU6*, *OsU6*, *AtUBQ*, *OsUBQ*, and *CaMV 35S* [86,91,95]. Similarly, the expression of the Cas9 endonuclease can be driven by either *EF1A*, *CaMV*, *UBO*, or *LTR* promoters. Among them, the *CaMV* 35S promoter has been used most often [96] to drive the expression of a single chimeric gRNA that has been shown to be more efficient than separate crRNA and tracrRNA components for site-targeted mutagenesis in plants [81,93]. Although, due to the differences in experimental setups, it is hard to compare transformation and detection methods; but in general, the targeting efficiency of the CRISPR/Cas9 system seems to be comparable to or exceeding that obtained with ZFNs and TALENs [97,98].

One of the criticisms of the CRISPR/Cas technology is the relatively high rate of off-target effects reported in studies on animals [78,99,100]. Similarly, the off-target mutagenesis was observed in rice in two separate studies by using the PCR/restriction enzyme assay [88,89]. At the same time, no off-target modifications have been observed in studies on

Arabidopsis, tobacco, sweet orange, and in a separate study on rice using different methods, including sequencing of PCR amplicons, the whole-genome sequencing, and the restriction enzyme loss method [81,87,101,102]. The reduced specificity of the CRISPR/Cas system in some of the previous reports is apparently due to the fact that only a fragment of 8–12 nt at the 3′-end (the seed sequence) is needed for target site recognition and cleavage [90,103]. In addition, multiple mismatches in the PAM-distal region can be tolerated, depending on a sequence [78,99,100]. It has been hypothesized that the reduced specificity of the CRISPR/Cas9 complex at nonseed positions in the crRNA spacer has evolved to decrease the escape of viruses with point mutations form the immune system of bacteria [104].

Overall, the limited data available thus far suggest that the off-target effects caused by the CRISPR/Cas system are rare in plants. Nevertheless, a careful selection of the specific gRNA sequences combined with the proper regulation of the CRISPR/Cas cassette expression should reduce the risk of unwanted genome modifications.

4.2 The Application of the CRISPR/Cas System in Crops

A simplicity in both design and assembly and an open access to the components of the CRISPR/Cas system made it highly applicable for the range of crops, including rice, sorghum [90], wheat [57,92], maize [98], tomato [105], and sweet orange [106]. Curiously, four independent reports have shown that the CRISPR/Cas9 technology is suitable for the introduction of biallelic or homozygous mutations directly in the first generation of stable transgenic rice and tomato plants [81,88,107,108]. More importantly, genetic crosses segregating the CRISPR/Cas cassettes away from the edited plants have allowed to obtain genome-edited but transgene-free rice [81]. These studies indicate an exceptionally high efficiency of the CRISPR/Cas system in agriculturally important crop species.

In 2014, the CRISPR/Cas9 technology was briefly characterized for its mutation efficiency in one of the most complicated sequenced genomes—bread wheat (*Triticum aestivum*) [57]. A stable transformation of the CRISPR/Cas9 cassette resulted in the recovery of mature plants with mutations at one of the three alleles of the *MILDEW RESISTANCE LOCUS* (*MLO-A1*) gene with a frequency of 5.6% comparable to that obtained by TALENs [57]. Future reports will demonstrate how efficient the CRISPR/Cas9 technology is for targeted mutagenesis of all alleles simultaneously in hexaploid wheat. The possibility of a relatively easy multiplexing and tolerance of the Cas9 enzyme to DNA methylation leaves very little doubt to suspect that the CRISPR/Cas9 system would not be efficient in the most complicated crop genomes.

4.3 Potential Limitations of the CRISPR/Cas System

One of the biggest concerns regarding the CRISPR/Cas system is its relatively high off-target mutagenesis reported previously in animals [78,99]. However, this seems to be not of a big concern for plants, possibly due to differences in the transformation efficiency, expression levels, and codon usage in plant systems. The optimization of Cas9 nuclease expression has been proposed as a way to control the specificity because high concentrations of Cas9 and gRNA components can cause off-target effects in animals [78,100,109]. Another approach is to carefully select target regions in the genome because the imperfectly matched spacer sequences can result in the cleavage at off-target positions. A comparison of several gRNAs targeting the same gene in human cells has revealed that the CRISPR/Cas system is less efficient at the sequences with an unusually high or low GC content as compared to those with an average GC level [57]. In addition, gRNAs designed to target a transcribed strand are less effective than those targeting a nontranscribed strand. Furthermore, the Cas9 enzyme preferentially binds to gRNAs containing purine residues in the last four positions of spacer sequence with a direct correlation between the affinity of Cas9 to gRNA and the cleavage activity. Although it still remains to be shown whether the same rules are applicable to plant systems, these examples can be taken into account for gRNA design in plants in order to increase the efficiency and reduce off-target effects of the CRISPR/Cas technology.

Unfortunately, the possibility of the CRISPR/Cas9 system to target a desired sequence may be limited by the availability of PAM sites [80]. The alteration of the PAM sequence greatly reduces but not abolishes the activity of the CRISPR/Cas system in plants, suggesting that although PAM is important, it is not absolutely required for the function of CRISPR/Cas [91]. A thorough examination of nuclear genome sequences in silico from eight representative plant species (*Arabidopsis thaliana*, *Medicago truncatula*, *G. max*, *Solanum lycopersicum*, *Brachpodium distachyon*, *Oryza sativa*, *Sorghum bicolor*, and *Zea mays*) using data from mammalian systems has revealed an occurrence of the PAM (NGG/NAG) site at the frequency of 5–12 times for every 100 bp [110]. The total number of PAMs correlated with genome size, and for all species except maize, it was possible to predict specific gRNAs to target 85.4–98.9% of the annotated transcript units. Since maize has the largest genome examined and the functional redundancy of some homologous genes with high sequence identity, only 30% of the transcription units could be targeted by specific gRNAs. It is, therefore, expected that similar challenges may occur for gRNA target prediction in wheat and barley that have even larger genomes than maize [110].

5. FUTURE PERSPECTIVES OF THE GENOME-EDITING TECHNOLOGY

The use of the ZFN, TALEN, and CRISPR/Cas technologies to target DSBs to the selected locus/loci has opened up the possibility of a precise, fast, and efficient genome editing both in prokaryotes and eukaryotes. The utilization of designed endonucleases will accelerate both functional genomics and applied crop improvement [3]. The connection between a particular gene and the resulted phenotype would be easy to establish for species in which a mutant is unknown or does not exist in nature. This, in turn, should speed up the efforts for the development of novel traits. Eventually, the products obtained by site-specific nucleases which do not contain a transgene cassette are expected to be regulated in the North America similar to conventionally bred genotypes, and thus be more cost effective to bring to market (http://www.genengnews.com/insight-and-intelligence/gene-editing-will-change-everything-just-not-all-at-one-time/77900351/). The use of designed endonucleases may remove a number of regulatory restrictions associated with transgenic plants. Although, the European regulatory organizations working on GM crops focus on the method and not the product (eg, plants produced by conventional mutagenesis and genome editing would be regulated differently under the current guidelines), there is a possibility that plants altered by the targeted mutagenesis using designed endonucleases would not be classified and regulated as GMOs [111,112].

The targeted, predicted integration of transgenes through a trait stacking approach can eliminate the resulting effect of an unintended disruption of host metabolism and/or production of toxic or allergenic compounds. By trait stacking, the plant cells can be efficiently engineered to act as a factory for the production of specific metabolites or proteins with a number of genes involved. This can be achieved by a careful examination of different loci in the plant genome for the influence of the chromatin and surrounding sequence on transgene expression. Eventually, a generic recipient line with a predetermined and characterized locus can be established for routine utilization of transgene insertion and strong expression, thus producing a high yield of the corresponding product [77].

Overall, it is expected that the CRISPR/Cas technology will advance more rapidly as compared to ZFNs and TALENs [113]. This is due to a combination of the few major factors: a simplicity in design and construction, a possibility of relatively easy multiplex targeting, tolerance to DNA methylation, and, most importantly, the open access policy of the CRISPR research community. Plasmids are freely available from the nonprofit repository (eg, Addgene), and the range of web tools have been developed for selecting gRNA sequences and predicting their specificity (eg, CRISPR-P, CRISPR-PLANT, and Cas-OFFinder) [77].The application of this tool in plants opens immense possibilities from the regulation of lignin biosynthesis in order to increase forage digestibility and kappa value in the pulping industry [44] to the generation of wheat-resistant cultivars by targeting the loss of susceptibility genes [114]. For these targets to be met in plants, the development of supporting technologies is required. In most of the cases, the limiting factor is the availability of an efficient transformation technique and a high-throughput molecular screening method for genome-editing analysis. Therefore, the improvement in cell and tissue culture together with the development of more efficient transformation techniques will continue to play an essential role in the further development of genome-editing technology in plants [3]. One of the promising approaches includes the utilization of a microspore culture together with the protein-mediated genome editing [115]. Overall, a brief overview of examples of targeted genome modification in plants mediated by designed endonucleases provides a clear indication that complex crop genomes can now be manipulated with a precision that far surpasses the conventional breeding practices. Therefore, it can be speculated that it is just a matter of time when genome-edited fruit, vegetable, and cereal crops will appear on the shelves of stores.

GLOSSARY

Clustered regularly interspaced short palindromic repeats/CRISPR-associated 9 An RNA-based genome-editing tool that consists of *Streptococcus pyogenes* Cas9 endonuclease and guided RNA.

Engineered endonucleases Artificial endonucleases that can be designed to digest a predetermined nucleotide sequence.

Genome editing A type of genome manipulation for the purpose of insertion, deletion, or replacement of the DNA sequence by using engineered endonucleases.

Homologous recombination repair A type of double-strand break repair in which the nucleotide sequences are exchanged between two identical or very similar molecules of DNA during genetic recombination.

Homing endonucleases/meganucleases Naturally occurring endonucleases characterized by a high specificity due to a long recognition site (12–40 bp).

Nonhomologous end joining repair A double-strand break repair pathway that involves a simple rejoining of the broken ends of the DNA molecule either in the presence or absence of the microhomology regions between broken ends.

Transcription activator-like effector nucleases The engineered endonucleases that contain a TALE DNA-binding domain at the amino terminus and a *Fok*I nuclease domain at the carboxyl terminus.

Zinc-finger nucleases The engineered endonucleases that consist of a zinc-finger protein at the N-terminus and a *Fok*I nuclease domain at the C-terminus.

LIST OF ACRONYMS AND ABBREVIATIONS

ABI4 An *ABAINSENSITIVE-4* gene
ADH1 An *ALCOHOL DEHYDROGENASE-1* gene
ALS An *ACETOLACTATE SYNTHASE* gene
CLV3 A *CLAVATA3* gene
crRNA CRISPR RNA
CRISPR/Cas Clustered regularly interspaced short palindromic repeats/CRISPR-associated
DCL1a A *DICER-LIKE1a* gene
DSB Double-strand break
EMN Engineered homing endonucleases/meganucleases
ES Embryonic stem cells
FAD2-1A* and *FAD2-1B Fatty desaturase genes
GFP GREEN FLUORESCENT PROTEIN
GMO Genetically modified organisms
gRNA Single guide RNA
GT Gene targeting
HEN1a *HUA ENHANCER 1a*gene
HR Homologous recombination
Indel Mutation caused either by insertion or deletion
IPK1 Gene that encodes the inositol-1,3,4,5,6-pentakisphosphate 2-kinase gene
MLO MILDEW-RESISTANCE LOCUS
NHEJ Nonhomologous end joining
NLS Nuclear localization signal
PAM Protospacer adjacent motif
PCR Polymerase chain reaction
PPOX *PROTOPORPHYRINOGEN OXIDASE* gene
PRO *PROCERA* gene
RDR6a *RNA-DEPENDENT RNA POLYMERASE 6a* gene
Rep Replication-initiation protein
RVD Repeat variable diresidue
SuRA* and *SuRB *SULFONYLUREA RECEPTOR* gene
SSA Single-strand annealing
TALEN Transcription activator-like effector nucleases
TLPs Trait landing pads
tracrRNA *Trans*-activating RNA
TT4 *TRANSPARENT TESTA-4* gene
VInv *VACUOLAR INVERTASE* gene
YFP Yellow fluorescent protein
ZFN Zinc-finger nucleases

REFERENCES

[1] Stadler LJ. Genetic effects of X-Rays in maize. Proc Natl Acad Sci USA 1928;14(1):69–75.
[2] Stadler LJ. Mutations in barley induced by X-Rays and radium. Science 1928;68(1756):186–7.
[3] Petolino JF. Genome editing in plants via designed zinc finger nucleases. In Vitro Cell Dev Biol Plant 2015;51(1):1–8.
[4] Bibikova M, Beumer K, Trautman JK, Carroll D. Enhancing gene targeting with designed zinc finger nucleases. Science 2003;300(5620):764.
[5] Puchta H. Gene replacement by homologous recombination in plants. Plant Mol Biol 2002;48(1–2):173–82.
[6] Sakuma T, Woltjen K. Nuclease-mediated genome editing: at the front-line of functional genomics technology. Dev Growth Differ 2014;56(1): 2–13.
[7] Zhang Y, Zhang F, Li X, Baller JA, Qi Y, Starker CG, et al. Transcription activator-like effector nucleases enable efficient plant genome engineering. Plant Physiol 2013;161(1):20–7.
[8] Terada R, Urawa H, Inagaki Y, Tsugane K, Iida S. Efficient gene targeting by homologous recombination in rice. Nat Biotechnol 2002;20(10): 1030–4.
[9] Hanin M, Volrath S, Bogucki A, Briker M, Ward E, Paszkowski J. Gene targeting in *Arabidopsis*. Plant J 2001;28(6):671–7.
[10] Choulika A, Perrin A, Dujon B, Nicolas JF. Induction of homologous recombination in mammalian chromosomes by using the I-SceI system of *Saccharomyces cerevisiae*. Mol Cell Biol 1995;15(4):1968–73.
[11] Cohen-Tannoudji M, Robine S, Choulika A, Pinto D, El Marjou F, Babinet C, et al. I-SceI-induced gene replacement at a natural locus in embryonic stem cells. Mol Cell Biol 1998;18(3):1444–8.

[12] Puchta H, Dujon B, Hohn B. Homologous recombination in plant cells is enhanced by in vivo induction of double strand breaks into DNA by a site-specific endonuclease. Nucleic Acids Res 1993;21(22):5034–40.

[13] Puchta H, Dujon B, Hohn B. Two different but related mechanisms are used in plants for the repair of genomic double-strand breaks by homologous recombination. Proc Natl Acad Sci USA 1996;93(10):5055–60.

[14] Gorbunova V, Levy AA. Non-homologous DNA end joining in plant cells is associated with deletions and filler DNA insertions. Nucleic Acids Res 1997;25(22):4650–7.

[15] Nakatsukasa T, Shiraishi Y, Negi S, Imanishi M, Futaki S, Sugiura Y. Site-specific DNA cleavage by artificial zinc finger-type nuclease with cerium-binding peptide. Biochem Biophys Res Commun 2005;330(1):247–52.

[16] Mahfouz MM, Piatek A, Stewart Jr CN. Genome engineering via TALENs and CRISPR/Cas9 systems: challenges and perspectives. Plant Biotechnol J 2014;12(8):1006–14.

[17] Osakabe Y, Osakabe K. Genome editing with engineered nucleases in plants. Plant Cell Physiol 2015;56(3):389–400.

[18] Gaj T, Gersbach CA, Barbas 3rd CF. ZFN, TALEN, and CRISPR/Cas-based methods for genome engineering. Trends Biotechnol 2013;31(7): 397–405.

[19] Enuameh MS, Asriyan Y, Richards A, Christensen RG, Hall VL, Kazemian M, et al. Global analysis of *Drosophila* Cys(2)-His(2) zinc finger proteins reveals a multitude of novel recognition motifs and binding determinants. Genome Res 2013;23(6):928–40.

[20] Persikov AV, Wetzel JL, Rowland EF, Oakes BL, Xu DJ, Singh M, et al. A systematic survey of the Cys2His2 zinc finger DNA-binding landscape. Nucleic Acids Res 2015;43(3):1965–84.

[21] Kim Y-G, Shi Y, Berg JM, Chandrasegaran S. Site-specific cleavage of DNA–RNA hybrids by zinc finger/FokI cleavage domain fusions. Gene 1997;203(1):43–9.

[22] Bitinaite J, Wah DA, Aggarwal AK, Schildkraut I. FokI dimerization is required for DNA cleavage. Proc Natl Acad Sci 1998;95(18):10570–5.

[23] Orlando SJ, Santiago Y, DeKelver RC, Freyvert Y, Boydston EA, Moehle EA, et al. Zinc-finger nuclease-driven targeted integration into mammalian genomes using donors with limited chromosomal homology. Nucleic Acids Res 2010;38(15):e152.

[24] Bibikova M, Golic M, Golic KG, Carroll D. Targeted chromosomal cleavage and mutagenesis in *Drosophila* using zinc-finger nucleases. Genetics 2002;161(3):1169–75.

[25] Shukla VK, Doyon Y, Miller JC, DeKelver RC, Moehle EA, Worden SE, et al. Precise genome modification in the crop species Zea mays using zinc-finger nucleases. Nature 2009;459(7245):437–41.

[26] Townsend JA, Wright DA, Winfrey RJ, Fu F, Maeder ML, Joung JK, et al. High-frequency modification of plant genes using engineered zinc-finger nucleases. Nature 2009;459(7245):442–5.

[27] Osakabe K, Osakabe Y, Toki S. Site-directed mutagenesis in *Arabidopsis* using custom-designed zinc finger nucleases. Proc Natl Acad Sci 2010;107(26):12034–9.

[28] Zhang F, Maeder ML, Unger-Wallace E, Hoshaw JP, Reyon D, Christian M, et al. High frequency targeted mutagenesis in *Arabidopsis thaliana* using zinc finger nucleases. Proc Natl Acad Sci 2010;107(26):12028–33.

[29] Curtin SJ, Zhang F, Sander JD, Haun WJ, Starker C, Baltes NJ, et al. Targeted mutagenesis of duplicated genes in soybean with zinc-finger nucleases. Plant Physiol 2011;156(2):466–73.

[30] Lloyd A, Plaisier CL, Carroll D, Drews GN. Targeted mutagenesis using zinc-finger nucleases in *Arabidopsis*. Proc Natl Acad Sci USA 2005;102(6):2232–7.

[31] De Pater S, Neuteboom LW, Pinas JE, Hooykaas PJJ, Van Der Zaal BJ. ZFN-induced mutagenesis and gene-targeting in Arabidopsis through Agrobacterium-mediated floral dip transformation. Plant Biotechnol J 2009;7(8):821–35.

[32] Tovkach A, Zeevi V, Tzfira T. A toolbox and procedural notes for characterizing novel zinc finger nucleases for genome editing in plant cells. Plant J 2009;57(4):747–57.

[33] Marton I, Zuker A, Shklarman E, Zeevi V, Tovkach A, Roffe S, et al. Nontransgenic genome modification in plant cells. Plant Physiol 2010;154(3):1079–87.

[34] Cai C, Doyon Y, Ainley WM, Miller J, DeKelver R, Moehle E, et al. Targeted transgene integration in plant cells using designed zinc finger nucleases. Plant Mol Biol 2009;69(6):699–709.

[35] Petolino J, Worden A, Curlee K, Connell J, Strange Moynahan T, Larsen C, et al. Zinc finger nuclease-mediated transgene deletion. Plant Mol Biol 2010;73(6):617–28.

[36] Qi Y, Li X, Zhang Y, Starker CG, Baltes NJ, Zhang F, et al. Targeted deletion and inversion of tandemly arrayed genes in *Arabidopsis thaliana* using zinc finger nucleases. G3 Genes|Genomes|Genetics 2013;3(10):1707–15.

[37] Haber JF, Partners and pathways. Trends Genet 2000;16(6):259–64.

[38] Pater S, Pinas JE, Hooykaas PJJ, Zaal BJ. ZFN-mediated gene targeting of the *Arabidopsis* protoporphyrinogen oxidase gene through Agrobacterium-mediated floral dip transformation. Plant Biotechnol J 2013;11(4):510–5.

[39] Petolino JF, Hopkins NL, Kosegi BD, Skokut M. Whisker-mediated transformation of embryogenic callus of maize. Plant Cell Rep 2000;19(8): 781–6.

[40] Ainley WM, Sastry-Dent L, Welter ME, Murray MG, Zeitler B, Amora R, et al. Trait stacking via targeted genome editing. Plant Biotechnol J 2013;11(9):1126–34.

[41] Kumar V, Jain M. The CRISPR-Cas system for plant genome editing: advances and opportunities. J Exp Bot 2015;66(1):47–57.

[42] Sander JD, Dahlborg EJ, Goodwin MJ, Cade L, Zhang F, Cifuentes D, et al. Selection-free zinc-finger-nuclease engineering by context-dependent assembly (CoDA). Nat Methods 2011;8(1):67–9.

[43] Chandrasekharan S, Kumar S, Valley CM, Rai A. Proprietary science, open science and the role of patent disclosure: the case of zinc-finger proteins. Nat Biotechnol 2009;27(2):140–4.

[44] Eudes A, Liang Y, Mitra P, Loque D. Lignin bioengineering. Curr Opin Biotechnol 2014;26:189–98.

[45] Christian M, Cermak T, Doyle EL, Schmidt C, Zhang F, Hummel A, et al. Targeting DNA double-strand breaks with TAL effector nucleases. Genetics 2010;186(2):757–61.

[46] Bogdanove AJ, Schornack S, Lahaye T. TAL effectors: finding plant genes for disease and defense. Curr Opin Plant Biol 2010;13(4):394–401.

[47] Boch J, Scholze H, Schornack S, Landgraf A, Hahn S, Kay S, et al. Breaking the code of DNA binding specificity of TAL-type III effectors. Science 2009;326(5959):1509–12.

[48] Moscou MJ, Bogdanove AJ. A simple cipher governs DNA recognition by TAL effectors. Science 2009;326(5959):1501.

[49] Weber E, Gruetzner R, Werner S, Engler C, Marillonnet S. Assembly of designer TAL effectors by Golden Gate cloning. PLoS One 2011;6(5):e19722.

[50] Mahfouz MM, Li L, Shamimuzzaman M, Wibowo A, Fang X, Zhu J-K. De novo-engineered transcription activator-like effector (TALE) hybrid nuclease with novel DNA binding specificity creates double-strand breaks. Proc Natl Acad Sci 2011;108(6):2623–8.

[51] Cermak T, Doyle EL, Christian M, Wang L, Zhang Y, Schmidt C, et al. Efficient design and assembly of custom TALEN and other TAL effector-based constructs for DNA targeting. Nucleic Acids Res 2011;39(12):e82.

[52] Christian M, Qi Y, Zhang Y, Voytas DF. Targeted mutagenesis of *Arabidopsis thaliana* using engineered TAL effector nucleases. G3 (Bethesda) 2013;3(10):1697–705.

[53] Forner J, Pfeiffer A, Langenecker T, Manavella P, Lohmann JU. Germline-transmitted genome editing in *Arabidopsis thaliana* Using TAL-effector-nucleases. PLoS One 2015;10(3):e0121056.

[54] Shan Q, Wang Y, Chen K, Liang Z, Li J, Zhang Y, et al. Rapid and efficient gene modification in rice and *Brachypodium* using TALENs. Mol Plant.6(4):1365–1368.

[55] Baltes NJ, Gil-Humanes J, Cermak T, Atkins PA, Voytas DF. DNA replicons for plant genome engineering. Plant Cell 2014;26(1):151–63.

[56] Haun W, Coffman A, Clasen BM, Demorest ZL, Lowy A, Ray E, et al. Improved soybean oil quality by targeted mutagenesis of the fatty acid desaturase 2 gene family. Plant Biotechnol J 2014;12(7):934–40.

[57] Wang Y, Cheng X, Shan Q, Zhang Y, Liu J, Gao C, et al. Simultaneous editing of three homoeoalleles in hexaploid bread wheat confers heritable resistance to powdery mildew. Nat Biotechnol 2014;32(9):947–51.

[58] Wendt T, Holm P, Starker C, Christian M, Voytas D, Brinch-Pedersen H, et al. TAL effector nucleases induce mutations at a pre-selected location in the genome of primary barley transformants. Plant Mol Biol 2013;83(3):279–85.

[59] Lor VS, Starker CG, Voytas DF, Weiss D, Olszewski NE. Targeted mutagenesis of the tomato PROCERA gene using transcription activator-like effector nucleases. Plant Physiol 2014;166(3):1288–91.

[60] Clasen BM, Stoddard TJ, Luo S, Demorest ZL, Li J, Cedrone F, et al. Improving cold storage and processing traits in potato through targeted gene knockout. Plant Biotechnol J 2015. n/a-n/a.

[61] Doyle EL, Booher NJ, Standage DS, Voytas DF, Brendel VP, Vandyk JK, et al. TAL effector-nucleotide targeter (TALE-NT) 2.0: tools for TAL effector design and target prediction. Nucleic Acids Res 2012;40(Web Server issue):W117–22.

[62] Hwang WY, Fu Y, Reyon D, Maeder ML, Tsai SQ, Sander JD, et al. Efficient genome editing in zebrafish using a CRISPR-Cas system. Nat Biotech 2013;31(3):227–9.

[63] Deng D, Yin P, Yan C, Pan X, Gong X, Qi S, et al. Recognition of methylated DNA by TAL effectors. Cell Res 2012;22(10):1502–4.

[64] Valton J, Dupuy A, Daboussi F, Thomas S, Marechal A, Macmaster R, et al. Overcoming transcription activator-like effector (TALE) DNA binding domain sensitivity to cytosine methylation. J Biol Chem 2012;287(46):38427–32.

[65] Holkers M, Maggio I, Liu J, Janssen JM, Miselli F, Mussolino C, et al. Differential integrity of TALE nuclease genes following adenoviral and lentiviral vector gene transfer into human cells. Nucleic Acids Res 2013;41(5):e63.

[66] Reyon D, Tsai SQ, Khayter C, Foden JA, Sander JD, Joung JK. FLASH assembly of TALENs for high-throughput genome editing. Nat Biotechnol 2012;30(5):460–5.

[67] Cermak T, Starker CG, Voytas DF. Efficient design and assembly of custom TALENs using the Golden Gate platform. Methods Mol Biol 2015;1239:133–59.

[68] Ishino Y, Shinagawa H, Makino K, Amemura M, Nakata A. Nucleotide sequence of the iap gene, responsible for alkaline phosphatase isozyme conversion in *Escherichia coli*, and identification of the gene product. J Bacteriol 1987;169(12):5429–33.

[69] Bolotin A, Quinquis B, Sorokin A, Ehrlich SD. Clustered regularly interspaced short palindrome repeats (CRISPRs) have spacers of extrachromosomal origin. Microbiology 2005;151(8):2551–61.

[70] Mojica FJ, Diez-Villasenor C, Garcia-Martinez J, Soria E. Intervening sequences of regularly spaced prokaryotic repeats derive from foreign genetic elements. J Mol Evol 2005;60(2):174–82.

[71] Pourcel C, Salvignol G, Vergnaud G. CRISPR elements in *Yersinia pestis* acquire new repeats by preferential uptake of bacteriophage DNA, and provide additional tools for evolutionary studies. Microbiology 2005;151(3):653–63.

[72] Barrangou R, Fremaux C, Deveau H, Richards M, Boyaval P, Moineau S, et al. CRISPR provides acquired resistance against viruses in prokaryotes. Science 2007;315(5819):1709–12.

[73] Brouns SJ, Jore MM, Lundgren M, Westra ER, Slijkhuis RJ, Snijders AP, et al. Small CRISPR RNAs guide antiviral defense in prokaryotes. Science 2008;321(5891):960–4.

[74] Deltcheva E, Chylinski K, Sharma CM, Gonzales K, Chao Y, Pirzada ZA, et al. CRISPR RNA maturation by trans-encoded small RNA and host factor RNase III. Nature 2011;471(7340):602–7.

[75] Garneau JE, Dupuis ME, Villion M, Romero DA, Barrangou R, Boyaval P, et al. The CRISPR/Cas bacterial immune system cleaves bacteriophage and plasmid DNA. Nature 2010;468(7320):67–71.

[76] Rousseau C, Gonnet M, Le Romancer M, Nicolas J. CRISPI: a CRISPR interactive database. Bioinformatics 2009;25(24):3317–8.

[77] Bortesi L, Fischer R. The CRISPR/Cas9 system for plant genome editing and beyond. Biotechnol Adv 2015;33(1):41–52.

[78] Hsu PD, Scott DA, Weinstein JA, Ran FA, Konermann S, Agarwala V, et al. DNA targeting specificity of RNA-guided Cas9 nucleases. Nat Biotechnol 2013;31(9):827–32.

[79] Qi Lei S, Larson Matthew H, Gilbert Luke A, Doudna Jennifer A, Weissman Jonathan S, Arkin Adam P, et al. Repurposing CRISPR as an RNA-guided platform for sequence-specific control of gene expression. Cell 2013;152(5):1173–83.

[80] Jinek M, Chylinski K, Fonfara I, Hauer M, Doudna JA, Charpentier E. A programmable dual-RNA-guided DNA endonuclease in adaptive bacterial immunity. Science 2012;337(6096):816–21.

[81] Zhou H, Liu B, Weeks DP, Spalding MH, Yang B. Large chromosomal deletions and heritable small genetic changes induced by CRISPR/Cas9 in rice. Nucleic Acids Res 2014;42(17):10903–14.

[82] Hale CR, Zhao P, Olson S, Duff MO, Graveley BR, Wells L, et al. RNA-guided RNA cleavage by a CRISPR RNA-Cas protein complex. Cell 2009;139(5):945–56.

[83] Cho SW, Kim S, Kim JM, Kim JS. Targeted genome engineering in human cells with the Cas9 RNA-guided endonuclease. Nat Biotechnol 2013;31(3):230–2.

[84] Jinek M, East A, Cheng A, Lin S, Ma E, Doudna J. RNA-programmed genome editing in human cells. Elife 2013;2:e00471.

[85] Mali P, Yang L, Esvelt KM, Aach J, Guell M, DiCarlo JE, et al. RNA-guided human genome engineering via Cas9. Science 2013;339(6121):823–6.

[86] Feng Z, Zhang B, Ding W, Liu X, Yang D-L, Wei P, et al. Efficient genome editing in plants using a CRISPR/Cas system. Cell Res 2013;23(10):1229–32.

[87] Nekrasov V, Staskawicz B, Weigel D, Jones JD, Kamoun S. Targeted mutagenesis in the model plant *Nicotiana benthamiana* using Cas9 RNA-guided endonuclease. Nat Biotechnol 2013;31(8):691–3.

[88] Shan Q, Wang Y, Li J, Zhang Y, Chen K, Liang Z, et al. Targeted genome modification of crop plants using a CRISPR-Cas system. Nat Biotech 2013;31(8):686–8.

[89] Xie K, Yang Y. RNA-guided genome editing in plants using a CRISPR–Cas system. Mol Plant 2013;6(6):1975–83.

[90] Jiang W, Zhou H, Bi H, Fromm M, Yang B, Weeks DP. Demonstration of CRISPR/Cas9/sgRNA-mediated targeted gene modification in *Arabidopsis*, tobacco, sorghum and rice. Nucleic Acids Res 2013;41(20):e188.

[91] Mao Y, Zhang H, Xu N, Zhang B, Gou F, Zhu JK. Application of the CRISPR-Cas system for efficient genome engineering in plants. Mol Plant 2013;6(6):2008–11.

[92] Upadhyay SK, Kumar J, Alok A, Tuli R. RNA-guided genome editing for target gene mutations in wheat. G3 (Bethesda) 2013;3(12):2233–8.

[93] Miao J, Guo D, Zhang J, Huang Q, Qin G, Zhang X, et al. Targeted mutagenesis in rice using CRISPR-Cas system. Cell Res 2013;23(10):1233–6.

[94] Vanyushin BF, Ashapkin VV. DNA methylation in higher plants: past, present and future. Biochim Biophys Acta 2011;1809(8):360–8.

[95] Fauser F, Schiml S, Puchta H. Both CRISPR/Cas-based nucleases and nickases can be used efficiently for genome engineering in *Arabidopsis thaliana*. Plant J 2014;79(2):348–59.

[96] Belhaj K, Chaparro-Garcia A, Kamoun S, Nekrasov V. Plant genome editing made easy: targeted mutagenesis in model and crop plants using the CRISPR/Cas system. Plant Methods 2013;9(1):39.

[97] Lozano-Juste J, Cutler SR. Plant genome engineering in full bloom. Trends Plant Sci 2014;19(5):284–7.

[98] Liang Z, Zhang K, Chen K, Gao C. Targeted mutagenesis in Zea mays using TALENs and the CRISPR/Cas system. J Genet Genomics 2014;41(2):63–8.

[99] Fu Y, Foden JA, Khayter C, Maeder ML, Reyon D, Joung JK, et al. High-frequency off-target mutagenesis induced by CRISPR-Cas nucleases in human cells. Nat Biotechnol 2013;31(9):822–6.

[100] Pattanayak V, Lin S, Guilinger JP, Ma E, Doudna JA, Liu DR. High-throughput profiling of off-target DNA cleavage reveals RNA-programmed Cas9 nuclease specificity. Nat Biotechnol 2013;31(9):839–43.

[101] Feng Z, Mao Y, Xu N, Zhang B, Wei P, Yang DL, et al. Multigeneration analysis reveals the inheritance, specificity, and patterns of CRISPR/Cas-induced gene modifications in *Arabidopsis*. Proc Natl Acad Sci USA 2014;111(12):4632–7.

[102] Jiang W, Yang B, Weeks DP. Efficient CRISPR/Cas9-mediated gene editing in *Arabidopsis thaliana* and inheritance of modified genes in the T2 and T3 generations. PLoS One 2014;9(6):e99225.

[103] Cong L, Ran FA, Cox D, Lin S, Barretto R, Habib N, et al. Multiplex genome engineering using CRISPR/Cas systems. Science 2013;339(6121):819–23.

[104] Semenova E, Jore MM, Datsenko KA, Semenova A, Westra ER, Wanner B, et al. Interference by clustered regularly interspaced short palindromic repeat (CRISPR) RNA is governed by a seed sequence. Proc Natl Acad Sci USA 2011;108(25):10098–103.

[105] Ron M, Kajala K, Pauluzzi G, Wang D, Reynoso MA, Zumstein K, et al. Hairy root transformation using *Agrobacterium rhizogenes* as a tool for exploring cell type-specific gene expression and function using tomato as a model. Plant Physiol 2014;166(2):455–69.

[106] Jia H, Wang N. Targeted genome editing of sweet orange using Cas9/sgRNA. PLoS One 2014;9(4):e93806.

[107] Brooks C, Nekrasov V, Lippman ZB, Van Eck J. Efficient gene editing in tomato in the first generation using the clustered regularly interspaced short palindromic repeats/CRISPR-associated9 system. Plant Physiol 2014;166(3):1292–7.

[108] Zhang H, Zhang J, Wei P, Zhang B, Gou F, Feng Z, et al. The CRISPR/Cas9 system produces specific and homozygous targeted gene editing in rice in one generation. Plant Biotechnol J 2014;12(6):797–807.

[109] Fujii W, Kawasaki K, Sugiura K, Naito K. Efficient generation of large-scale genome-modified mice using gRNA and CAS9 endonuclease. Nucleic Acids Res 2013;41(20):e187.

[110] Xie K, Zhang J, Yang Y. Genome-wide prediction of highly specific guide RNA spacers for CRISPR–Cas9-mediated genome editing in model plants and major crops. Mol Plant 2014;7(5):923–6.

[111] Hartung F, Schiemann J. Precise plant breeding using new genome editing techniques: opportunities, safety and regulation in the EU. Plant J 2014;78(5):742–52.

[112] Podevin N, Davies HV, Hartung F, Nogue F, Casacuberta JM. Site-directed nucleases: a paradigm shift in predictable, knowledge-based plant breeding. Trends Biotechnol 2013;31(6):375–83.

[113] Belhaj K, Chaparro-Garcia A, Kamoun S, Patron NJ, Nekrasov V. Editing plant genomes with CRISPR/Cas9. Curr Opin Biotechnol 2015;32: 76–84.

[114] O'Driscoll A, Kildea S, Doohan F, Spink J, Mullins E. The wheat-*Septoria* conflict: a new front opening up? Trends Plant Sci 2014;19(9):602–10.

[115] Bilichak A, Luu J, Eudes F. Intracellular delivery of fluorescent protein into viable wheat microspores using cationic peptides. Front Plant Sci 2015;6:666.

Chapter 13

Plant Genome Stability: General Mechanisms

A. Bilichak
Agriculture and Agri-Food Canada, Lethbridge, AB, Canada

Chapter Outline

1. INTRODUCTION

The maintenance of genome stability in every organism encompasses a complex of measures aimed at a precise replication of a native sequence or the repair of damaged DNA in order to avoid any alterations in the genetic material in somatic cells and in the progeny. Due to a wide number of intra- and extracellular genotoxic agents, different components of DNA can be damaged including the sugar residues, phosphodiester linkages, and purine and pyrimidine bases [1]. Hence, depending on DNA damage, different pathways are involved in the sensing of distinct lesions, their recognition and repair. It is believed that an efficient DNA repair is of a particular importance for plants because unlike most of the higher eukaryotes, they do not set aside gametes during early sporophytic development [1]. Thus, any stress factor encountered during plant sporophytic development can potentially affect the genome of the predecessors of gametes and be passed on to the progeny. Taking into account that plants are sessile organisms with an associated inability to initiate avoidance response during stress exposure, it is safe to assume that the genome maintenance mechanism has to be robust enough to cope with genotoxic factors for an intact passage of genetic information.

The typical external sources of DNA damage are UV-B, ozone, high temperatures, drought, air, and soil pollutants including heavy metals [2]. Reactive oxygen species (ROS) belong to the internal agents which may cause single-strand breaks (SSBs) either through the damage of deoxyribose units or covalent alterations of bases [2]. ROS are continuously generated in plant cells during normal oxidative cellular processes and possess a potential danger to the integrity of the plant genome even in the absence of environmental stressors. DNA damage can also occur spontaneously during DNA replication—the collapsed replication forks or replication through SSB [3]. In most of the cases, to mitigate the cytotoxic effects of DNA damage, the early detection, cell-cycle arrest, and the rapid repair of damaged regions have to take place. In this chapter, we provide a short review of the major DNA-repair pathways in plants that play a key role in the maintenance of genome stability.

2. DNA-DAMAGING AGENTS

Exogenous genotoxins as well as metabolic derivatives can react with DNA and cause a number of different base modifications and even SSBs. Extracellular DNA-damaging agents, such as salt, heavy metals, extreme alterations in the ambient

Genome Stability. http://dx.doi.org/10.1016/B978-0-12-803309-8.00013-6

temperature, water supply, pathogens, and elicitors can increase the level of DNA lesions and activate responsive pathways [4]. A common factor that links all of the stressors together is the generation of ROS that are also constantly produced in mitochondria and chloroplasts during respiration and photosynthesis, respectively [5]. The hydroxyl radical (•OH) is one of the most active ROS which effectively interacts with biomolecules at the diffusion-controlled rates [6]. Due to a high reactivity of •OH, it essentially reacts with biomolecules in the place of its generation in a reaction volume of less than 2 nm. •OH is mostly generated in the Fenton reaction as a product of the interaction of reduced redox-active metal ions (eg, Fe^{2+} and Cu^{+}) with intracellular hydrogen peroxide. The main targets of •OH are thymine nucleobases that eventually can give rise to *cis*- and *trans*-diastereomers of 5,6-dihydroxy-5,6-dihydrothymine. The oxidation of cytosine leads generates intermediate products which are highly unstable and give rise to analogues of uracil. In turn, the degradation products of guanine include 8-oxo-7,8-dihydroguanine (8-oxoGua) and 2,6-diamino-4-hydroxy-5-formamidopyrimidine (Fapy-Gua) with a higher efficiency of the intracellular generation of Fapy-Gua as compared to 8-oxoGua [7]. Similarly, the oxidation of adenine also leads to the formation of 8-oxoAde and Fapy-Ade. Singlet oxygen ($^1O_2$) is another ROS that is the main contributor of UV-A irradiation-induced oxidative damage and may cause DNA lesions by reacting selectively with guanine components [8]. As a result, 8-oxo-7,8-dihydro-20-deoxyguanosine (8-oxodGuo) is exclusively formed in the oxidation reaction. The oxidized guanine adducts are mainly associated with G-A and G-T mutations. In addition, the alkylation of guanine can lead to G-C mutation [9].

The artificial mutagenesis is also widely exploited in research by the utilization of ethyl methane sulfonate (EMS) to generate alkylation products, including O6-ethylguanine. Eventually, the DNA-replication machinery recognizes the modified base as adenine leading to G-A mutations, thus permanently altering the sequence of the genome [10,11].

3. SENSING DNA DAMAGE

The detection of DNA damage is the first step in the DNA-repair pathway. It is believed that there are few mechanisms for DNA lesions detection depending on the nature of damage. The two main proteins involved in the DNA-damage sensing in *Arabidopsis* are the phosphoinositide-3-kinase-related protein kinases ATM (ataxia telangiectasia mutated) and ATR (ataxia telangiectasia and Rad3-related) (Fig. 13.1). While the ATM protein is activated by double-strand breaks (DSBs), ATR responds to replication defects [12,13] (Table 13.1). ATR-deficient plants also demonstrate a hypersensitivity to UV-B radiation and show changes in G2-phase cell-cycle checkpoints [14,15]. The kinase activity of both proteins results in the phosphorylation of several hundred target proteins in animals, including H2AX, a histone 2A isoform, Nbs1, and the checkpoint-related protein kinases Chk1 and Chk2 [16]. The phosphorylation of H2AX in dividing root cells occurs very rapidly and demonstrates a peak of accumulation at10-min post-irradiation [17]. As a result of a wide variety of protein phosphorylation by ATM/ATR proteins, a rapid relocation of the DSB-repair

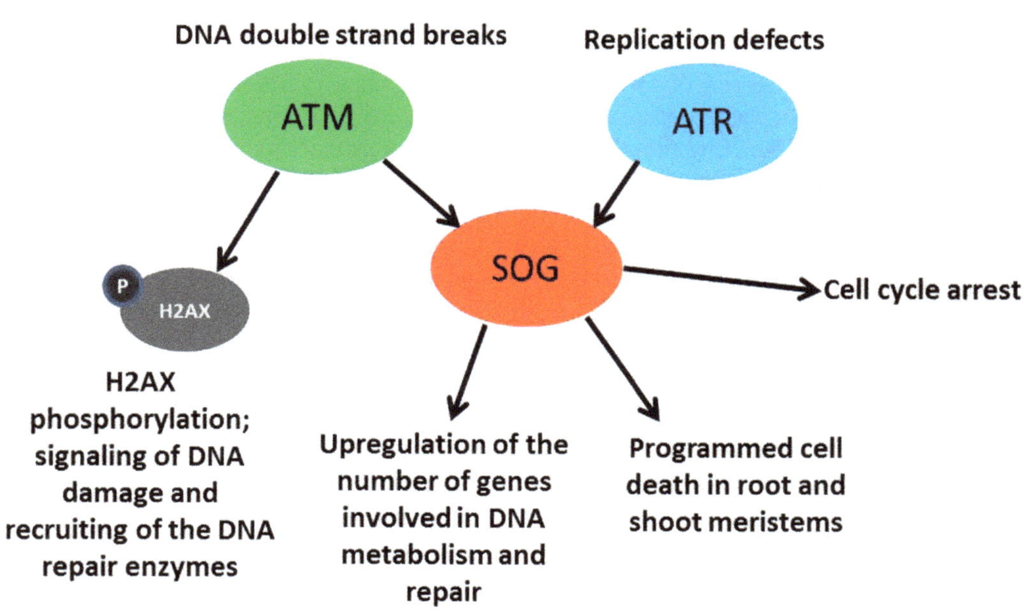

FIGURE 13.1 **DNA-damage signaling in plants.** The phosphoinositide 3-kinase-like protein kinases ATAXIA TELANGIECTASIA MUTATED (ATM), ATAXIA TELANGIECTASIA MUTATED, and RAD3-RELATED (ATR) are involved in the DNA-damage response in plants.

proteins occurs along with the activation of the cell-cycle checkpoint and initiation of DNA repair [18]. In addition, transcriptional changes in response to DNA damage play a vital role in lesion repair. While the ATM protein is required for the upregulation of genes involved in DNA metabolism, cell cycle, and homologous recombination (HR) repair, the ATR protein plays a minor role in the regulation of gene expression [19]. At the same time, NHEJ genes demonstrate a negligible transcriptional upregulation in response to DNA damage that is apparently due to a constitutive expression of genes in somatic tissues. In addition to the ATM/ATR proteins, a unique plant-specific transcription factor SUPPRESSOR OF GAMMA RESPONSE 1 (SOG1) has been demonstrated to play a role of a central regulator in the DNA-damage response [20]. It has been suggested to perform similar functions to the mammalian p53 protein involved in the cell-cycle checkpoint.

The UV-induced photoproducts are recognized by the DNA damage-binding (DDB) complex that consists of DDB1, DDB2 (XPE), Cullin4, and Rbx1 proteins [21]. In addition, in the case of DSBs, free DNA ends are detected by the KU70 and KU80 complex as well as by the MRN complex [12,22].

4. CHROMATIN ARCHITECTURE AND DNA REPAIR

Following the recognition of DNA lesions, the chromatin-remodeling enzymes provide an access of repair proteins to the damaged DNA. The evolutionarily conserved Chromatin assembly factor 1 (CAF-1) complex plays a key role in the deposition of H3 and H4 histones on to a newly synthesized DNA molecule [23] (Table 13.1). CAF1 is a chaperone complex consisting of FASCIATA 1 (FAS1), FAS2, and Multicopy suppressor of IRA1 (MSI1) subunits in *Arabidopsis* [24]. Mutations of CAF-1 components cause a hypersensitivity to genotoxins, the elevated level of DSBs, a constitutive activation of the DNA-damage response, including H2AX phosphorylation and RAD51 induction concomitant with a 40-fold increase in the rate of HR [25,26]. Similarly, in *Arabidopsis*, the overexpression of the RAD54a protein which is a member of the SWItch/Sucrose nonfermentable (SWI2/SNF2) superfamily and a chromatin-remodeling factor promotes HR and increases the rate of gene targeting by almost 30-fold [27]. In addition, screening of RNAi plants deficient in representatives of the SWI2/SNF2 subfamilies revealed their hypersensitivity to genotoxins, further highlighting the importance of chromatin-remodeling factors in the DNA repair [28].

In the process of DSB repair, especially during HR, the cohesion of sister chromatids plays a vital role in promoting the recombination. DNA damage–induced cohesion is stimulated by the large ATPases—structural maintenance of chromosomes (SMC) proteins [29,30]. *Arabidopsis* mutants deficient in the SMC6 homologue MIM (hypersensitive to MMS, Irradiation and MMC) demonstrate the reduced levels of DSB repair, including low rates of intrachromosomal HR and an increased sensitivity to a broad range of genotoxins [31,32].

In addition to DSB repair, the histone acetyltransferases HAM1 and HAM2 have been implemented in the repair of UV-B-induced DNA damage [33]. Similarly, the histone H3/H4 chaperone ANTI-SILENCING *FUNCTION1* (ASF1) is also involved in the repair of UV-B-induced DNA damage, thus further reinforcing the importance of chromatin modifiers in DNA-damage repair [34].

TABLE 13.1 Proteins Involved in the DNA Damage Sensing and Chromatin Composition in Plants

Pathway	Protein Name	Function
DNA-damage sensing and response	Ataxia telangiectasia mutated (ATM)	Activated by double-strand breaks and is required for the upregulation of genes involved in the DNA metabolism, cell cycle, and HR repair
	Ataxia telangiectasia and Rad3-related (ATR)	Responds to replication defects; plays a minor role in the regulation of gene expression
	Suppressor of gamma response 1 (SOG1)	Plays a role of the central regulator in the DNA-damage response
Chromatin composition	Chromatin assembly factor 1chaperone complex (CAF-1)	Involved in the deposition of H3 and H4 histones on to newly synthesized DNA
	Structural maintenance of chromosomes (SMC) proteins	The large ATPases that stimulate the DNA damage–induced cohesion
	Histone acetyltransferases HAM1 and HAM2	Involved in the repair of UV-B-induced DNA damage

5. PHOTOREACTIVATION

During their life cycle, plants are constantly exposed to the ultraviolet radiation (UVR). Solar UVR that reaches the Earth's surface is divided into three wavelength ranges: the UV-A (315–400 nm), UV-B (280–315 nm), and UV-C (200–280 nm) spectrum. While the first two types of UVR can reach the surface of the planet, the third type is strongly absorbed by oxygen and ozone in the atmosphere [1]. The UV-B radiation is also partially filtered through the stratospheric ozone layer and represents 1.5% of the total radiation that reaches the Earth's surface. At the same time, it is one of the most damaging types of solar radiation which reaches the surface [35]. More importantly, during the end of the 20th century, a decreasing layer of stratospheric ozone caused an increase in the amount of UV-B irradiation reaching the surface of our planet [36]. UVR can potentially damage almost all biomolecules; however, DNA damage is considered to be the most pronounced as compared to the damage of proteins and lipids [37]. UVR-induced alterations to the DNA structure include photosensitization reactions and dimer production between adjacent pyrimidine residues in the DNA strand caused either by UV-A/visible light or UV-B, respectively. UV-A/visible light can induce the generation of ROS, including $^1O_2$ which is very potent in damaging biomolecules [38]. In addition to pyrimidine dimers, UV-B light can produce the oxidized and hydrated bases, SSBs and cross-links (both DNA–protein and DNA–DNA) that can lead to the growth and development retardation in plants [39,40].

The two major types of pyrimidine dimers include the cyclobutane pyrimidine dimer (CPD) and the pyrimidine-pyrimidone (6-4) photoproduct (6-4PP). When induced by wavelengths longer than 290 nm, the last class of pyrimidine dimers may occasionally be converted to a Dewar isomer. Structurally, both CPD and 6-4PP can introduce distortions into the DNA double helix leading to either slight bending or even unwinding of the strands [40]. This in turn can impede the transcription and result in error-prone replication [41,42]. In plants, the occurrence of CPDs following UV-B exposure prevails as compared to 6-4PP. Although plants may tolerate a low level of CPDs in their genome [43,44], the pyrimidine dimers can jeopardize plant development due to the mutagenic and cytotoxic effects. Therefore, an efficient removal of pyrimidine dimers from the DNA structure is of a paramount importance for plant survival. A number of DNA-repair pathways can be engaged to restore the genome integrity following UV exposure, including photorepair (photoreactivation), base excision repair (BER), nucleotide excision repair (NER), or mismatch repair (MMR) [45]. Overall, depending on the light requirement, the pathways are divided into "dark" and "light" repair pathways. Dark repair pathways including BER, NER, MMR, and others are relatively inefficient in removal of UV-induced DNA lesions. At the same time, the photoreactivation process is considered more effective and depends on the wavelength of 350–450 nm [46]. The dimer splitting reaction is catalyzed by 450–550 amino acids-long monomeric enzymes called photolyases that upon binding to the UV-damaged DNA absorb UV-A light (350–450 nm) to induce cyclic electron transfer, split the CPD ring, and restore the bases to their normal state [47] (Table 13.2). The absorbance of photons is performed by cofactors, one of which is always a two-electron reduced form of flavin adenine dinucleotide ($FADH^-$), while another one is either a reduced pterin methenyltetrahydrofolate (MTHF) or 8-hydroxy-7,8-didemethyl-5-deazariboflavin (8-HDF). While the FAD cofactor is necessary for both a specific binding to the damaged DNA and for the reaction to occur, other chromophores are not essential for catalysis under normal light conditions and have no effect on enzyme-binding specificity [47]. At the same time, under the limiting light conditions, the second chromophore may increase the rate of DNA lesions repair by 10- to100-fold depending on the wavelength during catalysis. The dimer-splitting reaction includes two steps: in the "dark reaction," the photolyase binds to a pyrimidine dimer lesion, flips the dimers out of the double helix into the enzyme's catalytic site and forms a stable enzyme–substrate complex; in the light-dependent step, the chromophore absorbs a photon and transfers the excitation energy to $FADH^-$ followed by one-electron transfer to the enzyme-bound pyrimidine complex [1] (Fig. 13.2). Eventually, the covalent bond between two pyrimidine dimers is split, and the electron is transferred back to FADH. Similarly, 6-4PP is repaired by 6-4 photolyases, but with the inclusion of the thermal conversion step of 6-4PP to an unstable oxetane intermediate before the photochemical reversion steps. Distinct photolyases are responsible for the repair of different dimers and the photolyase that repairs one type of dimer cannot repair another. For instance, in *Arabidopsis*, whereas the *UVR2* gene encodes a photolyase (PHR1) that recognizes only CPDs, the *UVR3* gene encodes an enzyme that acts on 6-4PP [35]. Although the efficiency of CPD photorepair is lower as compared to 6-4PP, the quantum yield of photolyase repair of the first DNA lesion is higher as compared to that of the second one [47,48]. Similarly, the ATP-dependent NER pathway is 9.5–10.7 times more efficient in the repair of 6-4PP as compared to CDPs [47,49].

Both the induction and rate of photoproduct repair in plants are temperature dependent: they are the lowest at 0°C and the highest at around 30°C followed by either a further stabilization or decline of repair rate at higher temperatures [50,51]. In addition, the choice of the pathway for the repair of DNA lesions caused by UV-B irradiation depends on the plant's developmental stage as well as the severity of the damage. For instance, in alfalfa seedlings, whereas at the relatively low levels of pyrimidine dimers, the error-free photoreactivation mechanism is involved, at the higher DNA-damage levels, both the photoreactivation and excision-repair pathways repair DNA lesions [52]. Thus, when the level of pyrimidine dimers is

TABLE 13.2 Proteins Involved in the Photoreactivation and Base Excision Repair Pathways

Pathway	Protein Name	Function
Photoreactivation DNA-repair pathway	Photolyases	Catalyze the splitting of UV-induced pyrimidine dimers; photolyases absorb UV-A light to induce cyclic electron transfer to split cyclobutane pyrimidine dimer ring and restore the bases to normal state
Base excision repair (BER)	Glycosylases	Catalyze the recognition and excision of the damaged or incorrect base by hydrolysis of the *N*-glycosidic bond between the damaged base and the sugar
	Apurinic/apyrimidinic (AP) endonuclease	Cleaves the sugar–phosphate backbone at the 5′-region of the AP site leaving 3′-OH and blocking 5′-deoxyribose-5-phosphate (5′-dRP) termini
	DNA Pol λ	Possibly performs the gap filling in the short-patch repair
	Replicative DNA polymerase complex Pol δ/ε	Catalyzes DNA synthesis in the long-patch repair
	Flap endonuclease FEN1	Endonuclease which removes a "flap" structure generated by DNA polymerase complex Pol δ/ε in the long-patch repair
	Ligase I	Possibly involved in the repair of both single- and double-strand breaks in planta

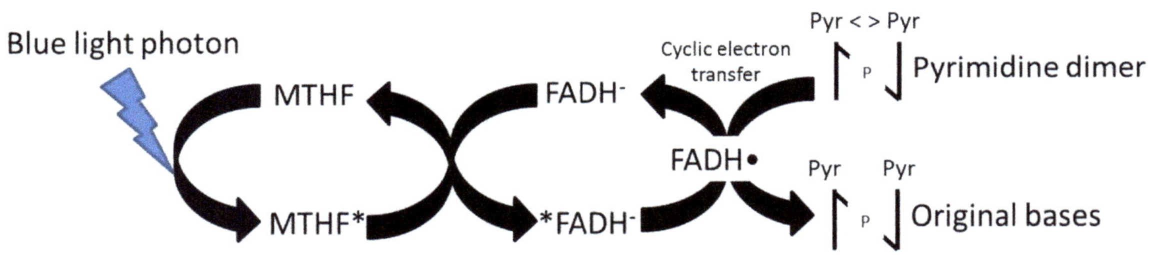

FIGURE 13.2 **A simplified scheme of the reaction catalyzed by the CPD photolyase enzyme.** The pterin methenyltetrahydrofolate (MTHF) chromophore absorbs a blue light, and the excited (MTHF*) transfers energy to the fully reduced flavin $FADH^-$. The last one, in its excited form ($*FADH^-$), induces a cyclic electron transfer step that leads to the splitting of the pyrimidine dimer into two pyrimidine monomers. The flavin coenzyme eventually is converted into the active two-electron fully reduced state. *Adapted from Bray CM, West CE. DNA repair mechanisms in plants: crucial sensors and effectors for the maintenance of genome integrity. New Phytol 2005;168(3):511–28.*

under 30 dimers per 10^6 bases, the external light energy is used for DNA repair, whereas when the DNA damage is more severe, the ATP-dependent pathway is involved.

The expression profile of genes involved in the photoreactivation and excision repair during the development has a clear tissue-dependent distribution in plants. For instance, the expression of genes involved in the excision-repair pathway is more profound in proliferating tissues. At the same time, the CPD photolyase is expressed in nonproliferating tissues, such as mature leaves and elongation zone of roots [53]. Furthermore, the DNA damage is not repaired efficiently in mature leaves in the dark, suggesting that the photoreaction is the major DNA-repair pathway for UV-induced DNA lesions in nonproliferating tissues. As expected, the examination of the photolyase protein distribution in *Arabidopsis* tissues using Western blot analysis revealed the highest level of both CPD and 6-4PP photolyases in aerial tissues including florets and leaves, and the level was very low in roots [51]. In addition, the highest level of 6-4PP photolyase was detected in siliques, and the protein is constantly expressed throughout the plant development. In contrast, the level of CPD photolyase was low in leaves of young *Arabidopsis* plants (7-day seedlings) with a further increase as plants get older (7–14 days), followed by a decrease in leaves of mature plants (6 weeks old). The low transcript level of CPD photolyase found in young leaves can apparently be compensated by a higher expression of excision-repair genes as compared to mature tissues found in *Arabidopsis* and wheat [46,53].

A more severe damage of bases and nucleotides is usually repaired through the excision repair pathways. Since many types of genotoxins introduce lesions only to a single strand of the double helix, the second strand can be used as a template for the repair. Depending on the DNA damage, the distinct but evolutionary conserved excision-repair pathways are involved that have a few common steps including: the recognition of a lesion and the excision of the damaged region, repair synthesis, and ligation [1].

6. BASE EXCISION REPAIR

BER is the major pathway for protecting DNA from genotoxic agents [54]. It is involved in the repair of DNA damages caused by ROS, alkylation, deamination, abasic (apurinic and/or apyrimidinic, AP) sites, and SSBs. The initial step of the classic BER pathway involves the recognition and excision of the damaged or incorrect base by hydrolysis of the *N*-glycosidic bond between the damaged base and the sugar catalyzed by glycosylases. The resulting site can become AP and acquire a single-stranded DNA break or a 1-nt gap flanked with a 5′-phosphate [55]. Different glycosylases are specific for every base adduct that they remove (Table 13.2). For instance, there are distinct glycosylases for the excision of either of 3-methyladenine or 8-oxoG adducts in plants [56,57]. The 8-oxoG residue is recognized as thymidine during replication, thus leading to G-T mutations. DNA glycosylases scan the DNA in search for damaged bases. During this process, the enzymes kink or bend the DNA double helix and flip the base into the catalytic site. The damaged bases usually destabilize the structure of DNA and are therefore more easily flipped out. In addition, the catalytic site of glycosylases is complementary to a specific structure and charge distribution on the damaged base, thus once in the catalytic pocket, the *N*-glycosidic bond is hydrolyzed [58]. The AP sites may also arise randomly by a spontaneous hydrolysis of the *N*-glycosylic bond [59]. Regardless of the cause of the appearance of AP sites, the following steps are parts of the SSB-repair pathway including an incision at the AP site, the formation of a gap, DNA-repair synthesis, and ligation. In the case of the monofunctional glycosylase, the AP site is further processed by an AP endonuclease which cleaves the sugar phosphate backbone at the 5′-region of the AP site leaving 3′-OH and blocking 5′-deoxyribose-5-phosphate (5′-dRP) termini [42,60]. The *Arabidopsis* genome encodes three homologues to human AP endonucleases, such as Arp, Ape1L, and Ape2. The enzymes have been suggested to act during embryo development and even programmed deletion of certain bases in gene promoters activated during the development process [61,62].

The bifunctional glycosylases, for instance, an 8-oxoG DNA glycosylase/AP lyase (OGG1), can perform both functions: the hydrolysis of *N*-glycosidic bond and the cleavage of the sugar–phosphate backbone. Other vital glycosylases which are responsible for the excision of oxidized purines and uracils are formamidopyrimidine DNA glycosylase (FPG) and uracil DNA glycosylase (UDG), respectively [61]. Curiously, while in bacteria, FPG is able to recognize and remove the 8-oxoG lesion, in plants, its homologue has structural differences that result in the minimal or no activity in DNA containing this oxidized base [63]. The appearance of uracil in DNA can arise due to a wrong incorporation of dUMP during the replication process as well as a hydrolytic deamination of cytosine [61].

The *Arabidopsis* genome encodes at least nine bifunctional glycosylases, the seven of which have been confirmed to have the AP lyase activity in vitro [54]. In addition, some of the glycosylases are actively involved in the DNA demethylation process by removing 5-methylcytosine [64].

A gap in the damaged strain may be filled in either by the insertion of a single nucleotide (short-patch repair, SP) or by DNA synthesis including a few nucleotides (long-patch repair, LP) [54,65] (Fig. 13.3). While in the former scenario, the gap filling is performed by DNA polymerase β (Pol β), in the latter case, DNA synthesis is done by the replicative DNA polymerase complex Pol δ/ε. The Pol β complex has an intrinsic deoxyribose lyase activity and is also capable of removing the sugar left by the monofunctional DNA glycosylase. Therefore, the polymerase releases the blocking 5′-dRP terminus, thus allowing for strand ligation by DNA ligase III. In the case of the LP pathway, the polymerases carry out the displacement of the strand containing the 5′-dRP terminus by 2–10 nucleotides at the 3′-region to the abasic site [1]. The generated flap structure is removed by the 5′-flap endonuclease FEN1 assisted by a proliferating cell nuclear antigen (PCNA) followed by ligation step by DNA ligase I. It seems that the choice of the BER-repair pathway depends on the nature of lesion as well as the type of glycosylase that initiates the repair [65]. The SP-repair pathway in plants seems to be missing distinct homologues of mammalian Pol β and DNA ligase III. At the same time, it has been shown that both the SP and LP pathways are active in *Arabidopsis* protein extracts and plants encode DNA ligase I that might be involved in the repair of both SSBs and DSBs in planta [54,66]. In addition, the *Arabidopsis* genome contains a gene which codes for DNA polymerase λ that contains N-terminal region that is similar to a human Pol λ. This polymerase has been shown to have the dRP-lyase activity in vitro, thus suggesting for its possible involvement in the BER pathway instead of Pol β [67].

In mammalian cells, in addition to Pol β, an X-ray repair cross-complementing protein 1 (XRCC1) is involved in the SP pathway. Although the *Arabidopsis* homologue of XRCC1 does not contain domains required for the interaction with Pol β and ligase III, it still contains the conserved BRCT domain responsible for the interaction with poly(ADP-ribose) polymerase (PARP). However, the XRCC1 homologue of rice has been hypothesized to interact with the plant Pol λ in the presence of the PCNA protein [68]. In *Arabidopsis*, the *PARP* gene has been shown to be responsive to DNA-damaging agents [69].

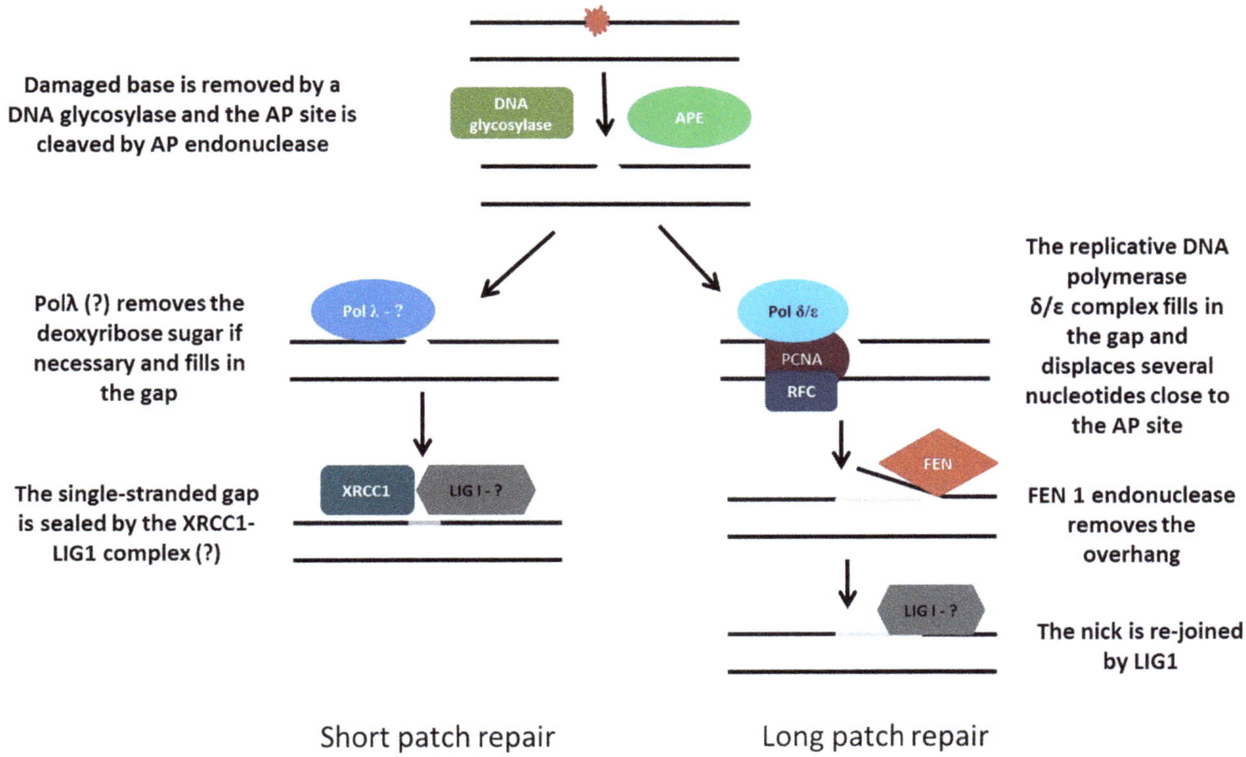

FIGURE 13.3 **Short- and long-patch repair of the base excision-repair (BER) pathway.** In the short-patch repair, the damaged base is cleaved off by either a DNA glycosylase or a bifunctional DNA glycosylase-endonuclease. Then an Apurinic/apyrimidinic endonuclease (APE) cleaves the sugar–phosphate backbone at the 5′-region of the AP site. Pol β removes the deoxyribose sugar if necessary and fills in the gap. In animals, the single-stranded gap is ligated by the XRCC1–LIG3 complex. Due to the absence of the definitive LIG3 homologue, this function is possibly performed by LIG1. In the long-patch repair, after base removal and APE nicking, the replicative DNA Pol δ/ε complex fills in the gap. During this process, a couple of nucleotides close to the AP site are displaced. The generated flap structure is removed by the 5′-flap endonuclease FEN1, and the nick is re-joined by LIG1. *Adapted from Bray CM, West CE. DNA repair mechanisms in plants: crucial sensors and effectors for the maintenance of genome integrity. New Phytol 2005;168(3):511–28.*

PARP1 has a high affinity for SSB intermediates produced during BER, but at the same time, it has been reported to have a negative effect on the rate of BER in animal cells and is not essential for an efficient completion of BER [70].

7. NUCLEOTIDE EXCISION REPAIR

The NER pathway is assumed to be less specific for a variety of DNA lesions because the enzymes involved in this pathway detect modifications broadly by conformational changes to the DNA double helix. NER is involved in the repair of DNA damage that causes significant distortions in the structure of the double helix, such as UV-photoproducts and bulky covalent lesions. Following the recognition, NER proteins remove a 24–32 oligonucleotide stretch with the altered nucleotide from the damaged strand, and the repair is completed by DNA synthesis and ligation [61]. The pathway includes two different modes, such as global genome repair (GGR) and transcription-coupled repair (TCR), with the difference between them being in the way of lesion recognition (Fig. 13.4). In plants, the pathway was discovered in the experiments involving EMS mutagenesis that resulted in a group of mutants which demonstrated a hypersensitivity to UV-C and γ-irradiation [71,72]. Thus, in addition to having an active photoreactivation pathway, plants rely on the NER pathway for the repair of UV adducts in DNA.

The recognition of DNA lesions in plants in the GGR pathway is performed by the multiprotein complex involving the xeroderma pigmentosum group C (XPC), RAD23, and centrin 2 (CEN2) [58]. The *Arabidopsis* homologue of the human XPC protein is *AtRAD4* [73]. Since the XPC protein is not capable of recognizing the UV-induced CPDs with a high efficiency, the DNA damage–binding (DDB) complex enhances its binding to the damaged DNA. The DDB complex consists of DDB1, DDB2 (XPE), Cullin4, Rbx1 proteins and is also complexed with an E3-ligase that targets specific E2-ubiquitin conjugating enzymes to other proteins including XPC. Upon the ubiquitination, the XPC protein acquires a higher affinity for the UV-damaged DNA. In addition, for the NER pathway to progress, the removal of nucleosomes around the damaged DNA region has to occur. In animal cells, it has been shown to be strictly related to the ubiquitination of H3 and H4

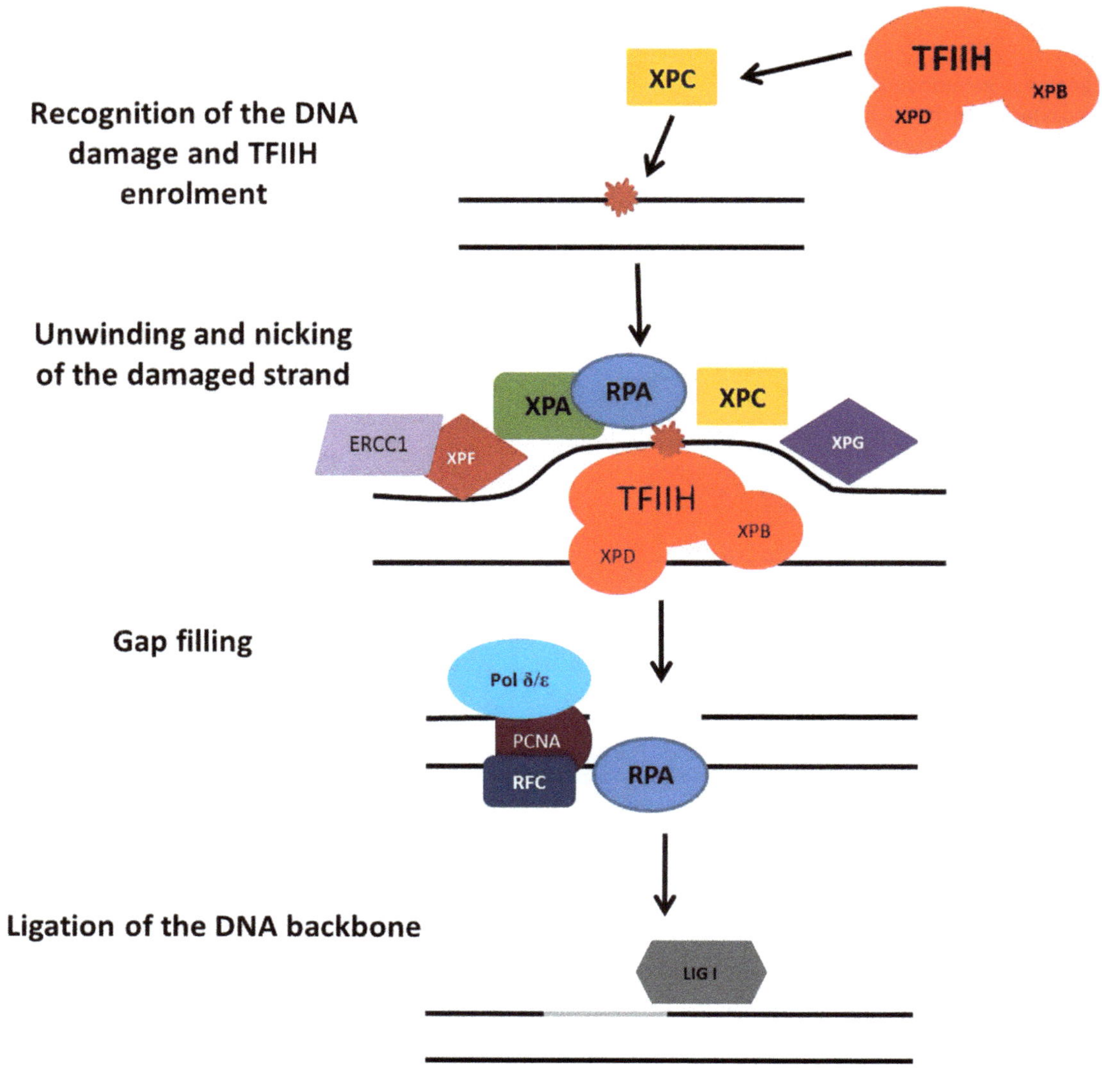

FIGURE 13.4 **Global genomic repair of the nucleotide excision-repair (NER) pathway.** The damaged DNA is recognized through binding by the XPC complex followed by recruiting the TFIIH complex. The complex contains nine subunits, including two helicases (XPB and XPD) which unwind the damaged region. The endonucleases XPG and XPF-ERCC1 allow a release of the single-stranded oligonucleotide. Following incision, a replication factor C (RFC) clamp loader adds PCNA at the 5′-site, and the replicative Pol δ/ε fill in the gap by rejoining the phosphodiester backbone by the DNA ligase I. *Adapted from Bray CM, West CE. DNA repair mechanisms in plants: crucial sensors and effectors for the maintenance of genome integrity. New Phytol 2005;168(3):511–28.*

histones that is performed by the components of an E3 ubiquitin ligase in response to UV exposure [74]. Mutation of the components involved in the DDB complex renders *Arabidopsis* plants hypersensitive to bulky DNA adducts caused by UV or cisplatin [21]. In line with these results, the overexpression of *DDB1A* and *DDB2* genes in *Arabidopsis* enhances UV-C tolerance in plants.

The recruitment of NER components following the stalling of the RNA polymerase II at the lesion site during transcription involves a Cockayne syndrome (CS) CSA and CSB proteins that may help remove the RNA polymerase complex and facilitate DNA repair. Although plants have two orthologues of CSA and one of CSB, their exact role remains uncharacterized [58].

Following the recognition of the DNA lesion, the XPC protein recruits a TFIIH complex that acts both as an RNA polymerase II transcription factor and a vital component involved in DNA repair. The complex contains nine subunits, including two helicases XPB and XPD which act in the 3′→5′- and 5′→3′-directions, respectively, and open the DNA helix around the lesions. Curiously, another transcription elongation factor II-S (TFIIS) stimulates RNA Pol II to bypass DNA regions containing specifically 8-oxoG in animal cells, thus preventing cell death due to the oxidative damage of DNA [75]. The TFIIS homologue was identified in plants, and the gene was shown to be responsive to stress in both the aerial parts and roots [76,77].

Additionally, the gene was up-regulated during seed imbibition that also requires the active DNA damage–repair process. Nevertheless, it still remains to be shown whether functions of the DNA lesion bypass pertain to the described homologue.

The *Arabidopsis* genome encodes two copies of XPB (*AtXPB1* and *AtXPB2*) which contain the conserved ATPase and helicase domains and also a homologue of XPD (*AtXPD*) [78]. While a complete knockout of the *AtXPB1*gene in *Arabidopsis* does not affect plant sensitivity to UV exposure, a mutation of the *AtXPD* gene is lethal. This apparently is due to a functional redundancy of *AtXPB1* and *AtXPB2* in the NER pathway. At the same time, both genes have been shown to be important during early stages of plant development because the *atxpb1* plants demonstrate a developmental delay, low seed viability, and a loss of germination synchrony.

The unwinding of the DNA helix around the lesion site exposes ssDNA and allows a replication protein A (RPA) to coat the strand [58]. In addition to the RPA protein, XPA is vital for the opening of the preincision complex followed by the recruitment of the XPF-ERCC1 complex by XPA. Although, plants appear to lack the orthologue of XPA, mutations of plant orthologues of XPF and ERCC1 render them hypersensitive to DNA-damaging agents. The cleavage at the 5′-end catalyzed by the XPF-ERCC1 nuclease complex releases the damaged DNA strand as a 24–32 oligonucleotide. The *Arabidopsis* genome encodes *AtRAD1*, a homologue of the XPF gene, the mutation of which causes a small reduction in the rate of CPD repair and hypersensitivity to DNA-damaging agents [79,80]. In addition, the RAD1/ERCC1 complex has been implemented in the removal of DNA flap structures generated during recombination. Following the incision, a replication factor C (RFC) clamp loader adds PCNA at the 5′-site, and replicative Pol δ/ε fill in the gap by the rejoining of the phosphodiester backbone by the DNA ligase (Table 13.3). The recognition of DNA adducts is the rate-limiting step in the NER pathway. This is due to the fact that the TCR pathway repairs different lesions at a constant rate, whereas the efficiency of GGR depends on the type of a lesion [81].

8. MISMATCH REPAIR

A mismatch is the mutagenic incorporation of an incorrect nucleotide that can occur during replication of both native and damaged DNA. An MMR pathway is involved in the efficient removal of erroneous nucleotides incorporated by the replicative DNA polymerase (Fig. 13.5). Although the proofreading function of the replicative polymerase limits the misincorporation of one nucleotide per 10^6–10^7 bp, this value is still high for the effective maintenance of genome integrity. Therefore, the error rate is further reduced by the MMR mechanism to one misincorporated base per 10^9–10^{10} nucleotides in the nascent DNA chain [61]. In addition, the MMR pathway may be also involved in eliminating mismatches at recombination sites. This allows to prevent recombination events which can cause inappropriate chromosome rearrangements [82]. Eukaryotic MMR involves the MutS HOMOLOGUE (MSH) proteins that are the evolutionarily conserved homologues of prokaryotic MutS. The *Arabidopsis* genome encodes seven MSH proteins, with MSH7 being unique to plants [83]. The heterodimeric protein complexes that include MSH2, 3, 6 and 7 are involved in the recognition of mismatches. Every dimer seems to be responsible for the recognition of specific lesions in the DNA sequence. For instance, the analysis of the in vitro produced MSH proteins revealed that an MSH2:MSH6 pair showed preference for a (T/G) base/base mispair and a one-nucleotide (+T) loop out [84]. At the same time, the heterodimer showed a minimal recognition of homoduplex (T/A) DNA, (C/C) heteroduplex, or to the three-nucleotide (+AAG) loop out. The recognition of the last DNA lesion, however, was best achieved by the MSH2:MSH3 heterodimer, although the binding to (C/C) and (T/G) mismatches was weak. Similarly to the MSH2:MSH6 pair, the MSH2:MSH7 heterodimer demonstrated an affinity for the (T/G) mispair and almost no binding to other lesions. While mutation of the *AtMSH2* gene in *Arabidopsis* is critical for the repair of mismatches in germline cells, it seems to be dispensable for MMR in somatic cells [83]. Similarly, both the *AtMSH4* and *AtMSH5* genes have been implemented in the DNA repair in gametes [85,86]. The expression of the genes is critical for floral organs, and their mutations cause a severe reduction in fertility due to meiotic defects. Moreover, the localization of AtMSH5 to the chromatin was compromised in the absence of AtMSH4.

The recognition of mismatches is linked to strand incision in eukaryotes by few orthologues of the prokaryotic MutL protein: MLH1, MLH2, MLH3, and PMS2. The *Arabidopsis* genome encodes orthologues of MLH1, PMS2, and a relatively distinct orthologue of MLH3 [83]. In animal cells, the MLH1:PMS2 pair plays a key role in the differentiation between the template and nascent DNA strands. Following the recognition step, the heterodimer catalyzes the excision of a stretch of the nascent strand DNA containing the erroneous nucleotide. The excision is terminated right after the mismatched base on the strand, and the DNA polymerase fills in the stretch of cleaved DNA followed by ligation of the strand by DNA ligase I [61] (Table 13.3).

A curious connection between MMR and an epigenetic pathway of the regulation of gene expression comes from 2015 studies on the *MSH1* gene [87,88]. MSH1 is a homologue of the yeast MSH1 protein which is absent in mammalian cells [89]. The protein is encoded in the nucleus but is localized to mitochondrial and chloroplast nucleoids and is involved in organelle genome stability [90]. Mutation of the *MSH1* gene increases the recombination rate of repeated sequences in the

TABLE 13.3 The Major Components of the Nucleotide Excision Repair and Mismatch Repair Pathways

Pathway	Protein Name	Function
Nucleotide excision repair (NER)	Multiprotein complex involving a Xeroderma pigmentosum group C (XPC), Rad23 and Centrin 2 (CEN2) proteins	Responsible for recognition of DNA lesions in plants during the global genome-repair pathway
	DDB complex consisting of DDB1, DDB2 (XPE), Cullin4, RBX1 proteins	Enhances binding of XPC protein to UV-induced cyclobutane pyrimidine dimers in the damaged DNA
	E3-ligase	Targets specific E2-ubiquitin conjugating enzymes to the XPC protein and others
	Cockayne syndrome (CS) CSA and CSB proteins	In plants, the exact role of orthologues remains uncharacterized. In animals, the proteins help to remove stalled RNA polymerase complex to facilitate the DNA repair
	TFIIH complex which contains nine subunits, including two helicases—XPB and XPD	Acts both as an RNA Pol II transcription factor and the vital component involved in the DNA repair
	Transcription elongation factor II-S (TFIIS)	Stimulates RNA Pol II to bypass DNA regions containing specifically 8-oxoG in animal cells; in plants, the homologue has been identified, but the function remains unknown
	Replication protein A (RPA)	Binds to ssDNA to stabilize it after unwinding of the DNA helix
	XPF-ERCC1 nuclease complex	Catalyzes the cleavage at the 5′-end that releases the damaged DNA strand as a 24–32 oligonucleotide; the *Arabidopsis* genome encodes a homologue of the XPF gene—*AtRAD1*
	Replicative Pol δ/ε	Fill in the gap
	LIGASE I	Possibly involved in the rejoining of DNA strands
Mismatch repair (MMR)	MutS homologue (MSH) proteins, such as heterodimeric protein complexes MSH2, 3, 6, and 7	Involved in the recognition of mismatches
	MSH1	The protein is encoded in nucleus but is localized to the mitochondrial and chloroplast nucleoids and is involved in organelle genome stability

mitochondrial genome, causes cytoplasmic male sterility, and a number of phenotypic abnormalities in plants [89]. This effect seems to be universal across plant kingdom and is referred to as developmental reprogramming [91]. Curiously, the *msh1*-associated phenotype can persist for multiple generations even when gene expression is restored to a wild-type level [92]. Furthermore, reciprocal crosses between the dwarf sorghum line with the *msh1*-associated phenotype (but with the WT genetic background) and the WT line resulted in the F1 progeny which demonstrated an enhanced vigor as compared to genuine WT plants. Similar results were obtained in other plant species including tomatoes, soybean, tobacco, and *Arabidopsis* [88,89]. A global analysis of a DNA methylation profile in *msh1* plants revealed alterations in the methylation level at CG and non-CG positions as compared to WT plants [87]. At the same time, no differences were detected in the DNA sequence itself that would argue against the developmental reprogramming phenotype caused by mutations in the genome. Since the gene is responsive to stress conditions in somatic tissues [91,93], it can be speculated that MSH1 is a novel component of the environmental sensing apparatus of plants which links the detection of alterations in ambient conditions through plastids to epigenetic responses of the whole plant [87].

9. DNA DOUBLE-STRAND BREAK REPAIR

An efficient repair of DSBs is of a particular importance for plant growth and development. The progression of DNA replication in the presence of DSBs can lead to the loss of chromosome fragments that can be detrimental in actively

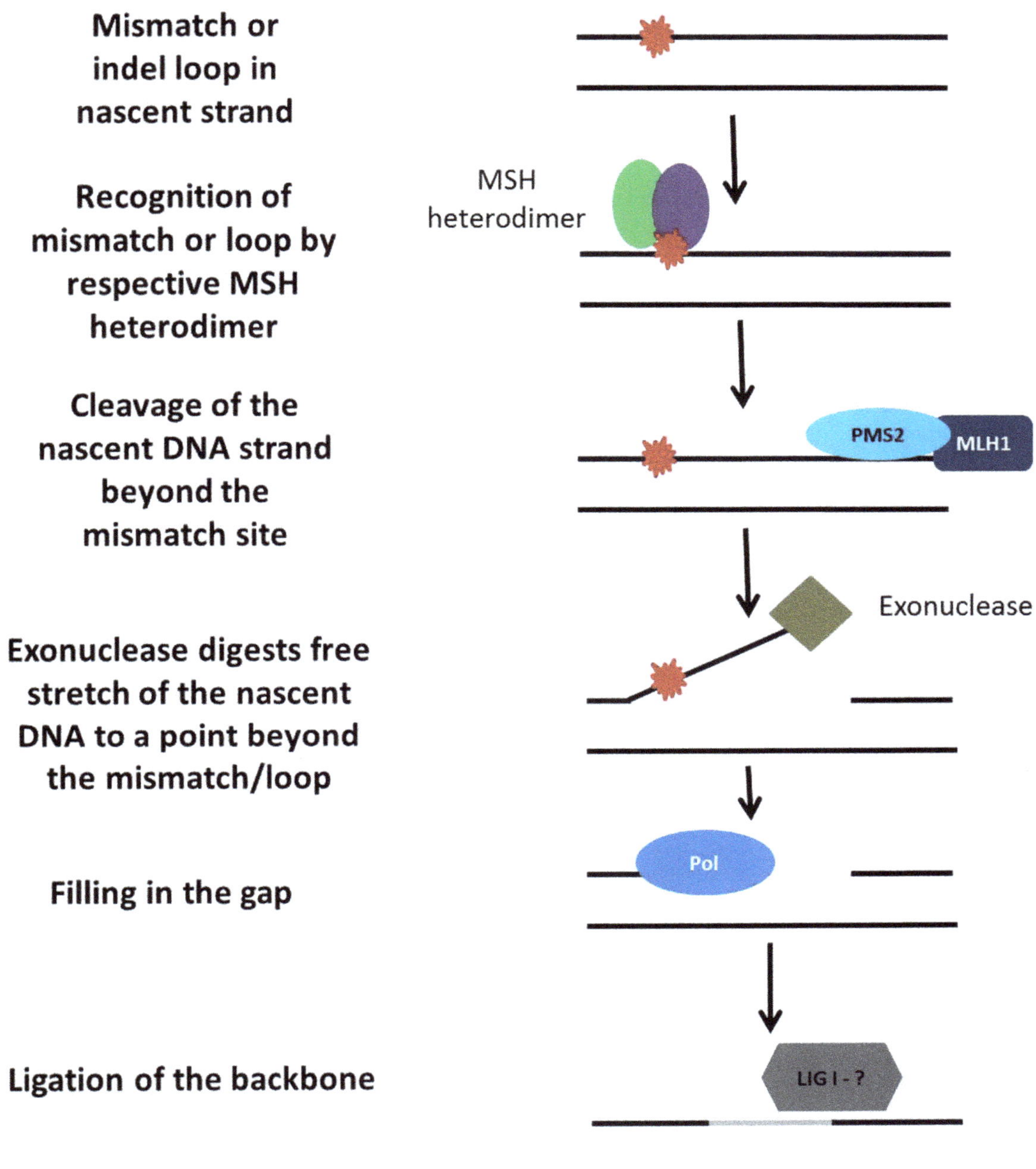

FIGURE 13.5 **A simplified scheme of the mismatch repair (MMR) pathway in plants.** A mismatch or loop is recognized by an appropriate MSH heterodimer (s) followed by cleavage of the nascent strand catalyzed by MLH1/PMS2. This promotes the unwinding of the DNA helix and digestion of the nascent strand to a point beyond the mismatch lesion/loop. The gap is filled in by DNA polymerase followed by backbone ligation. *Adapted from Bray CM, West CE. DNA repair mechanisms in plants: crucial sensors and effectors for the maintenance of genome integrity. New Phytol 2005;168(3):511–28.*

dividing cells [94]. In addition, an incorrect repair can cause chromosome fusions leading to dicentric chromosomes and anaphase bridges.

There are two major pathways involved in the repair of DNA DSBs (DSB) in living cells: homologous recombination (HR) and nonhomologous end-joining (NHEJ) or illegitimate recombination. The two pathways are considered to be responsible for maintaining a balance of genome stability versus genetic diversity, and the NHEJ pathway significantly prevails over HR in vascular plants [1]. The rate of the HR-mediated DSB repair in somatic cells is quite low and is about 1 in 10^3 repair events [95]. At the same time, when DSB occurs between tandem repeats, 30% of DSBs can be repaired by a single-strand annealing (SSA) and approximately 7%—by a synthesis-dependent strand annealing (SDSA) [96,97]. Similarly, the induced DSBs significantly increase the rate of HR that is probably concomitant with an increase in the frequency of repair through the NHEJ pathway. In addition, DNA-repair mutants, such as *uvr2–1* (CPD-photolyase), *atrad50*, and *atcen2* also demonstrate an increase in the rate of HR repair (Table 13.4).

9.1 Homologous Recombination

The HR pathway plays a vital role in both DSB repair in somatic cells and meiotic recombination during gametogenesis. The pathway includes RPA proteins and proteins in the RAD52 epistasis group, such as RAD51, RAD52, RAD54, and the MRN complex of RAD50, MRE11 and yeast XRS2/human NBS1. In addition, HR also requires the RAD51-like proteins. The *Arabidopsis* genome contains seven RAD51 homologues that are divided into two ancient groups: RADα and RADβ subfamilies [98]. While the former subfamily includes both RAD51 and DMC1 proteins, the latter one contains the RAD51B, RAD51C, RAD51D, XRCC2, and XRCC3 proteins. The RAD51, DMC1, RAD51C, and XRCC3 proteins have a unique role in meiotic HR and are necessary for a normal fertility [99]. In addition, the RAD51, RAD51C, and XRCC3 proteins are involved in DNA repair in somatic cells [1].

In yeast cells, Rad52 is a key protein involved in DBS repair and HR. It promotes DNA annealing and is involved in Rad51-mediated strand invasion. Eventually, Rad52 may participate in capturing the second DNA end followed by strand annealing to the D-loop and the formation of a Holliday junction [100]. The *Arabidopsis* genome also contains two homologues of the *RAD52* gene, the mutation of which causes a reduced fertility, a sensitivity to cross-linking drug mitomycin C, and the reduced level of intrachromosomal recombination compared to wild-type plants [101]. Similarly, null mutations of other components of the HR pathway, such as *AtRAD51*, *AtRAD50*, and *AtMRE11* result in a sterility due to severe meiotic defects [102–104].

Overall, three distinct models of HR are recognized: DSB repair (DSBR), SDSA, and SSA, with all three pathways being active in plants [105] (Fig. 13.6). The DSBR model is best described as a part of meiotic recombination [1]. The meiotic recombination occurs between homologous chromosomes rather than sister chromatids to stimulate the mixing of parental

TABLE 13.4 Proteins Involved in the DNA Double-Strand Break Repair

Pathway	Protein Name	Function
Homologous recombination	RAD51	Required for homology search
	The RAD51-like proteins: DMC1, RAD51B, RAD51C, RAD51D, XRCC2, and XRCC3	Involved in the steps of strand invasion and later on during the recombination
	RAD52	Promotes DNA annealing and is involved on RAD51-mediated strand invasion
	RAD54	Promotes chromatin remodeling and protein displacement from dsDNA and interacts with RAD51 protein that stimulates DNA strand exchange activity
	RECQ4A helicase	Plays a major role in replication fork regression
	FANCM helicase	Inhibits the formation of crossover recombinants during meiosis, thus favoring non-crossover resolutions
Nonhomologous end-joining	KU70 and KU80 heterodimer	Protects the DNA ends from exonuclease activity
	MRE11–RAD50–NBS1 (MRN)	Involved in the DNA-damage repair, DNA replication, meiosis, and telomere maintenance
	Poly(ADP-ribose) polymerase(PARP)	Involved in DNA-damage response in *Arabidopsis*
	XRCC1	A key protein in BER and single-strand break repair; acts as a scaffold for other DNA-repair proteins; involved in KU-independent alternative end-joining pathway that results in the large deletions at the joints
	DNA ligase IV	Is a specialized ligase that catalyzes a final step in the NHEJ pathway and together with its cofactor XRCC4 interacts with KU to seal a joint

genomes. The process starts by the creation of DSB and the formation of a Holliday junction followed by processing of ends to produce long 3′-tails. The single-stranded nucleoprotein filament mediates homology search and the invasion of the homologous chromosome. An in vitro study has shown that homology search requires a single RAD51 protein. The protein is also vital for plant reproduction because the *Arabidopsis atrad51* mutants are sterile [102]. At the same time, the mutant does not display any phenotypic abnormalities and is not hypersensitive to DNA-damaging agents. Additional RAD51-like proteins are

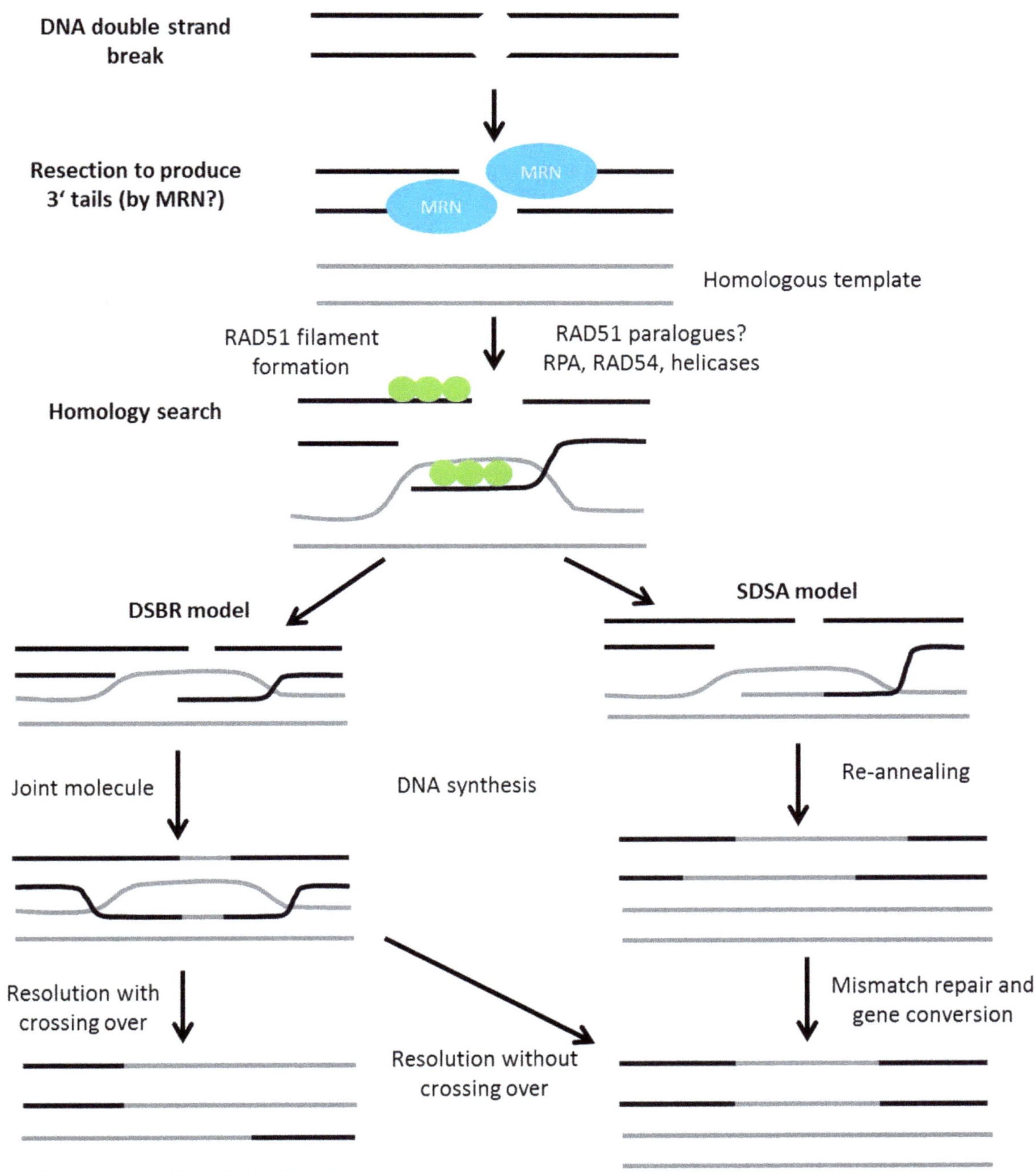

FIGURE 13.6 **The homologous recombination pathways of DNA double-strand break repair.** Three different models of homologous recombination (HR) pathways are recognized in plants, including double-strand break repair (DSBR), synthesis-dependent strand annealing (SDSA), and single-strand annealing (SSA). The former one is considered to be the most active pathway (not shown); it functions when a break occurs between the repeated sequences and results in the deletion of the intervening sequence. The SDSA model explains most of the recombination products observed in plants in the nonrepeat regions. The region copied during the recombination in this pathway remains unaltered. The DSBR model describes crossing over of chromosomes during meiosis. MRN is a complex of MRE11, RAD50, and NBS1. *Adapted from Waterworth WM, Drury GE, Bray CM, West CE. Repairing breaks in the plant genome: the importance of keeping it together. New Phytol 2011;192(4):805–22.*

also required in strand invasion steps and later on during the recombination [104]. In *Arabidopsis*, it has been speculated that the RAD51B, RAD51D, and XRCC2 proteins might interact with each other to form a protein complex that also cooperates with RAD51C in the HR pathway. Moreover, the RAD51B, RAD51D, and XRCC2 proteins are partially redundant because a triple mutant of the three genes demonstrates a higher sensitivity to bleomycin than single and double mutants [106].

The search for the homologous sequence will not be possible without alterations in the chromatin structure; therefore, the chromatin-remodeling enzymes and helicases assist in homology search. The invading strand may have differences in the sequence resulting in the occurrence of the heteroduplex DNA leading to the repair of mismatches by MMR [1]. The DNA synthesis is then initiated on the invaded strain by using a homologous DNA as a template. Eventually, the invading strand is ligated to the other side of DSB leading to the formation of two Holliday junctions that may undergo either a crossover or gene conversion to be resolved.

In the process of SDSA, a sister chromatid, the homologous chromosome, or an ectopic region of homology in the genome can be used as a recombination substrate. The repair mechanism is used both in meiotic and somatic cells, and similarly to DSBR, it is initiated by the generation of long single-stranded 3′-ends that invade a homologous stretch followed by the DNA synthesis. Unlike the DSBR pathway, SDSA rarely involves the formation of Holliday junctions and crossovers. This is achieved by annealing of the newly synthesized DNA to the other side of DSB to promote break repair, thus avoiding the formation of joint molecules [107]. The avoidance of crossovers during SDSA eliminates the possible mutagenic effect of recombination that can occur between ectopic regions of homology. At the same time, the Holliday junction can take place in the process when the recombination occurs between sister chromatids. SDSA is the main pathway of conservative HR repair in somatic cells in plants [108]. The involvement of SDSA over DSBR also plays a key role in the maintenance of genome stability. This is due to the contribution of crossovers to the DSBR process that if occurred in somatic cells at ectopic positions in chromosomes could potentially result in dicentric and acentric chromosomes, thus causing genome instability.

The SSA mechanism, in turn, can utilize tandem repeats that are arranged in a close proximity. In the genome, local duplication events as well as clustered ribosomal genes can become the preferred substrates for SSA. The DNA sequence between tandem repeats is removed during the recombination process, thus suggesting for an additional mechanism (NHEJ is the other one) that can cause DNA loss during the evolution and is mutagenic [1]. In the process of SSA following the DSB induction, the homologous regions are getting exposed during the resection. The partially complementary strands can directly anneal to each other, and the chimeric DNA-double helix can be formed. The extra 3′-overhangs can be trimmed or the single-stranded gaps can be filled in through the DNA synthesis. Eventually, the DNA backbone is joined together by the DNA ligase. The main difference between SSA and SDSA is that the latter pathway requires the strand exchange, whereas the former one does not. Unfortunately, no proteins have been characterized yet that are exclusively necessary for the SSA pathway. At the same time, the recombination process in the SDSA pathway requires homologues of the recombinase RecA, AtRAD51, AtXRCC3, AtRAD54, DNA helicases AtRECQ4A, and AtFANCM, as well as nucleases like AtMUS81 [108].

9.1.1 Replication-Associated HR

The HR pathway may be also involved in processing the stalled replication forks. A number of factors can cause stalling of the replication forks, including cross-linked DNA strands, the presence of the inhibitory modified base in the strand, and an increased tension of dsDNA due to supercoiling caused by an improper function or inactivation of topoisomerases [108]. Because the free ends of the double-stranded DNA are not available in the process, the repair cannot be done by the NHEJ and SSA pathways. Therefore, it is assumed that in animal cells, the pathway is more important during the S phase as compared to gap phases where the NHEJ and SSA pathways prevail. The repair can progress according to the two main scenarios: the replication bypass and the formation of one-sided DSB. In the case of the replication bypass, translational polymerases can synthesize the complementary strand besides the altered DNA bases. At the same time, if the DNA strand cannot be elongated due to a severe damage of the template strand, the second daughter strand can still be synthesized in the process known as "overshoot synthesis." The process resembles the steps involved in the SDSA mechanism and relies on the sister duplex strand for copying the information onto the shorter daughter strand, thus avoiding the damaged part at the template strand [108]. In the second scenario, DSB can form when the polymerase encounters a nick on one of the template strands or when the replication fork stalls due to the endonuclease activity. Since there is no second end present on another side of the break, the NHEJ pathway cannot be engaged in the repair. Thus, the one-sided DSB is repaired using the homologous sequence present on the sister chromatid.

Curiously, in the studies on *Arabidopsis* mutants, two inhibitors of the HR pathway were revealed, RECQA and FANCM. The genes seem to be involved in different mechanisms of HR suppression because the double mutant demonstrated higher spontaneous HR frequencies compared to the single mutants [108].

9.2 Nonhomologous End-Joining

Experiments involving a stable transformation of either somatic cells or gametes revealed that in most of the cases, the foreign DNA integrates randomly into the genome of higher plants without utilizing the regions of homology to the endogenous DNA. This and other observations have led to the conclusion that the majority of DSBs in plant cells are repaired through NHEJ in higher eukaryotes [1]. It is believed that there are at least two different mechanisms of NHEJ repair, such as classical and alternative NHEJ pathways (cNHEJ and aNHEJ, respectively) that differ in both the key players involved in the pathways and the final outcome of the repair process [109]. While in the cNHEJ pathway, DSB repair does not require the microhomology at the joints, the aNHEJ mechanism requires small homology regions. The ends in the first pathway are protected against the exonuclease activity by binding the KU70/80 heterodimer followed by ligation mediated by ligase IV. In addition, in yeast, if the ends of the damaged DNA do not carry the 5′-phosphate and 3′-hydroxyl group required for ligation, they can be processed by the MRE11–RAD50–XRS2 (MRX) complex [110] (Fig. 13.7). In plants, a similar complex—MRE11–RAD50–NBS1 (MRN)—is involved in DNA-damage repair, DNA replication, meiosis, and telomere maintenance. The complex recognizes DSBs by its ability to bind to the ends of DNA. Following binding, the complex unwinds and initiates the processing of the ends. The *mrnatr* double mutants demonstrate the growth retardation effect caused by the accumulation of DNA lesions and cell death [17]. Similarly, KU-deficient mutants are hypersensitive to DNA-damaging and alkylating agents, further suggesting the involvement of these genes in the DNA-repair process [111,112]. Additionally, mutants compromised in the KU70 protein demonstrate drastic telomere deregulation leading to an increase in the telomere length as compared to WT plants.

The aNHEJ mechanism resembles the SSA-repair pathway because similarly to the last mechanism, the 3′-resection of the broken ends occurs by a specific exonuclease enzyme complex followed by microhomology search, trimming and ligation of the broken ends by DNA ligase IV. In *Arabidopsis*, both PARP1 and XRCC1 are involved in the aNHEJ pathway, and it seems that the PARP1 protein competes with KU80 for DSBs [113]. The result of NHEJ repair is almost always a genomic change caused by either deletions of different sizes or insertions. In addition, if more than one break is induced simultaneously, the potential outcome can be a rearrangement, which leads to the generation of a new sequence combination. In most cases, the reshuffling of the genome would be detrimental to the viability of the progeny; therefore, it will not be propagated. In rare occasions, however, small rearrangements of chromosomes can be inherited, thus possibly affecting the speciation [108].

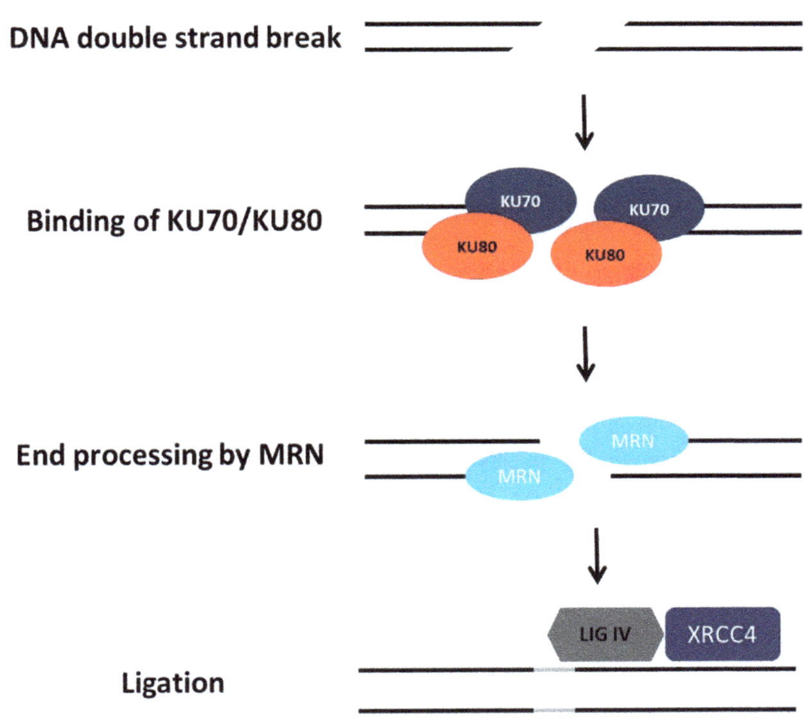

FIGURE 13.7 **The nonhomologous end-joining pathway of DNA double-strand break repair.** In the classical nonhomologous end-joining (NHEJ) pathway, the recognition of free DNA ends is performed by the KU70–KU80 complex. Later, the MRN complex (contains MRE11, RAD50, and NBS1 proteins) binds to DSB, and if the damaged ends do not carry the 5′-phosphate and 3′-hydroxyl group required for ligation, it unwinds the strands and initiates processing the ends. The DNA LIGASE 4–XRCC4 complex seals the phosphodiester backbone. In the absence of KU proteins, the repair can occur followed by microhomology search and requires the presence of either MRE11 or LIG4. *Adapted from Waterworth WM, Drury GE, Bray CM, West CE. Repairing breaks in the plant genome: the importance of keeping it together. New Phytol 2011;192(4):805–22.*

10. DNA REPAIR IN ORGANELLES

The maintenance of genome stability in plants also includes an efficient DNA-damage repair in chloroplasts and mitochondria which are the main factories of ROS in the cell. Unfortunately, the DNA-repair mechanisms in these organelles remain largely unexplored, although the 2014 studies provided evidence of resemblance of the HR pathway in chloroplasts to that observed in bacteria [114,115]. The HR repair in bacterial cells is performed by the RecA/RecBCD pathway [116]. RecA proteins that are targeted to mitochondria and chloroplasts were described in *Arabidopsis*, suggesting that the HR pathway is also active in organelles [115,117]. In addition, such components of the BER pathway as the endonuclease III homologues and the AP endonuclease were also identified in *Arabidopsis* chloroplasts [118]. Similarly, the BER pathway is also active in mitochondria because the uracil DNA glycosylase associated with mitochondrial membranes has been characterized in the previous study [119]. At the same time, substrates for the NER pathway are likely to be repaired by alternative pathways since there is little evidence for the presence of this pathway's components in organelles [61,119].

11. FUTURE PERSPECTIVE

The ability of plants to maintain the genome integrity in response to external and internal cues is detrimental for both the survival of an individual plant and the transmission of intact genetic information to future generations. At the same time, due to errors in DNA-repair pathways, genetic variation can lead to the genetic diversity, the appearance of altered traits, which occasionally can be beneficial in the new environment. Although our understanding of DNA-repair pathways in plants is far from complete, it is now clear that the utilization of DNA-repair components can benefit applied studies including those in genome editing and breeding for crop varieties with improved DNA-repair functions. In the case of genome editing, benefits of either the NHEJ or HR pathway are harnessed for the targeted gene disruption or insertion, respectively, by using designed endonucleases. Similarly, crop varieties with improved DNA-repair pathways may be more stress tolerant, whereas unrepaired DNA damage directly correlates with yield loss [61]. A deeper understanding of DNA-damage repair in organelles will also be essential for biotechnology applications since stable modifications of the organelle genome can be beneficial for both the improvement of photosynthesis and the avoidance of the spread of the modified genome through cross-pollination. Therefore, further elucidation of DNA-repair pathways in plants will be valuable for the generation of crops with improved traits.

GLOSSARY

Base excision repair DNA damage–repair process which is primarily responsible for the removal of small base lesions from the genome.

Homologous recombination repair Type of double-strand break repair in which nucleotide sequences are exchanged between two identical or very similar molecules of DNA during genetic recombination.

Mismatch The mutagenic incorporation of an incorrect nucleotide that can occur during replication and recombination of both native and damaged DNA.

Mismatch repair DNA-repair process involved in removing mismatches from the DNA structure.

Nonhomologous end-joining repair Double-strand break-repair pathway that involves simple rejoining of the broken ends of the DNA molecule either in the presence or absence of microhomology regions between the broken ends.

Nucleotide excision repair DNA damage–repair process which is responsible for removing bulky lesions in DNA that cause the disruption of the DNA helix.

Photoreactivation Repair process of UV-damaged DNA by photolyase enzymes with the utilization of the light source with the longer wavelengths.

LIST OF ACRONYMS AND ABBREVIATIONS

5′-dRP 5′-deoxyRibose-5-phosphate termini
6-4PP Pyrimidine-pyrimidone (6-4) photoproduct
8-HDF 8-hydroxy-7,8-didemethyl-5-deazariboflavin
8-oxoAde 8-oxo-7,8-dihydroadenine
8-oxoGua 8-oxo-7,8-dihydroguanine
8-oxodGuo 8-oxo-7,8-dihydro-20-deoxyadenosine
AP Apurinic/apyrimidinic
ASF1 *Anti-silencing function1* gene
ATM Ataxia telangiectasia-mutated protein
ATR Ataxia telangiectasia and Rad3-related protein
BER Base excision repair

CAF1 Chromatin assembly factor 1
CEN2 Centrin 2
Chk1 and Chk2 Checkpoint-related protein kinases
cNHEJ and aNHEJ Classical and alternative NHEJ, respectively
CPD Cyclobutane pyrimidine dimer
CSA Cockayne syndrome protein A
DDB DNA damage–binding complex
DSBR DSB repair by the HR pathway
EMS Ethyl methane sulfonate
FADH Flavin adenine dinucleotide
Fapy-Gua 2,6-diamino-4-hydroxy-5-formamidopyrimidine
FAS1 FASCIATA 1 protein
FEN1 5′-flap endonuclease
FPG Formamidopyrimidine-DNA glycosylase
GGR Global genome repair
H2AX Histone 2A isoform
HR Homologous recombination
LP Long-patch repair
MIM Hypersensitive to MMS, irradiation and MMC
MMR Mismatch repair
The MRX and MRN complexes MRE11–RAD50–XRS2 and MRE11–RAD50–NBS1 complexes, respectively.
MSH MutS Homologue
MSI1 Multicopy suppressor of IRA1
MTHF Pterin methenyltetrahydrofolate
Nbs1 Nijmegen breakage syndrome gene
NER Nucleotide excision repair
$^{1}O_2$ Singlet oxygen
OGG1 8-oxoG DNA glycosylase/AP lyase
·OH Hydroxyl radical
PARP Poly(ADP-ribose)polymerase
PCNA Proliferating cell nuclear antigen
Pol β DNA polymerase β
RFC Replication factor C clamp loader
ROS Reactive oxygen species
RPA Replication protein A
SDSA Synthesis-dependent strand annealing
SMC Structural maintenance of chromosome proteins
SOG1 Suppressor of gamma response 1
SP Short-patch repair
SSA Single-strand annealing
SSB Single-strand break
SWI2/SNF2 SWItch/Sucrose nonfermentable
TCR Transcription-coupled repair
TFIIH Transcription factor II H
TFIIS Transcription elongation factor II-S
UDG Uracil DNA glycosylase
UV Ultraviolet radiation
XPC Xeroderma pigmentosum group C
XRCC1 X-ray repair cross complementing protein 1

REFERENCES

[1] Bray CM, West CE. DNA repair mechanisms in plants: crucial sensors and effectors for the maintenance of genome integrity. New Phytol 2005;168(3):511–28.
[2] Roy S. Maintenance of genome stability in plants: repairing DNA double strand breaks and chromatin structure stability. Front Plant Sci 2014;5:487.
[3] Kuzminov A. Single-strand interruptions in replicating chromosomes cause double-strand breaks. Proc Natl Acad Sci USA 2001;98(15):8241–6.
[4] Boyko A, Kovalchuk I. Genome instability and epigenetic modification – heritable responses to environmental stress? Curr Opin Plant Biol 2011;14(3):260–6.

[5] Cadet J, Douki T, Gasparutto D, Ravanat J-L. Oxidative damage to DNA: formation, measurement and biochemical features. Mutat Res 2003;531(1–2):5–23.

[6] Cadet J, Wagner JR. DNA base damage by reactive oxygen species, oxidizing agents, and UV radiation. Cold Spring Harb Perspect Biol 2013;5(2).

[7] Pouget JP, Frelon S, Ravanat JL, Testard I, Odin F, Cadet J. Formation of modified DNA bases in cells exposed either to gamma radiation or to high-LET particles. Radiat Res 2002;157(5):589–95.

[8] Ravanat J-L, Saint-Pierre C, Di Mascio P, Martinez GR, Medeiros MHG, Cadet J. Damage to isolated DNA mediated by singlet oxygen. Helv Chim Acta 2001;84(12):3702–9.

[9] Jena NR. DNA damage by reactive species: mechanisms, mutation and repair. J Biosci 2012;37(3):503–17.

[10] Vidal A, Abril N, Pueyo C. DNA repair by Ogt alkyltransferase influences EMS mutational specificity. Carcinogenesis 1995;16(4):817–21.

[11] Greene EA, Codomo CA, Taylor NE, Henikoff JG, Till BJ, Reynolds SH, et al. Spectrum of chemically induced mutations from a large-scale reverse-genetic screen in *Arabidopsis*. Genetics 2003;164(2):731–40.

[12] Lee JH, Paull TT. Direct activation of the ATM protein kinase by the Mre11/Rad50/Nbs1 complex. Science 2004;304(5667):93–6.

[13] Sweeney PR, Britt AB, Culligan KM. The *Arabidopsis* ATRIP ortholog is required for a programmed response to replication inhibitors. Plant J 2009;60(3):518–26.

[14] Culligan K, Tissier A, Britt A. ATR regulates a G2-phase cell-cycle checkpoint in *Arabidopsis thaliana*. Plant Cell 2004;16(5):1091–104.

[15] Garcia V, Bruchet H, Camescasse D, Granier F, Bouchez D, Tissier A. AtATM is essential for meiosis and the somatic response to DNA damage in plants. Plant Cell 2003;15(1):119–32.

[16] Matsuoka S, Ballif BA, Smogorzewska A, McDonald 3rd ER, Hurov KE, Luo J, et al. ATM and ATR substrate analysis reveals extensive protein networks responsive to DNA damage. Science 2007;316(5828):1160–6.

[17] Amiard S, Charbonnel C, Allain E, Depeiges A, White CI, Gallego ME. Distinct roles of the ATR kinase and the Mre11-Rad50-Nbs1 complex in the maintenance of chromosomal stability in *Arabidopsis*. Plant Cell 2010;22(9):3020–33.

[18] Shiloh Y. The ATM-mediated DNA-damage response: taking shape. Trends Biochem Sci 2006;31(7):402–10.

[19] Culligan KM, Robertson CE, Foreman J, Doerner P, Britt AB. ATR and ATM play both distinct and additive roles in response to ionizing radiation. Plant J 2006;48(6):947–61.

[20] Yoshiyama K, Conklin PA, Huefner ND, Britt AB. Suppressor of gamma response 1 (SOG1) encodes a putative transcription factor governing multiple responses to DNA damage. Proc Natl Acad Sci USA 2009;106(31):12843–8.

[21] Molinier J, Lechner E, Dumbliauskas E, Genschik P. Regulation and role of *Arabidopsis* CUL4-DDB1A-DDB2 in maintaining genome integrity upon UV stress. PLoS Genet 2008;4(6):e1000093.

[22] Lieber MR. The mechanism of double-strand DNA break repair by the nonhomologous DNA end-joining pathway. Annu Rev Biochem 2010;79:181–211.

[23] Kirik A, Pecinka A, Wendeler E, Reiss B. The chromatin assembly factor subunit FASCIATA1 is involved in homologous recombination in plants. Plant Cell 2006;18(10):2431–42.

[24] Hennig L, Taranto P, Walser M, Schönrock N, Gruissem W. *Arabidopsis* MSI1 is required for epigenetic maintenance of reproductive development. Development 2003;130(12):2555–65.

[25] Endo M, Ishikawa Y, Osakabe K, Nakayama S, Kaya H, Araki T, et al. Increased frequency of homologous recombination and T-DNA integration in *Arabidopsis* CAF-1 mutants. EMBO J 2006;25(23):5579–90.

[26] Ramirez-Parra E, Gutierrez C. E2F regulates FASCIATA1, a chromatin assembly gene whose loss switches on the endocycle and activates gene expression by changing the epigenetic status. Plant Physiol 2007;144(1):105–20.

[27] Shaked H, Melamed-Bessudo C, Levy AA. High-frequency gene targeting in *Arabidopsis* plants expressing the yeast RAD54 gene. Proc Natl Acad Sci USA 2005;102(34):12265–9.

[28] Shaked H, Avivi-Ragolsky N, Levy AA. Involvement of the *Arabidopsis* SWI2/SNF2 chromatin remodeling gene family in DNA damage response and recombination. Genetics 2006;173(2):985–94.

[29] De Piccoli G, Cortes-Ledesma F, Ira G, Torres-Rosell J, Uhle S, Farmer S, et al. Smc5-Smc6 mediate DNA double-strand-break repair by promoting sister-chromatid recombination. Nat Cell Biol 2006;8(9):1032–4.

[30] Watanabe K, Pacher M, Dukowic S, Schubert V, Puchta H, Schubert I. The structural maintenance of chromosomes 5/6 complex promotes sister chromatid alignment and homologous recombination after DNA damage in *Arabidopsis thaliana*. Plant Cell 2009;21(9):2688–99.

[31] Mengiste T, Revenkova E, Bechtold N, Paszkowski J. An SMC-like protein is required for efficient homologous recombination in *Arabidopsis*. EMBO J 1999;18(16):4505–12.

[32] Kozak J, West CE, White C, da Costa-Nunes JA, Angelis KJ. Rapid repair of DNA double strand breaks in *Arabidopsis thaliana* is dependent on proteins involved in chromosome structure maintenance. DNA Repair (Amst) 2009;8(3):413–9.

[33] Campi M, D'Andrea L, Emiliani J, Casati P. Participation of chromatin-remodeling proteins in the repair of ultraviolet-B-damaged DNA. Plant Physiol 2012;158(2):981–95.

[34] Lario LD, Ramirez-Parra E, Gutierrez C, Spampinato CP, Casati P. ANTI-SILENCING FUNCTION1 proteins are involved in ultraviolet-induced DNA damage repair and are cell cycle regulated by E2F transcription factors in *Arabidopsis*. Plant Physiol 2013;162(2):1164–77.

[35] Nawkar GM, Maibam P, Park JH, Sahi VP, Lee SY, Kang CH. UV-Induced cell death in plants. Int J Mol Sci 2013;14(1):1608–28.

[36] Madronich S, McKenzie RL, Bjorn LO, Caldwell MM. Changes in biologically active ultraviolet radiation reaching the Earth's surface. J Photochem Photobiol B 1998;46(1–3):5–19.

[37] Britt AB. DNA damage and repair in plants. Annu Rev Plant Physiol Plant Mol Biol 1996;47:75–100.

[38] Alscher RG, Donahue JL, Cramer CL. Reactive oxygen species and antioxidants: relationships in green cells. Physiol Plant 1997;100(2):224–33.

[39] Britt AB. Molecular genetics of DNA repair in higher plants. Trends Plant Sci 1999;4(1):20–5.
[40] Tuteja N, Ahmad P, Panda BB, Tuteja R. Genotoxic stress in plants: shedding light on DNA damage, repair and DNA repair helicases. Mutat Res 2009;681(2–3):134–49.
[41] Hoeijmakers JH. Genome maintenance mechanisms for preventing cancer. Nature 2001;411(6835):366–74.
[42] Sancar A, Lindsey-Boltz LA, Unsal-Kacmaz K, Linn S. Molecular mechanisms of mammalian DNA repair and the DNA damage checkpoints. Annu Rev Biochem 2004;73:39–85.
[43] Britt AB, Chen JJ, Wykoff D, Mitchell D. A UV-sensitive mutant of *Arabidopsis* defective in the repair of pyrimidine-pyrimidinone(6-4) dimers. Science 1993;261(5128):1571–4.
[44] Britt A, Fiscus EL. Growth responses of *Arabidopsis* DNA repair mutants to solar irradiation. Physiol Plant 2003;118(2):183–92.
[45] Gill SS, Anjum NA, Gill R, Jha M, Tuteja N. DNA damage and repair in plants under ultraviolet and ionizing radiations. ScientificWorldJournal 2015;2015:250158.
[46] Taylor RM, Nikaido O, Jordan BR, Rosamond J, Bray CM, Tobin AK. Ultraviolet-B-induced DNA lesions and their removal in wheat (*Triticum aestivum* L.) leaves. Plant Cell Environ 1996;19(2):171–81.
[47] Sancar A. Structure and function of DNA photolyase and cryptochrome blue-light photoreceptors. Chem Rev 2003;103(6):2203–37.
[48] Jiang CZ, Yee J, Mitchell DL, Britt AB. Photorepair mutants of *Arabidopsis*. Proc Natl Acad Sci USA 1997;94(14):7441–5.
[49] Lo HL, Nakajima S, Ma L, Walter B, Yasui A, Ethell DW, et al. Differential biologic effects of CPD and 6-4PP UV-induced DNA damage on the induction of apoptosis and cell-cycle arrest. BMC Cancer 2005;5:135.
[50] Takeuchi Y, Murakami M, Nakajima N, Kondo N, Nikaido O. Induction and repair of damage to DNA in cucumber cotyledons irradiated with UV-B. Plant Cell Physiol 1996;37(2):181–7.
[51] Waterworth WM, Jiang Q, West CE, Nikaido M, Bray CM. Characterization of *Arabidopsis* photolyase enzymes and analysis of their role in protection from ultraviolet-B radiation. J Exp Bot 2002;53(371):1005–15.
[52] Quaite FE, Takayanagi S, Ruffini J, Sutherland JC, Sutherland BM. DNA damage levels determine cyclobutyl pyrimidine dimer repair mechanisms in alfalfa seedlings. Plant Cell 1994;6(11):1635–41.
[53] Kimura S, Tahira Y, Ishibashi T, Mori Y, Mori T, Hashimoto J, et al. DNA repair in higher plants; photoreactivation is the major DNA repair pathway in non-proliferating cells while excision repair (nucleotide excision repair and base excision repair) is active in proliferating cells. Nucleic Acids Res 2004;32(9):2760–7.
[54] Cordoba-Canero D, Morales-Ruiz T, Roldan-Arjona T, Ariza RR. Single-nucleotide and long-patch base excision repair of DNA damage in plants. Plant J 2009;60(4):716–28.
[55] Joldybayeva B, Prorok P, Grin IR, Zharkov DO, Ishenko AA, Tudek B, et al. Cloning and characterization of a wheat homologue of apurinic/apyrimidinic endonuclease Ape1L. PLoS One 2014;9(3):e92963.
[56] Santerre A, Britt AB. Cloning of a 3-methyladenine-DNA glycosylase from *Arabidopsis thaliana*. Proc Natl Acad Sci USA 1994;91(6):2240–4.
[57] Dany AL, Tissier A. A functional OGG1 homologue from *Arabidopsis thaliana*. Mol Genet Genomics 2001;265(2):293–301.
[58] Shu QY, Forster BP, Nakagawa H. Plant mutation breeding and biotechnology. Cambridge (MA): CABI International; 2012. p. 608.
[59] Lindahl T. Instability and decay of the primary structure of DNA. Nature 1993;362(6422):709–15.
[60] Babiychuk E, Kushnir S, Van Montagu M, Inze D. The *Arabidopsis thaliana* apurinic endonuclease Arp reduces human transcription factors Fos and Jun. Proc Natl Acad Sci USA 1994;91(8):3299–303.
[61] Balestrazzi A, Confalonieri M, Macovei A, Dona M, Carbonera D. Genotoxic stress and DNA repair in plants: emerging functions and tools for improving crop productivity. Plant Cell Rep 2011;30(3):287–95.
[62] Murphy TM, Belmonte M, Shu S, Britt AB, Hatteroth J. Requirement for abasic endonuclease gene homologues in *Arabidopsis* seed development. PLoS One 2009;4(1):e4297.
[63] Duclos S, Aller P, Jaruga P, Dizdaroglu M, Wallace SS, Doublie S. Structural and biochemical studies of a plant formamidopyrimidine-DNA glycosylase reveal why eukaryotic Fpg glycosylases do not excise 8-oxoguanine. DNA Repair (Amst) 2012;11(9):714–25.
[64] Zhu JK. Active DNA demethylation mediated by DNA glycosylases. Annu Rev Genet 2009;43:143–66.
[65] Fortini P, Dogliotti E. Base damage and single-strand break repair: mechanisms and functional significance of short- and long-patch repair subpathways. DNA Repair (Amst) 2007;6(4):398–409.
[66] Waterworth WM, Kozak J, Provost CM, Bray CM, Angelis KJ, West CE. DNA ligase 1 deficient plants display severe growth defects and delayed repair of both DNA single and double strand breaks. BMC Plant Biol 2009;9:79.
[67] Uchiyama Y, Kimura S, Yamamoto T, Ishibashi T, Sakaguchi K. Plant DNA polymerase λ, a DNA repair enzyme that functions in plant meristematic and meiotic tissues. Eur J Biochem 2004;271(13):2799–807.
[68] Uchiyama Y, Suzuki Y, Sakaguchi K. Characterization of plant XRCC1 and its interaction with proliferating cell nuclear antigen. Planta 2008;227(6):1233–41.
[69] Doucet-Chabeaud G, Godon C, Brutesco C, de Murcia G, Kazmaier M. Ionising radiation induces the expression of PARP-1 and PARP-2 genes in *Arabidopsis*. Mol Genet Genomics 2001;265(6):954–63.
[70] Ström CE, Johansson F, Uhlén M, Al-Khalili Szigyarto C, Erixon K, Helleday T. Poly (ADP-ribose) polymerase (PARP) is not involved in base excision repair but PARP inhibition traps a single-strand intermediate. Nucleic Acids Res 2010;39(8):3166–75.
[71] Jiang CZ, Yen CN, Cronin K, Mitchell D, Britt AB. UV- and gamma-radiation sensitive mutants of *Arabidopsis thaliana*. Genetics 1997;147(3):1401–9.
[72] Jenkins ME, Harlow GR, Liu Z, Shotwell MA, Ma J, Mount DW. Radiation-sensitive mutants of *Arabidopsis thaliana*. Genetics 1995;140(2):725–32.
[73] Liang L, Flury S, Kalck V, Hohn B, Molinier J. CENTRIN2 interacts with the *Arabidopsis* homolog of the human XPC protein (AtRAD4) and contributes to efficient synthesis-dependent repair of bulky DNA lesions. Plant Mol Biol 2006;61(1–2):345–56.

[74] Wang H, Zhai L, Xu J, Joo H-Y, Jackson S, Erdjument-Bromage H, et al. Histone H3 and H4 ubiquitylation by the CUL4-DDB-ROC1 ubiquitin ligase facilitates cellular response to DNA damage. Mol Cell 2006;22(3):383–94.

[75] Kuraoka I, Suzuki K, Ito S, Hayashida M, Kwei JS, Ikegami T, et al. RNA polymerase II bypasses 8-oxoguanine in the presence of transcription elongation factor TFIIS. DNA Repair (Amst) 2007;6(6):841–51.

[76] Grasser M, Kane CM, Merkle T, Melzer M, Emmersen J, Grasser KD. Transcript elongation factor TFIIS is involved in *Arabidopsis* seed dormancy. J Mol Biol 2009;386(3):598–611.

[77] Macovei A, Balestrazzi A, Confalonieri M, Buttafava A, Carbonera D. The TFIIS and TFIIS-like genes from *Medicago truncatula* are involved in oxidative stress response. Gene 2011;470(1–2):20–30.

[78] Morgante PG, Berra CM, Nakabashi M, Costa RM, Menck CF, Van Sluys MA. Functional XPB/RAD25 redundancy in *Arabidopsis* genome: characterization of AtXPB2 and expression analysis. Gene 2005;344:93–103.

[79] Gallego F, Fleck O, Li A, Wyrzykowska J, Tinland B. AtRAD1, a plant homologue of human and yeast nucleotide excision repair endonucleases, is involved in dark repair of UV damages and recombination. Plant J 2000;21(6):507–18.

[80] Liu Z, Hossain GS, Islas-Osuna MA, Mitchell DL, Mount DW. Repair of UV damage in plants by nucleotide excision repair: *Arabidopsis* UVH1 DNA repair gene is a homolog of *Saccharomyces cerevisiae* Rad1. Plant J 2000;21(6):519–28.

[81] Friedberg EC. Relationships between DNA repair and transcription. Annu Rev Biochem 1996;65(1):15–42.

[82] Wu SY, Culligan K, Lamers M, Hays J. Dissimilar mispair-recognition spectra of *Arabidopsis* DNA-mismatch-repair proteins MSH2*MSH6 (MutSalpha) and MSH2*MSH7 (MutSgamma). Nucleic Acids Res 2003;31(20):6027–34.

[83] Hays JB. *Arabidopsis thaliana*, a versatile model system for study of eukaryotic genome-maintenance functions. DNA Repair (Amst) 2002;1(8):579–600.

[84] Culligan KM, Hays JB. *Arabidopsis* MutS homologs-AtMSH2, AtMSH3, AtMSH6, and a novel AtMSH7-form three distinct protein heterodimers with different specificities for mismatched DNA. Plant Cell 2000;12(6):991–1002.

[85] Higgins JD, Armstrong SJ, Franklin FCH, Jones GH. The *Arabidopsis* MutS homolog AtMSH4 functions at an early step in recombination: evidence for two classes of recombination in *Arabidopsis*. Genes Dev 2004;18(20):2557–70.

[86] Higgins JD, Vignard J, Mercier R, Pugh AG, Franklin FC, Jones GH. AtMSH5 partners AtMSH4 in the class I meiotic crossover pathway in *Arabidopsis thaliana*, but is not required for synapsis. Plant J 2008;55(1):28–39.

[87] Virdi KS, Laurie JD, Xu YZ, Yu J, Shao MR, Sanchez R, et al. *Arabidopsis* MSH1 mutation alters the epigenome and produces heritable changes in plant growth. Nat Commun 2015;6:6386.

[88] Yang X, Kundariya H, Xu YZ, Sandhu A, Yu J, Hutton SF, et al. MutS HOMOLOG1-derived epigenetic breeding potential in tomato. Plant Physiol 2015;168(1):222–32.

[89] Xu YZ, Santamaria Rde L, Virdi KS, Arrieta-Montiel MP, Razvi F, Li S, et al. The chloroplast triggers developmental reprogramming when mutS HOMOLOG1 is suppressed in plants. Plant Physiol 2012;159(2):710–20.

[90] Abdelnoor RV, Yule R, Elo A, Christensen AC, Meyer-Gauen G, Mackenzie SA. Substoichiometric shifting in the plant mitochondrial genome is influenced by a gene homologous to MutS. Proc Natl Acad Sci USA 2003;100(10):5968–73.

[91] Xu Y-Z, Arrieta-Montiel MP, Virdi KS, de Paula WBM, Widhalm JR, Basset GJ, et al. MutS HOMOLOG1 is a nucleoid protein that alters mitochondrial and plastid properties and plant response to high light. Plant Cell 2011;23(9):3428–41.

[92] de la Rosa Santamaria R, Shao MR, Wang G, Nino-Liu DO, Kundariya H, Wamboldt Y, et al. MSH1-induced non-genetic variation provides a source of phenotypic diversity in *Sorghum bicolor*. PLoS One 2014;9(10):e108407.

[93] Shedge V, Davila J, Arrieta-Montiel MP, Mohammed S, Mackenzie SA. Extensive rearrangement of the *Arabidopsis* mitochondrial genome elicits cellular conditions for thermotolerance. Plant Physiol 2010;152(4):1960–70.

[94] Waterworth WM, Drury GE, Bray CM, West CE. Repairing breaks in the plant genome: the importance of keeping it together. New Phytol 2011;192(4):805–22.

[95] Gisler B, Salomon S, Puchta H. The role of double-strand break-induced allelic homologous recombination in somatic plant cells. Plant J 2002;32(3):277–84.

[96] Siebert R, Puchta H. Efficient repair of genomic double-strand breaks by homologous recombination between directly repeated sequences in the plant genome. Plant Cell 2002;14(5):1121–31.

[97] Orel N, Kyryk A, Puchta H. Different pathways of homologous recombination are used for the repair of double-strand breaks within tandemly arranged sequences in the plant genome. Plant J 2003;35(5):604–12.

[98] Lin Z, Kong H, Nei M, Ma H. Origins and evolution of the recA/RAD51 gene family: evidence for ancient gene duplication and endosymbiotic gene transfer. Proc Natl Acad Sci USA 2006;103(27):10328–33.

[99] Li W, Ma H. Double-stranded DNA breaks and gene functions in recombination and meiosis. Cell Res 2006;16(5):402–12.

[100] Mortensen UH, Lisby M, Rothstein R. Rad52. Curr Biol 2009;19(16):R676–7.

[101] Samach A, Melamed-Bessudo C, Avivi-Ragolski N, Pietrokovski S, Levy AA. Identification of plant RAD52 homologs and characterization of the *Arabidopsis thaliana* RAD52-like genes. Plant Cell 2011;23(12):4266–79.

[102] Li W, Chen C, Markmann-Mulisch U, Timofejeva L, Schmelzer E, Ma H, et al. The *Arabidopsis* AtRAD51 gene is dispensable for vegetative development but required for meiosis. Proc Natl Acad Sci USA 2004;101(29):10596–601.

[103] Puizina J, Siroky J, Mokros P, Schweizer D, Riha K. Mre11 deficiency in *Arabidopsis* is associated with chromosomal instability in somatic cells and Spo11-dependent genome fragmentation during meiosis. Plant Cell 2004;16(8):1968–78.

[104] Bleuyard J-Y, Gallego ME, Savigny F, White CI. Differing requirements for the *Arabidopsis* Rad51 paralogs in meiosis and DNA repair. Plant J 2005;41(4):533–45.

[105] Puchta H. The repair of double-strand breaks in plants: mechanisms and consequences for genome evolution. J Exp Bot 2005;56(409):1–14.

[106] Wang Y, Xiao R, Wang H, Cheng Z, Li W, Zhu G, et al. The *Arabidopsis* RAD51 paralogs RAD51B, RAD51D and XRCC2 play partially redundant roles in somatic DNA repair and gene regulation. New Phytol 2014;201(1):292–304.

[107] Paques F, Haber JE. Multiple pathways of recombination induced by double-strand breaks in *Saccharomyces cerevisiae*. Microbiol Mol Biol Rev 1999;63(2):349–404.

[108] Knoll A, Fauser F, Puchta H. DNA recombination in somatic plant cells: mechanisms and evolutionary consequences. Chromosome Res 2014;22(2):191–201.

[109] Mladenov E, Iliakis G. Induction and repair of DNA double strand breaks: the increasing spectrum of non-homologous end joining pathways. Mutat Res 2011;711(1–2):61–72.

[110] Zhang X, Paull TT. The Mre11/Rad50/Xrs2 complex and non-homologous end-joining of incompatible ends in *S. cerevisiae*. DNA Repair (Amst) 2005;4(11):1281–94.

[111] Riha K, Watson JM, Parkey J, Shippen DE. Telomere length deregulation and enhanced sensitivity to genotoxic stress in *Arabidopsis* mutants deficient in Ku70. EMBO J 2002;21(11):2819–26.

[112] West CE, Waterworth WM, Story GW, Sunderland PA, Jiang Q, Bray CM. Disruption of the *Arabidopsis* AtKu80 gene demonstrates an essential role for AtKu80 protein in efficient repair of DNA double-strand breaks in vivo. Plant J 2002;31(4):517–28.

[113] Osakabe K, Osakabe Y, Toki S. Site-directed mutagenesis in *Arabidopsis* using custom-designed zinc finger nucleases. Proc Natl Acad Sci USA 2010;107(26):12034–9.

[114] Nielsen BL, Cupp JD, Brammer J. Mechanisms for maintenance, replication, and repair of the chloroplast genome in plants. J Exp Bot 2010;61(10):2535–7.

[115] Rowan BA, Oldenburg DJ, Bendich AJ. RecA maintains the integrity of chloroplast DNA molecules in *Arabidopsis*. J Exp Bot 2010;61(10):2575–88.

[116] Cox MM. Regulation of bacterial RecA protein function. Crit Rev Biochem Mol Biol 2007;42(1):41–63.

[117] Shedge V, Arrieta-Montiel M, Christensen AC, Mackenzie SA. Plant mitochondrial recombination surveillance requires unusual RecA and MutS homologs. Plant Cell 2007;19(4):1251–64.

[118] Gutman BL, Niyogi KK. Evidence for base excision repair of oxidative DNA damage in chloroplasts of *Arabidopsis thaliana*. J Biol Chem 2009;284(25):17006–12.

[119] Boesch P, Ibrahim N, Paulus F, Cosset A, Tarasenko V, Dietrich A. Plant mitochondria possess a short-patch base excision DNA repair pathway. Nucleic Acids Res 2009;37(17):5690–700.

Section V

Genome Stability in Mammals

Chapter 14

Cell-Cycle Control and DNA-Damage Signaling in Mammals

V. Gomez, A. Hergovich
University College London, London, United Kingdom

Chapter Outline

1. INTRODUCTION

For the survival and normal functionality of a mammalian organism, it is essential to ensure the maintenance of genomic integrity and the accurate transmission of genetic information between daughter cells at the cellular level. To safeguard the integrity and functionality of their genomes, mammalian cells utilize molecular mechanisms to separate the replication of DNA in S phase from the equal distribution of genetic information during M phase. Very accurate cell cycle–control mechanisms ensure that a cell does not reduplicate chromosomes before sister chromatid separation. Conversely, cells do not enter into mitosis before DNA duplication is completed. These mechanisms are orchestrated by the oscillating activities of cyclin-dependent kinases (CDKs), the master regulators of cell-cycle progression [1,2]. CDKs are regulated by complex formation with different cell-cycle phase-specific cyclins, supporting kinase activity and substrate recruitment. The activities of CDK–cyclin complexes are regulated by controlling the synthesis and degradation of cyclins throughout the cell cycle. CDK activities are further modulated by specific activating or inhibiting phosphorylations and interactions with specific CDK inhibitors (CKIs). Collectively, these events enable mammalian cells to orderly progress through the cell cycle, and hence to preserve genomic stability.

Besides faulty genome replication and/or segregation, the integrity of mammalian genomes is also threatened by spontaneous endogenous DNA damage or exogenously induced DNA lesions. Therefore, cells utilize a fine-tuned DNA-damage response (DDR) to coordinate the detection, signaling, and repair of DNA damage. The cellular DDR was initially discovered in yeast, where it functions to arrest the cell cycle in order to allow DNA repair [3]. The mammalian DDR also couples DNA repair with cell-cycle progression and DNA replication, and possibly, when the DNA damage is beyond repair, in the commitment to undergo apoptosis or terminal differentiation through senescence [4–7]. Generally, the DDR is a kinase-based signaling network. DDR signaling plays a role in the coordination of the response to different types of DNA lesions, where the role of the DDR in DNA double-strand break (DSB) repair is best understood. In response to DSBs, as one of the earliest events in DDR signaling, the DNA-damage sensor and signal mediator complex MRE11/RAD50/NBS1 (MRN) is recruited to the sites of DNA lesions. The recruitment of the MRN complex assists the activation of ataxia

Genome Stability. http://dx.doi.org/10.1016/B978-0-12-803309-8.00014-8

telangiectasia–mutated kinase (ATM) by sequestering ATM at DSBs, which results in ATM-mediated formation of γH2AX in the vicinity of DNA breaks up to distances of megabases. Together MRN and γH2AX function as signal amplifiers that further enhance local ATM activation and recruit additional DDR mediators, effectors, and DNA-repair factors to DNA. In response to DNA damage, many molecular events take place involving posttranslational modifications, such as phosphorylation, ubiquitinylation, and others, which can help coordinate selective and regulatory protein–protein interactions of central DDR molecules [4–11].

By maintaining genomic integrity, the DDR is critical to prevent aging and defend against malignant transformation [5,12]. Therefore, it is not surprising that many commonly lost tumor-suppressor genes, including p53, $p16^{INK4A}$, RB, BRCA1/2, and ATM, play roles in the DDR which is among the most frequently compromised pathways in human cancers [13–20]. Defective DNA repair can also result in striking disease phenotypes. For example, deficiency in nonhomologous end joining (NHEJ) results in severe combined immunodeficiency (SCID) due to impaired V(D)J immunoglobulin arrangements [5]. Patients suffering from NHEJ defects are further predisposed to develop lymphoid tumors stemming from T and/or B cells with abnormal TCR and/or V(D)J recombination [5]. As another example, mutations in ATM or components of the MRN complex have been linked to a broad range of clinical features including progressive ataxis, telangiectasia, mild immunodeficiency, and cancer predisposition [19,21–26]. Generally, DDR defects can result in cancer predisposition, neurological disorders, premature aging, impaired immune biology, infertility, and other syndromes [5]. Specifically, since genome instability is a fundamental feature of cancer [27–29], and DNA-damaging agents, such as cisplatin, doxorubicin (Adriamycin), and ionizing radiation (IR) are routinely used in chemotherapy regimens [30,31], the bettering of our understanding of DNA-damage signaling has gained an extensive interest in order to improve DNA-damaging cancer therapies with the aim of reducing the frequency of cancer therapy resistance [32].

Here, we provide an overview of key molecular events in the control of the mammalian cell cycle with a particular emphasis on cell cycle–checkpoint activation and DNA-damage repair in response to DNA damage.

2. CELL-CYCLE PROGRESSION IN MAMMALIAN CELLS

To progress through the cell cycle in a coordinated fashion is a complex challenge for every eukaryotic cell, particularly in multicellular organisms with many specialized cell types. To ensure genomic integrity, a cell must be able to detect DNA lesions followed by signal integration to establish checkpoints which are specific for each cell-cycle stage, allowing sufficient time for various types of DNA repair.

2.1 Definition of Cell-Cycle Phases

The faithful transmission of genetic information between daughter cells is performed by two central processes: DNA replication and cell division. The mammalian cell cycle can be subdivided into four different stages occurring in the following order: G1, S, G2, and M phases (Fig. 14.1). G1, S, and G2 are referred to as interphase, while the M phase is also known as mitosis. During the first gap phase (G1), the cell prepares for a new round of duplication, ensuring that sufficient building blocks and the right environment/conditions are available for a successful cell multiplication. In case conditions are unfavorable, the cell halts in a transient G1 arrest, activates reversible cell cycle–exit mechanisms to enter G0 (quiescence), or commits to a permanent cell-cycle arrest (senescence). During the S phase, DNA synthesis takes place to completely and accurately replicate the double-stranded DNA molecules. During the second gap phase (G2), the cell prepares for the mitotic division. In mitosis, divided into prophase, prometaphase, metaphase, anaphase, telophase, and cytokinesis, the cell divides by segregating the chromosomes into two separate daughter cells. Collectively, during cell-cycle progression, a mammalian cell ensures to precisely replicate (copy) its genetic information in the S phase and to proper partition (distribute) chromosomes between daughter cells in the M phase. Equally important, the cell can identify and correct DNA lesions that can arise spontaneously or are induced exogenously. In this regard, the cell can activate cell-cycle checkpoints to arrest cells in the G1, S, and G2 cell-cycle phases (Fig. 14.1).

2.2 Molecular Regulation of Cell-Cycle Progression

The coordinated progression through all four phases of the cell cycle is dependent on protein phosphorylation [33,34]. The serine/threonine CDKs form catalytically active heterodimer complexes with cyclins which can be regulated by CKIs, such as p21 (CIP1/WAF1) or p27 (KIP1) [35]. The oscillating activities of CDKs are essential to safeguard the timely duplication and segregation of the genome (Fig. 14.1). Different CDKs contribute to cell-cycle progression: CDK2, CDK4, and CDK6 are active in G1, CDK2—during the S phase, and CDK1—during the G2 and M phases. The transition from G1

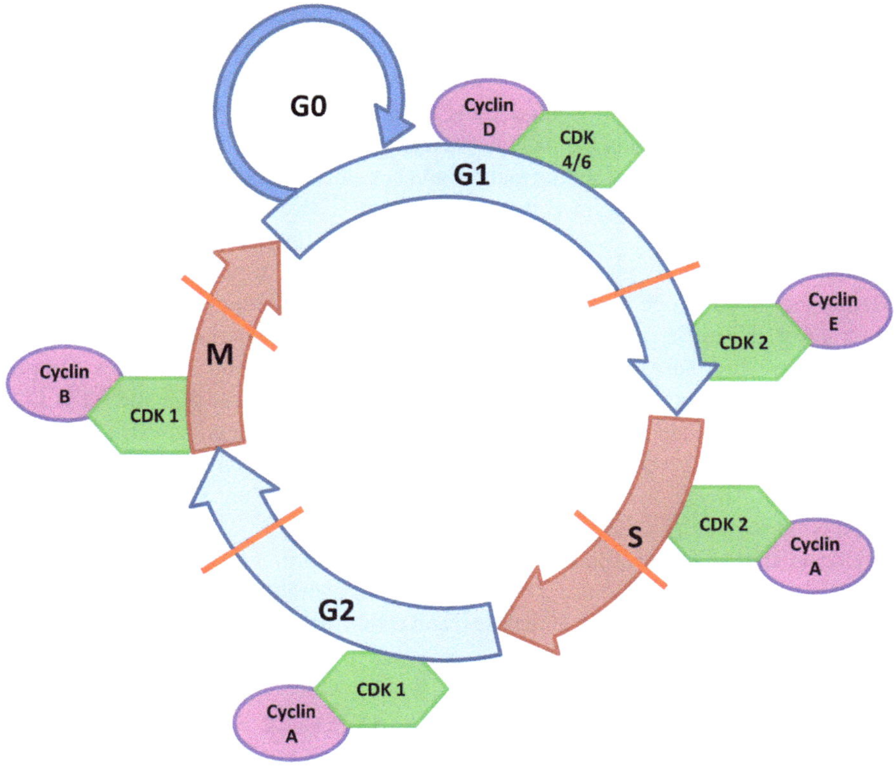

FIGURE 14.1 **Regulation of the mammalian cell cycle by CDK–cyclin complexes.** Different cyclin-dependent kinases (CDKs) play roles in regulating cell-cycle progression. In G1, cells can reversibly exit the cell cycle and enter into quiescence (G0) or permanently exit the cell cycle by entering into senescence. Alternatively, a mammalian cell can decide to commit to another cell cycle round. In this case, CDK4/6–cyclin D complexes in early G1 and CDK2–cyclin E complexes in late G1 phosphorylate pRB to promote the release of E2F which enables progression into the S phase. The progression through the S phase is promoted by phosphorylations performed by CDK2–cyclin E and CDK2–cyclin A complexes. In G2, the CDK1–cyclin A complex initially promotes cell-cycle progression, while in late G2, the CDK1–cyclin B complex takes over. The entry into mitosis (M) is triggered by a sharp increase in CDK1-cyclin B activity. Note that, if needed, mammalian cells can activate cell-cycle checkpoints in all four cell-cycle phases (indicated by *red lines*).

to S is initiated by phosphorylations performed by CDK4/6–cyclin D and CDK2–cyclin E complexes [33,36]. In normal conditions, the retinoblastoma protein (pRB) is phosphorylated by CDK4/6-cyclin D in early G1 and CDK2-cyclin E in late G1. Fully phosphorylated pRB then releases the E2F transcription factor which enables S-phase progression [36–42]. Progression through the S phase is subsequently coordinated by phosphorylations performed by CDK2–cyclin E and CDK2–cyclin A complexes. Protein phosphorylation is also crucial for the G2/M transition and in M-phase progression regulated by CDK1–cyclin A and CDK1–cyclin B complexes together with other mitotic kinases, such as Aurora, PLK, NEK, and Greatwall [43]. During the normal cell-cycle progression, the entry into mitosis is triggered by a sharp increase in CDK1–cyclin B activity, which requires the removal of inhibitory phosphorylations on CDK1 generated by Wee1 and Myt1 [44]. Members of the CDC25 family of phosphatases are responsible for the removal of these inhibitory phosphorylations. Significantly, the activity of CDC25 is also regulated by activating and inhibitory phosphorylations. Activating phosphorylations are added to CDC25 by PLK1, MAPKs, and CDK1, while inhibitory phosphorylations are added to CDC25 by CHK1, CHK2, and MK2 (MAPK-activated protein kinase 2) in response to DNA damage [8]. In particular, the CDC25A phosphatase needs to be tightly regulated in the context of the G1/S and intra-S-phase checkpoints (see later).

To prevent the inappropriate cell-cycle progression, the activity of CDKs and other cell-cycle kinases has to be tightly regulated at the G1/S and G2/M transitions. This is mainly achieved by specific interactions with CKIs and by a direct phosphorylation and/or dephosphorylation of kinases and regulatory cyclin subunits. CDKs are activated by T-loop phosphorylation mediated by CDK7-cyclin H. CDKs can be inhibited by Myt1- and Wee1-mediated phosphorylation which can be removed by CDC25 phosphatases [45–47]. Furthermore, two main protein families function as CKIs: (1) the INK4 family (composed of p15, p16, 18, and p19) inhibits CDK4/6 activity by preventing CDK4/6–cyclin D complex formation, and (2) the WAF proteins p21, p27, and p57 inhibit primarily CDK2–cyclin E and CDK4–cyclin D complexes. Another level of CDK regulation is subcellular localization. For example, the regulated translocation of CDK1–cyclin B complexes from the cytoplasm to the nucleus promotes mitosis [47,48]. E3 ubiquitin ligases also play key roles in cell-cycle progression. E3 ligases can mark their substrates by promoting the addition of polyubiquitin chains which targets the substrate for destruction by the proteasome. For example, CKIs p21 and p27 are regulated by phosphorylation-mediated degradation through the SCF^{SKP2} E3 complex, a major cell cycle–E3 ligase [49,50]. The APC/C, the second major cell cycle–E3 ligase, marks SKP2, cyclin A, and cyclin B for degradation to regulate CDK1 and CDK2 activities in the G1, G2, and M phases [51]. Specifically, the degradation of SKP2 counteracts the unscheduled degradation of SCF^{SKP2} substrates in G2, while the timely degradation of cyclin A and cyclin B is crucial for normal progression through G2/M and M. Upon G2/M transition, CDK2 activity is blocked by SCF^{FBW7}-mediated cyclin E degradation, and CDK1 activation is assisted by $SCF^{\beta TrCP}$-mediated degradation of Wee1.

Intriguingly, several of these regulatory events are linked to DNA-damage signaling by proteins that preferentially bind to phosphorylated serine and/or threonine residues. These phospho-binding proteins function together with cell cycle–checkpoint kinases to control cell-cycle progression [8]. In this regard, it is noteworthy that upon recovery from DNA-damage checkpoints, the polo-like kinase 1 (PLK1), the Wip1 phosphatase and other kinases and phosphatases play key roles [52,53]. Wip1 can suppress the phosphorylation of key DDR factors, such as p53 and ATM [52,53], while PLK1 is required for CDC25 activation to initiate mitosis as well as the progression through and exit from mitosis. In response to DNA damage, CHK2 inhibits PLK1 by phosphorylation. Conversely, upon completion of DNA repair, PLK1 shuts off ATM/ATR-CHK1/2 signaling to silence the G2/M checkpoint to allow mitotic entry. Thus, PLK1 plays a role as a downstream target and an upstream regulator of the DDR pathway.

In summary, in the basal state, CDK–cyclin complexes are ready to promote cell-cycle progression upon removal of their inhibitory modifications and/or interactions. CDK activation and consequently cell-cycle progression occur when the inhibitory phosphates are removed by CDC25 phosphatases, and inhibitory kinases and CKIs are inactivated, destroyed, or sequestered away from CDK–cyclin complexes.

3. DNA-DAMAGE SIGNALING AND REPAIR IN MAMMALS

Mammalian genomes are under constant assault from endogenous and exogenous DNA lesions [54,55] where endogenous DNA damage can be caused by defective/stalled DNA replication, reactive oxygen species, and other mechanisms [56]. DNA DSBs are believed to be the most harmful forms of DNA damage [57]. DSBs can occur accidentally during normal DNA metabolism [56] or after exposure to exogenous agents, such as IR or DNA-damaging chemotherapeutics during radio- or chemotherapy [31,58]. To counteract DNA damage, cells make use of an array of signaling and repair mechanisms, involving kinases, nucleases, topoisomerases, helicases, ligases, polymerases, glycosylases, and other enzymes. Once DNA lesions have been detected, DDR pathways can promote different outcomes depending on the severity of DNA damage and the cell type. Possible outcomes include slowing down or arresting cell-cycle progression, permanent cell-cycle arrest (senescence), or the initiation of apoptotic programs. In either case, the aim is to prevent the replication of damaged DNA and the inheritance of DNA damage by daughter cells.

Various factors involved in cell-cycle control, DDR signaling, and DNA-damage repair contain specific modular domains, including the phospho-binding BRCT and FHA modules that play important roles in regulatory protein–protein interactions in the DDR [8]. In this regard, the actions of protein kinases are linked to physiological events by controlling specific phosphorylation-dependent protein–protein interactions. In mammalian cells, the ATM-CHK2 and ATR-CHK1 kinase cascades are the two major types of DDR kinase signaling. The ATM and ATR (ATM and Rad3-related) kinases are members of the phosphatidylinositol 3-kinase-related kinases (PIKKs) family of serine/threonine protein kinases [59]. The DNA-dependent protein kinase (DNA-PK) is another member of the PIKK family [60] regulating DSB repair together with ATM (see later). Noteworthy here is, although ATM is the central DDR kinase, most ATM substrates are still phosphorylated to a certain degree in ATM-deficient cells in response to DNA damage, suggesting that other PIKKs, such as ATR and DNA-PK, are also important for DDR signaling [61].

3.1 DDR Signaling

Generally, DNA-damage signaling can be subdivided into four levels (Fig. 14.2): (1) DNA-damage detection by sensors, such as the MRN and Ku70/Ku80 complexes, (2) the formation/recruitment of mediators, such as γH2AX and MDC1, (3) the activation of signal transducers, such as the ATM and ATR kinases, (4) the activation of effectors, such as p53 and others (see later). In response to DNA damage, different factors, such as the MRN complex, PARP1, and Ku70/Ku80 bind rapidly to sites of DNA damage resulting in DDR activation [4–7,62]. In particular, DSBs trigger a range of signal transduction processes with the ATM-CHK2 pathway as a primary response, while the ATR-CHK1 pathway tends to be activated by exposure to ssDNA, stalled replication forks, and bulky DNA base adducts [4,5,7,61,63]. However, ATR can

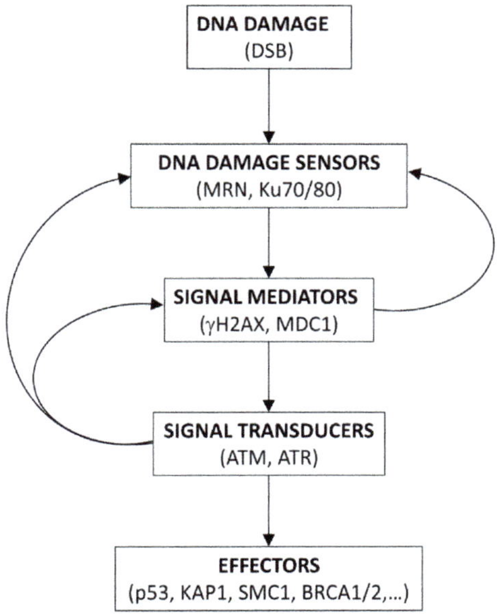

FIGURE 14.2 **Main steps of DNA damage–response signaling.** DNA-damage signaling can be subdivided into four levels: the detection of DNA damage (eg, DSBs) by DNA-damage sensors (eg, MRN and Ku70/Ku80 complexes), the formation/recruitment of signal mediators (eg, γH2AX and MDC1), the activation of signal transducers (eg, ATM and ATR kinases which are mainly activated by DNA breaks), and the activation of effectors (eg, p53 and others). Note that signal mediators can facilitate the amplification of the initial DNA-damage signal. Signal transducers can also act in positive-feedback loops by promoting the activation/recruitment of DNA-damage sensors and signal mediators, resulting in a further amplification of the initial DNA-damage signal. For a more complete overview of these DNA damage–signaling steps, please refer to the main text.

also be involved in DSB signaling [4–7]. Specifically, ATM-mediated DSB resection and the consequent single-stranded DNA (ssDNA) formation also promote ATR/CHK1 signaling in S and G2 [64,65]. Thus, the ATM response to DSBs is very rapid and cell-cycle independent, while the ATR response is generally slower and requires CDK-dependent resection, hence being restricted to S and G2. The general view is that the checkpoint kinases CHK1 and CHK2 are key effectors of ATR and ATM signaling. CHK2 is activated by ATM in a cell cycle–independent manner, while CHK1 is activated by ATR following ATM and MRN-mediated resection of DSBs in a cell cycle–dependent fashion. As a third main regulator of DNA damage–checkpoint activation, the p38MAPK/MK2 pathway can function downstream of ATM and ATR in response to DNA damage [66,67]. Here, we provide an overview of the key sensors, mediators, transducers, and effectors in mammalian DNA-damage signaling (Fig. 14.2).

The MRN complex functions as the initial DSB sensor tethering the broken DNA ends together [24–26]. The interaction between NBS1 and ATM [8] results in the initial sequestering of ATM to DSB sites [24–26]. Once activated by autophosphorylation [68], ATM phosphorylates many different substrates involved in DDR signaling [19,61]. On the one hand, the activated ATM phosphorylates the MRN component NBS1 to create a positive-feedback loop maintaining/amplifying ATM activity [61]. On the other hand, the ATM-mediated phosphorylation of the H2AX histone results in the formation of γH2AX which consequently accumulates in the vicinity of DSBs. The formation of γH2AX foci initiates the recruitment of other DDR factors to sites of DNA lesions. γH2AX formation is normally ATM dependent but can also be mediated by DNA-PK or ATR [61]. Note that in addition to phosphorylation, other posttranslational modifications, such as ubiquitylation and methylation are also involved in the early steps of DDR signaling [4–11]. In a nutshell, γH2AX formation recruits MDC1. MDC1 in turn interacts with MRN and ATM to tether MRN and ATM at DSBs, and also recruits the RNF8 (RING finger 8) and RNF168 E3 ligases which ubiquitylate H2A in the DSB vicinity. H2A ubiquitylation then influences the methylation status of other histones which can promote 53BP1 recruitment to DSBs, thereby influencing the choice of DSB repair (see later). Note that additional factors can be sequestered at DSBs in a cell cycle–dependent manner, including BRCA1 and other protein complexes [69]. However, since the phosphorylations mediated by the activated ATM are crucial for DDR signaling, we focus here on discussing key substrates of ATM in the DDR [61].

ATM-mediated γH2AX formation serves as a platform for the recruitment of DDR factors and as an amplifier of the initial signal. However, γH2AX deficiency has only subtle effects on cell-cycle checkpoints and DNA repair, indicating that γH2AX may only regulate repair and signaling of a portion of DSBs [61]. In addition, in mammalian cells, γH2AX-foci preferentially form in euchromatin, being mostly excluded from densely packed heterochromatin [70], suggesting that DNA lesions in heterochromatin are signaled by γH2AX-independent mechanisms.

The transcription factor p53 is a major effector of ATM signaling [36,61,71] mediating a G1 cell–cycle arrest mainly through the transcriptional upregulation of p21 [72]. Alternatively, if the DNA damage is too extensive, p53 triggers cell death through intrinsic and extrinsic pathways [73,74]. Generally, p53 as a downstream effector of ATM signaling acts as a major DNA damage–checkpoint regulator (see Chapter 15).

ATM also phosphorylates the checkpoint regulator CHK2. Once activated by ATM, CHK2 acts as an effector kinase by phosphorylating numerous downstream targets, including CDC25, p53, and BRCA1 [75]. Of note, ATM-mediated phosphorylation of CHK2 is followed by additional autophosphorylation events which are required for the full activation of CHK2 (summarized in [8]). In general, CHK2 is a key effector downstream of ATM signaling in the DDR.

ATM also phosphorylates the KAP1 (Krüppel-associated box (KRAB)-associated protein 1) [76] which allows the transient and localized relaxation of heterochromatin without affecting epigenetic marks in the vicinity of DSB [77–80]. Specifically, ATM-dependent KAP1 phosphorylation is essential for DSB repair within heterochromatin regions. This function is important since the connection between chromatin architecture and DNA-damage signaling can modulate the choice between different DNA-repair pathways (see later).

ATM further phosphorylates the structural maintenance of chromosomes 1 protein (SMC1) known to function in complex with SMC3 in sister chromatid cohesion and DNA recombination. This phosphorylation of SMC1 by ATM can play a role in the intra-S-phase checkpoint, hence being required for genomic stability and consequently cell survival in response to DNA damage [81–83].

Note that ATM also phosphorylates factors involved in the repair of DSBs, including CtIP, BRCA1, and RAD51 [61] which can influence DSB-repair pathway choice (see later). In this regard, it is most likely that in response to DNA damage, the ATM/ATR kinases may phosphorylate more than 700 different substrates in mammalian cells [84]. This underscores the central importance of ATM in the DDR, while highlighting the challenges that lie ahead to functionally decipher the importance of ATM- and/or ATR-mediated phosphorylation events. Last, but not least, one should also note that defective ATM activation is quantitative, not absolute, in MRN mutant cells [22,61,82,85,86], and MRN-mediated ATM activation can be dispensable for p53 and CHK2 phosphorylation [23,87]. This suggests, on the one hand, that ATM can be activated through different signaling routes, while, on the other hand, not all known ATM substrates may represent suitable readouts for ATM activity in certain settings.

3.2 DNA-Damage Repair

Tens of thousands of DNA-damaging events take place in every cell on a daily basis [54–56]. Thus, DNA lesions are recognized and processed by highly specialized DNA-repair systems to ensure a quick and accurate removal of DNA damage [4–7]. These systems are crucial since the persistence of DNA mutations can result in the altered gene functions potentially causing cancer development, tissue degeneration, and other human diseases [5,88]. Specifically, the defective DNA repair can lead to elevated mutation rates, which when occurring in tumor-suppressor genes or proto-oncogenes can cause cancer development, as, for example, is the case in mismatch repair (MMR) defects [89–92]. Defects in HRR can also predispose patients to cancer considering the data of heterozygous carriers with mutations in BRCA1/2, two HRR components that function as tumor-suppressor proteins [14]. In addition, human diseases can derive from deficiencies in DSB-repair pathways, exhibiting premature aging, cancer predisposition, and defects in neurobiology, immunology, and development [5,88,93–98].

Since DSBs represent very dangerous DNA lesions that will cause an uneven division of the genome during the M phase when not dealt with prior to mitotic entry, we focus our overview of DNA damage–repair mechanisms on DSB-repair pathways in the context of cell-cycle dependencies. Nonetheless, we also summarize other DNA-repair mechanisms, such as nucleotide excision repair (NER), base excision repair (BER), and MMR which are defined in more detail elsewhere [99]. Interstrand crosslink (ICL) repair is not discussed since it is primarily performed by a combination of NER, HRR, and other repair pathways.

3.2.1 Nucleotide Excision Repair

NER mainly fixes "bulky" DNA alterations including ultraviolet light (UV)-induced photoproducts, base adducts created by genotoxic agents, such as cisplatin, reactive oxygen species (ROS)-induced base modifications, and others [99]. NER occurs in four main steps: (1) DNA-damage recognition, (2) incision on both sides of the DNA lesion and removal of the damaged DNA fragment, (3) gap-filling DNA synthesis, and (4) ligation of open DNA ends. About 30 proteins function in the NER pathway, and defects in NER components are linked to human diseases, such as xeroderma pigmentosum, Cockayne syndrome, and others [99]. Generally, the NER pathway can be subdivided into two processes: (1) the global genome NER (GG-NER) functioning in a cell cycle–independent manner to remove UV-induced photoproducts and other "bulky" lesions, and (2) transcription-coupled NER (TC-NER) recognizing RNA polymerase stalled at "bulky" DNA lesions. GG-NER and TC-NER differ at the step of DNA recognition damage and share the remaining DNA-repair machinery [99].

3.2.2 Base Excision Repair

BER fixes nonbulky DNA base damage, abasic sites, and DNA single-strand breaks (SSBs) throughout all stages of the cell cycle [99]. BER occurs in five major steps: (1) recognition and excision of a damaged base, (2) incision at the abasic site, (3) replacement of the excised DNA nucleotide, (4) processing of DNA ends, and (5) sealing of the DNA nick. DNA glycosylases are responsible for the recognition and hydrolysis of DNA lesions followed by DNA polymerase β and XRCC1-ligase III-mediated nucleotide replacement and DNA nick sealing [99]. SSBs are detected by PARP1 which catalyzes the formation of poly-ADP-ribose (PAR) chains on itself and other proteins to facilitate the recruitment of specialized BER enzymes and DNA-repair factors, such as XRCC1, DNA polymerase β,Ligases I and III [99].

3.2.3 Mismatch Repair

MMR is responsible for the recognition and repair of base–base matches and insertion–deletion loops (IDLs) which are caused by faulty DNA replication and homologous recombination [99,100]. To preserve genome integrity, MMR must take place selectively on the newly synthesized DNA strand containing the error. MSH2-containing complexes recognize DNA lesions followed by the recruitment of MLH1/3 and PMS1/2 complexes, and then the endonucleases PMS2 and MLH3 make an incision at the site of the DNA lesion. Upon the marking of the appropriate strand by incision, the exonuclease Exo1 generates a multi-nucleotide gap which is filled and ligated by DNA polymerase δ and Ligase I [99,100]. Significantly, MMR increases DNA replication fidelity by about 100-fold.

3.2.4 DSB Repair

DSBs are severe lesions that can result in the acquisition of disease-promoting properties or premature cell death when not repaired properly [5,30,31,88]. To minimize the impact of DSBs, mammalian cells utilize different DSB-repair pathways [69,99]. The two major pathways are: more error-prone but fast DNA nonhomologous end joining (NHEJ) and error-free but slow HRR. Depending on the origin of DSBs, mammalian cells use different DSB-repair mechanisms. DSBs can arise during programmed DNA recombination (eg, V(D)J and class switch recombination upon immunoglobulin production) or accidental DNA breakage upon the arrest or stalling of DNA replication and exposure to DNA-damaging agents, such as IR or topoisomerase poisons. These events cause DSBs of a distinctive nature which are recognized and processed differently [69]. The regulated resection of DSBs represents a key step in the choice between NHEJ and HRR [11,101] where the accumulation of 53BP1 at DSBs can block the resection and consequently RAD51 loading, hence influencing DSB-repair pathway choice [11,102]. Thus, 53BP1 generally promotes NHEJ and restricts HRR. Specifically, 53BP1 recruitment of RIF1 promotes NHEJ, while by excluding 53BP1 from DSB sites, BRCA1 can promote HRR [103–109].

Mostly, NHEJ occurs independently of ATM signaling, and ATM signaling takes place independently of the NHEJ machinery. DNA-PK signaling is essential for NHEJ, and ATM signaling is essential for HRR. Note that in mammalian cells, about 80% of radiation-induced DSBs are repaired quickly, while 20% of DSBs are repaired slowly [110,111]. Considering further that ATM is required for the slow repair of DSBs [111], the majority of DSBs are repaired by NHEJ. Actually, NHEJ is recognized as the predominant DSB-repair pathway in G1 and G2 [69,112]. In G1 and G2, NHEJ promotes a fast DSB repair, while in G2, but not G1, a slow process occurs by HRR. In the current DSB-repair model, NHEJ makes the first attempt to repair DSBs, but when a rapid repair does not ensue, resection occurs, thereby committing to HRR as a slow process [69,113]. In this context, one should also note that the chromatin status is very likely to influence the choice of DSB repair since euchromatin is more accessible than highly compacted heterochromatin [69]. Therefore, ATM also regulates factors such as KAP1 to promote the repair of DSBs located in heterochromatin (see earlier). Not surprisingly, other regulators of chromatin assembly influence the efficiency of DSB detection and repair as well [69,114].

Generally, major determinants of DSB-repair pathway choice are the extent of DNA end processing and cell-cycle position. The chromatin structure also contributes to this choice, with studies since 2000 suggesting that HRR is predominantly used for DSB repair in the areas of heterochromatin [111–113]. Initially, the MRN and Ku70/Ku80 complexes recognize DSBs, but they trigger different repair mechanisms (see later).

3.2.4.1 Classical Nonhomologous End-Joining Repair

NHEJ functions throughout the cell cycle as a predominant DSB-repair mechanism in mammalian cells [69,115–117]. NHEJ occurs via three main steps: (1) DSB recognition, (2) processing of nonligatable DNA termini, and (3) joining of two suitable DSBs. Noteworthy here, NHEJ can also directly religate the broken DNA ends and does not require DNA end resection for repair initiation. Classical NHEJ (c-NHEJ) is mediated by the Ku70/Ku80 heterodimer which binds to DSBs within seconds and dictates NHEJ pathway choice [118]. This binding protects DSB from degradation and recruits DNA-PK [60] whose autophosphorylation is essential for c-NHEJ (summarized in [69]). After DSB end-processing, Ligase IV functions as an NHEJ-specific ligase supported by XRCC4 and XLF [69].

3.2.4.2 Homologous Recombination Repair

HRR is dependent on the availability of an undamaged DNA template to restore sequence information accurately; hence, HRR only functions in late S and G2 phases when a sister chromatid is available. HRR requires DNA end resection of the DNA break to initiate repair. HRR occurs via six different steps [69]: (1) generation of 3′ ssDNA—also known as 5′-3′ resection, (2) RPA coating of ssDNA, (3) BRCA2-assisted displacement of RPA by RAD51 to form RAD51 filaments, (4) strand invasion resulting in heteroduplex and Holliday junction formation, (5) branch migration, and (6) resolution. Current evidence suggests that resection (ssDNA generation) involves two steps, with CtIP and MRE11 functioning as initiators and other exonucleases performing the elongation [69,101]. The MRN/CtIP complex is not only important for the initiation of resection [101] but also represents the event that commits to HRR [113,119]. Noteworthy being, CDK-mediated phosphorylation of CtIP promotes DSB resection in a cell cycle–dependent manner [120], and BRCA2 phosphorylation by CDK inhibits the BRCA2/RAD51 interaction [121], suggesting that CDK activity can regulate different steps of HRR. CtIP function is also regulated by ATM phosphorylation [122], with CtIP-dependent DNA resection selectively promoting HRR while suppressing NHEJ [123].

3.2.4.3 Alternative DSB-Repair Mechanisms

Two other DSB-repair pathways are microhomology-mediated end joining (MMEJ) and single-strand annealing (SSA). Both processes require DNA end resection for repair initiation [69,99].

MMEJ can be dependent on c-NHEJ and alternative NHEJ (alt-NHEJ). MMEJ uses short homologous sequences (microhomologies) to align the broken ends prior to ligation. Alt-NHEJ does not utilize c-NHEJ factors, such as Ku70/Ku80 and Ligase IV, but rather utilizes PARP1, XRCC1, Ligase I/III, and potentially the MRN complex [69]. Sometimes alt-NHEJ and MMEJ are used mistakenly as synonyms [69]. However, alt-NHEJ most likely represents only one variant of different MMEJ processes [69], likely as a back-up mechanism for c-NHEJ since alt-NHEJ does not seem to be active unless c-NHEJ is defective [99]. In general, alt-NHEJ is more error-prone than c-NHEJ [99]. Nonetheless, a better understanding of the alt-NHEJ-repair pathway is likely to further improve our treatment options of cancers since the inhibition of the alt-NHEJ pathway selectively sensitizes leukemia cells to cytotoxic agents [124].

SSA repair can take place when the DSB is flanked by repetitive homologous nucleotide repeats on both sides [99]. This allows the resected DSB ends to anneal with each other instead of invading a homologous DNA sequence elsewhere, as it occurs in HRR. Thus, in contrast to HRR, SSA repair does not involve strand invasion and does not require sister chromatid exchange. Consequently, SSA repair functions independently of RAD51, while being facilitated by RPA and other HHR factors [99]. Overall, in contrast to HRR, SSA repair results in the deletion of DNA sequence and is therefore error prone.

4. CHECKPOINT CONTROL: DNA-DAMAGE SIGNALING AND THE MAMMALIAN CELL CYCLE

Healthy mammalian cells respond to DNA lesions by activating DNA-damage checkpoints which delay cell-cycle progression while promoting DNA-repair mechanisms [45,125,126]. Cell cycle–checkpoints function as regulators of the cell-cycle machinery in response to DNA damage and act as a guard against the propagation of damaged DNA [127]. Considering the importance of the cellular response to DSBs, the most toxic DNA lesions [57], we focus here on defining DNA damage–checkpoint signaling in response to DSBs [128].

Three different and major cell cycle–checkpoints function in the DDR: the G1/S, intra-S phase, and G2/M DNA-damage checkpoints (Fig. 14.3). These checkpoints are primarily controlled by the signal transduction pathways that coordinate DSB recognition and repair, namely the ATM/ATR kinases. In response to DSBs, ATM and ATR initiate phosphorylation cascades that result in cell-cycle arrest at DNA-damage checkpoints where many DNA-damage checkpoints are mediated by the activation of the p53 tumor-suppressor protein [33,125,129]. Generally, DNA-damage checkpoints can be subdivided into initiation, maintenance, and recovery phases. In particular, following DNA damage, the initiation and maintenance of DNA-damage checkpoints is tightly regulated by ATM/ATR-mediated signaling (Fig. 14.3). Of course, checkpoint release and cell-cycle reentry following DNA repair must also be coordinated accordingly [52,125].

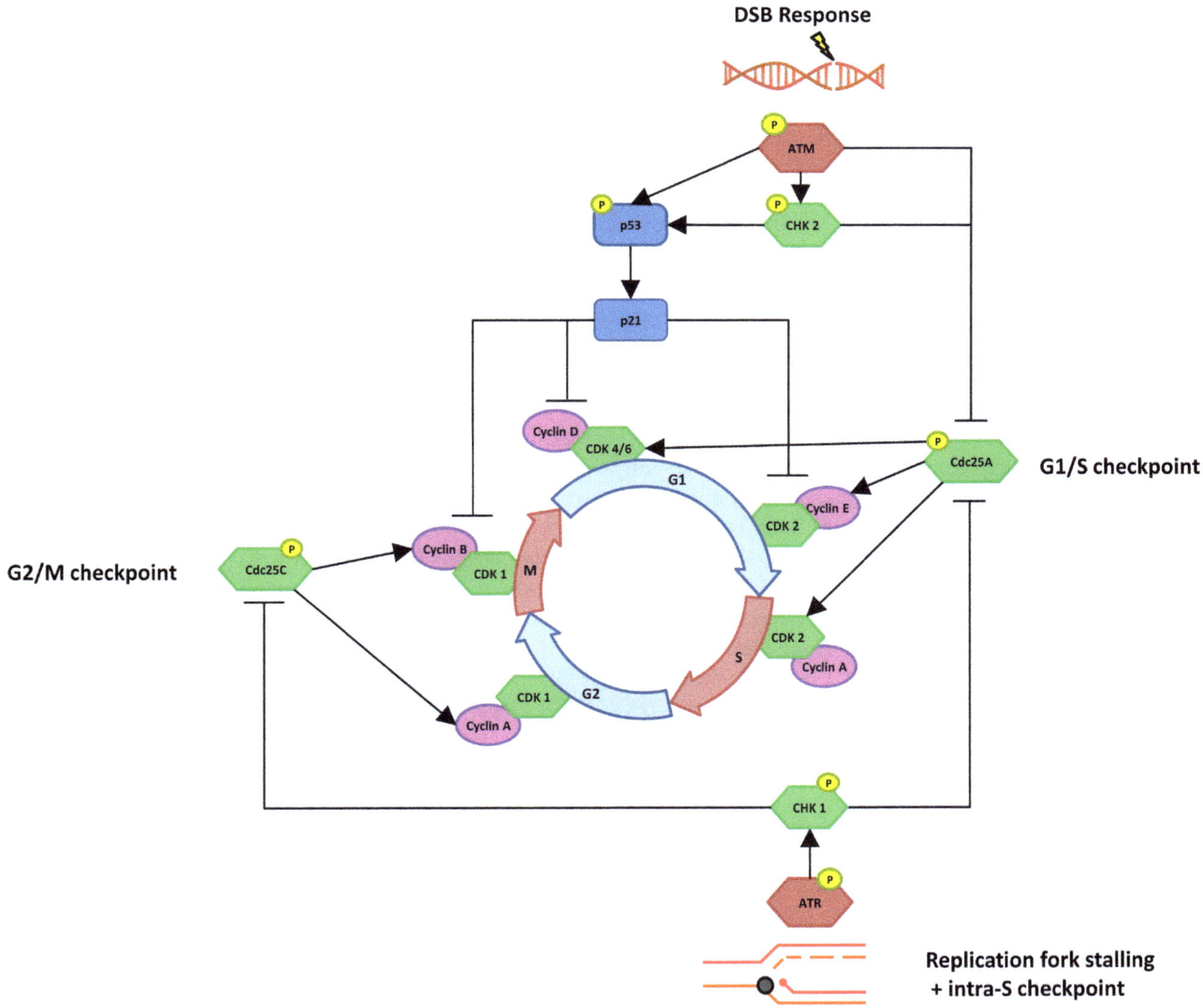

FIGURE 14.3 **The main molecular players in DNA-damage cell-cycle checkpoints.** In response to DNA damage, mammalian cells can activate three major DNA-damage checkpoints: G1/S, intra-S-phase, and G2/M checkpoints. DSBs trigger ATM-CHK2 kinase signaling which can block G1/S progression by promoting the p53–p21 axis and by inhibiting CDC25. The activated ATR-CHK1 kinase signaling can inhibit CDC25 as a part of the intra-S-phase checkpoint. ATM/ATR-CHK1/2 signaling can further halt cells at the G2/M checkpoint through different signaling routes as indicated. For a more complete summary of the functions of these key molecular players, please refer to the main text.

Here, we focus on summarizing the importance of ATM-CHK2 and ATR-CHK1 signaling in DNA-damage checkpoints. In this regard, one should note that the p38/MK2 pathway can complement the well-established ATM and ATR signaling nodes by converging on the regulation of CDC25 in the DNA-damage checkpoints [67,130,131]. Moreover, it is noteworthy that in response to DSBs, the efficiency of CDK1 activation as a part of DNA-damage checkpoints appears to be cell-cycle dependent since the DSB response is less efficient in G1 than in S/G2 [122].

4.1 The G1/S Cell-Cycle Checkpoint

The G1/S checkpoint is activated by very low numbers of DSBs, possibly one DSB being sufficient for the activation [132]. Two different processes primarily contribute to the activation of the G1/S checkpoint in response to DNA damage [75,126].

The first widely studied process involves the ATM–p53–p21 signaling axis [126,133]. Since this process requires the p53-dependent transcription of the CDKI p21, the entry into the S phase is only inhibited several hours after DNA damage has occurred [132]. The first G1/S checkpoint is mainly mediated by ATM, which results in the phosphorylation and activation of the p53 transcriptional activity. Consequently, the CKI p21 is upregulated in a p53-dependent manner which causes the p21-dependent inhibition of G1 and early S-phase CDK–cyclin complexes, finally resulting in the accumulation of hypo-phosphorylated pRB which can sequester E2F and consequently block S-phase initiation/progression [33,39].

In contrast, the second G1/S checkpoint does not fully prevent S-phase entry. Similar to the G2/M checkpoint (see later), it involves ATM/ATR activation of CHK1/2 which can result in CDK2 inhibition [132]. In this regard, the CDC25A phosphatase following phosphorylation by different kinases can be targeted for ubiquitin-mediated degradation by the $SCF^{\beta TrCP}$ E3 ligase in G1 (summarized in [8]).

In general, the limitation of the G1/S checkpoint is a slow and incomplete block of entry into the S phase since this cell-cycle arrest is mainly transcription dependent [132]. Consequently, some cells can escape from G1 to S phase with unrepaired DSBs, which can cause chromosome breakage in G2 [132].

4.2 The Intra-S-Phase Cell–Cycle Checkpoint

The intra-S-phase checkpoint is important to prevent the progression of DNA replication in the presence of DSBs [134]. In particular, the execution of CDK1-dependent events can be detrimental when the intra-S-phase checkpoint is not functional since the incompletely replicated chromosomes might be missegregated in mitosis. Therefore, similar to the G1/2 checkpoint, the intra-S-phase checkpoint can be divided into two distinct processes, involving, on the one hand, the arrest of ongoing replication fork progression, and on the other hand, the inhibition of late firing replication origins. More specifically, in the S phase, radiation-induced DNA damage slows down DNA synthesis by two ATM-dependent pathways: ATM-NBS1-SMC1 and ATM-CHK2-CDC25A signaling [33,135], with the latter pathway negatively regulating DNA replication by preventing loading of the replication factor CDC45 onto replication origins [136]. ATR-CHK1 signaling also prevents new replication origins from firing during the S phase [63]. Moreover, as part of the intra-S-phase checkpoint, the phosphorylation of CDC25A by CHK1 results in the rapid $SCF^{\beta TRCP}$-mediated proteasomal degradation of CDC25A which blocks the removal of inhibitory phosphorylations on CDKs [137–139].

4.3 The G2/M Cell–Cycle Checkpoint

The G2/M checkpoint has a defined threshold of sensitivity since it is estimated that at least 10 to 15 DSBs are required for the efficient checkpoint activation and maintenance [45,69]. Therefore, the G2/M checkpoint is considered inefficient to maintain genomic stability upon deregulation of the G1/S checkpoint [45].

In healthy cells, the activation of the G2/M checkpoint in response to DNA damage prevents the entry of cells into mitosis, thereby blocking the propagation of damaged DNA to daughter cells. The activation of the G2/M checkpoint involves the activation of CHK1/2 by ATM/ATR-mediated phosphorylation. The activated CHK1/2 phosphorylates CDC25 causing CDC25 inhibition by the cytoplasmic sequestration of CDC25. Consequently, CDK1 remains inactive by sustained inhibitory phosphorylation by Wee1, resulting in the rapid inhibition of entry into mitosis [125,137,140]. At first, ATM functions as the initiator of the G2/M checkpoint, but a delayed ATR response contributes to a sustained G2/M checkpoint response. Specifically, upon initiation of the G2/M checkpoint by DNA damage, the DNA end resection results in a switch from ATM dependency to ATR dependency for checkpoint control [48].

Another layer of control for the maintenance of the G2/M checkpoint is the p53 pathway [141] (see Chapter 15). Through transcriptional induction of p21, p53 can indirectly suppress CDK activity, which in turn enables the activation of the pRB tumor-suppressor pathway [33]. The activated pRB reduces the activity of E2F, thereby decreasing the pro-proliferative expression of E2F target genes, such as the APC/C inhibitor Emi1. This promotes the premature activation of APC/C in G2, resulting in the degradation of cyclins A and B and further supporting a G2/M cell–cycle arrest [142].

In summary, in response to DNA damage, mammalian cells can arrest at the G2/M checkpoint due to the inhibition and/or degradation of CDK1 activators, such as CDC25 and cyclin B combined with the activation of CDK1 inhibitors, such as Wee1. Note that, upon completion of DNA repair, PLK1 is activated, resulting in the phosphorylation-dependent degradation of Wee1 and the activation of CDC25 by nuclear accumulation.

4.4 DNA-Damage Checkpoints and Disease

DNA-damage checkpoints play important roles in diseases including cancer and aging [143]. They influence basic mechanisms of somatic and stem cell physiology, such as renewal, maintenance, and differentiation, but they also help prevent the development of cancer. Specifically, in mammals, at a young age, DNA-damage checkpoints can help to extend lifespan by promoting cancer resistance, while during aging, DNA-damage checkpoints may limit tissue integrity. On the one hand, DNA-damage checkpoints influence the self-renewal, maintenance, and quiescence of somatic mammalian stem cells, with a decreased responsiveness of ATM-p53-dependent checkpoints during aging [143]. On the other hand, the constant activation of DNA-damage checkpoints in the context of increased DNA damage can promote cellular transformation [143]. Likewise, defects in DNA-damage checkpoints can promote tumor progression by abrogating apoptotic and/or senescence programs, as is the case in patients suffering from ataxia telangectasia or Li–Fraumeni syndromes [19,144]. However, although the DDR and DNA damage–associated oncogene-induced senescence are activated in precancerous lesions, their activation in advanced cancers is rarely observed, indicating that malignant cancer cells find ways to bypass DNA-damage checkpoints and senescence [145].

In summary, DNA-damage checkpoints are certainly beneficial at a young age. However, their aging-associated decline associated with the accumulation of DNA damage in tissues has the potential to turn protective responses into damaging responses, possibly resulting in tissue dysfunction and selection of malignant cancer cells.

5. CONCLUSION

Taken together, mammalian cells rely on diverse signal transduction mechanisms to safeguard their genomic integrity. On the one hand, the accurate copying of genetic information in the S phase must be coupled with the precise and equal distribution of chromosomes between daughter cells in mitosis. On the other hand, mechanisms must be in place to detect and repair a broad range of DNA lesions that occur on a regular basis. Therefore, mammalian cells utilize DNA lesion–specific DNA damage–repair pathways to remove unwanted alterations of genetic information. Significantly, the repair of DNA lesions is synchronized with cell-cycle progression by the DDR. In response to DNA damage, a mammalian cell can respond with a transient cell-cycle arrest to allow the repair of DNA damage, or commit to a permanent cell-cycle arrest in the form of senescence, or initiate apoptosis in case the DNA damage is beyond repair. Generally, the DDR protects mammalian cells against the accumulation of DNA lesions which if not removed can cause human diseases including cancer, premature aging, and others. Thus, future research into bettering our understanding of DNA-damage checkpoints and DNA-repair mechanisms in health and disease is very likely to significantly expand our diagnosis, prediction, and treatment options in diverse human diseases.

GLOSSARY

Apoptosis The process of programmed cell death.

Ataxia telangiectasia A rare inherited disorder affecting the nervous system, immune system, and other body systems. Ataxia refers to a poor coordination and telangiectasia to small dilated blood vessels, two hallmarks of the disease.

Cell-cycle checkpoint Specific control mechanisms in eukaryotic cells ensuring a proper cell-cycle progression.

Cellular senescence An irreversible G1 cell–cycle arrest in which cells are refractory to growth factor stimulation.

Cytokinesis The separation of daughter cells by cytoplasmic division at the end of mitosis.

Cytotoxic Any process or agent that kills cells.

DNA damage–induced cell-cycle checkpoints (aka DNA-damage checkpoints) Cell-cycle checkpoints that are specifically activated upon the detection of DNA lesions.

DNA-damage response A complex network of cellular pathways responsible for the detection, signaling and repair of DNA lesions.

E3 ubiquitin ligase An enzyme that catalyzes the transfer of ubiquitin from the E2 ubiquitin–conjugating enzyme to specific protein substrates.

Endogenous DNA lesion A type of DNA damage that is a consequence of endogenous cellular processes.

Exogenous DNA lesion A type of DNA damage that is caused by exogenous genotoxic agents.

Genomic instability (aka genetic or genome instability) Defined as a high frequency of mutations within the genome where mutations can include changes in nucleic acid sequences, chromosomal rearrangements, and/or aneuploidy.

Genotoxic Damaging effects on a cell's genetic material.

Heterochromatin The tightly packed forms of DNA that play roles in the regulation of gene expression.

Li–Fraumeni syndrome A rare cancer predisposition hereditary syndrome associated with p53 mutations.

Malignant transformation The process by which cells acquire the properties of cancer.

Permanent cell-cycle arrest An irreversible exit from cell-cycle progression.

Quiescence The state of a cell when it is not dividing as a consequence of a reversible cell-cycle exit.

Transient cell-cycle arrest A fully reversible exit from cell-cycle progression.

Ubiquitin-mediated degradation E3 ubiquitin ligase–mediated marking of substrates by polyubiquitin chains, which targets substrates for destruction by the proteasome.

LIST OF ACRONYMS AND ABBREVIATIONS

ATM Ataxia telangiectasia mutated
ATR ATM and Rad3 related
BER Base excision repair
CDK Cyclin-dependent kinase
CHK1 Checkpoint kinase 1
CHK2 Checkpoint kinase 2
CKI CDK inhibitor
DDR DNA-damage response
DNA-PK DNA-dependent protein kinase
DSB DNA double-strand break
γH2AX H2AX phosphorylated on Ser139
HRR Homologous recombination repair
IR Ionizing radiation
MK2 MAPK-activated protein kinase 2
MMEJ Microhomology-mediated end joining
MMR Mismatch repair
MRN MRE11/RAD50/NBS1 complex
NER Nucleotide excision repair
NHEJ Nonhomologous end joining
PIKK Phosphatidylinositol 3-kinase–related kinase
PLK1 Polo-like kinase 1
pRB Retinoblastoma protein
SCF SKP1, Cullin1, F-box protein–containing E3 ligase complex
SSA Single-strand annealing
SSB DNA single-strand break
ssDNA Single-stranded DNA
UV Ultraviolet

ACKNOWLEDGMENTS

We apologize to all authors whose work we could not cite due to space limitations. We are very grateful to Joanna Lisztwan, Ahmad Sharif, and Nirmal Perera for their critical review of the manuscript, and also thank all members of the Hergovich laboratory for helpful discussions. The work of the Hergovich laboratory is supported by a Wellcome Trust grant (090090/Z/09/Z), a Cancer Research UK Centre Development Fund, and the National Institute for Health Research University College London Hospitals Biomedical Research Centre.

Authors' Contributions

Valenti Gomez and Alexander Hergovich researched the literature and wrote the manuscript together. Valenti Gomez created all figures. All authors read and approved the final manuscript.

REFERENCES

[1] Hochegger H, Takeda S, Hunt T. Cyclin-dependent kinases and cell-cycle transitions: does one fit all? Nat Rev Mol Cell Biol 2008;9(11):910–6.
[2] Malumbres M, Harlow E, Hunt T, Hunter T, Lahti JM, Manning G, et al. Cyclin-dependent kinases: a family portrait. Nat Cell Biol 2009;11(11):1275–6.
[3] Weinert T, Hartwell L. Control of G2 delay by the rad9 gene of *Saccharomyces cerevisiae*. J Cell Sci Suppl 1989;12:145–8.
[4] Ciccia A, Elledge SJ. The DNA damage response: making it safe to play with knives. Mol Cell 2010;40(2):179–204.
[5] Jackson SP, Bartek J. The DNA-damage response in human biology and disease. Nature 2009;461(7267):1071–8.
[6] Panier S, Durocher D. Regulatory ubiquitylation in response to DNA double-strand breaks. DNA Repair (Amst) 2009;8(4):436–43.
[7] Thompson LH. Recognition, signaling, and repair of DNA double-strand breaks produced by ionizing radiation in mammalian cells: the molecular choreography. Mutat Res 2012;751(2):158–246.
[8] Reinhardt HC, Yaffe MB. Phospho-Ser/Thr-binding domains: navigating the cell cycle and DNA damage response. Nat Rev Mol Cell Biol 2013;14(9):563–80.
[9] Bergink S, Jentsch S. Principles of ubiquitin and SUMO modifications in DNA repair. Nature 2009;458(7237):461–7.
[10] Jackson SP, Durocher D. Regulation of DNA damage responses by ubiquitin and SUMO. Mol Cell 2013;49(5):795–807.
[11] Panier S, Durocher D. Push back to respond better: regulatory inhibition of the DNA double-strand break response. Nat Rev Mol Cell Biol 2013;14(10):661–72.
[12] Bartek J, Bartkova J, Lukas J. DNA damage signalling guards against activated oncogenes and tumour progression. Oncogene 2007;26(56):7773–9.
[13] Bartkova J, Horejsi Z, Koed K, Kramer A, Tort F, Zieger K, et al. DNA damage response as a candidate anti-cancer barrier in early human tumorigenesis. Nature 2005;434(7035):864–70.

[14] Fackenthal JD, Olopade OI. Breast cancer risk associated with BRCA1 and BRCA2 in diverse populations. Nat Rev Cancer 2007;7(12):937–48.
[15] Frebourg T, Friend SH. Cancer risks from germline p53 mutations. J Clin Invest 1992;90(5):1637–41.
[16] Lavin MF, Shiloh Y. The genetic defect in ataxia-telangiectasia. Annu Rev Immunol 1997;15:177–202.
[17] Nevanlinna H, Bartek J. The CHEK2 gene and inherited breast cancer susceptibility. Oncogene 2006;25(43):5912–9.
[18] O'Driscoll M, Ruiz-Perez VL, Woods CG, Jeggo PA, Goodship JA. A splicing mutation affecting expression of ataxia-telangiectasia and Rad3-related protein (ATR) results in Seckel syndrome. Nat Genet 2003;33(4):497–501.
[19] Shiloh Y. ATM and related protein kinases: safeguarding genome integrity. Nat Rev Cancer 2003;3(3):155–68.
[20] Taylor AM, Groom A, Byrd PJ. Ataxia-telangiectasia-like disorder (ATLD)-its clinical presentation and molecular basis. DNA Repair (Amst) 2004;3(8–9):1219–25.
[21] Stewart GS, Maser RS, Stankovic T, Bressan DA, Kaplan MI, Jaspers NG, et al. The DNA double-strand break repair gene hMRE11 is mutated in individuals with an ataxia-telangiectasia-like disorder. Cell 1999;99(6):577–87.
[22] Waltes R, Kalb R, Gatei M, Kijas AW, Stumm M, Sobeck A, et al. Human RAD50 deficiency in a Nijmegen breakage syndrome-like disorder. Am J Hum Genet 2009;84(5):605–16.
[23] Lim DS, Kim ST, Xu B, Maser RS, Lin J, Petrini JH, et al. ATM phosphorylates p95/nbs1 in an S-phase checkpoint pathway. Nature 2000;404(6778):613–7.
[24] Rupnik A, Lowndes NF, Grenon M. MRN and the race to the break. Chromosoma 2010;119(2):115–35.
[25] Stracker TH, Petrini JH. The MRE11 complex: starting from the ends. Nat Rev Mol Cell Biol 2011;12(2):90–103.
[26] Williams GJ, Lees-Miller SP, Tainer JA. Mre11-Rad50-Nbs1 conformations and the control of sensing, signaling, and effector responses at DNA double-strand breaks. DNA Repair (Amst) 2010;9(12):1299–306.
[27] Hanahan D, Weinberg RA. Hallmarks of cancer: the next generation. Cell 2011;144(5):646–74.
[28] Negrini S, Gorgoulis VG, Halazonetis TD. Genomic instability – an evolving hallmark of cancer. Nat Rev Mol Cell Biol 2010;11(3):220–8.
[29] Stratton MR, Campbell PJ, Futreal PA. The cancer genome. Nature 2009;458(7239):719–24.
[30] Bouwman P, Jonkers J. The effects of deregulated DNA damage signalling on cancer chemotherapy response and resistance. Nat Rev Cancer 2012;12(9):587–98.
[31] Curtin NJ. DNA repair dysregulation from cancer driver to therapeutic target. Nat Rev Cancer 2012;12(12):801–17.
[32] Holohan C, Van Schaeybroeck S, Longley DB, Johnston PG. Cancer drug resistance: an evolving paradigm. Nat Rev Cancer 2013;13(10):714–26.
[33] Kastan MB, Bartek J. Cell-cycle checkpoints and cancer. Nature 2004;432(7015):316–23.
[34] Reinhardt HC, Yaffe MB. Kinases that control the cell cycle in response to DNA damage: Chk1, Chk2, and MK2. Curr Opin Cell Biol 2009;21(2):245–55.
[35] Sherr CJ, Roberts JM. CDK inhibitors: positive and negative regulators of G1-phase progression. Genes Dev 1999;13(12):1501–12.
[36] Harris SL, Levine AJ. The p53 pathway: positive and negative feedback loops. Oncogene 2005;24(17):2899–908.
[37] Delavaine L, La Thangue NB. Control of E2F activity by p21Waf1/Cip1. Oncogene 1999;18(39):5381–92.
[38] Ewen ME, Sluss HK, Sherr CJ, Matsushime H, Kato J, Livingston DM. Functional interactions of the retinoblastoma protein with mammalian D-type cyclins. Cell 1993;73(3):487–97.
[39] Ren B, Cam H, Takahashi Y, Volkert T, Terragni J, Young RA, et al. E2F integrates cell cycle progression with DNA repair, replication, and G(2)/M checkpoints. Genes Dev 2002;16(2):245–56.
[40] Shiyanov P, Bagchi S, Adami G, Kokontis J, Hay N, Arroyo M, et al. p21 Disrupts the interaction between cdk2 and the E2F-p130 complex. Mol Cell Biol 1996;16(3):737–44.
[41] Sherr CJ. Cancer cell cycles. Science 1996;274(5293):1672–7.
[42] Harbour JW, Luo RX, Dei Santi A, Postigo AA, Dean DC. Cdk phosphorylation triggers sequential intramolecular interactions that progressively block Rb functions as cells move through G1. Cell 1999;98(6):859–69.
[43] Ma HT, Poon RY. How protein kinases co-ordinate mitosis in animal cells. Biochem J 2011;435(1):17–31.
[44] Okamoto K, Sagata N. Mechanism for inactivation of the mitotic inhibitory kinase Wee1 at M phase. Proc Natl Acad Sci USA 2007;104(10):3753–8.
[45] Deckbar D, Jeggo PA, Lobrich M. Understanding the limitations of radiation-induced cell cycle checkpoints. Crit Rev Biochem Mol Biol 2011;46(4):271–83.
[46] Kousholt AN, Menzel T, Sorensen CS. Pathways for genome integrity in G2 phase of the cell cycle. Biomolecules 2012;2(4):579–607.
[47] Stark GR, Taylor WR. Control of the G2/M transition. Mol Biotechnol 2006;32(3):227–48.
[48] Lindqvist A, Rodriguez-Bravo V, Medema RH. The decision to enter mitosis: feedback and redundancy in the mitotic entry network. J Cell Biol 2009;185(2):193–202.
[49] Bornstein G, Bloom J, Sitry-Shevah D, Nakayama K, Pagano M, Hershko A. Role of the SCFSkp2 ubiquitin ligase in the degradation of p21Cip1 in S phase. J Biol Chem 2003;278(28):25752–7.
[50] Yu ZK, Gervais JL, Zhang H. Human CUL-1 associates with the SKP1/SKP2 complex and regulates p21(CIP1/WAF1) and cyclin D proteins. Proc Natl Acad Sci USA 1998;95(19):11324–9.
[51] Sivakumar S, Gorbsky GJ. Spatiotemporal regulation of the anaphase-promoting complex in mitosis. Nat Rev Mol Cell Biol 2015;16(2):82–94.
[52] Wang H, Zhang X, Teng L, Legerski RJ. DNA damage checkpoint recovery and cancer development. Exp Cell Res 2015;334(2):350–8.
[53] Lu X, Nguyen TA, Moon SH, Darlington Y, Sommer M, Donehower LA. The type 2C phosphatase Wip1: an oncogenic regulator of tumor suppressor and DNA damage response pathways. Cancer Metastasis Rev 2008;27(2):123–35.
[54] Lindahl T, Nyberg B. Rate of depurination of native deoxyribonucleic acid. Biochemistry 1972;11(19):3610–8.

[55] Loeb LA, Harris CC. Advances in chemical carcinogenesis: a historical review and prospective. Cancer Res 2008;68(17):6863–72.
[56] Lindahl T, Barnes DE. Repair of endogenous DNA damage. Cold Spring Harb Symp Quant Biol 2000;65:127–33.
[57] Khanna KK, Jackson SP. DNA double-strand breaks: signaling, repair and the cancer connection. Nat Genet 2001;27(3):247–54.
[58] Helleday T, Petermann E, Lundin C, Hodgson B, Sharma RA. DNA repair pathways as targets for cancer therapy. Nat Rev Cancer 2008;8(3):193–204.
[59] Bakkenist CJ, Kastan MB. Initiating cellular stress responses. Cell 2004;118(1):9–17.
[60] Lieber MR, Ma Y, Pannicke U, Schwarz K. Mechanism and regulation of human non-homologous DNA end-joining. Nat Rev Mol Cell Biol 2003;4(9):712–20.
[61] Shiloh Y, Ziv Y. The ATM protein kinase: regulating the cellular response to genotoxic stress, and more. Nat Rev Mol Cell Biol 2013;14(4):197–210.
[62] Polo SE, Jackson SP. Dynamics of DNA damage response proteins at DNA breaks: a focus on protein modifications. Genes Dev 2011;25(5):409–33.
[63] Cimprich KA, Cortez D. ATR: an essential regulator of genome integrity. Nat Rev Mol Cell Biol 2008;9(8):616–27.
[64] Jazayeri A, Falck J, Lukas C, Bartek J, Smith GC, Lukas J, et al. ATM- and cell cycle-dependent regulation of ATR in response to DNA double-strand breaks. Nat Cell Biol 2006;8(1):37–45.
[65] Sartori AA, Lukas C, Coates J, Mistrik M, Fu S, Bartek J, et al. Human CtIP promotes DNA end resection. Nature 2007;450(7169):509–14.
[66] Manke IA, Nguyen A, Lim D, Stewart MQ, Elia AE, Yaffe MB. MAPKAP kinase-2 is a cell cycle checkpoint kinase that regulates the G2/M transition and S phase progression in response to UV irradiation. Mol Cell 2005;17(1):37–48.
[67] Reinhardt HC, Aslanian AS, Lees JA, Yaffe MB. p53-deficient cells rely on ATM- and ATR-mediated checkpoint signaling through the p38MAPK/MK2 pathway for survival after DNA damage. Cancer Cell 2007;11(2):175–89.
[68] Bakkenist CJ, Kastan MB. DNA damage activates ATM through intermolecular autophosphorylation and dimer dissociation. Nature 2003;421(6922):499–506.
[69] Goodarzi AA, Jeggo PA. The repair and signaling responses to DNA double-strand breaks. Adv Genet 2013;82:1–45.
[70] Cowell IG, Sunter NJ, Singh PB, Austin CA, Durkacz BW, Tilby MJ. gammaH2AX foci form preferentially in euchromatin after ionising-radiation. PLoS One 2007;2(10):e1057.
[71] Kruse JP, Gu W. Modes of p53 regulation. Cell 2009;137(4):609–22.
[72] Vogelstein B, Lane D, Levine AJ. Surfing the p53 network. Nature 2000;408(6810):307–10.
[73] Fridman JS, Lowe SW. Control of apoptosis by p53. Oncogene 2003;22(56):9030–40.
[74] Vousden KH, Lu X. Live or let die: the cell's response to p53. Nat Rev Cancer 2002;2(8):594–604.
[75] Bartek J, Falck J, Lukas J. CHK2 kinase – a busy messenger. Nat Rev Mol Cell Biol 2001;2(12):877–86.
[76] White DE, Negorev D, Peng H, Ivanov AV, Maul GG, Rauscher 3rd FJ. KAP1, a novel substrate for PIKK family members, colocalizes with numerous damage response factors at DNA lesions. Cancer Res 2006;66(24):11594–9.
[77] Goodarzi AA, Jeggo P, Lobrich M. The influence of heterochromatin on DNA double strand break repair: getting the strong, silent type to relax. DNA Repair (Amst) 2010;9(12):1273–82.
[78] Goodarzi AA, Noon AT, Deckbar D, Ziv Y, Shiloh Y, Lobrich M, et al. ATM signaling facilitates repair of DNA double-strand breaks associated with heterochromatin. Mol Cell 2008;31(2):167–77.
[79] Noon AT, Shibata A, Rief N, Lobrich M, Stewart GS, Jeggo PA, et al. 53BP1-dependent robust localized KAP-1 phosphorylation is essential for heterochromatic DNA double-strand break repair. Nat Cell Biol 2010;12(2):177–84.
[80] White D, Rafalska-Metcalf IU, Ivanov AV, Corsinotti A, Peng H, Lee SC, et al. The ATM substrate KAP1 controls DNA repair in heterochromatin: regulation by HP1 proteins and serine 473/824 phosphorylation. Mol Cancer Res 2012;10(3):401–14.
[81] Kim ST, Xu B, Kastan MB. Involvement of the cohesin protein, Smc1, in Atm-dependent and independent responses to DNA damage. Genes Dev 2002;16(5):560–70.
[82] Kitagawa R, Bakkenist CJ, McKinnon PJ, Kastan MB. Phosphorylation of SMC1 is a critical downstream event in the ATM-NBS1-BRCA1 pathway. Genes Dev 2004;18(12):1423–38.
[83] Yazdi PT, Wang Y, Zhao S, Patel N, Lee EY, Qin J. SMC1 is a downstream effector in the ATM/NBS1 branch of the human S-phase checkpoint. Genes Dev 2002;16(5):571–82.
[84] Matsuoka S, Ballif BA, Smogorzewska A, McDonald 3rd ER, Hurov KE, Luo J, et al. ATM and ATR substrate analysis reveals extensive protein networks responsive to DNA damage. Science 2007;316(5828):1160–6.
[85] Theunissen JW, Kaplan MI, Hunt PA, Williams BR, Ferguson DO, Alt FW, et al. Checkpoint failure and chromosomal instability without lymphomagenesis in Mre11(ATLD1/ATLD1) mice. Mol Cell 2003;12(6):1511–23.
[86] Uziel T, Lerenthal Y, Moyal L, Andegeko Y, Mittelman L, Shiloh Y. Requirement of the MRN complex for ATM activation by DNA damage. EMBO J 2003;22(20):5612–21.
[87] Stracker TH, Morales M, Couto SS, Hussein H, Petrini JH. The carboxy terminus of NBS1 is required for induction of apoptosis by the MRE11 complex. Nature 2007;447(7141):218–21.
[88] Bunting SF, Nussenzweig A. End-joining, translocations and cancer. Nat Rev Cancer 2013;13(7):443–54.
[89] Poulogiannis G, Frayling IM, Arends MJ. DNA mismatch repair deficiency in sporadic colorectal cancer and Lynch syndrome. Histopathology 2010;56(2):167–79.
[90] Barnetson RA, Tenesa A, Farrington SM, Nicholl ID, Cetnarskyj R, Porteous ME, et al. Identification and survival of carriers of mutations in DNA mismatch-repair genes in colon cancer. N Engl J Med 2006;354(26):2751–63.

[91] Jacob S, Praz F. DNA mismatch repair defects: role in colorectal carcinogenesis. Biochimie 2002;84(1):27–47.

[92] Worthley DL, Walsh MD, Barker M, Ruszkiewicz A, Bennett G, Phillips K, et al. Familial mutations in PMS2 can cause autosomal dominant hereditary nonpolyposis colorectal cancer. Gastroenterology 2005;128(5):1431–6.

[93] D'Amours D, Jackson SP. The Mre11 complex: at the crossroads of DNA repair and checkpoint signalling. Nat Rev Mol Cell Biol 2002;3(5):317–27.

[94] Difilippantonio MJ, Petersen S, Chen HT, Johnson R, Jasin M, Kanaar R, et al. Evidence for replicative repair of DNA double-strand breaks leading to oncogenic translocation and gene amplification. J Exp Med 2002;196(4):469–80.

[95] Hoeijmakers JH. Genome maintenance mechanisms for preventing cancer. Nature 2001;411(6835):366–74.

[96] Lamb NE, Sherman SL, Hassold TJ. Effect of meiotic recombination on the production of aneuploid gametes in humans. Cytogenet Genome Res 2005;111(3–4):250–5.

[97] O'Driscoll M, Gennery AR, Seidel J, Concannon P, Jeggo PA. An overview of three new disorders associated with genetic instability: LIG4 syndrome, RS-SCID and ATR-Seckel syndrome. DNA Repair (Amst) 2004;3(8–9):1227–35.

[98] Sharpless NE, Ferguson DO, O'Hagan RC, Castrillon DH, Lee C, Farazi PA, et al. Impaired nonhomologous end-joining provokes soft tissue sarcomas harboring chromosomal translocations, amplifications, and deletions. Mol Cell 2001;8(6):1187–96.

[99] Iyama T, Wilson 3rd DM. DNA repair mechanisms in dividing and non-dividing cells. DNA Repair (Amst) 2013;12(8):620–36.

[100] Jiricny J. The multifaceted mismatch-repair system. Nat Rev Mol Cell Biol 2006;7(5):335–46.

[101] Symington LS, Gautier J. Double-strand break end resection and repair pathway choice. Annu Rev Genet 2011;45:247–71.

[102] Chapman JR, Taylor MR, Boulton SJ. Playing the end game: DNA double-strand break repair pathway choice. Mol Cell 2012;47(4):497–510.

[103] Callen E, Di Virgilio M, Kruhlak MJ, Nieto-Soler M, Wong N, Chen HT, et al. 53BP1 mediates productive and mutagenic DNA repair through distinct phosphoprotein interactions. Cell 2013;153(6):1266–80.

[104] Chapman JR, Barral P, Vannier JB, Borel V, Steger M, Tomas-Loba A, et al. RIF1 is essential for 53BP1-dependent nonhomologous end joining and suppression of DNA double-strand break resection. Mol Cell 2013;49(5):858–71.

[105] Daley JM, Sung P. RIF1 in DNA break repair pathway choice. Mol Cell 2013;49(5):840–1.

[106] Di Virgilio M, Callen E, Yamane A, Zhang W, Jankovic M, Gitlin AD, et al. Rif1 prevents resection of DNA breaks and promotes immunoglobulin class switching. Science 2013;339(6120):711–5.

[107] Escribano-Diaz C, Orthwein A, Fradet-Turcotte A, Xing M, Young JT, Tkac J, et al. A cell cycle-dependent regulatory circuit composed of 53BP1-RIF1 and BRCA1-CtIP controls DNA repair pathway choice. Mol Cell 2013;49(5):872–83.

[108] Feng L, Fong KW, Wang J, Wang W, Chen J. RIF1 counteracts BRCA1-mediated end resection during DNA repair. J Biol Chem 2013;288(16): 11135–43.

[109] Zimmermann M, Lottersberger F, Buonomo SB, Sfeir A, de Lange T. 53BP1 regulates DSB repair using Rif1 to control 5′ end resection. Science 2013;339(6120):700–4.

[110] Iliakis GE, Metzger L, Denko N, Stamato TD. Detection of DNA double-strand breaks in synchronous cultures of CHO cells by means of asymmetric field inversion gel electrophoresis. Int J Radiat Biol 1991;59(2):321–41.

[111] Riballo E, Kuhne M, Rief N, Doherty A, Smith GC, Recio MJ, et al. A pathway of double-strand break rejoining dependent upon ATM, Artemis, and proteins locating to gamma-H2AX foci. Mol Cell 2004;16(5):715–24.

[112] Beucher A, Birraux J, Tchouandong L, Barton O, Shibata A, Conrad S, et al. ATM and Artemis promote homologous recombination of radiation-induced DNA double-strand breaks in G2. EMBO J 2009;28(21):3413–27.

[113] Shibata A, Conrad S, Birraux J, Geuting V, Barton O, Ismail A, et al. Factors determining DNA double-strand break repair pathway choice in G2 phase. EMBO J 2011;30(6):1079–92.

[114] Lemaitre C, Soutoglou E. Double strand break (DSB) repair in heterochromatin and heterochromatin proteins in DSB repair. DNA Repair (Amst) 2014;19:163–8.

[115] DeFazio LG, Stansel RM, Griffith JD, Chu G. Synapsis of DNA ends by DNA-dependent protein kinase. EMBO J 2002;21(12):3192–200.

[116] Lees-Miller SP, Meek K. Repair of DNA double strand breaks by non-homologous end joining. Biochimie 2003;85(11):1161–73.

[117] Spagnolo L, Rivera-Calzada A, Pearl LH, Llorca O. Three-dimensional structure of the human DNA-PKcs/Ku70/Ku80 complex assembled on DNA and its implications for DNA DSB repair. Mol Cell 2006;22(4):511–9.

[118] Fattah F, Lee EH, Weisensel N, Wang Y, Lichter N, Hendrickson EA. Ku regulates the non-homologous end joining pathway choice of DNA double-strand break repair in human somatic cells. PLoS Genet 2010;6(2):e1000855.

[119] Shibata A, Moiani D, Arvai AS, Perry J, Harding SM, Genois MM, et al. DNA double-strand break repair pathway choice is directed by distinct MRE11 nuclease activities. Mol Cell 2014;53(1):7–18.

[120] Huertas P, Jackson SP. Human CtIP mediates cell cycle control of DNA end resection and double strand break repair. J Biol Chem 2009;284(14):9558–65.

[121] Esashi F, Christ N, Gannon J, Liu Y, Hunt T, Jasin M, et al. CDK-dependent phosphorylation of BRCA2 as a regulatory mechanism for recombinational repair. Nature 2005;434(7033):598–604.

[122] Li S, Ting NS, Zheng L, Chen PL, Ziv Y, Shiloh Y, et al. Functional link of BRCA1 and ataxia telangiectasia gene product in DNA damage response. Nature 2000;406(6792):210–5.

[123] Yun MH, Hiom K. CtIP-BRCA1 modulates the choice of DNA double-strand-break repair pathway throughout the cell cycle. Nature 2009;459(7245):460–3.

[124] Hahnel PS, Enders B, Sasca D, Roos WP, Kaina B, Bullinger L, et al. Targeting components of the alternative NHEJ pathway sensitizes KRAS mutant leukemic cells to chemotherapy. Blood 2014;123(15):2355–66.

[125] Bartek J, Lukas J. DNA damage checkpoints: from initiation to recovery or adaptation. Curr Opin Cell Biol 2007;19(2):238–45.

[126] Wahl GM, Linke SP, Paulson TG, Huang LC. Maintaining genetic stability through TP53 mediated checkpoint control. Cancer Surv 1997;29:183–219.

[127] Kerzendorfer C, O'Driscoll M. Human DNA damage response and repair deficiency syndromes: linking genomic instability and cell cycle checkpoint proficiency. DNA Repair (Amst) 2009;8(9):1139–52.

[128] Beishline K, Azizkhan-Clifford J. Interplay between the cell cycle and double-strand break response in mammalian cells. Methods Mol Biol 2014;1170:41–59.

[129] Riley T, Sontag E, Chen P, Levine A. Transcriptional control of human p53-regulated genes. Nat Rev Mol Cell Biol 2008;9(5):402–12.

[130] Morandell S, Reinhardt HC, Cannell IG, Kim JS, Ruf DM, Mitra T, et al. A reversible gene-targeting strategy identifies synthetic lethal interactions between MK2 and p53 in the DNA damage response in vivo. Cell Rep 2013;5(4):868–77.

[131] Reinhardt HC, Jiang H, Hemann MT, Yaffe MB. Exploiting synthetic lethal interactions for targeted cancer therapy. Cell Cycle 2009;8(19):3112–9.

[132] Deckbar D, Stiff T, Koch B, Reis C, Lobrich M, Jeggo PA. The limitations of the G1-S checkpoint. Cancer Res 2010;70(11):4412–21.

[133] Kastan MB, Zhan Q, el-Deiry WS, Carrier F, Jacks T, Walsh WV, et al. A mammalian cell cycle checkpoint pathway utilizing p53 and GADD45 is defective in ataxia-telangiectasia. Cell 1992;71(4):587–97.

[134] Grallert B, Boye E. The multiple facets of the intra-S checkpoint. Cell Cycle 2008;7(15):2315–20.

[135] Falck J, Mailand N, Syljuasen RG, Bartek J, Lukas J. The ATM-Chk2-Cdc25A checkpoint pathway guards against radioresistant DNA synthesis. Nature 2001;410(6830):842–7.

[136] Falck J, Petrini JH, Williams BR, Lukas J, Bartek J. The DNA damage-dependent intra-S phase checkpoint is regulated by parallel pathways. Nat Genet 2002;30(3):290–4.

[137] Donzelli M, Draetta GF. Regulating mammalian checkpoints through Cdc25 inactivation. EMBO Rep 2003;4(7):671–7.

[138] Jin J, Shirogane T, Xu L, Nalepa G, Qin J, Elledge SJ, et al. SCFbeta-TRCP links Chk1 signaling to degradation of the Cdc25A protein phosphatase. Genes Dev 2003;17(24):3062–74.

[139] Sorensen CS, Syljuasen RG, Falck J, Schroeder T, Ronnstrand L, Khanna KK, et al. Chk1 regulates the S phase checkpoint by coupling the physiological turnover and ionizing radiation-induced accelerated proteolysis of Cdc25A. Cancer Cell 2003;3(3):247–58.

[140] Bartek J, Lukas J. Chk1 and Chk2 kinases in checkpoint control and cancer. Cancer Cell 2003;3(5):421–9.

[141] Taylor WR, Stark GR. Regulation of the G2/M transition by p53. Oncogene 2001;20(15):1803–15.

[142] Lee J, Kim JA, Barbier V, Fotedar A, Fotedar R. DNA damage triggers p21WAF1-dependent Emi1 down-regulation that maintains G2 arrest. Mol Biol Cell 2009;20(7):1891–902.

[143] Sperka T, Wang J, Rudolph KL. DNA damage checkpoints in stem cells, ageing and cancer. Nat Rev Mol Cell Biol 2012;13(9):579–90.

[144] Malkin D, Li FP, Strong LC, Fraumeni Jr JF, Nelson CE, Kim DH, et al. Germ line p53 mutations in a familial syndrome of breast cancer, sarcomas, and other neoplasms. Science 1990;250(4985):1233–8.

[145] Larsson LG. Oncogene- and tumor suppressor gene-mediated suppression of cellular senescence. Semin Cancer Biol 2011;21(6):367–76.

Chapter 15

The Role of p53/p21/p16 in DNA-Damage Signaling and DNA Repair

Y. Kulaberoglu, R. Gundogdu, A. Hergovich
University College London, London, United Kingdom

Chapter Outline

1. INTRODUCTION

DNA damage as a mutagenic event threatens the integrity of genetic information. Thus, mammalian cells utilize a complex signaling network to detect, signal, and repair DNA damage with the aim to restore genomic stability. In case, the DNA damage is beyond repair, the damaged cell is eliminated from the proliferating cell pool by cell death or senescence. If DNA repair and cell death/senescence fail, mutations can accumulate in the genome, which can result in the deregulation of genes regulating cell growth, proliferation, and/or death, consequently increasing the risk for the development of cancer and other diseases [1,2]. Therefore, DNA mutations and genomic instability are the established hallmarks of cancer cells [3]. Furthermore, many commonly lost tumor-suppressor genes, including p53, $p16^{INK4A}$, RB, and BRCA1/2, play a role in the DDR and DNA-repair pathways which are among the most frequently compromised pathways in human cancers [4–6].

Given the biological significance of genomic integrity [2], the activation of cell-cycle checkpoints, apoptotic programs, and transcriptional changes are crucial end points of DDR signaling [7–9]. On the one hand, these cellular response mechanisms to DNA damage are crucial for tumor suppression by invoking cell-cycle arrest, senescence, and/or apoptosis [2,4,10,11]. On the other hand, the unrepaired DNA damage contributes to limitations in stem cell functionality and tissue homeostasis during aging [12] when the likelihood of cancer development and tissue dysfunction significantly increase with age because DNA damage accumulates gradually during cellular aging [13,14]. Different types of DNA damage can cause an acute or chronic DNA damage, which depending on the cellular context can result in reversible transient or irreversible permanent cell-cycle arrests. As an example of acute DNA damage, dysfunctional telomeres can cause a transient G1 cell–cycle arrest through the activation of p53 and its transcriptional target p21 [12], while continuous telomere dysfunction results in chronic DNA damage followed by the induction of p16 and pRB, thereby stabilizing senescence [12]. Thus, a detailed understanding of DDR involving p53, p21, and p16 also helps improve stem cell biology [12], in addition to their well-established actions in tumor suppression (see later).

Here, we summarize our current understanding of the key roles of p53, p21, and p16 in DNA-damage signaling and DNA-damage repair. In this regard, we are paying particular attention to the p53 tumor-suppressor protein and p21 as its effector protein, since p53 has been regarded as a prime example for the relationship between the DDR and tumor suppression [15–17].

Genome Stability. http://dx.doi.org/10.1016/B978-0-12-803309-8.00015-X

2. THE p53 TUMOR-SUPPRESSOR PROTEIN

Since the discovery and cloning of p53 in the late 1970s and early 1980s [17], the regulation and functions of p53 have been one of the most intensively studied areas of molecular cancer biology [16,18]. The p53 tumor suppressor is a central, versatile, and multifunctional player in the cellular DNA-damage response. Upon genotoxic stress, p53 is upregulated and induces transcriptional programs promoting transient cell–cycle arrest, permanent cell–cycle arrest in the form of senescence, DNA repair, and/or apoptosis [18–21]. The Tp53 gene is the most frequently mutated gene in cancer [18,22]. About 50% of all human cancers carry mutations in the p53 tumor-suppressor gene [23]. p53-deficient mice die with nearly 100% penetrance of cancer around 6 months of age [24–27], and patients suffering from Li–Fraumeni syndrome display an association with p53 mutations and cancer development [28,29]. Therefore, several researchers in the cancer research field have studied the consequences of the introduction of a normal p53 gene into tumor cells with mutant p53 which has the potential to restore their ability to undergo cell-cycle arrest, apoptosis, and/or the differentiation in response to DNA-damaging therapies [30]. The particular aim is to restore normal p53 function of the endogenous form with small molecules [30]. However, since p53 also acts in stem cell self-renewal and quiescence [12], these restoration approaches must be conducted with caution. Furthermore, p53 can exert pro- and anti-aging functions through the differential transcriptional regulation of apoptotic, senescence, and longevity target genes [31,32]. In this regard, a study of so-called "super-p53" mice showed that a moderately enhanced expression of p53 in combination with $p16^{INK4A}/p19^{ARF}$ can result in tumor-free life extension [33]. Further noteworthy, studies during 2010s challenged the view that the DNA damage–induced programs triggered by p53 are the sole and major mechanism by which p53 exerts its tumor-suppressor function [34–36].

2.1 p53 in the DNA-Damage Response

Already in the 1980s and 1990s, researchers observed that p53 was upregulated rapidly at the level of protein stabilization by genotoxic agents, including ultraviolet (UV) light, ionizing radiation (IR), and chemotherapeutics [37–40]. In response to DNA damage, p53 is extensively modified by phosphorylations and other posttranslational modifications [41]. Upstream regulators of p53 include the DDR kinases ATM, ATR, DNA-PK, CHK1, and CHK2. These modifications release p53 from the interaction with its negative regulator Mdm2 and promote the transcriptional activity of p53 [41], resulting in an increased transcription of p53 targets including the cyclin-dependent kinase (CDK) inhibitor (CKI) p21 [19,42]. Through the transcriptional induction of p21, p53 indirectly suppresses CDK activity, which in turn enables the activation of the pRB tumor-suppressor pathway [43]. The activated pRB reduces the activity of E2F, thereby decreasing the pro-proliferative expression of E2F target genes (Fig. 15.1).

Significantly, in response to DNA damage (and also other cellular stresses), the pure abundance of the p53 protein due to the inhibition of the Mdm2-mediated proteosomal degradation should not be considered as the sole determinant of p53 activity, although it normally can serve as a reliable readout of its activation [19]. As a transcription factor that can bind DNA in a sequence-specific manner [31], p53 is known to activate or repress the transcription of genes involved in cell-cycle arrest, DNA repair, apoptosis, metabolism, autophagy, and others [21,31], where depending on the cell system, p53 upregulation can trigger different biological effects [19,21]. Thus, monitoring a transcriptional readout is very suitable to complement measurements of p53 activity. Since in human cancers, most mutations in p53 affect the DNA-binding domain of p53 [44], it appears that the transcriptional function plays the main tumor-suppressive role of p53 in cancer cells, although p53 also displays nontranscriptional activities.

In case, the DNA damage is too extensive, p53 triggers cell death through intrinsic and extrinsic pathways [45,46]. In response to DNA damage, the upregulation of p53 can strongly induce apoptosis, but this function is cell type and DNA-damage dependent [32,39]. However, apoptosis can also take place without p53 [19]. Moreover, the extent of apoptosis in human lymphoma cells upon treatment with chemotherapeutics was dependent on the p53 status [47]. Collectively, these reports suggest that cell type, stress, and other signaling pathways determine whether p53 can induce apoptosis or not in response to DNA damage. Generally, p53 can promote apoptosis through three different routes: (a) transcriptional activation, (b) transcriptional repression, and (c) transcription-independent mechanisms. Considering the emphasis of this chapter, we refer the reader to an excellent summary of these three routes [19] for further information on the role of p53 in apoptosis induction. Note that various types of cellular stress, including DNA damage, can induce apoptosis by p53-dependent and p53-independent mechanisms, where apoptosis represents the primary response to DNA damage and telomere dysfunction in certain cell types [12,32].

Upon DNA damage, as an alternative to apoptosis, high p53 levels can induce a transient or permanent cell–cycle arrest [48–52]. In this context, p53 can facilitate either apoptosis or senescence in a cell type–dependent manner [49,53]. Although many factors have been shown to contribute to these different types of cell-cycle arrest (summarized in Ref. [19]),

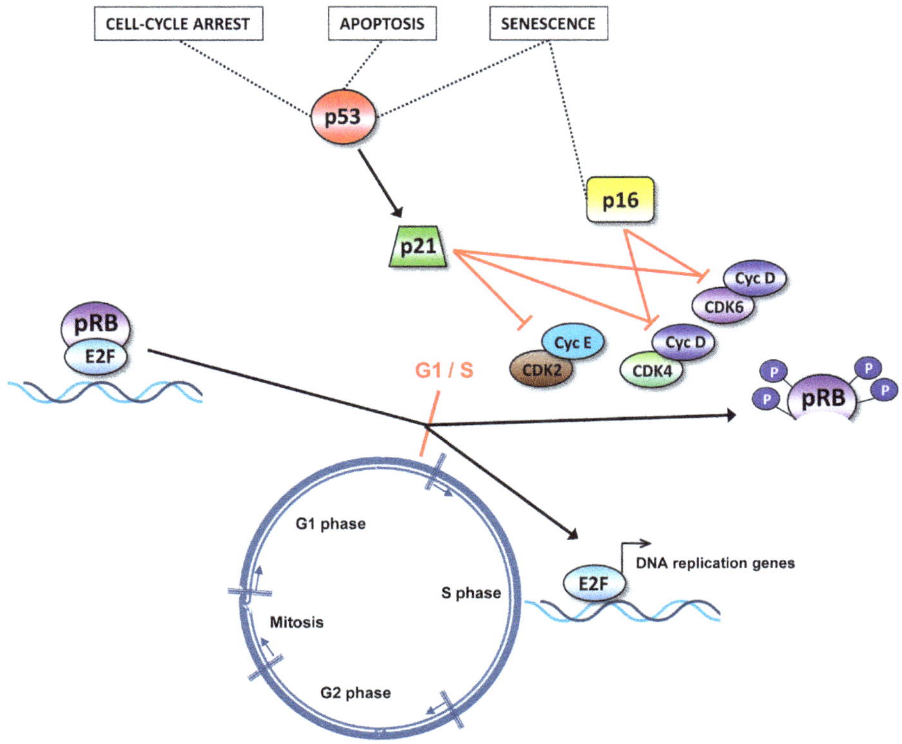

FIGURE 15.1 **p53/p21/p16 tumor suppressors in control of the G1/S cell–cycle checkpoint.** The tumor-suppressor proteins p53, p21, and p16 are major regulators of the G1/S cell–cycle checkpoint. Upon p53 stabilization, the transcription of p21 is induced, subsequently resulting in the inhibition of CDK4/6–cyclin D and CDK2–cyclin E complexes by p21. In response to senescence-inducing stimuli, p16 levels are elevated to inhibit the CDK4/6–cyclin D complexes. As a result of the inhibition of CDK–cyclin complexes, the hypo-phosphorylated form of pRB can accumulate, consequently inhibiting the transcription of S-phase promoting genes by E2F.

the parameters defining the choice between apoptosis, a transient cell–cycle arrest, and a permanent growth arrest (senescence) are not fully understood yet. In this regard, it is noteworthy that a sustained stabilization of p53 results in a permanent proliferation arrest through senescence, while pulses of stabilized p53 yield a transient, reversible cell–cycle arrest [54]. A misbalance between DNA damage and DNA repair normally results in higher p53 levels through pulses [55], causing a transient p53-dependent G1/S cell–cycle arrest [54]. This pulsatile behavior of p53 can be explained at least in part by ATM-induced activation of 53 followed by transactivation of Mdm2 and Wip1, two negative regulators of p53 [56]. This induction of Mdm2 and Wip1 counteracts p53 activity as a negative feedback loop. However, it still is important to fully understand these stabilization dynamics in the context of a variety of other distinct mechanisms that are known to stabilize p53 [41]. In this regard, the type of genotoxic stress also plays an important part since in contrast to radiation-induced DNA damage, UV-induced DNA damage triggers a sustained induction of p53 that does not seem to oscillate in waves [56].

Significantly, p53 can act in the G1/S and G2/M DNA-damage cell-cycle checkpoints (Figs. 15.1 and 15.2). These transient arrests prevent the amplification by replication and/or propagation by cell division of damaged DNA molecules. In particular, for the G1/S arrest in response to DNA damage, p53 is important. For example, p53-deficient mammalian cells display a lack of a G1/S arrest upon DNA-damage induction [57,58]. Considering that cells lacking p21 as a p53 effector display the same phenotype [59–61], it is well established that the p53-mediated upregulation of p21 is essential for the DNA damage–induced G1/S cell–cycle checkpoint, at least in mammalian tissue culture cells [62–64]. Mechanistically, p53 is a major effector of DDR kinase signaling [65], mediating a G1 cell–cycle arrest mainly through the transcriptional upregulation of the CKI p21 [16]. p21 subsequently inhibits the CDK2–cyclin E complex and consequently DNA replication, hence defining the p53/p21 pathway as a master regulator of the G1/S cell–cycle transition in the DDR [41].

Regarding the role of 53 in the G2/M DNA-damage checkpoint, one should note that the p53-mediated G2/M arrest involves the upregulation of various target genes with distinct functions which can negatively influence CDK1 activity (summarized in Ref. [19]). Although the induction of the G2/M arrest does not require p53 [38,42], p53 and its effector p21 seem to be required for the maintenance of the G2/M arrest [42,66]. In addition, p53 can directly and indirectly play a role in S-phase progression and DNA replication [19], where ATR and CHK1 can regulate the activation of p53 in response to DNA damage [67].

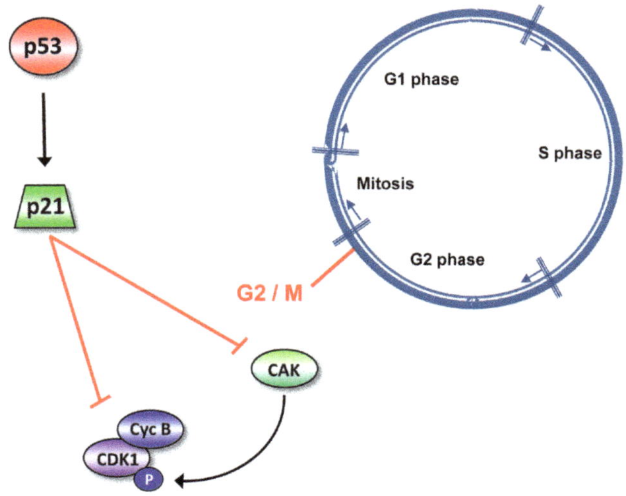

FIGURE 15.2 **p53/p21 in control of the G2/M cell–cycle checkpoint.** The p53 and p21 tumor-suppressor proteins can help to sustain a G2/M cell–cycle arrest in response to stress (eg, DNA damage). The stress-induced stabilization of p53 triggers an increased expression of p21. The elevated p21 levels can support a G2/M cell–cycle arrest through different routes, two of which are indicated here. On the one hand, p21 can inhibit the CDK-activating kinase (CAK) which interferes with the CAK-mediated activating phosphorylation of CDK1. Yet, on the other hand, p21 can directly inhibit the CDK1–cyclin B complex. For more details, please check the main text.

2.2 p53 in DNA-Damage Repair

p53-deficient cells display impaired nucleotide excision repair (NER) of UV-induced photoproducts [19]. p53 also appears to be involved in the base excision repair (BER) and mismatch repair (MMR) pathways (summarized in Ref. [19]). Moreover, p53 can regulate the repair of DNA double-strand breaks (DSBs) by homologous recombination (HRR) and nonhomologous end joining (NHEJ) [68]. Normal p53 can suppress HRR by binding to RAD51 and the BLM helicase [69–71], while mutant p53 can promote an increase in basal and DNA-damage induced RAD51 levels [72,73]. Thus, mutant p53 can cause a "hyper-recombination" phenotype [72,73] which potentially is related to the link between p53 and BRCA1 [74]. In the context of NHEJ repair and BER, it was reported that p53 can either suppress or promote these DNA-repair activities [19], suggesting that context- and cell system–dependent mechanisms must be carefully considered in this regard. Nonetheless, current evidence suggests that p53 facilitates DNA repair by at least three routes: (a) the transcription-dependent induction and maintenance of a transient cell–cycle arrest (mainly with p21/CIP1 upregulation to provide sufficient time for DNA repair), (b) the direct upregulation of the expression of DNA-repair genes (summarized in Ref. [19]), and (c) transcription-independent activities of p53 (see earlier and Ref. [19]). In addition, p53 can bind to damaged DNA, Holliday junctions, and heteroduplex joints in vitro [19], suggesting that p53 binding to abnormal DNA structures might also play a role in directing DNA repair, although these mechanisms are poorly understood.

2.3 p53 in Tumor Suppression and the DNA-Damage Response

Since the majority of mutations in p53 are likely to impair the transcriptional activity of p53 [44] and most cytotoxic clinical compounds induce the p53-mediated DDR, it was expected that the p53 status will have an influence on cancer progression and the outcome of cancer therapy. The initial xenograft experiments showed a clear correlation between the p53 status and apoptosis in response to DNA-damaging agents [75], but in spite of intensive efforts, the impact of the p53 status for a successful cancer therapy is yet to be fully understood (Ref. [19] and see later) with the current emphasis being on the development of small molecules that restore the normal p53 function of the endogenous form [30]. Before we close in the next subsection our summary of p53 with a discussion of p53 in the context DNA-damaging agents, we briefly summarize our current understanding of p53-mediated tumor suppression based on animal models. For an expert overview of p53-related animal models, we refer the reader to other reviews [76,77] since here, we focus on the role of p53-mediated DDR programs and tumor suppression.

As already mentioned, p53-deficient mice die of cancer with nearly 100% penetrance [24–27]. Radiation treatment of p53-deficient mice accelerated cancer formation even further [26]. However, follow-up studies revealed that the p53 status had no effect on cancer development triggered by radiation [78,79]. These studies rather revealed that p53 was required at later time points (after radiation treatment) to suppress tumor formation in mice [78,79]. Using mice expressing transcriptionally dead p53, it was subsequently shown that the transcriptional activity of p53 was essential for the tumor-suppressive function of p53 [34,80], where p53-mediated tumor suppression was most likely a result of the expression control of more than one target gene [19,76], with the apoptotic function(s) of p53 being an important factor [81]. In this regard, one should further note that in addition to apoptosis, cell-cycle arrest and senescence can also play a part in p53-mediated tumor suppression [19].

Last but not least, it is noteworthy that three different studies during 2010s challenged the view that p53-mediated apoptosis, cell cycle arrest, and senescence are the sole effectors of p53 in preventing cancer [34–36]. In the first study, although in response to DNA damage, cells carrying mutant p53 responded like p53-null cells regarding apoptosis, cell-cycle arrest, and senescence, the corresponding mice did not develop spontaneous tumors over a period of 16 months [35]. In the second study, mice carrying deletions of three crucial p53 target genes were studied, revealing that despite defective DNA damage–induced apoptosis and senescence, mutant mice did not develop spontaneous tumors [36]. In the third study, mice expressing a p53 mutant unable to trigger an acute DDR displayed a significant suppression of tumor formation [34]. Collectively, these studies suggest that classical p53-mediated DDR programs are dispensable for the suppression of spontaneous tumor formation. For sure, in response to DNA damage, the p53-mediated apoptotic cell-cycle arrest and senescence programs play important roles, but these studies illustrate that other effector processes, such as possibly the p53-mediated regulation of metabolic and autophagic processes and others, are contributing significantly to the tumor-suppressive functions of p53.

It is also noteworthy that it could be speculated that p53 inactivation should result in genomic instability based on the central roles of p53 in the DNA-damage checkpoint [82]. Nonetheless, p53 deficiency in mammalian cells does not lead to aneuploidy [83,84], and in human precancerous lesions, genomic instability can be observed before the detection of p53 mutations [4,85]. Collectively, these findings suggest that the loss of p53 is not sufficient to induce genomic instability.

2.4 p53 and Targeted DNA-Damaging Cancer Therapy

Genomic instability is a hallmark of tumor development and progression [3]. Consequently, cancer cells must acquire the ability to tolerate an increased amount of DNA damage when compared to untransformed cells. Frequently, this is achieved by dampening/suppressing one or more DNA damage–repair/signaling pathways, which allows cancer cells to function and proliferate in the presence of DNA damage [86]. On the other hand, these mutated or compromised DNA-damage pathways, although contributing to cancer development, also represent a potential Achilles' heel for cancer therapeutics. Many commonly lost tumor-suppressor genes, including p53, p16^{INK4A}, RB, BRCA1/2, and ATM, play a role in the DDR- and DNA-repair pathways which are among the most frequently compromised pathways in human cancers [4–6]. Thus, the targeting of oncogenic signaling pathways in combination with DNA-damaging chemotherapeutics may offer an improved and more selective efficacy than single treatments alone. Considering advances in radiotherapy in 2015 and that DNA damage is the most critical factor in radiation-induced cell death [87], these approaches are not limited to chemotherapy but potentially also applicable to radiotherapy.

The increased selectivity of cancer cells to DNA-damaging agents indicates that cancer cells have rewired their DDR- and DNA-repair pathways, and consequently established new dependencies between cell-cycle checkpoints and survival pathways that are much less pronounced in normal cells. For example, the p53-regulated G1/S checkpoint is a predominant DNA-damage checkpoint in normal mammalian cells which is lost due to p53 mutations in many cancer cell lines [16] generally associated with resistance to chemotherapeutics [18]. In this context, synthetic lethal strategies are gaining more and more interest [86]. Upon the loss of p53 function, cancer cells are often completely dependent on the intra-S and G2/M DNA–damage checkpoints to arrest cell cycle after genotoxic chemotherapeutic stress. Thus, the interference with the intra-S and G2/M checkpoints has become a promising strategy to sensitize G1 checkpoint-deficient cancer cells to DNA-damaging therapy [86].

Compared to genetically matched controls, mammalian mutant or null-p53 tumor cell lines can display the elevated cancer-specific sensitivities to compounds associated with DNA-damaging cancer therapy [86], although this p53 dependency was not observed in all tumor types analyzed [74]. For example, in combination with DNA-damaging agents, CHK1 inhibition causes a by-pass of both the intra-S and G2/M DNA–damage checkpoints in p53-deficient cells (summarized in Ref. [86]). A synthetic lethal interaction between ATR, the upstream activator of CHK1 and p53 deficiency has also been observed, suggesting that ATR or CHK1 inhibition in combination with DNA-damaging agents could be beneficial for the treatment of p53-deficient tumor cells [86]. Most promising in this context might be the discovery of a synthetic lethal interaction between p53 and the ATM–CHK2 pathway, with the loss of ATM or CHK2 together with p53 causing the loss of the G1/S and G2/M checkpoints, consequently driving cancer cells into a mitotic catastrophe in response to genotoxic DNA damage [86]. In this context, it was also observed that a single loss of ATM or p53 can promote resistance to DNA-damaging chemotherapy, while their combined loss can cause an increased chemosensitivity [88]. In this regard, the chemotherapy resistance of p53-proficient but ATM-deficient tumor cells may be reversible by DNA-PK inhibition. However, this will largely be dependent on the development of good pharmacological DNA-PK inhibitors [86,89]. As defined in more detail in Ref. [86], other kinases that can be used as interesting and promising clinical targets for the development of inhibitors that might display a synthetic lethality with p53 deficiency are the ATM- and ATR-activated kinases in the p38/MK2 pathway. The Wee1 kinase is also an attractive target since upon activation by CHK1, Wee1 inhibits the cell-cycle kinases CDK1 and CDK2 and thereby abrogates DNA damage–checkpoint activation [87,90]. Therefore, in ongoing and future clinical trials, the stratification of the p53 status together with functional assays of DNA-repair activities (eg, RAD51 foci formation as readout for HRR [91]) may provide valuable insights as stratification methods in the context of adjuvant chemotherapies including ATM and other kinase inhibitors [74]. However, these clinical analyses of the p53 status must include the determination of p53 single-point mutations since gain-of-function mutations have been described for p53 [92]. Specifically in the context of DNA-damage signaling, the most common p53 mutations, R248W and R273H, have been shown to exhibit gain-of-function properties by binding to MRE11, a key component of the DNA damage–sensing MRN complex [93], through which these p53 mutants can abolish early DDR signaling by negatively interfering with ATM activation in response to DSBs [94].

3. THE p21 TUMOR-SUPPRESSOR PROTEIN

In the 1990s, the p21^{CDKN1A} tumor-suppressor protein was independently isolated as CDK-interacting protein 1 (CIP1) [95,96], wild-type p53-activated factor 1 (WAF1) [64], and senescent cell–derived inhibitor 1 (SDI1) [97]. p21 is also known under the synonyms MDA6, CAP20, and PIC1. The official gene name for the gene encoding p21 is CDKN1A (cyclin-dependent kinase tumor-suppressor protein inhibitor 1A). p21 CIP1 was identified as the first CDK inhibitor (CKI) since it can inhibit various CDKs [96,98–100]. p21 can inhibit CDK activity by blocking the ATP-binding site of CDK [101], by interacting with CDK, or by interfering with CDK phosphorylation [102]. Particularly, p21 blocks CDK2 activity which is required for the inactivation of pRB to release E2F to prepare for DNA replication [103,104].

As a major transcriptional target of p53, p21 is important for the response of cells to DNA damage [63,64]. Depending on its subcellular localization, p21 can perform different functions in a mammalian cell. For example, the nuclear p21 can inhibit CDK1 and CDK2 kinase activities, thereby blocking G1/S and G2/M cell–cycle progression (see Figs. 15.1 and 15.2 and later). p21 is further required for the induction of senescence, and it has an anti-apoptotic function where the anti-apoptotic activity of p21 is related to the cytoplasmic pool of p21 [103]. As a target of p53, p21 can also promote stem cell quiescence [12].

p21 is regulated by p53-mediated transcription, posttranslational modifications, degradation, and subcellular localization [103–105] Since these regulatory mechanisms are complexly interlinked, it is beyond the scope of this chapter to discuss them in detail. Thus, we refer the reader to reviews discussing these somewhat complex aspects in more detail [103–105]. For example, during normal cell–cycle progression, p21 is regulated by proteolysis mediated by E3 ubiquitin ligases including SCF^{SKP2}, APC/C^{CDC20}, and CRL4^{CDT2}, while upon ATR activation, p21 can be degraded in a ubiquitin-independent manner (summarized in Refs. [103,104]).

Besides functioning in the DDR and DNA repair, p21 has additional functions in cell motility and transcriptional control which are summarized elsewhere [103–105]. Here, we focus on summarizing specific roles of p21 in the DDR and DNA repair.

3.1 p21 in the DNA-Damage Response

Upon encountering DNA damage, a mammalian cell relies on three fundamental processes: (a) cell-cycle arrest to allow sufficient time for DNA repair, (b) the repair of DNA lesions, and (c) the induction of apoptosis or senescence in case the DNA damage is beyond repair. Significantly, p21 can play essential roles in all these processes [103].

p21 is a key mediator of cell-cycle arrest in response to DNA damage. p21^{CDKN1A} is a DNA damage–inducible gene whose transcriptional induction can occur dependent on p53, but may also occur through p53-independent pathways. Thus, the induction of p21 expression is considered a paradigm of the cell response to genotoxic damage [106]. In response to DNA damage, the upregulation of p21 can cause a G1/S cell–cycle arrest, and it can also support the G2/M cell–cycle checkpoint [66,103] (Figs. 15.1 and 15.2).

In response to DNA damage, p53 accumulates in the nucleus driving the transcription of p21 and other response genes. Consequently, p21 accumulates in the nucleus causing the activation of the DNA-damage cell-cycle checkpoint in G1/S (Fig. 15.1). p21 functions as a potent inhibitor of CDKs which are major drivers of cell-cycle progression [96]. Particularly, p21 efficiently inhibits the CDK2–cyclin E and CDK2–cyclin A complexes which normally promote the G1/S cell–cycle transition [95]. In response to DNA damage, CDK2 is inhibited by increased levels of p21, resulting in the accumulation of hypo-phosphorylated pRB and thereby sequestering the transcription factor E2F whose activity is required for the entry into the S phase [43]. Since p21-deficient mammalian cells display the lack of G1/S arrest in response to DNA damage [59–61], it is well established that the upregulation of p21 is essential for the DNA damage–induced G1/S cell–cycle checkpoint, at least in mammalian tissue culture cells [62–64]. Generally, it is recognized that p21-deficient cells lack the ability to arrest at the G1/S DNA-damage checkpoint. In addition, p21 has been shown to influence the G1/S checkpoint by interacting with PCNA, a cofactor of the DNA polymerases δ and ε which are required for DNA replication and DNA repair [103].

In addition to its role in the DNA-damage checkpoint in G1/S, p21 also plays a crucial role in the G2/S checkpoint [42,107,108]. Although the induction of G2/M arrest does not require p53 [38,42], p53 and its effector p21 seem to be required for the maintenance of G2/M arrest [42,66] which possibly also involves the interaction of p21 with the CDK1–cyclin B complex [103,107,109,110]. In G2, p21 can also inhibit the CDK-activating kinase (CAK) and consequently block the activating phosphorylation of CDK1 [110]. p21 can further mediate cyclin B degradation [111]. Other targets of p21 in G2 are CDK1–cyclin A and CDK2–cyclin A complexes [112]. Furthermore, by suppressing CDK activity, p21 can indirectly activate pRB, consequently reducing E2F activity, thereby decreasing the pro-proliferative expression of E2F target genes, such as the APC/C inhibitor Emi1. This can promote the premature activation of APC/C in G2, resulting in the degradation of cyclins A and B followed by a G2/M cell–cycle arrest [113]. Collectively, the upregulation of p21 can sustain the G2/M cell–cycle arrest through different routes (Fig. 15.2).

The upregulation of p21 can also impact S-phase progression and DNA replication [103,104]. p21 has also an important role in the induction, but not maintenance, of replicative and stress-induced premature senescence, with p21 upregulation serving as the first marker of replicative senescence [114]. p21 also plays a fundamental role in reversible cell-cycle exit (quiescence) by regulating CDK2 activity [115]. Furthermore, p21 plays an active role in inhibiting apoptosis summarized in Refs. [103,104] (see also later). We also refer the reader to the same papers [103,104] to obtain an overview of the roles of p21 in transcriptional regulation in response to DNA damage.

3.2 p21 in DNA-Damage Repair

Besides inducing a cell-cycle arrest, p21 appears to also have distinct functions in DNA repair [103,104]. Initially, it was suggested that p21 may play a role in DNA damage repair since p21 can interact with PCNA, a cofactor of the DNA polymerases δ and ε, required for DNA replication and DNA repair. Several studies indicate that p21 is involved in major DNA-repair pathways including NER, BER, HRR, and NHEJ, where the interaction of p21 with PCNA seems to play a role in NER and BER. Considering that the involvement of p21 in NER is currently based on the contrasting results, we refer the reader to reviews which elegantly summarized these apparently opposing findings [103,104].

p21-deficient cells displayed elevated PARP1 activity, a defective BER, and increased sensitivity to DNA-alkylating agents [103]. In this regard, in vitro experiments suggest that p21 specifically interferes with the activity of DNA polymerase δ, and p21 has been shown to inhibit the DNA damage–sensor protein PARP1 which is important in BER [103,104]. Collectively, these results suggest that p21 acts as a regulator in BER. p21 can also play a role in the two main DSB-repair pathways: HRR and NHEJ [103,104]. p21 can colocalize with components of the DNA damage–sensing MRN complex, a promoter of HRR together with CtIP, and can be recruited to sites of DSBs [103,104]. Specifically, the spatiotemporal analysis of GFP-tagged p21 revealed that p21 is rapidly recruited to regions containing DNA damage. By inhibiting CDK activity, p21 can favor HRR. In NHEJ, p21 is recruited to DNA lesions independently of key NHEJ factors including DNA-PK and Ku70/Ku80. In addition, p21 may also contribute to MMR and other DNA-repair pathways, but the mechanistic role(s) of p21 in DNA-repair processes remains poorly understood [103,104].

3.3 p21 and Tumor Suppression

Through its roles in DNA damage–checkpoint signaling and DNA repair, p21 can protect mammalian cells against the accumulation of DNA damage and subsequent genome instability. However, p21-deficient mice are not really prone to cancer development and are only subtly sensitive to radiation-induced cancer formation [60,116,117], although the loss of p21 can promote tumor development in mice carrying a p53 loss-of-function mutation or oncogenic RAS [19].

As discussed previously, depending on its subcellular localization, p21 can perform different functions in a mammalian cell. Nuclear p21 promotes cell-cycle arrests in G1/S and G2/M and possibly DNA repair (see earlier), while cytoplasmic p21 can inhibit apoptosis induction in response to DNA damage through inhibitory binding to pro-apoptotic factors [103,104]. Different levels of DNA damage may direct the response controlled by p21. Low levels of DNA damage can stabilize p21 leading to cell-cycle checkpoint activation, while high levels of DNA damage promote the downregulation of p21 and consequently apoptosis. In this regard, although p21 can function as a tumor-suppressor protein by activating cell-cycle checkpoints (and possibly DNA repair), p21 can also play a role in tumor initiation by protecting the damaged cells from apoptosis. On the one hand, p21 deficiency enables the proliferation of cells carrying the damaged DNA promoting tumor progression. On the other hand, as anti-apoptotic factor p21 can act as an oncoprotein. In support of this notion, the elevated levels of p21 protein have already been described in different human cancer samples, frequently correlating with the invasiveness and malignancy of cancer [105]. Note that the deletion of p21 impairs the survival of leukemia stem cells, suggesting that in this context, p21 is required to maintain the self-renewal and quiescence capacities of cancer stem cells by protecting them from accumulating DNA damage and genomic instability, hence displaying an oncogenic activity [118]. Since p21 can also serve as an assembly factor for the formation of CDK4–cyclin D complexes [119,120] which may also contribute to the oncogenic properties of p21. Thus, it will be important to continue to decipher the context-dependent role of p21 in cancer prevention vs. initiation and maintenance in the context of cancer therapy [105,121]. In this regard, one should also note that p21 appears to have also context-dependent functions in stem cell biology, particularly in response to DNA damage and telomere dysfunction [12].

4. THE p16^{INK4A} TUMOR-SUPPRESSOR PROTEIN

Different stresses can induce senescence, including telomere dysfunction (eg, through replicative erosion), the induction of chronic or acute DNA damage (by UV radiation, IR, or genotoxic compounds), oncogene activation, and others [122,123]. Several signal transduction pathways are essential drivers of senescence, including p53–p21 signaling required for senescence induction and p16–pRB-signaling needed for senescence maintenance [123]. Here, we provide an overview of the roles of p16 in promoting senescence with a particular emphasis on DNA damage–signaling-induced senescence (see also Fig. 15.1).

p16 (also known as MTS1 and INK4A) was first discovered by yeast two-hybrid screens as a novel binding partner of CDK4 [124,125], and subsequently the full-length p16 was isolated [126]. Like p21 (see earlier), p16 also functions as CKI by specifically and directly binding to proto-oncogenic CDK4–cyclin D and CDK6–cyclin D complexes [125,127]. Consequently, pRB remains hypo-phosphorylated and keeps the S-phase initiating transcription factor E2F inactive, thereby stabilizing a G1/S cell–cycle arrest by activating the pRB checkpoint [43]. In support of this tumor-suppressive function, loss-of-function and overexpression studies showed that p16 functions as a tumor-suppressor protein [128]. Furthermore, it was observed that the CDKN2A locus which encodes the p16 protein is very frequently inactivated by deletions, point mutations, or promoter hypermethylation in melanoma, pancreatic carcinomas, leukemia, bladder cancer, head and neck carcinomas, and others [128]. However, in this context, one must note that the CDKN2A gene locus encodes for two independent tumor-suppressor genes, namely p16^{INK4A} and p14^{ARF}, through the use of alternative open reading frames [126]. Nonetheless, p16 inactivation by CDKN2A deletions in human cancers is likely the main event regarding tumor suppression since p16 functions in the regulation of CDK4, CDK6, and pRB [125,129]. Noteworthy, p16 can also be overexpressed in human cancers [130], in particular in the context of human papillomavirus (HPV)-transformed cells, where the elevated levels of p16 are considered a hallmark of HPV-positive cervical carcinoma and head and neck cancer [131]. Mechanistically, the HPV encoded oncoprotein E7 disrupts the function of pRB, consequently releasing pRB from its inhibitory role of E2F, hence allowing the entry into the S phase. Nevertheless, HPV-infected cells still upregulate p16 levels to block the proliferation of HPV-transformed cells, which, however, due to the deregulation of pRB by E7 is an unsuccessful attempt to stop the proliferation [132].

Generally, senescence is defined as an irreversible cell-cycle arrest that is associated with the secretion of a specific subset of growth factors, referred to as the senescence-associated secretory phenotype [133]. Mammalian cells undergoing senescence develop specific characteristics that distinguish them from other nondividing cell states, such as quiescence or terminal differentiation [134]. Senescence can be induced prematurely by DNA damage without telomere shortening, referred to as stress-induced premature senescence. In contrast, replicative senescence occurs as a response to telomere shortening. Moreover, senescence can be induced by the failure to repair DSBs. While replication-induced telomere erosion–dependent senescence is mainly triggered by the recognition of dysfunctional telomeres by the ATM-dependent DSB-signaling response driving p53-dependent mechanisms [135], DSBs have been observed in senescent cells independent of telomere erosion [13,14], suggesting that senescence can also result from nontelomeric DSBs, where the amount of unrepaired DSBs required to trigger senescence is dose-, damage-, and cell type–dependent [135].

Collectively, different stimuli can promote senescence, including telomere erosion which causes a permanent DDR and nontelomeric DNA–damage stress which can cause the persistent DDR activation through the misfired replication origins and replication fork collapses [135]. Thus, senescence can serve as an anticancer barrier [134]. In particular, in the context of oncogene-driven DNA damage, a specific type of stress-induced premature senescence has been recognized as a powerful barrier to the malignant transformation of pre-cancerous lesions [10,11,85]. For example, hyperactivated oncogenes can trigger chronic DSB signaling by causing error-prone DNA replication, resulting in the initiation of senescence [10]. This type of premature senescence is known as oncogene-induced senescence. Furthermore, most human tumors display inactivating mutations of the p53 and/or p16–pRB pathways, which are central components of the senescence response [135], further supporting the notion that senescence is an important tumor-suppressive mechanism. Thus, senescence induction is considered a possible mechanism for cancer therapy [136].

In mammals, cellular senescence is regulated by two major mechanisms: the p16–pRB pathway and the p53–p21 pathway [114,135]. While p21 plays a role in the initiation of senescence [114], the state of permanent cell–cycle arrest is maintained by p16 [137]. Specifically, the p16–pRB pathway seems to be essential for the maintenance of senescence since the senescence phenotype is not reversible once senescence arrest has been fully established by the p16–pRB pathway [138,139].

Significantly, chromatin alterations are also a key feature of senescence accompanied by the formation of senescence-associated heterochromatin foci (SAHF) [135]. The p16–pRB pathway is required for the formation of SAHF structures in response to oncogene-induced senescence [135], where SAHF formation can serve as a barrier to chronic DSB responses, with the potential to help cancer cells to bypass cell death induction by DSB signaling [140]. Most likely, SAHF surround an unresolved DSB and suppress it from contributing to DSB-response signaling, which could be beneficial for normal tissue functionality but detrimental in the context of cancer cell survival [135]. Thus, it is possible that SAHF formation might assist in preserving genomic integrity, in addition to muting the DSB response which normally would trigger cell death mechanisms, and thereby remove cells containing excessive DNA damage. Thus, considering that the p16–pRB pathway is needed for the maintenance of senescence as an anticancer barrier (see earlier) and is possibly required for the suppression of DSB signaling to promote cancer cell survival, it is very likely that p16 may play context-dependent roles in cancer as already defined for p21 (see earlier).

Last but not least, one should further note that the role of p16 in the context of DNA damage has been linked to stem cell biology and aging in addition to cancer (summarized in Ref. [12]). For example, cell-intrinsic DNA damage in tissues can result in systemic alterations of the blood, thus accelerating normally age-dependent functional defects in hematopoietic stem cell pools [12]. Moreover, the functional impairment of somatic stem cells due to accumulated DNA damage and the consequent DNA-damage checkpoint responses can lead to defects in tissue maintenance. In this regard, p16 expression is significantly increased in various tissues during mouse and human aging, which possibly involves a decline in ATM functionality combined with the accumulation of DNA damage during aging. Even more importantly, since the age-dependent accumulation of DNA damage also occurs in stem cells, stem cell maintenance can be improved by the deletion of p16 in murine cells. Taken together, the deregulated expression of p16 can have detrimental effects on the functionality of stem cells as well as somatic cells [12].

5. CONCLUSION

In summary, many commonly lost tumor-suppressor genes, including p53, $p16^{INK4A}$, and RB, play roles in the DDR- and DNA-repair pathways. As summarized in this chapter, p53 acts as a central, versatile and multifunctional player in the cellular DDR. In response to DNA damage, p53 protein levels are upregulated, thereby inducing diverse transcriptional programs that can promote a transient cell cycle arrest, a permanent cell–cycle arrest in the form of senescence, DNA repair, and/or apoptosis. Considering that the Tp53 gene is the most frequently mutated gene in human cancers and that p53-deficient mice display nearly 100% penetrance of cancer development at the young age, it is not surprising that over the past decades, very intensive research efforts have focused on deciphering the key tumor-suppressive functions of p53. In this regard, it is now fully established that p53 plays important roles in the G1/S and G2/M DNA–damage cell-cycle checkpoints. In response to DNA damage, p53 as a transcription factor is stabilized, allowing p53 to drive the transcription of the p21 tumor-suppressor gene and other targets. This p53–p21 axis is essential for the induction of the G1/S arrest, while "only" being required for the maintenance of the G2/M arrest. In addition, p53 directly and indirectly (through its effector p21) can play distinct roles in supporting DNA repair. p21 also plays an active role in inhibiting apoptosis, which in a context-dependent manner can have cancer-promoting effects, in contrast to the general tumor-suppressive role of p21. Moreover, the p53–p21 pathway acts in the initiation of senescence. Conversely, the maintenance of cellular senescence is promoted by the p16–pRB pathway. Considering that the p53–p21 and p16–pRB pathways are among the most frequently compromised pathways in human cancers, more research is now needed to decipher which of their cell biological functions are essential for tumor suppression in vivo. Research aiming to translate the p53, p21, and/or p16 status into clinical cancer settings may help improve (maybe even optimize) the prediction of responses to radio- and/or chemotherapies. In the context of clinically developing and testing selective DDR inhibitors, the analyses of p53, p21, and/or p16 levels and mutations may even help open up completely novel anticancer approaches.

GLOSSARY

Acute DNA damage Severe and temporally limited DNA damage.
Aneuploidy The presence of an abnormal number of chromosomes in a cell.
Apoptosis The process of programmed cell death.
Cell-cycle checkpoint Specific control mechanisms in eukaryotic cells ensuring proper cell-cycle progression.
Cellular senescence An irreversible G1 cell–cycle arrest in which cells are refractory to growth factor stimulation.
Chronic DNA damage A type of DNA damage that persists for a long time or that regularly recurs.
DNA-damage checkpoint A cell-cycle checkpoint that is specifically activated upon the detection of DNA lesions.
DNA-damage response A complex network of cellular pathways that is responsible for the detection, signaling, and repair of DNA lesions.
E3 ubiquitin ligase An enzyme that catalyzes the transfer of ubiquitin from the E2 ubiquitin-conjugating enzyme to specific protein substrates.
Genomic instability (aka genetic or genome instability) Defined as a high frequency of mutations within the genome, where mutations can include changes in nucleic acid sequences, chromosomal rearrangements, and/or aneuploidy.
Genotoxic A damaging effect on a cell's genetic material.
Malignant transformation The process by which cells acquire the properties of cancer.
Mitotic catastrophe A cellular event in which a cell is destroyed during mitosis.
Permanent cell–cycle arrest An irreversible exit from cell-cycle progression.
Quiescence The state of a cell when it is not dividing as a consequence of a reversible cell–cycle exit.
Synthetic lethal interaction (aka synthetic lethality) A type of genetic interaction where the cooccurrence of two genetic events results in organismal or cellular lethality.
Transient cell–cycle arrest A fully reversible exit from cell-cycle progression.

LIST OF ACRONYMS AND ABBREVIATIONS

ATM Ataxia telangiectasia mutated
ATR ATM and Rad3 related
BER Base excision repair
CDK Cyclin-dependent kinase
CDKN1A CDK tumor-suppressor protein inhibitor 1A
CDKN2A CDK tumor-suppressor protein inhibitor 2A
CHK1 Checkpoint kinase 1
CHK2 Checkpoint kinase 2
CIP1 CDK interacting protein 1
CKI CDK inhibitor
DDR DNA-damage response
DNA-PK DNA-dependent protein kinase
DSB DNA double-strand break
HPV Human papillomavirus
HRR Homologous recombination repair
IR Ionizing radiation
MK2 MAPK-activated protein kinase 2
MMR Mismatch repair
MRN MRE11/RAD50/NBS1 complex
NER Nucleotide excision repair
NHEJ Nonhomologous end joining
pRB Retinoblastoma protein
SAHF Senescence-associated heterochromatin foci
SDI1 Senescent cell–derived inhibitor 1
UV Ultraviolet
WAF1 Wild-type p53-activated factor 1

ACKNOWLEDGMENTS

We apologize to all authors whose work we could not cite due to space limitations. We are very grateful to Joanna Lisztwan, Ahmad Sharif, and Nirmal Perera for their critical review of the manuscript, and also thank all members of the Hergovich laboratory for helpful discussions. Y.K. and R.G. are supported by the Ministry of National Education of the Republic of Turkey. The work of the Hergovich laboratory is supported by a Wellcome Trust grant (090090/Z/09/Z), a Cancer Research UK centre development fund, and the National Institute for Health Research University College London Hospitals Biomedical Research Centre.

Author's Contributions

Yavuz Kulaberoglu, Ramazan Gundogdu, and Alexander Hergovich researched the literature and wrote the manuscript together. Yavuz Kulaberoglu and Ramazan Gundogdu created all figures. All authors read and approved the final manuscript.

REFERENCES

[1] Bunting SF, Nussenzweig A. End-joining, translocations and cancer. Nat Rev Cancer 2013;13(7):443–54.
[2] Jackson SP, Bartek J. The DNA-damage response in human biology and disease. Nature 2009;461(7267):1071–8.
[3] Hanahan D, Weinberg RA. Hallmarks of cancer: the next generation. Cell 2011;144(5):646–74.
[4] Bartkova J, Horejsi Z, Koed K, Kramer A, Tort F, Zieger K, et al. DNA damage response as a candidate anti-cancer barrier in early human tumorigenesis. Nature 2005;434(7035):864–70.
[5] Fackenthal JD, Olopade OI. Breast cancer risk associated with BRCA1 and BRCA2 in diverse populations. Nat Rev Cancer 2007;7(12):937–48.
[6] Frebourg T, Friend SH. Cancer risks from germline p53 mutations. J Clin Invest 1992;90(5):1637–41.
[7] Bartek J, Lukas J. DNA damage checkpoints: from initiation to recovery or adaptation. Curr Opin Cell Biol 2007;19(2):238–45.
[8] Deckbar D, Jeggo PA, Lobrich M. Understanding the limitations of radiation-induced cell cycle checkpoints. Crit Rev Biochem Mol Biol 2011;46(4):271–83.
[9] Wahl GM, Linke SP, Paulson TG, Huang LC. Maintaining genetic stability through TP53 mediated checkpoint control. Cancer Surv 1997;29:183–219.
[10] Di Micco R, Fumagalli M, Cicalese A, Piccinin S, Gasparini P, Luise C, et al. Oncogene-induced senescence is a DNA damage response triggered by DNA hyper-replication. Nature 2006;444(7119):638–42.
[11] Bartkova J, Rezaei N, Liontos M, Karakaidos P, Kletsas D, Issaeva N, et al. Oncogene-induced senescence is part of the tumorigenesis barrier imposed by DNA damage checkpoints. Nature 2006;444(7119):633–7.
[12] Sperka T, Wang J, Rudolph KL. DNA damage checkpoints in stem cells, ageing and cancer. Nat Rev Mol Cell Biol 2012;13(9):579–90.
[13] Sedelnikova OA, Horikawa I, Zimonjic DB, Popescu NC, Bonner WM, Barrett JC. Senescing human cells and ageing mice accumulate DNA lesions with unrepairable double-strand breaks. Nat Cell Biol 2004;6(2):168–70.
[14] Wang C, Jurk D, Maddick M, Nelson G, Martin-Ruiz C, von Zglinicki T. DNA damage response and cellular senescence in tissues of aging mice. Aging Cell 2009;8(3):311–23.
[15] Lane DP. Cancer. p53, guardian of the genome. Nature 1992;358(6381):15–6.
[16] Vogelstein B, Lane D, Levine AJ. Surfing the p53 network. Nature 2000;408(6810):307–10.
[17] Levine AJ, Oren M. The first 30 years of p53: growing ever more complex. Nat Rev Cancer 2009;9(10):749–58.
[18] Vousden KH, Lane DP. p53 in health and disease. Nat Rev Mol Cell Biol 2007;8(4):275–83.
[19] Speidel D. The role of DNA damage responses in p53 biology. Arch Toxicol 2015;89(4):501–17.
[20] Vousden KH. Outcomes of p53 activation–spoilt for choice. J Cell Sci 2006;119(Pt 24):5015–20.
[21] Vousden KH, Prives C. Blinded by the light: the growing complexity of p53. Cell 2009;137(3):413–31.
[22] Kandoth C, McLellan MD, Vandin F, Ye K, Niu B, Lu C, et al. Mutational landscape and significance across 12 major cancer types. Nature 2013;502(7471):333–9.
[23] Petitjean A, Mathe E, Kato S, Ishioka C, Tavtigian SV, Hainaut P, et al. Impact of mutant p53 functional properties on TP53 mutation patterns and tumor phenotype: lessons from recent developments in the IARC TP53 database. Hum Mutat 2007;28(6):622–9.
[24] Donehower LA, Harvey M, Slagle BL, McArthur MJ, Montgomery Jr CA, Butel JS, et al. Mice deficient for p53 are developmentally normal but susceptible to spontaneous tumours. Nature 1992;356(6366):215–21.
[25] Jacks T, Remington L, Williams BO, Schmitt EM, Halachmi S, Bronson RT, et al. Tumor spectrum analysis in p53-mutant mice. Curr Biol 1994;4(1):1–7.
[26] Kemp CJ, Wheldon T, Balmain A. p53-deficient mice are extremely susceptible to radiation-induced tumorigenesis. Nat Genet 1994;8(1):66–9.
[27] Purdie CA, Harrison DJ, Peter A, Dobbie L, White S, Howie SE, et al. Tumour incidence, spectrum and ploidy in mice with a large deletion in the p53 gene. Oncogene 1994;9(2):603–9.
[28] Malkin D, Li FP, Strong LC, Fraumeni Jr JF, Nelson CE, Kim DH, et al. Germ line p53 mutations in a familial syndrome of breast cancer, sarcomas, and other neoplasms. Science 1990;250(4985):1233–8.
[29] Srivastava S, Zou ZQ, Pirollo K, Blattner W, Chang EH. Germ-line transmission of a mutated p53 gene in a cancer-prone family with Li-Fraumeni syndrome. Nature 1990;348(6303):747–9.
[30] Khoo KH, Verma CS, Lane DP. Drugging the p53 pathway: understanding the route to clinical efficacy. Nat Rev Drug Discovery 2014;13(3):217–36.
[31] Menendez D, Inga A, Resnick MA. The expanding universe of p53 targets. Nat Rev Cancer 2009;9(10):724–37.
[32] Reinhardt HC, Schumacher B. The p53 network: cellular and systemic DNA damage responses in aging and cancer. Trends Genet 2012;28(3):128–36.
[33] Garcia-Cao I, Garcia-Cao M, Martin-Caballero J, Criado LM, Klatt P, Flores JM, et al. "Super p53" mice exhibit enhanced DNA damage response, are tumor resistant and age normally. EMBO J 2002;21(22):6225–35.
[34] Brady CA, Jiang D, Mello SS, Johnson TM, Jarvis LA, Kozak MM, et al. Distinct p53 transcriptional programs dictate acute DNA-damage responses and tumor suppression. Cell 2011;145(4):571–83.
[35] Li M, He Y, Dubois W, Wu X, Shi J, Huang J. Distinct regulatory mechanisms and functions for p53-activated and p53-repressed DNA damage response genes in embryonic stem cells. Mol Cell 2012;46(1):30–42.

[36] Valente LJ, Gray DH, Michalak EM, Pinon-Hofbauer J, Egle A, Scott CL, et al. p53 efficiently suppresses tumor development in the complete absence of its cell-cycle inhibitory and proapoptotic effectors p21, Puma, and Noxa. Cell Rep 2013;3(5):1339–45.

[37] Fritsche M, Haessler C, Brandner G. Induction of nuclear accumulation of the tumor-suppressor protein p53 by DNA-damaging agents. Oncogene 1993;8(2):307–18.

[38] Kastan MB, Onyekwere O, Sidransky D, Vogelstein B, Craig RW. Participation of p53 protein in the cellular response to DNA damage. Cancer Res 1991;51(23 Pt 1):6304–11.

[39] Lowe SW, Schmitt EM, Smith SW, Osborne BA, Jacks T. p53 is required for radiation-induced apoptosis in mouse thymocytes. Nature 1993;362(6423):847–9.

[40] Maltzman W, Czyzyk L. UV irradiation stimulates levels of p53 cellular tumor antigen in nontransformed mouse cells. Mol Cell Biol 1984;4(9):1689–94.

[41] Kruse JP, Gu W. Modes of p53 regulation. Cell 2009;137(4):609–22.

[42] Bunz F, Dutriaux A, Lengauer C, Waldman T, Zhou S, Brown JP, et al. Requirement for p53 and p21 to sustain G2 arrest after DNA damage. Science 1998;282(5393):1497–501.

[43] Kastan MB, Bartek J. Cell-cycle checkpoints and cancer. Nature 2004;432(7015):316–23.

[44] Olivier M, Hollstein M, Hainaut P. TP53 mutations in human cancers: origins, consequences, and clinical use. Cold Spring Harb Perspect Biol 2010;2(1):a001008.

[45] Fridman JS, Lowe SW. Control of apoptosis by p53. Oncogene 2003;22(56):9030–40.

[46] Vousden KH, Lu X. Live or let die: the cell's response to p53. Nat Rev Cancer 2002;2(8):594–604.

[47] Fan S, el-Deiry WS, Bae I, Freeman J, Jondle D, Bhatia K, et al. p53 gene mutations are associated with decreased sensitivity of human lymphoma cells to DNA damaging agents. Cancer Res 1994;54(22):5824–30.

[48] Agarwal ML, Agarwal A, Taylor WR, Stark GR. p53 controls both the G2/M and the G1 cell cycle checkpoints and mediates reversible growth arrest in human fibroblasts. Proc Natl Acad Sci USA 1995;92(18):8493–7.

[49] Di Leonardo A, Linke SP, Clarkin K, Wahl GM. DNA damage triggers a prolonged p53-dependent G1 arrest and long-term induction of Cip1 in normal human fibroblasts. Genes Dev 1994;8(21):2540–51.

[50] Stewart N, Hicks GG, Paraskevas F, Mowat M. Evidence for a second cell cycle block at G2/M by p53. Oncogene 1995;10(1):109–15.

[51] Sugrue MM, Shin DY, Lee SW, Aaronson SA. Wild-type p53 triggers a rapid senescence program in human tumor cells lacking functional p53. Proc Natl Acad Sci USA 1997;94(18):9648–53.

[52] Taylor WR, DePrimo SE, Agarwal A, Agarwal ML, Schonthal AH, Katula KS, et al. Mechanisms of G2 arrest in response to overexpression of p53. Mol Biol Cell 1999;10(11):3607–22.

[53] Clarke AR, Purdie CA, Harrison DJ, Morris RG, Bird CC, Hooper ML, et al. Thymocyte apoptosis induced by p53-dependent and independent pathways. Nature 1993;362(6423):849–52.

[54] Purvis JE, Karhohs KW, Mock C, Batchelor E, Loewer A, Lahav G. p53 dynamics control cell fate. Science 2012;336(6087):1440–4.

[55] Loewer A, Batchelor E, Gaglia G, Lahav G. Basal dynamics of p53 reveal transcriptionally attenuated pulses in cycling cells. Cell 2010;142(1):89–100.

[56] Batchelor E, Loewer A, Lahav G. The ups and downs of p53: understanding protein dynamics in single cells. Nat Rev Cancer 2009;9(5):371–7.

[57] Kastan MB, Zhan Q, el-Deiry WS, Carrier F, Jacks T, Walsh WV, et al. A mammalian cell cycle checkpoint pathway utilizing p53 and GADD45 is defective in ataxia-telangiectasia. Cell 1992;71(4):587–97.

[58] Lowe SW, Ruley HE, Jacks T, Housman DE. p53-dependent apoptosis modulates the cytotoxicity of anticancer agents. Cell 1993;74(6):957–67.

[59] Brugarolas J, Chandrasekaran C, Gordon JI, Beach D, Jacks T, Hannon GJ. Radiation-induced cell cycle arrest compromised by p21 deficiency. Nature 1995;377(6549):552–7.

[60] Deng C, Zhang P, Harper JW, Elledge SJ, Leder P. Mice lacking p21CIP1/WAF1 undergo normal development, but are defective in G1 checkpoint control. Cell 1995;82(4):675–84.

[61] Waldman T, Kinzler KW, Vogelstein B. p21 is necessary for the p53-mediated G1 arrest in human cancer cells. Cancer Res 1995;55(22):5187–90.

[62] Dulic V, Kaufmann WK, Wilson SJ, Tlsty TD, Lees E, Harper JW, et al. p53-dependent inhibition of cyclin-dependent kinase activities in human fibroblasts during radiation-induced G1 arrest. Cell 1994;76(6):1013–23.

[63] el-Deiry WS, Harper JW, O'Connor PM, Velculescu VE, Canman CE, Jackman J, et al. WAF1/CIP1 is induced in p53-mediated G1 arrest and apoptosis. Cancer Res 1994;54(5):1169–74.

[64] el-Deiry WS, Tokino T, Velculescu VE, Levy DB, Parsons R, Trent JM, et al. WAF1, a potential mediator of p53 tumor suppression. Cell 1993;75(4):817–25.

[65] Harris SL, Levine AJ. The p53 pathway: positive and negative feedback loops. Oncogene 2005;24(17):2899–908.

[66] Taylor WR, Stark GR. Regulation of the G2/M transition by p53. Oncogene 2001;20(15):1803–15.

[67] Cimprich KA, Cortez D. ATR: an essential regulator of genome integrity. Nat Rev Mol Cell Biol 2008;9(8):616–27.

[68] Menon V, Povirk L. Involvement of p53 in the repair of DNA double strand breaks: multifaceted Roles of p53 in homologous recombination repair (HRR) and non-homologous end joining (NHEJ). Subcell Biochem 2014;85:321–36.

[69] Kumari A, Schultz N, Helleday T. p53 protects from replication-associated DNA double-strand breaks in mammalian cells. Oncogene 2004;23(13):2324–9.

[70] Sengupta S, Linke SP, Pedeux R, Yang Q, Farnsworth J, Garfield SH, et al. BLM helicase-dependent transport of p53 to sites of stalled DNA replication forks modulates homologous recombination. EMBO J 2003;22(5):1210–22.

[71] Sturzbecher HW, Donzelmann B, Henning W, Knippschild U, Buchhop S. p53 is linked directly to homologous recombination processes via RAD51/RecA protein interaction. EMBO J 1996;15(8):1992–2002.

[72] Bertrand P, Rouillard D, Boulet A, Levalois C, Soussi T, Lopez BS. Increase of spontaneous intrachromosomal homologous recombination in mammalian cells expressing a mutant p53 protein. Oncogene 1997;14(9):1117–22.

[73] Saintigny Y, Rouillard D, Chaput B, Soussi T, Lopez BS. Mutant p53 proteins stimulate spontaneous and radiation-induced intrachromosomal homologous recombination independently of the alteration of the transactivation activity and of the G1 checkpoint. Oncogene 1999;18(24):3553–63.

[74] Beckta JM, Ahmad SF, Yang H, Valerie K. Revisiting p53 for cancer-specific chemo- and radiotherapy: ten years after. Cell Cycle 2014;13(5):710–3.

[75] Lowe SW, Bodis S, McClatchey A, Remington L, Ruley HE, Fisher DE, et al. p53 status and the efficacy of cancer therapy in vivo. Science 1994;266(5186):807–10.

[76] Bieging KT, Attardi LD. Deconstructing p53 transcriptional networks in tumor suppression. Trends Cell Biol 2012;22(2):97–106.

[77] Bieging KT, Mello SS, Attardi LD. Unravelling mechanisms of p53-mediated tumour suppression. Nat Rev Cancer 2014;14(5):359–70.

[78] Christophorou MA, Ringshausen I, Finch AJ, Swigart LB, Evan GI. The pathological response to DNA damage does not contribute to p53-mediated tumour suppression. Nature 2006;443(7108):214–7.

[79] Hinkal G, Parikh N, Donehower LA. Timed somatic deletion of p53 in mice reveals age-associated differences in tumor progression. PLoS One 2009;4(8):e6654.

[80] Jiang D, Brady CA, Johnson TM, Lee EY, Park EJ, Scott MP, et al. Full p53 transcriptional activation potential is dispensable for tumor suppression in diverse lineages. Proc Natl Acad Sci USA 2011;108(41):17123–8.

[81] Timofeev O, Schlereth K, Wanzel M, Braun A, Nieswandt B, Pagenstecher A, et al. p53 DNA binding cooperativity is essential for apoptosis and tumor suppression in vivo. Cell Rep 2013;3(5):1512–25.

[82] Kuerbitz SJ, Plunkett BS, Walsh WV, Kastan MB. Wild-type p53 is a cell cycle checkpoint determinant following irradiation. Proc Natl Acad Sci USA 1992;89(16):7491–5.

[83] Bunz F, Fauth C, Speicher MR, Dutriaux A, Sedivy JM, Kinzler KW, et al. Targeted inactivation of p53 in human cells does not result in aneuploidy. Cancer Res 2002;62(4):1129–33.

[84] Kang J, Ferguson D, Song H, Bassing C, Eckersdorff M, Alt FW, et al. Functional interaction of H2AX, NBS1, and p53 in ATM-dependent DNA damage responses and tumor suppression. Mol Cell Biol 2005;25(2):661–70.

[85] Gorgoulis VG, Vassiliou LV, Karakaidos P, Zacharatos P, Kotsinas A, Liloglou T, et al. Activation of the DNA damage checkpoint and genomic instability in human precancerous lesions. Nature 2005;434(7035):907–13.

[86] Morandell S, Yaffe MB. Exploiting synthetic lethal interactions between DNA damage signaling, checkpoint control, and p53 for targeted cancer therapy. Prog Mol Biol Transl Sci 2012;110:289–314.

[87] Morgan MA, Lawrence TS. Molecular pathways: overcoming radiation resistance by targeting DNA damage response pathways. Clin Cancer Res 2015;21(13):2898–904.

[88] Jiang H, Reinhardt HC, Bartkova J, Tommiska J, Blomqvist C, Nevanlinna H, et al. The combined status of ATM and p53 link tumor development with therapeutic response. Genes Dev 2009;23(16):1895–909.

[89] Postel-Vinay S, Vanhecke E, Olaussen KA, Lord CJ, Ashworth A, Soria JC. The potential of exploiting DNA-repair defects for optimizing lung cancer treatment. Nat Rev Clin Oncol 2012;9(3):144–55.

[90] Morandell S, Reinhardt HC, Cannell IG, Kim JS, Ruf DM, Mitra T, et al. A reversible gene-targeting strategy identifies synthetic lethal interactions between MK2 and p53 in the DNA damage response in vivo. Cell Rep 2013;5(4):868–77.

[91] Willers H, Taghian AG, Luo CM, Treszezamsky A, Sgroi DC, Powell SN. Utility of DNA repair protein foci for the detection of putative BRCA1 pathway defects in breast cancer biopsies. Mol Cancer Res 2009;7(8):1304–9.

[92] Brosh R, Rotter V. When mutants gain new powers: news from the mutant p53 field. Nat Rev Cancer 2009;9(10):701–13.

[93] Stracker TH, Petrini JH. The MRE11 complex: starting from the ends. Nat Rev Mol Cell Biol 2011;12(2):90–103.

[94] Song H, Hollstein M, Xu Y. p53 gain-of-function cancer mutants induce genetic instability by inactivating ATM. Nat Cell Biol 2007;9(5):573–80.

[95] Harper JW, Adami GR, Wei N, Keyomarsi K, Elledge SJ. The p21 Cdk-interacting protein Cip1 is a potent inhibitor of G1 cyclin-dependent kinases. Cell 1993;75(4):805–16.

[96] Xiong Y, Hannon GJ, Zhang H, Casso D, Kobayashi R, Beach D. p21 is a universal inhibitor of cyclin kinases. Nature 1993;366(6456):701–4.

[97] Noda A, Ning Y, Venable SF, Pereira-Smith OM, Smith JR. Cloning of senescent cell-derived inhibitors of DNA synthesis using an expression screen. Exp Cell Res 1994;211(1):90–8.

[98] Delavaine L, La Thangue NB. Control of E2F activity by p21Waf1/Cip1. Oncogene 1999;18(39):5381–92.

[99] Shiyanov P, Bagchi S, Adami G, Kokontis J, Hay N, Arroyo M, et al. p21 Disrupts the interaction between cdk2 and the E2F-p130 complex. Mol Cell Biol 1996;16(3):737–44.

[100] Garcia V, Phelps SE, Gray S, Neale MJ. Bidirectional resection of DNA double-strand breaks by Mre11 and Exo1. Nature 2011;479(7372):241–4.

[101] Russo AA, Jeffrey PD, Patten AK, Massague J, Pavletich NP. Crystal structure of the p27Kip1 cyclin-dependent-kinase inhibitor bound to the cyclin A-Cdk2 complex. Nature 1996;382(6589):325–31.

[102] Mandal M, Bandyopadhyay D, Goepfert TM, Kumar R. Interferon-induces expression of cyclin-dependent kinase-inhibitors p21WAF1 and p27Kip1 that prevent activation of cyclin-dependent kinase by CDK-activating kinase (CAK). Oncogene 1998;16(2):217–25.

[103] Cazzalini O, Scovassi AI, Savio M, Stivala LA, Prosperi E. Multiple roles of the cell cycle inhibitor p21(CDKN1A) in the DNA damage response. Mutat Res 2010;704(1–3):12–20.

[104] Dutto I, Tillhon M, Cazzalini O, Stivala LA, Prosperi E. Biology of the cell cycle inhibitor p21(CDKN1A): molecular mechanisms and relevance in chemical toxicology. Arch Toxicol 2015;89(2):155–78.

[105] Abbas T, Dutta A. p21 in cancer: intricate networks and multiple activities. Nat Rev Cancer 2009;9(6):400–14.

[106] Sakai R, Kondo C, Oka H, Miyajima H, Kubo K, Uehara T. Utilization of CDKN1A/p21 gene for class discrimination of DNA damage-induced clastogenicity. Toxicology 2014;315:8–16.

[107] Niculescu 3rd AB, Chen X, Smeets M, Hengst L, Prives C, Reed SI. Effects of p21(Cip1/Waf1) at both the G1/S and the G2/M cell cycle transitions: pRb is a critical determinant in blocking DNA replication and in preventing endoreduplication. Mol Cell Biol 1998;18(1):629–43.
[108] Waldman T, Lengauer C, Kinzler KW, Vogelstein B. Uncoupling of S phase and mitosis induced by anticancer agents in cells lacking p21. Nature 1996;381(6584):713–6.
[109] Charrier-Savournin FB, Chateau MT, Gire V, Sedivy J, Piette J, Dulic V. p21-Mediated nuclear retention of cyclin B1-Cdk1 in response to genotoxic stress. Mol Biol Cell 2004;15(9):3965–76.
[110] Smits VA, Klompmaker R, Vallenius T, Rijksen G, Makela TP, Medema RH. p21 inhibits Thr161 phosphorylation of Cdc2 to enforce the G2 DNA damage checkpoint. J Biol Chem 2000;275(39):30638–43.
[111] Gillis LD, Leidal AM, Hill R, Lee PW. p21Cip1/WAF1 mediates cyclin B1 degradation in response to DNA damage. Cell Cycle 2009;8(2):253–6.
[112] Dulic V, Stein GH, Far DF, Reed SI. Nuclear accumulation of p21Cip1 at the onset of mitosis: a role at the G2/M-phase transition. Mol Cell Biol 1998;18(1):546–57.
[113] Lee J, Kim JA, Barbier V, Fotedar A, Fotedar R. DNA damage triggers p21WAF1-dependent Emi1 down-regulation that maintains G2 arrest. Mol Biol Cell 2009;20(7):1891–902.
[114] Roninson IB. Oncogenic functions of tumour suppressor p21(Waf1/Cip1/Sdi1): association with cell senescence and tumour-promoting activities of stromal fibroblasts. Cancer Lett 2002;179(1):1–14.
[115] Spencer SL, Cappell SD, Tsai FC, Overton KW, Wang CL, Meyer T. The proliferation-quiescence decision is controlled by a bifurcation in CDK2 activity at mitotic exit. Cell 2013;155(2):369–83.
[116] Jackson RJ, Engelman RW, Coppola D, Cantor AB, Wharton W, Pledger WJ. p21Cip1 nullizygosity increases tumor metastasis in irradiated mice. Cancer Res 2003;63(12):3021–5.
[117] Martin-Caballero J, Flores JM, Garcia-Palencia P, Serrano M. Tumor susceptibility of p21(Waf1/Cip1)-deficient mice. Cancer Res 2001;61(16):6234–8.
[118] Viale A, De Franco F, Orleth A, Cambiaghi V, Giuliani V, Bossi D, et al. Cell-cycle restriction limits DNA damage and maintains self-renewal of leukaemia stem cells. Nature 2009;457(7225):51–6.
[119] LaBaer J, Garrett MD, Stevenson LF, Slingerland JM, Sandhu C, Chou HS, et al. New functional activities for the p21 family of CDK inhibitors. Genes Dev 1997;11(7):847–62.
[120] Welcker M, Lukas J, Strauss M, Bartek J. p21WAF1/CIP1 mutants deficient in inhibiting cyclin-dependent kinases (CDKs) can promote assembly of active cyclin D/CDK4(6) complexes in human tumor cells. Cancer Res 1998;58(22):5053–6.
[121] Rowland BD, Peeper DS. KLF4, p21 and context-dependent opposing forces in cancer. Nat Rev Cancer 2006;6(1):11–23.
[122] Bartek J, Bartkova J, Lukas J. DNA damage signalling guards against activated oncogenes and tumour progression. Oncogene 2007;26(56):7773–9.
[123] Campisi J, d'Adda di Fagagna F. Cellular senescence: when bad things happen to good cells. Nat Rev Mol Cell Biol 2007;8(9):729–40.
[124] Kamb A, Gruis NA, Weaver-Feldhaus J, Liu Q, Harshman K, Tavtigian SV, et al. A cell cycle regulator potentially involved in genesis of many tumor types. Science 1994;264(5157):436–40.
[125] Serrano M, Hannon GJ, Beach D. A new regulatory motif in cell-cycle control causing specific inhibition of cyclin D/CDK4. Nature 1993;366(6456):704–7.
[126] Quelle DE, Zindy F, Ashmun RA, Sherr CJ. Alternative reading frames of the INK4a tumor suppressor gene encode two unrelated proteins capable of inducing cell cycle arrest. Cell 1995;83(6):993–1000.
[127] Stein GH, Drullinger LF, Soulard A, Dulic V. Differential roles for cyclin-dependent kinase inhibitors p21 and p16 in the mechanisms of senescence and differentiation in human fibroblasts. Mol Cell Biol 1999;19(3):2109–17.
[128] Li J, Poi MJ, Tsai MD. Regulatory mechanisms of tumor suppressor P16(INK4A) and their relevance to cancer. Biochemistry 2011;50(25):5566–82.
[129] Matsushime H, Ewen ME, Strom DK, Kato JY, Hanks SK, Roussel MF, et al. Identification and properties of an atypical catalytic subunit (p34PSK-J3/cdk4) for mammalian D type G1 cyclins. Cell 1992;71(2):323–34.
[130] Romagosa C, Simonetti S, Lopez-Vicente L, Mazo A, Lleonart ME, Castellvi J, et al. p16(Ink4a) overexpression in cancer: a tumor suppressor gene associated with senescence and high-grade tumors. Oncogene 2011;30(18):2087–97.
[131] Mendenhall WM, Logan HL. Human papillomavirus and head and neck cancer. Am J Clin Oncol 2009;32(5):535–9.
[132] Nevins JR. E2F: a link between the Rb tumor suppressor protein and viral oncoproteins. Science 1992;258(5081):424–9.
[133] Rodier F, Coppe JP, Patil CK, Hoeijmakers WA, Munoz DP, Raza SR, et al. Persistent DNA damage signalling triggers senescence-associated inflammatory cytokine secretion. Nat Cell Biol 2009;11(8):973–9.
[134] Campisi J. Suppressing cancer: the importance of being senescent. Science 2005;309(5736):886–7.
[135] Klement K, Goodarzi AA. DNA double strand break responses and chromatin alterations within the aging cell. Exp Cell Res 2014;329(1):42–52.
[136] Lleonart ME, Artero-Castro A, Kondoh H. Senescence induction; a possible cancer therapy. Mol Cancer 2009;8:3.
[137] Shay JW, Roninson IB. Hallmarks of senescence in carcinogenesis and cancer therapy. Oncogene 2004;23(16):2919–33.
[138] Beausejour CM, Krtolica A, Galimi F, Narita M, Lowe SW, Yaswen P, et al. Reversal of human cellular senescence: roles of the p53 and p16 pathways. EMBO J 2003;22(16):4212–22.
[139] Jacobs JJ, de Lange T. Significant role for p16INK4a in p53-independent telomere-directed senescence. Curr Biol 2004;14(24):2302–8.
[140] Di Micco R, Sulli G, Dobreva M, Liontos M, Botrugno OA, Gargiulo G, et al. Interplay between oncogene-induced DNA damage response and heterochromatin in senescence and cancer. Nat Cell Biol 2011;13(3):292–302.

Chapter 16

Roles of RAD18 in DNA Replication and Postreplication Repair

C. Vaziri[1], S. Tateishi[2], E. Mutter-Rottmayer[1], Y. Gao[1]
[1]University of North Carolina at Chapel Hill, Chapel Hill, NC, United States; [2]Kumamoto University, Kumamoto, Japan

Chapter Outline

1. INTRODUCTION: THE DDR, DNA DAMAGE–TOLERANCE AND DNA DAMAGE–AVOIDANCE MECHANISMS

DNA damage poses a serious threat to genome stability and the S-phase of the cell cycle is particularly vulnerable to the detrimental effects of bulky replication fork-stalling DNA lesions. Cells have evolved an elaborate signaling network termed the DNA-damage response (DDR) that coordinates DNA replication and DNA repair with cell-cycle progression following genotoxic exposures. DNA damage acquired during S-phase elicits three important protective responses that are mediated at least in large part by the ATR and Chk1 checkpoint kinases [1]: Inhibition of initiation of DNA synthesis at unfired origins of replication and slowing of ongoing replication forks (a mechanism termed the "S-phase checkpoint") [1,2]; Stabilization of stalled replication forks, the crucial function of S-phase checkpoint signaling [2,3]; Inhibition of entry into mitosis in the presence of un-replicated DNA, a mechanism also termed the "replication checkpoint" [3,4]. It has become clear that attenuation of S-phase checkpoint signaling and recovery from DNA damage–induced cell-cycle delays is critically dependent on postreplication repair (PRR) mechanisms that facilitate resolution of stalled DNA replication forks and permit continued S-phase progression on damaged genomic DNA templates [5,6]. PRR of damaged DNA may proceed via trans-lesion synthesis (TLS), a DNA damage–tolerance process that uses error-prone Y-family DNA polymerases to synthesize daughter strand DNA using a damaged template (Fig. 16.1, left). Alternatively, cells may employ an error-free DNA damage–avoidance mechanism termed "template switching" (TS) that depends on the presence of a newly synthesized sister chromatid DNA template (Fig. 16.1, right). Collectively, TLS- and TS-mediated PRR mechanisms allow cells to survive exposure to a variety of genotoxins.

Genome Stability. http://dx.doi.org/10.1016/B978-0-12-803309-8.00016-1

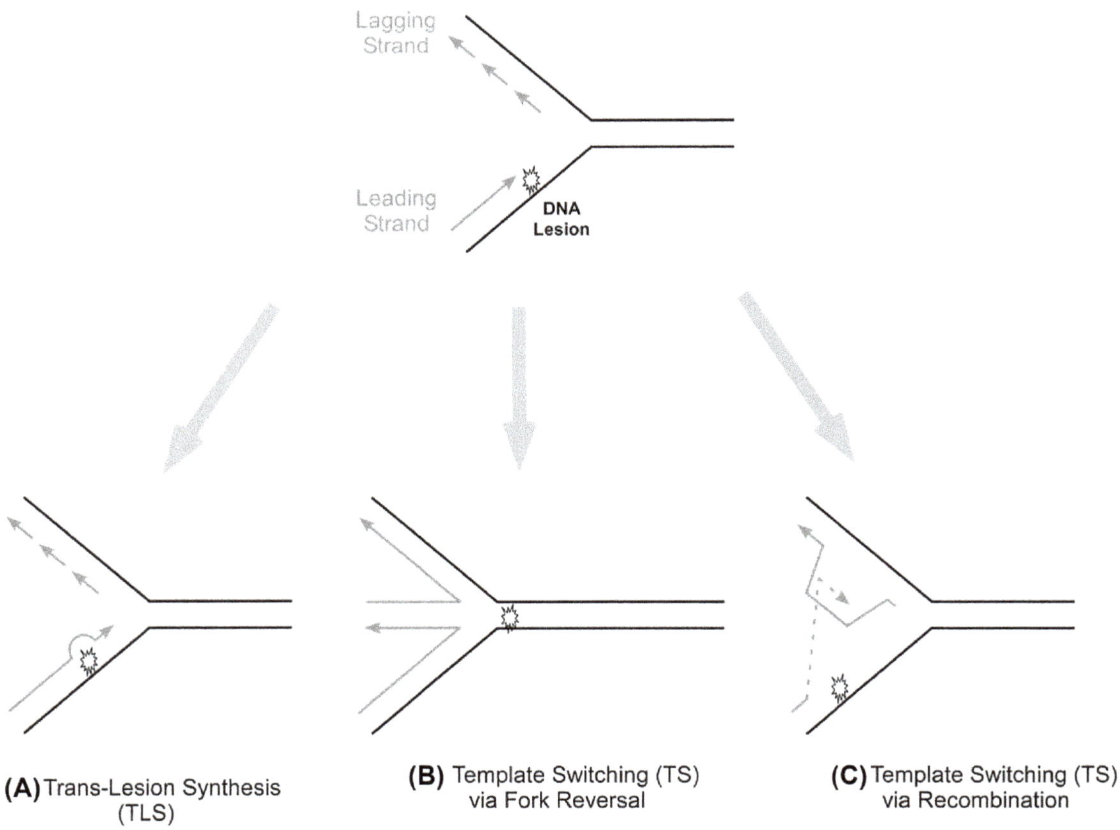

FIGURE 16.1 **Potential mechanisms of postreplication repair via TLS and TS.** During TLS (A), specialized DNA damage–tolerant Y-family DNA polymerases are recruited to stalled replication forks where they perform error-prone DNA synthesis using damaged templates. TS may proceed via fork reversal (B) or recombination-based (C) mechanisms, both using a newly synthesized undamaged sister chromatid as template for error-free DNA synthesis. See text for details.

TLS and TS are activated by ubiquitination of the DNA polymerase processivity factor Proliferating Cell Nuclear Antigen (PCNA). TLS relies on monoubiquitination of PCNA at Lysine 164 (K164), while TS is promoted by PCNA K164 polyubiquitination. RAD18 is the major PCNA K164-directed E3 ubiquitin ligase in eukaryotic cells. RAD18 exists as a complex with the E2 ubiquitin–conjugating enzyme RAD6 and is activated coincident with the S-phase checkpoint. Therefore, the RAD18–RAD6 complex represents a proximal activator of both TLS and TS pathways. Here we review the activation mechanisms of RAD18, and discuss the roles of its effector TLS and TS pathways in genome maintenance. In particular, we emphasize the basis for coordination of RAD18 with other elements of the DDR. Finally, we consider the potential impact of RAD18-mediated genome maintenance on development and disease.

2. IDENTIFICATION OF RAD18–RAD6 AS A MEDIATOR OF DNA DAMAGE TOLERANCE

The *Saccharomyces cerevisiae RAD18* and *RAD6* genes (encoding E3 ubiquitin ligase and E2 ubiquitin–conjugating enzymes RAD18 and RAD6, respectively) belong to the same epistasis group and were identified based on their roles in conferring tolerance of ultraviolet (UV) light and chemically induced DNA damage [7–9]. *rad18* and *rad6* mutant yeast have PRR defects and accumulate discontinuities in newly replicated DNA following genotoxin exposure [8,10]. Moreover, DNA damage–inducible mutagenesis is attenuated in *rad6* and *rad18* mutants. *S. cerevisiae* RAD6 is a ubiquitin-conjugating (E2) enzyme that can use histones H2A and H2B as substrates [11]. RAD18 associates directly with RAD6, has zinc finger domains that mediate nucleic acid binding [12], binds to ssDNA [13] and has ubiquitin-conjugating and ATP hydrolytic activities [14]. Prakash and colleagues first suggested that DNA-binding and nucleotide-binding activities might enable RAD18 protein to recognize damaged template DNA with high affinity [15]. Furthermore, these workers proposed that ubiquitination of replication factors may be required for activation of postreplicative bypass DNA-repair machinery [13,14].

Human *RAD18* was identified based on homology to the yeast *RAD18* gene. There are two human RAD6 homologues, RAD6A and RAD6B, both of which interact with RAD18 [16,17]. Human cells expressing hRAD18 protein with a "really-interesting gene" (RING) finger mutation are compromised for PRR [16]. Similarly, *Rad18*-knockout mouse embryonic stem cells generated by gene targeting are PRR-defective and hypersensitive to multiple DNA-damaging agents [18]. Mutation rates (measured by ouabain resistance) are similar between wild-type and *Rad18*-knockout cells. However, spontaneous sister chromatid exchanges (SCEs), random targeting of exogenous DNA into the genome, and gene targeting at the Oct3/4 locus are increased as a result of Rad18-deficiency, demonstrating that Rad18 represses illegitimate recombination events [18]. Increased SCE rates are also observed in *RAD18*$^{-/-}$ DT40 cells, indicative of a role for RAD18 in suppression of HR-mediated PRR [19]. Therefore, similar to *rad18* mutant yeast, *RAD18*-deficient vertebrate cells exhibit genome maintenance defects, indicating conservation of RAD18 function between species.

3. RAD18-MEDIATED PCNA MONOUBIQUITINATION AND THE TLS POLYMERASE SWITCH

PCNA is the critical target whose modification by RAD18-RAD6 directs PRR pathway activation [20]. Jentsch and colleagues showed that RAD18 recruits RAD6 to chromatin to promote PCNA monoubiquitination at K164. The ubiquitin-conjugating MMS2–UBC13 complex is recruited to chromatin by RAD5 (another RING-finger E3 ligase) leading to further K63-linked multi-ubiquitination of the monoubiquitinated PCNA. Thus, different PCNA modifications target for alternative functions in PRR. Stelter and Ulrich showed that PCNA monoubiquitination activates TLS via DNA polymerases eta and zeta, whereas PCNA polyubiquitination promotes error-free repair [21]. PCNA ubiquitination was also shown to be required for DNA damage–induced mutagenesis. Taken together these important studies demonstrated that PRR activation and the selection of error-prone TLS vs. error-free TS pathways are dependent upon posttranslational modifications of PCNA.

K164 is present in human PCNA, indicating that the mechanism of TLS pathway activation is conserved across species [20]. Lehman and colleagues demonstrated that UV irradiation induces PCNA monoubiquitination in a RAD18-dependent manner in human cells and that DNA polymerase eta (Polη, the mammalian homologue of yeast RAD30) associates preferentially with K164-monoubiquitinated PCNA [22]. Interestingly, RAD18 also has a noncatalytic role in regulating TLS via its interactions with Polη [23] (described in more detail later). RAD18-mediated PCNA monoubiquitination also promotes recruitment of DNA polymerase kappa (Polκ) [24], DNA polymerase iota (Polι) [25], and REV1 [26] to sites of replication fork stalling in genotoxin-treated cells. It is unclear whether Y-family polymerases other than Polη are regulated via direct interactions with RAD18. The presence of specialized ubiquitin-binding zinc finger (UBZ) and ubiquitin-binding motif (UBM) domains in the Y-family DNA polymerases provides the molecular basis for the association of Y-family TLS DNA polymerases with monoubiquitinated PCNA [27].

The extent to which PCNA-monoubiquitination is necessary for recruitment of Y-family polymerases to stalled replication forks has been controversial. In one study, PCNA ubiquitination did not disrupt Polδ–PCNA interactions or enhance the binding affinity of TLS DNA polymerases for PCNA, leading to the suggestion that K164 monoubiquitination displaces putative inhibitors of PCNA–TLS polymerase interactions [28]. A UBZ-deficient Polη mutant retaining the PCNA-interacting peptide ("PIP" domain) was able to complement UV-sensitivity defects of *xeroderma pigmentosum*–variant (XPV) cells which lack endogenous Polη [29], further suggesting that PCNA monoubiquitination is nonessential for TLS polymerase activation. However, in "knock-in" mouse cells harboring K164-mutated ubiquitination-resistant PCNA [30], Polη recruitment to stalled replication forks and TLS-mediated recovery from replication fork stalling are compromised but not completely attenuated. Most probably, therefore, monoubiquitinated PCNA does promote TLS but additional mechanisms (likely involving ubiquitin-independent PIP box interactions) contribute to stable association of Y-family TLS polymerases with PCNA.

4. RAD18 STRUCTURE, ACTIVATION, AND COORDINATION WITH THE DDR

4.1 RAD18 Structure

The domain organization of the 495 amino acid (AA) hRAD18 protein is shown in Fig. 16.2 and illustrates major conserved domains including the RING motif (AAs 25–63), a UBZ4-type zinc finger (AAs 201–225), the SAF-A/B, Acinus and PIAS (SAP) domain (AAs 248–282), the RAD6-binding domain (AAs 340–395), and a Polη-binding motif (AAs 401–445) [31]. A crystal structure for the RAD18–RAD6 complex is not yet available. However, biophysical studies indicate that RAD18 exists as an asymmetric heterotrimer consisting of two RAD18 molecules and a single molecule of RAD6 [32,33]. Multiple contacts between RAD18 and RAD6 are necessary for formation of the $[RAD18]_2$–RAD6 complex. The RAD18 RING domain is necessary for PCNA ubiquitination activity [16]. RING domains generally serve as interaction sites for E2

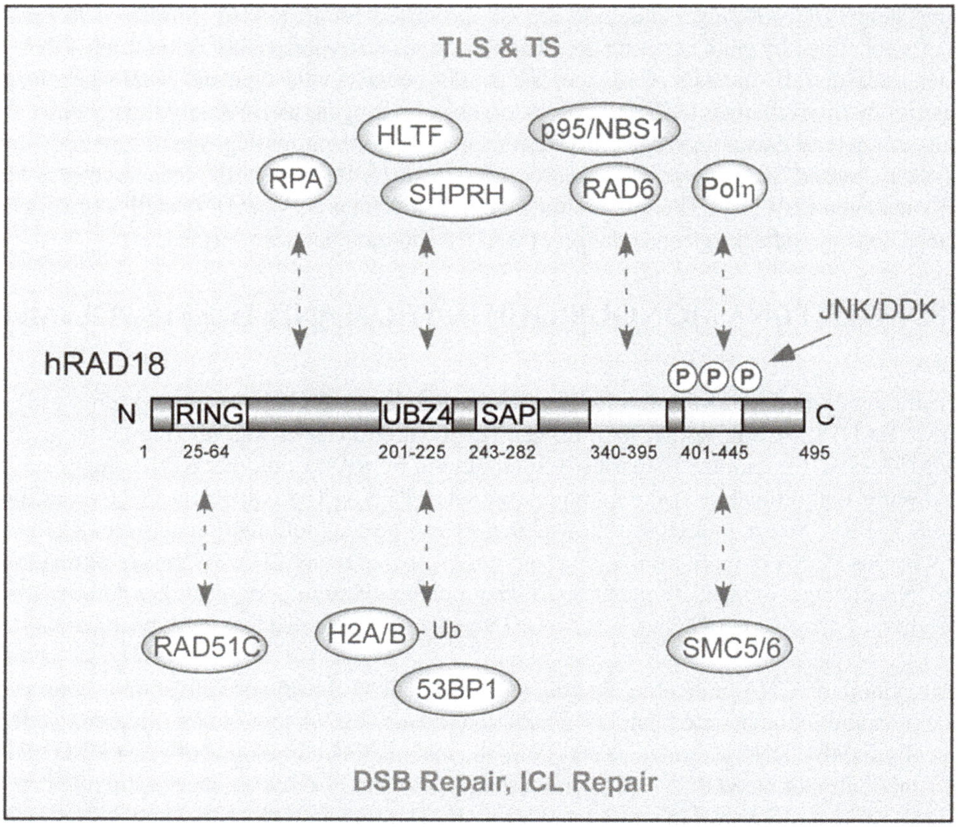

FIGURE 16.2 **Domain structure of hRAD18 indicating key domains involved in TLS, TS, and other genome maintenance activities.** The diagram shows relative locations of RING, UBZ, and SAP domains in the full-length (495 amino acid, AA) human RAD18 protein. Interaction sites for several key binding partners that mediate TLS and TS (top half of figure) and DSB/ICL repair (bottom half of figure) are indicated. The region spanning AAs 401–445 contains phosphorylation sites for JNK (serine 409) and a cluster of DDK sites (residing in serine residues 432–444). JNK and DDK-mediated phosphorylations are Chk1 dependent and promote associations with Polη (S409, S432–444) and with the SMC5/6 proteins (S432–444) to promote TLS and ICL repair, respectively. See text for details.

enzymes and bring substrates in proximity of the E2 to promote ubiquitination. Similar to other E3 ligases, the N-terminal RING domain of RAD18 contributes to E2 (RAD6) binding [16,31,34]. The RAD18 UBZ domain belongs to the UBZ4 subgroup that is also present in Polκ and WRIP1 [35] The UBZ4 domain is dispensable for RAD18–RAD6 complex formation, catalytic activity, and TLS [33], yet may facilitate DNA binding and may contribute to self-dimerization [16,34]. UBZ-mediated interactions between Rad18 and monoubiquitinated PCNA may also facilitate retention of Rad18 at sites of replication fork stalling, providing a feed-forward mechanism that amplifies the PCNA monoubiquitination response [36].

As discussed later (Section 7), RAD18 participates in DNA double-stranded breaks (DSB) repair independently of its role in TLS and the UBZ motif may facilitate RAD18 recruitment to DSB-flanking ubiquitinated histones [37]. The SAP domain [38] facilitates RAD18 recruitment into Polη-containing nuclear foci, PCNA monoubiquitination, and UV DNA-damage tolerance [39,40], yet is dispensable for the recruitment of RAD18 to DNA DSB [37]. Residues 401–445 of RAD18 interact with Polη and this association is necessary for efficient chaperoning of Polη to sites of replication stalling [23]. The importance of the RAD18–Polη interaction is demonstrated by the observation that Polη interaction–deficient RAD18 mutants that retains E3 ligase activity are compromised for DNA-damage tolerance [23,41]. As discussed later, the Rad18–Polη interaction also integrates TLS with the cell cycle and other genome maintenance pathways.

4.2 RAD18 Activation

DNA damage–induced accumulation of monoubiquitinated PCNA results both from inhibition of PCNA de-ubiquitination [42], and from increased PCNA ubiquitination by RAD18. The RAD18-inducible component of the overall PCNA ubiquitination seems to be a multistep process involving RAD18 recruitment to ssDNA in the vicinity of stalled DNA replication forks, followed by a Polη-mediated "hand-off" to PCNA, as described further on.

DNA damage–induced stalling of replicative DNA polymerases causes uncoupling of leading and lagging strand DNA synthesis and leads to single-stranded DNA (ssDNA) accumulation [43]. In *S. cerevisiae*, UV-induced replication stalling increases the length of replication-associated ssDNA tracts from about 100 to 200 bases [44]. ssDNA is the proximal trigger that activates several branches of the DDR including the ATR/Chk1-mediated S-phase checkpoint [45]. PCNA ubiquitination is selectively induced by genotoxins that generate ssDNA via uncoupling of replicative helicase and polymerase activities [46]. It has long been known that RAD18 has ssDNA-binding activity [13] and ssDNA generated during replication fork stalling is probably the basis for the initial recruitment of RAD18 to the vicinity of damaged DNA. Indeed, RAD18 preferentially recognizes synthetic ssDNAs that resemble replication fork intermediates [40]. ssDNA generated by stalled replication forks is coated by replication protein A (RPA), and RPA-ssDNA is a key mediator of ATR/Chk1 pathway activation. In *S. cerevisiae*, 95% degradation of temperature-sensitive rfa1 (the large subunit of yeast RPA) mutant sustains DNA replication yet abolishes PCNA monoubiquitination, indicative of a role for RPA-ssDNA accumulation in RAD18 activation [46]. Moreover, RAD18–RAD6 complex interacts with RFA1 and RFA2 subunits of yeast RPA, even in the absence of DNA. An N-terminal domain of yeast RAD18 confers RPA-binding activity while the SAP domain (necessary for ssDNA binding) is dispensable for RPA association. Therefore, recruitment of RAD18 to DNA at sites of replication stalling may require independent interactions of RAD18 with RPA and ssDNA, at least in yeast. An RPA-ssDNA-based mechanism of RAD18 activation explains the temporal correlation of PCNA ubiquitination and Chk1 phosphorylation in genotoxin-treated cells and provides a parsimonious mechanism for simultaneous activation of two major elements of the DDR (TLS and the S-phase checkpoint).

Although RPA-coated ssDNA might explain the initial recruitment of Rad18 to the local environment of stalled replication forks, this model does not explain how Rad18 associates with PCNA, its critical substrate in the TLS pathway. RAD18 lacks a PIP box or any known PCNA-interacting motifs. However, the RAD18–Polη interaction may facilitate association of RAD18 with PCNA: Polη interacts with PCNA via a PIP box, thereby providing a potential mechanism for targeting the Polη-bound RAD18 to PCNA. Indeed, Polη promotes association of RAD18 with PCNA *and* enhances PCNA monoubiquitination in vitro and in cultured human cells [47]. A catalytically inactive Polη mutant retains RAD18-binding activity, promotes PCNA monoubiquitination, and stimulates the recruitment of other TLS polymerases to PCNA [47]. Moreover, UV sensitivity of Polη-deficient cells is partially rescued by the expression of catalytically inactive Polη [48]. Therefore, Polη has a noncatalytic scaffolding role in promoting RAD18-mediated PCNA monoubiquitination and DNA-damage tolerance.

The RAD18-binding motif of Polη has not been mapped precisely, yet resides in a C-terminal domain (AAs 594–713) that is frequently deleted in XPV patients [49]. Therefore, genome instability in some XPV patients may result from defective Polη scaffold function and altered targeting of RAD18 to PCNA. The extent to which the other Y-family polymerases associate with RAD18 and promote PCNA monoubiquitination is unclear, although in a side-by-side comparison, Polκ fails to promote PCNA monoubiquitination as efficiently as Polη [50]. Interestingly, substitution of the Polκ PIP box with the Polη core PIP sequence plus PIP box-flanking residues confers increased PCNA monoubiquitination activity upon Polκ [47]. Therefore, the high affinity of the Polη PIP box for PCNA may explain why Polη supports RAD18-mediated PCNA ubiquitination preferentially when compared with other Y-family DNA polymerases.

RAD18 can perform sequential monoubiquitinations of multiple units of the PCNA homotrimer and the mono- and multi-monoubiquitinated PCNA trimers might activate distinct modes of DNA-damage tolerance [36]. Interestingly, trimeric PCNA complexes containing one or two K164-monoubiquitinated monomers are ubiquitinated more efficiently by RAD18 when compared with unmodified PCNA trimers [36]. That is, PCNA monoubiquitination appears to stimulate further ubiquitination of the other PCNA subunits. It is possible that the UBZ domain of RAD18 mediates its retention at monoubiquitinated PCNA, establishing a feed-forward mechanism for enhanced monoubiquitination of other PCNA monomers in the same trimer.

In addition to RPA-ssDNA and Polη, several other proteins may influence RAD18-mediated PCNA ubiquitination and TLS at sites of DNA replication stalling. For example, the orphan protein C1orph124 (also designated "Spartan") facilitates RAD18–PCNA association and modestly stimulates PCNA monoubiquitination [51]. Spartan/C1orf124 also interacts with the replicative DNA polymerase POLD3 and PDIP1 in the absence of DNA damage, but preferentially associates with Polη upon UV damage, perhaps indicating additional roles for Spartan in the polymerase switch [52]. Spartan may also promote accumulation of monoubiquitinated PCNA independently of its putative role in RAD18 activation by protecting against de-ubiquitination [53]. It must be noted, however, that the role of Spartan in TLS is not entirely clear since other studies indicate Spartan is not required for PCNA monoubiquitination, but instead interacts with p97 "segregase" to promote removal of Polη from sites of UV-induced DNA damage, thereby reducing mutagenesis [54]. Other reports indicate that Spartan depletion increases rates of mutagenesis [55]. Clearly therefore, the roles of Spartan in regulating TLS are complex and incompletely understood. Han and colleagues in 2014 identified the ARF-directed E3 ligase SIVA1 as another mediator that physically bridges chromatin-bound RAD18 and PCNA [56]. Therefore, SIVA1 may function as substrate receptor for RAD18 ubiquitin ligase that promotes PCNA ubiquitination.

Other proteins with known roles in distinct genome maintenance pathways have also been implicated in RAD18-mediated TLS. p95/NBS1 (mutated in Nijmegen breakage syndrome) interacts directly with the RAD6-binding domain of RAD18 [57] and promotes RAD18 distribution to sites of DNA replication stalling, stimulating PCNA monoubiquitination [57]. The BRCA1 (breast cancer 1) protein, a major component of the HR pathway, also recruits RPA, RAD18, Polη, and REV1 to damaged chromatin to promote TLS and template switching [58]. The participation of major DSB-sensing and repair factors in TLS is indicative of extensive crosstalk and coordination between genome maintenance pathways.

In summary, multiple factors (RPA, ssDNA, Polη, NBS1, BRCA1, SIVA1, and doubtless other proteins) associate with RAD18 and/or create a local environment that is permissive for PCNA monoubiquitination and TLS at stalled replication forks.

4.3 Transcriptional and Posttranslational Regulation of RAD18

Ectopic over-expression of RAD18 in cultured cells induces DNA damage–independent PCNA monoubiquitination, drives TLS polymerases to sites of DNA replication [24], and confers DNA-damage tolerance [59]. Therefore, stringent control of RAD18 expression is important for limiting error-prone DNA synthesis and maintaining genome stability. During the cell cycle, RAD18 protein levels are relatively low in G1, increase during S-phase, and decrease rapidly following mitosis [60]. Interestingly, the *RAD18* promoter is a target of the DNA damage–inducible E2F family member E2F3, which mediates transcriptional induction of RAD18 expression in genotoxin-treated cells [61]. Other mechanisms for transcriptional regulation of RAD18 expression have not been described. However, RAD18 protein levels are regulated via its ubiquitin-dependent proteolysis. RAD18 is polyubiquitinated (via auto-ubiquitination) and the polyubiquitinated species is targeted for proteasomal degradation [34]. A 2015 siRNA screen identified RAD18 as a target of the de-ubiquitinating enzyme USP7 [62]. Thus, USP7-mediated removal of polyubiquitin chains from RAD18 confers stability and represents an important mechanism for maintaining DNA-damage tolerance via TLS.

Integration of TLS with S-phase, checkpoint signaling, and stress kinase pathways is achieved through RAD18 phosphorylation [41,63]. The Polη-binding domain of hRAD18 contains a cluster of DBF4/DRF1-dependent kinase (DDK) phosphorylation sites (including the preferred DDK phosphorylation site at S434) embedded in an acidic region termed the "S-box" [41] and a c-Jun N-terminal kinase (JNK) phosphorylation site at S409 [63]. DDK is a critical protein kinase for the initiation of DNA synthesis [64] and JNK mediates signaling in response to diverse cellular stresses, including many genotoxic agents [65]. The JNK and DDK phosphorylation sites of RAD18 are conserved between species and serve to promote RAD18–Polη complex formation, contributing to DNA-damage tolerance. DBF4, the activating subunit of DDK binds RAD18 and likely directs CDC7 to RAD18 [66]. Interestingly, DBF4 might also promote PCNA monoubiquitination by facilitating RAD18 recruitment to damaged chromatin independently of its role in DDK-mediated RAD18 phosphorylation. RAD18 phosphorylation by JNK and DDK depends on Checkpoint Kinase 1 (CHK1), a key mediator of the S-phase checkpoint [67]. Therefore, RAD18 phosphorylation by DDK and JNK coordinates TLS with DNA replication and stress kinase signaling via the S-phase checkpoint.

ATR/CHK1 signaling promotes PCNA monoubiquitination [24,68], although the mechanism of Chk1-induced PCNA monoubiquitination is not known. CHK1-dependent formation of the RAD18–Polη complex (required for targeting RAD18 to PCNA) provides a plausible mechanism for the stimulatory effect of CHK1 on PCNA monoubiquitination [24,68]. The association of RAD18 with Polη also provides a basis for integrating RAD18-mediated PCNA monoubiquitination with p53 signaling. The *POLH* gene (encoding Polη) is a transcriptional target of p53 and Polη protein levels are induced by DNA damage [69]. In cultured cells, RAD18 protein is present in excess of Polη by about 100-fold [47], and consequently Polη levels are limiting for recruitment of RAD18 to PCNA. However, DNA damage–induced p53 activity stimulates Polη expression, increasing the availability of Rad18–Polη complexes that associate efficiently with PCNA.

In summary, we propose an integrated model for initiation of TLS (Fig. 16.3) in which the RAD18–RAD6–Polη complex is first recruited to the vicinity of stalled replication forks via interaction of RAD18 with RPA-coated ssDNA. Subsequent association of the RAD18 complex with PCNA is facilitated by Polη scaffolding activity, leading to K164 monoubiquitination of one PCNA subunit. Additional scaffolding proteins, such as Spartan, p95/NBS, and SIVA1 may facilitate the interaction of RAD18 with PCNA. USP7 and p53 contribute to maintaining RAD18 expression levels. DDK/JNK-mediated RAD18 phosphorylation preserves RAD18–Polη interactions and promotes PCNA monoubiquitination and TLS.

5. DNA REPLICATION–INDEPENDENT RAD18 ACTIVATION AND TLS

There is now considerable evidence that RAD18-mediated lesion bypass occurs postreplicatively and serves to fill ssDNA gaps remaining behind a newly-primed leading strand [70–72]. For example, TLS deficiency does not affect rates of leading

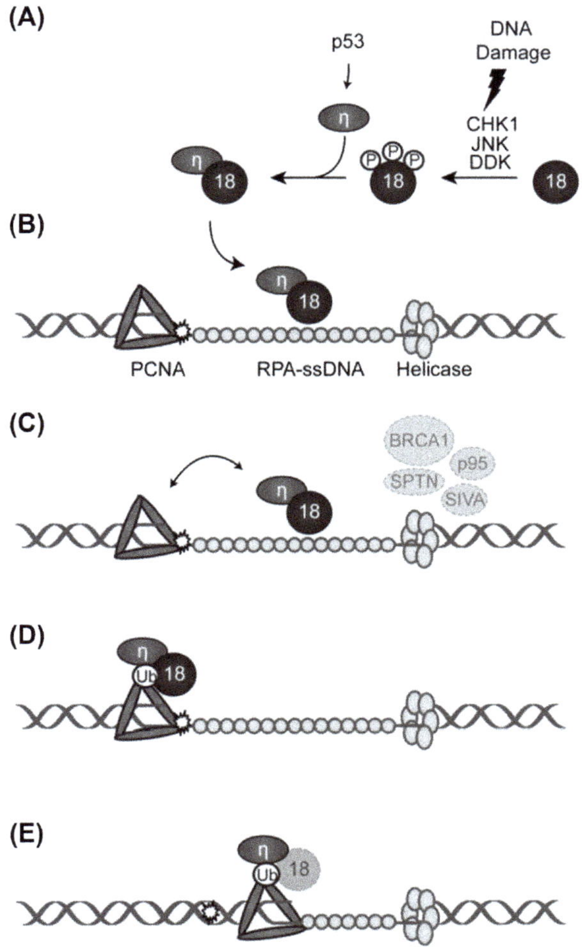

FIGURE 16.3 **Mechanisms of RAD18 recruitment to stalled replication forks.** (A) DNA damage induces RAD18 phosphorylation (by JNK and DDK), promoting its association with Polη. DNA damage also induces Polη expression via p53-dependent transcription, further contributing to the formation of RAD18–Polη complexes. (B) The RAD18–Polη complex is recruited to the vicinity of stalled replication forks via interactions between RAD18 and RPA-ssDNA. (C) Polη binds PCNA, thereby serving as a scaffold that mediates association of RAD18 with PCNA. Various other factors including BRCA1, p95/NBS1, Spartan, and SIVA may interact with core TLS proteins or create a local environment that facilitates RAD18 interactions with PCNA. (D) RAD18 monoubiquitinates PCNA leading to high affinity binding of Polη and other Y-family TLS polymerases. (E) Association of TLS polymerases with monoubiquitinated PCNA allows replicative bypass of DNA lesions.

strand synthesis on damaged templates, but instead leads to postreplicative gaps [72]. Limiting TLS to G2/M phase efficiently promotes lesion tolerance, fully consistent with the idea that TLS serves to fill ssDNA gaps behind newly re-primed replication forks [70]. In elegant experiments that visualized and quantified PRR tracts, TLS was temporally and spatially separable from global genomic DNA replication [71]. Thus, RAD18-mediated TLS is truly a PRR mechanism that operates distal to active replication forks.

Interestingly, several studies show that RAD18/TLS-mediated patch filling is not necessarily restricted to ssDNA behind replication forks, and also contributes to repair of ssDNA breaks (SSBs) that arise outside S-phase. For example, UV irradiation of quiescent (G0) human fibroblasts induces PCNA monoubiquitination and PCNA association of Polκ [73–75]. Polκ-deficient MEF exhibit reduced repair synthesis activity, particularly in the presence of the ribonucleotide reductase (RNR) inhibitor hydroxyurea (HU), suggesting that TLS polymerases participate in nucleotide excision repair (NER) when dNTP concentrations are limiting [76]. In nonproliferating cells, exonuclease 1 (EXO1) activity converts NER intermediates to long ssDNA gaps that are capable of activating the Chk1 pathway [77,78]. Similarly, it is likely that RPA-coated ssDNA generated at sites of NER could recruit RAD18, thereby initiating TLS independently of DNA replication. Indeed, recruitment of Polκ to monoubiquitinated PCNA is observed in nonreplicating wild-type but not XPA cells [73]. Therefore, DNA intermediates, such as ssDNA generated during the incision phase of NER are likely to initiate TLS outside S-phase.

RAD18-mediated PCNA monoubiquitination is also inducible by H_2O_2 (a source of oxidative DNA damage) in nonreplicating cells [74,79]. In contrast with UV-induced DNA damage (which induces PCNA ubiquitination via NER intermediates), H_2O_2-induced PCNA monoubiquitination depends on the MSH2–MSH6 complex (but not on MLH1) [79]. Thus, oxidative stress-induced clustered lesions evading repair by DNA glycosylases may activate MSH2–MSH6 to load an exonuclease (likely EXO1) that generates the ssDNA tracts needed to activate RAD18. RAD18-mediated PCNA monoubiquitination then facilitates recruitment of Polη, which contributes to repair synthesis. RAD18-mediated TLS is essential for facilitating completion of DNA replication and conferring cell survival after oxidative injury in S-phase [74]. Interestingly, however, the role of RAD18 in preventing H_2O_2-induced DSBs and lethality during G1 is nonessential owing to backup nonhomologous end joining (NHEJ)-mediated DSB repair [74]. Alkylating agents, such as MNNG also induce S-phase-independent PCNA monoubiquitination via noncanonical Mismatch Repair (MMR) [80]. While H_2O_2-induced PCNA monoubiquitination is MLH1 independent, MLH1 is necessary for PCNA monoubiquitination following exposure to MNNG. Therefore, noncanonical MMR in G1 may lead to MUTLα-induced endonucleolytic nicks and loading of EXO1, generating the ssDNA required for RAD18 activation and PCNA monoubiquitination. Extension of ssDNA tracts (by EXO1 and/or other exonucleases) likely represents a general mechanism for replication fork-independent recruitment of RAD18 to sites of NER or SSB repair. RAD18-mediated TLS can also repair ssDNA breaks persisting after replication in G2 and UV-induced PCNA ubiquitination is observed in synchronized metaphase-arrested cells [79]. Therefore, RAD18-mediated PCNA monoubiquitination occurs throughout the cell cycle. Why then would cells use error-prone TLS DNA polymerases in lieu of the error-free polymerases conventionally employed for NER or SSB repair? One possibility is that TLS polymerases may be required for SSB repair when clustered DNA lesions are generated on both strands. In addition, TLS DNA polymerases may be more efficient than high-fidelity DNA-repair polymerases when nucleotide concentrations are low (as is the case in G1 cells). With the realization that TLS is operational outside S-phase, RAD18 and its effector Y-family polymerases represent potential mediators of genome maintenance in diverse nonreplicating cell types including quiescent stem cells, postmitotic and differentiated neurons, and cardiomyocytes that experience high levels of oxidative stress.

6. RAD18 FUNCTIONS IN ERROR-FREE PRR VIA TEMPLATE SWITCHING

The error-free PRR pathway uses a newly synthesized daughter strand of the undamaged complementary sequence as a template for extending stalled leading strands [81]. The molecular basis of TS is not fully understood, but there is evidence for both fork reversal and recombination-mediated template-switching mechanisms (Fig. 16.1, right), as described further on.

In *S. cerevisiae*, error-free PRR involves the *RAD6* epistasis group genes *MMS2*, *UBC13*, and *RAD5* which prevent accumulation of daughter strand discontinuities opposite fork-stalling DNA lesions [82–84]. The ubiquitin-conjugating enzyme UBC13 and a noncanonical UBC variant MMS2 form a heteromeric complex with RAD5 [83]. RAD5 is an SWI/SNF ATPase family member [84] that contains a C3HC4 RING motif [85] and possesses DNA-dependent ATPase activity [84]. RAD5 recruits UBC13-MMS2 to damaged chromatin, to form a complex that cooperates with RAD6–RAD18 to polyubiquitinate PCNA at K164 [86].

There are two known mammalian RAD5 homologues, SHPRH and HLTF. Elegant biochemical studies have shown that purified HLTF and SHPRH cooperate with RAD18-RAD6 to polyubiquitinate PCNA, yet achieve PCNA polyubiquitination via distinct mechanisms. SHPRH polyubiquitinates PCNA via extension of monoubiquitinated K164 [87]. On the other hand, HLTF forms a thiol-linked Ub chain on UBC13 that is transferred to RAD6. RAD18 then transfers the pre-conjugated Ub chain to K164 of unmodified PCNA [20,88].

SHPRH mediates alkylating agent (MMS)-induced PCNA polyubiquitination and confers tolerance to MMS (but not to UV, 4-NQO, and MMC [89]), whereas HLTF mediates PCNA polyubiquitination and confers DNA-damage tolerance in response to bulky DNA lesions [90]. In UV-irradiated mammalian cells, HLTF enhances PCNA monoubiquitination and Polη recruitment, while inhibiting SHPRH function. Conversely, MMS promotes SHPRH–RAD18 interactions, while inducing HLTF degradation. Thus, HLTF and SHPRH promote error-free PRR in a DNA damage–specific manner [91].

It is hypothesized that polyubiquitinated PCNA generated via the concerted actions of RAD18 and RAD5 recruits the mediators of the TS pathway to stalled replication forks. ZRANB3 (Zn finger, RAN-binding domain containing 3, also known as Annealing Helicase two or AH2) is recruited to polyubiquitinated PCNA where it facilitates fork regression, replication fork restart, and DNA-damage tolerance [92–94]. Most likely, additional proteins remain to be identified whose docking at polyubiquitinated PCNA promotes template switching.

Biochemical studies in 2015 suggested a mechanism for HLTF in promoting fork reversal-based template switching [95,96]. Fork reversal occurs when the stalled replication fork is remodeled by pairing of newly synthesized chromatids to form a fourth regressed DNA duplex termed a Holliday junction (HJ). Fork reversal provides an opportunity for error-free DNA synthesis using the undamaged lagging strand as an alternative template (Fig. 16.1).

HLTF and RAD5 possess dsDNA translocase activity with 3′5′ polarity that catalyzes fork reversal and branch migration in an ATP-dependent fashion [97–99]. RAD5 and HLTF share a HIP116/HLTF RAD5 N-terminal (HIRAN) domain that is crucial for fork reversal activity [95,96]. The HIRAN domain is a unique "OB-fold" (a general nucleic acid–binding domain) that recognizes free 3′-ssDNA ends, thereby targeting HLTF and RAD5 to the 3′-end of the leading strand to direct fork remodeling and reversal [95,96]. Replication fork speed is globally increased in HLTF-deficient cells owing to the lack of fork reversal [95]. SHPRH lacks a HIRAN domain, indicating that additional mechanisms exist for recruiting RAD5 homologues to sites of TS.

HLTF can also promote D-loop formation in a Rad51-independent manner [100], possibly indicating dual roles in fork reversal and recombination-mediated modes of TS. Interestingly, ZRANB3 disrupts D-loops formed by strand invasion [92], perhaps suggesting that HLTF and ZRANB3 act in distinct early and late stages of TS, respectively.

Clearly, error-free and error-prone PRR act in opposition, with the RAD5 pathway preventing error-prone (mutagenic) TLS. It is not clear why cells would employ error-prone PRR (TLS) if an error-free (TS) pathway is available. It has been suggested that TS is employed when DNA damage is too severe to be processed via TLS and results in persistence of 3′-ends at stalled DNA replication forks [96]. Nevertheless, selection of error-free TS vs. error-prone TLS could profoundly influence genome stability and mechanisms of carcinogenesis. *HLTF* promoter methylation and loss of HLTF expression are observed in cancer [101] and may contribute to increased TLS and mutagenesis. In summary, RAD18 can direct both TS- and TLS-mediated PRR. The putative mechanisms that dictate the selection of RAD18-dependent TLS and TS remain to be determined.

7. TLS- AND TS-INDEPENDENT ROLES OF RAD18 IN GENOME MAINTENANCE

Although best known for its roles in error-prone TLS and TS, RAD18 participates in additional genome maintenance pathways, including DSB repair and ICL repair. A detailed discussion of noncanonical TLS/TS-independent RAD18 activities is beyond the scope of this review and roles of RAD18 in DSB and ICL repair are summarized very briefly.

In DT40 cells and mammalian cancer cells, RAD18 promotes homologous recombination [37,102]. RAD18 mediates HR by binding and chaperoning the RAD51C recombinase to "ionizing radiation induced foci" (ICRF, corresponding to sites of DSB repair) in the nucleus [37]. Association of the RAD18–RAD51C complex with IRIF depends upon RNF8, an E3 ligase which monoubiquitinates Histone H2A (and perhaps other chromatin components) in the vicinity of DSBs, and is mediated via the RAD18 UBZ domain. RAD18-mediated RAD51C chaperone activity does not require the SAP domain or E3 ubiquitin ligase activity. Therefore, the role of RAD18 in RAD51C regulation is fully separable from its PRR activities.

In addition to its role in HR, RAD18 may influence DSB repair via NHEJ.

RAD18 is recruited to X-ray-induced DSB in a 53BP1-dependent manner during G1. Moreover, RAD18 monoubiquitinates and promotes chromatin retention of 53BP1, conferring DNA-damage tolerance [103]. The RAD18 UBZ domain (which is dispensable for RAD18-mediated PCNA modification) is required for formation of 53BP1 IRIF. Therefore, mechanisms of RAD18-mediated PCNA and 53BP1 monoubiquitination are separable. 53BP1 plays important roles in the choice of DSB-repair mechanism, promoting NHEJ and inhibiting homology-directed repair (HDR) [104]. Therefore, RAD18–53BP1 signaling might promote DSB repair via NHEJ, although a direct role of RAD18 in NHEJ has not been formally demonstrated.

RAD18 is also implicated as a potential upstream activator of the Fanconi Anemia (FA) pathway. FA is an autosomal-recessive chromosomal instability syndrome characterized by developmental defects, bone marrow failure, and cancer propensity [105]. FA cells are hypersensitive to interstrand cross-link (ICL)-inducing agents including cisplatin and mitomycin C (MMC). There are at least 18 complementation groups of FA and the protein products of the *FANC* genes mutated in FA patients (termed "FANCA" through FANCT) function in a common ICL-repair pathway. When DNA replication forks encounter ICL, an FA "core complex" comprising "FANCs A, B, C, E, F, G, L, and M" functions as a multi-subunit E3 ubiquitin ligase to monoubiquitinate FANCD2 and FANCI. Monoubiquitinated FANCD2-FANCI is the presumed effector of the FA pathway and directs ICL repair, most likely promoting endolytic processing of cross-linked DNA [106].

RAD18 promotes FA pathway activation and FANCD2-dependent DNA-damage tolerance [107–111], although the mechanisms of RAD18-dependent FANCD2 ubiquitination are lesion specific. For bulky benzo[a]pyrene and cisplatin adducts and UV-induced DNA lesions, FA pathway activation requires PCNA monoubiquitination and Polη activation [107,110]. However, FA pathway activation in response to the Topoisomerase inhibitor camptothecin (CPT, which induces replication-dependent DSB) is RAD18 mediated but TLS independent [108]. Precisely how RAD18 facilitates FA pathway activation in response to DSB is unclear. However, catalytically inactive (C28>F-mutated) RAD18 does not support CPT-induced FANCD2 monoubiquitination, possibly indicating that an unidentified RAD18 substrate must be ubiquitinated to mediate FA pathway activation following Topoisomerase I inhibition. In addition to its proximal role(s) in FA pathway activation, RAD18 contributes to ICL repair by facilitating association of Structural Maintenance of Chromosome 5 and 6 (SMC5/6) to ubiquitinated histones in the vicinity of damaged chromatin [112]. RAD18 scaffold function in SMC5/6 recruitment and ICL repair is RAD6 independent and does not require E3 ubiquitin ligase activity. Similar to RAD18 function in RAD51C chaperoning, the recruitment of RAD18 to sites of ICL requires UBZ-mediated interactions with ubiquitinated chromatin. Interestingly, although the scaffolding role of RAD18 in ICL repair is TLS independent, the same DDK-mediated phosphorylations that promote RAD18–Polη [41] mediate SMC5/6 complex formation [112]. Therefore, DDK-dependent phosphorylation of RAD18 promotes both TLS and ICL repair, providing a common mechanism for S-phase-specific activation of two important genome maintenance pathways.

8. PHYSIOLOGICAL ROLES OF RAD18

Although numerous studies suggest roles for RAD18 in multiple genome maintenance pathways, physiological functions of RAD18 in vivo are poorly defined. Genome maintenance pathways often have enormous impact on development and tumorigenesis. The few known developmental roles of RAD18 and the potential impact of RAD18 on genome stability and tumorigenesis are considered briefly here.

8.1 Developmental Roles of RAD18

Rad18 (but not Polη) is expressed at high levels in mouse testes and localizes to undifferentiated spermatogonia and the XY body (a region containing transcriptionally silent unpaired XY chromosomes) [113], and to a subset of Spo11-induced meiotic DSB [114]. *Rad18*$^{-/-}$ mice are viable yet have decreased testes size and fertility defects upon aging. For example, while young (2-month old) *Rad18*$^{-/-}$ mice have normal spermatogenesis, 25% of the seminiferous tubules in aged animals (>12 month) lack germ cells, due to depletion of spermatogonial stem cells. Thus, Rad18 is important for long-term maintenance of spermatogenesis [115]. It is likely therefore that stem cells tolerate endogenous forms of DNA damage via Rad18-mediated DNA repair. However, the Rad18 effector pathways (TLS, FA, HR) required for maintenance of spermatogonial stem cells are not known. In stable *Rad18* knock-down (KD) mice, H3K4me2 is increased on the XY body (and elsewhere in the nucleus) and there is increased frequency of XY asynapsis when compared with WT mice [114]. Therefore, the roles of Rad18 in spermatogenesis and meiosis are probably TLS independent and involve DSB processing. Since FA patients and *Fanc*-deficient mice have fertility defects, it is possible that meiotic roles of Rad18 also involve the FA pathway. Indeed, the Spo11-induced redistribution of Fancd2 to the XY body is compromised in *Rad18*$^{-/-}$ mice [115a], consistent with a role for the Rad18-FA signaling axis in normal germ cell function. However, Rad18 mutant mice do not recapitulate baseline hematopoietic defects of FA patients and Fanc mutant mice [115a]. Therefore RAD18 is not an obligate component of the FA pathway in hematopoietic cells.

8.2 RAD18 Roles in Tumorigenesis

From cell culture studies, RAD18 clearly impacts many genome maintenance pathways: RAD18 has the potential to promote both error-free and mutagenic DNA-damage tolerance (via TS and TLS, respectively). RAD18 deficiency can generate DSB owing to defects in recovery from replication fork stalling. Moreover, RAD18 can promote DSB repair via error-free HR or perhaps stimulate indiscriminate genome-destabilizing NHEJ via 53BP1. Therefore, RAD18 could influence the fidelity or DNA replication/repair in ways that preserve genome stability (TS, HR) and suppress tumorigenesis or that cause mutations (via error-prone TLS or NHEJ) and drive tumorigenesis. Effects of *Rad18* on tumorigenesis in vivo have not been addressed experimentally. Nevertheless, the potential impact of Rad18 on mechanisms of genomic instability and carcinogenesis are considered further on.

Because Rad18 promotes Polη activity, *Rad18*$^{-/-}$ mice might recapitulate the UV-sensitivity and UV-induced skin cancer-propensity phenotypes of Polη-deficient mice [116,117]. Alternatively, *Rad18* deficiency and Polη deficiency could result in distinct phenotypes: UV-induced mutations in Polη-deficient cells result from error-prone compensatory lesion bypass by other Y-family DNA polymerases [118] whose activities are also RAD18 dependent. Therefore, it is possible that overall mutagenic bypass will be reduced when *Rad18* is absent—potentially leading to reduced carcinogenesis. On the other hand, because *Rad18*-deficiency in carcinogen-treated cells leads to of DSB [24,74], *Rad18*-deficienct cells could show reduced rates of point mutations (owing to reduced TLS), and increased translocations due to NHEJ-mediated DSB repair.

In addition to its potential roles in determining the balance between mutagenesis and gross chromosomal rearrangements, RAD18 might affect tumorigenesis by influencing tolerance of oncogenic stress. Oncogene expression in primary cells elicits "DNA replication stress" via diverse mechanisms including generation of genotoxic reactive oxygen species (ROS) [119–121], depletion of dNTP pools [122], and re-replication (repeated "firing" of replication origins every S-phase [123]). RAD18 is activated by many stresses commonly incited by oncogenes including ROS [74,79], dNTP depletion [124], and origin re-firing in geminin-depleted cells [125]. Importantly, RAD18 facilitates ongoing DNA synthesis in the face of excess ROS, dNTP shortage, and origin re-firing. Therefore, RAD18-mediated genome maintenance might enable proliferation and survival of neoplastic cells, thereby contributing to tumorigenesis. By analogy, the ATR-mediated S-phase checkpoint pathway (which is activated coincident with TLS) may in some instances promote survival of neoplastic cells and contribute to tumorigenesis [126]. Experiments with genetically engineered mice are required to elucidate the roles of Rad18 in tumorigenesis in response to different oncogenic drivers.

Cancer cells typically express very high levels of RAD18 and TLS polymerases when compared with primary untransformed cells—an observation that is potentially consistent with a selective advantage for TLS-proficient cells in oncogenic stress tolerance. Unfortunately, RAD18/TLS polymerase activity in cancer cells is likely to confer resistance to genotoxic therapeutic agents. Cisplatin is an important therapeutic agent for many cancers [127]. However, the success of cisplatin therapy is limited due to several mechanisms that confer cisplatin resistance including increased DNA-damage tolerance [128,129]. Polη allows replication of cisplatin-damaged DNA templates [130–137] and is a reliable marker of cisplatin resistance and poor outcome in patients with non-small cell lung cancer (NSLC) [138,139]. In cell culture studies, cancer cells lacking Polη [136,140,141] or RAD18 [19,142] fail to replicate cisplatin-damaged genomes and instead accumulate unfilled postreplicative gaps, collapsed replication forks, and lethal DNA DSBs. Therefore, RAD18-mediated TLS represents an appealing therapeutic target pathway whose inhibition may sensitize cells to cisplatin [143,144]. Cisplatin therapy also leads to serious side effects including ototoxicity, nephrotoxicity, and neurotoxicity [145–148]. Therefore, inhibition of RAD18-mediated TLS could lower the therapeutic dose of cisplatin and help minimize toxic side effects. Because RAD18 also participates in DSB repair [37,108], suppression of RAD18 function might also be a promising approach for sensitizing cancer cells to camptothecin or radiotherapy.

9. CONCLUSIONS AND PERSPECTIVES

The E3 ubiquitin ligase RAD18 is a major apical component of the DDR with important roles in both TLS and TS pathways of PRR, namely. RAD18 also has TLS/TS-independent roles in DSB repair and ICL repair. RAD18 functions in genome maintenance are integrated with the cell cycle, DNA replication, and checkpoint signaling via transcriptional and posttranslational mechanisms. RAD18 functions in genome maintenance have been identified mainly based on studies with cultured cell lines. However, *Rad18* is a nonessential gene (at least in mice) and *Rad18* deficiency does not result in any overt developmental defects or cancer propensity. Further work is necessary to define the physiological roles of Rad18 and to identify putative genes and pathways that may explain why *Rad18* is nonessential. We speculate that redundant genome maintenance mechanisms must be eliminated to reveal important roles of Rad18. Since RAD18 deficiency in cultured cells leads to DSBs, it is possible that back-up DSB-repair pathways compensate for Rad18 deficiency in vivo. In this regard, perhaps 2014 studies with *Caenorhabditis elegans* DNA-repair mutants are instructive: In *C. elegans* strains lacking Y-family TLS polymerases, DSBs are repaired via the A-family polymerase theta (PolQ, which mediates alternative NHEJ) [149]. It is possible that interesting genome maintenance defects will be revealed in mice harboring combined deficiencies in Rad18 and NHEJ or other DSB-repair genes. RAD18 deficiency sensitizes human cancer cells to therapeutic genotoxic agents. Therefore, understanding RAD18 signaling mechanisms in cancer cells may facilitate identification of synthetic lethalities and development of small molecule inhibitors that augment the anti-neoplastic effects of existing genotoxic therapies.

GLOSSARY

D-Loop A DNA structure formed during HR in which two strands of a double-stranded DNA molecule are separated for a stretch and held apart by a third invading strand of DNA.

Synthetic lethality Death resulting from combined mutations in two or more genes whose individual mutations do not compromise viability.

Template switch An error-free "DNA damage–avoidance" mechanism that allows continued DNA replication of damaged genomes by using a newly synthesized undamaged sister chromatid as a template.

LIST OF ABBREVIATIONS

CPT Camptothecin

DDR DNA-damage response

D-loop Displacement loop
DSB Double-stranded DNA break
FA Fanconi anemia
HDR Homology-directed repair
HJ Holliday junction
HR Homologous recombination
HU Hydroxyurea
ICL Interstrand crosslinker
IRIF Ionizing radiation-induced foci
MEF Mouse embryonic fibroblast
MMC Mitomycin C
MMR Mismatch repair
MNNG Methylnitronitrosoguanidine
NER Nucleotide excision repair
NHEJ Nonhomologous end joining
NSCLC Nonsmall cell lung cancer
PIP PCNA-interacting peptide
PRR Postreplication repair
RING Really interesting gene
RNR Ribonucleotide reductase
SCE Sister chromatid exchange
ssDNA Single-stranded DNA
TLS Trans-lesion synthesis
TS Template switching
UBM Ubiquitin-binding motif
UBZ Ubiquitin-binding zinc finger
UV Ultraviolet radiation
XPA *Xeroderma pigmentosum* complementation group A
XPV *Xeroderma pigmentosum* complementation group V

REFERENCES

[1] Nyberg KA, Michelson RJ, Putnam CW, Weinert TA. Toward maintaining the genome: DNA damage and replication checkpoints. Annu Rev Genet 2002;36:617–56.
[2] Lopes M, Cotta-Ramusino C, Pellicioli A, Liberi G, Plevani P, Muzi-Falconi M, et al. The DNA replication checkpoint response stabilizes stalled replication forks. Nature 2001;412(6846):557–61.
[3] Brown EJ. The ATR-independent DNA replication checkpoint. Cell Cycle 2003;2(3):188–9.
[4] Brown EJ, Baltimore D. Essential and dispensable roles of ATR in cell cycle arrest and genome maintenance. Genes Dev 2003;17(5):615–28.
[5] Lawrence C. The RAD6 DNA repair pathway in *Saccharomyces cerevisiae*: what does it do, and how does it do it? Bioessays 1994;16(4):253–8.
[6] Lehmann AR. Replication of damaged DNA in mammalian cells: new solutions to an old problem. Mutat Res 2002;509(1–2):23–34.
[7] Lawrence CW, Christensen R. UV mutagenesis in radiation-sensitive strains of yeast. Genetics 1976;82(2):207–32.
[8] Prakash L. The effect of genes controlling radiation sensitivity on chemical mutagenesis in yeast. Basic Life Sci 1975;5A:393–5.
[9] Prakash L. Effect of genes controlling radiation sensitivity on chemically induced mutations in *Saccharomyces cerevisiae*. Genetics 1976;83(2):285–301.
[10] Prakash L. Characterization of postreplication repair in *Saccharomyces cerevisiae* and effects of rad6, rad18, rev3 and rad52 mutations. Mol Gen Genet 1981;184(3):471–8.
[11] Sung P, Prakash S, Prakash L. The RAD6 protein of *Saccharomyces cerevisiae* polyubiquitinates histones, and its acidic domain mediates this activity. Genes Dev 1988;2(11):1476–85.
[12] Prakash L. The structure and function of RAD6 and RAD18 DNA repair genes of *Saccharomyces cerevisiae*. Genome 1989;31(2):597–600.
[13] Bailly V, Lamb J, Sung P, Prakash S, Prakash L. Specific complex formation between yeast RAD6 and RAD18 proteins: a potential mechanism for targeting RAD6 ubiquitin-conjugating activity to DNA damage sites. Genes Dev 1994;8(7):811–20.
[14] Bailly V, Lauder S, Prakash S, Prakash L. Yeast DNA repair proteins Rad6 and Rad18 form a heterodimer that has ubiquitin conjugating, DNA binding, and ATP hydrolytic activities. J Biol Chem 1997;272(37):23360–5.
[15] Jones JS, Weber S, Prakash L. The *Saccharomyces cerevisiae* RAD18 gene encodes a protein that contains potential zinc finger domains for nucleic acid binding and a putative nucleotide binding sequence. Nucleic Acids Res 1988;16(14B):7119–31.
[16] Tateishi S, Sakuraba Y, Masuyama S, Inoue H, Yamaizumi M. Dysfunction of human Rad18 results in defective postreplication repair and hypersensitivity to multiple mutagens. Proc Natl Acad Sci USA 2000;97(14):7927–32.
[17] Xin H, Lin W, Sumanasekera W, Zhang Y, Wu X, Wang Z. The human RAD18 gene product interacts with HHR6A and HHR6B. Nucleic Acids Res 2000;28(14):2847–54.

[18] Tateishi S, Niwa H, Miyazaki J, Fujimoto S, Inoue H, Yamaizumi M. Enhanced genomic instability and defective postreplication repair in RAD18 knockout mouse embryonic stem cells. Mol Cell Biol 2003;23(2):474–81.

[19] Yamashita YM, Okada T, Matsusaka T, Sonoda E, Zhao GY, Araki K, et al. RAD18 and RAD54 cooperatively contribute to maintenance of genomic stability in vertebrate cells. EMBO J 2002;21(20):5558–66.

[20] Hoege C, Pfander B, Moldovan GL, Pyrowolakis G, Jentsch S. RAD6-dependent DNA repair is linked to modification of PCNA by ubiquitin and SUMO. Nature 2002;419(6903):135–41.

[21] Stelter P, Ulrich HD. Control of spontaneous and damage-induced mutagenesis by SUMO and ubiquitin conjugation. Nature 2003;425(6954):188–91.

[22] Kannouche PL, Wing J, Lehmann AR. Interaction of human DNA polymerase eta with monoubiquitinated PCNA: a possible mechanism for the polymerase switch in response to DNA damage. Mol Cell 2004;14(4):491–500.

[23] Watanabe K, Tateishi S, Kawasuji M, Tsurimoto T, Inoue H, Yamaizumi M. Rad18 guides poleta to replication stalling sites through physical interaction and PCNA monoubiquitination. EMBO J 2004;23(19):3886–96.

[24] Bi X, Barkley LR, Slater DM, Tateishi S, Yamaizumi M, Ohmori H, et al. Rad18 regulates DNA polymerase kappa and is required for recovery from S-phase checkpoint-mediated arrest. Mol Cell Biol 2006;26(9):3527–40.

[25] Plosky BS, Vidal AE, Fernandez de Henestrosa AR, McLenigan MP, McDonald JP, Mead S, et al. Controlling the subcellular localization of DNA polymerases iota and eta via interactions with ubiquitin. EMBO J 2006;25(12):2847–55.

[26] Guo C, Sonoda E, Tang TS, Parker JL, Bielen AB, Takeda S, et al. REV1 protein interacts with PCNA: significance of the REV1 BRCT domain in vitro and in vivo. Mol Cell 2006;23(2):265–71.

[27] Bienko M, Green CM, Crosetto N, Rudolf F, Zapart G, Coull B, et al. Ubiquitin-binding domains in Y-family polymerases regulate translesion synthesis. Science 2005;310(5755):1821–4.

[28] Haracska L, Unk I, Prakash L, Prakash S. Ubiquitylation of yeast proliferating cell nuclear antigen and its implications for translesion DNA synthesis. Proc Natl Acad Sci USA 2006;103(17):6477–82.

[29] Acharya N, Yoon JH, Hurwitz J, Prakash L, Prakash S. DNA polymerase eta lacking the ubiquitin-binding domain promotes replicative lesion bypass in humans cells. Proc Natl Acad Sci USA 2010;107(23):10401–5.

[30] Hendel A, Krijger PH, Diamant N, Goren Z, Langerak P, Kim J, et al. PCNA ubiquitination is important, but not essential for translesion DNA synthesis in mammalian cells. PLoS Genet 2011;7(9):e1002262.

[31] Notenboom V, Hibbert RG, van Rossum-Fikkert SE, Olsen JV, Mann M, Sixma TK. Functional characterization of Rad18 domains for Rad6, ubiquitin, DNA binding and PCNA modification. Nucleic Acids Res 2007;35(17):5819–30.

[32] Huang A, Hibbert RG, de Jong RN, Das D, Sixma TK, Boelens R. Symmetry and asymmetry of the RING-RING dimer of Rad18. J Mol Biol 2011;410(3):424–35.

[33] Masuda Y, Suzuki M, Kawai H, Suzuki F, Kamiya K. Asymmetric nature of two subunits of RAD18, a RING-type ubiquitin ligase E3, in the human RAD6A-RAD18 ternary complex. Nucleic Acids Res 2012;40(3):1065–76.

[34] Miyase S, Tateishi S, Watanabe K, Tomita K, Suzuki K, Inoue H, et al. Differential regulation of Rad18 through Rad6-dependent mono- and polyubiquitination. J Biol Chem 2005;280(1):515–24.

[35] Hofmann K. Ubiquitin-binding domains and their role in the DNA damage response. DNA Repair (Amst) 2009;8(4):544–56.

[36] Kanao R, Masuda Y, Deguchi S, Yumoto-Sugimoto M, Hanaoka F, Masutani C. Relevance of simultaneous mono-ubiquitinations of multiple units of PCNA homo-trimers in DNA damage tolerance. PLoS One 2015;10(2):e0118775.

[37] Huang J, Huen MS, Kim H, Leung CC, Glover JN, Yu X, et al. RAD18 transmits DNA damage signalling to elicit homologous recombination repair. Nat Cell Biol 2009;11(5):592–603.

[38] Aravind L, Koonin EV. SAP – a putative DNA-binding motif involved in chromosomal organization. Trends Biochem Sci 2000;25(3):112–4.

[39] Nakajima S, Lan L, Kanno S, Usami N, Kobayashi K, Mori M, et al. Replication-dependent and -independent responses of RAD18 to DNA damage in human cells. J Biol Chem 2006;281(45):34687–95.

[40] Tsuji Y, Watanabe K, Araki K, Shinohara M, Yamagata Y, Tsurimoto T, et al. Recognition of forked and single-stranded DNA structures by human RAD18 complexed with RAD6B protein triggers its recruitment to stalled replication forks. Genes Cells 2008;13(4):343–54.

[41] Day TA, Palle K, Barkley LR, Kakusho N, Zou Y, Tateishi S, et al. Phosphorylated Rad18 directs DNA polymerase eta to sites of stalled replication. J Cell Biol 2010;191(5):953–66.

[42] Huang TT, Nijman SM, Mirchandani KD, Galardy PJ, Cohn MA, Haas W, et al. Regulation of monoubiquitinated PCNA by DUB autocleavage. Nat Cell Biol 2006;8(4):339–47.

[43] Byun TS, Pacek M, Yee MC, Walter JC, Cimprich KA. Functional uncoupling of MCM helicase and DNA polymerase activities activates the ATR-dependent checkpoint. Genes Dev 2005;19(9):1040–52.

[44] Sogo JM, Lopes M, Foiani M. Fork reversal and ssDNA accumulation at stalled replication forks owing to checkpoint defects. Science 2002;297(5581):599–602.

[45] Zou L, Elledge SJ. Sensing DNA damage through ATRIP recognition of RPA-ssDNA complexes. Science 2003;300(5625):1542–8.

[46] Davies AA, Huttner D, Daigaku Y, Chen S, Ulrich HD. Activation of ubiquitin-dependent DNA damage bypass is mediated by replication protein a. Mol Cell 2008;29(5):625–36.

[47] Durando M, Tateishi S, Vaziri C. A non-catalytic role of DNA polymerase eta in recruiting Rad18 and promoting PCNA monoubiquitination at stalled replication forks. Nucleic Acids Res 2013;41(5):3079–93.

[48] Ito W, Yokoi M, Sakayoshi N, Sakurai Y, Akagi J, Mitani H, et al. Stalled Poleta at its cognate substrate initiates an alternative translesion synthesis pathway via interaction with REV1. Genes Cells 2012;17(2):98–108.

[49] Broughton BC, Cordonnier A, Kleijer WJ, Jaspers NG, Fawcett H, Raams A, et al. Molecular analysis of mutations in DNA polymerase eta in xeroderma pigmentosum-variant patients. Proc Natl Acad Sci USA 2002;99(2):815–20.
[50] Hishiki A, Hashimoto H, Hanafusa T, Kamei K, Ohashi E, Shimizu T, et al. Structural basis for novel interactions between human translesion synthesis polymerases and proliferating cell nuclear antigen. J Biol Chem 2009;284(16):10552–60.
[51] Centore RC, Yazinski SA, Tse A, Zou L. Spartan/C1orf124, a reader of PCNA ubiquitylation and a regulator of UV-induced DNA damage response. Mol Cell 2012;46(5):625–35.
[52] Ghosal G, Leung JW, Nair BC, Fong KW, Chen J. Proliferating cell nuclear antigen (PCNA)-binding protein C1orf124 is a regulator of translesion synthesis. J Biol Chem 2012;287(41):34225–33.
[53] Juhasz S, Balogh D, Hajdu I, Burkovics P, Villamil MA, Zhuang Z, et al. Characterization of human Spartan/C1orf124, an ubiquitin-PCNA interacting regulator of DNA damage tolerance. Nucleic Acids Res 2012;40(21):10795–808.
[54] Davis EJ, Lachaud C, Appleton P, Macartney TJ, Nathke I, Rouse J. DVC1 (C1orf124) recruits the p97 protein segregase to sites of DNA damage. Nat Struct Mol Biol 2012;19(11):1093–100.
[55] Machida Y, Kim MS, Machida YJ. Spartan/C1orf124 is important to prevent UV-induced mutagenesis. Cell Cycle 2012;11(18):3395–402.
[56] Han J, Liu T, Huen MS, Hu L, Chen Z, Huang J. SIVA1 directs the E3 ubiquitin ligase RAD18 for PCNA monoubiquitination. J Cell Biol 2014;205(6):811–27.
[57] Yanagihara H, Kobayashi J, Tateishi S, Kato A, Matsuura S, Tauchi H, et al. NBS1 recruits RAD18 via a RAD6-like domain and regulates Pol eta-dependent translesion DNA synthesis. Mol Cell 2011;43(5):788–97.
[58] Tian F, Sharma S, Zou J, Lin SY, Wang B, Rezvani K, et al. BRCA1 promotes the ubiquitination of PCNA and recruitment of translesion polymerases in response to replication blockade. Proc Natl Acad Sci USA 2013;110(33):13558–63.
[59] Kermi C, Prieto S, van der Laan S, Tsanov N, Recolin B, Uro-Coste E, et al. RAD18 is a maternal limiting factor silencing the UV-dependent DNA damage checkpoint in *Xenopus* embryos. Dev Cell 2015;34(3):364–72.
[60] Masuyama S, Tateishi S, Yomogida K, Nishimune Y, Suzuki K, Sakuraba Y, et al. Regulated expression and dynamic changes in subnuclear localization of mammalian Rad18 under normal and genotoxic conditions. Genes Cells 2005;10(8):753–62.
[61] Varanasi L, Do PM, Goluszko E, Martinez LA. Rad18 is a transcriptional target of E2F3. Cell Cycle 2012;11(6):1131–41.
[62] Zlatanou A, Sabbioneda S, Miller ES, Greenwalt A, Aggathanggelou A, Maurice MM, et al. USP7 is essential for maintaining Rad18 stability and DNA damage tolerance. Oncogene February 25, 2016;35(8):965–76. http://dx.doi.org/10.1038/onc.2015.149. [Epub 2015 May 11].
[63] Barkley LR, Palle K, Durando M, Day TA, Gurkar A, Kakusho N, et al. c-Jun N-terminal kinase-mediated Rad18 phosphorylation facilitates Poleta recruitment to stalled replication forks. Mol Biol Cell 2012;23(10):1943–54.
[64] Labib K. How do Cdc7 and cyclin-dependent kinases trigger the initiation of chromosome replication in eukaryotic cells? Genes Dev 2010;24(12):1208–19.
[65] Davis RJ. Signal transduction by the JNK group of MAP kinases. Cell 2000;103(2):239–52.
[66] Yamada M, Watanabe K, Mistrik M, Vesela E, Protivankova I, Mailand N, et al. ATR-Chk1-APC/CCdh1-dependent stabilization of Cdc7-ASK (Dbf4) kinase is required for DNA lesion bypass under replication stress. Genes Dev 2013;27(22):2459–72.
[67] Heffernan TP, Simpson DA, Frank AR, Heinloth AN, Paules RS, Cordeiro-Stone M, et al. An ATR- and Chk1-dependent S checkpoint inhibits replicon initiation following UVC-induced DNA damage. Mol Cell Biol 2002;22(24):8552–61.
[68] Yang XH, Shiotani B, Classon M, Zou L. Chk1 and Claspin potentiate PCNA ubiquitination. Genes Dev 2008;22(9):1147–52.
[69] Liu G, Chen X. DNA polymerase eta, the product of the xeroderma pigmentosum variant gene and a target of p53, modulates the DNA damage checkpoint and p53 activation. Mol Cell Biol 2006;26(4):1398–413.
[70] Karras GI, Jentsch S. The RAD6 DNA damage tolerance pathway operates uncoupled from the replication fork and is functional beyond S phase. Cell 2010;141(2):255–67.
[71] Daigaku Y, Davies AA, Ulrich HD. Ubiquitin-dependent DNA damage bypass is separable from genome replication. Nature 2010;465(7300):951–5.
[72] Lopes M, Foiani M, Sogo JM. Multiple mechanisms control chromosome integrity after replication fork uncoupling and restart at irreparable UV lesions. Mol Cell 2006;21(1):15–27.
[73] Ogi T, Limsirichaikul S, Overmeer RM, Volker M, Takenaka K, Cloney R, et al. Three DNA polymerases, recruited by different mechanisms, carry out NER repair synthesis in human cells. Mol Cell 2010;37(5):714–27.
[74] Yang Y, Durando M, Smith-Roe SL, Sproul C, Greenwalt AM, Kaufmann W, et al. Cell cycle stage-specific roles of Rad18 in tolerance and repair of oxidative DNA damage. Nucleic Acids Res 2013;41(4):2296–312.
[75] Tsanov N, Kermi C, Coulombe P, Van der Laan S, Hodroj D, Maiorano D. PIP degron proteins, substrates of CRL4Cdt2, and not PIP boxes, interfere with DNA polymerase eta and kappa focus formation on UV damage. Nucleic Acids Res 2014;42(6):3692–706.
[76] Ogi T, Lehmann AR. The Y-family DNA polymerase kappa (pol kappa) functions in mammalian nucleotide-excision repair. Nat Cell Biol 2006;8(6):640–2.
[77] Giannattasio M, Follonier C, Tourriere H, Puddu F, Lazzaro F, Pasero P, et al. Exo1 competes with repair synthesis, converts NER intermediates to long ssDNA gaps, and promotes checkpoint activation. Mol Cell 2010;40(1):50–62.
[78] Sertic S, Pizzi S, Cloney R, Lehmann AR, Marini F, Plevani P, et al. Human exonuclease 1 connects nucleotide excision repair (NER) processing with checkpoint activation in response to UV irradiation. Proc Natl Acad Sci USA 2011;108(33):13647–52.
[79] Zlatanou A, Despras E, Braz-Petta T, Boubakour-Azzouz I, Pouvelle C, Stewart GS, et al. The hMsh2-hMsh6 complex acts in concert with monoubiquitinated PCNA and Pol eta in response to oxidative DNA damage in human cells. Mol Cell 2011;43(4):649–62.

[80] Pena-Diaz J, Bregenhorn S, Ghodgaonkar M, Follonier C, Artola-Boran M, Castor D, et al. Noncanonical mismatch repair as a source of genomic instability in human cells. Mol Cell 2012;47(5):669–80.

[81] Torres-Ramos CA, Prakash S, Prakash L. Requirement of RAD5 and MMS2 for postreplication repair of UV-damaged DNA in *Saccharomyces cerevisiae*. Mol Cell Biol 2002;22(7):2419–26.

[82] Broomfield S, Chow BL, Xiao W. MMS2, encoding a ubiquitin-conjugating-enzyme-like protein, is a member of the yeast error-free postreplication repair pathway. Proc Natl Acad Sci USA 1998;95(10):5678–83.

[83] Hofmann RM, Pickart CM. Noncanonical MMS2-encoded ubiquitin-conjugating enzyme functions in assembly of novel polyubiquitin chains for DNA repair. Cell 1999;96(5):645–53.

[84] Johnson RE, Henderson ST, Petes TD, Prakash S, Bankmann M, Prakash L. *Saccharomyces cerevisiae* RAD5-encoded DNA repair protein contains DNA helicase and zinc-binding sequence motifs and affects the stability of simple repetitive sequences in the genome. Mol Cell Biol 1992;12(9):3807–18.

[85] Deshaies RJ, Joazeiro CA. RING domain E3 ubiquitin ligases. Annu Rev Biochem 2009;78:399–434.

[86] Ulrich HD, Jentsch S. Two RING finger proteins mediate cooperation between ubiquitin-conjugating enzymes in DNA repair. EMBO J 2000;19(13):3388–97.

[87] Unk I, Hajdu I, Fatyol K, Szakal B, Blastyak A, Bermudez V, et al. Human SHPRH is a ubiquitin ligase for Mms2-Ubc13-dependent polyubiquitylation of proliferating cell nuclear antigen. Proc Natl Acad Sci USA 2006;103(48):18107–12.

[88] Masuda Y, Suzuki M, Kawai H, Hishiki A, Hashimoto H, Masutani C, et al. En bloc transfer of polyubiquitin chains to PCNA in vitro is mediated by two different human E2-E3 pairs. Nucleic Acids Res 2012;40(20):10394–407.

[89] Motegi A, Sood R, Moinova H, Markowitz SD, Liu PP, Myung K. Human SHPRH suppresses genomic instability through proliferating cell nuclear antigen polyubiquitination. J Cell Biol 2006;175(5):703–8.

[90] Unk I, Hajdu I, Fatyol K, Hurwitz J, Yoon JH, Prakash L, et al. Human HLTF functions as a ubiquitin ligase for proliferating cell nuclear antigen polyubiquitination. Proc Natl Acad Sci USA 2008;105(10):3768–73.

[91] Lin JR, Zeman MK, Chen JY, Yee MC, Cimprich KA. SHPRH and HLTF act in a damage-specific manner to coordinate different forms of postreplication repair and prevent mutagenesis. Mol Cell 2011;42(2):237–49.

[92] Ciccia A, Nimonkar AV, Hu Y, Hajdu I, Achar YJ, Izhar L, et al. Polyubiquitinated PCNA recruits the ZRANB3 translocase to maintain genomic integrity after replication stress. Mol Cell 2012;47(3):396–409.

[93] Weston R, Peeters H, Ahel D. ZRANB3 is a structure-specific ATP-dependent endonuclease involved in replication stress response. Genes Dev 2012;26(14):1558–72.

[94] Yuan J, Ghosal G, Chen J. The HARP-like domain-containing protein AH2/ZRANB3 binds to PCNA and participates in cellular response to replication stress. Mol Cell 2012;47(3):410–21.

[95] Kile AC, Chavez DA, Bacal J, Eldirany S, Korzhnev DM, Bezsonova I, et al. HLTF's ancient HIRAN domain binds 3′ DNA ends to drive replication fork reversal. Mol Cell 2015;58(6):1090–100.

[96] Hishiki A, Hara K, Ikegaya Y, Yokoyama H, Shimizu T, Sato M, et al. Structure of a novel DNA-binding domain of helicase-like transcription factor (HLTF) and its functional implication in DNA damage tolerance. J Biol Chem 2015;290(21):13215–23.

[97] Achar YJ, Balogh D, Haracska L. Coordinated protein and DNA remodeling by human HLTF on stalled replication fork. Proc Natl Acad Sci USA 2011;108(34):14073–8.

[98] Blastyak A, Hajdu I, Unk I, Haracska L. Role of double-stranded DNA translocase activity of human HLTF in replication of damaged DNA. Mol Cell Biol 2010;30(3):684–93.

[99] Blastyak A, Pinter L, Unk I, Prakash L, Prakash S, Haracska L. Yeast Rad5 protein required for postreplication repair has a DNA helicase activity specific for replication fork regression. Mol Cell 2007;28(1):167–75.

[100] Burkovics P, Sebesta M, Balogh D, Haracska L, Krejci L. Strand invasion by HLTF as a mechanism for template switch in fork rescue. Nucleic Acids Res 2014;42(3):1711–20.

[101] Moinova HR, Chen WD, Shen L, Smiraglia D, Olechnowicz J, Ravi L, et al. HLTF gene silencing in human colon cancer. Proc Natl Acad Sci USA 2002;99(7):4562–7.

[102] Saberi A, Hochegger H, Szuts D, Lan L, Yasui A, Sale JE, et al. RAD18 and poly(ADP-ribose) polymerase independently suppress the access of nonhomologous end joining to double-strand breaks and facilitate homologous recombination-mediated repair. Mol Cell Biol 2007;27(7):2562–71.

[103] Watanabe K, Iwabuchi K, Sun J, Tsuji Y, Tani T, Tokunaga K, et al. RAD18 promotes DNA double-strand break repair during G1 phase through chromatin retention of 53BP1. Nucleic Acids Res 2009;37(7):2176–93.

[104] Zimmermann M, de Lange T. 53BP1: pro choice in DNA repair. Trends Cell Biol February 2014;24(2):108–17. http://dx.doi.org/10.1016/j.tcb.2013.09.003. [Epub 2013 Oct 4].

[105] Kee Y, D'Andrea AD. Molecular pathogenesis and clinical management of Fanconi anemia. J Clin Invest 2012;122(11):3799–806.

[106] Knipscheer P, Raschle M, Smogorzewska A, Enoiu M, Ho TV, Scharer OD, et al. The Fanconi anemia pathway promotes replication-dependent DNA interstrand cross-link repair. Science 2009;326(5960):1698–701.

[107] Song IY, Palle K, Gurkar A, Tateishi S, Kupfer GM, Vaziri C. Rad18-mediated translesion synthesis of bulky DNA adducts is coupled to activation of the Fanconi anemia DNA repair pathway. J Biol Chem 2010;285(41):31525–36.

[108] Palle K, Vaziri C. Rad18 E3 ubiquitin ligase activity mediates Fanconi anemia pathway activation and cell survival following DNA Topoisomerase 1 inhibition. Cell Cycle 2011;10(10):1625–38.

[109] Williams SA, Longerich S, Sung P, Vaziri C, Kupfer GM. The E3 ubiquitin ligase RAD18 regulates ubiquitylation and chromatin loading of FANCD2 and FANCI. Blood 2011;117(19):5078–87.

[110] Geng L, Huntoon CJ, Karnitz LM. RAD18-mediated ubiquitination of PCNA activates the Fanconi anemia DNA repair network. J Cell Biol 2010;191(2):249–57.

[111] Park HK, Wang H, Zhang J, Datta S, Fei P. Convergence of Rad6/Rad18 and Fanconi anemia tumor suppressor pathways upon DNA damage. PLoS One 2010;5(10):e13313.

[112] Raschle M, Smeenk G, Hansen RK, Temu T, Oka Y, Hein MY, et al. DNA repair. Proteomics reveals dynamic assembly of repair complexes during bypass of DNA cross-links. Science 2015;348(6234):1253671.

[113] van der Laan R, Uringa EJ, Wassenaar E, Hoogerbrugge JW, Sleddens E, Odijk H, et al. Ubiquitin ligase Rad18Sc localizes to the XY body and to other chromosomal regions that are unpaired and transcriptionally silenced during male meiotic prophase. J Cell Sci 2004;117(Pt 21):5023–33.

[114] Inagaki A, Sleddens-Linkels E, Wassenaar E, Ooms M, van Cappellen WA, Hoeijmakers JH, et al. Meiotic functions of RAD18. J Cell Sci 2011;124(Pt 16):2837–50.

[115] Sun J, Yomogida K, Sakao S, Yamamoto H, Yoshida K, Watanabe K, et al. Rad18 is required for long-term maintenance of spermatogenesis in mouse testes. Mech Dev 2009;126(3–4):173–83.

[115a] Yang Y, Poe JC, Yang L, Fedoriw A, Desai S, Magnuson T, et al. Rad18 confers hematopoietic progenitor cell DNA damage tolerance independently of the Fanconi Anemia pathway in vivo. Nucleic Acids Res May 19, 2016;44(9):4174–88. http://dx.doi.org/doi:10.1093/nar/gkw072. PMID: 26883629 [Epub 2016 Feb 15].

[116] Lin Q, Clark AB, McCulloch SD, Yuan T, Bronson RT, Kunkel TA, et al. Increased susceptibility to UV-induced skin carcinogenesis in polymerase eta-deficient mice. Cancer Res 2006;66(1):87–94.

[117] Ohkumo T, Kondo Y, Yokoi M, Tsukamoto T, Yamada A, Sugimoto T, et al. UV-B radiation induces epithelial tumors in mice lacking DNA polymerase eta and mesenchymal tumors in mice deficient for DNA polymerase iota. Mol Cell Biol 2006;26(20):7696–706.

[118] Shachar S, Ziv O, Avkin S, Adar S, Wittschieben J, Reissner T, et al. Two-polymerase mechanisms dictate error-free and error-prone translesion DNA synthesis in mammals. EMBO J 2009;28(4):383–93.

[119] Weyemi U, Lagente-Chevallier O, Boufraqech M, Prenois F, Courtin F, Caillou B, et al. ROS-generating NADPH oxidase NOX4 is a critical mediator in oncogenic H-Ras-induced DNA damage and subsequent senescence. Oncogene 2012;31(9):1117–29.

[120] DeNicola GM, Karreth FA, Humpton TJ, Gopinathan A, Wei C, Frese K, et al. Oncogene-induced Nrf2 transcription promotes ROS detoxification and tumorigenesis. Nature 2011;475(7354):106–9.

[121] Vafa O, Wade M, Kern S, Beeche M, Pandita TK, Hampton GM, et al. c-Myc can induce DNA damage, increase reactive oxygen species, and mitigate p53 function: a mechanism for oncogene-induced genetic instability. Mol Cell 2002;9(5):1031–44.

[122] Bester AC, Roniger M, Oren YS, Im MM, Sarni D, Chaoat M, et al. Nucleotide deficiency promotes genomic instability in early stages of cancer development. Cell 2011;145(3):435–46.

[123] Di Micco R, Fumagalli M, Cicalese A, Piccinin S, Gasparini P, Luise C, et al. Oncogene-induced senescence is a DNA damage response triggered by DNA hyper-replication. Nature 2006;444(7119):638–42.

[124] de Feraudy S, Limoli CL, Giedzinski E, Karentz D, Marti TM, Feeney L, et al. Pol eta is required for DNA replication during nucleotide deprivation by hydroxyurea. Oncogene 2007;26(39):5713–21.

[125] Sekimoto T, Oda T, Kurashima K, Hanaoka F, Yamashita T. Both high-fidelity replicative and low-fidelity Y-family polymerases are involved in DNA rereplication. Mol Cell Biol 2015;35(4):699–715.

[126] Schoppy DW, Ragland RL, Gilad O, Shastri N, Peters AA, Murga M, et al. Oncogenic stress sensitizes murine cancers to hypomorphic suppression of ATR. J Clin Invest 2012;122(1):241–52.

[127] Masters JR, Koberle B. Curing metastatic cancer: lessons from testicular germ-cell tumours. Nat Rev Cancer 2003;3(7):517–25.

[128] DeVita VT, Hellman S, Rosenberg SA. Cancer: principles and practice of oncology. 6th ed. 2001.

[129] Johnson SW, Ferry KV, Hamilton TC. Recent insights into platinum drug resistance in cancer. Drug Resist Updat 1998;1(4):243–54.

[130] Kelland L. The resurgence of platinum-based cancer chemotherapy. Nat Rev Cancer 2007;7(8):573–84.

[131] Mamenta EL, Poma EE, Kaufmann WK, Delmastro DA, Grady HL, Chaney SG. Enhanced replicative bypass of platinum-DNA adducts in cisplatin-resistant human ovarian carcinoma cell lines. Cancer Res 1994;54(13):3500–5.

[132] Kunz BA, Straffon AF, Vonarx EJ. DNA damage-induced mutation: tolerance via translesion synthesis. Mutat Res 2000;451(1–2):169–85.

[133] Lehmann AR. Replication of damaged DNA by translesion synthesis in human cells. FEBS Lett 2005;579(4):873–6.

[134] Lord CJ, Ashworth A. The DNA damage response and cancer therapy. Nature;481(7381):287–94.

[135] Vaisman A, Masutani C, Hanaoka F, Chaney SG. Efficient translesion replication past oxaliplatin and cisplatin GpG adducts by human DNA polymerase eta. Biochemistry 2000;39(16):4575–80.

[136] Chen YW, Cleaver JE, Hanaoka F, Chang CF, Chou KM. A novel role of DNA polymerase eta in modulating cellular sensitivity to chemotherapeutic agents. Mol Cancer Res 2006;4(4):257–65.

[137] Bassett E, Vaisman A, Havener JM, Masutani C, Hanaoka F, Chaney SG. Efficiency of extension of mismatched primer termini across from cisplatin and oxaliplatin adducts by human DNA polymerases beta and eta in vitro. Biochemistry 2003;42(48):14197–206.

[138] Ceppi P, Novello S, Cambieri A, Longo M, Monica V, Lo Iacono M, et al. Polymerase eta mRNA expression predicts survival of non-small cell lung cancer patients treated with platinum-based chemotherapy. Clin Cancer Res 2009;15(3):1039–45.

[139] Teng KY, Qiu MZ, Li ZH, Luo HY, Zeng ZL, Luo RZ, et al. DNA polymerase eta protein expression predicts treatment response and survival of metastatic gastric adenocarcinoma patients treated with oxaliplatin-based chemotherapy. J Transl Med 2010;8:126.

[140] Alt A, Lammens K, Chiocchini C, Lammens A, Pieck JC, Kuch D, et al. Bypass of DNA lesions generated during anticancer treatment with cisplatin by DNA polymerase eta. Science 2007;318(5852):967–70.

[141] Albertella MR, Green CM, Lehmann AR, O'Connor MJ. A role for polymerase eta in the cellular tolerance to cisplatin-induced damage. Cancer Res 2005;65(21):9799–806.

[142] Wagner JM, Karnitz LM. Cisplatin-induced DNA damage activates replication checkpoint signaling components that differentially affect tumor cell survival. Mol Pharmacol 2009;76(1):208–14.

[143] Ummat A, Rechkoblit O, Jain R, Roy Choudhury J, Johnson RE, Silverstein TD, et al. Structural basis for cisplatin DNA damage tolerance by human polymerase eta during cancer chemotherapy. Nat Struct Mol Biol 2012;19(6):628–32.

[144] Zhao Y, Biertumpfel C, Gregory MT, Hua YJ, Hanaoka F, Yang W. Structural basis of human DNA polymerase eta-mediated chemoresistance to cisplatin. Proc Natl Acad Sci USA 2012;109(19):7269–74.

[145] Rybak LP, Mukherjea D, Jajoo S, Ramkumar V. Cisplatin ototoxicity and protection: clinical and experimental studies. Tohoku J Exp Med 2009;219(3):177–86.

[146] Park SB, Krishnan AV, Lin CS, Goldstein D, Friedlander M, Kiernan MC. Mechanisms underlying chemotherapy-induced neurotoxicity and the potential for neuroprotective strategies. Curr Med Chem 2008;15(29):3081–94.

[147] Sioka C, Kyritsis AP. Central and peripheral nervous system toxicity of common chemotherapeutic agents. Cancer Chemother Pharmacol 2009;63(5):761–7.

[148] Pasetto LM, D'Andrea MR, Rossi E, Monfardini S. Oxaliplatin-related neurotoxicity: how and why? Crit Rev Oncol Hematol 2006;59(2):159–68.

[149] Roerink SF, van Schendel R, Tijsterman M. Polymerase theta-mediated end joining of replication-associated DNA breaks in *C. elegans*. Genome Res 2014;24(6):954–62.

Chapter 17

Base Excision Repair and Nucleotide Excision Repair

T. Izumi, I. Mellon
University of Kentucky, Lexington, KY, United States

Chapter Outline

1. GENERAL OVERVIEW AND HISTORICAL PERSPECTIVES OF TWO DNA EXCISION-REPAIR PATHWAYS, BER AND NER

Base excision repair (BER) and nucleotide excision repair (NER) are two major DNA excision-repair pathways. They are conserved among eukaryotes from yeast to mammals, and prototype-repair systems exist in prokaryotes including *Escherichia coli*. It is well established that deficiencies in BER and NER can lead to mutations and cell death after exposure of cells to exogenous and endogenous forms of DNA-damaging agents. Biochemical, cell biological, and genetic studies unequivocally support the notion that BER and NER are pivotal for cells to survive exposure to different forms of DNA damage. If left unrepaired, mutations and cell death are unavoidable, and diseases arise in multicellular eukaryotes. Therefore, BER and NER have been intensively studied in molecular toxicology.

BER is capable of repairing small base damage, apurinic/apyrimidinic sites (AP-site, lacking a base), and DNA single-strand breaks [1–3]. NER, on the other hand, repairs relatively large ("bulky") adducts of DNA. These include

Genome Stability. http://dx.doi.org/10.1016/B978-0-12-803309-8.00017-3

photoproducts formed by ultraviolet irradiation and a multitude of base modifications produced by exposure to chemical carcinogens such as benzo(a)pyrenes and other aromatic hydrocarbons, aflatoxins, 2-acetylamiofluorenes, and chemotherapeutic agents such as platinum [4–8].

Four processes occur during both BER and NER: (1) recognition of damaged DNA, (2) excision of the damage, (3) DNA synthesis to fill the nucleotide(s) gap, and (4) the sealing of nicks (3′-OH and 5′-P pairs without gaps) in DNA. This simplified description will be expanded upon in greater detail later. Many BER and NER genes were identified and cloned by the mid-1990s, and we saw significant progress in understanding these core BER and NER reactions by using purified DNA-repair enzymes.

However, multicellular organisms conduct DNA repair in the context of the entire organism. Consequently, it is important to investigate and understand how DNA-repair proteins communicate with factors controlling cell-cycle checkpoints and apoptosis, and discern whether damage introduction and repair are influenced by other cellular processes such as transcription and DNA replication. These signaling activities are often referred to DDR (DNA-damage response), and they are currently extremely active topics of research. The sequencing of mammalian genomes and the development of new genomic approaches have required refinement of earlier studies of DNA repair to consider the consequences of the highly complex and dynamic DDR networks, to reveal the full scale of the cellular mechanisms needed to recover from DNA damage. We describe the basic mechanisms of BER and NER, and discuss recent advances in DDR that may functionally unite components of the BER and NER pathways.

2. MAMMALIAN BER

2.1 History and Overview of BER

Many BER proteins are relatively small, ranging from 20 to 60 kDa, and many enzymatic activities can be detected in biochemical assays without forming multi-subunit structures. This is in sharp contrast to many components of the NER pathway. In the late 1960s, enzymes functioning in BER were purified and characterized in studies using *E. coli*. An endonuclease that can recognize and cleave AP sites was biochemically isolated in the late 1960s and characterized in the 1970s. Also a uracil DNA glycosylase that recognizes and removes uracil in DNA to generate AP site was characterized by the early 1970s [9–11]. These studies helped scientists construct the concept of BER in the 1970s of a systematic DNA-repair pathway for small base damage [12]. This also had the important ramification for the understanding that cells are continuously attacked by not only exogenous DNA-damaging agents, but also by endogenously generated damage [13]. Since it was understood that the concept of "decaying DNA" was associated with mutation and genetic evolution, BER was then recognized as an essential cellular function. The identification, cloning, and characterization of many BER genes and recombinant proteins of *E. coli* occurred during the 1970s to 1980s. This was followed by similar advances in understanding BER in yeast and mammalian cells. Cloning the BER genes led to detailed genetic and biochemical characterization and the elucidation of X-ray crystal structures of many BER proteins [14,15]. By the end of the 1990s, a clear picture of the BER pathway was drawn. However, questions remained unsolved regarding the efficiencies of the recombinant BER proteins, particularly those of DNA glycosylases that carry out the first base removal steps. Purified DNA glycosylases exhibit very low catalytic activities when studied in vitro which questioned how BER proteins in cells succeeded in maintaining genomic integrity [16]. Based on studies carried out mostly in the 2000s, it became apparent that the BER reactions are coordinated to bring about efficient repair. A DNA–protein complex formed by a BER enzyme (eg, APE1) and the resulting cleaved DNA (eg, DNA cleaved by APE1) is in a conformation favored for interacting with a BER enzyme carrying out the next reaction (eg, DNA polymerase beta, Polβ) [17–19]. The BER coordination achieved by this "hand-off" mechanism is ensured by XRCC1, a BER scaffolding protein critical for facilitating the BER efficiency in vivo [20–24]. Understanding the coordination of BER that involves the scaffolding protein XRCC1 and the damage sensory protein poly(ADP-ribose) polymerases (PARPs) has greatly increased the possibility of modulation of BER in the intervention of diseases including cancer and neurodegeneration [25–27].

2.2 Types of DNA Damage Repaired by BER

We can define the BER pathway as a series of reactions by proteins that are capable of repairing abnormal bases, AP sites, and DNA single-strand breaks (SSBs).

2.2.1 Base Damage and DNA Single-Strand Breaks

DNA bases are vulnerable to alkylation, deamination, and oxidation.

2.2.1.1 Alkylation

A number of DNA-alkylating agents are known, including methyl methanesulfonate (MMS), 1-methyl-3-nitro-1-nitrosoguanidine (methylnitronitrosoguanidine; MNNG), and *N*-nitroso-*N*-methylurea (NMU) [1]. Temozolomide is an alkylating agent that is an FDA-approved chemotherapeutic drugs used for glioblastoma treatment [28,29]. Alkylation of purines may also occur endogenously with S-methyladenosine [30]. N7- and N3-alkyl purines are the major adducts in DNA caused by alkylating reagents; more than 80% of adducts produced by MMS are N7-alkylguanine, and about 10% are N3-alkyladenine [31]. Alkylated purines become highly unstable, and readily undergo depurination (loss of purine bases) in physiological conditions [32]. It is noted that O^6-methylguanine produced by alkylating agents is a highly mutagenic base adduct, but in mammals the lesion is repaired by a single enzyme MGMT (O^6-methylguanine-DNA methyltransferase) through a direct reversal mechanism [33].

2.2.1.2 Deamination

Exocyclic amino groups in the bases are subject to deamination. Deamination at N4 of cytosine results in the conversion of cytosine to uracil. Similarly, deamination of adenine at N6 changes the purine base to hypoxanthine. These reactions are mutagenic as uracil pairs with adenine in DNA, and hypoxanthine with cytosine. Another important deamination reaction occurs at N4 of 5-methylcytosine (5mC). 5mC is the result of methylation in CpG di-nucleotide in mammalian cells. Deamination of 5mC converts cytosine to thymine, and thus generates a G:T mispair which is mutagenic. These incorrect uracil and thymine bases are removed by uracil DNA glycosylase and thymine DNA glycosylase in the BER pathway (see Section 2.3.1 and Table 17.1).

2.2.1.3 Oxidation

Reactive oxygen species (ROS) are continuously generated in cells [3]. The mitochondrial respiratory chain is the major source of ROS, as the electron transport system in the inner mitochondrial membrane builds a necessary redox gradient, and electron leaks inevitably occur and are trapped by oxygens to produce superoxide (O_2^-) [34]. O_2^- are effectively scavenged by mitochondria-specific superoxide dismutase Mn-SOD (SOD2). SOD2 is an extremely efficient enzyme that can easily prevent O_2^- from accumulating inside cells. However, this reaction creates hydrogen peroxide, and in the presence of redox metals such as iron and copper, hydrogen peroxide may be further processed to hydroxyl radical (OH•) via the Haber–Weiss reaction [3]. OH• is highly reactive and readily attacks DNA to produce a plethora of different types of oxidative DNA base damage (reviewed in Hegde et al. [2] and Evans et al. [35]). Moreover, ROS also directly attacks the DNA backbone to produce SSBs [2]. SSBs produced by ROS often possess unusual 3′-end structures including 3′-phosphate and 3′-phosphoglycolate as the major products, and these have to be processed to 3′-OH termini in order for the repair process to be completed.

2.2.2 SSBs With Tyrosyl–DNA Covalent Linkage

Mammals possess three topoisomerases I, II, and III (TOP1, TOP2, and TOP3), that resolve higher-order supercoils and knot structures in DNA by introducing single-strand nicks in the DNA (TOP1 and TOP3), or DNA double-strand breaks (DSBs; TOP2) [36,37]. Topoisomerases form tyrosyl–DNA–phosphodiester covalent bonds as intermediate products during the reactions. The tyrosyl–DNA complex formation is transient and resolved in normal topoisomerase reactions. However, when steps of the reactions are inhibited or aborted, the covalent bonds become trapped. This can occur when the enzymes encounter sites of DNA damage such as AP sites and 8-oxoG, or when they are trapped by inhibitors of topoisomerases [38]. Tyrosine residues are trapped at either 3′- or 5′-termini depending on the type of topoisomerases. TOP1 generates a DNA 3′-tyrosyl-phosphodiester bond and a 5′-OH, while TOP2 generates a DNA 3′-OH and a 5′-tyrosyl-phosphodiester bond. In both cases the moieties that are formed block normal DNA synthesis and ligation, and thus they can be regarded as termini-blocking SSBs which require BER proteins repair them.

2.3 Mechanism of Mammalian BER

A model for the basic mechanism of mammalian BER was established by the mid-1990s. The entire BER pathway, the "single nucleotide gap-filling reaction" (SN-BER), could be reconstituted by five distinct reactions in vitro (the middle scheme in Fig. 17.1). (1) Base damage is recognized and removed by DNA glycosylases which leave AP sites. (2) AP sites are recognized by AP endonucleases and are incised, resulting in nicks in the DNA strand with a 3′-OH terminus and 5′-deoxyribose phosphate (dRP) structure [15]. (3) 5′-dRP is removed by DNA Polβ [39]. (4) Polβ fills a nucleotide in the

TABLE 17.1 Enzymes and Reactions in the BER Pathway

BER Sub-pathway	#	Reaction Description	Enzyme	Substrate	Product
1 nt-filling BER	A	Base removal	DNA glycosylases[a]	Abnormal bases	AP sites
	B	Incision upstream of AP sites	APE1	AP sites	SSB with 3′-OH/5′dRP
	C/C′	Incision downstream of dRP	Polβ(as a dRPase), DNA glycosylases[b]	dRP	SSB with 3′-OH/5′-P gap
	D	One nucleotide filling	Polβ	SSB with a 1 nt-gap	DNA with 3′-OH/5′-P nick
	E	DNA ligation	LigIIIα	DNA with 3′-OH/5′-P nick	Repaired DNA
Long-patch BER	F	Excision of flipped nucleotides	FEN1	Flipped strand breaks with 5′-dRP	SSB with multiple nucleotide gap
	G	Long-patch filling	Polβ, Polδ/Polε, PCNA	Multinucleotide gap	DNA with 3′-OH/5′-P nick
	H	DNA ligation	LigI	DNA with 3′-OH/5′-P nick	Repaired DNA
APE1-independent BER	I	δ-Elimination	NEIL1, NEIL2	3′-α,β-unsaturated aldehyde	SSB with 3′-P/5′-P gap
	J	3′-P removal	PNKP	SSB with 3′-P/5′-P gap	SSB with 3′-OH/5′-P gap
TDP[c]	K	3′-Tyrosyl-DNA-phosphodiesterase	TDP1	3′-Phosphotyrosyl linkage	3′-P
	L	Phosphate removal and addition	PNKP	3′-P and 5′-OH	3′-OH and 5′-P
	M	DNA ligation	LigIIIα (TDP1), LigIV (TDP2)	DNA with 3′-OH/5′-P nick	Repaired DNA
	N	5′-Tyrosyl-DNA-phosphodiesterase	TDP2	5′-Phosphotyrosyl linkage	3′-OH and 5′-P
Non-enzymatic reactions and enzymatic "mis"-reactions	1	Incision downstream of AP sites	Spontaneous β-elimination	AP sites	SSB with 3′-OH/5′-dRP gap
	2	Oxidation of AP site	Spontaneous oxidation	AP sites	Oxidized AP sites
	3	Stalled Topo I	Topoisomerase I	Normal DNA	3′-Phosphotyrosyl linkage
	4	Stalled Topo II	Topoisomerase II	Normal DNA	5′-Phosphotyrosyl linkage

Reactions are linked to the schemes (A–N) in Figs. 17.1 and 17.2.
[a]*DNA glycosylases without AP lyase activity: Methylpurine DNA glycosylase (MPG), uracil DNA glycosylases, MutY-homology (MYH), Thymine DNA glycosylase (TDG).*
[b]*DNA glycosylases with AP lyase activity: 8-oxoG DNA glycosylase (OGG1), EndoIII homology (NTH), EndoVIII-like 1 and 2 (NEIL1 and NEIL2). NEIL1 and NEIL2 also carry out βδ-elimination.*
[c]*The reactions do not involve DNA-repair synthesis.*

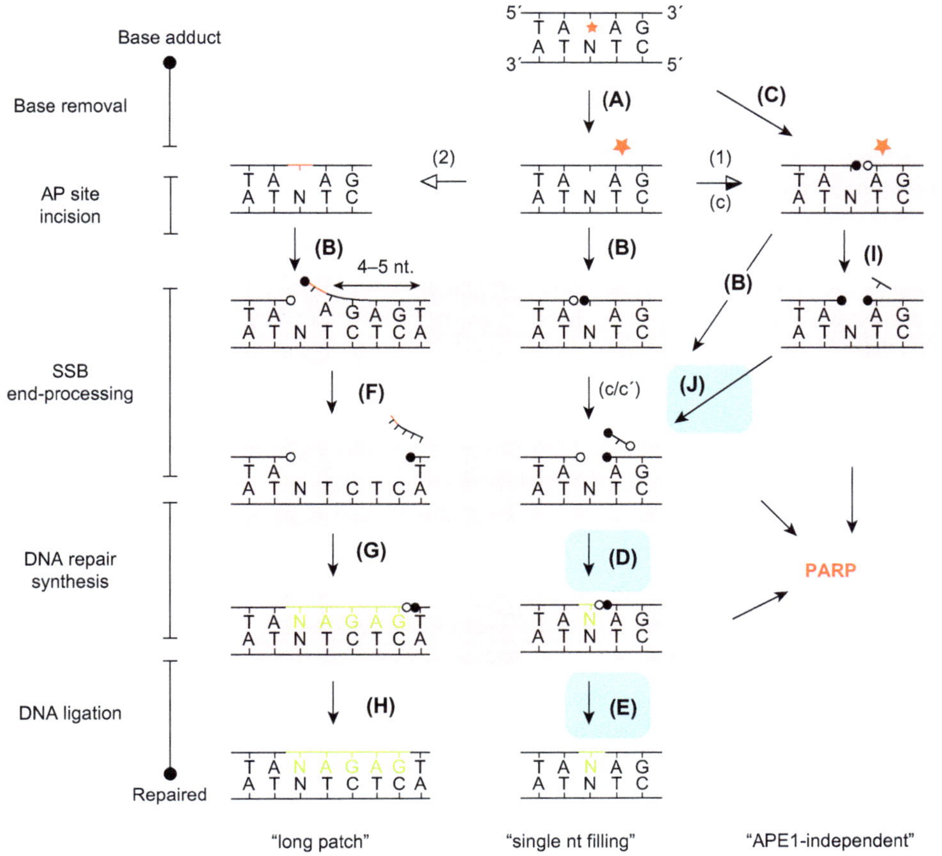

FIGURE 17.1 **DNA base excision repair.** The star in red represents abnormal bases including 8-oxoG and other oxidized and alkylated bases. An oxidized AP site after the reaction (2) is shown in red. Newly synthesized nucleotides are shown in green. Schemes (A–J) depict enzymatic reactions and (1) and (2) are spontaneously occurring reactions. The *open circles* at the 3′-end of SSBs denote 3′-OH termini, and the *filled circles* indicate 3′- or 5′-phosphate termini. Also see Table 17.1. XRCC1 and PARPs are not directly involved in the DNA processing but are pivotal for efficient BER in vivo. Reactions stimulated by XRCC1 (which is recruited to the DNA-damage sites by PARP1) are *colored by light blue*. PARP-activating DNA structures (ie, SSBs) are *encircled in red.*

gap and leaves a nick (3-OH and 5′-P without a gap). (5) The nick is sealed by DNA ligase IIIα [20,40]. Each step takes care of one type of DNA damage and leaves an intermediate lesion until the final nick-sealing reaction performed by DNA ligases occurs. Repair reactions may start at any of the intermediate lesions. For example, topoisomerase–DNA cross-links have been more recently characterized as forms of DNA damage, and repair of these trapped lesions does not follow the base removal step in the conventional BER pathway. Instead, resolution of the tyrosyl–DNA complex is followed by DNA end-processing reactions and by DNA synthesis and ligation, skipping the reactions described earlier as steps 1 and 2. This flexibility confers a versatility to BER and it can act on a plethora of different types of DNA damage that are generated endogenously or by exposure to exogenous DNA-damaging agents.

2.3.1 DNA Glycosylases

There are a total of 10 DNA glycosylases identified in mammals (Table 17.1). All DNA glycosylases cleave N-glycosylic bonds that link bases to the DNA-ribose backbone (Fig. 17.1, reaction A). This reaction creates AP sites that are processed further by an AP endonuclease (APE1). However, many DNA glycosylases further process the resulting AP sites using their intrinsic AP lyase activities (Fig. 17.1C). An AP lyase activity carries out a DNA strand–cleavage reaction through β- or βδ-elimination (Fig. 17.1B and I). The resulting 3′/5′-end structures are 3′-phospho-α,β-unsaturated aldehydes (3′-PUA)/5′-P by β-elimination and 3′-P/5′-P by βδ-elimination [15]. Importantly, these 3′-end structures cannot serve as primers for DNA-repair synthesis carried out by DNA polymerases, and require APE1 or polynucleotide kinase/phosphatase (PNKP) (Fig. 17.1B and J) to generate 3′-OH termini. Fig. 17.1 and Table 17.1 list the mammalian DNA glycosylases and summarize their reactions.

2.3.2 AP Endonuclease 1

Mammals appear to possess only a single active AP endonuclease; that is, APE1. APE1 not only incises AP sites to create 3′-OH/5′-dRP termini [15], but also it hydrolyzes 3′-phosphodiester bonds in 3′-PUA to generate 3′-OH (Fig. 17.1B) [40]. In both processes, APE1 generates 3′-OH ends which are absolutely required for DNA-repair synthesis carried out by DNA polymerases. Early studies of mice with homozygous knockouts of the Ape1 gene (Apex1) found that the gene disruptions resulted in early embryonic lethality [41,42], and APE1 was thought to be essential for cell viability [43,44]. However, in 2013, Masani et al. successfully created B cells defective in the APE1 gene [45]. Surprisingly, deleting the APE1 gene in the B cells did not affect the cell growth, although the cells exhibited a significant decrease in immunoglobulin class switch recombination [45,46], and were hypersensitive to treatment with MMS, an alkylating agent that produces AP sites. It needs to be determined whether cells other than B cells can survive without APE1, and this should be testable, given the advancement of the CRISPR gene–knockout technology. In a 2015 study, a particular mouse embryonic fibroblast cell line expressing APE1 at a level only 0.2% of normal cells was established [47]. While the cells with low APE1 grew normally, their mitochondrial respiratory activities and intracellular oxidative stress levels were greatly reduced. Thus, cells may be able to adapt to conditions with extremely low APE1 activity, which may explain why in previous studies that created an acute reduction in APE, the cells underwent apoptosis [43,44].

A second AP endonuclease, APE2, was identified based on amino acid sequence homology to APE1. However, its biological significance is not clear [45].

2.3.3 Enzymes That Process DNA Termini in BER

2.3.3.1 3′-End Cleaning (APE1 and PNKP)

As described earlier, for DNA-repair synthesis to be initiated, the 3′-terminal end used as a substrate for extension by DNA polymerases must possess a 3′-OH. However, when the AP lyase activity of a DNA glycosylase processes a damaged base instead of using the reaction of APE1 (Fig. 17.1B), a 3′-PUA is generated (Fig. 17.1, reactions 1 and C) and this differs from the requisite 3′-OH. Therefore, it is necessary for BER to process the "3′-blocking" damage. As depicted (Fig. 17.1B in the right scheme), APE1's phosphodiesterase activity has the capacity to remove these 3′-end structures including 3′-PUA and 3′-phosphoglycolate [40,48]. In contrast, APE1 has very weak activity on substrates containing 3′-phosphates (Fig. 17.1, product of I) [40]. Instead of APE1, PNKP has been shown to efficiently remove 3′-phosphate (Fig. 17.1J) [40,49,50].

2.3.3.2 5′-End Cleaning Enzymes

A 5′-phosphate is the end structure required for a DNA ligase reaction to occur with a 3′-OH (Fig. 17.1, prior to H and E reactions). Incision of AP sites by APE1 not only generates 3′-OH but also 5′-dRP which needs to be removed for the subsequent BER reaction (Fig. 17.1C). DNA glycosylases with intrinsic AP lyase activities (Table 17.1) remove the 5′-dRP moieties via β-elimination (Fig. 17.1C).

In addition to DNA glycosylases, DNA Polβ, the main DNA-repair DNA polymerase, has an intrinsic activity to remove 5′-dRP [39]. This "dRPase" reaction (Fig. 17.1C′) is catalyzed via hydrolysis and usually requires Mg^{2+} as a cofactor. The reaction leaves 5′-phosphate at the 5′-termini of the DNA strand breaks. While 5′-phosphate termini can also be generated by DNA glycosylases with intrinsic AP lyase activity (Fig. 17.1C), dRPase and AP lyase are different enzymatic reactions. Although the role of dRPases and AP lyases in cleaning up the 5′-termini is identical (Fig. 17.1C and C′), AP sites can be incised by AP lyases (Fig. 17.1C) but not by dRPases [51].

The dRP or AP sites may be oxidized or reduced in cells (Fig. 17.1, reactions 1 and 2). This modification makes it impossible for the AP lyase/dRPase to remove the sugar moiety [52]. When AP sites are modified by oxidation/reduction, FEN1 (flap structure-specific endonuclease 1) can recognize the 5′-flap end structure (Fig. 17.1F in the left scheme), and incise the nucleotide a few bases downstream of the 5′-dRP.

2.3.3.3 TDP1 and TDP2: Resolving Tyrosyl–DNA Cross-links

Tyrosyl–DNA phosphodiesterases (TDPs) are enzymes that can resolve tyrosyl–DNA cross-links formed during aberrant activities by topoisomerases (Fig. 17.2).

TDP1 can resolve this unique structure to resolve the tyrosyl-3′-phosphodiester cross-link (Fig. 17.2, reaction 3), and leave 3′-phosphate termini in the DNA via hydrolysis (Fig. 17.2K) [38]. Similar to APE1, TDP1 can also remove 3′-phosphoglycolate [53]. The 3′-phosphate groups remaining after TDP1 reactions are then processed by PNKP to generate 3′-OH similarly to the 3′-end cleaning process in Fig. 17.1J. Of note, TDP1 also reacts on 3′-phosphoglycolate to generate 3′-phosphate [53], which may be further processed by PNKP to 3′-OH.

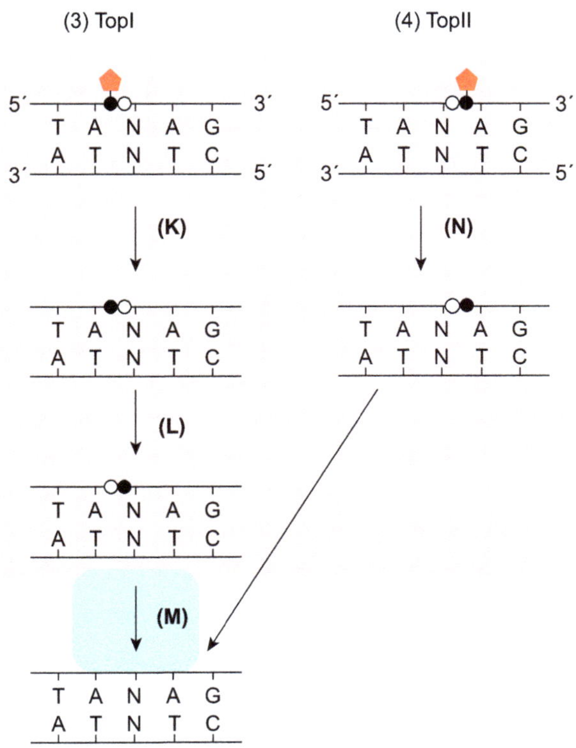

FIGURE 17.2 **Resolution of DNA–protein covalent linkage.** Stalled reactions on DNA by DNA topoisomerases I and II result in trapped 3′-phosphotyrosyl linkages which cause obstruction to transcription and replication, and result in cell death. Tyrosyl–DNA phosphodiesterase 1 and 2 (TDP1 and TDP2) resolve the stalled linkages followed by direct DNA ligation without DNA synthesis. In the case of TDP1, PNKP is required to generate proper 3′- and 5′-termini for DNA ligation. The *same labels and colored areas* as in Fig. 17.1 are used. Also see Table 17.1 for details.

Similar to TOP1, the catalytic Tyr-DNA intermediates of TOP2 may be trapped by TOP2 inhibitors such as etoposide [38]. Unlike TOP1, however, TOP2 incises the DNA to generate 3′-OH and 5′-P termini [37], and the Tyr residues form covalent cross-links to the 5′-phosphate termini (Fig. 17.2, reaction 4). TOP2 incises both strands and so it temporarily produces DSBs. When the TOP2 activity is inhibited and trapped, Tyr-5′-P intermediates are formed, DSB accumulate in DNA and become highly toxic. The trapped structure can be resolved by TDP2. The TDP2 reaction resolves the trapped linkage and releases the 5′-P termini in DNA (Fig. 17.2N).

2.3.4 Completion of an Entire BER Reaction: DNA Polymerases and DNA Ligases in Coordinated Reactions

The excision steps described earlier (Figs. 17.1A–C,F,I,J and 17.2K,L,N) are damage-specific BER reactions, and can only be processed by the enzymes that remove the particular lesions. In contrast, the DNA gap-filling (Fig. 17.1D and G) and sealing steps (Figs. 17.1E,H and 17.2M) do not involve damaged DNA, and thus theoretically any combinations of DNA polymerases and ligases should complete the processes. Although this may be the case in vitro, in cells there appears a stringent coordination that determines what DNA polymerases and ligases should follow each damage-specific BER process. It is believed that the coordination from the excision steps to the gap-filling/sealing reactions is to minimize the toxic effects of the intermediate DNA lesions. For example, gap-containing regions of DNA formed by BER might become even more toxic if they had to persist until they were randomly recognized by DNA polymerases. Instead, interactions of DNA polymerases and ligases with other BER proteins are known to improve the efficiency of the entire BER reaction. This is known as BER coordination [18,19,54,55], and it is controlled by interactions among the BER proteins and damaged DNA involved in each reaction step. The coordination of BER is further ensured by the presence of XRCC1 (see Section 2.3.5).

Thus, the DNA synthesis and ligation steps during BER should be viewed as sequential reactions that follow the damage-excision reactions. DNA polymerases beta, delta/epsilon, and lambda (Polβ, Polδ/ε, and Polλ) have been shown to function in DNA synthesis in BER, and DNA ligase I (LIGI) and III (LigIIIα) are the major DNA ligases in BER.

2.3.4.1 Single-Nucleotide Filling-BER

The involvement of Polβ has been studied since the 1980s. BER can be completed by Polβ with DNA LigIIIα in the simplest sub-pathway named single-nucleotide filling (SN)-BER which is shown in the middle column in Fig. 17.1A–E. As an example, uracil forms in DNA as the product of cytosine deamination (resulting in U:G mispair), and can be repaired by

SN-BER [14]. Uracil is removed by uracil DNA glycosylase (UNG) which produces AP site (Fig. 17.1A) [56]. APE1 then cleaves the DNA upstream of the AP site, and generates 3′-OH/5′-dRP termini (Fig. 17.1B). Polβ removes the dRP (Fig. 17.1C), fills the single nucleotide gap (Fig. 17.1D). This reaction leaves a nick with 3′-OH/5-P termini, which is sealed by LigIIIα tightly interacting with XRCC1 (Fig. 17.1E). Subsequent studies have shown that Polλ, whose amino acid sequence shows high homology to Polβ, can substitute for Polβ in this BER sub-pathway [57]. Interestingly, Polλ appears to be more critical than Polβ for cellular protection against oxidative DNA damage [58].

2.3.4.2 Long-Patch BER

An alternative repair pathway for oxidized (and reduced) AP sites was postulated that involves DNA Polδ and Polε (Fig. 17.1F–H) [52]. Oxidized AP sites (Fig. 17.1, reaction 2) are incised by APE1 in the same way as intact AP sites (Fig. 17.1B), but the resulting oxidized 5′-dRP cannot be removed by Polβ or by AP lyase-associated DNA glycosylases. Instead, flap structure–specific endonucleases (FEN1) remove the dRP-containing 5′-termini (Fig. 17.1F), leaving gaps spanning several nucleotides. DNA synthesis from these gapped DNA structures was shown to be specifically carried out by Polδ with PCNA as an essential elongation cofactor (Fig. 17.1G). Finally, DNA Lig I seals the nicked DNA to complete this BER sub-pathway (Fig. 17.1H).

2.3.4.3 APE1-Independent BER

As described earlier, two BER sub-pathways rely on APE1 to generate 3′-OH termini at damaged site in DNA, the essential primer for DNA polymerases. However, SSBs with 3′-phosphate termini are poor substrates for APE1, and thus the 3′-end cleaning step may become rate limiting.

The NEIL family of DNA glycosylases, NEIL1 and NEIL2, carry out β-elimination to generate 3′-PUA after the base damage is removed, and they further process PUA by δ-elimination to generate 3′-phosphate at the site (βδ-elimination) [59,60]. When PNKP was characterized for its pivotal role in SSB repair (SSBR) as a 3′-phosphatase/5′-kinase, Mitra and his colleagues examined the possibility of alternative BER sub-pathway that do not require APE1. Wiederhold et al. thus showed that AP sites can be processed to 3′-phosphate and 5′-phosphate by NEIL1 or NEIL2 (Fig. 17.1I), and then further processed by PNKP to generate 3′-OH (Fig. 17.1J) [40]. The concept that BER does not require an AP endonuclease has an important ramification in that APE1 can be dispensable in BER, and it also underscores the role of PNKP in BER.

2.3.5 Scaffolding Proteins in BER: Proteins That Do Not Directly Participate in DNA Processing

SSBs may be generated directly by DNA-damaging agents such as ROS or by enzymatic processing during BER. PARPs and XRCC1 play pivotal roles in SSBR. While PARPs and XRCC1 are not directly involved in DNA processing, they establish interactions with other BER enzymes for coordinated and efficient reactions. PARP1, the major PARP, binds to SSBs with a high affinity and protects the toxic DNA damage. PARP1 possesses an enzymatic activity that polymerizes ADP–ribosyl groups onto many cellular factors including itself. The PARylation activity of PARP1 is triggered by SSBs and by DSBs to some extent. PARP1 recruits XRCC1 which possesses a PAR-binding motif in its central domain [21] and thus interacts with PAR-modified PARP1 [61]. Auto-modification of PARP1 results in its decreased affinity for SSBs, and PARP1 is then dissociated from SSBs. XRCC1 then coordinates the BER-repair reactions by interacting with PNKP [23], Polβ [24], and LigIIIα [20]. XRCC1–LigIIIα interaction is essential for efficient SSBR. There are other BER proteins that reportedly XRCC1 interacts with to facilitate the whole BER pathway. These include PCNA, APE1, UNG, NEIL1, OGG1, MPG, NTL1, and NEIL2 [62–67]. However, XRCC1 is recruited on SSBs after PARP activation [21]. Although XRCC1 was shown to possess intrinsic affinity for DNA, SSBs are required for efficient interaction of XRCC1 with DNA [68,69]. Therefore, further studies should clarify how XRCC1 is recruited to DNA damage prior to the generation of SSBs to enhance the BER efficiency.

PARP1's role in BER has been studied for more than two decades, but new roles of PARP1 in enhancing BER are still being discovered [70,71]. This is partly because the PARylation reaction complicates cellular recovery from DNA damage. PARylation consumes cellular NAD^+, whose synthesis requires energy. Thus, the overactivation of PARP has long been known to deplete intracellular NAD^+ and ATP pools and cause cell death [72]. In addition, it was thought that a function of PARylation was to enhance the DNA-ligase reaction, given that PAR provides positive charges to the damaged sites, and this enhances activities of DNA ligases [73], particularly that of LigIIIα [74]. Intriguingly, in 2015, Weinfeld et al. reported that DNA LigIIIα, and not PARP1, is the SSB sensor and acts by recruiting XRCC1 and PNKP to affect the efficiency of SSB reactions in cells [75]. As PARP1 is involved in mitochondrial energy metabolism and apoptosis signaling [76,77], a definitive answer for PARP's role in BER needs additional investigation.

2.4 BER Gene Knockout in Mice and Cells

Many BER genes have been studied using genetic knockout approaches in mice to understand the roles of BER in normal physiology and how alterations impact risks for disease. Table 17.2 summarizes knockout studies of BER and NER genes. Homozygous deletions of many BER genes result in embryonic lethality in mice. Unsurprisingly, homozygous deletions of BER genes known to be required for essential activities in the cells (eg, DNA replication) result in embryonic lethality. Genes that belong to this category are Fen1, DNA ligases, and DNA polymerase genes required for DNA replication. However, many BER genes whose essential functions were not well defined can also result in embryonic lethality. These genes are Tdg, Apex1 (Ape1), Polb (Polβ), and Xrcc1. Notably, Parp1 homozygous knockout mice are viable, although cells lacking Parp1 are hypersensitive to many DNA-damaging agents, and double homozygous knockout of Parp1 and Parp2 result in embryonic lethality [78]. The deletion of individual DNA glycosylase genes does not produce serious phenotypic defects in mice except for the Tdg (thymine DNA glycosylase) gene. It is noted that TDG is required for demethylation of 5mC [79], and thus it is essential for the regulation of differentiation. Thus, losing this function is likely the cause of mouse embryonic lethality, rather than the deficiency in the repair of G:T mispairs in DNA [79]. These observations validate the belief that cells and the mammalian body cannot sustain the accumulation of endogenous DNA damage without BER, and they also underscore the role of BER in epigenetic DNA metabolism.

3. MAMMALIAN NER

3.1 History and Overview of NER

Excision-repair pathways involve the removal or "excision" of a stretch of DNA containing damaged DNA and the resulting gap is filled in by DNA replication using the undamaged DNA as a template. In the 1960s, several groups discovered key aspects of the NER pathway in bacteria and in mammalian cells. Paul Howard-Flanders, Richard Setlow, and their colleagues found that bacteria treated with UV light remove small fragments of DNA containing pyrimidine dimers [80,81]. At roughly the same time, Philip Hanawalt and David Pettijohn demonstrated that "DNA-repair synthesis" coincides with excision of fragments containing pyrimidine dimers in bacteria treated with UV light [82]. Robert Painter developed a novel technique to detect "DNA-repair synthesis" in mammalian cells treated with UV light [83]. This technique is still used today to measure "unscheduled DNA synthesis" or DNA synthesis that occurs outside of S phase as part of the NER pathway after cells are treated with a DNA-damaging agent. An additional seminal observation was made by James Cleaver, who working together with Robert Painter, found that cells from patients with the sun-sensitive and cancer-prone syndrome, xeroderma pigmentosum (XP), are deficient in NER [84]. This observation was groundbreaking for many reasons. It provided evidence that deficiencies in the NER pathway can predispose humans to the development of cancer. It also led to a cell complementation analysis of the clinically heterogeneous disease, XP and this paved the way to identifying many different genes involved in the NER pathway [85]. Seven genetic complementation groups have been identified in XP, designated XPA through XPG, that represent different genes required for the NER pathway [86–88].

Studies performed by many groups around the world during the 1980s and 1990s resulted in the cloning and biochemical characterization of many genes required for mammalian NER [89–93]. As suggested by the seminal observations made in the 1960s, the overall general strategy of NER in mammalian cells is similar to that found in bacteria. An initial step in the pathway involves DNA-damage recognition. This is followed by the introduction of two incisions in the damaged strand, one on each side of the damage. An oligonucleotide containing the DNA damage is removed, and this is followed by synthesis of new DNA to replace the excised, damaged DNA. Finally, there is ligation of the newly synthesized DNA to the parental DNA. While the overall strategy of NER has been conserved in mammals and bacteria, it has been estimated that NER in mammalian cells, in vivo, requires 30–50 different gene products, and hence it is much more complicated than that found in bacteria.

A perhaps unique characteristic of the NER pathway is that it can be coupled to the process of transcription (reviewed in Refs. [94,95]). This surprising aspect of NER was first documented by the investigation of DNA repair in specific regions of the genome. Using this approach, it was discovered that DNA damage can be preferentially removed from genes active in the transcription process [96,97], and this preferential repair is actually targeted to only the transcribed strand of an active gene while the nontranscribed stand is unaffected [98]. Subsequent investigations have provided evidence that many of the same genes are involved in NER and transcription-coupled NER (TC-NER), but the processes differ at the steps involving recognition of the DNA damage.

TABLE 17.2 Summary of Phenotypes of Homozygous Knockout Mice of BER and NER Genes

Pathway	Gene Symbol	MGI ID[a]	Reaction[b]	Homozygous knockout mouse phenotype	References
BER (Fig. 17.1)	Ung	109352	A	Viable; no significant phenotype	[195]
	Smug1	1918976	A	Viable; no significant phenotype	[196]
	Mpg (Aag)	97073	A	Viable; no significant phenotype	[197]
	Nthl1	1313275	C	Viable; no significant phenotype; increased tumors in Nthl1 Neil1 double knockout mice	[198,199]
	Mutyh	1917853	A≫C	Viable; increased intestinal tumors, particularly with exposure to $KBrO_3$	[200]
	Ogg1	1097693	A>C	Viable; $KBrO_3$ induces renal cancer	[201,202]
	Tdg	108247	A	Nonviable; critical to controlling epigenetic status	[79]
	Neil1	1920024	C, then I	Viable; reduced germinal B cell	[203]
	Neil2	2686058	C, then I	Viable; accumulation of oxidative DNA damage in transcriptionally active genes in aged mice	[204]
	Neil3	2384588	C?	Viable; reduced proliferation and sensitive to genotoxic stress	[205]
	Apex1 (Ape1)	88042	B	Nonviable; apoptotic	[41,42,206]
	Polb	97740	C′, then D	Neonatal lethality; immune deficiency	[207]
	Pold1	97741	G	Null likely nonviable; proofreading deficiency to elevated mutation and tumors; shortened longevity	[208]
	Pole	1196391	G	Null likely nonviable; proofreading deficiency to elevated mutation and tumors; shortened longevity	[209]
	Poll	1889000	D	Viable	[210]
	Fen1	102779	F	Nonviable; heterozygous knockout mice predisposed to adenocarcinoma; E359K mutation oncogenic	[211,212]
	Lig1	101789	H	Nonviable	[213]
	Lig3	109152	E	Nonviable	[214]
	Xrcc1	99137	–	Nonviable; increased spontaneous SSBs	[215]
	Parp1	1340806	–	Viable; cells sensitive to DNA damage; improved ischemic injury recovery; resistance to diabetes; Parp1 Parp2 double knockout embryonic lethal	[72,216–218]

NER (Fig. 17.3)	Ddb1 (Xpe)	1202384	A	Nonviable	[219]
	Ddb2 (Xpe)	1355314	A	Susceptible to UV-induced DNA damage and skin tumor	[220]
	Xpc (Rad4)	103557	A	High incidence of UV-induced skin tumors; high mutation frequency	[221]
	Rad23b	105128	A	Neonatal mortality; growth retardation and other abnormality; shortened life	[222]
	Ercc8 (Csa)	1919241	D	Increased skin tumor by UV irradiation	[223]
	Ercc6 (Csb)	1100494	C	UV sensitivity; increased skin and eye tumors; circling behavior; low body weight	[224]
	Ercc3 (Xpb)	95414	E, F	Null nonviable	[225]
	Ercc2 (Xpd)	95413	E, F	Null nonviable; a missense knock-in with brittle and graying hair, cachexia	[226,227]
	Xpa	99135	G, H	Predisposition to skin tumors induced by UV and other bulky DNA damage	[228]
	Ercc5 (Xpg)	103582	G, J	Postnatal mortality; hypersensitive to UV	[229]
	Ercc1 (Rad10)	95412	I	Growth and liver failure; postnatal death; early aging and sensitive to oxidative stress	[230]
	Ercc4 (Xpf)	1354163	I	Impaired growth; short life (~several weeks); cells hypersensitive to UV	[231]

[a]MGI, *Mouse Genome Informatics (http://www.informatics.jax.org).*
[b]*Reactions depicted in Fig. 17.1 (BER) and in Fig. 17.3 (NER).*

3.2 Types of DNA Damage Repaired by NER

The NER pathway is unusual, in that it recognizes and removes a wide spectrum of different types of DNA damage and the damage is usually formed by some covalent alteration or modification to one of the DNA bases [99–102]. It is generally held that the NER pathway actually recognizes a distortion in the localized structure of the DNA helix produced by the presence of a damaged base and it does not directly recognize the modified base in "a hand in clove" manner (described in Section 2.3.1 in more detail) [103,104]. Hence, NER can recognize and remove structurally unrelated base modifications including those formed by exposure to UV light, benzo(a)pyrenes and other aromatic hydrocarbons, aflatoxins, 2-acetylaminofluorenes and chemotherapeutic agents, such as platinum.

UV light has been used extensively to investigate the NER pathway. UV light results in the covalent linkage of adjacent pyrimidines and produces two predominant types of DNA damage; the cyclobutane pyrimidine dimer (CPD) and 6-4 photoproduct (6-4PP) [105,106]. Several organisms have developed additional strategies for removing UV photoproducts by the process of photoreactivation. However, humans and other placental mammals appear to lack photoreactivation pathways, and hence, NER is their sole means of removing CPDs and 6-4PPs. Left unrepaired, CPDs and 6-4PPs can produce mutations and contribute to the development of skin cancer. One of the hallmarks of the disease XP is an extremely elevated incidence of skin cancer. Many XP patients develop a form of skin cancer within the first decade of their life and develop many tumors in sun-exposed regions of their body. It is likely that UV photoproducts are formed in the skin of XP patients beginning early in life; however, since XP patients have a deficiency in NER, the photoproducts persist and lead to the formation of mutations, a driving force in cancer etiology and progression [107]. Efforts to protect XP patients from the harmful effects of sunlight and UV radiation are prolonging their lives but their deficiencies in NER appear to contribute to the development of other forms of cancer.

Thousands of compounds have been identified in the vapor and particulate phases of cigarette smoke and they include carcinogens, co-carcinogens, mutagens, and tumor promoters. About 70 of these compounds have been classified as carcinogens [108,109]. Different classes of carcinogens are present in tobacco smoke and include the polycyclic aromatic hydrocarbons (PAH) such as benzo(a)pyrene (BP), dibenz(a,h)anthracene, and dibenzo(a,i)pyrene. Metabolic activation of these and other chemical compounds found in tobacco smoke can create intermediates that react with DNA bases and produce DNA adducts that are substrates of NER. Hence, DNA adducts are likely continually formed in the lung tissues of people who smoke, and if they are not removed by DNA-repair processes, their persistence could lead to the formation of mutations and ultimately to lung cancer.

3.3 Mechanisms of Mammalian NER

Advances in the 1980s and 1990s led to the development of mammalian cell–free systems to investigate detailed mechanistic steps in NER [89,91,93,110]. During the mid-1990s, NER was reconstituted in vitro using the purified repair proteins: XPC-RAD23B, TFIIH (containing XPB and XPD), XPA, XPG, and ERCC1-XPF and the purified replication proteins: RPA, PCNA, and DNA Polδ [89,111]. Subsequent studies indicate that these and additional proteins function in the cell through an ordered and sequential assembly onto damaged DNA (reviewed in Refs. [99,101,107]).

3.3.1 DNA-Damage Recognition and Unwinding of the Damaged DNA Duplex

The properties that govern the ability of NER to recognize structurally diverse types of DNA damage were originally described in a model described as "bipartite recognition" [103,104]. In this model, the more favorable substrates for NER are those in which the DNA damage destabilizes the DNA helix and is bulky. Damage substrates that destabilize the helix can promote disruption of hydrogen bonding and bending of the DNA helix. The bipartite recognition model has been supported by studies that have compared the efficiency of NER on damaged substrates that differ in the degree to which they destabilize helix. Hence, it was discovered that 6-4PPs have a strong destabilizing effect on the DNA helix and are efficient substrates for NER, while CPDs do not and can be poorly repaired [112].

DNA-damage recognition in NER is achieved by the XPC protein (Fig. 17.3A) [113]. XPC binding to damaged DNA is promoted by destabilization of the DNA helix and XPC can even bind destabilized DNA in the absence of DNA damage as seen using substrates containing small loops or bubbles [114]. XPC resides in a complex with RAD23B and centrin-2, a member of the calmodulin family of calcium-binding proteins [113,115]. RAD23B stabilizes XPC and may help deliver it to the site of damage (Fig. 17.3A) [116]. The role of centrin-2 is less clear. Since the presence of CPDs in DNA do not promote disruption of hydrogen bonding or destabilizing of the helix, they are not efficiently recognized by XPC. An additional protein, UV DNA damage–binding protein 2 (UV-DDB2), is required for the removal of CPDs in cells and may directly and indirectly promote binding of XPC to CPDs (Fig. 17.3A) [115].

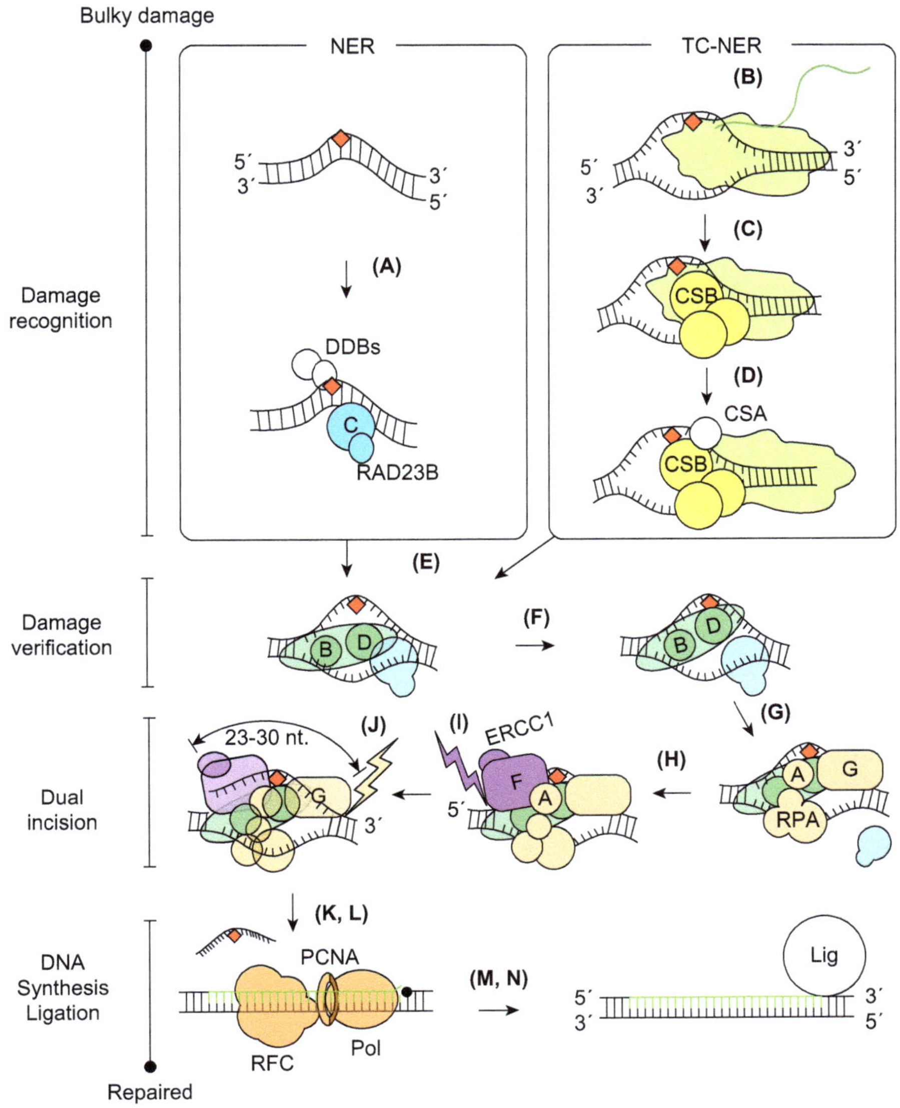

FIGURE 17.3 **DNA nucleotide excision repair. (DNA-damage recognition)** In the regular NER (A, global genome NER), the XPC/RAD23B complex is critical for the damage recognition. The XPC/RAD23B complex (blue) senses distortion in DNA structure containing damage. In the case of cyclobutane pyrimidine dimer (CPD), XPC/RAD23B requires UV–DDB1/2 (XPE) complex for DNA binding. (B–D) In TC-NER, RNA Pol II (light green) stalled by DNA damage (B) initiates NER. (C) CSB–UVSSA–USP7 complex (yellow) is recruited to the damage site, and (D) CSA–CSB complex formation facilitates RNA PolII backtracking to set up a NER platform. **(Damage verification)** (E) Transcription elongation factor TFIIH opens DNA double-strand at the damage site, and XPD plays a key role in damage verification step (F). **(DNA strand incision)** (G) XPA, RPA, and XPG bind to the open complex. (H) The ERCC1/XPF complex is recruited to the repair complex through interaction with XPA, (I) and XPF incises the damage-containing DNA strand at the 5′-upstream of the lesion. (J) The incision by XPF is immediately followed by downstream incision carried out by XPG. **(DNA-repair synthesis and ligation)** (K and L) Polδ/Polκ or Polε with PCNA and RFC fills the gap created by XPF and XPG, (M and N) and DNA LigIIIα/XRCC1 or DNA LigI seals the 3′-OH/5′-P DNA termini to complete the NER reaction. Also see Table 17.2.

TFIIH is a large complex that functions in both NER and transcription [117]. It is loaded onto sites of damaged DNA through interactions with XPC–RAD23B (Fig. 17.3E) [118–120]. It is comprised of 10 subunits that can be divided into the core complex which contains XPB and the cyclin-activated kinase (CAK) sub-complex which is not required for NER. XPD appears to serve as bridge between the core and CAK complexes. XPB and XPD are both helicases and ATPases and the roles of these activities in NER have been extensively studied. XPB helicase activity functions in 3′–5′-translocation and XPD helicase activity functions in 5′–3′-translocation (Fig. 17.3F, Table 17.3). In contrast to XPD, the ATPase activity of XPB is required for NER but not its helicase activity [121]. A major function of XPB in NER appears to be in disrupting the DNA helix, which assists in the loading of TFIIH onto damaged DNA. Once TFIIH is loaded onto DNA, XPD helicase activity results in its translocation along the DNA, unwinding the damaged duplex in the 5′–3′-direction until it encounters

TABLE 17.3 Enzymes and Reactions in the NER Pathway

NER Pathways	#	Reaction Description		Protein	Interaction With Preexisting Factor	DNA
NER	A	Damage recognition	Bulky damage recognition	XPC/Rad23B	DNA kink	Distortion in DNA due to bulky damage; CPD recognition requires UV-DDB
			CPD recognition[a]	DDB1/2	CPD	
				XPC/Rad23B	DDB 1/2	
TC-NER	B	Damage recognition	Transcription stalling	RNA Pol II	DNA damage	
	C		CSB recruitment	CSB/UVSSA/USP7	RNA Pol II	
	D		Backtracking RNA Pol II	CSA–CSB	CSB	
Downstream reactions common to NER/TC-NER	E	Strand opening		TFIIH		DNA strand opening
	F	Damage verification		XPB⇒XPD	Part of TFIIH	
	G	3′-Incision complex (XPG) formation		XPA-RPA-XPG		
	H	5′-Incision complex (ERCC1/XPF) formation		ERCC1/XPF	XPA	
	I	5′-Incision		ERCC1/XPF		Nick upstream of damage
	J	3′-Incision		XPG		Nick downstream of damage
	K	DNA synthesis		Polδ/Polκ/PCNA		22–30 nt incorporation
	L			Polε/PCNA		
	M	DNA ligation		LigIIIα/XRCC1		Repaired
	N			LigI		

Reactions (#) are linked to the schemes (A–N) in Fig. 17.3.
[a]CPD, *cyclobutane pyrimidine dimer;* CSA(B), *cockayne syndrome protein A (B);* RNA Pol II, *RNA polymerase II holoenzyme;* DDB1/2, *UV-damage DNA-binding protein 1 and 2;* UVSSA, *UV-stimulated scaffold protein A.*

a bulky covalent DNA base modification that results in blockage of additional translocation [114,121]. This blockage of XPD-mediated translocation of TFIIH at sites of damage is viewed as one step in DNA-damage verification which serves to prevent or reduce gratuitous NER at undamaged locations in the DNA (Fig. 17.3F, Table 17.3).

For many years it was held that XPA rather than XPC was involved in DNA-damage recognition. Instead, XPA appears to hold a central role in coordinating the loading of additional NER proteins at the site of damage and perhaps serves as an additional step in DNA-damage verification (Fig. 17.3G) [122–127]. When the translocation of XPD becomes stalled at a damaged site, XPC–HHR23B dissociates and XPA, RPA, and XPG bind the damaged site. A stable pre-incision complex is formed and comprised of TFIIH, XPA, RPA, and XPG. XPA serves an important role in assembling the pre-incision complex in its interaction with other NER proteins and single-stranded DNA. XPG binds through interactions with TFIIH.

3.3.2 Incision, Repair Synthesis and Ligation

Once the pre-incision complex is formed, XPA recruits ERCC1–XPF complex (Fig. 17.3H) [125,127]. ERCC1–XPF and XPG are junction-specific endonucleases that cleave DNA at junctions between double-stranded and single-stranded DNA. The unwinding of DNA by TFIIH and the assembly of XPA, RPA, XPG, and TFIIH produce a bubbled structure at the site of DNA damage. Evidence supports an ordered sequence of incisions; ERCC1–XPF makes the first incision on the 5′ side of the DNA damage (Fig. 17.3I) and XPG makes the second incision on the 3′ side of the damage (Fig. 17.3J) [128]. The oligonucleotide containing the DNA damage and TFIIH are released (Fig. 17.3K) [129]. Once the first incision is made by ERCC1–XPF, a free 3′-OH is formed that can be used by DNA polymerases in repair synthesis, and this could even occur before the second incision is made by XPG (Fig. 17.3L). The repair synthesis step in NER, about 25–30 nucleotides in length, was once assumed to be relatively straightforward. However, the discovery that the error-prone DNA polymerase kappa (Polκ), participates in repair synthesis during NER, in addition to Polδ and Polε suggests that this step in NER is complex [130–132]. Similarly, the ligation step which seals the final phosphodiester bond between the newly synthesized DNA and the parental DNA (Fig. 17.3N) appears more complicated than originally thought. It appears to be regulated by the proliferative state of the cell with DNA ligase I used in proliferative cells and DNA ligase IIIα used in quiescent and replicating cells (reviewed in Ref. [130]).

3.4 Transcription-Coupled NER

The process of TC-NER has been studied for several decades (reviewed in Refs. [94,95]). The existence of mechanisms that couple DNA repair to transcription was indicated many years ago by studies that followed the recovery of RNA synthesis and DNA-repair levels after cells were exposed to UV light [133]. It was found that RNA synthesis, which is initially inhibited by UV light, recovered before significant amounts of DNA damage were found to be removed from total cellular DNA. Subsequently, it was found that UV-induced CPDs were selectively removed from transcriptionally active genes in mammalian cells and the selective or preferential repair of DNA damage from active genes was due to selective repair of only the transcribed strands of the genes [96–98]. The selective repair of DNA damage from the transcribed strands of active genes was first documented in mammalian cells and then subsequently documented in *E. coli* and in yeast. These observations led to models of transcription-coupled repair in which recognition of DNA damage present in the transcribed strand of an active gene was a direct consequence of the stalling or arrest of RNA polymerase when it encountered the damage. This was supported by subsequent studies that found that certain types of bulky damage arrest elongation of RNA polymerase when they are located in the transcribed strand of an active gene but they do not block it when they are present in the nontranscribed strand (reviewed in Ref. [134]).

Investigating TC-NER in cell-free systems has been challenging and this is likely due to the combined complexities involved in the transcription elongation process, in NER and in chromatin structure. Biochemical and genetic studies indicate that damage recognition in TC-NER occurs through blockage or stalling of the RNA polymerase complex when it encounters damage in the transcribed strand (Fig. 17.3B). Many of the subsequent events, loading of TFIIH, XPA, RPA, ERCC1-XPF, and XPG, are likely similar to those found in global NER (Fig. 17.3E–N). However, a notable and major difference between NER and TC-NER involves processing of the RNA polymerase when it becomes stalled or arrested at DNA damage. Due to the large size of the RNA polymerase complex, some processing events are required to remove or displace it in order for the subsequent loading of essential NER proteins to occur. Different models for these processing events have been proposed and include the backward translocation of the RNA polymerase complex away from the damage (backtracking) and/or ubiquitin-mediated modification of damage-stalled RNA polymerase and subsequent degradation of the complex. It remains unclear how these processes occur in mammalian cells. However, genetic and biochemical studies support roles for Cockayne syndrome A (CSA), Cockayne syndrome B (CSB), UV-sensitive syndrome A (UVSSA), and XPA-binding protein 2 (XAB2) in TC-NER (Fig. 17.3C and D). Cell lines with defects in each of these genes exhibit deficiencies in TC-NER or recovery of RNA synthesis following treatment with DNA-damaging agents [118,135–138]. Studies since mid-2000s suggest that degradation of damage-stalled/arrested RNA polymerase complexes may not be a common event; instead, actual degradation of the RNA polymerase complex may serve as a less frequent method of simply clearing the polymerase from the damaged site which might then allow global NER to act at the damage. Similarly, in some rare or unusual instances, the RNA polymerase complex may actually bypass the damage [139]. However, for TC-NER to occur, the polymerase more likely backtracks or is transiently displaced or altered, and this movement of the polymerase serves as a mechanism for loading TFIIH and subsequent NER factors which ultimately results in DNA-damage removal, DNA-repair synthesis, and ligation [140–142].

Mutations in a gene required for NER or TC-NER generally renders cells more sensitive to treatment with agents that introduce bulky types of DNA damage. Their sensitivity to DNA damage can be severe to moderate depending on the gene that is mutated. Clearly, mutations in NER genes can predispose humans to the development of skin and other forms of cancer as illustrated by the disease XP and discussed in more detail later. However, mutations in genes specifically required for TC-NER such as CSA, CSB, and UVSS2 do not generally predispose humans or mice to cancer (reviewed in Ref. [107]). Instead, CS patients display complex phenotypes that include developmental and neurological abnormalities, growth arrest, mental retardation, and premature death. Both CS and UVSS2 patients show cutaneous sensitivity to UV irradiation. These observations together with biochemical and genetic studies may indicate that deficiencies in proteins required for the coupling of NER to transcription may lead to the persistence of RNA polymerase complexes arrested at sites of damage which in turn may trigger apoptotic events leading to cell death.

3.5 NER and Chromatin Structure

The recognition of DNA damage and the functions of many proteins involved in NER and TC-NER described earlier must take into consideration the packaging of DNA into chromatin when repair takes place in vivo. The presence of nucleosomes and the assembly of nucleosomes into higher-order chromatin structures likely impede DNA-damage recognition and NER. Hence, "an access, repair, restore" model proposes that chromatin and nucleosomes must be altered or displaced during DNA-damage recognition and repair, and this is followed by restoration of the nucleosome and chromatin structure following repair [143]. It is likely that this involves alterations in the posttranslational modifications of the histone tails such as by acetylation, alterations in the distribution of histone variants, and the recruitment of chromatin-remodeling complexes. Early studies indicated that nucleosomes become rearranged during NER and that the acetylation of histones stimulated NER. More studies conducted between 2012 and 2014 have provided more detailed mechanistic insights into how alterations in chromatin impact DNA-damage recognition and processing by NER and TC-NER (reviewed in Refs. [95,107,144–146]).

The access step which allows NER proteins to recognize and bind DNA damage appears to be influenced by many proteins. UV-DDB promotes ubiquitylation of core histones and associates with PARP1 to mediate PARylation of chromatin to open it up [116,147,148]. Histone acetylation by the histone acetyl transferases, p300 and GCN5, can also contribute to relaxing chromatin and the ATP-dependent chromatin-remodeling complexes, SWI/SNF and INO80, can promote repair by displacement of nucleosomes and by influencing the recruitment of XPC–RAD23B–centrin complex to the damage [149,150]. After repair is completed, the restore step to assemble the newly synthesized DNA into nucleosomes involves histone chaperones CCRF-associated factor (CAF1) and alternative splicing factor, ASF1 [151–153].

TC-NER occurs during the elongation stage of transcription since it serves and is signaled by RNA polymerase complex blocked at DNA damage. Hence, this state of chromatin is likely different from chromatin that is not transcriptionally active. For TC-NER, the chromatin has already been "opened" to allow transcription initiation and elongation. CSB is required for TC-NER and studies have found that it can remodel chromatin in vitro [154]. Whether it has chromatin remodeling functions during TC-NER is unclear. CSA and UVSSA can play different roles in targeting CSB for ubiquitylation and degradation. CSA promotes ubiquitylation of CSB, while UVSSA inhibits ubiquitylation of CSB [155,156]. CSA and CSB appear to promote the association of histone acetyl transferases and proteins that promote chromatin remodeling and chaperones to incorporate histones into newly reassembled nucleosomes [142].

3.6 Alterations in NER and Cancer Predisposition

It is clear that heritable mutations in NER genes can predispose individuals to the development of skin cancer and other forms of cancer. Many XP patients develop nonmelanoma skin cancer within the first decade of life. This is in sharp contrast to the development of nonmelanoma skin cancer in the general, non-XP population that occurs, on average, when people are well into their 60s. XP patients can also develop tumors in internal, non-UV-exposed organs including tumors of the brain and central nervous system and the lung [87]. Genetically modified mice with deficiencies in certain NER genes are also predisposed to UV-induced skin cancer and carcinogen-induced and spontaneous forms of lung cancer [157].

It is unclear how alterations in NER impact cancer etiology in the general, non-XP population. Deficiencies in NER could render an individual with a greater predisposition to the development of cancer and conversely, enhancement of NER capacity in an individual could render them less susceptible to the development of cancer. Alterations in individual repair capacity could also impact how an individual responds to treatment with chemotherapeutic agents that damage DNA. An individual's capacity to carry out NER could be influenced by the inheritance of polymorphic alleles of NER genes, by exposure to agents in the environment that impact NER efficiency, or by some combination of the two. These interactions are likely highly complex. There have been numerous studies that have investigated correlations between polymorphisms in NER genes and many different forms of cancer including those that occur in the lung, stomach, breast, skin, or blood. However, while linkages have been reported in some studies, many of these are either not supported or are found to be weak associations when studies are combined and subjected to meta-analyses [158–163]. In addition, while XP is a rare disease, the frequency of single mutant alleles is much greater, but it remains unclear if individuals containing only one mutant allele of an NER gene are more highly predisposed to the development of cancer.

4. BIOLOGICAL IMPLICATIONS BEYOND DNA DAMAGE AND REPAIR

Some functions of excision repair are indispensable for the organisms. The central theme in this section is the versatility of BER and that some proteins involved in BER can be utilized in fundamental cellular activities unrelated to DNA repair. As there are excellent reviews on these subjects [163,164], this section briefly describes recent studies investigating additional roles of proteins involved in BER.

4.1 Diversity of Immune Cells by Activation-Induced Deaminase

Somatic hypermutation (SHM) and class switch recombination (CSR) are necessary for antibody diversification in antigen-specific memory B cells, and both mechanisms require activation-induced deaminase (AID) [164–166]. Because AID deaminates cytosine to generate uracil in DNA, a well-known BER substrate, involvement of BER in this pathway is being established [45,166]. The canonical BER reactions depicted in Fig. 17.1 do not likely occur during SHM and CSR. Instead, when uracil is generated in DNA by AID, it serves as a flag to recruit error-prone bypassing DNA polymerases for SHM and components of DNA DSB repair cooperate with some of the BER enzymes to lead to CSR [46,164]. Continued understanding of the mechanisms of SHM and CSR involving BER, other DNA-repair and -signaling pathways should illuminate the sophisticated crosstalk among the DNA-repair pathways.

4.2 DNA Demethylation

Methylation of cytosine at CpG dinucleotides generates 5mC. 5mC is a major epigenetic DNA modification that controls gene expression. While abnormalities in the distribution of 5mC in the genome are a hallmark of cell transformation in cancer genomics, DNA methylation is pivotal in controlling normal cell differentiation during development.

Processes are required to regulate demethylation of DNA and this is necessary to remove 5mC and introduce cytosine. Studies in the past several years have established an essential role of the demethylation process not only in cell differentiation, particularly for the stem cell research, but also in cancer development [79,167,168].

An initial event in the demethylation process, described in detail in a previous review by Wu and Zhang [169], is the conversion of 5mC to 5-hydroxymethyl cytosine (5hmC) by Tet methylcytosine dioxygeneases (Tet1, Tet2, and Tet3; ten-eleven translocation 1, 2, 3 gene protein). The Tet proteins further process 5hmC to 5-formylcytosine (5fC) and then to 5-carboxylcytosine (5caC) [170]. Both 5fC and 5caC are processed by BER, as TDG recognizes and removes these unusual cytosine derivatives as its substrates, and leaves AP sites at these locations (Fig. 17.1A). The reactions that follow the generation of AP sites are not entirely clear. However, a 2010 study showed that the BER proteins including TDG, APE1, PARP1, and XRCC1 are upregulated during developmental stages in embryonic mice when whole-genome demethylation takes place [171]. In zygotic cells, PARP1 and XRCC1 were found to be physically associated with the paternal genome where demethylation takes place [171]. Several studies have found that the BER proteins, XRCC1, PARP1, and APE1, are utilized in the demethylation process in *Arabidopsis* and mammalian cells [167,172–175]. Because of the impact of demethylation on many study fields of study including stem cells, cell differentiation, cancer, and cancer stem cells, the advanced technology demonstrating the involvement of BER in the distribution of C, 5mC, 5hmC, 5fC, and 5caC [168,176,177] has broadened the role of BER beyond DNA repair and toward epigenetic maintenance.

5. INTERPLAY BETWEEN NER AND BER: THE KEY ROLE OF THE DNA-DAMAGE RESPONSE FOR PREVENTION OF CELLULAR DEGENERATION

5.1 Overlapping Substrate Specificity Between BER and NER

BER and NER enzymes may recognize the same types of DNA damage, and hence, this class of substrates could be repaired by either pathway. A role of NER in the repair of endogenously generated DNA damage has been suggested since it could explain the neurodegenerative phenotypes associated with some NER deficiencies. However, UV damage does not occur in neurons and hence the substrates for NER that may produce the neurodegenerative phenotype are an unsolved question.

Overlapping substrates for BER and NER were reported in *E. coli* and yeast [178–183]. Memisoglu et al. found that a deficiency in rad13, an NER protein in yeast, produced increased sensitivity to alkylating reagents, and they proposed that alkylated bases may be repaired directly by the NER pathway [180]. In 2010, the repair of AP sites was reported to be associated with TC-NER in yeast [184], and the investigators proposed a detailed mechanism for this observation.

Some studies have also reported overlapping roles of BER and NER in the removal of oxidative DNA damage in human cells. A biochemical study showed that human NER proteins could recognize and remove 8-oxoG in DNA [185]. Although it has been difficult to clearly show a role of NER in the removal of 8-oxoG in vivo, primary cells from XPC patients have been found to be hypersensitive to treatment with oxidizing reagents [186].

5.2 A Nuclear–Mitochondria Signaling Network as a Main Platform of BER/NER Interplay

An emerging field of study, how mitophagy is regulated by a DNA-damage response, is being formed that may finally delineate the crosstalk between BER and NER, and perhaps it also involves other DNA-repair pathways.

Mitophagy is a cellular process that degrades damaged mitochondria and facilitates generation of new mitochondria. Mitophagy and autophagy require common factors and reactions involving ubiquitin-dependent proteasome systems and LC3 conjugation [187,188]. However, compared to autophagy, mitophagy is a mechanism that provides a quality check for maintaining the integrity of mitochondria, and its biological role resembles that of apoptosis.

Mitophagy requires a ubiquitin ligase, Parkin [188]. Deficiencies in the gene for Parkin, PARK2, are a major cause of both early and late onset of Parkinson's disease. Parkin is a RING domain containing E3 ligase and it requires an essential cofactor, PINK1, which is also a Parkinson's disease–causative gene [189]. The astonishing finding that ubiquitin Ser65 phosphorylation regulates mitophagy was reported by several studies [190–193]. Impaired mitochondria lose the inner membrane potential, which induces phosphorylation at Ser65 of ubiquitin by PINK1. The Ser65-phosphoubiquitin facilitates Parkin's translocalization from the cytosol to the surface of mitochondria, and enhances its ubiquitination reaction. Therefore, cells proactively maintain the quality of mitochondria with Parkin, for which PINK1 plays a central regulatory role.

PINK1 is highly sensitive to proteolysis; the truncated PINK1 loses the kinase activity, and thus becomes incapable of activating Parkin and mitophagy. Although the sensitivity of PINK1 to proteolysis may provide an autoregulation of the Parkin–PINK1 protein degradation system, in 2014 it became apparent that PINK1 is susceptible to oxidative stress in causing its proteolysis [194]. An unexpected finding was that lack of XPA was associated with increased PINK1 cleavage, resulting in the impairment of the cellular function to check mitochondrial integrity [194]. This study proposed the following degenerative cellular events (Fig. 17.4): (1) Elevation of PARP1 activities in XPA-deficient cells causes insufficient NAD^+ concentration. (2) Low NAD^+ causes down-modulation of SIRT1 activity which in turn lowers PGC-1α activity. PGC-1α is the master regulator of mitochondrial regeneration and energy-generating activity in the cells. (3) Low PGC-1α activity causes lower UCP2 levels. Because UCP2 is an uncoupler that maintains the proper mitochondrial membrane potential, when UCP2 is increased, it is an inducer of mitophagy. In contrast, lower levels of UCP2 result in degradation of PINK1 and suppression of mitophagy. This effect is additive and results in the accumulation of damaged mitochondria. The unusually high mitochondrial and oxidative stresses are unsustainable and thus cause apoptosis. This phenotype and the novel link to mitochondria was not only found to be associated with deficiencies in XPA; it was also associated with Cockayne syndrome B (CSB) and ataxia telangiectasia-mutated (ATM) deficiencies. Surprisingly, it was not associated with deficiencies in XPC. Therefore, although both XPA and XPC are essential for the NER process, XPA appears to have an independent function in the maintenance of mitochondrial integrity.

These are remarkable discoveries in the field of BER and NER, and further studies may provide critical information as to why deficiencies in NER cause neurodegenerative diseases. However, an important question regarding the mechanism of NER has not been answered: Does XPA's direct involvement in repairing endogenous DNA damage help cells maintain intact mitophagy, or is XPA a signal transducer in this particular DNA-damage response? In other words, what initiates mitochondrial degeneration which is exacerbated by a deficiency in XPA? The unusually high oxidative stress caused by mitochondrial degeneration and UCP2 down-regulation may be a consequence rather than the cause of cellular degeneration. Similarly, PINK1 degradation may be induced by the elevated oxidative stress. The fundamental cause of these molecular events could be endogenous DNA damage (Fig. 17.4). Endogenous DNA damage is continuously generated under the normal physiology, and keeps PARPs at its equilibrium balance between the activated and dormant forms. It is hypothesized that, in a yet unidentified reaction scheme, the presence of XPA suppresses the activation of PARP, either by facilitating repair or by inhibiting PARP. Hence, a deficiency in XPA could result in the accumulation of active PARP, which results in a gradual, yet irreversible, degeneration of mitochondrial and an increase in oxidative stress that ultimately kills the cell. Crucial experiments remain to be carried out to identify interactions of XPA with molecules involved in the DNA-damage response, including endogenous DNA damage and DNA–repair intermediates (eg, AP sites, DNA-strand breaks, protein–DNA cross-links, and 5meC), and NER/BER proteins such as PARPs.

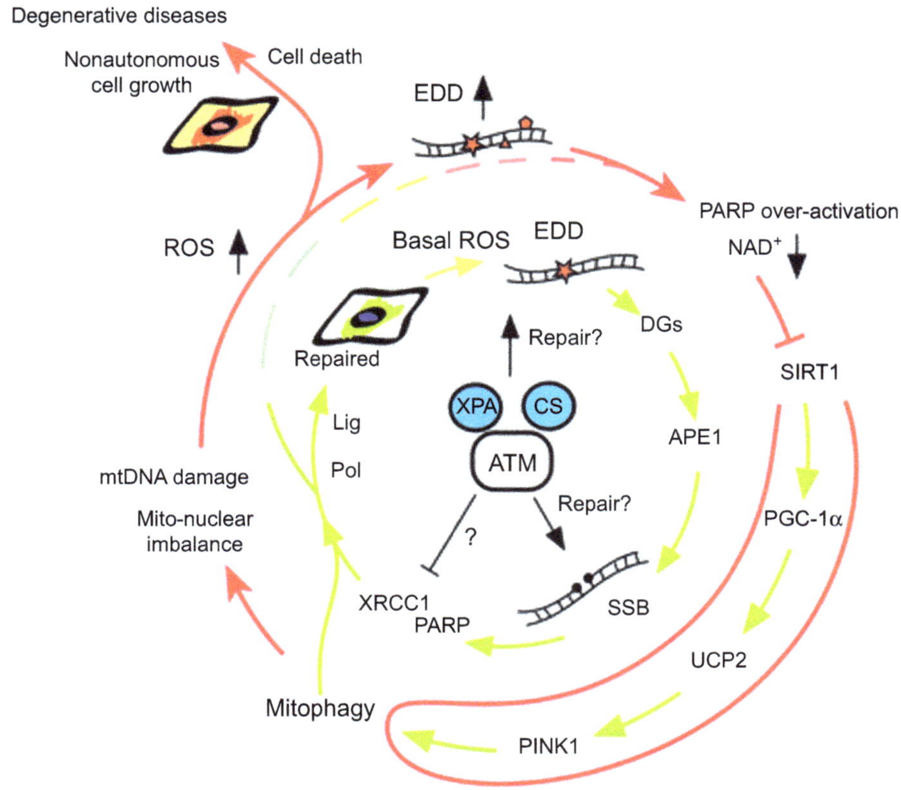

FIGURE 17.4 **A model of cellular degeneration caused by endogenous DNA damage.** (*Green, inner ring*) A normal cellular cycle in which EDD (endogenous DNA damages) are continuously generated but under the control of BER, and PARPs are activated to facilitate the repair process. (*Red, outer ring*) EDD generation at a high rate causes overactivation of PARP. Depletion of NAD^+ suppresses SIRT1-dependent PGC-1α activation, which abrogates mitochondrial quality check by Parkin/PINK1 [194]. Cells enter a vicious cycle involving ROS elevation and mitochondrial degeneration. XPA, CSB, and ATM, all appear to take part in keeping EDD at normal levels [194]. To determine what type of EDD exactly causes the PARP overactivation, and to understand the coordination of BER with XPA, CSB, and ATM will help improve precision medicine of degenerative diseases. *DGs*, DNA glycosylases.

6. CONCLUDING REMARKS

Here, we reviewed advances in understanding two mammalian DNA excision-repair mechanisms. We described the basic mechanisms of BER and NER, and reviewed recent studies regarding the interplay of BER and NER, revealing a novel role of NER which is independent of the repair of UV-induced DNA damage. The versatility of BER was also illustrated by describing how BER is also involved in processing modified DNA bases such as 5mC and this role in epigenetics is an indispensable function of BER. The stepwise reaction scheme of BER makes it a flexible pathway and thus an ideal DNA-modifying machinery that can adapt to different types of unusual bases.

Endogenous damage must be repaired by DNA-repair pathways to avoid pathophysiological conditions. NER has been studied mainly to understand its role in removing bulky DNA damage generated by UV radiation and by exposure to carcinogens. However, recent studies have led to the discovery of novel DNA-damage responses involving NER as well as BER that likely play roles in disease. One key to understanding the impact of these novel pathways on disease may be to identify the endogenous targets of NER.

GLOSSARY

3′-blocking damage Non-3′-OH termini at DNA strand breaks that cannot serve as DNA synthesis primers and therefore require 3′-end processing to generate 3′-OH termini. 3′-blocking damage includes 3′-phosphate, 3′-α,β-unsaturated aldehyde, 3′-phosphoglycolate.

AP lyase A lyase that catalyzes DNA-strand breakage at AP sites, and removes 5′-dRP from 5′-ends of DNA strand breaks. The AP lyase reaction occurs via β- or βδ-elimination through formation of Schiff base. Many DNA glycosylases possess an AP lyase activity. Also see dRPase.

AP sites (apurinic/apyrimidinic sites) A type of DNA damage where a base (either purine or pyrimidine) is removed. Also known as abasic sites.

Bipartate recognition of DNA damage during NER The efficient recognition and removal of DNA damage by NER generally requires that the damage to DNA possesses two important features. One feature is that the damage represents a covalent modification to the DNA. The second feature is that the presence of the covalent modification creates a significant alteration in the overall structure of the DNA helix.

Deamination Deamination may occur at exocyclic amino groups of cytosine, 5-methyl cytosine, adenine, and guanine, which are converted to uracil, thymidine, hypoxanthine, and xanthine, respectively. The bases resulted from deamination may form base pairs different from original pairs (U to A, T to A, HX to C) and thus potentially mutagenic.

Demethylation A process wherein 5-methyl cytosine is converted to cytosine. An active demethylation process in cells involves enzymes of BER. BER proteins that are shown to function in demethylation include TDG, PARP, XRCC1, and APE1.

DNA-alkylating agents Chemical compounds with electrophilic alkyl groups that attack nucleophilic groups in DNA. Commonly used alkylating agents in research include methyl methanesulfonate (MMS), methylnitronitrosoguanidine (MNNG), *N*-nitroso-*N*-methylurea (MNU), temozolomide.

dRPase An enzyme capable of removing 5′-dRP from 5′-ends of DNA strand breaks via hydrolysis. The term dRPase is often used to describe an AP lyase, due to the fact that their roles in the BER pathway in producing 5′-phosphate from 5′-dRP are identical. By definition, unlike AP lyases, dRPase does not incise AP sites. The difference between AP lyases and dRPases is described in detail by Piersen et al. [51].

Mitophagy An active process to digest damaged mitochondria involving protein degradation via ubiquitination catalyzed by Parkin and PINK1. It is a specialized autophagy for maintaining quality of mitochondria.

Oxidative DNA damage Bases and backbone of DNA can be oxidized spontaneously or induced by oxidizing reagents. These include 8-oxoguanine, thymine glycol, DNA strand breaks with 3′-blocking damage. Find details in Refs. [3,35].

RNA-polymerase backtracking Instead of moving along the template DNA strand in the 3′ to 5′ direction synthesizing new RNA, the RNA polymerase complex can translocate in the opposite direction and move backwards in the 5′ to 3′ direction.

TET Tet methylcytosine dioxygenase or ten-eleven translocation gene protein. TET enzymes catalyze base conversion reactions using 5-methylcytosine (5mC) as the starting substrate to generate 5-hydroxymethylcytosine (5hmC), then 5-formylcytosine (5fC), and finally 5-carboxycytosine (5caC). The converted cytosine derivatives are recognized and removed by TDG. Evidence indicates that AP sites, generated by TDG, are repaired by the traditional BER pathway involving APE1, PARP1, XRCC1, Polβ, and DNA ligase III. TET1, TET2, and TET3 belong to the TET enzyme family.

LIST OF ABBREVIATIONS

3′-PUA 3′-Phospho-α,β-unsaturated aldehydes
5caC 5-Carboxylcytosine
5fC 5-Formylcytosine
5hmC 5-Hydroxymethyl cytosine
5mC 5-Methylcytosine
6-4PP 6-4 Photoproduct
AID Activation-induced deaminase
APE1 AP Endonuclease 1
Apex1 Ape1 gene
AP-site Apurinic/apyrimidinic sites
ASF1 Alternative splicing factor
ATM Ataxia telangiectasia mutated
BER Base excision repair
BP Benzo(a)pyrene
CAF1 CCRF-associated factor
CAK Cyclin-activated kinase
CPD Cyclobutane pyrimidine dimer
CSA Cockayne syndrome A
CSB Cockayne syndrome B
CSR Class switch recombination
DDR DNA-damage response
dRP 2-Deoxyribose 5-phsphate
EDD Endogenous DNA damages
FEN1 Flap structure–specific endonucleases
LIGI DNA ligase
MGMT O^6-methylguanine-DNA methyltransferase
MMS Methanesulfonate
MNNG Methylnitronitrosoguanidine
MPG Methylpurine DNA glycosylase
MYH MutY homology
NEIL1 and NEIL2 EndoVIII-like 1 and 2
NER Nucleotide excision repair
NMU *N*-nitroso-*N*-methylurea
NTH EndoIII homology

O_2^- Superoxide
OGG1 8-oxoG DNA glycosylase
OH• Hydroxyl radical
PARPs Poly(ADP-ribose)polymerases
PNKP Polynucleotide kinase/phosphatase
Polβ DNA polymerase beta
Polδ/ε DNA polymerase delta/epsilon
Polλ DNA polymerase lambda
ROS Reactive oxygen species
SHM Somatic hypermutation
(SN)-BER Single-nucleotide filling base excision repair
SOD2 Manganese superoxide dismutase 2
SSBs Single-strand breaks
Tet1, Tet2, and Tet3 Tet methylcytosine dioxygeneases
TC-NER Transcription-coupled NER
Tdg Thymine DNA glycosylase
TDPs Tyrosyl–DNA phosphodiesterases
TOP1, TOP2, TOP3 Topoisomerases I, II, and III
UNG Uracil DNA glycosylase
UV-DDB2 UV DNA damage-binding protein 2
UVSSA UV-sensitive syndrome A
XAB2 XPA-binding protein 2
XP Xeroderma pigmentosum

ACKNOWLEDGMENT

This work was supported by NIH/NCI CA98664, the Markey Cancer Center Support Grant (CCSG), and Kentucky Lung Cancer Research Programs.

REFERENCES

[1] Singer B, Kusmierek JT. Chemical mutagenesis. Annu Rev Biochem 1982;51:655–93.
[2] Hegde ML, Izumi T, Mitra S. Oxidized base damage and single-strand break repair in mammalian genomes: role of disordered regions and post-translational modifications in early enzymes. Prog Mol Biol Transl Sci 2012;110:123–53.
[3] Breen AP, Murphy JA. Reactions of oxyl radicals with DNA. Free Radic Biol Med 1995;18(6):1033–77.
[4] Kriek E. Persistent binding of a new reaction product of the carcinogen *N*-hydroxy-*N*-2-acetylaminofluorene with guanine in rat liver DNA *in vivo*. Cancer Res 1972;32(10):2042–8.
[5] Weinstein IB, Jeffrey AM, Jennette KW, Blobstein SH, Harvey RG, Harris C, et al. Benzo(a)pyrene diol epoxides as intermediates in nucleic acid binding *in vitro* and *in vivo*. Science 1976;193(4253):592–5.
[6] Muench KF, Misra RP, Humayun MZ. Sequence specificity in aflatoxin B1–DNA interactions. Proc Natl Acad Sci USA 1983;80(1):6–10.
[7] Zamble DB, Lippard SJ. Cisplatin and DNA repair in cancer chemotherapy. Trends Biochem Sci 1995;20(10):435–9.
[8] Porter PC, Mellon I, States JC. XP-A cells complemented with Arg228Gln and Val234Leu polymorphic XPA alleles repair BPDE-induced DNA damage better than cells complemented with the wild type allele. DNA Repair (Amst) 2005;4(3):341–9.
[9] Ljungquist S, Lindahl T. A mammalian endonuclease specific for apurinic sites in double-stranded deoxyribonucleic acid. I. Purification and general properties. J Biol Chem 1974;249(5):1530–5.
[10] Lindahl T, Ljungquist S. Apurinic and apyrimidinic sites in DNA. Basic Life Sci 1975;5A:31–8.
[11] Ljungquist S. A new endonuclease from *Escherichia coli* acting at apurinic sites in DNA. J Biol Chem 1977;252(9):2808–14.
[12] Lindahl T. DNA glycosylases, endonucleases for apurinic/apyrimidinic sites, and base excision-repair. Prog Nucleic Acid Res Mol Biol 1979;22:135–92.
[13] Lindahl T. Instability and decay of the primary structure of DNA. Nature 1993;362(6422):709–15.
[14] Mol CD, Hosfield DJ, Tainer JA. Abasic site recognition by two apurinic/apyrimidinic endonuclease families in DNA base excision repair: the 3′ ends justify the means. Mutat Res 2000;460(3–4):211–29.
[15] Izumi T, Wiederhold LR, Roy G, Roy R, Jaiswal A, Bhakat KK, et al. Mammalian DNA base excision repair proteins: their interactions and role in repair of oxidative DNA damage. Toxicology 2003;193(1–2):43–65.
[16] Mitra S, Izumi T, Boldogh I, Bhakat KK, Hill JW, Hazra TK. Choreography of oxidative damage repair in mammalian genomes. Free Radic Biol Med 2002;33(1):15–28.
[17] Bennett RA, Wilson 3rd DM, Wong D, Demple B. Interaction of human apurinic endonuclease and DNA polymerase beta in the base excision repair pathway. Proc Natl Acad Sci USA 1997;94(14):7166–9.

[18] Mol CD, Izumi T, Mitra S, Tainer JA. DNA-bound structures and mutants reveal abasic DNA binding by APE1 and DNA repair coordination [corrected]. Nature 2000;403(6768):451–6.

[19] Wilson SH, Kunkel TA. Passing the baton in base excision repair. Nat Struct Biol 2000;7(3):176–8.

[20] Caldecott KW, McKeown CK, Tucker JD, Ljungquist S, Thompson LH. An interaction between the mammalian DNA repair protein XRCC1 and DNA ligase III. Mol Cell Biol 1994;14(1):68–76.

[21] Pleschke JM, Kleczkowska HE, Strohm M, Althaus FR. Poly(ADP-ribose) binds to specific domains in DNA damage checkpoint proteins. J Biol Chem 2000;275(52):40974–80.

[22] Iles N, Rulten S, El-Khamisy SF, Caldecott KW. APLF (C2orf13) is a novel human protein involved in the cellular response to chromosomal DNA strand breaks. Mol Cell Biol 2007;27(10):3793–803.

[23] Ali AA, Jukes RM, Pearl LH, Oliver AW. Specific recognition of a multiply phosphorylated motif in the DNA repair scaffold XRCC1 by the FHA domain of human PNK. Nucleic Acids Res 2009;37(5):1701–12.

[24] Cuneo MJ, London RE. Oxidation state of the XRCC1 N-terminal domain regulates DNA polymerase beta binding affinity. Proc Natl Acad Sci USA 2010;107(15):6805–10.

[25] Farez MF, Quintana FJ, Gandhi R, Izquierdo G, Lucas M, Weiner HL. Toll-like receptor 2 and poly(ADP-ribose) polymerase 1 promote central nervous system neuroinflammation in progressive EAE. Nat Immunol 2009;10(9):958–64.

[26] Bjorkhem I, Diczfalusy U, Olsson T, Russell DW, McDonald JG, Wang Y, et al. Detecting oxysterols in the human circulation. Nat Immunol 2011;12(7):577. author reply 577–8.

[27] Lord CJ, Tutt AN, Ashworth A. Synthetic lethality and cancer therapy: lessons learned from the development of PARP inhibitors. Annu Rev Med 2015;66:455–70.

[28] Singer B. N-nitroso alkylating agents: formation and persistence of alkyl derivatives in mammalian nucleic acids as contributing factors in carcinogenesis. J Natl Cancer Inst 1979;62(6):1329–39.

[29] Atkins RJ, Ng W, Stylli SS, Hovens CM, Kaye AH. Repair mechanisms help glioblastoma resist treatment. J Clin Neurosci 2015;22(1):14–20.

[30] Rydberg B, Lindahl T. Nonenzymatic methylation of DNA by the intracellular methyl group donor *S*-adenosyl-L-methionine is a potentially mutagenic reaction. EMBO J 1982;1(2):211–6.

[31] Friedberg EC. DNA repair. New York: W.H. Freeman; 1984. x, 614 pp.

[32] Gates KS, Nooner T, Dutta S. Biologically relevant chemical reactions of N7-alkylguanine residues in DNA. Chem Res Toxicol 2004;17(7):839–56.

[33] Mitra S. MGMT: a personal perspective. DNA Repair (Amst) 2007;6(8):1064–70.

[34] Scott TL, Rangaswamy S, Wicker CA, Izumi T. Repair of oxidative DNA damage and cancer: recent progress in DNA base excision repair. Antioxid Redox Signal 2014;20(4):708–26.

[35] Evans MD, Dizdaroglu M, Cooke MS. Oxidative DNA damage and disease: induction, repair and significance. Mutat Res 2004;567(1):1–61.

[36] Nitiss JL, Soans E, Rogojina A, Seth A, Mishina M. Topoisomerase assays. Curr Protoc Pharmacol 2012. [Chapter 3:Unit 3].

[37] Zeng Z, Sharma A, Ju L, Murai J, Umans L, Vermeire L, et al. TDP2 promotes repair of topoisomerase I-mediated DNA damage in the absence of TDP1. Nucleic Acids Res 2012;40(17):8371–80.

[38] Pommier Y, Huang SY, Gao R, Das BB, Murai J, Marchand C. Tyrosyl-DNA-phosphodiesterases (TDP1 and TDP2). DNA Repair (Amst) 2014;19:114–29.

[39] Matsumoto Y, Kim K. Excision of deoxyribose phosphate residues by DNA polymerase beta during DNA repair. Science 1995;269(5224):699–702.

[40] Wiederhold L, Leppard JB, Kedar P, Karimi-Busheri F, Rasouli-Nia A, Weinfeld M, et al. AP endonuclease-independent DNA base excision repair in human cells. Mol Cell 2004;15(2):209–20.

[41] Meira LB, Devaraj S, Kisby GE, Burns DK, Daniel RL, Hammer RE, et al. Heterozygosity for the mouse *Apex* gene results in phenotypes associated with oxidative stress. Cancer Res 2001;61(14):5552–7.

[42] Xanthoudakis S, Smeyne RJ, Wallace JD, Curran T. The redox/DNA repair protein, Ref-1, is essential for early embryonic development in mice. Proc Natl Acad Sci USA 1996;93(17):8919–23.

[43] Fung H, Demple B. A vital role for Ape1/Ref1 protein in repairing spontaneous DNA damage in human cells. Mol Cell 2005;17(3):463–70.

[44] Izumi T, Brown DB, Naidu CV, Bhakat KK, Macinnes MA, Saito H, et al. Two essential but distinct functions of the mammalian abasic endonuclease. Proc Natl Acad Sci USA 2005;102(16):5739–43.

[45] Masani S, Han L, Yu K. Apurinic/apyrimidinic endonuclease 1 is the essential nuclease during immunoglobulin class switch recombination. Mol Cell Biol 2013;33(7):1468–73.

[46] Xu J, Husain A, Hu W, Honjo T, Kobayashi M. APE1 is dispensable for S-region cleavage but required for its repair in class switch recombination. Proc Natl Acad Sci USA 2014;111(48):17242–7.

[47] Suganya R, Chakraborty A, Miriyala S, Hazra TK, Izumi T. Suppression of oxidative phosphorylation in mouse embryonic fibroblast cells deficient in apurinic/apyrimidinic endonuclease. DNA Repair (Amst) 2015;27C:40–8.

[48] Izumi T, Malecki J, Chaudhry MA, Weinfeld M, Hill JH, Lee JC, et al. Intragenic suppression of an active site mutation in the human apurinic/apyrimidinic endonuclease. J Mol Biol 1999;287(1):47–57.

[49] Jilani A, Ramotar D, Slack C, Ong C, Yang XM, Scherer SW, et al. Molecular cloning of the human gene, PNKP, encoding a polynucleotide kinase 3′-phosphatase and evidence for its role in repair of DNA strand breaks caused by oxidative damage. J Biol Chem 1999;274(34):24176–86.

[50] Weinfeld M, Mani RS, Abdou I, Aceytuno RD, Glover JN. Tidying up loose ends: the role of polynucleotide kinase/phosphatase in DNA strand break repair. Trends Biochem Sci 2011;36(5):262–71.

[51] Piersen CE, McCullough AK, Lloyd RS. AP lyases and dRPases: commonality of mechanism. Mutat Res 2000;459(1):43–53.

[52] Klungland A, Lindahl T. Second pathway for completion of human DNA base excision-repair: reconstitution with purified proteins and requirement for DNase IV (FEN1). EMBO J 1997;16(11):3341–8.

[53] Inamdar KV, Pouliot JJ, Zhou T, Lees-Miller SP, Rasouli-Nia A, Povirk LF. Conversion of phosphoglycolate to phosphate termini on 3′ overhangs of DNA double strand breaks by the human tyrosyl-DNA phosphodiesterase hTdp1. J Biol Chem 2002;277(30):27162–8.

[54] Liu Y, Prasad R, Beard WA, Kedar PS, Hou EW, Shock DD, et al. Coordination of steps in single-nucleotide base excision repair mediated by apurinic/apyrimidinic endonuclease 1 and DNA polymerase beta. J Biol Chem 2007;282(18):13532–41.

[55] Prasad R, Beard WA, Batra VK, Liu Y, Shock DD, Wilson SH. A review of recent experiments on step-to-step "hand-off" of the DNA intermediates in mammalian base excision repair pathways. Mol Biol (Mosk) 2011;45(4):586–600.

[56] Krokan HE, Standal R, Slupphaug G. DNA glycosylases in the base excision repair of DNA. Biochem J 1997;325(Pt 1):1–16.

[57] Braithwaite EK, Prasad R, Shock DD, Hou EW, Beard WA, Wilson SH. DNA polymerase lambda mediates a back-up base excision repair activity in extracts of mouse embryonic fibroblasts. J Biol Chem 2005;280(18):18469–75.

[58] Braithwaite EK, Kedar PS, Lan L, Polosina YY, Asagoshi K, Poltoratsky VP, et al. DNA polymerase lambda protects mouse fibroblasts against oxidative DNA damage and is recruited to sites of DNA damage/repair. J Biol Chem 2005;280(36):31641–7.

[59] Hazra TK, Kow YW, Hatahet Z, Imhoff B, Boldogh I, Mokkapati SK, et al. Identification and characterization of a novel human DNA glycosylase for repair of cytosine-derived lesions. J Biol Chem 2002;277(34):30417–20.

[60] Hazra TK, Izumi T, Boldogh I, Imhoff B, Kow YW, Jaruga P, et al. Identification and characterization of a human DNA glycosylase for repair of modified bases in oxidatively damaged DNA. Proc Natl Acad Sci USA 2002;99(6):3523–8.

[61] Okano S, Lan L, Caldecott KW, Mori T, Yasui A. Spatial and temporal cellular responses to single-strand breaks in human cells. Mol Cell Biol 2003;23(11):3974–81.

[62] Campalans A, Marsin S, Nakabeppu Y, O'Connor TR, Boiteux S, Radicella JP. XRCC1 interactions with multiple DNA glycosylases: a model for its recruitment to base excision repair. DNA Repair (Amst) 2005;4(7):826–35.

[63] Marsin S, Vidal AE, Sossou M, Menissier-de Murcia J, Le Page F, Boiteux S, et al. Role of XRCC1 in the coordination and stimulation of oxidative DNA damage repair initiated by the DNA glycosylase hOGG1. J Biol Chem 2003;278(45):44068–74.

[64] Hegde ML, Hegde PM, Arijit D, Boldogh I, Mitra S. Human DNA glycosylase NEIL1's interactions with downstream repair proteins is critical for efficient repair of oxidized DNA base damage and enhanced cell survival. Biomolecules 2012;2(4):564–78.

[65] Akbari M, Solvang-Garten K, Hanssen-Bauer A, Lieske NV, Pettersen HS, Pettersen GK, et al. Direct interaction between XRCC1 and UNG2 facilitates rapid repair of uracil in DNA by XRCC1 complexes. DNA Repair (Amst) 2010;9(7):785–95.

[66] Fan J, Otterlei M, Wong HK, Tomkinson AE, Wilson 3rd DM. XRCC1 co-localizes and physically interacts with PCNA. Nucleic Acids Res 2004;32(7):2193–201.

[67] Vidal AE, Boiteux S, Hickson ID, Radicella JP. XRCC1 coordinates the initial and late stages of DNA abasic site repair through protein-protein interactions. EMBO J 2001;20(22):6530–9.

[68] Marintchev A, Mullen MA, Maciejewski MW, Pan B, Gryk MR, Mullen GP. Solution structure of the single-strand break repair protein XRCC1 N-terminal domain. Nat Struct Biol 1999;6(9):884–93.

[69] London RE. The structural basis of XRCC1-mediated DNA repair. DNA Repair (Amst) 2015;30:90–103.

[70] Das BB, Huang SY, Murai J, Rehman I, Ame JC, Sengupta S, et al. PARP1-TDP1 coupling for the repair of topoisomerase I-induced DNA damage. Nucleic Acids Res 2014;42(7):4435–49.

[71] Prasad R, Dyrkheeva N, Williams J, Wilson SH. Mammalian base excision repair: functional partnership between PARP-1 and APE1 in AP-site repair. PLoS One 2015;10(5):e0124269.

[72] Eliasson MJ, Sampei K, Mandir AS, Hurn PD, Traystman RJ, Bao J, et al. Poly(ADP-ribose) polymerase gene disruption renders mice resistant to cerebral ischemia. Nat Med 1997;3(10):1089–95.

[73] Creissen D, Shall S. Regulation of DNA ligase activity by poly(ADP-ribose). Nature 1982;296(5854):271–2.

[74] Leppard JB, Dong Z, Mackey ZB, Tomkinson AE. Physical and functional interaction between DNA ligase IIIalpha and poly(ADP-Ribose) polymerase 1 in DNA single-strand break repair. Mol Cell Biol 2003;23(16):5919–27.

[75] Abdou I, Poirier GG, Hendzel MJ, Weinfeld M. DNA ligase III acts as a DNA strand break sensor in the cellular orchestration of DNA strand break repair. Nucleic Acids Res 2015;43(2):875–92.

[76] Mouchiroud L, Houtkooper RH, Moullan N, Katsyuba E, Ryu D, Canto C, et al. The NAD(+)/sirtuin pathway modulates longevity through activation of mitochondrial UPR and FOXO signaling. Cell 2013;154(2):430–41.

[77] Fouquerel E, Goellner EM, Yu Z, Gagne JP, Barbi de Moura M, Feinstein T, et al. ARTD1/PARP1 negatively regulates glycolysis by inhibiting hexokinase 1 independent of NAD^+ depletion. Cell Rep 2014;8(6):1819–31.

[78] Menissier de Murcia J, Ricoul M, Tartier L, Niedergang C, Huber A, Dantzer F, et al. Functional interaction between PARP-1 and PARP-2 in chromosome stability and embryonic development in mouse. EMBO J 2003;22(9):2255–63.

[79] Cortazar D, Kunz C, Selfridge J, Lettieri T, Saito Y, MacDougall E, et al. Embryonic lethal phenotype reveals a function of TDG in maintaining epigenetic stability. Nature 2011;470(7334):419–23.

[80] Setlow RB, Carrier WL. The disappearance of thymine dimers from DNA: an error-correcting mechanism. Proc Natl Acad Sci USA 1964;51:226–31.

[81] Boyce RP, Howard-Flanders P. Release of ultraviolet light-induced thymine dimers from DNA in *E. Coli* K-12. Proc Natl Acad Sci USA 1964;51:293–300.

[82] Pettijohn D, Hanawalt P. Evidence for repair-replication of ultraviolet damaged DNA in bacteria. J Mol Biol 1964;9:395–410.
[83] Rasmussen RE, Painter RB. Evidence for repair of ultra-violet damaged deoxyribonucleic acid in cultured mammalian cells. Nature 1964;203:1360–2.
[84] Cleaver JE. Defective repair replication of DNA in xeroderma pigmentosum. Nature 1968;218(5142):652–6.
[85] Bootsma D, De Weerd-Kastelein EA, Kleijer WJ, Keyzez W. Genetic complementation analysis of xeroderma pigmentosum. Basic Life Sci 1975;5B:725–8.
[86] Feltes BC, Bonatto D. Overview of xeroderma pigmentosum proteins architecture, mutations and post-translational modifications. Mutat Res Rev Mutat Res 2015;763:306–20.
[87] Kraemer KH, DiGiovanna JJ. Global contributions to the understanding of DNA repair and skin cancer. J Invest Dermatol 2014;134(e1):E8–17.
[88] Lehmann AR, McGibbon D, Stefanini M. Xeroderma pigmentosum. Orphanet J Rare Dis 2011;6:70.
[89] Mu D, Hsu DS, Sancar A. Reaction mechanism of human DNA repair excision nuclease. J Biol Chem 1996;271(14):8285–94.
[90] Boulikas T. Xeroderma pigmentosum and molecular cloning of DNA repair genes. Anticancer Res 1996;16(2):693–708.
[91] Guzder SN, Habraken Y, Sung P, Prakash L, Prakash S. Reconstitution of yeast nucleotide excision repair with purified Rad proteins, replication protein A, and transcription factor TFIIH. J Biol Chem 1995;270(22):12973–6.
[92] Bootsma D, Weeda G, Vermeulen W, van Vuuren H, Troelstra C, van der Spek P, et al. Nucleotide excision repair syndromes: molecular basis and clinical symptoms. Philos Trans R Soc Lond B Biol Sci 1995;347(1319):75–81.
[93] Wood RD, Robins P, Lindahl T. Complementation of the xeroderma pigmentosum DNA repair defect in cell-free extracts. Cell 1988;53(1):97–106.
[94] Spivak G, Ganesan AK. The complex choreography of transcription-coupled repair. DNA Repair (Amst) 2014;19:64–70.
[95] Vermeulen W, Fousteri M. Mammalian transcription-coupled excision repair. Cold Spring Harb Perspect Biol 2013;5(8):a012625.
[96] Mellon I, Bohr VA, Smith CA, Hanawalt PC. Preferential DNA repair of an active gene in human cells. Proc Natl Acad Sci USA 1986;83(23):8878–82.
[97] Bohr VA, Smith CA, Okumoto DS, Hanawalt PC. DNA repair in an active gene: removal of pyrimidine dimers from the DHFR gene of CHO cells is much more efficient than in the genome overall. Cell 1985;40(2):359–69.
[98] Mellon I, Spivak G, Hanawalt PC. Selective removal of transcription-blocking DNA damage from the transcribed strand of the mammalian DHFR gene. Cell 1987;51(2):241–9.
[99] Scharer OD. Nucleotide excision repair in eukaryotes. Cold Spring Harb Perspect Biol 2013;5(10):a012609.
[100] Kuper J, Kisker C. Damage recognition in nucleotide excision DNA repair. Curr Opin Struct Biol 2012;22(1):88–93.
[101] Sugasawa K. Multiple DNA damage recognition factors involved in mammalian nucleotide excision repair. Biochemisty (Mosc) 2011;76(1):16–23.
[102] Fuss JO, Tainer JA. XPB and XPD helicases in TFIIH orchestrate DNA duplex opening and damage verification to coordinate repair with transcription and cell cycle via CAK kinase. DNA Repair (Amst) 2011;10(7):697–713.
[103] Dip R, Camenisch U, Naegeli H. Mechanisms of DNA damage recognition and strand discrimination in human nucleotide excision repair. DNA Repair (Amst) 2004;3(11):1409–23.
[104] Hess MT, Schwitter U, Petretta M, Giese B, Naegeli H. Bipartite substrate discrimination by human nucleotide excision repair. Proc Natl Acad Sci USA 1997;94(13):6664–9.
[105] Richa, Sinha RP, Hader DP. Physiological aspects of UV-excitation of DNA. Top Curr Chem 2015;356:203–48.
[106] Pfeifer GP, Besaratinia A. UV wavelength-dependent DNA damage and human non-melanoma and melanoma skin cancer. Photochem Photobiol Sci 2012;11(1):90–7.
[107] Marteijn JA, Lans H, Vermeulen W, Hoeijmakers JH. Understanding nucleotide excision repair and its roles in cancer and ageing. Nat Rev Mol Cell Biol 2014;15(7):465–81.
[108] Hecht SS, Carmella SG, Stepanov I, Jensen J, Anderson A, Hatsukami DK. Metabolism of the tobacco-specific carcinogen 4-(methylnitrosamino)-1-(3-pyridyl)-1-butanone to its biomarker total NNAL in smokeless tobacco users. Cancer Epidemiol Biomarkers Prev 2008;17(3):732–5.
[109] Hecht SS. Lung carcinogenesis by tobacco smoke. Int J Cancer 2012;131(12):2724–32.
[110] Huang JC, Svoboda DL, Reardon JT, Sancar A. Human nucleotide excision nuclease removes thymine dimers from DNA by incising the 22nd phosphodiester bond 5′ and the 6th phosphodiester bond 3′ to the photodimer. Proc Natl Acad Sci USA 1992;89(8):3664–8.
[111] Aboussekhra A, Biggerstaff M, Shivji MK, Vilpo JA, Moncollin V, Podust VN, et al. Mammalian DNA nucleotide excision repair reconstituted with purified protein components. Cell 1995;80(6):859–68.
[112] Batty D, Rapic'-Otrin V, Levine AS, Wood RD. Stable binding of human XPC complex to irradiated DNA confers strong discrimination for damaged sites. J Mol Biol 2000;300(2):275–90.
[113] Sugasawa K, Ng JM, Masutani C, Iwai S, van der Spek PJ, Eker AP, et al. Xeroderma pigmentosum group C protein complex is the initiator of global genome nucleotide excision repair. Mol Cell 1998;2(2):223–32.
[114] Sugasawa K, Akagi J, Nishi R, Iwai S, Hanaoka F. Two-step recognition of DNA damage for mammalian nucleotide excision repair: directional binding of the XPC complex and DNA strand scanning. Mol Cell 2009;36(4):642–53.
[115] Sugasawa K, Okuda Y, Saijo M, Nishi R, Matsuda N, Chu G, et al. UV-induced ubiquitylation of XPC protein mediated by UV-DDB-ubiquitin ligase complex. Cell 2005;121(3):387–400.
[116] Bergink S, Toussaint W, Luijsterburg MS, Dinant C, Alekseev S, Hoeijmakers JH, et al. Recognition of DNA damage by XPC coincides with disruption of the XPC-RAD23 complex. J Cell Biol 2012;196(6):681–8.
[117] Compe E, Egly JM. TFIIH: when transcription met DNA repair. Nat Rev Mol Cell Biol 2012;13(6):343–54.
[118] Volker M, Mone MJ, Karmakar P, van Hoffen A, Schul W, Vermeulen W, et al. Sequential assembly of the nucleotide excision repair factors *in vivo*. Mol Cell 2001;8(1):213–24.
[119] Evans E, Moggs JG, Hwang JR, Egly JM, Wood RD. Mechanism of open complex and dual incision formation by human nucleotide excision repair factors. EMBO J 1997;16(21):6559–73.

[120] Yokoi M, Masutani C, Maekawa T, Sugasawa K, Ohkuma Y, Hanaoka F. The xeroderma pigmentosum group C protein complex XPC-HR23B plays an important role in the recruitment of transcription factor IIH to damaged DNA. J Biol Chem 2000;275(13):9870–5.

[121] Coin F, Oksenych V, Egly JM. Distinct roles for the XPB/p52 and XPD/p44 subcomplexes of TFIIH in damaged DNA opening during nucleotide excision repair. Mol Cell 2007;26(2):245–56.

[122] Saijo M, Takedachi A, Tanaka K. Nucleotide excision repair by mutant xeroderma pigmentosum group A (XPA) proteins with deficiency in interaction with RPA. J Biol Chem 2011;286(7):5476–83.

[123] Patrick SM, Turchi JJ. Xeroderma pigmentosum complementation group A protein (XPA) modulates RPA-DNA interactions via enhanced complex stability and inhibition of strand separation activity. J Biol Chem 2002;277(18):16096–101.

[124] Nocentini S, Coin F, Saijo M, Tanaka K, Egly JM. DNA damage recognition by XPA protein promotes efficient recruitment of transcription factor II H. J Biol Chem 1997;272(37):22991–4.

[125] Li L, Peterson CA, Lu X, Legerski RJ. Mutations in XPA that prevent association with ERCC1 are defective in nucleotide excision repair. Mol Cell Biol 1995;15(4):1993–8.

[126] Li L, Lu X, Peterson CA, Legerski RJ. An interaction between the DNA repair factor XPA and replication protein A appears essential for nucleotide excision repair. Mol Cell Biol 1995;15(10):5396–402.

[127] Li L, Elledge SJ, Peterson CA, Bales ES, Legerski RJ. Specific association between the human DNA repair proteins XPA and ERCC1. Proc Natl Acad Sci USA 1994;91(11):5012–6.

[128] Fagbemi AF, Orelli B, Scharer OD. Regulation of endonuclease activity in human nucleotide excision repair. DNA Repair (Amst) 2011;10(7): 722–9.

[129] Kemp MG, Reardon JT, Lindsey-Boltz LA, Sancar A. Mechanism of release and fate of excised oligonucleotides during nucleotide excision repair. J Biol Chem 2012;287(27):22889–99.

[130] Lehmann AR. DNA polymerases and repair synthesis in NER in human cells. DNA Repair (Amst) 2011;10(7):730–3.

[131] Araujo SJ, Tirode F, Coin F, Pospiech H, Syvaoja JE, Stucki M, et al. Nucleotide excision repair of DNA with recombinant human proteins: definition of the minimal set of factors, active forms of TFIIH, and modulation by CAK. Genes Dev 2000;14(3):349–59.

[132] Shivji MK, Podust VN, Hubscher U, Wood RD. Nucleotide excision repair DNA synthesis by DNA polymerase epsilon in the presence of PCNA, RFC, and RPA. Biochemistry 1995;34(15):5011–7.

[133] Mayne LV, Lehmann AR. Failure of RNA synthesis to recover after UV irradiation: an early defect in cells from individuals with Cockayne's syndrome and xeroderma pigmentosum. Cancer Res 1982;42(4):1473–8.

[134] Tornaletti S. Transcription arrest at DNA damage sites. Mutat Res 2005;577(1–2):131–45.

[135] Nakazawa Y, Sasaki K, Mitsutake N, Matsuse M, Shimada M, Nardo T, et al. Mutations in UVSSA cause UV-sensitive syndrome and impair RNA polymerase IIo processing in transcription-coupled nucleotide-excision repair. Nat Genet 2012;44(5):586–92.

[136] Kuraoka I, Ito S, Wada T, Hayashida M, Lee L, Saijo M, et al. Isolation of XAB2 complex involved in pre-mRNA splicing, transcription, and transcription-coupled repair. J Biol Chem 2008;283(2):940–50.

[137] Spivak G, Itoh T, Matsunaga T, Nikaido O, Hanawalt P, Yamaizumi M. Ultraviolet-sensitive syndrome cells are defective in transcription-coupled repair of cyclobutane pyrimidine dimers. DNA Repair (Amst) 2002;1(8):629–43.

[138] Mayne LV, Lehmann AR, Waters R. Excision repair in Cockayne syndrome. Mutat Res 1982;106(1):179–89.

[139] Marietta C, Brooks PJ. Transcriptional bypass of bulky DNA lesions causes new mutant RNA transcripts in human cells. EMBO Rep 2007;8(4): 388–93.

[140] Damsma GE, Alt A, Brueckner F, Carell T, Cramer P. Mechanism of transcriptional stalling at cisplatin-damaged DNA. Nat Struct Mol Biol 2007;14(12):1127–33.

[141] Brueckner F, Hennecke U, Carell T, Cramer P. CPD damage recognition by transcribing RNA polymerase II. Science 2007;315(5813):859–62.

[142] Fousteri M, Vermeulen W, van Zeeland AA, Mullenders LH. Cockayne syndrome A and B proteins differentially regulate recruitment of chromatin remodeling and repair factors to stalled RNA polymerase II *in vivo*. Mol Cell 2006;23(4):471–82.

[143] Smerdon MJ, Lieberman MW. Nucleosome rearrangement in human chromatin during UV-induced DNA-repair synthesis. Proc Natl Acad Sci USA 1978;75(9):4238–41.

[144] Dijk M, Typas D, Mullenders L, Pines A. Insight in the multilevel regulation of NER. Exp Cell Res 2014;329(1):116–23.

[145] Peterson CL, Almouzni G. Nucleosome dynamics as modular systems that integrate DNA damage and repair. Cold Spring Harb Perspect Biol 2013;5(9).

[146] Dinant C, Bartek J, Bekker-Jensen S. Histone displacement during nucleotide excision repair. Int J Mol Sci 2012;13(10):13322–37.

[147] Robu M, Shah RG, Petitclerc N, Brind'Amour J, Kandan-Kulangara F, Shah GM. Role of poly(ADP-ribose) polymerase-1 in the removal of UV-induced DNA lesions by nucleotide excision repair. Proc Natl Acad Sci USA 2013;110(5):1658–63.

[148] Luijsterburg MS, Lindh M, Acs K, Vrouwe MG, Pines A, van Attikum H, et al. DDB2 promotes chromatin decondensation at UV-induced DNA damage. J Cell Biol 2012;197(2):267–81.

[149] Jiang Y, Wang X, Bao S, Guo R, Johnson DG, Shen X, et al. INO80 chromatin remodeling complex promotes the removal of UV lesions by the nucleotide excision repair pathway. Proc Natl Acad Sci USA 2010;107(40):17274–9.

[150] Datta A, Bagchi S, Nag A, Shiyanov P, Adami GR, Yoon T, et al. The p48 subunit of the damaged-DNA binding protein DDB associates with the CBP/p300 family of histone acetyltransferase. Mutat Res 2001;486(2):89–97.

[151] Polo SE, Roche D, Almouzni G. New histone incorporation marks sites of UV repair in human cells. Cell 2006;127(3):481–93.

[152] Green CM, Almouzni G. Local action of the chromatin assembly factor CAF-1 at sites of nucleotide excision repair *in vivo*. EMBO J 2003;22(19):5163–74.

[153] Gaillard PH, Martini EM, Kaufman PD, Stillman B, Moustacchi E, Almouzni G. Chromatin assembly coupled to DNA repair: a new role for chromatin assembly factor I. Cell 1996;86(6):887–96.

[154] Citterio E, Van Den Boom V, Schnitzler G, Kanaar R, Bonte E, Kingston RE, et al. ATP-dependent chromatin remodeling by the Cockayne syndrome B DNA repair-transcription-coupling factor. Mol Cell Biol 2000;20(20):7643–53.

[155] Schwertman P, Lagarou A, Dekkers DH, Raams A, van der Hoek AC, Laffeber C, et al. UV-sensitive syndrome protein UVSSA recruits USP7 to regulate transcription-coupled repair. Nat Genet 2012;44(5):598–602.

[156] Groisman R, Kuraoka I, Chevallier O, Gaye N, Magnaldo T, Tanaka K, et al. CSA-dependent degradation of CSB by the ubiquitin-proteasome pathway establishes a link between complementation factors of the Cockayne syndrome. Genes Dev 2006;20(11):1429–34.

[157] Niedernhofer LJ. Nucleotide excision repair deficient mouse models and neurological disease. DNA Repair (Amst) 2008;7(7):1180–9.

[158] Xie F, Sun Q, Wu S, Xie X, Liu Z. Nucleotide excision repair gene ERCC1 19007T>C polymorphism contributes to lung cancer susceptibility: a meta-analysis. Genet Test Mol Biomarkers 2014;18(8):591–5.

[159] Mandal RK, Yadav SS, Panda AK. Meta-analysis on the association of nucleotide excision repair gene XPD A751C variant and cancer susceptibility among Indian population. Mol Biol Rep 2014;41(2):713–9.

[160] Huang D, Zhou Y. Nucleotide excision repair gene polymorphisms and prognosis of non-small cell lung cancer patients receiving platinum-based chemotherapy: a meta-analysis based on 44 studies. Biomed Rep 2014;2(4):452–62.

[161] Zhang L, Wang J, Xu L, Zhou J, Guan X, Jiang F, et al. Nucleotide excision repair gene ERCC1 polymorphisms contribute to cancer susceptibility: a meta-analysis. Mutagenesis 2012;27(1):67–76.

[162] Mei CR, Luo M, Li HM, Deng WJ, Zhou QH. DNA repair gene polymorphisms in the nucleotide excision repair pathway and lung cancer risk: a meta-analysis. Chin J Cancer Res 2011;23(2):79–91.

[163] Kiyohara C, Yoshimasu K. Genetic polymorphisms in the nucleotide excision repair pathway and lung cancer risk: a meta-analysis. Int J Med Sci 2007;4(2):59–71.

[164] Yousif AS, Stanlie A, Begum NA, Honjo T. Opinion: uracil DNA glycosylase (UNG) plays distinct and non-canonical roles in somatic hypermutation and class switch recombination. Int Immunol 2014;26(10):575–8.

[165] Arakawa H, Hauschild J, Buerstedde JM. Requirement of the activation-induced deaminase (AID) gene for immunoglobulin gene conversion. Science 2002;295(5558):1301–6.

[166] Kato L, Stanlie A, Begum NA, Kobayashi M, Aida M, Honjo T. An evolutionary view of the mechanism for immune and genome diversity. J Immunol 2012;188(8):3559–66.

[167] Okashita N, Kumaki Y, Ebi K, Nishi M, Okamoto Y, Nakayama M, et al. PRDM14 promotes active DNA demethylation through the ten-eleven translocation (TET)-mediated base excision repair pathway in embryonic stem cells. Development 2014;141(2):269–80.

[168] Shen L, Wu H, Diep D, Yamaguchi S, D'Alessio AC, Fung HL, et al. Genome-wide analysis reveals TET- and TDG-dependent 5-methylcytosine oxidation dynamics. Cell 2013;153(3):692–706.

[169] Wu H, Zhang Y. Mechanisms and functions of Tet protein-mediated 5-methylcytosine oxidation. Genes Dev 2011;25(23):2436–52.

[170] Ito S, Shen L, Dai Q, Wu SC, Collins LB, Swenberg JA, et al. Tet proteins can convert 5-methylcytosine to 5-formylcytosine and 5-carboxylcytosine. Science 2011;333(6047):1300–3.

[171] Hajkova P, Jeffries SJ, Lee C, Miller N, Jackson SP, Surani MA. Genome-wide reprogramming in the mouse germ line entails the base excision repair pathway. Science 2010;329(5987):78–82.

[172] Martinez-Macias MI, Cordoba-Canero D, Ariza RR, Roldan-Arjona T. The DNA repair protein XRCC1 functions in the plant DNA demethylation pathway by stimulating cytosine methylation (5-meC) excision, gap tailoring, and DNA ligation. J Biol Chem 2013;288(8):5496–505.

[173] Lee J, Jang H, Shin H, Choi WL, Mok YG, Huh JH. AP endonucleases process 5-methylcytosine excision intermediates during active DNA demethylation in Arabidopsis. Nucleic Acids Res 2014;42(18):11408–18.

[174] Li Y, Cordoba-Canero D, Qian W, Zhu X, Tang K, Zhang H, et al. An AP endonuclease functions in active DNA dimethylation and gene imprinting in arabidopsis. PLoS Genet 2015;11(1):e1004905.

[175] Yu H, Su Y, Shin J, Zhong C, Guo JU, Weng YL, et al. Tet3 regulates synaptic transmission and homeostatic plasticity via DNA oxidation and repair. Nat Neurosci 2015;18(6):836–43.

[176] Wheldon LM, Abakir A, Ferjentsik Z, Dudnakova T, Strohbuecker S, Christie D, et al. Transient accumulation of 5-carboxylcytosine indicates involvement of active demethylation in lineage specification of neural stem cells. Cell Rep 2014;7(5):1353–61.

[177] Nettersheim D, Heukamp LC, Fronhoffs F, Grewe MJ, Haas N, Waha A, et al. Analysis of TET expression/activity and 5mC oxidation during normal and malignant germ cell development. PLoS One 2013;8(12):e82881.

[178] Shah D, Kelly J, Zhang Y, Dande P, Martinez J, Ortiz G, et al. Evidence in *Escherichia coli* that N3-methyladenine lesions induced by a minor groove binding methyl sulfonate ester can be processed by both base and nucleotide excision repair. Biochemistry 2001;40(6):1796–803.

[179] Doetsch PW, Morey NJ, Swanson RL, Jinks-Robertson S. Yeast base excision repair: interconnections and networks. Prog Nucleic Acid Res Mol Biol 2001;68:29–39.

[180] Memisoglu A, Samson L. Contribution of base excision repair, nucleotide excision repair, and DNA recombination to alkylation resistance of the fission yeast *Schizosaccharomyces pombe*. J Bacteriol 2000;182(8):2104–12.

[181] Swanson RL, Morey NJ, Doetsch PW, Jinks-Robertson S. Overlapping specificities of base excision repair, nucleotide excision repair, recombination, and translesion synthesis pathways for DNA base damage in *Saccharomyces cerevisiae*. Mol Cell Biol 1999;19(4):2929–35.

[182] Guillet M, Boiteux S. Endogenous DNA abasic sites cause cell death in the absence of Apn1, Apn2 and Rad1/Rad10 in *Saccharomyces cerevisiae*. EMBO J 2002;21(11):2833–41.

[183] Saporito SM, Gedenk M, Cunningham RP. Role of exonuclease III and endonuclease IV in repair of pyrimidine dimers initiated by bacteriophage T4 pyrimidine dimer-DNA glycosylase. J Bacteriol 1989;171(5):2542–6.

[184] Kim N, Jinks-Robertson S. Abasic sites in the transcribed strand of yeast DNA are removed by transcription-coupled nucleotide excision repair. Mol Cell Biol 2010;30(13):3206–15.

[185] Reardon JT, Bessho T, Kung HC, Bolton PH, Sancar A. *In vitro* repair of oxidative DNA damage by human nucleotide excision repair system: possible explanation for neurodegeneration in xeroderma pigmentosum patients. Proc Natl Acad Sci USA 1997;94(17):9463–8.

[186] D'Errico M, Parlanti E, Teson M, de Jesus BM, Degan P, Calcagnile A, et al. New functions of XPC in the protection of human skin cells from oxidative damage. EMBO J 2006;25(18):4305–15.

[187] Tanida I, Ueno T, Kominami E. LC3 conjugation system in mammalian autophagy. Int J Biochem Cell Biol 2004;36(12):2503–18.

[188] Narendra D, Tanaka A, Suen DF, Youle RJ. Parkin-induced mitophagy in the pathogenesis of Parkinson disease. Autophagy 2009;5(5):706–8.

[189] Narendra DP, Jin SM, Tanaka A, Suen DF, Gautier CA, Shen J, et al. PINK1 is selectively stabilized on impaired mitochondria to activate Parkin. PLoS Biol 2010;8(1):e1000298.

[190] Kazlauskaite A, Kondapalli C, Gourlay R, Campbell DG, Ritorto MS, Hofmann K, et al. Parkin is activated by PINK1-dependent phosphorylation of ubiquitin at Ser65. Biochem J 2014;460(1):127–39.

[191] Kane LA, Lazarou M, Fogel AI, Li Y, Yamano K, Sarraf SA, et al. PINK1 phosphorylates ubiquitin to activate Parkin E3 ubiquitin ligase activity. J Cell Biol 2014;205(2):143–53.

[192] Koyano F, Okatsu K, Kosako H, Tamura Y, Go E, Kimura M, et al. Ubiquitin is phosphorylated by PINK1 to activate parkin. Nature 2014;510(7503):162–6.

[193] Wauer T, Swatek KN, Wagstaff JL, Gladkova C, Pruneda JN, Michel MA, et al. Ubiquitin Ser65 phosphorylation affects ubiquitin structure, chain assembly and hydrolysis. EMBO J 2015;34(3):307–25.

[194] Fang EF, Scheibye-Knudsen M, Brace LE, Kassahun H, SenGupta T, Nilsen H, et al. Defective mitophagy in XPA via PARP-1 hyperactivation and NAD(+)/SIRT1 reduction. Cell 2014;157(4):882–96.

[195] Nilsen H, Rosewell I, Robins P, Skjelbred CF, Andersen S, Slupphaug G, et al. Uracil-DNA glycosylase (UNG)-deficient mice reveal a primary role of the enzyme during DNA replication. Mol Cell 2000;5(6):1059–65.

[196] Kemmerich K, Dingler FA, Rada C, Neuberger MS. Germline ablation of SMUG1 DNA glycosylase causes loss of 5-hydroxymethyluracil- and UNG-backup uracil-excision activities and increases cancer predisposition of $Ung^{-/-}Msh2^{-/-}$ mice. Nucleic Acids Res 2012;40(13):6016–25.

[197] Engelward BP, Weeda G, Wyatt MD, Broekhof JL, de Wit J, Donker I, et al. Base excision repair deficient mice lacking the Aag alkyladenine DNA glycosylase. Proc Natl Acad Sci USA 1997;94(24):13087–92.

[198] Takao M, Kanno S, Shiromoto T, Hasegawa R, Ide H, Ikeda S, et al. Novel nuclear and mitochondrial glycosylases revealed by disruption of the mouse Nth1 gene encoding an endonuclease III homolog for repair of thymine glycols. EMBO J 2002;21(13):3486–93.

[199] Chan MK, Ocampo-Hafalla MT, Vartanian V, Jaruga P, Kirkali G, Koenig KL, et al. Targeted deletion of the genes encoding NTH1 and NEIL1 DNA *N*-glycosylases reveals the existence of novel carcinogenic oxidative damage to DNA. DNA Repair (Amst) 2009;8(7):786–94.

[200] Sakamoto K, Tominaga Y, Yamauchi K, Nakatsu Y, Sakumi K, Yoshiyama K, et al. MUTYH-null mice are susceptible to spontaneous and oxidative stress induced intestinal tumorigenesis. Cancer Res 2007;67(14):6599–604.

[201] Minowa O, Arai T, Hirano M, Monden Y, Nakai S, Fukuda M, et al. Mmh/Ogg1 gene inactivation results in accumulation of 8-hydroxyguanine in mice. Proc Natl Acad Sci USA 2000;97(8):4156–61.

[202] Arai T, Kelly VP, Minowa O, Noda T, Nishimura S. The study using wild-type and Ogg1 knockout mice exposed to potassium bromate shows no tumor induction despite an extensive accumulation of 8-hydroxyguanine in kidney DNA. Toxicology 2006;221(2–3):179–86.

[203] Mori H, Ouchida R, Hijikata A, Kitamura H, Ohara O, Li Y, et al. Deficiency of the oxidative damage-specific DNA glycosylase NEIL1 leads to reduced germinal center B cell expansion. DNA Repair (Amst) 2009;8(11):1328–32.

[204] Chakraborty A, Wakamiya M, Venkova-Canova T, Pandita RK, Aguilera-Aguirre L, Sarker AH, et al. Neil2-null mice accumulate oxidized DNA bases in the transcriptionally active sequences of the genome and are susceptible to innate inflammation. J Biol Chem 2015;290(41):24636–48.

[205] Rolseth V, Krokeide SZ, Kunke D, Neurauter CG, Suganthan R, Sejersted Y, et al. Loss of Neil3, the major DNA glycosylase activity for removal of hydantoins in single stranded DNA, reduces cellular proliferation and sensitizes cells to genotoxic stress. Biochim Biophys Acta 2013;1833(5):1157–64.

[206] Ludwig DL, MacInnes MA, Takiguchi Y, Purtymun PE, Henrie M, Flannery M, et al. A murine AP-endonuclease gene-targeted deficiency with post-implantation embryonic progression and ionizing radiation sensitivity. Mutat Res 1998;409(1):17–29.

[207] Gu H, Marth JD, Orban PC, Mossmann H, Rajewsky K. Deletion of a DNA polymerase beta gene segment in T cells using cell type-specific gene targeting. Science 1994;265(5168):103–6.

[208] Goldsby RE, Hays LE, Chen X, Olmsted EA, Slayton WB, Spangrude GJ, et al. High incidence of epithelial cancers in mice deficient for DNA polymerase delta proofreading. Proc Natl Acad Sci USA 2002;99(24):15560–5.

[209] Albertson TM, Ogawa M, Bugni JM, Hays LE, Chen Y, Wang Y, et al. DNA polymerase epsilon and delta proofreading suppress discrete mutator and cancer phenotypes in mice. Proc Natl Acad Sci USA 2009;106(40):17101–4.

[210] Bertocci B, De Smet A, Flatter E, Dahan A, Bories JC, Landreau C, et al. Cutting edge: DNA polymerases mu and lambda are dispensable for Ig gene hypermutation. J Immunol 2002;168(8):3702–6.

[211] Chung L, Onyango D, Guo Z, Jia P, Dai H, Liu S, et al. The FEN1 E359K germline mutation disrupts the FEN1-WRN interaction and FEN1 GEN activity, causing aneuploidy-associated cancers. Oncogene 2015;34(7):902–11.

[212] Kucherlapati M, Yang K, Kuraguchi M, Zhao J, Lia M, Heyer J, et al. Haploinsufficiency of Flap endonuclease (Fen1) leads to rapid tumor progression. Proc Natl Acad Sci USA 2002;99(15):9924–9.

[213] Bentley D, Selfridge J, Millar JK, Samuel K, Hole N, Ansell JD, et al. DNA ligase I is required for fetal liver erythropoiesis but is not essential for mammalian cell viability. Nat Genet 1996;13(4):489–91.

[214] Puebla-Osorio N, Lacey DB, Alt FW, Zhu C. Early embryonic lethality due to targeted inactivation of DNA ligase III. Mol Cell Biol 2006;26(10):3935–41.

[215] Tebbs RS, Flannery ML, Meneses JJ, Hartmann A, Tucker JD, Thompson LH, et al. Requirement for the Xrcc1 DNA base excision repair gene during early mouse development. Dev Biol 1999;208(2):513–29.

[216] Masutani M, Suzuki H, Kamada N, Watanabe M, Ueda O, Nozaki T, et al. Poly(ADP-ribose) polymerase gene disruption conferred mice resistant to streptozotocin-induced diabetes. Proc Natl Acad Sci USA 1999;96(5):2301–4.

[217] de Murcia JM, Niedergang C, Trucco C, Ricoul M, Dutrillaux B, Mark M, et al. Requirement of poly(ADP-ribose) polymerase in recovery from DNA damage in mice and in cells. Proc Natl Acad Sci USA 1997;94(14):7303–7.

[218] Wang ZQ, Auer B, Stingl L, Berghammer H, Haidacher D, Schweiger M, et al. Mice lacking ADPRT and poly(ADP-ribosyl)ation develop normally but are susceptible to skin disease. Genes Dev 1995;9(5):509–20.

[219] Cang Y, Zhang J, Nicholas SA, Bastien J, Li B, Zhou P, et al. Deletion of DDB1 in mouse brain and lens leads to p53-dependent elimination of proliferating cells. Cell 2006;127(5):929–40.

[220] Yoon T, Chakrabortty A, Franks R, Valli T, Kiyokawa H, Raychaudhuri P. Tumor-prone phenotype of the DDB2-deficient mice. Oncogene 2005;24(3):469–78.

[221] Cheo DL, Ruven HJ, Meira LB, Hammer RE, Burns DK, Tappe NJ, et al. Characterization of defective nucleotide excision repair in XPC mutant mice. Mutat Res 1997;374(1):1–9.

[222] Ng JM, Vrieling H, Sugasawa K, Ooms MP, Grootegoed JA, Vreeburg JT, et al. Developmental defects and male sterility in mice lacking the ubiquitin-like DNA repair gene mHR23B. Mol Cell Biol 2002;22(4):1233–45.

[223] van der Horst GT, Meira L, Gorgels TG, de Wit J, Velasco-Miguel S, Richardson JA, et al. UVB radiation-induced cancer predisposition in Cockayne syndrome group A (Csa) mutant mice. DNA Repair (Amst) 2002;1(2):143–57.

[224] van der Horst GT, van Steeg H, Berg RJ, van Gool AJ, de Wit J, Weeda G, et al. Defective transcription-coupled repair in Cockayne syndrome B mice is associated with skin cancer predisposition. Cell 1997;89(3):425–35.

[225] Andressoo JO, Weeda G, de Wit J, Mitchell JR, Beems RB, van Steeg H, et al. An Xpb mouse model for combined xeroderma pigmentosum and Cockayne syndrome reveals progeroid features upon further attenuation of DNA repair. Mol Cell Biol 2009;29(5):1276–90.

[226] de Boer J, Donker I, de Wit J, Hoeijmakers JH, Weeda G. Disruption of the mouse xeroderma pigmentosum group D DNA repair/basal transcription gene results in preimplantation lethality. Cancer Res 1998;58(1):89–94.

[227] Andressoo JO, Mitchell JR, de Wit J, Hoogstraten D, Volker M, Toussaint W, et al. An Xpd mouse model for the combined xeroderma pigmentosum/Cockayne syndrome exhibiting both cancer predisposition and segmental progeria. Cancer Cell 2006;10(2):121–32.

[228] Nakane H, Takeuchi S, Yuba S, Saijo M, Nakatsu Y, Murai H, et al. High incidence of ultraviolet-B-or chemical-carcinogen-induced skin tumours in mice lacking the xeroderma pigmentosum group A gene. Nature 1995;377(6545):165–8.

[229] Shiomi N, Kito S, Oyama M, Matsunaga T, Harada YN, Ikawa M, et al. Identification of the XPG region that causes the onset of Cockayne syndrome by using Xpg mutant mice generated by the cDNA-mediated knock-in method. Mol Cell Biol 2004;24(9):3712–9.

[230] Weeda G, Donker I, de Wit J, Morreau H, Janssens R, Vissers CJ, et al. Disruption of mouse ERCC1 results in a novel repair syndrome with growth failure, nuclear abnormalities and senescence. Curr Biol 1997;7(6):427–39.

[231] Tian M, Shinkura R, Shinkura N, Alt FW. Growth retardation, early death, and DNA repair defects in mice deficient for the nucleotide excision repair enzyme XPF. Mol Cell Biol 2004;24(3):1200–5.

Chapter 18

DNA Mismatch Repair in Mammals

M. Yang, P. Hsieh

National Institutes of Health, Bethesda, MD, United States

Chapter Outline

1. INTRODUCTION AND BRIEF HISTORY

Preserving genomic integrity is essential for all organisms to survive and reproduce. Ensuring high-fidelity replication is critical for maintaining genome stability as all organisms are frequently exposed to exogenous and endogenous sources of DNA damage. In eukaryotes, the replicative DNA polymerases Polδ and Polε select the correct nucleotide for incorporation with high precision; however, once every 10^4 to 10^5 nucleotides, an error is made [1]. These base–base mispairs, for example, T opposite G (T:G), are corrected by the exonuclease proofreading function of replicative polymerases when the abnormal geometry of a mismatched base pair slows the extension of DNA synthesis triggering the editing function. More problematic are insertion/deletion loops (IDLs) or indels. If left unrepaired, IDLs give rise to insertions or deletions and accompanying frameshift mutations. These arise in regions of nucleotide repeats, for example, a run of A's. During replication, the template strand can slip out of register with respect to the newly synthesized strand. These DNA loops generally escape the proofreading function of Polδ and Polε as they can be located away from the polymerase active site. DNA mismatch repair (MMR) targets both base–base mismatches that have escaped proofreading, and IDLs restoring the original parental sequence in an excision pathway referred to as post-replication repair. As such, MMR contributes between 100- and 1000-fold to the overall fidelity of replication. Given its central role in assuring replication fidelity, it is not surprising that MMR is extensively conserved. Unexpectedly, it is absent in *Actinobacteria*, many *Archaea*, *Helicobacter* and *Campylobacter* and most *Mollicutes* including *Mycobacterium tuberculosis* [2].

Loss of MMR confers a mutator phenotype in which the rate of spontaneous mutation is increased 50–1000-fold. In addition, the accumulation of IDLs leads to microsatellite instability (MSI) characterized by genomic expansion or contraction in regions containing 1–4 nucleotide repeat sequences. The central role of MMR in mutation avoidance and genome stability is underscored by the fact that inactivating mutations in several key MMR genes, most commonly *MSH2* and *MLH1*, but also *PMS2* and *MSH6*, are linked to Lynch syndrome, also known as hereditary nonpolyposis colorectal cancer [3,4]. Lynch syndrome, one of the most common hereditary cancers in humans, is inherited in an autosomal dominant fashion in which carriers are heterozygous for the germline-inactivating mutation. Loss of the functioning allele by epigenetic silencing or mutation results in colorectal carcinoma, endometrial carcinoma, and other cancers. Furthermore, epigenetic silencing of MMR genes, most commonly *MLH1*, is associated with a subset of sporadic tumors [5]. Lynch syndrome tumor cells frequently exhibit MSI due to the loss of MMR-induced IDL correction, as is the case in bacteria and fungi [6].

Genome Stability. http://dx.doi.org/10.1016/B978-0-12-803309-8.00018-5

This review focuses on MMR in mammalian cells—the molecular mechanisms that operate in post-replication repair, its regulation, and the role of MMR proteins in DNA-damage signaling. MMR proteins have important functions in a number of other cellular processes not discussed here, for example, (1) repairing DNA double-strand breaks (DSBs) and regulating homologous recombination in both meiosis and mitosis [7]; (2) promoting triplet repeat expansion in neurodegenerative diseases [8]; and (3) promoting somatic hypermutation in the variable regions of immunoglobulin genes in activated B cells [9,10]. The MMR literature is vast, and the reader is directed to several reviews on MMR [1,11,12]. Much of what is known about MMR has been learned from genetic, biochemical, and structural studies of MMR in unicellular organisms including its earliest description in studies of repair and recombination in bacteria and meiotic recombination in fungi. Mouse models have provided important information (reviewed in Ref. [13]), and studies in *Saccharomyces cerevisiae* inform virtually all aspects of MMR, although only a small subset are cited here. Reviews of bacteria and yeast MMR highlight ongoing advances in these experimentally tractable systems [1,12,14–17].

2. POST-REPLICATION MISMATCH REPAIR

2.1 Overview

MMR directed against replication errors has three main steps: (1) mismatch recognition in newly synthesized DNA; (2) DNA excision targeted to the newly synthesized DNA strand in the vicinity of the mismatch; and (3) high-fidelity, error-free DNA synthesis to fill the single-strand gap, thereby restoring an intact duplex with no errors (see Fig. 18.1). Excision exclusively on the newly synthesized strand is critical for mutation avoidance as indiscriminate excision on the template strand is mutagenic. The reconstitution of MMR in vitro with purified human proteins has facilitated dissection of the molecular pathway [18–20].

In discussing the key features of mammalian MMR, it is helpful to draw comparisons with MMR in *Escherichia coli* and fungi, particularly *S. cerevisiae*, that encompass many of the essential features of the mammalian pathway. Two key MMR proteins in *E. coli*, MutS and MutL, are homodimers encoded by single genes, *mutS* and *mutL*, respectively (Table 18.1). In eukaryotes multiple genes have arisen through gene duplication yielding MutS and MutL homologs that are heterodimeric proteins. This combinatorial aspect of eukaryotic MMR proteins facilitates their multifunctional roles mentioned earlier. MMR is initiated when *E. coli* MutS or eukaryotic MutS homologs (MSH) MutSα (MSH2–MSH6) and MutSβ (MSH2–MSH3) recognize and bind to mismatched DNA. *E. coli* MutS repairs seven of eight base–base mispairs and small IDLs; C:C mispairs are recalcitrant to MMR and are substrates for base excision repair (BER). MutSα and MutSβ have overlapping but distinct substrate specificities. MutSα targets all base–base mispairs except C:C as well as +1 IDLs and, to a lesser extent, +2 IDLs, whereas MutSβ targets 1–4 nt IDLs and a limited subset of base–base mispairs including C:C, A:A, and possibly G:G mispairs (see [21] and references cited therein). In addition to a mismatch-binding domain (MBD), MutS proteins have two nucleotide-binding domains (NBDs). ATP binding and hydrolysis license subsequent steps of MMR involving MutS and MutL proteins by modulating the binding interaction of MutS on DNA and its interaction with MutL in prokaryotes and, most frequently, MutLα (MLH1–PMS2) in human cells or Mlh1–Pms1 in *S. cerevisiae*. The designation PMS (post-meitoic segregation) derives from the earliest identification of MutL homologs in studies of meiotic recombination in fungi [16]. A second MutL homolog, MutLγ (MLH1–MLH3) is important for meiotic recombination, but also has a minor role in MMR based on genetic and biochemical data from *S. cerevisiae*, mice, and human cells (reviewed in Ref. [12]).

Activation of MutS and MutL proteins on mismatched DNA leads to nuclease excision exclusively on the newly synthesized strand. This is well understood for *E. coli* and a few closely related gamma-proteobacteria but not for most other organisms. Targeted excision of the newly synthesized strand in *E. coli* is mediated by the MutH protein, an endonuclease activated by MutL [15]. MutH nicks the unmethylated strand at hemimethylated GATC sequences that are substrates for the Dam methyltransferase and are present in newly replicated DNA that has not yet been methylated post replication. Thus, MutH nicking is restricted to the newly synthesized, transiently unmethylated strand. A 3′–5′-helicase, UvrD, also activated by MutL unwinds from the nick providing a single-strand substrate for multiple single-strand exonucleases possessing 5′–3′- or 3′–5′-directionality. In this way, MMR is bidirectional and can respond to strand discontinuities in the chromosome that are located on either side of the mismatch (reviewed in Ref. [22]). In most organisms including eukaryotes, no MutH homolog exists, and in eukaryotes, the only exonuclease known to function in MMR is EXO1, an obligate 5′–3′-exonuclease that can function on dsDNA. The C-terminus of scExo1 contains an Msh2-binding domain [23] and an Mlh1-interacting protein (MIP) box [24]. Thus, MMR can utilize EXO1 for 5′-nick-directed MMR, but must utilize a novel excision mechanism to insure strand specificity of repair and bidirectionality. The unexpected discovery of a latent endonuclease activity in the human MutL homolog PMS2 that is activated for cleavage via interactions with

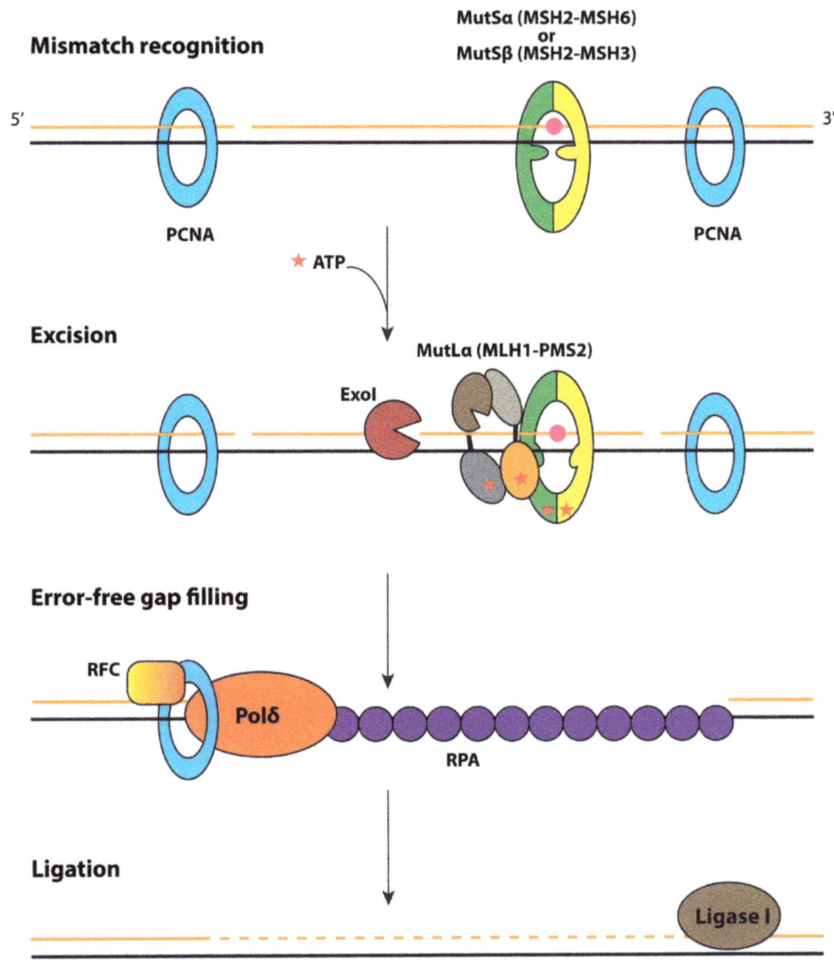

FIGURE 18.1 **Cartoon scheme for MMR in mammalian cells.** MMR is initiated when MutSα (MSH2–MSH6) or MutSβ (MSH2–MSH3) recognizes a mismatch in newly replicated DNA and forms a clamp structure. Nucleotide binding by MutSα or MutSβ induces a conformational switch allowing the recruitment of MutLα (MLH1–PMS2). MutSα or MutSβ can assume a sliding-clamp conformation. PCNA facilitates the recruitment of MMR proteins to the vicinity of the replication fork via a PIP motif on MSH3 and MSH6. ATP binding and hydrolysis at NBDs in both subunits of MutS and MutL homologs (indicated by *red star*) modulates protein–protein and protein–DNA interactions. Recruitment by MutSα and interaction with PCNA activate a latent endonuclease function in the PMS2 subunit of MutLα that nicks exclusively the newly synthesized strand. The nick provides an entry point for ExoI excision; alternatively, an ExoI-independent pathway requiring MutLα endonuclease activity is utilized (not shown). The resulting single-strand gapped DNA is protected by RPA. Error-free gap filling is carried out by replicative Polδ and DNA ligase I to restore the integrity of the duplex. See text for details.

MutSα, mismatched DNA, and PCNA provides an alternative pathway (see further on), though the molecular details are still being developed [25].

Formation of a single-strand gapped DNA in which the single-strand region is coated with *E. coli* single strand–binding (SSB) protein or eukaryotic replication protein A (RPA) yields a substrate for high-fidelity replicative polymerases, PolIII in *E. coli* or Polδ in eukaryotes. Ligation seals the nick resulting in an intact homoduplex devoid of mismatches. Clamp-like proteins that serve as processivity factors for DNA polymerases, bacterial β-clamp, and eukaryotic PCNA not only facilitate the gap-filling synthesis step, but also have critical albeit incompletely understood roles in recruiting MutS proteins to newly replicated DNA and, in the case of eukaryotes, activating and regulating an endonuclease activity that resides in the PMS2 subunit of MutLα (see further on).

2.2 MutS Homologs

Crystallographic studies of bacterial MutS proteins bearing short C-terminal truncations from *Thermus aquaticus* (Taq) or *E. coli* bound to a mismatched DNA containing a single unpaired T or a G:T mispair, respectively, and ADP provide important insights into MutS function (see Fig. 18.2A) [26–28]. In these structures, the two identical subunits forming the dimer each have five distinct structural domains separated by flexible linkers. domain I at the N-terminus is the MBD; domain II

TABLE 18.1 MMR Factors in *Escherichia coli* and *Homo sapiens*

Escherichia coli	*Homo sapiens*	Function
MutS–MutS	MSH2–MSH6 (MutSα) MSH2–MSH3 (MutSβ)	Mismatch recognition. Heterodimeric MutSα and MutSβ have distinct but overlapping mismatch specificities.
MutL–MutL	MLH1–PMS2 (MutLα)	Molecular matchmaker. *E. coli* MutL activates the MutH endonuclease. Human MutLα possesses an intrinsic endonuclease activity. Participates in excision termination in vitro.
	MLH1–PMS1 (MutLβ)	Unknown
	MLH1–MLH3 (MutLγ)	MutLγ can substitute for MutLα in a minor MMR role, but primary function is in meiotic recombination.
Dam methylase		Promotes N^6-adenine methylation at d(GATC) sites serving as strand discrimination signal in *E. coli*
MutH		Strand-specific endonuclease, nicks daughter strand
UvrD		DNA helicase II, promote excision reaction
RecJ, ExoVII		5′–3′-ssDNA exonuclease
ExoI, ExoVII, ExoX		3′–5′-ssDNA exonuclease
	ExoI	5′–3′-dsDNA exonuclease
β-Clamp	PCNA	DNA polymerase processivity factor; multiple MMR functions
γ-Complex	RFC	Loading of β-clamp/PCNA
SSB	RPA1–3	ssDNA-binding protein
DNA Pol III	Polδ	Replicative DNA polymerase that does gap filling
DNA ligase	Ligase I	Seal nicks after DNA resynthesis
	HMGB1	Accessory protein; stimulates excision

interacts with a second highly conserved MMR protein, MutL (see further on); domain III lies between the MBD and the NBD located in domain V. Long α helices or lever arms in domains III and IV propagate conformational changes between the MBD and NBDs of MutS that are separated by approximately 70Å. The two composite NBDs are members of the ABC (ATP-binding cassette) ATPase superfamily and are each comprised of residues from both subunits. They reside at the primary dimerization interface in domain V that also contains a conserved helix-turn-helix motif that promotes dimerization (see Ref. [29]). The structure of the short C-terminus that was deleted in earlier structural studies is essential for MMR at physiological levels of protein and may stabilize the dimer as well as help confer asymmetry of the NBDs as shown for hMutSβ (see further on Ref. [30]).

When bound to a mismatched DNA, MutS is a clamp in which the two previously identical subunits now exhibit asymmetry as only one MBD directly contacts the mismatched base, while the other MBD makes largely van der Waal and hydrogen bond contacts with flanking DNA. This structural and functional asymmetry in the bacterial proteins presages the heterodimeric nature of eukaryotic MMR proteins (see later). The DNA is sharply kinked at the mismatch by about 60° with widening of the minor groove at the mismatch to accommodate the MBD and corresponding narrowing of the opposing major groove.

Phe39 in a conserved Phe-X-Glu motif in domain I of Taq MutS was presumed to be in close proximity to the mismatched base, based on cross-linking studies of Taq MutS bound to a mismatch DNA containing a 5-iododeoxyuridine cross-linking moiety [29]. Mutation to alanine in the related *E. coli* MutS protein abolished mismatched DNA binding in vitro. The crystal structures confirmed that this Phe residue in one of the subunits approaches from the minor groove of the heteroduplex DNA and stacks with the unpaired base extruding it into the minor groove. A hydrogen bond between a carboxyl oxygen of a conserved Glu residue in the same subunit and the mismatched base is also observed [31]. Genetic and biochemical studies confirm that these two residues are essential for proper mismatch recognition in MutS and MutSα, but are notably absent in MutSβ (reviewed in Refs. [11,29]).

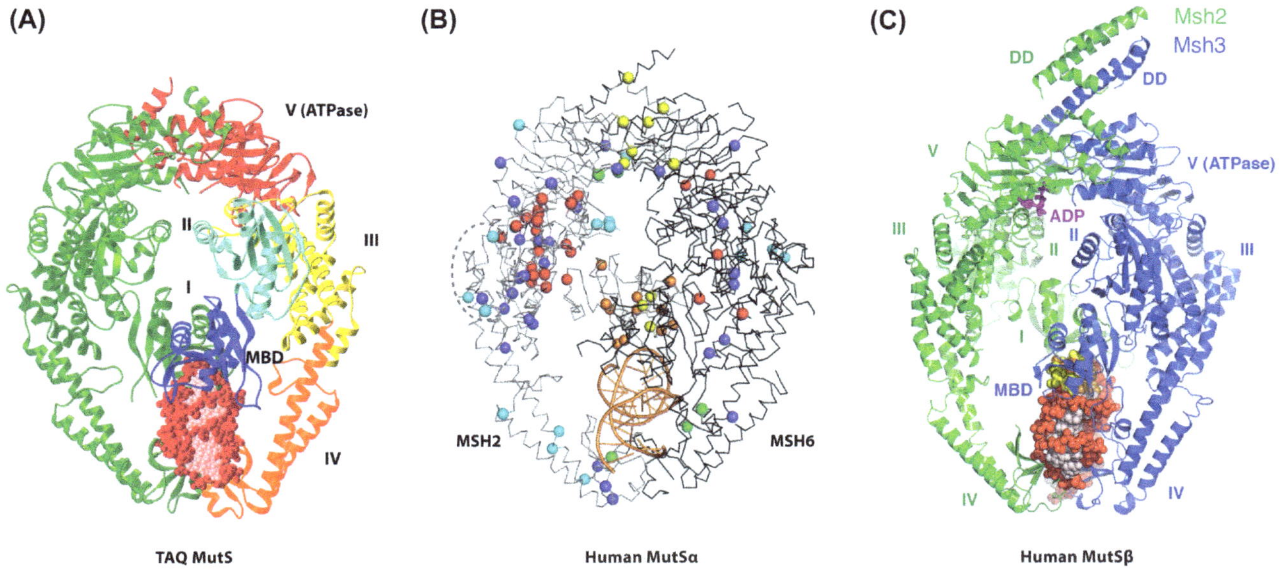

FIGURE 18.2 **Structural models for MutS homologs.** (A) *Thermus aquaticus* MutS bound to a +1IDL mismatched DNA (1EWQ.pdb). The two protein monomers are represented by ribbon diagrams. The DNA is shown in a space-filling model, in which the *backbone atoms are red* and *bases are pink*. In the A subunit, the five structural domains are colored—domain I (mispair-binding domain, MBD) is *blue*; II (connector domain) is *cyan*; III is *yellow*; IV is *pink*; V (nucleotide-binding domain, NBD) is *red*. The B subunit is *green (Reproduced with permission from Yang W, Junop MS, Ban C, Obmolova G, Hsieh P. DNA mismatch repair: from structure to mechanism. Cold Spring Harb Symp Quant Biol 2000;65:225–32)*. (B) Structural model for human MutSα with map of Lynch syndrome mutations (2O8B.pdb). MSH2 and MSh6 are shown as *light and dark gray* Cα chain traces, respectively. Mismatched G:T DNA is *orange*. Lynch syndrome (HNPCC) alleles are indicated by colored dots reflecting hypothetical function. *Cyan*—protein–protein interactions; *blue*—protein stability; *red*—stability/allostery; *yellow*—MSH2–MSH6 interface; *green*—nucleotide-binding sites *(Reproduced with permission from Warren JJ, Pohlhaus TJ, Changela A, Iyer RR, Modrich PL, Beese LS. Structure of the human MutSα DNA lesion recognition complex. Mol Cell 2007;26(4):579–92)*. (C) Ribbon diagram of the structure of human MutSβ, with *MSH2 in green* and *MSH3 in blue* bound to a +3IDL (3THY.pdb). The DNA is shown in a space-filling model with the backbone in *red*, bases in *light pink*, and the unpaired nucleotides in *yellow and orange*. ADP bound to MSH2 is shown in *purple sticks (Reproduced with permission from Gupta S, Gellert M, Yang W. Mechanism of mismatch recognition revealed by human MutSβ bound to unpaired DNA loops. Nat Struct Mol Biol 2012;19(1):72–8)*.

To a first approximation, the crystal structure of a hMutSα-ADP-G:T mismatch complex, comprising full-length MSH2 and a truncated MSH6 missing the first 340 residues, resembles the bacterial MutS structures [32]. Thus, MutSα forms a protein clamp with two channels, the larger one accommodating the kinked mismatched DNA (see Fig. 18.2B). Both MSH2 and MSH6 have five structured domains with the MBD and two composite NBDs containing the ABC ATPase motif at opposing ends of the molecule. Subunit asymmetry evident in the bacterial MutS structures is recapitulated in hMutSα. MSH6, containing the Phe-X-Glu motif (Phe432), contacts the mismatched base, while MSH2 contacts the flanking DNA accompanied by a 45° kink at the mismatch. Lynch syndrome alleles map to virtually all regions of hMutSα. While MSH2 is more or less colinear with bacterial MutS proteins, MSH6 and MSH3 have an additional N-terminal extension of several hundred amino acids in which resides the PCNA-interacting protein (PIP) motif that mediates physical association of MutSα and MutSβ with the processivity factor for replicative polymerases. This N-terminal domain has been deleted in the crystal structures of hMutSα and hMutSβ to aid in structural determination. However, small-angle X-ray scattering (SAXS) of yeast and human MutSα and hMutSβ provides information about the MutS homolog–PCNA interaction and reveals differences in the structure of this N-terminal extension that is disordered in yeast but a globular domain in human MutS homologs [33,34]. In addition, MSH6, but not MSH2 or MSH3, contains a PWWP domain that interacts with histone methylation marks (see further on).

The absence of a Phe-X-Glu mismatch-binding motif in MSH3 and the preference of MutSβ for both small and larger IDLs raise questions about how mismatch recognition is carried out by MutSβ. In addition to mismatch specificity and the architecture of the MBDs (see later), MutSα and MutSβ differ in other respects, for example, nucleotide-induced conformational changes and the ability of MutSα to interact simultaneously with MutLα and PCNA whereas PCNA and MutLα compete for binding to MutSβ [35]. *In vivo* studies of MMR in *S. cerevisiae* reveal that mismatch recognition differs in significant ways between Msh2–Msh3 and Msh2–Msh6 and suggest that mismatch recognition by Msh2–Msh3 requires DNA bending and strand separation at the mismatch using residues in the MBDs of both Msh2 and Msh3 [36–38].

An ensemble of crystal structures of a trimmed human MutSβ bound to IDLs of 2, 3, 4, or 6 nt and one ADP confirms the bending and strand separation model [39]. MutSβ injects a conserved tyrosine–lysine pair into the IDL site that, together with residues chiefly from the MBD of MSH3, but also from MSH2 in the case of larger IDLs, distort the sugar–phosphate backbone of IDLs to achieve sharp substrate bending and strand separation at the IDL (Fig. 18.2C). Importantly, MutSβ can accommodate IDLs of varying length by modulating the degree of DNA bending and allowing domains IV of MSH2 and MSH3 to move independently. The structural data in conjunction with extensive genetic and biochemical studies provide new insights into the roles of these proteins outside post-replication repair (reviewed in Ref. [12]). Single molecule approaches, such as atomic force microscopy (AFM), FRET, and SAXS are providing new insights into mismatch recognition (see Ref. [40]). DNA flexibility and base stacking influence recognition, as the propensity of a DNA lesion to bend and deform regulates access to the MBD [41–43].

The highly conserved composite ATP-binding sites of MutS are critical for MMR, as nucleotide binding regulates the interaction of MutS on DNA and its interaction with MutL and other proteins. The asymmetry of the two MutS subunits induced by mismatch binding is further elaborated by nucleotide-induced conformational changes; nucleotide binding and hydrolysis promote allosteric regulation as opposed to fueling a protein machine like a helicase. Because of its central role, much effort has been focused on understanding the role of ATP binding and hydrolysis in the context of MMR with multiple models under consideration at various times. See reviews in 2010s, as only a small portion of the original literature is cited here [11,12,44].

The structure of hMutSβ, including the extreme C-terminus that is absent in the bacterial MutS structures, defines the full dimerization domain of MutS proteins and provides a structural basis for the intimate connection between subunit dimerization and the NBDs. Several α helices from MSH2 and MSH3, including a previously described helix-turn-helix (HTH) domain found in all MutS proteins form a hydrophobic bundle that stabilizes the dimer [39,45]. The composite nature of the NBDs in which the N1, N3, and N4 nucleotide-binding motifs derive from one subunit, while the N2 motif is contributed by the other subunit underlie a complex structural arrangement in which nucleotide occupancy in one subunit can influence the ATP-binding site of the other. This is mediated by the dimerization domains that facilitate communication between the two NBDs. Thus, the HTH motif in the dimerization domain contacts the nucleotide-binding site directly through a conserved trio of amino acids and also interacts with the N2 nucleotide-binding motif contributed in trans by the partnering subunit. The MutSβ structural data also reveal for the first time motifs in the NBD that can bloc an MSH3 ATP-binding site providing a molecular framework for regulating asymmetric nucleotide binding and exchange regulated by mismatch binding [39].

The inherent asymmetry of the two protein subunits observed in the MBD is mirrored in the NBDs of MutS proteins (see Ref. [11]). Thus, MSH3 and MSH6 are more active ATPases than MSH2 in free MutSα and MutSβ. Biochemical, structural, and genetic experiments reveal that the NBDs of *E. coli* MutS and yeast and human Msh2 and Msh6 bind ATP with different affinities and kinetics and that nucleotide occupancy in one subunit influences the ATP-binding site of the partnering subunit [46–53]. In particular, mismatch binding strongly inhibits the ATPase activity of MSH3 and MSH6, but only weakly affects MSH2. Mismatch binding is correlated with broad movement of MSH3 and MSH6 domains leading to their intimate association, whereas the domains in MSH2 remain loosely associated. Collectively, genetic, biochemical, and structural data lead Gupta et al. to suggest that binding to a mismatch induces a conformational change in the ATPase domain and dimerization domain. Nucleotide binding is blocked in MSH3 and MSH6 but allowed in MSH2 resulting in a conformation that recruits MutLα and licenses MMR [39]. These structural studies provide a platform for testing molecular mechanism.

2.3 MutL Homologs

Less is known about how MutL proteins function compared to MutS homologs, but the endonuclease activity of most MutL proteins clearly plays a pivotal role in MMR and possibly other cellular functions as well. Thus, understanding how this class of MMR proteins works is paramount. Prokaryotic MutL proteins and eukaryotic MutL homologs are homodimers and heterodimers, respectively (reviewed in Ref. [54,55]). They belong to the GHKL (Gyrase b, Hsp90, Histidine kinases, and MutL homologs) superfamily of ATPases [56]. MutL proteins have a conserved N-terminal domain in which resides the four conserved motifs of the GHKL ATPase NBD and a C-terminal dimerization domain. The endonuclease activity found in some, but not all, MutL homologs [25], is located in the C-terminal domain (CTD). Separating these two domains is a flexible linker of varying lengths (see Fig. 18.3). The ATPase activity of MutL proteins is absolutely required for MMR in vivo; correspondingly, Lynch syndrome mutations cluster in this region. The recurring theme of nonequivalent subunits in the key MMR proteins is readily apparent in MutL proteins. Numerous studies reveal that "equivalent" point substitution mutations in the NBDs of MLH1 and PMS2 (or Pms1) do not yield equivalent phenotypes in vivo or biochemical properties

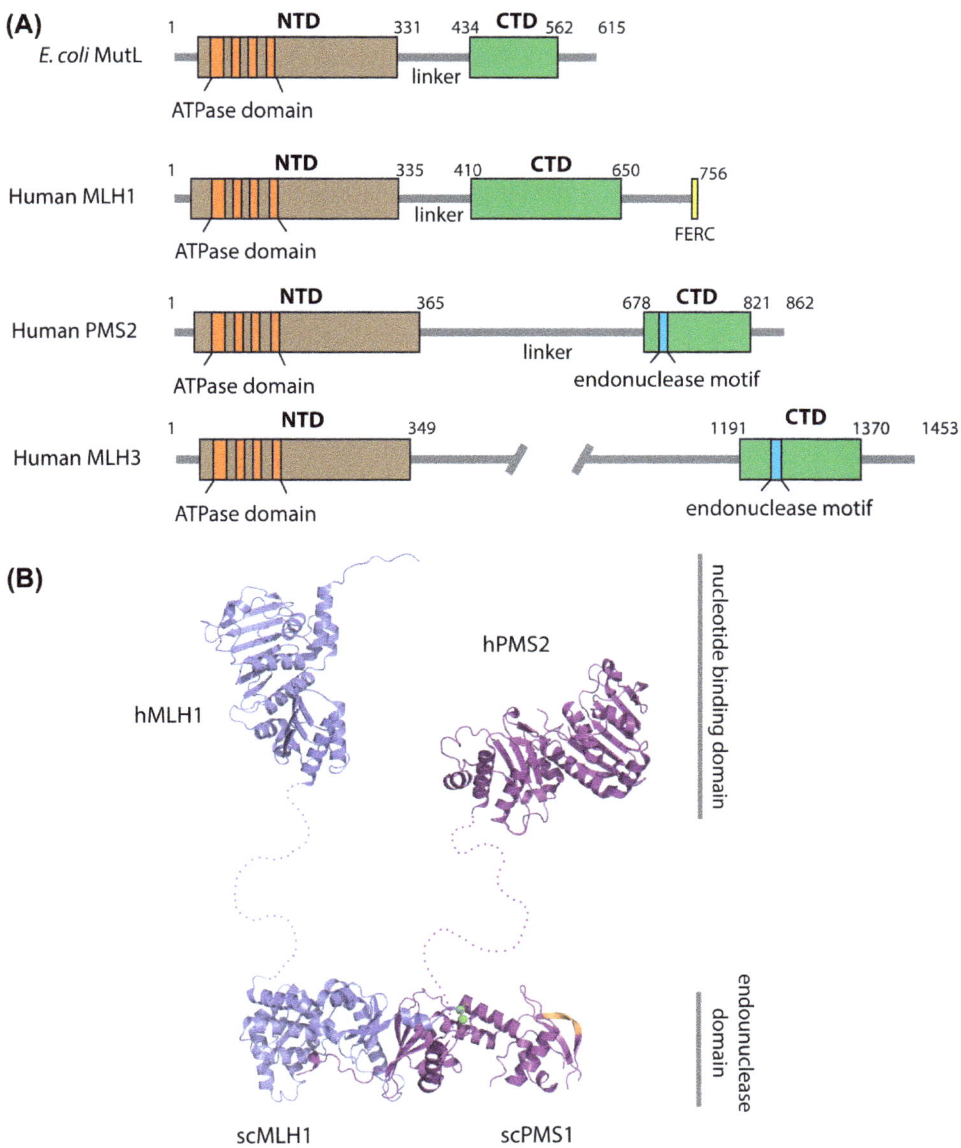

FIGURE 18.3 **Architecture of MutL homologs.** (A) Structural domains of *E. coli* MutL and human MutL homologs. The N-terminal domain (NTD) is indicated by a *tan box*, the C-terminal domain by a *green box*. The ATPase domain in the NTD consists of four highly conserved motifs (shown in *orange*). The endonuclease domain of PMS2 and MLH3 are in *blue*. The *yellow box* is the conserved FERC sequence of MLH1 (see text). (B) A hypothetical composite model for a eukaryotic MutLα (MLH1–PMS1/PMS2) based on available structures for human NTDs from MLH1 (PDB 4P7A) and PMS2 (PDB 1H7S) [59] shown as ribbon models and CTDs (endonuclease domain) from *S. cerevisiae* MLH1–PMS1 (PDB 4E4W) shown as ribbon models [24]. MLH1 is *blue*, and PMS2/PMS1 is *purple*. Green denotes two zinc atoms in the endonuclease active site. A putative PCNA interaction motif in PMS1 based on *B. subtilis* MutL is shown in *orange*. *Dotted line* represents the unstructured linker domain.

in vitro (see Ref. [54]). In eukaryotes, several MutL homologs exist that form heterodimers with sometimes overlapping but distinct functions. MutLα (MLH1–PMS2) or the equivalent scMlh1–Pms1 is the major MMR protein. Mutation of MLH1 accounts for a large fraction of Lynch syndrome alleles with a much smaller number attributable to loss of PMS2, and epigenetic silencing of the *MLH1* promoter at CpG islands occurs in spontaneous tumors. MutLβ (MLH1–PMS1) or scMlh1–Mlh2 is not thought to have a significant role in MMR, although data in 2014 on yeast suggest that it might act as an accessory factor [57]. Its role in human MMR, if any, is unknown. A third MutL protein, MLH1–MHL3, MutLγ, is important for meiotic recombination where it interacts with a meiotic-specific MutS homolog, Msh4–Msh5 (see Ref. [12]). It has a minor role in the repair of IDLs in yeast, and based on the cancer susceptibility and mutator phenotypes of knockout *MLH3* mice [58], probably contributes to MMR in mammals as well.

The N-terminal domain, about 330 residues, is multifunctional. In addition to the ATPase domain, it is responsible for both DNA binding and for interactions between MutL proteins and other repair partners including MutS proteins and PCNA.

Crystal structures have been solved for *E. coli* MutL–AMPPnP [56], hMLH1–ATP [59], hPMS2–ATPγS [60], and scPMS1–AMPPnP [61]. Structural and biochemical studies of the N-terminal domain from *E. coli* MutL reveal that ATP binding causes dimerization of the N-terminal domain with large numbers of conformational changes induced by a cycle of ATP binding, hydrolysis, and ADP release [60]. The structure also identifies a conserved DNA-binding groove, and establishes that binding to DNA stimulates the otherwise modest ATPase activity of MutL with attendant conformational changes. One consequence of ATP binding by MutL proteins is their conversion from open proteins to ring-like structures that have been shown to bind DNA (see later). AFM of yeast and human MutLα define at least four distinct conformations modulated by nucleotide binding, hydrolysis, and ADP release [62]. The proline-rich linker is poorly conserved but is thought to help mediate these large, asymmetric conformational changes. A working model based on these and other studies is that binding of ATP by MutL proteins results in dimerization of the N-terminal domains creating a DNA-binding groove that when occupied, results in a semicondensed state (reviewed in Ref. [54]). Efforts to relate these structural changes to MutL function are underway.

The CTD, about 200 residues, is the primary dimerization interface of MutL proteins. In prokaryotes with no MutH and in eukaryotes, the CTD of MutL or scPms1 (hPMS2) and Mlh3 harbors a conserved DQHA$(X)_2$E$(X)_4$E endonuclease motif as well as other conserved motifs that together constitute the endonuclease domain [25,55,63–65]. The CTD of MutLα also harbors the interaction domain with Exo1 [24,66]. Despite low sequence homology, the available structural information from CTDs of *E. coli*, *Bacillus subtilis*, and *S. cerevisiae* suggests overall conservation of topology [24,67,68]. In the structure of the CTD from scMutLα, each CTD is composed of distinct dimerization and regulatory domains [24]. The dimer interface between Mlh1 and Pms1 is extensive. The regulatory domain of Mlh1 contains an MIP-box motif that mediates interaction with other proteins including Exo1. The endonuclease site resides in a connector domain positioned at the dimerization interface between Mlh1 and Pms1. Given the importance of the endonuclease activity for MMR, its location at this critical interface is perhaps not surprising. It consists of the expected DQHA$(X)_2$E$(X)_4$E motif plus three other motifs that constitute the metal-binding site occupied by two zinc atoms. The C-terminal amino acid of Mlh1, Cys769, part of a conserved FERC motif, interacts with the metal-binding site in the crystal structure. The requirement for Cys769 in MMR in vivo is unclear as conflicting results are obtained in different mutator assays [24,64], and its role in human MutLα remained to be determined.

A longstanding question is the nature of the interaction between MutS and MutL proteins. When is MutL recruited to mismatched DNA in the presence of MutS, how long-lived is the interaction, and what is the stoichiometry of MutS and MutL proteins on mismatched DNA? Both ATP binding and mismatch bonding are required for the interaction of bacterial MutS with MutL and scMsh2–Msh6 with scMlh1–Pms1 ([69] and references cited therein). Hydrogen–deuterium exchange mass spectrometry in the presence of *E. coli* MutS, ATP, a mismatched DNA and *E. coli* MutL identifies a region in *E. coli* MutS "connector" domain II that exhibited decreased solvent accessibility in the presence of MutL [69]. Genetic and biochemical experiments confirm that Q211 or Q212 or both, mediate a MutS–MutL interaction in vitro and are required for MMR in vivo. The residues map to a structurally conserved region in scMsh2, but not scMsh6, and additional experiments confirm that the same region in Msh2, but not Msh6, is required for a MutSα–MutLα interaction in vitro and in vivo. Thus, the functional asymmetry evident in the MBDs of MutS proteins is also evident in the interaction with MutL, and the authors propose that ATP and mismatch recognition by MutS serve to present regions in domain II to MutL. Cross-linking and FRET studies also suggest an interaction between the N-terminal domain of *E. coli* MutL and domain II of only one MutS subunit [70]. In 2015, a crystal structure of the NTD of *E. coli* MutL site specifically cross-linked to *E. coli* MutS with a previously crystallized C-terminal truncation in the presence of AMP-PNP and a G:T mismatched DNA, reveals two interfaces [71]. One involves a region of the ATPase domain of MutL and a repositioned connector domain II of one MutS subunit consistent with deuterium exchange mass spectrometry [72]. The second involves an adjacent region of the ATPase domain of MutL and the ATPase and core domains of the other subunit of MutS including a conserved peptide loop in the core domain implicated in MutS–MutL interactions in *B. subtilis* MutS and human MutSβ [35,73]. Each MutL monomer is interacting with both subunits of the MutS dimer accompanied by large movements in multiple domains detected by FRET that are postulated to reflect an ATP-induced sliding-clamp conformation for MutS. Collectively, these advances highlight questions for further investigation.

2.4 Licensing Targeted Excision

An unresolved question is exactly how recognition of a mismatch by MutS proteins leads to recruitment of MutL proteins and licensing of the downstream excision step. In vitro MMR assays invariably require DNAs with a mismatch and a pre-existing nick even in the presence of MutLα with its latent nicking activity, and MMR-mediated excision is directed to the pre-nicked strand. *E. coli*, an outlier, depends on MutH for strand-specific nicking, but Dam GATC sites that are recognition sites for MutH can be several 1000 bp from a mismatch. MMR occurs on both the leading and lagging strands. In *S. cerevisae*, there is support for strand breaks in the Okazaki fragment-containing lagging strand serving as a strand discrimination signal, but the leading strand is thought to be comparatively barren of such breaks. It has been proposed that PCNA- and RNase H2-mediated removal of misincorporated ribonucleotides could provide strand breaks for MMR, though genetic data suggest it is a minor pathway (see Refs. [1,12]). In any case, MMR has to solve an action-at-a-distance problem.

Early models invoke a stationary complex of MutS and MutL at the mismatch and DNA bending or looping to bring distant sites together based on in vitro biochemical studies (see Ref. [29]) or ATP-powered translocation by MutS along the DNA to facilitate long-distance communication (reviewed in Ref. [22]). A nucleotide-switch or sliding-clamp model posits that ADP to ATP exchange upon mismatch binding induces conformational changes that convert MutS proteins from a clamp on the mismatch to a diffusing or sliding clamp that migrates along the DNA [74,75]. In this model, iterative rounds of MutS loading at the mismatch can occur leading to multiple MutS–MutL complexes on the DNA. Genetic, biochemical, and biophysical studies support a nucleotide switch in mismatch-bound MutS proteins and the formation of ATP-dependent sliding clamps in vitro, and mutations that disrupt this nucleotide switch disrupt MMR in vivo (reviewed in Refs. [11,12]). The sliding-clamp model is also consistent with in vitro experiments showing that a physical block between the mismatch and a DNA nick inhibits MMR [76,77]. Single-molecule fluorescence studies probe the movement of MutS on DNA and detect a corkscrew-like motion in the presence of ATP accompanied by distinct conformational changes in both the DNA and the protein that might facilitate a mismatch search [62,78–81].

What happens to a MutS–MutL complex? In a study of Q-dot-labeled MutSα and MutLα on λ DNA containing three GT mismatches, MutSα–MutLα complexes in a 1:1 ratio are seen to move along the DNA as sliding clamps [82] (see discussion in Ref. [40]). The final verdict is not in, however, as several studies suggest that the association of MutL with MutS, while ATP dependent, does not promote diffusion from the mismatch and involves a biased loading of MutL relative to MutS. A 2015 single-molecule FRET study proposes that Taq MutL traps MutS at the mismatch after MutS binds ATP and undergoes the first of multiple conformational changes [80]. Multiple loading of MutS proteins is observed that is inhibited by MutL, and the stoichiometry is consistent with a small excess of MutL over MutS. Visualization of fluorescently tagged and biologically active *E. coli* MutS and MutL in vivo [83] is also most consistent with colocalization at mismatches of MutS and MutL. MutL is found in several-fold excess over MutS and is thought to reflect multiple loading of MutL on the DNA extending from the mismatch towards a strand discrimination site in a manner possibly related to proposed catalytic loading of MutLα by MutSα in *S. cerevisiae* (see Refs. [12,84]). In reconstituted MMR assays, hMutLα helps to limit the extent of excision such that it terminates just beyond the mismatch suggesting that MutLα supplies mismatch positional information [19]. A requirement for more MutL than MutS might also explain why MutL is limiting for MMR in vivo and is consistent with previous *E. coli* MutS–MutL footprinting experiments (discussed in Refs. [40,83]).

Finally, even the existence of a stable ternary complex of MutS, MutL, and a mismatched DNA, particularly in the case of eukaryotic MMR proteins, is being questioned as studies in the 2010s suggest that the interactions between MutS and MutL may be transient in vivo. Attempts to isolate presumptive ternary complexes have required cross-linking or chemical trapping [70,71], and scMsh2 and scMlh1 foci do not always colocalize in vivo [84]. Furthermore, in vivo and in vitro studies indicate that MutSα and a mismatched DNA are not required to activate the endonuclease activity of MutLα per se, but may have roles in recruiting and/or retaining MutLα and PCNA to newly replicated and mismatched DNA so that PCNA can activate the MutLα endonuclease [84–86]. Much work remains to understand how MutS proteins find rare mismatches in a sea of genomic DNA and recruit MutL to license MMR.

2.5 Strand Discrimination

How is excision directly exclusively to the newly synthesized strand? *E. coli*, almost uniquely, exploits the transient undermethylation of newly synthesized DNA and incision by the MutH methyl-directed endonuclease. For virtually all other organisms, another mechanism(s) must be in play. In the case of the lagging strand containing Okazaki fragments, it is easy to envision EXO1 acting at transient breaks. In the case of the leading strand or in EXO1-independent excision, the mechanism is less obvious. It has been proposed that PCNA dictates the strand bias of MutLα nicking directing incision exclusively to the newly synthesized strand. The latent endonuclease activity of MutLα is activated in vitro by RFC and PCNA and utilizes the PIP motif in the PMS2 subunit for direct interaction. RFC and PCNA, but neither a mismatch nor MutSα, is required to activate MutLα in vitro and direct cleavage to the strand with a preexisting nick [85]. Because RFC loads PCNA with a fixed orientation preferentially at 3′-double-strand–single-strand junctions [87], PCNA serves as a de facto strand–discrimination signal. Perhaps a specific geometry of the MutLα–PCNA interaction imposes a strand bias on MutLα incision, but this remains unproven [88,89]. *In vivo*, the situation is more complicated as PCNA has been shown to bind to DNA structures that have single-strand characteristics, such as a small number of extruded triplet repeats leading, in principle, to error-prone repair if the template strand is indiscriminately nicked by MutLα [90]. Another possibility is that nicks introduced by RNaseH processing of ribonucleotides misincorporated into DNA may serve as a strand discrimination signal (see discussion in Ref. [1]).

However, *S. cerevisiae* strains missing RNaseH2 exhibit only a mild mutator phenotype suggesting that other mechanisms must operate to confer strand specificity (see discussion in Ref. [12]). An unambiguous mechanism for targeting the MutLα endonuclease to newly synthesized strands, particularly in the case of the leading strand, remains elusive.

A further wrinkle in the MMR excision step is the existence of Exo1-independent excision. Inactivation of *EXO1* in *S. cerevisiae* or mice confers only a weak mutator phenotype, and in mouse models fails to recapitulate the mutation or cancer spectrum of mutations in essential MMR genes. Low but detectable levels of MMR are observed in MMR assays using Exo1-deficient mouse cell extracts. A series of genetic tests in *S. cerevisiae* identified mutations that differentially affect Exo1-independent versus Exo1-dependent pathways (see Ref. [86]). In particular, loss of the Pms1 endonuclease activity (equivalent to hPMS2) conferred hypersensitivity in an *exo1Δ* strain as did certain mutations in the gene that encodes PCNA, *pol30* [91]. These results provide support for a critical role for MutLα endonuclease activity in an MMR excision pathway that does not involve Exo1 and suggests that recruitment of MutLα to mismatched DNA by MutSα and the activation of MutLα by PCNA are essential features of this pathway in vivo.

Imaging studies in *S. cerevisiae* identify Msh2–Msh6 foci that colocalize with replication factories. Mlh1–Pms1 foci are dependent on Msh2–Msh6, but they seldom colocalize with Msh2–Msh6 foci or replication machinery [84]. The authors propose a model in which MutSα (or MutSβ) bound to a mismatch catalytically loads multiple molecules of MutLα. Upon interaction with PCNA, these MutLα molecules can incise the newly synthesized strand providing access for Exo1 excising in a 5′–3′-fashion to create a gapped DNA intermediate. In an Exo1-independent scenario, several nonexclusive pathways for excision and processing of a MutLα-nicked heteroduplex may occur: (1) strand displacement synthesis by Polδ from the nick followed by flap cleavage and ligation; (2) additional DNA nicking by Mlh1–Pms1 followed by Polδ-dependent strand displacement and/or gap filling; (3) excision by the 3′–5′-proofreading exonuclease of replicative polymerases. In vitro MMR assays utilizing purified hMutSα, hMutLα, RFC, PCNA, RPA, and DNA Polδ yield no excision intermediates but support synthesis–driven strand displacement by Polδ in this EXO1-independent MMR system [92]. Confirmation in mammalian cells awaits as does a detailed study of the prevalence and kinetics of these pathways.

3. MISMATCH REPAIR AND THE DNA-DAMAGE RESPONSE

3.1 Alkylation Damage and Thiopurines

The MMR system is also implicated in the repair and cytoxicity of a subset of DNA lesions caused by S_N1 DNA alkylators, 6-thioguanine, fluoropyrimidines, cisplatin, UV light, and certain environmental carcinogens that form DNA adducts (reviewed in Refs. [4,29]). The S_N1 DNA alkylators, for example, *N*-methyl-*N′*-nitro-*N*-nitrosoguanidine (MNNG), methylnitrosourea, and the chemotherapy drug temozolomide, methylate all four DNA bases producing a variety of potentially cytotoxic lesions. Exposure to these alkylators induces a DNA-damage response resulting in cell cycle arrest and apoptosis that is dependent on MutSα and MutLα MMR proteins. Despite constituting a small fraction of total lesions, O^6-methylguanine (O^6me-G) is the key contributor to the mutagenic and cytotoxic effects of S_N1 alkylators. During replication, polymerases misincorporate opposite O^6me-G forming O^6me-G:T mispairs that, if unrepaired, lead to G to A transition mutations. O^6-methylguanine-DNA methyltransferase (MGMT) directly reverses O^6meG in cells and plays an important role in protecting against cytotoxic effects of S_N1 alkylators and preventing tumor formation in vivo [93]. Thiopurines, used in chemotherapy, are incorporated into DNA and undergo spontaneous methylation by endogenous *S*-adenosylmethionine to form structurally similar 6-thiomethylguanine. Tolerance to thiopurines is also tied to loss of MMR (see Ref. [94]).

First documented in *E. coli*, MMR-deficient mammalian cell lines also exhibit tolerance to alkylating agents and can be almost two orders of magnitude more resistant to killing than comparable MMR-proficient cells (reviewed in Ref. [94]). Low doses of MNNG induce a G_2/M cell cycle arrest in the second cell cycle after exposure that is dependent on MMR proteins (reviewed in Refs. [11,95]). In cells exposed to alkylating agents, ATM and Rad3-related (ATR) kinase undergoes autophosphorylation and, together with other proteins, such as ATR-interacting protein (ATRIP), an obligate ATR partner, Claspin and TopBP1, activates a signaling cascade ultimately leading to G_2/M cell cycle arrest mediated by downstream targets including the Chk1 checkpoint kinase and others (reviewed in Ref. [96]). Apoptosis directed in most cases by phosphorylated p53 also requires MutSα and MutLα. ATP–ATRIP is recruited to regions of ssDNA bound to RPA via an RPA-interacting motif in ATRIP [97]. Thus, persistent excision intermediates of MMR can activate ATR.

An explanation for the requirement of MMR proteins for cell killing by alkylating agents involves MMR processing of O^6me-G:T mispairs that are recognized by MutSα (reviewed in Ref. [22]). As discussed earlier, MMR-directed excision is targeted exclusively to the newly synthesized strand containing thymidine, whereas O^6meG remains in the DNA possibly triggering repeated cycles of MMR excision followed by resynthesis. Repeated rounds of excision could lead to DSBs. Aberrant or abortive MMR processing at sites of damage can lead to the accumulation of single-strand gaps visualized by EM [98]. Activation of ATR can occur when ATP–ATRIP is recruited to these regions of ssDNA bound to RPA. These single-strand intermediates that fail to engage the DNA synthesis machinery in the final step of MMR will give rise to broken chromosomes and damage signaling in the next round of replication explaining the delayed-damage response in cells exposed to alkylating agents. Ectopic expression of nuclease-dead EXO1 in mouse embryo fibroblasts in which endogenous EXO1 is absent restores a MSH2–CHK1 interaction and MNNG sensitivity providing support for the role of EXO1-mediated excision in a DNA-damage response [99].

An alternative model that remains to be proven involves direct recruitment of ATR to sites of damage by the MMR machinery (see Ref. [100]). MutSα and MutLα associate with ATR and other damage-signaling proteins, such as TopBP1 and Chk1 in multiprotein complexes in human cells [101,102]. In addition, MutSα and MutLα are required to recruit and activate ATR in the presence of O^6me-G:T-containing DNAs in an in vitro assay scoring for phosphorylation of Chk1 [103], and recruitment of ATR to sites of cisplatin damage is dependent on hMSH2 but not on RPA, Rad17, or the 9-1-1 complex [102].

3.2 Fluorouracil

Fluoropyrimidines, such as 5-fluorouracil (FU) are widely used in chemotherapy and evince a cytotoxic response that is dependent, in part, on MutSα and MutLα (reviewed in Ref. [104]). When FU is metabolized, thymidylate synthase, a key enzyme in de novo pyrimidine biosynthesis, is inhibited resulting in imbalances in nucleotide precursor pools and the incorporation of uracil and fluorouracil in DNA and RNA. Several lines of evidence indicate that incorporation of FdU into DNA is the primary pathway for cell killing. MutSα targets rare dFU:G mispairs resulting in the activation of MutSα ATPase activity [105]. Base excision repair (BER) also targets dFU (see Ref. [93]), and there is evidence that both BER and MMR contribute to the damage response and promote cell killing (see Refs. [106,107]).

3.3 Oxidative Damage and Noncanonical MMR

BER is the primary repair pathway for oxidative DNA damage in which specific glycosylase enzymes remove the damaged base followed by cleavage at the abasic site and gap repair. However, 7,8-dihydro-8-oxo-guanine (8-oxoG) templates 8-oxoG:A mispairs that often escape proofreading due to near normal geometry and are recognized by MutSα (see Refs. [1,11]). Interestingly, MutSα is implicated in a noncanonical MMR pathway that operates largely outside of S-phase in which MutSα recognizes clustered oxidative lesions leading to excision and monoubiquitination of PCNA [108]. This PCNA modification signals an error-prone polymerase, Polη, that carries out the gap-filling step in place of high-fidelity polymerases. Thus, MutSα can be recruited for a mutagenic process. A similar noncanonical MMR pathway dependent on MutSα, MutLα, monoubiquitinated PCNA, and Polη responds to S_N1-type alkylating agents in a variety of cell types and may explain the mutagenicity of alkylating agents (see Ref. [10]). In fact, mutagenic repair involving MMR may be more prevalent than previously thought. MMR-induced mutations are found flanking naturally occurring mismatches [109].

3.4 UV, Cisplatin, and DNA Cross-Links

UV causes cyclobutane pyrimidine dimers (CPDs) and genotoxic (6-4) pyrimidine pyrimidone dimers (6-4PP) in DNA. Although nucleotide excision repair is the primary repair pathway, *msh2*$^{-/-}$ mice exhibit an increased incidence of UV-induced skin tumors. Mammalian MSH2 in murine and human cells is implicated in cell cycle arrest and apoptosis induced by UV and its loss with increased mutagenesis (reviewed in Ref. [100]). In vitro, MutSα can bind to mismatched CPDs and 6-4PP. A novel pathway for MMR dependent, UV-induced mutagenesis, and DNA-damage signaling termed "post-TLS repair" invokes prior action at sites of UV damage by error-prone translesion synthesis (TLS) DNA polymerases that synthesize past the UV lesion residing in the template strand but introduce errors [110]. MutSα recognizes the mismatch products of TLS and initiates excision. If the single-strand gaps are not repaired, checkpoint induction occurs, but error-free filling of the gaps mitigates UVC mutagenicity. Deducing an explanation for organ tropism of tumors is oftentimes challenging. In Lynch syndrome, the rate of cellular proliferation is probably an important contributor, but it is unlikely to be the only one. Tsaalbi-Shtylik et al. suggest that loss of post-TLS repair and attendant elevated mutagenesis leads to disruption of multiple tumor-suppressing functions, and in combination with constant exposure to intestinal genotoxins, may explain the colorectal tropism of Lynch syndrome [110].

Cisplatin, a common chemotherapeutic drug, introduces intrastrand and lethal interstrand DNA cross-links (ICLs). There is a large literature on the effects of MMR on survival in cells treated with cisplatin with variable results (reviewed in Ref. [100]). MMR is unlikely to directly remove the cross-link; instead, multiple protein complexes involved in the Fanconi anemia pathway, homologous recombination, DSB repair, and NER converge on the ICL. MMR's role may be as a modulator of recombination or activity of the Fanconi anemia proteins. A FANCJ–MLH1 interaction suppresses MSH2 activity to promote restart at stalled replication forks [111,112], and MLH1 and PMS2 have been implicated in a p73-dependent apoptotic response to cisplatin [113] indicative of a pleiotropic role for MMR. MMR may modulate other repair pathways that target bulky DNA adducts formed by several environmental carcinogens, for example, benzo[c]phenanthrene dihydrodiol epoxide that modifies adenine residues or benzo[a]pyrene, a polycyclic aromatic hydrocarbon (see Refs. [4,100]).

4. REGULATION OF MMR

Spatiotemporal regulation of MMR is best exemplified by the close association of MMR with replication that confers several advantages. MMR can proceed efficiently, can utilize a transient open state of chromatin at the replication fork, and can collaborate with the replication and MMR machinery including PCNA, RPA, RFC, and replicases (reviewed in Ref. [1]). Correspondingly, expression of MMR genes is highest during S-phase though the increase is modest. Genome-wide assessments of mutational spectra in *S. cerevisiae* strains harboring mutations in MMR genes and/or replicases reveal that MMR is influenced by the replicase, leading versus lagging strand, mismatch composition and local sequence context. MMR corrects errors made by all three replicative polymerases.

How are MMR proteins recruited to newly synthesized DNA? Evidence from bacteria, yeast, and human cells point to a recruitment role for polymerase processivity factors like PCNA. Live cell imaging in *B. subtilis* and *S. cerevisiae* is revealing new details (reviewed in Ref. [12]). In *S. cerevisiae*, Msh2–Msh6 foci in S-phase colocalize with DNA polymerases, PCNA, and RPA; disruption of PCNA binding, for example, by mutating the PIP motif of Msh6 results in loss of the foci [84], and temporal coupling between MMR and replication in yeast is observed [114]. Similar interactions between *B. subtilis* and *E. coli* β-clamps and MutS have been reported (eg, Ref. [115]). The situation is likely more complicated, however, as loss of PCNA–Msh6 interactions in yeast only causes a partial loss of MMR in vivo supporting PCNA-independent pathways. Mlh1–Pms1 foci do not always colocalize with Msh2–Msh6 foci, consistent with the foci representing different MMR intermediates or events. Human MSH6 and MSH3 retain a PIP motif, and both MutSα and MutSβ interact with PCNA in vitro [116]. Furthermore, PCNA is an obligate partner in in vitro 5′-nick directed and bidirectional MMR assays utilizing human proteins and is required to activate the latent endonuclease activity of MutLα discussed previously [18–20,25,88]. Recently, Li and colleagues have reported that an epigenetic histone mark, trimethylation of histone H3K36, recruits MutSα to chromatin utilizing a PWWP recognition domain in MSH6 [117]. Epigenetic modification may serve as a general recruitment tool for MMR though MSH3 lacks a PWWP domain suggesting that other recruitment mechanisms exist.

Epigenetic marks on chromatin and chromatin architecture can modulate MMR. As discussed earlier, histone methylation at H3K36me3 may serve in general recruitment of MutSα to chromatin, an idea that is supported by the presence of MSI in SETD2 methylase-deficient cells [117]. Histone H3 acetylation is also suggested to modulate MMR [118]. Elements of chromatin structure can also inhibit MMR since DNA wrapped around nucleosome cores is generally less accessible, and nucleosomes block excision in in vitro systems (reviewed in Ref. [119]). Chromatin assembly factor 1 (CAF-1) promotes the assembly of nucleosomes on newly replicated DNA and protects lagging strands from excessive degradation in a reconstituted MMR system [120]. Thus, the coordination of MMR with nucleosome reassembly post-replication is critical. MMR delays nucleosome reassembly in vitro, and intriguingly, two key MMR players, MutSα and PCNA, interact with CAF-1 [121].

MMR can be inhibited by targeted degradation of MMR proteins by ubiquitin proteasomes. Histone deacetylase 6 (HDAC6) sequentially deacetylates and ubiquitinates MSH2 leading to the loss of MutSα and MutSβ in human cells [122]. Loss of MSH2 via an ubiquitination-dependent pathway also occurs in a subset of acute lymphoblastoid leukemia (ALL) cells that harbor inactivating chromosomal deletions in at least one of four genes that inhibit an MSH2 degradation pathway [123]. These primary ALL cells have low levels of MSH2 and exhibit MSI, and the loss of MMR may explain clinical tolerance to thiopurine therapy in this patient subpopulation. Another pathway for shutting off MMR is its indirect downregulation by the epidermal growth factor receptor (EGFR), a transmembrane receptor protein kinase that promotes cell growth, tumor progression, and metastasis. Following import into the nucleus, EGFR induces phosphorylation of PCNA at Y211 crippling its interaction with MMR proteins and inhibiting MMR [124]. Regulating MMR levels can also occur through changes in expression of microRNAs (miRs) that respond to DNA damage, with miR-422a, miR-21, and miR-155 being likely candidates ([125] and references cited therein).

5. FUTURE DIRECTIONS

The ubiquitous MMR system has been the focus of much attention in recent years as a critical player in genome stability and tumor suppression and as an important participant in numerous other diverse cellular processes. With respect to repair, a number of important questions remain. How is the MMR machinery recruited to newly replicated DNA and how is it positioned with respect to the advancing replisome? When do MutS and MutL proteins interact and when do they function separately? Exactly how is the MutL endonuclease activity targeted to the newly synthesized strand, and what is its biological scope? What is the mechanism of recruitment of replicative polymerases to single-strand gaps? In what contexts are error-prone polymerases employed instead and what are the consequences? How is MMR influenced by the higher-order architecture of chromatin and the nucleus? Finally, how can knowledge of MMR mechanism improve clinical diagnostics and therapeutic outcomes? The chapter on MMR is still being written.

GLOSSARY

Apoptosis It is also known as programmed cell death. A highly regulated and coordinated process that results in cell death preceded by characteristic changes including nuclear fragmentation, chromosomal DNA fragmentation, and mRNA decay. Apoptosis is part of normal developmental and differentiation processes and can also be triggered by DNA damage that blocks replication.

DNA excision repair Highly conserved molecular pathways that restore genome integrity after DNA damage by both endogenous and exogenous sources. Three pathways, base excision repair, nucleotide excision repair, and mismatch repair target distinct types of damage including oxidized or alkylated bases, UV photoproducts, and base mispairs by excising or enzymatically removing the damaged or incorrect bases and restoring the correct sequence using DNA polymerases and the undamaged strand as a template for correction. The 2015 Nobel Prize in Chemistry was awarded in recognition of basic advances in our understanding of these three excision repair pathways.

Epigenetic silencing Turning off gene expression by external or environmental factors such as DNA methylation at promoter sequences that inhibits transcription in contrast to changes in nucleotide sequence.

FRET Förster resonance energy transfer describes energy transfer between two light-sensitive molecules or chromophores, a donor and an acceptor. FRET is very sensitive to small changes in distance and is used to measure association/dissociation events and conformational changes in biological molecules bearing precisely positioned chromophores.

Indels The insertion or deletion of bases in the genome of an organism that accumulate in the absence of DNA mismatch repair; indels occur more frequently in microsatellite regions. In coding regions, indels that are not multiples of three will result in a frameshift mutation.

Microsatellite instability Contraction or expansion of a genomic region caused by loss of mismatch repair commonly in a region of mono- or dinucleotide repeats; its presence is strongly correlated with Lynch syndrome colorectal cancer.

Mutator phenotype Phenomenon whereby an organism exhibits a greatly elevated rate of spontaneous mutation, usually genome-wide, due to the genetic inactivation of a protective pathway (eg, DNA mismatch repair).

Posttranslational modification Covalent modification of proteins (eg, phosphorylation, acetylation, or ubiquitination), usually involving specialized enzymes, that occurs during or after protein synthesis by translating ribosomes on mRNA. Such modifications regulate many aspects of protein function, stability, and cellular localization.

Ubiquitin-mediated proteolysis In eukaryotes, the selective breakdown of proteins by a proteasome complex in response to the covalent addition of a small, 8.5 kDa regulatory protein, ubiquitin, by a group of ubiquitin-activating/conjugating/ligase enzymes.

LIST OF ABBREVIATIONS

6-4PP (6-4)Pyrimidine pyrimidone dimers
8-oxo-G 7,8-Dihydro-8-oxo-guanine
ATR ATM and Rad3 related
ATRIP TR-interacting protein
BER Base excision repair
CPDs Cyclobutane pyrimidine
CTD C-terminal domain
DSB Double-strand break
EGFR Epidermal growth factor receptor
FRET Förster resonance energy transfer
GHKL Gyrase b, Hsp90, histidine kinases and MutL homologs
HTH Helix-turn-helix
ICLs Interstrand cross-links
IDL Insertion/deletion loop
MBD Mismatch-binding domain of MutS
MGMT O^6-methylguanine-DNA methyltransferase
MIP MLH1-interacting protein
miR micro-RNA
MLH MutL homolog
MMR Mismatch repair
MNNG *N*-methyl-*N'*-nitro-*N*-nitrosoguanidine

MSH MutS homolog
MSI Microsatellite instability
NBD Nucleotide-binding domain
NTD N-terminal domain
***O*⁶me-G** *O*⁶-methylguanine
SAXS Small-angle X-ray scattering
TLS Translesion synthesis

ACKNOWLEDGMENTS

We thank Wei Yang for providing Figs. 18.2A and C, Lorena Beese for providing Fig. 18.2B, and Alba Guarné for the model in Fig. 18.3B and extend our apologies to our many colleagues whose work is not cited. The authors gratefully acknowledge the support of the Intramural Research Program of the National Institute of Diabetes and Digestive and Kidney Diseases.

REFERENCES

[1] Kunkel TA, Erie DA. Eukaryotic mismatch repair in relation to DNA replication. Annu Rev Genet 2015;49:291–313.
[2] Sachadyn P. Conservation and diversity of MutS proteins. Mutat Res 2010;694(1–2):20–30.
[3] Boland CR, Lynch HT. The history of Lynch syndrome. Fam Cancer 2013;12(2):145–57.
[4] Li GM. Mechanisms and functions of DNA mismatch repair. Cell Res 2008;18(1):85–98.
[5] Peltomaki P. Epigenetic mechanisms in the pathogenesis of Lynch syndrome. Clin Genet 2014;85(5):403–12.
[6] Strand M, Prolla TA, Liskay RM, Petes TD. Destabilization of tracts of simple repetitive DNA in yeast by mutations affecting DNA mismatch repair. Nature 1993;365(6443):274–6.
[7] Spies M, Fishel R. Mismatch repair during homologous and homeologous recombination. Cold Spring Harb Perspect Biol 2015;7(3).
[8] Iyer RR, Pluciennik A, Napierala M, Wells RD. DNA triplet repeat expansion and mismatch repair. Annu Rev Biochem 2015;84:199–226.
[9] Chahwan R, Edelmann W, Scharff MD, Roa S. AIDing antibody diversity by error-prone mismatch repair. Semin Immunol 2012;24(4):293–300.
[10] Pena-Diaz J, Jiricny J. Mammalian mismatch repair: error-free or error-prone? Trends Biochem Sci 2012;37(5):206–14.
[11] Jiricny J. Postreplicative mismatch repair. Cold Spring Harb Perspect Biol 2013;5(4):a012633.
[12] Reyes GX, Schmidt TT, Kolodner RD, Hombauer H. New insights into the mechanism of DNA mismatch repair. Chromosoma 2015;124(4).
[13] Edelmann L, Edelmann W. Loss of DNA mismatch repair function and cancer predisposition in the mouse: animal models for human hereditary nonpolyposis colorectal cancer. Am J Med Genet C Semin Med Genet 2004;129c(1):91–9.
[14] Lenhart JS, Pillon MC, Guarne A, Biteen JS, Simmons LA. Mismatch repair in Gram-positive bacteria. Res Microbiol 2015;167(1).
[15] Marinus MG. DNA mismatch repair. EcoSal Plus 2012;5(1).
[16] Harfe BD, Jinks-Robertson S. DNA mismatch repair and genetic instability. Annu Rev Genet 2000;34:359–99.
[17] Boiteux S, Jinks-Robertson S. DNA repair mechanisms and the bypass of DNA damage in *Saccharomyces cerevisiae*. Genetics 2013;193(4):1025–64.
[18] Dzantiev L, Constantin N, Genschel J, Iyer RR, Burgers PM, Modrich P. A defined human system that supports bidirectional mismatch-provoked excision. Mol Cell 2004;15(1):31–41.
[19] Zhang Y, Yuan F, Presnell SR, Tian K, Gao Y, Tomkinson AE, et al. Reconstitution of 5′-directed human mismatch repair in a purified system. Cell 2005;122(5):693–705.
[20] Constantin N, Dzantiev L, Kadyrov FA, Modrich P. Human mismatch repair: reconstitution of a nick-directed bidirectional reaction. J Biol Chem 2005;280(48):39752–61.
[21] Srivatsan A, Bowen N, Kolodner RD. Mispair-specific recruitment of the Mlh1-Pms1 complex identifies repair substrates of the *Saccharomyces cerevisiae* Msh2-Msh3 complex. J Biol Chem 2014;289(13):9352–64.
[22] Iyer RR, Pluciennik A, Burdett V, Modrich PL. DNA mismatch repair: functions and mechanisms. Chem Rev 2006;106(2):302–23.
[23] Tishkoff DX, Boerger AL, Bertrand P, Filosi N, Gaida GM, Kane MF, et al. Identification and characterization of *Saccharomyces cerevisiae* EXO1, a gene encoding an exonuclease that interacts with MSH2. Proc Natl Acad Sci USA 1997;94(14):7487–92.
[24] Gueneau E, Dherin C, Legrand P, Tellier-Lebegue C, Gilquin B, Bonnesoeur P, et al. Structure of the MutLα C-terminal domain reveals how Mlh1 contributes to Pms1 endonuclease site. Nat Struct Mol Biol 2013;20(4):461–8.
[25] Kadyrov FA, Dzantiev L, Constantin N, Modrich P. Endonucleolytic function of MutLα in human mismatch repair. Cell 2006;126(2):297–308.
[26] Obmolova G, Ban C, Hsieh P, Yang W. Crystal structures of mismatch repair protein MutS and its complex with a substrate DNA. Nature 2000;407(6805):703–10.
[27] Junop MS, Obmolova G, Rausch K, Hsieh P, Yang W. Composite active site of an ABC ATPase: MutS uses ATP to verify mismatch recognition and authorize DNA repair. Mol Cell 2001;7(1):1–12.
[28] Lamers MH, Perrakis A, Enzlin JH, Winterwerp HH, de Wind N, Sixma TK. The crystal structure of DNA mismatch repair protein MutS binding to a G×T mismatch. Nature 2000;407(6805):711–7.
[29] Hsieh P, Yamane K. DNA mismatch repair: molecular mechanism, cancer, and ageing. Mech Ageing Dev 2008;129(7–8):391–407.
[30] Mendillo ML, Putnam CD, Kolodner RD. *Escherichia coli* MutS tetramerization domain structure reveals that stable dimers but not tetramers are essential for DNA mismatch repair in vivo. J Biol Chem 2007;282(22):16345–54.

[31] Natrajan G, Lamers MH, Enzlin JH, Winterwerp HH, Perrakis A, Sixma TK. Structures of *Escherichia coli* DNA mismatch repair enzyme MutS in complex with different mismatches: a common recognition mode for diverse substrates. Nucleic Acids Res 2003;31(16):4814–21.

[32] Warren JJ, Pohlhaus TJ, Changela A, Iyer RR, Modrich PL, Beese LS. Structure of the human MutSα DNA lesion recognition complex. Mol Cell 2007;26(4):579–92.

[33] Shell SS, Putnam CD, Kolodner RD. The N terminus of *Saccharomyces cerevisiae* Msh6 is an unstructured tether to PCNA. Mol Cell 2007;26(4):565–78.

[34] Iyer RR, Pohlhaus TJ, Chen S, Hura GL, Dzantiev L, Beese LS, et al. The MutSα-proliferating cell nuclear antigen interaction in human DNA mismatch repair. J Biol Chem 2008;283(19):13310–9.

[35] Iyer RR, Pluciennik A, Genschel J, Tsai MS, Beese LS, Modrich P. MutLα and proliferating cell nuclear antigen share binding sites on MutSβ. J Biol Chem 2010;285(15):11730–9.

[36] Lee SD, Surtees JA, Alani E. *Saccharomyces cerevisiae* MSH2-MSH3 and MSH2-MSH6 complexes display distinct requirements for DNA binding domain I in mismatch recognition. J Mol Biol 2007;366(1):53–66.

[37] Shell SS, Putnam CD, Kolodner RD. Chimeric *Saccharomyces cerevisiae* Msh6 protein with an Msh3 mispair-binding domain combines properties of both proteins. Proc Natl Acad Sci USA 2007;104(26):10956–61.

[38] Dowen JM, Putnam CD, Kolodner RD. Functional studies and homology modeling of Msh2-Msh3 predict that mispair recognition involves DNA bending and strand separation. Mol Cell Biol 2010;30(13):3321–8.

[39] Gupta S, Gellert M, Yang W. Mechanism of mismatch recognition revealed by human MutSbeta bound to unpaired DNA loops. Nat Struct Mol Biol 2012;19(1):72–8.

[40] Erie DA, Weninger KR. Single molecule studies of DNA mismatch repair. DNA Repair 2014;20:71–81.

[41] Wang H, Yang Y, Schofield MJ, Du C, Fridman Y, Lee SD, et al. DNA bending and unbending by MutS govern mismatch recognition and specificity. Proc Natl Acad Sci USA 2003;100(25):14822–7.

[42] Mazurek A, Johnson CN, Germann MW, Fishel R. Sequence context effect for hMSH2-hMSH6 mismatch-dependent activation. Proc Natl Acad Sci USA 2009;106(11):4177–82.

[43] DeRocco VC, Sass LE, Qiu R, Weninger KR, Erie DA. Dynamics of MutS-mismatched DNA complexes are predictive of their repair phenotypes. Biochemistry 2014;53(12):2043–52.

[44] Fishel R. Mismatch repair. J Biol Chem 2015;290(44).

[45] Biswas I, Ban C, Fleming KG, Qin J, Lary JW, Yphantis DA, et al. Oligomerization of a MutS mismatch repair protein from *Thermus aquaticus*. J Biol Chem 1999;274(33):23673–8.

[46] Lamers MH, Georgijevic D, Lebbink JH, Winterwerp HH, Agianian B, de Wind N, et al. ATP increases the affinity between MutS ATPase domains. Implications for ATP hydrolysis and conformational changes. J Biol Chem 2004;279(42):43879–85.

[47] Antony E, Khubchandani S, Chen S, Hingorani MM. Contribution of Msh2 and Msh6 subunits to the asymmetric ATPase and DNA mismatch binding activities of *Saccharomyces cerevisiae* Msh2-Msh6 mismatch repair protein. DNA Repair 2006;5(2):153–62.

[48] Martik D, Baitinger C, Modrich P. Differential specificities and simultaneous occupancy of human MutSα nucleotide binding sites. J Biol Chem 2004;279(27):28402–10.

[49] Mazur DJ, Mendillo ML, Kolodner RD. Inhibition of Msh6 ATPase activity by mispaired DNA induces a Msh2(ATP)-Msh6(ATP) state capable of hydrolysis-independent movement along DNA. Mol Cell 2006;22(1):39–49.

[50] Hargreaves VV, Shell SS, Mazur DJ, Hess MT, Kolodner RD. Interaction between the Msh2 and Msh6 nucleotide-binding sites in the *Saccharomyces cerevisiae* Msh2-Msh6 complex. J Biol Chem 2010;285(12):9301–10.

[51] Zhai J, Hingorani MM. *Saccharomyces cerevisiae* Msh2-Msh6 DNA binding kinetics reveal a mechanism of targeting sites for DNA mismatch repair. Proc Natl Acad Sci USA 2010;107(2):680–5.

[52] Heinen CD, Cyr JL, Cook C, Punja N, Sakato M, Forties RA, et al. Human MSH2 (hMSH2) protein controls ATP processing by hMSH2-hMSH6. J Biol Chem 2011;286(46):40287–95.

[53] Monti MC, Cohen SX, Fish A, Winterwerp HH, Barendregt A, Friedhoff P, et al. Native mass spectrometry provides direct evidence for DNA mismatch-induced regulation of asymmetric nucleotide binding in mismatch repair protein MutS. Nucleic Acids Res 2011;39(18):8052–64.

[54] Guarne A. The functions of MutL in mismatch repair: the power of multitasking. Prog Mol Biol Transl Sci 2012;110:41–70.

[55] Guarne A, Charbonnier JB. Insights from a decade of biophysical studies on MutL: roles in strand discrimination and mismatch removal. Prog Biophys Mol Biol 2015;117(2–3).

[56] Ban C, Yang W. Crystal structure and ATPase activity of MutL: implications for DNA repair and mutagenesis. Cell 1998;95(4):541–52.

[57] Campbell CS, Hombauer H, Srivatsan A, Bowen N, Gries K, Desai A, et al. Mlh2 is an accessory factor for DNA mismatch repair in *Saccharomyces cerevisiae*. PLoS Genet 2014;10(5):e1004327.

[58] Chen PC, Dudley S, Hagen W, Dizon D, Paxton L, Reichow D, et al. Contributions by MutL homologues Mlh3 and Pms2 to DNA mismatch repair and tumor suppression in the mouse. Cancer Res 2005;65(19):8662–70.

[59] Guarne A, Junop MS, Yang W. Structure and function of the N-terminal 40 kDa fragment of human PMS2: a monomeric GHL ATPase. EMBO J 2001;20(19):5521–31.

[60] Ban C, Junop M, Yang W. Transformation of MutL by ATP binding and hydrolysis: a switch in DNA mismatch repair. Cell 1999;97(1):85–97.

[61] Arana ME, Holmes SF, Fortune JM, Moon AF, Pedersen LC, Kunkel TA. Functional residues on the surface of the N-terminal domain of yeast Pms1. DNA Repair 2010;9(4):448–57.

[62] Sacho EJ, Kadyrov FA, Modrich P, Kunkel TA, Erie DA. Direct visualization of asymmetric adenine-nucleotide-induced conformational changes in MutLα. Mol Cell 2008;29(1):112–21.

[63] Rogacheva MV, Manhart CM, Chen C, Guarne A, Surtees J, Alani E. Mlh1-Mlh3, a meiotic crossover and DNA mismatch repair factor, is a Msh2-Msh3-stimulated endonuclease. J Biol Chem 2014;289(9):5664–73.

[64] Smith CE, Mendillo ML, Bowen N, Hombauer H, Campbell CS, Desai A, et al. Dominant mutations in *S. cerevisiae* PMS1 identify the Mlh1-Pms1 endonuclease active site and an exonuclease 1-independent mismatch repair pathway. PLoS Genet 2013;9(10):e1003869.

[65] Erdeniz N, Nguyen M, Deschenes SM, Liskay RM. Mutations affecting a putative MutLα endonuclease motif impact multiple mismatch repair functions. DNA Repair 2007;6(10):1463–70.

[66] Schmutte C, Sadoff MM, Shim KS, Acharya S, Fishel R. The interaction of DNA mismatch repair proteins with human exonuclease I. J Biol Chem 2001;276(35):33011–8.

[67] Guarne A, Ramon-Maiques S, Wolff EM, Ghirlando R, Hu X, Miller JH, et al. Structure of the MutL C-terminal domain: a model of intact MutL and its roles in mismatch repair. EMBO J 2004;23(21):4134–45.

[68] Pillon MC, Lorenowicz JJ, Uckelmann M, Klocko AD, Mitchell RR, Chung YS, et al. Structure of the endonuclease domain of MutL: unlicensed to cut. Mol Cell 2010;39(1):145–51.

[69] Mendillo ML, Hargreaves VV, Jamison JW, Mo AO, Li S, Putnam CD, et al. A conserved MutS homolog connector domain interface interacts with MutL homologs. Proc Natl Acad Sci USA 2009;106(52):22223–8.

[70] Winkler I, Marx AD, Lariviere D, Manelyte L, Giron-Monzon L, Heinze RJ, et al. Chemical trapping of the dynamic MutS-MutL complex formed in DNA mismatch repair in *Escherichia coli*. J Biol Chem 2011;286(19):17326–37.

[71] Groothuizen FS, Winkler I, Cristovao M, Fish A, Winterwerp HH, Reumer A, et al. MutS/MutL crystal structure reveals that the MutS sliding clamp loads MutL onto DNA. eLife 2015;4:e06744.

[72] Mendillo ML, Putnam CD, Mo AO, Jamison JW, Li S, Woods Jr VL, et al. Probing DNA- and ATP-mediated conformational changes in the MutS family of mispair recognition proteins using deuterium exchange mass spectrometry. J Biol Chem 2010;285(17):13170–82.

[73] Lenhart JS, Pillon MC, Guarne A, Simmons LA. Trapping and visualizing intermediate steps in the mismatch repair pathway in vivo. Mol Microbiol 2013;90(4):680–98.

[74] Gradia S, Acharya S, Fishel R. The human mismatch recognition complex hMSH2-hMSH6 functions as a novel molecular switch. Cell 1997;91(7):995–1005.

[75] Gradia S, Subramanian D, Wilson T, Acharya S, Makhov A, Griffith J, et al. hMSH2-hMSH6 forms a hydrolysis-independent sliding clamp on mismatched DNA. Mol Cell 1999;3(2):255–61.

[76] Wang H, Hays JB. Mismatch repair in human nuclear extracts: effects of internal DNA-hairpin structures between mismatches and excision-initiation nicks on mismatch correction and mismatch-provoked excision. J Biol Chem 2003;278(31):28686–93.

[77] Pluciennik A, Modrich P. Protein roadblocks and helix discontinuities are barriers to the initiation of mismatch repair. Proc Natl Acad Sci USA 2007;104(31):12709–13.

[78] Gorman J, Chowdhury A, Surtees JA, Shimada J, Reichman DR, Alani E, et al. Dynamic basis for one-dimensional DNA scanning by the mismatch repair complex Msh2-Msh6. Mol Cell 2007;28(3):359–70.

[79] Cho WK, Jeong C, Kim D, Chang M, Song KM, Hanne J, et al. ATP alters the diffusion mechanics of MutS on mismatched DNA. Structure 2012;20(7):1264–74.

[80] Qiu R, Sakato M, Sacho EJ, Wilkins H, Zhang X, Modrich P, et al. MutL traps MutS at a DNA mismatch. Proc Natl Acad Sci USA 2015;112(35):10914–9.

[81] Lee JB, Cho WK, Park J, Jeon Y, Kim D, Lee SH, et al. Single-molecule views of MutS on mismatched DNA. DNA Repair 2014;20:82–93.

[82] Gorman J, Wang F, Redding S, Plys AJ, Fazio T, Wind S, et al. Single-molecule imaging reveals target-search mechanisms during DNA mismatch repair. Proc Natl Acad Sci USA 2012;109(45):E3074–83.

[83] Elez M, Radman M, Matic I. Stoichiometry of MutS and MutL at unrepaired mismatches in vivo suggests a mechanism of repair. Nucleic Acids Res 2012;40(9):3929–38.

[84] Hombauer H, Campbell CS, Smith CE, Desai A, Kolodner RD. Visualization of eukaryotic DNA mismatch repair reveals distinct recognition and repair intermediates. Cell 2011;147(5):1040–53.

[85] Kadyrov FA, Holmes SF, Arana ME, Lukianova OA, O'Donnell M, Kunkel TA, et al. *Saccharomyces cerevisiae* MutLα is a mismatch repair endonuclease. J Biol Chem 2007;282(51):37181–90.

[86] Goellner EM, Putnam CD, Kolodner RD. Exonuclease 1-dependent and independent mismatch repair. DNA Repair 2015;32.

[87] Johnson A, O'Donnell M. Cellular DNA replicases: components and dynamics at the replication fork. Annu Rev Biochem 2005;74:283–315.

[88] Pluciennik A, Dzantiev L, Iyer RR, Constantin N, Kadyrov FA, Modrich P. PCNA function in the activation and strand direction of MutLα endonuclease in mismatch repair. Proc Natl Acad Sci USA 2010;107(37):16066–71.

[89] Umar A, Buermeyer AB, Simon JA, Thomas DC, Clark AB, Liskay RM, et al. Requirement for PCNA in DNA mismatch repair at a step preceding DNA resynthesis. Cell 1996;87(1):65–73.

[90] Pluciennik A, Burdett V, Baitinger C, Iyer RR, Shi K, Modrich P. Extrahelical (CAG)/(CTG) triplet repeat elements support proliferating cell nuclear antigen loading and MutLα endonuclease activation. Proc Natl Acad Sci USA 2013;110(30):12277–82.

[91] Goellner EM, Smith CE, Campbell CS, Hombauer H, Desai A, Putnam CD, et al. PCNA and Msh2-Msh6 activate an Mlh1-Pms1 endonuclease pathway required for Exo1-independent mismatch repair. Mol Cell 2014;55(2):291–304.

[92] Kadyrov FA, Genschel J, Fang Y, Penland E, Edelmann W, Modrich P. A possible mechanism for exonuclease 1-independent eukaryotic mismatch repair. Proc Natl Acad Sci USA 2009;106(21):8495–500.

[93] Fu D, Calvo JA, Samson LD. Balancing repair and tolerance of DNA damage caused by alkylating agents. Nat Rev Cancer 2012;12(2):104–20.

[94] Karran P. Mechanisms of tolerance to DNA damaging therapeutic drugs. Carcinogenesis 2001;22(12):1931–7.

[95] Jiricny J. The multifaceted mismatch-repair system. Nat Rev Mol Cell Biol 2006;7(5):335–46.

[96] Zeman MK, Cimprich KA. Causes and consequences of replication stress. Nat Cell Biol 2014;16(1):2–9.

[97] Cortez D, Guntuku S, Qin J, Elledge SJ. ATR and ATRIP: partners in checkpoint signaling. Science 2001;294(5547):1713–6.

[98] Mojas N, Lopes M, Jiricny J. Mismatch repair-dependent processing of methylation damage gives rise to persistent single-stranded gaps in newly replicated DNA. Genes & Dev 2007;21(24):3342–55.

[99] Izumchenko E, Saydi J, Brown KD. Exonuclease 1 (Exo1) is required for activating response to S(N)1 DNA methylating agents. DNA Repair 2012;11(12):951–64.

[100] Li Z, Pearlman AH, Hsieh P. Mismatch repair and the DNA damage response. DNA Repair 2016;38:94–101.

[101] Wang Y, Qin J. MSH2 and ATR form a signaling module and regulate two branches of the damage response to DNA methylation. Proc Natl Acad Sci USA 2003;100(26):15387–92.

[102] Liu Y, Fang Y, Shao H, Lindsey-Boltz L, Sancar A, Modrich P. Interactions of human mismatch repair proteins MutSα and MutLα with proteins of the ATR-Chk1 pathway. J Biol Chem 2010;285(8):5974–82.

[103] Yoshioka K, Yoshioka Y, Hsieh P. ATR kinase activation mediated by MutSα and MutLα in response to cytotoxic O^6-methylguanine adducts. Mol Cell 2006;22(4):501–10.

[104] Li LS, Morales JC, Veigl M, Sedwick D, Greer S, Meyers M, et al. DNA mismatch repair (MMR)-dependent 5-fluorouracil cytotoxicity and the potential for new therapeutic targets. Br J Pharmacol 2009;158(3):679–92.

[105] Meyers M, Wagner MW, Mazurek A, Schmutte C, Fishel R, Boothman DA. DNA mismatch repair-dependent response to fluoropyrimidine-generated damage. J Biol Chem 2005;280(7):5516–26.

[106] Liu A, Yoshioka K, Salerno V, Hsieh P. The mismatch repair-mediated cell cycle checkpoint response to fluorodeoxyuridine. J Cell Biochem 2008;105(1):245–54.

[107] SenGupta T, Torgersen ML, Kassahun H, Vellai T, Simonsen A, Nilsen H. Base excision repair AP endonucleases and mismatch repair act together to induce checkpoint-mediated autophagy. Nat Commun 2013;4:2674.

[108] Zlatanou A, Despras E, Braz-Petta T, Boubakour-Azzouz I, Pouvelle C, Stewart GS, et al. The hMsh2-hMsh6 complex acts in concert with monoubiquitinated PCNA and Pol η in response to oxidative DNA damage in human cells. Mol Cell 2011;43(4):649–62.

[109] Chen J, Miller BF, Furano AV. Repair of naturally occurring mismatches can induce mutations in flanking DNA. eLife 2014;3:e02001.

[110] Tsaalbi-Shtylik A, Ferras C, Pauw B, Hendriks G, Temviriyanukul P, Carlee L, et al. Excision of translesion synthesis errors orchestrates responses to helix-distorting DNA lesions. J Cell Biol 2015;209(1):33–46.

[111] Peng M, Xie J, Ucher A, Stavnezer J, Cantor SB. Crosstalk between BRCA-Fanconi anemia and mismatch repair pathways prevents MSH2-dependent aberrant DNA damage responses. EMBO J 2014;33(15):1698–712.

[112] Cantor SB, Brosh Jr RM. What is wrong with Fanconi anemia cells? Cell Cycle 2014;13(24):3823–7.

[113] Shimodaira H, Yoshioka-Yamashita A, Kolodner RD, Wang JY. Interaction of mismatch repair protein PMS2 and the p53-related transcription factor p73 in apoptosis response to cisplatin. Proc Natl Acad Sci USA 2003;100(5):2420–5.

[114] Hombauer H, Srivatsan A, Putnam CD, Kolodner RD. Mismatch repair, but not heteroduplex rejection, is temporally coupled to DNA replication. Science 2011;334(6063):1713–6.

[115] Simmons LA, Davies BW, Grossman AD, Walker GC. β clamp directs localization of mismatch repair in *Bacillus subtilis*. Mol Cell 2008;29(3): 291–301.

[116] Kleczkowska HE, Marra G, Lettieri T, Jiricny J. hMSH3 and hMSH6 interact with PCNA and colocalize with it to replication foci. Genes & Dev 2001;15(6):724–36.

[117] Li F, Mao G, Tong D, Huang J, Gu L, Yang W, et al. The histone mark H3K36me3 regulates human DNA mismatch repair through its interaction with MutSα. Cell 2013;153(3):590–600.

[118] Kadyrova LY, Mertz TM, Zhang Y, Northam MR, Sheng Z, Lobachev KS, et al. A reversible histone H3 acetylation cooperates with mismatch repair and replicative polymerases in maintaining genome stability. PLoS Genet 2013;9(10):e1003899.

[119] Li GM. New insights and challenges in mismatch repair: getting over the chromatin hurdle. DNA Repair 2014;19:48–54.

[120] Kadyrova LY, Blanko ER, Kadyrov FA. CAF-I-dependent control of degradation of the discontinuous strands during mismatch repair. Proc Natl Acad Sci USA 2011;108(7):2753–8.

[121] Schopf B, Bregenhorn S, Quivy JP, Kadyrov FA, Almouzni G, Jiricny J. Interplay between mismatch repair and chromatin assembly. Proc Natl Acad Sci USA 2012;109(6):1895–900.

[122] Zhang M, Xiang S, Joo HY, Wang L, Williams KA, Liu W, et al. HDAC6 deacetylates and ubiquitinates MSH2 to maintain proper levels of MutSα. Mol Cell 2014;55(1):31–46.

[123] Diouf B, Cheng Q, Krynetskaia NF, Yang W, Cheok M, Pei D, et al. Somatic deletions of genes regulating MSH2 protein stability cause DNA mismatch repair deficiency and drug resistance in human leukemia cells. Nat Med 2011;17(10):1298–303.

[124] Ortega J, Li JY, Lee S, Tong D, Gu L, Li G-M. Phosphorylation of PCNA by EGFR inhibits mismatch repair and promotes misincorporation during DNA synthesis. Proc Natl Acad Sci USA 2015;112(18):5667–72.

[125] Mao G, Lee S, Ortega J, Gu L, Li GM. Modulation of microRNA processing by mismatch repair protein MutLα. Cell Res 2012;22(6):973–85.

Chapter 19

Repair of Double-Strand Breaks by Nonhomologous End Joining: Its Components and Their Function

P. Moskwa
Medical University of Greifswald, Greifswald, Mecklenburg-Vorpommern, Germany

Chapter Outline

1. INTRODUCTION

The genomic DNA is exposed to a continuous endogenous and exogenous damage through reactive oxygen species (ROS), chemicals, viral infections, replication errors, and ionizing radiation. The large variety of DNA-damaging agents causes a large variety of DNA damage, and thus, necessitates distinct and specialized DNA-repair pathways to guard genomic stability. This chapter will exclusively focus on double-strand breaks (DSBs) and their repair by nonhomologous end joining (NHEJ).

DSBs represent the most dangerous DNA damage and a single unrepaired break is sufficient to induce cell death. Nonetheless, hematopoietic cells undergo programmed DSBs during V(D)J and class switch recombination (CSR) to insure infinite variability of antibodies and T-cell receptors (TCRs) matching each invading pathogenic microorganism. Regardless whether DSBs are of toxic or programmed origin, inability to repair these breaks may cause either cell death or chromosomal rearrangements, which may lead to malignant transformation, often with deleterious consequences. Therefore, to minimize the chance of this catastrophic event, an immediate and faithful repair of DNA DSBs is demanded.

Dividing cells unpack and replicate genomic DNA making it particularly susceptible to previously mentioned damage. This poses a high risk of chromosomal aberration that could be passed to descending cells. Thus, to meet the cell cycle-specific requirements and preserve genomic stability, evolutionary two distinct DNA DSB-repair pathways have evolved. Homologous recombination (HR), a high-fidelity mechanism, which functions predominantly in the S and G2 phase of the cell cycle and utilizes the sister chromatid as a template to restore identical copy of the damaged DNA. In contrast, NHEJ is mainly prevalent in the G0 and G1 phase of the cell cycle and joins broken DNA ends without template. Recently, there are growing evidences that NHEJ consists of two pathways, the classical NHEJ (C-NHEJ) and the alternative NHEJ (A-NHEJ). The latter exhibits high mutagenic propensity, whereas the former is characterized by the ability to join DNA ends directly or with minimal processing.

Genome Stability. http://dx.doi.org/10.1016/B978-0-12-803309-8.00019-7

Here, the current knowledge of NHEJ will be summarized. It will be distinguished between the classical and alternative NHEJ pathway. Components of each pathway will be briefly introduced emphasizing their most important characteristics, such as structure, function, binding partners, and possible regulation. Common and distinct phenotypic changes in case of mutation or loss of NHEJ components will be pointed out. The role of C-NHEJ as an integral part of V(D)J and class switch recombination will be discussed.

2. CLASSICAL NHEJ

The C-NHEJ pathway of DSB repair is the major DSB-repair pathway in mammalian cells and has been already extensively studied for many years. The initial studies came from the laboratory of F. Alt, which described two nonlymphoid-specific genes involved in DNA repair, later identified as Ku80 und Xrcc4. Subsequently, over many years altogether eight components of C-NHEJ have been identified; four highly conserved Ku70, Ku80, Xrcc4, and Lig4 and three accessory units Artemis, XLF, and DNA-Pkcs (Fig. 19.1). In addition, PAXX protein, a paralog of XLF and Xrcc4, was reported in 2015 as potential component of C-NHEJ [1]. Defects in C-NHEJ lead to some characteristic phenotypic changes, such as radiosensitive severe combined immunodeficiency (RS-SCID), premature aging, microcephaly and growth retardation [2,3].

In undamaged cells, components of C-NHEJ are disassembled and a single DSB is sufficient to activate the assembly of a functional C-NHEJ complex at the site of damage to seal the break. The repair of a DSB is extremely complex and progresses through four successive phases: sensing and tethering, end processing, ligation of broken DNA ends, and finally dissociation of the C-NHEJ complex from resolved break (Fig. 19.1). In contrast, components mediating these steps functionally overlap these phases. For instance, the Ku70/80 heterodimer is not only able to sense and tether breaks, but it also removes abasic nucleotides from broken ends.

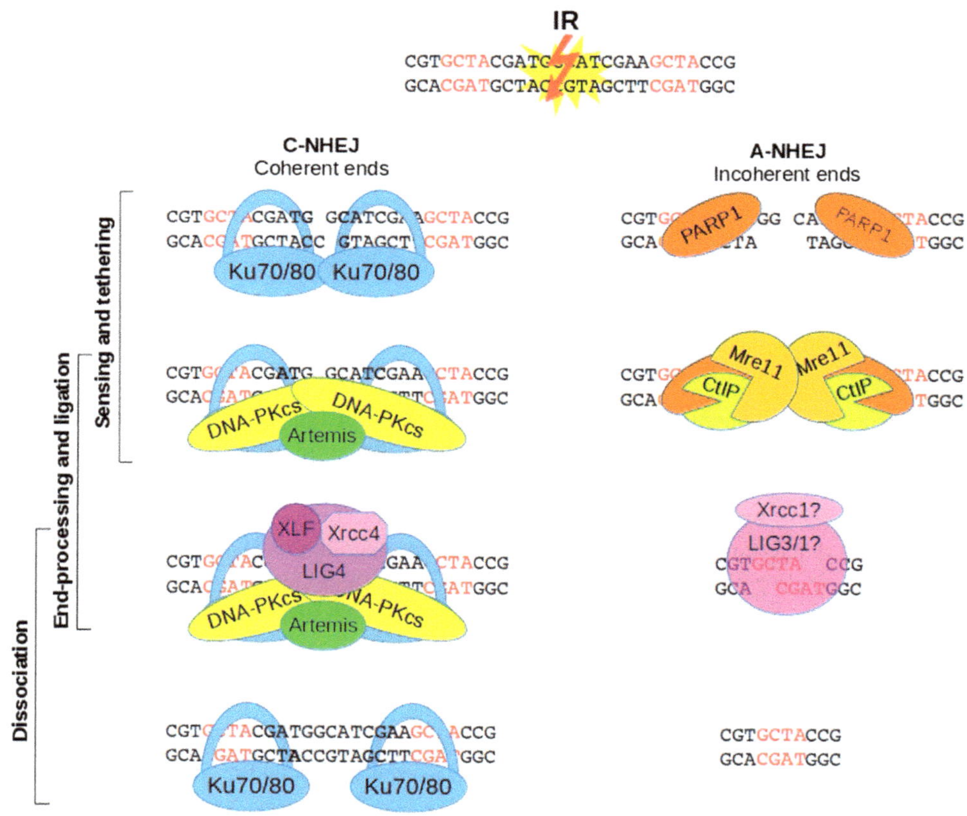

FIGURE 19.1 **Cartoon representing NHEJ-mediated repair.** Repair of coherent ends (on the left) is mediated by the classical NHEJ, whereas alternative NHEJ (on the right) mediates repair of non-coherent breaks. C-NHEJ: Ku70/80 heterodimers form a basket-like structure and instantly enclose broken DNA ends. Concurrently, Ku70/80 complex induces sequential assembly of functional C-NHEJ complex that ligates the breaks. Upon resolution of the break Ku70/80 complex remains trapped on linear DNA. The escape mechanism is not clear, degradation of Ku70 and Ku80 proteins is considered. A-NHEJ: PARP1 senses the DSB and possibly recruits Mre11 and CtIP nucleases to carry out limited resection (see also Fig. 19.4). This resection proceeds until microhomologies (highlighted with red) are found. The complementary overhangs hybridize and LIG3/1 seals the break. The phases are presented as overlapping to emphasize the multifaceted role of each component.

The first phase is initiated by Ku70/80 complex, which binds with high affinity to broken DNA ends and tethers them together. To execute this role, Ku70 and Ku80 form a ring structure suitable to accommodate DNA ends. Ku70/80 heterodimers associate with the tip of broken ends and anchor each other to hold opposite DNA ends in close vicinity. Concomitantly, Ku70/Ku80 complex serves as a docking platform for DNA-PKcs and other proteins involved in many aspects of DNA-damage response. Ku70/80 together with DNA-PKcs form a functional complex designated as DNA-PK [4].

Interaction of DNA-PKcs with Ku70/80 complex causes the latter to slide inwards on the DNA ends. The DNA-PKcs moves to the center of the synapsis and takes over the tethering of the DNA ends as well as stabilizes the entire DNA-PK/DNA synapsis. Reciprocally, this interaction induces conformational change of DNA-PKcs that activates the kinase domain. The catalytic activation of DNA-PKcs results in its autophosphorylation as well as phosphorylation of large array of proteins, among other C-NHEJ components [5,6].

Frequently, ends of DSBs are ragged and incompatible to undergo direct ligation. Thus, in the second phase, DNA ends are exposed to processing through specialized enzymes. To remove damaged or excessive nucleotides, nucleases trim DNA ends, whereas polymerases can fill in gaps, which arise when only partially complementary ends are annealed. Artemis is one of such nucleases and is recruited to DSB concurrently with DNA-PKcs. This enzyme cuts single-stranded DNA overhangs and opens hairpin structures. Polymerases lambda and mu (Pol λ and Pol μ, respectively) have been implicated in the gap filling during C-NHEJ-mediated repair, possibly they contribute to end retention [7,8].

The third phase is carried out by LIG4 complex consisting of DNA-ligase 4 (Lig4), Xrcc4, and XLF. This complex is recruited to the break by interaction with DNA-PK and restores the integrity of DNA molecule by covalently sealing two DNA ends. Once the break is resolved, in the fourth phase, C-NHEJ complex must dissociate from the DNA, a process that has been poorly studied. Possibly some of the components simply dissociate upon secondary modification, such as phosphorylation and may be recycled, whereas some others undergo ubiquitin-mediated degradation [7].

2.1 Components of Classical NHEJ

2.1.1 DNA-PK (DNA-Protein Kinase Catalytic Subunit/Ku70/Ku80) Complex

2.1.1.1 Ku70/80 Heterodimer

Ku70 and Ku80 make up the Ku70/80 heterodimer. They are encoded by the Xrcc5 and Xrcc6 genes, respectively, and are highly abundant in both prokaryotes and eukaryotes. The stability of these proteins depends on each other. Mice deficient in Ku70 show severely reduced expression of Ku80 and vice versa [9,10]. In eukaryotic cells, both Ku proteins consist of three domains (Fig. 19.2): an N-terminal von Willebrand A domain (vWA), a central DNA-binding domain, and a diverged C-terminal domain (CTD). Role of the vWA domain is poorly characterized but based on its homology to other proteins, it is thought to function as a protein–protein interaction site. The Ku80 CTD is only present in higher eukaryotes and is well established to interact with the DNA-PKcs, proximal to its kinase domain [5,11]. In lower eukaryotes lacking DNA-PKcs this domain is not present. The Ku70 CTD contains a distal SAP domain (SAF-A/B, Acinus, and PIAS motifs), which seems to increase binding to dsDNA and may also interact with other proteins [12,13].

Crystallographic structure of human Ku70/80 heterodimers has revealed that both proteins interact with each other through the central DNA-binding domain. These domains intertwine with each other to form an asymmetric ring structure, which avidly binds linear dsDNA as well as with lower-affinity hairpin ends in a sequence-independent manner. This is attributed to the inner structure of the ring that is lined with positively charged amino acid residues. These residues exclusively interact with dsDNA sugar–phosphate backbone and are able to accommodate about 14 base pair [14].

Simplified, Ku70/80 complex is a sensor of DSBs that initiates the assembly of C-NHEJ complex to restore integrity of a DNA molecule. In fact, the role of Ku70/80 proteins is highly complex and integrates different aspects of DNA repair. Besides sensing, Ku70/80 complex possibly tethers broken DNA ends keeping them in close vicinity for end-processing and subsequent ligation. Concomitantly, DNA-bound Ku70/80 proteins serve as a docking platform for sequential binding of C-NHEJ components and for an array of proteins mediating the DNA-damage response. Further, Ku70/80 have been reported to coordinate DSB repair with cell cycle arrest, pathway choice, and, if needed, apoptosis [4]. Interestingly, Roberts et al. published that Ku70 possesses 5′-dRP/AP lyase activity and is possibly involved in end-processing by removing 5′-apurinic and 5′-apyrimidinic (AP) sites [15].

Broken DNA ends are threaded through the eyelet-like structure formed by the Ku70/80 dimer and once the break is resolved, Ku70/80 remains trapped on the linear dsDNA molecule (Fig. 19.1). The escape mechanism of Ku70/80 complex from the DNA thread has been poorly studied and is not clear [4,7]. Based on the crystallographic structure, a conformational change leading to an opening of the eyelet is not possible [14]. However, there are some evidences for alternative

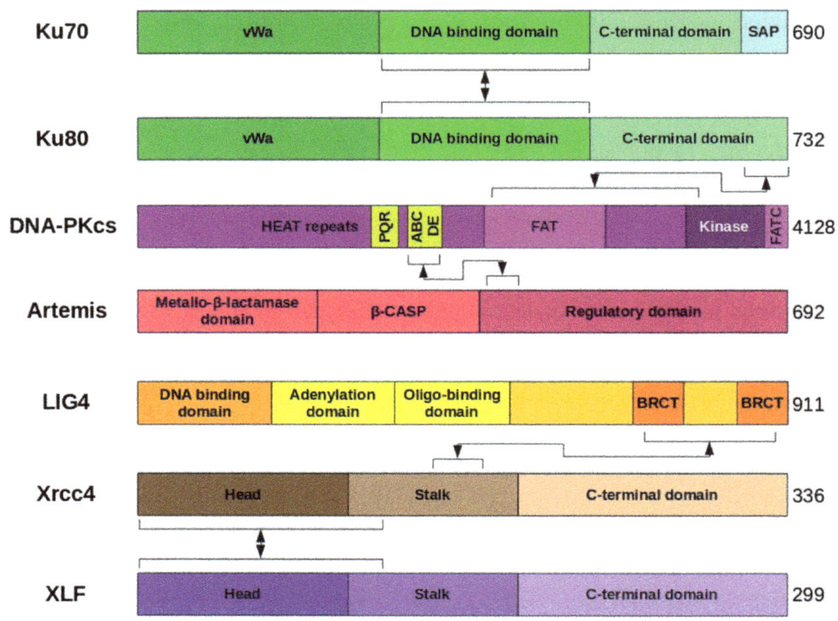

FIGURE 19.2 **Cartoon representing C-NHEJ components.** Indicated are main domains and possible interaction sites. *Summarized information is adopted from [4–6,24,36,38,45,55].*

mechanisms, such as DNA nicking or protein degradation. DNA nicking has been reported in yeast, but there are so far no studies supporting this mechanism in human. In contrast, RNF8 (RING finder protein 8) has been shown to ubiquitinate human Ku80 and to induce thereby its degradation. Depletion of RNF8 resulted in prolonged retention of Ku80 at the site of DNA damage and as a consequence NHEJ was impaired [4,16].

Mice deficient in Ku70 or Ku80 are viable, fertile, and show a strongly similar phenotype, such as growth retardation and SCID due to inability to rejoin V, D, and J segments. Surprisingly, in Ku70-deficient animals there is a residual development of T cells but it is not clear how these cells circumvent V(D)J recombination, a step that is absolutely indispensable for B- and T-cell development. Ku80 deficiency has no predisposition for tumorigenesis but in contrast, Ku70-deficient animals are significantly prone to develop thymic lymphomas. MEF cells isolated from Ku70- or Ku80-deficient animals exhibit radiosensitivity, intact DNA-damage checkpoints, and commit to premature senescence likely due to accumulation of DNA damage [9,10,17]. In humans, neither deficiency of Ku70 nor of Ku80 has been reported so far, most likely because deficiency is lethal. This speculation is further supported by the work of Wang et al., who demonstrated in human cells that Ku80 represses lethal telomere deletion [18]. The role of Ku70/80 in telomere maintenance is further supported in mice studies; however, these studies have provided conflicting results and the functional role in this process remains still to be determined [19,20].

2.1.1.2 DNA-PKcs (DNA-Dependent Protein Kinase Catalytic Subunit)

DNA-PKcs belongs together with ATM (ataxia telangiectasia related) and ATR (ataxia telangiectasia and Rad3 related) to the family of PIKK (phophatidyl inositol-3 kinase-like protein kinase) protein kinases, which mediate DNA-damage response and share overall a common structure. DNA-PKcs is a huge protein composed of 4128 amino acids corresponding to a molecular weight of about 465 kDa. The N-terminal domain is predicted to consist of HEAT repeats [Huntingtin, elongation factor 3 (EF3), protein phosphatase 2A (PP2A), yeast kinase TOR1] and encompasses two-third of the entire protein (Fig. 19.2). Within this region are located major phosphorylation sites: PQR and ABCDE cluster. The C-terminus contains the catalytic kinase domain and two FAT domains (FRAP, ATM, TRAP) [5].

DNA-PKcs is the catalytic unit of the heterotrimer DNA-PK consisting of Ku70, Ku80, and DNA-Pkcs. Assembly of DNA-PK requires the binding of Ku70/80 complex to linear dsDNA ends, otherwise DNA-PKcs stays dissociated [21]. In this complex DNA-PKcs fulfills multiple tasks. The interaction between the DNA-PKcs and the Ku70/80 complex causes the latter to slide inward on each broken dsDNA end bringing the DNA-PKcs in a central position able to tether the DSB [22,23]. Reciprocally, the interaction with Ku70/80 complex induces conformational change of DNA-PKcs resulting in its catalytic activation and enables recruitment of subsequent C-NHEJ components [5,24].

Acquiring of catalytic activity by DNA-PKcs is of particular importance in C-NHEJ, though the exact role of target phosphorylation is poorly understood. Deficiency of kinase activity causes radiosensitivity and inability to rejoin V, D,

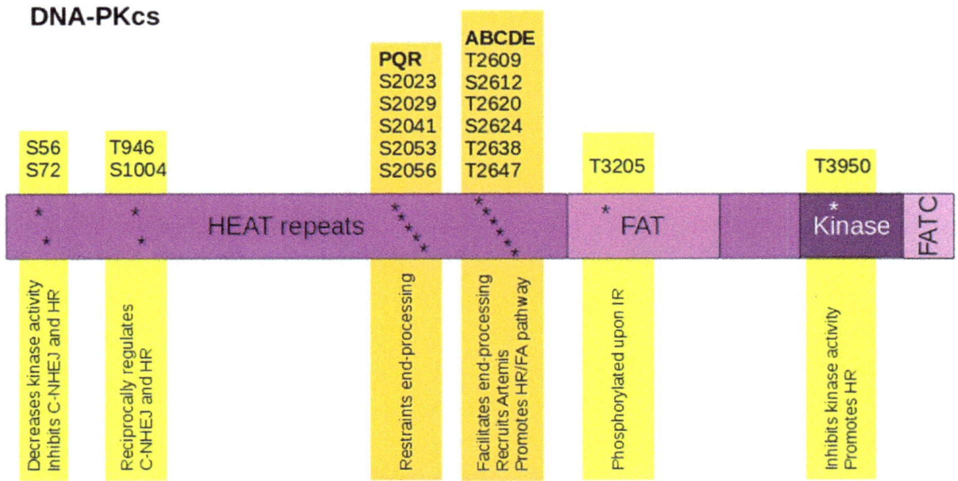

FIGURE 19.3 **Cartoon representing DNA-PKcs.** Highlighted are domains of DNA-PKcs and some phosphorylation sites. *Summarized information is adopted from reviews [4–6,24].*

and J segments [25]. However, phosphorylation of C-NHEJ components, such as Ku70, Ku80, Artemis, Xrcc4, and XLF seems to be redundant and does not affect C-NHEJ [24]. In contrast, phosphorylation of DNA-PKcs by the catalytic subunit itself and/or other kinases, such as ATM has been implicated in highly complex processes including regulation of end-processing, DNA-repair pathway choice, C-NHEJ complex dissociation, and auto-inactivation of kinase activity [5,7,24].

Altogether more than 40 sites of DNA-PKcs have been identified to undergo phosphorylation. Mutational analysis of both PQR and ABCDE cluster revealed that phosphorylation of these clusters do not affect the kinase activity but may reciprocally regulate end processing of dsDNA ends. The latter cluster seems to facilitate end processing and may promote a switch to the HR/FA (Fanconi) DNA-repair pathway, whereas PQC cluster is likely to counteract this process [24,26,27]. In contrast, phosphorylation of N-terminal cluster at serine 56 and 57 impacts kinase activity and blocks both C-NHEJ and HR. Phosphorylation of threonine 946 and serine 1004 does not affect kinase activity but it promotes HR. In line, phosphorylation of C-terminal threonine 3950 promotes HR and it also inhibits kinase activity. Threonine 3205 is phosphorylated in response to IR. Possibly differential phosphorylation of DNA-PKcs may be involved in fine-tuning of pathway choice (Fig. 19.3) [28,29].

DNA-PKcs knockout mice are viable but severely immunocompromised due to inability to carry out rejoining of V, D, and J segments and develop thymic lymphoblastic lymphomas. In contrast, mice expressing DNA-PKcs mutated in ABCDE cluster die prematurely from severe congenital bone marrow failure. This particularly severe phenotype is possibly attributed to the block of HR/FA (Fanconi) DNA-repair pathway resulting in apoptosis of hematopoietic stem cells [27,30,31].

At present, there are no reports on human deficient in DNA-PKcs but a missense mutation at position L3062R has been reported. In consequence, this patient exhibited a classical RS-SCID. On molecular basis, this was attributed to a functional failure in V(D)J recombination due to impaired activation of Artemis [32].

2.1.1.3 Artemis

Artemis was first identified in human patients suffering from RS-SCID. Alignment of its protein sequence revealed N-terminal catalytic domain encompassing metallo-β-lactamase and β-CASP subdomains and regulator CTD specific to Artemis (Fig. 19.2). Both N-terminal subdomains are conserved in nucleic acid processing enzyme belonging to the superfamily of metallo-β-lactamases [33,34]. Artemis is an endonuclease processing hairpin coding ends during V(D)J recombination and single-strand overhangs of dsDNA. As expected, Artemis deficiency in hematopoietic cells causes accumulation of unopened hairpins at the coding ends, whereas non-hematopoietic cells have increased sensitivity to ionizing radiation [35,36].

Recruitment of Artemis to a DSB as well as its subsequent catalytic activation is mediated through an interaction with DNA-PKcs. The interaction sites between both proteins are not well established. From the site of DNA-PKcs, the ABCDE cluster and its phosphorylation have be implicated in binding to the C-terminal region of Artemis. This site of Artemis is also phosphorylated by DNA-PKcs but its function remains to be determined [36,37].

2.1.2 LIG4 (DNA-Ligase 4/Xrcc4/XLF) Complex

LIG4 complex is composed of the DNA-Ligase 4, Xrcc4, and XLF and mediates the final step of the C-NHEJ repair. LIG4 is the catalytic unit of the complex capable of covalently sealing the ends of a DSB, whereas Xrcc4 and XLF do not show catalytic activity and rather play structural role. Both Xrcc4 and XLF form stable dimers by interaction between the head domains and proximal stalk regions (Fig. 19.2). These domains share structural but no sequence homology [38]. Xrcc4 is absolutely required for the stability of LIG4. Cells deficient in Xrcc4 do not exhibit LIG4 activity and the protein is not detectable suggesting rapid LIG4 degradation [39]. In contrast, XLF does not impact the stability of LIG4 or Xrcc4 [40,41]. The functional roles of both Xrcc4 and XLF proteins in LIG4 complex are not clear but both have been reported to stimulate LIG4 activity, possibly by promoting its adenylation [42,43]. Interestingly, XLF has been shown to particularly facilitate joining of non-cohesive ends [38,41,44].

In higher eukaryotes, there are three DNA ligases (LIG1, LIG3, and LIG4), which participate in different DNA-repair pathways [45]. They share a highly conserved DBD (DNA-binding domain) and a CD (catalytic domain) comprising of a nucleotidyl transferase and oligonucleotide/oligosaccharide-binding subdomain. Unique to LIG4 is the long C-terminal tail, which accommodates two BRCT (BRCA1 C-terminal) domains separated by a linker region. This region and the both BRCT domains are involved in binding of Xrcc4 [46,47]. Another unique feature to LIG4 is the catalytic ability to join non-cohesive ends containing gaps and the exclusive function only to participate in the C-NHEJ-mediated DNA repair [44]. In contrast, LIG1 and LIG3 are promiscuous and engage in distinct DNA-repair pathways and their function is shortly discussed in the context of A-NHEJ.

LIG4 catalyzes a multistep reaction requiring ATP to activate lysine K273 (in human) of the catalytic domain. The activated lysine carries AMP that is then transferred to 5′-PO4 of DNA to form activated 5′-AMP-DNA. Subsequently, the 3′-OH of a second DNA strand attacks the 5′-PO4 of the activated DNA releasing AMP and covalently sealing DNA strands. While this function is undisputed, the entire role of LIG4 complex in active NHEJ machinery is not clear [45].

Recruitment of LIG4 complex to the site of DNA damage is initiated by DNA-PK. But, there are emerging evidences that LIG4 complex itself may also contribute to assembly of a functional C-NHEJ complex serving as docking platform for end-processing enzymes. Studies carried out by Budman et al. clearly demonstrate that LIG4 complex is required for end processing of non-cohesive DNA ends [48]. The authors proposed a very attractive model, which implies that LIG4 complex binds to a break before end-processing enzymes, such as nucleases and polymerases and examines the nature of the break. Cohesive ends, which do not need end-processing, are directly ligated, whereas in case of non-cohesive ends, LIG4 complex recruits nucleases and polymerases. Once the ends are processed and made cohesive, LIG4 immediately seals the break protecting the DNA ends from further degradation [48,49].

As expected, targeted disruption of LIG4 or Xrcc4 in mice results in a strongly similar phenotype characterized by embryonic lethality due to massive neural apoptosis. Further, these animals suffer from arrested lymphocytosis through inability to rejoin V, D, and J segments and multiple other defects. MEF cells of these animals are markedly sensitive to IR [50,51]. Partially, the phenotype of LIG4 deficiency can be rescued by concomitant ablation of p53 or ATM. This is the case for embryonic lethality but not for arrested lymphocytosis. Rescue of lethality in LIG4/p53 or LIG4/ATM knockouts is not attributed to an improved or compensatory DNA-repair capacity, but rather due to attenuated response to DNA damage, and thus, diminished induction of neural apoptosis. In consequence, the cells bear to accumulate DNA damage but still fail to repair DSB or to rejoin V, D, and J segments [52,53]. In contrast, XLF null mice are viable, of normal size, and surprisingly, undergo almost unaffected V(D)J recombination, suggesting redundancy for this component in murine cells. Nonetheless, MEF cells show still increased sensitivity to IR [54].

In mouse, the LIG4 or Xrcc4 deficiency is not compatible with life and as expected, no deficiency in humans have been reported either. But, there are several reports of humans carrying mutations in LIG4, Xrcc4, and XLF genes. In case of LIG4, the severity of phenotype correlates with the residual LIG4 catalytic activity. These patients are generally characterized by sensitivity to radiation, features related to neural apoptosis, such as microcephaly and accordingly mental retardation, growth retardation, facial dysmorphisms, and skin abnormalities. Further, affected individuals show variable degrees of immunodeficiency (RS-SCID) [55]. Likewise, XLF or Xrcc4-affected patients show in general similar features, such as microcephaly, growth retardation, and radiosensitivity. But in striking contrast to LIG4 and XLF mutations, no immunodeficiency has been reported in patients carrying Xrcc4 mutations. This suggest that Xrcc4 may be redundant for V(D)J recombination in humans [40,41,56–61].

2.2 Programmed Double-Strand Breaks

To defend from continuously invading pathogens, higher vertebrates have developed an adaptive immune system composed of B and T cells. These cells have been equipped in a unique ability to tailor pathogen-specific immunoglobulins (Ig) and

TCRs. To do so, maturating B and T cells commit to highly dangerous programmed DSBs during V(D)J and CSR that, if not adequately repaired, result either in cell death or in chromosomal translocation [3,62].

V(D)J recombination occurs in both B and T cells and is initiated when the endonuclease RAG (recombination activating gene) binds to recombination signal sequences (RSS) flanking the variable (V), diverse (D), and joining (J) segments. RSS encompasses a heptamer of seven conserved base pairs (CACAGTG), a spacer region of 12 or 23 variable nucleotides and conserved nanomer (ACAAAAACC). Every V, D, and J segment is flanked with 12-RSS (12 base pair pacer) on one side and with 23-RSS (23 base pair spacer) on the other side. RAG creates single-strand DNA nicks within RSS regions and uses the reactive 3′-OH to disrupt the complementary DNA strand. The segment between 12-RSS and 23-RSS (12/23 rule) is permanently deleted from a chromosome and the ends are covalently joined (Fig. 19.4) [3,62].

While the RSS ends (deleted DNA sequence) are blunt and can be directly ligated, the chromosomal coding ends form hairpin structures that must undergo processing before ligation. The hairpin opening requires the endonuclease activity of Artemis. As expected, deficiency of this nuclease results in accumulation of unopened ends in the maturating B and T cells [35]. Once the coding ends are open, an array of enzymes, such as polymerases and possibly different nucleases act on these ends. A particular role is played by the TdT polymerase, which has a unique ability to attach random nontemplated nucleotides to the overhangs of coding ends. The extensive processing of ends is required to interspace nucleotide sequences, and thus, increase diversity of Ig and TCR [63].

The final ligation of coding ends is highly complex and is exclusively carried out by C-NHEJ machinery. This tunneling function is attributed to RAG proteins, which tether the coding ends and possibly concurrently block the access of other DNA-repair pathways. C-terminal mutation of murine RAG2 permits A-NHEJ to mediate this process [64]. Interestingly, XLF seems to be redundant for successful V(D)J recombination in mice and Xrcc4 in humans [54]. In contrast, knockout or hypomorphic mutations of any other C-NHEJ component results in RS-SCID due to inability to rejoin coding ends of V, D, and J segments in both mice and humans [3,62].

CSR is exclusive to peripheral antigen-stimulated B cells that express IgM and IgD on their surface. In similar fashion to the V(D)J recombination, first formation of DSB is initiated by AID (activation-induced cytidine deaminase). This enzyme deaminates cytidine in S-regions flanking the genes coding for IgM, IgD, IgG, IgE, and IgA. In multistep process, this lesion is converted to a DSB and the unwanted DNA segment is permanently deleted from the chromosome. The segments surrounding the break are brought together resulting in a switch of antibody isotype. In contrast, ligation of these segments is not absolutely dependent on C-NHEJ. Genetically engineered B cells to bypass V(D)J recombination undergo robust CSR in mice deficient in C-NHEJ key components, such as Ku70, LIG4, and Xrcc4. The joining junctions in these animals

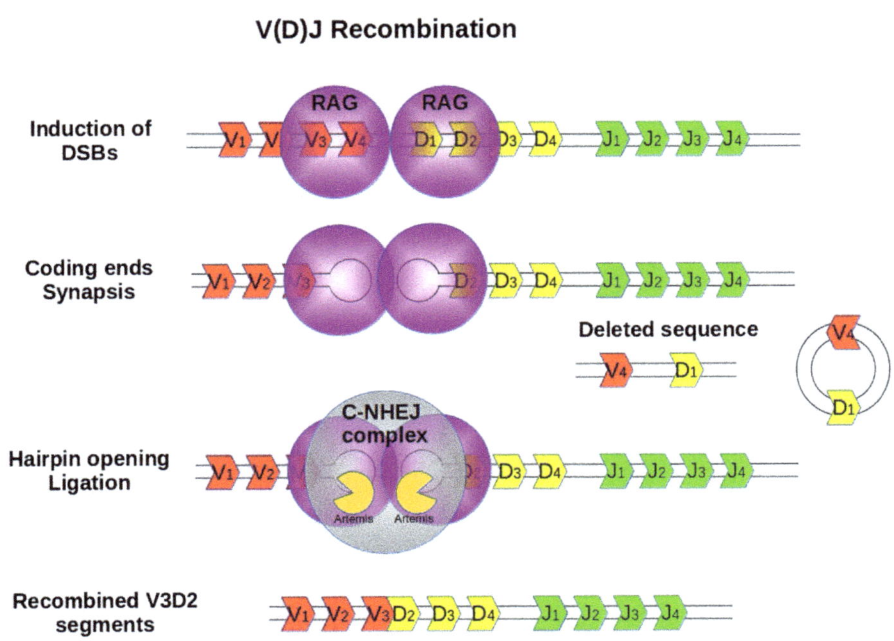

FIGURE 19.4 **Cartoon representing V(D)J recombination.** RAG induces DSBs in RSS regions in the vicinity of segments that are intended to be recombined and tethers the coding ends. The DNA sequence between these segments is permanently deleted from chromosome. The coding ends carry hairpin structure that must be opened before ligation. This is carried out by Artemis. The joining of coding ends is exclusively mediated by C-NHEJ and is tunnelled through RAG.

carry extended microhomology characteristic for A-NHEJ [65–67]. This implies that A-NHEJ has the ability to back up the C-NHEJ machinery in CSR but its physiological role is not understood.

3. ALTERNATIVE NHEJ

While the C-NHEJ has been intensively studied for many years and great knowledge has accumulated, the A-NHEJ, in contrast, has been gaining attention since 2010 as microhomologies (hallmark of A-NHEJ) became linked to chromosomal aberrations in murine and human cells [3,68].

The first evidence for existence of the A-NHEJ came from experiments, in which cells deficient in C-NEHJ component(s) still repair DSB in a template-independent manner [69–71]. Because its activity is mainly seen, when C-NHEJ is inactivated, A-NHEJ is also referred to as backup NHEJ [3,68]. Studies in 2015 further revealed that A-NHEJ may also be a backup mechanism for HR [72,73]. However, it is not clear what is the role of A-NHEJ when the both aforementioned mechanisms are functional. One possibility is, when for whatever reason C-NHEJ and/or HR fails to repair a DSB, then A-NHEJ engages and fixes this break. The fingerprint of A-NHEJ-mediated repair is microhomology with deletions at the repair junctions. This implies that initiation of A-NHEJ requires broken DNA ends to be resected until short homologous overhangs are uncovered (Fig. 19.5). Subsequently, these complementary ends anneal and the break is then directed to A-NHEJ machinery for repair. Due to these short homologies, A-NHEJ has been also designated as microhomology-mediated end joining (MHEJ).

Though the role of microhomology-mediated repair in DNA repair is well established, the nature of this pathway is still heavily disputed. It is not clear whether A-NHEJ is a single pathway or may be distinct pathways. Boboila et al. analyzed CSR junctions from murine B-cells deficient in LIG4 or Ku70 [67]. LIG4-deficient cells showed almost exclusively microhomology-mediated repair of junctions, but in contrast, Ku70-deficient cells carried a substantial fraction of direct junctions with no microhomology. Direct junctions are usually seen in WT cells. Based on these results, Ku70-dependent and Ku70-independent A-NHEJ pathway has been proposed, the latter able to mediate both direct and microhomology-mediated junctions [67]. However, it remains to be deciphered how these differences occur. Since the activity of A-NHEJ is mainly seen in cells deficient in C-NHEJ components, other scientists have suggested that what has been designated as A-NHEJ, is merely C-NHEJ, in which a redundant protein substitutes a missing component, for example, Lig3 substitutes for Lig4 [3,74].

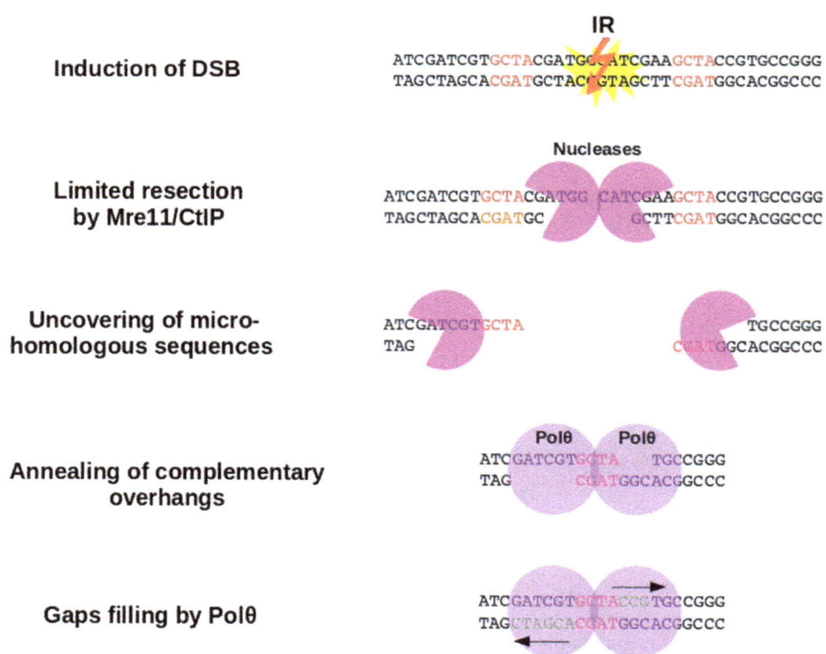

FIGURE 19.5 **Cartoon of limited end resection and gap filling.** Microhomology is highlighted with *red*. After double-strand break (DSB) induction, incoherent DNA ends undergo limited resection until homologous overhangs are uncovered. Pol θ stabilizes the hybridized overhangs and uses the opposite strand to fill in the gaps. *Arrows* indicate the direction of gap filling.

3.1 Components of A-NHEJ

Based on the literature many different proteins have been implicated to play a role in the A-NHEJ. Some of them are PARP1 (poly-ADP-ribose polymerase 1), Xrcc1, Lig1 and Lig3, Mre11, CtIP (CtBP-interacting protein), and Pol θ (polymerase theta). All of these components and others that are not mentioned here have been already assigned to distinct DNA-repair pathways. For instance, both CtIP and Mre11 have been assigned to HR-mediated repair and PARP1 was first allocated to base excision repair [68]. Thus, it makes it very difficult to define the A-NHEJ pathway.

Although it is not understood how and when cells choose to initiate A-NHEJ over C-NHEJ or HR-mediated DNA repair, there are supporting evidences that PARP1 and the end-processing nuclease CtIP and Mre11 may play a pivotal role in this process. Common step to each DNA-repair pathways is sensing of the break that is mediated by pathway-specific complex (sensor). Simplified, whereas Ku70/80 complex is the DSB sensor for C-NHEJ-mediated repair, MRN complex senses for HR [7,75]. Similarly, PAPR1 may be the sensor protein for A-NHEJ because it is able to bind single- and double-stranded DNA ends and is instantly recruited to DSBs. In addition, inhibition or knockdown of PARP1 counteracts activation of A-NHEJ. PARP1 interaction with broken DNA ends causes its heavy auto-poly-ADP-ribosylation that promotes recruitment of MRN, CtIP, and LIG3/Xrcc1 complex, possibly through direct interaction [76,77].

MRN complex is composed of three proteins Mre11, Rad50, and NBS and avidly binds to DNA DSBs [75]. Based on crystallographic structure, MRN complex tethers broken DNA ends and serves, on the other hand, as a docking platform for CtIP at the break [78,79]. Both Mre11 and CtIP bestow the MRN complex with nuclease activity and carry out initial limited end resection (Fig. 19.5). This resection is restricted to maximum of two to three hundred nucleotides and is needed to uncover microhomologous (few nucleotides long) complementary overhangs. Annealing of these overhangs is possibly required for initiation of A-NHEJ-mediated ligation. However, if cells choose to progress to HR-mediated repair, then there is a switch to an extensive resection over thousand nucleotides mediated by a different set of nucleases, such as EXO1 and BLM [80,81]. At present, it is not understood how the switch/pathway choice is made.

Another unanswered question is how the access of sensor proteins to DSBs is determined. It has been shown that Ku70/80, MRN, and PARP1 compete for binding to broken DNA ends. However, Ku70/80 complex seems to outcompete both MRN and PARP1 making C-NHEJ the default DNA-repair pathway [82–84]. Thus, it is understandable that Ku70 or Ku80 deficiency would favor the binding of PARP1 to a DSB and promote A-NHEJ, but how does deficiency of LIG4 or other C-NHEJ components promote A-NHEJ is not clear [66,67]. One possibility is that in the absence of C-NHEJ components, Ku70/80 binding to DNA ends cannot be stabilized and the complex falls off making the DNA ends again accessible for PARP1 or MRN complex [84,85].

Analog to the role of polymerases (Pol λ and Pol μ) in the C-NHEJ, in 2015, Pol θ was reported to play a role in A-NHEJ. Kent et al. presented in an in vitro assay that, once overhangs with microhomology are annealed, Pol θ stabilizes the hybridized sites of the overhangs and uses the opposing strand to fill in the gaps with complementary nucleotides (Fig. 19.5) [86]. In vivo studies have further supported the role of Pol θ in A-NHEJ and demonstrated its competitive nature with HR. Interestingly, cells compromised in HR due to BRCA1 deficiency rely on A-NHEJ as a backup mechanism and knockdown of Pol θ results in synthetic lethality making it an attractive chemotherapeutic target in a subset of cancers [72,73].

Multiple research groups have shown that the final step of ligation is mostly carried out by LIG3, possibly in complex with its binding partner Xrcc1. Depletion of this ligase in cells or cell extracts significantly reduced A-NHEJ-mediated events [87,88]. Della-Maria et al. provided an interesting observation that in wild-type (WT) cells, MRN and LIG3/Xrcc1 are associated [89]. Upon IR exposure, MRN and LIG3/Xrcc1 dissociate possibly to participate in distinct DNA-repair pathways. In contrast, MRN and LIG3/Xrcc1 remain associated in cells that are deficient in LIG4 or DNA-PKcs, suggesting redirection to A-NHEJ pathway [89]. In line, ablation of nuclear LIG3 decreased translocation rate and remaining translocation did not show bias toward microhomology. Concomitant knockdown of LIG1 but not of LIG4 further suppressed translocation rate making LIG1 the possible backup ligase for LIG3 [90].

The stability of LIG3 is dependent on its binding partner Xrcc1 that is analogous to LIG4/Xrcc4 complex. Ablation of Xrcc1 causes functional deficiency of LIG3 [45]. However, there are convicting studies showing that Xrcc1 may be redundant for A-NHEJ. Knockout of Xrcc1 in WT or in Xrcc4-deficient B cells did not affect A-NHEJ-mediated CSR and IgH/c-myc translocations. This is in line with the work of Soni et al. showing that Xrcc1 is not required for translocations in MEF cells [91]. Based on previous reports that in the absence of LIG3, LIG1 was still driving translocations, these studies further support an important role of LIG1 in A-NHEJ pathway and not necessary just as a backup ligase for LIG3 [65,91].

3.2 Role of A-NHEJ in Chromosomal Aberration

Translocations and other chromosomal aberrations are the hallmark of cancers and can determine the nature of a tumor as well as response to radiochemotherapy. In 2010s, microhomology signature was reported at the breakpoints of translocations and other chromosomal aberrations providing evidences for A-NHEJ as the executive mechanism for genomic instability under certain circumstances [2,3,92].

Animals deficient in Xrcc4/p53 or LIG4/p53 succumb uniformly from pro-B cell lymphomas that carry oncogenic translocations between chromosome 12 and 15 t(12; 15) resulting in IgH/c-myc fusion [93]. Analogously, non-hematopoietic mouse cells deficient in Ku70, Xrcc4, or LIG4 showed increased rate of translocations. Comparison of breakpoint junctions between WT and C-NHEJ component–deficient cells revealed similar characteristics, such as deletions, insertions, and microhomologies [90,94,95]. In line with the role of A-NHEJ in translocations, depletion of CtIP resulted in decreased translocation rate and reduced microhomologies in WT cells. All these studies support the role of A-NHEJ as a major mediator of translocations in mammalian cells. However, this concept was challenged in 2014 in human cells. Ghezraoui et al. found in multiple human cell lines including HCT116 that deficiency of Xrcc4 or LIG4 decreased translocation rate but remaining translocations carried microhomologies at breakpoint junctions and knocking down CtIP decreases translocation rate [96]. Thus, the authors concluded that C-NEHJ presents the major mechanism of translocation in humans and A-NHEJ plays only marginal role [96]. In contrast, Soni et al. presented opposite results supporting the finding in murine cells. In their hands, translocation rate was increased in the same LIG4-deficient HCT116 cells and frequency of the translocations was decreased by PARP inhibitor [91]. Intermediate results came from analysis of germline chromosomal rearrangement in human patients. Here, 31% breakpoint junctions disclosed microhomology [97]. Thus, eventual species-specific differences between humans and mice remain to be elucidated.

Interestingly, increased A-NHEJ activity has been demonstrated in leukemia cells expressing Bcr-abl or FLT3/ITD (FMS-like tyrosine kinase/internal tandem duplication). Bcr-abl is an oncogenic fusion protein resulting from t(9; 22) translocation and is considered to be the causative mechanism of CML (chronic myelogenic leukemia), whereas FLT/ITD is pathognostic for AML (acute myelogenic leukemia). Cells carrying either of these constitutively active kinases showed impaired DSB repair attributed to deregulation between C- and A-NHEJ pathways [98–100]. However, from these studies it is not clear whether impaired balance between these two mechanisms is a consequence of bcr-abl or FLT3/IDT expression or these aberrations were induced by preexisting increased A-NHEJ activity.

4. END PROCESSING

The nature of DNA-damaging agents determines the complexity of DSBs. Particularly, IR- and ROS-induced damage produce DNA ends that show highly complex structures and are not compatible for direct ligation. Such ends may contain abasic nucleotides, nucleotides missing 3′-OH or 5′-phosphate group, or overhangs with no or only partial complementarity. Ends with hairpin structures arise during V(D)J recombination (Fig. 19.4). Common to all these DNA ends is a requirement for adequate processing before the break can be sealed. On the other hand, coherent or blunt end that can be directly joined needs to be protected from enzymatic trimming [101].

To deal with the complexity of DNA DSBs, an array of regulatory and end-processing components has evolved. Both DNA-PKcs and LIG4 complex have been shown to be recruited to DSBs before the end-processing enzymes and may determine the accessibility to the DNA ends. Central role could be assigned to the ABCDE and PQR phosphorylation clusters of DNA-PKcs that reciprocally regulate end-processing, among others through interaction with Artemis [26,37,48,49]. In line, LIG4 complex serves as docking platform for end-processing enzymes and additionally, LIG4 alone possesses ability to ligate non-cohesive ends containing gaps [44,48,49]. Further, Ku70, a member of the DNA–PK complex, has been also implicated in the end-processing by removing abasic nucleotides [15]. Altogether, it seems that C-NHEJ complex possesses the ability to deal with less-complex breaks. Possibly, failure of C-NHEJ complex to adequately process the ragged ends may contribute to a switch to limited or extensive end resection revealing homologous sequences for A-NHEJ or HR pathway [2].

Another group of specialized enzymes including PNKP (polynucleotide kinase/phosphatase), aprataxin, and APLF (aprataxin and PNKP-like factor) have been implicated in trimming DNA ends during NHEJ. All three enzymes share a common FHA (fork-head associated) domain that mediates their binding to Xrcc4 or Xrcc1 in phosphorylation-dependent manner [102]. PNKP carries a 5′-kinase and 3′-phosphatase activity and is able to restore compatible 5′-phosphate and 3′-OH group at DNA ends [103]. Aprataxin was found to release AMP from the 5′-DNA ends of abortive DNA ligation intermediates, and thus, producing a 5′-phosphate that can undergo religation. Mutations in the gene coding for aprataxin cause neurological disorder, such as ataxia oculomotor apraxia-1, possibly through neuron death due to accumulation of DNA damage [104]. APLF interacts with Ku80 through the central domain and enhances NHEJ, possibly through its endonuclease activity [102].

A special situation arises when partially complementary ends anneal leaving open gaps that need to be filled in. Members of polymerases X family including Pol λ, Pol μ, and TdT (terminal deoxynucleotidyl transferase) have been implicated in filling in these gaps. Both Pol λ and Pol μ interact with Ku70/80 and LIG4/Xrcc4 complexes, whereas XLF seems to promote their activity. Deficiency of Pol λ and Pol μ during V(D)J recombination resulted in increased deletion of overhangs. Thus, these polymerases possibly contribute to retention of DNA overhangs, but their exact biological role remains to be determined [8]. TdT takes a special role among these polymerases as it is only expressed in hematopoietic B and T cells that undergo V(D)J recombination. During deletion of V, D, and J segments, TdT polymerase adds randomly nucleotides to overhangs of opened coding ends before they are religated. In consequence, the interspaced nucleotides increase the diversity of antigen-specific antibodies and TCRs [63].

5. CONCLUSIONS

Maintenance of genomic stability is of paramount importance for single cells, living organisms, and for conservation of the species. Inability to maintain genome stability has known deleterious consequences, such as cell death, malignant transformation causing cancer, and when germ cells are affected passing of diseases to subsequent generations. On the other hand, cells must accommodate certain level of mutability to ensure evolutionary adaptation. To deal with all these challenges, cells developed DNA-damage response, the ability to sense DNA damage and respond to it in adequate way. This involves several DNA-repair pathways, extensively reviewed in this book as well as activation of signaling pathways that synchronize the cell cycle, DNA replication, and cell metabolism with DNA repair and induce apoptosis, if repair fails.

Nowadays, after decades of intense studies addressing the DNA-repair mechanisms, the main DNA-repair pathways have been possibly discovered. The DNA pathways are tailored to instantly handle any kind of DNA damage at any location in the genome but differ in their fidelity to execute this function. In case of inability to repair the damage, cells are expected to undergo programmed cell death to prevent accumulation of unwanted aberrations and not to pass them to descending population of cells. Nonetheless, it is widely accepted that accumulation of chromosomal aberrations as a consequences of impaired DNA repair leads to malignant transformation, but the causative mechanisms are poorly understood.

A-NHEJ is considered as a backup mechanism of both C-NHEJ and possibly HR. HR is a high-fidelity mechanism that in most cases guarantees 100% faithful repair of DSBs. C-NHEJ, although considered error prone, can directly ligate coherent ends with no change in nucleotides sequence and incoherent ends are processed with minimal nucleotide loss. Cells tolerate well this low level of inaccuracy that may secure certain rate of mutability required to drive evolution. Both C- and A-NHEJ are active through the cell cycle but the latter shows great fluctuations. A-NHEJ peaks in G2 phase and almost vanishes in G1/0 phase of the cell cycle [101]. It is obscure why both C-NHEJ and HR pathway would need to be backed up by low-fidelity mechanism such as A-NHEJ that is blamed to cause genomic instability. It would be understandable that low-fidelity mechanism is activated in differentiated cells that will not commit to enter the cell cycle again such as circulating leukocytes or neurons but in contrast, A-NHEJ reaches its highest activity in G2 phase of dividing cells. May be under physiological conditions, A-NHEJ fulfills a different role and is not mutagenic at all. Merely the conditions, which are chosen to study its function, disclose the wrong nature of this pathway.

Possibly, better characterization of A-NHEJ at the molecular level and better understanding of pathway choice may provide some clarity to the earlier questions. However, the greatest limitation in understanding the complexity of DNA repair and any other cellular processes is attributed to the technics that are at present used in laboratories.

Cellular processes are highly dynamic; billions of molecules classified to hundreds of pathways are functioning at the same time in a single cell and carry out thousands of different reactions. But the vast majority of laboratory techniques enables just to take a snapshot of highly dynamic metabolism in a large number of cells. At present, live-time imaging is in its infancy and allows simultaneous tracing of few molecules at best with low resolution. Studying cellular process with current technics, it is like trying to reproduce and understand the dynamic life of New York City based on static pictures in "google maps." Only development of techniques that allow tracing hundreds or thousands molecules in real time at the resolution of single molecule in a single cell will provide better understanding of all these complex processes. Thus, it is a long way to go until we truly understand the DNA-repair pathways.

GLOSSARY

Chromosomal aberrations All unwanted changes in the genomic DNA.
Heterodimer Complex of two distinct proteins.
Heterotrimer Complex of three distinct proteins.
Microhomology Few nucleotides-long complementary DNA sequences.
Paralog A gene, which arose by duplication and evolved new function.
Processing Removing/adding of nucleotides at free DNA ends.
Radiosensitivity Increased rate of cell death to ionizing radiation.

Sensing To detect a double-strand break.
Synapsis Complex of proteins holding broken DNA ends together.
Tethering Holding two DNA ends together.

LIST OF ABBREVIATIONS

AID Activation-induced cytidine deaminase
APLF Aprataxin and PNKP-like factor
ATM Ataxia telangiectasia related
ATR Ataxia telangiectasia and Rad3 related
Bcr-abl Breakpoint cluster region-abl1 gene
BRCT BRCA1 C terminal
CD Catalytic domain
CSR Class switch recombination
CTD C-terminal domain
CtIP CtBP-interacting protein
DBD DNA-binding domain
DNA-PKcs DNA-protein kinase catalytic subunit
DSBs Double-strand breaks
FHA Fork-head associated
FLT3/ITD FMS-like tyrosine kinase/internal tandem duplication
HEAT Huntingtin, elongation factor 3 (EF3), protein phosphatase 2A (PP2A), yeast kinase TOR1
HR Homologous recombination
Ig Immunoglobulin
IR Ionizing radiation
MRN Mre11, Rad50, NBS
NBS Nijmegen breakage syndrome
NHEJ Nonhomologous end joining
PARP Poly-ADP-ribose polymerase
PIKK Phosphatidyl inositol-3 kinase-like protein kinase
PNKP Polynucleotide kinase/phosphatase
Pol θ Polymerase theta
RAG Recombination-activating gene
ROS Reactive oxygen species
RSS Recombination signal sequence
RS-SCID Radiosensitive severe combined immunodeficiency
SAP SAF-A/B, Acinus, and PIAS motifs
TCR T-cell receptor
TdT Terminal deoxynucleotidyl transferase
vWa von Willebrand A domain
Xrcc4 X-ray repair cross-complementing protein

REFERENCES

[1] Ochi T, Blackford AN, Coates J, Jhujh S, Mehmood S, Tamura N, et al. DNA repair. PAXX, a paralog of XRCC4 and XLF, interacts with Ku to promote DNA double-strand break repair. Science January 9, 2015;347(6218):185–8.
[2] Bunting SF, Nussenzweig A. End-joining, translocations and cancer. Nat Rev Cancer July 2013;13(7):443–54.
[3] Zhang Y, Gostissa M, Hildebrand DG, Becker MS, Boboila C, Chiarle R, et al. The role of mechanistic factors in promoting chromosomal translocations found in lymphoid and other cancers. Adv Immunol 2010;106:93–133.
[4] Fell VL, Schild-Poulter C. The Ku heterodimer: function in DNA repair and beyond. Mutat Res Rev Mutat Res March 2015;763:15–29.
[5] Jette N, Lees-Miller SP. The DNA-dependent protein kinase: a multifunctional protein kinase with roles in DNA double strand break repair and mitosis. Prog Biophys Mol Biol March 2015;117(2–3):194–205.
[6] Davis AJ, Chen BPC, Chen DJ. DNA-PK: a dynamic enzyme in a versatile DSB repair pathway. DNA Repair May 2014;17:21–9.
[7] Deriano L, Roth DB. Modernizing the nonhomologous end-joining repertoire: alternative and classical NHEJ share the stage. Annu Rev Genet November 23, 2013;47:433–55.
[8] Ramsden DA. Polymerases in nonhomologous end joining: building a bridge over broken chromosomes. Antioxid Redox Signal June 15, 2011;14(12):2509–19.
[9] Nussenzweig A, Chen C, da Costa Soares V, Sanchez M, Sokol K, Nussenzweig MC, et al. Requirement for Ku80 in growth and immunoglobulin V(D)J recombination. Nature August 8, 1996;382(6591):551–5.

[10] Gu Y, Seidl KJ, Rathbun GA, Zhu C, Manis JP, van der Stoep N, et al. Growth retardation and leaky SCID phenotype of Ku70-deficient mice. Immunity November 1997;7(5):653–65.

[11] Singleton BK, Torres-Arzayus MI, Rottinghaus ST, Taccioli GE, Jeggo PA. The C terminus of Ku80 activates the DNA-dependent protein kinase catalytic subunit. Mol Cell Biol May 1, 1999;19(5):3267–77.

[12] Hu S, Pluth JM, Cucinotta FA. Putative binding modes of Ku70-SAP domain with double strand DNA: a molecular modeling study. J Mol Model May 2012;18(5):2163–74.

[13] Cohen HY, Lavu S, Bitterman KJ, Hekking B, Imahiyerobo TA, Miller C, et al. Acetylation of the C terminus of Ku70 by CBP and PCAF controls Bax-mediated apoptosis. Mol Cell May 3, 2004;13(5):627–38.

[14] Walker JR, Corpina RA, Goldberg J. Structure of the Ku heterodimer bound to DNA and its implications for double-strand break repair. Nature August 9, 2001;412(6847):607–14.

[15] Roberts SA, Strande N, Burkhalter MD, Strom C, Havener JM, Hasty P, et al. Ku is a 5′-dRP/AP lyase that excises nucleotide damage near broken ends. Nature April 22, 2010;464(7292):1214–7.

[16] Feng L, Chen J. The E3 ligase RNF8 regulates KU80 removal and NHEJ repair. Nat Struct Mol Biol February 2012;19(2):201–6.

[17] Zhu C, Bogue MA, Lim D-S, Hasty P, Roth DB. Ku86-Deficient mice exhibit severe combined immunodeficiency and defective processing of V(D)J recombination intermediates. Cell August 9, 1996;86(3):379–89.

[18] Wang Y, Ghosh G, Hendrickson EA. Ku86 represses lethal telomere deletion events in human somatic cells. Proc Natl Acad Sci USA July 28, 2009;106(30):12430–5.

[19] Espejel S, Franco S, Rodríguez-Perales S, Bouffler SD, Cigudosa JC, Blasco MA. Mammalian Ku86 mediates chromosomal fusions and apoptosis caused by critically short telomeres. EMBO J May 1, 2002;21(9):2207–19.

[20] Samper E, Goytisolo FA, Slijepcevic P, van Buul PPW, Blasco MA. Mammalian Ku86 protein prevents telomeric fusions independently of the length of TTAGGG repeats and the G-strand overhang. EMBO Rep September 1, 2000;1(3):244–52.

[21] Gottlieb TM, Jackson SP. The DNA-dependent protein kinase: requirement for DNA ends and association with Ku antigen. Cell January 15, 1993;72(1):131–42.

[22] Yoo S, Dynan WS. Geometry of a complex formed by double strand break repair proteins at a single DNA end: recruitment of DNA-PKcs induces inward translocation of Ku protein. Nucleic Acids Res December 1, 1999;27(24):4679–86.

[23] DeFazio LG, Stansel RM, Griffith JD, Chu G. Synapsis of DNA ends by DNA-dependent protein kinase. EMBO J June 17, 2002;21(12):3192–200.

[24] Neal JA, Meek K. Choosing the right path: does DNA-PK help make the decision? Mutat Res June 3, 2011;711(1–2):73–86.

[25] Kienker LJ, Shin EK, Meek K. Both V(D)J recombination and radioresistance require DNA-PK kinase activity, though minimal levels suffice for V(D)J recombination. Nucleic Acids Res July 15, 2000;28(14):2752–61.

[26] Cui X, Yu Y, Gupta S, Cho Y-M, Lees-Miller SP, Meek K. Autophosphorylation of DNA-dependent protein kinase regulates DNA end processing and may also alter double-strand break repair pathway choice. Mol Cell Biol December 15, 2005;25(24):10842–52.

[27] Zhang S, Yajima H, Huynh H, Zheng J, Callen E, Chen H-T, et al. Congenital bone marrow failure in DNA-PKcs mutant mice associated with deficiencies in DNA repair. J Cell Biol April 18, 2011;193(2):295–305.

[28] Douglas P, Cui X, Block WD, Yu Y, Gupta S, Ding Q, et al. The DNA-dependent protein kinase catalytic subunit is phosphorylated in vivo on threonine 3950, a highly conserved amino acid in the protein kinase domain. Mol Cell Biol March 2007;27(5):1581–91.

[29] Neal JA, Dang V, Douglas P, Wold MS, Lees-Miller SP, Meek K. Inhibition of homologous recombination by DNA-dependent protein kinase requires kinase activity, is titratable, and is modulated by autophosphorylation. Mol Cell Biol April 15, 2011;31(8):1719–33.

[30] Jhappan C, Morse HC, Fleischmann RD, Gottesman MM, Merlino G. DNA-PKcs: a T-cell tumour suppressor encoded at the mouse scid locus. Nat Genet December 1997;17(4):483–6.

[31] Smith GCM, Jackson SP. The DNA-dependent protein kinase. Genes Dev April 15, 1999;13(8):916–34.

[32] van der Burg M, Ijspeert H, Verkaik NS, Turul T, Wiegant WW, Morotomi-Yano K, et al. A DNA-PKcs mutation in a radiosensitive T-B- SCID patient inhibits Artemis activation and nonhomologous end-joining. J Clin Invest January 2009;119(1):91–8.

[33] Ochi T, Wu Q, Blundell TL. The spatial organization of non-homologous end joining: from bridging to end joining. DNA Repair May 2014;17:98–109.

[34] Moshous D, Callebaut I, de Chasseval R, Corneo B, Cavazzana-Calvo M, Le Deist F, et al. Artemis, a novel DNA double-strand break repair/V(D)J recombination protein, is mutated in human severe combined immune deficiency. Cell April 20, 2001;105(2):177–86.

[35] Ma Y, Pannicke U, Schwarz K, Lieber MR. Hairpin opening and overhang processing by an Artemis/DNA-dependent protein kinase complex in nonhomologous end joining and V(D)J recombination. Cell March 22, 2002;108(6):781–94.

[36] Kurosawa A, Adachi N. Functions and regulation of Artemis: a goddess in the maintenance of genome integrity. J Radiat Res September 1, 2010;51(5):503–9.

[37] Jiang W, Crowe JL, Liu X, Nakajima S, Wang Y, Li C, et al. Differential phosphorylation of DNA-PKcs regulates the interplay between end-processing and end-ligation during nonhomologous end-joining. Mol Cell April 2, 2015;58(1):172–85.

[38] Mahaney BL, Hammel M, Meek K, Tainer JA, Lees-Miller SP. XRCC4 and XLF form long helical protein filaments suitable for DNA end protection and alignment to facilitate DNA double strand break repair. Biochem Cell Biol February 2013;91(1):31–41.

[39] Bryans M, Valenzano MC, Stamato TD. Absence of DNA ligase IV protein in XR-1 cells: evidence for stabilization by XRCC4. Mutat Res January 26, 1999;433(1):53–8.

[40] Buck D, Malivert L, de Chasseval R, Barraud A, Fondanèche M-C, Sanal O, et al. Cernunnos, a novel nonhomologous end-joining factor, is mutated in human immunodeficiency with microcephaly. Cell January 27, 2006;124(2):287–99.

[41] Ahnesorg P, Smith P, Jackson SP. Xlf interacts with the XRCC4-DNA ligase IV complex to promote DNA nonhomologous end-joining. Cell January 27, 2006;124(2):301–13.

[42] Modesti M, Hesse JE, Gellert M. DNA binding of Xrcc4 protein is associated with V(D)J recombination but not with stimulation of DNA ligase IV activity. EMBO J April 1, 1999;18(7):2008–18.

[43] Riballo E, Woodbine L, Stiff T, Walker SA, Goodarzi AA, Jeggo PA. XLF-Cernunnos promotes DNA ligase IV-XRCC4 re-adenylation following ligation. Nucleic Acids Res February 2009;37(2):482–92.

[44] Gu J, Lu H, Tippin B, Shimazaki N, Goodman MF, Lieber MR. XRCC4:DNA ligase IV can ligate incompatible DNA ends and can ligate across gaps. EMBO J February 21, 2007;26(4):1010–23.

[45] Ellenberger T, Tomkinson AE. Eukaryotic DNA ligases: structural and functional insights. Annu Rev Biochem 2008;77(1):313–38.

[46] Girard P-M, Kysela B, Härer CJ, Doherty AJ, Jeggo PA. Analysis of DNA ligase IV mutations found in LIG4 syndrome patients: the impact of two linked polymorphisms. Hum Mol Genet October 15, 2004;13(20):2369–76.

[47] Grawunder U, Zimmer D, Lieber MR. DNA ligase IV binds to XRCC4 via a motif located between rather than within its BRCT domains. Curr Biol July 16, 1998;8(15):873–6.

[48] Budman J, Kim SA, Chu G. Processing of DNA for nonhomologous end-joining is controlled by kinase activity and XRCC4/ligase IV. J Biol Chem April 20, 2007;282(16):11950–9.

[49] Budman J, Chu G. Processing of DNA for nonhomologous end-joining by cell-free extract. EMBO J February 23, 2005;24(4):849–60.

[50] Frank KM, Sekiguchi JM, Seidl KJ, Swat W, Rathbun GA, Cheng HL, et al. Late embryonic lethality and impaired V(D)J recombination in mice lacking DNA ligase IV. Nature November 12, 1998;396(6707):173–7.

[51] Gao Y, Sun Y, Frank KM, Dikkes P, Fujiwara Y, Seidl KJ, et al. A critical role for DNA end-joining proteins in both lymphogenesis and neurogenesis. Cell December 23, 1998;95(7):891–902.

[52] Frank KM, Sharpless NE, Gao Y, Sekiguchi JM, Ferguson DO, Zhu C, et al. DNA ligase IV deficiency in mice leads to defective neurogenesis and embryonic lethality via the p53 pathway. Mol Cell June 2000;5(6):993–1002.

[53] Sekiguchi J, Ferguson DO, Chen HT, Yang EM, Earle J, Frank K, et al. Genetic interactions between ATM and the nonhomologous end-joining factors in genomic stability and development. Proc Natl Acad Sci USA March 13, 2001;98(6):3243–8.

[54] Li G, Alt FW, Cheng H-L, Brush JW, Goff PH, Murphy MM, et al. Lymphocyte-specific compensation for XLF/cernunnos end-joining functions in V(D)J recombination. Mol Cell September 5, 2008;31(5):631–40.

[55] Chistiakov DA, Voronova NV, Chistiakov AP. Ligase IV syndrome. Eur J Med Genet November 2009;52(6):373–8.

[56] Guo C, Nakazawa Y, Woodbine L, Björkman A, Shimada M, Fawcett H, et al. XRCC4 deficiency in human subjects causes a marked neurological phenotype but no overt immunodeficiency. J Allergy Clin Immunol July 23, 2015;136(4).

[57] Bee L, Nasca A, Zanolini A, Cendron F, d'Adamo P, Costa R, et al. A nonsense mutation of human XRCC4 is associated with adult-onset progressive encephalocardiomyopathy. EMBO Mol Med 2015;7(7):918–29.

[58] Rosin N, Elcioglu NH, Beleggia F, Isgüven P, Altmüller J, Thiele H, et al. Mutations in XRCC4 cause primary microcephaly, short stature and increased genomic instability. Hum Mol Genet July 1, 2015;24(13):3708–17.

[59] de Bruin C, Mericq V, Andrew SF, van Duyvenvoorde HA, Verkaik NS, Losekoot M, et al. An XRCC4 splice mutation associated with severe short stature, gonadal failure, and early-onset metabolic syndrome. J Clin Endocrinol Metab May 2015;100(5):E789–98.

[60] Murray JE, van der Burg M, IJspeert H, Carroll P, Wu Q, Ochi T, et al. Mutations in the NHEJ component XRCC4 cause primordial dwarfism. Am J Hum Genet March 5, 2015;96(3):412–24.

[61] Dai Y, Kysela B, Hanakahi LA, Manolis K, Riballo E, Stumm M, et al. Nonhomologous end joining and V(D)J recombination require an additional factor. Proc Natl Acad Sci USA March 4, 2003;100(5):2462–7.

[62] Malu S, Malshetty V, Francis D, Cortes P. Role of non-homologous end joining in V(D)J recombination. Immunol Res December 2012;54(1–3):233–46.

[63] Gilfillan S, Benoist C, Mathis D. Mice lacking terminal deoxynucleotidyl transferase: adult mice with a fetal antigen receptor repertoire. Immunol Rev December 1995;148:201–19.

[64] Corneo B, Wendland RL, Deriano L, Cui X, Klein IA, Wong S-Y, et al. Rag mutations reveal robust alternative end joining. Nature September 27, 2007;449(7161):483–6.

[65] Boboila C, Oksenych V, Gostissa M, Wang JH, Zha S, Zhang Y, et al. Robust chromosomal DNA repair via alternative end-joining in the absence of X-ray repair cross-complementing protein 1 (XRCC1). Proc Natl Acad Sci USA February 14, 2012;109(7):2473–8.

[66] Boboila C, Jankovic M, Yan CT, Wang JH, Wesemann DR, Zhang T, et al. Alternative end-joining catalyzes robust IgH locus deletions and translocations in the combined absence of ligase 4 and Ku70. Proc Natl Acad Sci USA February 16, 2010;107(7):3034–9.

[67] Boboila C, Yan C, Wesemann DR, Jankovic M, Wang JH, Manis J, et al. Alternative end-joining catalyzes class switch recombination in the absence of both Ku70 and DNA ligase 4. J Exp Med February 15, 2010;207(2):417–27.

[68] Frit P, Barboule N, Yuan Y, Gomez D, Calsou P. Alternative end-joining pathway(s): bricolage at DNA breaks. DNA Repair May 2014;17:81–97.

[69] Boulton SJ, Jackson SP. *Saccharomyces cerevisiae* Ku70 potentiates illegitimate DNA double-strand break repair and serves as a barrier to error-prone DNA repair pathways. EMBO J September 16, 1996;15(18):5093–103.

[70] Kabotyanski EB, Gomelsky L, Han JO, Stamato TD, Roth DB. Double-strand break repair in Ku86- and XRCC4-deficient cells. Nucleic Acids Res December 1, 1998;26(23):5333–42.

[71] Wang H, Perrault AR, Takeda Y, Qin W, Wang H, Iliakis G. Biochemical evidence for Ku-independent backup pathways of NHEJ. Nucleic Acids Res September 15, 2003;31(18):5377–88.

[72] Ceccaldi R, Liu JC, Amunugama R, Hajdu I, Primack B, Petalcorin MIR, et al. Homologous-recombination-deficient tumours are dependent on Polθ-mediated repair. Nature February 12, 2015;518(7538):258–62.

[73] Mateos-Gomez PA, Gong F, Nair N, Miller KM, Lazzerini-Denchi E, Sfeir A. Mammalian polymerase θ promotes alternative NHEJ and suppresses recombination. Nature 2015;518(7538). Available from: http://www.nature.com.ezp-prod1.hul.harvard.edu/nature/journal/vaop/ncurrent/full/nature14157.html.

[74] Pannunzio NR, Li S, Watanabe G, Lieber MR. Non-homologous end joining often uses microhomology: implications for alternative end joining. DNA Repair May 2014;17:74–80.

[75] Lamarche BJ, Orazio NI, Weitzman MD. The MRN complex in double-strand break repair and telomere maintenance. FEBS Lett September 10, 2010;584(17):3682–95.

[76] Langelier M-F, Planck JL, Roy S, Pascal JM. Structural basis for DNA damage–dependent poly(ADP-ribosyl)ation by human PARP-1. Science May 11, 2012;336(6082):728–32.

[77] Beck C, Robert I, Reina-San-Martin B, Schreiber V, Dantzer F. Poly(ADP-ribose) polymerases in double-strand break repair: focus on PARP1, PARP2 and PARP3. Exp Cell Res November 15, 2014;329(1):18–25.

[78] Williams RS, Moncalian G, Williams JS, Yamada Y, Limbo O, Shin DS, et al. Mre11 dimers coordinate DNA end bridging and nuclease processing in double-strand-break repair. Cell October 3, 2008;135(1):97–109.

[79] Sartori AA, Lukas C, Coates J, Mistrik M, Fu S, Bartek J, et al. Human CtIP promotes DNA end resection. Nature November 22, 2007;450(7169):509–14.

[80] Symington LS. End resection at double-strand breaks: mechanism and regulation. Cold Spring Harb Perspect Biol August 2014;6(8).

[81] Truong LN, Li Y, Shi LZ, Hwang PY-H, He J, Wang H, et al. Microhomology-mediated End Joining and Homologous Recombination share the initial end resection step to repair DNA double-strand breaks in mammalian cells. Proc Natl Acad Sci USA May 7, 2013;110(19):7720–5.

[82] Cheng Q, Barboule N, Frit P, Gomez D, Bombarde O, Couderc B, et al. Ku counteracts mobilization of PARP1 and MRN in chromatin damaged with DNA double-strand breaks. Nucleic Acids Res December 2011;39(22):9605–19.

[83] Wang M, Wu W, Wu W, Rosidi B, Zhang L, Wang H, et al. PARP-1 and Ku compete for repair of DNA double strand breaks by distinct NHEJ pathways. Nucleic Acids Res 2006;34(21):6170–82.

[84] Fattah F, Lee EH, Weisensel N, Wang Y, Lichter N, Hendrickson EA. Ku regulates the non-homologous end joining pathway choice of DNA double-strand break repair in human somatic cells. PLoS Genet February 2010;6(2):e1000855.

[85] Zhang Y, Hefferin ML, Chen L, Shim EY, Tseng H-M, Kwon Y, et al. Role of Dnl4-Lif1 in nonhomologous end-joining repair complex assembly and suppression of homologous recombination. Nat Struct Mol Biol July 2007;14(7):639–46.

[86] Kent T, Chandramouly G, McDevitt SM, Ozdemir AY, Pomerantz RT. Mechanism of microhomology-mediated end-joining promoted by human DNA polymerase θ. Nat Struct Mol Biol 2015;22(3). Available from: http://www.nature.com.ezp-prod1.hul.harvard.edu/nsmb/journal/vaop/ncurrent/full/nsmb.2961.html.

[87] Liang L, Deng L, Nguyen SC, Zhao X, Maulion CD, Shao C, et al. Human DNA ligases I and III, but not ligase IV, are required for microhomology-mediated end joining of DNA double-strand breaks. Nucleic Acids Res June 2008;36(10):3297–310.

[88] Wang H, Rosidi B, Perrault R, Wang M, Zhang L, Windhofer F, et al. DNA ligase III as a candidate component of backup pathways of nonhomologous end joining. Cancer Res May 15, 2005;65(10):4020–30.

[89] Della-Maria J, Zhou Y, Tsai M-S, Kuhnlein J, Carney JP, Paull TT, et al. Human Mre11/human Rad50/Nbs1 and DNA ligase IIIα/XRCC1 protein complexes act together in an alternative nonhomologous end joining pathway. J Biol Chem September 30, 2011;286(39):33845–53.

[90] Simsek D, Brunet E, Wong SY-W, Katyal S, Gao Y, McKinnon PJ, et al. Dna ligase III promotes alternative nonhomologous end-joining during chromosomal translocation formation. PLoS Genet June 2, 2011;7(6):e1002080.

[91] Soni A, Siemann M, Grabos M, Murmann T, Pantelias GE, Iliakis G. Requirement for Parp-1 and DNA ligases 1 or 3 but not of Xrcc1 in chromosomal translocation formation by backup end joining. Nucleic Acids Res June 2014;42(10):6380–92.

[92] Hanahan D, Weinberg RA. Hallmarks of cancer: the next generation. Cell March 4, 2011;144(5):646–74.

[93] Zhu C, Mills KD, Ferguson DO, Lee C, Manis J, Fleming J, et al. Unrepaired DNA breaks in p53-deficient cells lead to oncogenic gene amplification subsequent to translocations. Cell June 28, 2002;109(7):811–21.

[94] Simsek D, Jasin M. Alternative end-joining is suppressed by the canonical NHEJ component Xrcc4-ligase IV during chromosomal translocation formation. Nat Struct Mol Biol April 2010;17(4):410–6.

[95] Weinstock DM, Brunet E, Jasin M. Formation of NHEJ-derived reciprocal chromosomal translocations does not require Ku70. Nat Cell Biol August 2007;9(8):978–81.

[96] Ghezraoui H, Piganeau M, Renouf B, Renaud J-B, Sallmyr A, Ruis B, et al. Chromosomal translocations in human cells are generated by canonical nonhomologous end-joining. Mol Cell September 18, 2014;55(6):829–42.

[97] Chiang C, Jacobsen JC, Ernst C, Hanscom C, Heilbut A, Blumenthal I, et al. Complex reorganization and predominant non-homologous repair following chromosomal breakage in karyotypically balanced germline rearrangements and transgenic integration. Nat Genet April 2012;44(4):390–7. S1.

[98] Sallmyr A, Tomkinson AE, Rassool FV. Up-regulation of WRN and DNA ligase IIIα in chronic myeloid leukemia: consequences for the repair of DNA double-strand breaks. Blood August 15, 2008;112(4):1413–23.

[99] Fan J, Li L, Small D, Rassool F. Cells expressing FLT3/ITD mutations exhibit elevated repair errors generated through alternative NHEJ pathways: implications for genomic instability and therapy. Blood December 9, 2010;116(24):5298–305.

[100] Li L, Zhang L, Fan J, Greenberg K, Desiderio S, Rassool FV, et al. Defective nonhomologous end joining blocks B-cell development in FLT3/ITD mice. Blood March 17, 2011;117(11):3131–9.

[101] Schipler A, Iliakis G. DNA double-strand-break complexity levels and their possible contributions to the probability for error-prone processing and repair pathway choice. Nucleic Acids Res September 2013;41(16):7589–605.

[102] Radhakrishnan SK, Jette N, Lees-Miller SP. Non-homologous end joining: emerging themes and unanswered questions. DNA Repair May 2014;17:2–8.

[103] Weinfeld M, Mani RS, Abdou I, Aceytuno RD, Glover JNM. Tidying up loose ends: the role of polynucleotide kinase/phosphatase in DNA strand break repair. Trends Biochem Sci May 2011;36(5):262–71.

[104] Ahel I, Rass U, El-Khamisy SF, Katyal S, Clements PM, McKinnon PJ, et al. The neurodegenerative disease protein aprataxin resolves abortive DNA ligation intermediates. Nature October 12, 2006;443(7112):713–6.

Chapter 20

Double-Strand Break Repair: Homologous Recombination in Mammalian Cells

C. Gelot, T. Le-Guen, S. Ragu, B.S. Lopez
Université Paris-Saclay, Villejuif, France

Chapter Outline

1. INTRODUCTION

DNA double–strand breaks (DSBs) are one of the most injurious lesions that can generate genomic rearrangements and challenge cell fate. DSBs are produced through exposure to exogenous treatments (such as ionizing radiation), the byproducts of endogenous cellular metabolism and arrested replication forks. DSB repair is essential for the maintenance of DNA integrity, but can also trigger profound genomic rearrangements. Conversely, DSBs can also generate genetic diversity in essential biological processes, such as meiosis and the establishment of the immune repertoire (discussed in Refs. [1,2]). Therefore, DSB repair must be tightly controlled.

Two major strategies are used to repair DSBs: homologous recombination (HR), which requires an intact homologous sequence and nonhomologous end joining (NHEJ), which joins the DNA double–stranded ends (DSEs) without requiring any extended homologous sequence [3], and is a prominent process in mammalian cells. The canonical NHEJ (C-NHEJ) pathway is Ku70/Ku80- and XRCC4-DNA ligase 4-dependent. During early 2000s, an additional highly mutagenic alternative end-joining pathway(s) (A-EJ) that is Ku70/Ku80-and XRCC4-DNA ligase 4-independent was described (for review, see Refs. [1,4]).

Here, we focus on HR in mammalian cells. HR is evolutionarily conserved in all organisms. The main roles of HR are the protection and reactivation of replication forks that have been blocked (reviewed in Refs. [2,5]), the gap filling of single-stranded DNA (ssDNA) and the repair of DSBs [3]. Therefore, HR also plays essential and pivotal roles in genome stability, diversity and plasticity.

Genome Stability. http://dx.doi.org/10.1016/B978-0-12-803309-8.00020-3

2. THE ROLE OF HR IN THE EQUILIBRIUM OF GENETIC STABILITY VERSUS DIVERSITY

The products of HR are gene conversions (GC, nonreciprocal exchange of genetic material) associated with or without crossing over (CO, reciprocal exchange of the adjacent sequences), which allows HR to generate new combinations of genetic material or eliminate mutations (Fig. 20.1A). This function combined with the ability to repair DNA breakage places HR at the heart of the equilibrium controlling the balance between genetic stability and variability. Therefore, HR is implicated in many fundamental biological processes (see Fig. 20.1B). Indeed, due to its versatility, HR is involved in essential biological processes ranging from molecular evolution to DNA repair and meiotic differentiation and is also relevant to the application of targeted gene replacement (Fig. 20.1B).

Other examples of the diverse roles of HR in genome plasticity are as follows (Fig. 20.1B):

- HR is a driving force for the evolution of multigene families. In some families of repeated genes, the duplicated genes co-evolved via a phenomenon called concerted evolution [6,7].
- During meiosis, HR favors allelic recombination between two homologous chromosomes. Because the two homologous chromosomes are not fully identical, this process ensures allele mixing and creates genetic diversity (for review, see Ref. [8]).
- HR participates in neurogenesis during embryonic and postnatal neural development [9]. Heterozygous mutations in *RAD51*, which is the central HR component, have been found in individuals with congenital mirror movements (CMM) [10].

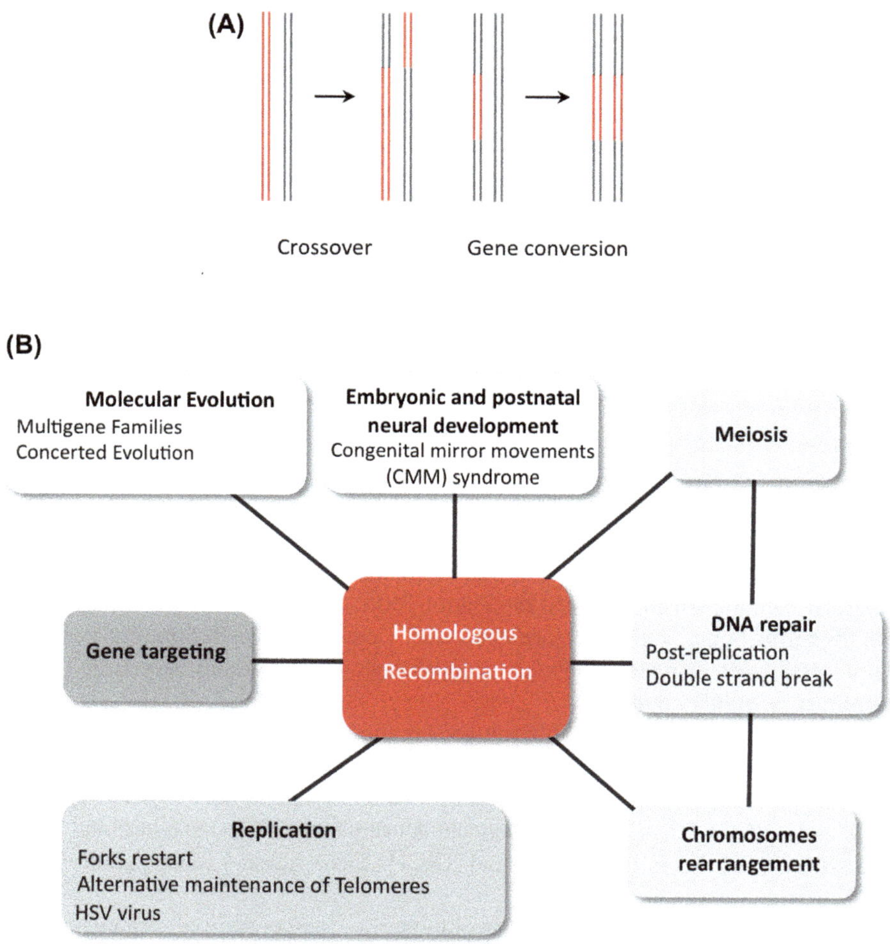

FIGURE 20.1 **The outcomes and roles of HR in mammalian cells.** (A) The products of HR. Crossover (left): Reciprocal exchange of the adjacent sequences. Gene conversion (right): Nonreciprocal transfer of genetic material.(B) The involvement of HR in many biological processes. HR participates in numerous fundamental processes controlled by the equilibrium between the stability and instability/diversity of the genome.

3. MOLECULAR MECHANISMS AND REGULATION OF HR

HR refers to different molecular mechanisms (Fig. 20.2). DSB repair by HR acts through several successive steps that need to be precisely coordinated to secure genome integrity. All of the different HR processes are initiated by a 5′–3′-single-strand break.

3.1 DSB Sensing and Chromatin Remodeling

The initial sensing of HR is mediated by the MRN complex (MRE11/RAD50/NBS1) in cooperation with the ATM (ataxia telangiectasia–mutated) kinase, which transduces the DNA-damage response (DDR). Inaccessible areas of the chromatin cannot be supported by the DDR. Thus, immediate changes in the DNA structure that result from the detection of the affected region are needed [11,12]. Indeed, the state of the chromatin changes after the recognition of DSBs, particularly through the phosphorylation of histone variant H2A.X in the vicinity of the lesion [13]. This change facilitates the accumulation of repair proteins at the damaged areas. The MDC1 mediator is recruited to the damage after its phosphorylation by ATM. MDC1 stabilizes the MRN complex, and its accumulation leads to remodeling of the chromatin by the ubiquitin ligases RNF8 and RNF168 (for review, see Ref. [2]).

3.2 Initiation of DNA Resection

After chromatin decondensation, the 53BP1 and Rap80–BRCA1 complex is recruited to the DSB. BRCA1 ("breast cancer type 1 susceptibility protein") is an HR mediator, and multiple roles of BRCA1 have been described in HR and DDR [14,15]. HR is initiated by the resection of the 5′-end toward the 3′-end to obtain ssDNA with a 3′-extension. This step is performed by nucleases and DNA helicases. One essential role of BRCA1 (in association with CtIP) during HR initiation is the removal of 53BP1 from the DNA ends, thereby making them accessible for resection initiation [16,17].

The resection occurs during two substeps and is modulated by the MNR complex. BRCA1 has a BRCT domain that enables interactions with phosphorylated proteins. BRCA1 forms a heterodimer with its BARD1 cofactor ("BRCA1-associated ring domain 1"), which possesses a RING domain [18] and gives the BRCA1/BARD1 complex E3 ubiquitin ligase activity. Thus, this complex allows the ubiquitination of the CtIP nuclease (exonuclease activity in the 3′- to 5′-direction) that cooperates with MRN to initiate the resection of the DSB [19–21]. Studies in 2015 showed that the MCM8–9 helicase complex was essential for DNA resection by the MRN complex at DSBs and was required for the proper localization of the MRN complex to the DSBs [22]. Then, the exonuclease 1 (Exo1) and/or the BLM/DNA2 complex ensures the elongation of the 3′-strand to generate a long 3′-overhang [23] (Fig. 20.2A); BLM is a member of the Rec Q helicase family and is mutated in Bloom syndrome [24]. Finally, the ssDNA is protected and stabilized by RPA.

3.3 Loading of RAD51 and Strand Exchange

BRCA2 ("breast cancer type 2 susceptibility protein") in association with Palb2 (which is also mutated in breast cancer familial cases) replaces RPA by RAD51, creating a presynaptic filament. RPA contributes to the polarity of the process. The RAD51/ssDNA filament promotes homologous pairing, through microhomologies scanning [24a] and strand invasion of a homologous duplex sequence, initiating then copy of the homologous matrix, and generating cruciform intermediates called Holliday junctions (HJs) (Fig. 20.2A). Rad54, which is a protein from the SWI2/SNF2 family, interacts with Rad51, thereby facilitating strand invasion [25]. Rad54 catalyzes the migration of the branches that takes place between the two strands.

The invading 3′-ssDNA allows the priming of DNA synthesis. This priming displaces the complementary strand creating a D-loop (displacing loop), which is then captured by the other broken DNA end. RAD54 also stabilizes this HR intermediate. This intermediate generates the cruciform HJs that are resolved by nucleases or dissolved, leading to the HR outcomes (Fig. 20.2A).

A family of six proteins (RAD51B, RAD51C, RAD51D, XRCC2, XRCC3, and RAD51AP1) known as the RAD51 paralogs (ie, proteins that share sequence homology with RAD51) has been identified. Genes encoding these paralogs may have derived from RAD51 gene duplication, and share at least 20% identity at the amino acid level with RAD51 and each other [26]. Two distinct complexes have been identified: RAD51B–RAD51C–RAD51D–XRCC2 (BCDX2) and RAD51C–XRCC3 (CX3) [28]. The early role of the RAD51 paralogs in HR is to promote the formation and stabilization of the RAD51 nucleoprotein filament, most likely by counteracting the disruption of the filament by the helicases. A 2004 work showed that the BCDX2 complex (but not the CX3 complex) was responsible

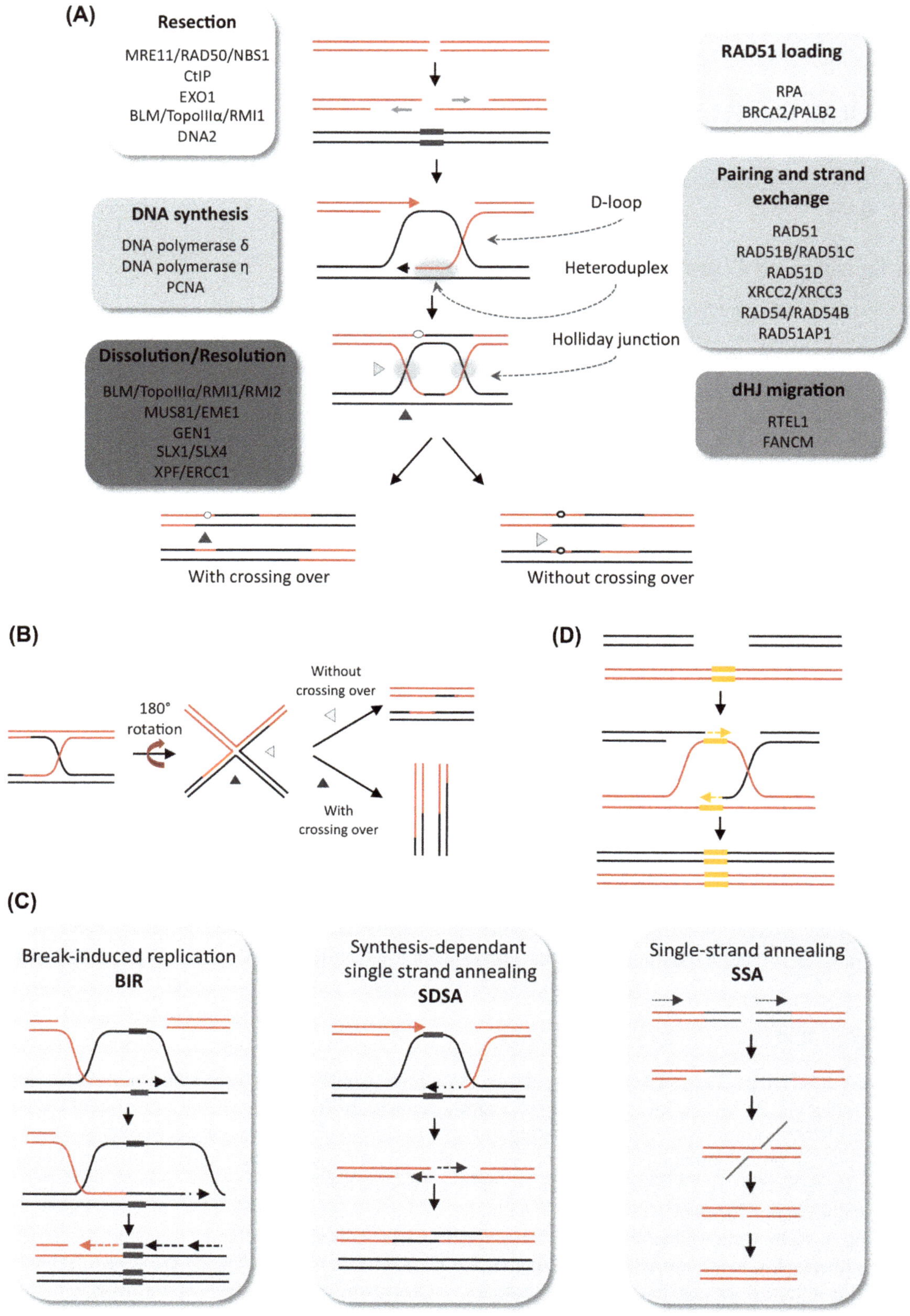

FIGURE 20.2 The different HR models. (A) The DSB repair model [100]. DSB resection (*gray arrows*) generates ssDNA tails that invade an intact homologous duplex DNA and initiate DNA synthesis (*black arrow*). This strand invasion displaces the complementary strand and creates a D-loop that anneals to the complementary strand of the recipient molecule. DNA synthesis fills in the gaps, and the processes results in two cruciform junctions (Holliday junctions) that can be resolved/dissolved or migrate. The final outcome is gene conversion associated with (*black arrow*) or without (*gray arrow*) crossing over. Mismatch forming heteroduplexe (*white circles*). The proteins involved in these steps are noted on both sides of the schematic. (B) Resolution of the Holliday junctions. The products of HR issue from a 180° rotation of the HJ followed by its resolution that leads to a crossing over (*black arrow*) or a non-crossing over (*white arrow*) event. (C) Other HR models. In the absence of resolution/dissolution of the dHJ, synthesis can be prolonged up to the end of the chromosome (left panel; BIR: break-induced replication). Alternatively, the invading strand can flip back to its parental molecule (middle panel; SDSA: synthesis-dependent single-strand annealing). Finally, another process results in the annealing of the complementary strand revealed by resection (right panel; SSA: single-strand annealing). SSA does not act through strand invasion of the duplex DNA, does not generate a dHJ, and is a nonconservative process because it leads to the deletion of the intervening sequence. (D) Gap filling. The broken molecule (black) invades an intact homologous molecule (red) containing a heterologous sequence (yellow). DNA synthesis initiated by HR copies the heterologous sequence (yellow) and transfers it to the acceptor molecule (black).

for RAD51 recruitment to DNA-damage sites in human cells. After RAD51-mediated strand invasion, the RAD51 paralogs influence GC tract length. Moreover, the RAD51 paralogs can bind to Y-shaped replication-like intermediates and synthetic HJ suggesting a role for the RAD51 paralogs in DNA repair during replication and the resolution of HR intermediary structures [29].

3.4 Resolution of the HJ and HR Outcomes

The nucleases GEN1 ("gen endonuclease homolog 1"), the heterodimer Mus81/Eme1 ("essential meiotic endonuclease 1"), and SLX1 and SLX4 [30–32] are among the factors that resolve HJ through cleavage. Topoisomerase III (TopoIII) and BLM resolve the double HJ (dHJ) substrate via the convergent migration of the two dHJs toward one another, leading to their collapse. TopoIII alpha recruits Rmi1 to catalyze dHJ dissolution. Finally, Rmi2 (an essential member of the "dissolvasome" complex) stimulates dHJ resolution [33] (Fig. 20.2A).

According to the orientation of the HJ resolution, the process may result in an exchange of adjacent sequences (Fig. 20.2B). Therefore, the products of HR are GCs with or without CO (depending on the resolution of the intermediate structure). However, the absence of HJ resolution will lead to break-induced replication (BIR) or synthesis-dependent strand annealing (SDSA) [3] (Fig. 20.2C).

Pairing and strand exchange of homologous DNA strands tolerate some differences between the molecules involved, thereby allowing the creation of a hybrid double-stranded DNA molecule called a heteroduplex that carries mismatches (Fig. 20.2A). Mismatch repair will result in the nonreciprocal transfer of genetic information from one DNA molecule to the other (ie, GC). In addition, the 3′-end of the invading strand allows DNA synthesis by copying the recipient molecule. In this process, a sequence absent from the invading strand can be copied, leading to the nonreciprocal transfer of genetic information from the invaded molecule to the invading molecule and therefore GC (Fig. 20.2D). Of note, all models are initiated by common steps beginning with the resection of the ssDNA, followed by the invasion and exchange of a homologous DNA strand that is a pivotal step in HR. These models are considered to be prominent for mitotic and meiotic recombination without crossover. Furthermore, BIR seems to be the mechanism underlying telomere maintenance by the ALT system in the absence of telomerase [34].

The last model (single-strand annealing or SSA) can occur between two sequences in tandem and is initialized by a single-strand resection. However, in contrast with the models described previously, this step is not followed by invasion of the DNA duplex and strand exchange. In fact, when the two sequences are in direct orientation, the ssDNA sequences revealed are complementary and can hybridize to form a branched structure. The following concerns should be noted in relation to the SSA model: it is a nonconservative process leading inevitably to a deletion of the intervening sequence; it cannot occur between inverted repeat sequences because the strands revealed by the resection are not complementary but identical; and finally, SSA can generate translocations if two breaks occur simultaneously in ectopic homologous sequences [35].

4. ROLES OF HR IN REPLICATION FORK REACTIVATION AND DSB REPAIR

4.1 Fork Stability/Restart by HR Upon Replication Stress

Replication fork progression is routinely challenged by diverse exogenous or endogenous stresses that ultimately lead to replication fork stalling, collapse, or breakage and trigger the DDR [5,36–39].

A crucial role for HR in genome stability maintenance is to escort replication fork progression. Indeed, HR is involved in the recovery of arrested replication forks (Fig. 20.3) [2,5,38,40].

Because the newly synthesized DNA strands produced by replication are complementary, reversion of the blocked fork can take place (Fig. 20.3) [41]. Notably, RAD51 participates in this replication fork reversion process [42].

The resumption of replication forks can be initiated by the loading of HR factors onto the single-strand DNA present at the stalled fork. Several different restart pathways have been proposed: (1) fork restart after repriming (ie, the loading of the replisome after a lesion) (Fig. 20.3A1); (2) restart after a fork reversion, when the newly synthesized DNA strand is homologous to the parental DNA downstream and creates a "chicken foot" structure (Fig. 20.3A2); and (3) restart using the ssDNA formed after fork regression (Fig. 20.3C) in a process analogous to BIR (Fig. 20.3C).

In some cases, single-ended DSBs are formed by either the passage of replication forks through a nick or an ssDNA gap (Fig. 20.3B) or the cleavage of the reversed forks by structure-specific endonucleases, such as MUS81 (Fig. 20.3A2). Then, HR can use the sister chromatid to prime DNA synthesis, thereby allowing the resumption of replication (Fig. 20.3).

In addition BRCA2 and RAD51 can protect the DNA ends of an arrested replication fork from resection by MRE, without leading to a recombination outcome [42a,42b,42c].

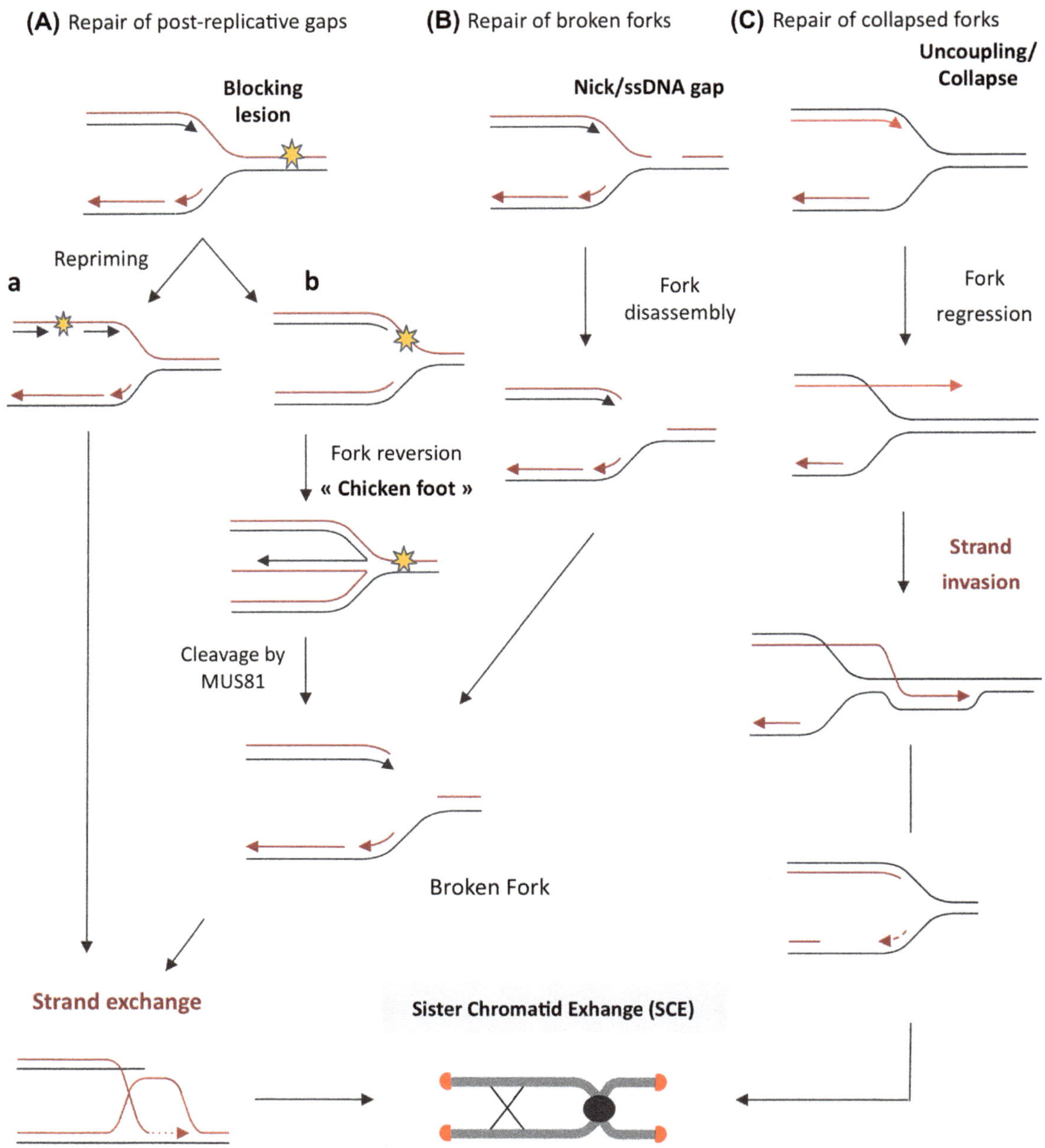

FIGURE 20.3 The role of HR in the reactivation of arrested replication forks. (A) Repair of post-replicative gaps. The replication fork (*arrows*) reaches a blocking DNA lesion (*yellow star*). DNA synthesis is primed downstream of the lesion to produce a single-strand gap bearing the lesion. This gap is filled in via sister chromatid exchange (SCE) by a copy of the intact sister chromatid. Reversion of the blocked fork leads to a "chicken foot" structure. Cleavage of this cruciform structure generates a DSB with only one end. SCE allows replication to resume. (B) Repair of broken forks. One replication fork reaching a ssDNA gap or nick is converted into a DSB. SCE can resume replication. (C) Repair of a collapsed fork. If a replication fork collapses, an uncoupling between the leading and lagging strand can occur generating a ssDNA strand that can invade the duplex matrix and restart DNA synthesis.

Because HR plays a pivotal role in the resumption of arrested replication forks, defects in HR lead to spontaneous slowed replication fork progression [43,44]. Replication defects in *HR*$^-$ cell lead to mitosis and chromosome defects, including anaphase bridges, common fragile sites, and supernumerary centrosomes, which result in multipolar mitosis and aneuploidy [44–50].

Thus, HR is an essential mechanism for the protection, recovery, and restart of replication forks. Consistent with the role in replication fork reactivation, HR-deficient cells are highly sensitive to agents that block the progression of replication forks, such as cisplatin or mitomycin C that generate interstrand cross-links in the DNA [51].

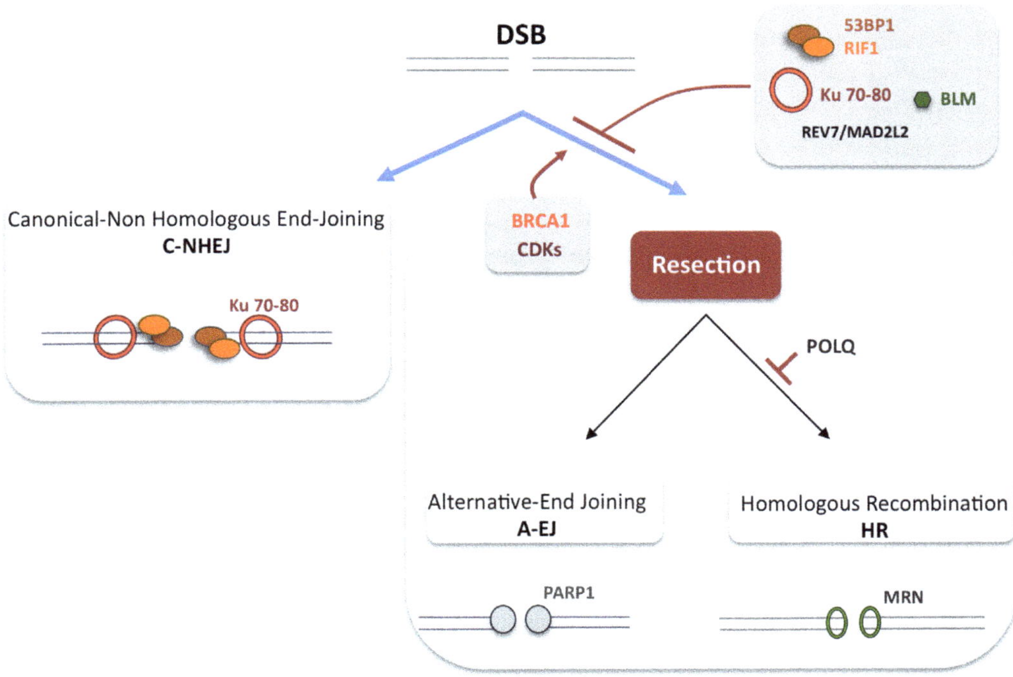

FIGURE 20.4 **The two-step model for the choice of the repair pathway.** The first alternative is the choice between C-NHEJ versus a resection producing ssDNA. The second alternative is competition between alternative end joining versus HR on the resected DNA. Some essential components of these processes are noted.

4.2 Competition for the DNA DSB–Repair Pathway Choice and Consequences for Meiosis and Genome Manipulation

4.2.1 Competition Between HR and End Joining for DSB Repair: A Two-Step Model

Several competing processes can repair DSBs: C-NHEJ, which is not initiated by DNA end resection, and HR and A-EJ, which are both initiated by resection involving common components, such as MRN and CtIP. Therefore, we propose a model in two steps [1,4,20] for the choice of the DSB-repair process (Fig. 20.4). First, competition occurs between C-NHEJ and resection. NHEJ is active throughout the cell cycle and resection is favored at S-phase entry, although A-EJ is also active throughout the cell cycle [52,53]. Second, when resection is initiated, competition between HR and A-EJ takes place. This competition can be modulated by the cell-cycle phase and the extent of the resection.

4.2.2 Meiosis

Defects in HR lead to sterility linked with meiotic division issues. The role of the meiotic program is to generate gametes with half of the chromosome content of the original progenitor cell. This task is accomplished by the occurrence of a single round of DNA replication, followed by two successive rounds of chromosome segregation. During meiosis, which aims to generate genetic diversity, sister-chromatid exchanges (SCEs) are repressed and HR between homologous chromosomes (which are not fully identical) is favored. HR plays a double role during meiosis division. The first is to assure the balance of the segregation of homologous chromosomes, and the second is to ensure the mixture of alleles to create genetic diversity. During reductional division, the cell must segregate the two homologous chromosomes into the two different daughter cells. However, these chromosomes do not have a physical link that would distinguish them (in mitosis, the chromosomes are linked by centromeres). This physical link is assured in meiosis by HR, which generates HJs; additionally, the generation of crossovers allows the rearrangement of alleles, thereby ensuring genetic diversity. Meiotic recombination is initiated by a DSB generated by the enzyme SPO11; the repair of this break essentially utilizes the same systems used for the repair of breaks induced by ionizing radiation (for review, see Refs. [8,54]).

4.2.3 Genome Manipulation

Due to its requirement for sequence homology, HR can be used for gene targeting (GT)—that is, the targeted modification (correction or insertional alteration) of a nuclear sequence by an exogenous sequence (Fig. 20.5). Drs. M. Cappecchi and

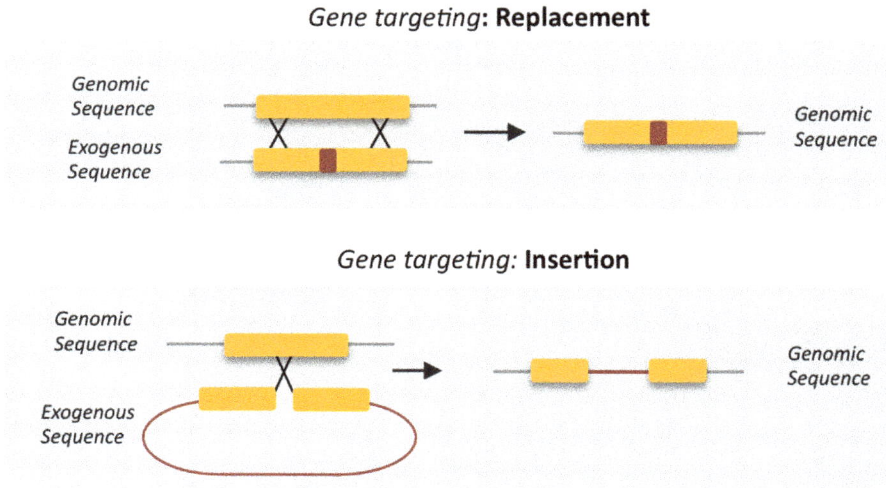

FIGURE 20.5 **Basic gene-targeting strategies via HR.** Upper panel: Replacement vector. Lower panel: Insertion vector. Yellow: Homologous sequences. Red: Modified sequence.

O. Smithies, who developed GT in mammalian cells, were awarded the Noble Prize in 2007. GT represents a promising strategy for gene therapy and the development of new biological models of interest for both academic and applied medical, biotechnological, and agronomic research. Importantly, GT allows the correction of a mutated nuclear gene, leading to the restoration of normal gene functions and thus targeted in situ gene therapy.

However, the efficiency of HR remains disappointing resulting in low efficiency of GT compared to the random integration of the correcting DNA. Because HR can repair DSBs, one promising strategy to increase the frequency of GT is to generate a DSB in the target sequence. The development of engineered sequence-specific nucleases (ie, a zinc finger, TALE, or sgRNA-Cas9 nuclease) allows the generation of the required DSBs in a given loci of the mammalian genome, thereby stimulating HR by several orders of magnitude [55]. Several studies have reported the feasibility of this ex vivo approach in human stem cells and primary cells [56] and in the liver of a hemophilia B mouse model [57], providing a proof of concept for the treatment of monogenic disease by genome editing with engineered nucleases.

5. THE DARK SIDE OF HR: PROMOTION OF GENOME INSTABILITY

HR contributes to the maintenance of genome stability/diversity through the combination of its different products and its ability to repair DNA. Because it copies an intact homologous DNA, HR is frequently classified as an error-free DNA-repair process. Indeed, HR-deficient cells exhibit increased genetic instability [58]. However, careful examination of the data can challenge this strict view (for review, see Ref. [2]):

1. CO between ectopic homologous sequences (nonallelic HR, NAHR) generates profound genome rearrangements leading to genetic instability (Fig. 20.6B). Moreover, BIR can also induce genome instability in mammalian cells. Indeed, it was reported in 2014 that replication stress induced by the overexpression of cyclin E in human cells leads to copy number alterations (CNAs). One-third of these genome alterations (duplications of less than 200 kb) have been attributed to BIR events. The authors propose that BIR repair of damaged replication forks may explain the presence of segmental genomic duplications in human cancers. The larger amplification (>200 kb) and deletion observed after the overexpression of cyclin E may arise from nonallelic HR [59].
2. GC with pseudo-genes can result in the extinction of the functional allele (Fig. 20.6A).
3. The accumulation of HR intermediates is toxic and can generate genetic instability [60].
4. The DNA synthesis initiated by HR is error prone, at least in yeast [60a].

6. PROTECTION AGAINST EXCESSIVE HR

HR plays an essential role in genome stability maintenance but can also jeopardize it (see earlier). Particularly, excess HR initiation can lead to the accumulation of HR intermediates, thereby generating genomic instability and cell death [60]. Thus, HR is a double-edged sword; on the one hand, it protects against genetic instability, but on the other hand, it can trigger cell lethality, profound genomic rearrangements, and point mutations. Therefore, HR should be tightly controlled to

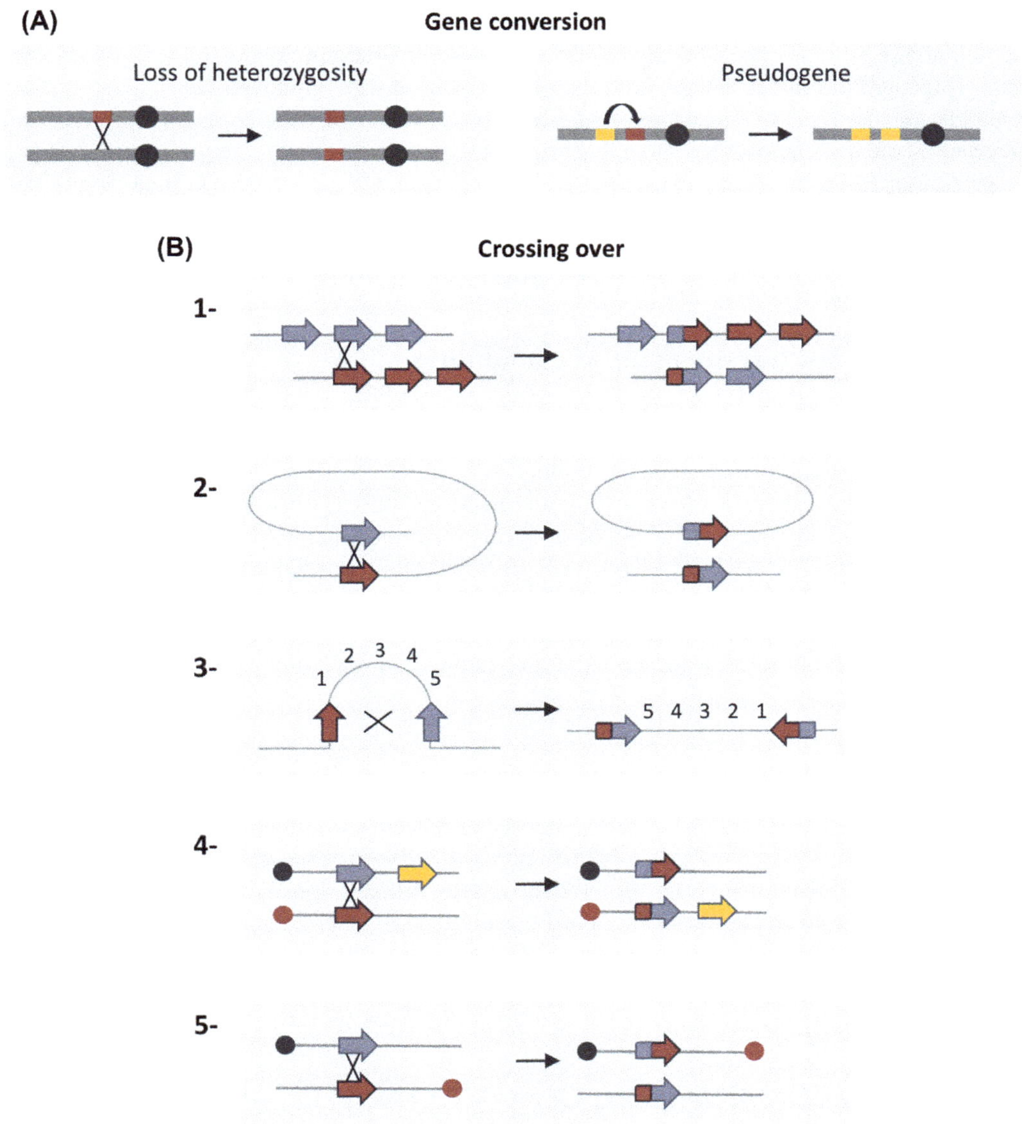

FIGURE 20.6 **Genetic instability induced by HR.** (A) By gene conversion. Nonreciprocal exchange of genetic information between two heteroalleles leads to a loss of heterozygosity (left panel). Gene conversion between a pseudogene (yellow), which often contains nonsense mutations, and a gene (red) transfers the stop codon, thereby inactivating gene expression (right panel). (B) By crossing over. Chromosomal rearrangements resulting from crossing over (CO) between repeat sequences. (1) Between repeat homologous sequences on two chromosomes or following unequal sister chromatid exchange on the same chromosome, resulting in the amplification of one molecule and the deletion of the other. (2) Intramolecular CO between two homologous sequences in a direct orientation, resulting in the excision of the intervening sequence. (3) Intramolecular CO between two homologous sequences in an inverted orientation, resulting in the inversion of the internal fragment. (4, 5) Interchromosomal CO. According to the orientation of the homologous sequences with respect to their centromeres (*gray or red circles*), this process generates translocation (4) or a dicentric and an acentric chromosome (5).

avoid unnecessary HR events. Excess HR can be controlled at several levels: initiation step, cell cycle, and destabilization of abortive HR intermediates through the action of helicases (reviewed in Refs. [61,62]). The fact that protective systems have evolved to counteract excess HR highlights the potential risks of this pathway.

6.1 Cell-Cycle Regulation

Prolonged blockage of replication forks leads to DSBs that can be addressed by HR or NHEJ [63]. However, unlike DSBs produced by enzymes, ionizing radiation, or endonucleases, breaks produced by replication stops have only one DSE. Ligation of two replication stress-induced DSEs involves distant DSEs, leading inexorably to a chromosomal rearrangement. During S phase, HR can take advantage of the intact sister chromatid to restore replication and avoid genetic instability.

Because the two chromatids have identical sequences, the genetic impact is minimal. The close proximity of the sister chromatids (particularly due to the cohesin complex) favors the use of sister chromatids for HR [64–66]. In addition, GC without crossover is favored in somatic cells to limit the risks associated with crossovers [64].

Sister chromatids are absent in G1. Therefore, should HR occur, the lack of sister chromatids would necessitate the use of sequences carried by other chromosomes, thereby jeopardizing genome stability. The maintenance of genome stability requires the restriction of HR to the S and G2 phases of the cell cycle when the sister chromatids are present. First, the CDK1/2-dependent phosphorylation of CtIP and EXO1 in S/G2 favors the initiation of resection and extension, respectively [67–69]. Second, resection is highly repressed by the loading of proteins on DSEs, such as 53BP1, RIF1, BLM, PTIP, the 2010 described REV7/MAD2L2 [70–75], and the C-NHEJ factors Ku70–80.

6.2 Protection Against HR Intermediate Accumulation

HR should be completed once initiated; otherwise, the accumulation of HR intermediates (RAD51 filaments or HJ) can generate genetic rearrangements and/or cell toxicity. The helicases from the RecQ family, which include *BLM*, *WRN*, *RECQ1*, and *RECQ5*, contribute to overall genome stability through the cleaning of abortive HR intermediates from the genome. RECQ5 can disrupt the RAD51 filament [76], and BLM and RECQ1 can melt D-loops. These helicases selectively dissociate recombination intermediates whose polarity can impair polymerase progression [77,78]. Additionally, two other helicases of the UvrD family (PARI and FBH1) can affect the stability of the RAD51 nucleoprotein filament. These helicases have been suggested to remove RAD51 from the ssDNA in a process that requires ATP hydrolysis by RAD51 [79–81]. Moreover, the ATP-dependent DNA helicase RTEL1 functions as an anti-recombinase that is dedicated to counteracting toxic recombination [82]. RTEL1 and FANCM also promote migration of the dHJ, thereby favoring SDSA and protecting against crossover events [83].

6.3 Repression of HR Initiation

Restricting the initiation of unscheduled HR has been proposed to prevent the accumulation of toxic HR intermediates. In mammalian cells, this protective role against excessive HR initiation has been proposed for Tp53, Bcl-2, and AKT1. Indeed, in these situations, essential HR components such as RAD51 or BRCA1 are trapped and mislocalized. Tp53 interacts with RAD51 and, in addition, may also affect the heteroduplexes resolution [84]; Bcl-2 interacts with BRCA1 and localizes it to the mitochondrial membrane [85]; and AKT1 activation BRCA1 and RAD51 are sequestrated into the cytoplasm leading to the inactivation of their nuclear functions [86]. In 2015, the POLQ polymerase was shown to play a role in the balance between HR and A-EJ after resection by inhibiting HR and triggering A-EJ at DSBs [87,88].

7. HOMOLOGOUS RECOMBINATION, GENOME STABILITY, AND CANCER

7.1 Misregulation of HR in Tumors

Both up- and down-regulation of HR have been described in oncogenic situations, highlighting the necessity for a precise equilibrium in HR regulation. Indeed, both decreased and increased HR can generate genetic instability (see earlier).

Defects in HR confer increased oncogenic risks. Most of the germ-line mutations involved in familial breast and ovary cancer affect genes directly involved in HR (the most frequently mutated genes BRCA1 and BRCA2, as well as Palb2, RAD51C, MRE11, NBS1, FANCJ, and FANCN), the HR/replication interface (claspin), or the regulation of HR (ATM, CHK2, and Tp53) [89]. A dominent negative form of RAD51 has been descibed in a subtype of Fanconi anemia, a syndrome associated with genetic instability and cancer predisposition [89a]. PTEN has also been shown to be mutated in familial breast cancer and to affect RAD51 expression [90]. Moreover, the oncogenic kinase AKT1 (which is antagonized by PTEN) has been demonstrated to be up-regulated in 40–60% of sporadic breast cancers and 40% of sporadic ovarian cancers. Importantly, AKT1 activation leads to HR repression through the cytoplasmic retention of BRCA1 and RAD51 [86]. Similarly, overexpression of the oncogene Bcl-2 results in HR down-regulation through the mitochondrial mislocalization of BRCA1 [85]. Finally, cells overexpressing a RAD51-dominant negative form that inhibits HR exhibit higher tumor development efficiency upon injection of nude mice [45].

Conversely, increased HR has also been described in oncogenic situation. Tp53 is the most frequently mutated gene in tumors. Importantly, Tp53 has been shown to inhibit HR, and *Tp53* cells exhibit increased HR activity. Bloom syndrome (BS) results from a mutation in BLM and is associated with high genetic instability and a cancer predisposition. Cells from BS patients exhibit high levels of SCE and hyper-recombination phenotypes (reviewed in ref. [24]). The fusion oncogene BCR/ABL from the translocation chromosome Philadelphia t(9:22) between BCR and ABL is present in chronic myelogenous leukemia (CML) and a number of other forms of acute lymphocytic leukemia (ALL). BCR/ABL expression results in constitutive tyrosine kinase activity that is responsible for resistance to drugs that generate DNA damage. Particularly, BCR/ABL expression causes the overexpression of RAD51 and the paralogs RAD51B, RAD51D, and XRCC2, resulting in increased HR and conferring resistance to cisplatin and mitomycin C [91]. More generally, components of the HR pathway are aberrantly expressed in many tumors [92,93], and the radioresistance of tumors exhibiting increased HR activity has been correlated with a poor prognosis. Thus, HR stimulation should fuel genome instability toward carcinogenic development, and confer resistance to anticancer treatments.

7.2 Anticancer Strategies

Many anticancer therapies are based on the induction of DSBs (ie, ionizing radiation or topoisomerases inhibitors) or DNA interstrand cross-links (ie, cisplatin or mitomycin C) [94]. Due to its role in DSB repair and the reactivation of arrested replication forks, HR represents a pharmacological target for the optimization of chemo- and radiotherapy.

Inhibitors targeting components of the HR pathway (MRN, RPA, and Rad51) [95], mediators/transducers of DSB signaling, and cell cycle–checkpoint regulation (ATM/ATR kinases, Chk1 and Chk2, Tp53, Wee1, and Cdc25) are in development.

ssDNA gaps or nicks are transformed into DSBs when the replication fork reaches them. Due to the roles of HR in both the replication stress response and DSB repair, HR-defective cells are highly sensitive to these treatments. PARP1 is involved in ssDNA gap and nick repair. Inhibition of PARP1 results in the accumulation of DNA alterations. PARP1 is also involved in DSB signaling and the competing/alternative DSB-repair pathway A-EJ. HR deficient–tumor cells (ie, BRCA1- or BRCA2-deficient cells) are highly sensitive to PARP1 inhibitors [96,97]. This strategy that consists of the induction of the inhibition of two metabolic pathways to generate cell death is called *synthetic lethality*. Similar conclusions have also been drawn for other DSB-repair pathways [98].

8. HR IN GENOMIC MOLECULAR EVOLUTION

Because HR plays a pivotal role in the balance between genetic stability and diversity and requires sequence homology, it is involved in the evolution (divergence versus co-evolution) of homologous sequences such as multigene families.

Following duplication, the two resulting sequences diverge during the course of evolution. However, in some families of repeated genes the two duplicated units do not evolve independently but co-evolve similarly in a process called *concerted evolution* [6,7]. During concerted evolution, one mutation present in one duplicated unit is transferred to the second duplicated unit by GC, which is the driving force behind the homogenization of duplicated sequences, and therefore concerted evolution. Sequences heterologies between the interacting DNA molecules impair HR and thus, represent barriers to concerted evolution. Introns can accumulate mutations without affecting the expression of the protein encoded by the gene. Consequently, introns have been proposed to serve as protective barriers against HR between repeated sequences and favor the maintenance of the genome organization [99]. Therefore, one can speculate that introns are antagonistic evolutionary forces to concerted evolution, routing evolution toward the divergence of repeated sequences.

Concerted evolution can occur between repeated α globin, histone, ribosomal, ubiquitin, and even mitochondrial genes, as well as between noncoding sequences ranging from α satellite sequences to dispersed repeat sequences.

9. CONCLUDING REMARKS

HR is a double-edged sword that can play opposite roles in the maintenance of genomic stability and also favoring genetic diversity up to genetic instability. Therefore, unscheduled excess HR can jeopardize genome stability and cell fate. Depending on the structure of the DNA partners involved in HR, GC, and CO are intrinsically capable to generate genetic variability/instability. In addition to cell-cycle regulation, which restricts HR at the S–G2 phase (and the tight cohesion of the sister chromatids that orientate exchange to equal SCE), several additional mechanisms repress HR: mismatch repair, helicases, and p53. Defects in these systems are associated with genome instability and cancer predisposition. Collectively, these capacities have been used by cell to generate beneficial genetic diversity. However, accidental HR can account for many pathological genome rearrangements.

Because of its main roles in DSB repair and reactivation of arrested replication forks, HR can be advantageously used in several applications: it is involved in gene targeted replacement; it represents a promising pharmacological target for cancer therapy. Therefore methods aiming at precise modulations of HR (either stimulation or repression) represent exciting challenges for future research and medical applications. Finally, HR plays a pivotal role in fundamental processes such as meiosis and molecular evolution of multigene families. Any novel knowledge on HR should thus benefit both academic and applied research.

GLOSSARY

Break-induced replication (BIR) A nonreciprocal recombination–dependent replication process.

Concerted evolution A process in which the paralogous genes within one species are more closely related to each other than to members of the same gene family in another species, even though the gene duplication event preceded the speciation event.

Crossing-over Reciprocal exchange of the adjacent sequences.

Displacement loop (D-loop) The single-stranded DNA formed when two strands of dsDNA are separated by the invasion of a third strand that anneals by base-pair complementation.

DNA helicase An enzyme that unwinds complementary duplex DNA.

DNA topoisomerase Enzymes that alter DNA topology by catalyzing strand passage.

Double Holliday junction (dHJ) Two adjacent Holliday junctions formed between four strands of DNA.

Double-strand breaks DNA damage that results in a break of both strands of DNA.

Gene conversion HR product, nonreciprocal exchange of genetic material.

Holliday junctions (HJ) Cross-strand exchange between two DNA molecules that results in a four-way junction.

Loss of heterozygosity Deletion, or mutation or recombination events that result in loss of the wild-type allele in a heterozygote.

MRN Mre11–Rad50–Nbs1 complex, responsible for recognizing and processing DNA ends.

Resection Degradation of one of the complementary strands of DNA specialized.

Sister-chromatid exchange (SCE) Reciprocal recombination between two sister chromatids degradation.

Synthetic lethality When the association of mutations in two or genes leads to cell death, whereas a mutation in only one of these genes is viable.

Synthesis-dependent strand annealing (SDSA) A recombination process that occurs when an extended strand is displaced and base paired with a complementary single strand to create a duplex without a crossover.

LIST OF ABBREVIATIONS

53BP1 p53-binding protein 1
A-EJ Alternative end joining
BIR Break-induced replication
C-NHEJ Canonical nonhomologous end joining
CDK Cyclin-dependent kinase
CO Crossing over
CMM Congenital mirror movements
DDR DNA-damage response
dHJ Holliday junctions
D-loop Displacement loop
DSB Double-strand break
DSE Double-strand end
EJ End joining
FANC Fanconi
GC Gene conversion
HR Homologous recombination
IR Ionizing radiation
MRN MRE11/RAD50/NBS1 complex
SCE Sister chromatid exchange
SDSA Synthesis-dependent strand annealing
ssDNA Single-strand DNA
T-SCE Telomere sister chromatid exchange

REFERENCES

[1] Betermier M, Bertrand P, Lopez BS. Is non-homologous end-joining really an inherently error-prone process? PLoS Genet 2014;10(1):e1004086.
[2] Guirouilh-Barbat J, Lambert S, Bertrand P, Lopez BS. Is homologous recombination really an error-free process? Front Genet 2014;5:175.
[3] Haber JE. Genome stability: DNA repair and recombination. 1st ed. Garland Science; 2014.
[4] Grabarz A, Barascu A, Guirouilh-Barbat J, Lopez BS. Initiation of DNA double strand break repair: signaling and single-stranded resection dictate the choice between homologous recombination, non-homologous end-joining and alternative end-joining. Am J Cancer Res 2012;2(3):249–68.
[5] Magdalou I, Lopez BS, Pasero P, Lambert SA. The causes of replication stress and their consequences on genome stability and cell fate. Semin Cell Dev Biol 2014;30C:154–64.
[6] Liao D. Concerted evolution: molecular mechanism and biological implications. Am J Hum Genet 1999;64(1):24–30.

[7] Arnheim N. Concerted evolution of multigene families. In: Koehn RK, Nei M, editors. Evol genes proteins. MA: Sinauer Assoc Sunderland; 1983. p. 38–61.
[8] Zickler D, Kleckner N. Recombination, pairing, and synapsis of homologs during meiosis. Cold Spring Harb Perspect Biol May 18, 2015;7(6). http://dx.doi.org/10.1101/cshperspect.a016626, pii: a016626.
[9] Frappart P-O, Lee Y, Lamont J, McKinnon PJ. BRCA2 is required for neurogenesis and suppression of medulloblastoma. EMBO J 2007;26(11): 2732–42.
[10] Depienne C, et al. RAD51 haploinsufficiency causes congenital mirror movements in humans. Am J Hum Genet 2012;90(2):301–7.
[11] Cohn MA, D'Andrea AD. Chromatin recruitment of DNA repair proteins: lessons from the fanconi anemia and double-strand break repair pathways. Mol Cell 2008;32(3):306–12.
[12] Lukas J, Lukas C, Bartek J. More than just a focus: the chromatin response to DNA damage and its role in genome integrity maintenance. Nat Cell Biol 2011;13(10):1161–9.
[13] Rogakou EP, Pilch DR, Orr AH, Ivanova VS, Bonner WM. DNA double-stranded breaks induce histone H2AX phosphorylation on serine 139. J Biol Chem 1998;273(10):5858–68.
[14] Moynahan ME, Chiu JW, Koller BH, Jasin M. Brca1 controls homology-directed DNA repair. Mol Cell 1999;4(4):511–8.
[15] Caestecker KW, Van de Walle GR. The role of BRCA1 in DNA double-strand repair: past and present. Exp Cell Res 2013;319(5):575–87.
[16] Bunting SF, et al. 53BP1 inhibits homologous recombination in Brca1-deficient cells by blocking resection of DNA breaks. Cell 2010;141(2): 243–54.
[17] Cao L, et al. A selective requirement for {53BP1} in the biological response to genomic instability induced by Brca1 deficiency. Mol Cell 2009;35(4):534–41.
[18] Wu LC, et al. Identification of a RING protein that can interact in vivo with the BRCA1 gene product. Nat Genet 1996;14(4):430–40.
[19] Yu X, Fu S, Lai M, Baer R, Chen J. BRCA1 ubiquitinates its phosphorylation-dependent binding partner CtIP. Genes Dev 2006;20(13):1721–6.
[20] Rass E, et al. Role of Mre11 in chromosomal nonhomologous end joining in mammalian cells. Nat Struct Mol Biol 2009;16(8):819–24.
[21] Bennardo N, Cheng A, Huang N, Stark JM. Alternative-NHEJ is a mechanistically distinct pathway of mammalian chromosome break repair. PLoS Genet 2008;4:e1000110.
[22] Lee KY, et al. MCM8-9 complex promotes resection of double-strand break ends by MRE11-RAD50-NBS1 complex. Nat Commun 2015;6:7744.
[23] Nimonkar AV, et al. BLM-DNA2-RPA-MRN and EXO1-BLM-RPA-MRN constitute two DNA end resection machineries for human DNA break repair. Genes Dev 2011;25(4):350–62.
[24] Chu WK, Hickson ID. RecQ helicases: multifunctional genome caretakers. Nat Rev Cancer 2009;9(9):644–54.
[24a] Qi Z, Redding S, Lee JY, Gibb B, Kwon Y, Niu H, et al. DNA sequence alignment by microhomology sampling during homologous recombination. Cell 2015;160:856–69.
[25] Sigurdsson S, Van Komen S, Petukhova G, Sung P. Homologous DNA pairing by human recombination factors Rad51 and Rad54. J Biol Chem 2002;277(45):42790–4.
[26] Sigurdsson S, et al. Mediator function of the human Rad51B-Rad51C complex in Rad51/RPA-catalyzed DNA strand exchange. Genes Dev 2001;15(24):3308–18.
[27] Deleted in review.
[28] Masson JY, et al. Identification and purification of two distinct complexes containing the five RAD51 paralogs. Genes Dev 2001;15(24):3296–307.
[29] Liu Y, Masson JY, Shah R, O'Regan P, West SC. RAD51C is required for Holliday junction processing in mammalian cells. Sci (80-) 2004;303(5655):243–6.
[30] Castor D, et al. Cooperative control of holliday junction resolution and DNA repair by the SLX1 and MUS81-EME1 nucleases. Mol Cell 2013;52(2):221–33. Available at: http://www.pubmedcentral.nih.gov/articlerender.fcgi?artid=3808987&tool=pmcentrez&rendertype=abstract.
[31] Munoz IM, et al. Coordination of structure-specific nucleases by human SLX4/BTBD12 is required for DNA repair. Mol Cell 2009;35(1):116–27.
[32] Adair GM, et al. Role of ERCC1 in removal of long non-homologous tails during targeted homologous recombination. EMBO J 2000;19(20): 5552–61.
[33] Singh TR, et al. BLAP18/RMI2, a novel OB-fold-containing protein, is an essential component of the Bloom helicase-double Holliday junction dissolvasome. Genes Dev 2008;22(20):2856–68.
[34] Dunham MA, Neumann AA, Fasching CL, Reddel RR. Telomere maintenance by recombination in human cells. Nat Genet 2000;26(4):447–50.
[35] Richardson C, Jasin M. Frequent chromosomal translocations induced by DNA double-strand breaks. Nature 2000;405(6787):697–700.
[36] Branzei D, Foiani M. Maintaining genome stability at the replication fork. Nat Rev Mol Cell Biol 2010;11(3):208–19.
[37] Hyrien O. Mechanisms and consequences of replication fork arrest. Biochimie 2000;82(1):5–17.
[38] Gelot C, Magdalou I, Lopez BS. Replication stress in Mammalian cells and its consequences for mitosis. Genes (Basel) 2015;6(2):267–98.
[39] Lambert S, Froget B, Carr AM. Arrested replication fork processing: interplay between checkpoints and recombination. DNA Repair 2007;6(7): 1042–61.
[40] Lambert S, Carr AM. Impediments to replication fork movement: stabilisation, reactivation and genome instability. Chromosoma 2013;122 (1–2):33–45.
[41] Higgins NP, Kato K, Strauss B. A model for replication repair in mammalian cells. J Mol Biol 1976;101(3):417–25.
[42] Zellweger R, et al. Rad51-mediated replication fork reversal is a global response to genotoxic treatments in human cells. J Cell Biol 2015;208(5): 563–79.
[42a] Hashimoto Y, Ray Chaudhuri A, Lopes M, Costanzo V. Rad51 protects nascent DNA from Mre11-dependent degradation and promotes continuous DNA synthesis. Nat Struct Mol Biol 2010;17:1305–11.

[42b] Schlacher K, Christ N, Siaud N, Egashira A, Wu H, Jasin M. Double-strand break repair-independent role for BRCA2 in blocking stalled replication fork degradation by MRE11. Cell 2011;145:529–42.
[42c] Ying S, Hamdy FC, Helleday T. Mre11-dependent degradation of stalled DNA replication forks is prevented by BRCA2 and PARP1. Cancer Res 2012;72:2814–21.
[43] Daboussi F, et al. A homologous recombination defect affects replication-fork progression in mammalian cells. J Cell Sci 2008;121(Pt 2):162–6. Available at: http://www.ncbi.nlm.nih.gov/entrez/query.fcgi?cmd=Retrieve&db=PubMed&dopt=Citation&list_uids=18089650.
[44] Wilhelm T, et al. Spontaneous slow replication fork progression elicits mitosis alterations in homologous recombination-deficient mammalian cells. Proc Natl Acad Sci USA 2014;111(2):763–8.
[45] Bertrand P, Lambert S, Joubert C, Lopez BS. Overexpression of mammalian Rad51 does not stimulate tumorigenesis while a dominant-negative Rad51 affects centrosome fragmentation, ploidy and stimulates tumorigenesis, in p53-defective CHO cells. Oncogene 2003;22(48):7587–92.
[46] Ingvarsson S, et al. Reduced Fhit expression in sporadic and BRCA2-linked breast carcinomas. Cancer Res 1999;59(11):2682–9.
[47] Turner BC, et al. The fragile histidine triad/common chromosome fragile site 3B locus and repair-deficient cancers. Cancer Res 2002;62(14): 4054–60.
[48] Deans B, Griffin CS, O'Regan P, Jasin M, Thacker J. Homologous recombination deficiency leads to profound genetic instability in cells derived from Xrcc2-knockout mice. Cancer Res 2003;63(23):8181–7.
[49] Wang RH, Yu H, Deng CX. A requirement for breast-cancer-associated gene 1 (BRCA1) in the spindle checkpoint. Proc Natl Acad Sci USA 2004;101(49):17108–13.
[50] Lahkim Bennani-Belhaj K, et al. The Bloom syndrome protein limits the lethality associated with RAD51 deficiency. Mol Cancer Res 2010;8(3): 385–94.
[51] Liu N, et al. XRCC2 and XRCC3, new human Rad51-family members, promote chromosome stability and protect against DNA cross-links and other damages. Mol Cell 1998;1(6):783–93.
[52] Rothkamm K, Kruger I, Thompson LH, Lobrich M. Pathways of DNA double-strand break repair during the mammalian cell cycle. Mol Cell Biol 2003;23(16):5706–15.
[53] Guirouilh-Barbat J, Huck S, Lopez BS. S-phase progression stimulates both the mutagenic KU-independent pathway and mutagenic processing of KU-dependent intermediates, for nonhomologous end joining. Oncogene 2008;27(12):1726–36.
[54] Buard J, de Massy B. Playing hide and seek with mammalian meiotic crossover hotspots. Trends Genet 2007;23(6):301–9.
[55] Gaj T, Gersbach CA, Barbas CF. ZFN, TALEN, and CRISPR/Cas-based methods for genome engineering. Trends Biotechnol 2013;31:397–405.
[56] Tebas P, et al. Gene editing of CCR5 in autologous CD4 T cells of persons infected with HIV. New Engl J Med 2014;370(10):901–10.
[57] Anguela XM, et al. Robust ZFN-mediated genome editing in adult hemophilic mice. Blood 2013;122(19):3283–7.
[58] Takata M, et al. The Rad51 paralog Rad51B promotes homologous recombinational repair. Mol Cell Biol 2000;20(17):6476–82.
[59] Costantino L, et al. Break-induced replication repair of damaged forks induces genomic duplications in human cells. Science 2014;343(6166): 88–91.
[60] Gangloff S, Soustelle C, Fabre F. Homologous recombination is responsible for cell death in the absence of the Sgs1 and Srs2 helicases. Nat Genet 2000;25(2):192–4.
[60a] Hicks WM, Kim M, Haber JE. Increased mutagenesis and unique mutation signature associated with mitotic gene conversion. Science 2010;329:82–5.
[61] Bernstein KA, Gangloff S, Rothstein R. The RecQ DNA helicases in DNA repair. Annu Rev Genet 2010;44:393–417.
[62] Vindigni A, Hickson ID. RecQ helicases: multiple structures for multiple functions? HFSP J 2009;3(3):153–64.
[63] Saintigny Y, et al. Characterization of homologous recombination induced by replication inhibition in mammalian cells. EMBO J 2001;20(14): 3861–70.
[64] Johnson RD, Jasin M. Sister chromatid gene conversion is a prominent double-strand break repair pathway in mammalian cells. EMBO J 2000;19(13):3398–407.
[65] Potts PR, Porteus MH, Yu H. Human SMC5/6 complex promotes sister chromatid homologous recombination by recruiting the SMC1/3 cohesin complex to double-strand breaks. EMBO J 2006;25(14):3377–88.
[66] Tittel-Elmer M, et al. Cohesin association to replication sites depends on rad50 and promotes fork restart. Mol Cell 2012;48(1):98–108.
[67] Huertas P, Jackson SP. Human CtIP mediates cell cycle control of DNA end resection and double strand break repair. J Biol Chem 2009;284(14): 9558–65.
[68] Chen X, et al. Cell cycle regulation of DNA double-strand break end resection by Cdk1-dependent Dna2 phosphorylation. Nat Struct Mol Biol 2011;18(9):1015–9.
[69] Tomimatsu N, et al. Phosphorylation of EXO1 by CDKs 1 and 2 regulates DNA end resection and repair pathway choice Nozomi. Nat Commun 2014;5:3561.
[70] Bothmer A, et al. Regulation of DNA end joining, resection, and immunoglobulin class switch recombination by 53BP1. Mol Cell 2011;42(3): 319–29.
[71] Grabarz A, et al. A role for BLM in double-strand break repair pathway choice: prevention of CtIP/Mre11-mediated alternative nonhomologous end-joining. Cell Rep 2013;5(1):21–8.
[72] Escribano-Diaz C, et al. A cell cycle-dependent regulatory circuit composed of 53BP1-RIF1 and BRCA1-CtIP controls DNA repair pathway choice. Mol Cell 2013;49(5):872–83.
[73] Feng L, Fong KW, Wang J, Wang W, Chen J. RIF1 counteracts BRCA1-mediated end resection during DNA repair. J Biol Chem 2013;288(16): 11135–43.
[74] Xu G, et al. REV7 counteracts DNA double-strand break resection and affects PARP inhibition. Nature 2015;521(7553):541–4.

[75] Boersma V, et al. MAD2L2 controls DNA repair at telomeres and DNA breaks by inhibiting 5′ end resection. Nature 2015;521:537–40.
[76] Hu Y, et al. RECQL5/Recql5 helicase regulates homologous recombination and suppresses tumor formation via disruption of Rad51 presynaptic filaments. Genes Dev 2007;21(23):3073–84.
[77] Wu L, et al. BLAP75/RMI1 promotes the BLM-dependent dissolution of homologous recombination intermediates. Proc Natl Acad Sci USA 2006;103(11):4068–73.
[78] Van Brabant AJ, Stan R, Ellis NA. DNA helicases, genomic instability, and human genetic disease. Annu Rev Genomics Hum Genet 2000;1(1): 409–59.
[79] Bacquin A, et al. The helicase FBH1 is tightly regulated by PCNA via CRL4(Cdt2)-mediated proteolysis in human cells. Nucleic Acids Res 2013;41(13):6501–13.
[80] Chu WK, et al. FBH1 influences DNA replication fork stability and homologous recombination through ubiquitylation of RAD51. Nat Commun 2015;6:5931.
[81] Moldovan G-L, et al. Inhibition of homologous recombination by the PCNA-interacting protein PARI. Mol Cell 2012;45(1):75–86.
[82] Barber LJ, et al. RTEL1 maintains genomic stability by suppressing homologous recombination. Cell 2008;135(2):261–71.
[83] Sun W, et al. The FANCM ortholog Fml1 promotes recombination at stalled replication forks and limits crossing over during DNA double-strand break repair. Mol Cell 2008;32(1):118–28.
[84] Bertrand P, Saintigny Y, Lopez BS. p53's double life: transactivation-independent repression of homologous recombination. Trends Genet 2004;20(6):235–43.
[85] Laulier C, et al. Bcl-2 inhibits nuclear homologous recombination by localizing BRCA1 to the endomembranes. Cancer Res 2011;71(10):3590–602. Available at: http://www.ncbi.nlm.nih.gov/entrez/query.fcgi?cmd=Retrieve&db=PubMed&dopt=Citation&list_uids=21444675.
[86] Plo I, et al. AKT1 inhibits homologous recombination by inducing cytoplasmic retention of BRCA1 and RAD51. Cancer Res 2008;68(22): 9404–12.
[87] Ceccaldi R, et al. Homologous-recombination-deficient tumours are dependent on Polθ-mediated repair. Nature 2015;518(7538):258–62.
[88] Mateos-gomez PA, et al. Mammalian polymerase θ promotes alternative NHEJ and suppresses recombination. Nature 2015;518:254–7.
[89] Walsh T, et al. Mutations in 12 genes for inherited ovarian, fallopian tube, and peritoneal carcinoma identified by massively parallel sequencing. Proc Natl Acad Sci USA 2011;108(44):18032–7.
[89a] Ameziane N, May P, Haitjema A, van de Vrugt HJ, van Rossum-Fikkert SE, Ristic D, et al. A novel Fanconi anaemia subtype associated with a dominant-negative mutation in RAD51. Nat Commun 2015;6:8829.
[90] Shen WH, et al. Essential role for nuclear PTEN in maintaining chromosomal integrity. Cell 2007;128(1):157–70.
[91] Slupianek A, et al. BCR/ABL regulates mammalian RecA homologs, resulting in drug resistance. Mol Cell 2001;8(4):795–806.
[92] Tennstedt P, et al. RAD51 overexpression is a negative prognostic marker for colorectal adenocarcinoma. Int J Cancer 2013;132(9):2118–26.
[93] Miyagawa K. Clinical relevance of the homologous recombination machinery in cancer therapy. Cancer Sci 2008;99(2):187–94.
[94] Jekimovs C, et al. Chemotherapeutic compounds targeting the DNA double-strand break repair pathways: the good, the bad, and the promising. Front Oncol 2014;4:86.
[95] Ward A, Khanna KK, Wiegmans AP. Targeting homologous recombination, new pre-clinical and clinical therapeutic combinations inhibiting RAD51. Cancer Treat Rev 2015;41(1):35–45.
[96] Bryant HE, et al. Specific killing of BRCA2-deficient tumours with inhibitors of poly(ADP-ribose) polymerase. Nature 2005;434(7035):913–7.
[97] Farmer H, et al. Targeting the DNA repair defect in BRCA mutant cells as a therapeutic strategy. Nature 2005;434(7035):917–21.
[98] Mladenov E, Magin S, Soni A, Iliakis G. DNA double-strand break repair as determinant of cellular radiosensitivity to killing and target in radiation therapy. Front Oncol 2013;3(May):113.
[99] Kricker MC, Drake JW, Radman M. Duplication-targeted DNA methylation and mutagenesis in the evolution of eukaryotic chromosomes. Proc Natl Acad Sci USA 1992;89(3):1075–9.
[100] Szostak JW, Orr-Weaver TL, Rothstein RJ, Stahl FW. The double-strand-break repair model for recombination. Cell 1983;33(1):25–35.

Chapter 21

Telomere Maintenance and Genome Stability

W. Hernandez-Sanchez[a], M. Xu[a], D.J. Taylor
Case Western Reserve University, Cleveland, OH, United States

Chapter Outline

1. INTRODUCTION

As early as the 1930s, it was noted that linear chromosomes possessed special ends that prevented them from being cleaved and fused together [1,2]. These data, along with similar discoveries, provided evidence that the free ends of the chromosomes contributed directly toward genome stability. The free, or natural, chromosome ends were eventually given the name "telomeres" [2]. Later, in the 1960s, Leonard Hayflick discovered that somatic cells could divide only a finite number of times before cell division was halted [3]. In the following decade, Alexei Olovnikov recognized that the ends of chromosomes could not be fully replicated and he suggested that the gradual shortening of chromosome DNA might contribute to the limited replicative capacity of the cell [4]. At around the same time, James Watson was investigating the so-called "end-replication problem" of DNA synthesis [5]. The end-replication problem refers to the inability of DNA polymerases to fully synthesize the ends of DNA on the lagging strand. As a consequence, the ends of chromosomes become shorter with each cell division. It was later demonstrated that, on average, chromosomes in human diploid cells lose between 100 and 150 base pairs per cell division, and most mammalian cells are able to divide about 40–60 times before entering a state of replicative senescence [6,7].

It was not until 1978 that telomere DNA sequence was identified in the ciliate protozoa, *Tetrahymena thermophila* [8]. In this work, it was discovered that the ends of chromosomes contain a repeating DNA sequence (TTGGGG in *T. thermophila*). Furthermore, it was demonstrated that *T. thermophila* telomeric sequence could be recognized in yeast, suggesting that a unique telomere replication process was conserved among distant organisms [9]. In 1985, such a specialized enzyme called "terminal transferase" and eventually named telomerase was identified as being responsible for the

a. These authors contributed equally to the preparation of this chapter.

Genome Stability. http://dx.doi.org/10.1016/B978-0-12-803309-8.00021-5

replication of telomere DNA [10]. The seminal work focusing on the discovery, composition, and maintenance of telomeres and telomerase eventually led to the awarding of the 2009 Nobel Prize in Physiology and Medicine to Elizabeth Blackburn, Jack Szostak, and Carol Greider.

We now know that telomere DNA extends for thousands of base pairs of double-stranded DNA (dsDNA) before ending with a G-rich single-stranded DNA (ssDNA) overhang that serves as the template for telomerase extension. One function of telomeres is to absorb the loss of DNA caused by the end-replication problem and to prevent the loss of genomic information. As such, telomere length in healthy, adult somatic cells is somewhat heterogeneous among individuals and populations; however, telomere lengths tend to gradually become shorter as part of the natural aging process. On average, telomere length in healthy, human adult cells ranges from 5 to 15kB, with the 3′ overhang extending for an additional 50–200 nucleotides [11,12]. While telomere sequence is conserved among mammals $(TTAGGG)_n$, telomere length varies among species. As an example, mice have extremely long telomeres (~30–150kb), which has complicated the use of mouse models to explore the processes involved in telomere regulation in humans.

Telomere DNA is bound and protected by a core group of six proteins that is collectively referred to as the shelterin complex [13]. Three shelterin proteins interact specifically with telomere DNA; telomeric repeat-binding factors 1 and 2 (TRF1 and TRF2) bind to the double-stranded region of the telomere and protection of telomere 1 (POT1) binds to the telomere ssDNA. RAP1 (the human ortholog of yeast repressor/activator protein 1) interacts directly with TRF2 to modulate its function. The two remaining shelterin proteins, TIN2 (TERF1-interacting nuclear factor 2) and TPP1 behave as a molecular conduit in the shelterin complex, interacting with TRF2, TRF1, and POT1 to form a direct, protein-mediated link between telomere dsDNA and ssDNA. The removal of individual shelterin proteins induces a complex set of DNA-damage responses, which includes traditional and nontraditional repair mechanisms [14–16]. As such, a primary function of shelterin is to bind telomere DNA and protect it from inadequate recognition of DNA damage–response machinery. However, data are emerging which suggest that the shelterin complex plays a much more versatile role in telomere maintenance and cell-signaling events.

Telomerase is a specialized and unique ribonucleoprotein (RNP) complex that is responsible for maintaining telomere length homeostasis [17]. Telomerase is minimally composed of a catalytic subunit that contains the telomerase reverse transcriptase (TERT) activity and a telomerase RNA component (TR or TERC), which serves as the template for telomeric DNA synthesis [17] (Fig. 21.1). In addition to nucleotide addition, telomerase translocates its RNA template after six nucleotides are synthesized, so that it may be reused as a template for the next set of six nucleotides to be synthesized. This mechanism, referred to as repeat addition processivity, is coordinated by multiple domains within TERT and TR to prevent dissociation from the telomere and to orchestrate realignment of the template RNA with the newly synthesized DNA product.

While TERT is responsible for telomerase catalytic functions, TR provides the RNA template needed to elongate telomeres. In addition to the nucleotides that are responsible for incorporation of the consensus sequence by Watson–Crick base pairing, the TR template contains additional nucleotides that are equally important for initial binding and for proper alignment. Upon recruitment, telomerase binds to the DNA 3′ end-flanking region by complementarity of those nucleotides that are adjacent to the coding template region (Fig. 21.1A). Nucleotides are then reverse transcribed into telomeric ssDNA until the end of the coding template is reached. At this point, telomerase translocates the DNA strand to realign the template region with freshly synthesized telomere DNA to repeat the entire process without primer dissociation.

2. TELOMERE LENGTH AND TELOMERASE REGULATION

In addition to age, telomere length generally correlates with cell function. Cells that exhibit a high proliferative rate (eg, during embryogenesis, in adult germline, and proliferative cells of tissue renewal) express telomerase to maintain longer telomeres and to prevent senescence [7]. In healthy somatic cells, telomerase activity is below detection limits and progressive telomere shortening is observed. Cells that express moderate amounts of telomerase, such as hematopoietic stem cells (HSCs), have the ability to maintain telomere length but not as efficiently as cells that constitutively express telomerase, as is the case in most cancer cells. Generally, cancer cells reactivate and/or upregulate telomerase to maintain telomeres, albeit at reduced lengths. Other evidence suggests a putative mechanism in which telomerase is activated in response to the detection of extremely short telomeres that are at a higher risk for inducing chromosome instability [18]. In these cases, activation of telomerase is sufficient for avoiding cell death mechanisms that would otherwise be initiated. The exact mechanism of how telomere length and telomerase expression is regulated, particularly during cancer progression, remains unclear. Nonetheless, there is a clear connection between telomere length, telomerase activity, and gene stability in a wide range of cell types.

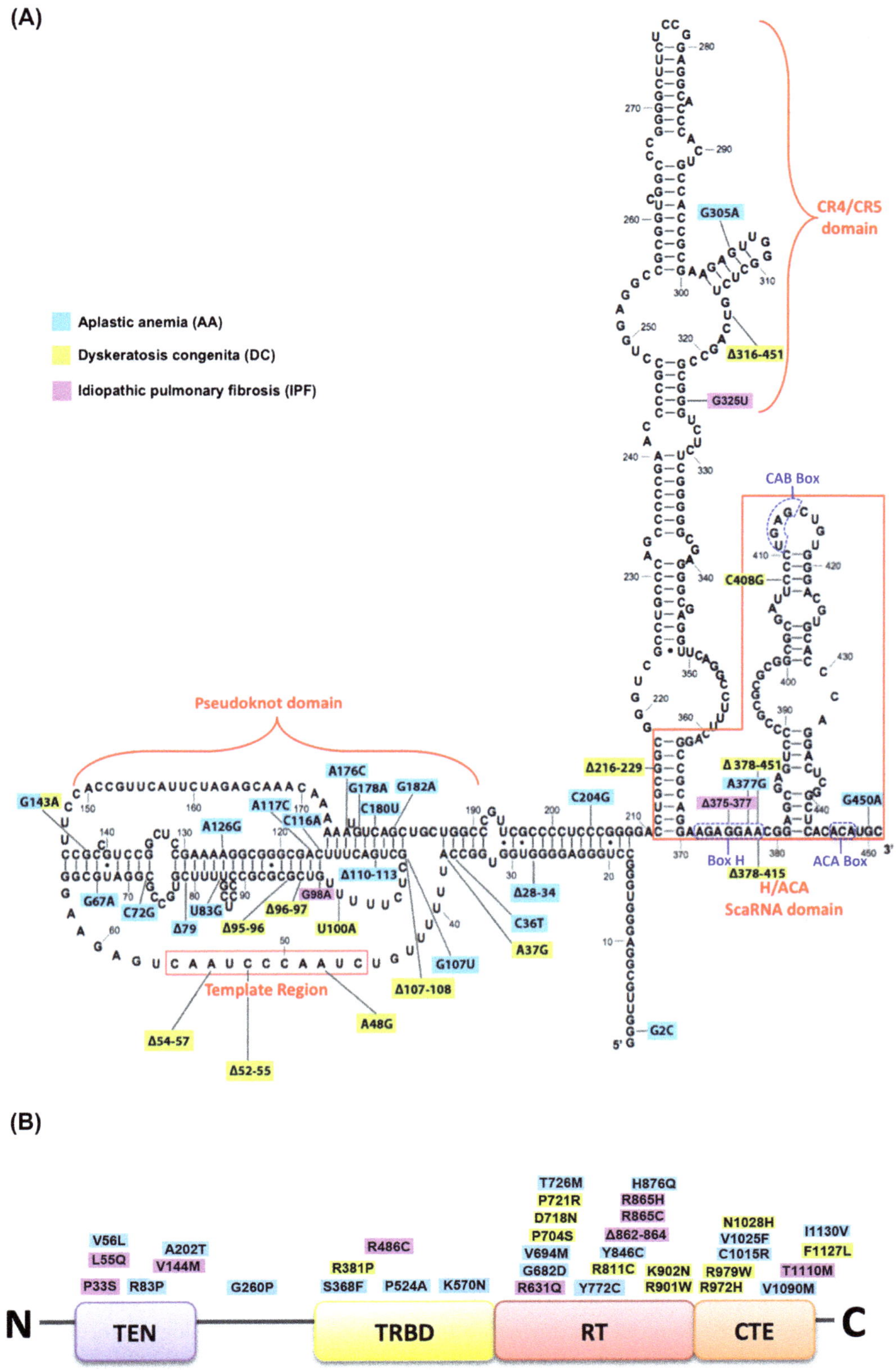

FIGURE 21.1 Telomerase (TERT and TR) domain topology. (A) Schematic representation of telomerase RNA structure. The four primary domains, pseudoknot domain, template region, CR4/CR5 domain, and H/ACA box, are labeled. Mutations associated with telomerase-associated disorders are individually identified and labeled by color code for each disease. (B) Ribbon diagram of TERT organization with N-terminal (TEN), RNA-binding domain (TRBD), reverse transcriptase (RT) domain, and C-terminal domain (CTE) domains labeled. Mutations related to telomerase-associated disorders are identified and labeled in color code for each disease. In both panels, *blue labels* are linked to AA, *green* for DC, and *purple* for IPF.

2.1 Germ Cells and Embryogenesis

In contrast to somatic cells, telomerase activity is detected in adult testes and ovaries. More precisely, the highly proliferative, immature germ cells (oocytes and spermatocytes) display high levels of telomerase activity [19]. Once the germ cells mature and become nonproliferative (eg, sperm and ovum), telomerase activity decreases. As expected, the telomere length in both female and male germline tends to be longer than those in somatic cells. Interestingly, the average telomere length of germ cells in female mice has been reported to decrease with age while maintaining a consistent length, or even increasing, in the germline of male mice [20]. The authors of this study speculate that telomerase activity is associated with fertility, and that telomerase expression and reproductive aging are intertwined. This hypothesis is supported by findings demonstrating that telomerase-negative mice (lacking TR) exhibit enhanced telomere erosion, defective spermatogenesis, decreased proliferation, and an increased incidence of apoptosis.

During embryo development, telomerase activity remains low until the blastocyte stage [21]. At the morula/blastocyte transition, telomeres are elongated significantly in a telomerase-dependent manner. Telomerase levels remains high until total histological differentiation of organs occurs. As an example, telomerase activity in human embryos becomes suppressed at 12 weeks gestation in heart cells when cardiac myoblasts differentiate to myocytes [19]. In mice, by 16 weeks gestation telomerase activity is heterogeneous across tissues, and by birth it diminishes to undetectable levels in brain, liver, skin, lung, muscle, and adrenal tissue [19].

2.2 Stem Cells

As is the case during development, telomerase activity tends to decrease with stem cell differentiation events. For example, when HSCs are stimulated with cytokines to induce differentiation, telomerase expression decreases simultaneously [22]. And although telomerase activity tends to be higher in stem cells than in somatic cells, it is still considerably lower than it is in germ cells, during embryogenesis, or in cancer [23]. One reason that may explain this difference could be due to the quiescent, or very slow growing, nature of some stem cells, which limits the rate of cell division and, thus, the need for telomerase activity. A comparison between HSCs and adult bone-marrow stem cells, and between stem cells from fetal liver and cord blood, revealed that a reduction in telomere length does occur in stem cells [24]. While there is agreement for elevated telomerase activity in HSCs, human mesenchymal stem cells (MSCs) do not have detectable levels of telomerase [25]. Also, MSC telomeres show attrition at a rate that is comparable to somatic cells (30–120 base pairs per cell population doubling) [25]. At least in mice, however, the removal of telomerase adversely affects MSC function and prevents differentiation [26]. Similarly, the proliferation and differentiation of stem cells is linked to telomere length and telomerase activity, further indicating that proper telomere maintenance is an important modulator of stem cell behavior [27]. As with somatic cells, these studies support a correlation between telomere length and telomerase activity in stem cells. Furthermore, they also highlight a critical role for telomerase in maintaining stem cell function, even when its activity is below current levels of detection.

3. ORGANIZATION AND FUNCTION OF TERT AND TR

3.1 TERT Organization

TERT is subdivided into four major domains: the essential N-terminal (TEN) domain, a TERT RNA-binding domain (TRBD), the reverse-transcriptase (RT) domain, and a C-terminal extension (CTE) (Fig. 21.1B) [17,28]. While each domain accounts for individual responsibilities, all four regions are necessary for TERT-orchestrated catalytic functions. The TEN domain facilitates telomerase recruitment and processivity, putatively by acting as a "clamp" to pull the freshly synthesized telomere DNA out of the telomerase active site to assist in translocation of the template, one hexamer at a time [29]. The TRBD domain is highly conserved among species and provides a high affinity-binding platform to facilitate specific interactions with TR that are important for holoenzyme assembly and function [30]. The RT region contains the active, catalytic site of the enzyme that ensures proper alignment of the telomerase RNA template and is responsible for nucleotide addition of the polymerase activity. The RT domain is also highly conserved and shares clear homology, particularly within its active site, with the RT domains of retrotransposons and retroviral RTs [28,31]. Despite sharing little sequence homology, the CTE is structurally analogous to the thumb domain of other RTs [32]. Functionally, the CTE stabilizes the telomerase–DNA complex and is required for telomerase-mediated nucleotide addition and processivity [33].

Although the full-length structure of TERT is yet to be determined, each of its domains have been solved individually or with other domains. The *T. thermophila* telomerase TEN domain represents a single, globular domain with a novel protein fold [34]. In addition to a C-terminal RNA-binding motif, the structure reveals a series of charged and conserved residues on its surface that comprise a putative DNA-binding groove. The structure of the *T. thermophila* RBD revealed that it contains two conserved (CP and T) motifs within its RNA-binding pocket [35]. The CP and T motifs form a wide hydrophilic pocket that is implicated in binding to the template boundary element of TR. The most complete structure of TERT is for that of the red flour beetle, *Tribolium castaneum* [32,36]. Although the protein lacks a TEN domain, the *T. castaneum* structure reveals how the RBD and RT domains interact to form a large cavity in the center of the protein to accommodate a DNA–RNA, primer–template duplex. The RT domain of TERT is structurally similar to other polymerases and RTs in that it includes a fingers, palm, and thumb topology for gripping nucleic acid. The TERT CTE is localized at the same position as the thumb domain with respect to the palm and finger subdomains of viral RNA and B-family DNA polymerases.

3.2 Telomerase RNA Component

In contrast to TERT, TR from different organisms varies in both sequence and length, ranging from 159 nt in *T. thermophila* and 451 nt in human to 1157 in *Saccharomyces cerevisiae* and 1213 nt in *Schizosaccharomyces pombe* [37]. Even though TR elements are highly divergent, several common motifs have been identified by phylogenetic and mutational studies among different species. These common elements include a template pseudoknot (T-PK) that contains the template used for telomeric DNA synthesis, conserved regions 4 and 5 (CR4/CR5), the vertebrate-specific hairpin-hinge-hairpin-ACA (H/ACA) and conserved region 7 (CR7) motifs (Fig. 21.1A). The T-PK and CR4/CR5 bind to TERT independently, and are the minimal TR elements required for reconstituting telomerase activity in vitro [38]. The H/ACA box, which is a small Cajal body-specific RNA at the 3′ end of vertebrate TR, is critical for telomerase holoenzyme biogenesis and regulation in vivo [39]. The CR7 domain is also located at the 3′ end of vertebrate TR and contains a conserved Cajal body (CAB box) localization element. The CAB box is a conserved feature that is responsible for targeting and localization of small Cajal body-specific RNAs (scaRNAs) to Cajal bodies where posttranscriptional processing occurs [40].

Although the expression of TR and TERT are the minimal requirements for reconstituting telomerase activity in vitro, there are a number of accessory proteins that are important for the composition and/or assembly of the telomerase holoenzyme. Several of these proteins, including dyskerin, NHP2, NOP10, and GAR1 comprise a major class of small nucleolar polypeptides that recognize sequence elements in RNA that include an H/ACA signature. The H/ACA box is a characteristic feature common in two classes of small RNAs—small nucleolar RNAs (snoRNAs) and scaRNAs. This motif generally guides the site-specific pseudouridylation of RNAs in the nucleolus and Cajal bodies, respectively [41]. The dyskerin pseudoduridine synthase, NHP2, NOP10, and GAR1 associate with H/ACA RNAs to guide site-specific pseudoridylation of ribosomal and spliceosomal RNAs, as well as other processes unique to eukaryotes and archaea [42]. In human TR, the H/ACA-motif (Fig. 21.1A) is bound by dyskerin, NHP2, NOP10, and GAR1 to modulate telomerase RNP biogenesis and stability [43]. In spite of lacking any pseudouridylation modifications, TR is classified as a scaRNA, primarily due to the presence and necessity of a CAB box and because of an essential requirement for localization to Cajal bodies for processing. The disruption of H/ACA protein interactions with TR, or abrogated Cajal body localization of TR, is an event that is associated with telomerase dysfunction and impaired telomere extension [39,44]. In addition to the H/ACA-binding proteins, another cofactor called TCAB1 (telomerase Cajal body protein 1; also known as WRAP53) mediates trafficking of telomerase to Cajal bodies for efficient biogenesis and subsequent telomere maintenance [45].

While it has yet to be determined for vertebrate TR biogenesis, the mature 3′-end of TR in fungi is generated by a spliceosome-mediated cleavage event [46,47]. The mature TR molecule in vertebrates contains a 2,2,7-trimethylguanosine cap with no polyA tail [48]. In human cells, mature telomerase exists in equilibrium with unassembled TR and TERT components [49]. Together, these data provide compelling evidence that implicates posttranscriptional processes, biogenesis, and assembly as important regulators in controlling telomerase activity.

4. TELOMERIC DNA STRUCTURE

Telomeric DNA is comprised of repetitive G-rich sequence motifs oriented 5′–3′ toward the chromosome end. The length and sequence of repeats varies among different species. It is represented by 4.5 repeats of (T_4G_4) sequence in the ciliate *Oxytricha nova*, 20–70 repeats of (T_2G_4) sequence in *T. thermophila*, 10–15 kb of (T_2AG_3) repeats in humans, and 20–50 kb in certain mouse and rat species [37]. Meanwhile, the telomeric DNA repeats in *S. cerevisae* is approximately 300 base pairs of a somewhat heterogeneous $(TG)_{1–4}G_{2–3}$ repeating sequence. In all organisms, telomeric DNA is composed mostly of dsDNA followed by a ssDNA overhang that serves as the substrate for telomerase-mediated extension.

Guanine-rich DNA is capable of forming very stable G-quadruplex (GQ) structures [50,51]. GQs are best characterized by the arrangement of planar arrays formed by four guanine bases held together by forming hydrogen-bonded Hoogsteen base-pairing interactions. GQ structures can form both inter- and intramolecularly and the morphology varies depending on several factors. For example, the orientation of the strands in the GQ can assemble as parallel or antiparallel, or a heterogeneous mix of both, strands. Similarly, the associated metal ion that stabilizes the GQ contributes to GQ topology and strand orientation. Finally, the sequence of nucleotides flanking the GQ structure can influence topology and stability. Multiple GQ topologies assembled using DNA with human telomere sequence have been characterized in molecular detail (Fig. 21.2). The K^+ containing structure of d[AGGG(TTAGGG)$_3$] determined by X-ray crystallography reveals a propeller-like structure with the strands oriented in a parallel fashion [52]. In this arrangement, the guanines are arranged into stacked G-quartets with the K^+ in the center and the TTA loops protruding away like the blades of a propeller. The structure of the same DNA sequence determined by NMR provides a different basket-type GQ conformation [53]. In this topology, all strands reside in an antiparallel orientation. While most of the current knowledge regarding GQ stability and structural polymorphism stems from biophysical experiments using isolated DNA, GQs have also been identified in the telomeres of human cells and in the macronuclei of ciliates and in frog oocytes [54–56].

Telomere loops (T-loop) describe another structure that has been characterized for telomeric DNA. To form a T-loop, the ssDNA overhang is predicted to invade the telomeric duplex DNA to form a lariat configuration [57]. Due to the elevated thermodynamic stability inherent to dsDNA, protein factors such as TRF2 are a requisite for T-loop assembly [57].

5. TELOMERE-INTERACTING PROTEINS

5.1 The Shelterin Complex

Vertebrate telomeres are capped by a multiple-protein complex called shelterin (Fig. 21.3) [13]. Two shelterin components, TRF1 and TRF2, localize specifically at telomeres (Fig. 21.3A). The two proteins are negative regulators of telomere length, as overexpression of either TRF1 or TRF2 leads to gradual telomere attrition in cancer cells [58,59]. TRF1 and TRF2 bind to the telomeric dsDNA as preformed homodimers, which interact through their N-terminal, TRF-homology (TRFH) domains [60]. The structure of the TRFH domain has been determined for both TRF1 and TRF2 and it resembles a twisted horseshoe-like structure with unique interface features to prevent heterodimerization [60]. For example, the amino acid sequence at the TRFH interface differs between TRF1 and TRF2 and the structures implicate these differences in inhibiting TRF1–TRF2 heterodimer formation. The protein–DNA interactions for TRF1 and TRF2 occur exclusively with

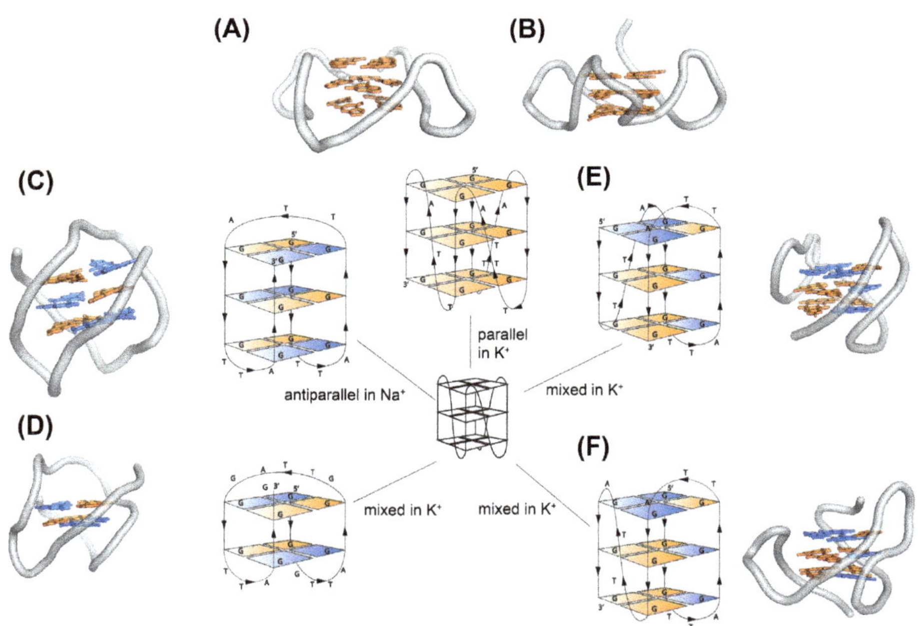

FIGURE 21.2 Three-dimensional and schematic structures of intramolecular G-quadruplexes formed by human telomeric sequences. (A) Parallel G-quadruplex observed for the sequence d[AGGG(TTAGGG)$_3$] in a K^+-containing crystal (PDB ID: 1KF1). (B) Parallel G-quadruplex observed by NMR for the sequence d[TAGGG(TTAGGG)$_3$] in a K^+-containing crowded solution (PDB ID: 2LD8). (C) antiparallel G-quadruplex observed by NMR for the sequence d[AGGG(TTAGGG)$_3$] in Na^+ -containing solution (PDB ID: 143D). (D) Hybrid G-quadruplex with two G-tetrad layers observed by NMR for the sequence d[GGG(TTAGGG)$_3$T] in K^+-containing dilute solution (PDB ID: 2KF8). (E) (3+1) form one hybrid G-quadruplex observed by NMR for the sequence d[TAGGG(TTAGGG)$_3$] in K^+-containing dilute solution (PDB ID: 2JSM). (F) (3+1) form 2 G-quadruplex observed by NMR for the sequence d[TAGGG(TTAGGG)$_3$TT] in K^+-containing dilute solution (PDB ID: 2JSL); Guanines in an anti-configuration are shown in *blue* and those in a syn-configuration are colored *yellow*.

telomere dsDNA and are orchestrated by conserved Myb domains that reside at the C-terminus of both proteins [61,62]. The tertiary structure of the Myb domain of TRFs is represented by three helices [62–64]. Notably, the third helix recognizes the core TAGGG sequence that resides in the major groove of the duplex, telomere DNA.

TIN2 is retained at the telomere through interactions that stabilize the TRF1 and TRF2 DNA-binding ability [65]. TIN2 comprises a central hub of the shelterin complex that maintains interactions with TRF1, TRF2, and TPP1 (Fig. 21.3B) [66–68]. Mutations in TIN2 that impair its binding with TRF1 or TRF2 destabilize telomeres and induce a DNA-damage response [69]. Meanwhile, interactions that reside between TIN2 and TPP1 are necessary for recruitment of the TPP1–POT1 heterodimer to the telomere to bind and protect the ssDNA overhang [70]. The removal of TIN2 protein in mice abrogates the localization of POT1–TPP1 protein at the telomere and triggers an ATR-mediated DNA-damage response [71]. These data suggest that TRF1 and TRF2 recruit TIN2 to the shelterin complex, which in turn recruits POT1–TPP1 to the telomere. In addition to forming interactions that keep the shelterin complex intact, the TRF1–TIN2 interaction prevents SCF^{Fbx4}-mediated ubiquitination and degradation of the TRF1 protein [72].

RAP1 is the most highly conserved shelterin protein with the least understood role in telomere biology. RAP1 forms a complex specifically with TRF2 to enhance its DNA-binding specificity (Fig. 21.3A) [73]. The RAP1–TRF2 interaction has been shown to protect telomeric DNA from nonhomologous end joining (NHEJ) [74]. The role of RAP1 in NHEJ has been controversial, however, as other data identify a role of RAP1 in suppressing homology-directed repair (HDR) at telomeres and not NHEJ, at least in cell lines devoid of KU70–KU80-signaling proteins [75]. Structurally, the RAP1 C-terminal domain forms a conserved module in proteins across species that guides interactions with TRF2 in humans, and is used to recruit SIR3 proteins to regulate gene silencing in budding yeast [76]. Interestingly, the removal of RAP1 from human cell lines has no affect on the other shelterin components or on telomere length homeostasis [77]. These data suggest that RAP1 may play a more crucial role in regulating transcription as opposed to a direct role in telomere maintenance.

The ssDNA overhang at the 3′ end of mammalian telomeres is bound and protected by POT1 protein (Fig. 21.3C) [78]. POT1 was originally thought to behave exclusively as a negative regulator of telomerase activity, but this interpretation gets complicated when POT1 functions with other shelterin proteins. For example, TPP1 is the binding partner of POT1 and the POT1-TPP1 heterodimer increases telomerase activity on telomere DNA [79,80]. Although TPP1 is not known to interact with telomere DNA directly, it increases the affinity of POT1 for telomere DNA substrates [79,80]. Furthermore,

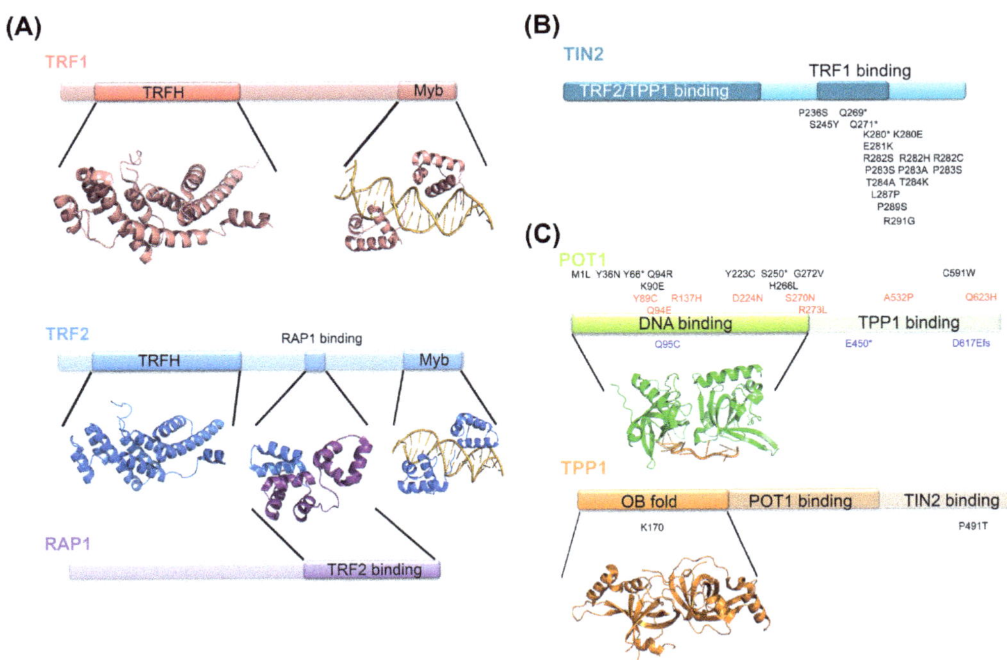

FIGURE 21.3 The structures of shelterin complex components. (A) The top of the panel shows the TRF homology (TRFH) domain (PDB ID: 1H6O) and double-stranded DNA (dsDNA)-bound Myb domain of TRF1; PDB ID: 1W0T). The bottom of the panel displays TRF2 structures, including the TRFH domain (PDB ID: 1H6P), dsDNA-bound Myb domain (PDB ID: 1W0U), and the RAP1–TRF2-interacting domains (PDB ID: 3K6G). (B) Schematic structure of TIN2 with mutations identified related to DC and other telomerase-associated disorders labeled. (C) The top panel shows POT1 organization and includes the structure of the DNA-binding domain bound to telomere DNA (PDB ID: 1XJV). The bottom panel shows TPP1 organization and includes the structure of an internal OB-fold domain (DB ID: 2I46). Mutations identified on POT1 are colored by cancer association: chronic lymphocytic leukemia are colored in black, melanoma in red, and glioma in blue. Mutations identified on TPP1 in Hoyeraal–Hreidarsson syndrome are shown.

TPP1 helps POT1 to discriminate between ssDNA and RNA substrates [81] and plays a central role in the shelterin complex by bridging TIN2, and thus the double-stranded region of the telomere, with POT1 and the ssDNA overhang [70,82,83].

Structurally, the N-terminal domain of POT1 folds into two oligonucleotide/oligosaccharide-binding (OB) domains, which interact intimately with telomeric DNA [84]. In general, the OB fold architecture is a conserved structure comprised of a β-barrel with an α-helix connecting two strands at the end of the barrel. In addition to POT1, a number of nucleic acid–binding proteins, including replication protein A, are represented by OB-fold motifs, indicating a universal role in the direct maintenance of genomic stability [85]. The central domain of TPP1 also represents an OB-fold [79]. However, instead of binding to nucleic acid, the OB-fold of TPP1 is responsible for interactions with telomerase [70,86,87]. Despite poor sequence identity, POT1 and TPP1 are structurally related to the *O. nova* telomere end binding α and β (TEBPα and TEBPβ) heterodimer [79]. TEBPα and TEBPβ are the first identified specific telomeric DNA-binding protein, and the structure of the TEBPα–TEBPβ–DNA complex has been solved by X-ray diffraction [88]. The structure reveals that TEBPα is represented as a series of three OB-fold domains, and TEBPβ is comprised of a single OB-fold motif. The two proteins interact with one another to clamp down on telomere DNA that resides in a groove that is formed between them.

5.2 CST Protein Complex

In addition to shelterin, another multiprotein, telomere-interacting assembly has been described as the CST complex. The CST complex was originally identified in budding yeast to assemble from Cdc13, Stn1, and Ten1 proteins to function in protecting telomeres from DNA degradation [89,90]. In mammalian cells, the homologs of Stn1 and Ten1 have been identified, and a conserved telomere-maintenance component 1 (CTC1) protein, which shares little sequence similarity with Cdc13, associates with STN1 and TEN1 at the telomere [91]. It was originally speculated that the CST complex functions by protecting telomere DNA and downregulating telomerase activity. However, only about 20% of mammalian CST complexes localize to the telomere, suggesting that it is involved in nontelomeric functions as well.

It remains unclear how shelterin and CST coordinate to regulate telomere maintenance. As is the case with shelterin proteins, defects or removal of CST components results in phenotypes that are consistent with telomere dysfunction. For example, the conditional knockdown of *CTC1* impairs efficient replication of telomere DNA [92]. CTC1 and STN1 have been copurified with DNA polymerase α and shown to increase polymerase processivity and assist in the synthesis of RNA–DNA Okazaki fragments [93]. Together, these findings reflect a general role of the CST complex in replication control of telomeres.

6. TELOMERE–TELOMERASE INTERACTIONS AND REGULATION

As mentioned, a primary function of shelterin proteins is to protect telomere DNA from illicit events, such as DNA degradation and end-to-end fusions of different chromosomes. However, the role of shelterin has expanded to include telomere-length maintenance as several of the proteins have been discovered to function in telomerase recruitment and regulation. For example, the POT1–TPP1 heterodimer enhances telomerase activity and processivity [79,80]. The enhancement of telomerase activity can be attributed to a direct protein–protein interaction that allows TPP1 to recruit telomerase to the telomere [70,86,87]. Independent of telomerase recruitment, the POT1–TPP1 heterodimer slows dissociation of telomerase from telomere DNA to assist translocation and enhances telomerase processivity [94]. A number of studies have identified posttranslational modifications to TPP1 that may provide a molecular switch between telomere protection and telomerase recruitment activities [95,96].

Several studies suggest that TIN2 facilitates the localization of POT1-TPP1 to telomeric ssDNA. The removal of TIN2 diminishes the amount of POT1 and TPP1 that localizes at the telomere [71]. Moreover, the depletion of TIN2 but not POT1 results in the failure of TPP1-dependent telomerase recruitment [70]. Together, these data present a scenario in which TRF1 and TRF2 nucleate assembly of the shelterin complex. Indeed, the deletion of both TRF1 and TRF2 in mouse cells results in a complete loss of detection for all six shelterin proteins at the telomere [16].

7. TELOMERE-ASSOCIATED DISEASES

7.1 Telomere Length Homeostasis and Related Diseases

Without question, telomere length homeostasis and the regulation of telomerase are important events to maintain proper genome stability. Short telomeres are associated with several degenerative and age-related disorders. Some examples include dyskeratosis congenita (DC), aplastic anemia (AA), and idiopathic pulmonary fibrosis (IPF). Many of these ailments are associated with single nucleotide polymorphisms within genes that code for TR or TERT (Fig. 21.1A). Several mutations related to disease, particularly for DC, have been identified in the dyskerin protein, which associates with TR. Still other afflictions, including in a wide range of human cancers, are associated with individual mutations in telomerase or one of the shelterin proteins. A comprehensive list of telomerase-associated mutations associated with human disease is maintained at http://telomerase.asu.edu/diseases.html. Within this communication, we focus on those that are the most prevalent and best understood.

7.1.1 Dyskeratosis Congenita

DC is a heritable, progressive telomerase-associated disorders that is inherited as autosomal dominant, autosomal recessive, or X-linked recessive. The clinical presentation of the disease, which includes hematopoietic deficiency, usually correlates with dysfunctional telomere maintenance. Several mutations in genes involved in telomere and telomerase assembly have been identified to correlate, particularly with X-linked DC, which is the most common form. The vast majority of mutations associated with DC (~40% cases) resides within the *DKC1* gene, which encodes the dyskerin protein [97]. While amino acid changes in DKC1 of DC patients are highly heterogeneous, the most common mutation is a missense change that occurs at position 1058 (C>G) of the gene resulting in an amino acid change at position 353 (A>V) on the protein level [98]. This particular *DKC1* mutation interferes with TR stability and pre-RNP assembly of telomerase [99].

Apart from *DKC1* there is a subset of genes involved in telomere maintenance that have been identified to contain mutations in patients with DC. Like dyskerin, *NHP2* and *NOP10* genes code for proteins that are important for assembly and biogenesis of the telomerase RNP and are associated with mutations in DC patients. DC-related point mutations to *NOP10* or *NHP2* impaired critical interactions between protein products and within dyskerin that are required for TR folding and biogenesis [99]. Mutations in *TCAB1* can abrogate the proper localization of TR to Cajal bodies, thus impairing telomerase holoenzyme assembly. Still other mutations associated with DC have been identified in the two primary components of telomerase, TERT and TR [100].

While these findings tend to agree that mutations linked to DC often result in improper TR assembly with TERT, the molecular pathway to this consequence can vary. For example, some of these mutations result in reduced telomerase catalytic activity, while others impair TR–TERT association, while still others cause loss in fidelity of telomeric repeat synthesis [100]. Regardless of the mechanism, it has become clear that telomerase-associated disorders and telomerase deficiency coincide with one another. It is likely that the telomerase deficiencies and shorter telomeres diminish the proliferative capacity of hematopoietic progenitors.

7.1.2 Aplastic Anemia

AA is another condition characterized by telomerase-associated disorders and associated with mutations in *TR* or *TERT*. Telomere length reduction can be detected in the peripheral blood leukocytes of one-third to one-half of these patients [101,102]. Telomere attrition in AA has been attributed to different mutations on telomerase (TERT and TR), as well as a reduction of shelterin complex proteins POT1 and TIN2 [102]. Like DC, mutations in *TERT* and *TR* found in AA are linked to loss or decrease of telomerase catalytic activity, reduction of telomerase repeat addition processivity, and/or TR instability [103,104]. As in DC, AA molecular mechanism linked to telomerase deficiency is based on haploinsufficiency of telomerase in bone-marrow stem cells that impairs telomere maintenance and proliferation. In fact, many DC patients also develop AA.

7.1.3 Idiopathic Pulmonary Fibrosis

IPF is a chronic lung disease characterized by fibroblast proliferation and excessive accumulation of extracellular matrix. IPF patients typically present with disrupted lung tissue architecture and respiratory difficulties. Although there is not a clear mechanism of the molecular pathway leading to this condition, IPF patients tend to have shorter telomeres in alveolar epithelial cells when compared to healthy individuals [105], suggesting that telomere length is an important feature for the pathophysiology of this disease. Mutations in *TERT* or *TR* are considered risk factors for IPF as they have been observed in 8–15% of familial cases [106]. Mutations in essential telomerase genes have been identified in a small subset of sporadic IPF cases as well [105].

As with DC and AA, mutations in the TERT and TR subunits have been identified in patients with familial IPF and are common genetic risk markers [106,107]. Two of the more characterized mutations in *TERT* (V144M and R865C/H) (Fig. 21.1B) that are associated with IPF patients have been shown to impair telomerase-mediated telomere extension [107]. While the V144M mutation did not impair telomerase catalytic activity in vitro, it exhibited impaired telomere synthesis in cultured human cells. In contrast, R865H or R865C mutations affected telomerase extension assays in vitro and impaired telomere synthesis in human cells. The R865 mutation likely affects proper nucleotide binding or incorporation, as that amino acid is within close proximity to the predicted TERT nucleotide-binding pocket.

Also similar to DC and AA, not all patients with IPF have mutations in telomere or telomerase-associated genes. However, a consistency of shorter telomeres among IPF patients suggests that dysfunctional telomere maintenance plays a key role in the molecular pathophysiology of this disease [108]. Mutations that impair telomerase activity or biogenesis commonly result in advanced shortening of telomeres, which limits the replicative potential of progenitor alveolar epithelial cells in IPF cases and can eventually induce apoptosis or senescence. The reduction of alveolar epithelial cell type 2 populations, which are key for the repairing of lung scarring tissue, would make the restructuring of alveolar tissue architecture more difficult in IPF patients with dysfunctional telomeres or telomerase.

7.2 Telomeres and Premature Aging Syndromes

Shortened telomeres are related to an onset of accelerated aging phenotype syndromes such as Werner's syndrome, Bloom's syndrome, and Hutchinson–Gilford progeria syndrome. All are characterized by an increased rate of senescence due to accelerated telomere shortening processes. On average, telomere lengths of the individuals afflicted with these syndromes are significantly shorter when compared to healthy individuals in a similar age group. While telomere shortening is the commonality among patients with advanced aging syndromes, all have different mechanisms leading to this consequence. In the cases of Werner's and Bloom's syndromes, the mechanism involves defects in the DNA helicases WRN and BLM, respectively, which generally increases the rate of transduction of nonfunctional proteins. Both DNA helicases belong to the RecQ family of DNA helicases and are necessary to resolve secondary structure within the telomere DNA (D-loop and GQ) and to unwind dsDNA for replication or DNA damage [109]. Dysfunctional WRN and BLM helicases are associated with incomplete replication of telomeres, recombination between sister chromatids, and inefficient replication of G-rich telomeric strands (WRN) [109,110]. In both cases, the overexpression of telomerase tends to rescue the extreme telomere-shortening phenotype.

As opposed to altered helicase function, Hutchinson–Gilford progeria syndrome is caused by a mutation in the lamin A gene (*LMNA*) at position 1824 (C->T) [111]. Lamin A belongs to a family of proteins that is involved in nuclear assembly, chromatin modifications, and nuclear structure. The C1824T mutation in *LMNA* increases the usage of an internal splicing site that creates a transcript that is missing 150 internal nucleotides. Translation of the transcript results in a protein, called progerin, which is an isoform of LMNA with the deletion of 50 internal amino acids [111]. Telomere shortening and senescent events in normal human fibroblasts activates progerin production and progressive telomere damage increases alternative splicing patterns in multiple genes [112]. In a circuitous cascade of harmful events, some studies have suggested that progerin accumulation leads to telomere deprotection by disrupting heterochromatin structure [112]. Together, these data demonstrate a clear relationship between progerin accumulation and telomere shortening.

7.3 Telomerase Activity in Cancer

While the aforementioned afflictions generally manifest with reduced or aberrant telomerase function, 85–90% of metastatic cancers display increased telomerase activity [113]. The regulation of telomerase activity is presumably governed by TERT expression, as TR and other components of the telomerase complex are constitutively expressed in adult somatic tissues as well as germ cells [114]. Conversely, TERT levels are virtually absent in most adult somatic tissues, but expressed at high levels in most cancer, reproductive, and stem cells where telomerase activity is also elevated [113]. The *TERT* promoter contains several response elements that are recognized by transcription factors, such as c-MYC and SP1 [115]. β-catenin is another protein that regulates gene transcription and interacts with KLF4 at *TERT* promoters to recruit a methyltransferase (Setd1A) to initiate *TERT* transcription [116]. Therefore, the overexpression of c-MYC and WNT/β-catenin that occurs in cancer, may explain the increased levels of TERT in those cells as well. Another explanation can be attributed directly to single-point mutations that have been identified in the *TERT* promoter site in a host of human cancers including skin, central nervous system, bladder, and thyroid [117]. These data indicate that sporadic mutations to regulatory sites might also be responsible for TERT activation in several types of cancer.

Mutations in the *TERT* promoter tend to result in higher *TERT* mRNA levels. One explanation for this correlation can be attributed to an enhanced binding motif that is created for transcription factors such as E-twenty-six (ETS) [117]. While the promoter clearly plays an important role in TERT transcription activation, a number of suppressors that include MAD1, E2F-1, and MZF-2, and PITX1, have been identified to contribute to the regulatory mechanism as well [118–121]. These suppressors exert repressor activity by binding directly to various sites within the *TERT* promoter. In addition, it has been speculated that TERT behaves as a transcriptional modulator of oncogenic genes that are critical for tumor proliferation

[122]. Other mechanisms such as alternative splicing events and epigenetic regulation are likely to be important modulators of TERT regulation as well.

7.4 Shelterin Mutations and Telomere-Related Diseases

Exome sequencing of familial glioma patients has identified inherited mutations in the *POT1* gene that are associated with this type of cancer (p.G95C, p.E450X, and p.D617Efs) [123]. The *POT1* G65C mutation is located within the DNA-binding groove and presumably disrupts interactions with telomere DNA. The *POT1* E450X introduces a premature STOP codon in the translated POT1 protein that would be predicted to lack its TPP1-interacting domain. Similar inherited mutations have been reported in the *POT1* gene of familial melanoma patients [124]. Most of the identified mutations are localized in the POT1 DNA-binding domain, which emphasizes an important relationship between POT1–DNA interactions and the development of familial cancer. Somatic mutations of *POT1* have also been detected in chronic lymphocytic leukemia (CLL) [125]. These studies suggest that mutant POT1 protein fails to localize at the telomere, leaving unprotected telomere ends that could lead to genome instability and tumorigenesis.

Besides POT1, mutations to the genes that code for TIN2 and TPP1 have been identified in patients with telomerase-associated disorders [126–128]. Most of the mutations identified in TIN2 have been identified in patients with DC. Mutations to TIN2 include missense changes as well as nonsense and frameshift mutations to the open reading frame. The corresponding mutations within the synthesized TIN2 protein likely interfere with critical interactions with other shelterin proteins. A 2014 study on Hoyeraal–Hreidarsson syndrome (HH)—a clinically severe variant of DC—revealed *ACD* (codes for TPP1) as a novel DC-related gene [127]. In this study, two mutations (ΔK170 and P491T) were identified at the protein level of TPP1, both of which are highly conserved in mammals. The first mutation is a single amino acid deletion of K170, which is located in a region of TPP1 that is responsible for conducting interactions with telomerase [87]. Interestingly, the ΔK170 mutation of TPP1 has been identified in patients with AA as well [128]. The second mutation is an amino acid substitution (P491T), which is located in the TIN2-interacting domain of TPP1. Together, these data further demonstrate that highly intricate protein–protein and protein–DNA interactions within the shelterin complex contribute to proper genome stability.

8. TELOMERES AS A DNA DAMAGE–PREVENTION SYSTEM

When telomeres become critically short, their capping function is compromised and a range of DNA damage–like responses are induced. In telomerase-deficient yeast cells, short telomeres are recognized as DNA damage and are arrested in G2/M [129]. Markers consistent with DNA-damage response are also triggered in human fibroblasts when telomeres reach a critically short threshold to invoke senescence. These signaling events are remarkably similar to those in cells bearing DNA double-stranded breaks (DSBs) and involve the activation of DNA damage–checkpoint kinases including CHK1 and CHK2. These findings, as well as others, have provided a clear connection between telomere-initiated senescence and innate DNA-damage responses.

In normal telomeres, the shelterin complex collaborates to repress at least six DNA-damage pathways that include ATM and ATR signaling, classical and alternative-nonhomologous end joining (alt-NHEJ), homologous recombination, and resection [16]. Single knockdown studies for each component of the shelterin complex have revealed similar and alternative mechanisms that explain how this protein complex functions to prevent telomeres from appearing as a DNA break in need of repair (Fig. 21.4). The deletion of a POT1 ortholog in mice results in telomere fusions and P53-dependent senescence [130]. POT1 knockdown experiments have provided evidence that it functions, at least in part, to prevent replication protein A (RPA) from binding to telomeric ssDNA, thereby preventing the RPA-induced activation of ATR-dependent DNA-damage responses [14,15,130]. Because of an expanded role in DNA-damage repair and DNA synthesis, RPA exists at a concentration that is much higher in the cell than that of POT1. Both POT1 and RPA display similar binding affinities for telomere DNA, yet physiological levels of POT1 are sufficient to prevent RPA from binding to the telomere. One explanation for this phenomenon can be attributed to a shelterin-related enhancement of POT1 localization and function. Interactions between POT1 and TPP1 with the rest of the shelterin complex effectively localize and concentrate POT1 protein exclusively at the telomere. Furthermore, the inclusion of TPP1 increases the binding affinity of POT1 to telomere DNA nearly 10 times over that of POT1 alone [79]. Highlighting the importance of TPP1 in localizing POT1 to protect ssDNA at the telomere, knockdown experiments revealed that the loss of TPP1 activates ATR-dependent DNA-damage response in a manner similar to that of POT1 removal [131]. Furthermore, TIN2 performs a similar role in recruitment, as its ablation prevents POT1-TPP1 localization at the telomere, thus allowing RPA binding and ATR signaling [71].

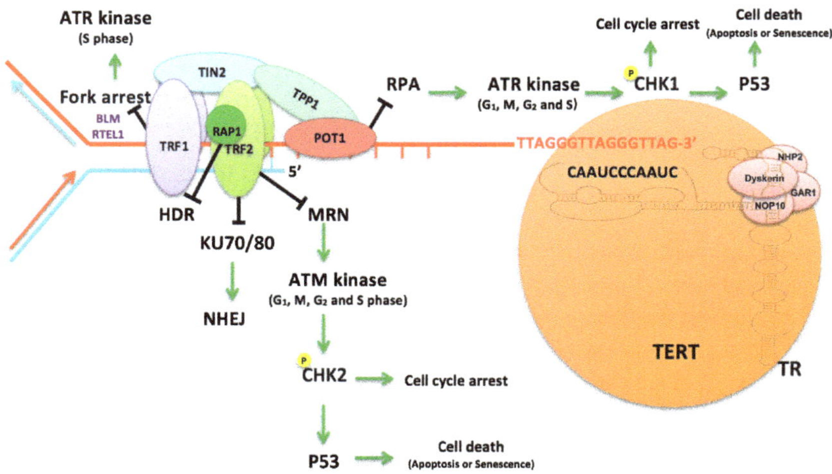

FIGURE 21.4 Shelterin prevents multiple DNA-damage responses at the telomere. The repression mechanism of DNA-damage response by shelterin components are indicated by *blunt-end arrows in black*. The telomerase main components, TR and TERT, are shown in *orange* and telomerase accessory ribonucleotide proteins (dyskerin, GAR1, NHP2, and NOP10) are colored in *pink*. Individual shelterin proteins and their role in preventing various DNA-damage responses are labeled accordingly.

At the double-stranded region of the telomere TRF1, TRF2, and RAP1 also shield telomere DNA from appearing as sites of breaks or damage. Knockdown studies of TRF1 revealed that it is essential for chromosome stability by limiting replicative stress. Mechanistically, TRF1 assists in the proper replication of telomeres by preventing ATR kinase activation and fork stalling [132]. TRF1 may function by coordinated interactions with essential helicases, such as BLM and RTEL1, to facilitate unwinding of GQ structures at the telomere and to avoid fork stalling. Mice deficient for TRF2 are early embryonic lethal. Knockdown of TRF2 in mouse embryonic fibroblast (MEF) cells causes telomere fusions and dsDNA break-like damage activation through MRN (MRE11, RAD50, and NBS1) recruitment and ATM activation [15]. Other studies in cell culture show that the removal of TRF2 allows the KU70–KU80 complex to load onto telomeres to initiate NHEJ DNA repair [133]. Although the TRF2-binding protein, RAP1, is dispensable for ATM activation and NHEJ events, it appears to be critical for repressing HDR at telomeres [75]. Strikingly, these studies demonstrate that HDR events in mouse embryonic fibroblasts lacking RAP1 occur at the telomere even in the absence of a DNA-damage signal. Cumulatively, these knockdown studies reveal a critical function of shelterin proteins in protecting against a range of DNA damage–response mechanisms.

9. CONCLUSIONS AND CLOSING REMARKS

Research relating DNA damage and repair pathways to telomere biology and genome stability will continue to emerge. Cellar phenotypes and molecular pathways have elucidated a role of shelterin and telomerase to function properly to prevent illicit events from occurring. Sequencing data have identified a series of mutations in shelterin proteins and in telomerase that are associated with telomerase-associated disorders and multiple forms of cancer. It will not be surprising to see this list of mutations expand significantly over the next several years. Continued work in telomere biology will also provide further insight into the details of how telomerase gets regulated during development and in cancer.

The basic research focused on telomere biology might lead to improved strategies for treating afflictions related to telomere dysfunction. For example, researchers have explored altering telomere integrity by changing the telomere DNA sequence that is reverse transcribed by telomerase. The overexpression of TR with mutations to its template region inhibits cell growth and induces apoptosis in cancer cells and mouse xenografts [134,135]. In mammals, the template region is 11 nucleotides long (3′-CAAUCCCAAUC-5′) and is localized between positions 46 and 56 of the TR RNA sequence. Studies have shown that different mutations on telomere sequence result in different telomere dysfunction behavior (Table 21.1). Although the usual trend of altering telomere sequence administers a toxic effect, it has been shown that certain sequences can be well tolerated by cells. An example of a nontoxic mutant telomere sequence is mutant TR TSQ1 (3′-CCAACGCCAAC-5′) that codifies telomeric sequence GTTGCG [136]. Studies have shown that TSQ1 mutation can be incorporated into telomeres and the cells are viable for several population doublings. Thus, future insight into what sequence is tolerated and how the telomere mutations coincide with shelterin mutations will provide additional, mechanistic insight into the precise requirements that are necessary for shelterin–telomere DNA interactions and those that result in DNA damage–response induction.

TABLE 21.1 Mutant Telomerase RNA Template

Mutant hTR	RNA Template Sequence	Predicted Sequence	Effect	References
Wild type	3′-CAAUCCCAAUC-5′	TTAGGG	Telomere stability	n/a
hTR-34, MuA, or 47A	3′-CAAACCCAAAC-5′	TTTGGG	Loss of cell viability, cell-cycle deregulation, alteration of nuclear morphology, telomere fusion, DSB-like damages ATM activation, and TRF2 repression	[136–139]
MuC	3′-CAACCCCAACC-5′	TTGGGG	Loss of cell viability, cell-cycle deregulation, alteration of nuclear morphology, and telomere fusion	[138]
MuD	3′-CAAUUCCAAUU-5′	TTAAGG	Loss of cell viability, cell-cycle deregulation, alteration of nuclear morphology, and telomere fusion	[138]
MuE	3′-CACUCCCACUC-5′	TGAGGG	Loss of cell viability, cell-cycle deregulation, alteration of nuclear morphology, and telomere fusion	[138]
AU5	3′-AAUAUAUAUAU-5′	TATATA	Loss of cell viability, DSB-like damages, ATM activation, and TRF2 repression	[134,139]
U11-hTer	3′-UUUUUUUUUUU-5′	AAAAAA	Decreased cell proliferation and cell viability, increased apoptosis, and decreased tumor growth rates	[137]
49A-hTer	3′-CAAUCCCA(AA)AUC-5′	T(TT)TAGGG	Decreased cell proliferation and cell viability, increased apoptosis, and decreased tumor growth rates	[134]
TSQ1	3′-CCAACGCCAAC-5′	GTTGCG	Well-tolerated sequence, do not seem to affect cell survival	[136]

Summary of different mutant TR-template region with their predicted telomeric sequence and genomic consequences. The mutations incorporated in nucleotide bases are underlined.

GLOSSARY

G-quadruplex Guanine-rich DNA capable of forming very stable structures. It is an arrangement of planar arrays formed by four guanine bases held together by forming hydrogen-bonded Hoogsteen base-pairing interactions.

Shelterin complex Group of six proteins (TRF1, TRF2, RAP1, TIN2, TPP1, POT1) that are bound to telomeric DNA. Its main function is to avoid telomere end-to-end fusion and to protect telomeres from being recognized from inappropriate DNA-damage response.

T loop Telomeric DNA structure characterized by the invasion of the ssDNA overhang into the telomeric duplex DNA to form a lariat configuration.

Telomerase Reverse transcriptase that uses its own RNA template to synthesize telomere DNA.

Telomere End cap of linear chromosomes. Its main function is to protect genomic information and control the lifespan of cells.

LIST OF ACRONYMS

AA Aplastic anemia
ACD Adrenocortical dysplasia
alt-NHEJ Alternative-nonhomologous end joining
ATM Ataxia telangiectasia mutated
ATR Ataxia telangiectasia and Rad3 related
BLM Bloom helicase
CAB box Cajal body localization element
CHK1 Checkpoint kinase 1
CHK2 Checkpoint kinase 2
CLL Chronic lymphocytic leukemia
CR4 Conserved region 2
CR5 Conserved region 5

CR7 Conserved region 7
CST Cdc13, Stn1, and Ten1 proteins
CTC1 Conserved telomere-maintenance component 1
CTE C-terminal extension
DC Dyskeratosis congenita
dsDNA Double-stranded DNA
ETS E-twenty-six
GAR1 H/ACA ribonucleoprotein complex subunit 1
GQ G-quadruplex
H/ACA Hairpin-hinge-hairpin-ACA
HDR Homology-directed repair
HSCs Hematopoietic stem cells
IPF Idiopathic pulmonary fibrosis
KLF4 Kruppel-like factor 4
LMNA Lamin A
MEF Mouse embryonic fibroblast
MRN MRE11, RAD50, and NBS1 complex
MSCs Mesenchymal stem cells
NHEJ Nonhomologous end joining
NHP2 H/ACA ribonucleoprotein complex subunit 2
NOP10 H/ACA ribonucleoprotein complex subunit 3
OB Oligonucleotide/oligosaccharide binding
PIP1 POT1-interacting protein 1
POT1 Protection of telomere 1
PTOP POT1- and TIN2-organizing protein
RAP1 Repressor/activator protein 1
RNP Ribonucleoprotein
RPA Replication protein A
RT Reverse transcriptase
RTEL1 Regulator of telomere elongation helicase 1
scaRNA Small Cajal body-specific RNA
SCF Skp, Cullin, F-box-containing complex
SIR3 Silent information regulator 3
snoRNA Small nucleolar RNA
SP1 Specificity protein 1
ssDNA Single-stranded DNA
TCAB1 Telomerase Cajal body protein 1
TEBP Telomere end-binding protein
TEN TERT N-terminal
TEN1 CST complex subunit TEN1
TERT Telomerase reverse transcriptase
TIN2 TERF1-interacting nuclear factor 2
TINT1 TIN2-interacting protein 1
T-loop Telomere loop
T-PK Template pseudoknot
TPP1 TINT1/PTOP/PIP1
TR or TERC Telomerase RNA component
TRBD TERT RNA-binding domain
TRF1 Telomeric repeat-binding factor 1
TRF2 Telomeric repeat-binding factor 2
TRFH TRF-homology domains
TSQ1 Tolerated sequence Q1
WRN Werner helicase

ACKNOWLEDGMENTS

We wish to thank Harry Scott for valuable suggestions and critical reading of this chapter. Research in the Taylor Lab is supported by NIH grants DP2 CA186571 and R21 CA169611, and an American Cancer Society Research Scholar Grant, RSG-13-211-01-DMC. W.H.S. is supported by NIH/NCI supplement DP2 CA186571-01S1.

REFERENCES

[1] McClintock B. The behavior in successive nuclear divisions of a chromosome broken at meiosis. Proc Natl Acad Sci USA 1939;25(8):405–16.
[2] Muller H. The re-making of chromosomes. Collect Net 1938;(13):181–98.
[3] Hayflick L. The limited in vitro lifetime of human diploid cell strains. Exp Cell Res 1965;37:614–36.
[4] Olovnikov AM. Principle of marginotomy in template synthesis of polynucleotides. Dokl Akad Nauk SSSR 1971;201(6):1496–9.
[5] Watson JD. Origin of concatemeric T7 DNA. Nat New Biol 1972;239(94):197–201.
[6] Huffman KE, Levene SD, Tesmer VM, Shay JW, Wright WE. Telomere shortening is proportional to the size of the G-rich telomeric 3′-overhang. J Biol Chem 2000;275(26):19719–22.
[7] Bodnar AG, Ouellette M, Frolkis M, Holt SE, Chiu CP, Morin GB, et al. Extension of life-span by introduction of telomerase into normal human cells. Science 1998;279(5349):349–52.
[8] Blackburn EH, Gall JG. Tandemly repeated sequence at termini of extrachromosomal ribosomal-RNA genes in Tetrahymena. J Mol Biol 1978;120(1):33–53.
[9] Szostak JW, Blackburn EH. Cloning yeast telomeres on linear plasmid vectors. Cell 1982;29(1):245–55.
[10] Greider CW, Blackburn EH. Identification of a specific telomere terminal transferase activity in Tetrahymena extracts. Cell 1985;43(2 Pt 1):405–13.
[11] Makarov VL, Hirose Y, Langmore JP. Long G tails at both ends of human chromosomes suggest a C strand degradation mechanism for telomere shortening. Cell 1997;88(5):657–66.
[12] de Lange T, Shiue L, Myers RM, Cox DR, Naylor SL, Killery AM, et al. Structure and variability of human chromosome ends. Mol Cell Biol 1990;10(2):518–27.
[13] de Lange T. Shelterin: the protein complex that shapes and safeguards human telomeres. Genes Dev 2005;19(18):2100–10.
[14] Guo X, Deng Y, Lin Y, Cosme-Blanco W, Chan S, He H, et al. Dysfunctional telomeres activate an ATM-ATR-dependent DNA damage response to suppress tumorigenesis. EMBO J 2007;26(22):4709–19.
[15] Denchi EL, de Lange T. Protection of telomeres through independent control of ATM and ATR by TRF2 and POT1. Nature 2007;448(7157):1068–71.
[16] Sfeir A, de Lange T. Removal of shelterin reveals the telomere end-protection problem. Science 2012;336(6081):593–7.
[17] Wyatt HD, West SC, Beattie TL. InTERTpreting telomerase structure and function. Nucleic Acids Res 2010;38(17):5609–22.
[18] Chin K, de Solorzano CO, Knowles D, Jones A, Chou W, Rodriguez EG, et al. In situ analyses of genome instability in breast cancer. Nat Genet 2004;36(9):984–8.
[19] Ulaner GA, Giudice LC. Developmental regulation of telomerase activity in human fetal tissues during gestation. Mol Hum Reprod 1997;3(9):769–73.
[20] Kalmbach KH, Fontes Antunes DM, Dracxler RC, Knier TW, Seth-Smith ML, Wang F, et al. Telomeres and human reproduction. Fertil Steril 2013;99(1):23–9.
[21] Xu J, Yang X. Telomerase activity in bovine embryos during early development. Biol Reprod 2000;63(4):1124–8.
[22] Chiu CP, Dragowska W, Kim NW, Vaziri H, Yui J, Thomas TE, et al. Differential expression of telomerase activity in hematopoietic progenitors from adult human bone marrow. Stem Cells 1996;14(2):239–48.
[23] Hiyama E, Hiyama K. Telomere and telomerase in stem cells. Br J Cancer 2007;96(7):1020–4.
[24] Vaziri H, Dragowska W, Allsopp RC, Thomas TE, Harley CB, Lansdorp PM. Evidence for a mitotic clock in human hematopoietic stem cells: loss of telomeric DNA with age. Proc Natl Acad Sci USA 1994;91(21):9857–60.
[25] Fehrer C, Lepperdinger G. Mesenchymal stem cell aging. Exp Gerontol 2005;40(12):926–30.
[26] Liu L, DiGirolamo CM, Navarro PA, Blasco MA, Keefe DL. Telomerase deficiency impairs differentiation of mesenchymal stem cells. Exp Cell Res 2004;294(1):1–8.
[27] Ferron SR, Marques-Torrejon MA, Mira H, Flores I, Taylor K, Blasco MA, et al. Telomere shortening in neural stem cells disrupts neuronal differentiation and neuritogenesis. J Neurosci 2009;29(46):14394–407.
[28] Lingner J, Hughes TR, Shevchenko A, Mann M, Lundblad V, Cech TR. Reverse transcriptase motifs in the catalytic subunit of telomerase. Science 1997;276(5312):561–7.
[29] Zaug AJ, Podell ER, Cech TR. Mutation in TERT separates processivity from anchor-site function. Nat Struct Mol Biol 2008;15(8):870–2.
[30] Lai CK, Mitchell JR, Collins K. RNA binding domain of telomerase reverse transcriptase. Mol Cell Biol 2001;21(4):990–1000.
[31] Nakamura TM, Morin GB, Chapman KB, Weinrich SL, Andrews WH, Lingner J, et al. Telomerase catalytic subunit homologs from fission yeast and human. Science 1997;277(5328):955–9.
[32] Gillis AJ, Schuller AP, Skordalakes E. Structure of the *Tribolium castaneum* telomerase catalytic subunit TERT. Nature 2008;455(7213):633–7.
[33] Hossain S, Singh S, Lue NF. Functional analysis of the C-terminal extension of telomerase reverse transcriptase. A putative "thumb" domain. J Biol Chem 2002;277(39):36174–80.
[34] Jacobs SA, Podell ER, Cech TR. Crystal structure of the essential N-terminal domain of telomerase reverse transcriptase. Nat Struct Mol Biol 2006;13(3):218–25.
[35] Rouda S, Skordalakes E. Structure of the RNA-binding domain of telomerase: implications for RNA recognition and binding. Structure 2007;15(11):1403–12.
[36] Mitchell M, Gillis A, Futahashi M, Fujiwara H, Skordalakes E. Structural basis for telomerase catalytic subunit TERT binding to RNA template and telomeric DNA. Nat Struct Mol Biol 2010;17(4):513–8.
[37] Nandakumar J, Cech TR. Finding the end: recruitment of telomerase to telomeres. Nat Rev Mol Cell Biol 2013;14(2):69–82.
[38] Mitchell JR, Collins K. Human telomerase activation requires two independent interactions between telomerase RNA and telomerase reverse transcriptase. Mol Cell 2000;6(2):361–71.

[39] Theimer CA, Jady BE, Chim N, Richard P, Breece KE, Kiss T, et al. Structural and functional characterization of human telomerase RNA processing and Cajal body localization signals. Mol Cell 2007;27(6):869–81.

[40] Richard P, Darzacq X, Bertrand E, Jady BE, Verheggen C, Kiss T. A common sequence motif determines the Cajal body-specific localization of box H/ACA scaRNAs. EMBO J 2003;22(16):4283–93.

[41] Kiss T, Fayet-Lebaron E, Jady BE. Box H/ACA small ribonucleoproteins. Mol Cell 2010;37(5):597–606.

[42] Mitchell JR, Cheng J, Collins K. A box H/ACA small nucleolar RNA-like domain at the human telomerase RNA 3′ end. Mol Cell Biol 1999;19(1):567–76.

[43] Cohen SB, Graham ME, Lovrecz GO, Bache N, Robinson PJ, Reddel RR. Protein composition of catalytically active human telomerase from immortal cells. Science 2007;315(5820):1850–3.

[44] Cristofari G, Adolf E, Reichenbach P, Sikora K, Terns RM, Terns MP, et al. Human telomerase RNA accumulation in Cajal bodies facilitates telomerase recruitment to telomeres and telomere elongation. Mol Cell 2007;27(6):882–9.

[45] Venteicher AS, Abreu EB, Meng Z, McCann KE, Terns RM, Veenstra TD, et al. A human telomerase holoenzyme protein required for Cajal body localization and telomere synthesis. Science 2009;323(5914):644–8.

[46] Box JA, Bunch JT, Tang W, Baumann P. Spliceosomal cleavage generates the 3′ end of telomerase RNA. Nature 2008;456(7224):910–4.

[47] Kannan R, Helston RM, Dannebaum RO, Baumann P. Diverse mechanisms for spliceosome-mediated 3′ end processing of telomerase RNA. Nat Commun 2015;6:6104.

[48] Gallardo F, Chartrand P. Telomerase biogenesis: the long road before getting to the end. RNA Biol 2008;5(4):212–5.

[49] Xi L, Cech TR. Inventory of telomerase components in human cells reveals multiple subpopulations of hTR and hTERT. Nucleic Acids Res 2014;42(13):8565–77.

[50] Williamson JR, Raghuraman MK, Cech TR. Monovalent cation-induced structure of telomeric DNA: the G-quartet model. Cell 1989;59(5):871–80.

[51] Sundquist WI, Klug A. Telomeric DNA dimerizes by formation of guanine tetrads between hairpin loops. Nature 1989;342(6251):825–9.

[52] Parkinson GN, Lee MP, Neidle S. Crystal structure of parallel quadruplexes from human telomeric DNA. Nature 2002;417(6891):876–80.

[53] Wang Y, Patel DJ. Solution structure of the human telomeric repeat d[AG3(T2AG3)3] G-tetraplex. Structure 1993;1(4):263–82.

[54] Biffi G, Tannahill D, McCafferty J, Balasubramanian S. Quantitative visualization of DNA G-quadruplex structures in human cells. Nat Chem 2013;5(3):182–6.

[55] Paeschke K, Juranek S, Simonsson T, Hempel A, Rhodes D, Lipps HJ. Telomerase recruitment by the telomere end binding protein-beta facilitates G-quadruplex DNA unfolding in ciliates. Nat Struct Mol Biol 2008;15(6):598–604.

[56] Hansel R, Lohr F, Foldynova-Trantirkova S, Bamberg E, Trantirek L, Dotsch V. The parallel G-quadruplex structure of vertebrate telomeric repeat sequences is not the preferred folding topology under physiological conditions. Nucleic Acids Res 2011;39(13):5768–75.

[57] Griffith JD, Comeau L, Rosenfield S, Stansel RM, Bianchi A, Moss H, et al. Mammalian telomeres end in a large duplex loop. Cell 1999;97(4):503–14.

[58] van Steensel B, de Lange T. Control of telomere length by the human telomeric protein TRF1. Nature 1997;385(6618):740–3.

[59] Smogorzewska A, van Steensel B, Bianchi A, Oelmann S, Schaefer MR, Schnapp G, et al. Control of human telomere length by TRF1 and TRF2. Mol Cell Biol 2000;20(5):1659–68.

[60] Fairall L, Chapman L, Moss H, de Lange T, Rhodes D. Structure of the TRFH dimerization domain of the human telomeric proteins TRF1 and TRF2. Mol Cell 2001;8(2):351–61.

[61] Broccoli D, Smogorzewska A, Chong L, de Lange T. Human telomeres contain two distinct Myb-related proteins, TRF1 and TRF2. Nat Genet 1997;17(2):231–5.

[62] Court R, Chapman L, Fairall L, Rhodes D. How the human telomeric proteins TRF1 and TRF2 recognize telomeric DNA: a view from high-resolution crystal structures. EMBO Rep 2005;6(1):39–45.

[63] Hanaoka S, Nagadoi A, Nishimura Y. Comparison between TRF2 and TRF1 of their telomeric DNA-bound structures and DNA-binding activities. Protein Sci 2005;14(1):119–30.

[64] Nishikawa T, Okamura H, Nagadoi A, Konig P, Rhodes D, Nishimura Y. Solution structure of a telomeric DNA complex of human TRF1. Structure 2001;9(12):1237–51.

[65] Ye JZ, Donigian JR, van Overbeek M, Loayza D, Luo Y, Krutchinsky AN, et al. TIN2 binds TRF1 and TRF2 simultaneously and stabilizes the TRF2 complex on telomeres. J Biol Chem 2004;279(45):47264–71.

[66] Ye JZ, Hockemeyer D, Krutchinsky AN, Loayza D, Hooper SM, Chait BT, et al. POT1-interacting protein PIP1: a telomere length regulator that recruits POT1 to the TIN2/TRF1 complex. Genes Dev 2004;18(14):1649–54.

[67] Liu D, O'Connor MS, Qin J, Songyang Z. Telosome, a mammalian telomere-associated complex formed by multiple telomeric proteins. J Biol Chem 2004;279(49):51338–42.

[68] Kim SH, Kaminker P, Campisi J. TIN2, a new regulator of telomere length in human cells. Nat Genet 1999;23(4):405–12.

[69] Kim SH, Beausejour C, Davalos AR, Kaminker P, Heo SJ, Campisi J. TIN2 mediates functions of TRF2 at human telomeres. J Biol Chem 2004;279(42):43799–804.

[70] Abreu E, Aritonovska E, Reichenbach P, Cristofari G, Culp B, Terns RM, et al. TIN2-tethered TPP1 recruits human telomerase to telomeres in vivo. Mol Cell Biol 2010;30(12):2971–82.

[71] Takai KK, Kibe T, Donigian JR, Frescas D, de Lange T. Telomere protection by TPP1/POT1 requires tethering to TIN2. Mol Cell 2011;44(4):647–59.

[72] Zeng Z, Wang W, Yang Y, Chen Y, Yang X, Diehl JA, et al. Structural basis of selective ubiquitination of TRF1 by SCFFbx4. Dev Cell 2010;18(2):214–25.

[73] Janouskova E, Necasova I, Pavlouskova J, Zimmermann M, Hluchy M, Marini V, et al. Human Rap1 modulates TRF2 attraction to telomeric DNA. Nucleic Acids Res 2015;43(5):2691–700.

[74] Bae NS, Baumann P. A RAP1/TRF2 complex inhibits nonhomologous end-joining at human telomeric DNA ends. Mol Cell 2007;26(3):323–34.

[75] Sfeir A, Kabir S, van Overbeek M, Celli GB, de Lange T. Loss of Rap1 induces telomere recombination in the absence of NHEJ or a DNA damage signal. Science 2010;327(5973):1657–61.

[76] Chen Y, Rai R, Zhou ZR, Kanoh J, Ribeyre C, Yang Y, et al. A conserved motif within RAP1 has diversified roles in telomere protection and regulation in different organisms. Nat Struct Mol Biol 2011;18(2):213–21.

[77] Zimmermann M, Kibe T, Kabir S, de Lange T. TRF1 negotiates TTAGGG repeat-associated replication problems by recruiting the BLM helicase and the TPP1/POT1 repressor of ATR signaling. Genes Dev 2014;28(22):2477–91.

[78] Baumann P, Cech TR. Pot1, the putative telomere end-binding protein in fission yeast and humans. Science 2001;292(5519):1171–5.

[79] Wang F, Podell ER, Zaug AJ, Yang Y, Baciu P, Cech TR, et al. The POT1-TPP1 telomere complex is a telomerase processivity factor. Nature 2007;445(7127):506–10.

[80] Xin H, Liu D, Wan M, Safari A, Kim H, Sun W, et al. TPP1 is a homologue of ciliate TEBP-beta and interacts with POT1 to recruit telomerase. Nature 2007;445(7127):559–62.

[81] Nandakumar J, Podell ER, Cech TR. How telomeric protein POT1 avoids RNA to achieve specificity for single-stranded DNA. Proc Natl Acad Sci USA 2010;107(2):651–6.

[82] O'Connor MS, Safari A, Xin H, Liu D, Songyang Z. A critical role for TPP1 and TIN2 interaction in high-order telomeric complex assembly. Proc Natl Acad Sci USA 2006;103(32):11874–9.

[83] Houghtaling BR, Cuttonaro L, Chang W, Smith S. A dynamic molecular link between the telomere length regulator TRF1 and the chromosome end protector TRF2. Curr Biol 2004;14(18):1621–31.

[84] Lei M, Podell ER, Cech TR. Structure of human POT1 bound to telomeric single-stranded DNA provides a model for chromosome end-protection. Nat Struct Mol Biol 2004;11(12):1223–9.

[85] Rajavel M, Mullins MR, Taylor DJ. Multiple facets of TPP1 in telomere maintenance. Biochim Biophys Acta 2014;1844(9):1550–9.

[86] Zhong FL, Batista LF, Freund A, Pech MF, Venteicher AS, Artandi SE. TPP1 OB-fold domain controls telomere maintenance by recruiting telomerase to chromosome ends. Cell 2012;150(3):481–94.

[87] Nandakumar J, Bell CF, Weidenfeld I, Zaug AJ, Leinwand LA, Cech TR. The TEL patch of telomere protein TPP1 mediates telomerase recruitment and processivity. Nature 2012;492(7428):285–9.

[88] Horvath MP, Schweiker VL, Bevilacqua JM, Ruggles JA, Schultz SC. Crystal structure of the *Oxytricha nova* telomere end binding protein complexed with single strand DNA. Cell 1998;95(7):963–74.

[89] Lin JJ, Zakian VA. The Saccharomyces CDC13 protein is a single-strand TG1-3 telomeric DNA-binding protein in vitro that affects telomere behavior in vivo. Proc Natl Acad Sci USA 1996;93(24):13760–5.

[90] Nugent CI, Hughes TR, Lue NF, Lundblad V. Cdc13p: a single-strand telomeric DNA-binding protein with a dual role in yeast telomere maintenance. Science 1996;274(5285):249–52.

[91] Miyake Y, Nakamura M, Nabetani A, Shimamura S, Tamura M, Yonehara S, et al. RPA-like mammalian Ctc1-Stn1-Ten1 complex binds to single-stranded DNA and protects telomeres independently of the Pot1 pathway. Mol Cell 2009;36(2):193–206.

[92] Surovtseva YV, Churikov D, Boltz KA, Song X, Lamb JC, Warrington R, et al. Conserved telomere maintenance component 1 interacts with STN1 and maintains chromosome ends in higher eukaryotes. Mol Cell 2009;36(2):207–18.

[93] Giraud-Panis MJ, Teixeira MT, Geli V, Gilson E. CST meets shelterin to keep telomeres in check. Mol Cell 2010;39(5):665–76.

[94] Latrick CM, Cech TR. POT1-TPP1 enhances telomerase processivity by slowing primer dissociation and aiding translocation. EMBO J 2010;29(5):924–33.

[95] Rai R, Li JM, Zheng H, Lok GT, Deng Y, Huen MS, et al. The E3 ubiquitin ligase Rnf8 stabilizes Tpp1 to promote telomere end protection. Nat Struct Mol Biol 2011;18(12):1400–7.

[96] Zhang Y, Chen LY, Han X, Xie W, Kim H, Yang D, et al. Phosphorylation of TPP1 regulates cell cycle-dependent telomerase recruitment. Proc Natl Acad Sci USA 2013;110(14):5457–62.

[97] Vulliamy TJ, Marrone A, Knight SW, Walne A, Mason PJ, Dokal I. Mutations in dyskeratosis congenita: their impact on telomere length and the diversity of clinical presentation. Blood 2006;107(7):2680–5.

[98] Knight SW, Heiss NS, Vulliamy TJ, Greschner S, Stavrides G, Pai GS, et al. X-linked dyskeratosis congenita is predominantly caused by missense mutations in the DKC1 gene. Am J Hum Genet 1999;65(1):50–8.

[99] Trahan C, Martel C, Dragon F. Effects of dyskeratosis congenita mutations in dyskerin, NHP2 and NOP10 on assembly of H/ACA pre-RNPs. Hum Mol Genet 2010;19(5):825–36.

[100] Robart AR, Collins K. Investigation of human telomerase holoenzyme assembly, activity, and processivity using disease-linked subunit variants. J Biol Chem 2010;285(7):4375–86.

[101] Ball SE, Gibson FM, Rizzo S, Tooze JA, Marsh JC, Gordon-Smith EC. Progressive telomere shortening in aplastic anemia. Blood 1998;91(10):3582–92.

[102] Wang T, Mei SC, Fu R, Wang HQ, Shao ZH. Expression of Shelterin component POT1 is associated with decreased telomere length and immunity condition in humans with severe aplastic anemia. J Immunol Res 2014;2014:439530.

[103] Yamaguchi H, Calado RT, Ly H, Kajigaya S, Baerlocher GM, Chanock SJ, et al. Mutations in TERT, the gene for telomerase reverse transcriptase, in aplastic anemia. N Engl J Med 2005;352(14):1413–24.

[104] Vulliamy T, Marrone A, Dokal I, Mason PJ. Association between aplastic anaemia and mutations in telomerase RNA. Lancet 2002;359(9324):2168–70.

[105] Alder JK, Chen JJ, Lancaster L, Danoff S, Su SC, Cogan JD, et al. Short telomeres are a risk factor for idiopathic pulmonary fibrosis. Proc Natl Acad Sci USA 2008;105(35):13051–6.

[106] Armanios MY, Chen JJ, Cogan JD, Alder JK, Ingersoll RG, Markin C, et al. Telomerase mutations in families with idiopathic pulmonary fibrosis. N Engl J Med 2007;356(13):1317–26.

[107] Tsang AR, Wyatt HD, Ting NS, Beattie TL. hTERT mutations associated with idiopathic pulmonary fibrosis affect telomerase activity, telomere length, and cell growth by distinct mechanisms. Aging Cell 2012;11(3):482–90.

[108] Alder JK, Barkauskas CE, Limjunyawong N, Stanley SE, Kembou F, Tuder RM, et al. Telomere dysfunction causes alveolar stem cell failure. Proc Natl Acad Sci USA 2015;112(16):5099–104.

[109] Kong CM, Lee XW, Wang X. Telomere shortening in human diseases. FEBS J 2013;280(14):3180–93.

[110] Crabbe L, Verdun RE, Haggblom CI, Karlseder J. Defective telomere lagging strand synthesis in cells lacking WRN helicase activity. Science 2004;306(5703):1951–3.

[111] De Sandre-Giovannoli A, Bernard R, Cau P, Navarro C, Amiel J, Boccaccio I, et al. Lamin a truncation in Hutchinson-Gilford progeria. Science 2003;300(5628):2055.

[112] Cao K, Blair CD, Faddah DA, Kieckhaefer JE, Olive M, Erdos MR, et al. Progerin and telomere dysfunction collaborate to trigger cellular senescence in normal human fibroblasts. J Clin Invest 2011;121(7):2833–44.

[113] Shay JW, Zou Y, Hiyama E, Wright WE. Telomerase and cancer. Hum Mol Genet 2001;10(7):677–85.

[114] Kim NW, Piatyszek MA, Prowse KR, Harley CB, West MD, Ho PL, et al. Specific association of human telomerase activity with immortal cells and cancer. Science 1994;266(5193):2011–5.

[115] Kyo S, Takakura M, Taira T, Kanaya T, Itoh H, Yutsudo M, et al. Sp1 cooperates with c-Myc to activate transcription of the human telomerase reverse transcriptase gene (hTERT). Nucleic Acids Res 2000;28(3):669–77.

[116] Hoffmeyer K, Raggioli A, Rudloff S, Anton R, Hierholzer A, Del Valle I, et al. Wnt/β-catenin signaling regulates telomerase in stem cells and cancer cells. Science 2012;336(6088):1549–54.

[117] Vinagre J, Almeida A, Populo H, Batista R, Lyra J, Pinto V, et al. Frequency of TERT promoter mutations in human cancers. Nat Commun 2013;4:2185.

[118] Gunes C, Lichtsteiner S, Vasserot AP, Englert C. Expression of the hTERT gene is regulated at the level of transcriptional initiation and repressed by Mad1. Cancer Res 2000;60(8):2116–21.

[119] Crowe DL, Nguyen DC, Tsang KJ, Kyo S. E2F-1 represses transcription of the human telomerase reverse transcriptase gene. Nucleic Acids Res 2001;29(13):2789–94.

[120] Fujimoto K, Kyo S, Takakura M, Kanaya T, Kitagawa Y, Itoh H, et al. Identification and characterization of negative regulatory elements of the human telomerase catalytic subunit (hTERT) gene promoter: possible role of MZF-2 in transcriptional repression of hTERT. Nucleic Acids Res 2000;28(13):2557–62.

[121] Qi DL, Ohhira T, Fujisaki C, Inoue T, Ohta T, Osaki M, et al. Identification of PITX1 as a TERT suppressor gene located on human chromosome 5. Mol Cell Biol 2011;31(8):1624–36.

[122] Li Y, Tergaonkar V. Noncanonical functions of telomerase: implications in telomerase-targeted cancer therapies. Cancer Res 2014;74(6):1639–44.

[123] Bainbridge MN, Armstrong GN, Gramatges MM, Bertuch AA, Jhangiani SN, Doddapaneni H, et al. Germline mutations in shelterin complex genes are associated with familial glioma. J Natl Cancer Inst 2015;107(1):384.

[124] Robles-Espinoza CD, Harland M, Ramsay AJ, Aoude LG, Quesada V, Ding Z, et al. POT1 loss-of-function variants predispose to familial melanoma. Nat Genet 2014;46(5):478–81.

[125] Ramsay AJ, Quesada V, Foronda M, Conde L, Martinez-Trillos A, Villamor N, et al. POT1 mutations cause telomere dysfunction in chronic lymphocytic leukemia. Nat Genet 2013;45(5):526–30.

[126] Walne AJ, Vulliamy T, Beswick R, Kirwan M, Dokal I. TINF2 mutations result in very short telomeres: analysis of a large cohort of patients with dyskeratosis congenita and related bone marrow failure syndromes. Blood 2008;112(9):3594–600.

[127] Kocak H, Ballew BJ, Bisht K, Eggebeen R, Hicks BD, Suman S, et al. Hoyeraal-Hreidarsson syndrome caused by a germline mutation in the TEL patch of the telomere protein TPP1. Genes Dev 2014;28(19):2090–102.

[128] Sung LY, Chang WF, Zhang Q, Liu CC, Liou JY, Chang CC, et al. Telomere elongation and naive pluripotent stem cells achieved from telomerase haplo-insufficient cells by somatic cell nuclear transfer. Cell Rep 2014;9(5):1603–9.

[129] IJpma A, Greider CW. Short telomeres induce a DNA damage response in *Saccharomyces cerevisiae*. Mol Biol Cell 2003;14(3):987–1001.

[130] Wu L, Multani AS, He H, Cosme-Blanco W, Deng Y, Deng JM, et al. Pot1 deficiency initiates DNA damage checkpoint activation and aberrant homologous recombination at telomeres. Cell 2006;126(1):49–62.

[131] Kibe T, Osawa GA, Keegan CE, de Lange T. Telomere protection by TPP1 is mediated by POT1a and POT1b. Mol Cell Biol 2010;30(4):1059–66.

[132] Sfeir A, Kosiyatrakul ST, Hockemeyer D, MacRae SL, Karlseder J, Schildkraut CL, et al. Mammalian telomeres resemble fragile sites and require TRF1 for efficient replication. Cell 2009;138(1):90–103.

[133] Doksani Y, Wu JY, de Lange T, Zhuang X. Super-resolution fluorescence imaging of telomeres reveals TRF2-dependent T-loop formation. Cell 2013;155(2):345–56.

[134] Kim MM, Rivera MA, Botchkina IL, Shalaby R, Thor AD, Blackburn EH. A low threshold level of expression of mutant-template telomerase RNA inhibits human tumor cell proliferation. Proc Natl Acad Sci USA 2001;98(14):7982–7.

[135] Li S, Rosenberg JE, Donjacour AA, Botchkina IL, Hom YK, Cunha GR, et al. Rapid inhibition of cancer cell growth induced by lentiviral delivery and expression of mutant-template telomerase RNA and anti-telomerase short-interfering RNA. Cancer Res 2004;64(14):4833–40.

[136] Diolaiti ME, Cimini BA, Kageyama R, Charles FA, Stohr BA. In situ visualization of telomere elongation patterns in human cells. Nucleic Acids Res 2013;41(18):e176.

[137] Marusic L, Anton M, Tidy A, Wang P, Villeponteau B, Bacchetti S. Reprogramming of telomerase by expression of mutant telomerase RNA template in human cells leads to altered telomeres that correlate with reduced cell viability. Mol Cell Biol 1997;17(11):6394–401.
[138] Guiducci C, Cerone MA, Bacchetti S. Expression of mutant telomerase in immortal telomerase-negative human cells results in cell cycle deregulation, nuclear and chromosomal abnormalities and rapid loss of viability. Oncogene 2001;20(6):714–25.
[139] Mahalingam D, Tay LL, Tan WH, Chai JH, Wang X. Mutant telomerase RNAs induce DNA damage and apoptosis via the TRF2-ATM pathway in telomerase-overexpressing primary fibroblasts. FEBS J 2011;278(19):3724–38.

Chapter 22

The Relationship Between Checkpoint Adaptation and Mitotic Catastrophe in Genomic Changes in Cancer Cells

L.H. Swift, R.M. Golsteyn
University of Lethbridge, Lethbridge, AB, Canada

Chapter Outline

1. CANCER AND ITS HALLMARKS

Cancer is a complex disease that was characterized by six hallmarks in 2000 [1]. These hallmarks are sustained proliferative signaling, resisting cell death, evading growth suppressors, activating invasion and metastasis, enabling replicative immortality, and inducing angiogenesis. Following a decade of research, these hallmarks were revisited in 2011 [1], and genomic instability remained as an overarching theme in cancer cell biology. Recent studies of checkpoint adaptation and DNA damage induced by mitosis suggest that these phenomena may perpetuate genome instability and in part contribute to the vast differences in genomes present in tumors.

Of the hallmarks of cancer, sustained proliferative signaling and resisting cell death are the most relevant to understanding the relationships between checkpoint adaptation and genomic instability. It is estimated that the majority of cancers have mutations in at least one of the genes that encode the tumor-suppressor retinoblastoma (Rb), p21 or p53 [2]. The Cancer Genome Atlas (TCGA) analyzed 3281 tumors from 12 cancer types (breast adenocarcinoma, lung adenocarcinoma, lung squamous cell carcinoma, uterine corpus endometrial carcinoma, glioblastoma multiforme, head and neck squamous cell carcinoma, colorectal carcinoma, bladder urothelial carcinoma, kidney renal clear cell carcinoma, ovarian serous carcinoma, and acute myeloid leukemia) for point mutations and small insertions/deletions and found that TP53 was mutated in 42% of samples, making it the most frequently mutated gene in human cancers [3]. Because cancer cells can lose the ability to regulate the cell cycle and to induce cell death, this has implications for cancer treatments. Many cancer treatments aim to inhibit cell division and induce cell death by causing irreparable amounts of DNA damage. Recent evidence about checkpoint adaptation and reexamination of older literature about the regulation of the cell cycle and genotoxic agents suggest that checkpoint adaptation has a role in contributing to genomic changes in cancer cells.

Genome Stability. http://dx.doi.org/10.1016/B978-0-12-803309-8.00022-7

2. THE CELL CYCLE

To divide, a cell must pass through four phases of the cell cycle: G1 (Gap 1) when cells grow and prepare for DNA synthesis; S (DNA synthesis) when cells replicate their DNA; G2 (Gap 2) when cells continue to grow and prepare for mitosis; and M (mitosis) when cells separate copies of their DNA as chromosomes and then undergo cytokinesis. To maintain the integrity of the genome during the cell cycle, several events must occur: DNA replication must be accurate; chromosomes must be distributed correctly during mitosis and cytokinesis; and damaged DNA must be detected and repaired [4].

During mitosis cells adopt a rounded morphology, known as mitotic cell rounding. Mitotic cell rounding is evolutionarily conserved and is nearly universal in metazoan and eukaryotic cells that lack a cell wall. Mitotic cell rounding is required for chromosome capture, spindle formation, and spindle stability and therefore has an important role in cell division [5]. To exhibit mitotic rounding, cells must disassemble focal adhesion complexes to decrease adhesion to their substrate and reorganize the actin cytoskeleton. The rounded morphology of mitotic cells can be used to visually distinguish mitotic cells from interphase cells. It can also be utilized to separate and collect mitotic cell populations from interphase cell populations during tissue culture by mechanical shake-off [6,7]. In experiments testing for checkpoint adaptation, cell rounding is a convenient feature that permits one to identify cells undergoing checkpoint adaptation (Fig. 22.1) and is used in addition to the detection of other mitotic markers such as histone H3 phosphorylated on serine-10 [8].

The cell cycle is highly regulated to prevent cells from transmitting damaged DNA to daughter cells. Cyclin-dependent kinases (Cdks) are the main regulators of the cell cycle and are highly conserved catalytic subunits of a family of serine/threonine kinases [9]. The mitotic cyclins, of which cyclin B is the prototype, are synthesized during S and G2 phases of the cell cycle [10]. Cyclin B binds to Cdk1 creating a Cdk1–cyclin B dimer, and the activity of this complex controls the transition between G2 phase and mitosis. Once Cdk1 is bound to cyclin B, Cdk1 is phosphorylated on threonine 161 by Cdk-activating kinase (CAK). This phosphorylation stabilizes the Cdk1–cyclin B complex and induces conformational changes necessary for kinase activity [10,11]. However, despite being bound to cyclin B, Cdk1-cyclin B is held inactive by phosphorylation on threonine 14 and tyrosine 15. Threonine 14 is phosphorylated by Myt1 kinase and tyrosine 15 is phosphorylated by Wee1 kinase [12]. These phosphate groups prevent Cdk1 from binding to and hydrolyzing adenosine triphosphate (ATP) [10], preventing the transfer of a phosphate group from ATP onto a Cdk1 substrate. The phosphate groups on threonine 14 and tyrosine 15 are removed by Cdc25 phosphatases, promoting entry into mitosis [13]. Once active, Cdk1-cyclin B phosphorylates the Cdk1 inhibitor Wee1 and the Cdk1-activating Cdc25 phosphatases through two different feedback loops [14]. Interestingly, the overexpression of only cyclin B and Cdc25 phosphatase is necessary to cause cells in G2 phase to prematurely enter mitosis [15].

The oscillating nature of cyclin B levels can be used to detect cells that are capable of entering mitosis. This is because cyclin B is synthesized through S and G2 phases of the cell cycle and then degraded at the metaphase–anaphase transition of mitosis [16]. Cdk1 phosphorylation is also a good measure of whether cells have active Cdk1 or not; if cells have high levels of tyrosine 15–phosphorylated Cdk1, then they do not contain enough active Cdk1 to be in mitosis [7,10]. Additionally, to detect Cdk1 activity, a specific Cdk1 activity assay can be performed. In 2013, Lewis et al. published a protocol to detect Cdk1 activity by using western blotting to quantify the amount of phosphorylation of threonine 320 on an artificial Cdk1 substrate consisting of glutathione *S*-transferase (GST) and amino acids 316–324 from the PP1-alpha catalytic subunit (PP1Cα) [17].

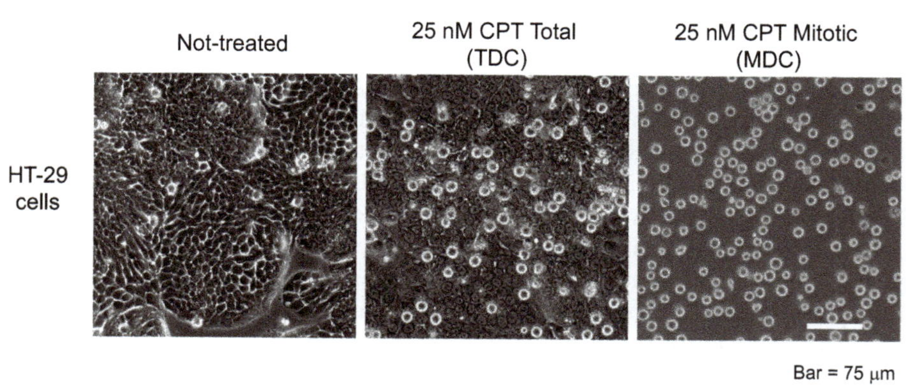

FIGURE 22.1 **Human cancer cells undergo checkpoint adaptation.** HT-29 human colon cancer cells were either not treated (left panel) or treated by 25 nM camptothecin (CPT) for 48 h (TDC; total and DNA-damaged cells, middle panel); 25 nM CPT is a cytotoxic concentration in which the cells die by 96 h. In a separate experiment, the rounded, mitotic cells (MDC; mitotic and DNA-damaged cells) from a TDC can be isolated by a mechanical shake-off, and investigated for survival properties and genomic changes. Scale bar represents 75 μm.

3. CELL-CYCLE CHECKPOINTS

Cell-cycle checkpoints enable a cell to ensure that important processes, such as DNA replication, are complete [18]. Cell-cycle checkpoints prevent the transmission of genetic errors to daughter cells. There exist three major cell-cycle checkpoints; the G1/S checkpoint, the G2/M checkpoint, and the spindle assembly checkpoint (SAC). The SAC ensures that chromosome segregation occurs correctly and is activated at the metaphase to anaphase transition in mitosis, in response to microtubule defects [19] or an erroneous kinetochore attachment [20]. Cells also arrest at the SAC when they enter mitosis with damaged DNA [21]. Inactivation of the SAC can lead to chromosome mis-segregation and aneuploidy [22].

The G1/S and G2/M checkpoints are initiated in response to DNA damage to prevent the transmission of damaged or incomplete chromosomes to daughter cells. The DNA-damage checkpoints provide cells with time to repair damaged DNA. If the DNA damage is irreparable, cells may initiate senescence (growth arrest) or cell death. The G1/S checkpoint prevents cells from replicating damaged DNA, whereas the G2/M checkpoint prevents cells from dividing with damaged DNA [18]. The G1/S checkpoint does not function when p53 or p21 are either absent or not functional [23]. This checkpoint is often defective in cancer cells because many of them have mutations in the genes that encode either p53, pRb, or p21 [2,19]. This means that the only DNA-damage checkpoint available to these cancer cells is the G2/M checkpoint [18].

The kinases ataxia telangiectasia mutated (ATM) and ATM and Rad3 related (ATR) are both involved in the initiation of the G2/M checkpoint, however ATR is the main effector kinase associated with G2/M arrest [24]. When single-stranded DNA (ssDNA) is present, it is bound by replication protein A (RPA) [25]. RPA recruits the ATR-interacting protein (ATRIP) in complex with ATR and the Rad9–Rad1–Hus1 (9-1-1) complex to ssDNA [26]. The 9-1-1 complex then recruits DNA topoisomerase-binding protein 1 (TOPBP1) which triggers the ATR-mediated phosphorylation of checkpoint kinase 1 (Chk1) [26]. The Rad17–replication factor C complex, the 9-1-1 complex, and the adaptor protein claspin are also required for Chk1 activation [25,27]. The Rad17–replication factor C complex acts as a clamp loader for the 9-1-1 complex [25] and claspin links ATR and Chk1, allowing for the phosphorylation of Chk1 on serine 317 and serine 345 [28]. Of these phosphorylation sites, serine 345 is essential for Chk1 activation, while serine 317 plays a contributory role [29].

Once active, Chk1 prevents the activation of Cdk1 by phosphorylating Cdc25A and Cdc25C, targeting them for cytoplasmic sequestration by the 14-3-3 proteins [30] or for ubiquitination and degradation by the proteasome [31]. This prevents the removal of inhibitory phosphates on threonine 14 and tyrosine 15 of Cdk1, preventing Cdk1 activity. Active Chk1 also stabilizes the Wee1 kinase, which is responsible for phosphorylating tyrosine 15 of Cdk1 [32].

It has been reported that Chk1, but not Chk2, is essential for the activation of the G2/M checkpoint. In 2000, Liu et al. generated an inducible Chk1 deficient line of murine embryonic stem cells [28]. They found that when this cell line was irradiated and Chk1 depleted, these cells abrogated the G2/M checkpoint [28]. It has also been demonstrated that H1299 human lung carcinoma cells treated with doxorubicin (a topoisomerase II inhibitor) and transfected with Chk1-silencing RNA (siRNA) abrogate the G2/M checkpoint [33]. Furthermore, when p53$^{-/-}$ HCT116 human colon carcinoma cells were treated with the genotoxic agent lidamycin, the cells transfected with Chk1 siRNA abrogated the G2/M checkpoint [34]. By contrast, the cells transfected with Chk2 siRNA remained arrested at the G2/M checkpoint [34]. Chk1 phosphorylation on serine 345 is a good measure of whether cells are arrested at the G2/M checkpoint or not. Loss of Chk1 serine 345 phosphorylation in mitotic cells collected by mechanical shake-off from interphase cells with Chk1 phosphorylated on serine 345 is a good indication that the cells have activated and abrogated the G2/M checkpoint, the second step of checkpoint adaptation (discussed in the following) [7].

4. GENOTOXIC AGENTS AS ANTICANCER DRUGS

Genotoxic agents are a mainstay of cancer therapy that cause cytotoxic levels of DNA damage. By inducing DNA damage, these agents cause cells to arrest at cell-cycle checkpoints leading to cell-cycle arrest and/or cell death. Genotoxic agents are widely used to treat cancer patients but are not always curative. Many of these agents are limited by the development of resistance to treatments, and genomic changes can be responsible for this acquisition of resistance to treatment [35]. For example, cisplatin is a widely used anticancer drug that has revolutionized the treatment of some types of cancers. Specifically, cisplatin and combination treatments with cisplatin have increased the cure rate for testicular cancer to 90%, if tumors are diagnosed early [36]. Cell lines and tumor samples that are resistant to cisplatin exhibit chromosomal abnormalities that distinguish them from cisplatin-sensitive cell lines and tumor samples. The genetic difference between cisplatin-sensitive and cisplatin-resistant cell lines and tumor samples has been studied using comparative genomic hybridization (CGH). This technique allows the detection of a change in chromosomal copy number [37]. In 1997, Wasenius et al. compared six ovarian carcinoma cell lines selected for resistance to cisplatin to two cisplatin-sensitive parental cell lines (2008 and A2780) using CGH [38]. They found that acquired resistance in the 2008 cells was associated with many chromosomal gains and losses by comparison to the parental cell line, and the average number of chromosome aberrations per resistant cell was 15. By contrast, acquired resistance in the A2780 cell line was only associated with five chromosomal aberrations, and all of these were losses [38]. These data suggest that acquired resistance to cisplatin is not associated with specific genetic changes. This has also been demonstrated in testicular germ-cell tumor (TGCT) cell lines using CGH. Three cisplatin-resistant TGCT cell lines (resistant GCT27, 833K, and Susa) were found to contain more gains and losses of chromosomal regions by comparison to the parental cell lines [39]. However, these losses and gains were different in each of the three resistant cell lines.

5. CELL DEATH

Cell death or permanent growth arrest are the desired outcomes of treating patients with anticancer genotoxic agents. It is often assumed (incorrectly) that cancer cells die by apoptosis following treatment with anticancer drugs, and apoptosis has been extensively studied with regards to cancer development and treatment. However, other modes of cell death and growth arrest exist, and an understanding of these may be equally or more important to understand how cancer cells respond to current cancer treatments. Three types of cell death frequently discussed in the literature with regard to cancer therapies are apoptosis, necrosis, and mitotic catastrophe (Fig. 22.2). Senescence is also a desired outcome of cancer treatment because senescent cells undergo permanent growth arrest.

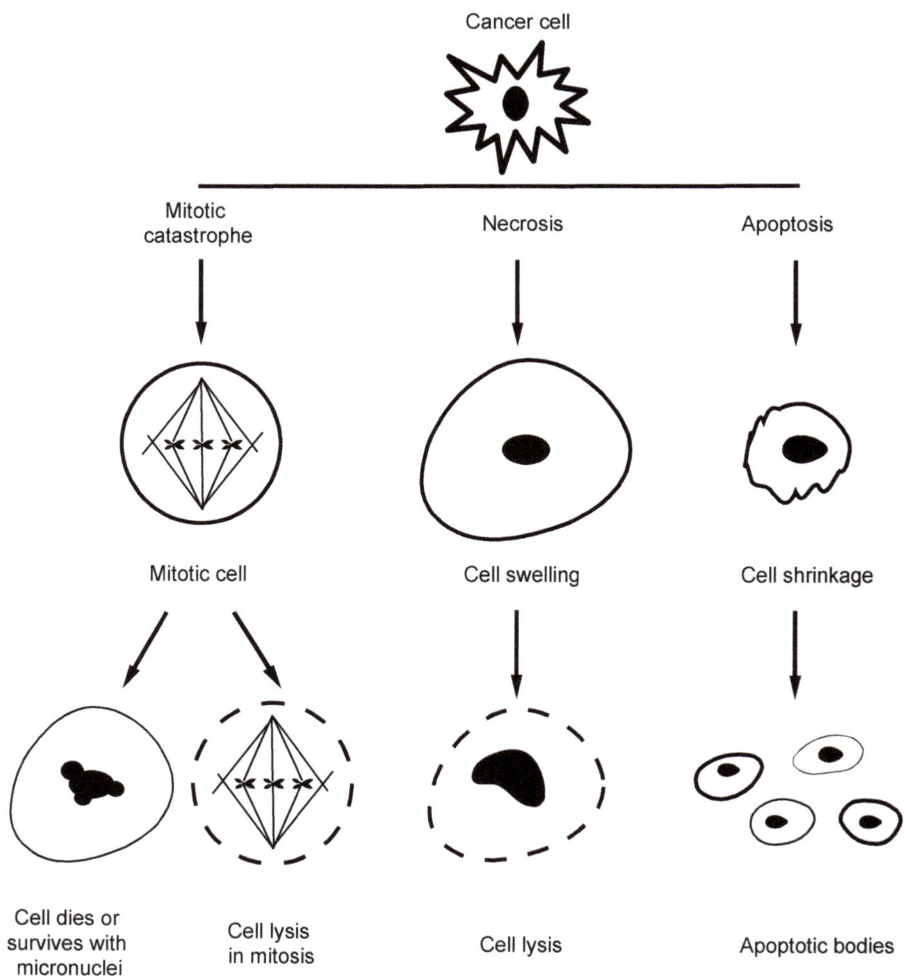

FIGURE 22.2 **An overview of three different modes of cell death.** Three of the major modes of cell death are apoptosis, necrosis, or mitotic catastrophe. Apoptosis is characterized by cell shrinkage and the formation of apoptotic bodies. Necrosis is characterized by cell swelling and lysis. Mitotic catastrophe is characterized by entry into mitosis prior to cell death, but cells may survive and contain micronuclei. Of these modes of cell death, mitotic catastrophe is not well understood, and it may be an outcome of different phenomena, including that of checkpoint adaptation. In this image, a *dashed line* indicates lysis and the mitotic cells are represented by a simple image of a mitotic spindle.

Apoptosis is considered to be the major mode of cell death in cancer cells treated with genotoxic agents. This is partly because the genes for many of the proteins that regulate apoptosis are mutated in human cancers [40]. However, there is increasing evidence that other cell death pathways have a major role in cancer cell death when solid tumors are treated with genotoxic agents. Inhibiting apoptosis is reported to have little or no effect on the clonogenic survival of cancer cells following treatment with anticancer drugs or ionizing radiation [41,42]. This has been demonstrated by several studies where the antiapoptotic protein Bcl-2 is overexpressed [43–48]. It was predicted that if Bcl-2 was overexpressed, then cells would be resistant to apoptosis and would therefore be less sensitive to treatment with genotoxic agents [42]. However, in these studies, although the overexpression of Bcl-2 prevented cells from undergoing apoptosis, it did not have a significant impact on clonogenic survival, indicating that the cells died by a mode of cell death other than apoptosis [43–48]. Wouters et al. also found that there was no difference in cell viability when apoptosis-proficient and apoptosis-deficient HCT116 cells were treated with either 5 μg/mL etoposide or 10 Gy ionizing radiation [49]. Ruth and Roninson made similar observations in cells engineered to express multidrug resistance protein 1 (MDR1), a P-glycoprotein that inhibits apoptosis [50]. HeLa-derived HtTA–MDR1 cervical adenocarcinoma cells and NIH 3T3 murine fibroblasts were treated with 9 Gy ionizing radiation and induced to express MDR1. Ruth and Roninson found that MDR1 expression prevented cells from undergoing apoptosis but did not change overall cell survival after treatment. Instead, the treated cells either died by mitotic catastrophe or initiated a senescence-like growth arrest [50].

6. MITOTIC CATASTROPHE

Mitotic catastrophe is a form of cell death related to mitosis; however, the exact definition of mitotic catastrophe is still debated [51]. It is debated whether mitotic catastrophe occurs as a direct result of a failed mitosis or if cells die by other cell death pathways following entry into mitosis. The induction of other cell death pathways due to a failed mitosis has three different consequences: (1) cell death during mitosis; (2) cell death once a cell has exited mitosis; and (3) senescence following exit from mitosis [51]. It is also debated whether mitotic catastrophe should be classified as a distinct form of cell death [41,52] or if cell death occurs by apoptosis or necrosis following aberrant mitosis [53,54].

It has long been known that cells treated with ionizing radiation enter mitosis. In 1961, Yamada and Puck found that the following irradiation with sub-cytotoxic concentrations of X-rays (either 0.3, 0.7, or 1.4 Gy), there was a decrease in the mitotic index of HeLa cells followed later by an increase in the mitotic index [55]. Because the cells delayed entry into mitosis, this suggests that they arrested at the G2/M checkpoint before entering mitosis. These cells could therefore have been undergoing checkpoint adaptation, however it is not known if they entered mitosis with damaged DNA or if they entered mitosis following the repair of damaged DNA.

By comparison to apoptosis, there are few distinguishing characteristics of mitotic catastrophe, other than mitosis itself. This means it is difficult to detect mitotic catastrophe as a form of cell death [52]. The main characteristic associated with mitotic catastrophe is the presence of micronuclei [51], but micronuclei only indicate that cells have undergone an aberrant mitosis. Some cells with micronuclei may not undergo mitotic catastrophe because they survive and do not die. Furthermore, cells may die directly in mitosis, and micronuclei cannot be used as a marker for this type of cell death. One of the best ways to study mitotic catastrophe is therefore time-lapse video microscopy, to observe cells in real time [56].

Mitotic catastrophe can be induced by DNA damage that directly affects the integrity of chromosomes by interference with the mitotic spindle [51] or by deficiencies in proteins and protein complexes involved in the process of mitosis itself [57]. Drugs such as the taxanes and vinca alkaloids induce mitotic catastrophe without damaged DNA by interfering with the mitotic spindle. This induces a mitotic arrest followed by cell death. The taxanes stabilize microtubules and induce a metaphase arrest, whereas the vinca alkaloids induce mitotic catastrophe by disrupting the dynamics of microtubule polymerization and depolymerization. Nocodazole is a compound that inhibits microtubule polymerization and is widely used as a positive control for mitotic cells in the laboratory because it arrests cells in mitosis.

Entry into mitosis with damaged DNA induces mitotic catastrophe through the SAC. HeLa cells treated with 1.5 μM aphidicolin entered mitosis with damaged DNA and arrested at metaphase [21]. Following this metaphase arrest, cells entered what the authors describe as a "precatastrophic phase" where chromosome segregation was attempted. This was followed by cell death. When either of the SAC proteins, Mad2 or BubR1, were depleted by siRNA in HeLa cells treated with aphidicolin, then the cells did not arrest at metaphase and continued with mitosis. This increased cell viability following treatment with aphidicolin by comparison to cells transfected with control siRNA [21].

Mitotic catastrophe has been observed in response to treatment with a variety of genotoxic agents that have different mechanisms of action (Table 22.1). These data demonstrate that mitotic catastrophe is an important and widely observed mode of death in response to treatment with genotoxic agents [58]. It is likely that these cells undergo checkpoint adaptation to enter mitosis with damaged DNA, but this was not addressed in the majority of these studies. Checkpoint adaptation

TABLE 22.1 A Table of Treatments That Induce Mitotic Catastrophe in Cell Lines

Treatment	Agent Type	Cell Type	Features of Mitotic Catastrophe	References
Aphidicolin	DNA replication inhibitor	HT0180 fibrosarcoma	Micronucleation	[103]
		P53$^{-/-}$ HCT116 colon carcinoma	Analysis of cell-cycle phase, increased mitotic index	[21]
Bleomycin	Radiomimetic, induces DSBs	DC-3F Chinese hamster lung fibroblast	Analysis of cell-cycle phase, micronucleation	[65]
Cisplatin	Cross-linking agent	CHO/UV41 Chinese hamster ovary	Rounded morphology, analysis of cell-cycle phase	[104]
		CHO	Micronucleation	[105]
		HCC metastatic hepatocellular carcinoma	Ser10 phospho-H3 positive, analysis of cell-cycle phase, micronucleation	[106]
		HT0180	Micronucleation	[103]
		SKOV-3 ovarian carcinoma	Lack of caspase activation, micronucleation	[63]
CPT	Topoisomerase I inhibitor	HT-29 human colon carcinoma	Checkpoint adaptation	[7]
		M059K	Checkpoint adaptation	
Cytarabine	Antimetabolite	HT0180	Micronucleation	[103]
Doxorubicin	Topoisomerase II inhibitor	HT0180	Micronucleation	[103]
		Huh-7 hepatocellular carcinoma (HCC)	Micronucleation, analysis of cell-cycle phase, lack of caspase activation	[64]
		SNU-354, -398, -449, -475 HCC	Micronucleation	
Etoposide	Topoisomerase II inhibitor	HT-29	Checkpoint adaptation	[7]
		HT0180	Micronucleation	[103]
5-FU	Antimetabolite	COLO320DM, HCT116, SW480	Analysis of cell-cycle phase, increased levels of cyclin B	[62]
		Colorectal adenocarcinoma		
Ionizing radiation	Physical agent that induces direct DSBs	U2OS osteosarcoma	Checkpoint adaptation	[99]
		MOLT4 leukemia	Checkpoint adaptation	[100]
		HeLa cervical adenocarcinoma	Analysis of cell-cycle phase, increased levels of cyclin B	[107]
		HT0180	Micronucleation	[103]
Oxaliplatin	Cross-linking agent	TE7 oesophageal adenocarcinoma	Analysis of cell-cycle phase, multinucleation	[108]
S23906	Atypical alkylating agent	HeLa	High levels of cyclin B, increased Cdk1 activity, Ser10 phospho-H3 positive	[58]
		HT-29		

may therefore be a common pathway that leads to cell death following treatment with genotoxic agents [59]. It is necessary to understand whether cells undergo checkpoint adaptation or not because checkpoint adaptation may contribute to cells surviving treatment with rearranged genomes. It may be possible to target the final step of checkpoint adaptation to prevent cells from entering mitosis with damaged DNA, preventing them from surviving treatment with rearranged genomes.

Cells that have undergone aberrant mitoses have also been observed in clinical samples. Micronuclei have been detected in clinical cervical [60] and oral carcinoma samples [61] after patients were treated with ionizing radiation. However, because mitotic catastrophe is difficult to detect, there is a lack of clinical data about mitotic catastrophe as a mode of cell death in vivo.

7. DUAL MODES OF CELL DEATH BY THE SAME GENOTOXIC AGENT

The mode of cell death may depend on different factors including the tissue of origin of a cell and the amount of DNA damage that a cell contains [62,63]. The effect of treatment concentration on the mode of cell death has been demonstrated in studies where the same cell lines treated with different concentrations of the same genotoxic agent died by different modes of cell death.

Three human colorectal adenocarcinoma cell lines, SW480, COLO320DM, and HCT116, were treated with relatively low and relatively high concentrations of the antimetabolite 5-fluorouracil (5-FU) [62]. SW480 and COLO320DM cells treated with 1000ng/mL 5-FU and HCT116cells treated with 100ng/mL 5-FU died by apoptosis, whereas SW480 and COLO320DM cells treated with 100ng/mL 5-FU and HCT116 cells treated with 10ng/mL 5-FU died by mitotic catastrophe [62]. The same results were observed when hepatocellular carcinoma cells were treated with doxorubicin [64]. Low-dose doxorubicin treatment (15–60ng/mL depending on the cell line) of five human hepatocellular carcinoma cell lines, Huh-7, SNU-354, -398, -449, and -475, induced entry into mitosis, followed by a senescence-like phenotype. By contrast, Huh-7 cells treated with a high dose of doxorubicin (10μg/mL) died by apoptosis [64]. Similarly, DC-3F Chinese hamster lung fibroblast cells treated with bleomycin died by mitotic catastrophe when treated with a low concentration of bleomycin (10nM), whereas DC-3F cells treated with a high concentration of bleomycin (10μM) died by apoptosis [65]. High concentrations of genotoxic agents likely induce high levels of damaged DNA. These results therefore suggest that the amount of damaged DNA affects which cell death pathway is activated following treatment.

Different modes of cell death were also observed when two different ovarian carcinoma cell lines (Caov-4 and SKOV-3) were treated with 33μM cisplatin [63]. Caov-4 cells died by apoptosis, whereas SKOV-3 cells died by entry into mitosis followed by necrosis-like lysis. However, the authors did not provide cytotoxicity data for Caov-4 and SKOV-3 cells treated with cisplatin, and so it may be that 33μM cisplatin was a high dose of cisplatin for Caov-4cells but not for SKOV-3 cells [63]. Overall, these studies have identified that the mode of cell death induced following treatment with genotoxic agents can be different and depends on either concentration of treatment or cell type. Furthermore, in all of the studies, cells underwent either cell death following entry into mitosis or cell death by apoptosis, highlighting the importance of these two cell death pathways in cells treated with genotoxic agents.

8. THE RELATIONSHIP BETWEEN ENTRY INTO MITOSIS WITH DAMAGED DNA AND GENOMIC INSTABILITY

Common types of genomic rearrangement are base substitutions, DNA insertions, DNA deletions, DNA translocations, and a change in copy number [66]. Genomic instability can occur following DNA damage, and two different events can induce this: (1) DNA-damage misrepair and (2) entry into mitosis with damaged DNA. DNA damage can be misrepaired during interphase inducing genomic rearrangements. For example, because non-homologous end joining does not require homologous DNA sequence to repair DNA double-strand breaks (DSBs), the ends of breaks from different chromosomes can be joined together, resulting in chromosomal translocations [67].

Entry into mitosis with damaged DNA can also be a source of genomic instability. In 2006, Nakada et al. found that when ATM-deficient primary fibroblast cells prematurely entered mitosis after treatment with etoposide, some cells survived with chromosomal translocations, including the 11q23 translocation associated with topoisomerase II inhibitor–induced secondary leukemia [68]. Entry into mitosis with damaged DNA induced by replication stress can also be a source of genomic instability. DNA damage induced by replication stress that occurred because of the overexpression of the oncogene E2F1 induced chromosome bridge formation and aneuploidy [69]. Replication stress can also occur in chromosomal instability positive (CIN^{+}) (aneuploid) human colorectal carcinoma cells lines and this was also found to induce structurally altered chromosomes that were subject to mis-segregation in mitosis, leading to genomic instability [70]. Cells treated with cisplatin can also enter mitosis with damaged DNA leading to the induction of chromosome aberrations [71].

Entry into mitosis with damaged DNA can lead to genomic instability by several different mechanisms. When DNA strand breaks occur, acentric fragments can be created. These fragments can be lost during cell division because they lack centromeres and are unable to attach to the mitotic spindle. These fragments can also be incorporated into micronuclei and either be lost or subjected to further genomic rearrangement [72], discussed later. The loss of genetic material following treatment with genotoxic agents has been detected experimentally. LA-9 murine cells containing a stable chromosome with integrated green fluorescent protein (GFP) were treated with either ionizing radiation or etoposide and assessed for loss of the GFP signal [73]. An increase in the percentage of non-fluorescent cells was observed when cells were either irradiated (3 or 5 Gy) or treated with etoposide (0.5 and 1 μM) by comparison to untreated cells [73].

DNA strand–break repair is inhibited in mitosis, once sites of damaged DNA induced by irradiation have been marked by the formation of histone γH2AX and mediator of DNA-damage checkpoint 1 (MDC1) foci [74]. Cdk1 activity is responsible for preventing DSB repair in mitosis by phosphorylating the key DSB-repair protein RNF8 (a ubiquitin ligase) at threonine 198, preventing it from interacting with MDC1 [74]. 53BP1 is also phosphorylated in mitosis, at threonine 1609 and serine 1618, preventing its recruitment to sites of damaged DNA. Because cells do not repair damaged DNA in mitosis, this means that when cells enter mitosis with damaged DNA, this damage can be transmitted to daughter cells. Instead of repairing damaged DNA in mitosis, mitotic cells progress to G1 where DNA-damage repair can occur. Although the repair of damaged DNA may occur in G1, it is possible that some genetic material could be lost or that micronuclei could form in cells in mitosis with damaged DNA, leading to genomic rearrangements. Furthermore, many cancer cells have a defective G1/S DNA-damage checkpoint. It is therefore plausible that a cancer cell can continue through a second cell cycle with damaged DNA following entry into mitosis with DNA damage (the final step of checkpoint adaptation), thus contributing to genomic instability.

Aberrant mitoses are also frequently associated with the formation of micronuclei and this can induce further genomic rearrangements in cells. In 2012, Crasta et al. demonstrated that micronuclei formed by errors in chromosome segregation during mitosis contribute to genomic instability [72]. Crasta et al. studied micronuclei in RPE-1-untransformed retinal pigment epithelial and U2OS osteosarcoma cells. They generated micronuclei in cells and then followed them through the cell cycle. The authors found that DNA contained in the micronuclei was damaged by DNA replication, and that 7.6% of chromosome spreads prepared from cells with micronuclei contained pulverized chromosomes. In addition, they reported that micronuclei persisted for several generations, and that the chromosomes contained in micronuclei could be reincorporated into the nuclei of daughter cells [72].

8.1 Chromothripsis

The chromosome shattering observed by Crasta et al. is called chromothripsis [75]. Chromothripsis describes a catastrophic event where tens to hundreds of genomic rearrangements are acquired in one or several regions of chromosomes [75]. A number of possibilities for how chromothripsis occurs have been suggested, including that chromothripsis occurs due to a high-energy ionizing radiation event during mitosis, that DNA fragmentation occurs as a result of aborted apoptosis, or that DSBs induced by genotoxic agents create dicentric fusions between sister chromatids that can be broken during mitosis [76]. However, the model for chromosome pulverization described by Crasta et al. is currently considered the most likely model for how chromothripsis arises [76]. This model of chromothripsis is supported by a 2015 study from the same group of researchers where live cell imaging and single-cell genome sequencing were used to characterize micronucleated cells [77]. Combining these techniques allowed the researchers to sequence cells where micronuclei were reincorporated into the main nucleus after one round of cell division. Zhang et al. (2015) used copy number analysis of the paired daughter cells present after the one round of cell division to determine which chromosomes were present in the micronuclei. They found that the mis-segregated chromosomes had a large number of genomic rearrangements in 8 of the 9 daughter cell pairs studied, by comparison to the normally segregated control chromosomes [77]. Till 2016, these are the only studies that provide experimental evidence for how chromothripsis can occur, although chromothripsis has been observed in a number of different cancer types including glioma [78], melanoma [75], multiple myeloma [79], medulloblastoma, acute myeloid leukemia [2], and breast cancer [80]. Additionally, chromothripsis has been detected in patients with congenital abnormalities [81]. Cells that undergo checkpoint adaptation enter mitosis with damaged DNA, and it is likely that this induces aberrant mitoses that may lead to genomic rearrangements and the induction of micronuclei, which can contribute to chromothripsis.

9. A HISTORY OF CHECKPOINT ADAPTATION

It is likely that the cells dying by mitotic catastrophe in the studies listed in Table 22.1 were undergoing checkpoint adaptation. However, this was not tested in the majority of the studies because checkpoint adaptation had not been identified in human cell lines at the time these studies were undertaken. Checkpoint adaptation is the process of entering mitosis with damaged DNA and is defined by three sequential steps: (1) a cell-cycle arrest induced by DNA damage; (2) overcoming this arrest; and (3) resuming the cell cycle with damaged DNA (Fig. 22.1) [82]. Checkpoint adaptation was first observed in 1993 by Sandell and Zakian [83]. DNA-damage repair-deficient *Saccharomyces cerevisiae* cells initiated and then overcame a G2 arrest following the loss of telomeric DNA from an extra dispensable chromosome [83].

Since the discovery that *S. cerevisiae* cells undergo checkpoint adaptation, several different research groups have explored this process in yeast cells. Because the DNA-damage response is highly conserved in eukaryotes, some of this research may provide an insight into how checkpoint adaptation is induced in other eukaryotic organisms such as humans. However, there are differences between G2/M checkpoint control in humans by comparison to those in *S. cerevisiae*. As described earlier, inhibitory phosphorylation of Cdk1 is maintained by the activation of Chk1 in human cells arrested at the G2/M checkpoint. This prevents cells from entering mitosis with damaged DNA. In *S. cerevisiae*, an arrest at the DNA-damage checkpoint does not require inhibitory phosphorylation of Cdk1, and two distinct pathways are involved in the activation of the DNA-damage checkpoint [84]. One pathway involves Chk1 and the other one involves Rad53, a second checkpoint kinase that is homologous to human Chk2. These pathways have different roles in the checkpoint following DNA damage; the Chk1 pathway acts pre-anaphase to prevent chromosome segregation, whereas the Rad53 pathway prevents mitotic exit [84]. Both of these kinases are activated by Mec1, the yeast homolog of ATR. Sanchez et al. (1999) found that Chk1 can prevent entry into anaphase by controlling phosphorylation and levels of Pds1 by preventing cohesion cleavage. Furthermore, they suggest that Rad53 induces cell-cycle arrest through inhibitory phosphorylation of its substrate Cdc5 and show that the overexpression of Cdc5 overrides checkpoint arrest [84]. Cdc5 is a polo-like kinase that induces mitotic exit by phosphorylating and inactivating proteins such as the Bfa1–Bub2 complex. Bfa1–Bub2 are part of the mitotic exit network (MEN), and they prevent mitotic exit until mitosis is complete [85].

That Cdc5 has a role in checkpoint adaptation in *S. cerevisiae* was first observed in 1997. Toczyski et al. (1997) identified two *S. cerevisiae* mutants that were checkpoint adaptation deficient in response to a single double-stranded DNA (dsDNA) break induced using the same *S. cerevisiae* model and method as Sandell and Zakian [82]. One of the mutants identified contained mutated CDC5 and the other mutated CKB2. Cdc5 is a member of the polo-like kinase (Plk) family of proteins. CKB2 encodes a nonessential subunit of casein kinase II (CKII), a serine–threonine kinase that is implicated in a number of pathways including the phosphorylation of the PP2-like phosphatase Ptc2 [86].

In 2001, Galgoczy and Toczyski used the Cdc5 mutant checkpoint adaptation-deficient *S. cerevisiae* strain to investigate the effect of checkpoint adaptation on cell viability and genomic instability [87]. They found that checkpoint adaptation increased cell viability when DNA damage was induced in a nonessential chromosome. Furthermore, they demonstrated that checkpoint adaptation proficient cells irradiated with 30 Gy of X-rays contained more chromosomal losses and translocations, by comparison to checkpoint adaptation-deficient cells. This indicates that checkpoint adaptation has a role in the induction of genomic instability in yeast. Pellicioli et al. also used *S. cerevisiae* checkpoint adaptation-proficient and -deficient cells to investigate checkpoint adaptation and found that the kinase activity of Rad53 was elevated for over 24 h in Cdc5 checkpoint adaptation deficient mutants, whereas Rad53 kinase phosphorylation and activity was lost in cells that underwent checkpoint adaptation [88]. Overexpression of Rad53 also prevented cells from undergoing checkpoint adaptation in response to the induction of a DSB. These data support the results from Sanchez et al., which demonstrated that Rad53 inhibited Cdc5, preventing mitotic entry [84].

Further studies have since confirmed that both Rad53 and Cdc5 have important roles in checkpoint adaptation in *S. cerevisiae*. Rad53 is dephosphorylated and inactivated by the PP2C-like phosphatase Ptc2, promoting checkpoint adaptation [86]. When the PTC2 gene was deleted, checkpoint adaptation proficient cells were unable to undergo checkpoint adaptation and when Ptc2 was overexpressed in checkpoint adaptation-deficient cells, the ability to undergo checkpoint adaptation was restored [86]. To dephosphorylate Rad53, Ptc2 must be phosphorylated on threonine 376 by CKII kinase [89]. This might explain the discovery of mutated CKII as a checkpoint mutant by Toczyski et al. [82]. In addition to Rad53 phosphorylation, Rad53 deacetylation also has a role in checkpoint adaptation. The deletion of histone deacetylase Rpd3 prevents checkpoint adaptation and leads to an increased level of acetylation on Rad53 [90]. Checkpoint adaptation is therefore promoted by deacetylation and inhibition of Rad53.

Till 2016, the precise biochemical pathway that induces checkpoint adaptation in *S. cerevisiae* has not been identified. However, it has been demonstrated that the checkpoint kinase Rad53 and the polo-like kinase Cdc5 have central roles in this process. This reflects the important roles of Rad53 and Cdc5 in the control of the cell cycle in budding yeast. Recently, proteins involved in several cellular responses that are not directly involved in checkpoint control have been identified as having a role in the biochemical pathway(s) that induce(s) checkpoint adaptation. Many of these studies have used checkpoint adaptation-deficient mutants to identify proteins that might be involved in checkpoint adaptation. However, these studies are not usually capable of elucidating the precise role of these proteins in checkpoint adaptation.

In 2015, Ghospurkar et al. identified that phosphorylation of replication factor A2 (Rfa2), the yeast homolog of RPA, induced *S. cerevisiae* cells to undergo checkpoint adaptation [91]. Cells with a phosphomimetic form of Rfa2 where all serine/threonines in the N-terminal domain (9 amino acids) were mutated to aspartic acid underwent checkpoint adaptation. Furthermore, checkpoint adaptation-deficient cells expressing these phosphomimetic proteins also underwent checkpoint adaptation. The authors therefore propose that the induction of checkpoint adaptation occurs when the Rfa proteins (Rfa1 and Rfa2) are modified following a prolonged arrest at the DNA-damage checkpoint [91].

Chromatin remodeling proteins are also involved in checkpoint adaptation. This was first demonstrated by Lee et al., who found that checkpoint adaptation-deficient *S. cerevisiae* cells contained mutated Rdh54/Tid1 [92]. In 2006, Papamichos-Chronakis et al. found that the chromatin remodeling protein Ino80 was required for checkpoint adaptation in *S. cerevisiae* following the induction of a DSB [93]. In 2012, Eapen et al. found that checkpoint-deficient *S. cerevisiae* cells contained mutant Fun30, a chromatin remodeling protein involved in homologous recombination (HR) [94]. The role of these chromatin remodeling proteins in checkpoint adaptation remains to be elucidated.

These *S. cerevisiae* studies indicate that checkpoint adaptation is an important area of research of interest to researchers worldwide. Because checkpoint adaptation increases the number of *S. cerevisiae* cells that survive a DNA-damaging event and also increases genomic instability [87], it has been suggested that checkpoint adaptation may be important in the development of tumorigenesis [95]. Additionally, these studies indicate that a polo-like kinase and checkpoint kinases are central to checkpoint adaptation in *S. cerevisiae*. These data provide starting points for elucidating the biochemical pathway(s) that induce(s) checkpoint adaptation in higher eukaryotes. However, these studies also highlight the complexity of checkpoint adaptation and the need for more research in this area. Furthermore, although proteins involved in the DNA-damage response and cell-cycle regulation are largely evolutionarily conserved across eukaryotes, the regulation of cell-cycle checkpoints in *S. cerevisiae* is not identical to the regulation of cell-cycle checkpoints in humans. To understand better the process of checkpoint adaptation in humans, it is therefore necessary to use human model systems such as human cancer cell lines.

Initially, it was proposed that checkpoint adaptation would only occur in unicellular organisms, as a last attempt to survive if DNA-damage repair was not successful [96]. This is because entry into mitosis with damaged DNA in multicellular organisms may have a detrimental effect on the survival of an organism as a whole, by increasing the risk of genomic instability. By contrast, one could rationalize that unicellular organisms have nothing to lose by attempting cell division when DNA damage is irreparable [96]. However, in 2004, Yoo et al. described checkpoint adaptation in *Xenopus* oocyte extracts [97]. Yoo et al. reported that when they blocked DNA replication with aphidicolin, cell-free extracts were arrested in interphase and then entered mitosis with only partially replicated chromosomes. Because *Xenopus* are multicellular organisms, this, for the first time, suggested that checkpoint adaptation may also occur in other metazoans such as humans [96,97]. However, *Xenopus* oocyte extracts are different from somatic cells because they are a cell-free system without intact cell membranes. They also rapidly alternate between S phase and mitosis without G1 or G2 phases of the cell cycle.

Checkpoint adaptation was next identified in plant cells in early 2006. The pathways involved in DNA-damage repair and cell-cycle checkpoints are largely conserved in all eukaryotes, including plants [98]. Root cells from *Allium cepa* were irradiated with 2.5–40 Gy X-rays and analyzed for entry into mitosis with damaged DNA. X-ray irradiation with either 5, 10, 20, or 40 Gy produced G2/M arrest in the cells. An increase of apoptotic cells was also observed when cells were treated with 20 and 40 Gy X-rays. The number of apoptotic cells relative to the not treated control cells increased at 16 h following treatment with 20 Gy and at 4 h following treatment with 40 Gy. However, some cells treated with both of these doses of X-rays still underwent checkpoint adaptation following treatment, albeit at later times by comparison to cells treated with 5 and 10 Gy X-rays. The authors also scored mitotic cells for aberrant mitoses. Broken chromatids and acentric chromosomal fragments (lacking a centromere) were observed in some cells that were in either metaphase, anaphase, or telophase. When the percentages of aberrant mitoses were quantified following treatment with either 5, 10, 20, or 40 Gy X-rays, between 20% and 90% of mitoses were aberrant. Furthermore, the percentage of aberrant mitoses depended on the dose of X-ray irradiation and time after treatment. Cells were fixed at 0, 2, and 4 h and then at 4 h intervals to 24 h. A consistent number of mitoses were aberrant when cells were treated with 5 Gy X-rays between 2 and 24 h following treatment (between 50% and 70%). By contrast, 90% of mitoses were aberrant 24 h after treatment with either 20 or 40 Gy. Cells treated with 20 and 40 Gy X-rays did not enter mitosis between 4 and 20 h following treatment, suggesting that they were arrested at the G2/M checkpoint for a long time before entering mitosis by comparison to cells treated with 10 Gy, which entered mitosis at 16 h, and cells treated with 5 Gy which entered mitosis at all times tested between 2 and 24 h [98]. This study demonstrates that chromosome aberrations are present when cells undergo checkpoint adaptation which may lead to genomic instability in cells that survive this process. Additionally, these results suggest that cells enter mitosis at different times when they have different levels of damaged DNA; cells treated with higher doses of X-rays required more time to enter mitosis with damaged DNA (the final step of checkpoint adaptation) by comparison to cells treated with lower doses of X-rays.

10. CHECKPOINT ADAPTATION IN HUMAN CELLS

Checkpoint adaptation was first observed in human cancer cells in 2006 [99]. Syljuåsen et al. treated U2OS cells with 6 Gy of ionizing radiation and cells accumulated in G2 phase of the cell cycle. Several of the key cellular features of checkpoint adaptation are described in Fig. 22.3. In the original report, cells were arrested at the G2/M checkpoint, as shown by flow cytometry and by Chk1 phosphorylation on serine 345 [99]. The arrested cells then began to enter mitosis with damaged DNA, observed by the detection of histone γH2AX and phosphorylated serine-10 histone H3 staining using immunofluorescence microscopy. The authors found that when Chk1 was inhibited in the irradiated cells by the Chk1 inhibitor UCN-01, more cells were in mitosis 18 h following treatment by comparison to cells treated with 6 Gy ionizing radiation alone [99]. Furthermore, cells that over-expressed Chk1 arrested in G2 for longer than cells with wild-type levels of Chk1 following treatment with 6 Gy ionizing radiation. Since the polo-like kinase Cdc5 has a role in checkpoint adaptation in *S. cerevisiae* [82] and polo-like kinase 1 (Plk1) was implicated in checkpoint adaptation in *Xenopus* egg extracts [97], Syljuåsen et al. tested if Plk1 has a role in checkpoint adaptation in human cancer cells. Plk1 was depleted by siRNA in U2OS cells which were then treated with ionizing radiation. These cells accumulated in G2/M 19 h after treatment, but did not enter mitosis, suggesting that Plk1 has a role in checkpoint adaptation in human cancer cells [99]. However, the authors did not determine whether checkpoint adaptation was prevented or just delayed as a result of inhibiting Plk1. Similar to the studies in *S. cerevisiae*, the results of Syljuåsen et al. implicated Chk1 and Plk1 in the pathway of checkpoint adaptation in human cells. However, it was unclear if this process was clinically relevant because humans are unable to tolerate a single dose of 6 Gy of ionizing radiation and are treated with fractions of ionizing radiation in the clinic.

The question of whether or not checkpoint adaptation occurs in cancer cells treated with fractionated doses of ionizing radiation was addressed in 2011. Rezacova et al. reported that 26% of MOLT4 leukemia cells treated with fractionated radiation initiated a G2/M arrest 48 h after treatment before entering mitosis with damaged DNA [100]. However, it was still unknown whether checkpoint adaptation occurred in response to treatments other than ionizing radiation.

In 2012, checkpoint adaptation was observed in HT-29 human colorectal adenocarcinoma cells treated with either camptothecin (CPT) or etoposide and in M059K human glioma cells treated with CPT [7]. Kubara et al. demonstrated that HT-29 cells treated with 25 nM CPT for 48 h were arrested in the G2/M phase by analyzing DNA content using flow cytometry and by the detection of serine 345–phosphorylated Chk1. These CPT-treated cells then entered mitosis with

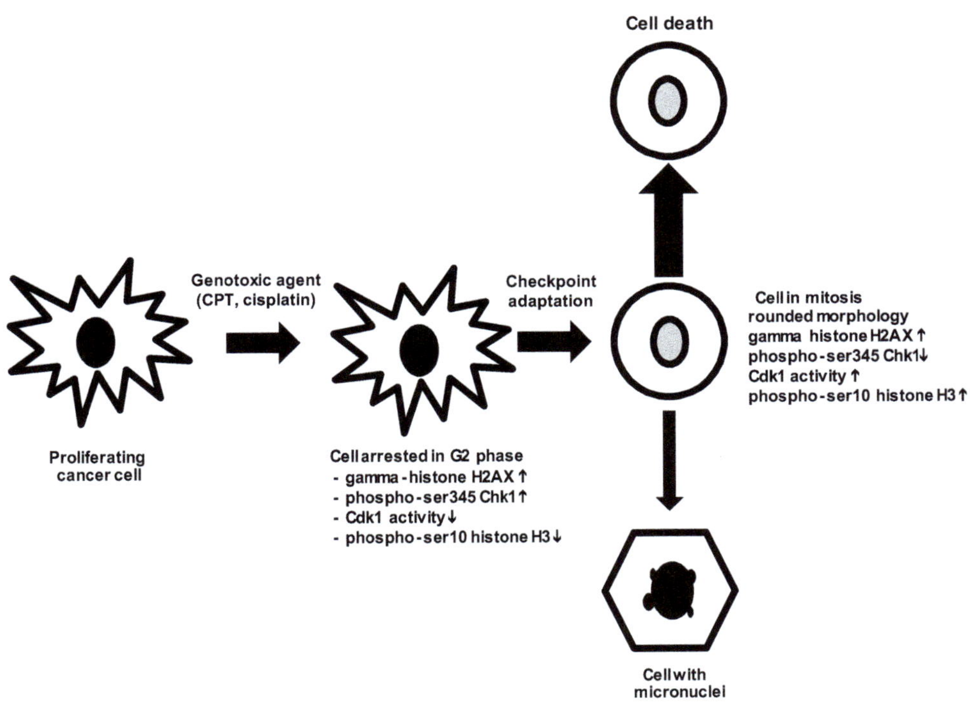

FIGURE 22.3 **A model of checkpoint adaptation.** In this figure we highlight several of the key cellular events that are characteristic of checkpoint adaptation. Proliferating cancer cells treated with genotoxic cancer drugs, such as camptothecin (CPT) or cisplatin, will arrest at the G2/M cell-cycle checkpoint with damaged DNA. The cells then enter mitosis with damaged DNA by undergoing checkpoint adaptation. The majority of cells die but some survive, likely with changes to their genome as demonstrated by micronuclei. The *arrows* list either increases or decreases in the biochemical and cellular events at each step in the process.

damaged DNA; they contained high levels of cyclin B1, had decreased levels of Cdk1 phosphorylated on tyrosine 15, and were positive for histone γH2AX and phosphorylated serine-10 histone H3 staining. Furthermore, Kubara et al. also demonstrated that entry into mitosis with damaged DNA could be prevented by co-treatment with the Cdk1 inhibitor CR8 [7]. Since HT-29 and M059K cells treated with CPT undergo checkpoint adaptation, CPT can be used as a positive control for checkpoint adaptation in these cell lines.

Kubara et al. (2012) also investigated the roles of Chk1 and Plk1 in checkpoint adaptation. HT-29 cells treated with CPT for 24 h and then with CPT and a Plk1 inhibitor for a further 24 h were partially prevented from entering mitosis with damaged DNA. However, there was only a slight decrease in the percentage of mitotic cells present in cells cotreated with CPT and a Plk1 inhibitor by comparison to cells treated with CPT alone [7]. This suggests that proteins other than Plk1 are involved in checkpoint adaptation in human cancer cells.

The role of Plk1 in human checkpoint adaptation remains unclear. In 2014, Liang et al. studied how single U2OS cells from a cell population treated with 1.5 Gy ionizing radiation responded to the activation of the G2/M checkpoint [101]. They found that different cells from the same population entered mitosis after a G2/M arrest at different times and with different levels of damaged DNA [101]. This entry into mitosis was reported to be dependent on levels of Plk1 and cyclin B. However, unpublished results from the laboratory of R.H. Medema indicate that inducing the expression of a constitutively active Plk1 mutant was not capable of overriding an established DNA damage–induced checkpoint [102]. The full molecular pathway(s) that are involved in the cellular induction of checkpoint adaptation remain to be elucidated.

11. THE CONSEQUENCES OF CHECKPOINT ADAPTATION

Checkpoint adaptation has several possible outcomes: (1) cells may die in mitosis; (2) cells may survive mitosis but die in subsequent phases of the cell cycle; or (3) cells may survive mitosis and divide with damaged DNA (Fig. 22.1) [7,99]. It has been demonstrated that the majority of human cancer cells that undergo checkpoint adaptation will die [7,99]. This is the desired outcome of treating cancer cells with a genotoxic agent. However, checkpoint adaptation is also a mechanism by which cells can transmit damaged DNA to daughter cells. It is therefore likely that checkpoint adaptation is a source of genomic instability in human cells that undergo this process (Fig. 22.3).

12. THE RELATIONSHIP BETWEEN CHECKPOINT ADAPTATION AND GENOMIC INSTABILITY

As described in Section 9, Galgoczy and Toczyski found that checkpoint adaptation in *S. cerevisiae* generated genomic rearrangements such as chromosome loss or chromosome rearrangement [87]. Additionally, these authors demonstrated that checkpoint adaptation increased yeast survival following DNA damage and that adaptation-deficient cells were less likely to survive a DNA-damaging event. These data suggest that, in yeast, checkpoint adaptation may contribute to cell survival following a genotoxic event and that cells surviving checkpoint adaptation are more likely to contain rearranged genomes. Some plant cells that enter mitosis following treatment with 5, 10, 20, and 40 Gy X-rays also contained broken chromatids, acentric chromosomal fragments, and chromosome bridges [98]. These data support the suggestion that checkpoint adaptation may induce genomic rearrangements in human cancer cells. However, these plant cells were fixed within 24 h of treatment, so it is unknown whether cells survived checkpoint adaptation with rearranged genomes or whether these chromosomal aberrations induced cell death.

To determine whether human cancer cells can survive checkpoint adaptation with rearranged genomes, it is first necessary to collect cells undergoing checkpoint adaptation and then culture these cells to assess cell viability. HT-29 cells can be used for this type of analysis because they display a pronounced rounded morphology in mitosis and spend a long time in mitosis following treatment. This allows cells that have undergone checkpoint adaptation and are in mitosis to be collected by mechanical shake-off (Fig. 22.1). These cells can then be assessed for cell survival by the clonogenic assay. Furthermore, once survival cells have been identified, these cells can be investigated to determine whether they contain rearranged genomes. Genomic rearrangements in these survival cells could be detected using cytogenetic techniques such as fluorescence in situ hybridization (FISH) to observe centromeres, telomeres, and chromosomal regions using fluorescent probes specific for these regions and by spectral karyotyping (SKY), which uses fluorescent probes of different colors to detect specific chromosomal regions. In addition to these techniques it will also be possible to sequence the genomes of individual cells surviving checkpoint adaptation when advances are made in single-cell DNA sequencing.

It has been reported that small numbers of HT-29 cells treated with CPT can survive entry into mitosis with damaged DNA [7]. Furthermore, preliminary data from our laboratory have suggested that HT-29 cells treated with CPT contain mitotic chromosomes that are shattered into different pieces (Rahman, Kernéis, and Golsteyn, unpublished data). CPT-treated HT-29 cells that survived checkpoint adaptation contained fewer chromosomes by comparison to untreated cells. Our laboratory has identified that entry into mitosis increases the number of micronuclei present in M059K cells treated with cisplatin. Checkpoint adaptation may thus facilitate the formation of survival cells with rearranged genomes, as a consequence of these cells surviving entry into mitosis with damaged DNA.

The investigation of checkpoint adaptation can provide an insight into the specific role of each step upon cell survival and genomic changes. For example, cancer cells appear to need to enter mitosis following treatment with a genotoxic agent to facilitate cell death in other phases of the cell cycle [96]. It is currently unknown whether cells that undergo checkpoint adaptation will still die if they are prevented from entering mitosis. If treated cancer cells can die when entry into mitosis with damaged DNA (the final step of checkpoint adaptation) is inhibited, then this could prevent cells from surviving treatment with rearranged genomes.

Current genotoxic anticancer treatments such as cisplatin are often limited by acquired resistance to treatment, which is induced by genomic changes [38,39]. By preventing treatment-induced genomic changes, it should be possible to prevent cells from acquiring resistance to treatment. These cancer cells would then be susceptible to cell killing induced by further rounds of treatment, thus improving the efficacy of genotoxic anticancer agents.

GLOSSARY

Mitotic cell rounding When cells enter mitosis they undergo a morphology change and display a rounded morphology instead of the flattened morphology of cells in interphase.

Chromothripsis Tens to hundreds of clustered genomic rearrangements are acquired in an event.

LIST OF ABBREVIATIONS

ATM Ataxia telangiectasia mutated
ATP Adenosine triphosphate
ATR ATM and Rad3 related
ATRIP ATR-interacting protein
CAK Cyclin-dependent kinase-activating kinase
Cdk Cyclin-dependent kinase
CGH Comparative genomic hybridization
Chk Checkpoint kinase
CIN Chromosomal instability
CKII Casein kinase II
CPT Camptothecin
DSB Double-strand breaks
dsDNA Double-stranded DNA
FISH Fluorescence in situ hybridization
GFP Green fluorescent protein
GST Glutathione *S*-transferase
HR Homologous recombination
MDC1 Mediator of DNA damage–checkpoint protein 1
MDR1 Multidrug-resistance protein 1
MEN Mitotic exit network
Plk1 Polo-like kinase 1
Rb Retinoblastoma
Rfa2 Replication factor A2
RPA Replication protein A
SAC Spindle assembly checkpoint
siRNA Silencing RNA
SKY Spectral karyotyping
ssDNA Single-stranded DNA
TCGA The Cancer Genome Atlas
TGCT Testicular germ cell tumor
TOPBP1 DNA topoisomerase binding protein 1
5-FU 5-Fluorouracil
9-1-1 Rad9–Rad1–Hus1 complex

ACKNOWLEDGMENTS

We thank members of the Cancer Cell Laboratory for valuable discussions. Research in the laboratory is supported by the University of Lethbridge ULRF, The School of Graduate Studies and Alberta Innovates Technology Futures.

REFERENCES

[1] Hanahan D, Weinberg RA. Hallmarks of cancer: the next generation. Cell 2011;144(5):646–74.
[2] Rausch T, Jones DT, Zapatka M, Stutz AM, Zichner T, Weischenfeldt J, et al. Genome sequencing of pediatric medulloblastoma links catastrophic DNA rearrangements with TP53 mutations. Cell 2012;148(1–2):59–71.
[3] Kandoth C, McLellan MD, Vandin F, Ye K, Niu B, Lu C, et al. Mutational landscape and significance across 12 major cancer types. Nature 2013;502(7471):333–9.
[4] Jackson SP, Bartek J. The DNA-damage response in human biology and disease. Nature 2009;461(7267):1071–8.
[5] Cadart C, Zlotek-Zlotkiewicz E, Le Berre M, Piel M, Matthews HK. Exploring the function of cell shape and size during mitosis. Dev Cell 2014;29(2):159–69.
[6] Terasima T, Tolmach LJ. Growth and nucleic acid synthesis in synchronously dividing populations of HeLa cells. Exp Cell Res 1963;30:344–62.
[7] Kubara PM, Kerneis-Golsteyn S, Studeny A, Lanser BB, Meijer L, Golsteyn RM. Human cells enter mitosis with damaged DNA after treatment with pharmacological concentrations of genotoxic agents. Biochem J 2012;446(3):373–81.
[8] Hendzel MJ, Wei Y, Mancini MA, Van Hooser A, Ranalli T, Brinkley BR, et al. Mitosis-specific phosphorylation of histone H3 initiates primarily within pericentromeric heterochromatin during G2 and spreads in an ordered fashion coincident with mitotic chromosome condensation. Chromosoma 1997;106(6):348–60.
[9] Malumbres M, Barbacid M. Mammalian cyclin-dependent kinases. Trends Biochem Sci 2005;30(11):630–41.
[10] Deibler RW, Kirschner MW. Quantitative reconstitution of mitotic CDK1 activation in somatic cell extracts. Mol Cell 2010;37(6):753–67.
[11] Larochelle S, Merrick KA, Terret ME, Wohlbold L, Barboza NM, Zhang C, et al. Requirements for Cdk7 in the assembly of Cdk1/cyclin B and activation of Cdk2 revealed by chemical genetics in human cells. Mol Cell 2007;25(6):839–50.
[12] Den Haese GJ, Walworth N, Carr AM, Gould KL. The Wee1 protein kinase regulates T14 phosphorylation of fission yeast Cdc2. Mol Biol Cell 1995;6(4):371–85.
[13] Millar JB, Russell P. The cdc25 M-phase inducer: an unconventional protein phosphatase. Cell 1992;68(3):407–10.
[14] Morgan DO. The cell cycle: principles of control. OUP/New Science Press; 2007.
[15] Karlsson C, Katich S, Hagting A, Hoffmann I, Pines J. Cdc25B and Cdc25C differ markedly in their properties as initiators of mitosis. J Cell Biol 1999;146(3):573–84.
[16] Pines J, Hunter T. Isolation of a human cyclin cDNA: evidence for cyclin mRNA and protein regulation in the cell cycle and for interaction with p34cdc2. Cell 1989;58(5):833–46.
[17] Lewis CW, Taylor RG, Kubara PM, Marshall K, Meijer L, Golsteyn RM. A western blot assay to measure cyclin dependent kinase activity in cells or in vitro without the use of radioisotopes. FEBS Lett 2013;587(18):3089–95.
[18] Kastan MB, Bartek J. Cell-cycle checkpoints and cancer. Nature 2004;432(7015):316–23.
[19] Schwartz GK, Shah MA. Targeting the cell cycle: a new approach to cancer therapy. J Clin Oncol 2005;23(36):9408–21.
[20] Nicklas RB. How cells get the right chromosomes. Science 1997;275(5300):632–7.
[21] Nitta M, Kobayashi O, Honda S, Hirota T, Kuninaka S, Marumoto T, et al. Spindle checkpoint function is required for mitotic catastrophe induced by DNA-damaging agents. Oncogene 2004;23(39):6548–58.
[22] Holland AJ, Cleveland DW. Boveri revisited: chromosomal instability, aneuploidy and tumorigenesis. Nat Rev Mol Cell Biol 2009;10(7):478–87.
[23] Deng C, Zhang P, Harper JW, Elledge SJ, Leder P. Mice lacking p21CIP1/WAF1 undergo normal development, but are defective in G1 checkpoint control. Cell 1995;82(4):675–84.
[24] Smith J, Tho LM, Xu N, Gillespie DA. The ATM-Chk2 and ATR-Chk1 pathways in DNA damage signaling and cancer. Adv Cancer Res 2010;108:73–112.
[25] Zou L, Liu D, Elledge SJ. Replication protein A-mediated recruitment and activation of Rad17 complexes. Proc Natl Acad Sci USA 2003;100(24):13827–32.
[26] Delacroix S, Wagner JM, Kobayashi M, Yamamoto K, Karnitz LM. The Rad9–Hus1–Rad1 (9-1-1) clamp activates checkpoint signaling via TopBP1. Genes Dev 2007;21(12):1472–7.
[27] Chini CC, Chen J. Human Claspin is required for replication checkpoint control. J Biol Chem 2003;278(32):30057–62.
[28] Liu Q, Guntuku S, Cui XS, Matsuoka S, Cortez D, Tamai K, et al. Chk1 is an essential kinase that is regulated by Atr and required for the G(2)/M DNA damage checkpoint. Genes Dev 2000;14(12):1448–59.
[29] Walker M, Black EJ, Oehler V, Gillespie DA, Scott MT. Chk1 C-terminal regulatory phosphorylation mediates checkpoint activation by de-repression of Chk1 catalytic activity. Oncogene 2009;28(24):2314–23.
[30] Lopez-Girona A, Furnari B, Mondesert O, Russell P. Nuclear localization of Cdc25 is regulated by DNA damage and a 14-3-3 protein. Nature 1999;397(6715):172–5.
[31] Ferguson AM, White LS, Donovan PJ, Piwnica-Worms H. Normal cell cycle and checkpoint responses in mice and cells lacking Cdc25B and Cdc25C protein phosphatases. Mol Cell Biol 2005;25(7):2853–60.
[32] Lee J, Kumagai A, Dunphy WG. Positive regulation of Wee1 by Chk1 and 14-3-3 proteins. Mol Biol Cell 2001;12(3):551–63.

[33] Chen Z, Xiao Z, Chen J, Ng SC, Sowin T, Sham H, et al. Human Chk1 expression is dispensable for somatic cell death and critical for sustaining G2 DNA damage checkpoint. Mol Cancer Ther 2003;2(6):543–8.
[34] Pan Y, Ren K-H, He H-W, Shao R-G. Knockdown of Chk1 sensitizes human colon carcinoma HCT116 cells in a p53-dependent manner to lidamycin through abrogation of a G2/M checkpoint and induction of apoptosis. Cancer Biol Ther 2014;8(16):1559–66.
[35] Shen DW, Pouliot LM, Hall MD, Gottesman MM. Cisplatin resistance: a cellular self-defense mechanism resulting from multiple epigenetic and genetic changes. Pharmacol Rev 2012;64(3):706–21.
[36] Jamieson ER, Lippard SJ. Structure, recognition, and processing of cisplatin-DNA adducts. Chem Rev 1999;99(9):2467–98.
[37] Weiss MM, Hermsen MA, Meijer GA, van Grieken NC, Baak JP, Kuipers EJ, et al. Comparative genomic hybridisation. Mol Pathol 1999;52(5): 243–51.
[38] Wasenius VM, Jekunen A, Monni O, Joensuu H, Aebi S, Howell SB, et al. Comparative genomic hybridization analysis of chromosomal changes occurring during development of acquired resistance to cisplatin in human ovarian carcinoma cells. Genes Chromosomes Cancer 1997;18(4): 286–91.
[39] Wilson C, Yang J, Strefford JC, Summersgill B, Young BD, Shipley J, et al. Overexpression of genes on 16q associated with cisplatin resistance of testicular germ cell tumor cell lines. Genes Chromosomes Cancer 2005;43(2):211–6.
[40] Brown JM, Wouters BG. Apoptosis, p53, and tumor cell sensitivity to anticancer agents. Cancer Res 1999;59(7):1391–9.
[41] Roninson IB, Broude EV, Chang BD. If not apoptosis, then what? Treatment-induced senescence and mitotic catastrophe in tumor cells. Drug Resist Updat 2001;4(5):303–13.
[42] Brown JM, Attardi LD. The role of apoptosis in cancer development and treatment response. Nat Rev Cancer 2005;5(3):231–7.
[43] Yin DX, Schimke RT. BCL-2 expression delays drug-induced apoptosis but does not increase clonogenic survival after drug treatment in HeLa cells. Cancer Res 1995;55(21):4922–8.
[44] Lock RB, Stribinskiene L. Dual modes of death induced by etoposide in human epithelial tumor cells allow Bcl-2 to inhibit apoptosis without affecting clonogenic survival. Cancer Res 1996;56(17):4006–12.
[45] Kyprianou N, King ED, Bradbury D, Rhee JG. bcl-2 over-expression delays radiation-induced apoptosis without affecting the clonogenic survival of human prostate cancer cells. Int J Cancer 1997;70(3):341–8.
[46] Elliott MJ, Stribinskiene L, Lock RB. Expression of Bcl-2 in human epithelial tumor (HeLa) cells enhances clonogenic survival following exposure to 5-fluoro-2′-deoxyuridine or staurosporine, but not following exposure to etoposide or doxorubicin. Cancer Chemother Pharmacol 1998;41(6):457–63.
[47] Elliott MJ, Murphy KM, Stribinskiene L, Ranganathan V, Sturges E, Farnsworth ML, et al. Bcl-2 inhibits early apoptotic events and reveals post-mitotic multinucleation without affecting cell cycle arrest in human epithelial tumor cells exposed to etoposide. Cancer Chemother Pharmacol 1999;44(1):1–11.
[48] Tannock IF, Lee C. Evidence against apoptosis as a major mechanism for reproductive cell death following treatment of cell lines with anti-cancer drugs. Br J Cancer 2001;84(1):100–5.
[49] Wouters BG, Denko NC, Giaccia AJ, Brown JM. A p53 and apoptotic independent role for p21waf1 in tumour response to radiation therapy. Oncogene 1999;18(47):6540–5.
[50] Ruth AC, Roninson IB. Effects of the multidrug transporter P-glycoprotein on cellular responses to ionizing radiation. Cancer Res 2000;60(10): 2576–8.
[51] Vitale I, Galluzzi L, Castedo M, Kroemer G. Mitotic catastrophe: a mechanism for avoiding genomic instability. Nat Rev Mol Cell Biol 2011;12(6):385–92.
[52] de Bruin EC, Medema JP. Apoptosis and non-apoptotic deaths in cancer development and treatment response. Cancer Treat Rev 2008;34(8): 737–49.
[53] Dewey WC, Ling CC, Meyn RE. Radiation-induced apoptosis: relevance to radiotherapy. Int J Radiat Oncol Biol Phys 1995;33(4):781–96.
[54] Castedo M, Perfettini JL, Roumier T, Andreau K, Medema R, Kroemer G. Cell death by mitotic catastrophe: a molecular definition. Oncogene 2004;23(16):2825–37.
[55] Yamada M, Puck TT. Action of radiation on mammalian cells. IV. Reversible mitotic lag in the S3 HeLa cell produced by low doses of X-rays. Proc Natl Acad Sci USA 1961;47:1181–91.
[56] Rello-Varona S, Kepp O, Vitale I, Michaud M, Senovilla L, Jemaa M, et al. An automated fluorescence videomicroscopy assay for the detection of mitotic catastrophe. Cell Death Dis 2010;1:e25.
[57] Broude EV, Loncarek J, Wada I, Cole K, Hanko C, Roninson IB, et al. Mitotic catastrophe in cancer therapy. In: Roninson IB, Brown JM, Bredesen DE, editors. Beyond apoptosis: cellular outcomes of cancer therapy. 1st ed. New York, USA: Informa Healthcare USA, Inc.; 2008. p. 307–20.
[58] Cahuzac N, Studeny A, Marshall K, Versteege I, Wetenhall K, Pfeiffer B, et al. An unusual DNA binding compound, S23906, induces mitotic catastrophe in cultured human cells. Cancer Lett 2010;289(2):178–87.
[59] Swift LH, Golsteyn RM. Genotoxic anti-cancer agents and their relationship to DNA damage, mitosis, and checkpoint adaptation in proliferating cancer cells. Int J Mol Sci 2014;15(3):3403–31.
[60] Widel M, Jedrus S, Owczarek S, Konopacka M, Lubecka B, Kolosza Z. The increment of micronucleus frequency in cervical carcinoma during irradiation in vivo and its prognostic value for tumour radiocurability. Br J Cancer 1999;80(10):1599–607.
[61] Bhattathiri VN. Amitotic cell divisions and tumour growth: an alternative model for cell kinetic compartments in solid tumours. Oral Oncol 2001;37(3):288–95.
[62] Yoshikawa R, Kusunoki M, Yanagi H, Noda M, Furuyama JI, Yamamura T, et al. Dual antitumor effects of 5-fluorouracil on the cell cycle in colorectal carcinoma cells: a novel target mechanism concept for pharmacokinetic modulating chemotherapy. Cancer Res 2001;61(3):1029–37.

[63] Vakifahmetoglu H, Olsson M, Tamm C, Heidari N, Orrenius S, Zhivotovsky B. DNA damage induces two distinct modes of cell death in ovarian carcinomas. Cell Death Differ 2008;15(3):555–66.

[64] Eom YW, Kim MA, Park SS, Goo MJ, Kwon HJ, Sohn S, et al. Two distinct modes of cell death induced by doxorubicin: apoptosis and cell death through mitotic catastrophe accompanied by senescence-like phenotype. Oncogene 2005;24(30):4765–77.

[65] Tounekti O, Pron G, Belehradek Jr J, Mir LM. Bleomycin, an apoptosis-mimetic drug that induces two types of cell death depending on the number of molecules internalized. Cancer Res 1993;53(22):5462–9.

[66] Stratton MR, Campbell PJ, Futreal PA. The cancer genome. Nature 2009;458(7239):719–24.

[67] Ferguson DO, Alt FW. DNA double strand break repair and chromosomal translocation: lessons from animal models. Oncogene 2001;20(40): 5572–9.

[68] Nakada S, Katsuki Y, Imoto I, Yokoyama T, Nagasawa M, Inazawa J, et al. Early G2/M checkpoint failure as a molecular mechanism underlying etoposide-induced chromosomal aberrations. J Clin Invest 2006;116(1):80–9.

[69] Ichijima Y, Yoshioka K, Yoshioka Y, Shinohe K, Fujimori H, Unno J, et al. DNA lesions induced by replication stress trigger mitotic aberration and tetraploidy development. PLoS One 2010;5(1):e8821.

[70] Burrell RA, McClelland SE, Endesfelder D, Groth P, Weller MC, Shaikh N, et al. Replication stress links structural and numerical cancer chromosomal instability. Nature 2013;494(7438):492–6.

[71] Edelweiss MI, Trachtenberg A, Pinheiro EX, da-Silva J, Riegel M, Lizardo-Daudt HM, et al. Clastogenic effect of cisplatin on Wistar rat bone marrow cells. Braz J Med Biol Res 1995;28(6):679–83.

[72] Crasta K, Ganem NJ, Dagher R, Lantermann AB, Ivanova EV, Pan Y, et al. DNA breaks and chromosome pulverization from errors in mitosis. Nature 2012;482(7383):53–8.

[73] Burns EM, Christopoulou L, Corish P, Tyler-Smith C. Quantitative measurement of mammalian chromosome mitotic loss rates using the green fluorescent protein. J Cell Sci 1999;112(Pt 16):2705–14.

[74] Orthwein A, Fradet-Turcotte A, Noordermeer SM, Canny MD, Brun CM, Strecker J, et al. Mitosis inhibits DNA double-strand break repair to guard against telomere fusions. Science 2014;344(6180):189–93.

[75] Stephens PJ, Greenman CD, Fu B, Yang F, Bignell GR, Mudie LJ, et al. Massive genomic rearrangement acquired in a single catastrophic event during cancer development. Cell 2011;144(1):27–40.

[76] Forment JV, Kaidi A, Jackson SP. Chromothripsis and cancer: causes and consequences of chromosome shattering. Nat Rev Cancer 2012;12(10): 663–70.

[77] Zhang CZ, Spektor A, Cornils H, Francis JM, Jackson EK, Liu S, et al. Chromothripsis from DNA damage in micronuclei. Nature 2015;522(7555): 179–84.

[78] Cohen A, Sato M, Aldape K, Mason CC, Alfaro-Munoz K, Heathcock L, et al. DNA copy number analysis of Grade II-III and Grade IV gliomas reveals differences in molecular ontogeny including chromothripsis associated with IDH mutation status. Acta Neuropathol Commun 2015;3:34.

[79] Magrangeas F, Avet-Loiseau H, Munshi NC, Minvielle S. Chromothripsis identifies a rare and aggressive entity among newly diagnosed multiple myeloma patients. Blood 2011;118(3):675–8.

[80] Przybytkowski E, Lenkiewicz E, Barrett MT, Klein K, Nabavi S, Greenwood CM, et al. Chromosome-breakage genomic instability and chromothripsis in breast cancer. BMC Genomics 2014;15:579.

[81] Bertelsen B, Nazaryan-Petersen L, Sun W, Mehrjouy MM, Xie G, Chen W, et al. A germline chromothripsis event stably segregating in 11 individuals through three generations. Genet Med 2015;18.

[82] Toczyski DP, Galgoczy DJ, Hartwell LH. CDC5 and CKII control adaptation to the yeast DNA damage checkpoint. Cell 1997;90(6):1097–106.

[83] Sandell LL, Zakian VA. Loss of a yeast telomere: arrest, recovery, and chromosome loss. Cell 1993;75(4):729–39.

[84] Sanchez Y, Bachant J, Wang H, Hu F, Liu D, Tetzlaff M, et al. Control of the DNA damage checkpoint by chk1 and rad53 protein kinases through distinct mechanisms. Science 1999;286(5442):1166–71.

[85] Hu F, Wang Y, Liu D, Li Y, Qin J, Elledge SJ. Regulation of the Bub2/Bfa1 GAP complex by Cdc5 and cell cycle checkpoints. Cell 2001;107(5): 655–65.

[86] Leroy C, Lee SE, Vaze MB, Ochsenbien F, Guerois R, Haber JE, et al. PP2C phosphatases Ptc2 and Ptc3 are required for DNA checkpoint inactivation after a double-strand break. Mol Cell 2003;11(3):827–35.

[87] Galgoczy DJ, Toczyski DP. Checkpoint adaptation precedes spontaneous and damage-induced genomic instability in yeast. Mol Cell Biol 2001;21(5):1710–8.

[88] Pellicioli A, Lee SE, Lucca C, Foiani M, Haber JE. Regulation of *Saccharomyces* Rad53 checkpoint kinase during adaptation from DNA damage-induced G2/M arrest. Mol Cell 2001;7(2):293–300.

[89] Guillemain G, Ma E, Mauger S, Miron S, Thai R, Guerois R, et al. Mechanisms of checkpoint kinase Rad53 inactivation after a double-strand break in *Saccharomyces cerevisiae*. Mol Cell Biol 2007;27(9):3378–89.

[90] Tao R, Xue H, Zhang J, Liu J, Deng H, Chen YG. Deacetylase Rpd3 facilitates checkpoint adaptation by preventing Rad53 overactivation. Mol Cell Biol 2013;33(21):4212–24.

[91] Ghospurkar PL, Wilson TM, Severson AL, Klein SJ, Khaku SK, Walther AP, et al. The DNA damage response and checkpoint adaptation in *Saccharomyces cerevisiae*: distinct roles for the replication protein A2 (Rfa2) N-terminus. Genetics 2015;199(3):711–27.

[92] Lee SE, Pellicioli A, Malkova A, Foiani M, Haber JE. The *Saccharomyces* recombination protein Tid1p is required for adaptation from G2/M arrest induced by a double-strand break. Curr Biol 2001;11(13):1053–7.

[93] Papamichos-Chronakis M, Krebs JE, Peterson CL. Interplay between Ino80 and Swr1 chromatin remodeling enzymes regulates cell cycle checkpoint adaptation in response to DNA damage. Genes Dev 2006;20(17):2437–49.

[94] Eapen VV, Sugawara N, Tsabar M, Wu WH, Haber JE. The *Saccharomyces cerevisiae* chromatin remodeler Fun30 regulates DNA end resection and checkpoint deactivation. Mol Cell Biol 2012;32(22):4727–40.

[95] Vidanes GM, Sweeney FD, Galicia S, Cheung S, Doyle JP, Durocher D, et al. CDC5 inhibits the hyperphosphorylation of the checkpoint kinase Rad53, leading to checkpoint adaptation. PLoS Biol 2010;8(1):e1000286.

[96] Lupardus PJ, Cimprich KA. Checkpoint adaptation; molecular mechanisms uncovered. Cell 2004;117(5):555–6.

[97] Yoo HY, Kumagai A, Shevchenko A, Shevchenko A, Dunphy WG. Adaptation of a DNA replication checkpoint response depends upon inactivation of Claspin by the Polo-like kinase. Cell 2004;117(5):575–88.

[98] Carballo JA, Pincheira J, de la Torre C. The G2 checkpoint activated by DNA damage does not prevent genome instability in plant cells. Biol Res 2006;39(2):331–40.

[99] Syljuåsen RG, Jensen S, Bartek J, Lukas J. Adaptation to the ionizing radiation-induced G2 checkpoint occurs in human cells and depends on checkpoint kinase 1 and Polo-like kinase 1 kinases. Cancer Res 2006;66(21):10253–7.

[100] Rezacova M, Rudolfova G, Tichy A, Bacikova A, Mutna D, Havelek R, et al. Accumulation of DNA damage and cell death after fractionated irradiation. Radiat Res 2011;175(6):708–18.

[101] Liang H, Esposito A, De S, Ber S, Collin P, Surana U, et al. Homeostatic control of polo-like kinase-1 engenders non-genetic heterogeneity in G2 checkpoint fidelity and timing. Nat Commun 2014;5:4048.

[102] Shaltiel IA, Krenning L, Bruinsma W, Medema RH. The same, only different—DNA damage checkpoints and their reversal throughout the cell cycle. J Cell Sci 2015;128(4):607–20.

[103] Chang BD, Broude EV, Dokmanovic M, Zhu H, Ruth A, Xuan Y, et al. A senescence-like phenotype distinguishes tumor cells that undergo terminal proliferation arrest after exposure to anticancer agents. Cancer Res 1999;59(15):3761–7.

[104] Demarcq C, Bunch RT, Creswell D, Eastman A. The role of cell cycle progression in cisplatin-induced apoptosis in Chinese hamster ovary cells. Cell Growth Differ 1994;5(9):983–93.

[105] Krishnaswamy G, Dewey WC. Cisplatin induced cell killing and chromosomal aberrations in CHO cells: treated during G1 or S phase. Mutat Res 1993;293(2):161–72.

[106] Wang XQ, Zhu YQ, Lui KS, Cai Q, Lu P, Poon RT. Aberrant Polo-like kinase 1-Cdc25A pathway in metastatic hepatocellular carcinoma. Clin Cancer Res 2008;14(21):6813–20.

[107] Ianzini F, Mackey MA. Delayed DNA damage associated with mitotic catastrophe following X-irradiation of HeLa S3 cells. Mutagenesis 1998;13(4):337–44.

[108] Ngan CY, Yamamoto H, Takagi A, Fujie Y, Takemasa I, Ikeda M, et al. Oxaliplatin induces mitotic catastrophe and apoptosis in esophageal cancer cells. Cancer Sci 2008;99(1):129–39.

Chapter 23

Chromatin, Nuclear Organization, and Genome Stability in Mammals

L. Boteva, N. Gilbert

The University of Edinburgh, Edinburgh, United Kingdom

Chapter Outline

1. INTRODUCTION

In mammalian cells, the long DNA molecules comprising the genome are wrapped with proteins to form a complex called chromatin. Chromatin fibers are then folded multiple times within the nucleus of the cell, allowing lengthy genomes to fit inside much smaller cells. Apart from overcoming space constraints, the folding of the genome also has a regulatory function, influencing fundamental processes such as gene expression, genome replication, and DNA-damage repair (DDR).

Chromatin is a complex structure with different levels of organization. At the most basic level, chromatin is made up of nucleosomes—147 base pairs (bp) of the DNA sequence wrapped around a protein octamer composed of eight histone proteins. Arrays of nucleosomes are then further folded to form a fiber measuring 30-nm in diameter; 30-nm fibers are then arranged into larger-scale structures forming domains with differing structural and functional properties and which ultimately form chromosomes—the largest units of chromatin organization.

A cell's genome is frequently exposed to factors that have the potential to introduce changes in the DNA sequence ranging from point mutations to chromosome structural aberrations and even chromosome gain or loss. Classically, threats to genome integrity were perceived to mainly come from external factors, such as drugs, chemical compounds, or UV radiation. A more current view is that internal factors and fundamental cellular processes such as transcription and replication also pose a risk to genome stability.

Whatever the source of the threat, chromatin is the context in which the genome is assaulted and then repaired. However, chromatin is more than just a passive bystander in the DNA-damage response. It forms a dynamic structure which plays an active role in a cell's response to genome damage and reacts to DNA damage with extensive changes to its structure and composition. The best accepted model describing chromatin dynamics upon induction of damage is the so-called "access, repair, restore" model [1]. It postulates that to fully repair a damaged locus, chromatin first has to be disrupted to allow access to the damaged template, followed by recruitment of factors that facilitate the repair process and finally, a reestablishment of the initial chromatin structure and eviction of the DNA-damage marks from the region. Failure in these processes can result in a serious predisposition to genomic damage and catastrophic consequences for the cell and the

Genome Stability. http://dx.doi.org/10.1016/B978-0-12-803309-8.00023-9

organism; therefore, our knowledge of mammalian DNA-damage response is incomplete without considering the contribution of chromatin context and the 3D organization of the genome.

2. HISTONES

Apart from DNA, chromatin contains numerous proteins with structural and regulatory functions. Among them, histone proteins are the most prominent. Core histones form nucleosomes—the basic repeating unit of chromatin, while linker histones provide the connections between nucleosomes. Histone proteins can be posttranslationally modified on their N-terminal tails, with different modifications exerting different effects on the chromatin fiber structure, adding a regulatory as well as a structural role to the range of histone functions. These posttranslational modifications (PTMs) include acetylation, methylation, and phosphorylation, as well as other, less well-characterized marks. In addition to the canonical histone proteins, the histone family also includes many histone variants, which can replace their classical counterparts in chromatin in a carefully regulated manner and in specific circumstances.

The histone proteins that form the nucleosome particle are called "core histones" and include H2A, H2B, H3, and H4, as well as their variants. Each nucleosome is an octamer consisting of two copies of each core histone, arranged as an H3/H4 tetramer and two H2A/H2B dimers. The core histones are positively charged proteins, rich in lysine and arginine residues. They bind to DNA noncovalently, through electrostatic interactions between positive charges on histones and the negatively charged DNA molecule. Nucleosomes are separated by linker DNA, whose length is not constant, but can vary from 10 to 100 bp between species and cell types. Histone proteins binding to this linker DNA are called linker histones and include H1 and its variant H5. Like core histones, H1 is also positively charged and is associated with both the linker DNA and the nucleosome particle. The H1 molecule consists of a globular domain and two tails, with the globular domain binding at the nucleosome dyad, while the tails contact the linker DNA and can drape along the chromatin fiber to stabilize the folding of nucleosomes into a 30-nm fiber structure.

2.1 Histone Variants

The classical histone molecules may be replaced in the nucleosomes and linker regions by histone variants—proteins with a high degree of sequence similarity to their common counterparts. H2A, H2B, H3, and H1, all have noncanonical variants which can replace canonical histones in the fiber in different circumstances. This replacement can have many effects, including a change in the fiber conformation (causing chromatin to become more or less tightly folded) or recruitment of regulatory proteins. Histone variant incorporation into chromatin can be independent of replication [2] and is carefully regulated by a class of proteins called histone chaperones.

An interesting example of a histone variant with an important role is CENP-A, an H3 variant present specifically at centromeres, deposited there by a histone chaperone called HJURP. Multiple studies have found evidence for a distinct chromatin structure at mammalian centromeres [3], which may be affected by the presence of CENP-A-containing nucleosomes. The presence of CENP-A is also important for recruitment of kinetochore components. Although the precise effects of CENP-A incorporation into nucleosomes in vivo are unclear, a 2011 study found that nucleosomal arrays containing CENP-A are more condensed compared to arrays containing canonical H3, suggesting that the presence of CENP-A helps to establish an unusual chromatin structure at centromeres [4].

As an important function of chromatin, locating and signaling DNA damage is also associated with a separate histone variant—H2AX, which comprises up to 32% of H2A throughout the genome. H2AX is phosphorylated at serine 139 as one of the primary events at sites of DNA damage and plays an essential role for DNA-damage signaling, detection, and repair, as discussed later in this chapter.

2.2 Histone Modifications

In addition to histone variants, chromatin fiber structure and composition can also be affected by PTMs, which are frequently present on the N-terminal tails of the histone proteins. These modifications include acetylation, methylation, phosphorylation, and ubiquitination and similarly to the presence of histone variants, can act by directly modifying chromatin structure or by recruiting regulatory factors recognizing specific posttranslational marks. Numerous posttranslational marks exist and their effects, both individual and combinatorial, are still under active investigation.

Acetylation of lysine residues in the N-terminal tails of H3 and H4 are marks often associated with active transcriptional states. Acetylation neutralizes the charge of the lysine residue, which is expected to weaken histone–histone and histone–DNA interactions, resulting in the opening of the chromatin fiber. However, a careful in vitro study performed on short arrays of nucleosomes reconstituted on a repetitive DNA sequence failed to demonstrate significant unfolding of the fiber, suggesting the effects of acetylation may depend on the wider chromatin context [5]. H3 and H4 can be acetylated at numerous positions, including H3K9, H3K14, H3K18, H4K5, H4K8, H4K12, and H4K16 providing binding sites for bromodomains that are present within some transcriptional activators and chromatin remodelers. The acetylation mark is put on histone molecules by a class of enzymes called histone acetyltransferases (HATs) and removed by histone deacetylases (HDACs). Interestingly, loss of HDAC function is associated with genome instability, including aneuploidy and lagging chromosomes [6].

Methylation occurs on lysine and arginine residues. Up to three methyl groups can be added on lysines, while arginines can only be mono- or dimethylated; unlike acetylation, methylation does not alter the charge of the residue affected. Lysines are methylated by lysine methyltransferases (HKMTs), which are very specific and only methylate-specific residues. Multiple HKMTs have been identified, all of which share a SET protein domain. Arginine residues are modified by arginine methyltransferases, also known as PRMTs, while removal of the methyl residues is done by demethylases. A few classes of lysine demethylases exist, including lysine-specific demethylase 1 (LSD1), which can demethylate different lysine residues depending on different accessory proteins and the jumonji domain demethylases, which act on trimethylated lysine residues. Methyl marks on H3 and H4 residues can be associated with active and inactive chromatin states. Examples include H3K9me3–a repressive mark which recruits the heterochromatin protein HP1 and H3K4me3–a mark present in actively transcribed regions. A direct relationship between appropriate methylation patterns and genome instability has been demonstrated via depletion of Suv39h, an H3K9 methyltransferase involved in establishing H3K9me3 at pericentromeric chromatin. Mice lacking Suv39h are prone to tumor formation, while embryonic fibroblasts derived from the animals have extremely unstable karyotypes [7]. While the H3K9 methylation mark probably exerts its effects on genomic stability through maintaining the structural state of certain genomic regions, another methylation mark, H3K79me, has been implicated in the DNA-damage response in a signaling manner. This mark is established by the DOT1 lysine methylase and is important for recruitment of 53BP1, a protein integral to the DDR, to a break site. 53BP1 recruitment by H379Kme is through a Tudor domain in the 53BP1 protein, a domain recognizing methylated residues; however, it is unclear whether the mark is established in response to a DNA break or whether chromatin changes in the vicinity of a break cause the mark to be exposed and recognized by 53BP1 [8].

Another important PTM, phosphorylation, can be added on serine, threonine, and tyrosine residues. This modification is placed by kinases and removed by phosphatases. Unlike acetylation and methylation, which are related to establishing chromatin domains with different properties, phosphorylation also has a major role in cell-cycle progression. The serine 10 position of H3 is phosphorylated genome-wide by the Aurora B kinase as cells progress through late G2 and into mitosis [9] in a manner that is interdependent on other histone modifications, such as H3K9me [10]. This modification is required for the mitotic condensation of chromosomes—a process in which chromosomes are compacted to facilitate chromosome separation and minimize entanglements during cell division.

Phosphorylation of the H2AX histone variant at serine 139 (phosho-H2AX or γ-H2AX) is the most widely studied DNA damage–associated histone modification. H2AX Serine 139 is rapidly phosphorylated in response to DNA damage and phosphorylation is dependent on the ATM, ATR, and DNA-PK kinases. The γ-H2AX mark spreads in large, megabase-sized domains surrounding the break region [11] and is essential for DNA-damage signaling and response. γ-H2AX–containing chromatin then serves as a platform for recruiting additional repair components, including 53BP1 and BRCA1 [12]. Interestingly, studies in which the H2AX phosphorylation site is disrupted indicate that lack of H2AX phosphorylation does not preclude initial recruitment of repair factors to the site (NBS1, BRCA1, and 53BP1); however, it affects their retention, suggesting that γ-H2AX provides a platform for maintaining factors necessary for repair. Following repair, H2AX phosphorylation is reversed by phosphatase complexes including PP2A and PP4 [13] and through histone exchange mediated by the FACT complex [14]. Mammalian cells lacking H2AX exhibit enhanced susceptibility to genomic instability and cancer [15]. Given the coordinated structural and signaling functions of other histone modifications, it is tempting to speculate that phosphorylation of H2AX may impact on chromatin fiber structure as well as via signaling in the DDR cascades. However, no such structural effects have been convincingly demonstrated and while changes in chromatin compaction are known to occur as a consequence of damage, they have been shown to be independent of the presence of γ-H2AX [16]. Other histone marks which may have a small role in the DNA-damage response include H2A ubiquitination [17], H2B phosphorylation at serine 14 [18], and H3 threonine 45 phosphorylation [19].

While the establishment of histone marks in response to DNA damage is well characterized, a 2015 publication by the Misteli Lab [20] explored the opposite idea—can certain histone PTMs predispose genomic regions to instability? Surprisingly, the study found enrichment of H3K4me1 and H3K27ac and depletion of the repressive H3K9me3 mark in genes frequently involved in translocations when compared to genes with similar expression patterns and levels. To demonstrate the correlation is causal, the authors tethered an H3K4 methyltransferase and an H3/H4 lysine acetyltransferase to a *LacO* array incorporating an artificially introduced unique restriction enzyme site. When the frequency of breaks was assessed, the authors found elevated rates in the presence of both the H3K4 methyltransferase and the H3/H4 lysine acetyltransferase, leading them to speculate that this created a more open chromatin environment making the genome more prone to instability. Interestingly, the H3K4me3 mark is also associated with the introduction of double-strand DNA breaks during V(D)J recombination in lymphocytes [21].

3. NUCLEOSOMES AND THE 30-NM FIBER

Independent of any variants or PTMs that may be present, core histones are invariably arranged in nucleosome structures, containing two H2A:H2B dimers and two H3:H4 dimers. One hundred and forty-seven base pairs of DNA are wrapped around each nucleosome with 10–100 bp "linker" DNA bound to histone H1, linking up different nucleosomes. With the help of linker histones, the arrays of nucleosomes fold into a fiber measuring 30 nm in diameter, the exact structure of which is still under intense debate (Fig. 23.1).

A number of models have been proposed for the arrangement of nucleosomes in the 30-nm fiber structure, including a solenoid model, where nucleosomes are organized in a helical array, a "zigzag" model with a zigzag arrangement of nucleosomes and an "irregular fiber" model with a disorganized arrangement and variable spacing of nucleosomes. Various techniques have been used since 1963 to try and resolve the structure of the 30-nm fiber, including variations of electron microscopy, X-ray diffraction, and in early 2010s, superresolution microscopy [22,23]. While successful observation of the 30-nm fiber structure is possible in chromatin reconstituted in situ and in some rare types of nuclei, it has proven impossible to resolve the fibers in intact nuclei, with chromatin appearing as a densely staining mass.

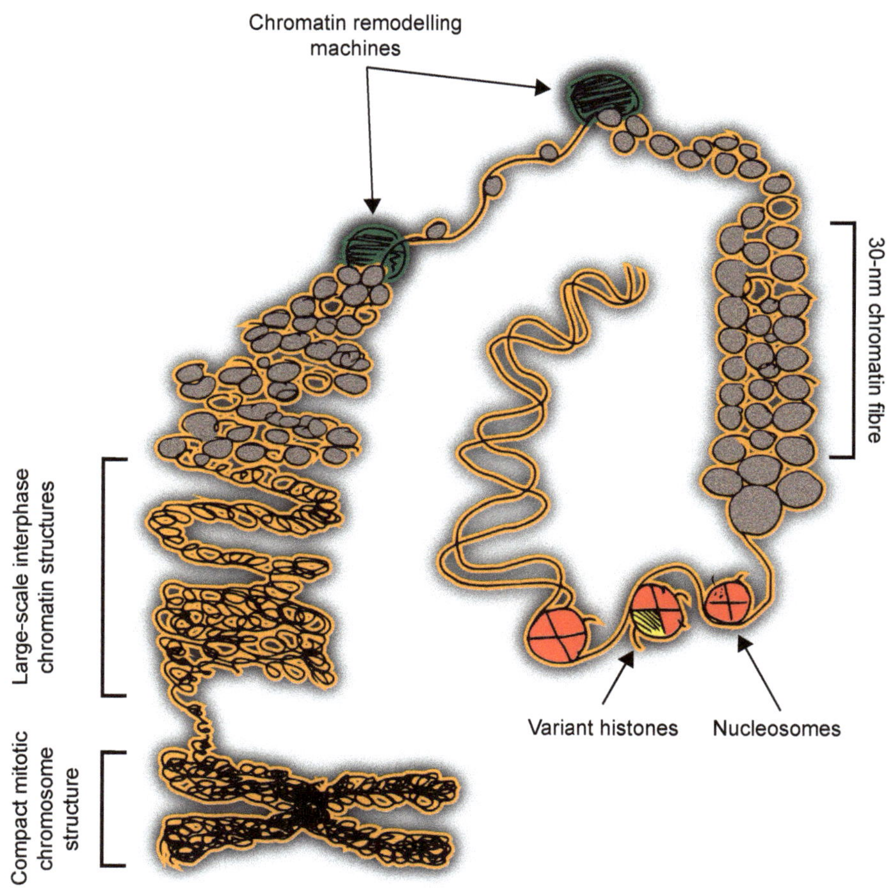

FIGURE 23.1 **Levels of chromatin organization.** At the primary level of chromatin folding, the DNA molecule is wrapped around histone octamers to form nucleosomes. Nucleosomes may contain core histones or histone variants and the N-terminal tails of the histones can carry various posttranslational marks. Interactions between nucleosomes cause further folding into a 30-nm fiber. The exact arrangement of nucleosomes within the 30-nm fiber is unknown, but it is likely not homogeneous and local disruptions caused by chromatin remodeling events are present. Larger-scale structures are formed by further folding of the 30-nm fibers. Interphase chromatin is additionally compacted for mitosis, giving rise to mitotic chromosomes.

In reality, as chromatin structures are very dynamic, it is likely the structure of the 30-nm fiber in living nuclei is not homogeneous, but instead is made up of a mixture of the models proposed with some regions being more compact and others more disrupted. In another illustration of the structure–function relationship, it has been shown that constitutively transcriptionally inactive parts of the genome (eg, centromeric heterochromatin) show a regular folding at the 30-nm level, while the bulk genome has a less regular conformation, interspersed with irregularities [3]. Nucleosomes can be moved and shuffled by proteins called chromatin remodelers to enable proteins such as transcription factors, replication-related proteins, and DNA-repair proteins to bind to the naked DNA template. It is easy to envisage how these movements of nucleosomes can introduce transient local disruptions in the chromatin fiber. A frequently used method to investigate nucleosome disruptions and 30-nm fiber structure is performed by testing the accessibility of the naked DNA by DNase I digestion, or a 2013 approach taking advantage of next-generation sequencing called ATAC-seq [24].

4. HIGHER-ORDER STRUCTURES

At a further level of chromatin organization, interactions between the 30-nm fibers give rise to so-called "large-scale" structures, an example of which are chromonema fibers measuring 100-nm in diameter, as observed by electron microscopy. The fine details of this level of organization are unknown (although looping of fibers is likely to be involved) and currently not many methods are available to investigate the mechanical composition of higher-order chromatin structures. Generally, these structures are organized into segments with differing functional properties, determined by a combination of sequence composition (AT:GC content), transcriptional state and the presence of different histone modifications and chromatin-bound proteins. A simplistic classical view is to split the genome into gene-rich segments with more open structures and gene-poor regions enriched in repeats and satellites where the folding of higher-order structures are more compact. However, a 2011 classification of the differing properties of chromatin types splits them into five categories based on the prevalence of histone modifications: yellow (constitutively transcriptionally active regions), red (tissue-specific active regions), blue (repressed development and differentiation-related regions), black (silenced regions containing genes), and green (constitutively inactive repeats and satellites) chromatin [25]. The first two categories contain the transcriptionally active portion of the genome, which is enriched in acetylated H3 and H4. The chromatin structure in such regions is likely to be enriched in disruptions at the 30-nm level, particularly at regulatory elements, for example, promoters and enhancers, while large-scale domains will be more unfolded, facilitating easy access of transcription, replication and DNA repair factors to the DNA template. In contrast, the chromatin structure within green regions is likely to be more compacted and less dynamic. Processes that require access to the DNA template such as replication and DNA repair may necessitate chromatin remodeling to open up chromatin in these regions of the genome. In fact, some 2008 data suggest that permanently silenced regions may act as a barrier to the DNA-damage response and that breaks within them may take longer to detect and repair [26].

The segmentation of the genome into higher-order domains with differing structures is essential for its correct function. A small number of human diseases related to perturbations of chromatin structure have been described, including ICF syndrome (immunodeficiency, centromeric instability, and facial anomalies syndrome) and Rett syndrome. ICF syndrome is caused by mutations in the *DNMT3B* gene, coding for a DNA methyltransferase, and patients show instability and breakage of the silenced, repeat-rich regions at the centromeres of chromosomes 1, 9, and 16 in lymphoblastoid cells [27].

5. CHROMATIN REMODELERS

Apart from histones, chromatin contains a range of other proteins with diverse roles, some of which function to prevent genomic instability and respond to DNA damage. One of the most important classes of nonhistone chromatin-associated proteins is the remodelers: proteins which can reposition and remove nucleosomes or change their composition in an ATP-dependent manner. Consequently, they introduce small-scale alterations in the state of the chromatin fiber and alter the accessibility of the DNA template. Chromatin remodelers are required for many nuclear processes, including transcription, replication, cell-cycle progression, and of course, DNA repair [28,29]. Numerous mammalian chromatin remodelers exist and they can be broadly divided into four families: SWI/SNF, ISWI, INO80, and CHD and their roles in DNA repair are summarized in Table 23.1.

Genes encoding chromatin remodelers of the SWI/SNF family are frequently mutated in cancer and components of the SWI/SNF members BAF and PBAF have been shown to localize to sites of DNA damage. PBAF subunit BAF180 has a role in silencing transcription at sites of DNA breaks [30], while Brg1, a subunit common to BAF and PBAF, is involved in sister chromatid decatenation at the G2/M boundary and its inhibition results in anaphase bridges and lagging chromosomes [31]. Hinting at the wide range of roles these remodelers have, the PBAF complex was also found to promote sister chromatid cohesion, especially at centromeres, with chromosomal breaks and abnormalities following its inhibition [32].

TABLE 23.1 Roles of Chromatin Remodelers in DNA-Damage Repair

Family	Features	Complexes	Role in DNA Repair	Subunit Implicated
SWI/SNF	Bromodomains	BAF	Decatenation of sister chromatids	Brg1
		PBAF	Silencing transcription at breaks Sister chromatid cohesion at centromeres	BAF180, Brg1
ISWI		ACF	Facilitates NHEJ	ACF1, SNF2H
INO80	Histone exchange	TIP60	Restores chromatin environment by removing gamma H2AX	p400
		INO80	Promotes HR repair	
CHD	Chromodomains	NuRD	Released from heterochromatin to promote relaxation	CHD3
			Recruited to heterochromatin to promote relaxation	CHD4

ACF-1, a component of two ISWI-type complexes, ACF and CHRAC, was also found to bind at laser-induced DNA breaks, colocalizing with γ-H2AX [33]. Cells depleted of ACF-1 are very sensitive to DNA damage, and the authors showed that ACF-1 facilitates the binding of NHEJ protein Ku at double-strand DNA breaks.

The CHD class of remodelers are characterized by the presence of chromodomains, which can read methyl marks on histones. An example is the nucleosome remodeling and deacetylase complex (NuRD), which promotes nucleosome compaction in heterochromatin. The CHD4 subunit of the NuRD complex is phosphorylated by ATM in response to genome damage [34] and is rapidly recruited to sites of damage [35]. In the same studies the authors observed increased rates of genomic breaks in CHD4-depleted cells, suggesting not only that CHD4 is essential for repair, but that its depletion might also make chromatin more susceptible to breaks. In contrast, NuRD complexes containing an alternative CHD3 isoform were released from heterochromatin upon treatment with ionizing radiation, promoting chromatin relaxation [36].

Mammalian cells depleted of the INO80 remodeler also exhibit DNA-repair problems, with homologous recombination (HR) specifically affected as INO80 seems to be involved with 5′- to 3′-DNA resection at break sites [37]. Depletion of p400, an INO80 component, primarily involved in the incorporation of the H2AZ variant at transcriptionally active regions, makes cells sensitive to DNA damage [38]. p400 also incorporates H2AZ at double-strand breaks, contributing to chromatin opening in the break region to facilitate access for repair proteins [39]. A further INO80 subunit, TIP60, which acetylates H2A and H4, has been implicated in restoring the chromatin environment following DNA-damage response by removing the phosphorylated H2AX from the affected regions [40]. An additional role for TIP60 also involves histone acetylation in heterochromatic breaks to potentiate chromatin relaxation before repair [41].

Overall, the study of chromatin remodelers in DNA repair is a very active field of research but complicated by the many functions of these enzymes. In addition, as chromatin remodelers tend to have a serious impact on gene expression, studies have to exclude indirect effects on genome instability due to altered transcription. This underlies the need for better and more representative in vitro chromatin models, which could be used to study the direct structural effects of the remodelers, separately from their other roles.

Apart from chromatin remodelers and the specialized repair factors described elsewhere in the book, a number of other nonhistone chromatin–associated proteins also assist with DNA repair, often as a secondary function; examples include topoisomerases, helicases, and structural scaffolding proteins. A particularly interesting example of this is the cohesin complex, a large molecular complex essential for sister chromatid cohesion and chromosome segregation. A ring-shaped structure, cohesin associates chromatin fibers not through direct binding, but rather topologically and contributes to the 3D organization of the genome [42]. It brings together the two sister chromatids following replication and functions in the HR pathway, ensuring proximity between the damaged chromatid and the repair template. Cohesin is recruited to DNA-damage sites following laser irradiation only in the S and G2 phases of the cell cycle through an interaction with Rad50 [43]. Other nonhistone proteins are implicated in maintenance of genomic stability indirectly by working to avoid the formation of DNA:RNA hybrids, conflicts between the transcription and replication machinery, and by rescuing stalled replication forks.

6. ACCESS, REPAIR, RESTORE

As illustrated by the extensive role of chromatin remodelers in the DNA-damage response, changes in chromatin conformation are essential for the repair process (Fig. 23.2).

There is some controversy about whether these changes are limited to the chromatin environment local to the break or whether they spread globally. Local changes have been demonstrated convincingly, using a variety of methods: HATs and HDACs are recruited to laser-induced tracks [44], while high-resolution imaging of chromatin in DNA repair foci shows chromatin in a state resembling a 10 nm fiber [45]. Consistent with this, a live cell imaging study utilizing the *SceI/LacO* system mentioned earlier demonstrated local chromatin remodeling in the proximity of a break [46]. In this study, authors used a photo-activated GFP fused to H2A, allowing them to induce damage and photoactivate chromatin within the damaged region simultaneously. They then measured changes in the H2A-GFP spot size and were able to show rapid expansion of the spot area lasting 1.5 min, followed by a recompaction phase lasting 15 min and then hyper-condensation beyond baseline level (20–30 min). A brief local decompaction, as demonstrated in this study, would enable access of the DDR

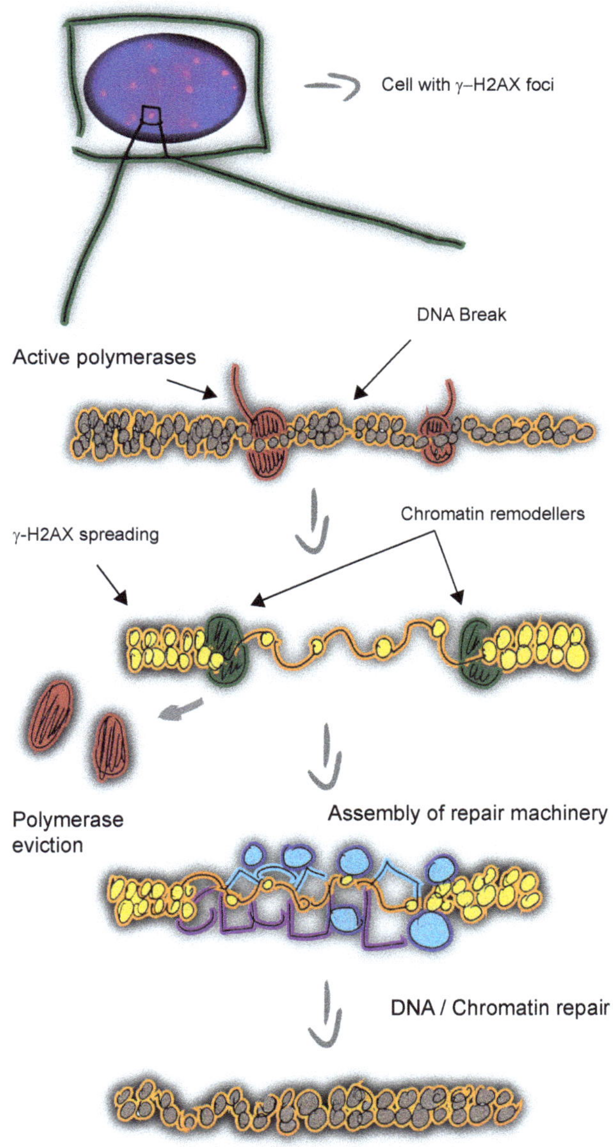

FIGURE 23.2 **Changes in local chromatin structure upon dsDNA breaks.** DNA breaks are accompanied by local changes in chromatin compaction and transcription. Upon breakage, chromatin remodelers alter the chromatin surrounding the break to be more accessible. Transcription stops and H2AX is phosphorylated (*yellow*) in a megabase-sized domain surrounding the break. The DDR components are recruited onto chromatin and retained there through γ-H2AX. Following repair, the normal chromatin environment is restored and the γ-H2AX mark is removed.

proteins to breaks. Alterations in the transcriptional activity of a locus in the vicinity of a DNA break also accompany local compaction changes. Ubiquitination of H2A at break sites was shown to correlate with transcriptional silencing near break regions [47] and recruitment of the SWI/SNF remodeler PBAF is found to contribute to this silencing [30]. A somewhat opposing finding was published in 2012, when Francia et al. [48] found evidence that transcription of small noncoding RNAs within a damaged region is required for the DNA-damage response. Whether the local changes in compaction and transcription spread globally is debatable. A 2006 study, using MNase digestion to assess genome-wide chromatin states, found evidence of global decondensation following DNA-damage induction [49]. However, a 2011 study from our lab found no evidence for global decompaction using the same approach or by sucrose gradient sedimentation to analyze the structure of soluble chromatin fibers [50].

Once the appropriate chromatin environment has been established, repair of the damage can proceed. The earliest step in the DDR involves rapid targeting of repair factors to the lesion and formation of DNA repair foci. The primary sensor is the MRN complex, composed of three different factors: MRE11, RAD50, and NBS1. The MRN complex activates ATM, which in turn phosphorylates H2AX at the damage site and the flanking chromatin up to a megabase away [11], amplifying the damage signal. An interesting question in the field is whether a full DDR is initiated only in response to DNA breaks. Surprisingly, not. Tethering of early repair components to genomic regions resulted in a full DNA-damage response and cell-cycle arrest, indicating that breaks are not needed beyond the initial recruitment of factors [51]. Consistently, treatment of cells with the HDAC-inhibitor TSA resulted in the activation of ATM raising the possibility that DDR can also be triggered by stimuli other than breaks, such as unusual chromatin structures [52].

Once the necessary factors have been recruited, repair can proceed. There are two main pathways for repair of double-strand DNA breaks—nonhomologous end joining (NHEJ) and HR, which are described in detail elsewhere in the book. Briefly, NHEJ works by joining the ends of the break together and is active throughout the cell cycle, while in HR, which is only possible in S and G2, the nondamaged homologous locus on the sister chromatid is used as a repair template. Interestingly, the 2014 evidence showed that breaks located in transcriptionally active segments of the genome are preferentially repaired with HR, while breaks in less-active regions are more frequently repaired vie NHEJ even as the cells transition into S and G2 [53]. The preferential recruitment of the HR machinery to breaks in transcribed regions is found to be dependent on an interaction between the H3K36me3 mark and LEDGF, a protein component of HR.

7. NUCLEAR ORGANIZATION OF CHROMATIN

Within cells, chromatin is contained within the nucleus—a complex organelle shaping the 3D organization of the genome. Positioning of the genome in the nucleus has important functional consequences; nuclear position is a significant characteristic of a locus, impacting on its transcriptional activity, replication timing, and proximity to other loci. Changes in the nuclear positioning of loci accompany development and differentiation, demonstrating the biological importance of nuclear organization.

The exact positioning of loci within the nucleus is probabilistic—it is not the same in every cell but is guided by a set of rules. With few exceptions, in mammalian cells, gene-rich, transcriptionally active regions of the genome are located toward the nuclear interior, while the gene-poor and heterochromatic regions are located toward the periphery. As a result, rather than having precisely defined locations, chromosomes have preferred radial positions in the nucleus. Centromeres also tend to be located toward the periphery [54], while telomeres are distributed through the nuclear volume.

The nuclear periphery is defined by its interaction with the nuclear lamina—a part of the inner nucleoplasmic membrane. The genomic regions that interact with the lamina are known as lamina-associated domains (LADs); they measure 0.1–10Mb in size and overlap with chromatin features such as low gene density and repressive histone marks. LADs can be divided into a *facultative* and a *constitutive* class. Facultative LADs are cell type specific, while constitutive LADs are shared between cell types. Interestingly, disruptions in the lamina structure have been associated with genome instability, as illustrated by a class of diseases known as laminopathies, caused by mutations in the genes coding for the proteins that make up the nuclear lamina. The best studied among them is the Hutchinson–Gilford progeria syndrome (HGPS), a rare premature aging syndrome caused by mutations in the *LMNA* gene. Cells from patients with HGPS show microscopically visible disruptions to the shape of the nuclear envelope, loss of the heterochromatic protein HP1 at the nuclear periphery, and altered histone modifications pattern. Although not deficient in any of the components of the DDR response, HGPS cells are sensitive to ionizing radiation and accumulate DNA damage when grown in culture [55]. They also display increased levels of γ-H2AX and ATR/ATM activation. In addition, Werner's syndrome, a disease that is phenotypically related to HGPS, is caused by mutations in the WRN protein—a DNA helicase that prevents DNA damage by resolving stalled replication forks. However, despite all the evidence that the lamina disruption observed in HGPS cells ultimately lead to increased DNA damage, the underlying molecular mechanisms are yet unknown.

8. CHROMOSOME TERRITORIES

Rather than being dispersed throughout the nucleus, each chromosome occupies a distinct volume, called a *chromosome territory*. This has been demonstrated by *chromosome painting*—a FISH-based technique where the genome is hybridized to a large number of chromosome-specific probes to allow visualization of individual chromosomes within the nucleus. The radial positioning of a chromosome is strongly influenced by its composition—gene-poor chromosomes tend to occupy positions closer to the nuclear periphery, while gene-rich chromosomes are more frequently located toward the interior [56]. This trend is illustrated by human chromosomes 18 and 19, which are very similar in size but have very different sequence composition: chromosome 18 is gene poor, while 19 is gene rich. The Bickmore lab used chromosome territory FISH to investigate the positions of the two chromosomes in the nucleus and found that chromosome 18 was consistently located closer to the nuclear periphery than chromosome 19 in both lymphoblastoid and fibroblast cell lines [57]. The radial positioning of chromosomes in the nucleus was also found to be tissue specific, with more closely related cell types exhibiting more similar chromosome positioning [58]. The human genome also contains five acrocentric chromosomes, containing rDNA sequences—chromosomes 13, 14, 15, 21, and 22 which are usually clustered around the nucleolus—the site of transcription and processing of ribosomal RNA.

The radial rule of chromosome positioning also influences the positioning of alternating gene-rich and gene-poor segments within chromosomes—in this case, gene-rich segments are located more centrally, while gene-poor regions occupy regions closer to the periphery. In addition, within chromosome territories, transcriptionally inactive segments are located internally and transcriptionally active segments are at the surface of the territory [59]. This arrangement allows transcriptionally active regions ready access to the transcription machinery and domains rich in mRNA metabolic factors such as SC-35 foci [60]. However, the fine-detail structure of chromosome territories is yet unclear, reflecting our lack of knowledge of the chromatin structures that shape them.

From a genome stability perspective, an important consequence of chromosome positioning patterns relates to translocations, the most frequent chromosomal abnormality seen within the human population. It is well established that the physical proximity of two chromosomes in the nucleus affects the probability of a translocation occurring between them (Fig. 23.3).

An analysis between the frequencies of different nonpathogenic translocations in the human population and the preferred radial positions of chromosomes in the nucleus found that chromosomes with similar nuclear positions form translocations

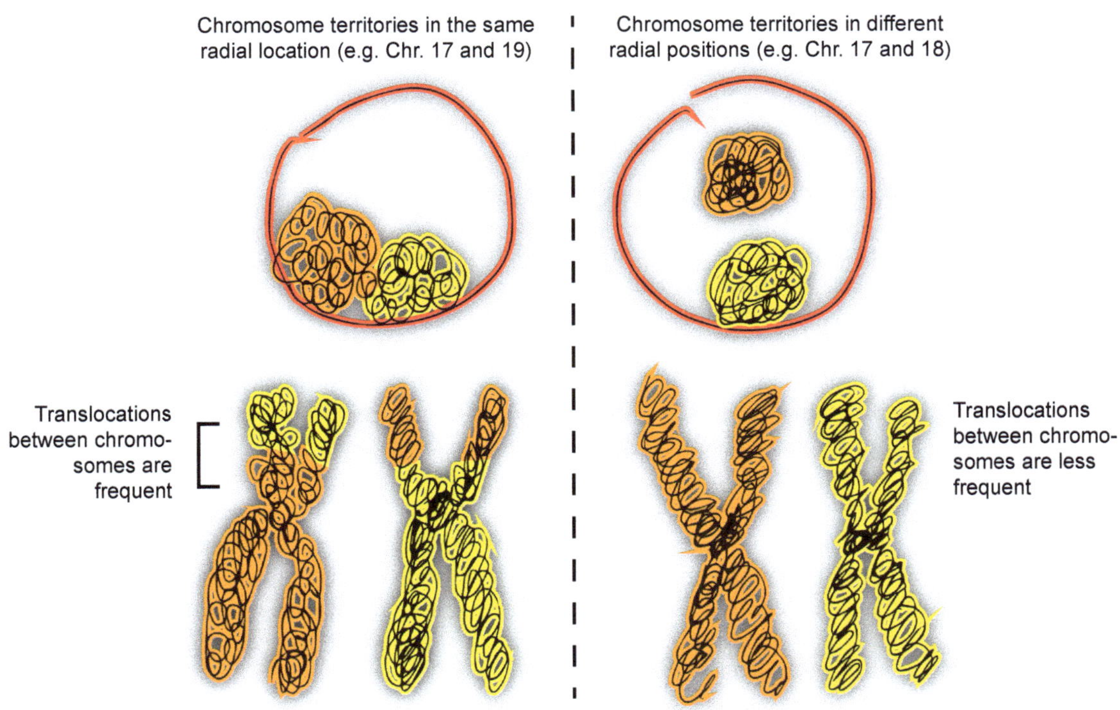

FIGURE 23.3 **Preferred positions of chromosomes in the nucleus influences translocation frequency.** Chromosomes with the same preferred radial position in the nucleus (eg, chromosomes 17 and 19) are more likely to be involved in translocations than chromosomes with different radial positions (eg, chromosomes 17 and 18).

more frequently than expected by chance [61]. Another study was able to demonstrate close proximity between the BCR and ABL loci, involved in the well-characterized t(9; 22) translocation forming a "Philadelphia" chromosome in chronic myeloid leukemia. The authors showed that the BCR and ABL loci were closer in B-lymphocytes than in hematopoietic progenitor cells, suggesting that cell type–specific aspects of nuclear organization may contribute to the association of certain translocations with particular cancer types. In 2013, the Misteli lab published a study [62] exploring the dynamics of double-strand breaks and subsequent translocation formation in an elegant system: NIH3T3duo cells encode a small number of *SceI* restriction enzyme sites integrated on different chromosomes, with some sites adjacent to a *LacO* array and other sites neighboring a *TetO* array. Upon break induction by the *SceI* enzyme, it was possible to track the breaks which were marked by fluorescently tagged Lac (LacR) and Tet (TetR)-repressor proteins; translocation formation was indicated by long-lasting, stable co-localization of the LacR and TetR signals. The authors were able to demonstrate that most translocations are formed by loci that are closely located prior to break induction (*contact-first model*), rather than as a result of a movement of double-strand breaks to proximal locations (*breakage-first model*).

Beyond methods for analysis of chromosome territories, two main complementary methods are used to study the 3D organization of the genome at the level of higher-order domain structure: FISH-based methods and chromosome confirmation capture methods [63]. FISH relies on hybridization of fluorescently labeled probes to visualize individual loci, defined portions of the genome or whole chromosomes. It provides a snapshot of nuclear structure at the single cell level, but disadvantages are that it is time-consuming and provides a limited amount of information at a low resolution. Chromatin conformation capture (3C) techniques rely on "freezing" the nuclear structure by cross-linking interactions within the nucleus, ligating DNA fragments held in proximity by the cross-links, followed by PCR or next-generation sequencing to identify hybrid DNA fragments, indicative of contacts. At the most sophisticated end, these techniques can theoretically identify all possible interactions throughout the genome, but there are also disadvantages. Unlike FISH, 3C techniques work on populations of cells rather than at a single cell level, producing a population average which may reflect a number of different contact configurations at the single cell level. Despite the caveats, 3C methodologies have been very influential in the field of 3D genome organization, contributing the concept of topologically associating domains (TADs). TADs are defined as regions measuring ~900 kb, where contact maps show increased interactions; FISH-based studies have shown that probes located within a TAD are physically closer than probes not located within the same TAD but separated by a similar "linear" genomic distance [63]. The full human genome is divided into approximately 2000 TADs which also overlap with the distribution of histone marks and other genomic features such as replication timing (described later). However, they are not cell-type specific and the question of what level of structural organization they reflect and their functional importance is still open to debate. Interestingly, the translocation frequency pattern seen with chromosome territories can be also traced to the TAD level of organization—a study conducted in B-cells found that the likelihood of translocation between two loci is strongly related to the contact frequency between them, as defined by chromosome confirmation capture-generated contact maps [64].

9. TRANSCRIPTION AND REPLICATION IN THE NUCLEUS

As we have seen earlier, the nucleus is a site of many correlations: radial position, gene density, histone mark enrichments, and transcriptional activity. Another correlation comes from the process of replication: the exact timing of replication of a locus also correlates with its nuclear position, as well as with its transcriptional activity. Replication proceeds in a well-controlled timely manner across the genome-alternating segments of chromosomes replicate at different times throughout S-phase, with gene-rich, transcriptionally active segments replicating early in S-phase and heterochromatic regions replicating last. These replication domains measure from 400 to 800 kb and control of replication timing is achieved by simultaneous firing of clusters of origins within the replication domains at defined times during S-phase. The correlation between replication timing and nuclear position is so strong that it gives rise to striking S-phase patterns visible in nuclei stained with markers of active replication (Fig. 23.4): early replicating cells show diffuse staining with markers excluded from the nuclear periphery; cells in mid-S have speckled patterns; and in nuclei in the latest stages of replication the staining overlaps with the nuclear periphery and heterochromatic regions. Replication timing domains partially overlap with TADs, however some replication domains are cell-type specific and change during development and differentiation, along with changes in transcription. About 80% of the genome has constant replication timing between cell types, with 50% showing development and differentiation-related changes.

A few studies to date have tried to separate out the effects of chromatin state, transcription and replication timing to investigate the real determinants of nuclear positioning. A 2014 study by the Bickmore Lab indicated that the chromatin compaction state may be the primary factor [65], with transcriptional activation influencing nuclear position, while replication timing was shown to be a consequence of transcriptional state. However, other studies have argued that replication plays a role in the

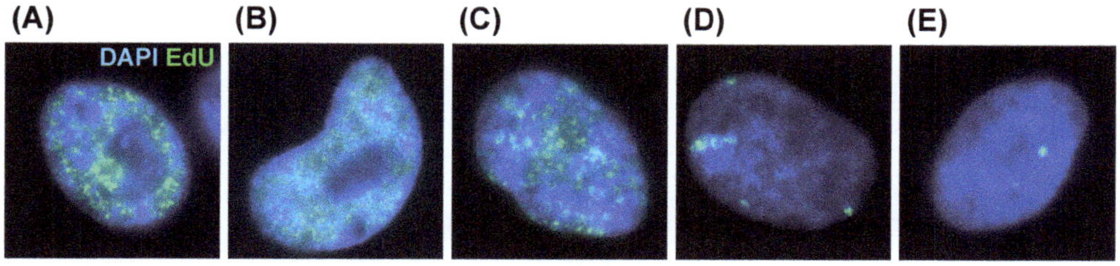

FIGURE 23.4 **Replication timing in the nucleus.** Correlation between replication timing and nuclear position gives rise to striking patterns in replicating cells, which discriminate between early (A,B), mid (C), and late (D,E) replicating cells. In this experiment, cells were pulsed with the thymidine analogue EdU which is incorporated into newly replicated DNA and can then be readily visualized (*green*). The nuclei are stained with the DNA dye 2-(4-amidinophenyl)-1H-indole-6-carboxamidine (DAPI, *blue*). EdU staining, indicating sites of active replication is diffuse in early S, speckly and close to the periphery in mid-S, and coinciding with heterochromatin in late-S.

establishment of nuclear organization. A 2015 chromatin confirmation capture study revealed that TAD structure is established during early G1, at the same time as the replication timing program [66]. Another 2015 study used high-throughput FISH to screen for factors affecting nuclear positioning of a small number of loci; it found that a number of replication-related proteins significantly affected positioning and also that replication was needed to maintain correct nuclear positioning [67].

The processes of replication and transcription have been at the heart of a conceptual shift in the field of genome stability since 2014. While historically research on the DNA-damage response was focused on external and severe mutagens such as UV light and carcinogenic drugs, recently it has become clear that DNA damage resulting from internal factors and fundamental cellular processes may be more physiologically relevant. A succession of recent studies has implicated replication and transcription as contributors to genome instability. For example, a study in 2015 determined that regions of very high mutation rates within the genome overlap with Okazaki fragment junctions; the underlying mechanism was found to be retention of short segments spanning the junctions synthesized by the error-prone DNA polymerase Pol-α [68]. An earlier study identified replication stress, physiologically present in cancer cells, as the root cause of structural and numerical chromosome instability in colorectal cancers with unstable karyotypes [69]. Transcription was implicated as a contributor to genomic instability in a publication by the Svejstrup Lab—the authors found that inhibition of a transcription-associated helicase caused transcription speed to increase, resulting in recurrent chromosomal rearrangements at particular genomic regions [70]. Another example is provided by the *RNU1*, *RNU2*, *RN5S*, and *PSU1* loci, all coding for tandemly repeated, highly transcribed small RNA sequences. These four loci exhibit fragility and appear as breaks on metaphase chromosomes upon either adenovirus infection or in the absence of the Cockayne syndrome group B (CSB) protein, which is mutated in Cockayne syndrome, a rare disorder characterized by neurological and developmental defects. As CSB functions as a transcription elongation factor, it has been speculated that its loss causes RNA polymerase stalling and blockage at the *RNU1*, *RNU2*, *RN5S*, and *PSU1* loci, which then interferes with chromosome condensation and consequently, the stability of the four regions [71].

Unlike external factor-mediated instability, which usually arises from stoichiometric interactions of the damage-inducing agents with DNA and results in predictable outcomes, internally mediated instability is stochastic: it is likely to result from a combination of factors, including the exact chromatin context at the location where problems arise. While in the past most common strategies for studying the role of chromatin in genome stability involve triggering DNA damage through methods such as irradiation, laser marks, or harsh damage-inducing agents such as hydroxyurea, it is clear that this new view of the field will require novel models and methods. A good model for how complex relationships between transcription, replication, and chromatin influence genome stability is presented by common fragile sites (CFS).

CFS are regions of the genome prone to instability in response to replication stress, manifesting as breaks, gaps, and constrictions on metaphase chromosomes. While it is known that CFS fragility is triggered by replication stress, the exact events leading up to genomic instability are unknown. As CFS fragility is cell-type specific—different genomic locations are fragile in different cell types—it is clear that factors beyond their sequence composition contribute to fragility; in particular, replication timing and transcription are considered important, while chromatin context is a promising but understudied potential contributor.

Three models have been proposed to explain how the induction of replication stress results in genomic instability in a locus-specific manner (Fig. 23.5).

The *replication fork collapse* model suggests that the AT-rich sequence of CFS makes them prone to forming secondary structures which contribute to replication fork stalling and collapse [72]. The *transcription–replication collisions* model is based on the observation that fragile sites frequently span long genes, raising the possibility that CFS instability can be the result of concomitant transcription and replication [73]. The *replication–initiation paucity* model explains CFS fragility as a consequence of cell type–specific features of replication timing [74].

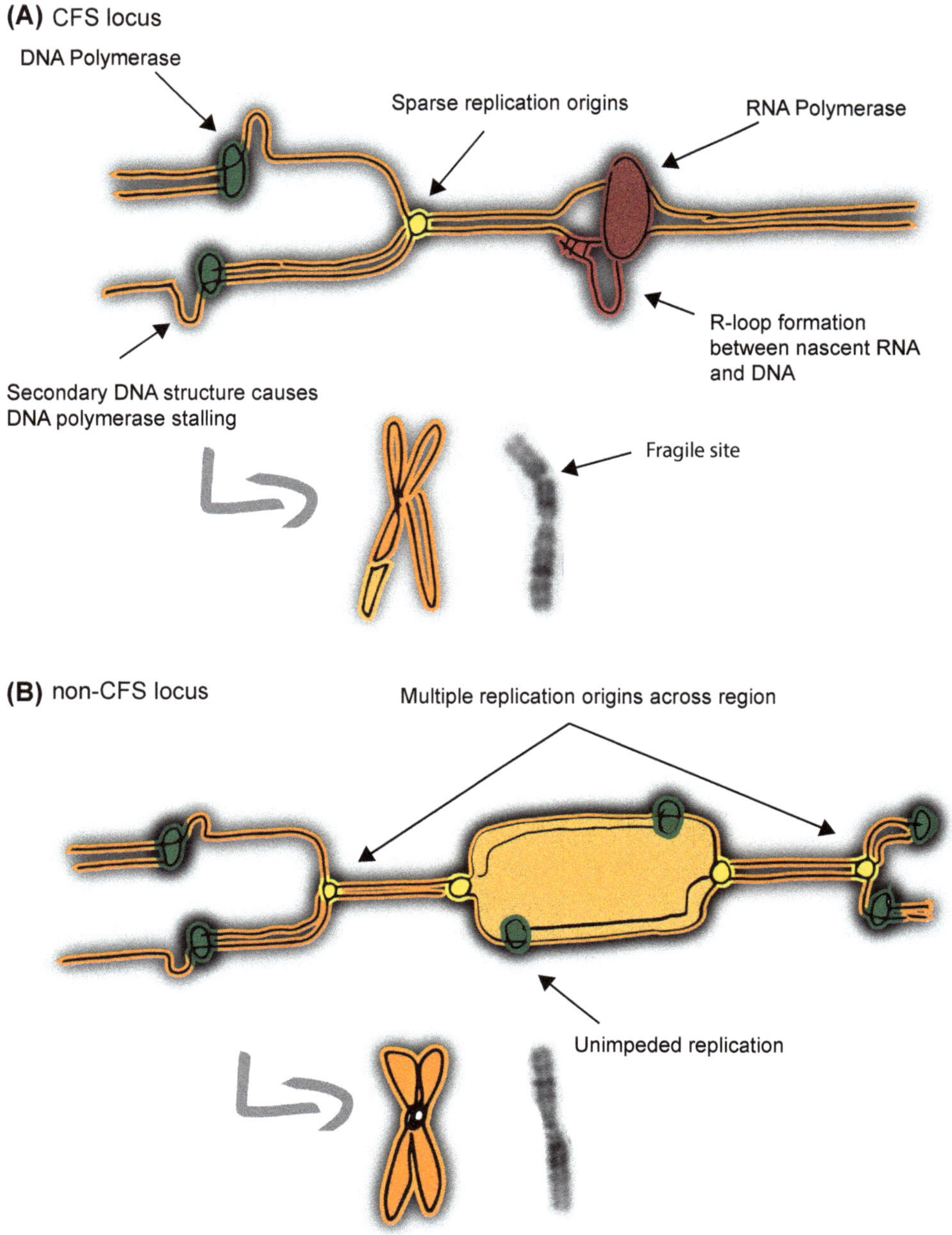

FIGURE 23.5 **Models of CFS formation.** Multiple models have been proposed to explain the cell type–specific fragility of CFS. (A) CFS region in a cell type–specific fragility inducing chromatin environment. (B) CFS region in a noninducing chromatin environment. In (A), the AT-rich sequence of fragile regions causes DNA polymerases to stall; transcription/replication encounters in the region result in the formation of R-loops; paucity of replication origins (*yellow*) means that the region is replicated very late in the cell cycle or remains unreplicated. In (B), the AT-rich sequence also causes stalling, but the lack of transcription in the vicinity of the CFS and frequently spaced replication origins across the region allow replication to proceed in time.

In support of the fork collapse model, two genetic disorders characterized by increased fragile site formation, Bloom syndrome, and Werner syndrome are caused by deficiencies of RecQ helicases specialized in resolving stalled replication intermediate structures [75]. Werner syndrome is caused by a deficiency of the Werner syndrome protein (WRN), an ATP-dependent helicase which efficiently unwinds structures resembling stalled replication bubbles such as Holliday Junctions (HJ). Cells derived from WRN-deficient patients form breaks at CFS spontaneously in the absence of aphidicolin treatment, while in wild-type cells, an increased frequency of CFS formation is observed following WRN depletion [76]. BLM syndrome is caused by a deficiency of the Bloom Syndrome protein (BLM) and is characterized by increased susceptibility to early onset cancers. BLM resolves structures that mimic replication and recombination intermediates, such as HJs, via homologous repair in a manner which does not result in a crossover and BLM has been shown to localize to stalled replication forks in vivo [77]. Cells from Bloom syndrome patients show an increased sensitivity to aphidicolin and an increased frequency of sister chromatid exchanges which could result from crossover-mediated repair of HJs by alternate nuclease complexes. Interestingly, in the absence of BLM and other Holliday junction dissolution mechanisms, extreme chromosome abnormalities resembling multiple fragile site breaks are observed [78]. Further evidence supporting the fork collapse model comes from observations that in the presence of aphidicolin, the replicative helicase complex becomes uncoupled from the replication machinery, giving rise to long stretches of single-stranded DNA [79]. Additional supporting evidence comes from a 2011 study demonstrating replication fork stalling at AT-rich sequences at the FRA16D fragile site [80]. A major disadvantage of this model however is that it fails to explain the cell-type specificity of CFS expression.

The tendency of fragile regions to encompass large genes has inspired a model suggesting that CFS instability results from collisions between the transcription and replication machinery. Large genes require longer times for transcription, sometimes exceeding the length of a full cell cycle, indicating that transcription might be ongoing during S-phase. Normally, S-phase transcription and replication are spatially separated in eukaryotic cells; most actively transcribed genes are early replicating and changes in transcription during development are accompanied by changes in replication timing [81]. In this model, aphidicolin treatment interferes with the temporal and spatial separation of replication and transcription at large genes, causing the occurrence of transcription and replication at fragile sites. The model speculates that concurrent transcription and replication can cause instability through the formation of *RNA–DNA (R-loop) hybrids* or through head-on collisions of the transcription machinery and the replication bubble, causing replication fork collapse. Efforts to correlate CFS fragility with gene expression in a cell type–specific manner have given conflicting results. A 2011 study showed a correlation between expression of the FHIT gene at the FRA3B fragile site and FRA3B fragility, accompanied by an increase in R-loop formation in the presence of aphidicolin [73]. However, a study from 2013 failed to find a correlation between expression and fragility on a more genome-wide scale [82]. Furthermore, breaks at CFS are not restricted to transcribed regions and can also occur at intergenic sequences. Therefore, unlike the RNU loci, active transcription is not required for induction of fragility at CFS, suggesting that the transcription–replication collision model does not fully explain CFS lesion formation.

In the *replication initiation paucity* model of CFS formation, instability is caused by a cell type–specific lack of initiation events across fragile regions, forcing the forks to travel long distances to replicate CFS loci and causing the regions to remain unreplicated at the end of S-phase in the presence of replication stress. Evidence supporting the model comes from a study demonstrating that a lack of initiation events across the well-studied FRA3B site correlates with its fragility in lymphoblastoid cells [74]; in contrast, initiation events across the site were observed in fibroblasts, where FRA3B is stable. In addition, the authors demonstrated increased use of origins in response to aphidicolin treatment at the flanking regions, but not the core of FRA3B, showing that a failure to utilize additional origins during replication stress may also contribute to fragility.

To date, no model has been found to exclusively explain the cell type–specific fragility or CFS loci and it is likely that aspects of all three models contribute to CFS instability. This complexity makes CFS a good model for studying how genomic instability develops in the complex landscape of the cell.

10. CONCLUSIONS

In summary, the study of the roles of chromatin and nuclear organization in maintaining genome stability is an active and developing field and the advance of novel technologies promises exciting new discoveries. CRISPR, the new genome-editing technology, will allow researchers to easily engineer specific mutations within chromatin-associated proteins and study their effects on genome stability; this technology can also be used to recruit proteins such as DDR components, repressors and activators, or fluorescent tags to endogenous genomic loci, in contrast with the *LacO/TetO* systems described earlier which are based on repeat arrays artificially integrated within the genome. Development of biologically faithful in vitro chromatin models is also a major aim of the chromatin field; such a system would be important both as a model to study chromatin structure and as a template which can be used to study how various components of the DDR interact with chromatin. The reduced cost of high-throughput sequencing has transformed the field, while high-throughput imaging approaches are also becoming more accessible. Still, numerous questions remain to be answered before our understanding of chromatin response to genomic instability is complete.

GLOSSARY

Breakage-first model Another model for the generation of translocations, in which the frequency of translocations between two chromosomes is independent of their preferred nuclear positions. This model hypothesizes that double-stranded breaks are brought together in the nucleus for repair, sometimes resulting in translocations. Recent data contradicts the breakage-first model.

Chromatin remodelers Proteins which can reposition and remove nucleosomes or change their composition in an ATP-dependent manner.

Chromonema A level of large-scale organization of the chromatin fiber measuring 100–130nm in diameter. Although chromonema fibers have been observed by electron microscopy, their fine-scale organization is unknown.

Chromosome painting A FISH-based technique based on hybridization of chromosome-specific probes which allows visualization of whole chromosome territories.

Chromosome territory The defined nuclear volume occupied by a chromosome.

Contact-first model A model for the generation of translocations which suggests that the likelihood of a translocation occurring between two chromosomes depends on their proximity in the nucleus prior to generation of double-stranded breaks.

Laminopathy A class of diseases resulting from defects in the structure of the lamina component of the nuclear envelope. Laminopathies are caused by mutations in genes encoding the components of the nuclear lamina, including *LMNA* and *LMNB2*, and have diverse phenotypic characteristics such as muscular dystrophy and premature aging.

RNA–DNA (R-loop) hybrids RNA:DNA hybrid structure which can occur during transcription if the nascent RNA hybridizes to the complement DNA strand and displaces the nontemplate DNA strand. When such structures are resolved, ds DNA breaks are generated, implicating R-loops in genomic instability.

Sister chromatid decatenation The process of separating entanglements and catenanes between sister chromatids following replication and prior to cell division.

LIST OF ABBREVIATIONS

ATP Adenosine triphosphate
CFS Common fragile sites
DDR DNA-damage repair
DNA Deoxyribonucleic acid
FISH Fluorescence in situ hybridization
GFP Green fluorescent protein
HAT Histone acetyltransferase
HDAC Histone deacetylase
HGPS Hutchinson–Gilford progeria syndrome
HJ Holliday junction
HKMT Lysine methyltransferase
HR Homologous recombination
ICF Immunodeficiency, centromeric instability, facial anomalies syndrome
LAD Lamina-associated domain
MNase Micrococcal nuclease
mRNA Messenger RNA
NHEJ Nonhomologous end joining
PRMT Arginine methyltransferases
PTM Posttranslational modification
rDNA Ribosomal DNA
RNA Ribonucleic acid
TAD Topologically associated domains
UV Ultraviolet

REFERENCES

[1] Green CM, Almouzni G. When repair meets chromatin. First in series on chromatin dynamics. EMBO Rep 2002;13:28–33.
[2] Henikoff S, Furuyama T, Ahmad K. Histone variants, nucleosome assembly and epigenetic inheritance. Trends Genet 2004;20:320–6.
[3] Gilbert N, Allan J. Distinctive higher-order chromatin structure at mammalian centromeres. Proc Natl Acad Sci USA 2001;98:11949–54.
[4] Panchenko T, Sorensen TC, Woodcock CL, Kan Z-Y, Wood S, Resch MG, et al. Replacement of histone H3 with CENP-A directs global nucleosome array condensation and loosening of nucleosome superhelical termini. Proc Natl Acad Sci USA 2011;108:16588–93.
[5] Neumann H, Hancock SM, Buning R, Routh A, Chapman L, Somers J, et al. A method for genetically installing site-specific acetylation in recombinant histones defines the effects of H3 K56 acetylation. Mol Cell 2009;36(1):153–63.
[6] Dovey OM, Foster CT, Conte N, Edwards SA, Edwards JM, Singh R, et al. Histone deacetylase 1 and 2 are essential for normal T-cell development and genomic stability in mice. Blood 2013;121:1335–44.
[7] Peters AHFM, O'Carroll D, Scherthan H, Mechtler K, Sauer S, Schöfer C, et al. Loss of the Suv39h histone methyltransferases impairs mammalian heterochromatin and genome stability. Cell 2001;107(3):323–37.
[8] Huyen Y, Zgheib O, Ditullio RA, Gorgoulis VG, Zacharatos P, Petty TJ, et al. Methylated lysine 79 of histone H3 targets 53BP1 to DNA double-strand breaks. Nature 2004;432:406–11.
[9] Wei Y, Yu L, Bowen J, Gorovsky MA, Allis CD. Phosphorylation of histone H3 is required for proper chromosome condensation and segregation. Cell 1999;97:99–109.
[10] Rea S, Eisenhaber F, O'Carroll D, Strahl BD, Sun ZW, Schmid M, et al. Regulation of chromatin structure by site-specific histone H3 methyltransferases. Nature 2000;406(6796):593–9.

[11] Rogakou EP, Boon C, Redon C, Bonner WM. Megabase chromatin domains involved in DNA double-strand breaks in vivo. J Cell Biol 1999;146(5):905–15.

[12] Misteli T, Soutoglou E. The emerging role of nuclear architecture in DNA repair and genome maintenance. Nat Rev Mol Cell Biol 2009;10:243–54.

[13] Chowdhury D, Keogh M-C, Ishii H, Peterson CL, Buratowski S, Lieberman J. gamma-H2AX dephosphorylation by protein phosphatase 2A facilitates DNA double-strand break repair. Mol Cell 2005;20(5):801–9.

[14] Heo K, Kim H, Choi SH, Choi J, Kim K, Gu J, et al. FACT-mediated exchange of histone variant H2AX regulated by phosphorylation of H2AX and ADP-ribosylation of Spt16. Mol Cell 2008;30(1):86–97.

[15] Celeste A, Difilippantonio S, Difilippantonio MJ, Fernandez-Capetillo O, Pilch DR, Sedelnikova OA, et al. H2AX haploinsufficiency modifies genomic stability and tumor susceptibility. Cell 2003;114(3):371–83.

[16] Kruhlak MJ, Celeste A, Dellaire G, Fernandez-Capetillo O, Müller WG, McNally JG, et al. Changes in chromatin structure and mobility in living cells at sites of DNA double-strand breaks. J Cell Biol 2006;172(6):823–34.

[17] Ui A, Nagaura Y, Yasui A. Transcriptional elongation factor ENL phosphorylated by ATM recruits polycomb and switches off transcription for DSB repair. Mol Cell May 2015;58(3):468–82.

[18] Fernandez-Capetillo O, Allis CD, Nussenzweig A. Phosphorylation of histone H2B at DNA double-strand breaks. J Exp Med 2004;199:1671–7.

[19] Lee J-H, Kang B-H, Jang H, Kim TW, Choi J, Kwak S, et al. AKT phosphorylates H3-threonine 45 to facilitate termination of gene transcription in response to DNA damage. Nucleic Acids Res May 19, 2015;43(9):4505–16.

[20] Burman B, Zhang ZZ, Pegoraro G, Lieb JD, Misteli T. Histone modifications predispose genome regions to breakage and translocation. Genes Dev July 1, 2015;29(13):1393–402.

[21] Stanlie A, Aida M, Muramatsu M, Honjo T, Begum NA. Histone3 lysine4 trimethylation regulated by the facilitates chromatin transcription complex is critical for DNA cleavage in class switch recombination. Proc Natl Acad Sci USA 2010;107(51):22190–5.

[22] Grigoryev SA, Woodcock CL. Chromatin organization – the 30 nm fiber. Exp Cell Res 2012;318(12):1448–55.

[23] Ricci MA, Manzo C, García-Parajo MF, Lakadamyali M, Cosma MP. Chromatin fibers are formed by heterogeneous groups of nucleosomes in vivo. Cell March 2015;160(6):1145–58.

[24] Buenrostro JD, Giresi PG, Zaba LC, Chang HY, Greenleaf WJ. Transposition of native chromatin for fast and sensitive epigenomic profiling of open chromatin, DNA-binding proteins and nucleosome position. Nat Methods 2013;10(12):1213–8.

[25] Van Steensel B. Chromatin: constructing the big picture. EMBO J 2011;30(10):1885–95.

[26] Goodarzi AA, Noon AT, Deckbar D, Ziv Y, Shiloh Y, Löbrich M, et al. ATM signaling facilitates repair of DNA double-strand breaks associated with heterochromatin. Mol Cell 2008;31:167–77.

[27] Xu GL, Bestor TH, Bourc'his D, Hsieh CL, Tommerup N, Bugge M, et al. Chromosome instability and immunodeficiency syndrome caused by mutations in a DNA methyltransferase gene. Nature 1999;402(6758):187–91.

[28] Brownlee PM, Meisenberg C, Downs JA. The SWI/SNF chromatin remodelling complex: its role in maintaining genome stability and preventing tumourigenesis. DNA Repair (Amst) 2015;32:127–33.

[29] Jeggo PA, Downs JA. Roles of chromatin remodellers in DNA double strand break repair. Exp Cell Res 2014;329(1):69–77.

[30] Kakarougkas A, Ismail A, Chambers AL, Riballo E, Herbert AD, Künzel J, et al. Requirement for PBAF in transcriptional repression and repair at DNA breaks in actively transcribed regions of chromatin. Mol Cell 2014;55:723–32.

[31] Dykhuizen EC, Hargreaves DC, Miller EL, Cui K, Korshunov A, Kool M, et al. BAF complexes facilitate decatenation of DNA by topoisomerase IIα. Nature 2013;497:624–7.

[32] Brownlee PM, Chambers AL, Cloney R, Bianchi A, Downs JA. BAF180 promotes cohesion and prevents genome instability and aneuploidy. Cell Rep 2014;6(6):973–81.

[33] Lan L, Ui A, Nakajima S, Hatakeyama K, Hoshi M, Watanabe R, et al. The ACF1 complex is required for DNA double-strand break repair in human cells. Mol Cell 2010;40(6):976–87.

[34] Polo SE, Kaidi A, Baskcomb L, Galanty Y, Jackson SP. Regulation of DNA-damage responses and cell-cycle progression by the chromatin remodelling factor CHD4. EMBO J September 15, 2010;29(18):3130–9.

[35] Larsen DH, Poinsignon C, Gudjonsson T, Dinant C, Payne MR, Hari FJ, et al. The chromatin-remodeling factor CHD4 coordinates signaling and repair after DNA damage. J Cell Biol 2010;190(5):731–40.

[36] Goodarzi AA, Kurka T, Jeggo PA. KAP-1 phosphorylation regulates CHD3 nucleosome remodeling during the DNA double-strand break response. Nat Struct Mol Biol 2011;18(7):831–9.

[37] Gospodinov A, Vaissiere T, Krastev DB, Legube G, Anachkova B, Herceg Z. Mammalian Ino80 mediates double-strand break repair through its role in DNA end strand resection. Mol Cell Biol 2011;31:4735–45.

[38] Courilleau C, Chailleux C, Jauneau A, Grima F, Briois S, Boutet-Robinet E, et al. The chromatin remodeler p400 ATPase facilitates RAD51-mediated repair of DNA double-strand breaks. J Cell Biol 2012;199(7):1067–81.

[39] Xu Y, Ayrapetov MK, Xu C, Gursoy-Yuzugullu O, Hu Y, Price BD. Histone H2A.Z controls a critical chromatin remodeling step required for DNA double-strand break repair. Mol Cell 2012;48(5):723–33.

[40] Kusch T, Florens L, Macdonald WH, Swanson SK, Glaser RL, Yates JR, et al. Acetylation by Tip60 is required for selective histone variant exchange at DNA lesions. Science 2004;306:2084–7.

[41] Fischle W. Tip60-ing the balance in DSB repair. Nat Cell Biol 2009;11:1279–81.

[42] Murayama Y, Uhlmann F. Biochemical reconstitution of topological DNA binding by the cohesin ring. Nature 2014;505(7483):367–71.

[43] Kim JS, Krasieva TB, LaMorte V, Malcolm A, Taylor R, Yokomori K. Specific recruitment of human cohesin to laser-induced DNA damage. J Biol Chem 2002;277(47):45149–53.

[44] Gong F, Miller KM. Mammalian DNA repair: HATs and HDACs make their mark through histone acetylation. Mutat Res 2013;750:23–30.
[45] Dellaire G, Kepkay R, Bazett-Jones DP. High resolution imaging of changes in the structure and spatial organization of chromatin, gamma-H2A.X and the MRN complex within etoposide-induced DNA repair foci. Cell Cycle 2009;8:3750–69.
[46] Burgess RC, Burman B, Kruhlak MJ, Misteli T. Activation of DNA damage response signaling by condensed chromatin. Cell Rep 2014;9(5): 1703–17.
[47] Shanbhag NM, Rafalska-Metcalf IU, Balane-Bolivar C, Janicki SM, Greenberg RA. ATM-Dependent chromatin changes silence transcription in cis to DNA double-strand breaks. Cell 2010;141(6):970–81.
[48] Francia S, Michelini F, Saxena A, Tang D, de Hoon M, Anelli V, et al. Site-specific DICER and DROSHA RNA products control the DNA-damage response. Nature 2012;488:231–5.
[49] Ziv Y, Bielopolski D, Galanty Y, Lukas C, Taya Y, Schultz DC, et al. Chromatin relaxation in response to DNA double-strand breaks is modulated by a novel ATM- and KAP-1 dependent pathway. Nat Cell Biol 2006;8(8):870–6.
[50] Hamilton C, Hayward RL, Gilbert N. Global chromatin fibre compaction in response to DNA damage. Biochem Biophys Res Commun 2011;414: 820–5.
[51] Soutoglou E, Misteli T. Activation of the cellular DNA damage response in the absence of DNA lesions. Science 2008;320:1507–10.
[52] Lee J-S. Activation of ATM-dependent DNA damage signal pathway by a histone deacetylase inhibitor, trichostatin A. Cancer Res Treat 2007;39: 125–30.
[53] Aymard F, Bugler B, Schmidt CK, Guillou E, Caron P, Briois S, et al. Transcriptionally active chromatin recruits homologous recombination at DNA double-strand breaks. Nat Struct Mol Biol 2014;21(4):366–74.
[54] Gilchrist S, Gilbert N, Perry P, Bickmore WA. Nuclear organization of centromeric domains is not perturbed by inhibition of histone deacetylases. Chromosom Res 2004;12(5):505–16.
[55] Musich PR, Zou Y. DNA-damage accumulation and replicative arrest in Hutchinson–Gilford progeria syndrome. Biochem Soc Trans 2011;39: 1764–9.
[56] Boyle S, Gilchrist S, Bridger JM, Mahy NL, Ellis JA, Bickmore WA. The spatial organization of human chromosomes within the nuclei of normal and emerin-mutant cells. Hum Mol Genet 2001;10(3):211–9.
[57] Croft JA, Bridger JM, Boyle S, Perry P, Teague P, Bickmore WA. Differences in the localization and morphology of chromosomes in the human nucleus. J Cell Biol 1999;145(6):1119–31.
[58] Parada LA, McQueen PG, Misteli T. Tissue-specific spatial organization of genomes. Genome Biol 2004;5(7):R44.
[59] Boyle S, Rodesch MJ, Halvensleben HA, Jeddeloh JA, Bickmore WA. Fluorescence in situ hybridization with high-complexity repeat-free oligonucleotide probes generated by massively parallel synthesis. Chromosom Res 2011;19(7):901–9.
[60] Shopland LS, Johnson CV, Byron M, McNeil J, Lawrence JB. Clustering of multiple specific genes and gene-rich R-bands around SC-35 domains: evidence for local euchromatic neighborhoods. J Cell Biol 2003;162(6):981–90.
[61] Bickmore WA, Teague P. Influences of chromosome size, gene density and nuclear position on the frequency of constitutional translocations in the human population. Chromosom Res 2002;10(8):707–15.
[62] Roukos V, Voss TC, Schmidt CK, Lee S, Wangsa D, Misteli T. Spatial dynamics of chromosome translocations in living cells. Science 2013;341(6146):660–4.
[63] Bickmore WA, van Steensel B. Genome architecture: domain organization of interphase chromosomes. Cell 2013;152:1270–84.
[64] Zhang Y, McCord RP, Ho YJ, Lajoie BR, Hildebrand DG, Simon AC, et al. Spatial organization of the mouse genome and its role in recurrent chromosomal translocations. Cell 2012;148(5):908–21.
[65] Therizols P, Illingworth RS, Courilleau C, Boyle S, Wood AJ, Bickmore WA. Chromatin decondensation is sufficient to alter nuclear organization in embryonic stem cells. Science 2014;346(6214):1238–42.
[66] Dileep V, Ay F, Sima J, Vera DL, Noble WS, Gilbert DM. Topologically associating domains and their long-range contacts are established during early G1 coincident with the establishment of the replication-timing program. Genome Res 2015;25(8):1104–13.
[67] Shachar S, Voss TC, Pegoraro G, Sciascia N, Misteli T. Identification of gene positioning factors using high-throughput imaging mapping. Cell 2015;162(4):911–23.
[68] Reijns MAM, Kemp H, Ding J, de Procé SM, Jackson AP, Taylor MS. Lagging-strand replication shapes the mutational landscape of the genome. Nature 2015;518(7540):502–6.
[69] Burrell RA, McClelland SE, Endesfelder D, Groth P, Weller MC, Shaikh N, et al. Replication stress links structural and numerical cancer chromosomal instability. Nature 2013;494:492–6.
[70] Saponaro M, Kantidakis T, Mitter R, Kelly GP, Heron M, Williams H, et al. RECQL5 controls transcript elongation and suppresses genome instability associated with transcription stress. Cell 2014;157:1037–49.
[71] Yu A, Fan HY, Liao D, Bailey AD, Weiner AM. Activation of p53 or loss of the Cockayne syndrome group B repair protein causes metaphase fragility of human U1, U2, and 5S genes. Mol Cell 2000;5(5):801–10.
[72] Durkin SG, Glover TW. Chromosome fragile sites. Annu Rev Genet 2007;41:169–92.
[73] Helmrich A, Ballarino M, Tora L. Collisions between replication and transcription complexes cause common fragile site instability at the longest human genes. Mol Cell 2011;44:966–77.
[74] Letessier A, Millot GA, Koundrioukoff S, Lachagès AM, Vogt N, Hansen RS, et al. Cell-type-specific replication initiation programs set fragility of the FRA3B fragile site. Nature 2011;470:120–3.
[75] Mohaghegh P, Karow JK, Brosh RM, Bohr VA, Hickson ID. The Bloom's and Werner's syndrome proteins are DNA structure-specific helicases. Nucleic Acids Res 2001;29:2843–9.

[76] Pirzio LM, Pichierri P, Bignami M, Franchitto A. Werner syndrome helicase activity is essential in maintaining fragile site stability. J Cell Biol 2008;180:305–14.
[77] Sengupta S, Linke SP, Pedeux R, Yang Q, Farnsworth J, Garfield SH, et al. BLM helicase-dependent transport of p53 to sites of stalled DNA replication forks modulates homologous recombination. EMBO J 2003;22:1210–22.
[78] Wechsler T, Newman S, West SC. Aberrant chromosome morphology in human cells defective for Holliday junction resolution. Nature 2011;471:642–6.
[79] Pacek M, Tutter AV, Kubota Y, Takisawa H, Walter JC. Localization of MCM2-7, Cdc45, and GINS to the site of DNA unwinding during eukaryotic DNA replication. Mol Cell 2006;21:581–7.
[80] Ozeri-Galai E, Lebofsky R, Rahat A, Bester AC, Bensimon A, Kerem B. Failure of origin activation in response to fork stalling leads to chromosomal instability at fragile sites. Mol Cell 2011;43:122–31.
[81] Hiratani I, Takebayashi S, Lu J, Gilbert DM. Replication timing and transcriptional control: beyond cause and effect–part II. Curr Opin Genet Dev 2009;19:142–9.
[82] Le Tallec B, Millot GA, Blin ME, Brison O, Dutrillaux B, Debatisse M. Common fragile site profiling in epithelial and erythroid cells reveals that most recurrent cancer deletions lie in fragile sites hosting large genes. Cell Rep 2013;4:420–8.

Chapter 24

Role of DNA Methylation in Genome Stability

D. Zhou, K.D. Robertson
Mayo Clinic, Rochester, MN, United States

Chapter Outline

1. INTRODUCTION TO THE CELLULAR FUNCTIONS OF DNA METHYLATION

Genomic information is inscribed within the DNA sequences and additional chemical modifications embedded in the chromatin structure. The orders given by such information to the particular cell, neighboring cells, and even the entire organism based on the underlying signal transduction and crosstalk sustain all basic and normal functionalities, guiding survival, reproduction, death, and ultimately biological evolution. Disturbing genome stability by intrinsic and/or extrinsic factors could disrupt growth or developmental trajectory, as well as regular cellular behaviors, leading to abnormal or even detrimental consequences. Both genetic and epigenetic mechanisms introduce DNA sequence–dependent and –independent changes, resulting in detrimental consequences in a genomic content. Genetic mutations, deletions, insertions, translocations, and chromosomal aneuploidy are well-recognized consequences resulting from genomic instability. Remarkably, epigenetic mechanisms, namely DNA-methylation and histone modifications, are established and acknowledged as contributing factors for maintaining genome integrity through regulating these genetic events during different cellular processes. In this chapter, we focus on the genomic instability triggered by epigenetic changes with a specific emphasis on the role of DNA methylation.

1.1 DNA-Methylation Dynamics

DNA methylation is dynamic and subject to alterations. To date, the DNA methyltransferase family DNMT1, 3A, 3B, and 3L, and the DNA demethylases, ten–eleven translocation enzymes (TETs) and thymine DNA glycosylase (TDG), have been identified in animals (Fig. 24.1). A clear division of labor exists in each family. DNMT1 binds specifically to the hemimethylated DNA double helix and faithfully maintains methylation patterns in the newly synthesized DNA

Genome Stability. http://dx.doi.org/10.1016/B978-0-12-803309-8.00024-0

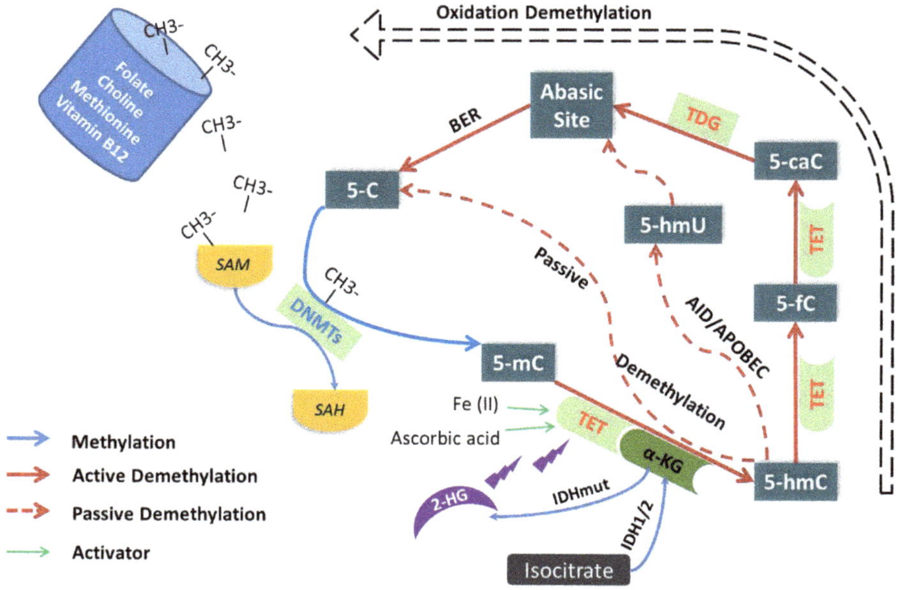

FIGURE 24.1 **DNA-methylation metabolic cycle and associated essential nutrients.** Methylation via DNMTs and active and passive demethylation via DNA methyltransferase (DNMT) and active and passive demethylation via ten–eleven translocation enzymes (TETs) and thymine DNA glycosylase (TDG) are depicted, along with natural resources of methyl donors, cofactors of TET, and effects of IDH1/2 mutations on demethylation enzymes. *Blue arrow*: S-adenosyl-L-methionine (SAM) assists DNMT activity by donating a methyl group to the DNA-methylation process, resulting in a methylated cytosine at position 5. *Red arrow*: Oxidation or active demethylation process takes place through the TET proteins, a dioxygenase protein family dependent on the availability of α-ketoglutarate and Fe^{2+}. TETs successively generate oxidized products 5-hydroxymethylC (5-hmC), 5-formylC (5-fC), and 5-carboxylC (5-caC). Highly oxidized 5-fC and 5-caC can be excised by TDG, forming an abasic site which can be repaired by base-excision repair (BER). *Red dash arrow*: Passive demethylation occurs in two possible pathways, one is through passive loss during replication, second is through AID/APOBEC-directed deamination process.

strand using the parental strand as a template [1]. This copy–paste process is essential for the inheritance of the biological information in the epigenomic structures to daughter cells during rapid cell proliferation. DNMT3A and DNMT3B are the de novo methyltransferases, capable of adding methyl-groups to the 5-position of unmodified cytosine, generating new patterns of DNA methylation [2]. DNMT3A and DNMT3B are extremely important in terms of establishing new DNA-methylation pattern during embryonic stem cell differentiation and tissue development [3]. Unlike stably expressed DNMT1, expression of DNMT3s is usually high in stem-like cells but reduced toward terminal differentiation. DNMT3L, lacking the C-terminal catalytic domain with the methyltransferase activity possessed by other DNMTs, mainly functions by influencing DNMT3A/3B activities to establish DNA-methylation markers [4]. De novo methylation drives the process of development and differentiation, and programs a cell with functional specificity [5]. It also creates dynamic DNA-methylation landscapes in response to intra- and extracellular signals, potentially contributing to environmental adaptation and evolutionary processes.

DNA demethylation is the process of removal or modification of a methyl (CH_3) group on DNA nucleotides. It can be achieved through both passive and active mechanisms. Passive demethylation could occur due to the absence of DNMT1 activity, with the newly synthesized DNA strands losing the methylation patterns such that upon several additional rounds of replication and division, this information will no longer be present in either strand. One of the examples of the passive loss of methylation is the passive demethylation upon inactivation of DNMT1 enzyme by 5-azacytidine [6]. Active DNA demethylation largely relies on the activity of TETs [7]. Unlike the relatively well-defined DNMTs, characterization of unique functions and possible redundancy of each TET are still underway. Oxidation of 5-methylcytosine (5-mC) takes place in a step-wise manner. In brief, TETs catalyze oxidation of an existing methyl group, yielding the first intermediate product 5-hydroxymethylC (5-hmC), which can be further oxidized into 5-formylC (5-fC) followed by 5-carboxylC (5-caC). Both 5-fC and 5-caC can be then replaced by an unmodified cytosine through TDG-mediated base-excision repair (BER). It worth noting that 5-hmC as well as the other two oxidation derivatives, 5-fC and 5-caC, are not recognized by DNMT1 during replication. Therefore, 5-hmC can be removed through active demethylation driven by TETs/TDG, or lost during replication by passive demethylation. Another proposed demethylation pathway involves a deamination process by cytidine deaminase (AID/APOBEC), which converts 5-hmC to 5-hydroxymethyluracil (5-hmU), generating an abasic site that can be removed by DNA glycosylase [8]. However, it is important to keep in mind that rather than just being intermediate products of a demethylation pathway, emerging studies show that 5-hmC and 5-fC, although present at fairly low levels in the genome, are stable DNA marks and may play important roles such as regulating gene transcription and cell proliferation [9,10].

Through the methionine cycle, the level of methylation intermediates S-adenosyl-L-methionine (SAM) and S-adenosyl-homocysteine (SAH) sustain DNA-methylation reactions in the body [11]. Dietary factors, especially some micronutrients such as folate, methionine, and choline, are essential methyl donors to one-carbon metabolism [12]. Methyl donors target DNA methylation through regulating the substrate availability. Micronutrients such as iron and ascorbate are important cofactors of demethylation enzymes and have been shown to generate health concerns when they are not provided in sufficient quantity. Adding ascorbic acid to the cells with proficient expression of TETs is capable of inducing 5-mC oxidation, leading to a substantial loss of 5-mC and gain of 5-hmC, 5-fC, and 5-caC [13].

1.2 Transcriptional Regulation by DNA Methylation

DNA methylation at promoters and gene bodies regulates transcriptional activity in different ways. In animals, DNA methylation occurs primarily at cytosine in a cytosine–phosphate–guanine context (CpG). Although DNA methylation can also occur in the context of CHG and CHH (where H represents a nucleotide other than guanine), gene-regulatory functions of this form of DNA methylation are less clear. Genomic regions can thus be classified according to CpG density. The most CG-rich regions of the genome, CpG islands, are a frequently studied feature for DNA-methylation regulation. CpG islands are defined using a moving window of 500 bp with CG content more than 60%. CpG island shores, by definition, are regions 2000 bp upstream and downstream of a CpG island. Both CpG islands and CpG island shores have been confirmed to possess key regulatory functions in the genome. Hypermethylation of CpG island(s) in promoters usually leads to gene silencing, whereas hypomethylation permits active transcription. About 70% of mammalian gene promoters bear a CpG island, including those associated with housekeeping genes, developmental genes, tumor suppressors, and cell-cycle genes [14]. Aberrant hypermethylation at promoters of these genes, such as $p16^{INK4a}$, Rb, BRCA1, MLH1, and MGMT, is frequently observed in cancer or other diseases [14]. However, promoter methylation is not the only factor determining gene activity. For example, MGMT expression is often inhibited due to promoter hypermethylation in glioblastoma, but in tumors that have developed temozolomide-resistance, MGMT is reactivated even in the presence of promoter hypermethylation [15], suggesting alternative mechanisms exist to promote MGMT expression. In contrast to promoter methylation, gene body methylation is often associated with active transcription. As seen in 5-aza-2′-deoxycytosine (5-aza-2′-dC) treated HCT116 cells, loss of methylation at gene bodies correlates with transcriptional repression in a large set of genes, whereas DNMT3B-mediated methylation at these regions reestablishes gene expression [16]. Genome-wide DNA-methylation mapping via high throughput sequencing revealed that methylation patterns at CpG island shores display lineage- and tissue-specific patterns, and associated strongly with gene expression. This is supported by findings showing that disrupted DNA methylation occurs most frequently at CpG island shores in colon cancer concentrated at the regions with tissue-specific methylation, and results in a loss of tissue-specific epigenetic signatures, suggesting a role for DNA methylation in sustaining cell identity [17].

2. MULTIFACETED REGULATION OF GENOME STABILITY BY DNA METHYLATION

Disrupting DNA-methylation patterns established during cell growth and development leads to loss of function, cell-cycle arrest, and can even be favorable to disease development and transformation. This is not only because DNA methylation regulates transcriptional activities of cell-cycle genes, oncogenes, and tumor suppressor genes, but also because it influences mutation frequencies when inappropriate methylation occurs to noncoding regions and DNA damage–repair processes. In Section 2.1, we focus on the role of DNA methylation in restricting the expansion of repeat elements and preventing abnormal homologous recombination (HR). Next, DNA-repair mechanisms that prevent and/or correct genetic errors incurred during replication and chromosomal rearrangement will be linked to DNA methylation in Section 2.2. In Section 2.3, we will also discuss the contribution of DNA methylation in maintaining nucleosome and heterochromatin structure.

2.1 Chromosomal Rearrangement and Changes in Nucleic Acid Sequences

2.1.1 *Instability of Repeat Elements*

Noncoding regions occupy about 98.5% of the human genome and are an important contributor to genome/chromosome stability. Repeat elements comprise nearly half of these noncoding sequences. Repeat elements in the human genome are classified into two groups, interspersed repeats mainly comprised of transposable elements (TEs), and tandem repeats ranging from a few bases to mega-bases [18]. Both classes of repeats are epigenetically modified, the status of which significantly contributes to genome stability and disease onset.

DNA-methylation mechanisms contribute to preserving stability of TEs (transposable elements) by silencing gene transcription, likely in a developmental stage–dependent manner. TEs are discrete mobile DNA segments capable of moving and integrating randomly within the genome. Depending on the nature of the element, transposition can be initiated by two different mechanisms, "cut and paste" and replicative transposition. DNA-only transposons are autonomous elements using a "cut and paste" mechanism initiated by transposes encoded within the transposon itself. The original repeat is directly relocated to the target site. In contrast, long terminal repeat (LTR) elements [19] and non-LTR elements are typical retrotransposons; they use replicative transposition requiring RNA synthesis and reverse transcription before the newly synthesized repeat sequence can be placed in the targeted site. Both long interspersed nuclear elements (LINEs) and short interspersed nuclear elements (SINEs) belong to the non-LTR family. Transposition induces genome instability in two ways. *First*, repeat element-directed recombination leads to intra- and interchromosomal rearrangement, dramatically increasing the frequency of deletions, duplications, and translocations. *Secondly*, transposition-associated mutations occur during RNA-based reverse transcription in autonomous retro-TEs, due to reduced processivity of the reverse transcriptase as compared to replicative polymerases. Both events are likely to have grave consequences on genome stability. In mouse embryos, transcription of intracisternal A-particle (IAP), a retrotransposon, is usually silenced due to a high degree of methylation in their LTR region. However in the absence of DNMT1, IAP transcript level increases by 50- to 100-fold, suggesting that DNMT1 plays an important role in maintaining the methylation level of IAPs [20]. In nondividing precursors of spermatogonial stem cells, deletion of DNMT3L disables the de novo methylation capacity at LTR and non-LTR retrotransposon elements of IAPs and thus enables active transcription [21]. Nonautonomous TEs (LINE and SINE) are critical components of heterochromatin enriched at regions flanking centromeres and telomeres [22]. DNA-methylation levels on these TEs are functionally relevant to the formation and stability of constitutive heterochromatin, delivering key messages to cell-cycle control and cell-fate decision markers. This is discussed further in Section 2.3.

Both DNMT1 protein and DNA methylation itself can stabilize tandem repeats. Tandem repeats are classified based on the size of the repeated sequence. A microsatellite comprises short tandem repeating units, usually less than 10 bp. Microsatellite instability (MSI) causes mutations through changes in length (expansion and contraction). It is therefore highly associated with hypermutation phenotypes in disease and contributes to lethal consequences in disorders like Huntington's, myotonic dystrophy, and a variety of cancers including hereditary nonpolyposis colorectal cancer (HNPCC). MSI is often a direct consequence of an impaired DNA mismatch repair (MMR) system. Accumulation of unrepaired DNA replication errors creates novel microsatellite fragments, or short tandem DNA repeats by definition, which are abundant in untranslated regions including introns. Mutations in DNMT1 result in increased instability of endogenous microsatellites and transgenic slippage reporter constructs without altering MMR components [23–25]. A novel function for *Dnmt1* in MMR was first assigned in genetic screening of *Blm*-deficient ES cells. This novel function of *Dnmt1* was confirmed in mouse ES cells in which cells with deficiency or homologous deletion of *Dnmt1* exhibited higher microsatellite slippage rate of a mononucleotide repeat carried by a reporter gene [24], as well as elevated frequencies of instability at endogenous microsatellite repeats [25]. More importantly, the flanking regions of mononucleotide repeats are always unmethylated regardless of *Dnmt1* expression level, suggesting that increased microsatellite slippage rate was not due to local DNA-methylation levels [24]. Other studies, however suggest that repeat stability is subject to DNA-methylation regulation at either local or adjacent regions. In human cells from myotonic dystrophy patients, inhibiting DNA methyltransferase through 5-aza-2′-dC treatment reproduces a similar consequence to that reported in the *Dnmt1*-deficient system, in which a 1000-fold increase in MSI was observed [26]. This effect could also be achieved through modifying CpG methylation of genes within or in the vicinity of microsatellites [27,28]. For example, expansion frequencies of CpG-free repeats, CAG.CTG, are highly affected by the CG content in the neighboring *cis*-sequence [26,28], suggesting methylation of CpGs at the flanking regions of microsatellite repeats also protects MSI from taking place. Trinucleotide repeats (TNRs) constitute a subset of microsatellites. Gain of methylation at CGG repeats artificially introduced into primate cells stabilizes these repeats [29]. Failure to maintain normal DNA-methylation patterns at repeat sequences during development contributes to the onset of genetic neurological disease. For example, fragile X syndrome (FXS) is caused by expansion of the CGG repeats at the fragile X mental retardation 1 (*FMR1*) gene on the X chromosome [30]. FMR1 mainly regulates dendritic protein synthesis, a class of proteins essential for synaptic strength. Normally, there are 30 CGG repeats at the 5′ untranslated region of FMR1. In patients with FXS, the number of repeats can be as high as 200 copies, resulting in hypermethylation of the entire repeat region and subsequent gene repression. A blockage in the AMPA-type glutamate receptor-signaling cascade arising from a lack of *FMR1* expression is primarily responsible for impaired learning and memory process [31]. Furthermore, $(CGG)_n$ repeat amplification is observed in the germline of male FXS patients carrying unmethylated repeats [32]. Another kind of TNR disorder, triggered by CAG repeat expansion, also accounts for a myriad of neurodegenerative disorders [33] such as Huntington's disease, which is manifested by an increase of more than 35 consecutive CAG repeats on the gene encoding huntingtin [34]. Even though CAG repeats are devoid of CpG sites, studies show that DNA methylation of adjacent sequences is associated with CAG repeat stability. In a *Dnmt1*-knockout mouse model, intergenerational expansion of CAG repeats is observed at the spinocerebellar ataxia type 1 (*Sca1*) locus and is associated with aberrant DNA methylation at regions adjacent to the repeat tract [35]. Moreover, it appears that DNMT1 knockdown induces CAG repeat contraction through activating CAG repeat transcription [36]. This controversial observation suggests DNMT1 carries some complex functions remaining to be discovered. Interaction between DNMT1 and histone modifiers may also contribute to microsatellite stability. This is because DNMT1 interacts with histone deacetylases (HDAC1 and HDAC2) at microsatellites [37], promoting deacetylation so as to constrain the chromatin structure from being accessible to the transcriptional machinery. However, overexpressing HDACs does not necessarily reduce the frequency of MSI. In a human colorectal cancer (CRC) tissue survey, an inverse correlation was found between expression of *SIRT1*, a class III histone deacetylase, and incidence of MSI [38].

2.1.2 Chromosomal Recombination

DNA hypomethylation is generally associated with an elevated frequency of gene rearrangements and chromosomal translocations as a consequence of increased HR. HR occurs regularly during meiosis, naturally increasing the biological diversity within a species. It also occurs occasionally in somatic cells. The chances of HR increase significantly during transcription, when single-stranded DNA is exposed, spatially facilitating HR. It often occurs between DNA regions sharing extensive sequence identity (eg, sister chromatids) or highly similar (eg, two homologs) sequences. HR between repeated sequences leads to chromosome rearrangement, including deletions, duplications, and translocations of large DNA segments with disastrous consequences. Multiple lines of evidence showed that DNA methylation negatively affects HR in mammals [39–42]. These studies show that V(D)J recombination rate is significantly reduced by CpG methylation using minichromosome substrates [39], and that DNA hypomethylation at peri-centromeric satellite DNA is associated with increased rates of peri-centromeric chromosomal rearrangements [41]. Transcriptional silencing mediated by DNA methylation inhibits HR from taking place [43]. Studies in several mouse models, where genomic hypomethylation induced by a deficiency of *Dnmt1* resulted in an increase of HR [43] and loss of heterozygosity [44], suggest that *Dnmt1* contributes to repression of HR. Similarly, DNMT1 and DNMT3 recruitment to peri-centromeric and centromeric regions is believed to protect these loci against unlicensed HR [45]. Extensive DNA hypomethylation significantly increases mutation rates potentially through increasing the rate of mitotic chromosomal recombination. In ES cells carrying nullizygous *Dnmt1*, two specific genes, endogenous hypoxanthine phosphoribosyltransferase (*Hprt*) and a viral thymidine kinase (*tk*) transgene, show large increases in locus-specific deletions and mutations [40]. In 2011, a study using genome-wide sequencing identified that mutation rate varies across the genome [46] and is inversely correlated with DNA-methylation levels [47]. In particular, within CpGs sites, low (20–40%) to intermediate (40–60%) methylated CpG sites are prone to accumulate more mutations based on the density of single nucleotide polymorphisms (SNP) [47], again indicating that mutation rates are negatively correlated with methylation levels. The effects of aberrant DNA methylation on repeat elements are depicted in Fig. 24.2.

2.2 DNA-Damage Repair

DNA replication and chromosomal rearrangements are the most likely processes to yield mutations. The spontaneous error rate of mammalian DNA polymerase is about 10^{-5} to 10^{-6} per base pair, whereas the true mutation rate is only about 10^{-9} to 10^{-10}. This considerable reduction of final mutation rate is attributed to the polymerase proofreading system and DNA-damage repair. Depending on the type of DNA errors, different repair mechanisms will become active. Single nucleotide damage can be repaired by BER, nucleotide-excision repair (NER), MMR, and atypical modification of a specific nucleobase, such as 3-methyladenine and 8-oxoguanine, is corrected by BER through DNA glycosylase activity [48]. NER also responds when large and complex types of damage are found on DNA, such as intrastrand and DNA-protein cross-links, and bulky adduct formations [49]. Nucleotide misincorporation generated during DNA replication that escapes proofreading is resolved by MMR, as well as strand slippage- and recombination-resulted erroneous insertions and deletions at repeated DNA sequences (tandem repeats, microsatellites). MMR also corrects abnormally modified nucleotides including O^6-methylguanine (O^6-meG), 8-oxoguanine, and DNA adducts formed between DNA and carcinogenic chemicals through covalent bonds [50]. Repairing single nucleotide damage using an MMR mechanism requires the other strand as a template. DNA breaks can attack either one or both DNA strands. Single-strand breaks (SSB) result in discontinuity in one DNA strand and are often accompanied with loss of a single nucleotide. Filling the gap introduced by SSB requires the unbroken strand as template [51]. Double-strand breaks (DSBs) employ two mechanisms to repair, HR and nonhomologous end

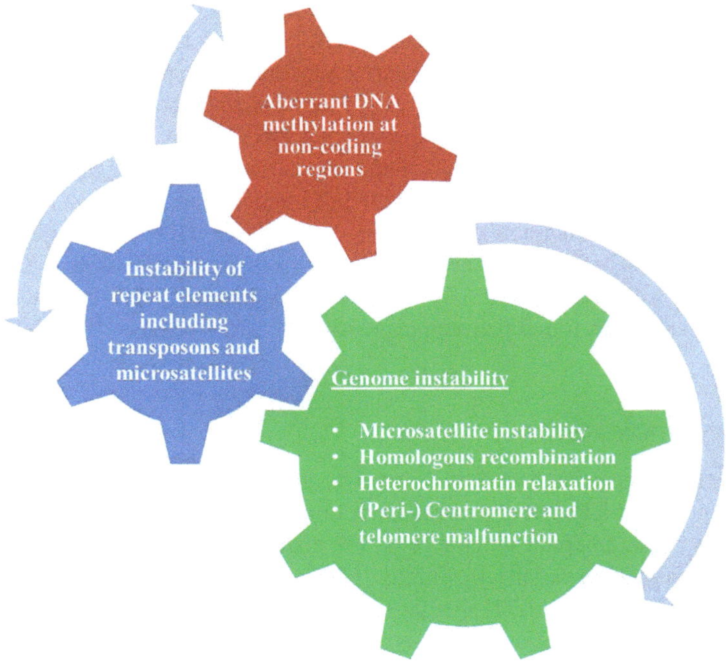

FIGURE 24.2 **A chain reaction induced by DNA methylation at repeat elements.** Aberrant DNA methylation at noncoding repeats destabilizes transposons and microsatellites, which result in microsatellite instability, increased rate of homologous recombination, heterochromatin structure change, and (peri-)centromere and telomere malfunction.

joining (NHEJ). HR takes advantage of the existence of a (nearly) identical sequence and uses it as a template for repair. NHEJ is mutagenic and therefore a less preferred mechanism as it usually results in point mutations and deletions of various size during repair [52]. The enzymes involved in each repairing process and their molecular functions are summarized in Table 24.1.

2.2.1 The Role of DNMT1 and DNA Methylation in DNA-Damage Repair

DNMT1 is an essential protein participating at the replication fork. Recruitment of DNMT1 to the replication fork requires interaction with proliferating cell nuclear antigen (PCNA), a cofactor of DNA polymerase delta (Polδ) and a component of DNA replication forks [53], and ubiquitin-like with PHD and ring finger domains 1 (UHRF1), a protein of unclear function that specifically recognizes hemi-methylated DNA and targets DNMT1 to such foci through a unique SET and ring-associated (SRA) domain [54]. UHRF1 and the complex formed by DNMT1 and G9a, euchromatic histone-lysine N-methyltransferase, colocalize with H3K9me2 at replication foci, enhancing the fidelity of DNA and histone methylation [55,56]. Depletion of DNMT1 at the replication fork leads to activation of checkpoint kinases 1 and 2 (CHK1 and CHK2, key effector kinases of the ATM/ATR-mediated DNA damage–response pathway), followed by degradation of cell division control protein 25a (CDC25a) and formation of γ H2A.X foci (H2A Histone family member X, a hallmark of DSBs), and eventually replication arrest [57]. This intra-S-phase replication arrest is not dependent on DNA demethylation as treating cells with 5-aza-dC, a nucleoside analogue trapping DNMT1 at the progressing replication fork, does not produce the same result [58]. Indeed, neither DNA demethylation by 5-aza-dC nor loss of catalytic activity of DNMT1 can stimulate a damage response similar to DNMT1 depletion [57]. Therefore, it appears that DNMT1 depletion triggers a protective mechanism to genome integrity through intra-S-phase replication arrest. It prevents global demethylation and epigenetic information loss by activating checkpoint pathways while being physically absent from the replication fork.

Accumulation of DNMT1 at DNA-damage sites and its association with MMR processes have been identified in a number of studies. The basic protein components of mammalian MMR are MutS (mutator S) α, MutSβ, MutLα, exonuclease 1 (EXO1), replication factor C (RFC), PCNA, replication protein A (RPA), DNA Polδ, and DNA ligase [59]. The MutS complex comprises a heterodimer of MSH2/MSH6 (MutS α) and MSH2/MSH3 (MutSβ), whereas the MutL complex consists of a heterodimer of MLH1/postmeiotic segregation increased 2 (PMS2) [60]. The principle of MMR resides in the nature of the DNA replication process, in which daughter strands should be faithfully synthesized using the parental sequence as the sole template. Therefore, upon receiving mismatching signals, three key actions are taken: *first*, recognition of the

TABLE 24.1 Key Enzymes of DNA-Repair Pathways

	DNA Glycosylase Scanning System				**Endonuclease**				**DNA Polymerase**	**DNA Ligase**	**Reference**
BER	UNG	OGG-1	NTHL1	NEIL1-3	APE1				Polβ	Lig1	[48]
	DNA Damage–Detection Complex				**Exonuclease Excision Complex**						
NER	XPC	XPG	RAD23B	ERCC6	XPA	XPG	RPA	ERCC1,3,4			[49]
	Mismatch-Recognition Complex				**Repair Machinery**		**Exonuclease**				
MMR	MutS	MutL			PCNA	RFC	EXO-1				[50]
	Tool Belt				**Approximation Process**						
NHEJ	Ku70	Ku80			Mre11–Rad50–Nbs1 (MRN)				–	Lig4	[52]
	DSB Processing		**Homologous Pairing and DNA Strand Invasion**					**Endonuclease**			
HR	MRN	Exo1	RAD51	RPA	BRAC2	XRCC2	XRCC3	Mus81–Eme1	Rev1, 3, 7	E3	[119]

mismatched base pair by MutSα complex and recruitment of MutLα, *secondly*, cleavage of the incorrectly placed nucleotide on the daughter strand by EXO1, and *lastly*, resynthesis of the damaged region by the PCNA/Polδ complex using the parental strand as a template [61]. A key premise of the MMR process is to distinguish between parental and daughter strands under the guidance of DNMT1. DNMT1 binds specifically to hemimethylated DNA during replication. MMR takes advantage of the hemimethylated state, identifies the parental strand, and then immediately digests the region containing the mismatched nucleotide (a short oligonucleotide spanning the mismatch site) on the new strand, allowing DNA polymerase to resynthesize the strand fragment [62]. In addition to being associated with PCNA at replication sties, DNMT1 also interacts with PCNA at DNA-damage sites [63] where MLH1 is also recruited [64], to methylate the new strand, demonstrating another aspect of DNMT1's role in MMR [63]. A protein–protein interaction between MLH1 and DNMT1 is possibly achieved through methyl-CpG binding domain 4 (MBD4), which binds MLH1 at its C-terminal glycosylase domain and DNMT1 via its N-terminal MBD domain [64]. Colocalization of DNMT1, MBD, and MLH1 occurs at heterochromatic regions and DNA-damage sites. In fact, DNMT1 deficiency impairs MMR function. Knockdown of DNMT1 in immortalized human fibroblasts yields resistance to the drug 6-thioguanine and a 10-fold increase of mutation rates at a CA_{17} microsatellite reporter gene, two hallmarks of MMR defects [65]. MMR defects in this study also appeared to be mediated by the reduction of steady-state protein levels of MSH2, MSH6, and PMS2. An important interaction between the MSH2/MSH6 heterodimer and DNMT1 was established in 2015 in a study of oxidation-induced DNA damage [66]. This study showed that oxidative damage triggered by hydrogen peroxide exposure reduces transcription of genes with promoter CpG islands. This repression is effectively blocked by knocking down MSH6 or DNMT1, suggesting accumulation of DNMT1 at the damaged site serves to prevent transcription from interfering with the repair process. An early study demonstrates that PCNA binds MSH6 and MSH3 at the replication fork during S phase [67], suggesting that accumulation of DNMT1 with MSH6/3 at the replication fork is likely through PCNA. In addition, DNMT1 contributes to DSB repair through interaction with both PCNA and ATR effector kinase CHK1. Immediately after laser microirradiation-induced DSBs, colocalization of DNMT1/PCNA/γH2A.X is observed at damage sites. The interaction between DNMT1 and PCNA or CHK1 is responsible for the recruitment of DNMT1 to the damage site, but is independent of its catalytic activity. This transient localization of DNMT1 to regions of DSBs modulates the rate of DSBs repair [68,69], again suggesting a methylation-independent role of DNMT1 in the DNA-repair process. MSI is a common mechanism for tumor development and can be driven by defective MMR. For example, knockdown of MMR components MSH2 or MSH3 inhibit contraction of CAG repeats, whereas depletion of MLH1 or PMS2 elevates contraction frequency [36]. Epigenetic silencing and mutations of MMR genes, including MLH1, MSH2, and PMS2, occur in many MSI tumors such as sporadic and hereditary colorectal and endometrial carcinomas [70]. This correlation between MMR and MSI has been brought into clinical application. Specially, the MSI phenotype is determined by MMR immunohistochemistry and is used to predict the risk of Lynch syndrome in patients with endometrial carcinomas [71]. These interactions between DNMT1 and other protein factors at replication fork and DNA-damaged site are illustrated in Fig. 24.3.

DNA damage could also be introduced by inappropriate DNA methylation. Exposure to alkylating agents, for example, results in the formation of O^6-meG, 1-methyladenine (1-meA), and 3-methylcytosine (3-meC). These aberrantly modified nucleotides form adducts, disrupt normal replication and transcription, and induce cell-cycle checkpoints and apoptosis [72]. Long-term accumulation of alkylation damage is prone to induce site-specific mutation (G to A) [73]. Direct reversal repair (DR) is involved to correct this type of DNA damage by employing two types of protein, O^6-methylguanine-DNA methyltransferase (MGMT or AGT) and the ALKBH family of Fe (II)/α-ketoglutarate-dependent dioxygenases (FeKGDs). Unlike BER or MMR, MGMT and ALKBH remove alkylation damage at DNA base-paring sites in a template-independent manner, and correct DNA base damage by directly accepting the methyl group [74]. The promoter of MGMT contains a CpG island, methylation of which usually remains low to ensure the proper expression of MGMT. Methylation of cytosine

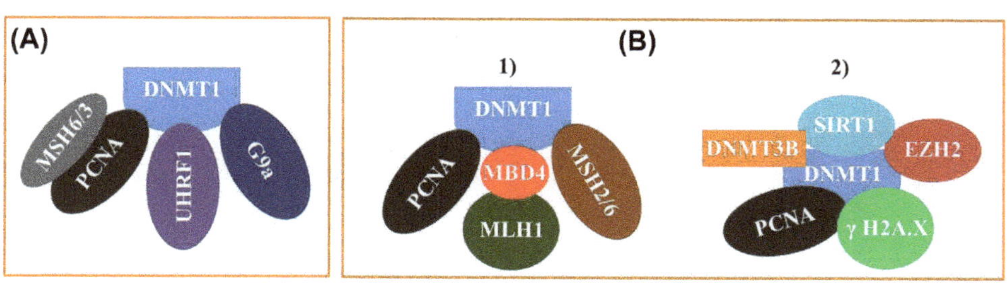

FIGURE 24.3 **Protein–protein interactions between DNA methyltransferase (DNMT1) and DNA replication and repair proteins.** Illustration of protein complex assembly at replication fork (A) and DNA-damaged sites (B). Accumulation of DNA mismatch repair (MMR) pathway-induced (1) and double-strand breaks (DSBs)-associated (2) DNA-repair protein is depicted, respectively.

is also mutagenic as it causes C to T transition mutations through deamination. The deamination product is mainly removed by thymine-DNA glycosylase, a key enzyme discussed in the context of the DNA-demethylation pathway.

Some interplay has been shown between HR-directed DNA damage repair, large DNA fragment exchange, and DNA methylation. HR serves as a means for repairing DSBs, resulting in gene conversion or loss of heterozygosity. Homologously recombined gene segments are often silenced through epigenetic mechanisms, involving DNA hypermethylation [75]. This event was induced at damaged site to repress local transcription from taking place [76,77], and achieved mainly through recruiting DNMT1 and the DNMT3s and introducing repressive histone modifications including H3K9me2/3 and H3K27me3 at the repair site [75,77]. Such epigenetic remodeling could either be transient or heritable, resulting in temporary or permanent gene silencing, respectively.

2.2.2 Transcriptional Regulation of DNA Damage–Repair Genes by DNA Methylation

MLH1 is an MMR protein that forms a complex with DNA-repair protein PMS2, and coordinates the other DNA-repair protein effectors to repair mismatches arising during DNA replication. Promoter hypermethylation of the MLH1 gene is highly associated with repressed expression, and is observed in many cancer types, including gastric cancer, nonsmall cell lung cancer, ovarian cancer, HNPCC, and CRC [60]. The frequency of MLH1 promoter hypermethylation however varies among cancer types and specimens, ranging from 1% to 66.9% in sporadic CRC, or from 0% to 21.4% in LS-CRC [78]. MLH1 promoter hypermethylation was observed in a subset of CRC with hypermethylation at a large number of CpG islands (termed CpG island methylator phenotype, or CIMP). In CIMP-positive CRC and gastric cancer, hypermethylation of MLH1 leads to a dysfunctional MMR pathway, resulting in an MSI phenotype [79,80]. This connection is supported by early evidence that MMR deficiency results in strong repression of a transgenic reporter gene through DNA hypermethylation [81].

Promoter methylation of the MGMT gene is a key factor determining the therapeutic efficacy in treating glioblastoma multiforme (GBM), one of most common and aggressive brain tumors. MGMT corrects the mutagenic DNA lesion O^6-meG in the DR pathway. During replication and transcription, O^6-meG mispairs with thymine. Thymine pairs with adenine in the next round of replication giving rise to permanent nucleotide alterations. Mutation or epigenetic silencing of MGMT is observed frequently in CRC [82]. Temozolomide (TMZ), an alkylating agent applied widely in chemotherapy, achieves better therapeutic effects when the MGMT promoter is hypermethylated [83,84]. This is because TMZ induces widespread N-7 or O-6 guanine methylation, which results in DNA damage accumulation and triggers cell death, but only when MGMT is not expressed. However, MGMT expression is not solely determined by promoter methylation. In GBM, long-term treatment of TMZ leads to drug resistance. In many cases of TMZ resistance, expression of MGMT is reactivated even with a hypermethylated promoter [15,85], suggesting that alternative gene-regulatory mechanisms exist.

As part of the BER pathway, TDG corrects G/T mismatches arising from the 5-mC deamination process. TDG interacts with deaminase AID and the damage response protein GADD45a (TDG is essential for active DNA demethylation by linked deamination BER). Promoter hypermethylation inhibits TDG expression. In multiple myeloma, epigenetic silencing of TDG contributes to genomic instability as it reduces DNA-repair efficiency [86]. Overexpression of TDG in cancer cell lines partially restores this DNA-repair pathway. Moreover, methylation-associated gene deregulation is found in many other DNA-repair genes, including XPC in bladder cancer, ERCC1 in GBM, and RAD23B in myeloma [60].

DNA methylation also regulates the transcription of genes involved in HR-directed DNA repair and NHEJ. HR promotes error-free repair by employing the sister chromatid as a template. Decreased rates of HR reduce DNA-repair efficiency, which is also carcinogenic. Cells deficient in breast cancer susceptibility gene 1 or 2 (BRCA1 or BRCA2) display reduced HR rate by at least sixfold in the presence of a DSB [87–89]. This is partially explained by the finding that both BRCA1 and BRAC2 interact with the RAD51 protein, which catalyzes the primary reaction in HR [90]. Epigenetic silencing of BRCA1 and BRCA2 genes by promoter hypermethylation is observed in breast cancer and several other cancer types [60]. In the NHEJ pathway, the XRCC5 gene that encodes the KU80 protein is also silenced by promoter hypermethylation, although this does not seem to be the only silencing mechanism in cancers like non-small-cell lung carcinoma where the gene is frequently down-regulated [91].

2.3 DNA Methylation and Heterochromatin Stability

2.3.1 Nucleosome Positioning and Packaging

Nucleosome structure and packaging are also influenced by DNA methylation. Studies reported during 2012–15 have used fluorescence resonance energy transfer (FRET) to monitor histone binding while modifying CpG methylation in a given DNA sequence. These studies revealed that CpG methylation of a DNA sequence tightened the association between double stranded DNA and core histone proteins, increased histone content within this region, and eventually expedited the formation of more compact and rigid nucleosome structures [92,93]. Using the same method, another study showed that 5-hmC increases DNA binding to histones, but is more likely to keep the nucleosome in an open state for active transcription [94]. Interestingly, some contradictory results were raised from a current study using a nanopore-based force spectroscopy approach. In this method, the binding affinity between nucleosomal DNA and histone core proteins was examined by giving constant or time-varying force [95]. The result showed that nucleosome stability is more sequence dependent, rather than methylation dependent, as displacing DNA from the associated nucleosome required equal force regardless of methylation status.

2.3.2 Heterochromatin Instability

DNA-methylation patterns across the entire genome are responsible for establishing condensed heterochromatin domains or loose euchromatin domains. Two major types of heterochromatin are present in eukaryotic cells, *constitutive heterochromatin* that is enriched for tandemly repeated sequences and forms (peri-)centromeres or telomeres containing discrete satellite DNA, and *facultative heterochromatin* that comprises LINE-type repeats and silenced gene clusters that reversibly transition to euchromatin in the presence of developmental stage–dependent cellular cues [96]. Heterochromatin is tightly packed and localizes to the periphery of the nucleus. Maintenance of heterochromatin relies heavily on epigenetic landmarks, including nonrandom deposition of heterochromatin protein 1 (HP1) together with H3K9me3 and DNA methylation. HP1 keeps heterochromatin tightly packed and transcriptionally repressed. Interaction between HP1 and the nuclear membrane protein, lamin B receptor, contributes to heterochromatin localization. H3K9me3 recruits HP1 at constitutive heterochromatin [97], whereas H3K27me3 is mainly enriched at facultative heterochromatin. HP1 then attracts DNMT3B to the locus and stabilizes the region in heterochromatin by seeding DNA methylation. In addition, UHRF1, which facilitates DNMT1 recruitment, also specifically binds to H3K9me3 [98]. Methyl CpG binding protein 2 (MeCP2) recruits HDACs, which serve as an additional mechanism to maintain transcriptional inactivity and heterochromatin stability. In addition to its association with DNA replication sites during S phase, *Dnmt1* is also localized to constitutive heterochromatin during G2 and M phase [99]. Interestingly, this association exists independent of other heterochromatic marks like H3K9me3, suppressor of variegation 3-9 homolog 1 (Suv39H1) and HP1, suggesting a separate mechanism of establishing stable heterochromatin domains and maintenance of DNA methylation [99].

Heterochromatin at different chromosomal locations performs specific functions [100]. An inability to restrain heterochromatin territories by DNA-methylation or histone marks leads to malfunction and heterochromatin spreading [96]. During mitotic processes, chromosomal rearrangements may place an euchromatic region next to a heterochromatic region or remove the original boundaries protecting this euchromatic region, resulting in heterochromatin invasion into adjacent euchromatin and inactivation of gene clusters residing in this region. Alternatively, disrupting heterochromatin boundaries also leads to heterochromatin spreading, accompanied by DNA-methylation gains outside of the original regions [101]. Conversely, losing hallmarks of heterochromatin leads to deconstruction of heterochromatin structure. *Suv39h1/2*-deficient mouse embryonic fibroblasts exhibit severe chromosome mis-segregation and increased aneuploidy, suggesting a key role for *Suv39h* in maintaining genome stability [102]. Massive reduction in H3K9me3 and significant increase in transcription of peri-centromeric satellite 2 (Sat2) and centromeric α-satellite (α-Sat) are observed following loss of H3K9 methylation by inactivating Suv39H1. Both loss of H3K9me3 and transcriptional activation of satellite repeats are indicative of heterochromatin relaxation in this case [103]. Interestingly, DNA demethylation may induce a similar effect in that it is able to diminish H3K9me3 at the same loci [104]. Occupation of H3K9me3/HP1 usually prevents recruitment of the PRC1/2 complex. In the absence of DNA methylation at these loci, H3K27me3 level increases due to polycomb-group (PcG) protein binding. This colocalization pattern of H3K9me3 and H3K27me3 suggests that switching from constitutive to facultative heterochromatin requires an absence of DNA methylation.

Both centromeric and peri-centromeric heterochromatin serves as the structural basis for chromosome condensation and cohesion between sister chromatids, assisting proper segregation of mitotic chromosomes. Therefore, appropriate heterchromatinization at the peri-centromere satellites is a prerequisite for centromere function. DNA methylation is well known for its role in maintaining the integrity of peri-centromeric heterochromatin structure. For example, DNMT1 facilitates accumulation of H3S10P foci and Aurora-B targeting at peri-centromeres [105], whereas DNMT3B enables centromeric heterochromatin formation and chromosomal condensation [106]. Establishing DNA methylation at peri-centromeric heterochromatin also requires the Suv39H1/2 anchoring H3K9me3 marker [97]. DNMT3A and DNMT3B interact with HP1 via its chromodomain [97]. At peri-centromeric satellite repeats, coexistence of both DNA methylation and H3K9me3 has proved to be essential. *Suv39h1/2* double knockout in mouse cells profoundly reduced DNA methylation and *Dnmt1* binding at peri-centromeric heterochromatin [97]. An additional link between DNA methylation and centromere stability lies in the interaction between DNMTs and centromere proteins (CENPs). Both CENP-B [107] and CENP-C [108] are important kinetochore proteins essential for ensuring proper kinetochore assembly during mitosis. CENP-B is crucial for centromere identity as it binds to unmethylated regions within the centromere to prevent the formation of multiple centromeres, while also promoting DNA methylation to maintain heterochromatin structure [107]. Colocalization of CENP-C and DNMT3B at centromeric regions is required for HP1 recruitment and kinetochore formation; loss of either mark results in a compromised association of the other to targeted sites, reduced DNA methylation, and impaired chromosomal segregation [109]. Some level of peri-centromeric repeat transcription has been shown to occur in most cells, but the underlying biological significance of these transcripts remains elusive. Although the exact role of DNA methylation in regulating this event is not yet clear, it is known that hypomethylation at this region in tumor cells results in transcriptional activation at peri-centromeric loci.

DNA methylation may also be responsible for maintaining telomere integrity through indirect regulation. Telomeres in most metazoans are comprised of a short DNA repeat sequence (5′-TTAGGG-3′) and are enriched for H3K9me3. Although these repeats do not appear to be directly affected by DNA methylation, an inverse relationship between sub-telomeric DNA methylation and telomere length and recombination was observed in a DNMT-deficient mouse model, which exhibited increased telomeric recombination and telomere-length changes [110]. A study reported in 2014 also suggested that DNA methylation at a subset of gene promoters is highly associated with telomere length in human leukocytes [111]. On the other hand, human telomerase gene expression can be activated following 5-aza-2′-dC treatment, suggesting that DNA methylation plays a role in regulating hTERT expression [112]. In addition, although positive correlations between telomere length and DNA methylation at LINE-1 and sub-telomeric regions in patients with dyskeratosis congenital were identified [113], this correlation was not stably observed across all research settings but rather was related to transcriptional and mutational landscapes [114]. For example, a study examining DNA methylation in human cancer cell lines showed no significant correlation between sub-telomeric methylation and telomere length [115]. Thus, whether and how DNA methylation affects telomeres is still controversial.

Immunodeficiency, centromere instability, facial anomalies (ICF) syndrome is a rare autosomal recessive immune disorder characterized by deficiency of serum immunoglobulin levels due to maturation blockage of naive B cells [116] and facial abnormalities. Different mutations have been mapped and are grouped into ICF subclasses, with type I (~50% of all cases) ICF carrying germline hypomorphic mutation in DNMT3B, type II (~30%) zinc-finger and BTB domain containing 24 (ZBTB24) mutations, type III cell division cycle associated 7 (CDCA7) mutations, and type IV lymphoid-specific helicase (LSH, or HELLS) mutations [117]. Hypomethylation at juxtacentromeric heterochromatin repeats accounts for the major pathogenic epigenetic mechanism that characterizes the genomic instability in ICF syndrome patients. In eukaryotes, integrity of centromeric heterochromatin is key for proper construction of cohesion and the kinetochore during mitosis. Centromeric regions and juxtacentromeric satellites are enriched with compact heterochromatin structures, methylation of which is usually maintained at a high level so as to maintain these DNA domains condensed, constrained, and silenced for transcription. Even though DNA methylation is not indispensable to heterochromatin formation, heavily methylated CpGs are believed to stabilize the heterochromatin structure. Extensive hypomethylation of constitutive heterochromatin regions results in loss of heterochromatin structure and consequential loss of mitosis-related functions. In ICF patients, classical satellite DNA is exclusively unmethylated in all tissue types, accompanied by chromosomal decondensation, frequent regional breakage, and rejoining taking place at satellite 2 regions of chromosomes 1 and 16, and satellite 3 regions of chromosome 9 [118]. It has been known that DNMT3B, through interaction with CENP-C, localizes specifically to the centromeric and peri-centromeric heterochromatin regions [109]. Mutation of DNMT3B in type I ICF leads to a hypomethylation phenotype and consequential abnormally arranged chromosome structure [109]. Although how mutations in ZBTB24, CDCA7, and HELLS contribute to the common epigenetic abnormalities and clinical manifestations in all ICF subclass remains to be answered, it is apparent that marked loss of methylation at (peri-)centromere regions is directly or indirectly attributed to these mutations.

3. CONCLUSIONS AND FUTURE DIRECTION

This chapter summarizes the essential functions of DNA methylation and DNA methyltransferases, especially DNMT1, in maintaining genome stability. DNA methylation at noncoding regions, including repeat sequences and heterochromatin regions such as centromeres and telomeres, inhibits spurious transcription and unlicensed HR, thus ensuring the proper functions of centromeres and telomeres at different cell stages. Disruption of these structures leads to mutations or genome rearrangements, which are monitored and repaired by a number of DNA-repair mechanisms. Proper performance of DNA-repair mechanisms requires both DNA methylation and DNMT1. DNA-methylation levels are involved in mediating expression of repair genes, whereas incorporating DNMT1 into replication forks and DNA damage sites through interacting with protein components of the DNA-repair machinery sustains the DNA-repair processes. Taken together, these results outline the indispensable role of epigenetics, especially DNA methylation, in maintaining genome stability. Meanwhile, some intriguing questions are raised for future research: (1) The exact mechanisms of how and why DNMT1, independent of its methyltransferase function, regulates the DNA damage–repair processes are not fully defined. (2) The methylation modifications identified in 2009, 5-hmC, 5-fC, and 5-caC, although present at relatively low amount in the genome, regulate gene transcription and enhancer activities. However, less is known about their functional relevance to genetic stability and disease development. (3) As more evidence accumulates to define the relationships between DNA marks, chromatin modifiers, and their associated histone modifications, a cooperative epigenetic pattern may arise, that is particularly essential for sustaining heterochromatin integrity and DNA-repair functions. (4) Epigenetic mechanisms are an integral part of the etiologies for many types of cancer, as well as neurological and immune disorders. These diseases are often manifested by co-occurrence of genetic mutations and epigenetic modulations. Therefore, it is becoming increasingly important to assign "driver" and "passenger" roles to these events, so that effective therapeutic approaches can be implemented. This said, as of 2016, indirect connections between genetic mutations and epigenetic perturbations are constantly being discovered. For example, in type II, III, and IV ICF syndrome, mutations of the zinc-finger and BTB domain containing 24 (ZBTB24), cell division cycle associated 7 (CDCA7), and lymphoid-specific helicase (LSH, or HELLS) have been identified [117]. Given that the epigenetic abnormalities (marked loss of methylation at peri-centromeres and centromeres) are common to these subclasses, ICF syndrome provides a platform for discovering new epigenetic regulators. More importantly, in light of the fast-developing and increasingly applied high-throughput sequencing technologies, epigenetic research is no longer restricted to a limited number of gene loci; rather it becomes a genome-wide approach to understand the comprehensive gene-regulatory network in cell type–specific and developmental stage–dependent manners. Coupling genome-wide mapping for mutations, transcriptomes, copy number variations (CNVs), and SNPs, research connecting epigenetics to genetic stability is becoming broader, but also revealing previously unknown relationships between these pathways and the machinery that mediates them.

GLOSSARY

Chromosome rearrangement Abnormal structural change occurs to native chromosome resulting in deletions, duplications, inversions, and translocations.

Heterozygosity A genotype where two different alleles of a gene are present at the same locus of homologous chromosome.

Homologous recombination A process in which two similar or identical fragments of DNA exchange their genetic location.

Mismatch repair A strand-specific process that can recognize and repair errors arising from DNA replication- and recombination-induced insertions, deletions and nucleotide misincorporation, and certain other types of DNA damage.

Nullizygous A genome type in which both alleles lose function for the same gene but due to different means of mutation.

Retrotransposon A DNA transposon element transposes itself through a retroviral-like mechanism, in which an RNA template is transcribed and then reverse transcribed into a new DNA element for insertion.

Satellite DNA Large arrays of tandem repeats, mostly enriched at centromeric regions.

Tandem repeats A DNA fragment that contains multiple and adjacent copies of a sequence of two or nucleotides.

Transposase An enzyme that binds to the end of a transposon, cuts and then transports the transposon element to a different genetic location.

LIST OF ACRONYMS AND ABBREVIATIONS

5-aza-2′-dC 5-aza-2′-deoxycytidine

5-caC 5-Carboxylcytosine

5-fC 5-Formylcytosine

5-hmC 5-Hydroxymethylcytosine

5-mC 5-Methylcytosine

ATM/ATR Ataxia telangiectasia mutated, a serine/threonine protein kinase; ataxia telangiectasia and Rad3-related protein, a serine/threonine-protein kinase

CHK1/CHK2 Checkpoint kinase 1/2, two serine/threonine-specific protein kinases

CIMP CpG island methylator phenotype

CRC Colorectal cancer

DNMT DNA methyltransferase

DSBs Double-strand breaks

ERCC1 Excision repair cross-complementation group

FMR1 Fragile X mental retardation one

FXS Fragile X syndrome

GBM Glioblastoma multiforme

H3K9me2/3 Di-/trimethylated histone H3-lysine nine

H3K27me3 Trimethylated histone H3-lysine 27
HDAC Histone deacetylase
HR Homologous recombination
IAP Intracisternal A-particle
ICF Immunodeficiency, centromeric instability, and facial anomalies
LINE Long interspersed nuclear element
LTR Long terminal repeat
MBD Methyl CpG-binding domain
MGMT O^6-alkylguanine DNA alkyltransferase
MLH1 MutL homolog 1
MMR DNA mismatch repair
MSI Microsatellite instability
PCNA Proliferating cell nuclear antigen
PMS2 Postmeiotic segregation increased 2
SAM S-adenosylmethionine
SINE Short interspersed nuclear element
TDG Thymine-DNA glycosylase
TEs Transposon elements
TET Ten–eleven translocation enzymes
TMZ Temozolomide
TNR Trinucleotide repeats
XPC Xeroderma pigmentosum, complementation group C

REFERENCES

[1] Leonhardt H, Page AW, Weier HU, Bestor TH. A targeting sequence directs DNA methyltransferase to sites of DNA replication in mammalian nuclei. Cell 1992;71(5):865–73.

[2] Liang G, Chan MF, Tomigahara Y, Tsai YC, Gonzales FA, Li E, et al. Cooperativity between DNA methyltransferases in the maintenance methylation of repetitive elements. Mol Cell Biol 2002;22(2):480–91.

[3] Huang K, Fan G. DNA methylation in cell differentiation and reprogramming: an emerging systematic view. Regen Med 2010;5(4):531–44.

[4] Suetake I, Shinozaki F, Miyagawa J, Takeshima H, Tajima S. DNMT3L stimulates the DNA methylation activity of Dnmt3a and Dnmt3b through a direct interaction. J Biol Chem 2004;279(26):27816–23.

[5] Okano M, Bell DW, Haber DA, Li E. DNA methyltransferases Dnmt3a and Dnmt3b are essential for de novo methylation and mammalian development. Cell 1999;99(3):247–57.

[6] Chen ZX, Riggs AD. DNA methylation and demethylation in mammals. J Biol Chem 2011;286(21):18347–53.

[7] Li D, Guo B, Wu H, Tan L, Lu Q. TET family of dioxygenases: crucial roles and underlying mechanisms. Cytogenet Genome Res 2015;146(3):171–80.

[8] Wu H, Zhang Y. Mechanisms and functions of Tet protein-mediated 5-methylcytosine oxidation. Genes Dev 2011;25(23):2436–52.

[9] Bachman M, Uribe-Lewis S, Yang X, Burgess HE, Iurlaro M, Reik W, et al. 5-Formylcytosine can be a stable DNA modification in mammals. Nat Chem Biol 2015;11(8):555–7.

[10] Bachman M, Uribe-Lewis S, Yang X, Williams M, Murrell A, Balasubramanian S. 5-Hydroxymethylcytosine is a predominantly stable DNA modification. Nat Chem 2014;6(12):1049–55.

[11] Sibani S, Melnyk S, Pogribny IP, Wang W, Hiou-Tim F, Deng L, et al. Studies of methionine cycle intermediates (SAM, SAH), DNA methylation and the impact of folate deficiency on tumor numbers in Min mice. Carcinogenesis 2002;23(1):61–5.

[12] Shorter KR, Felder MR, Vrana PB. Consequences of dietary methyl donor supplements: is more always better? Prog Biophys Mol Biol 2015;118(1–2):14–20.

[13] Yin R, Mao SQ, Zhao B, Chong Z, Yang Y, Zhao C, et al. Ascorbic acid enhances Tet-mediated 5-methylcytosine oxidation and promotes DNA demethylation in mammals. J Am Chem Soc 2013;135(28):10396–403.

[14] Esteller M. CpG island hypermethylation and tumor suppressor genes: a booming present, a brighter future. Oncogene 2002;21(35):5427–40.

[15] Kitange GJ, Carlson BL, Mladek AC, Decker PA, Schroeder MA, Wu W, et al. Evaluation of MGMT promoter methylation status and correlation with temozolomide response in orthotopic glioblastoma xenograft model. J Neurooncol 2009;92(1):23–31.

[16] Yang X, Han H, De Carvalho DD, Lay FD, Jones PA, Liang G. Gene body methylation can alter gene expression and is a therapeutic target in cancer. Cancer Cell 2014;26(4):577–90.

[17] Irizarry RA, Ladd-Acosta C, Wen B, Wu Z, Montano C, Onyango P, et al. The human colon cancer methylome shows similar hypo- and hypermethylation at conserved tissue-specific CpG island shores. Nat Genet 2009;41(2):178–86.

[18] Ahmed M, Liang P. Transposable elements are a significant contributor to tandem repeats in the human genome. Comp Funct Genomics 2012;2012:947089.

[19] Dudley KJ, Sloboda DM, Connor KL, Beltrand J, Vickers MH. Offspring of mothers fed a high fat diet display hepatic cell cycle inhibition and associated changes in gene expression and DNA methylation. PLoS One 2011;6(7):e21662.

[20] Walsh CP, Chaillet JR, Bestor TH. Transcription of IAP endogenous retroviruses is constrained by cytosine methylation. Nat Genet 1998;20(2):116–7.
[21] Bourc'his D, Bestor TH. Meiotic catastrophe and retrotransposon reactivation in male germ cells lacking Dnmt3L. Nature 2004;431(7004):96–9.
[22] Schueler MG, Sullivan BA. Structural and functional dynamics of human centromeric chromatin. Annu Rev Genomics Hum Genet 2006;7:301–13.
[23] Guo G, Wang W, Bradley A. Mismatch repair genes identified using genetic screens in Blm-deficient embryonic stem cells. Nature 2004;429(6994):891–5.
[24] Kim M, Trinh BN, Long TI, Oghamian S, Laird PW. Dnmt1 deficiency leads to enhanced microsatellite instability in mouse embryonic stem cells. Nucleic Acids Res 2004;32(19):5742–9.
[25] Wang KY, James Shen CK. DNA methyltransferase Dnmt1 and mismatch repair. Oncogene 2004;23(47):7898–902.
[26] Gorbunova V, Seluanov A, Mittelman D, Wilson JH. Genome-wide demethylation destabilizes CTG.CAG trinucleotide repeats in mammalian cells. Hum Mol Genet 2004;13(23):2979–89.
[27] Nichol Edamura K, Leonard MR, Pearson CE. Role of replication and CpG methylation in fragile X syndrome CGG deletions in primate cells. Am J Hum Genet 2005;76(2):302–11.
[28] Brock GJ, Anderson NH, Monckton DG. Cis-acting modifiers of expanded CAG/CTG triplet repeat expandability: associations with flanking GC content and proximity to CpG islands. Hum Mol Genet 1999;8(6):1061–7.
[29] Nichol K, Pearson CE. CpG methylation modifies the genetic stability of cloned repeat sequences. Genome Res 2002;12(8):1246–56.
[30] Landgrave-Gomez J, Mercado-Gomez O, Guevara-Guzman R. Epigenetic mechanisms in neurological and neurodegenerative diseases. Front Cell Neurosci 2015;9:58.
[31] Gos M. Epigenetic mechanisms of gene expression regulation in neurological diseases. Acta Neurobiol Exp (Wars) 2013;73(1):19–37.
[32] Reyniers E, Vits L, De Boulle K, Van Roy B, Van Velzen D, de Graaff E, et al. The full mutation in the FMR-1 gene of male fragile X patients is absent in their sperm. Nat Genet 1993;4(2):143–6.
[33] McMurray CT. Mechanisms of trinucleotide repeat instability during human development. Nat Rev Genet 2010;11(11):786–99.
[34] Walker FO. Huntington's disease. Lancet 2007;369(9557):218–28.
[35] Dion V, Lin Y, Hubert Jr L, Waterland RA, Wilson JH. Dnmt1 deficiency promotes CAG repeat expansion in the mouse germline. Hum Mol Genet 2008;17(9):1306–17.
[36] Lin Y, Wilson JH. Diverse effects of individual mismatch repair components on transcription-induced CAG repeat instability in human cells. DNA Repair (Amst) 2009;8(8):878–85.
[37] Robertson KD, Ait-Si-Ali S, Yokochi T, Wade PA, Jones PL, Wolffe AP. DNMT1 forms a complex with Rb, E2F1 and HDAC1 and represses transcription from E2F-responsive promoters. Nat Genet 2000;25(3):338–42.
[38] Nosho K, Shima K, Irahara N, Kure S, Firestein R, Baba Y, et al. SIRT1 histone deacetylase expression is associated with microsatellite instability and CpG island methylator phenotype in colorectal cancer. Mod Pathol 2009;22(7):922–32.
[39] Hsieh CL, Lieber MR. CpG methylated minichromosomes become inaccessible for V(D)J recombination after undergoing replication. EMBO J 1992;11(1):315–25.
[40] Chen RZ, Pettersson U, Beard C, Jackson-Grusby L, Jaenisch R. DNA hypomethylation leads to elevated mutation rates. Nature 1998;395(6697):89–93.
[41] Ji W, Hernandez R, Zhang XY, Qu GZ, Frady A, Varela M, et al. DNA demethylation and pericentromeric rearrangements of chromosome 1. Mutat Res 1997;379(1):33–41.
[42] Miniou P, Jeanpierre M, Blanquet V, Sibella V, Bonneau D, Herbelin C, et al. Abnormal methylation pattern in constitutive and facultative (X inactive chromosome) heterochromatin of ICF patients. Hum Mol Genet 1994;3(12):2093–102.
[43] Dominguez-Bendala J, McWhir J. Enhanced gene targeting frequency in ES cells with low genomic methylation levels. Transgenic Res 2004;13(1):69–74.
[44] Eden A, Gaudet F, Waghmare A, Jaenisch R. Chromosomal instability and tumors promoted by DNA hypomethylation. Science 2003;300(5618):455.
[45] Jaco I, Canela A, Vera E, Blasco MA. Centromere mitotic recombination in mammalian cells. J Cell Biol 2008;181(6):885–92.
[46] Mugal CF, Ellegren H. Substitution rate variation at human CpG sites correlates with non-CpG divergence, methylation level and GC content. Genome Biol 2011;12(6):R58.
[47] Xia J, Han L, Zhao Z. Investigating the relationship of DNA methylation with mutation rate and allele frequency in the human genome. BMC Genomics 2012;13(Suppl. 8):S7.
[48] Kim YJ, Wilson 3rd DM. Overview of base excision repair biochemistry. Curr Mol Pharmacol 2012;5(1):3–13.
[49] Dijk M, Typas D, Mullenders L, Pines A. Insight in the multilevel regulation of NER. Exp Cell Res 2014;329(1):116–23.
[50] Iyer RR, Pluciennik A, Burdett V, Modrich PL. DNA mismatch repair: functions and mechanisms. Chem Rev 2006;106(2):302–23.
[51] Caldecott KW. Mammalian single-strand break repair: mechanisms and links with chromatin. DNA Repair (Amst) 2007;6(4):443–53.
[52] Grundy GJ, Moulding HA, Caldecott KW, Rulten SL. One ring to bring them all–the role of Ku in mammalian non-homologous end joining. DNA Repair (Amst) 2014;17:30–8.
[53] Chuang LS, Ian HI, Koh TW, Ng HH, Xu G, Li BF. Human DNA-(cytosine-5) methyltransferase-PCNA complex as a target for p21WAF1. Science 1997;277(5334):1996–2000.
[54] Sharif J, Muto M, Takebayashi S, Suetake I, Iwamatsu A, Endo TA, et al. The SRA protein Np95 mediates epigenetic inheritance by recruiting Dnmt1 to methylated DNA. Nature 2007;450(7171):908–12.
[55] Liu X, Gao Q, Li P, Zhao Q, Zhang J, Li J, et al. UHRF1 targets DNMT1 for DNA methylation through cooperative binding of hemi-methylated DNA and methylated H3K9. Nat Commun 2013;4:1563.
[56] Esteve PO, Chin HG, Smallwood A, Feehery GR, Gangisetty O, Karpf AR, et al. Direct interaction between DNMT1 and G9a coordinates DNA and histone methylation during replication. Genes Dev 2006;20(22):3089–103.

[57] Unterberger A, Andrews SD, Weaver IC, Szyf M. DNA methyltransferase 1 knockdown activates a replication stress checkpoint. Mol Cell Biol 2006;26(20):7575–86.

[58] Milutinovic S, Zhuang Q, Niveleau A, Szyf M. Epigenomic stress response. Knockdown of DNA methyltransferase 1 triggers an intra-S-phase arrest of DNA replication and induction of stress response genes. J Biol Chem 2003;278(17):14985–95.

[59] Constantin N, Dzantiev L, Kadyrov FA, Modrich P. Human mismatch repair: reconstitution of a nick-directed bidirectional reaction. J Biol Chem 2005;280(48):39752–61.

[60] Jin B, Robertson KD. DNA methyltransferases, DNA damage repair, and cancer. Adv Exp Med Biol 2013;754:3–29.

[61] Pena-Diaz J, Jiricny J. Mammalian mismatch repair: error-free or error-prone? Trends Biochem Sci 2012;37(5):206–14.

[62] Hare JT, Taylor JH. One role for DNA methylation in vertebrate cells is strand discrimination in mismatch repair. Proc Natl Acad Sci USA 1985;82(21):7350–4.

[63] Mortusewicz O, Schermelleh L, Walter J, Cardoso MC, Leonhardt H. Recruitment of DNA methyltransferase I to DNA repair sites. Proc Natl Acad Sci USA 2005;102(25):8905–9.

[64] Ruzov A, Shorning B, Mortusewicz O, Dunican DS, Leonhardt H, Meehan RR. MBD4 and MLH1 are required for apoptotic induction in xDNMT1-depleted embryos. Development 2009;136(13):2277–86.

[65] Loughery JE, Dunne PD, O'Neill KM, Meehan RR, McDaid JR, Walsh CP. DNMT1 deficiency triggers mismatch repair defects in human cells through depletion of repair protein levels in a process involving the DNA damage response. Hum Mol Genet 2011;20(16):3241–55.

[66] Ding N, Bonham EM, Hannon BE, Amick TR, Baylin SB, O'Hagan HM. Mismatch repair proteins recruit DNA methyltransferase 1 to sites of oxidative DNA damage. J Mol Cell Biol 2015.

[67] Umar A, Buermeyer AB, Simon JA, Thomas DC, Clark AB, Liskay RM, et al. Requirement for PCNA in DNA mismatch repair at a step preceding DNA resynthesis. Cell 1996;87(1):65–73.

[68] Ha K, Lee GE, Palii SS, Brown KD, Takeda Y, Liu K, et al. Rapid and transient recruitment of DNMT1 to DNA double-strand breaks is mediated by its interaction with multiple components of the DNA damage response machinery. Hum Mol Genet 2011;20(1):126–40.

[69] Palii SS, Van Emburgh BO, Sankpal UT, Brown KD, Robertson KD. DNA methylation inhibitor 5-Aza-2′-deoxycytidine induces reversible genome-wide DNA damage that is distinctly influenced by DNA methyltransferases 1 and 3B. Mol Cell Biol 2008;28(2):752–71.

[70] Laghi L, Bianchi P, Malesci A. Differences and evolution of the methods for the assessment of microsatellite instability. Oncogene 2008; 27(49):6313–21.

[71] McConechy MK, Talhouk A, Li-Chang HH, Leung S, Huntsman DG, Gilks CB, et al. Detection of DNA mismatch repair (MMR) deficiencies by immunohistochemistry can effectively diagnose the microsatellite instability (MSI) phenotype in endometrial carcinomas. Gynecol Oncol 2015;137(2):306–10.

[72] Drablos F, Feyzi E, Aas PA, Vaagbo CB, Kavli B, Bratlie MS, et al. Alkylation damage in DNA and RNA–repair mechanisms and medical significance. DNA Repair (Amst) 2004;3(11):1389–407.

[73] Ellison KS, Dogliotti E, Connors TD, Basu AK, Essigmann JM. Site-specific mutagenesis by O^6-alkylguanines located in the chromosomes of mammalian cells: influence of the mammalian O^6-alkylguanine-DNA alkyltransferase. Proc Natl Acad Sci USA 1989;86(22):8620–4.

[74] Verbeek B, Southgate TD, Gilham DE, Margison GP. O^6-Methylguanine-DNA methyltransferase inactivation and chemotherapy. Br Med Bull 2008;85:17–33.

[75] Cuozzo C, Porcellini A, Angrisano T, Morano A, Lee B, Di Pardo A, et al. DNA damage, homology-directed repair, and DNA methylation. PLoS Genet 2007;3(7):e110.

[76] Morano A, Angrisano T, Russo G, Landi R, Pezone A, Bartollino S, et al. Targeted DNA methylation by homology-directed repair in mammalian cells. Transcription reshapes methylation on the repaired gene. Nucleic Acids Res 2014;42(2):804–21.

[77] O'Hagan HM, Mohammad HP, Baylin SB. Double strand breaks can initiate gene silencing and SIRT1-dependent onset of DNA methylation in an exogenous promoter CpG island. PLoS Genet 2008;4(8):e1000155.

[78] Li X, Yao X, Wang Y, Hu F, Wang F, Jiang L, et al. MLH1 promoter methylation frequency in colorectal cancer patients and related clinicopathological and molecular features. PLoS One 2013;8(3):e59064.

[79] Weisenberger DJ, Siegmund KD, Campan M, Young J, Long TI, Faasse MA, et al. CpG island methylator phenotype underlies sporadic microsatellite instability and is tightly associated with BRAF mutation in colorectal cancer. Nat Genet 2006;38(7):787–93.

[80] Shigeyasu K, Nagasaka T, Mori Y, Yokomichi N, Kawai T, Fuji T, et al. Clinical significance of MLH1 methylation and CpG island methylator phenotype as prognostic markers in patients with gastric Cancer. PLoS One 2015;10(6):e0130409.

[81] Lengauer C, Kinzler KW, Vogelstein B. Genetic instability in colorectal cancers. Nature 1997;386(6625):623–7.

[82] Shima K, Morikawa T, Baba Y, Nosho K, Suzuki M, Yamauchi M, et al. MGMT promoter methylation, loss of expression and prognosis in 855 colorectal cancers. Cancer Causes Control 2011;22(2):301–9.

[83] Kim YS, Kim SH, Cho J, Kim JW, Chang JH, Kim DS, et al. MGMT gene promoter methylation as a potent prognostic factor in glioblastoma treated with temozolomide-based chemoradiotherapy: a single-institution study. Int J Radiat Oncol Biol Phys 2012;84(3):661–7.

[84] Hasina R, Surati M, Kawada I, Arif Q, Carey GB, Kanteti R, et al. O-6-methylguanine-deoxyribonucleic acid methyltransferase methylation enhances response to temozolomide treatment in esophageal cancer. J Carcinog 2013;12:20.

[85] Kitange GJ, Carlson BL, Schroeder MA, Grogan PT, Lamont JD, Decker PA, et al. Induction of MGMT expression is associated with temozolomide resistance in glioblastoma xenografts. Neuro Oncol 2009;11(3):281–91.

[86] Peng B, Hurt EM, Hodge DR, Thomas SB, Farrar WL. DNA hypermethylation and partial gene silencing of human thymine- DNA glycosylase in multiple myeloma cell lines. Epigenetics 2006;1(3):138–45.

[87] Moynahan ME, Pierce AJ, Jasin M. BRCA2 is required for homology-directed repair of chromosomal breaks. Mol Cell 2001;7(2):263–72.

[88] Moynahan ME, Chiu JW, Koller BH, Jasin M. Brca1 controls homology-directed DNA repair. Mol Cell 1999;4(4):511–8.

[89] Xia F, Taghian DG, DeFrank JS, Zeng ZC, Willers H, Iliakis G, et al. Deficiency of human BRCA2 leads to impaired homologous recombination but maintains normal nonhomologous end joining. Proc Natl Acad Sci USA 2001;98(15):8644–9.
[90] Cox MM. Recombinational DNA repair in bacteria and the RecA protein. Prog Nucleic Acid Res Mol Biol 1999;63:311–66.
[91] Lee MN, Tseng RC, Hsu HS, Chen JY, Tzao C, Ho WL, et al. Epigenetic inactivation of the chromosomal stability control genes BRCA1, BRCA2, and XRCC5 in non-small cell lung cancer. Clin Cancer Res 2007;13(3):832–8.
[92] Lee JY, Lee TH. Effects of DNA methylation on the structure of nucleosomes. J Am Chem Soc 2012;134(1):173–5.
[93] Lee JY, Lee J, Yue H, Lee TH. Dynamics of nucleosome assembly and effects of DNA methylation. J Am Chem Soc 2015;290(7):4291–303.
[94] Mendonca A, Chang EH, Liu W, Yuan C. Hydroxymethylation of DNA influences nucleosomal conformation and stability in vitro. Biochim Biophys Acta 2014;1839(11):1323–9.
[95] Langecker M, Ivankin A, Carson S, Kinney SR, Simmel FC, Wanunu M. Nanopores suggest a negligible influence of CpG methylation on nucleosome packaging and stability. Nano Lett 2015;15(1):783–90.
[96] Tamaru H. Confining euchromatin/heterochromatin territory: jumonji crosses the line. Genes Dev 2010;24(14):1465–78.
[97] Lehnertz B, Ueda Y, Derijck AA, Braunschweig U, Perez-Burgos L, Kubicek S, et al. *Suv39h*-mediated histone H3 lysine 9 methylation directs DNA methylation to major satellite repeats at pericentric heterochromatin. Curr Biol 2003;13(14):1192–200.
[98] Rothbart SB, Krajewski K, Nady N, Tempel W, Xue S, Badeaux AI, et al. Association of UHRF1 with methylated H3K9 directs the maintenance of DNA methylation. Nat Struct Mol Biol 2012;19(11):1155–60.
[99] Easwaran HP, Schermelleh L, Leonhardt H, Cardoso MC. Replication-independent chromatin loading of Dnmt1 during G2 and M phases. EMBO Rep December 2004;5(12):1181–6.
[100] Saksouk N, Simboeck E, Dejardin J. Constitutive heterochromatin formation and transcription in mammals. Epigenetics Chromatin 2015;8:3.
[101] Honda S, Lewis ZA, Huarte M, Cho LY, David LL, Shi Y, et al. The DMM complex prevents spreading of DNA methylation from transposons to nearby genes in *Neurospora crassa*. Genes Dev 2010;24(5):443–54.
[102] Peters AH, O'Carroll D, Scherthan H, Mechtler K, Sauer S, Schofer C, et al. Loss of the Suv39h histone methyltransferases impairs mammalian heterochromatin and genome stability. Cell 2001;107(3):323–37.
[103] Wang D, Zhou J, Liu X, Lu D, Shen C, Du Y, et al. Methylation of SUV39H1 by SET7/9 results in heterochromatin relaxation and genome instability. Proc Natl Acad Sci USA 2013;110(14):5516–21.
[104] Saksouk N, Barth TK, Ziegler-Birling C, Olova N, Nowak A, Rey E, et al. Redundant mechanisms to form silent chromatin at pericentromeric regions rely on BEND3 and DNA methylation. Mol Cell 2014;56(4):580–94.
[105] Monier K, Mouradian S, Sullivan KF. DNA methylation promotes Aurora-B-driven phosphorylation of histone H3 in chromosomal subdomains. J Cell Sci 2007;120(Pt 1):101–14.
[106] Geiman TM, Sankpal UT, Robertson AK, Chen Y, Mazumdar M, Heale JT, et al. Isolation and characterization of a novel DNA methyltransferase complex linking DNMT3B with components of the mitotic chromosome condensation machinery. Nucleic Acids Res 2004;32(9):2716–29.
[107] Okada T, Ohzeki J, Nakano M, Yoda K, Brinkley WR, Larionov V, et al. CENP-B controls centromere formation depending on the chromatin context. Cell 2007;131(7):1287–300.
[108] Kwon MS, Hori T, Okada M, Fukagawa T. CENP-C is involved in chromosome segregation, mitotic checkpoint function, and kinetochore assembly. Mol Biol Cell 2007;18(6):2155–68.
[109] Gopalakrishnan S, Sullivan BA, Trazzi S, Della Valle G, Robertson KD. DNMT3B interacts with constitutive centromere protein CENP-C to modulate DNA methylation and the histone code at centromeric regions. Hum Mol Genet 2009;18(17):3178–93.
[110] Gonzalo S, Jaco I, Fraga MF, Chen T, Li E, Esteller M, et al. DNA methyltransferases control telomere length and telomere recombination in mammalian cells. Nat Cell Biol 2006;8(4):416–24.
[111] Buxton JL, Suderman M, Pappas JJ, Borghol N, McArdle W, Blakemore AI, et al. Human leukocyte telomere length is associated with DNA methylation levels in multiple subtelomeric and imprinted loci. Sci Rep 2014;4:4954.
[112] Dessain SK, Yu H, Reddel RR, Beijersbergen RL, Weinberg RA. Methylation of the human telomerase gene CpG island. Cancer Res 2000;60(3):537–41.
[113] Gadalla SM, Katki HA, Shebl FM, Giri N, Alter BP, Savage SA. The relationship between DNA methylation and telomere length in dyskeratosis congenita. Aging Cell 2012;11(1):24–8.
[114] Ng LJ, Cropley JE, Pickett HA, Reddel RR, Suter CM. Telomerase activity is associated with an increase in DNA methylation at the proximal subtelomere and a reduction in telomeric transcription. Nucleic Acids Res 2009;37(4):1152–9.
[115] Lee ME, Rha SY, Jeung HC, Chung HC, Oh BK. Subtelomeric DNA methylation and telomere length in human cancer cells. Cancer Lett 2009;281(1):82–91.
[116] Blanco-Betancourt CE, Moncla A, Milili M, Jiang YL, Viegas-Pequignot EM, Roquelaure B, et al. Defective B-cell-negative selection and terminal differentiation in the ICF syndrome. Blood 2004;103(7):2683–90.
[117] Thijssen PE, Ito Y, Grillo G, Wang J, Velasco G, Nitta H, et al. Mutations in CDCA7 and HELLS cause immunodeficiency-centromeric instability-facial anomalies syndrome. Nat Commun 2015;6:7870.
[118] Gisselsson D, Shao C, Tuck-Muller CM, Sogorovic S, Palsson E, Smeets D, et al. Interphase chromosomal abnormalities and mitotic missegregation of hypomethylated sequences in ICF syndrome cells. Chromosoma 2005;114(2):118–26.
[119] Li X, et al. Homologous recombination in DNA repair and DNA damage tolerance. Cell Res January 2008;18(1):99–113.

Chapter 25

Noncoding RNAs in Genome Integrity

I. Kovalchuk
University of Lethbridge, Lethbridge, AB, Canada

Chapter Outline

1. INTRODUCTION

Noncoding RNAs (ncRNA) are the RNA molecules that are not encoding for any protein and have diverse functions in the cell. There is a great variety of ncRNAs in terms of size, mechanisms of biogenesis, and functions. Among these ncRNAs, there are well-known ribosomal RNAs (rRNAs) and transfer RNAs (tRNAs) that are involved in the process of translation as well as small nuclear RNAs (smRNAs) and small nucleolar RNAs (snoRNAs) playing an essential role in the maturation of mRNA and rRNA [1]. Many other ncRNAs are involved in the regulation of gene expression at all levels, including chromatin, DNA, and RNA. Several ncRNAs are also implicated in the regulation of genome stability or/and DNA repair itself. Such regulation can be indirect (by targeting and changing the expression level of genes involved in DNA repair) or direct (by interfering or aiding the process of DNA repair in a direct manner). A direct role for ncRNAs in DNA repair is supported by the fact that several of them are being able to interact with DNA-repair proteins such as 53BP1 [2], BRCA1 [3], and Ku70 [4]. Moreover, RNA-binding proteins have been shown to be recruited to the site of DNA damage and influence the repair efficiency [5,6]. In this chapter, we describe the involvement of various ncRNAs in the regulation of genome integrity in various organisms, with more emphasis on eukaryotes.

2. TARGETING BACTERIOPHAGE GENOMES BY CRISPR/CAS9

In bacteria, the CRISPR (clustered regularly interspaced short palindromic repeats)/Cas9 (CRISPR-associated protein 9) system effectively protects bacteria from various bacteriophages by incorporating portions of the bacteriophage genome into the bacterial genome and then produces a single-guide RNA (sgRNA) from the newly incorporated regions. sgRNAs

Genome Stability. http://dx.doi.org/10.1016/B978-0-12-803309-8.00025-2

guide the cleavage of bacteriophage genomes in a sequence-specific manner. Details of the function of the CRISPR/Cas9 system can be found in Chapter 6. The ability of the CRISPR/Cas9 system to recognize a specific sequence and to generate a double-strand break (DSB) is nowadays used for high-precision genome editing (see Chapter 6 for details).

3. DNA ELIMINATION IN CILIATES

In ciliates, various classes of small and long ncRNAs known as scan RNAs (scnRNAs) are involved in the programmed DNA elimination and DNA rearrangement. One of the ncRNA classes, PIWI (P-element–induced wimpy testis)-interacting RNAs (piRNAs), is involved in somatic genome rearrangements in *Tetrahymena* and *Oxytricha*. In *Tetrahymena*, piRNAs arise from the germline and target germline-specific sequences of the developing somatic macronucleus for elimination. In contrast, piRNAs in *Oxytricha* stem from the parental somatic macronucleus; they direct the retention of somatic genes in the mature somatic macronucleus [7]. More detailed information about RNA-directed DNA elimination can be found in Chapter 7.

4. TELOMERASE RNA AND TELOMERE LENGTH

The number of cell divisions for a given cell is limited, and one of the limitation factors is the shortening of chromosome ends—telomeres. With each cell division, the chromosome ends get shorter due to the inability of the polymerase to replicate the leading strand from the first nucleotide (due to the use of RNA primer). The telomerase enzyme also known as telomere terminal transferase adds the missing DNA using telomerase RNA. More details about the role of telomerase RNA in the maintenance of genome stability can be found in Chapter 21.

5. ROLE OF MICRO-RNAS IN THE REGULATION OF DNA REPAIR AND GENOME STABILITY

5.1 A Brief Overview of Micro-RNA Biogenesis

Micro-RNAs or miRNAs are the most abundant and perhaps the most well-described class of ncRNAs generated in plants and animals. Although the mechanism of biogenesis differs in plants and animals, in general, miRNAs are typically transcribed by RNA polymerase II (Pol II) from the loci with a well-defined gene structure [8]. The initial mRNA transcript containing single or multiple miRNAs is 5′-capped and 3′-polyA-tailed; it is typically folded to form the single or multiple hairpin structures with an imperfect pairing. These initial primary transcripts are called pri-miRNA; they are processed to precursor miRNA (pre-miRNA) by RNAse III enzyme DROSHA in the nucleus and once more by DICER in the cytoplasm generating double-stranded RNA with 2-nt overhangs at the 3′-ends [9]. In humans, the DICER complex is associated with two different double-stranded RNA (dsRNA)-binding proteins, a protein activator of PKR (PACT), and a transactivation-response RNA-binding protein (TRBP). Lee et al. demonstrated that PACT together with DICER can inhibit the processing of pre-miRNA substrates [10]. They also showed that PACT and TRBP are nonredundant in the generation of miRNAs with different sizes, the so-called miRNA isoforms (isomiRs), which may have different targets. TRBP protein in humans is phosphorylated by the MAPK/ERK protein kinase and the phosphorylation of TRBP increases the stability of the DICER complex enhancing miRNA production (Fig. 25.1) [11]. In plants, a double cleavage event occurs in the nucleus with the help of a dicer-like 1 (DCL1) protein. In addition, in plants, the generated dsRNAs are methylated (2′-O-methylation) at the 3′-termini by HEN1 methyltransferase [8]. iRNAs regulate gene expression at the posttranscriptional level either by degrading target mRNAs in plants or by interfering with translation (polyA shortening, preventing ribosome loading, and so on, in animals). The degradation in plants requires a perfect homology between the designated miRNA and its target, whereas translation inhibition in animals relies on an imperfect homology, which allows that a single miRNA targets multiple mRNAs or a single mRNA is targeted by multiple miRNAs.

5.2 Indirect Impact of miRNAs on Genome Stability

Being able to regulate multiple independent mRNA targets, miRNAs undoubtedly can have a significant impact on the rate and efficiency of DNA-damage repair. A potential indirect role of miRNAs in DNA-damage response has been demonstrated by the observation that exposure to 2 Gy of radiation in human cells results in an overall decrease in the expression of miRNAs in the first 30 min of exposure, which is paralleled by an increase in the expression of their mRNA targets [12]. Exposure to UV has been shown to change the expression of several miRNAs in human fibroblasts at 4 h time point [13].

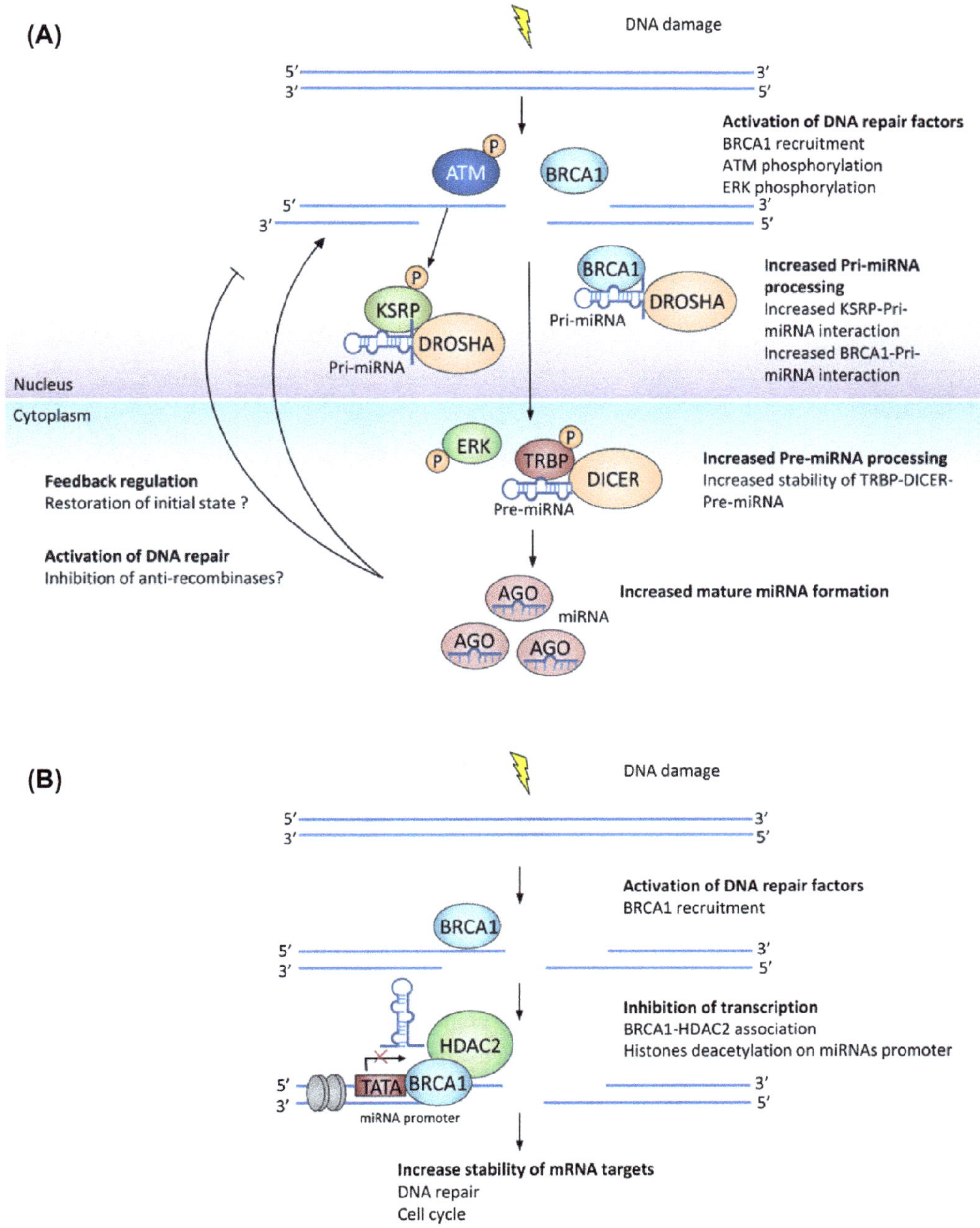

FIGURE 25.1 **Interplay of miRNAs and the DDR.** (A) *DNA damage–induced miRNA biogenesis.* DNA damage activates a signaling cascade which activates the processing of miRNA precursors. DNA damage–induced ATM phosphorylates KSRP and enhances its ability to recruit pri-miRNAs to DROSHA [16]. BRCA1 directly interacts with both pri-miRNAs and the DROSHA complex [18]. Processing by the DROSHA complex allows cytoplasmic export of pre-miRNAs. The MAPK ERK is also phosphorylated after DNA damage [92]. ERK phosphorylates TRBP and phospho-TRBP stabilizes the TRBP-DICER complex to promote pre-miRNAs processing in the cytoplasm [11]. Increased levels of mature miRNAs could play a role in the DNA-damage response by (1) decreasing the levels of anti-repair genes (such as the anti-recombinases, Srs2, PARI, RTEL1 [93]) and (2) down-regulate DDR proteins through a feedback regulation loop to restore pre-DNA-damage levels. (B) *DNA damage–induced repression of miRNA transcription.* BRCA1 associates with HDAC2 which deacetylates histone H2A and H3 on miR-155 promoter, leading to miR-155 transcriptional repression [19]. Transcriptional repression of miRNAs could contribute to the DDR by allowing increased expression of target proteins that are involved in DNA-repair and checkpoint control. (C) *miRNAs impacting DSB repair–pathway choice.* There is interplay of NHEJ and HR pathways during the course of the cell cycle, and this is critical for cell health. NHEJ is known to be active throughout the cell cycle phases and HR activity is maximum in S phase and gradually decreases over G2 phase [93,94]. In G1, NHEJ is promoted by 53BP1 and H2AX which prevent CtIP-mediated resection of the broken end [95]. Resection at a DSB impedes NHEJ and allows HR. In S-phase HR is active and BRCA1 is a key player in the recruitment of HR proteins to DSBs, thereby excluding NHEJ factors like 53BP1 [96]. CtIP promotes ends resection to allow formation of RPA-coated ssDNA at a DSB [97]. BRCA2 is the mediator protein that is essential for replacing RPA with RAD51 and the formation of the RAD51-ssDNA nucleoprotein filament. In normal cells, miRNAs maintain optimal expression of DNA-repair factors allowing efficient repair of DSBs. However, when miRNAs are aberrantly expressed it disrupts the correct choice of DSB-repair pathway. For example, overexpression of miRNAs targeting H2AX (indicated by bold arrow) may allow CtIP-mediated resection in G1 preventing NHEJ. HR-mediated repair in G1 is detrimental to cell health as it would lead to the loss of heterozygosity. Conversely, in S-phase overexpression of miRNAs targeting BRCA1 will impede HR and allow factors such as 53BP1 to direct the DSB to the NHEJ mediated–repair pathway, which in turn leads to higher mutation rates and chromosomal instability. *Reproduced from Chowdhury D, Choi YE, Brault ME. Charity begins at home: non-coding RNA functions in DNA repair. Nat Rev Mol Cell Biol 2013;14(3):181–9, with permission.*

Continued

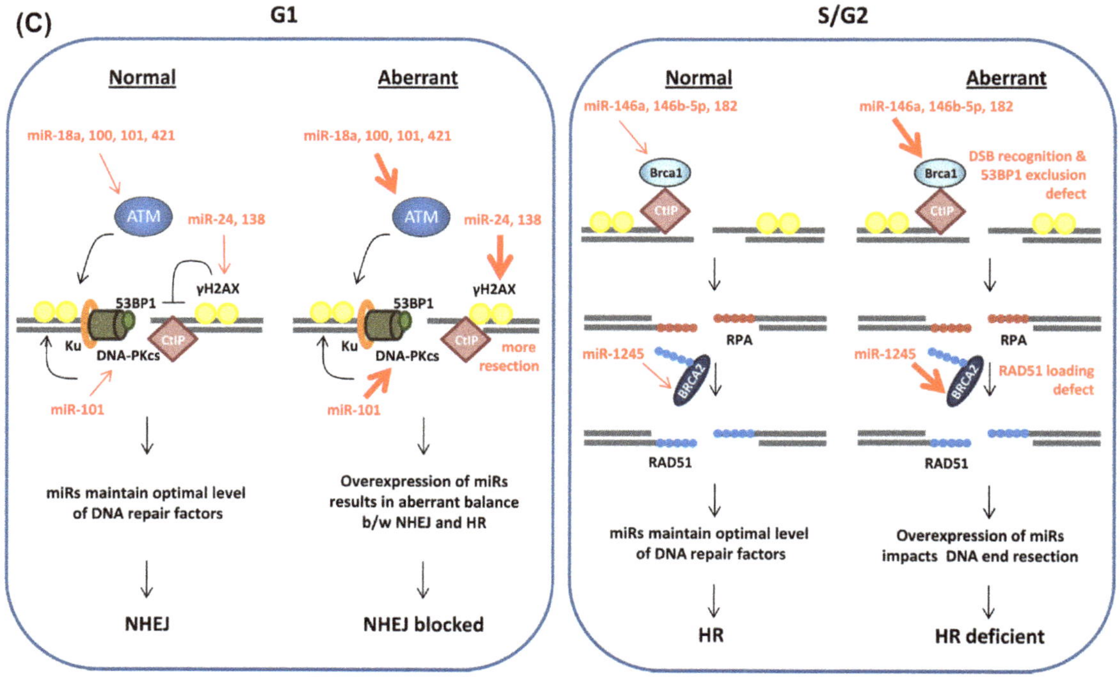

FIGURE 25.1 Cont'd

Similarly, 6 Gy of ionizing radiation (IR) result in an over threefold change in the expression of 22 miRNAs in prostate tumor lines [14]. Among 22 miRNAs, there was miR-521 that targets Cockayne syndrome protein A (CSA) involved in transcription-coupled repair [14].

5.3 DNA-Repair Factors Can Affect miRNA Biogenesis in Response to Stress

DNA-repair proteins can directly regulate biogenesis of miRNAs in response to IR (Fig. 25.1A) [15]. Specifically, in mice, the radiomimetic drug neocarzinostatin induces over 70 miRNAs in an ATM-dependent manner [16]. Biogenesis of miRNAs is dependent on the activity of KH-type splicing-regulatory protein (KSRP), one of the components of DROSHA and DICER complexes, and a direct substrate of ATM phosphorylation (Fig. 25.1A). KSRP phosphorylation significantly increases its activity by allowing a more efficient recruiting of pri-miRNAs to DROSHA for further processing in the nucleus [17]. It remains to be demonstrated whether the same occurs at the level of the pre-miRNA and Dicer complex in the cytoplasm. Therefore, it appears that ATM does not alter the transcription of miRNAs, but rather impacts their biogenesis in the step of conversion of pri-miRNAs.

Another DNA-repair protein that interacts with proteins processing miRNAs is BRCA1. BRCA1 binds DROSHA and several specific pri-miRNAs by regulating their biogenesis in a positive and negative manner (Fig. 25.1A) [18]. In addition, BRCA1 can regulate the expression of specific miRNAs at the level of transcription. For example, BRCA1 represses miR-155 transcription via its association with the histone deacetylase HDAC2 and the deacetylation of promoter region (Fig. 25.1B) [19]. Also, the anti-apoptotic transcription factor NF-Kβ is recruited to miR-21 promoter upon DNA damage, and via its interaction with the signal transducer and activator of transcription 3 (STAT3), it enhances the transcription of miR-21 [20].

5.4 Regulation of the Activity of DNA-Damage Sensors and Effectors by miRNAs

5.4.1 Regulation of Sensors

DNA strand breaks are sensed by several groups of proteins, such as the Mre11–Rad50–Nbs1 (MRN) complex, Ku70/80, and 53BP1. Proteins like ATM and γH2AX (the phosphorylated form of H2AX protein) also play an essential role in the initial damage recognition and signaling because H2AX is one of the first immediate targets of ATM phosphorylation. The repair choice is influenced by this initial binding (see Chapter 14). Therefore, the regulation of the abundance of one or several components of these sensors may significantly influence DNA-repair choice and outcomes.

Two component proteins involved in sensing strand breaks, Nbs1 and Ku80, may likely be regulated by miRNAs as they both contain the long 3′-UTRs with a high number of miRNA binding sites that can serve as a potential target for translation inhibition. Indeed, a 2015 work showed that Ku80 expression could indeed be affected by hsa–miR–526b in nonsmall-cell lung carcinoma (NSCLC) [21]. Hsa–miR–526b was found to be downregulated and Ku80 upregulated in the NSCLC cells compared to healthy tissues. No experimental data exist for Nbs1, but an association study demonstrated that NBS1 as well as Mre11 were likely to be regulated by miRNA; a case–control study revealed the association between the presence of SNPs in binding of several miRNAs at the 3′-UTR of these genes with an increased risk of breast cancer development [22]. Similar data for Nbs1 were observed in case-control studies involving colorectal cancer [23].

The expression of ATM is also regulated by miRNA at the posttranslational level; in neuroblastoma and HeLa cells, miR-421 downregulates ATM activity by modulating cell-cycle checkpoints and changing cell sensitivity to IR [24]. Similarly, miR-100 [25], miR-101 [26], and miR-18a [27] are also likely to regulate ATM because all of them were shown to target the 3′-UTR of ATM and downregulate it. Details of miRNA impact on various steps of DSB repair are shown in Fig. 25.1C [15].

5.4.2 Regulation of Effectors

The expression of effector proteins such as H2AX, BRCA1 and BRCA2, MDC1, RAD51, RAD52, and others, may also be affected by miRNAs. The formation of γH2AX loci, the result of the association of the phosphorylated H2AX histone with DSB lesion, may also be affected by miRNA expression. Several miRNAs that inhibit γH2AX foci formation have been identified [28]. Among them, there is miR-138 that appears to target 3′-UTR of H2AX mRNA by decreasing the number of the formed γH2AX foci and inducing chromosomal instability upon DNA damage [28]. In addition, miR-138 overexpression severely inhibits homologous recombination and increases cell sensitivity to various DNA-damaging agents, including cisplatin and IR.

BRCA1 is one of the essential DSB-repair proteins; its decreased expression is observed in more than 90% of all breast cancers [29]. Several miRNAs, including miR-182, miR-146a, and miR-146b-5p, are known to regulate BRCA1 expression. The overexpression of miR-182 alters homologous recombination and the sensitivity of breast tumors to DNA-damaging agents [30].

Effector proteins Rad51 and Rad52 are also likely to be regulated at the posttranscriptional level; an association study similar to the earlier mentioned one for Nbs1 and Mre11 revealed the correlation between the presence of SNPs in the predicted binding sites for several miRNAs and the chance to get breast cancer [22]. Also, miR-96 expressed from the same polycistronic transcript as miR-182 regulates the expression of RAD51 [31].

The DNA-dependent protein kinase catalytic subunit (DNA-PKcs) is recruited to the damaged DNA by the Ku heterodimer. DNA-PKcs and ATM1 are both targeted by the same miRNA—miR-101; the protein amounts were substantially reduced upon miR-101 overexpression sensitizing cancer cells to DNA damage [26].

More details about various miRNAs targeting the essential DNA-repair components can be found elsewhere [15,32].

6. THE ROLE OF PIWI-INTERACTING RNA IN THE MAINTENANCE OF GENOME STABILITY IN THE GERMLINE

The animal genome contains a great number of transposable elements, and many of them are active and expressed in somatic cells. Many such transposons are also expressed in germ cells where their excision or a "copy-and-paste" mechanism can result in genome instability and inheritance of an increased number of such elements. In *Drosophila*, for example, TAHRE, TART, HetA, copia, and the I element are expressed in the germline [33–36], whereas gypsy, ZAM, and idefix are expressed also in somatic cells of the ovary [37–39]. Therefore, controlling the activity of such elements is an important task.

An important discovery was made in 2006 by several independent groups—a specific class of small RNAs was found to be abundant in mouse testes [40–43]. These RNAs were named piRNAs and later on were proposed to control the genome stability in the animal's germline.

piRNAs were first discovered in *Drosophila* as ncRNAs that were involved in silencing transposable elements [44]. Later on, they have been found in multiple metazoan species, including worm, frog, zebrafish, mice, rats, and humans. piRNAs sequences are not conserved among different species and are different even within the same species. Moreover, piRNAs also differ from other small RNAs such as micro-RNAs (miRNAs) and small-interfering RNAs (siRNAs) because they do not have a double-stranded RNA precursor and are not processed by Dicer. In particular, in *Drosophila* and vertebrates, piRNAs are about 26–30-nt long, have a preference for a 5′-uracil, and are 2′-O-methylated at the 3′-(sugar)-end [45]. *Caenorhabditis elegans* piRNAs are rather similar, with the exception of the size—they are 21-nt in length.

6.1 piRNAs in *Drosophila*

PIWI protein was originally identified in *Drosophila* during the analysis of the results of an enhancer trap screen using a P-element. Male *Drosophila* mutants containing the P-element insertion at specific genomic locations in the germline cells had severe defects in spermatogenesis leading to sterility [46]. Further research led to the discovery of three PIWI proteins: PIWI, AUBERGINE (AUB), and ARGONAUTE3 (AGO3). The PIWI protein is localized in the nuclei of somatic follicle and germ cells in the ovary; it functions as a regulator of heterochromatic gene silencing in the transposon regions, and its function requires an interaction with heterochromatin protein 1a (HP1a) and methylation of the DNA at the target locus [47]. AUB protein is expressed in the cytoplasm of germ cells with a partial localization to the nuage, an electron-dense cytoplasmic region located around the nucleus that plays a prominent role in piRNA function. AUB protein is involved in silencing of the repetitive *Stellate* locus via sequence-specific antisense piRNAs [44]. Aub deficiency leads to a loss of anterior–posterior and dorsal–ventral patterning in embryos, which is likely the consequence of double-stranded DNA breaks occurring in the oocyte in the absence of Aub. Finally, AGO3 strictly localizes to the nuage region of the cytoplasm of germ cells [47].

In *Drosophila*, more than 80% of piRNA sequence reads map to the genomic regions harboring transposons [48]. As to the size, two sets of piRNAs are produced in *Drosophila*: 24–28 nt in length associated with AUB and AGO3 proteins and the 29–31 nt–long ones mostly associated with PIWI [49]. piRNAs in *Drosophila* derive from unidirectional or bidirectional clusters of repetitive elements located in the pericentromeric and telomeric regions where piRNAs are encoded by one or two strands, respectively [47]. In the case of bidirectional clusters, convergent transcription may result in stalling of polymerases at the site of conversion, thus contributing to the production of aberrant transcripts [50]. In *Drosophila*, transcripts formed by bidirectional transposon clusters are recognized by the HP1 homolog Rhino (Rhi) [51] and its colocalization partner UAP56 (Fig. 25.2A) [45]. During transcription, Rhi binds the chromatin repressive mark H3K9me3 on bidirectional clusters and recruits the protein deadlock and the transcription termination cofactor cutoff [45]. It is proposed that Cutoff binds the uncapped 5′-end of the piRNA precursor, preventing the degradation or/and splicing of the precursor [52]. In contrast, the transcription of unidirectional clusters occurs through the normal POL II transcription process involving the defined promoters, termination sequences, 5′-capping, and polyadenylation [52].

Both types of transcripts are likely transported through a nuclear pore with the aid of UAP56 [53], and then processed with the mitochondrial surface protein Zucchini (Zuc); Zuc likely trims the 5′- and possibly 3′-ends of the piRNA precursors, although very little detail is known about this process. Among other factors that seem to be involved in the process of biogenesis is CG2183 (Gasz); this protein colocalizes with Zuc and is believed to function as an adapter protein that recruits Piwi proteins to mitochondria for further piRNA maturation [54]. Next step of piRNA maturation requires 2′-O-methylation with the piRNA methyltransferase Pimet, the homolog of *Arabidopsis* methyltransferase HEN1 [55].

Mature piRNAs then bind to PIWI, AUB, and AGO3 proteins that initiate piRNA amplification via a cytoplasmic ping-pong mechanism. The participating proteins have a preference for a different set of small RNAs. While PIWI and AUB have a preference for transposon-derived antisense piRNAs with a 5′-uridine (U), AGO3 preferentially binds sense transposon piRNAs with no enrichment for 5′-U [47]. The binding of antisense piRNAs associated with AUB to a complementary transposon transcript (a length of 10 nt is typically sufficient) results in the endonucleolytic cleavage of the target between the 10th and 11th nt position of the piRNA. The product of this cleavage is a new piRNA with a 10 nt 5′-overlap in the antiparallel orientation (the sense strand) with the initial antisense piRNA; these new piRNAs have an adenosine residue at position 10 complementary to uridine at the initial antisense piRNA [45]. The 3′-end of a newly formed piRNAs is further processed, modified and picked up by AGO3. AGO3-associated piRNAs target primary piRNA cluster transcripts [47]. The cleaved piRNA transcript (antisense) is then picked up by the pair of PIWI/AUB proteins that further trim the transcript. These antisense piRNAs are identical to original antisense piRNAs produced by the initial binding and processing by PIWI/AUB. The complex of antisense piRNA/PIWI/AUB is then ready to enter a new ping-pong cycle or is available for the regulation of translation either in a negative manner by cleavage of target transposon transcripts (which is most common) or in a positive manner by the activation of transcription (which is much less frequent). The earlier mentioned ping-pong mechanism is also found in mice and zebrafish [56,57].

piRNAs in *Drosophila* function both in the cytoplasm and in the nucleus. Studies in *Drosophila* have revealed that the nuclear localization of the Piwi protein is essential for the establishment of chromosomal marks, whereas a slicer activity of the Piwi protein is not [58]. Moreover, loss of Piwi in *Drosophila* results in a substantial loss of the repressive H3K9me3 histone mark and an increased occupancy of POL II at transposable elements [59]. The overexpression of piRNAs against a specific genomic locus leads to the accumulation of the H3K9me3 mark, a decreased POL II occupancy and the recruitment of the heterochromatin protein HP1 [59]. Based on this information, it can be suggested that as a part of the piRNA-induced–silencing complex (pi-RISC), Piwi together with piRNAs translocate to the nucleus where they interact with DNA

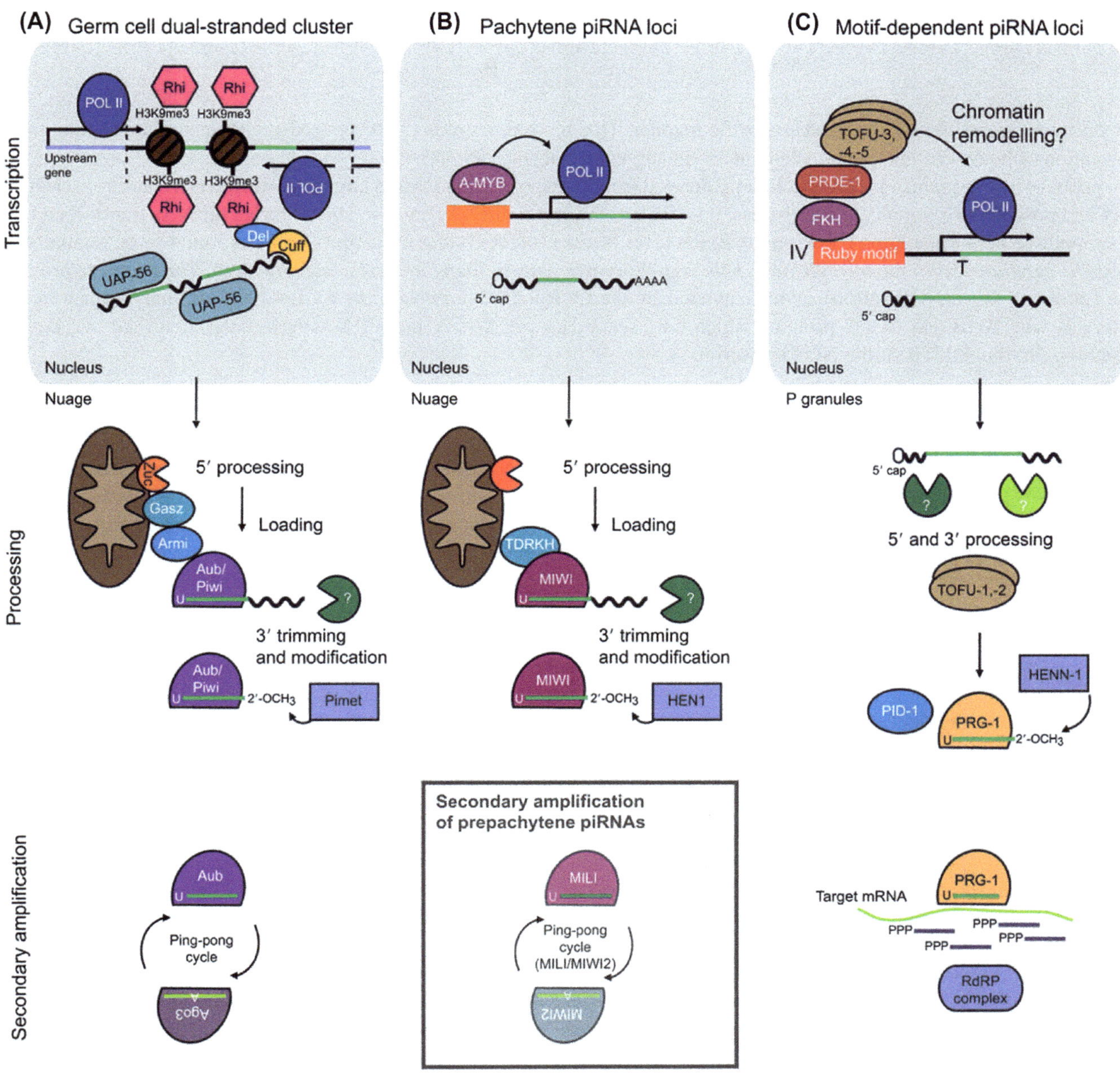

FIGURE 25.2 **Mechanisms of piRNA biogenesis in different organisms.** (A–C) Models of piRNA generation from dual-stranded clusters in *D. melanogaster* (A), from pachytene piRNA loci in *Mus musculus* (B), and from Ruby motif-containing loci in *C. elegans* (C). (A) Convergent transcription from neighboring genic loci generates piRNA precursors from dual-stranded clusters upon binding of the heterochromatin protein Rhino (Rhi) to H3K9me3 on cluster loci. Rhi in turn associates with Deadlock (Del) and Cutoff (Cuff), the latter of which is thought to protect the 5′-end of the noncanonical precursor transcript from degradation. Nuclear export of the piRNA, mediated by UAP-56, is followed by 5′-end processing, likely mediated by mitochondria-associated nuclease Zucchini (Zuc). Additional factors (eg, CG2183 (Gasz) and Armitage (Armi)) lead to Piwi protein recruitment, piRNA loading, and 3′-end processing, which likely involves an unknown trimmer activity as well as the action of methyltransferase Pimet. Extensive secondary piRNA amplification occurs via the ping-pong cycle, which takes place in *Drosophila* germ cells. (B) The transcription factor A-MYB (MYBL1) binds to a canonical promoter motif and initiates piRNA precursor transcription by POL II while simultaneously inducing expression of piRNA pathway genes (eg, Miwi and Mitopld). The precursor transcripts are 5′-capped and poly-A-tailed and, after export from the nucleus, processed by the murine homolog of Zuc, MITOPLD. Loading onto MIWI is likely followed by 3′-end trimming of the precursor and 2′-O-methylation by murine HEN1. For conceptual comparison of secondary amplification mechanisms in different organisms, the MILI-MIWI2 ping-pong cycle occurring only for prepachytene piRNAs is included here (inset). This process does not take place as part of the biogenesis of pachytene MIWI-bound piRNAs shown here. (C) The conserved Ruby motif is bound by Forkhead proteins (FKH) and possibly additional factors, and transcription of 5′-capped 28 or 29 nt precursors is initiated. Transcription and/or stability of these precursors depend on PRDE-1, TOFU-3, TOFU-4, and TOFU-5. After 5′- and 3′-end processing of the precursor, a process that may be mediated by TOFU-1 and TOFU-2, 2′-O-methylation of the 3′ end of the piRNA by HENN-1 takes place. PID-1 is another novel factor involved in piRNA biogenesis or stabilization, possibly acting at the same level as the *C. elegans* Piwi protein PRG-1. PRG-1:piRNA:target RNA interaction leads to the generation of secondary 22G-RNAs carrying a 5′-triphosphate (indicated as PPP) by a multi-protein machinery containing RNA-dependent RNA polymerases (RdRP). These small RNAs are incorporated into a secondary Argonaute and mediate target silencing. Question marks indicate unknown factors or functions; green lines represent piRNA sequences; and blue lines represent upstream sequences. Although the role of *D. melanogaster* UAP-56 in the targeting of piRNA precursors for nuclear export has been described, analogous mechanisms in mice and *C. elegans* have not yet been discovered; therefore, the sequence of events showing export of a long precursor from the nucleus to the cytoplasm in these organisms is speculative. Biogenesis of uni-stranded clusters in *D. melanogaster*, which occurs concomitantly with dual-strand cluster expression in germ cells and is the only mode of piRNA generation in somatic follicle cells, may be similar to canonical POL II transcription of protein-coding genes. It is currently less well studied and not depicted here. The same is the case for primary biogenesis of prepachytene piRNAs, which are expressed in the fetal germline in *M. musculus*, and for Ruby motif-independent piRNAs, which make up a small proportion of the overall piRNA population in *C. elegans* adults. Ago3, Argonaute3; Aub, Aubergine; RdRP, RNA-dependent RNA polymerase; TDRKH, a tudor domain protein. *Reproduced from Weick EM, Miska EA. piRNAs: from biogenesis to function. Development 2014;141(18):3458–71, with permission.*

or a primary transcript in a sequence-specific manner. This binding recruits other repressive elements, thus leading to the formation of heterochromatin and silencing of the target region (Fig. 25.3A).

piRNAs in *Drosophila* are predominantly derived from transposon regions and target transposon elements transcriptionally and posttranscriptionally. Nevertheless, it is likely that certain piRNAs are also able to target mRNAs transcribed from nonrepetitive genomic regions. For example, the level of the protein-coding transcript Fas3 was found to be regulated by piRNAs generated from the 3′-UTR [60]. This regulatory mechanism likely includes deadenylation. The evidence provides the following facts: NOS mRNA is deadenylated by the CCR4–NOT complex; the CCR4–NOT deadenylation complex interacts with AUB and AGO3 proteins which bind secondary piRNAs; some piRNAs have been identified that contain homology to the 3′-UTR of the NOS transcript.

6.2 piRNAs in Mammals

In contrast to *Drosophila*, in an adult mouse, about 93% of piRNA sequence reads are mapped to a single defined site in the genome. They appear to originate from unidirectional clusters; they are transcribed either from single strands or from two nonoverlapping strands. The chromatin immunoprecipitation analysis showed that piRNAs are derived from RNA Pol II transcripts (Fig. 25.2B) [48].

In mammals, piRNAs are also processed by the PIWI clade proteins: MIWI (PIWIL1), MILI (PWIL2), and MIWI2 (PWIL4). Each of these proteins has a very specific expression pattern during the development of male gametes. The expression of MILI starts at 12.5 days of embryonic development and persists into adulthood, whereas the expression of MIWI2 starts at about 14 days of embryonic development and persists until 3 days postpartum. Finally, the expression of MIWI does not start until 14 days postpartum, which coincides with the beginning of the pachytene stage of meiosis of male gamete development.

In accordance with PIWI protein expression, in germ cells of mice, two different populations of piRNAs have been developed, pre-pachytene and pachytene piRNAs. The expression of pre-pachytene piRNAs in early stages of spermatogenesis originates from transposon and gene-derived sequences; their expression coincides with the expression of two out of three PIWI family proteins—MIWI2 and MILI. MILI-bound piRNAs are 26 or 27 nt, whereas MIWI2-bound piRNAs are 28 nt. In contrast, pachytene piRNAs are 30 nt in length, and their production depends on MAEL and MIWI proteins [61]; they arise from intergenic loci unrelated to repeat sequences [48,62]. Pre-pachytene piRNAs are involved in silencing of transposon regions through de novo DNA methylation, thus contributing toward the stabilization of the genome.

The function of pachytene piRNAs was not clear for a long time. It was demonstrated in 2014 that pachytene piRNAs play a critical role during spermatogenesis by targeting and eliminating large amounts of mRNAs in spermatids. Together with MIWI and the catalytic subunit of the CCR4-CAF1-NOT deadenylation complex CAF1, pachytene piRNAs form pi-RISC. The function of pi-RISC is to deadenylate and degrade specific mRNA targets. A critical importance of MIWI and CAF1 has been demonstrated by knocking down these proteins in the elongating spermatids (ES); about 5000 genes are upregulated in cells with knockdown, with about 90% of them being the genes known to be regulated by MIWI and CAF1 [63]. The immunoprecipitation analysis has shown that over 60% of all mRNAs found in ES cells are physically associated with MIWI, likely as a part of pi-RISC. The fact that the majority of mRNAs physically associated with MIWI have a counterpart match in the form of a specific set of piRNAs is therefore not surprising. Moreover, the levels of these mRNAs inversely correlate with the expression of piRNAs with a sequence-specific match.

Similarly to piRNA-mediated silencing of transposons initially documented in *Drosophila melanogaster*, piRISC is also formed in mice by MILI and MIWI2 [64]. As MIWI2 localizes both in the cytoplasm and the nucleus, it has been proposed to bind secondary piRNAs in the nuage into the cytoplasm and shuttle them into the nucleus in the form of piRISC (Fig. 25.3B).

piRNAs are predominantly mapped to transposon regions, and the depletion of the PIWI protein results in a massive enrichment of transposon mRNAs. piRNAs are most abundant in germline tissues of animals, and they appear to be important for fertility; animals lacking Piwi exhibit various fertility defects [46,57,65]. At the same time, several studies demonstrated an essential role of piRNAs in various somatic tissues, especially during early embryogenesis. In mammals, an evidence for the existence of piRNA in somatic tissues is not yet certain, although the expression of the PIWI protein has been clearly demonstrated, and it has been shown to be even elevated in several human cancers (reviewed in [7]).

6.3 piRNAs in *C. elegans*

Unique features of piRNAs in *C. elegans* include the fact that piRNAs are 21-nt in size and that *C. elegans* has a single PIWI homolog—PRG-1. Because of their unique size and a bias in the location of U at the 5′-end, piRNAs in *C. elegans* are also referred to as 21U-RNAs. Curiously, these RNAs do not match to transposon elements. Most of the primary piRNA

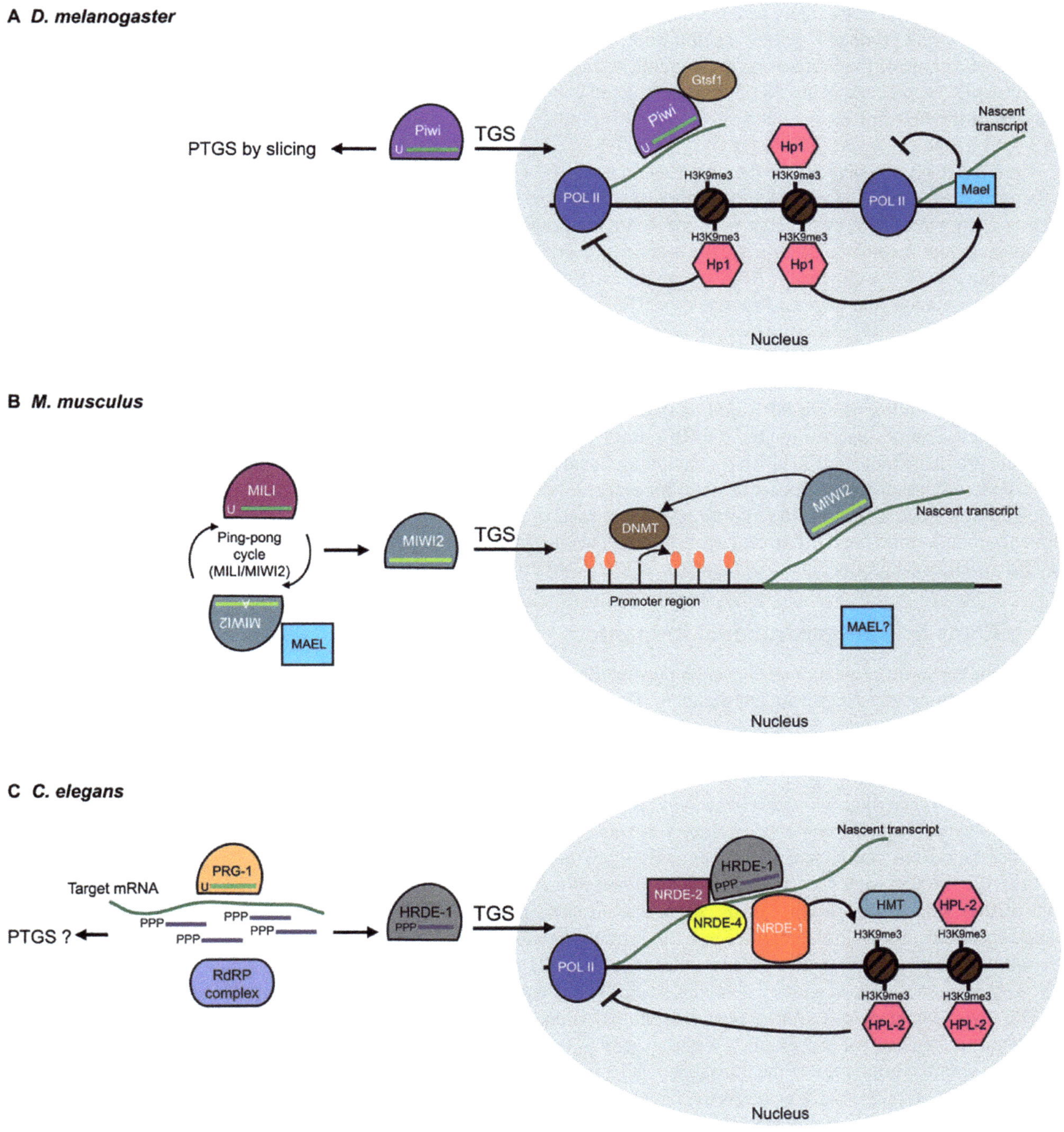

FIGURE 25.3 Mechanisms of piRNA-mediated transcriptional silencing. (A) In *D. melanogaster*, Piwi localizes to the nucleus and initiates repressive histone H3K9 trimethylation and RNA polymerase II stalling. Whether Piwi interacts with the nascent transcript or directly with DNA is not understood. The zinc-finger protein Gtsf1 likely directly interacts with Piwi, whereas the heterochromatin protein Hp1 binds to H3K9me3. Mael acts downstream of H3K9me3 methylation and is required for POL II repression; however, its mechanism of action also remains to be determined. In parallel to transcriptional gene silencing (TGS), posttranscriptional gene silencing (PTGS; ie, slicing) plays a well-defined role in *D. melanogaster* piRNA-mediated transposon silencing. (B) In *Mus musculus*, MIWI2 engages in the ping-pong cycle with MILI and translocates to the nucleus where it initiates CpG methylation of promoter elements upstream of transposon loci by DNA methyltransferase (DNMT) action. The murine MAEL homolog is found in the cytoplasm at MIWI2 sites; a role for this protein in the nucleus analogous to that described in *D. melanogaster* remains to be determined. Interaction of MIWI2 with the nascent transcript is speculative. (C) Secondary siRNAs are generated upon PRG-1:piRNA:target RNA interaction by RNA-dependent RNA polymerase (RdRP). These small RNAs are incorporated into the germline secondary Agonaute HRDE-1 which translocates to the nucleus to initiate H3K9me3 methylation and POL II stalling, likely by interacting with pre-mRNA and nuclear RNAi (NRDE) factors. For clarity, this model depicts establishment of repressive histone methylation marks by histone methyltransferases (HMT) as preceding POL II repression; the actual order of events remains to be determined experimentally and the reverse may also be the case. In *C. elegans*, target slicing is not essential for piRNA function; however, as of 2016, other mechanisms of PTGS have not been experimentally investigated. HPL, a H3K9me3-binding protein. *Reproduced from Weick EM, Miska EA. piRNAs: from biogenesis to function. Development 2014;141(18):3458–71, with permission.*

transcripts in *C. elegans* are likely generated by POL II from two major clusters containing over 16,000 genes. A large fraction of piRNAs are produced by transcription from specific promoters (motifs), since 2006 has been referred to as the Ruby motif (by the name of the first author of the publication describing such motifs) (Fig. 25.2C) [45,66].

These piRNA precursors are 28–29 nt in length and contain a 5′-cap. The processing of these precursors occurs through cleavage of the first 2 nt at the 5′-end and about 4–5 nt at the 3′-end [45]. The cleaved precursors are further methylated at the 3′-end by the *C. elegans* HEN1 ortholog HENN-1.

The PRG-1 protein may not be the only protein needed for biogenesis of piRNAs in nematodes since the *prg-1* mutant still contains low levels of piRNAs. Nevertheless, homologs of murine (or *Drosophila*) proteins involved in piRNA biogenesis are not yet found in *C. elegans*. Moreover, a ping-pong mechanism of the generation of secondary piRNAs is also not documented in nematodes. Instead, the secondary piRNAs (siRNAs) of 22 nt in size in worms, known as 22G-RNAs, are generated by RNA-dependent RNA polymerases (RdRPs) (Fig. 25.3C). The secondary siRNAs are picked up by the Argonaute homolog, the nuclear Argonaute NRDE-3 (which is RNAi deficient), and this complex (somewhat similar to pi-RISC in mice) shuttles into the nucleus where it is involved in transcriptional gene silencing. The complex binds pre-mRNAs in a sequence-specific manner and recruits two more proteins, NRDE-2 and NRDE-1. These proteins are involved in the establishment of the repressive H3K9me3 methylation mark that is further reinforced by the *C. elegans* homolog of the H3K9me3-binding protein HP1, HPL-2 [67,68] at the target site (Fig. 25.3C). The additional mechanism includes a direct suppression of transcription by POL II stalling during the elongation phase of transcription.

There are many targets of 21U-RNAs, including transposable elements and many protein-coding genes. Transcriptional gene silencing in nematodes occurs through imperfect binding to pre-mRNAs, allowing targeting of many different transcripts. Distinct classes of piRNAs may target different sets of proteins. piRNAs produced from promoters that lack specific sequence motifs (motif-less piRNAs) target mRNAs encoding immune-response genes. This suggests that there might be a specific mechanism producing a specific class of 21U-RNAs that regulate specific biological mechanisms.

6.4 piRNAs in Transgenerational Response

Several studies in different animals have demonstrated an interesting feature of piRNAs—the capacity to induce silencing across generations—transgenerational silencing. The first phenomenon of transgenerational transmission of information that later on was proved to have piRNAs as an essential component was described in *Drosophila*. Crosses between wild-caught males and laboratory-strain females resulted in infertile animals, whereas reciprocal crosses gave fertile flies. This phenomenon referred to as hybrid dysgenesis was believed to occur due to the presence of *P*-element or *I*-element transposons in wild animals. Since no sterility was observed in wild females, it was suggested that the suppressing factor was of a cytoplasmic origin and was transmitted by female gametes. It was later on demonstrated that the maternally deposited piRNAs provide the antisense piRNA component of the ping-pong loop, mounting an active defense response against transposons [69].

In 2012, piRNAs were shown to mediate the repression of the LacZ transgene cluster that has lasted for over 50 generations of flies without the presence of the allele that initiates silencing [70]. The trans-silencing effect (TSE) was achieved using repeat clusters of P-element-derived LacZ transgenes by exposing the nonsilencing cluster to the cytoplasm of a *Drosophila* strain carrying a cluster that exhibited strong TSE activities. Such type of silencing based on the transient interaction between the silenced and active alleles is similar to a paramutation phenomenon reported in plants and mice [71,72].

The involvement of piRNAs in transgenerational silencing has also been well documented in *C. elegans*. Feeding transgenic *C. elegans* that express the GFP transgene in the germline with *E. coli* that overexpress double-stranded RNAs (dsRNAs) with the homology to the transgene results in transgene silencing that has been observed in the germline for four consecutive generations [67]. Curious is the fact that once established, the transgenerational memory becomes independent of the piRNA trigger, although it still requires the nuclear RNAi/chromatin modification machinery described earlier. Reestablishing silencing in each generation requires piRNA transmission across generations via the germline followed by the amplification mechanism in somatic tissues.

In 2015, it was demonstrated that double-stranded RNA (dsRNA) generated in somatic tissues (neurons) of *C. elegans* can be transported to germline cells and trigger silencing of genes of matching sequence in germline cells [73]. Silencing within the germline can persist for over 25 generations.

Despite the similarity in the role of piRNAs in triggering transgenerational silencing in *Drosophila* and worms, the maintenance of silencing across generations may be different. Silencing in *Drosophila* is triggered by piRNAs deposited maternally and is maintained by using maternal piRNAs. In contrast, in worms, the maintenance phase can occur in both sexes. However, it should also be noted that all transgenerational effects of silencing were demonstrated by using transgenes and by exposing animals to silencer alleles. It remains to be shown whether the effects are equally strong in the naturally occurring silencing effects.

7. THE ROLE OF SMALL INTERFERING RNAs IN THE MAINTENANCE OF GENOME STABILITY

siRNAs, also referred to as short interfering RNAs, are a group of double-stranded ncRNAs that are usually perfectly complementary, 20–24 nt in size, have a 2-nt overhang, and contain the phosphorylated 5′-ends and the hydroxylated 3′-ends. They are typically processed from either larger endogenous double-stranded RNA molecules, hairpin RNAs, or exogenous (viral) double-stranded RNA molecules. siRNAs play the major role in transcriptional and posttranscriptional gene silencing via the establishment of a repressive chromatin state (via DNA methylation or repressive histone marks, predominantly in plants) or translational inhibition (mostly in animals) or transcript cleavage and degradation (mostly in plants).

The involvement of siRNAs in the RNA interference (RNAi) process has been first reported in plants as a phenomenon of cosuppression of both the integrated transgene and the endogenous homologous sequence [74]. Later on, RNAi has been documented in many different species in response to transgene overexpression, the activation of transposon elements or infection with viruses [75–77]. Therefore, RNAi is believed to be one of the ancient immune mechanisms of protection for the cell and genome, in particular against foreign genetic elements.

7.1 siRNAs in *Neurospora crassa*

In *Neurospora crassa*, siRNAs are involved in quelling, the posttranscriptional gene-silencing mechanism that occurs in response to the insertion of multiple copies of a transgene. Since transgenes often integrate as multiple, frequently truncated, and inverted copies, they are likely treated as repetitive transposable elements. *Neurospora* does not appear to have many copies of repetitive elements, and therefore RNAi against transgenes is considered to be a defense mechanism. Indeed, RNAi has also been shown to suppress the proliferation of multiple copies of LINE-like transposons [78]. *Neurospora* has several other RNAi-like genome defense mechanisms functioning during the sexual cycle. These are repeat-induced point mutation (RIP) and meiotic silencing by unpaired DNA (MSUD, also known as meiotic silencing) [79]. RIP causes C→A mutations in repetitive sequences, whereas MSUD triggers sequence-specific silencing when unpaired DNA is present and two haploid nuclei are fused into a diploid nucleus.

The quelling pathway requires the function of quelling deficient-1 (QDE-1), a protein with the activity of DNA-dependent RNA polymerase (DdRP) and RNA-dependent RNA polymerase (RdRP) [80] (Fig. 25.4). First, RNA Pol II transcribes complex transgenes producing aberrant RNAs. Then, QDE-1 converts aRNAs into dsRNA [80]. These dsRNAs are further processed by Dicer, and the resulting siRNAs of about 25 nt in size are loaded onto the Argonaute homolog QDE-2 to target homologous RNAs and induce posttranscriptional gene silencing in the form of RISC [81]. dsRNAs consist of two strands, the guide strand and passenger strand. While the former one guides the cleavage of mRNA targets, the latter one has to be degraded to activate RISC. The activation of RISC requires two steps: first, the QDE-2 protein nicks the passenger strand, and second, the QIP exonuclease degrades it. An active RISC consists of the guide siRNA strand and the QDE-2 protein.

7.1.1 qiRNA: DNA Damage–Induced siRNAs in Neurospora

DSB repair is a complex process that depends on the concerted action of various DNA-repair and chromatin-modifying proteins. Evidence suggests that the repair process also requires a new subclass of siRNAs. In *Neurospora*, they are produced in response to DNA damage; while studying the regulation of QDE-2, Lee et al. [82] observed that adding histidine rather than any other amino acids to the growth medium increased *qde-2* expression. Since histidine is known to induce DNA damage and mutation rates, it was hypothesized that DNA damage is likely the cause of the induction of *qde-2* expression. Indeed, treatment with other DNA-damaging agents like ethyl methanesulfonate, hydroxyurea, or methyl methanesulfonate also induced *qde-2* expression. The exact mechanism of qiRNA production is not clear, but it is possible that the presence of tandem repetitive DNA sequences within rDNA loci results in the formation of aberrant DNA structures that are recognized by the *recQ* DNA helicase QDE-3. However, a 2015 work demonstrated that the presence of a tandem repeat itself is not sufficient for the generation of qiRNAs and that a strand break in the region of tandem repeats is required for the initiation of aRNA production [83]. QDE-3 unwinds dsDNA to produce ssDNA, and RPA proteins stabilize the structure (Fig. 25.4). QDE-1 is then recruited to this complex to produce aRNA and convert it to dsRNA. RNA polymerase I (Pol I), which is normally required for the transcription from rDNA, is not needed for the generation of DNA damage–induced aRNAs because they are still produced in an RNA Pol I mutant [82]. Similarly, Pol II and III are also not needed for aRNA generation. These qiRNAs are shorter in size than those produced during quelling, they are about 20–21 nt in length and have a strong preference for uridine (93%) at the 5′-end and for A (49%) at the 3′-end. Most of the qiRNAs (86%) were stemming from the nucleolus organizer region (NOR) formed by about 200 copies of ribosomal DNA (rDNA) repeats. In addition, qiRNAs originate from the external and internal transcribed spacer regions of rDNA loci as well as from the intergenic

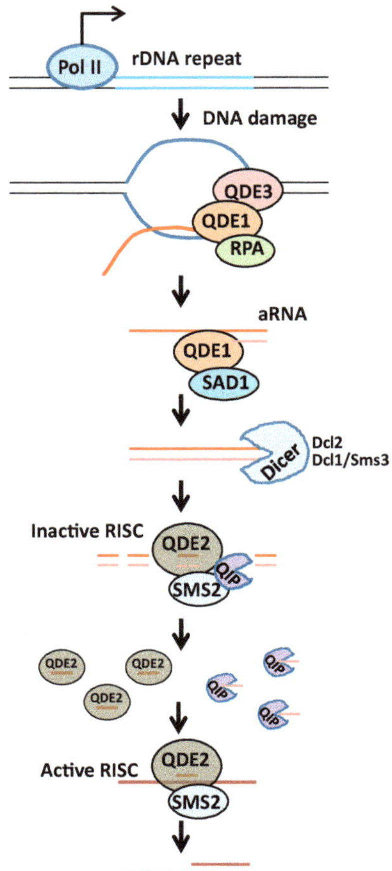

FIGURE 25.4 **Mechanisms of quelling and qiRNA biogenesis.** Mechanism of biogenesis of siRNAs in quelling and qiRNAs appears to be the same. Locus containing repetitive sequences of transgenes or transposons in quelling and rDNA in qiRNA generation is transcribed by QDE-1 protein, polymerase with high affinity to both DNA and RNA. QDE-3 is a helicase that unwinds dsDNA and stabilizes transcription. DNA damage is an essential component for generation of qiRNAs but not siRNAs in quelling. RPA protein likely aids QDE-1 in transcription by recruiting to the locus and by blocking the formation of DNA/RNA hybrids. QDE-1 then uses its RdRP activity to synthesize the second RNA strand using aRNA as a template. QDE-1 may be assisted by SAD-1 (suppressor of ascus dominance 1), a putative RdRP protein. Dicer protein (DCL2 or DCL1, although known as Sms3) dices dsRNA into small duplexes. RISC complexes consist of Argonaute homolog QDE-2, exonuclease QIP, and possibly Sms-2 (suppressor of meiotic silencing-2) protein, another Argonaute homolog, although it might only function in sexual stage of *Neurospora* development. RISC is activated by removing the passenger strand of the siRNA duplex by QDE-2 cleaving the passenger strand and QIP degrading the cleaved remnants. The guide strand is then loaded onto RISC and executes silencing by binding mRNA and cleaving it in a sequence-specific manner.

regions (6.57%), open reading frames (ORFs, 4.37%), and tRNAs (1.45%). Despite differences in size of siRNAs produced in quelling and upon DNA damage, it is believed that the mechanism of biogenesis is the same (Fig. 25.4).

The role of DNA damage–induced qiRNAs is not very clear, but it is possible that they inhibit rRNA biogenesis and/or the translation process after DNA damage. Indeed, protein synthesis is inhibited upon DNA damage in *Neurospora*, and this inhibition is partially removed in *qde-1* and *qde-3* mutants. The authors who discovered qiRNAs proposed that the production of qiRNAs is one of the mechanisms employed by fungi functioning similarly to DNA-damage checkpoints. Indeed, since the G2 phase requires intensive protein synthesis, its inhibition delays cell division, thus giving more time for DNA repair. Later on, it has been shown that qiRNA production requires homologous recombination (HR) [84]; the introduction of a site-specific DNA break initiates a homologous recombination event. HR at repetitive elements may generate recombination intermediates that are recognized by QDE-3. The QDE-3 helicase may function as a resolvase, by resolving the recombination intermediates into ssDNA. Therefore, it is possible that HR at the site of DNA damage functions to produce qiRNAs only from the repetitive DNA loci, thus distinguishing between genomic regions formed by transposon replication and other nonrepetitive genomic regions.

7.2 DNA Strand Break–Induced Small RNAs or diRNAs Are Involved in DSB Repair

DNA strand breaks—single stranded and double stranded (SSB and DSB, respectively)—occur as a result of oxidative damage due to the normal metabolic activity or as a result of a response to environmental pressures. Regardless of whether

DNA damage is direct or indirect, the outcome is the disruption of the nucleotide chain that has to be fixed. Two major repair pathways are competing for strand break nonhomologous end joining (NHEJ) and homologous recombination (HR). While the former one is a frequent but error-prone repair, the latter one is relatively rare and mostly error free. Cells utilize NHEJ in most of the cases, whereas they use HR mostly in the late S phase and G2 phase largely to avoid the loss of heterozygosity of a given deleterious mutation when sister chromatids are not available in the G1 phase. NHEJ plays a very important role in quiescent cells (G0) by taking care of most of the generated strand breaks.

The HR-repair pathway consists of several major steps. First, the MRN complex consisting of Mre11, Rad50, and Nbs1 proteins binds the region of strand breaks, and the broken DNA strands are unwound by helicases and resected by nucleases to expose long stretches of 3′-single-stranded DNA (ssDNA). Second, the exposed ssDNAs are covered by RPA. In parallel, the protein kinase ataxia telangiectasia (ATR) that is recruited to the region phosphorylates the proximal histone variant H2AX generating γH2AX; H2AX phosphorylation spreads distally to allow the exposure of a large stretch of DNA sequence [85]. Third, RAD51 displaces RPA from the ssDNA by forming long filaments. The RAD51-ssDNA filament searches for a homologous sequence template located typically on a sister chromatid or (less frequently) a homologous chromosome. The invading ssDNA forms a stretch of dsDNA with a template ssDNA, initiating DNA synthesis to replace the DNA surrounding the former strand break [85]. Fourth, the formed D-loop is resolved via the migration of double Holliday junction intermediate or by simple dissociation of one of the invading strands. Details of the type of HR repair and proteins involved can be found in Chapter 20.

The DNA damage–induced production of ncRNAs was reported in *Neurospora* (see qiRNAs in the preceding section) (Figs. 25.4 and 25.5). The DSB-induced production of ncRNAs has also been demonstrated in other species, including plants, flies, and mammals [86–88].

A class of 21 nt–long small RNAs induced by DSBs that originates in the proximity of the break (named DSB-induced small RNAs or diRNAs) was discovered in 2012 [86]. The authors used a special transgenic *Arabidopsis* recombination reporter line DGU.US consisting of the disrupted version of the GUS gene encoding the beta-glucuronidase enzyme. The transgene contained a recognition site for the rare cutter endonuclease I-SceI. When transgenic plants are crossed with plants carrying this endonuclease, a nick is generated in the transgene, creating a strand break, and the repair of this strand break results in the HR-dependent restoration of the transgene structure, which leads to its expression; its expression can be monitored via a histochemical assay that allows to score the number of recombination events [89]. The authors crossed these lines with various DCL mutants that are known to be involved in siRNA biogenesis (*dcl2*, *dcl3*, and *dcl4*) and demonstrated that the recombination frequency in mutants was decreased by 42%, 90%, and 44%, respectively. The results obtained allowed authors to hypothesize that siRNAs are likely involved in recombination repair. The Northern blot analysis with a probe spanning a region of about 450 nt flanking the I-SceI recognition site showed that indeed there were siRNAs produced from this region; siRNAs were significantly enhanced when the I-SceI endonuclease was introduced by crosses between plants overexpressing the endonuclease and plants containing the recognition site. The deep sequencing analysis confirmed the presence of these siRNAs and demonstrated that they were produced from both strands, thus allowing us to suggest that they were likely produced via dsRNA intermediates. The analysis of mutants showed that siRNAs were depleted in the *atr* mutant and all three DCL mutants [86]. In particular, the production of diRNAs was reduced by 98% in the *dcl3* mutant, which was consistent with the data demonstrating that in these plants, the HR frequency is reduced by about 90%. Furthermore, it was demonstrated that RdRPs (RDR2 and RDR6) and Pol IV (RNA-dependent DNA methylation or RdDM components, see [90]) were found to be involved in diRNA biogenesis; mutations in RDR2 and RDR6 caused 87% and 82% reductions in the levels of diRNAs, respectively. The analysis of other components of RdDM, including AGO4 and DRM2 (the de novo DNA methyltransferase), showed that they were not involved in the diRNA-mediated repair of DSBs. The authors also tested whether other components of RISC are involved in diRNA biogenesis/function. They found that AGO2 expression is induced by radiation; using immunoprecipitation followed by Northern blotting or deep sequencing, they demonstrated that radiation results in an increase in the number of diRNAs associated with AGO2 [86]. To further deduce the function of diRNAs, the authors analyzed whether diRNAs are required for the induction of H2AX phosphorylation and the formation of γH2AX foci; the analysis showed that diRNAs are likely not needed for these steps, and they are functioning downstream of γH2AX focus formation.

This work was extended from *Arabidopsis* to humans using the reporter line similar to the one used in *Arabidopsis*, but with GFP replacing GUS. The analysis of diRNA formation in response to radiation in transgenic human cell lines confirmed that they were induced, and their production was dependent on the presence of Dicer and Ago2 [86].

Finally, it was demonstrated that other components of diRNA biogenesis, such as Ago2, Dicer or Drosha/DGCR8, impaired HR to a similar degree as the knockdown of Rad51 [91]. They also showed that another DSB-repair pathway, NHEJ, did not benefit from diRNAs and was not regulated by them; the NHEJ activity was not altered in Dicer and Ago2 mutants. Their work showed that Ago2 formed a complex with Rad51 and that Rad51 accumulation at DSB loci was

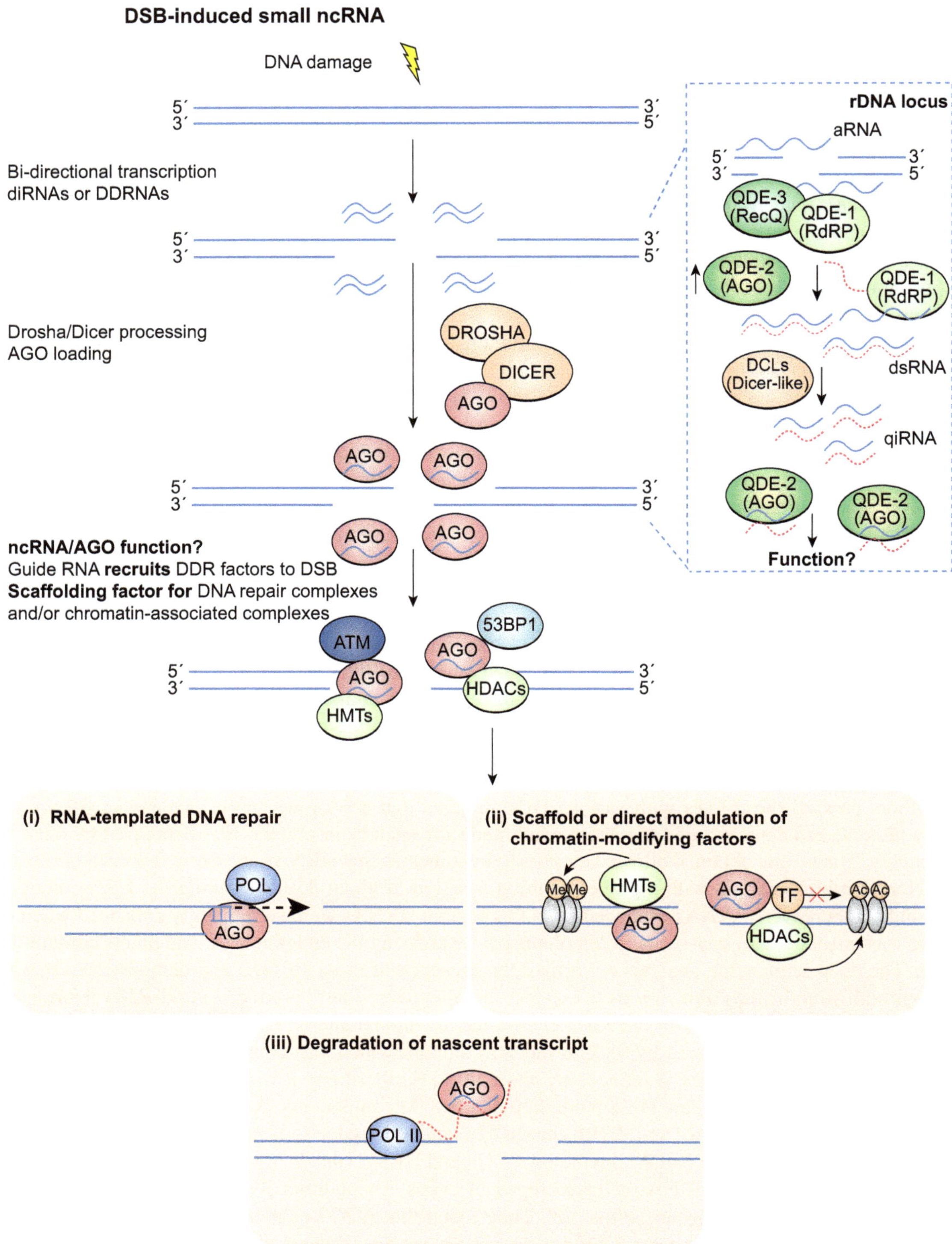

FIGURE 25.5 Other noncoding RNAs (diRNAs, qiRNAs) in DSB repair. DSBs can trigger the production of short ncRNA at the site of the DNA lesion. The inset is a schematic of the DNA damage–induced RdRP-dependent (QDE1) production of qiRNAs in *Neurospora crassa* [82]. qiRNAs were identified in complex with QDE2 (AGO homolog), but their role in DDR remains unknown. In higher eukaryotes lacking RdRPs, it is postulated that antisense transcripts lead to the formation of dsRNAs that are processed by the miRNA biogenesis machinery (DROSHA/DICER). The AGO-bound ncRNAs localizes at the DSB, and potentially facilitates the recruitment of DDR factors (53BP1, *p*-ATM) to the DSB site. It is feasible that the homology of the ncRNA to sequences proximal to the DSB allows it to serve as a "guide" for recruiting chromatin-modifying proteins/DDR factors to the DSB [98]. Alternatively, the ncRNA/AGO complex may also serve as a stable scaffold for maintaining the DNA-repair foci and facilitating the process of repair [99]. After the recruitment of the DNA-repair machinery to the DSB sites the precise role of the short ncRNAs is not known. It is speculated that the short ncRNAs could serve as RNA template to fill-in resected DNA during homologous recombination mediated DSB repair. Induced RNAs might play a role in mediating chromatin silenced state at the breaks either by recruiting associated chromatin factors at the breaks [100], or directly by interacting and modulating chromatin-associated factor in cis [101]. The short ncRNA could also act in the conventional siRNA pathway to degrade nascent RNA at the breaks in order to prevent deregulated expression of compromised genes. *Reproduced from Chowdhury D, Choi YE, Brault ME. Charity begins at home: non-coding RNA functions in DNA repair. Nat Rev Mol Cell Biol 2013;14(3):181–9, with permission.*

dependent on the catalytic activity and small RNA-binding capability of Ago2. Other steps of DSB repair, such as DSB resection, RPA and Mre11 loading, were not affected by Ago2 or Dicer depletion. Importantly, the authors demonstrated a direct role of diRNAs in HR repair; HR repair could be restored in Dicer-depleted cells by adding ncRNAs purified from human cells transfected with the I-SceI enzyme [91].

Francia et al. [87] demonstrated a similar diRNA activity in humans, mice, and zebrafish. They induced a strand break at a specific genomic site and by sequencing ncRNAs, they found ncRNA enrichment from the vicinity of the break; these ncRNAs were named DDRNAs. It was further shown that the formation of DDR foci depended on the production of DDRNAs, and DICER and DROSHA were the required components. Similar to the work by Gao et al. [91], a work by Francia et al. [87] showed that DDRNAs, either synthesized chemically or generated in vitro by DICER cleavage, were sufficient to restore the DNA-repair activity in the absence of other cellular RNAs.

Finally, Michalik et al. [88] showed a similar generation of diRNAs in *Drosophila*. They found that the structure of DSB ends (either blunt or with overhangs) did not influence diRNA production. The authors have reported an interesting novel finding. First, they have found that diRNAs are formed more actively when breaks occur in active transcribed genes. Second, they have shown that a DSB within a transcribed gene initiates transcription away from the break site [88]. The formation of such antisense transcripts results in a quick and efficient formation of dsRNA, the activation of RNAs, and a more efficient repair process.

Taking into consideration all the earlier mentioned results, a model for diRNA generation (qiRNAs, DDRNAs) can be proposed. Sequence-specific diRNAs are produced from the regions with DSBs. They are mainly processed by DCL3 (in cooperation with DCL2 and DCL4 in plants and Dicers in animals) from dsRNA intermediates, are picked up by AGO2 (in plants), and are incorporated into diRISC. Being incorporated in diRISC, diRNAs guide Ago2 to promote Rad51 recruitment and/or retention at DSBs to facilitate repair by HR [15].

8. CONCLUSION

Here, we provided a relatively short summary of ncRNAs that are directly or indirectly involved in the regulation of genome stability and in DNA repair itself. The role of ncRNAs is essential in prokaryotes where they are involved in the regulation of transcription, DNA repair, and chromatin accessibility to potential DNA lesions as well as in ncRNA-mediated degradation of foreign genomes (phages or other bacteria). It is even more critical in eukaryotes where they are involved in the targeted DNA elimination, the regulation of transposon activity, the regulation of transcription and chromatin structure, the regulation of the activity of DNA-repair components, and their recruitment to DNA lesions, and likely in many other processes. Many questions remain unanswered. It will have to be demonstrated that diRNAs, qiRNAs, and other damage-induced ncRNAs are actually actively involved in DNA repair, rather than being the byproduct of DNA damage. It will be important to establish whether there are ncRNAs that are produced specifically in response to a certain type of DNA damage, or whether they influence (aid) a certain type of DNA repair (although it was shown that diRNAs do not influence NHEJ but rather HR repair). It will also be important to develop an in vitro system that allows to study the kinetics of biogenesis of these ncRNAs and their role in the repair of certain type of lesions. Finally, it is important to demonstrate similarities and differences in the generation of various types of ncRNAs in different organisms.

GLOSSARY

Bidirectional clusters Genetic clusters representing genes (transgenes, transposons, and so on) located on both strands of DNA.

Convergent transcription The simultaneous transcription of bidirectional clusters containing overlapping gene sequences.

diRNAs DSB-induced siRNAs produced in the vicinity of DNA damage; observed in plants and animals.

Hybrid dysgenesis High rates of mutations in germline cells of *Drosophila* obtained upon crossing males carrying autonomous P elements with females without these elements.

MSUD Meiotic silencing by unpaired DNA, also known as meiotic silencing.

Nuage A perinuclear organelle, the site of piRNA processing in *Drosophila*.

Ping-pong mechanism The mechanism of generation of piRNAs using transposon-derived RNAs as a template.

qiRNAs DSB-induced siRNAs typically derived from rDNA loci in the *Neurospora* genome.

Quelling The mechanism of posttranscriptional gene silencing observed in *Neurospora* in response to transgene integration in a clustered repetitive manner.

Transgenerational memory A memory of exposure to certain environmental stimuli, manifesting itself in the form of changes in DNA methylation, histone modifications, DNA-repair efficiency, tolerance to stress, changes in transcriptomes, and so on.

Transgenerational silencing Silencing of transgenes or endogenes observed in the progeny in response to stimuli that occurred in parents.

LIST OF ABBREVIATIONS

CRISPR/Cas9 Clustered regularly interspaced short palindromic repeats/CRISPER-associated protein 9
CSA Cockayne syndrome protein A
DdRP DNA-dependent RNA polymerase
diRNA DSB-induced small RNAs
DNA-PKcs DNA-dependent protein kinase catalytic subunit
DSB double-strand break
isomiRs miRNA isoforms
miRNA Micro-RNA
MRN Mre11–Rad50–Nbs1
ncRNA noncoding RNA
PACT protein activator of PKR
piRNAs (P-element-induced wimpy testis)-interacting RNAs
pre-miRNA Precursor miRNA
PTGS Posttranscriptional gene silencing
qiRNA QDE-2-interacting small RNAs
RdDM RNA-dependent DNA methylation
RdRP RNA-dependent RNA polymerase
RNAi RNA interference
RPA Replication protein A
rRNA Ribosomal RNA
scnRNAs Scan RNAs
sgRNA Single-guide RNA
siRNA Small interfering RNA
smRNA Small nuclear RNA
snoRNA Small nucleolar RNAs (snoRNAs)
SSB Single-strand break
STAT3 Signal transducer and activator of transcription 3
TGS Transcriptional gene silencing
TRBP Trans-activation response RNA-binding protein
tRNA Transfer RNA
TSE Transsilencing effect
Zuc Zucchini

REFERENCES

[1] Kowalczyk MS, Higgs DR, Gingeras TR. Molecular biology: RNA discrimination. Nature 2012;482(7385):310–1.
[2] Pryde F, Khalili S, Robertson K, Selfridge J, Ritchie AM, Melton DW, et al. 53BP1 exchanges slowly at the sites of DNA damage and appears to require RNA for its association with chromatin. J Cell Sci 2005;118(Pt. 9):2043–55.
[3] Ganesan S, Silver DP, Greenberg RA, Avni D, Drapkin R, Miron A, et al. BRCA1 supports XIST RNA concentration on the inactive X chromosome. Cell 2002;111(3):393–405.
[4] Yoo S, Dynan WS. Characterization of the RNA binding properties of Ku protein. Biochemistry 1998;37(5):1336–43.
[5] Adamson B, Smogorzewska A, Sigoillot FD, King RW, Elledge SJ. A genome-wide homologous recombination screen identifies the RNA-binding protein RBMX as a component of the DNA-damage response. Nat Cell Biol 2012;14(3):318–28.
[6] Polo SE, Blackford AN, Chapman JR, Baskcomb L, Gravel S, Rusch A, et al. Regulation of DNA-end resection by hnRNPU-like proteins promotes DNA double-strand break signaling and repair. Mol Cell 2012;45(4):505–16.
[7] Ross RJ, Weiner MM, Lin H. PIWI proteins and PIWI-interacting RNAs in the soma. Nature 2014;505(7483):353–9.
[8] Millar AA, Waterhouse PM. Plant and animal microRNAs: similarities and differences. Funct Integr Genomics 2005;5(3):129–35.
[9] Ha M, Kim VN. Regulation of microRNA biogenesis. Nat Rev Mol Cell Biol 2014;15(8):509–24.
[10] Lee HY, Zhou K, Smith AM, Noland CL, Doudna JA. Differential roles of human dicer-binding proteins TRBP and PACT in small RNA processing. Nucleic Acids Res 2013;41(13):6568–76.
[11] Paroo Z, Ye X, Chen S, Liu Q. Phosphorylation of the human microRNA-generating complex mediates MAPK/Erk signaling. Cell 2009;139(1):112–22.
[12] Maes OC, An J, Sarojini H, Wu H, Wang E. Changes in MicroRNA expression patterns in human fibroblasts after low-LET radiation. J Cell Biochem 2008;105(3):824–34.
[13] Pothof J, Verkaik NS, van IW, Wiemer EA, Ta VT, van der Horst GT, et al. MicroRNA-mediated gene silencing modulates the UV-induced DNA-damage response. EMBO J 2009;28(14):2090–9.
[14] Josson S, Sung SY, Lao K, Chung LW, Johnstone PA. Radiation modulation of microRNA in prostate cancer cell lines. Prostate 2008;68(15):1599–606.

[15] Chowdhury D, Choi YE, Brault ME. Charity begins at home: non-coding RNA functions in DNA repair. Nat Rev Mol Cell Biol 2013;14(3):181–9.
[16] Zhang X, Wan G, Berger FG, He X, Lu X. The ATM kinase induces microRNA biogenesis in the DNA damage response. Mol Cell 2011;41(4): 371–83.
[17] Trabucchi M, Briata P, Garcia-Mayoral M, Haase AD, Filipowicz W, Ramos A, et al. The RNA-binding protein KSRP promotes the biogenesis of a subset of microRNAs. Nature 2009;459(7249):1010–4.
[18] Kawai S, Amano A. BRCA1 regulates microRNA biogenesis via the DROSHA microprocessor complex. J Cell Biol 2012;197(2):201–8.
[19] Chang S, Wang RH, Akagi K, Kim KA, Martin BK, Cavallone L, et al. Tumor suppressor BRCA1 epigenetically controls oncogenic microRNA-155. Nat Med 2011;17(10):1275–82.
[20] Niu J, Shi Y, Tan G, Yang CH, Fan M, Pfeffer LM, et al. DNA damage induces NF-kappaB-dependent microRNA-21 up-regulation and promotes breast cancer cell invasion. J Biol Chem 2012;287(26):21783–95.
[21] Zhang ZY, Fu SL, Xu SQ, Zhou X, Liu XS, Xu YJ, et al. By downregulating Ku80, hsa-miR-526b suppresses non-small cell lung cancer. Oncotarget 2015;6(3):1462–77.
[22] Wu Z, Wang P, Song C, Wang K, Yan R, Li J, et al. Evaluation of miRNA-binding-site SNPs of MRE11A, NBS1, RAD51 and RAD52 involved in HRR pathway genes and risk of breast cancer in China. Mol Genet Genomics 2015;290(3):1141–53.
[23] Li JT, Zhong BY, Xu HH, Qiao SY, Wang G, Huang J, et al. Associations between NBS1 Polymorphisms and Colorectal Cancer in Chinese Population. PLoS One 2015;10(7):e0132332.
[24] Hu H, Du L, Nagabayashi G, Seeger RC, Gatti RA. ATM is down-regulated by N-Myc-regulated microRNA-421. Proc Natl Acad Sci USA 2010;107(4):1506–11.
[25] Ng WL, Yan D, Zhang X, Mo YY, Wang Y. Over-expression of miR-100 is responsible for the low-expression of ATM in the human glioma cell line: M059J. DNA Repair (Amst) 2010;9(11):1170–5.
[26] Yan D, Ng WL, Zhang X, Wang P, Zhang Z, Mo YY, et al. Targeting DNA-PKcs and ATM with miR-101 sensitizes tumors to radiation. PLoS One 2010;5(7):e11397.
[27] Song L, Lin C, Wu Z, Gong H, Zeng Y, Wu J, et al. miR-18a impairs DNA damage response through downregulation of ataxia telangiectasia mutated (ATM) kinase. PLoS One 2011;6(9):e25454.
[28] Wang Y, Huang JW, Li M, Cavenee WK, Mitchell PS, Zhou X, et al. MicroRNA-138 modulates DNA damage response by repressing histone H2AX expression. Mol Cancer Res 2011;9(8):1100–11.
[29] Mueller CR, Roskelley CD. Regulation of BRCA1 expression and its relationship to sporadic breast cancer. Breast Cancer Res 2003;5(1):45–52.
[30] Moskwa P, Buffa FM, Pan Y, Panchakshari R, Gottipati P, Muschel RJ, et al. MiR-182-mediated downregulation of BRCA1 impacts DNA repair and sensitivity to PARP inhibitors. Mol Cell 2011;41(2):210–20.
[31] Wang Y, Huang JW, Calses P, Kemp CJ, Taniguchi T. MiR-96 downregulates REV1 and RAD51 to promote cellular sensitivity to cisplatin and PARP inhibition. Cancer Res 2012;72(16):4037–46.
[32] Tessitore A, Cicciarelli G, Del Vecchio F, Gaggiano A, Verzella D, Fischietti M, et al. MicroRNAs in the DNA damage/repair network and cancer. Int J Genomics 2014;2014:820248.
[33] Brennecke J, Aravin AA, Stark A, Dus M, Kellis M, Sachidanandam R, et al. Discrete small RNA-generating loci as master regulators of transposon activity in *Drosophila*. Cell 2007;128(6):1089–103.
[34] Chambeyron S, Popkova A, Payen-Groschene G, Brun C, Laouini D, Pelisson A, et al. piRNA-mediated nuclear accumulation of retrotransposon transcripts in the *Drosophila* female germline. Proc Natl Acad Sci USA 2008;105(39):14964–9.
[35] Shpiz S, Kwon D, Uneva A, Kim M, Klenov M, Rozovsky Y, et al. Characterization of *Drosophila* telomeric retroelement TAHRE: transcription, transpositions, and RNAi-based regulation of expression. Mol Biol Evol 2007;24(11):2535–45.
[36] Malone CD, Brennecke J, Dus M, Stark A, McCombie WR, Sachidanandam R, et al. Specialized piRNA pathways act in germline and somatic tissues of the *Drosophila* ovary. Cell 2009;137(3):522–35.
[37] Desset S, Buchon N, Meignin C, Coiffet M, Vaury C. In *Drosophila melanogaster* the COM locus directs the somatic silencing of two retrotransposons through both Piwi-dependent and -independent pathways. PLoS One 2008;3(2):e1526.
[38] Mevel-Ninio M, Pelisson A, Kinder J, Campos AR, Bucheton A. The flamenco locus controls the gypsy and ZAM retroviruses and is required for *Drosophila* oogenesis. Genetics 2007;175(4):1615–24.
[39] Pelisson A, Payen-Groschene G, Terzian C, Bucheton A. Restrictive flamenco alleles are maintained in *Drosophila melanogaster* population cages, despite the absence of their endogenous gypsy retroviral targets. Mol Biol Evol 2007;24(2):498–504.
[40] Aravin A, Gaidatzis D, Pfeffer S, Lagos-Quintana M, Landgraf P, Iovino N, et al. A novel class of small RNAs bind to MILI protein in mouse testes. Nature 2006;442(7099):203–7.
[41] Girard A, Sachidanandam R, Hannon GJ, Carmell MA. A germline-specific class of small RNAs binds mammalian Piwi proteins. Nature 2006;442(7099):199–202.
[42] Grivna ST, Beyret E, Wang Z, Lin H. A novel class of small RNAs in mouse spermatogenic cells. Genes Dev 2006;20(13):1709–14.
[43] Watanabe T, Takeda A, Tsukiyama T, Mise K, Okuno T, Sasaki H, et al. Identification and characterization of two novel classes of small RNAs in the mouse germline: retrotransposon-derived siRNAs in oocytes and germline small RNAs in testes. Genes Dev 2006;20(13):1732–43.
[44] Aravin AA, Naumova NM, Tulin AV, Vagin VV, Rozovsky YM, Gvozdev VA. Double-stranded RNA-mediated silencing of genomic tandem repeats and transposable elements in the *D. melanogaster* germline. Curr Biol 2001;11(13):1017–27.
[45] Weick EM, Miska EA. piRNAs: from biogenesis to function. Development 2014;141(18):3458–71.
[46] Lin H, Spradling AC. A novel group of pumilio mutations affects the asymmetric division of germline stem cells in the *Drosophila* ovary. Development 1997;124(12):2463–76.

[47] Brower-Toland B, Findley SD, Jiang L, Liu L, Yin H, Dus M, et al. *Drosophila* PIWI associates with chromatin and interacts directly with HP1a. Genes Dev 2007;21(18):2300–11.
[48] Li XZ, Roy CK, Moore MJ, Zamore PD. Defining piRNA primary transcripts. Cell Cycle 2013;12(11):1657–8.
[49] Kim VN, Han J, Siomi MC. Biogenesis of small RNAs in animals. Nat Rev Mol Cell Biol 2009;10(2):126–39.
[50] Hobson DJ, Wei W, Steinmetz LM, Svejstrup JQ. RNA polymerase II collision interrupts convergent transcription. Mol Cell 2012;48(3):365–74.
[51] Klattenhoff C, Xi H, Li C, Lee S, Xu J, Khurana JS, et al. The *Drosophila* HP1 homolog Rhino is required for transposon silencing and piRNA production by dual-strand clusters. Cell 2009;138(6):1137–49.
[52] Mohn F, Sienski G, Handler D, Brennecke J. The rhino-deadlock-cutoff complex licenses noncanonical transcription of dual-strand piRNA clusters in *Drosophila*. Cell 2014;157(6):1364–79.
[53] Zhang F, Wang J, Xu J, Zhang Z, Koppetsch BS, Schultz N, et al. UAP56 couples piRNA clusters to the perinuclear transposon silencing machinery. Cell 2012;151(4):871–84.
[54] Czech B, Preall JB, McGinn J, Hannon GJ. A transcriptome-wide RNAi screen in the *Drosophila* ovary reveals factors of the germline piRNA pathway. Mol Cell 2013;50(5):749–61.
[55] Saito K, Sakaguchi Y, Suzuki T, Suzuki T, Siomi H, Siomi MC. Pimet, the *Drosophila* homolog of HEN1, mediates 2′-O-methylation of Piwi-interacting RNAs at their 3′ ends. Genes Dev 2007;21(13):1603–8.
[56] Aravin AA, Hannon GJ, Brennecke J. The Piwi-piRNA pathway provides an adaptive defense in the transposon arms race. Science 2007;318(5851):761–4.
[57] Houwing S, Kamminga LM, Berezikov E, Cronembold D, Girard A, van den Elst H, et al. A role for Piwi and piRNAs in germ cell maintenance and transposon silencing in Zebrafish. Cell 2007;129(1):69–82.
[58] Klenov MS, Sokolova OA, Yakushev EY, Stolyarenko AD, Mikhaleva EA, Lavrov SA, et al. Separation of stem cell maintenance and transposon silencing functions of Piwi protein. Proc Natl Acad Sci USA 2011;108(46):18760–5.
[59] Le Thomas A, Rogers AK, Webster A, Marinov GK, Liao SE, Perkins EM, et al. Piwi induces piRNA-guided transcriptional silencing and establishment of a repressive chromatin state. Genes Dev 2013;27(4):390–9.
[60] Saito K, Inagaki S, Mituyama T, Kawamura Y, Ono Y, Sakota E, et al. A regulatory circuit for piwi by the large Maf gene traffic jam in *Drosophila*. Nature 2009;461(7268):1296–9.
[61] Castaneda J, Genzor P, van der Heijden GW, Sarkeshik A, Yates 3rd JR, Ingolia NT, et al. Reduced pachytene piRNAs and translation underlie spermiogenic arrest in maelstrom mutant mice. EMBO J 2014;33(18):1999–2019.
[62] Aravin AA, Sachidanandam R, Bourc'his D, Schaefer C, Pezic D, Toth KF, et al. A piRNA pathway primed by individual transposons is linked to de novo DNA methylation in mice. Mol Cell 2008;31(6):785–99.
[63] Gou LT, Dai P, Yang JH, Xue Y, Hu YP, Zhou Y, et al. Pachytene piRNAs instruct massive mRNA elimination during late spermiogenesis. Cell Res 2014;24(6):680–700.
[64] Kuramochi-Miyagawa S, Watanabe T, Gotoh K, Totoki Y, Toyoda A, Ikawa M, et al. DNA methylation of retrotransposon genes is regulated by Piwi family members MILI and MIWI2 in murine fetal testes. Genes Dev 2008;22(7):908–17.
[65] Carmell MA, Girard A, van de Kant HJ, Bourc'his D, Bestor TH, de Rooij DG, et al. MIWI2 is essential for spermatogenesis and repression of transposons in the mouse male germline. Dev Cell 2007;12(4):503–14.
[66] Ruby JG, Jan C, Player C, Axtell MJ, Lee W, Nusbaum C, et al. Large-scale sequencing reveals 21U-RNAs and additional microRNAs and endogenous siRNAs in *C. elegans*. Cell 2006;127(6):1193–207.
[67] Ashe A, Sapetschnig A, Weick EM, Mitchell J, Bagijn MP, Cording AC, et al. piRNAs can trigger a multigenerational epigenetic memory in the germline of *C. elegans*. Cell 2012;150(1):88–99.
[68] Shirayama M, Seth M, Lee HC, Gu W, Ishidate T, Conte Jr D, et al. piRNAs initiate an epigenetic memory of nonself RNA in the *C. elegans* germline. Cell 2012;150(1):65–77.
[69] Brennecke J, Malone CD, Aravin AA, Sachidanandam R, Stark A, Hannon GJ. An epigenetic role for maternally inherited piRNAs in transposon silencing. Science 2008;322(5906):1387–92.
[70] de Vanssay A, Bouge AL, Boivin A, Hermant C, Teysset L, Delmarre V, et al. Paramutation in *Drosophila* linked to emergence of a piRNA-producing locus. Nature 2012;490(7418):112–5.
[71] Chandler V, Alleman M. Paramutation: epigenetic instructions passed across generations. Genetics 2008;178(4):1839–44.
[72] Rassoulzadegan M, Grandjean V, Gounon P, Vincent S, Gillot I, Cuzin F. RNA-mediated non-mendelian inheritance of an epigenetic change in the mouse. Nature 2006;441(7092):469–74.
[73] Devanapally S, Ravikumar S, Jose AM. Double-stranded RNA made in *C. elegans* neurons can enter the germline and cause transgenerational gene silencing. Proc Natl Acad Sci USA 2015;112(7):2133–8.
[74] Napoli C, Lemieux C, Jorgensen R. Introduction of a chimeric chalcone synthase gene into petunia results in reversible co-suppression of homologous genes in trans. Plant Cell 1990;2(4):279–89.
[75] Sijen T, Plasterk RH. Transposon silencing in the *Caenorhabditis elegans* germ line by natural RNAi. Nature 2003;426(6964):310–4.
[76] Siomi MC, Saito K, Siomi H. How selfish retrotransposons are silenced in *Drosophila* germline and somatic cells. FEBS Lett 2008;582(17):2473–8.
[77] Wang X, Hsueh YP, Li W, Floyd A, Skalsky R, Heitman J. Sex-induced silencing defends the genome of *Cryptococcus neoformans* via RNAi. Genes Dev 2010;24(22):2566–82.
[78] Nolan T, Braccini L, Azzalin G, De Toni A, Macino G, Cogoni C. The post-transcriptional gene silencing machinery functions independently of DNA methylation to repress a LINE1-like retrotransposon in *Neurospora crassa*. Nucleic Acids Res 2005;33(5):1564–73.

[79] Dang Y, Yang Q, Xue Z, Liu Y. RNA interference in fungi: pathways, functions, and applications. Eukaryot Cell 2011;10(9):1148–55.

[80] Lee HC, Aalto AP, Yang Q, Chang SS, Huang G, Fisher D, et al. The DNA/RNA-dependent RNA polymerase QDE-1 generates aberrant RNA and dsRNA for RNAi in a process requiring replication protein A and a DNA. Helicase PLoS Biol 2010;8(10).

[81] Catalanotto C, Pallotta M, ReFalo P, Sachs MS, Vayssie L, Macino G, et al. Redundancy of the two dicer genes in transgene-induced posttranscriptional gene silencing in *Neurospora crassa*. Mol Cell Biol 2004;24(6):2536–45.

[82] Lee HC, Chang SS, Choudhary S, Aalto AP, Maiti M, Bamford DH, et al. qiRNA is a new type of small interfering RNA induced by DNA damage. Nature 2009;459(7244):274–7.

[83] Yang Q, Ye QA, Liu Y. Mechanism of siRNA production from repetitive DNA. Genes Dev 2015;29(5):526–37.

[84] Zhang Z, Chang SS, Zhang Z, Xue Z, Zhang H, Li S, et al. Homologous recombination as a mechanism to recognize repetitive DNA sequences in an RNAi pathway. Genes Dev 2013;27(2):145–50.

[85] Yamanaka S, Siomi H. diRNA-Ago2-RAD51 complexes at double-strand break sites. Cell Res 2014;24(5):511–2.

[86] Wei W, Ba Z, Gao M, Wu Y, Ma Y, Amiard S, et al. A role for small RNAs in DNA double-strand break repair. Cell 2012;149(1):101–12.

[87] Francia S, Michelini F, Saxena A, Tang D, de Hoon M, Anelli V, et al. Site-specific DICER and DROSHA RNA products control the DNA-damage response. Nature 2012;488(7410):231–5.

[88] Michalik KM, Bottcher R, Forstemann K. A small RNA response at DNA ends in *Drosophila*. Nucleic Acids Res 2012;40(19):9596–603.

[89] Orel N, Kyryk A, Puchta H. Different pathways of homologous recombination are used for the repair of double-strand breaks within tandemly arranged sequences in the plant genome. Plant J 2003;35(5):604–12.

[90] Herr AJ, Jensen MB, Dalmay T, Baulcombe DC. RNA polymerase IV directs silencing of endogenous DNA. Science 2005;308(5718):118–20.

[91] Gao M, Wei W, Li MM, Wu YS, Ba Z, Jin KX, et al. Ago2 facilitates Rad51 recruitment and DNA double-strand break repair by homologous recombination. Cell Res 2014;24(5):532–41.

[92] Tang D, Wu D, Hirao A, Lahti JM, Liu L, Mazza B, et al. ERK activation mediates cell cycle arrest and apoptosis after DNA damage independently of p53. J Biol Chem 2002;277(15):12710–7.

[93] Chapman JR, Taylor MR, Boulton SJ. Playing the end game: DNA double-strand break repair pathway choice. Mol Cell 2012;47(4):497–510.

[94] Karanam K, Kafri R, Loewer A, Lahav G. Quantitative live cell imaging reveals a gradual shift between DNA repair mechanisms and a maximal use of HR in mid S phase. Mol Cell 2012;47(2):320–9.

[95] Helmink BA, Tubbs AT, Dorsett Y, Bednarski JJ, Walker LM, Feng Z, et al. H2AX prevents CtIP-mediated DNA end resection and aberrant repair in G1-phase lymphocytes. Nature 2011;469(7329):245–9.

[96] Bunting SF, Callen E, Wong N, Chen HT, Polato F, Gunn A, et al. 53BP1 inhibits homologous recombination in Brca1-deficient cells by blocking resection of DNA breaks. Cell 2010;141(2):243–54.

[97] Yun MH, Hiom K. CtIP-BRCA1 modulates the choice of DNA double-strand-break repair pathway throughout the cell cycle. Nature 2009;459(7245):460–3.

[98] Verdel A, Jia S, Gerber S, Sugiyama T, Gygi S, Grewal SI, et al. RNAi-mediated targeting of heterochromatin by the RITS complex. Science 2004;303(5658):672–6.

[99] Tattermusch A, Brockdorff N. A scaffold for X chromosome inactivation. Hum Genet 2011;130(2):247–53.

[100] Grosshans H, Johnson T, Reinert KL, Gerstein M, Slack FJ. The temporal patterning microRNA let-7 regulates several transcription factors at the larval to adult transition in *C. elegans*. Dev Cell 2005;8(3):321–30.

[101] Wang X, Arai S, Song X, Reichart D, Du K, Pascual G, et al. Induced ncRNAs allosterically modify RNA-binding proteins in cis to inhibit transcription. Nature 2008;454(7200):126–30.

Section VI

Human Diseases Associated With Genome Instability

Chapter 26

Human Diseases Associated With Genome Instability

B.C. Feltes[a], J. de Faria Poloni[a], K.N. Miyamoto[a], D. Bonatto[a]
Federal University of Rio Grande do Sul, Porto Alegre, Brazil

Chapter Outline

1. INTRODUCTION

The main focus of the following section (Section 2) of this review is to introduce rare genetic diseases associated with different aspects and pathways of DNA repair. Molecular aspects regarding the connection among different aspects of the repair pathways are also considered. In Section 3, we address the main genetic alterations that drive cells to genome instability resulting in acquiring cancerous phenotypes. In addition, we discuss why understanding these phenomena are useful in oncological clinical care. Finally, in Section 4 we discuss epigenetic mechanisms that influence the cell-cycle regulation and the DNA-repair response. Furthermore, the topic also contemplates the most common abnormalities in the epigenetic-regulation mechanisms and their impact on the cell-fate acquisition.

2. RARE GENETIC DISEASES ASSOCIATED WITH DNA REPAIR

The DNA molecule is constantly threatened by a wide range of exogenous and endogenous mutagenic agents, such as reactive oxygen species (ROS), chemical pollutants, drugs, and radiation such as ultraviolet (UV) light [1,2]. However, during evolution, cells have established molecular mechanisms to protect and repair the DNA molecule. These include, but are not limited to, compacting the DNA in the form of chromatin, lowering its contact with the cellular environment, and developing repair mechanisms like the nucleotide excision repair (NER), base excision repair (BER), homologous recombination (HR), DNA interstrand cross-link repair (ICLR), double-strand break (DSB), and mismatch repair (MMR) [1,3]. However, if the damage is not repaired, the cell can undergo apoptosis, senescence, or can lose control of its mitosis and can start an abnormal proliferation and become a tumor [1].

Different consequences can arise from nonrepaired DNA mutations caused by defects in the repair mechanisms, the so-called genetic diseases. In the next two sections, we focus on rare diseases related to defects in the repair machinery: xeroderma pigmentosum (XP), Cockayne syndrome (CS), trichothiodystrophy (TTD), and Fanconi anemia (FA). All four diseases are associated with defects in genes that encode proteins related to DNA repair—XP, TTD, and CS phenotypes

a. All authors contributed equally to this work.

Genome Stability. http://dx.doi.org/10.1016/B978-0-12-803309-8.00026-4

are derived from mutations in genes that act on the NER pathway, whereas FA is a result from mutations in genes on the ICLR pathway [1,4].

We then focus on the diseases affected by mutations in RECQ family genes, which are Bloom syndrome (BS), Rothmund–Thomson syndrome (RTS), and Werner syndrome (WS).

In the last section, we discuss genetic diseases that are not specific to a single pathway, such as ataxia telangiectasia (AT) and Hutchinson–Gilford progeria syndrome (HGPS). Explanation of the whole spectrum of outcomes and molecular pathways of the listed diseases are complex, hence we focus on the major differences and how they are associated.

2.1 NER-Related Diseases: Xeroderma Pigmentosum, Trichothiodystrophy, and Cockayne Syndrome

In order to comprehend the complexity of CS, TTD, and XP, it is required to understand the functionality of the NER pathway. NER is specialized in removing UV-induced DNA damage, where 6,4-photoproducts (6,4-PP) and cyclobutane pyrimidine dimers (CPDs) are the most common lesions, although there are other types of UV-induced lesions [1]. NER is divided in two sub-pathways: the global genome NER (GG-NER), which is responsible for the removal of DNA lesion in nonactive genes, heterochromatin, and transcribed strands of active genes, and the transcription-coupled NER (TC-NER), responsible for removing DNA damage only from transcribed strands of active genes [5,6]. Molecularly, the main difference in both sub-pathways is that in TC-NER RNA polymerase is hindered in the lesion site with the aid of specific factors, the DNA-dependent ATPases (CSA, CSB) and the pre-mRNA splicing factor XAB2 that bind to the lesion where RNA polymerase is stalled, whereas in GG-NER the lesion is recognized by the *xeroderma pigmentosum, complementation group C* (XPC)-HR23B (RAD23B) heterodimer or the DDB-complex (composed by the DNA-binding proteins DDB1 and DDB2, and the ubiquitin–ligase complex CUL4A and ROC1). XPC-HR23B have a high affinity for 6,4-PP lesions and the DDB complex for CPD lesions, but it is known that the DDB complex recruits XPC–HR23B to the site once the damage is recognized [5]. After the damage recognition, both pathways follow a core NER reaction of damage excision as follows: (1) the recruitment of the TFIIH helicase complex to open the damaged site; (2) recruitment of XPA–RPA heterodimer to form a platform of protein–protein interaction; (3) DNA-damage excision by the endonucleases XPF and XPG; and (4) synthesis of a new DNA strand [5] (Fig. 26.1). XPF forms a heterodimer with the ERCC1 protein, and it is still debatable whether XPG is recruited to the excision complex or it is a subunit of the TFIIH complex [7].

CS, TTD, and XP are autosomal diseases characterized by hypersensitivity to sunlight, premature aging, and a shorter life span, but differ in the extension of other symptoms. XP was the first NER-related discovered disease, described in 1874 by Moriz Kaposi [1]. XP affects 1:250,000 individuals in Western countries and 1:45,000 in Japan and North Africa, where individuals show severe risk to develop skin cancer and sunburns, in which skin neoplasms can appear during childhood [1,2]. They also present ocular degeneration in the lids, cornea, and conjunctiva [8]. Neurological symptoms are less common, but can appear in some cases [1,2].

Moreover, CS was the second NER-related disease, discovered 62 years later, in 1936, by Edward Alfred Cockayne, and 44 years later, in 1980, TTD was described by Price [1]. In contrast to XP, CS, and TTD individuals commonly present cognitive impairments and neurological degeneration, cachectic dwarfism, skeletal and muscular defects as well for a facial characteristic called "bird-like" face, defined by deep sunken eyes and preeminent ears [2,9,10]. Some cases of CS can develop cerebro-oculo-facial-skeletal (COFS) syndrome, a disorder that can cause neurological and visual deficiencies, whereas TTD patients can present decreased fertility and osteosclerosis, combined with more aggressive neurological symptoms, such as tremors, low IQ, and incomplete myelination of nervous fibers [2,5,9]. It is also interesting to highlight that CS individuals do not exhibit skin cancer predisposition, although they are hypersensitive to sunlight, indicating that the TC-NER pathway is not required to prevent skin cancer.

The differences in each syndrome are related to the genes affected. As seen in Fig. 26.1, three proteins are specific for TC-NER, CSA, CSB, and XAB2, where mutations in CSA and CSB are responsible for the CS phenotypes. Mutations in CSA are related to Type I (classical) form of CS, where manifestations occur around the first years of life, and to Type III (mild), where individuals show a greater life span than other types and retain basic cognitive function such as walking and speaking [10]. On the other hand, mutations in CSB can manifest themselves as any type of CS, including Type II (severe), where individuals have a maximum life span of 7 years and display strong mental retardation and loss of basic cognitive functions [10].

In XP, the differences lie on which XP gene was compromised. There are seven XP genes in NER (XPA–XPG) (Fig. 26.1); mutations in any of those genes provoke an XP phenotype and in case of XPB, XPD, XPF, and XPG, some manifestations show a CS-like characteristic, such as neurological abnormalities [2,10].

Different from XP and CS, the core origin of the TTD phenotype lies in mutations in the helicases XPB and XPD, with XPD mutations being the major cause [9]. Thus, TTD phenotype appears to be related to the TFIIH complex activity more

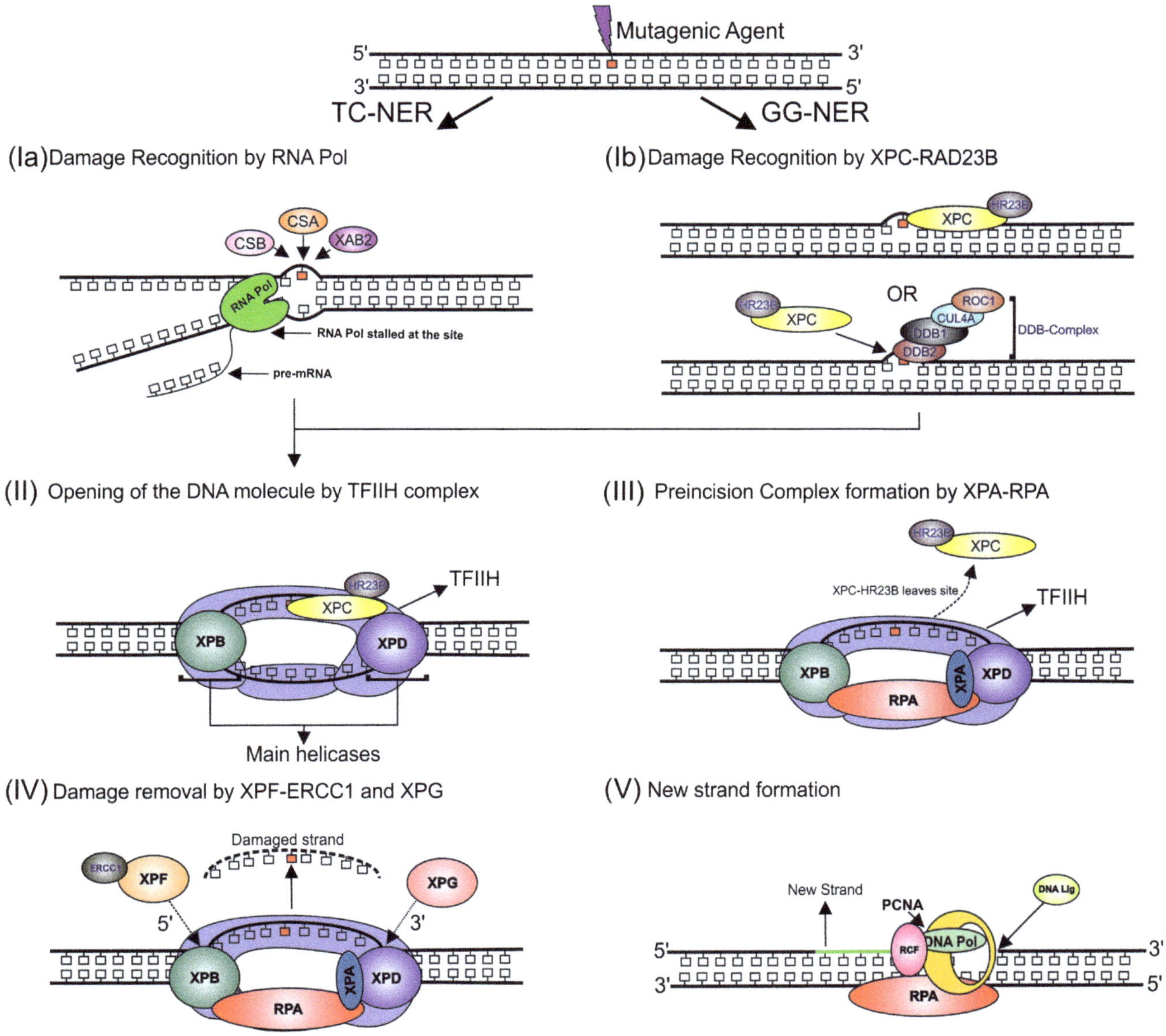

FIGURE 26.1 **The NER pathway divided by its fundamental steps.** It begins after DNA damage and one of the two sub-pathways is triggered. (Ia) TC–NER pathway is triggered if the damage site is in an active gene that is currently being transcribed. RNA Pol II is stalled at the lesion with the aid of CSA and CSB, that bind to the lesion and help in the recruitment of other factors, such as XAB2. (Ib) GG–NER can be triggered in any case, since it can act on heterochromatin, nonactive genes, and euchromatin. In this case, the damage is recognized by the XPC–HR23B heterodimer or the DDB complex. This separation is required to understand that it is a case of affinity: both complexes can recognize different damages, but they have higher specificity for a given substrate. XPC–HR23B have higher affinity for 6,4-PP while the DDB complex has it for CPDs. Nonetheless, XPC–HR23B is recruited by the DDB complex after it recognized the damage. After the initial recognition, both pathways converge in the core NER steps. (II) XPC–HR23B recruits the helicase complex TFIIH, where the 3′-helicase XPB and the 5′-helicase XPD act on opening the damage site. (III) XPC–HR23B leaves the site and the XPA–RPA heterodimer binds to unwounded DNA to allow a protein–protein platform. (IV) The 3′-endonuclease XPF and the 5′-endonuclease XPG are recruited to excise the damaged strand. (V) Finally, all proteins leave the site except for RPA which is required for the final polymerization step that recruits PCNA, DNA Polδ, and RCF to create a new strand of DNA. DNA ligase connects the new strand with the ends of the old strand.

than any other molecular aspect. The severity of TTD manifestation depends on what residue was mutated in the XPD protein, where the R112H, R592P, D673G, and R722W mutations are the cause of a severe phenotype, and R658C, R658H, and A725P result in a mild form of TTD [9].

It is interesting to observe that XP, TTD, and CS share the core NER reaction of damage excision, and the mutations in the XP genes related to this step can provoke similar phenotypes. Moreover, although all three diseases are associated with NER, the regular phenotypes are distinct, indicating that the proteins derived from the mutated genes are probably acting on other pathways beside DNA repair.

One explanation is that XP proteins, shared in both pathways, are interacting with a broad range of other proteins [11]. Taking the XP genes related to XP/CS/TTD phenotypes for example: XPB and XPD helicases are part of the TFIIH complex, where both exert structural role in interconnecting other subunits of the complex [11]. The TFIIH helicase complex is also required for regular transcription, and its deregulation can affect a broad range of cellular processes. In addition, XPB and XPD interact with proteins related to DNA repair such as p53, and to RAD52, where they would have a role in HR pathway [11]. Additionally, XPG is associated with the BER pathway by interacting with the protein NTH1. NTH1 plays a role in repairing thymine glycol mutations, and its affinity for the lesion is increased by XPG [11]. Finally, XPF interacts with RAD51 and RAD52 of the HR-repair pathway and with TRF2, a telomere elongation factor and with the Fanconi anemia, complementation group A (FANCA) protein, which is discussed later in more detail [11]. These relations clearly indicate that affecting XP proteins may disrupt a variety of mechanisms besides NER.

2.2 Fanconi Anemia

FA was discovered by Guido Fanconi in 1927. FA is distinct from XP, TTD, and CS, since the genes involved in the FA are mainly connected to the ICLR pathway instead of NER. ICLs are DNA lesions that covalently links paired strands of DNA, preventing the separation of the strands and the formation of the replication/transcription fork [12].

FA is mainly a hematopoietic disease that ultimately causes bone-marrow failure [12]. Individuals do not show most of the symptoms that appear in XP or CS, although they can present short stature and facial deformities [13]. Although the whole ICLR pathway consists of more than 30 genes, there are 16 FA genes related to ICLR, where 8 of them compose a multisubunit ubiquitin E3 ligase complex (FA core), and mutation in any of those 16 genes leads to FA [4,12–14]. A summary of the ICLR pathway can be found in Fig. 26.2. Broadly, the ICLR pathway can be divided into five stages: (1) damage recognition, (2) FA core recruitment, (3) complex assembly, (4) translesion polymerase activation, and (5) HR pathway triggering [4,12–16] (Fig. 26.2).

One interesting aspect of this pathway is its link to NER, since XPF is one of the main proteins that act on the pathway, and one of the responsible proteins for the XP/CS phenotype [2,10,15]. Remarkably, a 2013 work from Kashiyama et al. [17] described a patient who showed phenotypes associated with XP, CS, and FA [17]. The XPF–ERCC1 heterodimer is extremely important to proper removal of damaged sites, and it is known that ERCC1 mutant mice display neurodegeneration, whereas mutation in XPF results in a genetic disease called XPE progeroid syndrome, which is characterized by premature aging and aging related-diseases [15,17]. These studies indicate that there is a connection between ICLR and NER that may result in a combined phenotype of the three diseases, although this association is yet to be established. It is possible that this association was evolutionarily selected to enhance the response to DNA damage.

2.3 RECQ-Related Diseases: Rothmund–Thomson Syndrome, Werner Syndrome, and Bloom Syndrome

Another syndrome associated with the DNA-repair pathways, like BS, RTS, and WS arise from mutations in the RECQ helicase family. In this sense, before we discuss each syndrome individually, a broad view of the RECQ role on DNA repair is necessary [18].

The human RECQ helicases family consists of five proteins, RECQL1, WRN, BLM, RECQL4, and RECQL5, all of which play a crucial role in DNA damage–sensing and –repair pathways, either for helping other proteins to assemble the repair machinery or to recognize and unwind specific DNA rearrangements (Fig. 26.3) [18]. They interact with DSB repair–pathway proteins at different stages, and when DSB repair is initiated by HR, they are important at the initial DSB recognition, further disassembly of RAD51–ssDNA nucleoprotein filaments during recombination, and in resolving double Holliday junctions (DHJ) (Fig. 26.3-IV) at the branch migration phase [18]. On the other hand, in nonhomologous end joining (NHEJ) pathway, RECQ helicases act by modulating protein complexes like the DNA-dependent protein kinase catalytic subunit (DNA-PKCS), the Ku70/80 heterodimer, which detect DNA damage, and the XRCC4/ligase IV, responsible for DNA end ligation [18]. RECQ family, especially WRN, also mediates base lesion targeting in BER pathways [19], which is discussed later. Additionally, the role of RECQ helicases in the NER pathway remains poorly understood. Finally, RECQ proteins are also important to replication events; they are recruited at stalled or collapsed replication forks, interact with replication repair–machinery proteins, mainly RPA, guiding the DNA-damage fixing, and further replication restart [19]. This is an interesting fact, since RPA is also a close partner of XPA during NER (Fig. 26.1-III) [1,2].

Since the RECQ family is essential for genome maintenance, mutations in these genes could cause defects in many repair pathways, leading to genome instability. For this reason, diseases like WS, BS, and RTS, all caused by mutations in members of the RECQ family, are characterized by a wide range of symptoms and cancer development. We address each syndrome individually.

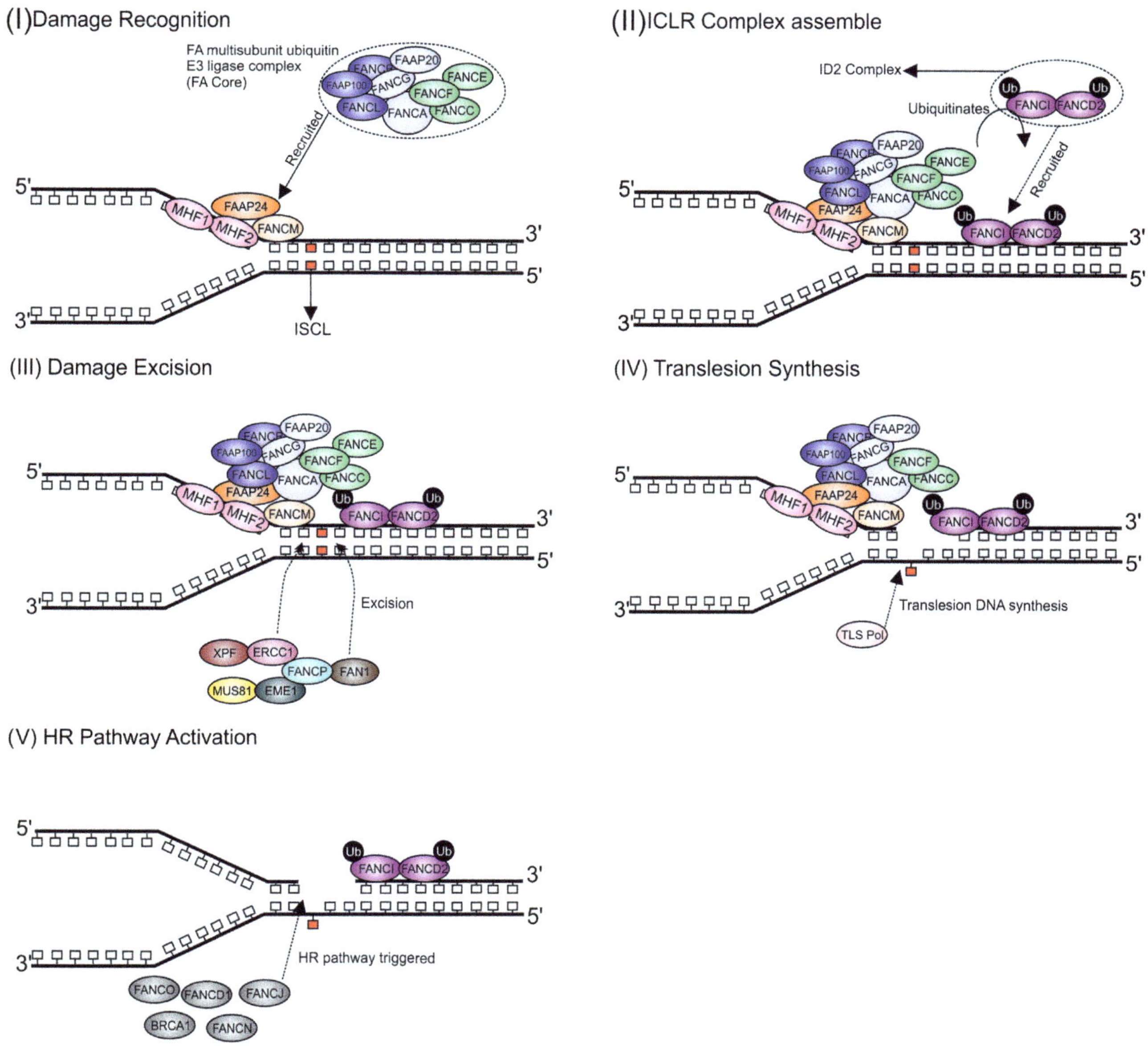

FIGURE 26.2 **The ICLR pathway divided by its fundamental steps.** (I) It begins with the ICL recognition by the proteins FANCM, MHF1–MHF2, and FAAP24 that bind to unwounded DNA and recruit the FA core. The FA core is composed by three subcomplexes, one composed of FANCL, FAAP100, and FANCB (blue); the other composed of FANCG, FANCA, and FAAP20 (light gray); and the third composed of FANCF, FANCC, and FANCE (green). (II) The FA core then ubiquitinates the ID2 complex, which is composed of FANCI and FANCD2 and binds to the unwounded DNA. With the complex formed, the excision machinery composed of XPF–ERCC1, MUS81–EME1, FANCP, and FAN1 excise the damaged region. (III) TLS polymerase than adds nucleotides to the removed strand. (IV) Finally, the HR machinery is triggered by FANCO, BRCA1, BRCA2, FANCD1, and FANCJ.

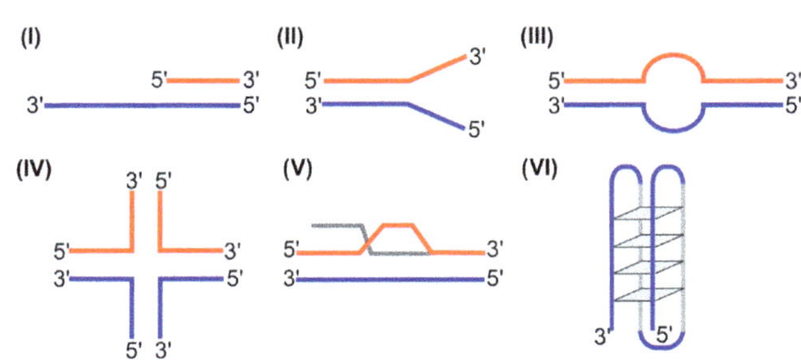

FIGURE 26.3 **Types of DNA arrangements that are substrates of the RECQ family.** (I) 3′-tailed DNA. (II) Forked DNA. (III) "Bubble" structured DNA. (IV) Holliday junction. (V) D-loop. (VI) G-quadruplex.

RTS is an autosomal recessive disease first described in 1868 by August von Rothmund, and then redescribed, with addition of phenotype variances by Matthew Sydney Thomson in 1926 [20]. However, the term Rothmund–Thompson was just coined in 1967 by William Taylor [20]. RTS major symptoms are epidermis-related tissue malformation (eg, hair, skin, and nails), short stature derived from skeletal malformations, cataracts, and cancer predisposition, especially osteosarcomas and spinocellular carcinomas [20].

RTS is caused by mutations in the RECQL4 protein, and ATP-dependent helicase, that is the only RECQ helicase present in both nucleus and mitochondria [21]. RECQL4 is necessary for the initiation of DNA replication and studies with RTS patients have shown that RECQL4 is also associated with sister chromatid separation, DSB, and BER repair pathways and telomere replication [18,20,21]. This protein can unwind forked duplexes, Holliday junctions, G-quadruplex structures, "bubble" structures, and D-loops, but cannot unwind normal duplex DNA (Fig. 26.3) [22]. Interestingly, RECQL4 shows little DNA-unwinding activity, when compared to other RECQ proteins, and seems to be more prone to anneal DNA [21]. Another fact to be observed is that mutations in RECQL4 are also related to two other syndromes: (1) Baler–Gerold syndrome (BGS), characterized by craniosynostosis and radial hypoplasia, together with short stature, and (2) the radial hypoplasia, patella hypoplasia and cleft or arched palate, diarrhea and dislocated joints, little size and limb malformation, slender nose and normal intelligence (RAPADILINO) syndrome [20].

Since RECQL4 is the only of the RECQ family to be present in the nucleus and the mitochondria, further studies focusing on understanding of RTS relationship with mitochondrial function are necessary. Additionally, the fact that RECQL4 appears to be colocalized with XPA after UV irradiation in the nucleus suggests a possible role for RECQL4 in NER that might show promising explanations for the RTS phenotype and possibly for XP [22].

Another autosomal recessive disease related to defects in genes of the RECQ family is Bloom syndrome, which was discovered by David Bloom in 1954 [23]. BS individuals present morphological abnormalities, such as long limbs, short stature, and the bird-like features similar to CS and TTD, as well as low subcutaneous fat content and dermatological conditions like photosensitivity and poikilorderma [24]. In addition, patients show high predisposition to cancers (eg, breast, larynx, skin, and colorectal cancers), lymphoma, and leukemia [24].

BS is another disease caused by a mutation in a gene of the RECQ family—in this case, the ATP-dependent RECQL2 known as BLM [18,25]. BLM can unwind the same DNA structures as RECQL4 plus 3′-end of the DNA, but shows preference for unwinding G-quadruplex DNA (Fig. 26.3), [25]. Similarly to RECQL4, BLM also has an ss-DNA annealing capacity, although the full mechanism by which it can promote strand annealing is not fully understood [25]. BLM also repairs centers of collapsed or stalled replication forks, where it appears to promote fork regression [18]. Thus, the loss of BLM function is related to a broad range of chromosomal aberrations and cancer formation [24,25]. It is not a surprise that BS individuals show high levels of sister chromatids exchanges, chromosomal breakage, translocation, and chromosomal quadri-radials [24,25].

One interesting molecular aspect of BLM is the fact that it appears to interact with FANCM, one of the proteins that comprises the FANC complex [24], showing that there might be a connection between the molecular pathways that lead to BS and FA. FANCM-deficient cells show high levels of sister chromatids exchange, similar to BS-derived cells [26]. A study made by Hoadly et al. [26] suggests that FANCM binds to the damage site and recruits the BLM complex (composed of BLM, topoisomerase IIIα, and the RMI1/RMI2 heterodimer) through its interaction with the RMI1/RMI2 heterodimer [26].

Chromosomal maintenance is a complex process that involves a wide variety of proteins and pathways, and understanding the interplay between BLM and these other mechanisms might help improve the knowledge about BS.

Finally, the last disease related to mutations in the RECQ family is WS, an autosomal recessive disease originated from mutations in WRN gene, which encodes for RECQL3/WRN protein. WS was described for the first time in 1904 by Carl Wilhelm Otto Werner. Its clinical manifestations are extensive, although most of the presented symptoms are aging related, such as atherosclerosis, diabetes mellitus type 2, osteoporosis, and cataracts, among others [27]. Nonaging-like symptoms include hypogonadism, reduced fertility, low height, and others. In addition, WS patients are also susceptible to development of tumors, especially sarcomas [28]. However, even though this could represent an important risk factor, most WS-affected individuals decease by a myocardial infarction between their fourth and fifth decades of life [28].

WRN protein has a helicase domain which has specificity for certain DNA structures (Fig. 26.3), especially G-quadruplex (Fig. 26.3-VI.) and Holliday junction (Fig. 26.3-IV.), DNA structures found mainly in telomeric DNA and in the recombination process, respectively [19,29]. In addition, WRN has a unique 3′–5′ exonuclease activity that digests 3′-recessed termini or blunt DNA duplexes that contain structures like bubbles (Fig. 26.3-III), forked duplexes (Fig. 26.3-II), Holliday junctions, and DNA–RNA heteroduplexes. This exonuclease acts coordinately with a helicase domain through the DNA duplexes size reduction, allowing proper helicase-unwinding role, although both domains have also independent functions [28].

WRN interacts with many regulating proteins related to DNA-repair pathways. For example, in the long-patch BER (LP-BER) pathway, WRN interacts with NEIL1, a formamidopyrimidine lesion glycosylase, DNA polymerase β, responsible for base replacement at the lesion site, and with FEN1, an endonuclease that removes the 5′-overhanging flap [19]. In DSB repair, WRN plays an active role in NHEJ by interacting with KU70/80, DNA-PKCS and XRCC4/ligase IV complexes. During HR, WRN interacts with proteins like RAD51 and RAD52, which are fundamental during strand invasion and annealing [28]. Moreover, WRN is important in telomere replication and maintenance pathways, recognizing D-loops (Fig. 26.3-V), G-quadruplex structures in telomeric DNA, and interacting with the shelterin complex proteins, such as TRF1, TRF2, and POT1 [29]. Conversely, WS cells present characteristics that are directly associated with impaired DNA repair, including telomeric erosions, oxidative DNA damage, and a defective DNA interstrand cross-link removal, which can contribute to develop aging-related symptoms and tumorigenic processes like sarcomas in WS patients [29].

2.4 Ataxia Telangiectasia

Different from all the previously discussed diseases, AT is not caused by defects in a specific DNA-repair pathway. AT is a rare autosomal recessive disorder described originally in 1926, but the term "ataxia telangiectasia" was suggested by Elena Boder and Robert P. Sedgewick in 1957 [30]. The clinical characteristics of AT disorder are progressive neurological dysfunctions, which includes oculocutaneous telangiectasia and cerebellar ataxia. In addition, individuals with this disorder show cancer predisposition, susceptibility to bronchopulmonary disease, and multisystem abnormalities, such as immunodeficiency, radiosensitivity, infertility, and endocrine dysfunctions [30,31].

AT is caused by mutations in ataxia telangiectasia–mutated gene (ATM), located at chromosome 11q22–23, which encodes to an ATM serine/threonine kinase [30]. The ATM protein is a member of the phosphoinositide 3-kinase (PI3K)-related protein kinase (PIKK) family, which is able to induce a DNA-damage response [30,31].

ATM is activated by different biological processes, such as cell-cycle checkpoint and DNA damage [30]. Indeed, many ATM substrates are cell-cycle regulators with important roles in DNA-damage response, such as p53, CHK2, and BRCA1 [31,32]. During DNA-damage repair, a complex composed of MRE11–RAD50–NBS1 (MRN complex) recognizes DSB and leads to ATM activation, as well as performs an adaptor role to subsequent phosphorylation of downstream ATM substrates [30–32]. In response to the MRN-complex signaling, ATM undergoes autophosphorylation at serine 1981, and is converted from inactive multimeric to an active monomeric kinase [30,31]. Once activated, ATM orchestrates a signaling cascade in response to DSB that coordinates cell-cycle arrest, DNA repair, or the cell-apoptosis process [30,31]. The response capacity of ATM to DNA damage is the primary in vivo function of this kinase and is intrinsically related to phenotypes of AT disorder [31].

In this sense, the role of ATM in DNA repair involves the phosphorylation of specific repair factors, like KRAB-associated protein 1 (KAP-1), which relaxes chromatin structure and allow the accessibility of repair proteins (the chromatin role in DNA repair is discussed later) [31]. Another target of ATM phosphorylation is the FANCD2, a protein whose defects leads to FA; in response to DNA damage, FANCD2 is phosphorylated at Ser222 [32].

2.5 Hutchinson–Gilford Progeria Syndrome

HGPS is a progeroid syndrome described at first time by Jonathan Hutchinson in 1886 and by Hastings Gilford in 1897 [33]. However, this syndrome was described in greater detail in 2003, when the molecular basis of the disease was discovered [33,34].

Despite HGPS patients born with normal appearance and weight, the clinical symptoms appear within 12 months and progress rapidly [35]. The best described characteristic of HGPS patients is the development of age-related diseases, such as cardiovascular pathologies, prominent superficial veins, skin complications, and alopecia. In addition, these individuals show disturbed growth, lipodystrophy, joint abnormalities, and osteolysis [35]. This disease is caused by a single nucleotide substitution on the gene *LMNA* that encodes the A-type nuclear lamin proteins [33,34]. Differential alternative splicing generates A-type lamin proteins, with the most abundant being lamins A and C; however, mutated *LMNA* leads to aberrant splicing that results in the deletion of 50 amino acid residues from C-terminal region of prelamin A [33,34,36]. This aberrant splicing produces a mutant protein called "progerin," which accumulates in a farnesylated form, affecting the nuclear organization, chromatin dynamics, epigenetic regulation, and gene expression, causing genomic instability, premature senescence, and telomeres disruptions [34,36].

It was observed that progerin modifies the composition and mechanical properties of nuclear lamina, which are related to abnormal nuclear morphology [34,36]. This occurs due to the high affinity of progerin to nuclear envelope and by immobilization of A-type lamins in the nuclear lamina induced by progerin. In addition, high levels of γH2AX phosphorylation are an indicative of activated DNA-damage response and cellular accumulation of DNA damage [34,36]. Furthermore, HPGS cells show reduced survival and proliferation, besides sensitivity to DSB and delayed recruitment of repair proteins [34,36].

Among the proteins involved in this impaired recruitment are the components of MRN complex, which are crucial for HR [36]. Furthermore, another DNA-repair defect observed in HPGS cells is the mislocalization of XPA to DSB that can be associated with the delay activation of DNA-repair proteins, such as the MRN complex [36,37]. Thus, HGTS molecular pathways have an interplay with the ATM related–molecular mechanisms that may lead to AT, as well as with NER-related proteins, such as XPA.

2.6 Rare Genetic Diseases: Summary

In conclusion, the understanding of DNA-repair diseases is crucial to boost the knowledge of the DNA-repair machinery and the consequences of its defects. Moreover, the phenotypes that result from diseases described earlier go from a broad range of anatomical abnormalities to neurodegeneration and cancer development, indicating that they regulate much more than DNA repair. For example, XP proteins are also targets of different post-translational modifications and have different protein–protein interaction sites that may answer how they are regulated and to what proteins they may be connected [5]. Other examples are CSA and CSB, that are also known to regulate cellular redox balancing, where cells lacking CSA and CSB show increased level of ROS [38]. These proteins are also connected to BER and the maintenance of the stability of mitochondrial DNA [38]. CSB is associated with biological processes, such as cell growth, angiogenesis, proliferation, and cell death [39]. The same logic goes for the proteins related to FA, BS, RTS, WS, AT, and HGTS which comprise large complexes and are connected to multiple proteins.

3. CANCER AND GENOME INSTABILITY

It is impossible to discuss genetic diseases without mentioning cancer, since it is intimately associated with DNA-repair defects and is a common outcome of the rare genetic diseases described before. Cancer onset begins when precancerous cells acquire uncontrollable growth, sustain angiogenesis, and become able to invade different tissues [40]. These are the main factors that contribute to the extent of cancer malignancy. However, since the human organism has redundant and self-regulating pathways to maintain homeostatic conditions, an extensive set of genes must be affected to reach conditions necessary to carcinogenesis. This change in genome profile can be achieved if DNA damage–repair systems and/or replicating machineries work improperly, or when cells are exposed to mutagenic or genotoxic agents, such as tobacco smoke, UV light, and ionizing radiations, among others. The increased mutation rate that changes drastically the genome landscape deregulates basal expression and surpasses genome integrity, and cell-cycle surveillance promotes cells transformation. It is important to reinforce the idea that genome instability makes cells more susceptible to carcinogenesis [41]. Since there are many pathways associated with cancer development, the focus of this section is to describe the basis of the different types of genome instability, as well as its relationship with cancer development and how understanding these phenomena can be useful in clinical practice and therapy.

Genome instabilities can vary from a single nucleotide mutation to a whole chromosome structure modification (clastogenesis). These changes alter cell homeostasis in many ways, depending on which genes are affected. Single nucleotide mutations arise usually due to the high cell exposure to DNA-damaging agents or when DNA-repair genes involved—for example in NER, BER, and MMR—are mutated, although a nucleotide-deficient environment could also promote such imbalance (Fig. 26.4-I) [42,43]. For example, XP patients are highly susceptible to sunlight UV-induced carcinogenesis (about 2000–10,000 times higher than a healthy person) due to the accumulation of nucleotide mutations that are caused by a deficient NER from one or more mutated XP proteins [8]. Some regions of the genome, containing repeated nucleotide sequences—one to six nucleotides repeated multiple times—called microsatellites, suffer more extensive modifications caused mainly by an inefficient MMR system, leading to insertions and/or deletions (indels) in these regions during the S phase of cell cycle (Fig. 26.4-II). This microsatellite instability (MSI) promotes frameshifts in coding sequences of genes resulting in truncated or nonfunctional proteins. MSI is present in some cancers, mainly in colorectal tumors, which corresponds to 15% of these cases [44].

Chromosomal instability (CIN), especially chimeric chromosomes and/or aneuploidies, on the other hand, is a common genome instability in most cancers. Since the discovery of Philadelphia chromosome, formed by the chromosome 9 and 22 translocation, much has been done to understand the importance of CIN events in tumor progression. In this sense, defective DNA-repair mechanisms can induce multiple chromosomal fragmentations. For example, DSB-repair failures contribute to CIN generation, creating a chromothripsis phenomenon characterized by multiple chromosome breaks and rearrangements (Fig. 26.5-I) [45]. In addition, other factors also contribute to the acquisition of an abnormal karyotype

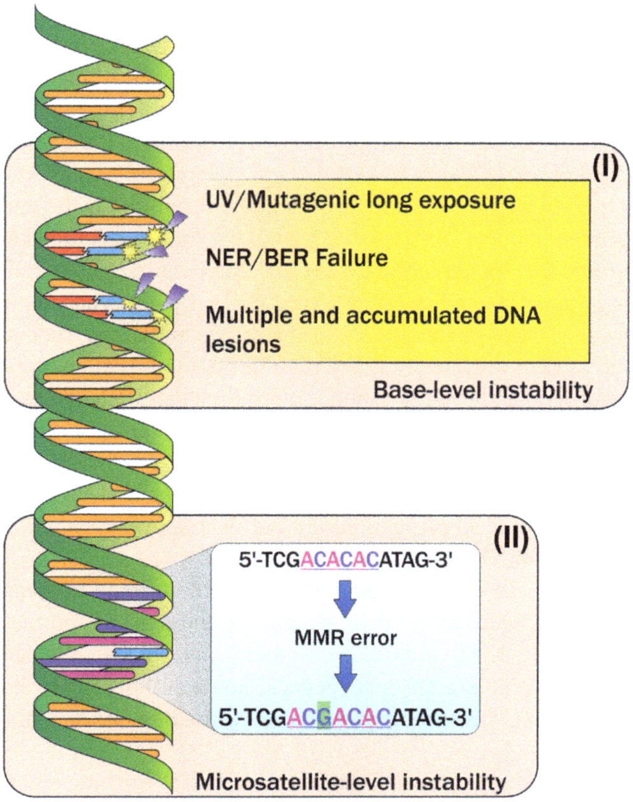

FIGURE 26.4 **Cancer-related DNA mutations that cause genome instability.** (I) Base-level accumulated mutations, caused by high exposure to mutagenic agents and/or defective NER/BER pathways. (II) Indel events in microsatellite regions caused by defective MMR systems.

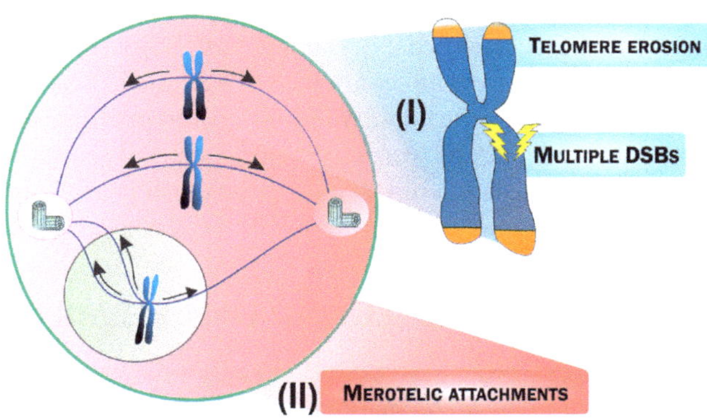

FIGURE 26.5 **Cancer-related events that lead to chromosome instability and clastogenic phenomenon.** (I) Chromosomal breakage and rearrangements by the progressive telomere shortening and/or double-strand breaks. (II) Multiple microtubules attached to one kinetochore (merotelic attachments) that causes spindle pole asymmetry and incorrect chromosome segregation.

in cancer cells. Telomere shortening or loss is one of the main driving forces of chromosome fragmentation. These six tandem repeated sequences at the end of chromosomes, along the protein complex called shelterins, are crucial to protect chromosomes from fusing to each other during cell cycle and generating aberrant chromosomes (Fig. 26.5-I). When cells naturally cease to express telomerase, they die due to a telomeric erosion condition in which cells either enter a replicative senescence state or begin to generate multiple chromosome fusions [46]. This process is tightly controlled in the cell and when they are impaired, the cells are able to proliferate, leading to uncontrolled cell growth. Curiously, tumor cells can even restabilize their genome through the reexpression of telomerase or via a homologous recombination alternative lengthening of telomere (ALT) mechanism, although the triggering mechanisms remain unclear, especially for ALT [47–49].

Finally, another CIN-inducing event is the incorrect segregation of chromosomes during cell cycle. Mutated proteins responsible for chromosome organization and cell structure can compromise the sister chromatids separation (karyokinesis) or cells separation (cytokinesis) resulting in aneuploidies [50]. Centrosome dynamics during mitosis, for example, is a very coordinated process that requires a myriad of signaling proteins. Thus, any change in the velocity of chromatids separation during the M phase of the cell cycle can induce CIN. This could make the chromosome's kinetochore attach to microtubules coming from both spindle poles, instead of just one of them, forming merotelic attachments (Fig. 26.5-II). This creates a spindle asymmetry that compromises the correct chromosome segregation and creates a lagging chromosome that will result in aneuploidic daughter cells [51].

One of many hypotheses that tried to explain the origins of genome instability is the mutator phenotype, which suggests that mutations in genes involved in the genome-maintenance pathways, known as caretaker genes, makes cells more prone to DNA lesions or replication failures that facilitate the cancer development [52]. In hereditary cancers, mutations in caretaker genes mostly drive cells to genome instability, as observed in the Lynch syndrome (in which MSI caused by MMR failure triggers oncogenesis) as well as other inherited DNA-repair gene mutations, such as mutations in the FANC family and in the BRCA1 gene. However, as demonstrated in many high throughput–sequencing studies, these types of mutations are unlikely to occur in sporadic cancers, and even if they do, it requires that both alleles must be affected to drive cells to genome instability [40]. In these nonhereditary cancers, the most accepted hypothesis is that the same altered pathways that activated oncogenes also drives deacceleration or stalling of replication fork progression, especially in regions denominated as common fragile sites, creating chromosome breakage at these locations. This favors the selection of defective tumor-suppressor genes, such as TP53, and genome amplification of other oncogenes, leading to cancer development through the escape from apoptosis and senescence [40,53–55].

Understanding how complex patterns of genome instability events contribute to cancer has many implications in the clinical health care. In cancer diagnosis, detection of chromosome instabilities is important to determine tumor aggressiveness and patient's prognosis. For example, MSI-containing colorectal cancers are considered to have more favorable prognosis as compared to stable microsatellite colorectal cancer types. Some authors hypothesized that translation frameshifts caused by MSI generate novel peptides at C-terminus region that are immunogenic and stimulate an inflammatory response against tumor cells [44].

Also, the comprehension of how cancer begins and develops is crucial for new chemotherapeutics drug design. One strategy is inducing mitotic catastrophe by small molecules that act on the kinetochore and spindle poles assembly proteins, such as aurora kinase inhibitors [56]. However, since many of these potential chemotherapeutics are toxic to the bone marrow, many types of DNA damage–response inhibitors were tested as adjuvants to maximize genomic instability in cancer cells, promoting mitotic catastrophe and apoptosis and avoiding potential drug resistance [57,58].

Therefore, cancer development can be accelerated by genome instability. High-proliferative capacity, sustained angiogenesis, cell cycle–checkpoint evasion, for example, are part of so-called "Cancer Hallmarks," and most of these features are acquired by multiple events of genome instability [59,60]. Although more studies are required to understand the complex relationship between cancer and genome rearrangements, especially in sporadic tumors, there is still plenty of information available that can help oncologists in clinical care to establish patient prognosis and to search for potential anticancer targets.

4. EPIGENETIC REGULATION OF CELL CYCLE AND DNA REPAIR IN CANCER

The transformation of healthy cell toward a cancerous cell occurs gradually by a series of factors, including genetic and epigenetic modifications. Proper maintenance of epigenetic marks is essential to healthy cells and is associated with cell-fate acquisition [61–63]. Epigenetic alterations can change chromatin structure to loose state, which is transcriptionally active (called euchromatin) or to compact state, resulting in a transcriptionally inactive configuration (called heterochromatin) (Fig. 26.6-I) [64]. In this sense, chromatin alterations changes DNA accessibility and are responsible for modulation of the gene expression by affecting the interaction of DNA with transcriptional complexes, resulting in activation or inhibition of different signaling pathways (Fig. 26.6-I) [61,63]. Furthermore, histone modifications may affect DNA–histone or histone–histone interactions, or recruit nonhistone proteins to chromatin, creating a binding site for specific proteins that can act as regulatory factors [62,65]. In addition, different biological processes such as transcription, cell cycle, DNA repair, and replication are regulated by posttranslational histone modifications [62,65].

During cell cycle, checkpoints are surveillance systems that have the capacity to interrupt cell-cycle progression [66]; however, abnormalities in checkpoints and signaling pathways associated with proliferation are commonly observed in cancerous cells [66]. An example of signaling pathway disturbed in cancerous cells that is addressed in this chapter involves the retinoblastoma tumor-suppressor protein (RB) [66]. RB is a tumor suppressor whose activity is associated with different biological processes, such as differentiation, apoptosis, DNA-damage response and repair, DNA replication, and cell cycle [67]. During cell cycle, RB binds to the transcription factor E2F and prevents the transcriptional activation of E2F target

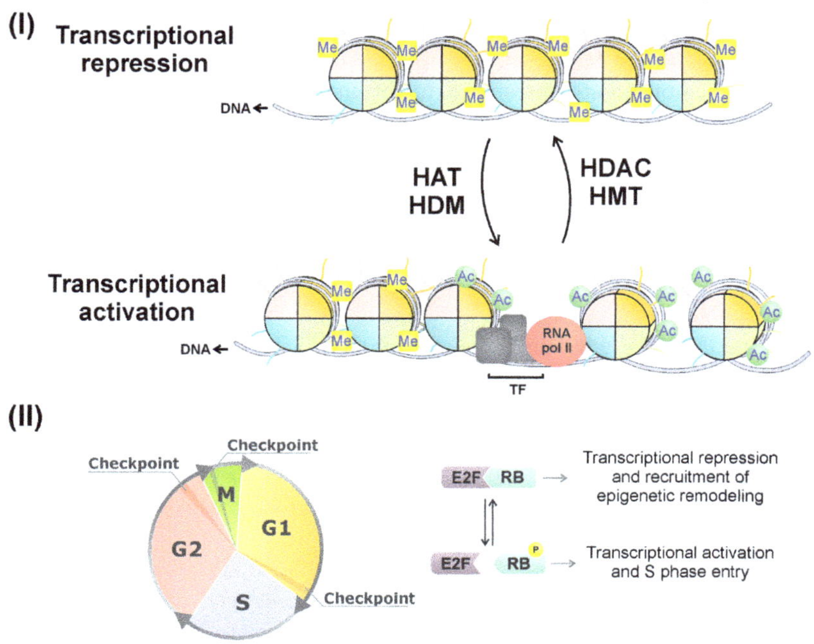

FIGURE 26.6 **(I) Simplified schematic of a transcriptionally inactive configuration of chromatin changing to an active chromatin configuration.** This transition is regulated by modifications in histone tails, where methylation and acetylation are most commonly observed. The proteins responsible for this transition include histone acetyltransferases (HAT), histone demethylases (HDM), histone methyltransferases (HMT), and histone deacetylases (HDAC). (II) Cell-cycle representation and RB-associated transcriptional-regulation mechanism. RB interacts with E2F, preventing the transcription of E2F target genes and recruiting chromatin remodelers to lead the transcriptional repression of genes associated with the cell-cycle progression. Phosphorylated RB is unable to interact with E2F, allowing the transcription and cell-cycle progression.

genes (Fig. 26.6-II) [67]. E2F is a transcription factor that regulates the expression of different genes associated with DNA synthesis and cell-cycle progression from G1 to S phase [66,68]. The RB activity is linked to the inhibition of cell cycle by interacting with E2F, leading to the down-regulation of specific cell cycle–related genes (Fig. 26.6-II) [66,68]. To make this inhibition more efficient, RB also recruits chromatin remodelers (Fig. 26.6-II), such as the co-repressor SIN3 transcription regulator family member B (SIN3B) that promotes lysine deacetylation from histone tails by recruiting histones deacetylases 1 and 2 (HDAC1 and 2) [68]. Lysine acetylation is correlated to transcriptional activation, and its deacetylation leads to a more compacted chromatin structure and consequently, transcriptional repression of E2F-target gene promoters [61,68]. Furthermore, histone methyltransferases (HMT), DNA methyltransferase 1 (DNMT1), and heterochromatin protein 1 (HP1) also are chromatin remodelers recruited by RB stimulation, promoting methylation in promoter region of genes regulated by E2F, contributing to its transcriptional repression [67]. Accordingly, RB is a crucial tumor suppressor, and it is necessary to ensure proper cell-cycle progression, one that promotes the silencing of genes that regulate cell-cycle progression and DNA replication [67].

RB mutations have been associated with reduced H4K20 trimethylation on the N-terminal tail [67,68]. Methylation state of H4K20 has an important role in cell cycle, and has been associated with cell-cycle progression, transcription, chromosome condensation, and origin firing for DNA replication [68]. It is observed that mono- and dimethylation are related to DNA repair and DNA replication, whereas trimethylation of H4K20 is associated with silenced heterochromatin formation and with cell-cycle arrest [68,69]. In addition, aberrant methylation in RB promoter leads to decrease in RB gene expression, and these abnormalities have been observed in different cancers, such as retinoblastoma, bladder cancer, neuroblastoma, and gastric carcinoma, among others [70].

Cell growth and division are both processes regulated tightly by a set of coordinated proteins that monitor cell-cycle progression and DNA integrity. The loss of cell-cycle control and DNA-damage propagation has emerged as the main inducer of cancer and other diseases. In this sense, modifications in the DNA sequence may alter gene products or lead to a loss of gene function. To prevent such genetic deregulation, DNA-damage response is activated, leading subsequently to cell-cycle arrest, recruitment of DNA-repair machinery, and damage correction or apoptosis [71]. Defective activity of epigenetic regulators can also lead to gene expression deregulation and, consequently, cell transformation [72]. As a result, proto-oncogenes expression can be activated by promoter hypomethylation, whereas the expression of tumor suppressors may be silenced by its promoter hypermethylation [72].

Thus, proper DNA repair is necessary for the coordination between chromatin modifications, cell cycle, and DNA-repair machineries [65]. In this sense, different proteins are able to mediate the communication among chromatin and repair, such as ATM, whose activation occurs in response to chromatin structure changes, like formation of DSB [65]. In addition to ATM, DNA-dependent protein kinase (DNA-PK) and/or Rad3-related protein (ATR) mediate the phosphorylation on the variant histone H2AX, which creates a binding motif to mediator of DNA damage–checkpoint protein (MDC1), that recruits other DNA-repair proteins, such as E3 ubiquitin–protein ligase (RNF8) and Nijmegen breakage syndrome 1 (NBS1) [65]. H2AX phosphorylation is the histone modification in response to a DNA break, but it is viewed to act as a broad signal in response to DNA damage, although primarily in the form of DSBs, as well as a triggering pathway in response to stalled replication forks (Fig. 26.7-I) [73].

At DNA-damage sites, the ubiquitin ligase Rnf20/Rnf40 mediates the ubiquitylation of H2B [65]. H2BK123 ubiquitylation is necessary to H3K4 and H3K79 methylation, being these modifications are required to alter the chromatin structure and allow the access of proteins involved in DNA repair (Fig. 26.7-I) [65,73].

Nonetheless, in DNA-damage region, methylated H4K20 (Fig. 26.7-I) acts as a binding platform to the P53-binding protein (53BP1), providing a stable 53BP1–chromatin association [69]. In human cancer–derived cells, the decreasing H4K20 trimethylation has been proposed as a common hallmark related to cell transformation [74]. This transformation can be associated with the fact that low H4K20 methylation avoids the repression of genes that regulate cell-cycle progression [74].

Another histone modification is the acetylation of H3K56 (Fig. 26.7-I) that occurs in response to replication fork damage [73]. During DSB repair, H3K56 acetylation is necessary for Rad52-dependent repair and promotes sister chromatids recombination [73].

Furthermore, in response to UV irradiation, H3K9 (Fig. 26.7-I) is acetylated during the NER process [73]. H3K9 acetylation regulates two pathways, the recruitment of histone acetyltransferase GCN5 to DNA lesions and the coordination of the activity of tumor suppressor p53 and acetyltransferase p300 [73]. In addition, ING2 activity is required to enhance the p53 and p300 interaction to induce the histone acetylation and to mediate a relaxation in chromatin structure; it is also required to recruit the XPA protein to the DNA lesions caused by UV irradiation [62,75]. In this sense, ING2 promotes the chromatin remodeling, which is adequate to DNA-damage repair and to the proper NER [75]. Many aspects of tumor biology, including cancer invasion and metastasis, are associated with deficiency in ING activity [75].

The events described earlier illustrate the importance of chromatin modifications and remodeling, acting in the DNA damage–response signaling and allowing formation of the complexes that mediate the DNA repair. In this sense, according to histone modifications, different chromatin–protein interactions are allowed, interfering with the propagation of repair signaling and turnover of factors involved in repair signaling. In addition, once repair process finishes, histone modifications are frequently reversed, allowing to establish a "prior to DNA damage" chromatin state.

Chromatin remodeling during cell cycle and DNA repair is a mechanistic step that allows or impairs the access to the specific DNA regions. Different chromatin remodelers and histone modifications act as signaling messengers, promoting the recruitment of proteins responsible for different DNA-repair pathways or cell-cycle progression. In addition, epigenetic inactivation of different genes is associated with increased genetic instability and with abnormal cell growth. In this sense, interference with the establishment of histone modifications, changes in chromatin accessibility, or silencing of chromatin are intrinsically associated with tumorigenesis. Thus, epigenetics has become an area of increased interest for the development of therapy and clinical strategies against cancer.

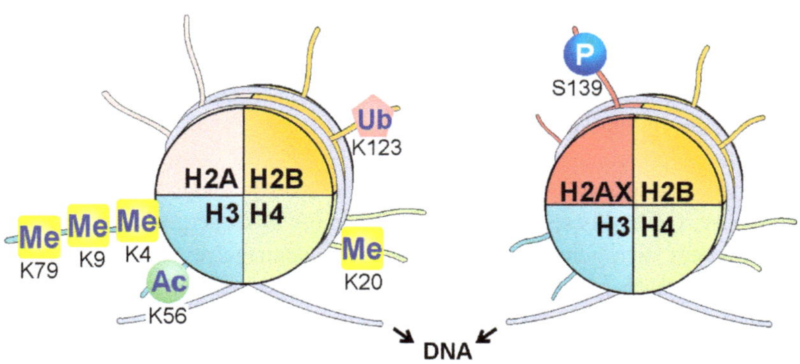

FIGURE 26.7 **Nucleosome representation with the most described modifications in histone tails related to DNA-repair response.**

GLOSSARY

Ataxia telangiectasia (AT) Is a disorder caused by mutations in the ATM gene. This disease is characterized by neurological dysfunctions, cancer predisposition, immunodeficiency, radiosensitivity, infertility and endocrine dysfunctions, among others.

Bloom syndrome (BS) Autosomal recessive genetic disease characterized by the lack of the BLM protein activity. Individuals affected by this disease show morphological abnormalities, photosensitivity, poikiloderma, and high predisposition to cancers.

Cerebro-oculo-facial-skeletal (COFS) syndrome Autosomal recessive genetic disease characterized by an intrinsic inability of the global genome nucleotide excision repair machinery to remove DNA lesions. Individuals affected by this disease present developmental delay, facial abnormalities, microcephaly, cataracts, and microphthalmia, among other symptoms.

Chromothripsis An event of multiple chromosomal rearrangements in a single event.

Clastogenesis A process defined by the loss, addition, or any rearrangement of chromosomes.

Cockayne syndrome (CS) Autosomal recessive genetic disease characterized by an intrinsic inability of the transcription-coupled nucleotide excision repair machinery to remove DNA lesions. Individuals affected by this disease present impaired neurodevelopment photosensitivity progeria, among other symptoms. However, unlike XP, individuals bearing the CS phenotype do not display high predisposition to skin cancers.

Common fragile sites Encountered in the majority of individuals, they are specific locations that are prone to chromosomal rearrangements.

Fanconi anemia Inherited blood disorder caused by intrinsic defects on the interstrand cross-link–repair machinery. Individuals affected by this disease shows severe predisposition to develop myelogenous leukemia and bone-marrow failure, in addition to numerous morphological abnormalities.

Hutchinson–Gilford progeria syndrome (HGTS) Genetic disease caused by a mutation in the LMNA gene. It is a progeroid syndrome characterized by age-related diseases, such as cardiovascular pathologies, skin complications, alopecia, lipodystrophy, joint abnormalities, and osteolysis.

Indels Base insertion and/or deletion.

Lynch syndrome (formerly known as hereditary nonpolyposis colorectal cancer) An inherited condition that makes carriers susceptible to develop certain types of cancer (especially colorectal cancers). They are characterized by mutations in mismatch repair genes.

Merotelic attachments Characterized when a centromere is attached to microtubules coming from both spindle poles.

Mutator phenotype A carcinogenesis hypothesis which postulates that cells acquire cancerous features (see Cancer Hallmarks) due to defective genes responsible for the maintenance of the genome stability.

Philadelphia chromosome A translocation of chromosomes 9 and 22, resulting in a shorter 22 chromosome and a BCR–ABL gene fusion. This phenomenon is present in some hematological malignancies such as acute lymphoblastic leukemia and chronic myelogenous leukemia.

Rothmund–Thomson syndrome (RTS) Autosomal recessive genetic disease characterized by the lack of the RECQL4 protein activity. Individuals affected by this disease show epidermis-related tissue malformation, morphological abnormalities, and cancer predisposition.

Trichothiodystrophy Autosomal recessive genetic disease characterized by an intrinsic inability of the global-genome nucleotide excision repair machinery to remove DNA lesions. Individuals affected by this disease present neurological impairments, brittle hair, and short stature, but do not show photosensitivity.

Werner syndrome (WS) A rare progeroid autosomal recessive disease defined by mutations in WRN gene. Clinical manifestations are aging-like symptoms such as diabetes mellitus type-2, osteoporosis, and cataracts, among others. The affected individuals are also more susceptible to develop cancers (especially sarcomas) and cardiovascular diseases.

Xeroderma pigmentosum (XP) Autosomal recessive genetic disease characterized by an intrinsic inability of the global-genome nucleotide excision repair machinery to remove DNA lesions. Individuals affected by this disease present with high predisposition of skin cancers, skin hyperpigmentation, and, in some cases, can develop neurological impairments, progeria, and cataracts.

LIST OF ACRONYMS AND ABBREVIATIONS

53BP1 P53-binding protein
6,4-PP 6,4-Photoproducts
ALT Alternative lengthening of telomeres
AT Ataxia telangiectasia
ATM Ataxia telangiectasia mutated
ATR Rad3-related protein
BER Base excision repair
BGS Baler–Gerold syndrome
BLM/RECQL3 Bloom syndrome, RecQ helicase-like
BRCA1 Breast cancer gene 1
BS Bloom syndrome
CHK2 Checkpoint kinase 2
CIN Chromosome instability
CPD Cyclobutane pyrimidine dimmers
COFS Cerebro-oculo-facial-skeletal syndrome
CSA Cockayne syndrome WD repeat protein CSA
CSB Cockayne syndrome protein CSB

CS Cockayne syndrome
CUL4A Cullin 4A
DDB1 and DDB2 Damage-specific DNA-binding protein 1 and 2
DNA-PK DNA-dependent protein kinase
DNA Lig DNA ligase IV
DNMT1 DNA methyltransferase
DNA Pol DNA polymerase (delta)
DSB Double-strand break
E2F E2F transcription factor
ERCC1 Excision repair cross-complementation group 1
FA Fanconi anemia
FANC family Fanconi anemia complementation group (composed by many proteins)
FEN1 Flap structure-specific endonuclease 1
GCN5 Histone acetyltransferase GCN5
GG-NER Global genome-nucleotide excision repair
H2AFX H2A histone family, member X
H2BK123 Lysine 123 of histone H2B
H3K4 Lysine 4 of histone H3
H3K9 Lysine 20 of histone H3
H3K79 Lysine 79 of histone H3
H4K20 Lysine 20 of histone H4
HAT Histone acetyltransferase
HDAC1 and 2 Histones deacetylase 1 and 2
HDM Histone demethylase
HGPS Hutchinson–Gilford progeria syndrome
HMT Histone methyltransferase
HP1 Heterochromatin protein 1
HR Homologous recombination
HR23B/RAD23B XP-C repair complementing protein
ICLR Interstrand cross-link repair
ING2 Inhibitor of growth family, member 2
KAP1 KRAB-associated protein 1
KU70 and KU80 X-ray repair cross-complementing protein 6 and 5
LP-BER Long-patch base-excision repair
LMNA Lamin
MDC1 Mediator of DNA-damage checkpoint
MMR Mismatch repair
MRE11 Meiotic recombination 11 homolog 1
MSI Microsatellite instability
NBS1 Nijmegen breakage syndrome 1
NEIL1 Nei endonuclease VIII-like 1 (*Escherichia coli*)
NER Nucleotide excision repair
NHEJ Nonhomologous end joining
NTH1 Nth endonuclease III-like 1
P53 Tumor protein p53
P300 E1A-binding protein P300
PCNA Proliferating cell nuclear antigen
PIKK Phosphoinositide 3-kinase(PI3K)-related protein kinase family
POT1 Protection of telomere 1
PR-Set7 Lysine N-methyltransferase
RAD51 RAD51 recombinase
RAD52 RAD52 homolog (*Saccharomyces cerevisiae*)
RAD53 Serine/threonine-protein kinase RAD53
RAPADILINO *RA*dial hypoplasia, *P*atella hypoplasia and cleft or *A*rched palate, *DI*arrhea and dislocated joints, *LI*ttle size and limb malformation, slender *N*ose and n*O*rmal intelligence syndrome
RB Retinoblastoma tumor suppressor protein
RCF Replication Factor C
RECQL4 RecQ protein-like 4
RMI1/RMI2 RecQ-mediated genome instability 1 and 2
RNA Pol RNA polymerase II

Rnf20/Rnf40 Ring finger protein 20/40 complex
RNF8 Ring finger protein 8
ROC1 Regulator of cullins 1
ROS Reactive oxygen species
RPA Replication protein A
RTS Rothmund–Thomson syndrome
SIN3A SIN3 transcription regulator family member A
SIN3B SIN3 transcription regulator family member B
ssDNA Single-stranded DNA
TC-NER Transcription-coupled-nucleotide excision repair
TFIIH Transcription factor II human
TRF1 and 2 Telomeric repeat binding factor 1 and 2
TTD Trichothiodystrophy
UV Ultraviolet light
WRN/RECQL2 Werner syndrome, RecQ helicase like
WS Werner syndrome
XP Xeroderma pigmentosum
XPA to XPG Xeroderma pigmentosum complementation group A to G
XAB2 XPA-binding protein 2
XRCC4 X-ray repair cross-complementing protein 4

REFERENCES

[1] Gillet LC, Schärer OD. Molecular mechanisms of mammalian global genome nucleotide excision repair. Chem Rev 2006;106(2):253–76.
[2] Menck CF, Munford V. DNA repair diseases: what do they tell us about cancer and aging? Genet Mol Biol 2014;37:220–33.
[3] Palomera-Sanchez Z, Zurita M. Open, repair and close again: chromatin dynamics and the response to UV-induced DNA damage. DNA Repair (Amst) 2011;10(2):119–25.
[4] Kim H, D'Andrea AD. Regulation of DNA cross-link repair by the Fanconi anemia/BRCA pathway. Genes Dev 2012;26(13):1393–408.
[5] Feltes BC, Bonatto D. Overview of xeroderma pigmentosum proteins architecture, mutations and post-translational modifications. Mutat Res Rev Mutat Res 2015;763:306–20.
[6] Schärer OD. Nucleotide excision repair in eukaryotes. Cold Spring Harb Perspect Biol 2013;5(10):a012609.
[7] Egly JM, Coin F. A history of TFIIH: two decades of molecular biology on a pivotal transcription/repair factor. DNA Repair (Amst) 2011;10(7):714–21.
[8] Lehmann AR, et al. Xeroderma pigmentosum. Orphanet J Rare Dis 2011;6:70.
[9] Bergmann E, Egly JM. Trichothiodystrophy, a transcription syndrome. Trends Genet 2011;17(5):279–86.
[10] Jaarsma D, et al. Cockayne syndrome pathogenesis: lessons from mouse models. Mech Ageing Dev 2013;134(5–6):180–95.
[11] Shell SM, Zou Y. Other proteins interacting with XP proteins. Adv Exp Med Biol 2008;637:103–12.
[12] Clauson C, et al. Advances in understanding the complex mechanisms of DNA interstrand cross-link repair. Cold Spring Harb Perspect Biol 2013;5(10):a012732.
[13] Longerich S, et al. Stress and DNA repair biology of the Fanconi anemia pathway. Blood 2014;124(18):2812–9.
[14] Zhang J, Walter JC. Mechanism and regulation of incisions during DNA interstrand cross-link repair. DNA Repair (Amst) 2014;19:135–42.
[15] Mouw KW, D'Andrea AD. Crosstalk between the nucleotide excision repair and Fanconi anemia/BRCA pathways. DNA Repair (Amst) 2014;19:130–4.
[16] Su X, Huang J. The Fanconi anemia pathway and DNA interstrand cross-link repair. Protein Cell 2011;(9):704–11.
[17] Kashiyama K, et al. Malfunction of nuclease ERCC1-XPF results in diverse clinical manifestations and causes Cockayne syndrome, xeroderma pigmentosum, and Fanconi anemia. Am J Hum Genet 2013;92(5):807–19.
[18] Singh DK, et al. RecQ helicases in DNA double strand break repair and telomere maintenance. Mutat Res 2012;736(1–2):15–24.
[19] Bohr VA. Rising from the RECQ-age: the role of human RECQ helicases in genome maintenance. Trends Biochem Sci 2008;33(12):609–20.
[20] Larizza L, et al. Rothmund-Thomson syndrome. Orphanet J Rare Dis 2010;5:1–16.
[21] Croteau DL, et al. RECQL4 in genomic instability and aging. Trends Genet 2012;28(12):624–31.
[22] Fan W, Luo J. RecQ4 facilitates UV light-induced DNA damage repair through interaction with nucleotide excision repair factor xeroderma pigmentosum group A (XPA). J Biol Chem 2008;283(43):29037–44.
[23] Bloom D. Congenital telangiectatic erythema resembling lupus erythematosus in dwarfs; probably a syndrome entity. AMA Am J Dis Child 1954;88(6):754–8.
[24] Arora H, et al. Bloom syndrome. Int J Dermatol 2014;53(7):798–802.
[25] Cheok CF, et al. Roles of the Bloom's syndrome helicase in the maintenance of genome stability. Biochem Soc Trans 2005;33(Pt 6):1456–9.
[26] Hoadley KA, et al. Defining the molecular interface that connects the Fanconi anemia protein FANCM to the Bloom syndrome dissolvasome. Proc Natl Acad Sci USA 2012;109(12):4437–42.
[27] Ozgenc A, Loeb LA. Current advances in unraveling the function of the Werner syndrome protein. Mutat Res Fund Mol M 2005;577(1–2 Spec. Iss.):237–51.

[28] Opresko PL, et al. Werner syndrome and the function of the Werner protein; what they can teach us about the molecular aging process. Carcinogenesis 2003;24(5):791–802.

[29] Rossi ML, et al. Roles of Werner syndrome protein in protection of genome integrity. DNA Repair 2010;9(3):331–44.

[30] Lavin MF. Ataxia-telangiectasia: from a rare disorder to a paradigm for cell signalling and cancer. Nat Rev Mol Cell Biol 2008;9(10):759–69.

[31] McKinnon PJ. ATM and the molecular pathogenesis of ataxia telangiectasia. Annu Rev Pathol 2012;7:303–21.

[32] Shiloh Y. ATM and related protein kinases: safeguarding genome integrity. Nat Rev Cancer 2003;3(3):155–68.

[33] Ghosh S, Zhou Z. Genetics of aging, progeria and lamin disorders. Curr Opin Genet Dev 2014;26:41–6.

[34] Kudlow BA, et al. Werner and Hutchinson-Gilford progeria syndromes: mechanistic basis of human progeroid diseases. Nat Rev Mol Cell Biol 2007;8(5):394–404.

[35] Parreno J, Cruz AV. Accelerated aging in patients with Hutchinson-Gilford progeria syndrome: clinical signs, molecular causes, treatments, and insights into the aging process. UBC Med J 2011;3:8–12.

[36] Gonzalo S, Kreienkamp R. DNA repair defects and genome instability in Hutchinson-Gilford progeria syndrome. Curr Opin Cell Biol 2015;34:75–83.

[37] Liu Y, et al. Involvement of xeroderma pigmentosum group A (XPA) in progeria arising from defective maturation of prelamin A. FASEB J 2008;22(2):603–11.

[38] D'Errico M, et al. The role of CSA and CSB protein in the oxidative stress response. Mech Ageing Dev 2013;134(5–6):261–9.

[39] Vélez-Cruz R, Egly JM. Cockayne syndrome group B (CSB) protein: at the crossroads of transcriptional networks. Mech Ageing Dev 2013;134(5–6):234–42.

[40] Negrini S, et al. Genomic instability – an evolving hallmark of cancer. Nat Rev Mol Cell Bio 2010;11(3):220–8.

[41] Lengauer C, et al. Genetic instabilities in human cancers. Nature 1998;396(6712):643–9.

[42] Abbas T, et al. Genomic instability in cancer development. Cold Spring Harb Perspect Biol 2013;5:1–18.

[43] Bester AC, et al. Nucleotide deficiency promotes genomic instability in early stages of cancer development. Cell 2011;145(3):435–46.

[44] Kloor M, et al. Clinical significance of microsatellite instability in colorectal cancer. Langenbeck Arch Surg 2014;399(1):23–31.

[45] Forment JV, et al. Chromothripsis and cancer: causes and consequences of chromosome shattering. Nat Rev Cancer 2012;12(10):663–70.

[46] Desmaze C, et al. Telomere-driven genomic instability in cancer cells. Cancer Lett 2003;194(2):173–82.

[47] Günes C, Rudolph KL. The role of telomeres in stem cells and cancer. Cell 2013;152(3):390–3.

[48] Hanahan D, Weinberg RA. Hallmarks of cancer: the next generation. Cell 2011;144(5):646–74.

[49] Murnane JP. Telomere dysfunction and chromosome instability. Mutat Res-Fund Mol M 2012;730(1–2):28–36.

[50] Ricke RM, et al. Whole chromosome instability and cancer: a complex relationship. Trends Genet 2008;24(9):457–66.

[51] Nam HJ, et al. Centrosome dynamics as a source of chromosomal instability. Trends Cell Biol 2015;25(2):65–73.

[52] Loeb LA. A mutator phenotype in cancer. Cancer Res 2001;61(8):3230–9.

[53] Halazonetis TD, et al. An oncogene-induced DNA damage model for cancer development. Science 2008;319(5868):1352–5.

[54] Mazouzi A, et al. DNA replication stress: causes, resolution and disease. Exp Cell Res 2014;329(1):85–93.

[55] Ozeri-Galai E, et al. The complex basis underlying common fragile site instability in cancer. Trends Genet 2012;28(6):295–302.

[56] Kollareddy M, et al. Aurora kinase inhibitors: progress towards the clinic. Invest New Drug 2012;30(6):2411–32.

[57] Curtin NJ. DNA repair dysregulation from cancer driver to therapeutic target. Nat Rev Cancer 2012;12(12):801–17.

[58] Vitale I, et al. Mitotic catastrophe: a mechanism for avoiding genomic instability. Nat Rev Mol Cell Bio 2011;12(6):385–92.

[59] Geigl JB, et al. Defining "chromosomal instability". Trends Genet 2008;24(2):64–9.

[60] Nowak MA, et al. The role of chromosomal instability in tumor initiation. P Natl Acad Sci USA 2002;99(25):16226–31.

[61] Sharma S, et al. Epigenetics in cancer. Carcinogenesis 2010;31(1):27–36.

[62] Li S. Implication of posttranslational histone modifications in nucleotide excision repair. Int J Mol Sci 2012;13(10):12461–86.

[63] Ma Y, et al. How the cell cycle impacts chromatin architecture and influences cell fate. Front Genet 2015;6:19.

[64] Lahtz C, Pfeifer GP. Epigenetic changes of DNA repair genes in cancer. J Mol Cell Biol 2011;3(1):51–8.

[65] Johnson DG, Dent SY. Chromatin: receiver and quarterback for cellular signals. Cell 2013;152(4):685–9.

[66] McLaughlin F, et al. The cell cycle, chromatin and cancer: mechanism-based therapeutics come of age. Drug Discov Today 2003;8(17):793–802.

[67] Gonzalo S, Blasco MA. Role of Rb family in the epigenetic definition of chromatin. Cell Cycle 2005;4(6):752–5.

[68] Corney DC, Coller HA. On form and function: does chromatin packing regulate the cell cycle? Physiol Genomics 2014;46(6):191–4.

[69] Jorgensen S, et al. Histone H4 Lysine 20 methylation: key player in epigenetic regulation of genomic integrity. Nucleic Acids Res 2013;41(5): 2797–806.

[70] Mastrangelo D, et al. Retinoblastoma as an epigenetic disease: a proposal. J Cancer Ther 2011;2(3):362–71.

[71] Cao J, Yan Q. Histone ubiquitination and deubiquitination in transcription, DNA damage response, and cancer. Front Oncol 2012;2:26.

[72] Henrique R, et al. The epigenetics of renal cell tumors: from biology to biomarkers. Front Genet 2012;3:94.

[73] House NCM, et al. Chromatin modifications and DNA repair: beyond double-strand breaks. Front Genet 2014;5:296.

[74] Evertts AG, et al. H4K20 methylation regulates quiescence and chromatin compaction. Mol Biol Cell 2013;24(19):3025–37.

[75] Jafarnejad SM, Li G. Regulation of p53 by ING family members in suppression of tumor initiation and progression. Cancer Metastasis Rev 2012;31(1–2):55–73.

Chapter 27

Cancer and Genomic Instability

W. Wei[1], Y. Cheng[2,4], B. Wang[3]

[1]*Massachusetts General Hospital Cancer Center, Boston, MA, United States;* [2]*University of British Columbia, Vancouver, BC, Canada;* [3]*University of Lethbridge, Lethbridge, AB, Canada;* [4]*Xiamen University, Xiamen, P.R. China*

Chapter Outline

1. INTRODUCTION

Cancer is a complex disease that is often associated with genomic instability. Since 1975 when an important role of genomic (chromosome) instability was first postulated in the development of lymphoid malignancies [1], it has been recognized as a key hallmark of cancer due to its characteristics in most human cancers [2]. In 2000s, many new findings have highlighted that certain DNA-repair pathways and cell cycle–control processes have important consequences for genomic stability and cancer cell biology. The deficiency of DNA-repair pathways may therefore result in genomic instability and cancer development.

2. DNA-REPAIR PATHWAYS

A large body of evidence has demonstrated that the genomic DNA of eukaryotic cells is constantly challenged by genotoxic stresses arising from either physiological metabolism or environmental exposure or both which frequently result in numerous DNA lesions (Table 27.1). It is estimated that every day thousands of DNA lesions are induced in each human cell [3]. However, multiple important DNA-repair pathways, including base excision repair (BER), nucleotide excision repair (NER), mismatch repair (MMR), nonhomologous end joining (NHEJ), and homology-directed repair (HDR) pathways, have evolved to remove the damaged regions from genomic DNA, thus preventing a key molecule from the deleterious consequences of lesion accumulation.

Genome Stability. http://dx.doi.org/10.1016/B978-0-12-803309-8.00027-6

TABLE 27.1 DNA Lesions and Repair Pathways

DNA-Damage Source	Type of Lesion	DNA-Repair Pathway
ROS, IR, alkylating agents	Altered base, abasic site, SSBs	BER
Ultraviolet light, cisplatin intrastrand adducts, polyaromatic hydrocarbons	Intrastrand cross-links, bulky DNA adducts	NER
IR, alkylating agents, cisplatin	Interstrand cross-links, DSBs	DSB repair (NHEJ/HR)
Replication errors	Base mismatches, insertions and deletions	Mismatch repair

This table was drawn based on that summarized elsewhere Reed SH, et al. Nucleotide excision repair in chromatin: damage removal at the drop of a HAT. DNA Repair 2011;10(7):734–42.

2.1 Base Excision–Repair Pathway

It has been indicated that small, nonhelix-distorting base lesions induced by oxidization, alkylation, and deamination in genomic DNA are the primary target of the BER pathway [4,5]. Generally, the BER process can be divided into four basic steps: apurinic/apyrimidinic (AP) site formation and excision, DNA end processing, gap filling, and DNA ligation (Fig. 27.1) [5]. It is believed that the BER pathway is initiated by DNA glycosylase that binds and removes the damaged bases, forming an abasic site–containing intermediate [4,5]. The apurinic endonuclease 1 (APE1) has the AP endonuclease activity and cleaves at the 5′-side of the AP-site ribose, producing either a 5′-deoxyribose phosphate (dRP) causing nonoxidized damages or a 5′-ribose phosphate causing oxidized damages. This process generates a temporary DNA single-strand break (SSB)—one of the most frequent lesions in genomic DNA. The 5′-dRP intermediate is further processed by DNA polymerase β (Polβ) which possesses the 5′-deoxyribose phosphate (dRP) lyase activity removing the 5′-dRP and forming a single-nucleotide gap that can be filled up by DNA Polβ. Other proteins, such as APE1, polynucleotide kinase 3′-phosphotase (PNKP), aprataxin (APTX), and the X-ray-repair cross-complementing protein 1 (XRCC1), may also be involved in the repair of the damaged termini. The nicked DNA is then ligated by DNA ligase 3 (LIG3). This pathway is termed short-patch (1 nt) BER. However, the SSBs induced by the oxidized ribose phosphate can be repaired through the long-patch BER pathway. Poly(ADP-ribose) polymerase 1 (PARP1), a chromatin-associated enzyme that catalyzes protein ADP-ribosylation and functions as an SSB sensor, primarily binds and is rapidly activated by DNA strand breaks. As a result, DNA end-processing enzymes, such as APE1, PNKP, APTX, and PARP1, get modified with branched poly(ADP-ribose) chains that repair the damaged 3′-termini. Simultaneously, flap endonuclease 1 (FEN1) that is activated by proliferating-cell nuclear antigen (PCNA) and PARP1 remove the damaged 5′-terminus of two or more nucleotides (2–12 nt). The left gap is then filled up with Polβ and Polδ/ε, and the nicked DNA is rejoined by DNA ligase 1 (LIG1) (Fig. 27.1).

2.2 Nucleotide Excision–Repair Pathway

NER is the major pathway for removing bulky, conformation-distorting DNA lesions induced by exogenous genotoxic agents such as UV irradiation, environmental mutagens, and certain chemotherapeutic drugs [6,7]. As it is demonstrated, two NER sub-pathways contribute to the removal of these DNA lesions. The global genome NER (GG-NER or GGR) has evolved to repair the lesions throughout the whole genome, while transcription-coupled NER (TC-NER or TCR) specifically removes DNA lesions in the transcribed strand of the activated genes.

Upon UV-induced DNA damage, GG-NER is triggered by the UV-damaged DNA-binding protein (UV-DDB) and XPC/RAD23B/CETN2 complexes that recognize and bind UV-induced DNA photolesions such as cyclobutane pyrimidine dimers (CPD) and 6–4 photoproducts (6–4 PP) (Fig. 27.2 left). Once the XPC complex is bound to the lesion, the component RAD23B dissociates from the complex. On the other hand, TC-NER is initiated by the stalled RNA polymerase II (RNA PolII) during transcription elongation (Fig. 27.2 right). It has been demonstrated that RNA PolII arrested by DNA damage interacts closely with the Cockayne syndrome B protein (CSB), the UV-stimulated scaffold protein A (UVSSA), and the ubiquitin specific–processing protease 7 (USP7). Upon stalling at the lesion site, the Cockayne syndrome WD repeat protein CSA and CSB complex is formed and causes RNA PolII to move backwards (also termed backtracking), which makes the lesion accessible for repair. After lesion recognition, the transcription initiation factor IIH (TFIIH) complex in which XPB, XPD, and other five subunits form a core complex and the single-strand DNA-binding protein RPA (the replication protein A) are recruited to the lesion site, hence they form the pre-incision complex in both the GG-NER and TC-NER pathways. XPB and XPD may help create transcription bubble because of their helicase and ATPase activities.

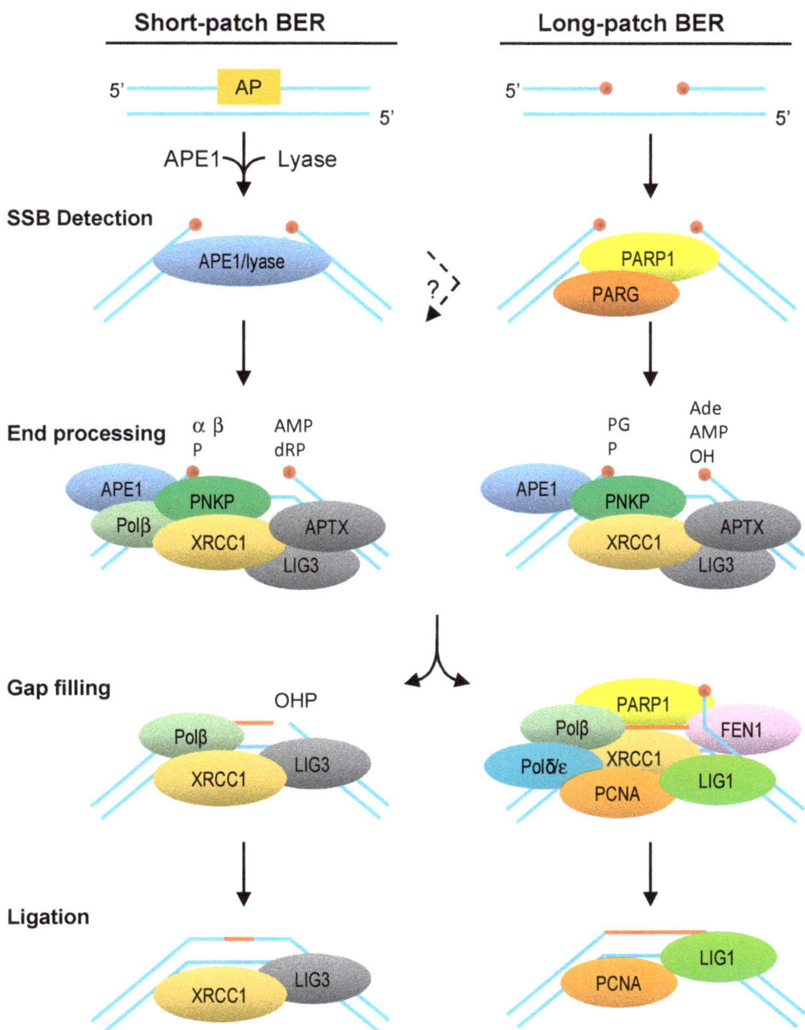

FIGURE 27.1 **A model for the base excision-repair pathway.** Single-strand breaks arisen indirectly from enzymatic incision at the AP site during BER or directly from disintegration of oxidized deoxyribose are primarily removed via the BER pathway which is generally divided into short-patch and long-patch BERs based on the size of nucleotide patch length. Numerous proteins overlap in the "four-step" repair processes. *Red circles* indicate damaged termini. This model was redrawn according to that proposed by Caldecott [5]. *We thank Nature Publishing Group's for permission, http://www.nature.com/nrg/journal/v9/n8/full/nrg2380.html.*

The XPG endonuclease also binds to the pre-incision NER complex. After dissociation with the CAK (CDK-activating kinase) subcomplex, the helicase activity of the TFIIH complex further unwinds the double helix, thus facilitating XPA to recruit and activate XPF/ERCC1. Once the 5′-side of the lesion was incised by the XPF/ERCC1 heterodimer, XPG is activated and incises the damaged strand at the 3′-side of the lesion releasing a 22–30 nt–long strand. PCNA then recruits DNA polymerase δ, κ, or ε to fill up the gap, and DNA ligase I or III rejoins the nick ends to complete the repair process.

2.3 Mismatch-Repair Pathway

Mismatch repair (MMR) is a major pathway that removes mismatch lesions generated during DNA replication, which represents a considerable threat to the genomic integrity [8,9]. The initiation of the MMR pathway can be attributed to the recognition of mismatches by either MutSα (MSH2–MSH6 heterodimer) or MutSβ (MSH2–MSH3 heterodimer) which recruits MutLα to form an ATP dependent–ternary complex (Fig. 27.3), thus facilitating this complex to undergo an ATP driven–conformational change and form a sliding clamp. This clamp can track the strand discontinuity in either direction along DNA. Replication factor C (RFC) acting as a clamp loader binds the 3′-terminus of the strand break and helps load PCNA onto the DNA. The clamps that slid to the upstream RFC loaded at the 5′-terminus of the strand break lead to RFC disassociation with DNA, facilitating the recruitment of EXO1 that mediates the degradation of the mismatch-containing

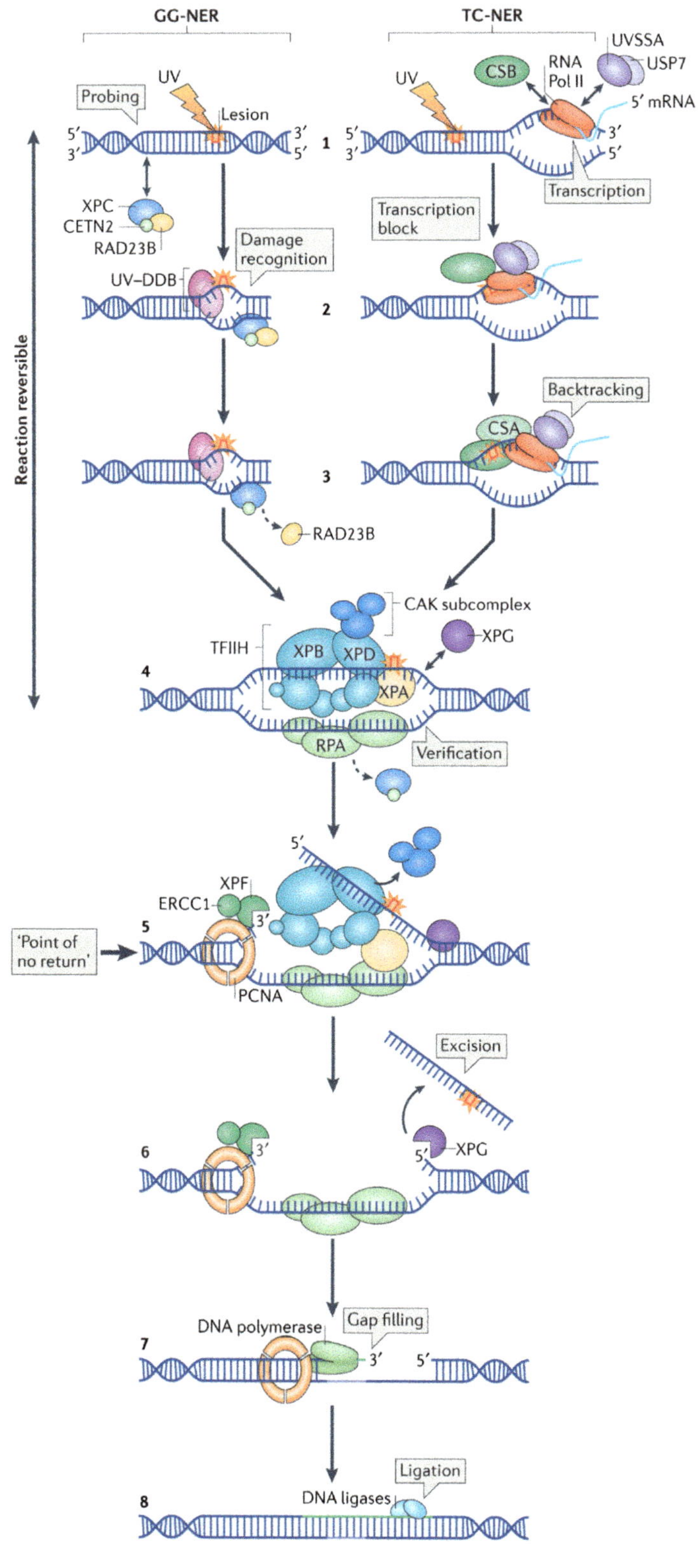

FIGURE 27.2 **A model for the nucleotide excision–repair pathway.** Bulky DNA lesions induced by exogenous genotoxic agents are mainly repaired by NER which is divided into global genome NER (GG-NER) and transcription-coupled NER (TC-NER) sub-pathways. *Adapted from Marteijn JA, et al. Understanding nucleotide excision repair and its roles in cancer and ageing. Nat Rev Mol Cell Biol 2014;15(7):465–81. We thank Nature Publishing Group's for permission, http://www.nature.com/nrm/journal/v15/n7/full/nrm3822.html.*

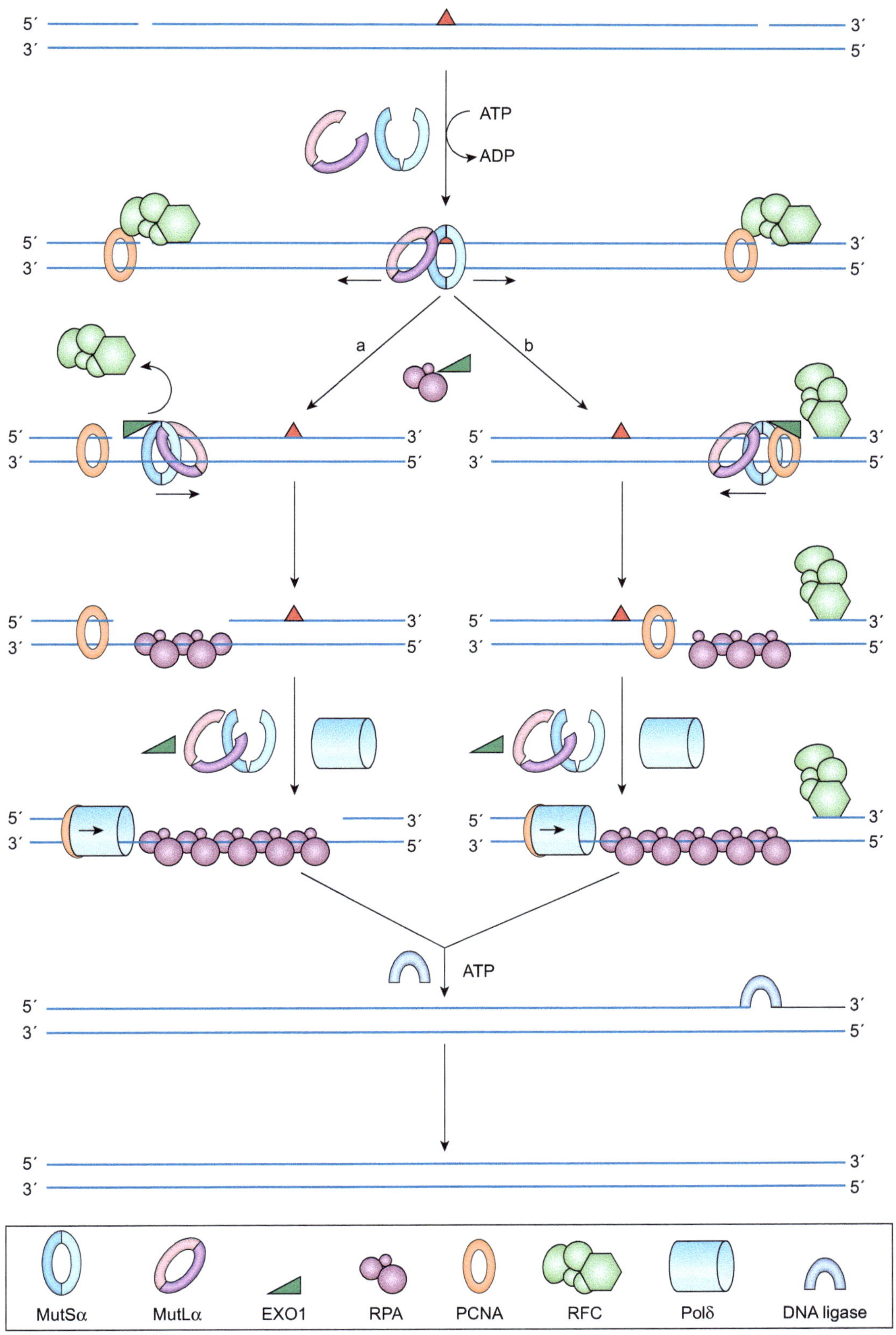

FIGURE 27.3 **A model for the mismatch-repair pathway.** Mismatch lesions arisen during DNA replication are predominantly removed via the mismatch-repair pathway. *Adapted from Jiricny J, et al. The multifaceted mismatch-repair system. Nat Rev Mol Cell Biol 2006;7(5):335–46. We thank Nature Publishing Group's for permission, http://www.nature.com/nrm/journal/v7/n5/full/nrm1907.html. Copyright © 2006 Nature Publishing Group.*

strand in a 5′→3′-direction (Fig. 27.3A). RPA then binds and stabilizes a single-strand gap. Once the mismatch lesion is removed, the EXO1 activity is suppressed by MutLα and released from the DNA. DNA Polδ then loads at the 3′-terminus of the discontinuity that is bound with PCNA homotrimer which is essential for Polδ-mediated DNA synthesis. This DNA Polδ resynthesizes a new DNA to fill the gap, and DNA ligase I seals the remaining nick. The clamps that track the downstream PCNA homotrimer bind at the 3′-terminus of the strand break (Fig. 27.3B) recruiting and/or activating EXO1 that triggers the degradation of the strand containing a mismatch in a 3′→5′-direction. After the removal of the mismatch lesion, Polδ is responsible to fill the gap, and DNA ligase I seals the remaining nick to complete the process of MMR.

2.4 Repair of DNA Double-Strand Breaks

Canonical nonhomologous end joining (C-NHEJ) and HDR are two major pathways engaged to repair DNA double-strand breaks (DSB). The initiation of C-NHEJ is triggered by heterodimer Ku (Ku70 and Ku80) that binds to the ends of DSB and recruits the DNA-dependent protein kinase catalytic subunit (DNA-PKcs) to form the DNA-PK holoenzyme [10]. The active kinase complex phosphorylates other repair-related proteins and promotes Artemis-mediated end processing. DNA integrity is then restored by rejoining of the DNA ends by the Lig4/XRCC4/XLF complex. C-NHEJ is a fast but error prone–repair machinery.

In contrast to C-NHEJ, homologous recombination HDR is an error-free DSB-repair pathway that is primarily governed by homologous sister chromatids in higher eukaryotes. HDR preferentially monitors the DSB that occurs in the late S and G2 phases of the cell cycle [11]. It is believed that the HDR-repair pathway is initiated by binding of ATM (the ataxia-telangiectasia-mutated gene) to the DSB [12] that recruits repair proteins such as BRCA1 (breast cancer 1 gene) and MRN complex (Mre11, Rad50, and Nbs1) to process the ends of DSB, thus generating a long 3′-single-stranded DNA (ssDNA) overhanging on both sides of the break [10]. RPA then binds and stabilizes ssDNAs. However, the binding of the recombinase RAD51 results in the release of RPA from DNA. RAD51 acts together with RAD54 to search for DNA homology, the ATPase activity of RAD54 helps unwind DNA facilitating strand invasion [13]. After displacement loop (D-loop) formation and strand invasion, a DNA polymerase extends the end of the invading 3′-strand by synthesizing a new DNA. After DNA synthesis, two sub-pathways—double-strand break repair (DSBR) and synthesis-dependent strand annealing (SDSA) may be engaged [10]. Both 3′-overhangs are involved in DSBR to generate a double Holliday junction which is then converted into recombination products by nicking endonucleases, resulting in crossover (common) or non-crossover products. In SDSA, only one 3′-overhang participates to form a single Holliday junction, resulting in non-crossover products.

3. GENOMIC INSTABILITY IN HEREDITARY CANCER

As suggested by the mutator phenotype hypothesis, tumors arise as a consequence of the accumulation of mutations and the corresponding dysregulation of key cellular functions [14]. However, the natural mutation rates in human cells are insufficient to account for the genomic instability observed in tumor cells. In addition, no more than 150 genes have been identified as cancer drivers [15], while most of the mutations observed in cancer cells do not present a significant functional impact on the neoplastic phenotype. Therefore, to confer oncogenic transformation, pre-neoplastic cells have to incur the increased rates of mutagenesis which will aggregate as a consequence of initial transformation and aggravate throughout the whole process of neoplastic progression.

The genomic maintenance involves the integration of multiple DNA damage–repair mechanisms to remove molecular lesions resulted from various intrinsic and extrinsic challenges. The identification of mutations in genes responsible for sensing and repairing DNA damages has provided a strong support for the mutator hypothesis. In contrast to somatic mutations which sporadically occur in individual cells, germline mutations may have been present since birth in every cell in an individual's body. Cells in individuals carrying germline mutations of genes that primarily function to maintain the genomic stability may present higher intrinsic mutagenesis rates due to haplodeficiency which makes precancerous cells more susceptible to genotoxic challenges. For example, the loss of the remaining wild-type allele will completely abrogate the gene function, thus resulting in uncontrolled mutation accumulation that will eventually trigger the neoplastic transformation, as predicted by the mutator hypothesis. Therefore, individuals with an inherent germline mutation of genes that encode DNA damage–response proteins frequently elicit a number of genomic instability syndromes, and these disorders often result in a heightened predisposition to cancer, as seen in familial cancer syndromes.

DNA-repair defects are a common cause of inherited cancer susceptibility, and many examples have now been recognized. Some of these are autosomal recessive conditions such as ataxia telangiectasia, Fanconi anemia (FA), and xeroderma pigmentosum (XP). Hereditary nonpolyposis colon cancer and breast cancer susceptibility due to mutations of MMR cluster *BRCA1/2*, respectively, are examples of autosomal dominant cancer susceptibility syndromes due to inherited alterations in genes that are involved in DNA MMR and HDR.

3.1 Li–Fraumeni Syndrome and TP53

The Li–Fraumeni syndrome (LFS) is an autosomal dominantly inherited disorder primarily determined by heterozygous germline mutations in the tumor-suppressor gene *TP53* (located at chromosome 17p13) [16]. *TP53* encodes a transcription factor activated in response to various stress signals, including DNA-damage signaling, and is thus implicated in the maintenance of genomic stability. The activation of TP53 prevents the proliferation of cells with damaged DNA by inducing cell-cycle arrest to allow DNA repair or directly trigger programmed cell death upon irreversible DNA damages. Because of its comprehensive role in DNA-damage response, TP53 is also defined as "the guardian of the genome." LFS individuals carry missense mutations of *TP53*, most frequently within exons 5–8 coding for the DNA-binding domain of the protein. Such mutation nullifies the TP53 transcriptional activity resulting in haploinsufficiency with reduced protein expression and consequently impairs the genomic protective role of TP53. Interestingly, the spectrum of *TP53* germline mutations found in LFS patients reflects those found in sporadic tumors, suggesting the importance of TP53 inactivation in tumorigenesis [17].

As one of the most well-recognized cancer predisposition syndromes, LFS has been characterized by a strikingly increased risk of early-onset breast cancer, sarcomas, brain tumors, adrenocortical carcinoma, and other neoplasms [18]. Particularly, about 50% of patients with LFS develop the first tumor by the age of 40, compared with 1% in the general population, and 90% of the carriers are diagnosed with cancer by the age of 60 [19]. In a retrospective study of 200 cancer-affected LFS individuals who survived childhood malignancies, 15% developed multiple primary cancers over their lifetimes [20].

Although the molecular basis of tumor predisposition in LFS is still a matter of debate, several potential mechanisms have been described since the identification of *TP53* mutation in 1990. In the setting of LFS, although not exclusively, the risk of tumor is specifically and significantly related to germline mutation of *TP53*, regardless of tumor type [16]. Germline *TP53* mutation–associated cancer eventually develops in 73% of men and almost 100% of women who carry such mutations, with the higher penetrance in the latter predominantly attributable to breast cancer. Analyses of genomic instability in fibroblasts established from LFS carriers have revealed that TP53 loss of function may increase spontaneous mutation activities which could lead to high rates of chromosomal instability and an allelic loss of key tumor-suppressor genes because these LFS fibroblasts spontaneously immortalize [21]. In 2006, *TP53* germline mutation carriers have been found to have a higher prevalence to carry single-nucleotide polymorphism (SNP) 309 (T>G variation) in the murine double-minute 2 (*MDM2*) gene [22]. MDM2 is a key negative regulator of TP53 that increases protein turnover through proteasomal degradation. The SNP 309T>G variation located in the first intron of *MDM2* increases Sp1 transcription factor binding and consequently the MDM2 expression levels. This augmented negative feedback loop further aggravates the haploinsufficiency of TP53, thus making the LFS patients who have SNP T309G in *MDM2* suffer an even earlier onset of tumor formation. Conversely, a higher occurrence rate in TP53-negative patients with the *MDM2* SNP 309G allele also suggests the loss of TP53-related protection against potential genotoxic mutations in LFS patients.

In addition, *TP53* mutation in LFS has also been demonstrated to be associated with severe chromosomal aberrations and aneuploidy. TP53 is involved in the very processes known to give rise to copy number variations (CNVs), which has a 100–10,000 times higher natural occurrence rate than point mutation in the human genome [23]. Defective TP53 has been linked to an increased CNV and genomic instability in tumors. Similarly, LFS fibroblasts have a tendency of a loss of chromosomal regions containing genes involved in cell-cycle control or senescence [24]. In addition, large-scale comparative genomic hybridization studies have revealed that the CNV frequency is remarkably similar among healthy individuals but is significantly increased in LFS patients with germline *TP53* mutation [25]. LFS patients with *TP53* germline mutation present a broad spectrum of copy number alterations affecting multiple loci, including exceptionally large deletions or duplications [26]. The CNVs in p53 mutation carriers themselves frequently encompass cancer genes. In 2012, a large *BRCA1* intragenic deletion related to germline *TP53* mutation was reported in breast cancer patients previously diagnosed with LFS [27].

3.2 MYH-Associated Polyposis and Deficiency in Base Excision Repair

One of the major roles of BER in the maintenance of genome stability is to correct subtle modifications of DNA induced by reactive oxygen species (ROS)-related DNA damage which generates 8-oxoguanine products (8-oxoG). A misincorporated oxidized guanine can mismatch with adenine (Hoogsteen base pair), resulting in G:C to T:A transversion and a consequent point mutation. The codon GAA in particular is more susceptible to such mutational event which will lead to a stop codon (TAA) on condition that BER is nonfunctional. In humans, the repair of misincorporated 8-oxoG is primarily orchestrated by three proteins. The DNA glycosylase OGG1removes the 8-oxoG from 8-oxoG:C base pairs in duplex DNA; the MYH

(MUTYH) DNA glycosylase excises adenine residues mismatching unrepaired 8-oxoG replication, while MTH1 is an 8-oxo-dGTPase which hydrolyzes 8-oxo-dGTP preventing its re-incorporation into a newly synthesized DNA during the patching step of BER [4].

In 2002, biallelic mutation of the *MYH* gene (alias: *MUTYH* located on chromosome 1p32.1–p34.3) was linked to an autosomal recessive CRC predisposition syndrome associated with multiple colonic polyps [28]. More than 80 germline variants have been reported with the deleterious missense mutations Y165C and G382D account for more than 80% of mutations occurring in affected individuals with a Caucasian background. Excess risk of colon cancer occurs in biallelic *MYH* mutation carriers, of whom about 70% carry both Y165C and G382D mutations. Jenkins et al. reported a 3-fold increase in risk for colorectal cancer in monoallelic carriers and a 50-fold increase in risk in biallelic carriers (the 8% and 80% cumulative risk to age 70 years, respectively) [29]. The *APC* gene has been suggested as a major target of MYH mutation-related BER deficiency, probably due to its rich in GAA codon [28]. The rate of spontaneous G:C to T:A transversions in the *APC* gene is significantly higher in colorectal tumor cells with biallelic *MYH* mutation compared to tumor cells with intact *MYH* [30]. Concordantly, many MYH-associated polyposis patients elicit a comparable phenotype to those suffering classic or attenuated familial adenomatous polyposis (FAP) which is hallmarked by the mutated *APC* gene and hyperactivation of the β-catenin/TCF pathway [31]. In addition, G:C to T:A transversions have also been described in the proto-oncogene K-Ras, resulting in a point mutation of G12C in K-Ras. Such K-Ras mutations have been associated with a poor prognosis in colorectal cancer patients [32].

3.3 Xeroderma Pigmentosum and a Deficiency in Nucleotide Excision Repair

XP is an autosomal recessive disorder characterized by an extreme sensitivity to ultraviolet light, hyperpigmentation of the sun-exposed area, neurodegeneration, and a greatly elevated incidence of skin cancers. Compared to the general population, XP individuals have a 10,000-fold increase in the incidence of developing nonmelanoma tumors and a 2,000-fold increase in the incidence of melanoma before the age of 20 in the sun-exposed tissues such as the skin, eyes, lips, and the tip of the tongue [33]. The median age at first diagnosis of skin neoplasm for XP individuals is 8 years, nearly 50 years younger than that found in the general population. XP also manifests a 10–20-fold increase in the risk of internal neoplasms such as lung cancer [34], which may be attributed to a higher susceptibility to environmental carcinogens such as tobacco.

Although XP presents itself as heterogeneous clinical features with eight different subtypes, most of the molecular background of XP indicates defects in NER. The DNA molecule has a strong absorption of both UV-B and UV-C which have a wavelength ranging from 280 to 320 nm and 240 to 280 nm, respectively. Upon exposure to UV radiation, DNA in epidermal cells tends to form two major photoproducts: cyclobutane pyrimidine dimers (CPD) and pyrimidine (6–4) pyrimidone photoproducts (6–4 PP) that involve both T and C pyrimidines. These pyrimidine dimer-related DNA lesions require an excision of single-stranded regions of DNA by the NER system rather than BER that shows a high efficiency for small-nucleotide adducts such as methylation or incorporation of apurinic/apyrimidinic nucleotides. Large DNA adducts such as CPD and 6–4 PP, if left unrepaired, will induce DNA conformational changes and subsequently transcription and replication failure which will lead to cell senescence, apoptosis as well as mutation accumulation to promote carcinogenesis [35]. Indeed, XP presents a unique model for analyzing the effects of unrepaired DNA lesions in skin carcinogenesis. The skin cancer predisposition observed in XP patients has been attributed to eight genes, of which each point missense mutation resulted in an XP subtype. Seven of these genes, namely *XPA* (9q22.3), *XPB/ERCC3* (2q21), *XPC* (3p25), *XPD/ERCC2* (19q13.2–q13.3), *XPE/DDB2* (11p11–12), *XPF/ERCC4* (16p13), and *XPG/ERCC5* (13q33), are required to remove UV damage from the DNA. The eighth one (XPV/POLH or DNA Polη) is required to replicate the DNA containing unrepaired damage. Eukaryotic NER includes two major branches, transcription-coupled repair and global genome repair [6,7]. Transcription-coupled NER specifically induced by the damaged DNA that blocks the progression of RNA PolII and efficiently repairs DNA lesions present on the transcribed strand of actively expressed genes. Conversely, global genome repair is an RNA transcription-independent, random process that inspects the entire genome for damage and is therefore critical in protecting the nontranscribed–regulatory DNA sequences from UC-induced damages. The NER pathway eliminates DNA legions by a multi-step "cut and patch" reactions, including the recognition of helix distortions (by XPE and XPC), DNA-damage verification (by XPA), unwinding of the DNA from around the lesion (by XPB and XPD along with the TFIIH protein to prime excision of damaged DNA), DNA cut (by XPF and XPG for 5′ and 3′, respectively), and finally resynthesis for gap filling (by DNA Polδ, DNA Polκ, or DNA Polε) and DNA ligation (such as DNA ligase III).

It is clear that UV exposure is one if not the foremost risk factors for all three major subtypes of skin cancer, including a basal cell carcinoma, squamous cell carcinoma, and melanoma. The photochemistry of DNA actually leaves a characteristic fingerprint on the mutation spectrum known as UV-signature mutation. In general, dipyrimidine sites, particularly the 3′C of a TC or CC site, are most susceptible to UV-induced mutation due to the inclination of deamination in the C–C photo-adduct leaving a C>T transition. Indeed, the analysis of type A to G XP tumors that have the defective NER capacity shows that the unrepaired DNA lesions result in higher levels of C to T and the UV signature tandem mutation CC to TT. Conversely, although the mechanism has not been fully revealed, mutation frequencies from UV damage are also increased in cells that lack POLH/XPV which encodes the class-Y polymerase capable of bypassing dipyrimidine DNA lesion to continue DNA replication. This explains the phenotype of Type-V XP which is hallmarked by defects in *POLH/XPV* gene but still can have an intact NER pathway.

UV signature mutations in several oncogenes and tumor-suppressive genes have been demonstrated to be implicated in XP-related tumor predisposition. The examination of mutations in the *TP53* gene in tumors from XP individuals reveals the C>T UV-signature mutations at a rate of more than 90% [36]. In contrast, the *TP53* mutation rate observed in sporadic skin tumors is 50%. In addition, $p16^{INK4a}$ and $p14^{ARF}$, both of which participate in the p53-related control of cell cycle, are also susceptible to UV-induced deleterious mutations in XP cells [37]. On the other hand, active mutations in all three Ras oncogenes, in particular in *N-ras*, have been observed at an approximately doubled rate in XP-related skin cancer compared to skin cancer in non-XP population (50% vs. 25%) [36]. Finally, it is important to note that the hyperactivation of the mitogenic sonic hedgehog (SHH) pathway has been associated to the initiation of skin basal cell carcinoma (BCC) in patients with XP background [38]. Germline mutations of the *PTCH* gene, one of the major components of a negative regulator of the SHH pathway and therefore a tumor-suppressor gene, have been identified to have UV-induced mutation in about 90% of XP-related BCC, which is significantly higher than the somatic mutation rate of 10–40% in non-XP BCC. In addition, significantly higher rates of UV-signature mutation in BCC from XP individuals have also been observed in two positive regulators of the SHH pathway, namely Shh (the ligand, ~20% in XP-BCC vs. <1% in sporadic BCC) and Smo (GLI1/2 transcription factor activator, ~30% in XP-BCC vs 10–15% in sporadic BCC) [36].

3.4 Hereditary Cancers Associated With Defects in DNA Mismatch Repair

Hereditary nonpolyposis colon cancer (HNPCC), also known as Lynch syndrome, is one of the commonest forms of inherited predisposition to colorectal cancer (CRC) accounting for 2–5% of all CRCs. CRC in individuals with Lynch syndrome differs from sporadic CRC by an earlier age of diagnosis (in the mid-40s). Individuals with Lynch syndrome have a probability of 70% to develop CRC by the age of 70 years. In contrast, the risk in the general population is around 5%. Moreover, Lynch syndrome–affected individuals are at an increased risk of a number of extra-colonic malignancies. The lifetime risk of endometrial adenocarcinoma is 30–60% (<3% in the general population) diagnosed at an average age of 40 years. Gastric, hepatobiliary tract, urinary tract, ovarian, and small bowel carcinoma have also been documented as Lynch syndrome–related malignancies with the cumulative lifetime risk at 19%, 18%, 10%, 9%, and 4%, respectively [39]. In addition, multiple sebaceous skin tumors occur in a subgroup of HNPCC, the Muir-Torre syndrome [40].

Lynch syndrome is primarily attributed to deleterious germline mutations in MMR genes, including *MLH1* at 3p21.3, *MSH2* at 2p21-22, *MSH3* at 5q11-12, *PMS1* at 2q31-33, *PMS2* at 7p22, and *MSH6* at 2p16, which lead to the loss of expression of one or more of the MMR proteins. In HNPCC, the most frequently affected genes are *MSH2* (52%), *MLH1* (22%), MSH6 (13%), and *PMS2* (9%), *with PMS1 and MSH3 gene being* occasionally involved [41]. In another study, *MLH1* and *MSH2* account for more than 90% of the MMR mutations in Lynch syndrome families [42]. The mutational profile of MMR genes spans from point missense or nonsense mutations to large deletions and rearrangements which compromise 5–10% of *MLH1* mutations and more than 20% of *MSH2* mutations in HNPCC [43]. Several studies have revealed correlation between the MMR gene involved and the spectrum of cancer risks. Carriers of *MSH2* mutations appear to be at a higher risk of cancer than *MLH1* mutation carriers, especially for extra-colonic cancers [44]. *MSH6* germline mutation has been reported to have a particularly strong association with endometrial cancer (a lifetime risk of about 70%), yet a lower penetrance for other Lynch syndrome–related malignancies including CRC [45]. In most of cases, one mutated allele with the affected MMR gene is inherited, the loss of heterozygosity (LOH) then happens somatically to inactivate the second allele in the form of mutation, methylation, or a combination of both. It is worth to note that one mutated allele is sufficient to confer an increased risk of cancer, thus making Lynch syndrome a dominant hereditary disease. In rare cases where both inherited alleles are mutated leading to constitutional mismatch repair deficiency (CMMR-D), patients will have a phenotype resembling neurofibromatosis type 1 with an onset of a broad spectrum of malignancies during childhood, in contrast to individuals with Lynch syndrome harboring a heterozygous mutant MMR gene allele at the age of 40–50 years [46]. In 2009, another intriguing mechanism of *MSH2* inhibition has been identified through mutation of *EPCAM,* a non-MMR gene. Germline deletions of the 3′-region of *EPCAM* cause transcriptional read-through which results in silencing of *MSH2* by hypermethylation and, subsequently, the development of a MMR-deficient phenotype [47].

The strong mutator phenotype and high-frequency microsatellite instability (MSI-H) resulted from MMR deficiency has been demonstrated to be the primary source of tumor predisposition within Lynch syndrome. Microsatellites are short tandem (1–6 base pairs) repeated DNA sequences such as $[A]_n$ or $[CA]_n$ that are present in large numbers in both noncoding and coding sequences in the eukaryotic genome. During DNA synthesis, the primer and template strands in a microsatellite (sometimes termed as "slippery DNA") can occasionally dissociate and re-anneal incorrectly [48]. This gives rise to heteroduplex DNA molecules presenting a different number of tandem-repeats in a newly synthesized strand and the template strand. These heterogeneities are known as insertion/deletion loops (IDLs) due to the aberrant conformation of unpaired nucleotides. Together with base:base mismatches arisen from sporadic escaping the proofreading function of DNA polymerase, IDLs are corrected by the MMR system which excises erroneous nucleotides of the newly synthesized strand and therefore provides the DNA polymerase with another chance to generate an accurate copy of the template sequence. In the absence of MMR, replication errors are left uncorrected and accumulated, thus leading to the creation of novel microsatellite that can be readily detected by PCR-based assays for the evidence of MSI.

Several tumor-suppressor genes contain simple repeated sequences or microsatellites in their coding sequence, including receptor kinases (eg, transforming growth factor beta receptor II/TGFBR2, activin receptor 2/ACVR2, and ephrin receptor B2/EPHB2), cell proliferation regulators (PTEN, GRB1, TCF-4, and WISP3), apoptosis inducers (eg, BAX, BCL-10, FAS, and APAF1), and DNA repair (MLH3, MSH3, MSH6, MRE11, RAD50, BLM, MBD4, and CHK1) [49,50]. Coding MSI in genes implicated in tumorigenesis causes frameshift mutations and functional inactivation of affected proteins (eg, through the generation of premature translation termination codons), thereby providing a selective growth advantage to MMR-deficient cells. Remarkably, every human MMR gene except *MLH1* includes a mononucleotide repeat. It is thus conceivable that the deficiency of MMR could be exacerbated with cumulative losses of components on the system [51]. Nevertheless, few studies have clarified which mutations of genes potentially affected by MSI-H are of a functional significance in an early tumor onset in Lynch syndrome individuals.

Several genes have been identified to be relevant for the initiation and/or progression of MSI-H-mediated tumorigenesis. The first gene in such category is *TGFBR2* in CRCs. The mutational inactivation of *TGFBR2* occurs in about 30% of colon cancers and promotes the formation of colon cancer by abrogating the antiproliferative activity of the TGF-β signaling pathway. Human colon cancer cell lines with high rates of MSI were found absent of TGFBR2 expression which is critical for TGF-β ligands to exert their antiproliferative functions [52]. The most frequent mutation is the 1-bp deletion in a tract of 10 constitutive adenines in the third exon of the gene, resulting in a truncated nonfunctional TGFBR2 protein. Clinically, *TGFBR2* mutation occurs in more than 90% of CRC cases showing MSI-H arisen from either hereditary or sporadic mutation of MMR genes. The pro-apoptotic gene *BAX* that has a long repeat of eight guanines is another gene that is commonly mutated in MSI-H CRC or endometrioid tumors [53]. Frameshift mutation of *BAX* has been detected in more than 50% of HNPCCs [54]. Activated by p53, BAX exerts its effects in the process of apoptosis to remove cells with extensive genomic legions. In HNPCC cells, the loss of function in BAX genes can result in the accelerated cancer development, even with wild-type p53 [54].

3.5 Hereditary Cancers Associated With Defects of DNA Double-Strand Break Repair

3.5.1 Ataxia–Telangiectasia (Louis–Bar Syndrome) and ATM

Ataxia-telangiectasia (A–T) is an autosomal recessive disorder that is characterized by progressive neurodegeneration, oculocutaneous blood vessels, and immunodeficiency due to a disrupted maturation of T and B cells, hypersensitivity to ionizing radiation, and a marked predisposition to malignancies. One-third of all A–T patients will develop cancer, and 15% will die of cancer. Overall, an A–T patient has a 50-fold to 150-fold excess risk to develop cancer (non-Hodgkin lymphoma or leukemia in particular) than individuals in the general population. Increased rates of breast cancer, gastric cancer, medulloblastomas, basal cell carcinomas, gliomas, ovarian cancer, and uterine cancer have been reported [55]. The A–T phenotype is caused by mutations of both alleles of the ATM gene (located in chromosome 11q22–23) that frequently result in a truncated nonfunctional gene product. Heterozygous carriers of an altered ATM allele appear to be clinically normal but are reported to be at an increased risk of developing cancer, especially in the breast [56].

A–T has been categorized as a hereditary genomic instability syndrome. This disorder involves a marked defect in sensing and responding to DNA DSBs, the most severe type of DNA damage. DNA DSBs can be generated by physiological processes such as meiotic recombination and V(D)J and class switch recombination during lymphocyte maturation, ionizing radiation, free radicals as well as genotoxic reagents that generate nicks, adducts, or intercalation leading to a collapse of stalled replication forks. Such genotoxic reagents include DNA topoisomerase inhibitors (eg, camptothecin and doxorubicin), DNA intrastrand and interstrand cross-linking reagents (eg, hydroxyurea, mitomycin, cisplatin, and nitrogen mustards), and PARP inhibitors (through the accumulation of SSBs). DNA DSBs represent a major disruption in the integrity of the genome. If not repaired correctly, DSBs can cause deletions, translocations, and fusions in the DNA. It is therefore conceivable that the ATM-deficient A–T immature lymphocytes are incapable to repair DNA DSBs that are generated during V(D)J recombination, resulting in unaccomplished T-cell and B-cell ontogeny and a consequent immunodeficiency [57].

The major known role of ATM is to participate in the responses to DNA DSBs for DNA repair and cell-cycle checkpoint activation. Upon DSB, ATM is rapidly recruited at the site of break through interacting with the MRN (MRE11-RAD50-NBS1) complex as well as the regions that flank the break. In the flanking regions, the activated ATM phosphorylates p53 and CHK2 so that G1/S checkpoint or S-phase checkpoint can be initiated to facilitate DNA repair. Notably, the activation of p53 is defective in A–T cells, resulting in failure to activate the G1/S checkpoint upon irradiation-induced DSB [58]. In addition, ATM also signals to the DNA-repair machinery to assist in the repair of DNA DSBs.

3.5.2 Hereditary Cancers and the FA/BRCA Pathway

Hereditary breast and ovarian cancer (HBOC) and FA are two inherited syndromes arising primarily from defects in DNA repair. Proteins encoded by the BRCA and FA genes form a conserved DNA-repair pathway known as the FA/BRCA pathway which removes interstrand cross-links and DSBs by HDR [59,60]. So far, 15 genes have been found being mutated in FA patients, including *FANCA, B, C, D1* (*BRCA2*), *D2, E, F, G, I, J* (*BRIP1*), *L, M, N* (*PALB2*), *O* (*RAD51C*), and *P* (*SLX4*). Upon replicative stresses, seven of the 15-gene-encoded FA proteins, including FANCA, B, C, E, F, G, L, and the FA-associated protein (FAAP100) assemble the FA core complex [61]. Upon stalled DNA replication, the interstrand cross-link (ICL) is recognized by FANCM and the associated proteins, then the latter will recruit the FA core complex that monoubiquitinates other two FA proteins, FANCD2 and FANCI. These ubiquitinated FANCD2 and FANCI proteins in chromatin recruit endonucleases (eg, Fanconi-associated nuclease 1, FAN1) and DNA polymerases for translesion DNA synthesis (Polε, Polκ, Polι, and Polν) that are required for the repair process. In coupled with translesion DNA synthesis, the interstrand adduct was removed sequentially from the antisense and sense strand by nucleases involved in NER, leaving a temporary DNA DSB. Next, the ubiquitinated FANCD2-FANCI recruits three downstream FA proteins (FANCD1/BRCA2, FANCN/PALB2, FANCJ/BRIP1, and BRCA1) to initiate homology-directed DNA repair (HDR) to remove the DSB. Finally, FANCM recruits the Blm helicase, and together they resolve the intermediate structure and reinstate DNA replication [61]. It is worth to note that, in addition to participating in resolving the intermediate DSB in DNA interstrand cross-linking repair as a positive regulator of DNA homologous recombination, FANCD1/BRCA2, FANCN/PALB2, FANCJ/BRIP1, and BRCA1 are also crucial in repairing DNA replication-independent DSBs, including ionizing radiation–induced DNA lesions. Deficiencies of these genes have been associated with an increased susceptibility of neoplastic transformation such as in hereditary breast and ovarian cancer.

Germline mutations in *BRCA1* and *BRCA2* genes are responsible for only 2–5% (up to 10%) of all breast and ovarian cancers [62]. However, these mutations account for about 50% and 45% of all hereditary breast and ovarian cancers, respectively [59]. A 2003 meta-analysis using pooled pedigree data from 22 studies involving 8139 index cases indicated that the average cumulative risk for *BRCA1*-mutation carriers to develop breast or ovarian cancer by the age 70 was 65% (44–78%) and 39% (18–54%), respectively [63]. The corresponding analysis for *BRCA2*-mutation carriers showed a 45% (31–56%) lifetime risk of developing breast cancer and a11% (2.4–19%) lifetime risk of developing ovarian cancer. In 2008, an analysis using data from 155 *BRCA1* and 164 *BRCA2* mutation carrier families in Spain showed the similar average cumulative risks of breast and ovarian cancers [64]. Furthermore, a significantly increased risk of other cancers has also been noted in *BRCA*-mutation carriers, such as stomach, pancreas, prostate, colon, and hematologic cancers [62,65]. For example, loss-of-function mutations of *BRCA* genes have also been associated with an increased risk of developing leukemia and lymphoma (up to 2000-fold) [65]. Interestingly, *BRCA1* and *BRCA2* mutation carriers are eventually characterized by tumor phenotypes. The majority of breast cancers that arise in *BRCA1* mutation carriers are triple-negative loss of function in estrogen receptor (ER), progesterone receptor (PR), and tyrosine kinase-type cell surface receptor HER2 (HER2/ERBB2) [66]. However, tumors arising in *BRCA2* mutation carriers are most frequently ER- and PR-positive. Although a number of studies showed a better prognosis and a 5-year survival in patients with breast or ovarian cancer arising in *BRCA* mutation carriers compared to sporadic cancers, growing evidence also indicated no difference in survival rate between *BRCA* mutation carriers and noncarriers [67], suggesting that further studies are essential to evaluate whether *BRCA* mutation could act as an independent favorable prognostic factor for breast cancer clinical outcome. It is believed that the functional status of DNA-repair pathways may impact the cancer cell response to anticancer chemotherapy. As demonstrated by several lines of evidence from both clinical and laboratory-based studies, a hereditary or sporadic loss of function in BRCA1, BRCA2, PALB2, and BRIP proteins (termed as BRCAness) implicates a susceptibility to platinum and other genotoxic agents in breast and ovarian cancers due to an inadequate homology-directed DNA repair [68,69]. Moreover, the additional inhibition of other DNA-repair pathways may potentiate the extent of DSB and consequently lead to synthetic lethality in HDR-deficient cells. This has been employed in the treatment of "BRCAness" breast and ovarian cancers with PARP inhibitors and revealed promising results.

FA is one of the rare hereditary bone marrow failure syndromes with an increased susceptibility to cancer, including leukemia [70], due to the biallelic mutation in FA genes [60]. The incidence rate of FA is about 1 in 100,000 births [71]. Generally, the majority of FA patients are identified as young children due to the presence of physical anomalies, such as short stature, café au lait spots and hyper/hypopigmentation, cytopenias, abnormal thumbs, microcephaly, micro-ophthalmia, and so on. However, up to a quarter of patients with the FA gene mutations display a normal phenotype [72]. Yet the mechanisms involved in FA have not been completely understood. Biallelic germline loss-of-function mutations of genes in the FA/BRCA pathway may contribute predominantly to the initiation and progress of FA. The knockdown of FANCA and FANCD2 in human embryonic stem cells using RNAi results in defects in the early development of hematopoietic lineage [73]. FA has been linked to cancer which is one of the most frequent causes of FA-related death. The most common malignancies arising in FA patients are acute myeloid leukemia (AML), head and neck squamous cell carcinoma (HNSCC), vaginal squamous cell carcinoma, liver and brain tumors [72], representing a 600-fold increase in risk of AML, a 500-fold increase in risk of HNSCC, and a 3000-fold increase in risk of vaginal squamous cell carcinoma, compared to the general population. Although the FA pathway has been demonstrated to be functionally linked to DNA-repair pathways, yet many details remain poorly understood. A better understanding of this pathway and more detailed information on its connection to DNA-repair pathways may offer opportunities to identify diagnostic biomarkers, prognostic indicators, and therapeutic targets for the HBOC and FA syndromes.

4. GENOMIC INSTABILITY IN SPORADIC CANCERS

The vast majority of cancers are nonhereditary "sporadic cancers." Almost all of the human sporadic cancers are characterized by genomic instability, especially chromosomal instability (CIN). The heterogeneity in the extent and type of genomic instability, including nucleotide, microsatellite, or chromosomal instabilities, has been shown both within the same neoplastic tissue and between cancer types [74]. Genomic instability refers to a transient or persistent state with an increased frequency of mutations within the cancer genome or a cellular lineage. These mutations include nonsynonymous alterations in nucleic acid sequences, chromosomal rearrangements, or changes in chromosome numbers, resulting in the continuous modification of tumor cell genomes, the subsequent acquisition of additional DNA alterations, clonal evolution, and tumor heterogeneity. Therefore, genomic instability which has been observed in a range of malignant stages [75,76] has been considered as a driving force of tumorigenesis [77]—from pre-neoplastic lesions to advanced tumors.

4.1 CIN in Sporadic Cancers

Chromosomal instability is the predominant form of genomic instability that leads to changes in both chromosome numbers and structure [75]. Numerical CIN is a high rate of either gain or loss of whole chromosomes, also called aneuploidy. Normal cells make errors in chromosome segregation in about 1% of cell divisions, whereas cells with CIN increase the error rate to 20% of cell divisions [78]. By contrast, structural CIN is the rearrangement of parts of chromosomes and amplifications or deletions within a chromosome. Almost all solid tumors show CIN, and about 90% of human cancers exhibit chromosomal abnormalities and aneuploidy [79]. The features of CIN tumor include global aneuploidy, loss of heterozygosity, homozygous deletions, translocation, and chromosomal changes such as deletions, insertions, inversions, and amplifications.

CIN leads to karyotypic instability and the simultaneous growth of multiple tumor subsets, resulting in inter- or intro-tumor heterogeneity [80]. Epithelial tumors generally exhibit a greater degree of genomic instability than hematologic and mesenchymal malignancies [74]. This is supported by a finding that epithelial-derived cancers such as breast, melanoma, lung, and prostate cancers have more somatic mutations than blood cancers [81]. Notably, distinct instability phenotypes could be displayed in the same cancer type. For example, lung cancer in smokers and nonsmokers shows a different extent of segmental alterations and genome instability [82]. Moreover, lung adenocarcinoma and squamous cell carcinoma exhibit a distinct type of genomic changes. Even within lung adenocarcinomas, a greater genomic instability has been found in the magnoid subtype compared to other subtypes of adenocarcinoma [83].

Studies in 2008 suggested that CIN is associated with a poor clinical outcome in solid tumors [84]. By developing a computational model, Carter et al. have identified a chromosomal instability signature that predicts a poor survival in 12 data-sets representing six cancer types, including breast, lung, and brain tumors as well as mesothelioma, glioma, and lymphoma [85]. One explanation of the clinical relevance of CIN is that CIN may be related to an increased tumor cell heterogeneity, thereby enhancing the ability of tumors to adapt to environmental stresses [86]. Moreover, preclinical studies have shown that CIN is associated with the intrinsic multidrug resistance both *in vitro* and *in vivo* [87].

4.2 Hypothesis of the Mechanisms of CIN

Although the fundamental importance of CIN in cancer biology has been recognized for decades, the molecular basis of CIN in sporadic cancers remains unclear. This is primarily due to the heterogeneous nature of CIN in sporadic cancers; numerous genes have been uncovered to contribute to CIN, including, but not limited to, those involved in chromosome condensation and segregation (*STAG2*) [88], sister chromatid cohesion (hSecutin) [89], cytokinesis (*MOS*, *RAS*, *RAF*) [90], telomere function (*TRF1* and tankyrase) [91], DNA damage (*TP53* and *ATM*) [92], and mitotic checkpoint (*SAC*, *BUB1*, *MAD2*) [93–95]. It is challenging to unify these often conflicting mechanisms into one general mechanism to explain CIN in human sporadic cancers.

Studies to explore the presence of genomic instability in sporadic cancers have led to three prevailing hypotheses. The first is the mutator hypothesis which states that mutations in caretaker genes (refer to the genes that primarily function to maintain genomic stability) [96] cause genomic instability in precancerous lesions and drive tumor development by increasing the spontaneous mutation rate. The second hypothesis is the oncogene-induced DNA replication stress model which demonstrates that CIN in sporadic cancers is the consequence of the oncogene-induced collapse of DNA replication forks, which in turn leads to DNA DSBs and genomic instability. The third theory states that telomere erosion and dysfunction give rise to CIN [97].

Caretaker genes encode proteins that stabilize the genome, thus mutations in caretaker genes lead to genomic instability. The classical caretaker genes are majorly of two types: DNA-repair genes and mitotic checkpoint genes. It is important to note that although the TP53 gene can be considered as a caretaker gene, TP53 is subject to selective pressure for the inactivation in cancer, whereas other classical caretaker genes are not. Therefore, when we refer to caretaker genes, TP53 is not included.

The mutator model is generally considered as the major mechanism responsible for the presence of genomic instability in hereditary cancers. Attempts to identify mutations in caretaker genes in sporadic cancers were unsuccessful in the past few years [98]. For example, in 2004, Wang et al. analyzed the sequences of 100 cell-cycle checkpoint and DNA-repair genes in early passage human colon cancer cell lines, yet they identified very few mutations [99]. Furthermore, Cahill and his colleagues generated chromosome instability in the experimental model by mutating the mitotic checkpoint gene budding uninhibited by benzimidazoles 1 (*BUB1*). However, sequencing studies showed that *BUB1* mutations were rare in human cancers [93]. To date, whole cancer genome sequencing has failed to identify putative caretaker genes that are frequently mutated in human cancer, and only 3–31% of sporadic cancers harbor at least one mutation of a caretaker gene [100]. These studies suggest that mutations in caretaker genes are not the cause, or at least a major cause, of genomic instability in sporadic cancers. Moreover, few studies convincingly support telomere erosion hypothesis. Therefore, an oncogene-induced DNA stress model is currently a generally accepted hypothesis to explain CIN in sporadic cancers, which we discuss further.

4.3 High-Throughput Sequencing Studies on CIN in Various Cancers

The development of high-throughput sequencing technologies enables researchers to investigate the genetic profiles of human cancers in a much more efficient manner. Here, we highlighted several important high-throughput sequencing studies that help us understand the molecular mechanism of CIN in several common types of human cancer such as breast, colorectal, lung cancers, and glioblastoma.

To determine the spectrum and extent of somatic mutations in human cancers, Kinzler and his colleagues sequenced 18,191 genes in genomic DNA isolated from 11 breast and 11 colorectal tumors [101]. Mutations were found in 1137 genes from breast cancers, and 848 genes from colorectal cancers. But in the additional validation screening of 24 breast and 24 colon cancers, only 167 of 1137 (14.69%) gene mutations and 183 of 848 (21.58%) gene mutations were detected. Then the same group of researchers analyzed sequences of 23,219 transcripts, representing 20,661 protein-coding genes in 24 pancreatic cancers [102]. Of 20,661 genes analyzed by sequencing, 1327 were mutated in at least one sample, and 148 were mutated more than twice among 24 cancers examined. In the subsequent validation screening, 39 genes that were mutated in more than one of the 24 cancers in the discovery set were sequenced in the additional 90 pancreatic cancers. The results of the analysis showed that 255 nonsilent somatic mutations occurred in 23 genes. Deletions and amplifications of these cancer genomes were also examined using SNP. The classical tumor-suppressor genes *CDKN2A* (encodes p16INK4A and p14ARF), *SMAD4* and *TP53,* were found mutated, whereas the small GTPase *KRAS* was found to be one of the most frequently mutated oncogenes in pancreatic cancer.

The same 20,661 protein-coding genes were also sequenced in 22 human glioblastomas [103]. The analysis results showed that 3.4% of the 20,661 genes (698 genes) were mutated at least once. Most of the mutations were single-base substitutions (94%), whereas the others were small insertions, deletions, or duplications. A set of 21 mutated genes were selected and evaluated in a second screen comprising an additional 83 glioblastomas. Nonsilent somatic mutations were identified in 16 of 21 genes in the additional tumor samples, and the mutation rates were significantly increased from 1.5 mutations per Mb in the discovery screen to 23 mutations per Mb in the consequent validation screen. The most frequently changed genes were *CDKN2A* (altered in 50% of samples); *TP53*, *EGFR*, and *PTEN* (altered in 30–40% of samples); NF1, *CDK4*, and *RB1* (altered in 12–15% of samples); and *PIK3CA* and *PIK3R1* (altered in 8–10% of samples).

Another unbiased genomic study on glioblastomas was conducted by The Cancer Genome Atlas Research Network [104]. A total of 91 matched tumor–normal pairs including 72 untreated and 19 treated cases were examined for mutations in 601 cancer-relevant genes. The results uncovered 453 validated nonsilent somatic mutations in 223 genes, and 79 of these genes were mutated twice or more. Interestingly, the frequency of mutations was remarkably different between untreated and treated tumors (98 events among 72 untreated cases versus 111 among 19 treated cases, $P<10^{-21}$). This difference was primarily driven by seven hypermutated samples, six of which harbored mutations in at least one of the MMR genes *MLH1*, *MSH2*, *MSH6*, or *PMS2*, suggesting that mutations in caretaker genes and a decreased DNA-repair activity may subsequently occur spontaneously during tumor progression or due to genotoxic therapeutic reagents, rather than primarily contribute to the initiation of the tumor.

Lung cancer was also studied by the Cancer Genome Atlas Research Network. DNA sequencing of 623 cancer-related genes in 188 human lung adenocarcinomas discovered more than 1000 somatic mutations across samples [105]. This study identified 26 frequently mutated genes that include tumor-suppressor genes *NF1*, *APC*, *RB1*, and *ATM*. Also, the NHEJ DNA-repair gene *PRKDC* and the MMR gene *MSH6* were mutated in six and four cases, respectively. Mutations in homologous recombination-repair genes *BRCA1*, *BRCA2*, the BRCA1-associated RING domain 1 (*BARD1*), *BAP1*, the BER gene *XRCC1*, the NER gene *XPD*, and the mitotic checkpoint genes *BUB1* and *STK12* were also detected in lung adenocarcinomas.

A similar genome-wide study of the molecular basis of CIN was conducted in hematologic cancer. Whole-genomic sequencing of four representative chronic lymphocytic leukemia (CLL) cases identified 46 somatic mutations, and the further screening of 363 CLL patients found four recurrent mutations of notch1 (*NOTCH1*), exportin 1 (*XPO1*), myeloid differentiation primary response gene 88 (*MYD88*) and kelch-like 6 (*KLHL6*) [106]. The subsequent functional and clinical analyses indicated that recurrent mutations are oncogenic and associated with a poor clinical outcome.

In 2013, the mutation spectra of esophageal adenocarcinoma (EAC) have been investigated by both whole-exome (149 EAC tumor–normal pairs) and whole-genome sequencing (15 of 149 pairs) [107]. Chromatin-modifying factors and candidate contributors were found significantly mutated in EAC, including *SPG20*, *TLR4*, *ELMO1*, and *DOCK2*. A more comprehensive study on the genomic instability patterns in esophagus, stomach, and colon adenocarcinomas showed that a significant recurrent amplification and deletion of genes were distinct to various gut-derived adenocarcinomas [108].

High-throughput sequencing studies provided deeper insights on the molecular basis of CIN in human cancers. These unbiased genome-wide studies indicated that mutations in the caretaker genes were infrequent. The frequency of caretaker gene mutations ranged from 14 to 31%, depending on the tumor type [100]. Collectively, 69–97% of cancers did not have mutations in caretaker genes, suggesting that the inactivation of caretaker genes is not the major cause of genomic instability in many sporadic cancers.

4.4 Oncogenes Induce CIN

The hypothesis that the most frequently mutated or deregulated oncogenes in response to DNA damage in sporadic cancers are responsible for CIN seems to be a favorable mechanism based on the high-throughput studies mentioned earlier. The frontier studies have provided evidence for this hypothesis. In 1996, Mai et al. increased the level of c-Myc protein in Rat1A–MycER cells and investigated chromosomal aberrations [109]. This early study indicated that the upregulation of the c-Myc protein which is an important cell-cycle regulator would lead to chromosome numerical changes, chromosome breakage, chromosome fusions, and other chromosomal abnormalities. In 1999, Charles et al. showed that the overexpression of cyclin E, another important positive cell-cycle regulator and oncogene, led to CIN in both rat embryo fibroblasts and human breast epithelial cells [110]. Later studies in 2000 have suggested that the disrupted expression of cyclin E in karyotypically stable colorectal cancer cells may lead to increased CIN and consequently a more malignant phenotype [111]. This phenotype attributed to a defect in the execution of metaphase and subsequent transmission of chromosomes caused by aberrant accumulation of cyclin E. In 2006, Di Micco et al. showed that the activation of oncogene H-RasV12 in normal cells was able to trigger cell proliferation, DNA-damage response, and cancerous transformation, both *in vitro* and *in vivo* [112]. Mounting evidences are supportive to the hypothesis that oncogenes trigger CIN.

It is important to note that CIN and neoplastic formation induced by oncogenes requires inactivation of the p53 pathway. Richard and Randy showed that oncogenic H-RasV12 was unable to induce CIN in the mouse embryonic fibroblasts (MEFs) with wild-type p53 ($p53^{+/+}$), but was capable of inducing CIN in the $p53^{-/-}$ MEFs cultured in the same conditions [113]. Furthermore, using *in vivo* mouse models, this group of investigators also found that the loss of p53 function was critical for cells to create a permissive environment allowing cancerous transformation.

Oncogene-induced CIN hypothesis highlighted the importance of the p53 protein and distinguished this molecule from other tumor suppressors. As the product of TP53, p53 has been dubbed "the guardian of the genome," and it controls cell fate after DNA damage such as DNA repair and survival or programmed death. Not surprisingly, TP53 is one of the most frequently mutated genes (>50%) in human cancer [114]. As TP53 is a DNA damage–checkpoint protein, its inactivation was expected to induce genomic instability. However, numerous studies have proved that the deletion of the TP53 gene in mouse model and human cell lines does not induce aneuploidy [115].

According to the oncogene-induced DNA replication stress model, CIN is mainly caused by DNA damage or other forms of DNA replication stress. Specific genomic loci called common fragile sites that preferentially exhibit gaps and breaks on metaphase chromosomes have been demonstrated to be more sensitive to the inhibition of DNA synthesis [116]. Previous CIN studies on human precancerous lesions and different experimental systems showed that these common fragile sites were prone to genomic instability induced by oncogenes [117].

4.5 Chromothripsis

Sporadic cancer is driven by gradually accumulated genomic alterations, such as somatic point mutations, chromosome rearrangements, and numerical and structural changes in chromosomes. Recent studies using next-generation DNA sequencing and SNP array analyses and bioinformatics methods have uncovered a new form of genomic chaos called chromothripsis (from Greek for "chromosome" (chromo) and "shattering into pieces" (thripsis)) [118]. Chromothripsis can be defined by three major features: remarkable chromosomal rearrangements in localized regions; a low degree of chromosomal gain or loss showing haploid (heterozygous deletion) or diploid across the rearranged region; and the preservation of heterozygosity.

Chromothripsis was first described by Stephens and his colleagues [119]. Using paired-end sequencing of the genome of a chronic lymphocytic leukemia (CLL) sample, they identified one patient who had 42 somatically acquired genomic rearrangements involving the long arm of chromosome 4. Although most of the rearrangements occurred in the region of chromosome 4, some of them also included segments from the regions of chromosome 1, 12, and 15. Further studies indicated that the stamp of chromothripsis can be seen in at least 2–3% of diverse cancer cell types. Moreover, chromothripsis is more prevalent in bone cancers; about 25% of osteosarcoma and chordoma exhibit features of this phenomenon.

Rearrangements generated in a single genomic crisis have shown their effects on multiple cancer-related genes. Chromothripsis in a chordoma patient showed the loss of tumor-suppressor genes *CDKN2A* (cyclin-dependent kinase inhibitor 2A), *WRN* (Werner syndrome ATP-dependent helicase), and *FBXW7* (F-box and WD-40 domain containing 7), indicating that several tumor-promoting events might occur concurrently.

Although further work is needed to fully understand these mechanisms and the prevalence of chromothripsis, some current studies have already shown the promise of its clinical implications. Studies on multiple myeloma and neuroblastoma patients have revealed the correlations of chromothripsis and a poor patient survival, suggesting that the examination of hallmarks of chromothripsis might be a novel way to identify high-risk cancer patients [120].

4.6 Microsatellite Instability in Sporadic Cancer

MSI is another form of genomic instability in sporadic cancer which is generally caused by deletions or random insertions of microsatellites [121]. Microsatellites are repetitive DNA sequences ranging from 2 to 5 base pairs that occur at thousands of locations in the human genome [121]. Microsatellites are characterized by high mutation rates and diversity in the population. MSI results from impaired MMR system, especially alterations of the *MLH1*, *MSH2*, *MSH6,* and *PMS2* genes.

MSI has been reported in numerous types of cancers, including gastric, endometrial, ovarian, lung, and colorectal cancers (CRC) [122–125]. Colorectal cancer is the first cancer type where MSI has been described and extensively studied. MSI occurs in about 15% of CRC cases that show a poorer clinical outcome compared to MSI-negative CRCs [126]. MSI is present in both hereditary and sporadic CRCs through different mechanisms. Hereditary nonpolyposis colon cancer (HNPCC, also known as Lynch syndrome) is characterized by inactivating germline mutations of *MSH2*, *MSH6*, *PMS2*, or *MLH1*, whereas sporadic CRCs with MSI are associated with hypermethylation (CpG island methylator phenotype referred to CIMP) and the loss of MLH1 expression [127]. CIMP is characterized by simultaneous methylation of multiple CpG islands of tumor-suppressor genes, such as *MLH1*, *CDKN2A*, and *THBS1* [127,128].

Genomic instability has been accepted as the major cause that plays a central role in sporadic cancer development, although our understanding of the molecular basis of genomic instability is very limited. It is contradictory about the process, timing, and extent of genetic alterations that occur in cancer: tumorigenesis could be the result of the accumulation of several genetic errors; or it could be due to multiple genetic alterations in a single catastrophic event, chromothripsis. Compared to other forms of genetic alterations, CIN is the dominant phenotype that may result from the gain-of-function alteration, rather than the inactivation of caretaker genes. The importance of CIN in sporadic cancer should make it a top priority in cancer biology research that provides promise for the improved therapeutic strategies.

5. TRIGGERING EXCESSIVE GENOMIC INSTABILITY BY TARGETING DNA-REPAIR PATHWAYS AS A STRATEGY FOR CANCER THERAPY

In the last three decades, one of the significant advances in our understanding of the mechanisms involved in the maintenance of genome integrity is the discovery of members of the poly(ADP-ribose) polymerase (PARP) gene family, such as *PARP1*. Proteins encoded by *PARP* genes have been demonstrated to be crucial in the repair of DNA SSBs resulted from direct DNA oxidization, BER, or erroneous topoisomerase activity at a rate of three orders of magnitude more frequently than DSBs. PARP proteins exert their function by directly binding to SSBs and promoting a rapid access by, and the accumulation of, downstream repair factors such as XRCC1 [129]. To date, as many as 17 mammalian PARP family members have been identified [130], while only a few, such as PARP1-4, TNKS (TRF1-interacting, ankyrin-related ADP-ribose polymerase), and TNKS2, display an enzymatic activity that catalyzes the poly(ADP-ribosyl)ation of nuclear proteins. The well-defined structure of PARP1 is primarily composed of three functional domains, including a DNA-binding domain (DBD), an automodification domain (AD), and a catalytic domain (CD) (Fig. 27.4A) [131].

Once activated by DNA damage, PARP1 forms a homodimer that can recognize a DNA lesion via its DBD domain (Fig. 27.4B). Using NAD^+ as a substrate, PARP1 catalyzes the poly(ADP-ribosyl)ation on acceptor proteins, such as histones and PARP1 itself. PARP1 loses its DNA-binding affinity due to the net negative charge of poly(ADP) ribose (pADPr, yellow beads); as a result, pADPr recruits other repair proteins to the site of the damaged DNA (blue and purple circles, Fig. 27.4B) [131], for instance, XRCC1, histone H1, Ku70, and Ku80, thus allowing them to repair the damaged DNA. The pADPr chain is then hydrolyzed into ADP-ribose units by poly(ADP-ribose) glycohydrolase (PARG) and/or ADP-ribose hydrolase 3 (ARH3). The ADP-ribose is further converted into AMP by pyrophosphohydrolase NUDIX, increasing AMP:ATP ratios and consequently leading to the activation of AMP-activated protein kinase (AMPK) for further adaptive stress responses such as metabolic control and pro-survival autophagy [131].

In contrast to other components of the DNA-repair machinery, such as BRCA1/2 mediating the homology-directed DNA DSB repair, that are frequently inactivated during the initiation and progression of hereditary and sporadic cancers, PARP1, the key player in the BER pathway for repairing DNA SSBs has been demonstrated to be upregulated in various human malignancies, such as breast cancer [132]. PARP1 overexpression has been associated with a higher tumor grade, estrogen independence, and a worse metastasis-free survival, implicating an oncogenic role of PARP1 in breast cancer. The oncogenic characteristics of PARP1 may contribute primarily to cancer cell survival via DNA repair to maintain the cellular homeostasis by removing DNA lesions resulted from intrinsic replication errors, metabolic damage, and, more importantly, extrinsic stresses through genotoxic therapeutic reagents because excessive genomic instability is detrimental and can usually lead to cell death [133].

The development of therapeutic reagents that specifically target malignant cells has been proved difficult. However, the concept of synthetic lethality has begun to shed some light on targeting certain malignancies with defined genetic defects [134]. Synthetic lethality is known as a phenomenon upon which cell death can be induced by combined genomic abnormalities (either a gain of function or a loss of function) of two distinct pathways, whereas a dysfunction in either pathway alone has no significant effect on viability. One of the most studied models for synthetic lethality is the particular susceptibility to PARP inhibitors in cells lacking HDR-mediated DNA double-strand repair. Several independent studies have demonstrated that the deficiency of key HDR genes, such as BRCA1 and BRCA2, profoundly sensitizes tumor cells to PARP inhibitors [135], resulting in chromosomal instability, cell-cycle arrest, and apoptosis. The potential mechanism of the hypersensitivity to PARP inhibitors in HDR-deficient cells may be attributed to the persistence and accumulation of DNA DSBs derived from unrepaired SSBs during DNA replication (Fig. 27.5). Conversely, cells with functional HDR elicit a final safeguard to efficiently prevent the cytotoxic DSB accumulation when treated with PARP inhibitors, allowing damaged DNA to be replicated and repaired by error-prone DNA polymerases. Considering the prevalence of germline and

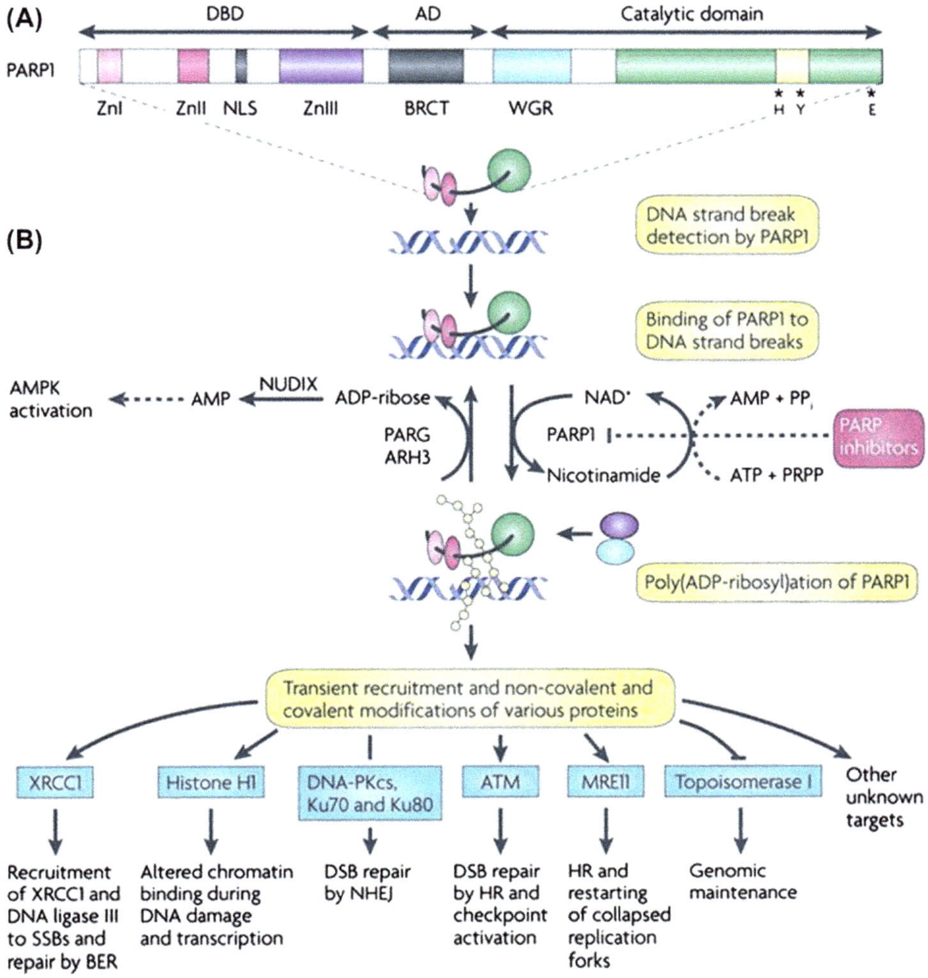

FIGURE 27.4 **Structural and functional features of poly(ADP-ribose) polymerase 1 (PARP1).** (A) PARP1 structure. (B) Biological functions of PARP1. The asterisk indicates essential residuals for NAD^+ binding (H, histone; Y, tyrosine) and polymerase activity (E, glutamic acid). *Adapted from Rouleau M, et al. PARP inhibition: PARP1 and beyond. Nat Rev Cancer 2010;10(4):293–301. We thank Nature Publishing Group's for permission, http://www.nature.com/nrc/journal/v10/n4/full/nrc2812.html.*

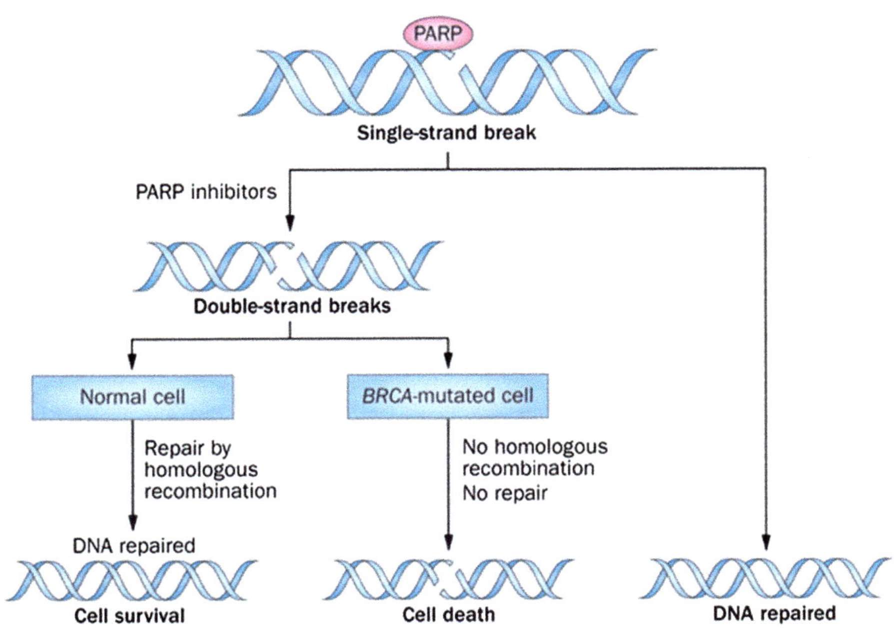

FIGURE 27.5 **Synthetic lethality induced by PARP inhibitor.** *Adapted from Sonnenblick, A, et al. An update on PARP inhibitors—moving to the adjuvant setting. Nat Rev Clin Oncol 2015;12(1):27–41. We thank Nature Publishing Group's for permission, http://www.nature.com/nrclinonc/journal/v12/n1/full/nrclinonc.2014.163.html.*

sporadic mutations of HDR-related genes such as *BRCA1* and FA pathway–related genes in various types of malignancies, the synthetic lethality may therefore represent a new direction in anticancer drug development [136].

Owing to the unique antitumor features, the development of selective and effective PARP inhibitors has become an active area in drug discovery. To date, more than 30 clinical trials have been reported to evaluate PARP inhibitors in different phases in treatment of a variety of carcinomas [137,138]. PARP inhibitors have been used as either monotherapeutic agents to induce synthetic lethality in HDR-deficient cancers such as breast, ovarian, and prostate cancers, or they can be combined with other DSB-inducing chemo- and/or radio-therapies to potentiate the therapeutic efficacy in glioblastoma, melanoma, head and neck cancers [138].

However, it should be noted that in the majority of *BRCA1*-mutant cancers (>80%), *p53* was also mutated [139]. The mutated p53 may nullify the apoptotic pathway mediated by p53 that could account for synthetic lethality of most BRCA1-deficient carcinomas, unless the PARP inhibitors induce a p53-independent apoptosis. In addition, although PARP1 and PARP2 are primarily involved in SSB repair, diverse functions of the PARP family members have been demonstrated in a wide range of biological processes [137]. Further investigations are required for better understanding of the molecular mechanisms underlying PARP inhibition.

6. CONCLUSION

Genomic instability has been accepted as the major cause that plays a central role in cancer progression, although our understanding of the molecular basis of genomic instability is very limited. DNA-repair pathways are pivotal processes in the maintenance of genomic stability. Thus, it comes as no surprise that any events leading to a deficiency in these pathways will increase susceptibility to cancer. Although the molecular basis of genomic instability is well defined in inherited cancers owing to the established relationship between mutations in DNA-repair genes and tumorigenesis, it is poorly understood in sporadic cancers. However, the importance of CIN in sporadic cancer should make it a top priority in cancer biology research and provide promise for the improved therapeutic strategies. The direct evidence showing that the genomic instability is the driving force for cancer development is scanty, which is partially due to the lethal effect of germline mutations, especially tumor suppressors, on the development of model animals before tumor formation. However, the recently developed conditional knockout technology may effectively overcome these challenges. Although targeting DNA-repair pathways has been shown as a novel and promising strategy for cancer therapy, further studies in different types of human cancer are required to better understand the underlying mechanisms of drug resistance and refractoriness, so that the therapeutic efficacy can be further potentiated.

GLOSSARY

BRCAness refers to germline or sporadic mutation of BRCA1, BRCA2, or other Fanconi anemia genes which results in deficiency in homology-directed DNA repair and cancer susceptibility. Tumor cells with BRCAness are particularly sensitive to genotoxic anticancer reagent such as platinum and PARP inhibitors.

Haploinsufficiency refers to a situation in diploid organisms that a single copy of wild-type gene is by itself incapable to maintain normal function due to the reduced expression of corresponding transcripts. Haploinsufficiency is one of the major causes of certain dominant inherited diseases, as a heterozygosity or hemizygosity elicits significant phenotypic impacts.

Slippery DNA refers to certain DNA sequences rich in single-nucleotide repeats or tandem repeats (microsatellites) at which the replicating DNA polymerase is error prone to potentiate microsatellites instability if replicative errors are left unfixed by the DNA mismatch repair.

LIST OF ABBREVIATIONS

ACVR2 Activin Areceptor type 2A
AD Automodification domain
AML Acute myeloid leukemia
AMPK AMP-activated protein kinase
AP Apurinic/apyrimidinic
APAF1 Apoptotic protease activating factor 1
APC APC gene
APE1 AP endonuclease 1
APTX Aprataxin
ARH3 ADP-ribosyl hydrolase 3
ATM Ataxia-telangiectasia-mutated gene

BAP1 BRCA1-associated protein 1
BARD1 BRCA1-associated ring domain 1
BAX BCL2-associated X protein
BCC Basal cell carcinoma
BCL10 B-cell CLL/lymphoma 10
BER Base excision repair
BLM BLM gene
BRCA1 Breast cancer 1 gene
BRCA2 BRCA2 gene
BRIP1 BRCA1-interacting protein 1
BUB1 Budding uninhibited by benzimidazoles 1
CAK CDK-activating kinase
CD Catalytic domain
CDK4 Cyclin-dependent kinase 4
CDKN2A Cyclin-dependent kinase inhibitor 2A
CETN2 Centrin, EF-hand protein 2
CHK1 Checkpoint, *Schizosaccharomyces pombe*, homolog of, 1
CIN Chromosomal instability
CLL Chronic lymphocytic leukemia
CMMR-D Constitutional mismatch repair deficiency
C-NHEJ Canonical NHEJ
CNVs Copy number variations
CPD Cyclobutane pyrimidine dimers
CRC Colorectal cancer
CSA Cockayne syndrome A
CSB Cockayne syndrome B
DBD DNA-binding domain
D-loop Displacement loop
DNA-PK DNA-dependent protein kinase
DNA-PKcs DNA-dependent protein kinase catalytic subunit
DOCK2 Dedicator of cytokinesis 2
dRP Deoxyribose phosphate
DSB Double-strand break
DSBR Double-strand break repair
EAC Esophageal adenocarcinoma
EGFR Epidermal growth factor receptor
ELMO1 Engulfment and cell motility gene 1
EPCAM Epithelial cellular adhesion molecule
EPHB2 Ephrin receptor EphB2
ER Estrogen receptor
ERCC1 Excision repair, complementing defective, in Chinese hamster, 1
EXO1 Exonuclease 1, *Saccharomyces cerevisiae*, homolog of
FA Fanconi anemia
FAAP100 Fanconi anemia–associated protein, 100-kD subunit
FAP Familial adenomatous polyposis
FAS FAS cell surface death receptor
FBXW7 F-box and WD-40 domain containing 7;
FEN1 Flap endonuclease 1
GG-NER Global genome NER
GRB1 Phosphatidylinositol 3-kinase-associated p85-alpha
HBOC Hereditary breast and ovarian cancer
HER2 Tyrosine kinase-type cell surface receptor HER2
HNPCC Hereditary nonpolyposis colon cancer
HNSCC Head and neck squamous cell carcinoma
HR Homologous recombination
IDLs Insertion/deletion loops
IR Irradiation
KLHL6 Kelch-like 6
KRAS V-Ki-ras2 Kirsten rat sarcoma viral oncogene homolog

Ku70 Ku antigen, 70 kD subunit
Ku80 Ku antigen, 80 kD subunit
LFS Li–Fraumeni syndrome
LIG1 DNA ligase 1
LIG3 DNA ligase 3
LIG4 DNA ligase 4
LOH Loss of heterozygosity
MAD2 Mitotic arrest-deficient 2
MBD4 Methyl-CpG-binding domain protein 4
MDM2 Murine double minute 2
MEFs Embryonic fibroblasts
MLH1 MutL, *Escherichia coli*, homolog of, 1
MLH3 MutL, *Escherichia coli*, homolog of, 3
MMR Mismatch repair
MOS v-MOS Moloney murine sarcoma viral oncogene homolog
MRE11 Meiotic recombination 11, *Saccharomyces cerevisiae*, homolog of, A
MSH2 MutS, *Escherichia coli*, homolog of, 2
MSH3 MutS, *Escherichia coli*, homolog of, 3
MSH6 MutS, *Escherichia coli*, homolog of, 6
MSI Microsatellite instability
MSI-H High-frequency microsatellite instability
MutLα MLH1-PMS2 heterodimer
MutSα MSH2–MSH6 heteroduplex
MutSβ MSH2–MSH3 heteroduplex
MUTYH MutY, *Escherichia coli*, homolog of
MYD88 Myeloid differentiation primary response gene 88
NAD Nicotinamide adenine dinucleotide
Nbs1 Nibrin
NER Nucleotide excision repair
NF1 Neurofibromin 1
NHEJ Nonhomologous end joining
NOTCH1 NOTCH, *Drosophila*, homolog of, 1
OGG1 8-oxoguanine DNA glycosylase
8-oxoG 8-oxoguanine products
PADPr Poly(ADP) ribose
PALB2 Partner and localizer of BRCA2
PARG Poly(ADP-ribose) glycohydrolase
PARP1 Poly(ADP-ribose) polymerase 1
PARP2 Poly(ADP-ribose) polymerase 2
PCNA Proliferating-cell nuclear antigen
PIK3CA Phosphatidylinositol 3-kinase, catalytic, alpha
PIK3R1 Phosphatidylinositol 3-kinase, regulatory subunit 1
PMS1 Postmeiotic segregation increased, *S. cerevisiae*, 1
PMS2 Postmeiotic segregation increased, *S. Cerevisiae*, 2
PNKP Polynucleotide kinase 3′-phosphotase
Polβ DNA polymerase β
Polδ/ε DNA polymerase δ/ε
POLH Polymerase, DNA, eta
Polκ DNA polymerase κ
Polν DNA polymerase ν
6–4PP 6–4 photoproducts
PR Progesterone receptor
PRKDC Protein kinase, DNA-activated, catalytic subunit
PTCH Patched, *Drosophila*, homolog of
PTEN Phosphatase and tensin homolog
RAD23B RAD23 homolog B
RAD50 RAD50, *Saccharomyces cerevisiae*, homolog of
RAD51 RAD51, *S. cerevisiae*, homolog of
RAD54 RAD54, *S. cerevisiae*, homolog of

RAF Proto-oncogene RAF
RAS Oncogene RAS
RB1 RB1 gene
RFC Replication factor C
RNA PolII RNA polymerase II
ROS Reactive oxygen species
RPA Replication protein A
SAC Soluble adenylyl cyclase
SDSA Synthesis-dependent strand annealing
SHH Sonic hedgehog
SLX4 SLX4, *Saccharomyces cerevisiae*, homolog of
SMAD4 SMA- and MAD-related protein 4
SNP Single-nucleotide polymorphism
Sp1 Transcription factor Sp1
SPG20 SPG20 gene
SSB Single-strand break
SsDNA Single-stranded DNA
STAG2 Stromal antigen 2
STK12 Serine/threonine protein kinase 12
TCF4 Transcription factor 4
TC-NER Transcription-coupled NER
TFIIH Transcription initiation factor IIH
TGF-β Transforming growth factor, beta-1
TGFBR2 Transforming growth factor beta receptor 2
THBS1 Thrombospondin 1
TLR4 Toll-like receptor 4
TNKS TRF1-interacting, ankyrin-related ADP-ribose polymerase
TP53 Tumor protein p53
TRF1 Telomeric repeat-binding factor 1
USP7 Ubiquitin-specific processing protease 7
UV Ultraviolet
UV-DDB UV-damaged DNA-binding protein
UVSSA UV-stimulated scaffold protein A
WISP3 WNT1-inducible signaling pathway protein 3
WRN Werner syndrome ATP-dependent helicase
XLF XRCC4-like factor
XP Xeroderma pigmentosum
XPA XPA gene
XPB Xeroderma pigmentosum, complementation group B
XPC XPC gene
XPD XPD gene
XPE Xeroderma pigmentosum, complementation group E
XPF Xeroderma pigmentosum, complementation group F
XPG Xeroderma pigmentosum, complementation group G
XPO1 Exportin 1
XRCC1 X-ray repair, complementing defective, in Chinese hamster, 1
XRCC4 X-ray repair, complementing defective, in Chinese hamster, 4

REFERENCES

[1] McCaw BK, et al. Somatic rearrangement of chromosome 14 in human lymphocytes. Proc Natl Acad Sci USA 1975;72(6):2071–5.
[2] Hanahan D, et al. Hallmarks of cancer: the next generation. Cell 2011;144(5):646–74.
[3] Reed SH. Nucleotide excision repair in chromatin: damage removal at the drop of a HAT. DNA Repair 2011;10(7):734–42.
[4] Maynard S, et al. Base excision repair of oxidative DNA damage and association with cancer and aging. Carcinogenesis 2009;30(1):2–10.
[5] Caldecott KW. Single-strand break repair and genetic disease. Nat Rev Genet 2008;9(8):619–31.
[6] Schärer OD. Nucleotide excision repair in eukaryotes. Cold Spring Harb Perspect Biol 2013;5(10):a012609.
[7] Marteijn JA, et al. Understanding nucleotide excision repair and its roles in cancer and ageing. Nat Rev Mol Cell Biol 2014;15(7):465–81.
[8] Jiricny J. The multifaceted mismatch-repair system. Nat Rev Mol Cell Biol 2006;7(5):335–46.
[9] Peña-Diaz J, et al. Mammalian mismatch repair: error-free or error-prone? Trends Biochem Sci 2012;37(5):206–14.

[10] Dueva R, et al. Alternative pathways of non-homologous end joining (NHEJ) in genomic instability and cancer. Transl Cancer Res 2013;2(3):163–77.

[11] Takata M, et al. Homologous recombination and non-homologous end-joining pathways of DNA double-strand break repair have overlapping roles in the maintenance of chromosomal integrity in vertebrate cells. EMBO 1998;17(18):5497–508.

[12] Menon V, et al. Involvement of p53 in the repair of DNA double strand breaks: multifaceted roles of p53 in homologous recombination repair (HRR) and non-homologous end joining (NHEJ). Subcell Biochem 2014;85:321–36.

[13] Mazin AV, et al. A novel function of Rad54 protein. Stabilization of the Rad51 nucleoprotein filament. J Biol Chem 2003;278(16):14029–36.

[14] Loeb LA, et al. Multiple mutations and cancer. Proc Natl Acad Sci USA 2003;100(3):776–81.

[15] Vogelstein B, et al. Cancer genome landscapes. Science 2013;339(6127):1546–58.

[16] McBride KA, et al. Li-Fraumeni syndrome: cancer risk assessment and clinical management. Nat Rev Clin Oncol 2014;11(5):260–71.

[17] Varley JM, et al. Characterization of germline TP53 splicing mutations and their genetic and functional analysis. Oncogene 2001;20(21):2647–54.

[18] Palmero EI, et al. Tumor protein 53 mutations and inherited cancer: beyond Li-Fraumeni syndrome. Curr Opin Oncol 2010;22(1):64–9.

[19] Birch JM, et al. Cancer phenotype correlates with constitutional TP53 genotype in families with the Li-Fraumeni syndrome. Oncogene 1998;17(9):1061–8.

[20] Hisada M, et al. Multiple primary cancers in families with Li-Fraumeni syndrome. J Natl Cancer Inst 1998;90(8):606–11.

[21] Liu PK, et al. Analysis of genomic instability in Li-Fraumeni fibroblasts with germline p53 mutations. Oncogene 1996;12(11):2267–78.

[22] Ruijs MW, et al. The single-nucleotide polymorphism 309 in the MDM2 gene contributes to the Li-Fraumeni syndrome and related phenotypes. Eur J Hum Genet 2007;15(1):110–4.

[23] Lupski JR. Genomic rearrangements and sporadic disease. Nat Genet 2007;39(7 Suppl.):S43–7.

[24] Burt EC, et al. Genomic alterations associated with loss of heterozygosity for TP53 in Li-Fraumeni syndrome fibroblasts. Br J Cancer 2000;83(4):467–72.

[25] Shlien A, et al. Excessive genomic DNA copy number variation in the Li-Fraumeni cancer predisposition syndrome. Proc Natl Acad Sci USA 2008;105(32):11264–9.

[26] Silva AG, et al. The profile and contribution of rare germline copy number variants to cancer risk in Li-Fraumeni patients negative for TP53 mutations. Orphanet J Rare Dis 2014;9:63.

[27] Silva AG, et al. Li-Fraumeni-like syndrome associated with a large BRCA1 intragenic deletion. BMC Cancer 2012;12:237.

[28] Al-Tassan N, et al. Inherited variants of MYH associated with somatic G:C-->T:A mutations in colorectal tumors. Nat Genet 2002;30(2):227–32.

[29] Jenkins MA, et al. Risk of colorectal cancer in monoallelic and biallelic carriers of MYH mutations: a population-based case-family study. Cancer Epidemiol Biomarkers Prev 2006;15(2):312–4.

[30] Sieber OM, et al. Multiple colorectal adenomas, classic adenomatous polyposis, and germ-line mutations in MYH. N Engl J Med 2003;348(9):791–9.

[31] Aretz S, et al. MUTYH-associated polyposis: 70 of 71 patients with biallelic mutations present with an attenuated or atypical phenotype. Int J Cancer 2006;119(4):807–14.

[32] Cerottini JP, et al. The type of K-ras mutation determines prognosis in colorectal cancer. Am J Surg 1998;175(3):198–202.

[33] Kraemer KH, et al. The role of sunlight and DNA repair in melanoma and nonmelanoma skin cancer. The xeroderma pigmentosum paradigm. Arch Dermatol 1994;130(8):1018–21.

[34] Kiyohara C, et al. Genetic polymorphisms in the nucleotide excision repair pathway and lung cancer risk: a meta-analysis. Int J Med Sci 2007;4(2):59–71.

[35] Moriwaki S, et al. Xeroderma pigmentosum—bridging a gap between clinic and laboratory. Photodermatol Photoimmunol Photomed 2001;17(2):47–54.

[36] Daya-Grosjean L, et al. The role of UV induced lesions in skin carcinogenesis: an overview of oncogene and tumor suppressor gene modifications in xeroderma pigmentosum skin tumors. Mutat Res 2005;571(1–2):43–56.

[37] Soufir N, et al. Association between INK4a-ARF and p53 mutations in skin carcinomas of xeroderma pigmentosum patients. J Natl Cancer Inst 2000;92(22):1841–7.

[38] Couvé-Privat S, et al. Functional analysis of novel sonic hedgehog gene mutations identified in basal cell carcinomas from xeroderma pigmentosum patients. Cancer Res 2004;64(10):3559–65.

[39] Aarnio M, et al. Cancer risk in mutation carriers of DNA-mismatch-repair genes. Int J Cancer 1999;81(2):214–8.

[40] Mangold E, et al. A genotype-phenotype correlation in HNPCC: strong predominance of msh2 mutations in 41 patients with Muir-Torre syndrome. J Med Genet 2004;41(7):567–72.

[41] Hampel H, et al. Screening for the Lynch syndrome (hereditary nonpolyposis colorectal cancer). N Engl J Med 2005;352(18):1851–60.

[42] Peltomaki P, et al. Mutations associated with HNPCC predisposition – update of ICG-HNPCC/INSiGHT mutation database. Dis Markers 2004;20(4–5):269–76.

[43] Woods MO, et al. A new variant database for mismatch repair genes associated with Lynch syndrome. Hum Mutat 2007;28(7):669–73.

[44] Pande M, et al. Cancer spectrum in DNA mismatch repair gene mutation carriers: results from a hospital based Lynch syndrome registry. Fam Cancer 2012;11(3):441–7.

[45] Ramsoekh D, et al. Cancer risk in MLH1, MSH2 and MSH6 mutation carriers; different risk profiles may influence clinical management. Hered Cancer Clin Pract 2009;7(1):17.

[46] Wimmer K, et al. Constitutional mismatch repair-deficiency syndrome. Haematologica 2010;95(5):699–701.

[47] Niessen RC, et al. Germline hypermethylation of MLH1 and EPCAM deletions are a frequent cause of Lynch syndrome. Genes Chromosom Cancer 2009;48(8):737–44.

[48] Kunkel TA. Nucleotide repeats. Slippery DNA and diseases. Nature 1993;365(6443):207–8.

[49] Duval A, et al. Mutations at coding repeat sequences in mismatch repair-deficient human cancers: toward a new concept of target genes for instability. Cancer Res 2002;62(9):2447–54.

[50] Woerner SM, et al. Microsatellite instability of selective target genes in HNPCC-associated colon adenomas. Oncogene 2005;24(15):2525–35.

[51] Perucho M. Microsatellite instability: the mutator that mutates the other mutator. Nat Med 1996;2(6):630–1.

[52] Markowitz S, et al. Inactivation of the type II TGF-beta receptor in colon cancer cells with microsatellite instability. Science 1995;268(5215):1336–8.

[53] Rampino N, et al. Somatic frameshift mutations in the BAX gene in colon cancers of the microsatellite mutator phenotype. Science 1997;275(5302):967–9.

[54] Yagi OK, et al. Proapoptotic gene BAX is frequently mutated in hereditary nonpolyposis colorectal cancers but not in adenomas. Gastroenterology 1998;114(2):268–74.

[55] Swift M, et al. Incidence of cancer in 161 families affected by ataxia-telangiectasia. N Engl J Med 1991;325(26):1831–6.

[56] Easton DF. Cancer risks in A–T heterozygotes. Int J Radiat Biol 1994;66(6 Suppl):S177–82.

[57] Bredemeyer AL, et al. ATM stabilizes DNA double-strand-break complexes during V(D)J recombination. Nature 2006;442(7101):466–70.

[58] Kastan MB, et al. A mammalian cell cycle checkpoint pathway utilizing p53 and GADD45 is defective in ataxia-telangiectasia. Cell 1992;71(4):587–97.

[59] Stecklein SR, et al. Identifying and exploiting defects in the Fanconi anemia/BRCA pathway in oncology. Transl Res 2012;160(3):178–97.

[60] Kottemann MC, et al. Fanconi anaemia and the repair Watson and Crick DNA crosslinks. Nature 2013;493(7432):356–63.

[61] Deans AJ, et al. DNA interstrand crosslink repair and cancer. Nat Rev Cancer 2011;11(7):467–80.

[62] Kobayashi H, et al. Hereditary breast and ovarian cancer susceptibility genes (Review). Oncol Rep 2013;30(3):1019–29.

[63] Antoniou A, et al. Average risks of breast and ovarian cancer associated with BRCA1 or BRCA2 mutations detected in case series unselected for family history: a combined analysis of 22 studies. Am J Hum Genet 2003;72(5):1117–30.

[64] Milne RL, et al. The average cumulative risks of breast and ovarian cancer for carriers of mutations in BRCA1 and BRCA2 attending genetic counseling units in Spain. Clin Cancer Res 2008;14(9):2861–9.

[65] Friedenson B. The BRCA1/2 pathway prevents hematologic cancers in addition to breast and ovarian cancers. BMC Cancer 2007;7:152.

[66] Lakhani SR, et al. The pathology of familial breast cancer: predictive value of immunohistochemical markers estrogen receptor, progesterone receptor, HER2, and p53 in patients with mutations in BRCA1 and BRCA2. J Clin Oncol 2002;20(9):2310–8.

[67] Rennert G, et al. Clinical outcomes of breast cancer in carriers of BRCA1 and BRCA2 mutations. N Engl J Med 2007;357(2):115–23.

[68] Kennedy RD, et al. The role of BRCA1 in the cellular response to chemotherapy. J Natl Cancer Inst 2004;96(22):1659–68.

[69] Tassone P, et al. BRCA1 expression modulates chemosensitivity of BRCA1-defective HCC1937 human breast cancer cells. Br J Cancer 2003;88(8):1285–91.

[70] Alter BP. Fanconi anemia and the development of leukemia. Best Pract Res Clin Haematol 2014;27(3–4):214–21.

[71] Rosenberg PS, et al. How high are carrier frequencies of rare recessive syndromes? Contemporary estimates for Fanconi anemia in the United States and Israel. Am J Med Genet A 2011;155A(8):1877–83.

[72] Shimamura A, et al. Pathophysiology and management of inherited bone marrow failure syndromes. Blood Rev 2010;24(3):101–22.

[73] Tulpule A, et al. Knockdown of Fanconi anemia genes in human embryonic stem cells reveals early developmental defects in the hematopoietic lineage. Blood 2010;115(17):3453–62.

[74] Pikor L, et al. The detection and implication of genome instability in cancer. Cancer Metastasis Rev 2013;32(3–4):341–52.

[75] Nowell PC. The clonal evolution of tumor cell populations. Science 1976;194(4260):23–8.

[76] Lengauer C, et al. Genetic instability in colorectal cancers. Nature 1997;386(6625):623–7.

[77] Gorgoulis VG, et al. Activation of the DNA damage checkpoint and genomic instability in human precancerous lesions. Nature 2005;434(7035):907–13.

[78] McGranahan N, et al. Cancer chromosomal instability: therapeutic and diagnostic challenges. EMBO Rep 2012;13(6):528–38.

[79] Lengauer C, et al. Genetic instabilities in human cancers. Nature 1998;396(6712):643–9.

[80] Gagos S, et al. Chromosome instability in neoplasia: chaotic roots to continuous growth. Int J Biochem Cell Biol 2005;37(5):1014–33.

[81] Loeb LA. Human cancers express mutator phenotypes: origin, consequences and targeting. Nat Rev Cancer 2011;11(6):450–7.

[82] Thu KL, et al. Lung adenocarcinoma of never smokers and smokers harbor differential regions of genetic alteration and exhibit different levels of genomic instability. PLoS One 2012;7(3):e33003.

[83] Wilkerson MD, et al. Differential pathogenesis of lung adenocarcinoma subtypes involving sequence mutations, copy number, chromosomal instability, and methylation. PLoS One 2012;7(5):e36530.

[84] Walther A, et al. Association between chromosomal instability and prognosis in colorectal cancer: a meta-analysis. Gut 2008;57(7):941–50.

[85] Carter SL, et al. A signature of chromosomal instability inferred from gene expression profiles predicts clinical outcome in multiple human cancers. Nat Genet 2006;38(9):1043–8.

[86] Chandhok NS, et al. A little CIN may cost a lot: revisiting aneuploidy and cancer. Curr Opin Genet Dev 2009;19(1):74–81.

[87] Swanton C, et al. Chromosomal instability determines taxane response. Proc Natl Acad Sci USA 2009;106(21):8671–6.

[88] Solomon DA, et al. Mutational inactivation of STAG2 causes aneuploidy in human cancer. Science 2011;333(6045):1039–43.

[89] Jallepalli PV, et al. Securin is required for chromosomal stability in human cells. Cell 2001;105(4):445–57.

[90] Fukasawa K, et al. Synergy between the Mos/mitogen-activated protein kinase pathway and loss of p53 function in transformation and chromosome instability. Mol Cell Biol 1997;17(1):506–18.

[91] O'Sullivan RJ, et al. Telomeres: protecting chromosomes against genome instability. Nat Rev Mol Cell Biol 2010;11(3):171–81.

[92] Levitt NC, et al. Caretaker tumour suppressor genes that defend genome integrity. Trends Mol Med 2002;8(4):179–86.

[93] Cahill DP, et al. Mutations of mitotic checkpoint genes in human cancers. Nature 1998;392(6673):300–3.
[94] Hanks S, et al. Constitutional aneuploidy and cancer predisposition caused by biallelic mutations in BUB1B. Nat Genet 2004;36(11):1159–61.
[95] Li Y, et al. Identification of a human mitotic checkpoint gene: hsMAD2. Science 1996;274(5285):246–8.
[96] Kinzler KW, et al. Cancer-susceptibility genes. Gatekeepers and caretakers. Nature 1997;386(6627):761–3.
[97] Hackett JA, et al. Balancing instability: dual roles for telomerase and telomere dysfunction in tumorigenesis. Oncogene 2002;21(4):619–26.
[98] Rajagopalan H, et al. Aneuploidy and cancer. Nature 2004;432(7015):338–41.
[99] Wang Z, et al. Three classes of genes mutated in colorectal cancers with chromosomal instability. Cancer Res 2004;64(9):2998–3001.
[100] Negrini S, et al. Genomic instability–an evolving hallmark of cancer. Nat Rev Mol Cell Biol 2010;11(3):220–8.
[101] Wood LD, et al. The genomic landscapes of human breast and colorectal cancers. Science 2007;318(5853):1108–13.
[102] Jones S, et al. Core signaling pathways in human pancreatic cancers revealed by global genomic analyses. Science 2008;321(5897):1801–6.
[103] Parsons DW, et al. An integrated genomic analysis of human glioblastoma multiforme. Science 2008;321(5897):1807–12.
[104] Cancer Genome Atlas Research Network. Comprehensive genomic characterization defines human glioblastoma genes and core pathways. Nature 2008;455(7216):1061–8.
[105] Ding L, et al. Somatic mutations affect key pathways in lung adenocarcinoma. Nature 2008;455(7216):1069–75.
[106] Puente XS, et al. Whole-genome sequencing identifies recurrent mutations in chronic lymphocytic leukaemia. Nature 2011;475(7354):101–5.
[107] Dulak AM, et al. Exome and whole-genome sequencing of esophageal adenocarcinoma identifies recurrent driver events and mutational complexity. Nat Genet 2013;45(5):478–86.
[108] Dulak AM, et al. Gastrointestinal adenocarcinomas of the esophagus, stomach, and colon exhibit distinct patterns of genome instability and oncogenesis. Cancer Res 2012;72(17):4383–93.
[109] Mai S, et al. Genomic instability in MycER-activated Rat1A-MycER cells. Chromosome Res 1996;4(5):365–71.
[110] Spruck CH, et al. Deregulated cyclin E induces chromosome instability. Nature 1999;401(6750):297–300.
[111] Rajagopalan H, et al. Inactivation of hCDC4 can cause chromosomal instability. Nature 2004;428(6978):77–81.
[112] Di Micco R, et al. Oncogene-induced senescence is a DNA damage response triggered by DNA hyper-replication. Nature 2006;444(7119):638–42.
[113] Woo RA, et al. Activated oncogenes promote and cooperate with chromosomal instability for neoplastic transformation. Genet Dev 2004;18(11):1317–30.
[114] Surget S, et al. Uncovering the role of p53 splice variants in human malignancy: a clinical perspective. Onco Targets Ther 2013;7:57–68.
[115] Bunz F, et al. Targeted inactivation of p53 in human cells does not result in aneuploidy. Cancer Res 2002;62(4):1129–33.
[116] Durkin SG, et al. Chromosome fragile sites. Annu Rev Genet 2007;41:169–92.
[117] Tsantoulis PK, et al. Oncogene-induced replication stress preferentially targets common fragile sites in preneoplastic lesions. A genome-wide study. Oncogene 2008;27(23):3256–64.
[118] Forment JV, et al. Chromothripsis and cancer: causes and consequences of chromosome shattering. Nat Rev Cancer 2012;12(10):663–70.
[119] Stephens PJ, et al. Massive genomic rearrangement acquired in a single catastrophic event during cancer development. Cell 2011;144(1):27–40.
[120] Molenaar JJ, et al. Sequencing of neuroblastoma identifies chromothripsis and defects in neuritogenesis genes. Nature 2012;483(7391):589–93.
[121] Thibodeau SN, et al. Microsatellite instability in cancer of the proximal colon. Science 1993;260(5109):816–9.
[122] Risinger JI, et al. Genetic instability of microsatellites in endometrial carcinoma. Cancer Res 1993;53(21):5100–3.
[123] Rhyu MG, et al. Microsatellite instability occurs frequently in human gastric carcinoma. Oncogene 1994;9(1):29–32.
[124] Kim WS, et al. Microsatellite instability(MSI) in non-small cell lung cancer(NSCLC) is highly associated with transforming growth factor-beta type II receptor(TGF-beta RII) frameshift mutation. Anticancer Res 2000;20(3A):1499–502.
[125] Fishel R, et al. The human mutator gene homolog MSH2 and its association with hereditary nonpolyposis colon cancer. Cell 1993;75(5):1027–38.
[126] Goldstein J, et al. Multicenter retrospective analysis of metastatic colorectal cancer (CRC) with high-level microsatellite instability (MSI-H). Ann Oncol 2014;25(5):1032–8.
[127] Toyota M, et al. CpG island methylator phenotype in colorectal cancer. Proc Natl Acad Sci USA 1999;96(15):8681–6.
[128] Ward R, et al. Microsatellite instability and the clinicopathological features of sporadic colorectal cancer. Gut 2001;48(6):821–9.
[129] Amé JC, et al. The PARP superfamily. Bioessays 2004;26(8):882–93.
[130] Wang Z, et al. The role of PARP1 in the DNA damage response and its application in tumor therapy. Front Med 2012;6(2):156–64.
[131] Rouleau M, et al. PARP inhibition: PARP1 and beyond. Nat Rev Cancer 2010;10(4):293–301.
[132] Goncalves A, et al. Pol (ADP-ribose) polymerase-1 mRNA expression in human breast cancer: a meta-analysis. Breast Cancer Res Treat 2011;127(1):272–81.
[133] Abbas T, et al. Genomic instability in cancer. Cold Spring Harb Perspect Biol 2013;5(3):a012914.
[134] Tucker CL, et al. Lethal combinations. Nat Gent 2003;35(3):204–5.
[135] Farmer H, et al. Targeting the DNA repair defect in BRCA mutant cells as a therapeutic strategy. Nature 2005;434(7035):917–21.
[136] Iglehart JD, et al. Synthetic lethality – a new direction in cancer-drug development. N Engl J Med 2009;361(2):189–91.
[137] Sonnenblick A, et al. An update on PARP inhibitors—moving to the adjuvant setting. Nat Rev Clin Oncol 2015;12(1):27–41.
[138] Mangerich A, et al. How to kill tumor cells with inhibitors of poly (ADP-ribosyl) ation. Int J Cancer 2011;128(2):251–65.
[139] Rosen EM, et al. Targeting the BRCA1/2 tumor suppressors. Curr Drug Targets 2014;15(1):17–31.

Chapter 28

Chromatin Modifications in DNA Repair and Cancer

M. Renaud-Young, K. Riabowol, J. Cobb

University of Calgary, Calgary, AB, Canada

Chapter Outline

1. INTRODUCTION

The DNA of eukaryotic organisms is bound and organized by numerous proteins. DNA and these proteins are collectively known as "chromatin." Gaining access to damaged DNA in the context is necessary for repair. Protein complexes can modify the structure of chromatin to alter the geometry of DNA and allow repair of damaged regions [2,3]. The pathway utilized for repair depends on the type of DNA lesion, the cell-cycle stage, and the chromatin environment in which the damage is detected [4]. It is generally accepted that all in vivo DNA transactions, including repair occur within the context of chromatin and its remodelers.

Cellular processes that are critical to survival and division are evolutionarily conserved, and the diverse pathways that modulate chromatin, to alter protein–DNA interactions during repair can be found in organisms ranging from humans to the budding yeast *Saccharomyces cerevisiae*. Furthermore, the factors that modulate chromatin in the inherently different processes of transcription, DNA replication, and repair have underlying similarities as each requires disruption of internucleosomal interactions, nucleosome disassembly to enable the process, and reassembly after it is complete. In this chapter, we review the hierarchical pathways of chromatin modifications in DNA replication, and repair, with emphasis on DNA-repair processes, focusing on the replication associated DNA-damage tolerance (DDT) pathway in the model system, *S. cerevisiae*.

2. INTERRELATIONSHIP OF DNA AND CHROMATIN

In eukaryotes, DNA is organized within the nucleus by octamers of histones that are assembled into nucleosomes, the basic units of chromatin [5]. Canonical nucleosomes contain two sets of core histones that preferentially form head-to-tail

Genome Stability. http://dx.doi.org/10.1016/B978-0-12-803309-8.00028-8

heterodimers of H2A–H2B and H3–H4 when they are not part of a nucleosome. The process of nucleosome construction occurs by stepwise assembly of H3–H4 dimers into $(H3–H4)_2$ tetramer. This half nucleosome is partially incorporated into DNA as two sets of H2A–H2B dimers are added to opposite sides of the tetramer (reviewed in Ref. [6]). Nucleosomes containing about 147 base pairs (bp) of DNA are spaced apart by between 10 and 70 bp of linker DNA [7]. The linker DNA can remain nucleosome free or become bound by the monomeric histone H1.

Normal expression of the core histones is highly regulated with respect to the DNA-replication cycle, and expression is maintained at stoichiometrically consistent levels for each dimer set, such that H2A and H2B are cotranscribed, and the same with H3 and H4 [6]. Misregulation causing an upset to the balance of histones can lead to cell-cycle arrest and cell death [8–10]. Histone chaperones are critical regulators of nucleosome-free histone pools, performing an essential role in nucleosome dynamics that permits transcription, replication, and repair, which require nucleosome disassembly, reassembly, and eviction [6,11].

Histones are bound by the chaperones Nap1 and Asf1 immediately after translation and these proteins facilitate transport into the nucleus. There are preferential associations of chaperones for different dimers: Nap1 binds H2A–H2B, while Asf1 binds H3–H4. After nuclear import, the chaperones may transfer the dimers to other chaperones. For example, H3–H4 dimers are transferred from Asf1 to the replication and DNA damage–specific chaperone CAF-1 (chromatin assembly factor-1), which incorporates nucleosomes into newly synthesized DNA [12]. Similarly, histone chaperones Spt6 and FACT permit RNA polymerase II (Pol II) passage during transcription. Nucleosomes are disassembled by sequential H2A–H2B removal by FACT and H3–H4 interaction with Spt6 ahead of RNA polymerase II (RNA Pol II) progression. Following the polymerase passage, FACT is instrumental to the incorporation of recycled H3–H4 into DNA [13].

Histone chaperones are critical to the dynamic placement of nucleosomes in chromatin and facilitate the changes that coincide with replication, transcription, and repair.

3. HISTONE MODIFICATIONS AND CHROMATIN REMODELERS

Some of the earliest biochemical characterizations of histones describe acetylation of H3 and H4 [14,15]. It was later found that histones are extensively modified, primarily by acetylation, methylation, citrullination, ubiquitination, sumoylation, and phosphorylation (reviewed in Refs. [16,17]) (Fig. 28.1). The posttranslational modifications (PTMs) serve several important functions. Many of the modifications occur on lysine (K) residues in the amino termini of histones, altering the net charge and topology, and changing the interactions of nucleosomes with each other. These modifications also serve as a signaling platform for chromatin modifiers, transcription factors, and polymerases. In addition to surface modification, nucleosomes can be translocated along the DNA by chromatin remodelers. For many DNA processes, the nucleosome presents a functional barrier that needs to be moved or evicted. In several cases, there are examples of hierarchical interactions, where a PTM facilitates a binding interaction with another complex leading to a downstream function [26,27]. Different combinations of posttranslational modifications can recruit downstream effectors that promote transcriptional activation or a type of DNA damage–repair pathway.

The combinatorial effect of histone PTMs on downstream function inspired the popular model known as the "histone code" [28]. Histone "code writers" add a modification to one or more histones of the complex and "code readers" recognize the PTM and initiate the subsequent signal propagation step or modification step. The enzymes that modify histones are usually components of multimeric protein complexes that are comprised of code readers, writers, and erasers [16]. As an example, the NuA3 complex has two methyl lysine recognizing subunits, Yng1 and Pdp3 [26,27], which allow the complex to recognize trimethylated H3–K4 and H3–K36 in promoter regions. Once bound, Sas3, another subunit and acetyltransferase, acetylates H3 at lysine 14. Additionally, there are cases where the same proteins can be found in complexes of different enzymatic function, such as Arp4 being a component of NuA4, SWR-C, and Ino80 [29].

Although there are several types of PTMs, not all histones are targeted equally for a type of modification, as shown in Fig. 28.1. For example, the majority of acetylations occur on histones H3 and H4. In *S. cerevisiae*, methylation is restricted to H3, and ubiquitination occurs on H2B. This may be a reflection of the asymmetry of histone placement within the nucleosome as well as binding constraints of the modifying complexes.

3.1 Histone Acetyltransferases and Deacetylases

Acetyltransferase proteins catalyze the addition of an acetyl group to the terminal amine of a lysine residue. This creates the effect of neutralizing the positively charged lysine (Fig. 28.2A), and has been proposed to weaken internucleosomal interactions to create more flexibility and access to chromatin [30,31] (Fig. 28.2B). Support for this model is the observation of an abundance of acetylated histones in highly transcribed euchromatic regions of the genome, as well as in the region

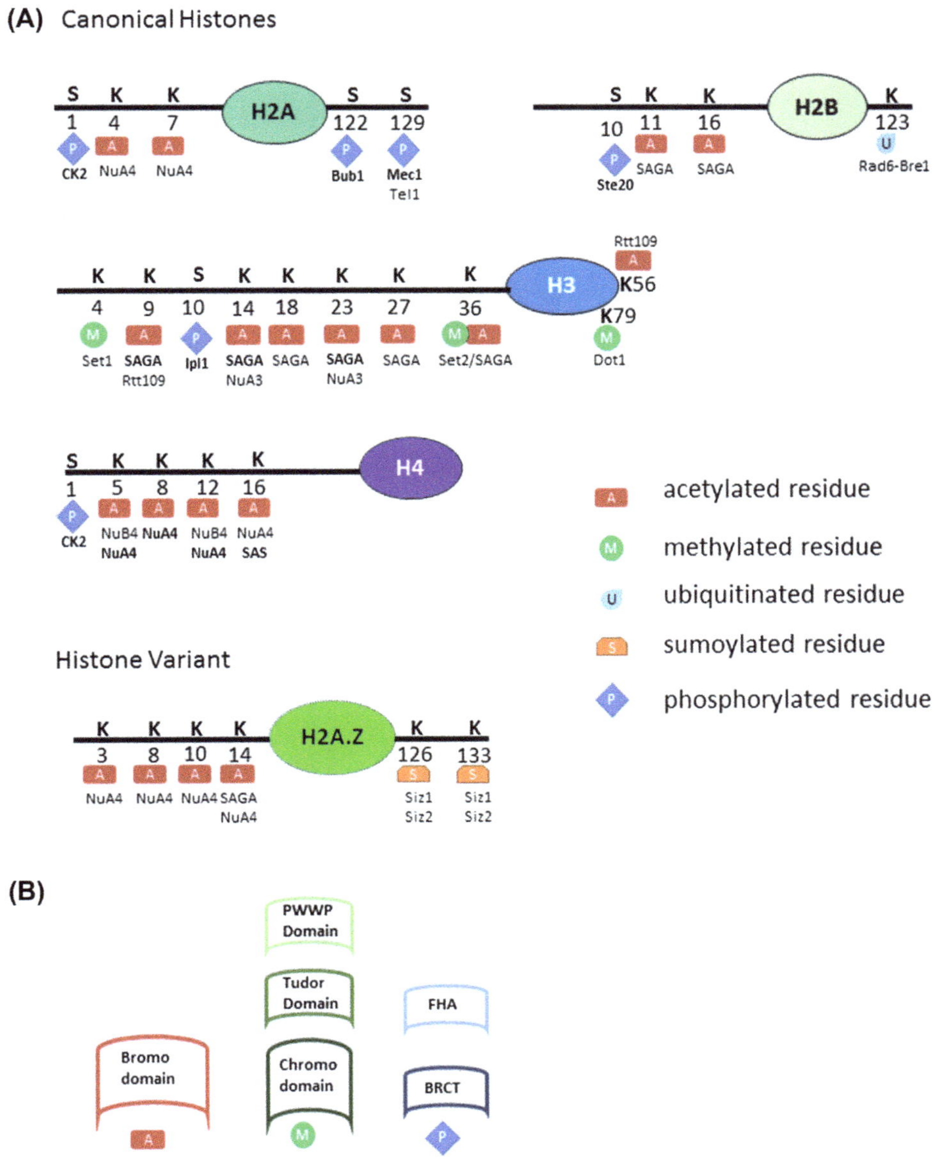

FIGURE 28.1 **Posttranslational modifications (PTMs) to histones in *Saccharomyces cerevisiae*.** (A) Histones are extensively posttranslationally modified by acetylation, methylation, ubiquitination, and sumoylation on lysine (K), and phosphorylation on serine (S) or threonine (T) [2,18–23]. While some modifications can only be made by a single complex, others are redundantly modified by several complexes (depicted beneath). With other modifications, such as sumoylation, all histones can be observed to be sumoylated at several K residues by Siz1 and Siz2 SUMO ligases [24]. There are many abundant PTMs observed in higher eukaryotes, which do not always occur on the exact residues that are observed with yeast. (B) PTMs often enable recognition by protein interaction motifs. The following are examples of common recognition motifs: bromodomains bind acetylated lysines; chromodomains, Tudor domains, and PWWP domains bind methylated lysines; BRCT repeats and FHA domains mediate phospho-serine and threonine interactions [25]. In many processes, the application of PTMs allows the histone tails to serve as signaling platforms for downstream processes.

directly adjacent to damaged DNA [32,33]. In addition to charge modulation, acetylation can also permit interactions with proteins that contain bromodomains. The RSC and SWR-C chromatin–remodeling complexes more strongly interact with acetylated histones through bromodomain-containing proteins [34–36].

Although there are many histone acetyltransferases (HATs) that redundantly acetylate histones, two stand out as being the most important to acetylating bulk chromatin proteins: Gcn5 and Esa1. Gcn5 is part of the Spt-Ada-Gcn5-acetyltransferase (SAGA) complex, SAGA-like (SLIK) complex and the smaller subcomplex ADA, and it acetylates H2B and H3 [37,38]. Similarly, Esa1 is a component of NuA4, the yeast homolog of Tip60, and the subcomplex piccolo NuA4. Esa1 acetylates H2A, H2A.Z, and H4 [18,39,40]. HAT activity is not restricted to histones or even other nuclear proteins, with new targets continually being discovered [41–43].

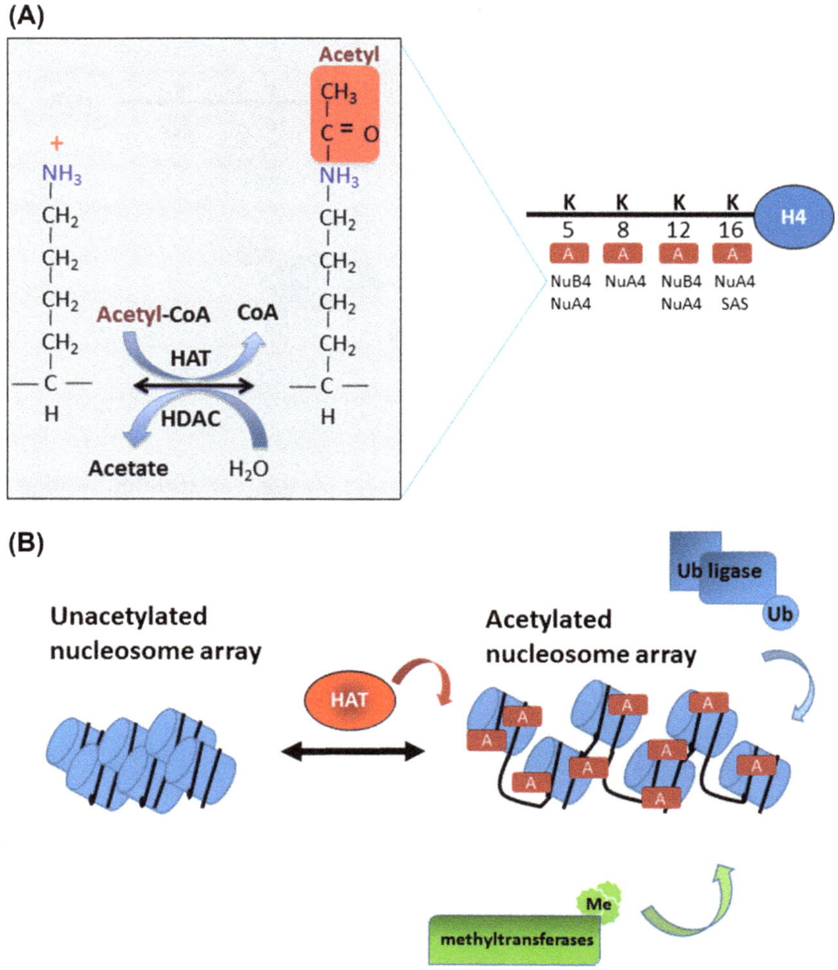

FIGURE 28.2 **Acetylation/deacetylation of lysine residues by histone acetyltransferase (HAT) and histone deacetylase (HDAC) complexes.** (A) HAT complex catalyzes the addition of acetyl groups to lysine residues, while HDACs catalyze the reverse reaction *(Adapted from Yang XJ, Seto E. The Rpd3/Hda1 family of lysine deacetylases: from bacteria and yeast to mice and men. Nat Rev Mol Cell Biol 2008;9:206–18.)*. Acetylation neutralizes the basicity of the lysine, which in the case of histones changes the chromatin environment, lessening the interactions of the nucleosomes with DNA. (B) Model depicting how abundant acetylation of histones creates more loosely associated nucleosomes. The "opening" of chromatin makes more downstream histone-modifying interactions more amenable.

Histone deacetylases (HDACs) catalyze the removal of the acetyl group from acetylated lysine residues and restore the lysine to its premodified state. There are four classes of HDACs, although the RPD3 and HDA complexes perform the vast majority of transcription-associated deacetylation [44–46]. An exception is Sir2 in budding yeast, together with Sir3 and Sir4, having a specialized role in deacetylating H4–K16 to augment the spread of silent chromatin [47]. As with the acetyltransferases, HDACs have targets that extend beyond histones and chromatin into every cellular system [42]. Much of RPD3 and HDA targeting is mediated by other proteins in the complex, which stabilize its localization to specific genomic sites including intergenic regions and certain promoters [46]. It appears that with regard to catalyzing deacetylation, RPD3 and HDA do not target specific lysines of histone tails but have a broad recognition of acetylated residues [44,47].

3.2 Histone Lysine Methyltransferases and Demethylases

Methylation of lysine residues differs from acetylation in that it functions primarily as a binding platform for signal transduction [25] (Fig. 28.1). Histone lysine methyltransferases (HKMT) have relatively fewer targets compared to acetyltransferases. In *S. cerevisiae*, the most described and biologically relevant HKMTs are Set1, Set2, and Dot1, which catalyze trimethylation of H3–K4, H3–K36, and H3–K79, respectively [48]. Generally, methylated lysines are binding targets for proteins that contain chromodomains, plant homeodomains (PHD), PWWP, and Tudor domains [49]. Unlike acetylation, where a single acetyl group can be added, up to three methyl groups can be added to lysine. HKMTs can be very precise as to what modification they catalyze. An enzyme that adds a single methylation mark often will not di- or trimethylate that residue (reviewed in Ref. [50]). This variation creates a level of plasticity in which reader complexes will target a specific methylation mark, as they have a preference for certain lysine-modification signatures. An example of this targeted interaction is the H3–K4 trimethylation (H3–K4me3) mark, which recruits proteins with a PHD motif. Although these PHDs will bind dimethylated and monomethylated forms, it is to a much lesser degree than its preferential trimethyl target [26]. Downstream effects of methylation include: (1) regulating DNA repair, in the case of H3–K79me3; (2) distinguishing chromatin regions for transcriptional functions; and (3) marking boundaries between active chromatin and subtelomeric silent chromatin [51,52]. As with acetylation, methylation is a reversible modification, mediated by the Jumanji domain–containing histone lysine demethylases (HKDM).

3.3 Histone Ubiquitination and Sumoylation

Histone ubiquitinatination and sumoylation result from an enzymatic cascade where the small proteins ubiquitin or small ubiquitin-like modifier (SUMO) are covalently linked to a lysine residue [53]. Both ubiquitin and SUMO modifications produce bulkier alterations than acetylation or methylation, and may distort the nucleosomal conformation to make it more accessible for further modification [25], or compete with and block other modifications at the same, or nearby lysine residues. Ubiquitination of H2B at K123 is associated with the coding section of transcribed regions [19]. This modification plays a critical role in permitting downstream nucleosome methylation by Set1 (COMPASS complex) and Dot1 to methylate H3–K4 and K79, respectively [54,55]. Like acetylation and methylation, ubiquitin modifications can be removed by ubiquitin proteases [56].

Histone sumoylation is dependent on Siz1 and Siz2 E3 ligases, and generally corresponds to repression of transcription. Transcription induction produces a loss of SUMO and gain of acetylation. Therefore, sumoylation is proposed to compete with transcription-promoting modifications and help maintain transcriptional quiescence at inducible genes [24].

3.4 Histone Phosphorylation

Phosphorylation of histones is associated with signaling events that are extremely relevant to genome stability, such as chromatin condensation in mitosis and the response to DNA damage. Like acetylation, phosphorylation that is proximal to lysines can nullify the highly basic charge of histones. In damage and mitosis signaling, its primary function is to create interaction sites for phosphorylation-binding motifs of downstream targets. As part of the DNA damage and signaling cascade, H2A is phosphorylated on S129 in yeast (abbreviated to γH2A) [20] (Fig. 28.1). In this circumstance, it is bound by BRCT repeat or forkhead-associated domain (FHA)-containing proteins that are part of the signal cascade leading to activation of cell-cycle checkpoint. Coordination of mitotic processes is strongly regulated by histone phosphorylation/dephosphorylation events. Phosphorylation of H3–S10 by the kinase Ipl1 (homolog of Aurora B) [20] is important to establishment and maintenance of chromatin condensation. After mitosis, it is removed by PP1 phosphatase Glc7. Phosphorylation of H2A–S122 occurs as part of the spindle assembly checkpoint (SAC) in the event of DNA damage and misaligned spindles [57].

3.5 Nucleosome Exchangers and Remodelers

Histone modifiers often act cooperatively with histone exchangers and remodelers, to alter nucleosome composition, positioning, and occupancy. Chromatin remodelers can be grouped into six families: Snf2-like (Chd1, SWI/SNF, RSC, ISWI); Swr1-like (SWR-C, INO80); SMARCAL1-like (Fun30); Rad54-like; Rad5/16-like; and ERCC6/SSO1653-like [58–60]. Remodelers have different capabilities, some can slide histones relative to DNA and some can completely evict octamers; however, they all require ATP hydrolysis for functionality. Many remodelers contain histone code readers for targeting to the genomic regions that they alter. The Chd1 protein associates directly with the SAGA acetyltransferase complex, acting cooperatively with acetylation to promote transcription progression [61]. Additionally, Chd1 and other CHD family members contain tandem chromodomains at their N-termini that facilitate interaction with H3–K4 [62].

Remodelers can be grouped according to specific functional roles. Members of the Snf2-like and Swr1-like families tend to promote transcription and processes that open up the chromatin and make it more accessible, through both nucleosome movement, histone exchange, and eviction [63]. However, ISWI and CHD family members can behave in the opposite manner, altering internucleosomal spacing to remove nucleosome-free regions (NFRs) and shorten internucleosomal gaps to such a degree that further shortening is not possible [60,62,64].

3.6 Histone Variants

Histone variants share many of the structural features of their canonical nucleosomal counterparts but have variation in sequence and transcriptional regulation that is independent of core histones. Variants also associate with chaperones; however, unlike "bulk histones" their incorporation into chromatin is replication independent. A variant that is essential and conserved in all eukaryotes is a version of H3 that designates the chromatin region of the centromere (CenH3). CenH3 (also CENP-A or Cse4) is incorporated by its chaperone Scm3, and is critical for stable formation of the kinetochore in mitosis (reviewed in Choy [65]; and Henikoff, & Furuyama [66]).

The variant H2A.Z (Htz1) functions in transcription, boundary formation between silenced DNA and active genes, and the response to DNA damage [67]. H2A.Z shares only 60% sequence homology with canonical H2A, varying most in the linker regions within the histone fold and C-terminus [68,69]. The unique C-terminal region is required to allow the SWR–C complex to interact with it and insert it into the DNA. The difference in linker regions alters the interaction of H2A.Z with H3–H4, and is predicted to create a less stable nucleosome [70].

Like H2A, H2A.Z-H2B dimers are bound and imported into the nucleus by the chaperone Nap1. When in the nucleus, the H2A.Z–H2B-specific chaperone, Chz1, participates in correctly targeting H2A.Z to promoters and telomeres [71–73]. The incorporation of H2A.Z is mediated by the Swi2/Snf2-related chromatin-remodeling complex SWR-C [74–76], which exchanges canonical H2A–H2B in chromatin for the variant dimer H2A.Z–H2B. Biochemical experiments indicate that H2A.Z nucleosomes may be heterotypic, where there is one H2A.Z and one H2A, or homotypic, where there are two H2A.Zs [77]. Because of differences in linker region 1 (L1) length around the histone fold of the protein, it is predicted that a heterotypic H2A.Z nucleosome would be highly unstable [70]. This instability may account for observations of H2A.Z presence at transcription start sites reducing the barrier function that most nucleosomes present to RNA Pol II [78].

Although H2A.Z incorporation has diverse functions, this variant is often placed in regions that require dynamic regulation, and are proximal to NFRs. At transcription start sites, a 70–120 bp NFR precedes the coding region of the gene. The NFR is followed by the so-called "+1 nucleosome," which is the first nucleosome encountered by RNA Pol II. In most genes, regardless of activity, this region is bound by SWR-C and INO80, both of which are recruited by the NFR at that site (Swc2 of SWR-C and Nhp10, Arp8 and Ies5 of INO80) [79]. SWR-C localization is further augmented by NuA4-mediated acetylation of H4 nucleosomes (Bdf1 of SWR-C binds these) [36,79]. In a study measuring the processivity rate of RNA Pol II in transcription, it was observed that the nucleosomal barrier to transcription could be alleviated by H2A.Z occupying the +1 position [78,80]. The model of how the combined activity of these complexes cooperates with transcription is that SWR-C incorporates H2A.Z while INO80 removes it from the +1 nucleosome, in a dynamic fashion [79]. This disruption of the nucleosome eases progression of the polymerase. Along with affecting its incorporation, H2A.Z acetylation by NuA4 and Gcn5 is associated with both highly transcribed genes, and retention of the variant by inhibiting INO80-mediated H2A.Z to H2A exchange [81–83]. Additionally, Rtt109 enhances removal of H2A.Z through acetylation of H3–K56, permitting SWR-C to catalyze the reverse exchange of incorporation: H2A for H2A.Z [84].

H2A.Z nucleosomes are also localized in proximity to replication origins [85], and sites of DNA damage [86], which are also surrounded by NFRs [87]. Therefore, a general mechanism of H2A.Z incorporation may be to relieve the nucleosomal barrier to polymerases at transcription start sites and to replisomes of newly fired replication origins.

4. CHROMATIN MODIFIERS IN GENOME STABILITY

The activities of various chromatin modifiers and remodelers in the functions of replication and DNA-damage response are critical in the maintenance of genome stability. There are many instances of cooperative and antagonistic interactions that create or deny access to certain genomic regions.

4.1 DNA Replication

The process of DNA replication is a highly coordinated event where bidirectionally oriented replication complexes or replisomes unwind and duplicate DNA from several genomic loci (reviewed in Bell, & Dutta [88]). The intersection of unreplicated duplex DNA with a single replisome and the replicated daughter strands is referred to as the replication fork. The replication fork is comprised of a leading strand, which is polymerized continuously and the lagging strand, which is discontinuously synthesized as Okazaki fragments [88].

During S phase, chromatin is duplicated along with the DNA template and requires both histone recycling and the synthesis of new histones to have enough octamers for both the newly generated strands. The assembly of nucleosomes after fork passage is rapid (30 s), and it was found that nucleosomes were present on the leading and lagging strands 125–300 bp from the replication complex (reviewed in Annunziato [89]). Therefore, addition of nucleosomes from recycled parental and newly synthesized histone pools progresses with similar timing to the synthesis of DNA. However, the presence of nucleosomes on nascent DNA did not follow the same kinetics as restoring all chromatin marks (chromatin maturity), which took about 15 min following replication [89]. This "maturity" time could be increased by the addition of sodium butyrate, an HDAC inhibitor, which provided some of the first evidence that newly translated H4 is acetylated [89].

It has since been found that newly translated histones H3 and H4 are acetylated in the cytoplasm by chaperone-associated acetyltransferase complexes. NuB4, which is comprised of the Hat1/Kat1, Hat2, and the chaperone Hif1, acetylates newly synthesized cytoplasmic H4 on K5, and K12 [90–92]. In a similar fashion, Rtt109 acetylates newly synthesized H3 on K9 and K56, associating with the chaperone Asf1 for H3–K56 and both Vps75 and Asf1 for H3–K9 [93]. Prior to nuclear import, H3 and H4 form dimers and are imported by Asf1 [92]. Experimental evidence suggests that H4 acetylation is important for nuclear import of these dimers, because this process is compromised in an H4–K5,12R allele, producing a defect of prolonged S phase [94]. Asf1 then transfers the H3–H4 dimers to the replication associated chaperones: CAF-1 and Rtt106. Although both are associated with the complex, CAF-1 incorporates the newly synthesized H3–H4, and this selectivity may be due to preferential association with the acetylated forms of the proteins [95,96]. CAF-1 and Rtt106 accept H3–H4 dimers from Asf-1 to form tetramers and place them into nascent DNA, while H2A–H2B dimers are added afterward.

The acetylation of newly translated histones is transient, and removal of these marks is evident in late S phase to G2 of the cell cycle [97]. As early studies of nascent chromatin indicate, this deacetylation event coincides with the appearance of compacted, nuclease-resistant chromatin [89]. Compaction is largely mediated by the remodeler Iswi2, which is associated with nucleosome assembly after replication [98]. Iswi2 preferentially interacts with unmodified H4 [99–101] and the Rpd3L HDAC complex [102] and is important for establishing correct spacing and nucleosome density after replication and transcription [64].

In replication, acetylation of nascent nucleosomes also provides continued access to underlying, newly replicated DNA. Open chromatin is more permissive to DNA-repair proteins, permitting access to damaged DNA and stalled replication forks. In support of this theory, H3–K56ac is associated with regions of DNA-damage and -replication stress that subsides only after repair has occurred [103]. This mark is positioned at the entry/exit point where DNA wraps around the nucleosome [104], and is proposed to create a loose interaction that inhibits compaction of nucleosome arrays and facilitates unwrapping. Therefore, the requirement of open chromatin may be the reason for H3–K56ac association with newly incorporated nucleosomes, highly transcribed genes, and sites of damage.

4.2 DNA-Damage Response and Repair

When studying DNA-damage processes, chromatin is typically viewed as an obstruction to repair. In the "access, repair, and restore" model [105], chromatin modifiers clear the way for repair processes and restore chromatin to the predamage state following repair completion. This model has evolved to acknowledge that the predamage chromatin state and chromatin modifiers that participate in the repair process can exert some control over the repair outcome [3]. Although there are many sources of DNA damage, the types of damage they create can be grouped into categories of DNA base lesions, mismatches, and breaks in the DNA backbone leading to single-strand breaks (SSBs) and double-strand breaks (DSBs) (Fig. 28.3). Base excision repair (BER) and nucleotide excision repair (NER) resolve base lesions, while mismatch repair (MMR) resolves incorrect base pairs. DNA DSBs are repaired by homologous recombination (HR) and nonhomologous end joining (NHEJ). These types of damage create a different type of lesion when encountered by replication complexes. Base lesions obstruct replication polymerases because they cannot incorporate bases across from damaged DNA. The DDT pathway permits replication to continue in spite of damage. In addition to pathways that mediate repair and damage tolerance during replication, checkpoint pathways coordinate repair with cell-cycle stage and prevent transitions until repair is complete. In this section we review how the chromatin environment and modifiers participate in these pathways with emphasis on replication and DDT.

4.2.1 Repair of DNA Base Damage

Chemical and environmental agents, including ultraviolet radiation (UV), reactive oxygen species (ROS), and chemotherapeutic agents such as alkylating agents cause base lesions. Repair of these requires excision of the damaged nucleotide, or a patch surrounding it, and use of the unaffected strand as a template to synthesize DNA to fill in the space [1] (Fig. 28.3). In the BER pathway, proteins that detect a lesion also initiate the repair process. DNA N-glycosylases translocate throughout the genome, providing a general surveillance of base damage. When a base lesion is recognized, the DNA glycosylase

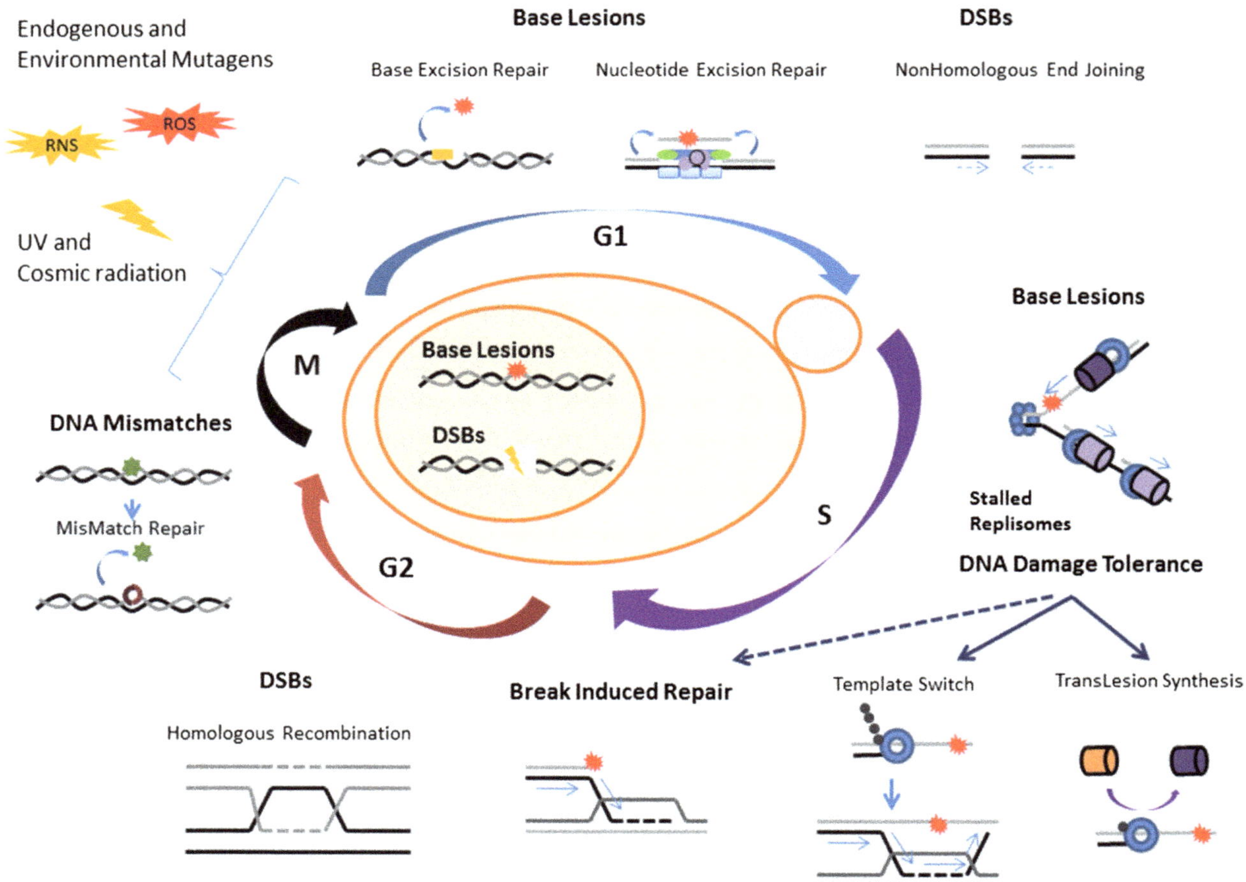

FIGURE 28.3 **DNA-damage repair and damage bypass pathways in different phases of the cell cycle.** The repair pathways that are utilized for different types of damage depend on the agent that caused the damage and the stage of the cell cycle [1,2,4]. Endogenous base lesions may be caused by reactive oxygen species (ROS) or reactive nitrogen species (RNS). Cosmic radiation or UV light cause base lesions and, more severely, DSBs. In G1, base lesions are repaired by BER and NER, while a DSB is repaired by NHEJ. In S phase, bypass of base lesions occurs via the DDT pathway, resulting in TLS or template switch. BIR occurs if the replisome encounters a nick in the DNA backbone or the replisome collapses. HR is the preferred repair pathway of DSBs during S and G2 of the cell cycle. Lastly, following replication, the newly replicated DNA is scanned for mismatches by the MMR pathway.

releases the damaged base from the deoxyribose sugar. This is followed by binding of an AP-endonuclease/lyase, which nicks the DNA backbone at the apyrimidinic/apurinic (AP) site. The nicked, AP deoxyribose is then recognized by a 3′- or 5′-phosophodiesterase, which removes the sugar leaving an empty space where the damage was. This gap is then filled by a DNA polymerase and the nick is sealed by DNA ligase I (reviewed in Boiteux, & Jinks-Robertson [4]). There is evidence, however, that BER can be inhibited at intermediate repair stages, after the DNA backbone is nicked, if the original DNA lesion resides is within a nucleosome. In an in vitro study of BER processes, it was found that repair of lesions near the center of the nucleosome was not completed because DNA Polβ could not access the abasic site [106]. Retention of repair intermediates at abasic sites can lead to genome instability as replication forks encountering nicked DNA can lead to DSBs in the DNA backbone [107]. Both in vitro and in vivo studies have identified a role for the RSC complex in supporting BER completion by moving the lesion to more accessible sites [108,109].

The NER pathway primarily repairs damage from UV and reactive oxygen-induced lesions. There are two types of NER: global genomic (GG-NER) and transcription coupled (TC-NER). GG-NER detects structural aberrations of the DNA helix resulting from a lesion that is often in untranscribed regions of the genome. Chromatin-modifying events in GG-NER are important to form the recognition platform and excision complex. The earliest steps in the pathway involve lesion recognition and localization of the heterodimers Rad4–Rad23 and Rad16–Rad7. Nucleosomes adjacent to the damage are repositioned by Rad4–Rad23-dependent recruitment of the chromatin remodeler SWI/SNF [110], and Rad16–Rad7 facilitates the recruitment of SAGA [111].

Histone acetylation by SAGA/Gcn5 and the recruitment of additional factors, Rad14 and Rad1–Rad10 (structure-specific endonuclease), assist in the subsequent incorporation of H2A.Z [112]. The initial GG-NER recognition factors eventually become targeted for Ub-mediated proteasomal degradation, while the NER pre-incision complex is formed.

This complex, which includes the essential RNA-Pol II–associated TFIIH, aggregates around the lesion and opens up the DNA, allowing binding of RPA to stabilize the undamaged ssDNA strand [4]. Following this, endonucleases are recruited to nick the DNA backbone 5′ and 3′ to the damaged region. The patch with the damage is then removed from the DNA and the gap is filled in by a DNA polymerase, Polε or Polδ, with DNA ligase I sealing the nick. TC-NER-damage removal from the pre-incision complex to completion of the repair is the same as with GG-NER, but the detection mechanism is different. The damage sensor is RNA-Pol II, which activates TC-NER when a lesion is encountered on the DNA template during transcription [4]. The chromatin remodeler RSC is required for efficient repair for both TC and GG-NER [113]. Although it is somewhat counterintuitive, there are some genes associated with TC-NER that are not a part of GG-NER. For example, Rad26 (ortholog of CSB), an Snf2-related DNA/RNA helicase, is proposed to function exclusively with TC-NER because it associates with RNA Pol II when it stalls during transcription elongation [114].

4.2.2 Repair of DNA Double-Stranded Breaks

DNA DSBs result from a variety of endogenous and exogenous sources. Hydroxyl radical attack and ionizing radiation can break the phosphodiester bond of the DNA backbone, replication stress can lead to fork collapse, and chemotherapeutic agents that poison topoisomerase II, can cause DSBs. DNA DSBs are considered to be the most dangerous DNA lesions, because they are the most likely to result in mutation or caused unresolvable lesions that may lead to cell death [115]. Repair of DSBs can be through two mechanisms: HR, in which newly replicated DNA serves as the template of repair, and NHEJ, in which broken DNA ends are directly annealed to each other (Fig. 28.3). It is generally felt that NHEJ is more error prone as it does not follow a template, and in the process of DSB formation and processing, some sequence between broken strands may be lost. Of these two repair pathways, HR is only selected between S phase and mitosis of the cell cycle, when there is a sister chromatid available to repair from. Because NHEJ does not use a template, it occurs during all cell-cycle phases though its function is most critical for cell survival in G1 and in mammalian cells, and G0 as well. As with base lesions, repair of DSBs involves several chromatin-modifying complexes, some of which are outlined in Fig. 28.4.

4.2.2.1 Nonhomologous End Joining

One of the most important factors for the initiation of NHEJ is the stable association of DNA ends with the damage-recognition heterodimer yKu70/yKu80 [119]. These proteins serve two important functions: inhibiting 3′ to 5′ resection of DNA ends, and serving as the recruiting platform for downstream factors that will stabilize and form a protein bridge between broken DNA ends. In yeast cells, the MRX complex serves the cross-bridging function between ends, while the phosphatidyl inositol kinase-like kinase (PIKK) family member DNA-dependent protein kinase (DNA-PK) [120], is instrumental for this in mammalian cells. MRX is a highly conserved heterotrimeric complex comprised of the Mre11 exonuclease, the SMC family member Rad50 and scaffold protein Xrs2. Although it is not an essential complex in yeast, DNA-damage signaling and repair are severely compromised in MRX mutants [121]. In addition to cross-bridging, the ends need to be processed to become completely blunt ended, either through removal of overhangs or utilizing a polymerase, Pol4, to fill in the space. Ligation of the DNA ends in yeast is performed by Dnl4 ligase with the help of Nej1 and Lif1 (reviewed in Daley [119] and Mathiasen & Lisby [122]). In addition to the canonical NHEJ pathway, alternative NHEJ pathways, such as microhomology-mediated end joining (MMEJ), can anneal regions with short sections of conserved sequence; however, MMEJ is highly error prone [123–125].

4.2.2.2 Checkpoint Activation and DNA Resection at a DSB

There are extensive chromatin modifications in the region of a DSB and the phosphorylation of H2A at S129 is central for many downstream events. The cell-cycle checkpoint is a protein kinase-mediated signal transduction cascade, which is initiated in response to DNA damage, preventing transition to the next stage of the cell cycle until repair is completed. The damage checkpoint and the DSB-repair pathways are activated primarily through signal amplification from the binding of the MRX complex to DNA ends (Fig. 28.4). MRX binds at the DSB and acts as a recruiting platform and activator of the damage signal transduction PIKK kinase Tel1. PIKKs (reviewed in Jackson [126]) are central to the DNA-damage response in all eukaryotic organisms. Two proteins of this family that are critical to damage signal transduction in *S. cerevisiae* are Tel1 and Mec1, the orthologs of ATM and ATR, respectively. They are considered the main transducers of a damage signal because their activation is between proteins that directly sense damage and the amplification of kinase cascades to activate cell cycle–checkpoint proteins Chk1 and Rad53 and phosphorylate other targets proximal to the break.

During S and G2 of the cell cycle, when HR can occur, MRX recruits the Sae2 exonuclease to initiate resection of DNA from the break, creating a 3′ ssDNA overhang [127,128]. Resection is initiated by Sae2 (mammalian CtIP), and continued by the unwinding activities of the RecQ helicase, Sgs1, and the exonucleases, Exo1 and Dna2, which catalyze

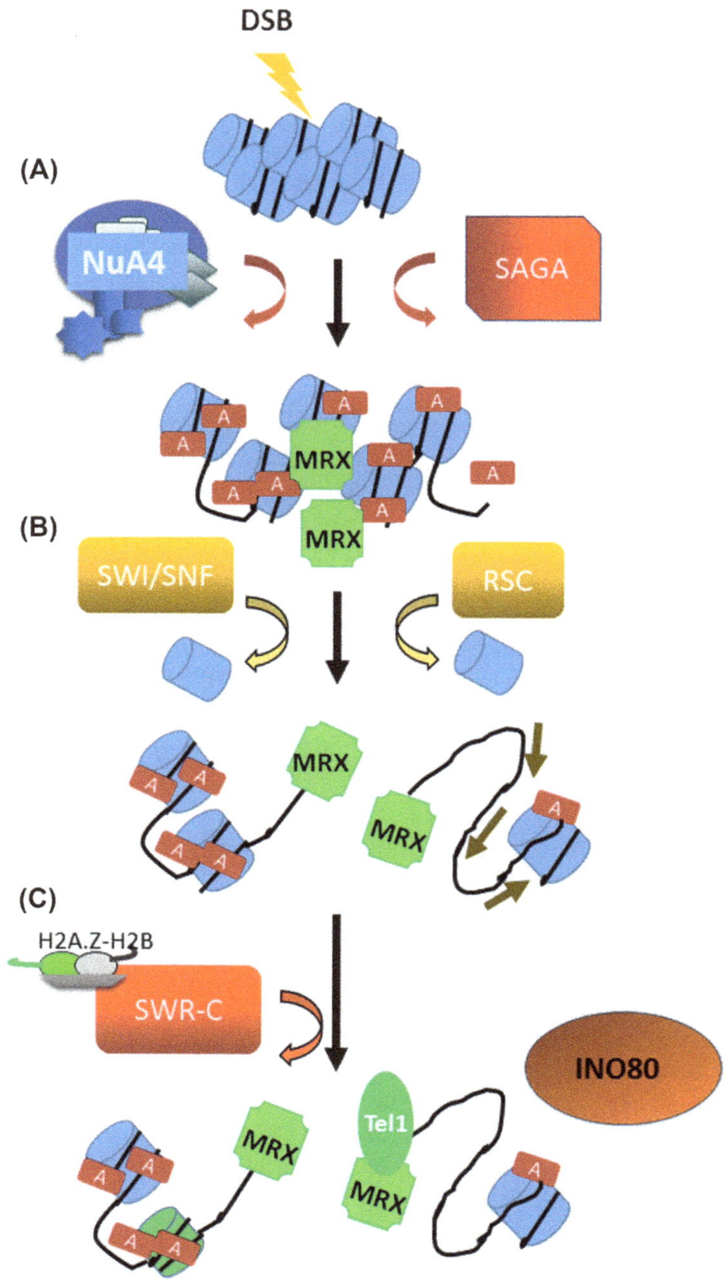

FIGURE 28.4 **Activity of chromatin modifiers and remodelers at a DSB.** An early event after damage induction is acetylation of histones H4 and H3 by NuA4 and SAGA, respectively [116,117]. Chromatin remodelers SWI/SNF and RSC recruit to the site of DSB, to slide nucleosomes and evict them from the region of the DSB [63]. SWR-C localizes to a DSB and incorporates H2A.Z into nucleosomes proximal to the break [86]. INO80 also recruits to a DSB, and is important in mediating downstream functions of break movement within the nucleus [118].

further resection. As the exonucleases resect DNA, the ssDNA-binding protein RPA accumulates and creates a platform for Mec1–Ddc2(ATRIP) binding. Mec1 kinase activity can be stimulated by interactions with Ddc1 (9-1-1 complex) and RPA interactions with both Ddc2 and Ddc1 function to tether Mec1–Ddc2 and 9-1-1 to the RPA-ssDNA [129], coupling Mec1 to its activator, the 9-1-1 complex.

Maintenance and amplification of checkpoint signaling is driven by a chromatin-based signaling platform that is initiated by Mec1- and Tel1-dependent γH2A formation. γH2A recruits Rad9 (homolog of 53BP1), a BRCT-containing scaffold protein that also binds the effector kinases Rad53 and Chk1 [122]. Once proximal to the break, Mec1/Tel1-dependent phosphorylation of Rad9 promotes its oligomerization via its BRCT repeats [130], and the further binding of Mec1 kinase-activating proteins, Ddc1 and Dpb11 [131]. Binding all of these proteins close to breaks serves to colocalize Mec1 with its activators, amplifying the signal by recruiting more Mec1–Ddc2, and activating its effector kinases, Rad53 and Chk1. Once activated, effector kinases are then able to perform their functions, which include preventing cell-cycle progression and inducing transcription of repair factors.

4.2.2.3 Homology Search and Repair

Following resection, the ssDNA-binding protein Rad51 displaces RPA on the 3′ overhang, to align with regions of the sister-chromatid template in the search for a homologous region to copy for repair (Fig. 28.3). Sequence search and verification is mediated partially by the activities of Rad52, and other members of the HR complementation group (Rad54, Rad55, and the Shu complex) [132]. When a complementary region is found, the 3′ resected strand invades the undamaged duplex, displacing the other strand of the donor sequence. The donor sequence strand that is displaced is then used as a template for filling in the gap from the other strand of the break to form the Holliday junction. After the DNA polymerase has passed the region of the break, the junction is resolved by nucleases that separate the connections between the donor and invading strands [133].

4.2.2.4 Histone Modifications and Remodelers Associated With DSBs

Along with phosphorylation of histones [21], histone acetylation at H4 by NuA4, and H3 by SAGA occur as an early event in the DNA damage response [32,33,41] (Fig. 28.4). This initiating event is important to opening chromatin to facilitate remodeling [2,134]. Additionally, the acetylation marks are recognized by the bromodomain-containing chromatin remodeler SWI/SNF [116], which enhances γH2A signaling at the break [86]. The remodelers SWR-C, INO80, and RSC also recruit to the DSB [117], and H2A.Z has been reported to be incorporated in nucleosomes for several kilobases flanking the break. The localization of H2A.Z and remodelers at a DSB is associated with extensive resection at the break [86]. However, the trend of extensive processing of a DSB does not agree with events that would promote NHEJ. Therefore, it is unlikely that the very different repair processes of HR and NHEJ would utilize all the same mechanisms in nucleosome restructuring. HR would require extensive reorganization to facilitate DNA resection, while NHEJ would require inhibition of those modifications to prevent it.

A caveat to many of the studies that have characterized histone modifications and remodelers at a DSB is that they did not restrict their observations to discrete cell-cycle phases [135]. A 2013 study sought to observe if there were differences in levels of chromatin modifications and remodelers between G1 and G2/M of the cell cycle, which they were able to do by using mating pheromone to synchronize cells in G1. Their findings revealed that within a chromatin context, there are mechanisms in place to influence repair choice between NHEJ and HR [117].

There are discrete differences among cell-cycle phases in recruitment of NuA4, INO80, SWR-C, SWI/SNF, and RSC. Generally, all of these complexes abundantly localized to a DSB in G2/M, in support of previous publications (reviewed in Papamichos-Chronakis & Peterson [2]; Price & D'Andrea [134]), but were not present at a DSB in G1, in which γH2A and yKu70/yKu80 were strongly enriched [117]. This result strongly differed from previous publications where γH2A was previously identified as being a recruiting mark for complexes containing Arp4, which includes NuA4, SWR-C, and INO80 [32]. Furthermore, mutation of H2A, so that it cannot be phosphorylated at S129, had no negative influence on recruitment of NuA4, INO80, SWR-C, SWI/SNF, and RSC to a DSB.

Subsequent exploration of the inverse relationship between γH2A and chromatin-modifier recruitment revealed that inhibiting DNA resection and HR strongly reduced recruitment of NuA4, INO80, SWR-C, SWI/SNF, RSC, and RPD3L to DSBs, while it increased γH2A levels [117]. Therefore, end processing and resection of DNA at a break is required to initiate remodeler activity and inhibit γH2A. At the same time, they observed that yKu70/80 strongly influenced remodeler and γH2A enrichment. NuA4, INO80, SWR-C, SWI/SNF, and RSC localized to a DSB in an *yku70Δ* in G1, and additionally, γH2A signaling was lost [117]. This supports that extensive modifications and remodeling are required to create access to the damage to facilitate early HR steps of resection.

This study raises many questions about the regulation of factors that promotes or inhibits resection at the break. As previously mentioned, Rad9 interacts with H3–K79me3 and γH2A to provide a binding site for this scaffold at sites of damage in G1 and G2 [136,137]. Rad9 forms oligomers via its BRCT repeats interactions with phosphorylated residues from Mec1/Tel1, and that this ability is important for damage-signal propagation and checkpoint signaling [130]. However, there is a model that suggests that Rad9 and associated proteins form an obstacle to nucleases, helicases, and chromatin modifiers that is proposed to inhibit resection [138,139]. To circumvent this obstacle, the SMARCAL1-like family chromatin remodeler, Fun30, is necessary to promote resection in the presence of Rad9 [138]. Fun30 localization at a DSB depends on proteins that facilitate resection (Sgs1, Exo1, and MRX). Additionally, Fun30 coimmunoprecipitates with RPA, Exo1, and Dna2 in damaged cells.

The Rad9 obstacle is likely to vary with the cell cycle, as it forms a complex at damage with Dpb11, 9-1-1, and Mec1 that is CDK dependent and uniquely occurring in G2 of the cell cycle [131]. CDK activity also has an impact on the capacity of Rad9 to regulate resection in G2 [140]. Therefore, the dynamic assembly of Rad9 at damage likely creates a different kind of relationship to downstream factors that promote resection and modify chromatin such as Fun30, causing Rad9 to be a barrier to resection that is dependent on the cell-cycle phase.

Another set of papers that supports different chromatin events happening in HR vs. NHEJ are from studies in *Schizosaccaromyes pombe* (fission yeast) and human cells [141,142]. In these studies, an intriguing relationship was identified where acetylation and methylation at H3–K36 compete to either enable or block resection of a DSB. Methylation of H3–K36 by Set2 occurs primarily in G1, in association with transcription, preventing resection and thus HR. However, acetylation of H3–K36 by Gcn5 in S and G2 allows HR to occur.

The observation that different chromatin modifiers and remodelers are present for G1 and S/G2 phases are intuitive for what would be expected at a DSB, based on the extreme difference in HR- and NHEJ-pathway mechanisms that are required for repair.

5. REPLICATION STRESS, ACTIVATION OF THE S-PHASE CHECKPOINT AND DNA-DAMAGE TOLERANCE

The S-phase checkpoint integrates the DNA-damage response with the replication process, enabling the cell to proceed through replication in the presence of damage and replication fork obstruction. Obstruction to replisome progression causes RPA-ssDNA accumulation, which results in a stressful configuration that can cause the replisome to disassemble, the replication fork to collapse, and the formation of a DSB (reviewed in Friedel et al. [143]; Yekezare et al. [144]; Yoshida et al. [145]). As with activation of checkpoint due to a DSB, RPA-ssDNA is again the recruiting platform for proteins that sense replication stress, most critical of which are Ddc2 (Mec1–Ddc2) and Ddc1 (9-1-1 complex). Binding of Ddc1 and Dpb11 to Mec1 activates Mec1 kinase activity, leading to phosphorylation of many components of the replisome to slow fork progression in a controlled manner and directly recruit and activate the checkpoint-kinase Rad53 [143].

The main repair pathways active during replication are MMR, when a base is misincorporated in the daughter strand, DDT pathway, when replication past a damaged lesion must occur, and HR when there is a DNA-DSB [4]. MMR is similar to BER and NER in that one half of the helix is the repair template for the mismatched base. MMR occurs only after DNA replication, proofreading the newly synthesized DNA for mismatches, insertions, and deletions using the MutS homologs (MSHs). When MSH proteins identify mismatches, the MutL homologs (MLH) proteins interact and begin to mediate the excision of the misincorporated daughter nucleotide and replacement with a correct one [4]. Much of what is known about MMR was derived from in vitro studies with "naked" DNA. Purified MMR proteins, however, are unable to perform the repair in nucleosome-DNA arrays, implying that like BER and NER, access to the lesion is important.

A current model is that MMR complex components are recruited to nascent DNA by association to modified histones, either by localizing to marks that are placed prior to replication and nucleosome disassembly, or to newly incorporated nucleosomes following the replication fork [146,147]. There is evidence in *S. cerevisiae* that H3–K56Ac plays a role in MMR [148]. As previously mentioned, H3–K56Ac is a mark of newly synthesized H3 and replication associated incorporation of new nucleosomes [93]. Association with H3–K56Ac would also localize MMR proteins to nascent DNA and the replication fork. Additionally, H3–K56Ac creates more flexibility in nucleosome/DNA interactions because the modification is at the entry/exit point of where DNA interacts with the nucleosome [104]. Association with modifications that are indicative of immature chromatin may indicate that MMR proteins interact with DNA while nucleosomal positioning is flexible, creating the ability to scan newly synthesized DNA for mismatches.

5.1 DNA-Damage Tolerance

The base lesion repair pathways of BER and NER are not usually sufficient to repair all damage while DNA is being replicated in S phase [4]. A replication fork will stall at a base lesion because it is unable to incorporate a base opposite the one that is damaged [149]. In this instance, the replication bypass pathway known as DDT or post-replication repair (PRR) is utilized (Fig. 28.5). This allows DNA replication to occur through the damage and permit NER or BER to do the actual repair after replication has occurred. DDT occurs by two mechanisms considered error free (EF) and translesion synthesis (TLS). When EF is selected, the replication template is switched to the undamaged strand in a process that utilizes many of the proteins and mechanisms of HR [150,151]. The template is switched back once replication has bypassed the region that was damaged. In TLS, the high fidelity–replication polymerase is switched to a lower-fidelity

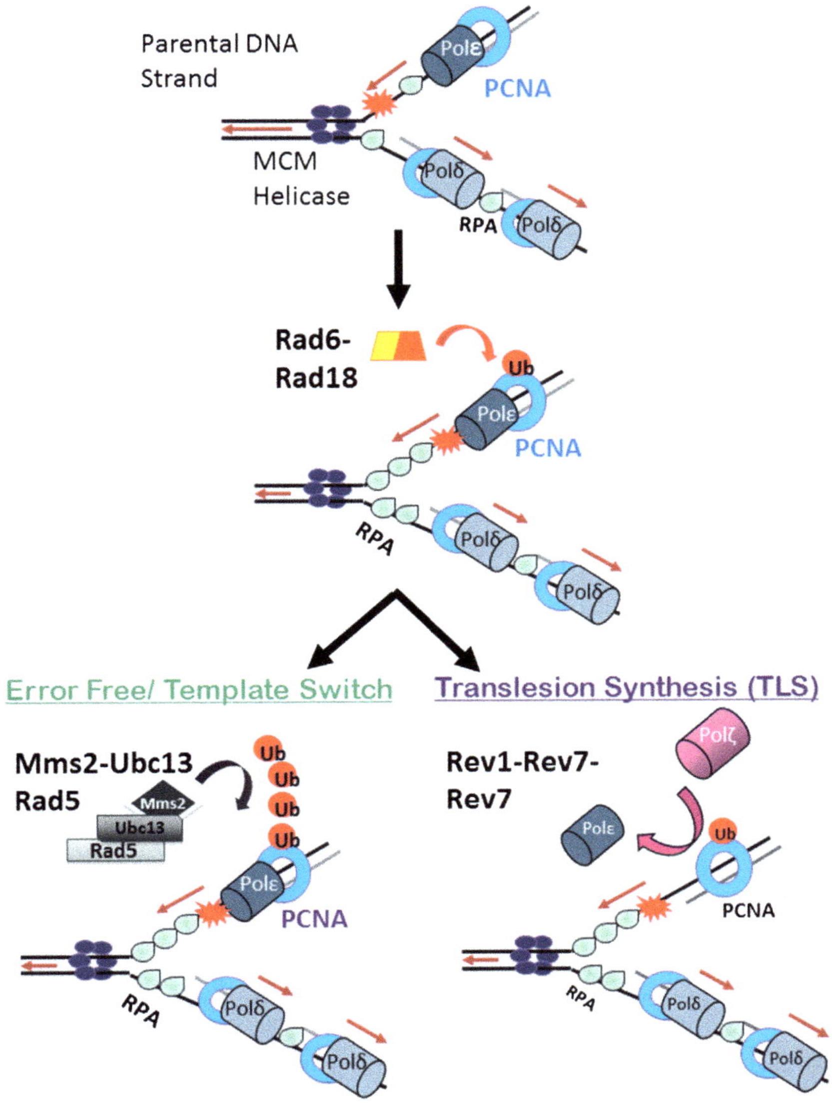

FIGURE 28.5 **Activation of the DNA-damage tolerance (DDT) pathway.** DDT is a means of bypassing base damage that is encountered by a replisome [4]. When a progressing replisome encounters base damage, the complex stalls because the high-fidelity polymerase cannot incorporate a nucleotide across from a damaged base. PCNA becomes ubiquitinated on K164 by Rad6–Rad18. If PCNA remains monoubiquitinated, bypass will proceed by TLS, where a low-fidelity polymerase (Polζ: Rev1, Rev3, and Rev7) replicates past the damaged base. If PCNA is polyubiquitinated by Mms2–Ubc13 and Rad5, bypass will be mediated by EF-DDT.

polymerase that is able to incorporate a base across from the damage, though it may be mutagenic. Once damage is bypassed, the high fidelity–replication polymerase continues DNA replication (reviewed in Sale et al. [149]; Moldovan, et al. [152]; Fu et al. [153]).

5.2 PCNA Modification and DNA-Damage Tolerance

The upstream mechanism that initiates utilization of EF vs. TLS bypass is through the posttranslational modification of the replication sliding–clamp PCNA [154,155] (Fig. 28.5). PCNA is a homotrimeric complex that encircles the DNA duplex and is essential to the function of DNA polymerases. Replication stalling occurs when the polymerase encounters a base lesion, resulting in an accumulation of ssDNA. Rad18, an E3 ubiquitin ligase, is recruited to ssDNA and sumoylated PCNA [156,157], and together with the E2 ubiquitin ligase Rad6, can ubiquitinate PCNA at K164 [154,155,158,159]. If PCNA remains monoubiquitinated, then bypass will proceed by the TLS pathway. However, if PCNA is polyubiquitinated on K164, then the EF pathway will be selected.

5.3 EF-Damage Tolerance

EF-DDT is selected by polyubiquitination of PCNA on K164 [160]. A K63 polyubiquitination linkage is added to PCNA by the E2 ligases Mms2–Ubc13, and E3 ligase Rad5 [158,161–164] (Fig. 28.5). Because of the similarity between EF-DDT and HR, many of the proteins utilized in homology search and replication using the sister template are also critical for template switch. Among the proteins needed for both pathways are Rad51, Rad52, Exo1, the Shu complex, Sgs1, and the 9-1-1 complex [151,165–167].

5.4 Translesion Synthesis–Damage Tolerance

TLS is selected when PCNA remains monoubiquitinated on K164 (Fig. 28.5) [155]. An error-prone polymerase assumes replication for sufficient number of bases to permit the damage to be bypassed and is then displaced by the higher fidelity–replication polymerase (reviewed in Makarova & Burgers [168]). Rev3 is from the same protein family as Pol3 of the Polδ lagging–strand polymerase, and like Pol3, Rev3–Rev7 interacts with the accessory subunits Pol31 and Pol32 [169,170]. The in vivo complex of Polζ also includes Rev1 [171]. Rev1's function when associated with Polζ is independent of its catalytic function, and it appears that acting as a scaffold via association with PCNA and DNA is the critical function of the protein for TLS [172–174]. Because it operates as a trimer with Rev1, Polζ is sometimes dubbed "the mutasome" [168]. Mutasome activity is generally restricted to G2 of the cell cycle, with Rev1 levels being regulated by proteolytic targeting [173,175]. Additionally, the activity of Rev1 has been shown to be regulated by Mec1 phosphorylation [176].

5.5 Checkpoint Signaling, Chromatin, and DDT

Mec1 and Rad53 function in the DDT pathway by different mechanisms, in that Mec1 promotes the TLS pathway, while Rad53 promotes template switch [177,178]. A potential mechanism for this is that Mec1 modifies targets that are distinct from the Rad53 checkpoint kinase that is downstream of it. To differentiate the function of Mec1 and one of its most important kinase targets, Rad53, it is intuitive that Mec1 is affecting processes that are independent of checkpoint activation. This has been demonstrated in its role in the localization and activity of Rev1 in DDT, at a DSB and potentially also with NER [176,179–181]. Genetic analysis of the interaction of the Mec1 target γH2A with the DDT pathway suggests that γH2A also supports TLS [182]. Also, given that Rad9 association with chromatin functions to activate the checkpoint and influence resection, it is possible that a role for Mec1 and Rad9, independent of Rad53 activity, is to restrict resection and favor TLS over EF [136,140]. In budding yeast, EF is preferentially used during S phase of the cell cycle [183]. The mechanisms mediating this bias appear to be chromatin based and could be dependent on the DNA damage–associated scaffold protein Rad9.

The signaling cascade involving ubiquitination of H2B–K123 and methylation of H3–K79 biases DDT away from TLS [184,185]. Histone H2B is ubiquitinated by the E2 and E3 ligases Rad6 and Bre1, respectively. If H2B–K123ub is present on nucleosomes bound by Dot1, its methyltransferase activity is directed toward H3–K79 [186,187]. Genetic interactions of *DOT1* support a role for it as promoting template switch DDT [184,185]. Consistent with a function in EF-DDT, loss of *DOT1* produces an increase in the rate of spontaneous mutagenesis, which is used as an indirect measurement of TLS polymerase activity. H3–K79me3 may be bound by Rad9 in the event of DNA damage via its Tudor domains [188] (Fig. 28.6). The interaction of *DOT1* with *RAD9* is dependent on the type of DNA damage, in that they are additively sensitive when treated with IR, but epistatic when treated with the base lesion causing agent methyl methanesulfonate (MMS). This indicates that Rad9 is associated with H3–K79me3 to mediate the survival promoting response to base lesion–induced damage. Support for Rad9 cooperating with EF is that a deletion mutant has a high rate of spontaneous mutagenesis which is dependent on functional TLS polymerases [189]. Therefore, a model for H3–K79me3 promoting EF could be that *dot1Δ* mutants combined with repair defects do not accumulate damage intermediates due to their mutated repair pathways because they undergo unregulated TLS bypass of the damage due to the inability of Rad9 to block TLS (Fig. 28.6). Dot1/H3–K79me3-mediated recruitment of Rad9 is an important activator of Rad53, and initiation of resection, which, by genetic interactions, is shown to support EF-DDT. Mec1 and γH2A promote TLS. It may be that Mec1 and the γH2A alter the chromatin platform of DDR to create an environment more conducive to TLS by inhibition of Rad9.

Other chromatin components have been found to have a role in DDT using the trinucleotide repeat screen. The DDT pathway is important for accurate replication of genomic regions that are difficult to replicate due to repeats of CAG/CTG sequences. Trinucleotide repeats are fragile sites within the genome that are vulnerable to expansion and contraction due to their tendency to spontaneously form DNA hairpins when unwound [190]. Errors in DNA repair, namely BER [191–194], lead to replication fork stalling at repeat expansions [195]. Defects in genes associated with HR and EF have a higher level

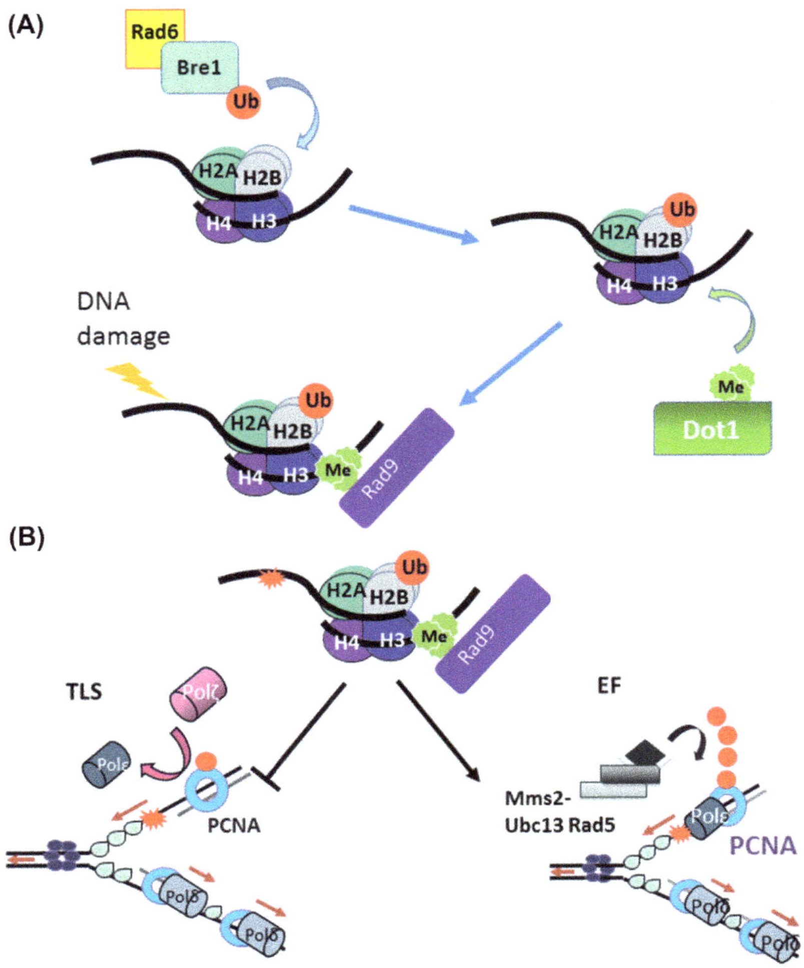

FIGURE 28.6 **The histone H3–K79me3 modification biases DDT toward EF-mediated bypass.** Rad9 is able to interact with the H3–K79me3 chromatin modification via its Tudor domains [186,187] (A). This signal cascade biases DDT toward the EF pathway by unknown mechanisms (as of 2016). Loss of DOT1 rescues mutants of the EF pathway and other damage-repair pathways by permitting unregulated TLS to occur [184,185] (B).

of repeat instability [195]. Contrary to EF-associated genes, TLS genes can lead to a greater rate of errors when other genes important for repeat stability are mutated, such as the *MPH1* helicase [196] or the replicative polymerases, Pol2 (ε) and Pol3 (δ) [197]. Additionally, TLS polymerases and template switch genes may cooperatively introduce repeat expansions at stalled replication structures [198].

Hmo1 (high mobility group protein 1) localizes to repeat tracts and reduces instability [199], a function consistent with its endogenous role of binding to the gene bodies of rDNA to prevent DNA-hairpin formation during transcription [200]. It was later shown that Hmo1 binding with DNA predisposes the pathway choice toward the EF pathway with a predicted model in which Hmo1 bends ssDNA to favor the initiation of strand invasion into the duplicated sister strand for template switch [201].

The acetyltransferase Rtt109, which acetylates H3K56 and strongly contributes to nucleosome–DNA interaction flexibility has been shown to have a role in suppressing repeat contractions [202]. In a synthetic genetic analysis screen, Rtt109 has been shown to favor TLS DDT [203]. A 2014 publication demonstrates that the acetyltransferase complex NuA4 and chromatin remodeler RSC suppress Rad5-dependent trinucleotide repeat expansions [204]. Without NuA4-acetylating histones which then recruit RSC to the region, EF bypass is highly error prone. Together, these reports indicate the need for manipulation of nucleosome–DNA interactions to either prevent hairpins from forming or cooperatively resolve them when they do occur.

A study that more comprehensively explored the role of NuA4 in DDT revealed that the primary role of the complex was toward the process of TLS [205]. Mutants of NuA4 reduced the rate of spontaneous mutagenesis in addition to increasing the sensitivity of mutants of EF to MMS. In addition to NuA4, its downstream effectors H4 and H2A.Z demonstrated

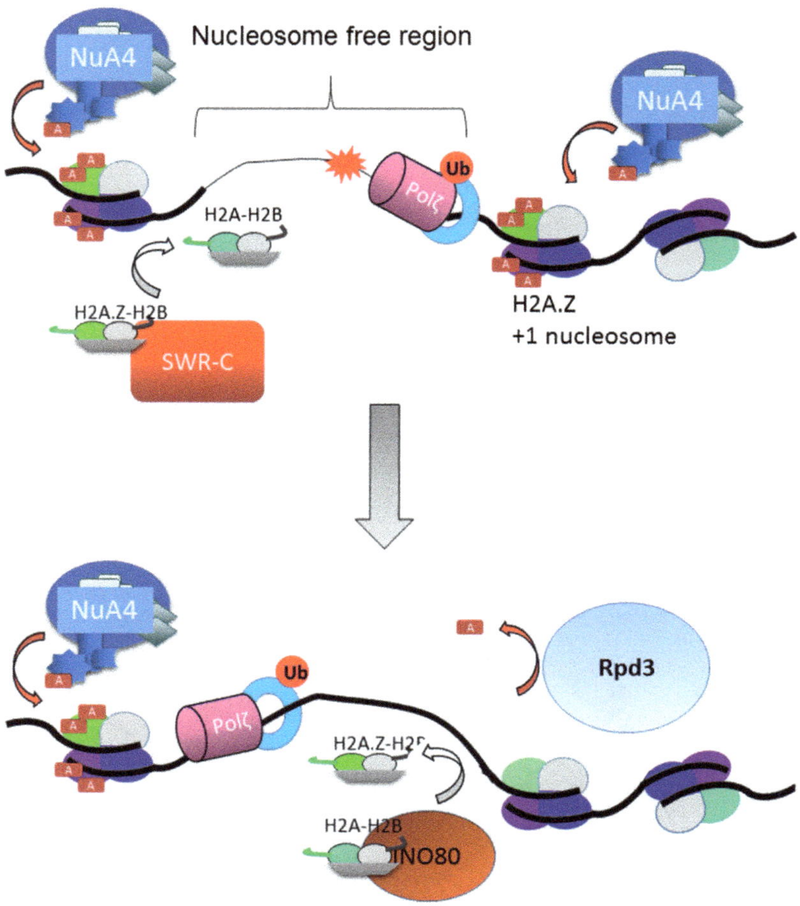

FIGURE 28.7 **Model of NuA4 and other regulators of H2A.Z in promoting function of TLS polymerases.** Regulators of H2A.Z function to promote TLS. TLS preferentially occurs in G2 after the bulk of replication completion, where remaining lesions in the genome are gaps of ssDNA [183]. NuA4 acetylates H4 proximal to damaged DNA, stabilizing SWR-C at the region, and leading to H2A.Z incorporation at regions proximal to damage. Acetylation of H2A.Z may stabilize its retention further by protecting it from INO80-mediated removal [83]. The presence of H2A.Z creates a boundary to an NFR surrounding the lesion and ssDNA gap [79,206]. Following TLS of the lesion, INO80 as well as SWR-C and Rtt109 facilitate removal of H2A.Z from the damage site to restore the chromatin environment after damage [84].

genetic interactions with the DDT pathway. An interpretation of this is that complexes that control H2A.Z incorporation and removal perform a function that is important to allow TLS (Fig. 28.7). NuA4 acetylates H4 to create binding sites for the bromodomains of SWR-C. NuA4 further stabilizes H2A.Z at specific loci by acetylation. Rtt109 facilitates removal of H2A.Z by acetylation of H3–K56, permitting SWR-C to remove this variant.

A function for H2A.Z may be to keep the chromatin environment accessible until replication over the damaged region has occurred. H2A.Z nucleosomes are considered to be less stable than canonical nucleosomes where a dynamic instability exists at transcription start sites, yet they can form stable barriers to the spread of heterochromatin. Perhaps the most critical function of H2A.Z behind the fork is to maintain a flexible chromatin environment that is accessible to TLS polymerases. When damage is present, NuA4 mediated acetylation of the N-terminus of H2A.Z may be needed to protect it from removal by INO80 and following TLS, H2A.Z would be deacetylated, allowing its removal by INO80 or SWR-C.

6. OVERVIEW: THE RELATIONSHIP BETWEEN CHROMATIN AND REPAIR CHOICE

The modification of chromatin is essential for optimal transcription, replication, and DNA repair. In spite of the inherent differences in these DNA processes, there are consistent steps in histone-modification pathways to enable these functions.

Chromatin opening is facilitated by posttranslational modifications of nucleosomes. Acetylation is often the first modification step to create open chromatin at transcription start sites, newly translated nucleosomes, and regions of damage. The hierarchical modifications of acetylation are often followed by the activity of remodelers such as SWR-C, RSC, SWI/SNF, and INO80, many of which have bromodomains that allow interactions with acetylated nucleosomes. NuA4- and

SAGA-mediated nucleosome acetylation followed by SWR-C-mediated histone exchange of H2A for H2A.Z, is associated with transcriptional start sites, heterochromatin boundaries, and TLS-DDT. Common pathways are also involved in chromatin condensation following replication, transcription, and repair. HDAC complexes such as RPD3 deacetylate histones and ISWI alters spacing between nucleosomes.

The hierarchy of interactions is also dependent on what the process requires and the phase of the cell cycle. Throughout the cell cycle, histone modifiers and remodelers participate in repair processes of BER and NER to shift nucleosomes and allow repair machinery to access DNA. In the case of the DSB-repair processes of NHEJ and HR, very different modifications around the break occur, in a cell cycle–dependent manner. In the case of HR, which occurs in S and G2, and requires resection to create access to large tracts of DNA, the acetyltransferases SAGA and NuA4, as well as the chromatin remodelers RSC, SWI/SNF, SWR-C, and INO80 localize to DSBs. However, for NHEJ which is restricted mostly to G1, resection is inhibited, and therefore there is a low level of recruitment of HATs and remodelers, but an abundance of γH2A and yKu70/yK80 [117]. The differences in chromatin modifications required for these two processes strongly illustrates the interplay between chromatin-modification complexes and repair pathways.

The use of the budding yeast model has been critical in gaining understanding of the involvement of chromatin modifications in processes associated with DNA transcription, replication, and repair. The goal of this chapter was to highlight the consistencies of chromatin-modifying activities in these processes, so as to illustrate cooperative behaviors among complexes and enable a better understanding of how they function in repair and genome stability. Indeed, some chromatin modifications play a role in repair that is so extensive as to determine repair-pathway outcome, such as with the replication-damage pathway of DDT. A full understanding of the many chromatin modifications needed for efficient DNA repair are clearly relevant to understanding human disease. This is particularly evident given that many human cancers likely originate from errors in DNA replication accumulated during the many cycles that stem cells need to undergo throughout the lifetime, explaining, in part, the strong linkage between age and the incidence of all forms of cancer [207].

GLOSSARY

Mutasome Protein complex allowing error-prone translesion synthesis.

LIST OF ABBREVIATIONS

BER Base excision repair
DDT DNA-damage tolerance
DSB Double-strand break
EF Error free
FHA Forkhead-associated domain
GG Global genome
HDAC Histone deacetylase
HKDM Histone lysine demethylases
HKMT Histone lysine methyltransferases
HR Homologous recombination
MMEJ Microhomology-mediated end joining
MMR Mismatch repair
NER Nucleotide excision repair
NHEJ nonhomologous end joining
NRFs Nucleosome-free regions
PHD Plant homeodomains
PRR Postreplication repair
TC Transcription coupled
TLS Translesion synthesis

REFERENCES

[1] Ciccia A, Elledge SJ. The DNA damage response: making it safe to play with knives. Mol Cell 2010;40:179–204.
[2] Papamichos-Chronakis M, Peterson CL. Chromatin and the genome integrity network. Nat Rev Genet 2013;14:62–75.
[3] Soria G, Polo SE, Almouzni G. Prime, repair, restore: the active role of chromatin in the DNA damage response. Mol Cell 2012;46:722–34.
[4] Boiteux S, Jinks-Robertson S. DNA repair mechanisms and the bypass of DNA damage in *Saccharomyces cerevisiae*. Genetics 2013;193:1025–64.
[5] Richmond TJ, Davey CA. The structure of DNA in the nucleosome core. Nature 2003;423:145–50.

[6] Elsasser SJ, D'Arcy S. Towards a mechanism for histone chaperones. Biochim Biophys Acta 2013;1819:211–21.

[7] Luger K, Mader AW, Richmond RK, Sargent DF, Richmond TJ. Crystal structure of the nucleosome core particle at 2.8 A resolution. Nature 1997;389:251–60.

[8] Meeks-Wagner D, Hartwell LH. Normal stoichiometry of histone dimer sets is necessary for high fidelity of mitotic chromosome transmission. Cell 1986;44:43–52.

[9] Kim UJ, Han M, Kayne P, Grunstein M. Effects of histone H4 depletion on the cell cycle and transcription of *Saccharomyces cerevisiae*. EMBO J 1988;7:2211–9.

[10] Eriksson PR, Ganguli D, Nagarajavel V, Clark DJ. Regulation of histone gene expression in budding yeast. Genetics 2012;191:7–20.

[11] Laskey RA, Honda BM, Mills AD, Finch JT. Nucleosomes are assembled by an acidic protein which binds histones and transfers them to DNA. Nature 1978;275:416–20.

[12] Winkler DD, Zhou H, Dar MA, Zhang Z, Luger K. Yeast CAF-1 assembles histone (H3-H4)2 tetramers prior to DNA deposition. Nucleic Acids Res 2012;40:10139–49.

[13] Jamai A, Puglisi A, Strubin M. Histone chaperone spt16 promotes redeposition of the original h3-h4 histones evicted by elongating RNA polymerase. Mol Cell 2009;35:377–83.

[14] Riggs MG, Whittaker RG, Neumann JR, Ingram VM. n-Butyrate causes histone modification in HeLa and Friend erythroleukaemia cells. Nature 1977;268:462–4.

[15] Boffa LC, Vidali G, Mann RS, Allfrey VG. Suppression of histone deacetylation in vivo and in vitro by sodium butyrate. J Biol Chem 1978;253:3364–6.

[16] Shahbazian MD, Grunstein M. Functions of site-specific histone acetylation and deacetylation. Annu Rev Biochem 2007;76:75–100.

[17] Gurard-Levin ZA, Almouzni G. Histone modifications and a choice of variant: a language that helps the genome express itself. F1000Prime Rep 2014;6:76. eCollection 2014.

[18] Clarke AS, Lowell JE, Jacobson SJ, Pillus L. Esa1p is an essential histone acetyltransferase required for cell cycle progression. Mol Cell Biol 1999;19:2515–26.

[19] Li B, Carey M, Workman JL. The role of chromatin during transcription. Cell 2007;128:707–19.

[20] Rossetto D, Avvakumov N, Cote J. Histone phosphorylation: a chromatin modification involved in diverse nuclear events. Epigenetics 2012;7:1098–108.

[21] Cheung WL, et al. Phosphorylation of histone H4 serine 1 during DNA damage requires casein kinase II in *S. cerevisiae*. Curr Biol 2005;15:656–60.

[22] Ahn SH, et al. Sterile 20 kinase phosphorylates histone H2B at serine 10 during hydrogen peroxide-induced apoptosis in *S. cerevisiae*. Cell 2005;120:25–36.

[23] Kawashima S, et al. Global analysis of core histones reveals nucleosomal surfaces required for chromosome bi-orientation. EMBO J 2011;30:3353–67.

[24] Nathan D, et al. Histone sumoylation is a negative regulator in *Saccharomyces cerevisiae* and shows dynamic interplay with positive-acting histone modifications. Genes Dev 2006;20:966–76.

[25] Bannister AJ, Kouzarides T. Regulation of chromatin by histone modifications. Cell Res 2011;21:381–95.

[26] Pena PV, et al. Molecular mechanism of histone H3K4me3 recognition by plant homeodomain of ING2. Nature 2006;442:100–3.

[27] Gilbert TM, et al. A PWWP domain-containing protein targets the NuA3 acetyltransferase complex via histone H3 lysine 36 trimethylation to coordinate transcriptional elongation at coding regions. Mol Cell Proteomics 2014;13:2883–95.

[28] Strahl BD, Allis CD. The language of covalent histone modifications. Nature 2000;403:41–5.

[29] Lu PY, Levesque N, Kobor MS. NuA4 and SWR1-C: two chromatin-modifying complexes with overlapping functions and components. Biochem Cell Biol 2009;87:799–815.

[30] Chahal SS, Matthews HR, Bradbury EM. Acetylation of histone H4 and its role in chromatin structure and function. Nature 1980;287:76–9.

[31] Korolev N, Yu H, Lyubartsev AP, Nordenskiold L. Molecular dynamics simulations demonstrate the regulation of DNA-DNA attraction by H4 histone tail acetylations and mutations. Biopolymers 2014;101:1051–64.

[32] Downs JA, et al. Binding of chromatin-modifying activities to phosphorylated histone H2A at DNA damage sites. Mol Cell 2004;16:979–90.

[33] Tamburini BA, Tyler JK. Localized histone acetylation and deacetylation triggered by the homologous recombination pathway of double-strand DNA repair. Mol Cell Biol 2005;25:4903–13.

[34] Hassan AH, Neely KE, Workman JL. Histone acetyltransferase complexes stabilize swi/snf binding to promoter nucleosomes. Cell 2001;104:817–27.

[35] Kasten M, et al. Tandem bromodomains in the chromatin remodeler RSC recognize acetylated histone H3 Lys14. EMBO J 2004;23:1348–59.

[36] Altaf M, et al. NuA4-dependent acetylation of nucleosomal histones H4 and H2A directly stimulates incorporation of H2A.Z by the SWR1 complex. J Biol Chem 2010;285:15966–77.

[37] Sterner DE, et al. Functional organization of the yeast SAGA complex: distinct components involved in structural integrity, nucleosome acetylation, and TATA-binding protein interaction. Mol Cell Biol 1999;19:86–98.

[38] Grant PA, et al. Expanded lysine acetylation specificity of Gcn5 in native complexes. J Biol Chem 1999;274:5895–900.

[39] Chang CS, Clarke A, Pillus L. Suppression analysis of esa1 mutants in *Saccharomyces cerevisiae* links NAB3 to transcriptional silencing and nucleolar functions. G3 (Bethesda) 2012;2:1223–32.

[40] Torres-Machorro AL, Pillus L. Bypassing the requirement for an essential MYST acetyltransferase. Genetics 2014;197:851–63.

[41] Lin YY, et al. A comprehensive synthetic genetic interaction network governing yeast histone acetylation and deacetylation. Genes Dev 2008;22:2062–74.

[42] Kaluarachchi Duffy S, et al. Exploring the yeast acetylome using functional genomics. Cell 2012;149:936–48.

[43] Mitchell L, et al. mChIP-KAT-MS, a method to map protein interactions and acetylation sites for lysine acetyltransferases. Proc Natl Acad Sci USA 2013;110:E1641–50.

[44] Rundlett SE, et al. HDA1 and RPD3 are members of distinct yeast histone deacetylase complexes that regulate silencing and transcription. Proc Natl Acad Sci USA 1996;93:14503–8.

[45] Kurdistani SK, Robyr D, Tavazoie S, Grunstein M. Genome-wide binding map of the histone deacetylase Rpd3 in yeast. Nat Genet 2002;31:248–54.

[46] Robyr D, et al. Microarray deacetylation maps determine genome-wide functions for yeast histone deacetylases. Cell 2002;109:437–46.

[47] Kurdistani SK, Grunstein M. Histone acetylation and deacetylation in yeast. Nat Rev Mol Cell Biol 2003;4:276–84.

[48] Kouzarides T. Chromatin modifications and their function. Cell 2007;128:693–705.

[49] Greer EL, Shi Y. Histone methylation: a dynamic mark in health, disease and inheritance. Nat Rev Genet 2012;13:343–57.

[50] Cheng X, Collins RE, Zhang X. Structural and sequence motifs of protein (histone) methylation enzymes. Annu Rev Biophys Biomol Struct 2005;34:267–94.

[51] Nguyen AT, Zhang Y. The diverse functions of Dot1 and H3K79 methylation. Genes Dev 2011;25:1345–58.

[52] Kim W, Choi M, Kim JE. The histone methyltransferase Dot1/DOT1L as a critical regulator of the cell cycle. Cell Cycle 2014;13:726–38.

[53] Passmore LA, Barford D. Getting into position: the catalytic mechanisms of protein ubiquitylation. Biochem J 2004;379:513–25.

[54] Wood A, et al. Bre1, an E3 ubiquitin ligase required for recruitment and substrate selection of Rad6 at a promoter. Mol Cell 2003;11:267–74.

[55] Shilatifard A. The COMPASS family of histone H3K4 methylases: mechanisms of regulation in development and disease pathogenesis. Annu Rev Biochem 2012;81:65–95.

[56] Schulze JM, et al. Splitting the task: Ubp8 and Ubp10 deubiquitinate different cellular pools of H2BK123. Genes Dev 2011;25:2242–7.

[57] Kawashima SA, Yamagishi Y, Honda T, Ishiguro K, Watanabe Y. Phosphorylation of H2A by Bub1 prevents chromosomal instability through localizing shugoshin. Science 2010;327:172–7.

[58] Gerhold CB, Gasser SM. INO80 and SWR complexes: relating structure to function in chromatin remodeling. Trends Cell Biol 2014;24:619–31.

[59] Seeber A, Dion V, Gasser SM. Remodelers move chromatin in response to DNA damage. Cell Cycle 2014;13:877–8.

[60] Bartholomew B. Regulating the chromatin landscape: structural and mechanistic perspectives. Annu Rev Biochem 2014;83:671–96.

[61] Pray-Grant MG, Daniel JA, Schieltz D, Yates 3rd JR, Grant PA. Chd1 chromodomain links histone H3 methylation with SAGA- and SLIK-dependent acetylation. Nature 2005;433:434–8.

[62] Clapier CR, Cairns BR. The biology of chromatin remodeling complexes. Annu Rev Biochem 2009;78:273–304.

[63] Chambers AL, Downs JA. The RSC and INO80 chromatin-remodeling complexes in DNA double-strand break repair. Prog Mol Biol Transl Sci 2012;110:229–61.

[64] Smolle M, et al. Chromatin remodelers Isw1 and Chd1 maintain chromatin structure during transcription by preventing histone exchange. Nat Struct Mol Biol 2012;19:884–92.

[65] Choy JS, Mishra PK, Au WC, Basrai MA. Insights into assembly and regulation of centromeric chromatin in *Saccharomyces cerevisiae*. Biochim Biophys Acta 2012;1819:776–83.

[66] Henikoff S, Furuyama T. The unconventional structure of centromeric nucleosomes. Chromosoma 2012;121:341–52.

[67] Billon P, Cote J. Precise deposition of histone H2A.Z in chromatin for genome expression and maintenance. Biochim Biophys Acta 2013;1819:290–302.

[68] Jackson JD, Falciano VT, Gorovsky MA. A likely histone H2A.F/Z variant in *Saccharomyces cerevisiae*. Trends Biochem Sci 1996;21:466–7.

[69] Jackson JD, Gorovsky MA. Histone H2A.Z has a conserved function that is distinct from that of the major H2A sequence variants. Nucleic Acids Res 2000;28:3811–6.

[70] Zlatanova J, Thakar A. H2A.Z: view from the top. Structure 2008;16:166–79.

[71] Luk E, et al. Chz1, a nuclear chaperone for histone H2AZ. Mol Cell 2007;25:357–68.

[72] Wan Y, et al. Histone chaperone Chz1p regulates H2B ubiquitination and subtelomeric anti-silencing. Nucleic Acids Res 2010;38:1431–40.

[73] Liu H, Luo K, Zhou Z, Mu Y, Wan Y. Histone chaperone Chz1 facilitates the disfavouring property of Spt16 to H2A.Z-containing genes in *Saccharomyces cerevisiae*. Biochem J 2014;460:387–97.

[74] Krogan NJ, et al. A Snf2 family ATPase complex required for recruitment of the histone H2A variant Htz1. Mol Cell 2003;12:1565–76.

[75] Kobor MS, et al. A protein complex containing the conserved Swi2/Snf2-related ATPase Swr1p deposits histone variant H2A.Z into euchromatin. PLoS Biol 2004;2:E131.

[76] Mizuguchi G, et al. ATP-driven exchange of histone H2AZ variant catalyzed by SWR1 chromatin remodeling complex. Science 2004;303:343–8.

[77] Luk E, et al. Stepwise histone replacement by SWR1 requires dual activation with histone H2A.Z and canonical nucleosome. Cell 2010;143:725–36.

[78] Weber CM, Ramachandran S, Henikoff S. Nucleosomes are context-specific, H2A.Z-modulated barriers to RNA polymerase. Mol Cell 2014;53:819–30.

[79] Ranjan A, et al. Nucleosome-free region dominates histone acetylation in targeting SWR1 to promoters for H2A.Z replacement. Cell 2013;154:1232–45.

[80] Teves SS, Weber CM, Henikoff S. Transcribing through the nucleosome. Trends Biochem Sci 2014;39:577–86.

[81] Mehta M, et al. Individual lysine acetylations on the N terminus of *Saccharomyces cerevisiae* H2A.Z are highly but not differentially regulated. J Biol Chem 2010;285:39855–65.

[82] Millar CB, Xu F, Zhang K, Grunstein M. Acetylation of H2AZ Lys 14 is associated with genome-wide gene activity in yeast. Genes Dev 2006;20:711–22.

[83] Papamichos-Chronakis M, Watanabe S, Rando OJ, Peterson CL. Global regulation of H2A.Z localization by the INO80 chromatin-remodeling enzyme is essential for genome integrity. Cell 2011;144:200–13.

[84] Watanabe S, Radman-Livaja M, Rando OJ, Peterson CL. A histone acetylation switch regulates H2A.Z deposition by the SWR-C remodeling enzyme. Science 2013;340:195–9.

[85] Dhillon N, Oki M, Szyjka SJ, Aparicio OM, Kamakaka RT. H2A.Z functions to regulate progression through the cell cycle. Mol Cell Biol 2006;26:489–501.

[86] Kalocsay M, Hiller NJ, Jentsch S. Chromosome-wide Rad51 spreading and SUMO-H2A.Z-dependent chromosome fixation in response to a persistent DNA double-strand break. Mol Cell 2009;33:335–43.

[87] Eaton ML, Galani K, Kang S, Bell SP, MacAlpine DM. Conserved nucleosome positioning defines replication origins. Genes Dev 2010;24:748–53.

[88] Bell SP, Dutta A. DNA replication in eukaryotic cells. Annu Rev Biochem 2002;71:333–74.

[89] Annunziato AT. Assembling chromatin: the long and winding road. Biochim Biophys Acta 2013;1819:196–210.

[90] Poveda A, et al. Hif1 is a component of yeast histone acetyltransferase B, a complex mainly localized in the nucleus. J Biol Chem 2004;279:16033–43.

[91] Ai X, Parthun MR. The nuclear Hat1p/Hat2p complex: a molecular link between type B histone acetyltransferases and chromatin assembly. Mol Cell 2004;14:195–205.

[92] Li Y, et al. Hat2p recognizes the histone H3 tail to specify the acetylation of the newly synthesized H3/H4 heterodimer by the Hat1p/Hat2p complex. Genes Dev 2014;28:1217–27.

[93] Fillingham J, et al. Chaperone control of the activity and specificity of the histone H3 acetyltransferase Rtt109. Mol Cell Biol 2008;28:4342–53.

[94] Ejlassi-Lassallette A, Mocquard E, Arnaud MC, Thiriet C. H4 replication-dependent diacetylation and Hat1 promote S-phase chromatin assembly in vivo. Mol Biol Cell 2011;22:245–55.

[95] Kaufman PD, Kobayashi R, Kessler N, Stillman B. The p150 and p60 subunits of chromatin assembly factor I: a molecular link between newly synthesized histones and DNA replication. Cell 1995;81:1105–14.

[96] Verreault A, Kaufman PD, Kobayashi R, Stillman B. Nucleosome assembly by a complex of CAF-1 and acetylated histones H3/H4. Cell 1996;87:95–104.

[97] Celic I, et al. The sirtuins hst3 and Hst4p preserve genome integrity by controlling histone h3 lysine 56 deacetylation. Curr Biol 2006;16:1280–9.

[98] Vincent JA, Kwong TJ, Tsukiyama T. ATP-dependent chromatin remodeling shapes the DNA replication landscape. Nat Struct Mol Biol 2008;15:477–84.

[99] Corona DF, Clapier CR, Becker PB, Tamkun JW. Modulation of ISWI function by site-specific histone acetylation. EMBO Rep 2002;3:242–7.

[100] Clapier CR, Nightingale KP, Becker PB. A critical epitope for substrate recognition by the nucleosome remodeling ATPase ISWI. Nucleic Acids Res 2002;30:649–55.

[101] Bowman GD. Mechanisms of ATP-dependent nucleosome sliding. Curr Opin Struct Biol 2010;20:73–81.

[102] Yadon AN, Tsukiyama T. DNA looping-dependent targeting of a chromatin remodeling factor. Cell Cycle 2013;12:1809–10.

[103] Masumoto H, Hawke D, Kobayashi R, Verreault A. A role for cell-cycle-regulated histone H3 lysine 56 acetylation in the DNA damage response. Nature 2005;436:294–8.

[104] Tessarz P, Kouzarides T. Histone core modifications regulating nucleosome structure and dynamics. Nat Rev Mol Cell Biol 2014;15:703–8.

[105] Smerdon MJ. DNA repair and the role of chromatin structure. Curr Opin Cell Biol 1991;3:422–8.

[106] Rodriguez Y, Smerdon MJ. The structural location of DNA lesions in nucleosome core particles determines accessibility by base excision repair enzymes. J Biol Chem 2013;288:13863–75.

[107] Kidane D, Murphy DL, Sweasy JB. Accumulation of abasic sites induces genomic instability in normal human gastric epithelial cells during *Helicobacter pylori* infection. Oncogenesis 2014;3:e128.

[108] Menoni H, Shukla MS, Gerson V, Dimitrov S, Angelov D. Base excision repair of 8-oxoG in dinucleosomes. Nucleic Acids Res 2012;40:692–700.

[109] Czaja W, Mao P, Smerdon MJ. Chromatin remodelling complex RSC promotes base excision repair in chromatin of *Saccharomyces cerevisiae*. DNA Repair (Amst) 2014;16:35–43.

[110] Gong F, Fahy D, Smerdon MJ. Rad4-Rad23 interaction with SWI/SNF links ATP-dependent chromatin remodeling with nucleotide excision repair. Nat Struct Mol Biol 2006;13:902–7.

[111] Yu S, Teng Y, Waters R, Reed SH. How chromatin is remodelled during DNA repair of UV-induced DNA damage in *Saccharomyces cerevisiae*. PLoS Genet 2011;7:e1002124.

[112] Yu Y, Deng Y, Reed SH, Millar CB, Waters R. Histone variant Htz1 promotes histone H3 acetylation to enhance nucleotide excision repair in Htz1 nucleosomes. Nucleic Acids Res 2013;41:9006–19.

[113] Srivas R, et al. A UV-induced genetic network links the RSC complex to nucleotide excision repair and shows dose-dependent rewiring. Cell Rep 2013;5:1714–24.

[114] Malik S, et al. Rad26p, a transcription-coupled repair factor, is recruited to the site of DNA lesion in an elongating RNA polymerase II-dependent manner in vivo. Nucleic Acids Res 2010;38:1461–77.

[115] Lees-Miller SP, Meek K. Repair of DNA double strand breaks by non-homologous end joining. Biochimie 2003;85:1161–73.

[116] Bennett G, Peterson CL. SWI/SNF recruitment to a DNA double-strand break by the NuA4 and Gcn5 histone acetyltransferases. DNA Repair (Amst) 2015;30:38–45.

[117] Bennett G, Papamichos-Chronakis M, Peterson CL. DNA repair choice defines a common pathway for recruitment of chromatin regulators. Nat Commun 2013;4:2084.

[118] Dion V, Kalck V, Horigome C, Towbin BD, Gasser SM. Increased mobility of double-strand breaks requires Mec1, Rad9 and the homologous recombination machinery. Nat Cell Biol 2012;14:502–9.

[119] Daley JM, Palmbos PL, Wu D, Wilson TE. Nonhomologous end joining in yeast. Annu Rev Genet 2005;39:431–51.

[120] Williams GJ, et al. Structural insights into NHEJ: building up an integrated picture of the dynamic DSB repair super complex, one component and interaction at a time. DNA Repair (Amst) 2014;17:110–20.

[121] Symington LS. Role of RAD52 epistasis group genes in homologous recombination and double-strand break repair. Microbiol Mol Biol Rev 2002;66:630–70. table of contents.

[122] Mathiasen DP, Lisby M. Cell cycle regulation of homologous recombination in *Saccharomyces cerevisiae*. FEMS Microbiol Rev 2014;38:172–84.

[123] McVey M. RPA puts the brakes on MMEJ. Nat Struct Mol Biol 2014;21:348–9.

[124] Deng SK, Gibb B, de Almeida MJ, Greene EC, Symington LS. RPA antagonizes microhomology-mediated repair of DNA double-strand breaks. Nat Struct Mol Biol 2014;21:405–12.

[125] McVey M, Lee SE. MMEJ repair of double-strand breaks (director's cut): deleted sequences and alternative endings. Trends Genet 2008;24:529–38.

[126] Jackson SP. Sensing and repairing DNA double-strand breaks. Carcinogenesis 2002;23:687–96.

[127] Mimitou EP, Symington LS. Sae2, Exo1 and Sgs1 collaborate in DNA double-strand break processing. Nature 2008;455:770–4.

[128] Clerici M, Mantiero D, Lucchini G, Longhese MP. The *Saccharomyces cerevisiae* Sae2 protein promotes resection and bridging of double strand break ends. J Biol Chem 2005;280:38631–8.

[129] Xu X, et al. The basic cleft of RPA70N binds multiple checkpoint proteins, including RAD9, to regulate ATR signaling. Mol Cell Biol 2008;28:7345–53.

[130] Usui T, Foster SS, Petrini JH. Maintenance of the DNA-damage checkpoint requires DNA-damage-induced mediator protein oligomerization. Mol Cell 2009;33:147–59.

[131] Granata M, et al. Dynamics of Rad9 chromatin binding and checkpoint function are mediated by its dimerization and are cell cycle-regulated by CDK1 activity. PLoS Genet 2010;6. http://dx.doi.org/10.1371/journal.pgen.1001047.

[132] Krejci L, Altmannova V, Spirek M, Zhao X. Homologous recombination and its regulation. Nucleic Acids Res 2012;40:5795–818.

[133] Maloisel L, Fabre F, Gangloff S. DNA polymerase delta is preferentially recruited during homologous recombination to promote heteroduplex DNA extension. Mol Cell Biol 2008;28:1373–82.

[134] Price BD, D'Andrea AD. Chromatin remodeling at DNA double-strand breaks. Cell 2013;152:1344–54.

[135] Holmes A, Haber JE. Physical monitoring of HO-induced homologous recombination. Methods Mol Biol 1999;113:403–15.

[136] Wysocki R, et al. Role of Dot1-dependent histone H3 methylation in G1 and S phase DNA damage checkpoint functions of Rad9. Mol Cell Biol 2005;25:8430–43.

[137] Javaheri A, et al. Yeast G1 DNA damage checkpoint regulation by H2A phosphorylation is independent of chromatin remodeling. Proc Natl Acad Sci USA 2006;103:13771–6.

[138] Chen X, et al. The Fun30 nucleosome remodeller promotes resection of DNA double-strand break ends. Nature 2012;489:576–80.

[139] Ferrari M, et al. Functional interplay between the 53BP1-ortholog Rad9 and the Mre11 complex regulates resection, end-tethering and repair of a double-strand break. PLoS Genet 2015;11:e1004928.

[140] Lazzaro F, et al. Histone methyltransferase Dot1 and Rad9 inhibit single-stranded DNA accumulation at DSBs and uncapped telomeres. EMBO J 2008;27:1502–12.

[141] Pai CC, et al. A histone H3K36 chromatin switch coordinates DNA double-strand break repair pathway choice. Nat Commun 2014;5:4091.

[142] Pfister SX, et al. SETD2-dependent histone H3K36 trimethylation is required for homologous recombination repair and genome stability. Cell Rep 2014;7:2006–18.

[143] Friedel AM, Pike BL, Gasser SM. ATR/Mec1: coordinating fork stability and repair. Curr Opin Cell Biol 2009;21:237–44.

[144] Yekezare M, Gomez-Gonzalez B, Diffley JF. Controlling DNA replication origins in response to DNA damage – inhibit globally, activate locally. J Cell Sci 2013;126:1297–306.

[145] Yoshida K, Poveda A, Pasero P. Time to be versatile: regulation of the replication timing program in budding yeast. J Mol Biol 2013;425:4696–705.

[146] Li GM. New insights and challenges in mismatch repair: getting over the chromatin hurdle. DNA Repair (Amst) 2014;19:48–54.

[147] Li F, et al. The histone mark H3K36me3 regulates human DNA mismatch repair through its interaction with MutSalpha. Cell 2013;153:590–600.

[148] Kadyrova LY, et al. A reversible histone H3 acetylation cooperates with mismatch repair and replicative polymerases in maintaining genome stability. PLoS Genet 2013;9:e1003899.

[149] Sale JE, Lehmann AR, Woodgate R. Y-family DNA polymerases and their role in tolerance of cellular DNA damage. Nat Rev Mol Cell Biol 2012;13:141–52.

[150] Merrill BJ, Holm C. The RAD52 recombinational repair pathway is essential in pol30 (PCNA) mutants that accumulate small single-stranded DNA fragments during DNA synthesis. Genetics 1998;148:611–24.

[151] Xu X, et al. The yeast Shu complex utilizes homologous recombination machinery for error-free lesion bypass via physical interaction with a Rad51 paralogue. PLoS One 2013;8:e81371.

[152] Moldovan GL, Pfander B, Jentsch S. PCNA, the maestro of the replication fork. Cell 2007;129:665–79.

[153] Fu D, Calvo JA, Samson LD. Balancing repair and tolerance of DNA damage caused by alkylating agents. Nat Rev Cancer 2012;12:104–20.

[154] Hoege C, Pfander B, Moldovan GL, Pyrowolakis G, Jentsch S. RAD6-dependent DNA repair is linked to modification of PCNA by ubiquitin and SUMO. Nature 2002;419:135–41.

[155] Stelter P, Ulrich HD. Control of spontaneous and damage-induced mutagenesis by SUMO and ubiquitin conjugation. Nature 2003;425:188–91.

[156] Bailly V, Lamb J, Sung P, Prakash S, Prakash L. Specific complex formation between yeast RAD6 and RAD18 proteins: a potential mechanism for targeting RAD6 ubiquitin-conjugating activity to DNA damage sites. Genes Dev 1994;8:811–20.

[157] Parker JL, Ulrich HDA. SUMO-interacting motif activates budding yeast ubiquitin ligase Rad18 towards SUMO-modified PCNA. Nucleic Acids Res 2012;40:11380–8.

[158] Ulrich HD, Jentsch S. Two RING finger proteins mediate cooperation between ubiquitin-conjugating enzymes in DNA repair. EMBO J 2000;19:3388–97.

[159] Xiao W, Chow BL, Broomfield S, Hanna M. The *Saccharomyces cerevisiae* RAD6 group is composed of an error-prone and two error-free postreplication repair pathways. Genetics 2000;155:1633–41.

[160] Xiao W, et al. Genetic interactions between error-prone and error-free postreplication repair pathways in *Saccharomyces cerevisiae*. Mutat Res 1999;435:1–11.

[161] Hofmann RM, Pickart CM. Noncanonical MMS2-encoded ubiquitin-conjugating enzyme functions in assembly of novel polyubiquitin chains for DNA repair. Cell 1999;96:645–53.

[162] Torres-Ramos CA, Prakash S, Prakash L. Requirement of RAD5 and MMS2 for postreplication repair of UV-damaged DNA in *Saccharomyces cerevisiae*. Mol Cell Biol 2002;22:2419–26.

[163] Broomfield S, Chow BL, Xiao W. MMS2, encoding a ubiquitin-conjugating-enzyme-like protein, is a member of the yeast error-free postreplication repair pathway. Proc Natl Acad Sci USA 1998;95:5678–83.

[164] Brusky J, Zhu Y, Xiao W. UBC13, a DNA-damage-inducible gene, is a member of the error-free postreplication repair pathway in *Saccharomyces cerevisiae*. Curr Genet 2000;37:168–74.

[165] Ball LG, Zhang K, Cobb JA, Boone C, Xiao W. The yeast Shu complex couples error-free post-replication repair to homologous recombination. Mol Microbiol 2009;73:89–102.

[166] Bernstein KA, et al. Sgs1 function in the repair of DNA replication intermediates is separable from its role in homologous recombinational repair. EMBO J 2009;28:915–25.

[167] Karras GI, Jentsch S. The RAD6 DNA damage tolerance pathway operates uncoupled from the replication fork and is functional beyond S phase. Cell 2010;141:255–67.

[168] Makarova AV, Burgers PM. Eukaryotic DNA polymerase zeta. DNA Repair (Amst) 2015;29.

[169] Acharya N, Johnson RE, Pages V, Prakash L, Prakash S. Yeast Rev1 protein promotes complex formation of DNA polymerase zeta with Pol32 subunit of DNA polymerase delta. Proc Natl Acad Sci USA 2009;106:9631–6.

[170] Makarova AV, Stodola JL, Burgers PM. A four-subunit DNA polymerase zeta complex containing Pol delta accessory subunits is essential for PCNA-mediated mutagenesis. Nucleic Acids Res 2012;40:11618–26.

[171] Waters LS, Walker GC. The critical mutagenic translesion DNA polymerase Rev1 is highly expressed during G_2/M phase rather than S phase. Proc Natl Acad Sci USA 2006;103:8971–6.

[172] D'Souza S, Walker GC. Novel role for the C terminus of *Saccharomyces cerevisiae* Rev1 in mediating protein-protein interactions. Mol Cell Biol 2006;26:8173–82.

[173] D'Souza S, Waters LS, Walker GC. Novel conserved motifs in Rev1 C-terminus are required for mutagenic DNA damage tolerance. DNA Repair (Amst) 2008;7:1455–70.

[174] Kosarek JN, et al. Comparative analysis of in vivo interactions between Rev1 protein and other Y-family DNA polymerases in animals and yeasts. DNA Repair (Amst) 2008;7:439–51.

[175] Wiltrout ME, Walker GC. Proteasomal regulation of the mutagenic translesion DNA polymerase, *Saccharomyces cerevisiae* Rev1. DNA Repair (Amst) 2011;10:169–75.

[176] Sabbioneda S, Bortolomai I, Giannattasio M, Plevani P, Muzi-Falconi M. Yeast Rev1 is cell cycle regulated, phosphorylated in response to DNA damage and its binding to chromosomes is dependent upon MEC1. DNA Repair (Amst) 2007;6:121–7.

[177] Gangavarapu V, Santa Maria SR, Prakash S, Prakash L. Requirement of replication checkpoint protein kinases Mec1/Rad53 for postreplication repair in yeast. MBio 2011;2:e00079–11.

[178] Gallego-Sanchez A, Ufano S, Andres S, Bueno A. Analysis of the tolerance to DNA alkylating damage in MEC1 and RAD53 checkpoint mutants of *Saccharomyces cerevisiae*. PLoS One 2013;8:e81108.

[179] Hirano Y, Sugimoto K. ATR homolog Mec1 controls association of DNA polymerase zeta-Rev1 complex with regions near a double-strand break. Curr Biol 2006;16:586–90.

[180] Pages V, Santa Maria SR, Prakash L, Prakash S. Role of DNA damage-induced replication checkpoint in promoting lesion bypass by translesion synthesis in yeast. Genes Dev 2009;23:1438–49.

[181] Kozmin SG, Jinks-Robertson S. The mechanism of nucleotide excision repair-mediated UV-induced mutagenesis in nonproliferating cells. Genetics 2013;193:803–17.

[182] Renaud-Young M. Interaction of the NuA4 complex and chromatin modifications with the DNA damage tolerance pathway, vol. 1. 2015.

[183] Huang D, Piening BD, Paulovich AG. The preference for error-free or error-prone postreplication repair in *Saccharomyces cerevisiae* exposed to low-dose methyl methanesulfonate is cell cycle dependent. Mol Cell Biol 2013;33:1515–27.

[184] Conde F, San-Segundo PA. Role of Dot1 in the response to alkylating DNA damage in *Saccharomyces cerevisiae*: regulation of DNA damage tolerance by the error-prone polymerases Polzeta/Rev1. Genetics 2008;179:1197–210.

[185] Levesque N, Leung GP, Fok AK, Schmidt TI, Kobor MS. Loss of H3 K79 trimethylation leads to suppression of Rtt107-dependent DNA damage sensitivity through the translesion synthesis pathway. J Biol Chem 2010;285:35113–22.

[186] Schulze JM, et al. Linking cell cycle to histone modifications: SBF and H2B monoubiquitination machinery and cell-cycle regulation of H3K79 dimethylation. Mol Cell 2009;35:626–41.

[187] Takahashi YH, et al. Dot1 and histone H3K79 methylation in natural telomeric and HM silencing. Mol Cell 2011;42:118–26.

[188] Grenon M, et al. Docking onto chromatin via the *Saccharomyces cerevisiae* Rad9 Tudor domain. Yeast 2007;24:105–19.

[189] Murakami-Sekimata A, Huang D, Piening BD, Bangur C, Paulovich AG. The *Saccharomyces cerevisiae* RAD9, RAD17 and RAD24 genes are required for suppression of mutagenic post-replicative repair during chronic DNA damage. DNA Repair (Amst) 2010;9:824–34.

[190] McMurray CT. Mechanisms of trinucleotide repeat instability during human development. Nat Rev Genet 2010;11:786–99.

[191] Xu M, et al. Base excision repair of oxidative DNA damage coupled with removal of a CAG repeat hairpin attenuates trinucleotide repeat expansion. Nucleic Acids Res 2014;42:3675–91.

[192] Schermerhorn KM, Delaney S. A chemical and kinetic perspective on base excision repair of DNA. Acc Chem Res 2014;47:1238–46.

[193] Budworth H, McMurray CT. Bidirectional transcription of trinucleotide repeats: roles for excision repair. DNA Repair (Amst) 2013;12:672–84.

[194] Goula AV, Merienne K. Abnormal base excision repair at trinucleotide repeats associated with diseases: a tissue-selective mechanism. Genes (Basel) 2013;4:375–87.

[195] Sundararajan R, Gellon L, Zunder RM, Freudenreich CH. Double-strand break repair pathways protect against CAG/CTG repeat expansions, contractions and repeat-mediated chromosomal fragility in *Saccharomyces cerevisiae*. Genetics 2010;184:65–77.

[196] Schmidt KH, et al. Defects in DNA lesion bypass lead to spontaneous chromosomal rearrangements and increased cell death. Eukaryot Cell 2010;9:315–24.

[197] Shah KA, et al. Role of DNA polymerases in repeat-mediated genome instability. Cell Rep 2012;2:1088–95.

[198] Northam MR, et al. DNA polymerases zeta and Rev1 mediate error-prone bypass of non-B DNA structures. Nucleic Acids Res 2014;42:290–306.

[199] Kim H, Livingston DM. A high mobility group protein binds to long CAG repeat tracts and establishes their chromatin organization in *Saccharomyces cerevisiae*. J Biol Chem 2006;281:15735–40.

[200] Albert B, et al. Structure-function analysis of Hmo1 unveils an ancestral organization of HMG-Box factors involved in ribosomal DNA transcription from yeast to human. Nucleic Acids Res 2013;41:10135–49.

[201] Gonzalez-Huici V, et al. DNA bending facilitates the error-free DNA damage tolerance pathway and upholds genome integrity. EMBO J 2014;33:327–40.

[202] Yang JH, Freudenreich CH. The Rtt109 histone acetyltransferase facilitates error-free replication to prevent CAG/CTG repeat contractions. DNA Repair (Amst) 2010;9:414–20.

[203] Guenole A, et al. Dissection of DNA damage responses using multiconditional genetic interaction maps. Mol Cell 2013;49:346–58.

[204] House NC, Yang JH, Walsh SC, Moy JM, Freudenreich CH. NuA4 initiates dynamic histone H4 acetylation to promote high-fidelity sister chromatid recombination at postreplication gaps. Mol Cell 2014;55:818–28.

[205] Renaud-Young M, et al. The NuA4 complex promotes translesion synthesis (TLS)-mediated DNA damage tolerance. Genetics 2015;199(4).

[206] Yen K, Vinayachandran V, Pugh BF. SWR-C and INO80 chromatin remodelers recognize nucleosome-free regions near +1 nucleosomes. Cell 2013;154:1246–56.

[207] Tomasetti C, Vogelstein B. Cancer risk: role of environment-response. Science 2015;347:729–31.

[208] Yang XJ, Seto E. The Rpd3/Hda1 family of lysine deacetylases: from bacteria and yeast to mice and men. Nat Rev Mol Cell Biol 2008;9:206–18.

Chapter 29

Genomic Instability and Aging: Causes and Consequences

C. Sidler
ETH Zürich, Zürich, Switzerland

Chapter Outline

1. INTRODUCTION

Aging is associated with the progressive functional decline of the body's tissues and organs, resulting in an increasing chance of death at any time point. As such it is a major risk factor for developing age-related pathologies including cancer, cardiovascular diseases, autoimmune diseases, and neurodegenerative diseases. The increasing fraction of the elderly within human populations has made aging a primary health concern. Studies on aging in various model systems, from unicellular organisms (such as yeast) to mammals (such as mice) have revealed several common molecular traits that are associated with aging. These include altered epigenetic profile, mitochondrial dysfunction, altered protein homeostasis, cellular senescence, reduced stem cell function, changes in inter- and intracellular signaling, genomic instability, and telomere shortening [1].

While there is evidence that supports roles for all of these in aging, there is also extensive interplay between the different processes. For instance, mitochondrial dysfunction is characterized by reduced mitochondrial biogenesis, increased respiratory rates, or dysfunction of the electron transport chain, resulting in increased generation of mitochondrial reactive oxygen species (ROS) with age. ROS can in turn cause oxidative damage to the macromolecules within the cell [2].

The "somatic mutation accumulation theory of aging" postulates that accumulation of mutations with age results in functional decline and ultimately leads to an increasing chance of death at any given time point [3,4], and was later modified

Genome Stability. http://dx.doi.org/10.1016/B978-0-12-803309-8.00029-X

into the "DNA-damage accumulation theory of aging." As mentioned previously, organismal aging is caused by a complex interplay of different molecular changes affecting the various tissues within the organism and resulting in their functional deterioration. While keeping this in mind, there is extensive evidence that indicates a central role of DNA-damage accumulation and resulting genomic instability in aging.

2. AGE-RELATED ACCUMULATION OF DNA DAMAGE AND GENOMIC INSTABILITY

The "somatic mutation accumulation theory of aging" was originally put forward by Failla to explain the increasing death rate within the male population of New York with increasing age [3], and by Szilard as a testable theory to explain why organisms age [4]. The central idea of these theories is that the genetic material acquires mutations at a steady rate. While these mutations are random, the risk to accumulate an amount of mutations that is no longer compatible with survival increases with age. According to this theory, aging would result in increased cell death due to mutation accumulation and would thereby promote functional decline. As the role of DNA as genetic material became better understood, this theory evolved into the "DNA-damage theory of aging," which considers the role of DNA damage and its molecular and cellular consequences in the process of aging. In agreement with this theory, different types of DNA damage including DNA double-strand breaks (DSBs) and oxidative DNA damage accumulate with age in various model organisms [5].

When considering the amount of DNA lesions that are detected in a cell at a given time point, it is important to keep in mind that this represents a reflection of the steady state. While DNA damage is reversible, products of faulty DNA repair and replication are irreversible and promote genomic instability. Genomic instability is commonly triggered at sites of single-stranded DNA gaps and DSBs and frequently results in point mutations, microsatellite contractions or expansions, copy number variation, loss of heterozygosity, or large genome rearrangements. However, their detection in the context of a whole genome is not entirely straightforward, as these events are rare at the scale of a whole tissue.

2.1 Accumulation of Point Mutations, Insertions, and Deletions

Most of the evidence for the accumulation of smaller somatic mutations, such as point mutations, small insertions, and deletions, was obtained by using reporter assays. These are commonly based on altered phenotypes caused by mutation of either an endogenous gene (such as the hypoxanthine phosphoribosyl transferase (*HPRT*) locus [6]) or a transgene (such as a *lacZ* reporter [7]). While scoring mutation frequencies using an endogenous reporter system relies on the suitability of the cells under study for cultivation, transgene-based scoring of mutation frequencies can be performed by excising the transgene and determining the mutation frequency in *Escherichia coli*. Results from such studies have provided evidence for increasing mutation rates with age [8].

2.2 Accumulation of Large Chromosomal Aberrations

Due to their easier detection, large chromosomal abnormalities have been observed in aging cells quite early on. Large chromosomal abnormalities accumulate in proliferating as well as in postmitotic cells with increasing age [9]. For instance, human brain cells accumulate high levels of aneuploidy with increasing age [10].

3. CAUSES OF AGE-DEPENDENT ACCUMULATION OF GENOMIC INSTABILITY

The accumulation of mutations with age is the result of a balance between lifelong exposure to DNA-damaging agents and subsequent repair of the lesions. DNA is exposed to various intrinsic and extrinsic sources of DNA damage, including chemicals, radiation, pathogens, ROS, hydrolysis, and DNA replication and repair errors. The predictability of patterns of aging symptoms points to a central role of intrinsic factors in the process. Therefore, intrinsic factors that affect the level of genomic instability in a cell will be discussed in the following sections, while keeping in mind that extrinsic damaging factors may accelerate the accumulation of DNA damage additionally.

3.1 Oxidative Stress

Among the intrinsic damaging agents, ROS are considered central contributors to human aging as formulated in the "free radical theory of aging" [11]. This theory is based on the observation that metabolic activity negatively correlates with life span, which may be mediated by increased ROS production and consequent oxidative damage.

The most prominent ROS are superoxide, hydrogen peroxide, hydroxyl radicals, and singlet oxygen, which are mainly produced during oxidative phosphorylation in the inner mitochondrial membrane.

Exposure of DNA to ROS can result in the oxidation of bases, and formation of abasic sites, DNA single-strand breaks (SSBs), and DSBs. Unlike damaged proteins or lipids, damaged DNA cannot simply be replaced. Therefore, the cell needs efficient repair mechanisms; otherwise, these lesions can cause replication stress or result in point mutations through repair errors. For instance, 8-oxo-guanine, which is one of the most frequent types of oxidative damage, can result in GC→TA transversions through mispairing during DNA replication [12].

Under normal physiological conditions, cells employ various strategies for detoxification of ROS, including enzymes, such as catalases, glutathione peroxidases, and sodium dismutases, and molecules with antioxidant properties, such as vitamins and glutathione.

Nonetheless, several lines of evidence support a role of oxidative damage in the aging process. First, several studies reported an accumulation of oxidative DNA damage with age [5]. Second, mitochondrial ROS generation increases with age, due to increased respiratory rates or dysfunction of the electron transport chain [2]. On the other hand, studies on the effects of mutations affecting antioxidant enzymes on life span have yielded contradictory results.

As previously mentioned, many tissues in different species have been shown to accumulate oxidative DNA damage with increasing age. An analysis of urinary excretion rates of oxidized nucleotides has shown decreasing rates of excretion with increasing age, suggesting that the rate of oxidation decreases with the decreasing metabolic rate during aging [13]. However, cells in the aging organism still accumulate oxidative DNA damage at steady rates, indicating reduced capacity for repair of oxidative damage with increasing age. In line with this, the repair of oxidative damage is more efficient in cells from young when compared to aged subjects [14] (see Section 3.4).

Although the exact role of oxidative damage in aging is not completely clear, it is unarguably an important internal source of DNA damage contributing to the age-associated accumulation of DNA damage and genomic instability.

3.2 Depurination, Depyrimidination, and Deamination

In addition to ROS, DNA is abundantly exposed to water, which can induce spontaneous hydrolysis of the glycosylic bond, resulting in abasic sites. This occurs at up to 10,000 sites per cell per day [15]. If not repaired before DNA replication, these sites can be subject to mispairing and are therefore potentially mutagenic. However, whether this contributes significantly to age-associated genomic instability remains to be tested.

While not that frequent, deamination also contributes to the accumulation of point mutations. In particular, deamination of 5-methylcytosine to thymine creates potentially mutagenic G-T base pairs. Since deamination of 5-methylcytosine creates thymine, which is a normal nucleotide, it is not as easily detected as damage. Therefore, 5-methylcytosine deamination constitutes a significant source of mutation [16]. Due to the important role of CpG methylation in epigenetic regulation, such deamination may not only affect the DNA sequence per se, but also its regulation. While a role of this mutagenicity in cancer is well established, no significant accumulation of C→T transitions has been detected in mouse livers with increasing age [17]. This is likely due to the activity of a thymine glycosylase (MBD4) that specifically recognizes and repairs G-T mispairs that are preferably located in a CpG sequence context [18].

Thus, while both spontaneous hydrolysis of glycosylic bonds and deamination of bases are internal sources of DNA damage, their role in human aging is unclear.

3.3 Replication Errors and Replication Stress

In general, genomic DNA is most vulnerable during DNA replication and is subject to replication errors and replication stress that can cause secondary damage.

While DNA polymerases replicate DNA with very high fidelity, faulty incorporation of nucleotides occurs in about 1 in 100,000 bases, of which 99% are fixed by proofreading. The remaining mismatches rely on repair by mismatch repair (MMR) [19]. In addition to mismatches, replication slippage contributes to length variations of repetitive sequences, which is a result of misalignment of the template and the newly synthesized DNA strand, thus introducing either small insertions or deletions [20].

While the fidelity of DNA replication seems to be unaltered with age [21], changes in the activity of MMR may potentially contribute to the accumulation of mutations caused by replication errors (see Section 3.4).

In addition, several conditions can result in slowing or stalling of replication forks, which is termed replication stress. These include reduced nucleotide pool, unrepaired DNA damage, frequency of initiation of DNA replication, impaired de novo nucleosome assembly, and mutations in genes required for replication [22]. Upon prolonged fork stalling, the replication fork can collapse or regress, resulting in the formation of mutagenic Holliday junctions [23], and can induce several types of genomic instability, including SSBs and DSBs. The major repair mechanism to process DSBs generated at stalled replication forks is homologous recombination (HR), but in the absence of HR, nonhomologous end joining (NHEJ) can repair the lesion, potentially resulting in genomic instability in the form of large genome rearrangements [24]. Processing of breaks generated at stalled replication forks can further result in sister chromatid exchange, and chromosome loss or fragility. Thus, this process is very tightly coordinated with cell cycle regulation and damage checkpoints.

Mice with defects in Mcm2 exhibit reduced licensing of DNA-replication origins, severe problems in proliferative cells, such as stem cells, and have a drastically reduced life span [25]. In addition, several human premature aging syndromes, including Werner syndrome (WS), Bloom syndrome (BS), and Rothmund–Thomson syndrome (RTS), are caused by mutations in DNA helicases of the RECQ family, which play a role in stabilizing stalled replication forks, checkpoint activation, and preventing and resolving mutagenic intermediate structures at stalled replication forks [26]. Taken together, this indicates that DNA-replication stress may contribute to aging by contributing to genomic instability, particularly in proliferating cells.

3.3.1 Werner Syndrome

WS is the human premature aging syndrome that recapitulates the most traits that are also associated with normal human aging, including increased risk for age-related diseases, such as atherosclerosis, osteoporosis, diabetes, and cancer, as well as other symptoms of aging, such as hair loss and cataracts [27] (Table 29.1). It is caused by a mutation in RECQ-like DNA helicase and exonuclease, which is involved in various processes including DNA replication and recombination, with a major role in the reinitiation of stalled replication forks [28]. In line with this, cells from WS patients accumulate DNA

TABLE 29.1 Human Premature Aging Syndromes and Associated Impairments in Genome Maintenance

Syndrome	Mutation	Role in Genome Maintenance	Traits of Aging	References
Trichothiodystrophy (TTD)	*TFIIH, XPB, XPD*	TC-NER, transcription	Neurologic and skeletal degeneration, osteoporosis, ichthyosis, early graying of hair, infertility, and brittle hair and nails	[46]
Cockayne syndrome (CS)	*CSA* or *CSB*	TC-NER	Cachexia, neuronal degeneration, loss of retinal cells, poor growth, cataracts, photosensitivity, atherosclerosis, diabetes, hypertension	[46]
Xeroderma pigmentosum (XP)	*XPA-XPG*	NER	CS symptoms and in addition: hypersensitivity to UV exposure, pigment alterations, and high incidence of skin cancer	[46]
Ataxia telangiectasia (AT)	*ATM*	DDR	Progressive cerebellar degeneration, severe ataxia, growth retardation, dilated blood vessels, immunologic defects, and cancer	[51]
Rothmund–Thomson syndrome (RTS)	*RECQL4*	DNA repair	Growth deficiency, gray hair, cataracts, poikiloderma, osteosarcomas, and skin cancers	[28]
Werner syndrome (WS)	*WRN*	Telomere maintenance, DNA recombination and repair	Atrophic skin, thin gray hair, osteoporosis, type II diabetes, autoimmunity, skin and muscle atrophy, poor wound healing, cataracts, atherosclerosis, hypogonadism, and cancer	[27,28]
Bloom syndrome (BS)	*BLM*	Mitotic recombination	Growth retardation, sun sensitivity, immune deficiency, genomic instability, cancer, and diabetes	[28]
Hutchinson–Gilford progeria syndrome (HGPS)	*LMNA*	Nuclear lamina function	Alopecia, sarcopenia, atherosclerosis, osteolysis, prominent scalp veins, loss of subcutaneous fat, vascular problems, limited sexual development, and high-pitched voice	[59]
Dyskeratosis congenita (DC)	*DKC1*	Telomere maintenance	Growth retardation, microcephaly, cerebellar hypoplasia, mental retardation, progressive combined immune deficiency, and aplastic anemia	[72]

damage, such as DSBs, and genomic instability in the form of instability of repetitive loci, chromosomal aberrations and mutations, and telomere instability [29].

3.3.2 Bloom Syndrome

BS is caused by mutations in *BLM*, which also belongs to the family of RECQ helicases (Table 29.1). BLM helicase resolves Holliday junction-like recombination intermediates, blunt-ended DNA duplexes with internal bubbles, and G-quadruplexes that are prevalent within telomeric DNA [28]. BS cells are characterized by the accumulation of chromosomal aberrations, including sister chromatid exchanges, polycentric chromosomes, breaks, and translocations [30].

3.3.3 Rothmund–Thomson Syndrome

RTS is triggered by a mutation in a RecQ-like helicase *RECQL4* [31]. The molecular functions of RECQL4 are less understood, but it has been associated with function in DSB repair, DNA replication, and telomere maintenance [28]. Further, RTS cells also exhibit chromosomal aberrations that mainly include large chromosomal rearrangements and isochromosome formation [32]. However, although patients present with signs of premature aging, most patients seem to have a normal life span (Table 29.1).

3.4 Deterioration of Genome-Maintenance Mechanisms

In addition to mechanisms that generate DNA damage, there is evidence that DNA-repair pathways deteriorate with increasing age, mainly due to the reduced expression and/or activity of several key enzymes. However, the role of deteriorating genome-maintenance mechanisms in aging is still controversial. The observation that polymorphisms in DNA-repair genes, such as *ATM* and *XPD*, are associated with longevity in human populations [33,34] supports a role for genome maintenance in preventing functional deterioration.

On the other hand, experimental reduction of DNA damage does not consistently result in life span extension. Therefore, defects in single DNA-repair pathways may have tissue-specific effects rather than affecting organismal aging.

3.4.1 Mismatch Repair

MMR mainly repairs DNA lesions caused by faulty DNA replication or repair, resulting in mismatches or small insertion and deletion loops, or deamination of 5-methylcytosine. Failure of MMR is often associated with point mutations or microsatellite instability (Fig. 29.1). Both elevated rates of microsatellite instability with increasing age [35,36], and the reduced capacity of cell extracts from old donors to repair induced mismatches [37], indicate an age-dependent decline in MMR activity.

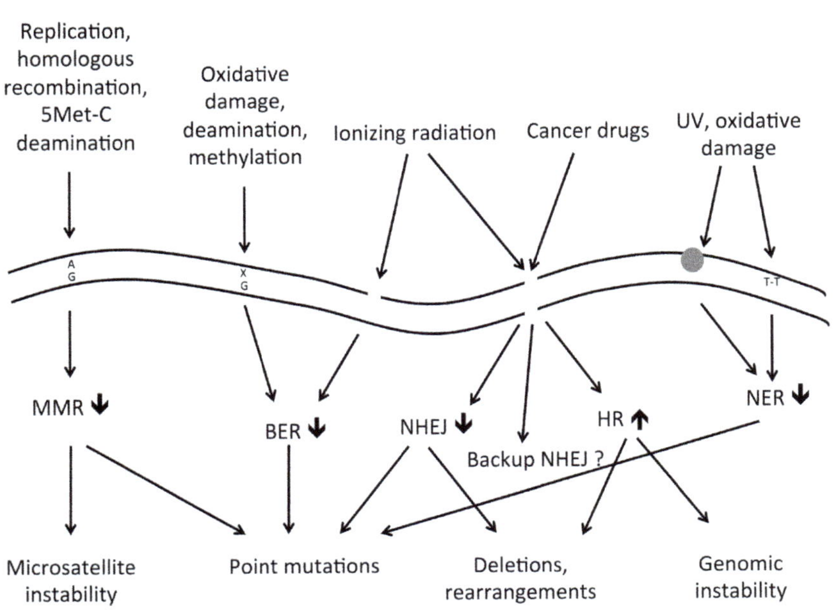

FIGURE 29.1 DNA damage and age-related changes in DNA repair. Different DNA-damaging agents that cause different types of DNA damage are listed on top, the repair pathways responsible to fix them in the middle and possible consequences of their dysfunction on the bottom. *Arrows next to the repair pathways* indicate functional decline with age. The *gray circle* indicates DNA damage in the form of bulky adducts. *Modified from Hoeijmakers JH. Genome maintenance mechanisms for preventing cancer. Nature 2001;411(6835):366–74.*

Further, mutations in MMR genes in humans are associated with cancer susceptibility [38]. This may indicate a role for age-dependent decline in MMR in the increasing cancer susceptibility with increasing age.

3.4.2 Base Excision Repair

Base excision repair (BER) is responsible for fixing damaged DNA bases, such as products of oxidative damage, deamination, and SSBs. Failure to repair these types of lesions can result in point mutations (Fig. 29.1). BER activity decreases with age in several organs, including the brain, likely due to the reduced activity of several DNA glycosylases and DNA polymerase β [39].

Assessing the effect of mutations in BER genes on life span in mammalian model organisms is difficult, since most of them are embryonic lethal. Milder defects in BER genes, on the other hand, result in elevated cancer incidence—for instance, in mice with haploinsufficiency of DNA polymerase β [40]. Although, when tested in yeast, mutations in BER genes result in life span shortening and this effect is cumulative [41], suggesting that functional BER is required to prevent aging.

Since BER is an important repair pathway to repair oxidative damage, cells with high metabolic activity and ROS generation may be particularly sensitive to declining BER activity with age. In line with this, BER deficiency has been detected in brains of Alzheimer's disease (AD) patients [42], and mutations in AP endonucleases were associated with amyotrophic lateral sclerosis (ALS) [43]. Thus, increased oxidative stress paralleled with impaired repair of oxidative damage seems to contribute to age-related neurodegeneration and neurodegenerative disorders.

3.4.3 Nucleotide Excision Repair

Nucleotide excision repair (NER) excises and repairs nucleotides that are modified by bulkier adducts, including oxidative damage and cyclobutane pyrimidine dimers caused by UV exposure. Defective NER contributes to the accumulation of point mutations (Fig. 29.1).

There is evidence that NER activity decreases with age, based on the observation that the repair of UV-induced damage in normal skin fibroblasts declines and becomes more mutagenic with increasing donor age [44]. A possible reason may be the reduced expression of NER genes in older individuals [45].

Mutations in several NER genes cause premature aging syndromes in humans, indicating an important role of NER in preventing age-dependent functional decline. These include trichothiodystrophy (TTD), which is caused by a mutation in the TFIIH helicase that is involved in NER, Cockayne syndrome (CS), which is caused by mutations in the *CSA* or *CSB* genes that function in the transcription-coupled NER pathway, and xeroderma pigmentosum (XP), which is caused by mutations in the *XPA-G* genes that are also involved in the NER pathway (Table 29.1) (Chapter 25) (for review see Ref. [46]). CS cells exhibit extensive chromosomal instability, however this does not make patients more prone to developing cancer, possibly due to their higher propensity to undergo apoptosis in response to UV damage [47]. In contrast, XP cells also exhibit defects in global genome NER, thus resulting in the accumulation of genome-wide mutations. XP patients show dramatically increased the rate of skin cancer and accelerated aging limited to areas of the body that are exposed to the sun [48]. Further, the effect of mutations in NER genes on aging is cumulative, indicating that functional NER is required to prevent functional decline.

3.4.4 Double-Strand Break Repair

DSBs are particularly toxic lesions that can be generated as a result of exposure to ionizing radiation, oxidative damage, or during replication, and if left unrepaired, they may result in loss of chromosomal segments. Errors in repair of DSBs can also be detrimental as they can result in translocations, ring chromosomes, and end fusions.

There are two major pathways involved in DSB repair: homologous recombination (HR) and NHEJ. HR uses stretches of extensive sequence homology as templates for repair and is therefore considered a precise repair mechanism. NHEJ, on contrary, rejoins the ends at a DSB with little or no consideration for homology and is therefore considered error prone.

Relative pathway usage depends on cell-cycle stage and sequence context. HR is mostly limited to S and G2 phases of the cell cycle; therefore, NHEJ is the main pathway for DSB repair during G1 phase including in nonproliferating and senescent cells. HR frequency increases if DSBs occur in close proximity to repetitive sequences [49]. Overall, the use of NHEJ exceeds the use of HR by orders of magnitude.

The initial step in both mechanisms involves DNA-damage signaling and rapid recruitment of ataxia telangiectasia mutated (ATM) [50], which induces downstream repair. The importance of this in maintaining genome integrity and preventing aging is accentuated by the fact that loss of function of ATM causes ataxia telangiectasia (AT), which shares several features of normal aging. For instance, cells from AT patients have unstable telomeres and enter premature cellular senescence. This may be due to the function of ATM in the formation of the telomeric T-loop that protects chromosome ends [51].

In addition to the DNA-damage response, the repair mechanisms that repair DSBs are also affected by aging, as is discussed in the following sections.

3.4.4.1 Nonhomologous End Joining

NHEJ ligates two broken ends together, thereby frequently resulting in small insertions and deletions. However, postmitotic cells rely on NHEJ for repairing DSBs. Failure to faithfully repair DSBs can result in point mutations, deletions, and large genome rearrangements (Fig. 29.1). However, NHEJ activity and fidelity decline with age [52], which may be in part due to altered expression levels, activity, and distribution of key repair enzymes [53]. Further, mutations in NHEJ genes including *Ku70* and *Ku80* have been associated with shortened life spans in mice [54]. In addition, defects in DNA-PKcs resulted in impaired telomere maintenance and shortened life span in mice [55]. Taken together, these lines of evidence suggest that NHEJ plays an important role in preventing age-related increase in genomic instability and functional decline.

3.4.4.2 Homologous Recombination

HR is a precise mechanism of DSB repair that makes use of homologous sequences to repair the lesion. However, the recombination of misaligned sequences can result in chromosome rearrangement and copy number variants. In fact, the frequency of such nonallelic recombination resulting in increasing frequency of genomic rearrangements increases with age [56]. This increase in HR seemed to be tissue specific, due to limited capacity of cells that underwent rearrangements to clonally expand, for instance, in skin [57].

In summary, most genome-maintenance mechanisms have been shown to become less active and/or less precise with increasing age, contributing to age-associated genomic instability. Moreover, genetic impairments in DNA-repair pathways are often associated with shortened life span and increased cancer susceptibility, further indicating that faithful and efficient DNA repair is crucial in preventing age-related disease and functional decline.

3.5 Altered Nuclear Architecture

The role of the nuclear architecture and organization in aging has only become appreciated in recent years. By segregating genomic DNA into regions of euchromatin and heterochromatin, active transcription and repression, and sites of active DNA repair, the overall nuclear organization has profound effects on gene expression patterns, but also on DNA stability.

Perturbation of the nuclear architecture caused by a mutation in the lamin A (*LMNA*) gene is a hallmark of the Hutchinson–Gilford progeria syndrome (HGPS). The most common *LMNA* mutation results in missplicing of the *LMNA* transcript, resulting in the accumulation of a truncated protein, which is called progerin [58]. Nuclei from HGPS patients are characterized by the loss of interaction of heterochromatin with the nuclear lamina as well as general perturbation of heterochromatin, which results in induction of transcription of pericentric satellites III. HGPS cells further exhibit impairments in the DNA-damage response, recruitment of repair proteins to DNA-damage sites, and DNA repair, which is reflected in elevated genomic instability. The *LMNA* splicing defect also leads to inappropriate localization of telomeres within the nucleus, resulting in a loss of heterochromatin at telomeres, telomere shortening and genomic instability, thereby contributing to the establishment of premature cellular senescence (for review see Ref. [59]). Taken together these observations indicate that the integrity of the nuclear membrane and its role in nuclear organization are crucial to maintaining genome integrity and preventing aging.

Changes in nuclear architecture have also been observed during normal human aging. For instance, progerin also accumulates in human cells during normal ageing [60] and is associated with similar changes to nuclear organization as in HGPS cells. In addition, senescing cells accumulate senescence-associated heterochromatin foci (SAHF), which are formed de novo for instance at E2F-target promoters and silence proliferative genes, thereby promoting cell-cycle exit [61]. SAHF formation is promoted by senescence-dependent reorganization of the nuclear lamina and associated chromatin [62]. This is paralleled with heterochromatin relaxation at perinuclear repetitive DNA sequences, which may promote instability of these regions.

Further, since the DNA-damage response promotes alterations in chromatin that can extend up to megabases around the break site [63], the accumulation of unrepaired DNA damage during aging may also contribute to age-related changes in chromatin structure. This can affect expression patterns of genes around the damaged site. For instance, persistent oxidative damage in promoter regions is associated with gene repression in the cortex of the human brain [64], suggesting that break-induced changes in gene expression may contribute to age-related alterations of the gene-expression profile and thus contribute to functional changes in cells and tissues.

In summary, it seems that altered nuclear organization and chromatin structure play a crucial role in maintaining genome integrity and preventing age-associated changes, while, on the other hand, accumulation of DNA damage affects chromatin structure, potentially resulting in age-associated changes in gene-expression profiles. A better mechanistic understanding of the establishment and role of altered nuclear organization in aging will clarify the significance of epigenetic changes in genomic instability during the aging process.

3.6 Selection

Evidence presented in the previous paragraphs indicates that the accumulation of unrepaired and misrepaired DNA damage and resulting genomic instability is promoted by several active mechanisms that either generate or fail to repair DNA lesions. In addition, passive accumulation through a shift in selection that limits survival of cells in the presence of DNA damage may occur as well.

Selection requires a quality check system, which in the case of proliferating cells is given by cell-cycle regulation and checkpoints that monitor genomic stability and ensure that a cell only propagates in the absence of DNA damage.

Early in life, when a large amount of cell divisions is required to form tissues and organs of the growing organism, DNA damage also accumulates [65]. Unrepaired DNA damage leads to checkpoint activation and cells with erroneous genomes either repair the damage or get eliminated, which can be detrimental to organismal survival. This is evidenced by the observation that the majority of spontaneously aborted embryos carry chromosomal abnormalities [66]. On the other hand, in the older organism many tissues contain large number of postmitotic cells that do not undergo the same quality control any longer and therefore passively accumulate unrepaired damage and genomic instability [67]. Thus, while the accumulation of DNA damage early in life can be detrimental, the accumulation later in life is more tolerated and interferes with tissue function rather than with organismal survival.

In summary, numerous internal sources of DNA damage, in particular oxidative damage, along with lifelong exposure to external sources of DNA damage, contribute to the accumulation of DNA damage throughout life. While several genome-maintenance mechanisms are in place to cope with the damage, these deteriorate with increasing age. In addition, alterations in nuclear architecture affect chromatin organization, which in turn influences the appropriate localization, stability, and regulation of genome regions. Lastly, nonproliferative cells with unrepaired DNA damage may accumulate in tissues of older organisms due to the lack of selective mechanisms that eliminate them. Taken together, this can explain the observed accumulation of point mutations and larger chromosomal abnormalities observed in tissues of aging organisms.

4. GENOMIC REGIONS WITH VARIOUS SUSCEPTIBILITY TO GENOMIC INSTABILITY

When considering the consequences such accumulation of DNA damage and resulting genomic instability may have, this depends on the affected sequence and the type of instability. In addition to protein-coding sequences, the genome consists of a large fraction of repetitive sequences, which are inherently more difficult to replicate and repair, and are therefore common targets for genomic instability.

Moreover, genomic DNA consists of nuclear DNA, which encodes the large majority of all protein-coding genes and exists in one diploid copy per cell, and mitochondrial DNA (mtDNA), which encodes 37 genes that mostly encode mitochondrial components and exists as two to five copies of circular, supercoiled DNA in hundreds to thousands of mitochondria per cell. The unique challenges these different sequence contexts present to the maintenance of genome integrity and their consequences for the aging process are discussed in the following sections.

4.1 Nuclear DNA

Nuclear DNA encodes the majority of genes that are required for life. The nuclear DNA consists of protein-coding sequences, which account for 2% of the total DNA, and noncoding sequences that include RNA-coding sequences, structural components, regulatory sequences, but also extensive repetitive DNA sequences. While a point mutation affecting a protein-coding gene or its regulatory elements may affect the function of this particular protein and the molecular network it plays a role in, larger aberrations like rearrangements may affect the functionality of a larger number of genes and thereby have further reaching physiological consequences. If a mutation is introduced at a specific locus in a specific cell within a tissue, this may not have a significant effect on the tissue. However, clonal expansion of cells that carry mutations can lead to an amplification of the phenotypic outcome and affect tissue function to a greater extent [68].

On the other hand, about 30% of the nuclear DNA consists of highly repetitive sequences, such as telomeres, ribosomal DNA (rDNA), microsatellites, and minisatellites that are difficult to replicate and repair. Thus, they are more susceptible to the accumulation of genomic instability and seem to play more specific roles in cellular senescence and aging, as will be discussed in the following sections.

4.1.1 Telomeric DNA

Human telomeres consist of TTAGGG repeats that can extend more than 10 kilobases at the linear ends of DNA. Telomeric DNA is protected by the shelterin complex to prevent nucleolytic degradation, constant activation of DNA-damage signaling, and unscheduled DNA repair at the unprotected ends, which could otherwise result in end-to-end fusions [69].

However, during the replication of linear DNA, 50 to 200 base pairs are lost from the ends in each cycle. Thus, in the absence of mechanisms that resynthesize them, telomeres shorten with every cell division. When telomeres become critically short and exposed, DNA-damage signaling is activated and triggers growth arrest [70].

In addition, due to their G-rich sequence, telomeres form G-quadruplex structures, which can interfere with normal DNA-replication fork progression [71] and contribute to replication stress and genomic instability.

While a role of telomere shortening in cellular senescence is widely accepted, there is also ample support for a role of telomere shortening in organismal aging. For instance, mutations that result in compromised telomere maintenance are associated with human premature aging syndromes, such as WS, HGPS, and dyskeratosis congenita (DC). DC patients carry mutations in *DKC1*, which is a structural part of RNP complexes including telomerase [72] (Table 29.1).

Further, telomere shortening also occurs during normal aging. For instance, leukocyte telomere length decreases over time in most people [73] and reduced telomere length correlates with the development of several age-related deficiencies, such as atherosclerosis and risk for development of cardiovascular disease [74].

As telomeres shorten with each cell division, cells that are highly proliferative cells, such as stem cells and immune cells, are particularly sensitive to telomere shortening [75]. Thus, telomere shortening may play a role in age-associated functional decline through inducing senescence in highly proliferative tissues.

4.1.2 Ribosomal DNA

rDNA constitutes another large region of repetitive DNA sequence within nuclear DNA. Due to its repetitive nature, rDNA is susceptible to recombination and as a consequence to deletion/insertions, which makes rDNA one of the largest fragile sites of the genome. In addition, rDNA constitutes a common site for replication fork arrest due to the presence of multiple replication fork barriers [76], which serve to prevent collisions between DNA replication and transcription.

The role of rDNA instability in aging is well characterized in yeast (for review see Ref. [77]). Yeast senescence is associated with increased recombination within rDNA, resulting in rDNA circle excision, which is likely caused by the relaxation of heterochromatin in this region. However, mutant yeast that exhibit rDNA instability without accumulating rDNA circles have a shorter life span than wildtype, suggesting that rDNA instability is sufficient to promote aging. In line with this, rDNA instability was identified as one of the major reasons that affected life span in genetically diverse yeast strains. Based on this, Kobayashi has put forward the "rDNA theory of aging," which postulates that rDNA is the region within the genome that is most sensitive to age-dependent accumulation of DNA damage and thereby may act as a DNA-damage sensor within the genome [78].

In contrast to yeast cells, human somatic cells do not express telomerase for the maintenance of telomeres. Therefore, telomere instability seems to play a more prominent role in human cellular senescence and aging than rDNA instability. However, some evidence for increased rDNA instability with age in humans exists. For instance, nondividing cells such as nerve, heart, and skeletal muscle tissues exhibit extensive loss of rDNA with increasing age [79]. Similarly, the accumulation of extrachromosomal rDNA with age has also been described in normal human cells [80]. In addition, cells from BS patients show extensive rDNA instability due to aberrant recombination [81], which may contribute to the increased cancer susceptibility. However, while rDNA also exhibits instability during human aging and in cells from patients with premature aging disorders, its role in human aging is still unclear.

4.1.3 DNA Repeats

Other DNA repeats, such as retrotransposons, micro- and minisatellites, are also prone to genomic instability. Such repetitive sequences are prone to random expansions and deletions due to faulty replication or repair, or in the case of retrotransposons to mutagenesis by excision and integration.

Transposable elements make a considerable fraction of the total genomic DNA sequence of humans. There are three major families of retrotransposons—L1, Alu, and SVA, and they make up for about 50% of the human genome. These sequences are usually silenced by heterochromatin; however, it was shown that, for instance, during differentiation of brain cells, retrotransposons can be activated and integrate into protein-coding genes, thereby modulating their expression [82]. Further, as cells become senescent, heterochromatin in regions of constitutive heterochromatin is increasingly reduced and results in the expression of transposable elements and ultimately in retrotransposition [83]. In addition to modifying expression levels, retrotransposons have also been shown to contribute to genomic instability by causing the DSBs [84], and due to their high frequency within the genome, they can also provide substrates for unequal recombination, resulting in sequence loss, inversions, or duplication [85].

Although less abundant, micro- and minisatellites and satellite DNA sequences constitute about 3% of the genomic DNA sequence in humans. Microsatellites are tandemly repeated sequences, which consist of units that are 1–6 base pairs long. Repeats of longer units are classified as microsatellites or satellites, such as centromeric tandem repeats. It was shown that the mutation rate within microsatellite regions increases with age in humans [86]. Mutation of microsatellite sequences is often in the form of small expansions or deletions, which are the result of DNA-replication slippage. Increased microsatellite instability was also detected in hematopoietic stem and progenitor cells and T-cell clones from human subjects with increasing age and correlated with reduced expression of MMR gene *MLH1* [36], and may contribute to replicative senescence and tumorigenesis.

4.2 Mitochondrial DNA

In contrast to nuclear DNA, mtDNA is not associated with histones and is therefore much more accessible to DNA damage. Due to the close proximity to the respiratory chain, mtDNA is also highly exposed to oxidative damage. This is countered by a much less efficient repair system than that in place at the nuclear DNA, as mtDNA repair is limited to BER [87] and MMR [88]. Similar to nuclear BER, mitochondrial BER activity also declines with age [89], and mtDNA has also been shown to accumulate small deletions with increasing age. In addition, replication errors have also been determined as a significant source of point mutations in mtDNA, and mitochondrial deletions are thought to occur in a replication-dependent manner through mispairing between direct repeats within mtDNA. However, DSBs may also promote mtDNA deletions by unknown molecular mechanisms (for review see Ref. [90]).

On the other hand, every cell contains several hundreds to thousands of mitochondrial genomes, possibly allowing for complementation [91]. Also, every single mitochondrial genome can tolerate a certain extent of mutations and deletions before inducing mitochondrial dysfunction. Thus, while being more prone to the accumulation of genomic instability, mitochondrial genomes also seem to be very resistant to its consequences.

However, several specific point mutations reach very high copy numbers in individuals of increasing age. Clonally expanded mitochondrial mutations of the *COX* gene accumulate in single muscle fibers with age, leading to functional deterioration, and in neurons within the substantia nigra, resulting in impaired respiration within the affected neurons [90]. The absence of a human progeroid syndrome that is characterized by mtDNA instability suggests that mtDNA instability is not a central cause of human aging. However, since clonally expanded mtDNA mutations affect cells within tissues with high respiratory requirements by resulting in respiratory impairments, they may contribute to the functional decline of these tissues.

5. ROLE OF GENOMIC INSTABILITY IN AGING?

As discussed thus far, there is ample evidence for DNA-damage accumulation with increasing age, which is a result of lifelong exposure to DNA-damaging agents, including intrinsic exposure to oxidative stress, and deteriorating genome-maintenance mechanisms coupled with altered selection. If this constitutes a driving force to human aging, this must translate into functional deterioration of tissues and organs in a somewhat predictable way.

Since the function of cells, tissues, organs, and organisms relies on complex regulatory networks, mutations that affect any point of these networks are expected to impede the appropriate function of the entire network. Therefore, many different mutations may result in a similar phenotype. Such mutations occurring in highly differentiated postmitotic tissues are very likely to negatively affect cell function and thereby contribute to the functional decline of the tissue. On the other hand, genomic instability also interferes with cell physiology; for instance, the accumulation of unrepaired DNA damage can result in cell death or cellular senescence and interfere with tissue homeostasis.

5.1 Effect of Genomic Instability on the Gene Expression Profile

Aging is characterized by extensive changes to the gene expression profile. Some of these changes have been linked to oxidative damage–induced repression of affected promoters. For instance, DNA damage–induced epigenetic silencing of promoters of genes required for cognitive function may play a role in cognitive decline associated with aging [64]. Such DNA damage–induced silencing is usually reversed upon faithful repair of the lesion; however, in a fraction of cells the lesion remains unrepaired resulting in permanent silencing of the locus [92].

5.2 Physiological Consequences

Further, in cells that divide actively, unrepaired DNA damage or telomere dysfunction triggers DNA-damage signaling and activation of DNA-damage checkpoints. Depending on the severity of the damage, cells can enter a terminal cell–cycle arrest or undergo apoptosis, both of which can interfere with tissue homeostasis and contribute to the functional decline.

While apoptosis contributes to the loss of functional cells, the consequences of cells entering senescence likely depend on the cells or cell types undergoing senescence. While senescence is a feature of the differentiated cells in most organs, senescence of the stem cells that are responsible for tissue renewal and repair is likely detrimental to the function of the tissue. In line with this, hematopoietic stem cells were shown to accumulate DNA damage with increasing age, which contributes to dysfunction of stem cells and the functional decline of the hematopoietic tissue [93]. However, while a decreasing functionality of stem cells with age is observed, it is still unclear whether this is due to stem cell aging or whether it is a result of the changing environment for stem cells within the aging tissue. A 2014 study supports an active role of stem cell aging in the age-dependent functional decline, by showing that adult quiescent muscle stem cells switch to irreversible senescence in muscles of geriatric mice [94].

The accumulation of senescent cells in tissues can result in altered microenvironment through altered secretion profiles [95], which can also disrupt tissue homeostasis and even stimulate proliferation of premalignant cells [96], or interfere with tissue regeneration [97]. Thus, the altered microenvironment within aging tissues may contribute both to the functional decline of the tissue as well as to the increased cancer incidence with increasing age.

On the other hand, immune cells are differentiated cells that rely on proliferation in order to perform their function. After several rounds of clonal selection, T-cells can become replicatively senescent in vitro, and T-cells with senescence-like features are also found in vivo [98]. This is thought to contribute to immunosenescence—the age-related deterioration of the immune system.

6. CONCLUSION

Genomic instability has long been considered to be a central driving force in aging, and extensive evidence has accumulated over the years supporting DNA-damage accumulation with age and its role in genomic instability. The findings that compromised genome-maintenance results in shortened life spans of model organisms or in human premature-aging syndromes suggest that genome maintenance is indeed a central antiaging mechanism.

However, aging of an organism is a very complex process, during which different tissues gradually functionally deteriorate. Since each tissue has its own characteristics and challenges, it is likely that their functional deterioration also follows individual patterns. Vijg and Dolle have put forward a model, according to which aging is not a clonal phenomenon, but rather arises from increasing heterogeneity of the cells in a tissue [99]. At the level of cells, accumulation of random mutations can contribute to this heterogeneity. At the level of tissues, different challenges to genomic integrity may promote aging. For instance, neural cells, which have high respiratory requirements, may be more susceptible to the accumulation of oxidative damage and cell death, whereas T cells that rely on continued proliferation may be more susceptible to senescence caused by telomere shortening.

If genomic instability is considered as a mechanism driving this age-associated mosaicism by contributing accumulation of mutations and promoting downstream outcomes such as altered gene expression, cell-cycle arrest, or cell death (Fig. 29.2), a better mechanistic understanding of what triggers age-related deterioration of genome-maintenance mechanisms or changes in nuclear architecture may provide valuable insights into the role of genomic instability in aging. In addition, a better understanding of interactions between cells and tissues in the aging organism will help in determining a hierarchy of age-related changes.

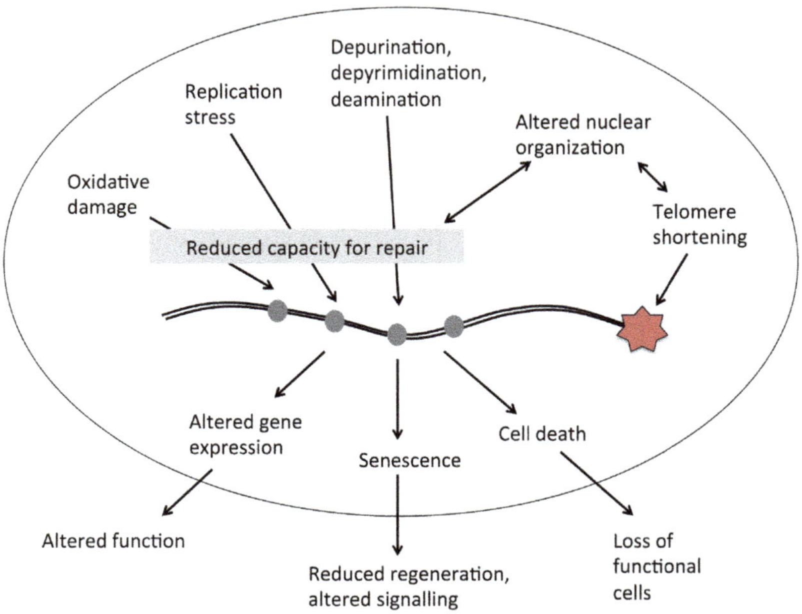

FIGURE 29.2 **Mechanisms contributing to genomic instability and their role during aging.** *Gray bubbles* represent DNA lesions, *arrows* indicate contributions of processes to the given outcomes, and *two-directional arrows* indicate interplay between two processes.

GLOSSARY

Cellular senescence Terminal cell–cycle arrest of normal diploid cells.
Chromosomal abnormalities Missing, additional, or irregular DNA sequence within chromosomes, including aberrations in chromosome number.
G-quadruplex Four-stranded structure in DNA or RNA formed through hydrogen bonds between four guanines.
Haploinsufficiency The presence of only one functional copy of a gene is not sufficient to produce the wild-type phenotype.
Isochromosome Chromosome with two identical arms, either two short arms or two long arms.
Large chromosomal rearrangements Large structural changes to chromosomes including duplications, deletions, inversions, and translocations.
Microsatellites Short sequence of tandem repeats of a 2–5 base pair–long motif.
Minisatellites Repetitive sequence consisting of repeats of a 10–60 base pair–long motif.
Replication origin licensing Assembly of required factors that allows replication origin to start DNA replication.
Replication stress Slowing or stalling of replication forks caused by several conditions including unrepaired DNA damage.
T-loop Structure formed at the telomere ends by looping back of single-stranded overhangs and annealing with double-stranded telomeric sequence.

LIST OF ABBREVIATIONS

AD Alzheimer's disease
ALS Amyotrophic lateral sclerosis
AT Ataxia telangiectasia
ATM Ataxia telangiectasia mutated
BER Base excision repair
BS Bloom syndrome
CS Cockayne syndrome
DKC Dyskeratosis congenita
DSB Double-strand break
HGPS Hutchinson–Gilford progeria syndrome
HR Homologous recombination
MMR Mismatch repair
nDNA Nuclear DNA
NER Nucleotide excision repair
NHEJ Nonhomologous end joining
PD Parkinson's disease
rDNA Ribosomal DNA
ROS Reactive oxygen species
RTS Rothmund–Thomson syndrome

SAHF Senescence-associated heterochromatin foci
SSB Single-strand break
TTD Trichothiodystrophy
WS Werner syndrome
XP Xeroderma pigmentosum

REFERENCES

[1] Lopez-Otin C, Blasco MA, Partridge L, Serrano M, Kroemer G. The hallmarks of aging. Cell 2013;153(6):1194–217.

[2] Lee HC, Wei YH. Mitochondria and aging. Adv Exp Med Biol 2012;942:311–27.

[3] Failla G. The aging process and cancerogenesis. Ann N Y Acad Sci 1958;71(6):1124–40.

[4] Szilard L. On the nature of the aging process. Proc Natl Acad Sci USA 1959;45(1):30–45.

[5] Moskalev AA, Shaposhnikov MV, Plyusnina EN, Zhavoronkov A, Budovsky A, Yanai H, et al. The role of DNA damage and repair in aging through the prism of Koch-like criteria. Ageing Res Rev 2013;12(2):661–84.

[6] Albertini RJ. HPRT mutations in humans: biomarkers for mechanistic studies. Mutat Res 2001;489(1):1–16.

[7] Gossen JA, Vijg J. A selective system for lacZ- phage using a galactose-sensitive *E. coli* host. Biotechniques 1993;14(3):326, 330.

[8] Dolle ME, Snyder WK, Gossen JA, Lohman PH, Vijg J. Distinct spectra of somatic mutations accumulated with age in mouse heart and small intestine. Proc Natl Acad Sci USA 2000;97(15):8403–8.

[9] Jacobs PA, Court Brown WM, Doll R. Distribution of human chromosome counts in relation to age. Nature 1961;191:1178–80.

[10] Rehen SK, Yung YC, McCreight MP, Kaushal D, Yang AH, Almeida BS, et al. Constitutional aneuploidy in the normal human brain. J Neurosci 2005;25(9):2176–80.

[11] Harman D. Aging: a theory based on free radical and radiation chemistry. J Gerontol 1956;11(3):298–300.

[12] Shibutani S, Takeshita M, Grollman AP. Insertion of specific bases during DNA synthesis past the oxidation-damaged base 8-oxodG. Nature 1991;349(6308):431–4.

[13] Loft S, Poulsen HE. Cancer risk and oxidative DNA damage in man. J Mol Med (Berl) 1996;74(6):297–312.

[14] Chevanne M, Calia C, Zampieri M, Cecchinelli B, Caldini R, Monti D, et al. Oxidative DNA damage repair and parp 1 and parp 2 expression in Epstein-Barr virus-immortalized B lymphocyte cells from young subjects, old subjects, and centenarians. Rejuvenation Res 2007;10(2):191–204.

[15] Lindahl T, Nyberg B. Rate of depurination of native deoxyribonucleic acid. Biochemistry 1972;11(19):3610–8.

[16] Cooper DN, Youssoufian H. The CpG dinucleotide and human genetic disease. Hum Genet 1988;78(2):151–5.

[17] Singhal RP, Mays-Hoopes LL, Eichhorn GL. DNA methylation in aging of mice. Mech Ageing Dev 1987;41(3):199–210.

[18] Hendrich B, Hardeland U, Ng HH, Jiricny J, Bird A. The thymine glycosylase MBD4 can bind to the product of deamination at methylated CpG sites. Nature 1999;401(6750):301–4.

[19] Kunkel TA. Evolving views of DNA replication (in)fidelity. Cold Spring Harb Symp Quant Biol 2009;74:91–101.

[20] Streisinger G, Okada Y, Emrich J, Newton J, Tsugita A, Terzaghi E, et al. Frameshift mutations and the genetic code. This paper is dedicated to Professor Theodosius Dobzhansky on the occasion of his 66th birthday. Cold Spring Harb Symp Quant Biol 1966;31:77–84.

[21] Silber JR, Fry M, Martin GM, Loeb LA. Fidelity of DNA polymerases isolated from regenerating liver chromatin of aging *Mus musculus*. J Biol Chem 1985;260(2):1304–10.

[22] Burhans WC, Weinberger M. DNA replication stress, genome instability and aging. Nucleic Acids Res 2007;35(22):7545–56.

[23] Sogo JM, Lopes M, Foiani M. Fork reversal and ssDNA accumulation at stalled replication forks owing to checkpoint defects. Science 2002;297(5581):599–602.

[24] Aguilera A, Garcia-Muse T. Causes of genome instability. Annu Rev Genet 2013;47:1–32.

[25] Pruitt SC, Bailey KJ, Freeland A. Reduced Mcm2 expression results in severe stem/progenitor cell deficiency and cancer. Stem Cells 2007;25(12):3121–32.

[26] Bennett RJ, Keck JL. Structure and function of RecQ DNA helicases. Crit Rev Biochem Mol Biol 2004;39(2):79–97.

[27] Goto M. Hierarchical deterioration of body systems in Werner's syndrome: implications for normal ageing. Mech Ageing Dev 1997;98(3):239–54.

[28] Croteau DL, Popuri V, Opresko PL, Bohr VA. Human RecQ helicases in DNA repair, recombination, and replication. Annu Rev Biochem 2014;83:519–52.

[29] Ariyoshi K, Suzuki K, Goto M, Watanabe M, Kodama S. Increased chromosome instability and accumulation of DNA double-strand breaks in Werner syndrome cells. J Radiat Res 2007;48(3):219–31.

[30] German J, Archibald R, Bloom D. Chromosomal breakage in a rare and probably genetically determined syndrome of man. Science 1965;148(3669):506–7.

[31] Kitao S, Shimamoto A, Goto M, Miller RW, Smithson WA, Lindor NM, et al. Mutations in RECQL4 cause a subset of cases of Rothmund–Thomson syndrome. Nat Genet 1999;22(1):82–4.

[32] Lindor NM, Devries EM, Michels VV, Schad CR, Jalal SM, Donovan KM, et al. Rothmund–Thomson syndrome in siblings: evidence for acquired in vivo mosaicism. Clin Genet 1996;49(3):124–9.

[33] Chen T, Dong B, Lu Z, Tian B, Zhang J, Zhou J, et al. A functional single nucleotide polymorphism in promoter of ATM is associated with longevity. Mech Ageing Dev 2010;131(10):636–40.

[34] Polosak J, Roszkowska-Gancarz M, Kurylowicz A, Owczarz M, Dobosz P, Mossakowska M, et al. Decreased expression and the Lys751Gln polymorphism of the XPD gene are associated with extreme longevity. Biogerontology 2010;11(3):287–97.

[35] Krichevsky S, Pawelec G, Gural A, Effros RB, Globerson A, Yehuda DB, et al. Age related microsatellite instability in T cells from healthy individuals. Exp Gerontol 2004;39(4):507–15.

[36] Neri S, Gardini A, Facchini A, Olivieri F, Franceschi C, Ravaglia G, et al. Mismatch repair system and aging: microsatellite instability in peripheral blood cells from differently aged participants. J Gerontol A Biol Sci Med Sci 2005;60(3):285–92.

[37] Annett K, Duggan O, Freeburn R, Hyland P, Pawelec G, Barnett Y. An investigation of DNA mismatch repair capacity under normal culture conditions and under conditions of supra-physiological challenge in human CD4+T cell clones from donors of different ages. Exp Gerontol 2005;40(12):976–81.

[38] Fishel R, Lescoe MK, Rao MR, Copeland NG, Jenkins NA, Garber J, et al. The human mutator gene homolog MSH2 and its association with hereditary nonpolyposis colon cancer. Cell 1993;75(5):1027–38.

[39] Cabelof DC, Raffoul JJ, Yanamadala S, Ganir C, Guo Z, Heydari AR. Attenuation of DNA polymerase beta-dependent base excision repair and increased DMS-induced mutagenicity in aged mice. Mutat Res 2002;500(1–2):135–45.

[40] Cabelof DC, Ikeno Y, Nyska A, Busuttil RA, Anyangwe N, Vijg J, et al. Haploinsufficiency in DNA polymerase beta increases cancer risk with age and alters mortality rate. Cancer Res 2006;66(15):7460–5.

[41] Maclean MJ, Aamodt R, Harris N, Alseth I, Seeberg E, Bjoras M, et al. Base excision repair activities required for yeast to attain a full chronological life span. Aging Cell 2003;2(2):93–104.

[42] Weissman L, Jo DG, Sorensen MM, de Souza-Pinto NC, Markesbery WR, Mattson MP, et al. Defective DNA base excision repair in brain from individuals with Alzheimer's disease and amnestic mild cognitive impairment. Nucleic Acids Res 2007;35(16):5545–55.

[43] Olkowski ZL. Mutant AP endonuclease in patients with amyotrophic lateral sclerosis. Neuroreport 1998;9(2):239–42.

[44] Moriwaki S, Ray S, Tarone RE, Kraemer KH, Grossman L. The effect of donor age on the processing of UV-damaged DNA by cultured human cells: reduced DNA repair capacity and increased DNA mutability. Mutat Res 1996;364(2):117–23.

[45] Goukassian D, Gad F, Yaar M, Eller MS, Nehal US, Gilchrest BA. Mechanisms and implications of the age-associated decrease in DNA repair capacity. FASEB J 2000;14(10):1325–34.

[46] Lehmann AR. DNA repair-deficient diseases, xeroderma pigmentosum, Cockayne syndrome and trichothiodystrophy. Biochimie 2003;85(11):1101–11.

[47] Ljungman M, Zhang F. Blockage of RNA polymerase as a possible trigger for UV light-induced apoptosis. Oncogene 1996;13(4):823–31.

[48] Kraemer KH, Lee MM, Scotto J. Xeroderma pigmentosum. Cutaneous, ocular, and neurologic abnormalities in 830 published cases. Arch Dermatol 1987;123(2):241–50.

[49] Siebert R, Puchta H. Efficient repair of genomic double-strand breaks by homologous recombination between directly repeated sequences in the plant genome. Plant Cell 2002;14(5):1121–31.

[50] Andegeko Y, Moyal L, Mittelman L, Tsarfaty I, Shiloh Y, Rotman G. Nuclear retention of ATM at sites of DNA double strand breaks. J Biol Chem 2001;276(41):38224–30.

[51] Di Domenico EG, Romano E, Del Porto P, Ascenzioni F. Multifunctional role of ATM/Tel1 kinase in genome stability: from the DNA damage response to telomere maintenance. Biomed Res Int 2014;2014:787404.

[52] Seluanov A, Mittelman D, Pereira-Smith OM, Wilson JH, Gorbunova V. DNA end joining becomes less efficient and more error-prone during cellular senescence. Proc Natl Acad Sci USA 2004;101(20):7624–9.

[53] Seluanov A, Danek J, Hause N, Gorbunova V. Changes in the level and distribution of Ku proteins during cellular senescence. DNA Repair (Amst) 2007;6(12):1740–8.

[54] Li H, Vogel H, Holcomb VB, Gu Y, Hasty P. Deletion of Ku70, Ku80, or both causes early aging without substantially increased cancer. Mol Cell Biol 2007;27(23):8205–14.

[55] Espejel S, Martin M, Klatt P, Martin-Caballero J, Flores JM, Blasco MA. Shorter telomeres, accelerated ageing and increased lymphoma in DNA-PKcs-deficient mice. EMBO Rep 2004;5(5):503–9.

[56] Flores M, Morales L, Gonzaga-Jauregui C, Dominguez-Vidana R, Zepeda C, Yanez O, et al. Recurrent DNA inversion rearrangements in the human genome. Proc Natl Acad Sci USA 2007;104(15):6099–106.

[57] Wiktor-Brown DM, Olipitz W, Hendricks CA, Rugo RE, Engelward BP. Tissue-specific differences in the accumulation of sequence rearrangements with age. DNA Repair (Amst) 2008;7(5):694–703.

[58] De Sandre-Giovannoli A, Bernard R, Cau P, Navarro C, Amiel J, Boccaccio I, et al. Lamin a truncation in Hutchinson–Gilford progeria. Science 2003;300(5628):2055.

[59] Gonzalo S, Kreienkamp R. DNA repair defects and genome instability in Hutchinson–Gilford Progeria Syndrome. Curr Opin Cell Biol 2015;34:75–83.

[60] Scaffidi P, Misteli T. Lamin A-dependent nuclear defects in human aging. Science 2006;312(5776):1059–63.

[61] Narita M, Nunez S, Heard E, Lin AW, Hearn SA, Spector DL, et al. Rb-mediated heterochromatin formation and silencing of E2F target genes during cellular senescence. Cell 2003;113(6):703–16.

[62] Sadaie M, Salama R, Carroll T, Tomimatsu K, Chandra T, Young AR, et al. Redistribution of the Lamin B1 genomic binding profile affects rearrangement of heterochromatic domains and SAHF formation during senescence. Genes Dev 2013;27(16):1800–8.

[63] Rogakou EP, Boon C, Redon C, Bonner WM. Megabase chromatin domains involved in DNA double-strand breaks in vivo. J Cell Biol 1999;146(5):905–16.

[64] Lu T, Pan Y, Kao SY, Li C, Kohane I, Chan J, et al. Gene regulation and DNA damage in the ageing human brain. Nature 2004;429(6994):883–91.

[65] Vanneste E, Voet T, Le Caignec C, Ampe M, Konings P, Melotte C, et al. Chromosome instability is common in human cleavage-stage embryos. Nat Med 2009;15(5):577–83.

[66] Lebedev IN, Ostroverkhova NV, Nikitina TV, Sukhanova NN, Nazarenko SA. Features of chromosomal abnormalities in spontaneous abortion cell culture failures detected by interphase FISH analysis. Eur J Hum Genet 2004;12(7):513–20.

[67] Rehen SK, McConnell MJ, Kaushal D, Kingsbury MA, Yang AH, Chun J. Chromosomal variation in neurons of the developing and adult mammalian nervous system. Proc Natl Acad Sci USA 2001;98(23):13361–6.

[68] Piotrowski A, Bruder CE, Andersson R, Diaz de Stahl T, Menzel U, Sandgren J, et al. Somatic mosaicism for copy number variation in differentiated human tissues. Hum Mutat 2008;29(9):1118–24.

[69] Sfeir A, de Lange T. Removal of shelterin reveals the telomere end-protection problem. Science 2012;336(6081):593–7.

[70] Herbig U, Jobling WA, Chen BP, Chen DJ, Sedivy JM. Telomere shortening triggers senescence of human cells through a pathway involving ATM, p53, and p21(CIP1), but not p16(INK4a). Mol Cell 2004;14(4):501–13.

[71] Edwards DN, Machwe A, Wang Z, Orren DK. Intramolecular telomeric G-quadruplexes dramatically inhibit DNA synthesis by replicative and translesion polymerases, revealing their potential to lead to genetic change. PLoS One 2014;9(1):e80664.

[72] Mitchell JR, Wood E, Collins K. A telomerase component is defective in the human disease dyskeratosis congenita. Nature 1999;402(6761):551–5.

[73] Chen W, Kimura M, Kim S, Cao X, Srinivasan SR, Berenson GS, et al. Longitudinal versus cross-sectional evaluations of leukocyte telomere length dynamics: age-dependent telomere shortening is the rule. J Gerontol A Biol Sci Med Sci 2011;66(3):312–9.

[74] Willeit P, Willeit J, Brandstatter A, Ehrlenbach S, Mayr A, Gasperi A, et al. Cellular aging reflected by leukocyte telomere length predicts advanced atherosclerosis and cardiovascular disease risk. Arterioscler Thromb Vasc Biol 2010;30(8):1649–56.

[75] Lee HW, Blasco MA, Gottlieb GJ, Horner 2nd JW, Greider CW, DePinho RA. Essential role of mouse telomerase in highly proliferative organs. Nature 1998;392(6676):569–74.

[76] Little RD, Platt TH, Schildkraut CL. Initiation and termination of DNA replication in human rRNA genes. Mol Cell Biol 1993;13(10):6600–13.

[77] Ganley AR, Kobayashi T. Ribosomal DNA and cellular senescence: new evidence supporting the connection between rDNA and aging. FEMS Yeast Res 2014;14(1):49–59.

[78] Kobayashi T. A new role of the rDNA and nucleolus in the nucleus–rDNA instability maintains genome integrity. Bioessays 2008;30(3):267–72.

[79] Johnson R, Strehler BL. Loss of genes coding for ribosomal RNA in ageing brain cells. Nature 1972;240(5381):412–4.

[80] Kunisada T, Yamagishi H, Ogita Z, Kirakawa T, Mitsui Y. Appearance of extrachromosomal circular DNAs during in vivo and in vitro ageing of mammalian cells. Mech Ageing Dev 1985;29(1):89–99.

[81] Killen MW, Stults DM, Adachi N, Hanakahi L, Pierce AJ. Loss of Bloom syndrome protein destabilizes human gene cluster architecture. Hum Mol Genet 2009;18(18):3417–28.

[82] Baillie JK, Barnett MW, Upton KR, Gerhardt DJ, Richmond TA, De Sapio F, et al. Somatic retrotransposition alters the genetic landscape of the human brain. Nature 2011;479(7374):534–7.

[83] De Cecco M, Criscione SW, Peckham EJ, Hillenmeyer S, Hamm EA, Manivannan J, et al. Genomes of replicatively senescent cells undergo global epigenetic changes leading to gene silencing and activation of transposable elements. Aging Cell 2013;12(2):247–56.

[84] Gasior SL, Wakeman TP, Xu B, Deininger PL. The human LINE-1 retrotransposon creates DNA double-strand breaks. J Mol Biol 2006;357(5):1383–93.

[85] Fitch DH, Bailey WJ, Tagle DA, Goodman M, Sieu L, Slightom JL. Duplication of the gamma-globin gene mediated by L1 long interspersed repetitive elements in an early ancestor of simian primates. Proc Natl Acad Sci USA 1991;88(16):7396–400.

[86] Coolbaugh-Murphy MI, Xu J, Ramagli LS, Brown BW, Siciliano MJ. Microsatellite instability (MSI) increases with age in normal somatic cells. Mech Ageing Dev 2005;126(10):1051–9.

[87] LeDoux SP, Wilson GL, Beecham EJ, Stevnsner T, Wassermann K, Bohr VA. Repair of mitochondrial DNA after various types of DNA damage in Chinese hamster ovary cells. Carcinogenesis 1992;13(11):1967–73.

[88] de Souza-Pinto NC, Mason PA, Hashiguchi K, Weissman L, Tian J, Guay D, et al. Novel DNA mismatch-repair activity involving YB-1 in human mitochondria. DNA Repair (Amst) 2009;8(6):704–19.

[89] Imam SZ, Karahalil B, Hogue BA, Souza-Pinto NC, Bohr VA. Mitochondrial and nuclear DNA-repair capacity of various brain regions in mouse is altered in an age-dependent manner. Neurobiol Aging 2006;27(8):1129–36.

[90] Kujoth GC, Bradshaw PC, Haroon S, Prolla TA. The role of mitochondrial DNA mutations in mammalian aging. PLoS Genet 2007;3(2):e24.

[91] Ono T, Isobe K, Nakada K, Hayashi JI. Human cells are protected from mitochondrial dysfunction by complementation of DNA products in fused mitochondria. Nat Genet 2001;28(3):272–5.

[92] O'Hagan HM, Mohammad HP, Baylin SB. Double strand breaks can initiate gene silencing and SIRT1-dependent onset of DNA methylation in an exogenous promoter CpG island. PLoS Genet 2008;4(8):e1000155.

[93] Rossi DJ, Bryder D, Seita J, Nussenzweig A, Hoeijmakers J, Weissman IL. Deficiencies in DNA damage repair limit the function of haematopoietic stem cells with age. Nature 2007;447(7145):725–9.

[94] Sousa-Victor P, Gutarra S, Garcia-Prat L, Rodriguez-Ubreva J, Ortet L, Ruiz-Bonilla V, et al. Geriatric muscle stem cells switch reversible quiescence into senescence. Nature 2014;506(7488):316–21.

[95] Coppe JP, Patil CK, Rodier F, Sun Y, Munoz DP, Goldstein J, et al. Senescence-associated secretory phenotypes reveal cell-nonautonomous functions of oncogenic RAS and the p53 tumor suppressor. PLoS Biol 2008;6(12):2853–68.

[96] Krtolica A, Parrinello S, Lockett S, Desprez PY, Campisi J. Senescent fibroblasts promote epithelial cell growth and tumorigenesis: a link between cancer and aging. Proc Natl Acad Sci USA 2001;98(21):12072–7.

[97] Conboy IM, Conboy MJ, Smythe GM, Rando TA. Notch-mediated restoration of regenerative potential to aged muscle. Science 2003;302(5650):1575–7.

[98] Effros RB. From Hayflick to Walford: the role of T cell replicative senescence in human aging. Exp Gerontol 2004;39(6):885–90.

[99] Vijg J, Dolle ME. Large genome rearrangements as a primary cause of aging. Mech Ageing Dev 2002;123(8):907–15.

[100] Hoeijmakers JH. Genome maintenance mechanisms for preventing cancer. Nature 2001;411(6835):366–74.

Chapter 30

Nucleolar Contributions to DNA-Damage Response and Genomic (In)Stability in the Nervous System

M. Hetman
University of Louisville, Louisville, KY, United States

Chapter Outline

1. INTRODUCTION

While cancer is recognized as a genomic instability disease, neurodegenerative disorders may also involve disturbed maintenance of the genome. For instance, DNA damage is extensively documented in various neurodegenerative conditions including Alzheimer's disease [1]. Moreover, mutations in various components of DNA-repair machinery produce neurodegenerative phenotypes in animals and humans [2,3]. In addition, despite being postmitotic cells, neurons are highly sensitive to various DNA-damaging agents [4]. Lastly, genomic instability may produce disease-causing mutations of specific loci triggering neurodegeneration [5]. Except these latter cases, role of DNA damage/genomic instability in neurodegeneration is not clear. In particular, neurotoxic mechanisms of DNA damage that are relevant for degeneration of mature neurons are poorly understood.

In proliferating cells, the nucleolus is a prominent subnuclear structure whose best recognized role is ribosomal biogenesis [6]. The nucleolus is not limited by a membrane but, instead, is formed around transcribed genes for nucleolar rRNA (47S rRNA in humans, rDNA). The rDNA is organized as clusters of repeated genes which in a human haploid genome include about 300 copies in 5 clusters that are located on acrocentric chromosomes. RNA-polymerase-1 (Pol1) transcribes rDNA initiating ribosomal biogenesis. The nucleolus is also a site of further steps of this critical process including rRNA processing and assembly of ribosomes. Inhibition of Pol1 disrupts ribosomal biogenesis and nucleolar localization of many proteins [7]. Such a nucleolar stress also leads to activation of the stress-activated transcription factor p53 [8]. Hence, the nucleolus may serve as a sensor for such stressors that inhibit Pol1. In addition, nucleolar accumulation of various components of ribonucleoprotein biogenesis machinery underlies its involvement in such nonribosomal roles as spliceosome assembly or generation of the signal recognition particle (SRP) [9]. By regulated sequestration of various important

Genome Stability. http://dx.doi.org/10.1016/B978-0-12-803309-8.00030-6

regulatory proteins of the cell cycle and/or differentiation programs, nucleolus affects these processes [9]. Finally, the nucleolus appears to play a structural role in genome organization [10–12].

In this review, the literature is discussed that suggests a role for the nucleolus in DNA-damage response (DDR) and genome maintenance in the brain. In the first part, nucleolar involvement in sensing DNA damage is presented. In the second part, the case for the nucleolus as a site of neurodegeneration-associated genomic instability is laid out.

2. NUCLEOLUS AS A SENSOR OF NEURONAL DNA DAMAGE

2.1 Effects of DNA Damage on the Nucleolus

The stress-sensing function of the nucleolus relies on the requirement of Pol1 activity for the maintenance of the nucleolar structure [8]. The Pol1-dependent nucleolar compartmentalization of various proteins provides a number of potential stress mediators which can be released upon Pol1 inhibition [8]. Importantly, the stress-sensing function of the nucleolus is able to operate in a wide range of cells provided that they contain an active nucleolus. Of note, many types of mature neurons including those highly prone to neurodegeneration (motoneurons, dopaminergic neurons of the substantia nigra, cerebellar Purkinje neurons) contain prominent nucleoli. Hence, the nucleolus-based stress-sensing mechanism could remain active in many types of differentiated cells including postmitotic neurons in the adult brain.

Consistent with such a concept, the structure of the neuronal nucleolus is sensitive to various types of injuries including DNA-damaging anticancer drugs, ionizing irradiation, hypoxia, and oxidants [13–18]. In many of these instances, it is likely that nucleolar disruption is a consequence of DNA lesions that directly interfere with rDNA transcription [4]. Alternatively, stress-activated signaling cascades including hydrogen peroxide-activated JNK2 or DNA double-strand break (DSB)-activated ATM may inhibit Pol1-driven transcription to induce the nucleolar stress [19–21].

2.2 Nucleolar Stress-Mediated Responses to Neuronal DNA Damage

A DNA damage–sensing role of Pol1 inhibition and the subsequent nucleolar stress have been proposed in developing neurons that are challenged with the DNA-damaging anticancer drug camptothecin (CPT) [22]. Camptothecin inhibits DNA topoisomerase-1 (Topo1) and induces DNA single-strand breaks (SSBs) as well as DNA-Topo1 adducts [23]. In cultured rat cortical neurons that were isolated from newborn animals, CPT blocked Pol1 activity and triggered nucleoplasmic translocation of the nucleolar marker protein nucleophosmin/B23 (Npm) [22]. Nmp translocation is a convenient marker of Pol1 inhibition in many cell types including proliferating cancer cell lines, cultured primary neurons and mature neurons in whole animal brain [17,24,25]. The CPT-induced nucleolar stress was followed by activation of the DNA damage–regulated pro-apoptotic transcription factor p53 and the p53-dependent apoptosis. Importantly, nucleoplasmic Npm translocation that indicates nucleolar stress also occurred if p53 and/or apoptosis were blocked [22]. Inhibition of Pol1 by knockdown of the specific Pol1 coactivator TIF1A was sufficient to induce p53-dependent neuronal apoptosis. Such a response required protein synthesis suggesting that the p53-mediated transcription of protein-coding genes was involved. This conclusion is also supported by observations that neuronal apoptosis was induced by the transcriptional inhibitor actinomycin D if it were applied at low concentrations which were relatively selective against Pol1 than the mRNA-transcribing RNA-polymerase-2 (Pol2). Of note, conditional knockout of TIF1A that was limited to the neuroprogenitor cells of the developing mouse brain was also followed by p53-dependent apoptosis [26]. Therefore, in immature neurons that are challenged with DNA damage that blocks Pol1-driven transcription, the p53-dependent apoptosis seems to be a direct response to the nucleolar stress.

One should emphasize the apparent specificity of the nucleolar stress response to only certain types of DNA damage. For instance, in cultured rat cortical neurons etoposide that blocks DNA toposisomerase-2 (Topo2) and produces SSBs, DSBs, and, Topo2-DNA adducts triggers nucleolar stress only at high concentrations that exceed 10 μM [25]. Conversely, the low concentration of 1 μM that is sufficient to induce strong DSB response does not induce nucleolar stress [25]. In addition, induction of DSBs within the 28S rRNA-coding region of rDNA using the *Physarum* endonuclease I–Ppo1 was insufficient to induce Pol1 inhibition and disrupt the nucleolus [25]. Such findings could be interpreted as evidence that in neurons, nucleolar stress is not a consequence of DSBs. Instead, SSBs and/or Topo–DNA adducts that accumulate in sufficient amounts after CPT and/or high concentration etoposide treatments are capable of triggering nucleolar stress, most likely by direct damage to rDNA and the subsequent interference with Pol1-driven transcription.

Interestingly, unlike neurons, proliferating cells were reported to show signs of nucleolar stress after DSB induction [20,21]. Such a response appeared to be caused by inhibition of Pol1 and required the DSB-activated DNA damage-signaling kinase ATM. Insensitivity of the neuronal nucleolus to DSBs may be part of the mechanism that underlies the well-established tolerance of such lesions by mature neurons [27]. Conversely, single-strand breaks and DNA adducts that interfere with Pol1 activity and trigger nucleolar stress appear to be highly neurotoxic (for review see Ref. [4]). For instance, a 2014 study has documented the role of unrepaired Topo–DNA adducts as a likely trigger for neurodegeneration that is a major phenotypic manifestation of ATM deficiency in humans [28]. Moreover, it has been proposed that such a phenotype is directly related to transcriptional inhibition. Likewise, accumulation of DNA adducts was proposed as a trigger for neurodegeneration that is associated with deficiencies in nucleotide excision repair including Cockayne syndrome or xeroderma pigmentosum [3]. Finally, SSBs and various forms of oxidative adducts are present in the degenerating brain [1]. Therefore, these transcription-interfering lesions may trigger a nucleolar stress response in brain cells.

2.3 DNA Damage–Induced Nucleolar Stress in Intact Brain

Some forms of the DDR are developmentally restricted. For instance, DNA damage–induced apoptosis occurs in neuroblasts and early postmitotic neurons but not in mature neurons [25,29]. However, no developmental restriction was observed for nucleolar stress as it was present in brain neurons of either neonate or adult rats that received intracerebroventricular- or intracarotid etoposide injections, respectively [25]. Therefore, the nucleolus may participate in sensing at least some forms of DNA damage in both developing and mature neurons. Such a role is supported by observations of dispersed NPM in dopaminergic neurons of substantia nigra of adult mice that received the pro-oxidant neurotoxin MPTP [17]. Likewise, in human postmortem samples from that region, NPM dispersion was associated with Parkinson's diseases in which oxidative DNA damage is also documented [17,30].

2.4 Mediators of the Nucleolar Stress Response

The p53-mediated neuronal apoptosis is among consequences of DNA damage–induced nucleolar stress [22] (Fig. 30.1). Hence, the key question is that of the identity of the nucleolar stress–specific activators of the p53 pathway that trigger this response. While there are no published reports on such mediators in neurons, there is abundant literature on the mechanisms linking nucleolar stress and the p53 pathway in proliferating cells. Most attention in this respect was focused on regulation of the p53 ubiquitin ligase MDM2 (as reviewed in Ref. [31]). Under basal conditions, this protein promotes degradation of p53 preventing its accumulation and the p53-dependent responses including apoptosis (Fig. 30.1). Nucleolar stress triggers nucleoplasmic release of several proteins that may bind to MDM2 inhibiting its negative effects on p53. As a result, p53 becomes stabilized and capable of activating apoptosis and/or cell cycle arrest. Various nucleolar proteins were demonstrated to bind and inhibit MDM2 including Npm, nucleolin, and several ribosomal proteins (RPs). RPs are highly abundant components of the ribosomes. However, they also occur in relatively smaller quantities as free proteins. Such a free pool of

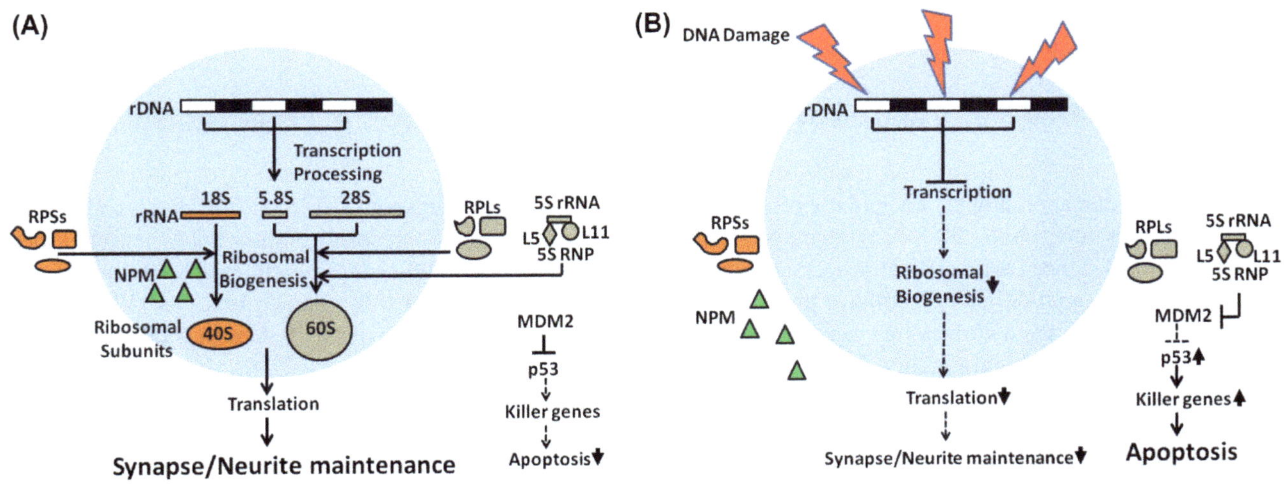

FIGURE 30.1 **A hypothetical model illustrating role of the nucleolar stress in neuronal DNA-damage response.** (A) Under normal conditions, nucleolus is maintained by ongoing transcription of the active copies of rDNA (white boxes) that initiates ribosomal biogenesis. Ribosomal proteins (RPSs and RPLs) as well as 5SrRNA are used for that process; under such conditions, MDM2 suppresses the p53 pathway including apoptosis. In addition, sufficient supply of ribosomes supports translation-mediated maintenance of neuronal structure and function. (B) Damage of rDNA inhibits its transcription disrupting ribosomal biogenesis. Ribosomal components including the 5S ribonucleoprotein (5S RNP) that is composed of 5S rRNA, RPL5, and RPL11 are no longer used for ribosome assembly. Thus, the 5S RNP inhibits MDM2 producing activation of p53. In immature neurons, that leads to the p53-mediated apoptosis. In the absence of apoptosis, chronic nucleolar stress may compromise neuronal translation and negatively affect structural as well as functional integrity of the neuron.

RPs is concentrated in the nucleolus and may be released when ribosomal biogenesis is blocked. It appears that L5 and L11 are two RPs playing a central role in p53 activation [31] (Fig. 30.1).

Together with 5S rRNA, these RPLs form the 5S ribonucleoprotein complex (5S RNP) that may inhibit MDM2. The MDM2 region that binds RPL11 has been identified and MDM2 knock-in mice were generated with a MDM2 point mutation (C305F) that selectively disrupts the RPL11 interaction [32]. In these mice, MDM2 ability to sense blockage of ribosomal biogenesis is impaired [32]. At least proliferating fibroblasts as well splenocytes and thymocytes from such animals did not appear to have any issues with activating p53 in response to various DNA-damaging agents including the Topo-1 inhibitor doxorubicin or SSB/DSB-inducing γ-irradiation. However, their p53 response to inhibition of ribosomal biogenesis was impaired. As neurons are postmitotic cells, their ability to engage the cell cycle–checkpoint-based sensing of DNA damage may be reduced. Therefore, it is tempting to speculate in these cells, relative contribution of the nucleolar stress to DNA damage–mediated activation of p53 may be greater than in fibroblasts. Indeed, preliminary observations suggest that L11 knockdown reduces neuronal apoptosis in response to not only TIF1A knockdown, but also to the Topo1 inhibitor CPT (M. Hetman and J. Hallgren, unpublished observations). Clearly, future studies are needed to determine which components of the nucleolus including specific RPs are involved in the neuronal responses to the DNA damage–induced nucleolar stress and whether p53 activation in such cells involves inhibition of MDM2.

Beyond p53, nucleolar stress may also engage other effector mechanisms to modulate neuronal survival. For instance, Npm has been shown to act as a chaperone for the pro-apoptotic protein Bax [33]. Moreover, in an experimental model of stroke, mitochondrial translocation of Npm has been proposed to increase Bax abundance/activity at this critical location for its death-promoting effects. Of note, in 2013, beneficial effects of a peptide that blocks Npm/Bax interactions was demonstrated using renal ischemia model [34]. PARP is another nucleolar protein which is released after nucleolar stress [35]. Also, in nonneuronal cells, PARP release from the nucleolus, sensitized cells to DNA damage [35]. It is tempting to speculate that a similar PARP-mediated sensitization may occur in neurons that are challenged with DNA damage that induces nucleolar stress. Lastly, Npm was shown to limit pro-excitotoxic activity of the GAPDH–SIAH1 complex in NMDA-treated cortical neurons [36]. Hence, one can consider a possibility that nucleolar stress may also promote neuronal survival.

In this context, one should note that the p53-dependent apoptosis is a consequence of nucleolar stress that is restricted to developing neurons [22,25,26]. Moreover, adult mouse neurons with conditional deletion of TIF1A survive for months despite absence of Pol1 activity, inhibition of mTOR signaling, mitochondrial impairment, and presence of oxidative stress [17,26,37]. Likewise, no apoptosis was found after transient disruption of neuronal nucleoli in adult rats that were treated with etoposide [25]. Hence, nucleolar stress may also activate a survival program that allows mature neurons to cope with unfavorable conditions. Finally, under hypoxia and/or oxidative damage, a transient inhibition of Pol1 may support survival by conserving energy [38]. Taken together, nucleolar stress may mediate neuronal responses to DNA damage that are not limited to cell death. In particular, one can consider a possibility that the nucleolar stress helps neurons to survive DNA damage.

2.5 Ribosomal Deficiency and Neurodegeneration as Consequences of Persistent Nucleolar Stress

While the DNA damage–induced nucleolar stress may activate stress-signaling response, it is also tempting to consider a possibility that if unrepaired, DNA damage may chronically impair ribosomal biogenesis compromising neuronal translation (Fig. 30.1). Since many critical processes including synaptic plasticity and synapse maintenance require translation, chronic ribosomal deficits could have profound negative effects on neuronal structure and function [39,40]. Thus, an interesting possibility emerges that neurodegeneration-associated accumulation of DNA damage induces neuronal atrophy and impairs neuronal function by inducing chronic deficiency of ribosomal biogenesis (Fig. 30.1). Indeed, conditional deletions of TIF1A in various populations of mature neurons produced chronic neurodegeneration [17,26,37]. Moreover, dendritic atrophy was observed after inhibition of ribosomal biogenesis in cultured hippocampal neurons with established dendritic trees [40a]. In that case, dendritic degeneration was accompanied by appearance of RNA stress granules which mark translational inhibition and are often found in various neurodegenerative diseases including AD [41]. Likewise, a causative connection has been proposed between oxidative DNA damage, nucleolar disruption, ribosomal deficits, and neurodegeneration that have been documented in cerebellar Purkinje neurons of the *pcd*-mutant mice [18].

Interestingly, reduced nucleolar size indicative of Pol1 inhibition has been reported in neurons from brain regions that are affected by AD [42,43]. Conversely, in PD, reduced nucleolar size and nucleolar stress have been observed in the *substantia nigra* [17,44].

Importantly, analysis of postmortem samples of AD brain revealed reduced numbers of ribosomes and/or extensive oxidation of rRNA in the AD-affected regions including in hippocampus and the parietal cortex [45,46]. Such deficits were associated with decreased protein synthesis [45,46]. Oxidation of rRNA as well as ribosomal defects have been found in mild cognitive impairment (MCI) which often represents an early stage AD [46]. AD-associated hypermethylation of the rDNA promoter was also observed [47]. Such epigenetic silencing of the nucleolar rRNA genes was most pronounced at the early stages of AD.

However, causative relationship between DNA damage, nucleolar deficits, and neurodegeneration remains to be established. Although, currently, such a relationship is not certain, one can consider an interesting possibility that rescue of nucleolar defects can improve structural integrity and function of the brain regions that are challenged with neurodegenerative pathologies such as AD. Indeed, nucleolar hypertrophy has been found in cortical and hippocampal neurons from asymptomatic patients with AD pathology [43].

3. NEURODEGENERATION-ASSOCIATED INSTABILITY OF rDNA

Various consequences of aging including proliferative senescence, cancer, and neurodegeneration have been proposed to be caused by accumulation of DNA damage and subsequent genomic instability [27,48]. For instance, loss of telomeres as well as de-regulated homologous recombination (HR) have been associated with aging-related disorders [48,49]. In AD, telomere shortening that in proliferating cells triggers DDR and cell cycle arrest may have complex effects including neuroprotective reduction of inflammation but also increased neuronal dysfunction [50]. The nucleolar rDNA is another site of genomic instability whose consequences started to emerge only recently. In addition, instability of rDNA has also been documented in human neurodegeneration [47,51]. Here, potential mechanisms and consequences of rDNA instability in the degenerating brain will be presented.

3.1 Consequences of rDNA Instability in Nonneuronal Systems

In early 1970s, it has been proposed that in nondividing mammalian cells including neurons, aging is associated with reduced number of rDNA copies [52]. It has been further hypothesized that such a reduction may underlie aging-associated decline in protein synthesis by reducing ribosomal biogenesis due to insufficient number of rRNA genes. However, other studies revealed no aging-associated changes in rDNA copy number in mouse or human brain [53,54]. Moreover, in chickens, fruit flies, and yeast, it was demonstrated that relatively few rRNA genes are fully sufficient to meet all the ribosomal biogenesis needs including periods of rapid proliferative growth which requires high rates of ribosomal generation [55–57].

In 1990s, the concept of rDNA instability as a component of aging got new support from work in a yeast model of replicative senescence. In that model, a correlation has been established between excessive rDNA recombination and the senescence-associated inability by the mother cell to produce offspring [58]. Moreover, accumulation of extrachromosomal rDNA circles (ERCs) rather than deficient ribosomal biogenesis has been proposed as a toxic mechanism underlying such a phenotype [58].

Subsequent studies confirmed that instability of rDNA contributes to yeast replicative aging (for review see Ref. [12]). However, such an effect appeared to be a direct consequence of rDNA loss rather than the secondary accumulation of ERCs [59]. While deficient ribosomal biogenesis prolongs replicative life span of yeast [60], a hypothesis has been proposed that loss of rDNA induces aging not by impairment of ribosome generation but by activation of the DDR [61]. According to this hypothesis, DDR activation would be a consequence of insufficient buffering capacity of DDR mediators that is provided by rDNA and would be similar to the DDR activation following loss of telomeres. In addition, structural role of rDNA in stabilization of yeast genome has been also proposed. Such an effect could be mediated by promoting chromatid cohesion to enable recombination-mediated amplification countering rDNA loss. Thus, DDR but not ribosomal insufficiency would lead to growth arrest and cell senescence. In support of such a possibility, yeast strains with decreasing content of rDNA were shown to have increased sensitivity to DNA damage [57]. Interestingly, such sensitivity was dependent on increased transcription of the few remaining rDNA genes as the high rate of transcription interfered with rDNA repair.

These and other yeast studies prompted a proposal of the rDNA hypothesis of aging [12]. This hypothesis claims that due to its high transcriptional activity, rDNA is hypersensitive to aging-associated DNA damage which in turn triggers its instability due to activation of DNA repair. Then, the rDNA instability becomes a direct cause for the aging-associated DDR and the replicative senescence.

Fruit fly is another experimental system in which consequences of rDNA instability have been directly investigated. By stimulating rDNA recombination with the rDNA-specific endonuclease I-CreI, a series of fly strains was generated with a defined content of rDNA [56]. While ribosomal supply defects were obvious when rDNA copy number was below the minimally required threshold for appropriate levels of ribosomal biogenesis, additional nonribosomal effects were also identified in those strains with adequate number of ribosomes. For instance, changes in rDNA copy number affected the general content of heterochromatin [10]. Specifically, high or low rDNA content was associated with high or low levels of heterochromatin markers, respectively. Likewise, the strength of heterochromatin-mediated regulation of gene expression was directly correlated with the amount of rDNA.

Effects of variation in rDNA copy number were also observed in the euchromatin [62]. For instance, relatively small but widespread changes of gene expression were observed in fly strains with different rDNA content. Similar effects were observed not only in the engineered mutants of rDNA, but also strains with natural variation in rDNA copy number. The strongest influence of rDNA copy number was noted on genes involved in mitochondrial function and lipid metabolism.

Finally, there are some correlative data that suggest role of rDNA in heterochromatin maintenance in mammals. Thus, in a mouse cell line, loss of heterochromatin coincided with loss of rDNA [63]. Such effects were observed after perturbing rDNA silencing by knocking down the critical component of the nucleolar repressive complex (NoRC), Tip5/Baz2b [63]. Interestingly, loss of constitutive heterochromatin including that at centromere and telomere regions was also observed in human cell lines with knock down of NoRC [11]. Growth arrest, defects in chromosome segregation, and genomic instability were among outcomes of such a deficiency. In addition, in human cancer cell line HeLa, analysis of nucleolus-associated chromatin revealed enrichment for repetitive elements as well as gene families with monoallelic expression limited to highly specialized cells including immunoglobulin receptors, T-cell antigen receptors, and odorant receptors [64]. Genes that are expressed during brief periods of embryonic development were also overrepresented suggesting that nucleolar association characterizes silent chromatin. Indeed, this nucleolus-associated domain (NAD) was enriched in repressive histone marks suggesting its heterochromatic nature [64]. Of note, differences have been observed between NAD and another well-established heterochromatin region, the lamina-associated domain (LAD). For instance, NAD also contained transcriptionally active RNA genes that are transcribed by RNA-Polymerase-3 (Pol3) including tRNA and 5S RNA. Hence, in mammals, rDNA may be of particular importance for maintenance of specific domain of heterochromatin that combines silenced and active genes. It remains to be determined whether such role of rDNA is by providing a nearby source of NoRC activity for the maintenance of this domain, or by supplying noncoding RNAs that are required for heterochromatin formation or by structural effects including interactions with cohesins.

3.2 Mechanisms of rDNA Instability in Nonneuronal Systems

In yeast, instability of rDNA is mediated by HR [65]. These mechanisms operate during DNA replication and involve unequal sister chromatid exchange leading to rDNA loss or rDNA expansion. The rDNA loss and expansion appear to balance each other in nonsenescent yeast cells [61].

The key determinants of rDNA instability include DNA sequences in the noncoding portion of the yeast rDNA unit (the intergenic spacer, IGS) (for review see Refs. [12,61]). First, the replication fork block (RFB) site binds the nuclease Fob1 that induces a recombinogenic DSB to stimulate HR. Second, RNA-Polymerase-2-mediated transcription from the E-Pro promoter displaces cohesins. It has been proposed that such events lead to unequal sister chromatid recombination leading to expansion of rDNA. In addition, the noncoding IGS transcript appears to promote unequal sister chromatid recombination by forming the recombinogenic RNA:DNA hybrid (R loop) which may collide with the replication fork [66]. Stability of rDNA is, therefore, regulated by the activity of another IGS site with the replication origin activity [12]. Its strength is closely correlated with rDNA stability and lifespan of yeast strains. In addition, Sir2 stabilizes rDNA by silencing the E-Pro-driven transcription. Likewise, the yeast ataxin-2 Pbp-1 prevents R-loop formation stabilizing rDNA [66]. Thus, loss of Fob1 or Sir2 and Pbp1 extends or shortens yeast life span, respectively [12,66]. In addition, deletions of HR components counteract increased DNA damage sensitivity in yeast strains with low copy number of rDNA copies [57]. Noncoding antisense transcription of rRNA has been also shown to be a source unequal sister chromatid recombination within rDNA [67]. Such a recombinogenic activity was inhibited by Dicer [67]. Finally, rDNA stabilizing role of the yeast chromatin structure is illustrated by instability of rDNA after disruption of its tethering to the nuclear membrane [68].

Critical role of HR control within rDNA is further illustrated by compartmentalization of HR machinery. Thus, its critical component, Rad52 is generally excluded from the nucleolus where the rDNA is located [69]. Rad52 interacts with rDNA DSBs only after a transient exit of the affected rDNA region from the nucleolus. Such a compartmentalization is at least in part due to sumoylation of Rad52. Conversely, anti-recombinogenic regulators including DNA helicase Srs2 have been found in proximity of rDNA [69].

Mechanisms of rDNA instability in higher eukaryotes are not clear. However, RFB activity was demonstrated in human rDNA IGS [70]. As in yeast, loss of HR control appears to destabilize human rDNA as its remarkable instability has been reported in cell lines from patients with the defective human DNA helicase BLM [71]. Like Srs2, BLM inhibits HR.

Moderate instability of human rDNA that was present in ATM-deficient cells may be also a result of increased HR as unrepaired DSBs may provide an additional stimulus for that process [71]. Finally, a study conducted in 2015 using targeted DSBs inside rDNA revealed similar HR-based repair compartmentalization as that observed in yeast [21]. Therefore, HR-based instability of mammalian rDNA appears to be suppressed by anti-recombinogenic activity, efficient DSB repair, and restricted access of HR to transcriptionally active regions of the nucleolus. Last but not least, epigenetic silencing and the resulting changes in chromatin structure appear as important factors promoting stability of mammalian rDNA.

Interestingly, nutrient availability affects rDNA stability of the Y-chromosome-linked rDNA cluster of fruit flies [72]. Thus, diet enriched in yeast extract has been shown to induce rDNA instability resulting in rDNA loss. Such changes were persistent and occurred both in germ line and in somatic cells. Interestingly, mTOR was required for this effect. Increased insulin receptor signaling mimicked effect of enriched diet on rDNA stability. Lastly, inhibition of rDNA transcription with actinomycin D improved rDNA stability. Such observations suggest that HR-based destabilization of rDNA is promoted when high nutrient supply increases Pol1 activity. In addition, it has been proposed that rDNA stabilization may at least partially contribute to antiaging effects of caloric restriction.

3.3 Evidence of rDNA Instability in the Brain

Early evidence that suggested rDNA loss in the brains of aged dogs, mice, and people has not been confirmed by follow-up studies [53,54]. Moreover, in 2014 a study using quantitative real-time PCR methodology revealed no apparent loss of rDNA in human parietal cortex [51]. Therefore, unlike in yeast, brain rDNA seems to be stable during normal aging. There are, however, some limitations to this conclusion. First, the number of individuals that were included in this study was relatively low including 14 young and 9 old individuals. While such sample size allows excluding big effects of aging on rDNA content, smaller changes may require greater number of cases to reach statistical significance. Indeed, far larger group ($n=120$) was investigated to demonstrate modest age-associated declines in rDNA content of human adipose tissue [73]. In addition, age-associated rDNA loss may be limited to only some cell types such as neurons. In such a case, tissue homogenate-based assay would miss and/or underestimate rDNA loss as in mature human brain neurons are outnumbered by glia. Future studies using DNA from selectively dissected neurons and glia could unequivocally solve the issue of aging-associated rDNA instability in these cells.

However, the qPCR-based analysis of total genomic DNA from cerebrocortical tissue revealed instability of rDNA in two different forms of age-associated neurodegeneration. AD is the most common form of dementia [74,75]. Alzheimer's disease (AD) is associated with excessive production of amyloid-β leading to tau pathology, synapse loss, neuronal atrophy, and subsequent neuronal death. Such changes occur throughout the forebrain including the hippocampus and the cortex. AD pathology appears to develop first in the temporal lobe, then in the parieto-occipital cortex and finally in other cortical areas including the prefrontal cortex. In postmortem cerebrocortical samples from AD patients, increased genomic content of the rDNA 18S-coding region was observed in both the parietal and the prefrontal cortex [47]. In addition, similar expansion of rDNA was also found in parietal cortex samples from patients with mild cognitive impairment (MCI), which, in most cases, represents early stages of AD. The relative increase of rDNA ranged between 1.5- and 2.4-fold of age-matched controls as compared to the genomic content of the *tRNA-K* (CTT anticodon) gene whose 17 copies are dispersed throughout several human chromosomes (http://gtrnadb.ucsc.edu/).

To determine whether such a change may be unique for AD, rDNA content was analyzed in dementia with Lewy bodies (DLB) [51]. DLB is the second most common type of dementia after AD [76] (http://www.omim.org/entry/127750). It is associated with intraneuronal accumulation of α-synuclein-containing inclusions, the Lewy bodies. In DLB, Lewy bodies accumulate in the cerebral cortex including parietal cortex. Progressive neurodegeneration appears to be a consequence of such a cortical synucleinopathy as α- or β-synuclein mutations have been associated with rare familial cases of DLB. As the causative relationship between the synuclein pathology and neurodegeneration is also present in Parkinson's disease (PD), DLB and PD may represent the same neurodegenerative process that affects different regions of the brain. Interestingly, PD and DLB may also coincide with pathology found in both the nigral and cortical neurons. Therefore, although DLB affects similar brain areas as AD, the underlying pathologies of these two forms of dementia are different.

In postmortem parietal cortex samples of DLB patients, increased genomic content of rDNA was observed using three different rDNA amplicons that probed 18S-, 5.8S-, or 28S-coding regions of rDNA [51]. Importantly, similar results were obtained when a multiplied- or a single locus gene was used as a total genomic DNA normalizer (*tRNA-K*^{ctt} or *ALB*, respectively). The magnitude of rDNA amplification was estimated between 1.5 and 2.3-fold of age-matched controls. Surprisingly, when the DLB pathology-free cerebellum was analyzed, nearly 50% reduction of rDNA content was found in DLB patients using 5.8S and 28S amplicons. Therefore, similarly to AD, DLB appears to be associated with rDNA instability.

As in AD, rDNA expanded in the pathology-affected cerebral cortex. In contrast, rDNA loss was found in the pathology-free cerebellum. These findings suggest that at least in DLB, rDNA becomes unstable in neurodegeneration-affected and unaffected brain regions. In the degenerating areas, the rDNA expansions would be the predominant product of such a destabilization; in the nondegenerating areas, rDNA instability would result in rDNA loss. Similar divergence has been reported for another unstable region of the genome, the telomere. Telomere contraction has been found in the peripheral blood leukocytes in AD, and DLB [77–79]. However, at least in AD, telomere expansion has been observed in the hippocampus that is affected by AD pathology at the early stages of this disease [78].

Although somatic rDNA instability appears to be the most probable source of the observed differences between the DLB and the control group, one cannot formally exclude a possibility that in addition to rDNA instability, the observed effects of neurodegeneration on brain rDNA content are also affected by germ line-derived rDNA copy number polymorphisms that are associated with an increased risk of AD or DLB. Thus, expanding the sample sizes and including material from other nonbrain tissues could help to examine such a possibility. However, at least for DLB, the demonstrated divergence between the cerebral cortex and the cerebellum argues for somatic instability as a major source of the observed variation. Indeed, mitotic recombination of human rDNA has been previously reported in normal individuals [80]. Interestingly, mitotic instability of rDNA has been also found in at least 50% of human colon and lung cancer samples [81].

3.4 Potential Mechanisms and Significance of Neurodegeneration-Associated Instability of rDNA

Because HR has been identified as a major mechanism of rDNA recombination [61,65], findings of neurodegeneration-associated instability in this region suggest that HR becomes activated in the brain. HR is thought to occur during or after DNA replication peaking in the S phase of the cell cycle [82]. Hence, the observed rDNA instability may be localized to reactive glia that underwent divisions in response to neuronal loss [83]. At least in AD-affected brain regions, neuronal cell–cycle reentry has been also documented including DNA synthesis [84]. Therefore, neurons that reenter the cell cycle could also activate the HR.

As human rDNA is present in five clusters on five distinct chromosomes that all reside in close physical proximity within the nucleolus and the perinucleolar heterochromatin, rDNA may also undergo HR in the absence of the replication-generated sister chromatids. Such a "nonreplicative" recombination could engage homologous rDNA units within the same chromatid or from different chromosomes [49]. Years 2012 and 2013 reports have documented HR-like activity of nonmitotic human somatic cells in G_0 [85,86]. Because such activity required the recombining DNA to be transcribed, this noncanonical HR may be possible at the active rDNA units. In addition, in G_1 HeLa cells, rDNA DSBs can trigger activation of HR and unscheduled DNA synthesis in the perinucleolar region where such lesions are translocated [21]. Therefore, HR of rDNA may also occur in cells with unreplicated genomes including most neurons.

Although major components of the HR machinery are not expressed in normal mature mammalian brain (http://www.brain-map.org/), there are no data on their expression/activity in the degenerating brain. Moreover, the reported expansion of telomeres in the AD hippocampus suggests HR is active in that tissue as HR is presumed to be a major positive regulator of telomere length in most somatic cells in which telomerase is not expressed [78,82].

While rDNA instability is a direct result of rDNA damage, recombinogenic potential of this region may be regulated by changes in chromatin structure, Pol1 activity, and noncoding RNAs as discussed in Section 3.2. Thus, in the neurodegeneration-affected human brain tissue, altered epigenetic status of rDNA and/or changes to Pol1 activity and/or altered ncRNA expression/processing may co-operate with DNA damage to destabilize rDNA.

Loss of heterochromatin was reported in AD and proposed to drive neurodegeneration in a fruit fly tauopathy model [87]. However, rDNA methylation analysis revealed increased presence of this repressive chromatin mark in rDNA from AD brains and no changes in DLB [47,51]. Thus, it remains to be tested whether neurodegeneration-associated instability of rDNA is due to loss of perinucleolar heterochromatin. At least in AD, increased rDNA promoter methylation and ribosomal deficits could suggest transcriptional insufficiency of Pol1 as a possible trigger of instability of rDNA [45–47]. Conversely, nucleolar hypertrophy that was observed in some cells of AD brain suggests that in some cells Pol1 activity increases [43]. Such an increase could also promote rDNA instability. Finally, as RNA metabolism changes are well documented in various neurodegenerative diseases including AD, excessive R-loop formation in rDNA could contribute to rDNA instability in the degenerating brain [41].

Recombination of rDNA is initiated by DNA damage [65,82]. DSBs that are the major recombinogenic form of DNA damage have been reported in normal mouse forebrain neurons following periods of increased physiological neuronal activity [88]. Moreover, DSB repair appeared to be impaired in a mouse model of AD-like amyloidosis. In addition, bulky DNA adducts and single-strand DNA gaps that are generated by the adduct removal via the nucleotide excision repair may also activate the HR pathway [82,89]. Therefore, in the degenerating brain, instability of rDNA may be initiated by excessive DNA damage (Fig. 30.2).

What may be possible consequences of rDNA destabilization in human brain? The reported changes in rDNA content are unlikely to affect ribosomal biogenesis as only a fraction of rDNA units is required for efficient ribosome supply even during accelerated growth in development [55,56]. Thus, one can consider the ribosome-unrelated effects of altered rDNA content as discussed in Section 3.1. For instance, rDNA amplification may increase structural support for cohesion- and/or NoRC-mediated stabilization of the genome (Fig. 30.2). Indeed, the recently suggested AD-associated loss of heterochromatin could promote defense mechanisms supporting its expansion (Fig. 30.2). Effects of rDNA content on euchromatin gene expression are another interesting possibility of rDNA interaction with the neurodegenerative pathology [62]. Enrichment of rDNA-regulated genes that affect mitochondrial function and lipid metabolism fit well with the AD- and DLB-associated dysfunction of the mitochondria and/or dysregulation of brain lipid homeostasis [90–93]. Last but not least, yeast studies led to a proposition that the inactive copies of rDNA sequester mediators of the cytotoxic DDR pathway [57,61]. Of note, in mammalian cells, stress-induced sequestration of various proteins has been demonstrated by the rRNA noncoding regions of rDNA [94]. Therefore, changes in rDNA copy number may affect cellular stress response including that to DNA damage.

A working model may be proposed that increasing content of rDNA would be beneficial for the brain cells by promoting heterochromatin formation, tighter regulation of euchromatic genes and stronger capacity to regulate the DDR. The opposite may be true if rDNA content declines. Therefore, increases of rDNA in the degenerating cerebral cortex may support cell survival. Such an effect could explain the apparent discrepancy between the cerebellar- and the cerebrocortical rDNA content in DLB. As in DLB, the cerebellum is not confronted with the pathology/cell loss, there is no selection factor against suboptimal genomic arrangements resulting in overall reduction of rDNA content. In the cortex, cell death would produce enrichment of most resistant cells with expanded rDNA. One could speculate that similar mechanisms may promote rDNA copy number variability in cancer cells helping them to adapt to changes in cellular environment including therapy resistance.

4. CONCLUDING REMARKS

The nucleolus-based process of ribosomal biogenesis emerges as an important participant in the neuronal DDR acting both as a sensor and a source of transducers for various DNA damage signals. In addition, changes of rDNA content my further modulate brain tissue response to stress including DNA damage and neurodegeneration. Thus, the nucleolus may exert multilayer influence on the nervous system that is coping with an injury. Such an influence may involve its most canonical function of making ribosomes and its core structure of rDNA. The key challenge for the future research is to directly evaluate contributions of the nucleolar stress pathway as well as rDNA instability to the nervous system maintenance in physiological and pathological conditions. Thus, identification of nucleolar stress mediators could set a stage for experiments to directly evaluate role of nucleolar stress in animal models of neurodegenerative diseases. Moreover, transgenic technologies could be harnessed to determine how changes in rDNA content determine outcome of neurodegeneration. As such studies address the fundamental question of the mechanisms that underlie age-associated neurodegenerative diseases, they deserve significant attention of the research community.

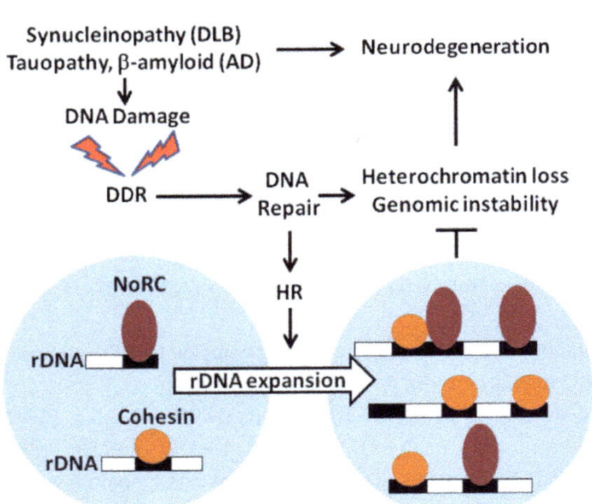

FIGURE 30.2 A hypothetical model presenting potential mechanisms and consequences of the neurodegeneration-associated rDNA expansion. The rDNA expansion that occurs in Alzheimer's diseases (AD) and in dementia with Lewy bodies (DLB) may be a consequence of increased DNA damage that activates the DNA-damage response (DDR) including rDNA repair via homologous recombination (HR). Additional copies of rDNA may provide stronger structural support for the chromatin. Such a support could include cohesion-mediated inter- and intrachromosomal interactions and/or increased supply of silencing complexes such as the NoRC. As in yeast, structural role of rDNA may be fulfilled by the inactive copies of rDNA (black). Thus, expanded rDNA would promote genomic stability and heterochromatin maintenance counteracting cytotoxic consequences of the neurodegenerative pathologies.

GLOSSARY

Extrachromosomal rDNA circles This is a form of extrachromosomal DNA that is observed in yeast. It is made of circular rDNA fragments that are generated by homologous recombination and which replicate during cell divisions.

Mild cognitive impairment (MCI) A brain function disorder involving cognitive impairments beyond those expected based on the age of the individual, but which are not severe enough to interfere with daily activities. It often represents a transitional stage between normal aging and dementia.

rDNA Ribosomal DNA, rRNA genes.

LIST OF ABBREVIATIONS

5S RNP 5S ribonucleoprotein
AD Alzheimer's disease
CPT Camptothecin
DDR DNA-damage response
DLB Dementia with Lewy bodies
DSB Double-strand break
HR Homologous recombination
mTOR Mammalian target of rapamycin
NoRC Nucleolar-repressive complex
Npm Nucleophosmin
PD Parkinson's disease
Pol1 RNA polymerase-1
Pol2 RNA polymerase-2
rDNA Ribosomal DNA
RP Ribosomal protein
SSB Single-strand break
TIF1A Transcription initiation factor-1A
Topo1 DNA topoisomerase-1
Topo2 DNA topoisomerase-2

ACKNOWLEDGMENTS

This work was supported by NIH (NS073584 and 8P30GM103507), NSF (IOS1021860), and the Commonwealth of Kentucky Challenge for Excellence Fund.

REFERENCES

[1] Markesbery WR, Lovell MA. DNA oxidation in Alzheimer's disease. Antioxid Redox Signal 2006;8(11–12):2039–45.
[2] Katyal S, McKinnon PJ. DNA repair deficiency and neurodegeneration. Cell Cycle 2007;6(19):2360–5. Epub 2007/08/19.
[3] Niedernhofer LJ. Nucleotide excision repair deficient mouse models and neurological disease. DNA Repair (Amst) 2008;7(7):1180–9. Epub 2008/02/15.
[4] Hetman M, Vashishta A, Rempala G. Neurotoxic mechanisms of DNA damage: focus on transcriptional inhibition. J Neurochem 2010;114(6):1537–49. Epub 2010/06/19.
[5] Budworth H, McMurray CT. A brief history of triplet repeat diseases. Methods Mol Biol 2013;1010:3–17. Epub 2013/06/12.
[6] Drygin D, Rice WG, Grummt I. The RNA polymerase I transcription machinery: an emerging target for the treatment of cancer. Annu Rev Pharmacol Toxicol 2010;50:131–56. Epub 2010/01/09.
[7] Andersen JS, Lam YW, Leung AK, Ong SE, Lyon CE, Lamond AI, et al. Nucleolar proteome dynamics. Nature 2005;433(7021):77–83.
[8] Boulon S, Westman BJ, Hutten S, Boisvert FM, Lamond AI. The nucleolus under stress. Mol Cell 2010;40(2):216–27. Epub 2010/10/23.
[9] Pederson T, Tsai RY. In search of nonribosomal nucleolar protein function and regulation. J Cell Biol 2009;184(6):771–6. Epub 2009/03/18.
[10] Paredes S, Maggert KA. Ribosomal DNA contributes to global chromatin regulation. Proc Natl Acad Sci USA 2009;106(42):17829–34. Epub 2009/10/12.
[11] Postepska-Igielska A, Krunic D, Schmitt N, Greulich-Bode KM, Boukamp P, Grummt I. The chromatin remodelling complex NoRC safeguards genome stability by heterochromatin formation at telomeres and centromeres. EMBO Rep 2013;14(8):704–10. Epub 2013/06/26.
[12] Ganley AR, Kobayashi T. Ribosomal DNA and cellular senescence: new evidence supporting the connection between rDNA and aging. FEMS Yeast Res 2014;14(1):49–59. Epub 2014/01/01.
[13] Tomiwa K, Nolan C, Cavanagh JB. The effects of cisplatin on rat spinal ganglia: a study by light and electron microscopy and by morphometry. Acta Neuropathol Berl 1986;69(3–4):295–308.

[14] Bigotte L, Olsson Y. Degeneration of trigeminal ganglion neurons caused by retrograde axonal transport of doxorubicin. Neurology 1987;37(6):985–92.
[15] Jamieson SM, Liu J, Connor B, McKeage MJ. Oxaliplatin causes selective atrophy of a subpopulation of dorsal root ganglion neurons without inducing cell loss. Cancer Chemother Pharmacol 2005;56(4):391–9. Epub 2005/05/11.
[16] Casafont I, Palanca A, Lafarga V, Berciano MT, Lafarga M. Effect of ionizing radiation in sensory ganglion neurons: organization and dynamics of nuclear compartments of DNA damage/repair and their relationship with transcription and cell cycle. Acta Neuropathol 2011;122(4):481–93. Epub 2011/09/15.
[17] Rieker C, Engblom D, Kreiner G, Domanskyi A, Schober A, Stotz S, et al. Nucleolar disruption in dopaminergic neurons leads to oxidative damage and Parkinsonism through repression of mammalian target of rapamycin signaling. J Neurosci 2011;31(2):453–60. Epub 2011/01/14.
[18] Baltanas FC, Casafont I, Weruaga E, Alonso JR, Berciano MT, Lafarga M. Nucleolar disruption and cajal body disassembly are nuclear hallmarks of DNA damage-induced neurodegeneration in Purkinje cells. Brain Pathol 2011;21(4):374–88. Epub 2010/11/09.
[19] Mayer C, Bierhoff H, Grummt I. The nucleolus as a stress sensor: JNK2 inactivates the transcription factor TIF-IA and down-regulates rRNA synthesis. Gen Dev 2005;19(8):933–41. Epub 2005/04/05.
[20] Kruhlak M, Crouch EE, Orlov M, Montano C, Gorski SA, Nussenzweig A, et al. The ATM repair pathway inhibits RNA polymerase I transcription in response to chromosome breaks. Nature 2007;447(7145):730–4.
[21] van Sluis M, McStay B. A localized nucleolar DNA damage response facilitates recruitment of the homology-directed repair machinery independent of cell cycle stage. Gen Dev 2015;29(11):1151–63. Epub 2015/05/29.
[22] Kalita K, Makonchuk D, Gomes C, Zheng JJ, Hetman M. Inhibition of nucleolar transcription as a trigger for neuronal apoptosis. J Neurochem 2008;105(6):2286–99. Epub 2008/03/05.
[23] Covey JM, Jaxel C, Kohn KW, Pommier Y. Protein-linked DNA strand breaks induced in mammalian cells by camptothecin, an inhibitor of topoisomerase I. Cancer Res 1989;49(18):5016–22. Epub 1989/09/15.
[24] Rubbi CP, Milner J. Disruption of the nucleolus mediates stabilization of p53 in response to DNA damage and other stresses. EMBO J 2003;22(22):6068–77.
[25] Pietrzak M, Smith SC, Geralds JT, Hagg T, Gomes C, Hetman M. Nucleolar disruption and apoptosis are distinct neuronal responses to etoposide-induced DNA damage. J Neurochem 2011;117(6):1033–46. Epub 2011/04/27.
[26] Parlato R, Kreiner G, Erdmann G, Rieker C, Stotz S, Savenkova E, et al. Activation of an endogenous suicide response after perturbation of rRNA synthesis leads to neurodegeneration in mice. J Neurosci 2008;28(48):12759–64.
[27] Katyal S, McKinnon PJ. DNA strand breaks, neurodegeneration and aging in the brain. Mech Age Dev 2008;129(7–8):483–91. Epub 2008/05/06.
[28] Katyal S, Lee Y, Nitiss KC, Downing SM, Li Y, Shimada M, et al. Aberrant topoisomerase-1 DNA lesions are pathogenic in neurodegenerative genome instability syndromes. Nat Neurosci 2014;17(6):813–21. Epub 2014/05/06.
[29] Herzog KH, Chong MJ, Kapsetaki M, Morgan JI, McKinnon PJ. Requirement for Atm in ionizing radiation-induced cell death in the developing central nervous system. Science 1998;280(5366):1089–91.
[30] Alam ZI, Jenner A, Daniel SE, Lees AJ, Cairns N, Marsden CD, et al. Oxidative DNA damage in the parkinsonian brain: an apparent selective increase in 8-hydroxyguanine levels in substantia nigra. J Neurochem 1997;69(3):1196–203.
[31] Bursac S, Brdovcak MC, Donati G, Volarevic S. Activation of the tumor suppressor p53 upon impairment of ribosome biogenesis. Biochim Biophys Acta 2014;1842(6):817–30. Epub 2014/02/12.
[32] Macias E, Jin A, Deisenroth C, Bhat K, Mao H, Lindstrom MS, et al. An ARF-independent c-MYC-activated tumor suppression pathway mediated by ribosomal protein-Mdm2 Interaction. Cancer Cell 2010;18(3):231–43. Epub 2010/09/14.
[33] Kerr LE, Birse-Archbold JL, Short DM, McGregor AL, Heron I, Macdonald DC, et al. Nucleophosmin is a novel Bax chaperone that regulates apoptotic cell death. Oncogene 2007;26(18):2554–62. Epub 006 Oct 30.
[34] Wang Z, Gall JM, Bonegio R, Havasi A, Illanes K, Schwartz JH, et al. Nucleophosmin, a critical Bax cofactor in ischemia-induced cell death. Mol Cell Biol 2013;33(10):1916–24. Epub 2013/03/06.
[35] Rancourt A, Satoh MS. Delocalization of nucleolar poly(ADP-ribose) polymerase-1 to the nucleoplasm and its novel link to cellular sensitivity to DNA damage. DNA Repair (Amst) 2009;8(3):286–97. Epub 2009/01/16.
[36] Lee SB, Kim CK, Lee KH, Ahn JY. S-nitrosylation of B23/nucleophosmin by GAPDH protects cells from the SIAH1-GAPDH death cascade. J Cell Biol 2012;199(1):65–76. Epub 2012/10/03.
[37] Kreiner G, Bierhoff H, Armentano M, Rodriguez-Parkitna J, Sowodniok K, Naranjo JR, et al. A neuroprotective phase precedes striatal degeneration upon nucleolar stress. Cell Death Differ 2013;20(11):1455–64. Epub 2013/06/15.
[38] Murayama A, Ohmori K, Fujimura A, Minami H, Yasuzawa-Tanaka K, Kuroda T, et al. Epigenetic control of rDNA loci in response to intracellular energy status. Cell 2008;133(4):627–39.
[39] McCann CM, Nguyen QT, Santo Neto H, Lichtman JW. Rapid synapse elimination after postsynaptic protein synthesis inhibition in vivo. J Neurosci 2007;27(22):6064–7.
[40] Klann E, Sweatt JD. Altered protein synthesis is a trigger for long-term memory formation. Neurobiol Learn Mem 2008;89(3):247–59. Epub 2007 Oct 4.
[40a] Slomnicki LP, Pietrzak M, Vashishta A, Jones J, Lynch N, Elliot S, et al. Requirement of neuronal ribosome synthesis for growth and maintenance of the dendritic tree. J Biol Chem March 11, 2016;291(11):5721–39.
[41] Wolozin B. Physiological protein aggregation run amuck: stress granules and the genesis of neurodegenerative disease. Discov Med 2014;17(91):47–52. Epub 2014/01/15.
[42] Mann DM, Marcyniuk B, Yates PO, Neary D, Snowden JS. The progression of the pathological changes of Alzheimer's disease in frontal and temporal neocortex examined both at biopsy and at autopsy. Neuropathol Appl Neurobiol 1988;14(3):177–95.

[43] Iacono D, O'Brien R, Resnick SM, Zonderman AB, Pletnikova O, Rudow G, et al. Neuronal hypertrophy in asymptomatic Alzheimer disease. J Neuropathol Exp Neurol 2008;67(6):578–89. Epub 2008/06/04.

[44] Mann DM, Yates PO. Pathogenesis of Parkinson's disease. Arch Neurol 1982;39(9):545–9. Epub 1982/09/01.

[45] Honda K, Smith MA, Zhu X, Baus D, Merrick WC, Tartakoff AM, et al. Ribosomal RNA in Alzheimer disease is oxidized by bound redox-active iron. J Biol Chem 2005;280(22):20978–86. Epub 2005 Mar 14.

[46] Ding Q, Markesbery WR, Chen Q, Li F, Keller JN. Ribosome dysfunction is an early event in Alzheimer's disease. J Neurosci 2005;25(40):9171–5.

[47] Pietrzak M, Rempala G, Nelson PT, Zheng J-J, Hetman M. Epigenetic silencing of nucleolar rRNA genes in Alzheimer's disease. PLoS One 2011;6(7):e22585.

[48] Lombard DB, Chua KF, Mostoslavsky R, Franco S, Gostissa M, Alt FW. DNA repair, genome stability, and aging. Cell 2005;120(4):497–512. Epub 2005/03/01.

[49] Moynahan ME, Jasin M. Mitotic homologous recombination maintains genomic stability and suppresses tumorigenesis. Nat Rev Mol Cell Biol 2010;11(3):196–207. Epub 2010/02/24.

[50] Rolyan H, Scheffold A, Heinrich A, Begus-Nahrmann Y, Langkopf BH, Holter SM, et al. Telomere shortening reduces Alzheimer's disease amyloid pathology in mice. Brain 2011;134(Pt 7):2044–56. Epub 2011/06/16.

[51] Hallgren J, Pietrzak M, Rempala G, Nelson PT, Hetman M. Neurodegeneration-associated instability of ribosomal DNA. Biochim Biophys Acta 2014;1842(6):860–8. Epub 2014/01/07.

[52] Johnson R, Strehler BL. Loss of genes coding for ribosomal RNA in ageing brain cells. Nature 1972;240(5381):412–4. Epub 1972/12/15.

[53] Gaubatz J, Prashad N, Cutler RG. Ribosomal RNA gene dosage as a function of tissue and age for mouse and human. Biochim Biophys Acta 1976;418(3):358–75. Epub 1976/02/05.

[54] Peterson CR, Cryar JR, Gaubatz JW. Constancy of ribosomal RNA genes during aging of mouse heart cells and during serial passage of WI-38 cells. Arch Gerontol Geriatr 1984;3(2):115–25. Epub 1984/07/01.

[55] Delany ME, Muscarella DE, Bloom SE. Effects of rRNA gene copy number and nucleolar variation on early development: inhibition of gastrulation in rDNA-deficient chick embryos. J Hered 1994;85(3):211–7. Epub 1994/05/01.

[56] Paredes S, Maggert KA. Expression of I-CreI endonuclease generates deletions within the rDNA of Drosophila. Genetics 2009;181(4):1661–71. Epub 2009/01/28.

[57] Ide S, Miyazaki T, Maki H, Kobayashi T. Abundance of ribosomal RNA gene copies maintains genome integrity. Science 2010;327(5966):693–6. Epub 2010/02/06.

[58] Sinclair DA, Guarente L. Extrachromosomal rDNA circles–a cause of aging in yeast. Cell 1997;91(7):1033–42. Epub 1998/01/15.

[59] Ganley AR, Ide S, Saka K, Kobayashi T. The effect of replication initiation on gene amplification in the rDNA and its relationship to aging. Mol Cell 2009;35(5):683–93. Epub 2009/09/15.

[60] Steffen KK, McCormick MA, Pham KM, MacKay VL, Delaney JR, Murakami CJ, et al. Ribosome deficiency protects against ER stress in *Saccharomyces cerevisiae*. Genetics 2012;191(1):107–18. Epub 2012/03/02.

[61] Kobayashi T. How does genome instability affect lifespan?: roles of rDNA and telomeres. Genes Cells 2011;16(6):617–24. Epub 2011/05/25.

[62] Paredes S, Branco AT, Hartl DL, Maggert KA, Lemos B. Ribosomal DNA deletions modulate genome-wide gene expression: "rDNA-sensitive" genes and natural variation. PLoS Genet 2011;7(4):e1001376. Epub 2011/05/03.

[63] Guetg C, Lienemann P, Sirri V, Grummt I, Hernandez-Verdun D, Hottiger MO, et al. The NoRC complex mediates the heterochromatin formation and stability of silent rRNA genes and centromeric repeats. EMBO J 2010;29(13):2135–46. Epub 2010/02/20.

[64] Nemeth A, Conesa A, Santoyo-Lopez J, Medina I, Montaner D, Peterfia B, et al. Initial genomics of the human nucleolus. PLoS Genet 2010;6(3):e1000889. Epub 2010/04/03.

[65] Eckert-Boulet N, Lisby M. Regulation of rDNA stability by sumoylation. DNA Repair (Amst) 2009;8(4):507–16. Epub 2009/03/06.

[66] Salvi JS, Chan JN, Szafranski K, Liu TT, Wu JD, Olsen JB, et al. Roles for Pbp1 and caloric restriction in genome and lifespan maintenance via suppression of RNA-DNA hybrids. Dev Cell 2014;30(2):177–91. Epub 2014/07/30.

[67] Castel SE, Ren J, Bhattacharjee S, Chang AY, Sanchez M, Valbuena A, et al. Dicer promotes transcription termination at sites of replication stress to maintain genome stability. Cell 2014;159(3):572–83. Epub 2014/11/25.

[68] Mekhail K, Seebacher J, Gygi SP, Moazed D. Role for perinuclear chromosome tethering in maintenance of genome stability. Nature 2008;456(7222):667–70. Epub 2008/11/11.

[69] Torres-Rosell J, Sunjevaric I, De Piccoli G, Sacher M, Eckert-Boulet N, Reid R, et al. The Smc5-Smc6 complex and SUMO modification of Rad52 regulates recombinational repair at the ribosomal gene locus. Nat Cell Biol 2007;9(8):923–31. Epub 2007/07/24.

[70] Akamatsu Y, Kobayashi T. The human RNA polymerase I transcription terminator complex acts as a replication fork barrier that coordinates the progress of replication with rRNA transcription activity. Mol Cell Biol 2015;35(10):1871–81. Epub 2015/03/18.

[71] Killen MW, Stults DM, Adachi N, Hanakahi L, Pierce AJ. Loss of Bloom syndrome protein destabilizes human gene cluster architecture. Hum Mol Genet 2009;18(18):3417–28. Epub 2009/06/19.

[72] Aldrich JC, Maggert KA. Transgenerational inheritance of diet-induced genome rearrangements in Drosophila. PLoS Genet 2015;11(4):e1005148. Epub 2015/04/18.

[73] Zafiropoulos A, Tsentelierou E, Linardakis M, Kafatos A, Spandidos DA. Preferential loss of 5S and 28S rDNA genes in human adipose tissue during ageing. Int J Biochem Cell Biol 2005;37(2):409–15. Epub 2004/10/12.

[74] Serrano-Pozo A, Frosch MP, Masliah E, Hyman BT. Neuropathological alterations in Alzheimer disease. Cold Spring Harb Perspect Med 2011;1(1):a006189. Epub 2012/01/10.

[75] Huang Y, Mucke L. Alzheimer mechanisms and therapeutic strategies. Cell 2012;148(6):1204–22. Epub 2012/03/20.

[76] McKeith I, Mintzer J, Aarsland D, Burn D, Chiu H, Cohen-Mansfield J, et al. Dementia with Lewy bodies. Lancet Neurol 2004;3(1):19–28. Epub 2003/12/25.
[77] Cai Z, Yan LJ, Ratka A. Telomere shortening and Alzheimer's disease. Neuromolecular Med 2013;15(1):25–48. Epub 2012/11/20.
[78] Thomas P, O' Callaghan NJ, Fenech M. Telomere length in white blood cells, buccal cells and brain tissue and its variation with ageing and Alzheimer's disease. Mech Age Dev 2008;129(4):183–90. Epub 2008/02/05.
[79] Kume K, Kikukawa M, Hanyu H, Takata Y, Umahara T, Sakurai H, et al. Telomere length shortening in patients with dementia with Lewy bodies. Eur J Neurol 2012;19(6):905–10. Epub 2012/02/01.
[80] Stults DM, Killen MW, Pierce HH, Pierce AJ. Genomic architecture and inheritance of human ribosomal RNA gene clusters. Genome Res 2008;18(1):13–8. Epub 2007/11/21.
[81] Stults DM, Killen MW, Williamson EP, Hourigan JS, Vargas HD, Arnold SM, et al. Human rRNA gene clusters are recombinational hotspots in cancer. Cancer Res 2009;69(23):9096–104. Epub 2009/11/19.
[82] Krejci L, Altmannova V, Spirek M, Zhao X. Homologous recombination and its regulation. Nucleic Acids Res 2012;40(13):5795–818. Epub 2012/04/03.
[83] Robel S, Berninger B, Gotz M. The stem cell potential of glia: lessons from reactive gliosis. Nat Rev Neurosci 2011;12(2):88–104. Epub 2011/01/21.
[84] Yang Y, Mufson EJ, Herrup K. Neuronal cell death is preceded by cell cycle events at all stages of Alzheimer's disease. J Neurosci 2003;23(7): 2557–63. Epub 2003/04/10.
[85] Gandhi M, Evdokimova VN, Cuenco KT, Bakkenist CJ, Nikiforov YE. Homologous chromosomes move and rapidly initiate contact at the sites of double-strand breaks in genes in G(0)-phase human cells. Cell Cycle 2013;12(4):547–52. Epub 2013/02/02.
[86] Gandhi M, Evdokimova VN, Cuenco KT, Nikiforova MN, Kelly LM, Stringer JR, et al. Homologous chromosomes make contact at the sites of double-strand breaks in genes in somatic G0/G1-phase human cells. Proc Natl Acad Sci USA 2012;109(24):9454–9. Epub 2012/05/31.
[87] Frost B, Hemberg M, Lewis J, Feany MB. Tau promotes neurodegeneration through global chromatin relaxation. Nat Neurosci 2014;17(3):357–66. Epub 2014/01/28.
[88] Suberbielle E, Sanchez PE, Kravitz AV, Wang X, Ho K, Eilertson K, et al. Physiologic brain activity causes DNA double-strand breaks in neurons, with exacerbation by amyloid-beta. Nat Neurosci 2013;16(5):613–21. Epub 2013/03/26.
[89] Ma W, Westmoreland JW, Resnick MA. Homologous recombination rescues ssDNA gaps generated by nucleotide excision repair and reduced translesion DNA synthesis in yeast G2 cells. Proc Natl Acad Sci USA 2013;110(31):E2895–904. Epub 2013/07/17.
[90] Cutler RG, Kelly J, Storie K, Pedersen WA, Tammara A, Hatanpaa K, et al. Involvement of oxidative stress-induced abnormalities in ceramide and cholesterol metabolism in brain aging and Alzheimer's disease. Proc Natl Acad Sci USA 2004;101(7):2070–5. Epub 2004/02/19.
[91] Reddy PH, Beal MF. Are mitochondria critical in the pathogenesis of Alzheimer's disease? Brain Res 2005;49(3):618–32. Epub 2005/11/05.
[92] Bosco DA, Fowler DM, Zhang Q, Nieva J, Powers ET, Wentworth Jr P, et al. Elevated levels of oxidized cholesterol metabolites in Lewy body disease brains accelerate alpha-synuclein fibrilization. Nat Chem Biol 2006;2(5):249–53. Epub 2006/03/28.
[93] Navarro A, Boveris A, Bandez MJ, Sanchez-Pino MJ, Gomez C, Muntane G, et al. Human brain cortex: mitochondrial oxidative damage and adaptive response in Parkinson disease and in dementia with Lewy bodies. Free Radic Biol Med 2009;46(12):1574–80. Epub 2009/03/21.
[94] Audas TE, Jacob MD, Lee S. Immobilization of proteins in the nucleolus by ribosomal intergenic spacer noncoding RNA. Mol Cell 2012;45(2): 147–57. Epub 2012/01/31.

Section VII

Effect of Environment on Genome Stability

Chapter 31

Diet and Nutrition

L.R. Ferguson
University of Auckland, Auckland, New Zealand

Chapter Outline

1. INTRODUCTION

While much media publicity points to exogenous causes of cancer such as cigarette smoking or occupational hazards, the evidence may be somewhat weaker as to the role of diet and nutrition. The monographs of the International Agency for Research on Cancer (IARC) provide excellent summaries, provided by international working groups, definitively showing that tobacco exposure [1], certain office working environments, and various other occupational exposures are unequivocally associated with human cancer risk [2]. DNA damage can occur through unintentional exposure to genotoxic chemicals in the diet, which may induce oxidation, DNA alkylation, cross-linking, dimerization, and strand breaks. As such, repair of this DNA damage (or protection against its formation) is essential to preserving genome stability.

Dietary factors play a well-established role in increasing cancer risk, through enhancing genomic instability. Vromman et al. ranked food components and environmental contaminants of food in terms of their potential hazard [3]. They considered arsenic and lead to be of high concern, while cadmium, methylmercury, dioxins, polychlorinated biphenyls, and toxaphene were ranked as medium priority. Although posing some risk at high levels, polybrominated biphenyls, chlordane, heptachlor, dichlorodiphenyl-trichloroethane, hexachlorocyclohexane, polychlorophenols, and their salts were classed as lower priority. Many of the reported exposures to such compounds may be inadvertent, through environmental pollutants that are accumulated by plants and animals eaten by humans [4], through mycotoxin formation on badly stored foods [5], or exposure to various cooked food mutagens/carcinogens [6,7]. However, because we eat a mixed diet, definitive human proof of many of these effects may not be as easily obtained. Furthermore, it is unethical to continue exposure to a putative mutagen/carcinogen in the expectation of providing definitive evidence of human harm. Thus, molecular evidence of genomic instability as a biomarker of cancer risk or likely cancer protection may be both more desirable and more readily obtained [8].

Dietary factors also play an essential role in protection against genomic instability, and there is increasing evidence for this [9]. In the long term, exploiting such beneficial dietary items and encouraging their increase in the diet may be the most constructive approach to protecting against human genomic instability and cancer initiation and progression.

Genome Stability. http://dx.doi.org/10.1016/B978-0-12-803309-8.00031-8

2. DIETARY CAUSES OF GENOMIC INSTABILITY

2.1 Dietary Excess (Obesity)

Obesity results from excess weight accumulation, generally considered to be caused by an excess of caloric energy intake in comparison with the amounts used in metabolism or consumed through exercise (Fig. 31.1). Increased oxidative stress is distinctive of obesity [10]. This condition occurs where there is an excessive production of reactive oxygen species (ROS), in comparison with the level of natural antioxidants. Enhanced ROS production is associated with a mitochondrial dysfunction in these individuals [11], and may affect the regulation of DNA methylation [12]. Global hypomethylation in repetitive sequences of the genome provides an important mechanism by which cells develop genomic instability [13]. Hypomethylation may be especially important where it occurs on the promoter of oncogenes. Delgado-Cruzata and coworkers studied the effects of weight loss on global DNA methylation in Hispanic, African-American, and Afro-Caribbean breast cancer survivors [14]. They found that DNA methylation of long interspersed nucleotide element 1 (LINE-1) was statistically significantly elevated after the intervention. Conversely, excess weight accumulation is associated with lower DNA-methylation levels.

Various posttranslational modifications alter the function of histones [15]. For example, acetylation of the lysine residues at the N-terminus of histone proteins leads to a reduction in the affinity between histones and DNA, enabling the access of RNA polymerase and transcription factors to gene-promoter regions [16]. In general, transcription is enhanced by histone acetylation and repressed by histone deacetylation. Histone ubiquitination modifies DNA-repair capacity, leading to chromatin structures conducive to the assembly of nucleotide excision repair (NER) complexes on damaged DNA [17]. Histone phosphorylation is required for efficient DNA repair. The net impact of perturbation of epigenetic mechanisms contributes significantly to genomic instability. Obesity has been found to have significant (adverse) effects on the function of histones [18,19].

Mitochondrial (mt) DNA alterations lead to oxidative phosphorylation and the generation of adenosine triphosphate (ATP) and ROS. Not only somatic mtDNA mutations, but also changes in mtDNA copy number have been shown to lead to mitochondrial dysfunction and increased genomic instability [20,21]. Mitochondrial dysfunctions have often been related to obesity and/or lipid imbalances in the diet, since mitochondrial membranes are lipid based and their maintenance is essential to the effective functioning of the mitochondria [22,23].

2.2 Alcohol

High alcohol consumption has been associated with increased carcinogenicity of the upper gastrointestinal tract. The primary mechanism of this has been suggested as through the formation of acetaldehyde, a metabolite of ethanol which can form DNA adducts [24].

2.3 Red Meats

Red meats include beef, veal, pork, lamb, mutton, horse, or goat meat. High red meat intake has been related to an increased risk of cancer, and this appears at least partly to associate with cooking processes (Fig. 31.2). Heterocyclic amines (HCAs)

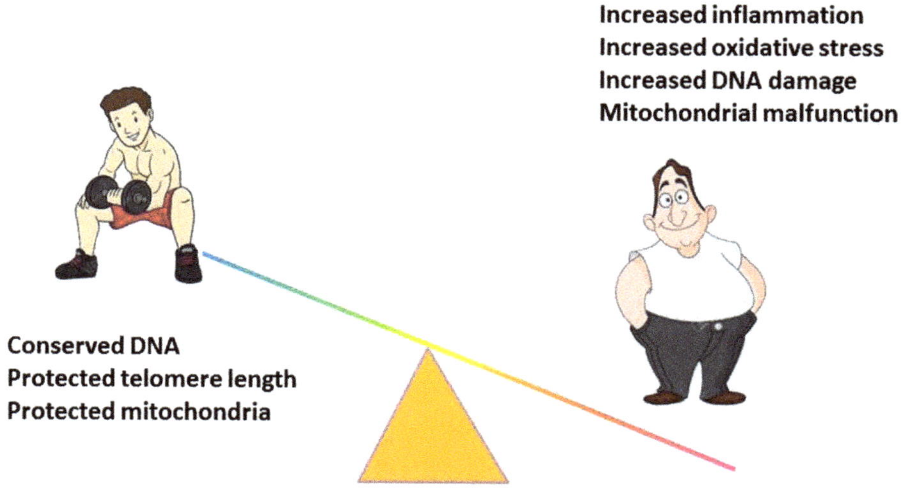

FIGURE 31.1 The significance of obesity in decreasing genomic stability, and of adequate diet and exercise in preventing this development.

formed by high-temperature cooking can generate reactive oxygen species. Carvalho and coworkers [25] correlated high temperature–cooking processes of such meats with cancer risk, and by measuring oxidative stress as malondialdehyde concentration in the plasma found suggestive evidence of a causal relationship in their Sao Paulo population. DNA-reactive polycyclic aromatic hydrocarbons (PAHs) are also formed during cooking of such meats, especially following direct heat exposure [6,26]. A 2015 IARC evaluation has concluded that well done cooked red meats enhance oxidative stress and other measures of genomic instability, at least in model systems, and are possible human carcinogens [27,28].

Advanced glycation end products such as N(E)-(carboxymethyl)lysine are present in both cooked and uncooked foods, leading to oxidative stress, aberrant cell signaling, and genomic instability, and have been associated with at least one type of cancer [29]. These reactive metabolites are produced as a byproduct of sugar metabolism [30], and appear to be related to various socioeconomic and risk factors linked to cancer susceptibility. While these are present at low levels in unprocessed red meats, they increase significantly upon cooking, but this process is reduced by previous marination of the meat [29,30].

2.4 Mutagens Formed During Food Processing

The 2015 IARC evaluation concluded that the evidence for carcinogenesis by processed meat was significantly stronger than for unprocessed meats, and the former should be considered as human carcinogens [27,28]. Processed meats are those which have been modified by salting, curing, fermentation, or other processes to enhance flavor or preservation. N-nitroso compounds in particular are DNA reactive, and are often formed during processing of red meats [31,32]. Smoking of salmon was shown to lead to the formation of various types of PAHs [33]. In a study from Taiyuan, China, PAHs were also shown to be formed during the cooking of vegetables, wheat flour, and fruits [8].

2.5 Mutagens Formed During Storage of Foods

Styrene has been widely used in food storage and also food preparation, as well as being released in various industrial settings. In workers exposed to this chemical, there is evidence of genotoxicity in the form of DNA adducts and strand breaks [34]. Styrene intake from various sources has also been associated with increased risk of invasive breast cancer in a population study in Texas [35]. Inappropriate storage containers of various food and drinks may themselves create a hazard. For example, an alcoholic beverage (cachaça) was found to be contaminated with PAHs when stored in a polyethylene tank, but this contamination was much less of a problem when storage was in a glass container [36]. Refrigeration has been found to be an important factor in food storage that helps to protect against the formation of various fungal toxins [37,38]. Various aflatoxins including aflatoxin B1 and aflatoxin M1, as well as ochratoxin A and fumonisin B1, are examples of important fungal secondary metabolites on badly stored nuts, grains, and other plant foods that cause DNA damage and promote genomic instability [37,39–41].

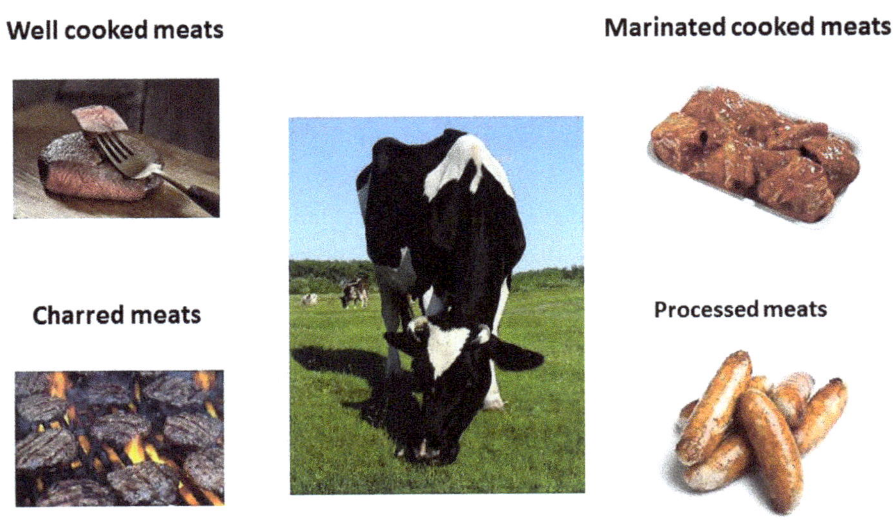

FIGURE 31.2 Some of the various ways in which red meats can enhance genomic instability. Heterocyclic amines are common in well-cooked meats, and polycyclic aromatic hydrocarbons in charred meats, both leading to DNA damage including the formation of DNA adducts. Advanced glycation end products are present in low concentrations in uncooked meats, but increase upon cooking, which process may be reduced by marinating the meats. Processing of meats to various products including sausages increases the concentration of mutagenic N-nitroso compounds.

FIGURE 31.3 The way in which certain food plants such as broccoli may function to either increase or reduce genomic instability. Glucosinolates such as 1-MIM glucosinate are innocuous unless activated by the release of myrosinase enzymes, usually contained in separate cells, when they release a DNA-reactive component. In contrast, various phytochemicals from broccoli, including sulforaphane, act to protect the integrity of DNA.

2.6 Accumulation of Environmental Pollutants in Animal Flesh

Various chemical toxins may accumulate in the flesh of grazing animals or fish in polluted estuarine regions [42,43]. These may include pesticides [3,44] and heavy metals such as mercury, cadmium, or lead [45,46]. The latter have been found at high levels in the flesh of slaughter-house animals, and have recognized genotoxic effects [43].

2.7 Natural Pesticides in Food Plants

While much publicity focuses around red meat and adverse effects associated with cooked or processed animal flesh, it is also important to record that many food plants contain natural pesticides, or are able to release such products under some conditions. An apparently enigmatic example is provided by broccoli (Fig. 31.3). While this contains a number of important and generally beneficial phytochemicals, as will be discussed in the next section, it is also able to release toxins in response to tissue damage caused by insect or other pests [47]. For example, 1-methoxy-3-indoylmethyl (1-MIM) glucosinate is found at high levels in cruciferous vegetables such as broccoli and cabbage. This forms DNA adducts in vitro, and is mutagenic following activation by the myrosinase enzyme [48]. It is noteworthy that these two plant components (1-MIM glucosinate and myrosinase) are typically found in separate cells. But after pest-induced cell damage, the two components can combine to form a DNA-reactive end product, in vitro and in vivo [48].

3. DIETARY PROTECTION AGAINST GENOMIC INSTABILITY

3.1 Classic Nutrients

Biomarkers relevant to genomic stability, including telomere length and mtDNA deletions, have been utilized in establishing recommended daily intakes for nutrients [49,50]. Accumulating evidence shows that genome integrity is highly sensitive to nutrient status, and that optimal levels may differ among individuals. Many investigations to date are limited by considering only the effects of single nutrients, without looking at the potential interactions among these, and of nutrients with toxicants in the diet. For example, Fenech has suggested that it is inappropriate to consider single nutrients, but we should be looking at nutrient combinations, using what he describes as a nutriome [51]. Nevertheless, it is clear that different nutrients and classes of nutrients have some important functional differences in the maintenance of genomic stability. Examples of some important nutrients and bioactives are given in Table 31.1.

3.1.1 Lipids

As described earlier, high intakes of saturated fats from animal products have been associated with obesity, and consequently detrimental effects on genomic stability [52]. Obesity and/or higher caloric intake also had a marked effect in promoting telomere shortening. While too high a fat intake overall may be detrimental, some fats may play an important

TABLE 31.1 Important Nutrients and Phytochemicals in the Maintenance of Genomic Stability

Processes Affected	Nutrients or Bioactive Substances
DNA oxidation	Vitamins (C, D, and E), Se, DHA, EPA, genistein, curcumin, RSV
DNA synthesis	Folate, vitamin B12, zinc, magnesium
DNA repair	Niacin, zinc, folate
DNA methylation	Vitamins A and D, folate
Other epigenetic effects	Vitamin D, RSV, EGCG, sulforaphane
Necrosis/apoptosis	Vitamins A, C, D, K12, niacin, zinc, DHA, EPA, curcumin
Chromosome segregation	Vitamin A, folate, magnesium
Telomere length	Vitamin D, niacin, folate, Se, DHA, EPA, curcumin, RSV

Data from references identified in the text.

protective role in the maintenance of genomic stability. Those for which the most information is available are the omega-3 and 6 (n-3 and n-6) polyunsaturated fatty acids (PUFAs). Within the n-3 PUFA family, the two long-chain PUFA eicosapentanoic acid (EPA) and docosahexaenoic acid (DHA) have been demonstrated to affect various key events that may protect against DNA oxidation and leukocyte telomere shortening, while promoting apoptosis and reduction of damaged cells [53–57]. These two PUFA also play key roles in protection against inflammation, reducing the possibility that an overreaction to immune stimulation may lead to DNA damage and genomic instability [22,58–61]. What may be almost as important as the individual PUFA here are the ratios between n-6 PUFA (which may promote inflammation) and long-chain n-3 PUFA.

There is considerable interest in mechanisms associated with the loss of function of the phosphatase and tensin homolog deleted on chromosome 10 (PTEN) tumor-suppressor gene in cancer [62–65]. The PTEN gene has lipid phosphatase activity and acts in the nucleus to promote genomic stability and DNA repair. Consequently, loss of this function leads to increased genomic instability [62]. In a PTEN-null mouse model, it was possible to demonstrate the importance of lipid-modifying enzymes in converting saturated fatty acids to monosaturated fatty acids, and also the negative implications of an increased ratio of long-chain omega-6 PUFA to omega-3 PUFA in genomic stability and cancer risk [65].

3.1.2 Vitamins

3.1.2.1 Carotenoids

Peto et al. [66] originally suggested a cancer-preventive role for β-carotene, based on a number of cross-sectional or case–control studies. Most such studies showed a negative correlation between blood carotenoid levels and various biomarkers of DNA damage. However, some placebo-controlled carotenoid intervention trials using disease and mortality as outcomes have suggested a significant increase rather than decrease in mortality associated with vitamin A, β-carotene, or vitamin E supplements [67]. It is possible that this depends upon the concentration used in the supplement, and also the population tested. Pro-vitamin A carotenoids include α- and β-carotene, β-cryptoxanthin, retinoic acid, retinal, and retinol, while non-vitamin A carotenoids include lycopene, lutein, astaxanthin, and zeaxanthin. A number of tissue culture studies have involved cotreatment with a DNA-damaging agent and various carotenoids [5,68–77]. While the non-vitamin A carotenoids usually decreased the DNA damage, thereby promoting genomic stability, the pro-vitamin A carotenoids had little or no effect at low concentrations but increased genomic instability at higher concentrations.

3.1.2.2 Other Vitamins

Various B vitamins have beneficial effects on the stability of both nuclear and mitochondrial genomes. These include niacin (vitamin B3), folate (vitamin B9), and vitamin B12. Folate is an essential factor in one-carbon metabolism, acting to supply the methyl units for DNA methylation. Folate deficiency, especially in the presence of suboptimal levels of vitamin B6 and vitamin B12, may have significant effects on chromosomal fragility, resulting in chromosome breaks and mtDNA deletions, as well as reduced telomere length [78,79]. Folate is a key component of a number of root vegetables, including pulses such as red kidney beans, chickpeas, and lentils [79]. The B vitamin class also includes choline, which also interacts with folate biosynthesis [80,81]. A deficiency of this nutrient can lead to DNA hypomethylation and an accumulation of strand breaks [80].

Vitamin C has been considered to be an antioxidant. In human studies, the effects of vitamin C supplementation on various markers of genome stability depend on individual responses to vitamin C levels in the diet, and on concomitant exposure to oxidative stresses [82]. Vitamin C also protects against DNA damage, DNA strand breakage, and chromosomal aberrations [69,82].

Vitamin D is also critical in the maintenance of genome stability, preventing oxidative stress, chromosomal aberrations, telomere shortening, and inhibition of telomerase activity [83–86]. There is reason to believe that a primary function of vitamin D is in preventing DNA damage, while a secondary effect is in the regulation of cellular growth [83].

Other vitamins such as biotin (or vitamin H) and the vitamin-like co-enzyme Q10 are also important in the maintenance of genomic stability [87,88]. It is important to recognize that there are considerable interindividual differences in the ability to absorb and metabolize all of these vitamins [89]. Recognizing the optimal amount is of considerable importance.

3.1.3 Minerals

While a number of minerals are typically considered as toxicants, some of these are essential micronutrients, albeit usually with a narrow window of efficacy as compared with toxicity. These include iron [90], selenium (Se) [91], and zinc [92]. Se provides a useful illustration of these complexities, since the population generally shows a "U"-shaped response curve, with both low and high selenium levels increasing genomic instability. The optimal form of Se at the optimal level may protect against DNA or chromosome breakage, chromosome gain or loss, damage to mtDNA, and detrimental effects on telomere length and function [93]. However, the optimal level of Se differs among individuals, and also with the form incorporated into the diet [91,94]. Various genetic polymorphisms may affect both the uptake and utilization of selenium among individuals [94].

3.2 Bioactive Food Components

Bioactives, sometimes called phytochemicals, have been defined as "constituents in foods or dietary supplements, other than those needed to meet human nutritional needs, which are responsible for changes in health status" [95]. This group includes various polyphenols, defined as having several hydroxyl groups on one or more aromatic rings, and divided into various groups according to chemical structure [96]. There is compelling evidence that a considerable range of polyphenols may stabilize genomic DNA, through various processes, including effects on DNA methylation [96].

Both genistein and dadzein are soy-derived phytoestrogens that bind to estrogen receptors and have both weak estrogenic and weak antiestrogenic effects [75,97,98]. Genistein has been shown to have antioxidant effects and may act in concert with other nutrients such as β-carotene in beneficially affecting genomic stability [75]. Genistein also showed beneficial effects in combination with the DNA-damaging agent, bisphenol A [99]. However, in common with other such compounds, genistein has been found to have adverse effects in combination with such compounds as diethylstilbestrol [100].

Curcumin is a polyphenol that is also the active ingredient in the spice, turmeric. In a rodent model of colorectal cancer, curcumin treatment led to downregulation of telomerase activity, and this effect was associated with cell-cycle arrest and induction of apoptosis [101]. Protection against genomic instability has also been shown by curcumin in combination with certain genotoxic agents. For example, in human hepatocyte LO2 cells, curcumin was able to protect against the genotoxicity of quinocetone (QCT), a controversial compound which has been used as an antimicrobial feed additive in China. Curcumin pretreatment significantly attenuated the formation of ROS, DNA fragmentation, and micronucleus formation [102]. However, in a different tissue culture model using Raji cells, curcumin increased ROS and cell-cycle arrest, leading to structural chromosome abnormalities [103].

Resveratrol (RSV) is another polyphenol which is considered to be the beneficial component in red wine. High intakes of RSV have usually been considered beneficial to human health, including cancer-protective and antiaging effects. For example, it is generally considered to be an antioxidant, and has shown a chemopreventive effect in different mouse cancer models [97,104,105]. In HeLa S3 mammalian cells, RSV has effects on gene expression leading to the induction of telomere-maintenance factors, without effects on cell proliferation. That is, it can protect against changes in telomere length [106]. However, in the HeLa colon cancer cell model, RSV has also induced DNA damage through pro-oxidant effects, leading to apoptosis [105].

Indole-3-carbinol and epigallocatechin–3-gallate (EGCG) from green tea are both examples of polyphenols that show strong evidence of modulating genomic stability through various epigenetic mechanisms [9,95].

Some phytochemicals may have complementary activities in protection against genomic stability. For example, in broccoli (Fig. 31.3), the isothiocyanate, sulforaphane, and the polyphenol, quercetin, may complement one another in their epigenetic actions [107]. Duthie [108] suggested that the evidence is particularly strong for berry phytochemicals, specifically anthocyanins (a class of flavonoids), which modulate various biomarkers of DNA damage and carcinogenesis, in both in vitro and in vivo animal studies. However, evidence for cancer-preventive effects of any of these phytochemicals in human studies is currently weak.

Tumor-promoting inflammation is inhibited by all of the compounds except vitamin B [79,93,109–121], while only vitamin D, carotenoids, and RSV prevent tumor cells from evading the immune system [122–124].

4. THE SIGNIFICANCE OF GENETIC POLYMORPHISMS

There is no question but that genetic polymorphisms in various genes such as breast cancer 1 early onset (BRCA1) and breast cancer 2 early onset (BRCA2) affect cancer susceptibility, independent of nutrition [125]. However, it is also increasingly clear that the risk of developing cancer depends upon a complex interplay among genetic susceptibility, lifestyle, and diet, and that a number of the important genes are associated with nutrient uptake, transport, metabolism, and excretion [9]. While general population recommendations for nutrients are clearly of benefit, these do not necessarily indicate the optimal diet for an individual [50,51,89,126]. Folate provides an excellent example of a vitamin for which there is strong influence of the interplay between the nutrient intake, and also certain genetic polymorphisms [51]. There is also considerable interest in vitamin D, where there have been several hundred genes reported, which may affect uptake and function of the nutrient in various ways [84]. The minerals Se, zinc, and iron are also required at different concentrations according to genotype [90,92,94].

5. CONCLUSIONS

Genomic instability plays a critical role in cancer initiation and progression. The fidelity of the genome is protected at every stage of the cell cycle. In cancer, the presence of aneuploid or tetraploid cells indicates the failure of one or many of these safety nets. The resultant genomic heterogeneity may offer the cancer cells a selection advantage against the selective nature of emerging therapies. Understanding these protective mechanisms, and how they are bypassed in cancer cells, may highlight new and more specific mechanisms for therapeutic attack and/or cancer prevention.

While much work has focused on the development of new cancer drugs, this review makes it clear that focusing on nutrition, both in terms of preventing cancer development and also its progression, may be more fruitful. Vitamins (such as B, C, and D), minerals (such as Se), and phytochemicals (such as RSV, sulforaphane, and EGCG) have shown remarkable potential for diminishing tumor risk and tumor progression. In addition to their protective properties against genomic instability, these compounds are known to inhibit proliferative signaling [119,127,128], attenuate oncogenic metabolism [126,129–134], and block inflammation [79,93,109–121].

Despite progress in antitumor therapies, the death rates from cancer remain alarming [135,136]. However, diet and lifestyle are increasingly being shown for their potential in reducing cancer risks and/or slowing tumor progression. In particular, antioxidants are critical for the prevention of DNA damage that enables cancer initiation and growth. Growing evidence shows that vitamins, minerals, and other dietary factors have profound and protective effects against cancer cells, whether they are grown in the laboratory, in animals, or studied in human populations. A better understanding of the effects and synergy of these dietary factors in the prevention and treatment of genomic instability is critical to the future reduction of mortality associated with cancer.

GLOSSARY

Anthocyanins Water-soluble vacuolar pigments that may appear red, purple, or blue depending on the pH, belonging to a parent class of molecules called flavonoids.

Bioactives Constituents in foods or dietary substances, other than those required to meet nutritional needs, which are responsible for health status.

Nutriome The combination of nutrients and their doses that optimizes genomic stability for an individual.

Polyphenols A structural class of mainly natural, but also synthetic or semisynthetic, organic chemicals characterized by the presence of large multiples of phenol structural units.

Phytoestrogen Plant-derived xenoestrogens not generated within the endocrine system but consumed by eating phytoestrogenic plants.

LIST OF ABBREVIATIONS

1-MIM 1-Methoxy-3-indoylmethyl
ATP Adenosine triphosphate
BRCA1 Breast cancer 1 early onset

BRCA2 Breast cancer 2 early onset
DHA Docosahexaenoic acid
EGCGE Pigallocatechin gallate
EPA Eicosapentaenoic acid
HCA Heterocyclic amine
IARC International Agency for Research on Cancer
LINE-1 Long interspersed nucleotide element 1
mtDNA Mitochondrial DNA
n-3 Omega-3
n-6 Omega-6
NER Nucleotide excision repair
PAH Polycyclic aromatic hydrocarbon
PTEN Phosphatase and tensin homolog
PUFA Polyunsaturated fatty acid
QCT Quinocetone
ROS Reactive oxygen species
RSV Resveratrol
Se Selenium

REFERENCES

[1] Mishra A, Chaturvedi P, Datta S, Sinukumar S, Joshi P, Garg A. Harmful effects of nicotine. Indian J Med Paediatr Oncol 2015;36(1):24–31.
[2] Cogliano VJ, Baan R, Straif K, Grosse Y, Lauby-Secretan B, El Ghissassi F, et al. Preventable exposures associated with human cancers. J Natl Cancer Inst 2011;103(24):1827–39.
[3] Vromman V, Maghuin-Rogister G, Vleminckx C, Saegerman C, Pussemier L, Huyghebaert A. Risk ranking priority of carcinogenic and/or genotoxic environmental contaminants in food in Belgium. Food Addit Contam Part A Chem Anal Control Expo Risk Assess 2014;31(5):872–88.
[4] Yazici Z, Sisman T. Genotoxic effects of water pollution on two fish species living in Karasu River, Erzurum, Turkey. Environ Monit Assess 2014;186(11):8007–16.
[5] Reddy L, Odhav B, Bhoola K. Aflatoxin B1-induced toxicity in HepG2 cells inhibited by carotenoids: morphology, apoptosis and DNA damage. Biol Chem 2006;387(1):87–93.
[6] Behsnilian D, Butz P, Greiner R, Lautenschlaeger R. Process-induced undesirable compounds: chances of non-thermal approaches. Meat Sci 2014;98(3):392–403.
[7] Levine H, Berman T, Goldsmith R, Goen T, Spungen J, Novack L, et al. Urinary concentrations of polycyclic aromatic hydrocarbons in Israeli adults: demographic and life-style predictors. Int J Hyg Environ Health 2015;218(1):123–31.
[8] Nie J, Shi J, Duan X, Wang B, Huang N, Zhao X. Health risk assessment of dietary exposure to polycyclic aromatic hydrocarbons in Taiyuan, China. J Environ Sci 2014;26(2):432–9.
[9] Ferguson LR, Chen H, Collins AR, Connell M, Damia G, Dasgupta S, et al. Genomic instability in human cancer: molecular insights and opportunities for therapeutic attack and prevention through diet and nutrition. Semin Cancer Biol December 2015;35(Suppl.):S5–24.
[10] Bonnefont-Rousselot D. Obesity and oxidative stress: potential roles of melatonin as antioxidant and metabolic regulator. Endocr Metab Immune Disord Drug Targets 2014;14(3):159–68.
[11] Chattopadhyay M, Khemka VK, Chatterjee G, Ganguly A, Mukhopadhyay S, Chakrabarti S. Enhanced ROS production and oxidative damage in subcutaneous white adipose tissue mitochondria in obese and type 2 diabetes subjects. Mol Cell Biochem 2015;399(1–2):95–103.
[12] Yara S, Lavoie JC, Levy E. Oxidative stress and DNA methylation regulation in the metabolic syndrome. Epigenomics 2015;7(2):283–300.
[13] Robertson KD. DNA methylation and human disease. Nat Rev Genet 2005;6(8):597–610.
[14] Delgado-Cruzata L, Zhang W, McDonald JA, Tsai WY, Valdovinos C, Falci L, et al. Dietary modifications, weight loss, and changes in metabolic markers affect global DNA methylation in Hispanic, African American, and Afro-Caribbean breast cancer survivors. J Nutr 2015;145(4):783–90.
[15] Sproul D, Gilbert N, Bickmore WA. The role of chromatin structure in regulating the expression of clustered genes. Nat Rev Genet 2005;6(10): 775–781. Epub 2005/09/15.
[16] Turner BM. Histone acetylation and an epigenetic code. Bioessays 2000;22(9):836–45. Epub 2000/08/17.
[17] Luijsterburg MS, Goedhart J, Moser J, Kool H, Geverts B, Houtsmuller AB, et al. Dynamic in vivo interaction of DDB2 E3 ubiquitin ligase with UV-damaged DNA is independent of damage-recognition protein XPC. J Cell Sci 2007;120(15):2706–16.
[18] Chang CC, Wu MJ, Yang JY, Camarillo IG, Chang CJ. Leptin-STAT3-G9a signaling promotes obesity-mediated breast cancer progression. Cancer Res 2015;75(11):2375–86.
[19] Horsburgh S, Robson-Ansley P, Adams R, Smith C. Exercise and inflammation-related epigenetic modifications: focus on DNA methylation. Exerc Immunol Rev 2015;21:26–41.
[20] Chatterjee A, Dasgupta S, Sidransky D. Mitochondrial subversion in cancer. Cancer Prev Res 2011;4(5):638–54. Epub 2011/05/06.
[21] Dasgupta S, Shao C, Keane TE, Duberow DP, Mathies RA, Fisher PB, et al. Detection of mitochondrial deoxyribonucleic acid alterations in urine from urothelial cell carcinoma patients. Int J Cancer J Int du Cancer 2012;131(1):158–64. Epub 2011/08/10.

[22] Ferramosca A, Zara V. Dietary fat and hepatic lipogenesis: mitochondrial citrate carrier as a sensor of metabolic changes. Adv Nutr 2014;5(3): 217–25.

[23] Monteiro JP, Morais CM, Oliveira PJ, Jurado AS. Mitochondrial membrane lipids in life and death and their molecular modulation by diet: tuning the furnace. Curr Drug Targets 2014;15(8):797–810.

[24] Balbo S, Brooks PJ. Implications of acetaldehyde-derived DNA adducts for understanding alcohol-related carcinogenesis. Adv Exp Med Biol 2015;815:71–88.

[25] Carvalho AM, Miranda AM, Santos FA, Loureiro AP, Fisberg RM, Marchioni DM. High intake of heterocyclic amines from meat is associated with oxidative stress. Br J Nutr 2015;113(8):1301–7.

[26] Abid Z, Cross AJ, Sinha R. Meat, dairy, and cancer. Am J Clin Nutr 2014;100(Suppl. 1):386S–93S.

[27] Bouvard V, Loomis D, Guyton KZ, Grosse Y, Ghissassi FE, Benbrahim-Tallaa L, et al. Carcinogenicity of consumption of red and processed meat. Lancet Oncol December 2015;16(16):1599–600.

[28] IARC Working Group L-S. Consumption of red meat and processed meat. IARC monogr eval carcinog risks hum, vol. 114. International Agency for Research on Cancer; October 2015:6–13.

[29] Jiao L, Stolzenberg-Solomon R, Zimmerman TP, Duan Z, Chen L, Kahle L, et al. Dietary consumption of advanced glycation end products and pancreatic cancer in the prospective NIH-AARP Diet and Health Study. Am J Clin Nutr 2015;101(1):126–34.

[30] Turner DP. Advanced glycation end-products: a biological consequence of lifestyle contributing to cancer disparity. Cancer Res 2015;75(10):1925–9.

[31] Catsburg CE, Gago-Dominguez M, Yuan JM, Castelao JE, Cortessis VK, Pike MC, et al. Dietary sources of N-nitroso compounds and bladder cancer risk: findings from the Los Angeles bladder cancer study. Int J Cancer 2014;134(1):125–35.

[32] Dellavalle CT, Xiao Q, Yang G, Shu XO, Aschebrook-Kilfoy B, Zheng W, et al. Dietary nitrate and nitrite intake and risk of colorectal cancer in the Shanghai Women's Health Study. Int J Cancer 2014;134(12):2917–26.

[33] Motorykin O, Santiago-Delgado L, Rohlman D, Schrlau JE, Harper B, Harris S, et al. Metabolism and excretion rates of parent and hydroxy-PAHs in urine collected after consumption of traditionally smoked salmon for Native American volunteers. Sci Total Environ 2015;514:170–7.

[34] Henderson LM, Speit G. Review of the genotoxicity of styrene in humans. Mutat Res 2005;589(3):158–91.

[35] Coyle YM, Hynan LS, Euhus DM, Minhajuddin AT. An ecological study of the association of environmental chemicals on breast cancer incidence in Texas. Breast Cancer Res Treat 2005;92(2):107–14.

[36] de RMAM, Cardoso M, Dorea HS, Emidio ES, Silva MM, dos Anjos JP, et al. Contamination of cachaca by PAHs from storage containers. Food Chem 2014;146:65–70.

[37] Felizardo RJ, Camara NO. Hepatocellular carcinoma and food contamination: aflatoxins and ochratoxin A as a great prompter. World J Gastroenterol 2013;19(24):3723–5.

[38] Graham S, Haughey B, Marshall J, Brasure J, Zielezny M, Freudenheim J, et al. Diet in the epidemiology of gastric cancer. Nutr Cancer 1990;13(1–2): 19–34.

[39] Sun LH, Lei MY, Zhang NY, Gao X, Li C, Krumm CS, et al. Individual and combined cytotoxic effects of aflatoxin B1, zearalenone, deoxynivalenol and fumonisin B1 on BRL 3A rat liver cells. Toxicon 2015;95:6–12.

[40] Ogah CO, Coker HA, Adepoju-Bello AA. Pesticide residue levels in maize samples from markets in Lagos State, Nigeria. Niger Q J Hosp Med 2011;21(2):169–74.

[41] Tomerak RH, Shaban HH, Khalafallah OA, El Shazly MN. Assessment of exposure of Egyptian infants to aflatoxin M1 through breast milk. J Egypt Public Health Assoc 2011;86(3–4):51–5.

[42] Jinadasa BK, Edirisinghe EM. Cadmium, lead and total mercury in Tilapia sp. in Sri Lankan reservoirs. Food Addit Contam Part B-Surveillance 2014;7(2):90–4.

[43] Fatima M, Usmani N, Firdaus F, Zafeer MF, Ahmad S, Akhtar K, et al. In vivo induction of antioxidant response and oxidative stress associated with genotoxicity and histopathological alteration in two commercial fish species due to heavy metals exposure in northern India (Kali) river. Comparative Biochemistry & Physiology. Toxicol Pharmacol 2015;176-7:17–30.

[44] L'Heritier F, Marques M, Fauteux M, Gaudreau L. Defining molecular sensors to assess long-term effects of pesticides on carcinogenesis. Int J Mol Sci 2014;15(9):17148–61.

[45] Itoh H, Iwasaki M, Sawada N, Takachi R, Kasuga Y, Yokoyama S, et al. Dietary cadmium intake and breast cancer risk in Japanese women: a case-control study. Int J Hyg Environ Health 2014;217(1):70–7.

[46] Adetunji VO, Famakin IO, Chen J. Lead and cadmium levels in cattle muscle and edible tissues collected from a slaughter slab in Nigeria. Food Addit Contam Part B-Surveil 2014;7(2):79–83.

[47] Ames BN, Profet M, Gold LS. Dietary pesticides (99.99% all natural). Proc Natl Acad Sci USA 1990;87(19):7777–81.

[48] Schumacher F, Engst W, Monien BH, Florian S, Schnapper A, Steinhauser L, et al. Detection of DNA adducts originating from 1-methoxy-3-indolylmethyl glucosinolate using isotope-dilution UPLC-ESI-MS/MS. Anal Chem 2012;84(14):6256–62.

[49] Otten JJ, Hellwig JP, Meyers LD. DRI, dietary reference intakes: the essential guide to nutrient requirements. Washington, USA: National Academies Press; 2006.

[50] Fenech MF. Dietary reference values of individual micronutrients and nutriomes for genome damage prevention: current status and a road map to the future. Am J Clin Nutr 2010;91(5):1438S–54S.

[51] Fenech MF. Nutriomes and personalised nutrition for DNA damage prevention, telomere integrity maintenance and cancer growth control. Cancer Treat Res 2014;159:427–41.

[52] Ferreira P, Cravo M, Guerreiro CS, Tavares L, Santos PM, Brito M. Fat intake interacts with polymorphisms of Caspase9, FasLigand and PPARgamma apoptotic genes in modulating Crohn's disease activity. Clin Nutr 2010;29(6):819–23.

[53] Kiecolt-Glaser JK, Epel ES, Belury MA, Andridge R, Lin J, Glaser R, et al. Omega-3 fatty acids, oxidative stress, and leukocyte telomere length: a randomized controlled trial. Brain Behav Immun 2013;28:16–24.

[54] Murray M, Hraiki A, Bebawy M, Pazderka C, Rawling T. Anti-tumor activities of lipids and lipid analogues and their development as potential anticancer drugs. Pharmacol Ther 2015;150:109–28.

[55] O'Callaghan N, Parletta N, Milte CM, Benassi-Evans B, Fenech M, Howe PR. Telomere shortening in elderly individuals with mild cognitive impairment may be attenuated with omega-3 fatty acid supplementation: a randomized controlled pilot study. Nutrition 2014;30(4):489–91.

[56] Borghetti G, Yamaguchi AA, Aikawa J, Yamazaki RK, de Brito GA, Fernandes LC. Fish oil administration mediates apoptosis of Walker 256 tumor cells by modulation of p53, Bcl-2, caspase-7 and caspase-3 protein expression. Lipids health Dis 2015;14:94.

[57] Kim N, Jeong S, Jing K, Shin S, Kim S, Heo JY, et al. Docosahexaenoic acid induces cell death in human non-small cell lung cancer cells by repressing mTOR via AMPK activation and PI3K/akt inhibition. BioMed Res Int 2015;2015. 239764.

[58] Flock MR, Skulas-Ray AC, Harris WS, Gaugler TL, Fleming JA, Kris-Etherton PM. Effects of supplemental long-chain omega-3 fatty acids and erythrocyte membrane fatty acid content on circulating inflammatory markers in a randomized controlled trial of healthy adults. Prostagl Leukot Essent Fat Acids 2014;91(4):161–8.

[59] Janakiram NB, Rao CV. Role of lipoxins and resolvins as anti-inflammatory and proresolving mediators in colon cancer. Curr Mol Med 2009;9(5):565–79.

[60] Park JM, Kwon SH, Han YM, Hahm KB, Kim EH. Omega-3 polyunsaturated Fatty acids as potential chemopreventive agent for gastrointestinal cancer. J Cancer Prev 2013;18(3):201–8.

[61] Serhan CN, Chiang N, Dalli J. The resolution code of acute inflammation: novel pro-resolving lipid mediators in resolution. Semin Immunol 2015;27(3):200–15.

[62] Dillon LM, Miller TW. Therapeutic targeting of cancers with loss of PTEN function. Curr Drug Targets 2014;15(1):65–79.

[63] Goel A, Nagasaka T, Arnold CN, Inoue T, Hamilton C, Niedzwiecki D, et al. The CpG island methylator phenotype and chromosomal instability are inversely correlated in sporadic colorectal cancer. Gastroenterology 2007;132(1):127–38. Epub 2006/11/08.

[64] Milella M, Falcone I, Conciatori F, Cesta Incani U, Del Curatolo A, Inzerilli N, et al. PTEN: multiple functions in human malignant tumors. Front 2015;5:24.

[65] Muir K, Hazim A, He Y, Peyressatre M, Kim DY, Song X, et al. Proteomic and lipidomic signatures of lipid metabolism in NASH-associated hepatocellular carcinoma. Cancer Res 2013;73(15):4722–31.

[66] Peto R, Doll R, Buckley JD, Sporn M. Can dietary beta-carotene materially reduce human cancer rates? Nature 1981;290(5803):201–8.

[67] Bjelakovic G, Nikolova D, Gluud LL, Simonetti RG, Gluud C. Mortality in randomized trials of antioxidant supplements for primary and secondary prevention: systematic review and meta-analysis. JAMA J Am Med Assoc 2007;297(8):842–57.

[68] Zhou G-D, Richardson M, Fazili IS, Wang J, Donnelly KC, Wang F, et al. Role of retinoic acid in the modulation of benzo (a) pyrene-DNA adducts in human hepatoma cells: implications for cancer prevention. Toxicol Appl Pharm 2010;249(3):224–30.

[69] Bergström T, Bergman J, Möller L. Vitamin A and C compounds permitted in supplements differ in their abilities to affect cell viability, DNA and the DNA nucleoside deoxyguanosine. Mutagenesis 2011;26(6):735–44.

[70] Lorenzo Y, Azqueta A, Luna L, Bonilla F, Domínguez G, Collins AR. The carotenoid β-cryptoxanthin stimulates the repair of DNA oxidation damage in addition to acting as an antioxidant in human cells. Carcinogenesis 2009;30(2):308–14.

[71] Rosignoli P, Fabiani R, De Bartolomeo A, Fuccelli R, Pelli MA, Morozzi G. Genotoxic effect of bile acids on human normal and tumour colon cells and protection by dietary antioxidants and butyrate. Eur J Nutr 2008;47(6):301–9.

[72] Yurtcu E, Iseri OD, Sahin FI. Effects of ascorbic acid and β-carotene on HepG2 human hepatocellular carcinoma cell line. Mol Biol Rep 2011;38(7):4265–72.

[73] van Helden YG, Keijer J, Heil SG, Picó C, Palou A, Oliver P, et al. Beta-carotene affects oxidative stress-related DNA damage in lung epithelial cells and in ferret lung. Carcinogenesis 2009;30(12):2070–6.

[74] van Helden YG, Keijer J, Knaapen AM, Heil SG, Briedé JJ, Van Schooten FJ, et al. β-Carotene metabolites enhance inflammation-induced oxidative DNA damage in lung epithelial cells. Free Radic Biol Med 2009;46(2):299–304.

[75] Shiau R-J, Chen K-Y, Wen Y-D, Chuang C-H, Yeh S-L. Genistein and β-carotene enhance the growth-inhibitory effect of trichostatin A in A549 cells. Eur J Nutr 2010;49(1):19–25.

[76] Kontek R, Drozda R, Śliwiński M, Grzegorczyk K. Genotoxicity of irinotecan and its modulation by vitamins A, C and E in human lymphocytes from healthy individuals and cancer patients. Toxicol Vitro 2010;24(2):417–24.

[77] Jang SH, Lim JW, Kim H. Mechanism of β-carotene-induced apoptosis of gastric Cancer cells: involvement of ataxia-telangiectasia-mutated. Ann NY Acad Sci 2009;1171(1):156–62.

[78] Bull CF, Mayrhofer G, O'Callaghan NJ, Au AY, Pickett HA, Low GK, et al. Folate deficiency induces dysfunctional long and short telomeres; both states are associated with hypomethylation and DNA damage in human WIL2-NS cells. Cancer Prev Res 2014;7(1):128–38.

[79] Sharp L, Carsin AE, Cantwell MM, Anderson LA, Murray LJ, Group FS. Intakes of dietary folate and other B vitamins are associated with risks of esophageal adenocarcinoma, Barrett's esophagus, and reflux esophagitis. J Nutr 2013;143(12):1966–73. Epub 2013/10/18.

[80] Lu L, Ni J, Zhou T, Xu W, Fenech M, Wang X. Choline and/or folic acid deficiency is associated with genomic damage and cell death in human lymphocytes in vitro. Nutr Cancer 2012;64(3):481–7.

[81] Ferrari A, de Carvalho AM, Steluti J, Teixeira J, Marchioni DML, Aguiar Jr S. Folate and nutrients involved in the 1-carbon cycle in the pretreatment of patients for colorectal cancer. Nutrients 2015;7(6):4318–35.

[82] Sram RJ, Binkova B, Rossner Jr P. Vitamin C for DNA damage prevention. Mutat Res 2012;733(1):39–49.

[83] Nair-Shalliker V, Armstrong BK, Fenech M. Does vitamin D protect against DNA damage? Mutat Res 2012;733(1–2):50–7.

[84] Ferguson IR, Laing B, Marlow G, Bishop KS. The role of vitamin D in reducing gastrointestinal disease risk and assessment of individual dietary intake needs: focus on genetic and genomic technologies. Mol Nutr Food Res January 2016;60(1):119–133.

[85] Kalman B, Toldy E. Genomic binding sites and biological effects of the vitamin D–VDR complex in multiple sclerosis (corrected). Neuromol Med 2014;16(2):265–79.

[86] Moen EL, Mariani CJ, Zullow H, Jeff-Eke M, Litwin E, Nikitas JN, et al. New themes in the biological functions of 5-methylcytosine and 5-hydroxymethylcytosine. Immunol Rev 2015;263(1):36–49.

[87] Zempleni J, Teixeira DC, Kuroishi T, Cordonier EL, Baier S. Biotin requirements for DNA damage prevention. Mutat Res 2012;733(1–2):58–60.

[88] Du Vigneaud V, Melville DB, Gyorgy P, Rose CS. On the identity of vitamin H with biotin. Science 1940;92(2377):62–3.

[89] van Ommen B, El-Sohemy A, Hesketh J, Kaput J, Fenech M, Evelo CT, et al. The Micronutrient Genomics Project: a community-driven knowledge base for micronutrient research. Genes & Nutr 2010;5(4):285–96.

[90] Prá D, Franke SIR, Henriques JAP, Fenech M. Iron and genome stability: an update. Mutat Res 2012;733(1):92–9.

[91] Ferguson LR, Karunasinghe N, Zhu S, Wang AH. Selenium and its' role in the maintenance of genomic stability. Mutat Res 2012;733(1):100–10.

[92] Sharif R, Thomas P, Zalewski P, Fenech M. The role of zinc in genomic stability. Mutat Res 2012;733(1):111–21.

[93] Barrett C, Singh K, Motley A, Lintel M, Matafonova E, Bradley A, et al. Dietary selenium deficiency exacerbates DSS-induced epithelial injury and AOM/DSS-induced tumorigenesis. PLoS One 2013;8(7).

[94] Karunasinghe N, Han DY, Zhu S, Yu J, Lange K, Duan H, et al. Serum selenium and single-nucleotide polymorphisms in genes for selenoproteins: relationship to markers of oxidative stress in men from Auckland, New Zealand. Genes & Nutr 2012;7(2):179–90.

[95] Lupton JR, Atkinson SA, Chang N, Fraga CG, Levy J, Messina M, et al. Exploring the benefits and challenges of establishing a DRI-like process for bioactives. Eur J Nutr 2014;53(Suppl. 1):1–9.

[96] Ferguson LR. Role of plant polyphenols in genomic stability. Mutat Res 2001;475(1–2):89–111.

[97] N'Soukpoe-Kossi CN, Bourassa P, Mandeville JS, Bekale L, Tajmir-Riahi HA. Structural modeling for DNA binding to antioxidants resveratrol, genistein and curcumin. J Photochem Photobiol B – Biol 2015;151:69–75.

[98] Mattison DR, Karyakina N, Goodman M, LaKind JS. Pharmaco- and toxicokinetics of selected exogenous and endogenous estrogens: a review of the data and identification of knowledge gaps. Crit Rev Toxicol 2014;44(8):696–724.

[99] Bernardo BD, Brandt JZ, Grassi TF, Silveira LTR, Scarano WR, Barbisan LF. Genistein reduces the noxious effects of in utero bisphenol A exposure on the rat prostate gland at weaning and in adulthood. Food & Chem Toxicol 2015;84:64–73.

[100] Jeng Y-J, Kochukov M, Nauduri D, Kaphalia BS, Watson CS. Subchronic exposure to phytoestrogens alone and in combination with diethylstilbestrol - pituitary tumor induction in Fischer 344 rats. Nutr Metab 2010;7:40.

[101] Rana C, Piplani H, Vaish V, Nehru B, Sanyal SN. Downregulation of telomerase activity by diclofenac and curcumin is associated with cell cycle arrest and induction of apoptosis in colon cancer. Tumour Biol 2015;36(8):5999–6010.

[102] Dai C, Tang S, Li D, Zhao K, Xiao X. Curcumin attenuates quinocetone-induced oxidative stress and genotoxicity in human hepatocyte L02 cells. Toxicol Mech Methods 2015;25(4):340–6.

[103] Sharma V, Jha AK, Kumar A, Bhatnagar A, Narayan G, Kaur J. Curcumin-mediated reversal of p15 gene promoter methylation: implication in anti-neoplastic action against acute lymphoid leukaemia cell line. Folia Biol 2015;61(2):81–9.

[104] Bagul PK, Deepthi N, Sultana R, Banerjee SK. Resveratrol ameliorates cardiac oxidative stress in diabetes through deacetylation of NFkB-p65 and histone 3. J Nutr Biochem 2015;26(11):1298–307.

[105] Demoulin B, Hermant M, Castrogiovanni C, Staudt C, Dumont P. Resveratrol induces DNA damage in colon cancer cells by poisoning topoisomerase II and activates the ATM kinase to trigger p53-dependent apoptosis. Toxicol Vitro 2015;29(5):1156–65.

[106] Uchiumi F, Watanabe T, Hasegawa S, Hoshi T, Higami Y. Tanuma S-i. The effect of resveratrol on the Werner syndrome RecQ helicase gene and telomerase activity. Curr Aging Sci 2011;4(1):1–7.

[107] Appari M, Babu KR, Kaczorowski A, Gross W, Herr I. Sulforaphane, quercetin and catechins complement each other in elimination of advanced pancreatic cancer by miR-let-7 induction and K-ras inhibition. Int J Oncol 2014;45(4):1391–400.

[108] Duthie SJ. Berry phytochemicals, genomic stability and cancer: evidence for chemoprotection at several stages in the carcinogenic process. Mol Nutr Food Res 2007;51(6):665–74.

[109] Hopkins MH, Owen J, Ahearn T, Fedirko V, Flanders WD, Jones DP, et al. Effects of supplemental vitamin D and calcium on biomarkers of inflammation in colorectal adenoma patients: a randomized, controlled clinical trial. Cancer Prev Res 2011;4(10):1645–54. Epub 2011/07/05.

[110] Krishnan AV, Moreno J, Nonn L, Swami S, Peehl DM, Feldman D. Calcitriol as a chemopreventive and therapeutic agent in prostate cancer: role of anti-inflammatory activity. J Bone Min Res 2007;22(Suppl. 2):V74–80. Epub 2008/03/20.

[111] Zhang X-H, Ma J, Smith-Warner S, Lee J, Giovannucci E. Vitamin B6 and colorectal cancer: current evidence and future directions. World J Gastroenterol 2013;19(7):1005–10.

[112] Kristal A, Arnold K, Neuhouser M, Goodman P, Platz E, Albanes D, et al. Diet, supplement use, and prostate cancer risk: results from the prostate cancer prevention trial. Am J Epidemiol 2010;172(5):566–77.

[113] Ozten N, Horton L, Lasano S, Bosland M. Selenomethionine and alpha-tocopherol do not inhibit prostate carcinogenesis in the testosterone plus estradiol-treated NBL rat model. Cancer Prev Res Phila Pa 2010;3(3):371–80.

[114] Trejo-Solís C, Pedraza-Chaverrí J, Torres-Ramos M, Jiménez-Farfán D, Cruz Salgado A, Serrano-García N, et al. Multiple molecular and cellular mechanisms of action of lycopene in cancer inhibition. Evidence-Based Comp Altern Med 2013;2013. 705121.

[115] Wertz K. Lycopene effects contributing to prostate health. Nutr cancer 2009;61(6):775–83.

[116] Swindall A, Stanley J, Yang E. PARP-1: friend or foe of DNA damage and repair in tumorigenesis? Cancers 2013;5(3):943–58.

[117] Masutani M, Fujimori H. Poly(ADP-ribosyl)ation in carcinogenesis. Mol Aspects Med 2013;34(6):1202–16.

[118] Chen MC, Chang WW, Kuan YD, Lin ST, Hsu HC, Lee CH. Resveratrol inhibits LPS-induced epithelial-mesenchymal transition in mouse melanoma model. Innate Immun 2012;18(5):685–93. Epub 2012/02/22.
[119] Zhang M, Zhou X, Zhou K. Resveratrol inhibits human nasopharyngeal carcinoma cell growth via blocking pAkt/p70S6K signaling pathways. Int J Mol Med 2013;31(3):621–7. Epub 2013/01/15.
[120] Fahey J, Wehage S, Holtzclaw W, Kensler T, Egner P, Shapiro T, et al. Protection of humans by plant glucosinolates: efficiency of conversion of glucosinolates to isothiocyanates by the gastrointestinal microflora. Cancer Prev Res Phila Pa 2012;5(4):603–11.
[121] Li R, Li C, Wang T. Transcriptomic alterations in human prostate cancer cell LNCaP tumor xenograft modulated by dietary phenethyl isothiocyanate. Mol Carcinog 2013;52(6):426–37.
[122] Kusmartsev S, Cheng F, Yu B, Nefedova Y, Sotomayor E, Lush R, et al. All-trans-retinoic acid eliminates immature myeloid cells from tumor-bearing mice and improves the effect of vaccination. Cancer Res 2003;63(15):4441–9.
[123] Najjar Y, Finke J. Clinical perspectives on targeting of myeloid derived suppressor cells in the treatment of cancer. Front Oncol 2013;3:49.
[124] Sayeed A, Luciani-Torres G, Meng Z, Bennington JL, Moore DH, Dairkee SH. Aberrant regulation of the BST2 (Tetherin) promoter enhances cell proliferation and apoptosis evasion in high grade breast cancer cells. PLoS One 2013;8(6):e67191. Epub 2013/07/11.
[125] Castilla LH, Couch FJ, Erdos MR, Hoskins KF, Calzone K, Garber JE, et al. Mutations in the BRCA1 gene in families with early-onset breast and ovarian cancer. Nat Genet 1994;8(4):387–91. Epub 1994/12/01.
[126] Ferguson L, Schlothauer R. The potential role of nutritional genomics tools in validating high health foods for cancer control: broccoli as example. Mol Nutr food Res 2012;56(1):126–46.
[127] Liu W, Asa SL, Fantus IG, Walfish PG, Ezzat S. Vitamin D arrests thyroid carcinoma cell growth and induces p27 dephosphorylation and accumulation through PTEN/akt-dependent and -independent pathways. Am J Pathol 2002;160(2):511–9. Epub 2002/02/13.
[128] Amiri F, Zarnani AH, Zand H, Koohdani F, Jeddi-Tehrani M, Vafa M. Synergistic anti-proliferative effect of resveratrol and etoposide on human hepatocellular and colon cancer cell lines. Eur J Pharmacol 2013;718(1–3):34–40. Epub 2013/09/24.
[129] Chen H, Landen C, Li Y, Alvarez R, Tollefsbol T. Epigallocatechin gallate and sulforaphane combination treatment induce apoptosis in paclitaxel-resistant ovarian cancer cells through hTERT and Bcl-2 down-regulation. Exp Cell Res 2013;319(5):697–706.
[130] El-Bayoumy K. The protective role of selenium on genetic damage and on cancer. Mutat Res 2001;475(1–2):123–39.
[131] Lamy E, Herz C, Lutz-Bonengel S, Hertrampf A, Márton M-R, Mersch-Sundermann V. The MAPK pathway signals telomerase modulation in response to isothiocyanate-induced DNA damage of human liver cancer cells. PLoS One 2013;8(1).
[132] Moon D-O, Kang S-H, Kim K-C, Kim M-O, Choi Y, Kim G-Y. Sulforaphane decreases viability and telomerase activity in hepatocellular carcinoma Hep3B cells through the reactive oxygen species-dependent pathway. Cancer Lett 2010;295(2):260–6.
[133] Donkena KV, Yuan H, Young CY. Vitamin Bs, one carbon metabolism and prostate cancer. Mini Rev Med Chem 2010;10(14):1385–92. Epub 2010/10/13.
[134] Lee SC, Chan JY, Pervaiz S. Spontaneous and 5-fluorouracil-induced centrosome amplification lowers the threshold to resveratrol-evoked apoptosis in colon cancer cells. Cancer Lett 2010;288(1):36–41. Epub 2009/07/21.
[135] Seeta Rama Raju G, Benton L, Pavitra E, Yu JS. Multifunctional nanoparticles: recent progress in cancer therapeutics. Chem Commun 2015;51(68):13248–59.
[136] Siegel RL, Miller KD, Jemal A. Cancer statistics, 2015. CA Cancer J Clin 2015;65(1):5–29.

Chapter 32

Chemical Carcinogens and Their Effect on Genome and Epigenome Stability

O. Kovalchuk
University of Lethbridge, Lethbridge, AB, Canada

Chapter Outline

1. INTRODUCTION

Cancer is a disease that is characterized by the uncontrolled growth, proliferation, and spread of cells. This deadly disease is projected to affect one in four Americans and Canadians. Development of cancer (the process of carcinogenesis) is a long, multistep transformation of normal cells into malignant cells, which includes initiation, promotion, and progression stages from single initiated cells to carcinoma in situ and further to large-scale carcinoma and metastasis [1–3]. For a long time, cancer was thought to be a genetic disease, whereby pathological changes were caused by a progressive accumulation of a multitude of genetic and cytogenetic alterations. Since the first cancer-promoting mutation in a human *RAS* oncogene was reported in 1983, a large body of evidence has accumulated on various polymorphisms, point mutations, deletions, insertions, and translocations associated with cancer development [2].

Induction and development of some cancers can be genetically predetermined via inherited mutations that are passed from generation to generation, such as those in BRCA1, MSH2, MLH1, and other genes, albeit those direct gene defects account only for 5–10% of cases.

A large proportion of cancers is induced by the harmful influence of deleterious cancer-causing factors—carcinogens that can cause mutations or alter the proper function and stability of the cellular genome, leading to a loss of cellular growth controls. Carcinogens can be of a chemical, physical (UV radiation, X-rays, and magnetic fields), or biological (bacterial, viral, or altered metabolism) nature. Chemical carcinogens account for a lion's share of all induced cancers and can be divided into genotoxic and non-genotoxic ones based on their mechanisms of action and, especially, based on their capability to alter the DNA sequence [4]. Historically, the term *genotoxic carcinogen* has been used to define a chemical that is "capable of producing cancer by directly altering the genetic material of target cells." A *non-genotoxic carcinogen* refers to "a chemical capable of producing cancer by some secondary mechanism not related to direct gene damage" [5].

As such, most genotoxic carcinogens interact with DNA or produce metabolites that can react with DNA, and these direct reactions with DNA alter DNA and the chromosome structure or chromosome number. Some genotoxic carcinogens (eg, ethyl methane sulfate) act directly, while others require metabolic activation (2-acetylaminofluorene). Genotoxic carcinogens cause various types of DNA damage, including base alkylation, oxidation, base loss, formation of DNA adducts, interstrand cross-links, DNA–protein cross-links, and breaks in DNA. As a main repository of genetic information, DNA is the only cellular molecule that is repaired; the rest of them are simply replaced.

Genome Stability. http://dx.doi.org/10.1016/B978-0-12-803309-8.00032-X

To combat these attacks on the genome, cells have evolved a response system that induces cell-cycle arrest, allowing sufficient time for specialized groups of proteins to repair the incurred damage. To protect themselves, cells harbor elaborate and highly effective DNA-repair machinery that includes more than 100 proteins and a precise cell cycle–control system that induces cell-cycle arrest to allow time for repairs. Consequently, in the vast majority of cases, DNA damage is repaired. Moreover, in the case of irreparable damage, the cellular DNA damage–response system induces apoptosis. If the damage is unrepaired or misrepaired, this can lead to mutation and result in the phenomenon of genome instability that manifests as an elevated accumulation of mutations that are persistent in the cellular lineage. Genomic instability can facilitate the process of cancer initiation and/or progression [6], and indeed, the loss of genomic stability is believed to be a hallmark of many cancers, as well as an important prerequisite for cancer formation [7–9].

Non-genotoxic carcinogens do not affect DNA directly, but rather cause epigenetic changes, affect gene expression, and thus cause metabolic changes, increase peroxisome proliferation, disrupt cellular structures, change the rate of cell proliferation, or foster other processes that are responsible for cellular homeostasis. Disruption of those processes may in turn predispose cells to indirect DNA damage and can lead to carcinogenesis.

Yet, even though numerous spontaneous and carcinogen-induced cancer-causing mutations have been identified and cataloged, they cannot explain, by far, a wide variety of different malignant tumors (cancers and soft tissue tumors) and their relationship with environmental factors or some of the puzzling patterns of tumor predisposition and inheritance. Additionally, numerous lines of evidence have suggested that cancer can arise due to aberrant gene expression and regulation [2]. As such, it is now well accepted that cancer is both a genetic and an epigenetic disease [1,2]. Additionally, research conducted in 2009 indicates that genetic and epigenetic mechanisms can mediate the toxicity of various environmental chemicals, both genotoxic and non-genotoxic ones [4].

2. EPIGENETIC REGULATORS

Modern science defines *epigenetics* as mechanisms that establish and maintain mitotically and meiotically stable and heritable patterns of gene expression and regulation and occur without changes in DNA sequence. Epigenetic processes impact gene expression and chromatin structure and include DNA methylation, histone modifications, chromatin remodeling, and noncoding RNAs [2,10].

2.1 DNA Methylation

Among epigenetic regulators, cytosine DNA methylation was the first epigenetic mark identified, and it is one of the most widely studied epigenetic phenomena. DNA methylation is a covalent modification of DNA, in which a methyl group from S-adenosyl-L-methionine is added to the carbon 5 position of cytosine, yielding 5-methylcytosine (5 mC) in DNA. Genomic DNA methylation refers to the overall content of methylated cytosine (5 mC) in the genome [2,11].

DNA methylation is a key regulator of gene expression and genome stability. It is crucial for the proper functioning of normal cells and tissues. In normal cells, it governs the regulation of cell-type and tissue-specific gene expression, the silencing of parasitic and highly repetitive sequences, X-chromosome inactivation, the correct organization of active and inactive chromatin, and genomic imprinting [10,12] (Fig 32.1). In mammals, including humans, DNA methylation occurs mainly in the context of CpG dinucleotides that are methylated to 70–90% [11]. The highest frequency of CpGs is in the CpG island areas, which are often located in the 5′-end control regions of genes [11,12].

DNA methylation is accomplished by DNA methyltransferase enzymes [11]. In mammals, three DNA methyltransferases (DNMT1, DNMT3a, and DNMT3b) are responsible for establishing and maintaining DNA-methylation patterns at CpG sites [10]. DNMT1 is the major enzyme involved in the maintenance of DNA-methylation patterns after DNA replication. It is localized at replication forks where, in collaboration with de novo methyltransferases and methyl-binding proteins, it directly modifies nascent DNA strands after replication and thereby maintains DNA-methylation patterns. DNMT3a and DNMT3b are de novo methyltransferases that target unmethylated and hemimethylated sites and that initiate and establish DNA methylation. Deregulation of methyltransferases may lead to altered methylation patterns [11]. DNA methylation is known to be associated with the inactive chromatin state and, in most cases, with repressed gene expression activity, while the loss of DNA methylation oftentimes correlates with elevated gene expression.

The altered patterns of genomic DNA methylation constitute a well-known characteristic of cancer cells [1,2,11]. Both hypermethylation and hypomethylation alterations occur in cancer (Fig. 32.1). The DNA-methylation profile of cancer cells is characterized by global genome DNA hypomethylation, cancer-associated gene-specific hypomethylation, and concurrent hypermethylation of CpG islands within the gene promoters of tumor suppressors [1,2,11,12].

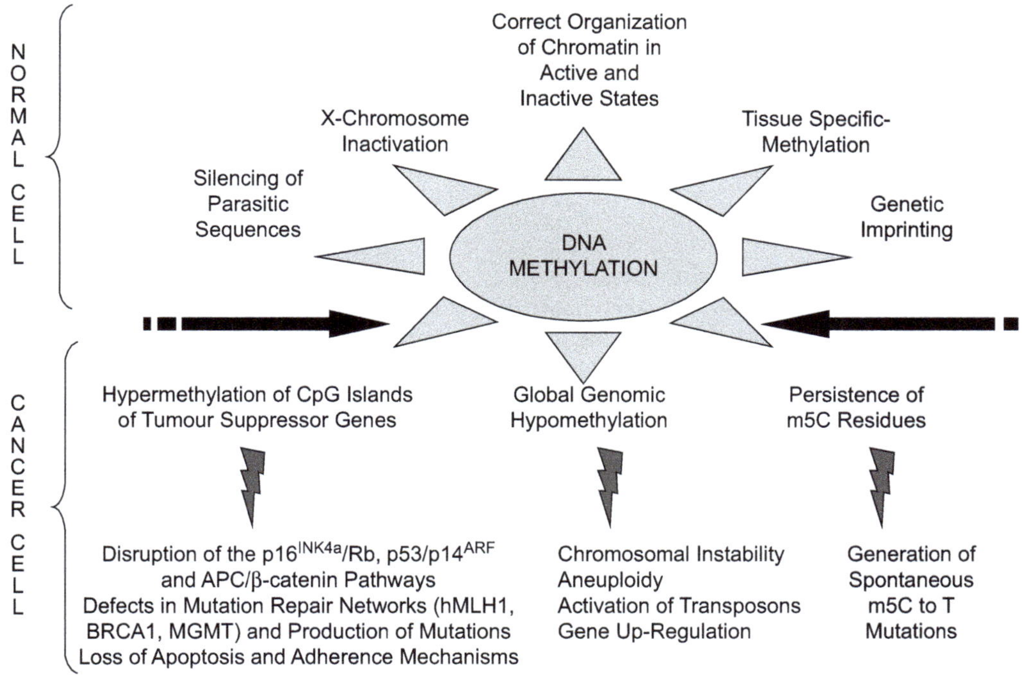

FIGURE 32.1 DNA methylation at the center of the normal and malignant behavior of the cell. *Adopted with permission from Esteller M, Herman JG. Cancer as an epigenetic disease: DNA methylation and chromatin alterations in human tumours. J Pathol 2002;196(1):1–7.*

The global loss of genomic DNA methylation has been linked to the activation of transposable elements, such as long and short interspersed nucleotide elements (LINEs and SINEs), retroviral intracisternal (A) particles (IAPs), and other elements located in the centromeric, pericentromeric, and subtelomeric chromosomal regions. It is also associated with elevated levels of chromosome breakage, aneuploidy, and increased mutation rates. All of these are signs of global genomic instability and the hallmarks of carcinogenesis [11].

Alongside global genomic DNA hypomethylation, cancer cells exhibit gene-specific hypomethylation. This is one of the prominent and potent mechanisms for the reactivation of oncogenes. Up to now, several hypomethylated tumor-promoting genes have been identified in major cancers. These include proto-oncogenes, plasminogen activators, urokinase (UPA), heparanase (HPA), and many others [1,2,11,12].

Hypermethylation, the gain of methylation at sites that are normally undermethylated, is another characteristic feature of cancer cells. Hypermethylation is the most extensively studied epigenetic change in cancer. Aberrant promoter methylation leads to silencing of a large number of protein-coding genes as well as genes coding for small RNAs [11]. Genes affected by DNA hypermethylation include crucial tumor-suppressor genes, such as the key gatekeeper p53 and retinoblastoma, cyclin-dependent kinase inhibitor 2A (p16INK4A), adenomatous polyposis coli (APC) gene, and Ras association (RalGDS/AF-6) domain family member 1 (RASSF1A), just to name a few. Hypermethylation may also affect DNA-repair genes, such as breast cancer 1 and 2 genes (BRCA1 and BRCA2), MutL homolog 1 (MLH1), and genes involved in apoptosis control [2,11].

In sum, in cancer, aberrant DNA methylation causes altered gene expression and regulation and leads to the deregulation of key processes that are critical for tumor initiation and progression, such as cell growth and persistent proliferative signaling, replicative immortality, resistance to apoptosis, signal transduction, inflammation, angiogenesis, and invasion [1,2,11,12] (Fig. 32.1).

2.2 Histone Modifications

The basic structure that comprises the chromatin is the nucleosome or beads on a string. It is essentially comprised of a strand of DNA being wrapped around an octamer of histone proteins containing a tetramer of H3–H4 histone proteins with an H2A–H2B dimer situated on either side. These four proteins are classified as core histones. A fifth histone, H1 (or linker histone), secures this structure and is involved in higher-order chromatin packing. These proteins are made up of a high proportion of positively charged amino acids, such as lysine and arginine, which allows them to interact electrostatically with net negatively charged DNA. The interactions between the nucleic acids and histone proteins in this structure function to organize and pack DNA and can have important implications for gene expression. Furthermore, histone proteins are often subjected to modifications, particularly on their N-terminal tails, which may alter their interactions with the associated DNA strand and contribute to the differential regulation of gene expression. Included in these alterations are acetylation, deacetylation, methylation, phosphorylation, and ubiquitination [13,14].

2.3 RNA-Induced Effects

This field was first initiated in the 1990s with the discovery of transgene silencing, first observed in *Petunia hybrida* in 1990 [15]. At first, the mechanism behind this phenomenon was unknown, but it was determined upon a later study that it was actually the transgene-derived RNA mediating the sequence-specific silencing. RNA-induced silencing was first seen in animals by Fire and Mello (1998) using *Caenorhabditis elegans*. Since these initial findings, the knowledge base surrounding RNA-mediated regulation of gene expression has expanded to include several groups of small RNA, including microRNA (miRNA), small interfering RNA (siRNA), and piwi-interacting RNA (piRNA) [16]. All are distinguished by their small size (ranging from 20 to 30 nucleotides), their role as a guiding agent for members of the Argonaute (Ago) family of proteins, and their involvement in reducing the expression of target genes [16]. Among those, miRNAs are key major negative regulators of expression of the vast majority of mammalian protein-coding genes [17]. miRNAs are key players in carcinogenesis and important biomarkers of cancer predisposition, development, and treatment outcomes [18–21].

Numerous environmental factors have been linked to aberrant changes in epigenetic pathways, both in experimental and epidemiological studies. In addition, epigenetic mechanisms may mediate specific mechanisms of toxicity and responses to certain chemicals [4].

3. EFFECTS OF METALS

Several studies have established an association between DNA methylation, and environmental metals and metalloids, including nickel, chromium, lead, cadmium, and particularly arsenic. Metals can cause oxidative stress because of the elevated production of reactive oxygen species (ROS) via redox signaling. Oxidative DNA damage has been shown to alter the activity of DNA methyltransferases and to affect their ability to interact with DNA, thus leading to aberrant DNA-methylation patterns.

Among toxic metals, cadmium is associated with several cancers, such as lung, kidney, uterine, and ovarian cancer [22–25], albeit it has very low direct mutagenic potential [26]. Cadmium exposure stems from tobacco smoke, air pollution, and diet, as well as from some occupational exposures [27]. Mechanistically, cadmium exposure has been shown to cause elevated ROS and reduced global genome DNA methylation via the noncompetitive inhibition of DNA methyltransferases [4,28,29]. Along with decreased global DNA methylation, cadmium also leads to hypomethylation and the aberrant expression of proto-oncogenes, thus inducing cellular proliferation and transformation [29]. Cadmium exposure has been shown to cause hypermethylation and decreased expression of the tumor-suppressor genes *RASSF1A* and *P16*, along with the overexpression and increased activity of DNMT activity [30,31]. Cadmium exposures cause aneuploidy that may be mediated by global DNA hypermethylation, which may be one of the mechanisms of cadmium-induced carcinogenesis [32]. Along with hypermethylation, cadmium can cause DNA hypomethylation of LINE1-transposable elements. Loss of LINE1 methylation is a common epigenetic event in malignancies and may also be important for cadmium-induced carcinogenesis [33].

Nickel, one of the most abundant metals, is found in coins, jewelry, stainless steel, batteries, and medical devices. Occupational nickel exposures occur during refinery, plating, and welding operations, and have been significantly associated with liver, lung, nasal, and pharyngeal cancers [34], although the precise mechanisms of nickel carcinogenicity are largely unknown [25]. Nickel exposures induce DNA damage, oxidative stress, and epigenetic alterations [35,36]. Studies during 1995 and 1998 showed that nickel exposure leads to promoter hypermethylation and increased global DNA methylation [37,38]. Nickel also causes significant posttranslational histone modification effects [27,39,40]. Nickel exposures has also been shown to cause hypermethylation and repression of several tumor-suppressor and DNA-repair genes, such as O6-methylguanine DNA methyltransferase, *MGMT*, *RARβ2*, *RASSF1A*, and the *P16* promoter [41,42].

Over the past few decades, environmental lead exposures have significantly decreased due to regulations banning or decreasing the use of lead in gasoline and paint [27,43], although the general population is still exposed due to tobacco smoke [44]. Lead exposures increase the risk of cardiovascular, kidney, and neurocognitive diseases and cancer [27]. Lead exposure has been reported to cause changes in DNA methylation and global gene expression, specifically, global hypomethylation [45–47], but the precise mechanisms of lead-induced changes in DNA methylation need to be further delineated.

Hexavalent chromium (Cr (VI)) is another well-known human and animal carcinogen [48]. People are environmentally exposed to chromium by drinking chromium-contaminated water or by using chromium-containing products, such as dyes, paints, inks, and plastics. Occupational chromium exposure has been associated with the elevated incidence of lung, stomach, liver, and kidney cancer [49–52]. Even though Cr (VI) is a known mutagen, it exerts its carcinogenicity largely through epigenetic mechanisms [25]. Several previous studies have highlighted the potential epigenetic effects of chromium (Cr(VI)), especially DNA-methylation effects [53]. Chromate Cr (VI) exposure has been associated with the hypermethylation of certain genes involved in the cell cycle control and DNA repair, such as *MLH1*, *p16*, *and APC* genes [54–56].

Arsenic is a common metalloid element and a human carcinogen [25]. Arsenic exposure exerts acute and chronic toxicity and has been associated with skin, lung, liver, and bladder cancers, as well as cardiovascular and neurological diseases [57–59]. Altered DNA methylation is a known sign of arsenic-induced carcinogenesis, and arsenic has been shown to cause both DNA hypermethylation and hypomethylation. Several studies have associated arsenic with gene-specific hypermethylation, whereby arsenic causes hypermethylation of the *P53* and *P16* genes, as well as the death-associated protein kinase (*DAPK*) gene. Additionally, arsenic causes an increase in global 5mC, indicative of global DNA hypermethylation [60]. Along with hypermethylation, global DNA hypomethylation is an early event in some cancers and occurs in response to arsenic exposure [61]. Moreover, arsenic exposures also affect other epigenetic mechanisms, such as histone modifications.

Overall, epigenetic changes play intricate roles in the regulation of gene expression upon metal exposures. The unique methylome alterations have been displayed in cancer cells after exposure to carcinogenic metals, such as nickel, lead, arsenic, cadmium, and chromium (VI). The metal-stimulated deviations to the methylome are possible mechanisms for metal-induced carcinogenesis and may provide potential biomarkers for cancer detection. These mechanisms are discussed in depth in an elegant review by Brocato and Costa (2013) [25].

Moreover, in-depth systematic analyses carried out in 2015 established that miRNA changes are caused by exposures to toxic chemicals and may be sensitive biomarkers of toxicant exposure [62,63]. Also, miRNAs are important players in arsenic-induced carcinogenesis. Luo et al., in 2014, demonstrated that arsenic-induced malignant transformation of lung HBE cells is associated with an increased expression of oncogenic miR-21 [64]. This miRNA has been found to be overexpressed in virtually all human cancers and is implicated in the carcinogenic process through the regulation of cell proliferation, genome instability, inflammation, evading apoptosis, invasion and metastasis, and angiogenesis [63]. Arsenic exposure can lead to upregulated miR-190 and result in downregulation of the PH domain leucine-rich repeat protein phosphatase (PHLPP) through the direct interaction of miR-190 with the 3′-UTR of the PHLPP mRNA, leading to activation of the AKT-signaling pathway [65].

Similarly to HBE cells, miR-21 can be upregulated by arsenic in immortalized human keratinocytes [66]. Additionally, arsenic exposure can lead to the reduction of let-7a, let-7b, let7c, and miR-34a during malignant transformation [63,67]. In arsenic-induced bladder carcinogenesis, arsenic treatment has resulted in the downregulation of the anti-EMT miRNAs, miR-200a, miR-200b, and miR-200c. In transformed human prostate epithelial cells, arsenic exposure has been shown to cause the downregulation of miR-134, miR-138, miR-155, miR-181c, miR-181d, miR-205, miR-373, and let-7 [68]. Decreased expression of the aforementioned miRNAs has been associated with the overexpression of target genes, RAN, RAB22A, RAB27A, and KRAS, which are key regulators of carcinogenesis. Additionally, arsenic exposure can cause the pronounced upregulation of miR-9, a small RNA that targets the miRNA-processing enzyme DICER1 [69].

Nickel exposure causes dose-dependent elevated miRNA-21 expression and promotes lung tumorigenesis [70]. Furthermore, the level of miR-152 can be significantly downregulated in nickel-transformed cells compared to passage-matched control cells. Additionally, in experimental nickel-induced carcinogenesis, miR-222 was upregulated, and miR-203 was downregulated [71–73].

Chromium exposure causes decreased levels of miR-143. He and colleagues suggested that the downregulation of miR-143 promotes chromium-induced cell transformation through increasing the expression of the insulin-like growth factor-1 receptor and the insulin receptor substrate-1 and through further activation of the ERK/hypoxia-induced factor 1α/NF-κB-signaling pathway [74].

In comparison to arsenic, little is known about the role of miRNA deregulation in cadmium-induced carcinogenesis. An analysis made in 2015 and 2016 showed that cadmium exposure results in the significant deregulation of multiple miRNAs that are predicted to target cell-cycle regulation, p53-signaling, and Wnt-signaling pathways [75,76].

4. TAMOXIFEN EFFECTS

Tamoxifen is a selective nonsteroidal anti-estrogen that has been used in the treatment of breast cancer since mid-1980s. It has been used lately as an effective chemopreventive agent for breast cancer in women who have a high risk of developing breast cancer [77,78]. While tamoxifen has proven to be beneficial for preventing the occurrence or recurrence of breast cancer [79], the IARC has classified it as a known human carcinogen, since it has been shown to increase the incidence of endometrial cancer [80,81]. Additionally, tamoxifen is a potent hepatocarcinogen in rodents, where it exhibits both cancer-initiating and cancer-promoting properties [82–85].

Mechanistically, tamoxifen-induced hepatic tumors in rats occur, at least in part, due to a genotoxic mechanism resulting from the formation of tamoxifen–DNA adducts [86–90]. Along with genotoxic mechanisms, tamoxifen causes profound gene expression and epigenetic changes [91–94].

High-throughput microarray technology has allowed researchers to establish the gene expression profiles in liver tissues during the early stages of tamoxifen-induced rat hepatocarcinogenesis. Global gene expression profiling of the liver tissues of rats treated with tamoxifen for 12 or 24 weeks indicated that the early stages of tamoxifen-induced liver carcinogenesis are characterized by alterations in several major cellular pathways, such as those involved in drug metabolism, lipid metabolism, cell cycle, apoptosis, and cell-proliferation control. Tamoxifen exposure can cause significant, progressive, and sustained increases in expression of the Pdgfc, Calb3, Ets1, and Ccnd1 genes, accompanied by an elevated level of the PI3K, p-PI3K, Akt1/2, Akt3, and cyclin B, D1, and D3 proteins [93].

An analysis of tamoxifen-induced epigenetic changes revealed pronounced global genomic DNA demethylation and altered activity and expression maintenance DNA methyltransferase (DNMT1) and de novo (DNMT3a and DNMT3b) DNA methyltransferases. Tamoxifen-induced DNA hypomethylation was paralleled by the progressive loss of histone H4K20me3 in liver tissues of tamoxifen-treated animals [92] and the loss of global histone H4 lysine 20 trimethylation [92]. Tamoxifen exposure also caused accumulation of DNA lesions in the liver tissues of tamoxifen-treated female F344 rats. Moreover, long-term exposure of female F344 rats to tamoxifen led to a substantial and progressive loss of CpG methylation in the regulatory sequences of LINE-1 and a subsequent pronounced increase in the levels of expression of LINE-1 elements and the c-myc proto-oncogene, that are known manifestations of genome instability. The accumulation of tamoxifen-induced changes was accompanied by the decreased level of key DNA-repair proteins, Rad51, Ku70, and DNA polymerase beta, which are very important for the maintenance of genome stability. Molecular and epigenetic changes were paralleled by increased regenerative cell proliferation, and taken together, the data showed that exposure of animals to tamoxifen led to the emergence of cancer-related epigenetic phenotypes prior to tumor formation [91].

In another study, the treatment of Fisher 344 rats to tamoxifen for 24 weeks caused substantial changes in the expression of miRNA genes in the liver. In this study, tamoxifen exposure caused a significant upregulation of known oncogenic miRNAs, such as the 17–92 cluster, miR-106a, and miR-34, and miRNA changes resulted in corresponding changes in the expression of proteins targeted by these miRNAs, including cell-cycle regulators, chromatin modifiers, and the expression regulators implicated in the regulation of genome stability and carcinogenesis. Moreover, the observed tamoxifen-induced miRNA changes occur prior to tumor formation and are not merely a consequence of a transformed state [94].

5. EFFECTS OF 1,3-BUTADIENE

Environmental contamination by numerous industrial chemicals is becoming a serious global problem. The gaseous olefin 1,3-butadiene is one such industrial chemical that is widely used in the production of plastic, rubber, and resins. This highly volatile chemical constitutes a key component of industrial and automobile exhaust and is also found in cigarette smoke. Furthermore, it is commonly found in urban ambient air and industrial complexes [95]. The International Agency for Research on Cancer (IARC) has classified 1,3-butadiene as a known human and rodent carcinogen that is associated with lung, liver, and hematopoietic system cancers [95–97].

1,3-Butadiene is a well-known genotoxic carcinogen, and the main mechanism of tumor induction by 1,3-butadiene exposure is the formation of its highly reactive metabolic epoxides (1,2-epoxy-3-butene, 1,2:3,4-diepoxybutane, and 3,4-epoxy-1,2-butanediol), which cause the formation of DNA adducts [96]. While its genotoxic potential has been established, this potential does not fully explain all of the carcinogenic mechanisms of 1,3-butadiene effects. Studies during 2011 and 2014 by the Pogribny and Rusyn laboratories have established that 1,3-butadiene profoundly alters gene expression and epigenetic processes in the affected cells and tissues [98–100].

In 2011, in a series of elegant studies, murine exposure to 1,3-butadiene also resulted in profound changes in global gene expression patterns, causing altered expression of Hmox1, Nqo1, Car3, Srebf1, and Lgr5, all of which are indicative of liver injury. The short-term inhalation of 1,3-butadiene in C57BL/6J mice caused, along with DNA adduct formation, extensive epigenetic changes such as a significant decrease in global DNA methylation, marked hypomethylation of repetitive elements, and a pronounced loss of histone H3K9, H3K27, and H4K20 trimethylation that are signs of genome instability in the liver tissue [99].

The methylation of lysine residues 9 and 27 at histone H3 and lysine 20 at histone H4 is crucially important for heterochromatin formation and maintenance as well as for transcriptional repression and general stability of the genome.

Therefore, the observed 1,3-butadiene exposure-induced loss of H3K9 and H4K20 trimethylation may in turn contribute to chromatin relaxation and overall genome instability. Indeed, a 2011 report further proved the decondensation of chromatin and activation of repetitive elements in the livers of 1,3-butadiene-exposed C57BL/6J mice [98,99].

Such an open chromatin structure may in turn increase the vulnerability of DNA to the influence of genotoxic DNA-reactive metabolites of 1,3-butadiene and contribute to adduct formation. Analysis also revealed significant interstrain differences in genetic and epigenetic responses to the inhalational of 1,3-butadiene in murine liver tissues, and these strain differences are associated with differences in histone H3K9, H3K27, and H4K20 methylation levels and alterations in chromatin structure [98].

Furthermore, epigenetic changes may underlie the tissue specificity of 1,3-butadiene-induced genome instability and tumorigeneses. A 2014 study by the Rusyn Laboratory analyzed 1,3-butadiene-induced changes in the kidney, liver, and lung tissues of mice that had had inhalational exposure. They noted that while 1,3-butadiene exposure caused DNA damage in all three tissues, epigenetic changes varied between the kidney, liver, and lung tissues of the exposed animals. 1,3-Butadiene-induced epigenetic changes indicative of genome instability included demethylation of repetitive DNA sequences and alterations in histone–lysine acetylation levels observed in the liver and lung tissues of the exposed mice. On the other hand, no DNA-methylation changes were seen in the kidneys of the exposed mice. Moreover, the histone marks of condensed heterochromatin and transcriptional silencing (histone–lysine trimethylation) were increased in kidney tissue, suggesting genome-stabilization effects. Therefore, epigenetic in-tissue differences may help to explain the differences in cancer predisposition. These modifications may represent a potential mechanistic explanation for the tissue specificity of cancer predisposition upon exposure to 1,3-butadiene [100]. As of 2016, no studies have analyzed the effects of 1,3-butadiene exposure on tissue microRNAome, and such analysis may provide more insight into the mechanisms of 1,3-butadiene-induced carcinogenesis.

6. INFLUENCE OF POLYCYCLIC AROMATIC HYDROCARBONS

Polycyclic aromatic hydrocarbons (PAHs) are the most widespread organic pollutants found in the environment. They are extensively present in crude oil, coal, and tar deposits and originate from the burning of fossil fuels and forest fires. Other significant PAH exposures stem from automobile exhaust, cigarette smoke, industrial exposure at coal-tar production plants, and municipal trash incinerators [101]. PAHs also come from dietary fats and overused cooking oils [102,103]. Human exposure to PAHs is associated with a wide array of diseases and conditions such as asthma, obstructive lung disease, heart disease, as well as lung, bladder, and other cancers [101,103]. PAH exposure significantly affects the in utero development of children, having been associated with cognitive defects and fetal growth impairment. One of the most sensitive periods for PAH exposure is during early-embryonic development. A large-scale longitudinal cohort study involving 700 children revealed significant epigenetic effects from transplacental PAH exposure [104]. Further studies have shown that maternal PAH exposure leads to aberrant global DNA methylation and gene-specific methylation, as well as the accumulation of DNA adducts in cord blood [103,105].

PAH exposures reportedly cause formation of DNA adducts, and they have also been associated with increased methylation levels of *Alu* and *LINE-1* and abnormal DNA-methylation patterns at specific sequences of the *p53* gene promoter, which correlate with the levels of PAH exposure seen in chronically exposed industrial coke–oven workers [106]. In occupationally exposed firefighters, PAH exposure has led to a higher level of *DUSP22* promoter hypomethylation in blood DNA, when compared with unexposed controls [107].

The carcinogenic activity of PAHs has often been associated with the induction of genotoxic and non-genotoxic alterations that both lead to genome instability [63]. Among those, exposure to the PAH benzo[*a*]pyrene has led to very pronounced formation of *anti*-7β, 8α-dihydroxy-9α, 10α-epoxy-7, 8, 9,10-tetrahydrobenzo[*a*]pyrene adducts. Interestingly, these occurred at the major hot-spot mutation codons 157, 248, and 273 of the *P53* gene and at codon 14 in the human *KRAS* gene—key genes implicated in a wide array of cancers [108,109]. Along with adducts, PAH benzo[*a*]pyrene exposures have also been shown to cause alterations in global and gene-specific DNA-methylation status and to affect genome stability in cancer-targeted tissues and lines [103,110].

PAHs, and especially PAH benzo[*a*]pyrene (BP) exposure, have been shown to cause changes in the expression of several miRNAs that may contribute to genome instability and malignant cell transformation. To this effect, a study by Shen and colleagues showed that benzo[*a*]pyrene exposure caused neoplastic transformation of normal human bronchial epithelial cells, which they associated with the profound upregulation of 45 and the downregulation of nine miRNAs [111]. Also upregulated were oncogenic miRNAs such as miR-17-5p, miR-20a, miR-92, miR-106a, miR-129, miR-320, miR-494, and miR498. Conversely, miR-10a, miR-363*, and miR-493-5p were significantly downregulated [111].

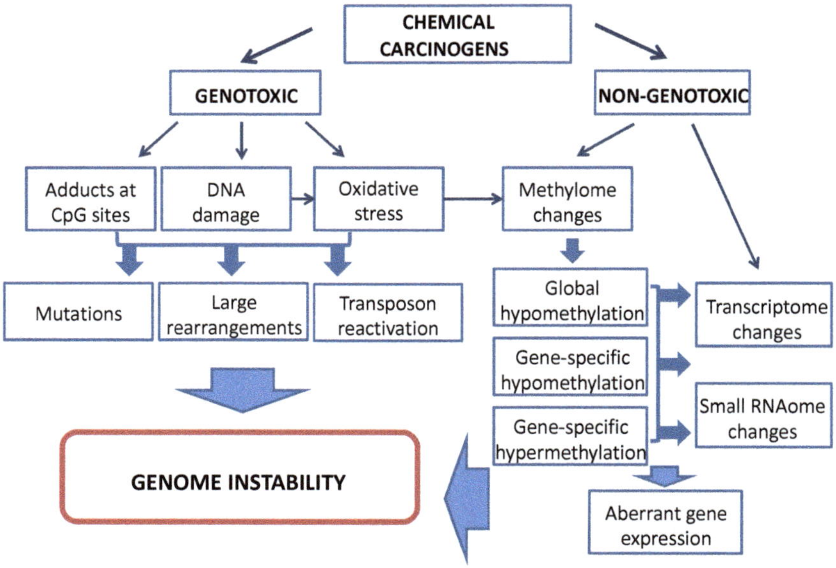

FIGURE 32.2 **Genetic and epigenetic effects of carcinogens on genome stability.** Exposure to genotoxic chemical carcinogens causes direct or indirect DNA damage, as well as effects on methylome, and subsequent aberrant global gene expression. Additionally, genotoxic effects on the methylome can lead to the increased levels of carcinogen–DNA-adduct formation at methylated CpG sites. Unrepaired or misrepaired damage can lead to genome instability. Non-genotoxic carcinogens affect methylome, cause oxidative stress, and cause global loss of DNA methylation, gene-specific hypermethylation, and gene-specific hypomethylation. Moreover, DNA hypomethylation can also lead to elevated mutation rates and genome instability. *Modified after Pogribny IP, Beland FA. DNA methylome alterations in chemical carcinogenesis. Cancer Lett 2013;334(1):39–45.*

BP-induced upregulation of miR-17-92 and miR-106a was important in the BP-induced malignant transformation of HBE cells. Increasing miR-106a levels in the transformed cells caused increased proliferation and inhibited apoptosis. Contrarily, inhibition of oncogenic miR-106a in the BP-transformed cells inhibited cell proliferation, induced cell arrest and apoptosis, and also inhibited growth of tumor xenografts in nude mice [112].

Several other studies reported BP-induced upregulation of miR-22, miR-494, and miR-638 and downregulation of miR-10a, miR-34c, and miR-506 during BP-induced carcinogenesis.

MicroRNAs were also deregulated in peripheral lymphocytes of PAH-exposed workers; among those, upregulation occurred in miR-638 [113] and other miRNAs that target tumor-suppressor proteins such as miR-20a, miR-17-5p, miR-106a, miR-494, miR-22, and miR-34c.

Several miRNAs were downregulated by BP exposure; among those, miR-10a was one of the most downregulated miRNAs in BP-transformed human 16HBE cells and in the lungs of rats exposed to BP-containing tobacco smoke [111,114].

7. CONCLUSIONS

Numerous environmental toxicants and chemicals have been reported to cause both genetic and epigenetic effects that play a key role in chemical-induced genome instability and carcinogenesis. These include endocrine disruptors—important environmental xenobiotics that interfere with the normal development and functioning of male and female reproductive systems and cause numerous health effects. The most studied endocrine disruptors include vinclozolin, methoxychlor, and other pesticides, as well as plasticiser bisphenol A, all of which act at different levels of epigenetic control and cause transgenerational effects [115–119].

Mycotoxins are toxic compounds of fungal origin that are common contaminants in human and animal food products. Among these, aflatoxin B_1, fumonisin B1, and ochratoxin are known to be possible human carcinogens [120,121]. While they are genotoxic carcinogens, they have also been reported to cause a complex network of epigenetic alterations [122–124]. Both genetic and epigenetic mechanisms of toxicity and carcinogenicity were described for alcohol exposure, cigarette smoke, chemotherapy agents, and many other toxicants. Many of these were described in several elegant review articles published in 2015 [63,125–131].

In sum, epigenetic modifications are influenced by the environment and environmental toxicants, including chemical carcinogens. The majority of genotoxic and non-genotoxic carcinogens affect genome stability and cause cancer by perturbing epigenetic processes in the cells (Fig. 32.2). These carcinogen-induced epigenetic changes are stable and can be used as important exposure biomarkers. On the other hand, epigenetic changes are pliable and reversible, and, therefore, analysis of chemical carcinogens-induced epigenetic alterations may uncover novel mechanism-based approaches for cancer treatment and cancer prevention. Furthermore, in the future, the incorporation of epigenetic technologies in carcinogenesis analysis and cancer-risk assessment will enhance the efficiency of carcinogenicity testing.

GLOSSARY

Genotoxic carcinogen A chemical that is capable of producing cancer by directly altering the genetic material of target cells.
Non-genotoxic carcinogen A chemical capable of producing cancer by some secondary mechanism not related to direct gene damage.

LIST OF ABBREVIATIONS

IAPs Intracisternal (A) particles
LINEs Long interspersed nucleotide elements
miRNA MicroRNA
PAHs Polycyclic aromatic hydrocarbons
piRNA piwi-interacting RNA
SINEs Short interspersed nucleotide elements
siRNA Small interfering RNA

REFERENCES

[1] You JS, Jones PA. Cancer genetics and epigenetics: two sides of the same coin? Cancer Cell 2012;22(1):9–20.
[2] Sandoval J, Esteller M. Cancer epigenomics: beyond genomics. Curr Opin Genet Dev 2012;22(1):50–5.
[3] Pogribny IP, Rusyn I, Beland FA. Epigenetic aspects of genotoxic and non-genotoxic hepatocarcinogenesis: studies in rodents. Environ Mol Mutagen 2008;49(1):9–15.
[4] Baccarelli A, Bollati V. Epigenetics and environmental chemicals. Curr Opin Pediatr 2009;21(2):243–51.
[5] Hayashi Y. Overview of genotoxic carcinogens and non-genotoxic carcinogens. Exp Toxicol Pathol 1992;44(8):465–71.
[6] Cheng KC, Loeb LA. Genomic instability and tumor progression: mechanistic considerations. Adv Cancer Res 1993;60:121–56.
[7] Goldberg Z. Clinical implications of radiation-induced genomic instability. Oncogene 2003;22(45):7011–7.
[8] Little JB. Genomic instability and radiation. J Radiol Prot 2003;23(2):173–81.
[9] Loeb LA, Loeb KR, Anderson JP. Multiple mutations and cancer. Proc Natl Acad Sci USA 2003;100(3):776–81.
[10] Jaenisch R, Bird A. Epigenetic regulation of gene expression: how the genome integrates intrinsic and environmental signals. Nat Genet 2003;33(Suppl.):245–54.
[11] Pogribny IP, Beland FA. DNA methylome alterations in chemical carcinogenesis. Cancer Lett 2013;334.
[12] Esteller M, Herman JG. Cancer as an epigenetic disease: DNA methylation and chromatin alterations in human tumours. J Pathol 2002;196(1):1–7.
[13] Jenuwein T, Allis CD. Translating the histone code. Science 2001;293(5532):1074–80.
[14] McGowan PO, Meaney MJ, Szyf M. Diet and the epigenetic (re)programming of phenotypic differences in behavior. Brain Res 2008;1237:12–24.
[15] Napoli C, Lemieux C, Jorgensen R. Introduction of a Chimeric Chalcone Synthase gene into Petunia results in reversible Co-Suppression of homologous genes in trans. Plant Cell 1990;2(4):279–89.
[16] Ghildiyal M, Zamore PD. Small silencing RNAs: an expanding universe. Nat Rev Genet 2009;10(2):94–108.
[17] Guo H, Ingolia NT, Weissman JS, Bartel DP. Mammalian microRNAs predominantly act to decrease target mRNA levels. Nature 2010;466(7308):835–40.
[18] Croce CM. Causes and consequences of microRNA dysregulation in cancer. Nat Rev Genet 2009;10(10):704–14.
[19] Heyns M, Kovalchuk O. Non-coding RNAs including miRNAs, piRNAs, and tRNAs in human cancer. Oncotarget 2015;6(27):23055–7.
[20] Koturbash I, Zemp FJ, Pogribny I, Kovalchuk O. Small molecules with big effects: the role of the microRNAome in cancer and carcinogenesis. Mutat Res 2011;722(2):94–105.
[21] Krishnan P, Ghosh S, Wang B, Li DP, Narasimhan A, Berendt R, et al. Next generation sequencing profiling identifies miR-574-3p and miR-660-5p as potential novel prognostic markers for breast cancer. BMC Genomics 2015;16.
[22] Beveridge R, Pintos J, Parent ME, Asselin J, Siemiatycki J. Lung cancer risk associated with occupational exposure to nickel, chromium VI, and cadmium in two population-based case-control studies in Montreal. Am J Ind Med 2010;53(5):476–85.
[23] Chen C, Xun P, Nishijo M, He K. Cadmium exposure and risk of lung cancer: a meta-analysis of cohort and case-control studies among general and occupational populations. J Expo Sci Environ Epidemiol 2016.
[24] Song JK, Luo H, Yin XH, Huang GL, Luo SY, Lin DR, et al. Association between cadmium exposure and renal cancer risk: a meta-analysis of observational studies. Sci Rep-UK 2015;5.
[25] Brocato J, Costa M. Basic mechanics of DNA methylation and the unique landscape of the DNA methylome in metal-induced carcinogenesis. Crit Rev Toxicol 2013;43(6):493–514.
[26] Filipic M, Hei TK. Mutagenicity of cadmium in mammalian cells: implication of oxidative DNA damage. Mutat Res-Fund Mol M 2004;546(1–2):81–91.
[27] Ruiz-Hernandez A, Kuo CC, Rentero-Garrido P, Tang WY, Redon J, Ordovas JM, et al. Environmental chemicals and DNA methylation in adults: a systematic review of the epidemiologic evidence. Clin Epigenetics 2015;7(1):55.
[28] Huang DJ, Zhang YM, Qi YM, Chen C, Ji WH. Global DNA hypomethylation, rather than reactive oxygen species (ROS), a potential facilitator of cadmium-stimulated K562 cell proliferation. Toxicol Lett 2008;179(1):43–7.

[29] Takiguchi M, Achanzar WE, Qu W, Li GY, Waalkes MP. Effects of cadmium on DNA-(Cytosine-5) methyltransferase activity and DNA methylation status during cadmium-induced cellular transformation. Exp Cell Res 2003;286(2):355–65.

[30] Benbrahim-Tallaa L, Waterland RA, Dill AL, Webber MM, Waalkes MP. Tumor suppressor gene inactivation during cadmium-induced malignant transformation of human prostate cells correlates with overexpression of de novo DNA methyltransferase. Environ Health Perspect 2007;115(10):1454–9.

[31] Jiang G, Xu L, Song S, Zhu C, Wu Q, Zhang L, et al. Effects of long-term low-dose cadmium exposure on genomic DNA methylation in human embryo lung fibroblast cells. Toxicology 2008;244(1):49–55.

[32] Inglot P, Lewinska A, Potocki L, Oklejewicz B, Tabecka-Lonczynska A, Koziorowski M, et al. Cadmium-induced changes in genomic DNA-methylation status increase aneuploidy events in a pig Robertsonian translocation model. Mutat Res 2012;747(2):182–9.

[33] Hossain MB, Vahter M, Concha G, Broberg K. Low-level environmental cadmium exposure is associated with DNA hypomethylation in Argentinean women. Environ Health Persp 2012;120(6):879–84.

[34] Chen K, Liao QL, Ma ZW, Jin Y, Hua M, Bi J, et al. Association of soil arsenic and nickel exposure with cancer mortality rates, a town-scale ecological study in Suzhou, China. Environ Sci Pollut Res Int 2015;22(7):5395–404.

[35] Pool-Zobel BL, Lotzmann N, Knoll M, Kuchenmeister F, Lambertz R, Leucht U, et al. Detection of genotoxic effects in human gastric and nasal mucosa cells isolated from biopsy samples. Environ Mol Mutagen 1994;24(1):23–45.

[36] Arita A, Costa M. Epigenetics in metal carcinogenesis: nickel, arsenic, chromium and cadmium. Metallomics 2009;1(3):222–8.

[37] Lee YW, Broday L, Costa M. Effects of nickel on DNA methyltransferase activity and genomic DNA methylation levels. Mutat Res 1998;415(3):213–8.

[38] Lee YW, Klein CB, Kargacin B, Salnikow K, Kitahara J, Dowjat K, et al. Carcinogenic nickel silences gene expression by chromatin condensation and DNA methylation: a new model for epigenetic carcinogens. Mol Cell Biol 1995;15(5):2547–57.

[39] Yan Y, Kluz T, Zhang P, Chen HB, Costa M. Analysis of specific lysine histone H3 and H4 acetylation and methylation status in clones of cells with a gene silenced by nickel exposure. Toxicol Appl Pharmacol 2003;190(3):272–7.

[40] Sorahan T, Williams SP. Mortality of workers at a nickel carbonyl refinery, 1958-2000. Occup Environ Med 2005;62(2):80–5.

[41] Zhang J, Zhang J, Li M, Wu Y, Fan Y, Zhou Y, et al. Methylation of RAR-beta2, RASSF1A, and CDKN2A genes induced by nickel subsulfide and nickel-carcinogenesis in rats. Biomed Environ Sci 2011;24(2):163–71.

[42] Govindarajan B, Klafter R, Miller MS, Mansur C, Mizesko M, Bai XH, et al. Reactive oxygen-induced carcinogenesis causes hypermethylation of p16(Ink4a) and activation of MAP kinase. Mol Med 2002;8(1):1–8.

[43] Muntner P, Menke A, DeSalvo KB, Rabito FA, Batuman V. Continued decline in blood lead levels among adults in the United States – the National Health and Nutrition Examination Surveys. Arch Intern Med 2005;165(18):2155–61.

[44] Apostolou A, Garcia-Esquinas E, Fadrowski JJ, McLain P, Weaver VM, Navas-Acien A. Secondhand tobacco smoke: a source of lead exposure in US children and adolescents. Am J Public Health 2012;102(4):714–22.

[45] Bihaqi SW, Zawia NH. Alzheimer's disease biomarkers and epigenetic intermediates following exposure to Pb in vitro. Curr Alzheimer Res 2012;9(5):555–62.

[46] Senut MC, Sen A, Cingolani P, Shaik A, Land SJ, Ruden DM. Lead exposure disrupts global DNA methylation in human embryonic stem cells and alters their neuronal differentiation. Toxicol Sci 2014;139(1):142–61.

[47] Dosunmu R, Alashwal H, Zawia NH. Genome-wide expression and methylation profiling in the aged rodent brain due to early-life Pb exposure and its relevance to aging. Mech Ageing Dev 2012;133(6):435–43.

[48] Schnedl W, Dev VG, Tantravahi R, Miller DA, Erlanger BF, Miller OJ. 5-Methylcytosine in heterochromatic regions of chromosomes: chimpanzee and gorilla compared to the human. Chromosoma 1975;52(1):59–66.

[49] Gibb HJ, Lees PSJ, Pinsky PF, Rooney BC. Lung cancer among workers in chromium chemical production. Am J Ind Med 2000;38(2):115–26.

[50] Koh DH, Kim TW, Jang SH, Ryu HW. Cancer mortality and incidence in cement industry workers in Korea. Saf Health Work 2011;2(3):243–9.

[51] Halasova E, Matakova T, Kavcova E, Musak L, Letkova L, Adamkov M, et al. Human lung cancer and hexavalent chromium exposure. Neuro Endocrinol Lett 2009;30(Suppl. 1):182–5.

[52] Linos A, Petralias A, Christophi CA, Christoforidou E, Kouroutou P, Stoltidis M, et al. Oral ingestion of hexavalent chromium through drinking water and cancer mortality in an industrial area of Greece – an ecological study. Environ Health 2011;10.

[53] Lou J, Wang Y, Chen J, Ju L, Yu M, Jiang Z, et al. Effects of soluble and particulate Cr(VI) on genome-wide DNA methylation in human B lymphoblastoid cells. Mutat Res Genet Toxicol Environ Mutagen 2015;792:12–8.

[54] Kondo K, Takahashi Y, Hirose Y, Nagao T, Tsuyuguchi M, Hashimoto M, et al. The reduced expression and aberrant methylation of p16(INK4a) in chromate workers with lung cancer. Lung Cancer-J Iaslc 2006;53(3):295–302.

[55] Takahashi Y, Kondo K, Ishikawa S, Uchihara H, Fujino H, Sawada N, et al. Microscopic analysis of the chromium content in the chromium-induced malignant and premalignant bronchial lesions of the rat. Environ Res 2005;99(2):267–72.

[56] Ali AH, Kondo K, Namura T, Senba Y, Takizawa H, Nakagawa Y, et al. Aberrant DNA methylation of some tumor suppressor genes in lung cancers from workers with chromate exposure. Mol Carcinog 2011;50(2):89–99.

[57] Jensen TJ, Novak P, Eblin KE, Gandolfi AJ, Futscher BW. Epigenetic remodeling during arsenical-induced malignant transformation. Carcinogenesis 2008;29(8):1500–8.

[58] Tseng CH, Huang YK, Huang YL, Chung CJ, Yang MH, Chen CJ, et al. Arsenic exposure, urinary arsenic speciation, and peripheral vascular disease in blackfoot disease-hyperendemic villages in Taiwan. Toxicol Appl Pharmacol 2005;206(3):299–308.

[59] Wasserman GA, Liu X, Parvez F, Ahsan H, Factor-Litvak P, van Geen A, et al. Water arsenic exposure and children's intellectual function in Araihazar, Bangladesh. Environ Health Perspect 2004;112(13):1329–33.

[60] Majumdar S, Chanda S, Ganguli B, Mazumder DN, Lahiri S, Dasgupta UB. Arsenic exposure induces genomic hypermethylation. Environ Toxicol 2010;25(3):315–8.

[61] Nohara K, Baba T, Murai H, Kobayashi Y, Suzuki T, Tateishi Y, et al. Global DNA methylation in the mouse liver is affected by methyl deficiency and arsenic in a sex-dependent manner. Arch Toxicol 2011;85(6):653–61.

[62] Vrijens K, Bollati V, Nawrot TS. MicroRNAs as potential signatures of environmental exposure or effect: a systematic review. Environ Health Persp 2015;123(5):399–411.

[63] Pogribny IP, Beland FA, Rusyn I. The role of microRNAs in the development and progression of chemical-associated cancers. Toxicol Appl Pharmacol 2015.

[64] Luo F, Ji J, Liu Y, Xu Y, Zheng G, Jing J, et al. MicroRNA-21, up-regulated by arsenite, directs the epithelial-mesenchymal transition and enhances the invasive potential of transformed human bronchial epithelial cells by targeting PDCD4. Toxicol Lett 2014;232(1):301–9.

[65] Beezhold K, Liu J, Kan H, Meighan T, Castranova V, Shi X, et al. miR-190-mediated downregulation of PHLPP contributes to arsenic-induced Akt activation and carcinogenesis. Toxicol Sci 2011;123(2):411–20.

[66] Lu XL, Luo F, Liu Y, Zhang AH, Li J, Wang BR, et al. The IL-6/STAT3 pathway via miR-21 is involved in the neoplastic and metastatic properties of arsenite-transformed human keratinocytes. Toxicol Lett 2015;237(3):191–9.

[67] Herbert KJ, Holloway A, Cook AL, Chin SP, Snow ET. Arsenic exposure disrupts epigenetic regulation of SIRT1 in human keratinocytes. Toxicol Appl Pharmacol 2014;281(1):136–45.

[68] Ngalame NN, Tokar EJ, Person RJ, Xu Y, Waalkes MP. Aberrant microRNA expression likely controls RAS oncogene activation during malignant transformation of human prostate epithelial and stem cells by arsenic. Toxicol Sci 2014;138(2):268–77.

[69] Leucci E, Zriwil A, Gregersen LH, Jensen KT, Obad S, Bellan C, et al. Inhibition of miR-9 de-represses HuR and DICER1 and impairs Hodgkin lymphoma tumour outgrowth in vivo. Oncogene 2012;31(49):5081–9.

[70] Chiou YH, Liou SH, Wong RH, Chen CY, Lee H. Nickel may contribute to EGFR mutation and synergistically promotes tumor invasion in EGFR-mutated lung cancer via nickel-induced microRNA-21 expression. Toxicol Lett 2015;237(1):46–54.

[71] Zhang J, Zhou Y, Ma L, Huang S, Wang R, Gao R, et al. The alteration of miR-222 and its target genes in nickel-induced tumor. Biol Trace Elem Res 2013;152(2):267–74.

[72] Zhang J, Zhou Y, Wu YJ, Li MJ, Wang RJ, Huang SQ, et al. Hyper-methylated miR-203 dysregulates ABL1 and contributes to the nickel-induced tumorigenesis. Toxicol Lett 2013;223(1):42–51.

[73] Ji W, Yang L, Yuan J, Yang L, Zhang M, Qi D, et al. MicroRNA-152 targets DNA methyltransferase 1 in NiS-transformed cells via a feedback mechanism. Carcinogenesis 2013;34(2):446–53.

[74] He J, Qian X, Carpenter R, Xu Q, Wang L, Qi Y, et al. Repression of miR-143 mediates Cr (VI)-induced tumor angiogenesis via IGF-IR/IRS1/ERK/IL-8 pathway. Toxicol Sci 2013;134(1):26–38.

[75] Liu Q, Zheng CJ, Shen HY, Zhou ZH, Lei YX. MicroRNAs-mRNAs expression profile and their potential role in malignant transformation of human bronchial epithelial cells induced by cadmium. Biomed Res Int 2015;2015.

[76] Humphries B, Wang Z, Yang C. The role of microRNAs in metal carcinogen-induced cell malignant transformation and tumorigenesis. Food Chem Toxicol 2016.

[77] Dunn BK, Wickerham DL, Ford LG. Prevention of hormone-related cancers: breast cancer. J Clin Oncol 2005;23(2):357–67.

[78] Jordan VC. Tamoxifen: a most unlikely pioneering medicine. Nat Rev Drug Discov 2003;2(3):205–13.

[79] Hutchings O, Evans G, Fallowfield L, Cuzick J, Howell A. Effect of early American results on patients in a tamoxifen prevention trial (IBIS). International Breast Cancer Intervention Study. Lancet 1998;352(9135):1222.

[80] White IN. The tamoxifen dilemma. Carcinogenesis 1999;20(7):1153–60.

[81] Cuzick J, Powles T, Veronesi U, Forbes J, Edwards R, Ashley S, et al. Overview of the main outcomes in breast-cancer prevention trials. Lancet 2003;361(9354):296–300.

[82] Greaves P, Goonetilleke R, Nunn G, Topham J, Orton T. Two-year carcinogenicity study of tamoxifen in Alderley Park Wistar-derived rats. Cancer Res 1993;53(17):3919–24.

[83] Dragan YP, Fahey S, Nuwaysir E, Sattler C, Babcock K, Vaughan J, et al. The effect of tamoxifen and two of its non-isomerizable fixed-ring analogs on multistage rat hepatocarcinogenesis. Carcinogenesis 1996;17(3):585–94.

[84] Williams GM, Iatropoulos MJ, Karlsson S. Initiating activity of the anti-estrogen tamoxifen, but not toremifene in rat liver. Carcinogenesis 1997;18(11):2247–53.

[85] Wogan GN. Review of the toxicology of tamoxifen. Semin Oncol 1997;24(1 Suppl. 1):S1-87–97.

[86] Beland FA, McDaniel LP, Marques MM. Comparison of the DNA adducts formed by tamoxifen and 4-hydroxytamoxifen in vivo. Carcinogenesis 1999;20(3):471–7.

[87] Gamboa da Costa G, McDaniel-Hamilton LP, Heflich RH, Marques MM, Beland FA. DNA adduct formation and mutant induction in Sprague-Dawley rats treated with tamoxifen and its derivatives. Carcinogenesis 2001;22(8):1307–15.

[88] Phillips DH. Understanding the genotoxicity of tamoxifen? Carcinogenesis 2001;22(6):839–49.

[89] Phillips DH, Carmichael PL, Hewer A, Cole KJ, Poon GK. Alpha-Hydroxytamoxifen, a metabolite of tamoxifen with exceptionally high DNA-binding activity in rat hepatocytes. Cancer Res 1994;54(21):5518–22.

[90] Carthew P, Lee PN, Edwards RE, Heydon RT, Nolan BM, Martin EA. Cumulative exposure to tamoxifen: DNA adducts and liver cancer in the rat. Arch Toxicol 2001;75(6):375–80.

[91] Tryndyak VP, Kovalchuk O, Muskhelishvili L, Montgomery B, Rodriguez-Juarez R, Melnyk S, et al. Epigenetic reprogramming of liver cells in tamoxifen-induced rat hepatocarcinogenesis. Mol Carcinog 2007;46(3):187–97.

[92] Tryndyak VP, Muskhelishvili L, Kovalchuk O, Rodriguez-Juarez R, Montgomery B, Churchwell MI, et al. Effect of long-term tamoxifen exposure on genotoxic and epigenetic changes in rat liver: implications for tamoxifen-induced hepatocarcinogenesis. Carcinogenesis 2006;27(8):1713–20.
[93] Pogribny IP, Bagnyukova TV, Tryndyak VP, Muskhelishvili L, Rodriguez-Juarez R, Kovalchuk O, et al. Gene expression profiling reveals underlying molecular mechanisms of the early stages of tamoxifen-induced rat hepatocarcinogenesis. Toxicol Appl Pharmacol 2007;225(1):61–9.
[94] Pogribny IP, Tryndyak VP, Boyko A, Rodriguez-Juarez R, Beland FA, Kovalchuk O. Induction of microRNAome deregulation in rat liver by long-term tamoxifen exposure. Mutat Res 2007;619(1–2):30–7.
[95] Humans IWGotEoCRt. 1,3-butadiene, ethylene oxide and vinyl halides (vinyl fluoride, vinyl chloride and vinyl bromide). In: IARC monographs on the evaluation of carcinogenic risks to humans, vol. 97. World Health Organization, International Agency for Research on Cancer; 2008. p. 3–471.
[96] Walker VE, Walker DM, Meng Q, McDonald JD, Scott BR, Seilkop SK, et al. Genotoxicity of 1,3-butadiene and its epoxy intermediates. Res Rep 2009;(144):3–79.
[97] Melnick RL, Sills RC. Comparative carcinogenicity of 1,3-butadiene, isoprene, and chloroprene in rats and mice. Chemico Biol Interact 2001;135-136:27–42.
[98] Koturbash I, Scherhag A, Sorrentino J, Sexton K, Bodnar W, Swenberg JA, et al. Epigenetic mechanisms of mouse interstrain variability in genotoxicity of the environmental toxicant 1,3-butadiene. Toxicol Sci 2011;122(2):448–56.
[99] Koturbash I, Scherhag A, Sorrentino J, Sexton K, Bodnar W, Tryndyak V, et al. Epigenetic alterations in liver of C57BL/6J mice after short-term inhalational exposure to 1,3-butadiene. Environ Health Perspect 2011;119(5):635–40.
[100] Chappell G, Kobets T, O'Brien B, Tretyakova N, Sangaraju D, Kosyk O, et al. Epigenetic events determine tissue-specific toxicity of inhalational exposure to the genotoxic chemical 1,3-butadiene in male C57BL/6J mice. Toxicol Sci 2014;142(2):375–84.
[101] Moorthy B, Chu C, Carlin DJ. Polycyclic aromatic hydrocarbons: from metabolism to lung cancer. Toxicol Sci 2015;145(1):5–15.
[102] Simon R, Gomez Ruiz JA, von Holst C, Wenzl T, Anklam E. Results of a European inter-laboratory comparison study on the determination of EU priority polycyclic aromatic hydrocarbons (PAHs) in edible vegetable oils. Anal Bioanal Chem 2008;391(4):1397–408.
[103] Ho SM, Johnson A, Tarapore P, Janakiram V, Zhang X, Leung YK. Environmental epigenetics and its implication on disease risk and health outcomes. ILAR J/Natl Res Counc Inst Laboratory Animal Resour 2012;53(3–4):289–305.
[104] Perera F, Tang WY, Herbstman J, Tang D, Levin L, Miller R, et al. Relation of DNA methylation of 5′-CpG island of ACSL3 to transplacental exposure to airborne polycyclic aromatic hydrocarbons and childhood asthma. PLoS One 2009;4(2):e4488.
[105] Tang WY, Levin L, Talaska G, Cheung YY, Herbstman J, Tang D, et al. Maternal exposure to polycyclic aromatic hydrocarbons and 5'-CpG methylation of interferon-gamma in cord white blood cells. Environ health Perspect 2012;120(8):1195–200.
[106] Pavanello S, Bollati V, Pesatori AC, Kapka L, Bolognesi C, Bertazzi PA, et al. Global and gene-specific promoter methylation changes are related to anti-B[a]PDE-DNA adduct levels and influence micronuclei levels in polycyclic aromatic hydrocarbon-exposed individuals. Int J Cancer 2009;125(7):1692–7.
[107] Ouyang B, Baxter CS, Lam HM, Yeramaneni S, Levin L, Haynes E, et al. Hypomethylation of dual specificity phosphatase 22 promoter correlates with duration of service in firefighters and is inducible by low-dose benzo[a]pyrene. J Occup Environ Med/Am Coll Occup Environ Med 2012;54(7):774–80.
[108] Denissenko MF, Pao A, Tang M, Pfeifer GP. Preferential formation of benzo[a]pyrene adducts at lung cancer mutational hotspots in P53. Science 1996;274(5286):430–2.
[109] Hu W, Feng Z, Tang MS. Preferential carcinogen-DNA adduct formation at codons 12 and 14 in the human K-ras gene and their possible mechanisms. Biochemistry 2003;42(33):10012–23.
[110] Damiani LA, Yingling CM, Leng S, Romo PE, Nakamura J, Belinsky SA. Carcinogen-induced gene promoter hypermethylation is mediated by DNMT1 and causal for transformation of immortalized bronchial epithelial cells. Cancer Res 2008;68(21):9005–14.
[111] Shen YL, Jiang YG, Greenlee AR, Zhou LL, Liu LH. MicroRNA expression profiles and miR-10a target in anti-benzo[a] pyrene-7, 8-diol-9, 10-epoxide-transformed human 16HBE cells. Biomed Environ Sci 2009;22(1):14–21.
[112] Jiang Y, Wu Y, Greenlee AR, Wu J, Han Z, Li X, et al. miR-106a-mediated malignant transformation of cells induced by anti-benzo[a]pyrene-trans-7,8-diol-9,10-epoxide. Toxicol Sci 2011;119(1):50–60.
[113] Li D, Wang Q, Liu C, Duan H, Zeng X, Zhang B, et al. Aberrant expression of miR-638 contributes to benzo(a)pyrene-induced human cell transformation. Toxicol Sci 2012;125(2):382–91.
[114] Izzotti A, Calin GA, Arrigo P, Steele VE, Croce CM, De Flora S. Downregulation of microRNA expression in the lungs of rats exposed to cigarette smoke. FASEB J 2009;23(3):806–12.
[115] Casati L, Sendra R, Sibilia V, Celotti F. Endocrine disrupters: the new players able to affect the epigenome. Front Cell Dev Biol 2015;3:37.
[116] Ferreira LL, Couto R, Oliveira PJ. Bisphenol A as epigenetic modulator: setting the stage for carcinogenesis? Eur J Clin Invest 2015;45:32–6.
[117] Collotta M, Bertazzi PA, Bollati V. Epigenetics and pesticides. Toxicology 2013;307:35–41.
[118] Anway MD, Skinner MK. Epigenetic transgenerational actions of endocrine disruptors. Endocrinology 2006;147(6 Suppl.):S43–9.
[119] Skinner MK, Manikkam M, Guerrero-Bosagna C. Epigenetic transgenerational actions of endocrine disruptors. Reprod Toxicol 2011;31(3):337–43.
[120] Ferguson LR, Philpott M. Nutrition and mutagenesis. Annu Rev Nutr 2008;28:313–29.
[121] Wild CP, Gong YY. Mycotoxins and human disease: a largely ignored global health issue. Carcinogenesis 2010;31(1):71–82.
[122] Zhang YJ, Ahsan H, Chen Y, Lunn RM, Wang LY, Chen SY, et al. High frequency of promoter hypermethylation of RASSF1A and p16 and its relationship to aflatoxin B1-DNA adduct levels in human hepatocellular carcinoma. Mol Carcinog 2002;35(2):85–92.
[123] Zhang YJ, Chen Y, Ahsan H, Lunn RM, Lee PH, Chen CJ, et al. Inactivation of the DNA repair gene O6-methylguanine-DNA methyltransferase by promoter hypermethylation and its relationship to aflatoxin B1-DNA adducts and p53 mutation in hepatocellular carcinoma. Int J Cancer 2003;103(4):440–4.

[124] Su H, Zhao J, Xiong Y, Xu T, Zhou F, Yuan Y, et al. Large-scale analysis of the genetic and epigenetic alterations in hepatocellular carcinoma from Southeast China. Mutat Res 2008;641(1–2):27–35.
[125] Marrone AK, Beland FA, Pogribny IP. Noncoding RNA response to xenobiotic exposure: an indicator of toxicity and carcinogenicity. Expert Opin Drug Metab Toxicol 2014;10(10):1409–22.
[126] Marrone AK, Beland FA, Pogribny IP. The role for microRNAs in drug toxicity and in safety assessment. Expert Opin Drug Metab Toxicol 2015;11(4):601–11.
[127] Pogribny IP, Beland FA. DNA hypomethylation in the origin and pathogenesis of human diseases. Cell Mol Life Sci 2009;66(14):2249–61.
[128] Pogribny IP, Beland FA. DNA methylome alterations in chemical carcinogenesis. Cancer Lett 2013;334(1):39–45.
[129] Pogribny IP, Rusyn I. Environmental toxicants, epigenetics, and cancer. Adv Exp Med Biol 2013;754:215–32.
[130] Secretan B, Straif K, Baan R, Grosse Y, El Ghissassi F, Bouvard V, et al. A review of human carcinogens-Part E: tobacco, areca nut, alcohol, coal smoke, and salted fish. Lancet Oncol 2009;10(11):1033–4.
[131] Koturbash I, Beland FA, Pogribny IP. Role of microRNAs in the regulation of drug metabolizing and transporting genes and the response to environmental toxicants. Expert Opin Drug Metab Toxicol 2012;8(5):597–606.

Chapter 33

Environmental Sources of Ionizing Radiation and Their Health Consequences

A.A. Goodarzi, A. Anikin, D.D. Pearson
University of Calgary, Calgary, AB, Canada

Chapter Outline

1. INTRODUCTION

As unstable atoms decay over time, they release radiation in the form of electromagnetic waves and subatomic particles. Some forms of radiation have sufficient energy to detach electrons (ie, ionize the atomic structure of the substances they pass through), and are thus called ionizing radiation (IR). Common types of IR include alpha particles, beta particles, neutrons, X-rays, and gamma rays (Table 33.1). The molecular, cellular, and physiological effects of IR on human beings are myriad, and depend on the source, quality, and dose of IR, the mode of exposure and the genetic background of the individual in question. The two main physical outcomes of IR exposure are stochastic effects, representing genome instability and consequent cancers (whose likelihood increases with IR dose but is not guaranteed), and deterministic effects, encompassing immediate and predictable effects with severity relating to the dose. The sources of IR exposure that are to be discussed in this chapter include, from the rarest to most commonly documented types: nuclear attack or disaster, space and aeronautical high-altitude exposure, radiotherapy, diagnostic medical imaging, and radon gas inhalation (Fig. 33.1).

2. THE MOLECULAR EFFECTS OF IR IN CELLS

IR exposure inflicts various types of damage to genomic DNA, including (but not limited to) single-strand breaks (SSBs), double-strand breaks (DSBs), base and sugar–backbone damage and DNA:DNA or DNA:protein cross-linkages. Such damage occurs either via the direct ionization of the DNA molecule or through the production of reactive oxygen species (ROS) via the ionization of intracellular water, which in turn react with the DNA (Fig. 33.2). A DSB forms when two SSBs on opposing strands of the DNA double helix occur in close enough proximity for the base pairing of intervening sequence to fail. DSBs are among the most serious of IR-induced lesions as, if unrepaired, they will lead to chromosomal fragmentation and potentially premature cell aging or death. IR can also induce intra- and interstrand DNA cross-links (ICLs), severe lesions that can distort DNA helices or block strand separation [1]. For IR to trigger the formation of an ICL, there is a requirement of three to five mismatched bases at the site of ionization [2], increasing the likelihood of these lesions in regions of non-hybridized DNA such as telomere D-loops, DNA-replication forks, and transcription bubbles, as well as in other non-β DNA structures such as slipped DNA, hairpins, and tetrahelical structures [3]. Even though ICLs will most

Genome Stability. http://dx.doi.org/10.1016/B978-0-12-803309-8.00033-1

TABLE 33.1 The Types of Ionizing Radiation (IR), Their Source, and Common Use

Name	Type of Ionizing Radiation (IR)	Source of IR
Gamma rays	High-energy photons with low linear energy transfer (LET) through matter and great penetrance	Cancer therapeutics optimizing gamma rays to shrink tumor size
X-rays	Charged photons with low-LET IR	Diagnostic imaging
Alpha radiation	Heavy particle with high-LET through matter and poor penetrance	Byproduct of uranium decay and its progeny
Beta radiation	Light-charged particles high-LET IR	^{90}Strontium industrial usage
Neutrons	High-LET IR particles with no charge but strong penetrance ability	Cancer therapy, for example, boron capture therapy

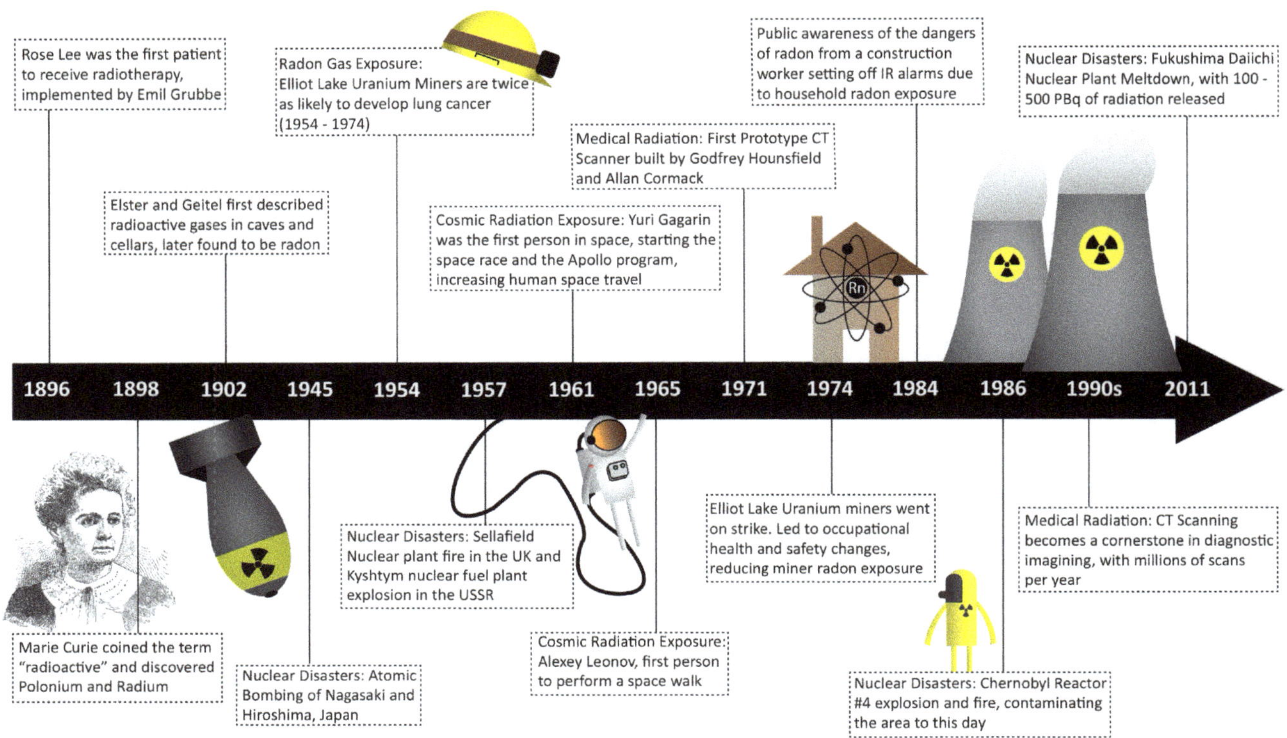

FIGURE 33.1 **A historical timeline of human exposure to sources of ionizing radiation.**

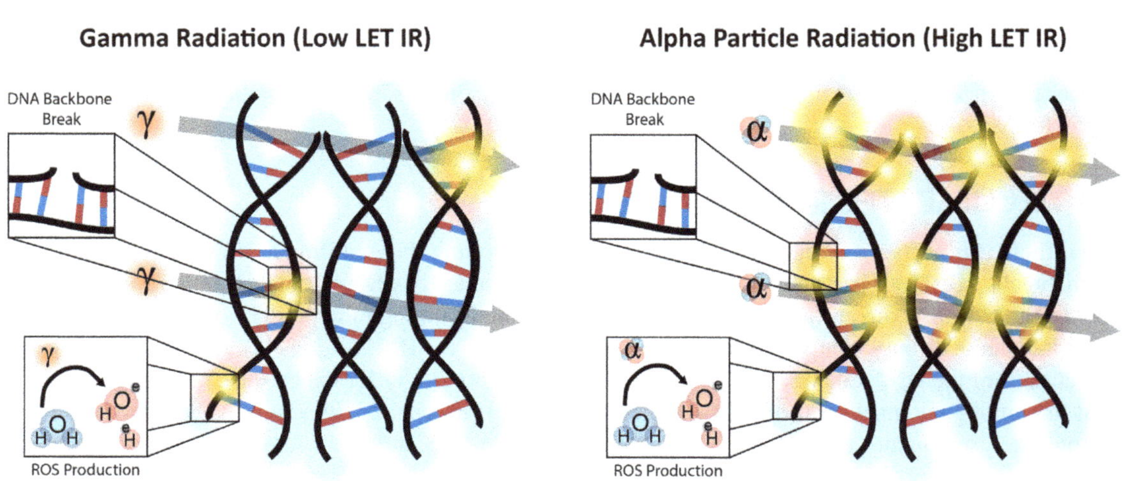

FIGURE 33.2 **The impact of high- versus low-LET ionizing radiation exposure on DNA.** Left panel: The direct effect of gamma rays, a type of low-LET IR that produces widely spaced DNA-damaged sites, both by direct ionization and indirect damage to the phosphodiester backbone by reactive oxygen species (ROS). Right panel: The direct effect of alpha particles, a form of high-LET IR whose high proximity ionization events produces clustered DNA-damage sites, defined as multiple lesions within few nanometers in a DNA molecule.

likely occur at a lower frequency than DSBs following IR, on a per lesion basis they are probably as or even more toxic due to the fact they occur commonly in regions of DNA-regulating key cellular processes [2].

In response to IR-induced DNA damage, life has evolved highly effective DNA-repair mechanisms [4]. IR induces SSBs more abundantly than any other type of lesion, although DSBs are considered to be the major factor responsible for cell death post exposure. If DSBs go unrepaired or are improperly repaired, there is a high chance of chromosomal loss, gain, or translocation events leading to a plethora of human diseases, including cancer [5]. In mammalian cells, formation of a DSB triggers a global cellular DNA-damage response (DDR) [6]. One of the earliest events to occur locally at the DSB site is phosphorylation of histone H2AX (at S139, to form "γH2AX foci"), which may be monitored by immunofluorescence microscopy and is often used as a tool to measure DSB induction and repair and diagnose radiation sensitivity [6,7]. The steady-state amount of DSBs observed by γH2AX foci enumeration will reflect a combination of IR exposure dosage and a cell's innate DSB-repair capacity, which can differ considerably between persons and tissues. DSBs are repaired generally by homologous recombination (HR) or nonhomologous end joining (NHEJ), depending on the cell-cycle stage of a cell when damage has occurred (HR, requiring an undamaged, copied sister chromatid for the repair process, is restricted to S and G2 phases). In cases where both NHEJ and HR fail or are unavailable, seen often in cancer cells, the so-called "alternative NHEJ" (alt-NHEJ) pathway can take place, although this generally produces large genomic alterations exacerbating genomic instability [8]. The choice between these pathways is a topic of intense research focus, at present, and is influenced not only by cell-cycle phase, but also the chromatin context within which the DNA damage occurs and the type of radiation from which the damage originates [8].

3. RADIATION DOSAGE AND LINEAR ENERGY TRANSFER

The emission of IR from a source is measured in becquerels (Bq), equating with one radioactive disintegration per second. The absorption of IR by a living cell is measured in gray (Gy), equating to 1 J of energy per kilogram of cells. One gray of IR is considered to produce about 20 DSBs per G0/G1 phase human cell [9], increasing (through S phase) toward 40 DSBs/Gy IR in the case of a G2 phase (which has twice as much DNA) [10]. The Sievert (Sv) is often equivalent to Gy, and represents the dose of IR absorbed by a cell/tissue/body that also has a measurable biological effect. When examining the impact of IR on cells and DNA, an important property to take into account is its linear energy transfer (LET) [11]. LET describes the rate at which energy is released by a radioactive source over a fixed distance. High-LET IR (such as alpha particles) emits more energy over the same relative distance compared to low-LET IR sources (such as gamma and X-rays) [12], and actually encompasses the majority of annual human IR exposure (see Section VI). Dose for dose, high-LET IR is much more lethal (at a cellular level) and carcinogenic than low-LET IR [13,14]. This is largely because high-LET IR produces DSBs and other lesions in very close proximity, which are both harder for the DNA-repair machinery to fix and have a greater chance of causing detrimental mutations, chromosomal translocations, cell death, or cancer [15]. Such clustered DNA damage, or multiple damaged sites (MDS), are specific types of DNA lesions generated by a single track of IR [16], including more than two individual lesions within one or two helical turns of DNA. The lesion found within clustered damage sites can be of several types: base modifications, abasic or apurinic/apyrimidinic sites, and oxidized purines or pyrimidines or SSBs in very close proximity to each other, as well as DSBs and ICLs. The complexity of IR-induced clustered damage correlates directly to ionization density; in fact, it is believed that clustered damage containing DSBs is mainly due to energy depositions of at least two to three ionization events localized within a 1–4 nm range [17]. Ninety percent of DSBs induced by high-LET IR is considered clustered, and it has been demonstrated experimentally that this is much more problematic for the cellular DDR to repair efficiently or accurately [18]. By contrast, the majority (99.9%) of DNA damage caused by low-LET IR is believed to occur indirectly via the formation of localized ROS, such as hydroxyl radical (·OH) from (relatively) isolated ionization events [19]. Consequently, they produce much more sporadic DSBs which are spread out across the nucleus and are repaired with faster kinetics compared to lesions formed by high-LET IR [9] (Fig. 33.2).

High-LET alpha particles, depositing energy at 100–150 keV/μm, produce highly linear tracks when traversing through a human cell. The microscopic pattern of energy deposition by high-LET IR represents complexity in terms of the nature of the radiation field, and the concepts of core and penumbra are important for understanding the physical characteristics of a charged particle track [20]. The core region is based on the Bohr adiabatic principle with a radius of around 0.0015 μm; within the core, excitation of all medium molecules occurs. In addition, some energy is deposited by extremely low-energy secondary electrons (δ-rays) that cannot successfully exit the core, and this defines the penumbra region. High-LET IR sources deposit their energy in two ways: 50% comes from within the core region from direct ionization and excitation of medium molecules, while δ-rays (the other 50%) are emitted from any collisions and extend for hundreds of microns outward [20]. A high-LET alpha particle passing through DNA will deposit large amounts of energy (300–500 eV) with a high probability of causing clustered lesions: about 20 DSBs per 10 μm track, with very few DSBs forming outside the track.

For high-LET IR delivered at the same dose as low-LET IR, there are also about three additional SSBs and three damaged bases produced near each DSB [20,21].

4. NUCLEAR MILITARY ATTACKS AND CIVILIAN NUCLEAR DISASTERS

While thankfully rare, nuclear attacks and disasters have provided a great deal of information on the effects of IR on human populations. These dramatic events, while undeniably tragic cases of mass, often whole-body irradiation of thousands of individuals, represent ideal scientific conditions for studying large populations with strong statistical significance. One of the largest studies of this kind is the Life Span Study (LSS) of Japanese survivors of the atomic bombing events in Hiroshima and Nagasaki towards the end of World War II [22]. The LLS examined 120,000 irradiated survivors and nonirradiated individuals, and was one of the first large studies to conclude the linkages between solid cancer incidence and IR dose. Additionally, nuclear power plant meltdowns, such as the 1986 meltdown of the Chernobyl nuclear reactor in the former USSR, provided data on initial high-dose irradiation as well as continual irradiation due to nuclear fallout. By examining the results of these studies, the initial and long-term health consequences of mass irradiation can be compared.

Irradiation from an explosive nuclear event may be broken up into two parts, the initial burst of neutrons, alpha particles, gamma- and X-rays, followed by the slower release of radioactive elements (fallout). The initial, high-dose IR bursts leads to strong deterministic effects such as acute radiation sickness (ARS). Using data obtained from the Chernobyl disaster, a correlation between the increasing severity of ARS to the increasing IR dose was observed, with the most severe ARS seen at estimated radiation doses between 6.5 and 16 Gy [23]. About 600 workers were at the Chernobyl nuclear plant immediately after the explosion, of which 237 were detectably symptomatic, 134 were officially diagnosed with ARS and 28 died [23,24]. ARS symptoms include a compromised immune system, vomiting, nausea, and diarrhea, and are associated with an overwhelming number of DNA-damaging events that kill cells outright and result in acute tissue damage [25], with proliferating cells being the most affected (particularly bone marrow and gastrointestinal cells). At exceptionally high IR doses, cerebrovascular and pulmonary dysfunction syndromes are noted, as even nonproliferating cells are affected [26,27]. Unfortunately, for those individuals close enough to the atomic event to receive massive doses, most die shortly thereafter due to bone marrow aplasia [28].

As distance from the disaster hypocenter ("ground zero") increases, estimated IR doses decrease significantly, following an inverse squared relation of dose to distance. For example, individuals within the 2 km radius of the Hiroshima/Nagasaki atomic bomb detonations received >0.5 Gy, whereas those 2–4 km from the hypocenter received acute doses of about 50 mGy [22,29]. At these doses, cases of ARS are rarer and individuals manifest stochastic effects such as cancer, with solid malignancies increasing at a rate of 26/10,000 cases per Gy IR [22], and deterministic effects such as fetal developmental abnormalities (including growth and mental retardation) [30]. Following initial high-dose irradiation, the resulting fallout of radioactive elements leads to continual low-dose irradiation over a much greater area than the initial blast radius. This generally represents the deposition of radioactive elements such as Iodine 131 and Cesium 137, which have half-lives of 8 days and 30 years, respectively, and can contaminate an area making it unsuitable for human population for decades [31]. In the case of Chernobyl, an estimated quintillion Bq (>1 EBq) of radioisotopes were released, with the initial evacuation zone being 3 km^2 but later expanded to 30 km^2 to account for spreading nuclear fallout [23], with evidence of contamination found over 100 km away from the Chernobyl hypocenter [32]. The Fukushima Daiichi Power Plant meltdown on March 11, 2011 in Japan released an estimated 100–500 PBq (1 PBq = a quadrillion Bq) of 131 Iodine and 6–20 PBq of 137 Cesium [33].

With extensive fallout, the documented incidences of thyroid, stomach, lung, liver, and blood cancers rise considerably following nuclear disaster with studies determining an inverse correlation of age at exposure to that of lifetime risk of cancer [22]. Indeed, the LSS (described earlier) demonstrated a 29% increase in the likelihood of solid cancers per decade decrease in age of exposure—meaning, the younger an individual is at the time of irradiation, the greater their likelihood of developing a tumor [22]. This fits with our understanding of IR-induced DNA damage, where genomic instability triggered by elevated DNA damage increases the chances of a mutagenic event that either ablates a tumor-suppressor gene or modifies and activates an oncogene. The longer the irradiated individual has left on their "natural life span," the more likely such mutations will occur within the same cell to trigger tumorigenesis. In addition, the highly proliferative tissue of younger humans is more susceptible to potentially cancer-causing mutations due to the vulnerability of DNA-replication processes to IR-induced DNA damage [34].

Radioactive fallout–induced cancers occur frequently in the thyroid gland due to its iodine-sequestering ability, wherein ingested Iodine 131 is progressively concentrated and increases the effective radiation dose by 1000–2000 times [24]. Lessons learned from the Chernobyl disaster have mitigated the uptake of nuclear fallout in Japan where, 6 months after the March 2011 nuclear meltdown, only 3286 of 9498 residents of the Fukushima prefecture had detectable radioisotope levels at an average of 11.4 Bq/kg, compared to Chernobyl's 49 Bq/kg still seen up to a decade afterward [35]. The Chernobyl incident marked a 30-fold increase in childhood thyroid cancers, with 98% of these tumors derived from the papillary cells versus the 67% commonly seen in thyroid tumors in nonirradiated populations [36,37]. This bias suggested a pathology specific for IR damage, which when further investigated revealed that 50–90% of the papillary cancers had a *RET* rearrangement leading to a replacement of the RET tyrosine kinase ligand-binding domain with a coiled-coil region, stimulating uncontrolled papillary cell growth [38,39]. Interestingly, the *RET* rearrangement alone was insufficient to induce tumorigenesis, an additional point mutation was required and, the earlier the individual was exposed to radiation, the higher was the likelihood of inducing cancer [40]. Follow-up animal studies have suggested that other IR-induced thyroid cell cancers are possible, such as medullary carcinomas, indicating that IR damage does not have a specific pathology per se, only that there may yet be an influx of other thyroid cancers from nuclear fallout victims [24,41]. In addition to thyroid cancer [37,38,42], the other highest incidence form of radiation-induced malignancy is leukemia [43–45]. Indeed, data from nuclear industry workers indicate that the excess relative risk (ERR), which compares the level of risk for an exposed person to that of the risk in a nonexposed person, is 2.18/Sv for all types of leukemia (except chronic lymphocytic leukemia) [43]. Mechanistically, the mutations accrued in radiation-induced leukemias do not share as striking a homology as the papillary thyroid cancers, but show random chromosomal aberrations in hemopoietic stem cells, ultimately leading to a higher likelihood of cancer [25,28]. Indeed, these rearrangements have shown to inactivate key proteins such as p53 and ATM, whose inactivation have been linked to several cancers [46–48].

Other than cancers, nuclear disasters have been implicated in acute or systemic health conditions of the blood, heart, brain, and circulatory and respiratory systems, as well as psychological disorders [22,49,50]. Although the direct mechanism of action is unknown, evidence has indicated that cellular death, endothelial changes, or microvascular damage may contribute to the occurrence of heart disease and stroke [50]. High-dose irradiation can also leave survivors in an immune-compromised state with bone marrow ablation, T-cell apoptosis, and a host of other immune system effects, increasing their susceptibility to infections such as pneumonia and influenza [26,27,49,51]. In the case of the civilian nuclear meltdowns, the permanent physical evacuation of individuals from contaminated areas has resulted in obesity, hypertension, and polycythemia due to the psychological stress of being displaced from their lives [49,52]. Taking all these factors into consideration, the health effects of a nuclear disaster are serious but, in large population health terms, thankfully rare [53].

5. AEROSPACE TRAVEL

Removed from the full protection of Earth's atmosphere and magnetosphere, astronauts experience greater doses of high-LET radiation from high atomic number and energy (HZE) particles, and cosmic radiation [54]. Even airplane flight staff, who spend significant amounts of time closer to space versus the general population, experience significantly elevated HZE bombardment over their careers [55]. HZE ions are a component of galactic cosmic rays (GCR) and are also emitted by individual solar proton events, particle storms of massively accelerated protons that are emitted by Earth's Sun during solar flares or during coronal mass ejection shock waves. HZE particles are a significant health concern for astronauts and individuals who spend a great deal of time traveling by air [56,57]. In outer space, the effects of HZE particles are so pronounced that for the Apollo space crew members, who left Earth's protective magnetosphere between 1968 and 1972, HZE particles and proton interactions with the retina were perceived as flashes of light [58]. Not only do HZE particles induce DNA damage in a similar manner to alpha particles, more importantly, HZE particles can ionize other atoms inside the body to become sources of DNA damage, effectively multiplying their deleterious effects [59]. It is well established that HZE atoms and alpha particles have greater overall relative biological effectiveness (RBE) than either X- or gamma rays, with the complexity of induced DNA damage being directly related to the total energy of the radiation source [21]. RBE is a term referring to the ratio of biological effectiveness of one IR type in comparison to another, given the same quantity of absorbed energy. Current animal studies demonstrate that HZE nuclei have a greater carcinogenic effect compared to low-LET gamma radiation. The values of RBE measured in rodents for multiple tumors such as those of the skin and mammary gland are as high as 24–40, even with low doses of HZE ions [60].

The frequency of HZE radiation and our current inability to effectively shield astronauts from these types of radiation is what makes space travel implicitly dangerous. Space missions venturing outside Earth's protective magnetic field for significant periods of time, such as manned missions to Mars, which could last for decades, would place astronauts at significant risk. One of the most immediate health concerns is the increased susceptibility to infection [61], with HZE particle irradiation–associated decreases in B-cell and T-cell counts, and IL-2 secretion of the spleen [62]. Immune-suppression in outer space is so significant that about 50% of Apollo mission astronauts contracted either bacterial, viral, or fungal infections [61]. Even latent viral infections, such as herpes and Epstein–Barr, have been observed to reemerge in space crews [62,63]. One method to mitigate immune-suppression is to provide preemptive broad-spectrum antibiotics; however, this strategy is flawed as it enables the development of antibiotic-resistant bacteria and does not address viral or fungal infections [61]. Of course a major risk of long-term exposure to HZE particles is cancer [56]. The likelihood of death at age 40 years due to cancer from a deep space mission (eg, Mars), calculated for lung, colon, stomach, bladder, bone marrow, breast, and ovarian malignancies, is estimated to be 4.2% for men and 5.1% for women [56]. For conventional airline pilots and cabin crew, a meta-analysis of 266,431 individuals conducted in 2015 determined crew to have about doubled rates of skin melanoma compared to the general population [64]. One likely contributing factor to all these cancers is that HZE particles have a 30-fold greater efficacy in causing interchromosomal exchanges compared to low-LET IR [56]. This suggests the HZE particles are able to elicit greater genome instability, as indicated by the presence of increased chromosomal truncations in cells exposed to HZE particles [65]. Increased cancer risk within this population would also be exacerbated by immune-suppression, as the destruction of nascent cancer cells by the immune system would be impaired [66].

Another consequence of continual exposure to low doses of high-LET radiation is reduced cognitive abilities associated with lowered neural plasticity [67–69]. Animal studies demonstrate that low IR doses alter neuronal gene regulation by suppressing five distinct genes: GNAS (a G-protein), GRIA3 (an AMPA glutamate receptor), SLC1A1 (a glutamate transporter), PRKCB1 (protein kinase C), and MEF2C (a transcription factor) [67]. These genes are also downregulated in aging and Alzheimer's disease, possibly indicative of the accelerated neuronal aging impact of irradiation. Learning, memory, and cognition capacity would certainly all be lowered, increasingly disabling spacecraft crews required to function at a high level to maintain their environment [69]. Fortunately, preliminary work has shown that antioxidants, such as α-lipoic acid reduce, albeit not completely, the effects of IR on cognition [68]. Long-term HZE particle exposure is also linked to cataracts [70–72], with several studies indicating that space and airline flight crews have a higher incidence than the general population [71–73]. The most likely causes have been associated with the production of hydroxykynurenine from the interaction of tryptophan residues with UV light, loss of antioxidant capabilities due to accelerated aging, and lenticular cell changes [71]. There also appears to be a distinction in the spatial localization of IR-induced cataracts versus age-related cataracts, with the majority of the former appearing in the posterior capsule of the lens and the latter dominating the lens nucleus [71,72]. For airline flight crews, cataracts may be mitigated through surgery; however, they pose a problem for deep space missions where optical surgery may not be possible.

6. MEDICAL RADIATION (RADIOTHERAPY AND MEDICAL IMAGING)

Each year, about 64,000 Canadian and nearly 1,000,000 American cancer patients undergo radiotherapy (RT), because this is considered one of the most effective antitumor treatments [74]. Delivered in targeted, short-spaced, very high-dose fractions, RT is very effective at killing cancer cells; however, it can also elicit tissue damage to normal cells resulting in deterministic effects (such as cataracts, heart disease, stroke, erythema, ulcers, telangiectasia, dermal atrophy, or cognitive dysfunction) or stochastic effects like secondary cancers [75–78]. As mentioned earlier, rapidly proliferating cells are particularly sensitive to the killing effects of IR, explaining the efficacy of RT as an antitumor agent and underlying the off-target side effects such as hair-loss, nausea, anemia, and delayed wound healing. Additionally, at sublethal doses, IR-induced DNA damage increases the likelihood of carcinogenesis [79]. Hence, normal cells not killed but damaged sufficiently by RT to alter their DNA can transform and become a secondary cancer years later [78]. On top of this, mild-to-severe over-responses to RT (mostly manifesting as grade 2–4 toxicity) are noted in 1–3% of adult cases (ie, >30,000 cancer patients in North America per year), likely due to undiagnosed radiosensitivity, multiplying the deleterious effects of radiation.

Children who undergo RT are thought to be especially at risk of adverse health effects due to their added life span and the elevated proliferation status of their normal tissue. Along with increased secondary cancers, it has been well documented that radiation has a profound effect on the developing cognitive abilities of children [76,80–82]. From a study of pediatric patients with acute lymphocytic leukemia (ALL) or medulloblastoma/posterior fossa primitive neural ectodermal tumor (PNET), a direct relation between cranial radiation dose and IQ lose was seen. Here, patients were given IQ tests before and after treatment and for those who received a total of 18 Gy, their scores were an average of 12.3 IQ points higher in follow-up compared to those who received 36 Gy [80]. Additionally, age was a large factor in the observed IQ decline, where patients <3 years old were predicted to lose an average of 11.9 more IQ points than for those 3–10 years old [80]. In fact, the cognitive effects are so prominent that those affected by radiation have ended up in special education or institutionalized [81]. Mechanistically, this cognitive decline results from decreased neurogenesis, where rat models have shown a 62% reduction in hippocampal neural stem/precursor cell proliferation as well as near ablation of differentiation into neurons or glia [81]. It is suggested that radiation induces mitotic stress, preventing in vitro cells from proliferating after 2–3 divisions, and the increased inflammation response alters the differentiation pathway choice [76,81]. Notably, there has been some work demonstrating reduced memory loss and increased neurogenesis with the use of nonsteroidal anti-inflammatory drugs (NSAIDs) [77].

The effects of medical radiation are exaggerated in utero, where time of exposure and dose will alter the effects of radiation in the developing fetus, causing congenital malformations, growth retardation, cancer, or death [30,83]. It is known that if doses of about 100 mGy are received during the 8–15-week gestation period, then subsequent IQ is lowered, while about 1000 mGy will cause severe mental retardation [30,83]. Indeed, the relative risk of developing cancers when exposed in utero is 1.4 times higher than nonexposed with a dose of only 10 mGy [30]. Although RT has proved effective in many cases, the side effects of treatment are numerous and have dictated the path of many treatment regimes.

The most common man-made source of IR exposure, by a large margin, is modern medical imaging with a mean effective dose estimated to be 1–3 mSv per person per year [75]. To put medical imaging into perspective, close to 100 million CT examinations are performed in Canada and the United States annually, with diagnosed and potential cancer patients representing a huge part of this population. Each whole-body CT scan represents exposure to about 10 mGy IR, a significant dose equating with one DSB for every five cells or, when considered on a whole-body level, a staggering 7.44 trillion DSBs per person [84,85]. Although widely prescribed, given these doses, it is not surprising that CT scanning is recognized as significantly increasing cancer risk, with the NCI estimating that 29,000 new US cancer cases were caused directly by the 72 million CT scans that took place in 2007 alone [86]. It should be noted that there is some dispute over the risks associated with medical imaging, as the precedence for relative risk of cancers is based upon the LSS of Hiroshima and Nagasaki. As mentioned previously, the LSS indicated a 26/10,000 increase in solid cancers per Gy of exposure; however, this trend is only seen for doses >100 mSv [22,87]. Based on this data, a linear through zero model was created where low doses also induce an observable increase in cancers, yet the evidence for this is minimal [87]. Due to the limited evidence, it has been suggested that the lower doses of radiation may not increase the likelihood of carcinogenesis [87]; however, what has not been taken into account in these suggestions is the stochastic nature of induced mutations. Although a direct correlation is difficult to draw with low-dose exposure, the fact remains that IR induces DNA damage which can lead to tumorigenesis [88]. A higher dose simply leads to a greater number of DSBs, which would increase the chances of activating an oncogene or deactivating a tumor suppressor. Given the latency period associated with IR-induced cancers (from causative event to diagnosis), and since the overwhelming majority of CT scans have taken place only since the late 1990s [89], it is possible (even likely) that most cancers resulting from CT scan usage have yet to be documented.

7. RADON GAS

The largest natural causes of IR-induced genetic damage are radioisotopes of the odorless and colorless noble gas radon, which is highly enriched within soil gases of uranium-rich geologies and can accumulate in homes via basements. In 1904, while at McGill University in Canada and based on landmark experiments conducted two years earlier by Elster and Geitel [90], Ernest Rutherford wrote [91]:

> *There can thus be little doubt that the abnormal activity observed in caves and cellars is due to a radioactive emanation, present within the earth, which gradually diffuses to the surface and collects in places where the air is not disturbed … [this activity] decays to half value in about 3.3 days, while the activity of the radium emanation decays to half value in an interval of 3.7 to 4 days. Considering the difficulty of making accurate determinations of these quantities, the rates of decay of the activity of the emanations from the earth and from radium agree within the limits of experimental error.*

Of course, Rutherford, while not fully knowing it at the time, was talking about radon gas—the "radium emanation." Radon is now recognized as second only to tobacco smoking as a direct cause of lung cancer, with over 10% of all cases worldwide attributed to its exposure [92,93]. The most common isotope of radon, ^{222}Rn, has a short half-life of 3.8 days, and will emit high-LET alpha particles before decaying rapidly to isotopes of polonium, bismuth, and lead. Radon's carcinogenic properties are attributed to the fact that every decaying atom of ^{222}Rn will emit four high-LET IR alpha particles, the first emission resulting in a transition from the gaseous state to a precipitated solid which becomes irreversibly embedded within lung tissue. These radon "daughter" radioisotopes will continue to emit alpha radiation within the lungs, with a cumulative half-life of just over 22 years before eventually decaying into solid lead. It is unsurprising, given this series of events, that radon is classified by the United Nations (UN) World Health Organization (WHO) as a Class I carcinogen in the same hazard category as benzene, mustard gas, and asbestos [94]. For most healthy individuals not involved in IR-prone occupations, radon inhalation represents the largest single source (37%) of annual radiation exposure throughout their lives [95].

Initial evidence for the significance of radon as a carcinogen came from observations of uranium mine workers who, after being exposed to high levels of radon progeny as a byproduct of ^{238}Uranium decay, developed lung cancer at substantially higher rates relative to equivalent professions [96]. Between the 1950s and 1970s, Ontario uranium miners in Canada were exposed to extraordinarily high levels of radon gas in poorly ventilated mines. By the mid-1970s miners were being diagnosed with lung cancer at twice the expected rates, such that by 1984 a total of 285 miners had already died of lung cancer [97]. Thankfully, this eventually led to fundamental changes within the mining industry and, through adequate ventilation, such mines were rendered safe for workers [98]. In the decades that followed, radon accumulation in residential homes and workplaces was additionally recognized as a means of cancer-causing exposure and, in 2009, following several decades of research within the medical and scientific community, the UN WHO released a "Handbook on Indoor Radon," declaring radon to be "a major contributor to the ionizing radiation dose received by the general population" and "that the lung cancer risk increases proportionally with increasing radon exposure." The consolidated studies summarized by the WHO indicated that lifetime risk of lung cancer increases 16% for every 100 Bq/m^3 of radon inhaled within the domestic or workplace environment over the long term, and recommended this level as the maximum acceptable limit for human environments, with the ideal levels being as low as achievable.

Since radon is odorless, colorless, and has no immediate, detectable impact on human respiration (unlike carbon monoxide), it usually goes unnoticed and becomes a problem when it is concentrated within well-insulated homes and offices. Actual residential levels of radon will vary between buildings due to the amount of radon produced in the underlying geological substrate and factors such as the presence or absence of ventilation, insulation, or heating systems. Heating a poorly ventilated (eg, a home sealed during cold winter months) but highly insulated (eg, energy efficient) home will hyperconcentrate radon in dwellings over certain geology, as the thermal stack effect (hot air rising, creating a pressure differential) actively draws radon-laden gases up through the foundations into the structure. This may easily be countered, thankfully, through relatively straightforward and moderately inexpensive radon testing and mitigation technologies that can accurately measure ambient levels of radon gas over the long term, and prevent accumulation permanently (often via sub-slab depressurization or increased ventilation) [99]. Unsurprisingly, exposure to the high-LET IR from radon during childhood increases significantly the risk of developing lung cancer later in life. Indeed, childhood (ages 0–17 years) exposure to even moderately high radon concentrations (400 Bq/m^3) is equivalent to a lifetime exposure at 100 Bq/m^3 radon concentration; less than 2 years in a home with 4000 Bq/m^3 is sufficient to achieve the same level of risk [100]. Thus, while it is advisable for anyone to test their homes and workplaces for radon (and mitigate if a problem is detected), any homes, schools, and daycares where small children and young adults spend a great deal of time should become a priority for radon elimination.

Curiously, perhaps alarmingly, radon inhalation has been and is commercially advertised to provide beneficial health effects in some locations. For example, there are so-called "radon spas" in Austria and the United States where people pay to be exposed up to 80,000 Bq/m^3 to provide relief from chronic pain, inflammatory diseases, and dermatological conditions, ostensibly by modulating the immune system and the DNA-repair machinery [101]. While the occasional visitor to such spas might experience no significant long-term health effects (from a probably short exposure window, few days per year), permanent employees in these facilities would be expected to experience the same increase in lung cancer risk as has been documented extensively for uranium miners prior to adequate mine ventilation; hence, we would advise that these facilities should be approached with some informed caution.

Considering that radon inhalation and subsequent high-LET alpha particle irradiation of lung tissue is, overwhelmingly, the most common mode of radiation exposure encountered by humanity, we know surprisingly little about genetic risk factors for radon-induced cancer. Genetic polymorphisms of factors participating in the detoxifying (often antioxidant) process of environmental carcinogens can modulate the risk of lung cancer dramatically, but few studies have investigated the association between residential radon exposure and different cancer-susceptibility genes [102]. In 2014, studies on animals indicated that lung cancer risk from radon is higher in mice lacking genes encoding microsomal glutathione S-transferase proteins *mu* GSTM1 and *theta* GSTT1; this is likely because alpha particles, like all IR, will generate ROS that damages DNA within lung epithelia and GST proteins scavenge ROS [102]. Interestingly, studies carried out in mice exposed to high-LET IR ^{56}Fe (iron ions), showed no overall differential expression in liver-metabolizing genes; however, looking at the lung epithelium revealed a reduction in the expression of O^6-methylguanine-DNA-methyl transferase (MGMT), which is crucial in maintaining genomic instability by reversing mutagenic O^6-methylguanine back to guanine, which otherwise could trigger mismatch error and ultimately contribute to carcinogenesis [103]. Further studies exposing mice to radon gas showed differential expression of genes involved in carcinogenesis: an upregulation of E-cadherin mRNA (involved in both carcinogenesis and metastasis), downregulation of protein A1 (associated with chromosomal rearrangements), and casein kinase delta (involved in apoptosis as well as chromosomal segregation) [104].

Logically, the mutation of most factors involved in the IR-induced DNA-damage response would also be a risk factor for radon-induced lung cancer, and further work is required to confirm whether this is the case. An important consideration is that gene–environment interactions would have the greatest impact at relatively low indoor radon concentrations (50–200 Bq/m^3) most commonly observed in homes. Although some countries and agencies have started to become proactive in mitigating radon levels >300 Bq/m^3, relatively little attention has been paid to the effects of low radon concentrations <200 Bq/m^3, which is what the bulk of the population are exposed to and are responsible for the majority of radon-related deaths. Chronic, low-dose irradiation has qualitatively distinct biological consequences to acute irradiation and often fails to trigger cell-cycle checkpoints [25]. Under these conditions, difficult-to-repair DNA damage (as would be caused by high-LET IR) would persist and be more likely to enter into DNA replication or cell division, where risks of chromosomal fragmentation or mutation would increase dramatically [92]. Any genetic polymorphisms associated with even mild DNA-repair delay would exacerbate this effect, and potentially render such individuals particularly at risk of radon-induced genetic mutation and thus cancer risk.

8. CONCLUSION

Although varied in source, features, and consequences, DNA damage underlies nearly all IR-induced human diseases. Whether from acute, high-dose exposure, chronic low-dose exposure, or IR of varying LET, it is the extent of damage and the type of cells impacted that dictates the outcome on health. At lethal doses, necrosis or apoptosis occurs and the immediate effects are seen in ARS due to tissue failure. By contrast, if a cell can survive IR-induced DNA damage, the chances of mutation are increased since repair mechanisms may erroneously repair DSBs, enabling chromosome translocations, rearrangements, and/or point mutations; however, carcinogenesis is not guaranteed. Through several studies of IR exposure survivors, a correlation for the stochastic effects of radiation has been made where the higher the IR dose at the younger the age of exposure, the more likely a cancer is to develop. This also holds true for deterministic effects of radiation, with ARS increasing in severity as the IR dose increases. Based on all of this, a simple conclusion can be drawn, radiation is a health concern whose exposure needs to be minimized and its resulting effects need to be mitigated through continued study and further understanding.

GLOSSARY

Alpha particle A helium nucleus (two protons and two neutrons).
Becquerel (Bq) One radioactive disintegration per second.
Beta particles A high-speed and energy electron or position.
Clustered damage Results from high-LET ionizing radiation, with large amounts of damage accrued over a defined distance.
Core (relating to high LET) High-LET radiation track structure consisting of energy deposited close to the high-LET particle trajectory. This arises from the excitation and collective oscillations of atoms very close to the track.
Daughter radioisotopes Radioactive elements decay to form progeny products.
Deterministic effects The immediate and predictable effects of ionizing radiation with severity relating to the dose received.
Excess relative risk This is the measure for the relative chance of developing a disorder (largely used for cancer risk assessment) for an exposed person, compared to the risk in a nonexposed person—that is, a person is 2 times as likely to develop a certain cancer per Sv of radiation received compared to a person who has not been exposed to radiation.
Fallout The radioactive elements released into an environment due to a nuclear incident.
Gamma rays Photons that have energies that can overlap with X-rays and range between 100 keV and 10 MeV, but are primarily defined as originating from atomic nuclear decay.
Gray (Gy) One Joule of ionizing radiation energy per kilogram of cells.
H2AX Histone 2A variant X (encompassing about 10% of the total histone 2A population).
γH2AX H2AX phosphorylated at serine 139 in response to DNA double-strand breaks.
Homologous recombination The error-free repair of damaged DNA double-strand breaks using template DNA (such as a sister chromatid).
LET The rate at which energy is released by a radioactive source over a fixed distance.
Neutrons Noncharged subatomic particle.
Nonhomologous end joining The error-prone repair of DNA double-strand break, not relying on template DNA.
Pneumbra (relating to high LET) Refers to high-LET radiation track structure when energy is deposited away from the trajectory of the high-LET particle—that is, scatter deposition of energy at some distance adjacent from the particle trajectory.
Relative biological effectiveness Is a term referring to the ratio of biological effectiveness of one IR type in comparison to another, given the same quantity of absorbed energy.
Sievert (Sv) Often equivalent to gray, and represents the dose of ionizing radiation absorbed by a cell/tissue/body that also has a measurable biological effect.
Stochastic effects The nonpredictable effects of ionizing radiation on health, such as genomic instability and cancer.
X-rays Photons with an energy of 100 eV–300 keV.

LIST OF ABBREVIATIONS

Alt-NHEJ Alternative NHEJ
ALL Acute lymphocytic leukemia
ARS Acute radiation sickness
ATM Ataxia telangiectasia mutated

CT Computed tomography
DDR DNA-damage response
DNA Deoxyribonucleic acid
DSBs DNA double-stranded breaks
ERR Excess relative risk
GCR Galactic cosmic rays
HR Homologous recombination
HZE High atomic number and energy
ICL Interstrand cross-link
IR Ionizing radiation
LET Linear energy transfer
LSS Life span study
MDS Multiple-damage sites
MGMT Methylguanine-DNA-methyl transferase
NHEJ Nonhomologous end joining
NSAIDs Nonsteroidal anti-inflammatory drugs
·OH Hydroxyl radical
SSBs DNA single-stranded breaks
RBE Relative biological effectiveness
ROS Reactive oxygen species
RT Radiotherapy
UN United Nations
UV Ultraviolet
WHO World Health Organization

REFERENCES

[1] Noll DM, Mason TM, Miller PS. Formation and repair of interstrand cross-links in DNA. Chem Rev 2006;106(2):277–301.

[2] Dextraze M-E, Gantchev T, Girouard S, Hunting D. DNA interstrand cross-links induced by ionizing radiation: an unsung lesion. Mutat Res January 2010;704(1–3):101–7. [cited 2015 June 3] Available from: http://www.sciencedirect.com/science/article/pii/S1383574210000074.

[3] Freudenreich CH. Chromosome fragility: molecular mechanisms and cellular consequences. Front Biosci 2007;12:4911–24.

[4] Kumar R, Horikoshi N, Singh M, Gupta A, Misra HS, Albuquerque K, et al. Chromatin modifications and the DNA damage response to ionizing radiation. Front Oncol 2013;2: 214.

[5] Jackson SP, Bartek J. The DNA-damage response in human biology and disease. Nature 2009;461(7267):1071–8.

[6] Rogakou EP, Pilch DR, Orr AH, Ivanova VS, Bonner WM. DNA double-stranded breaks induce histone H2AX phosphorylation on serine 139. J Biol Chem 1998;273(10):5858–68.

[7] Kinner A, Wu W, Staudt C, Iliakis G. γ-H2AX in recognition and signaling of DNA double-strand breaks in the context of chromatin. Nucleic Acids Res. Oxford University Press; October 4, 2008;36(17):5678–94. Available from: http://www.ncbi.nlm.nih.gov/pmc/articles/PMC2553572/.

[8] Frit P, Barboule N, Yuan Y, Gomez D, Calsou P. Alternative end-joining pathway(s): bricolage at DNA breaks. DNA Repair (Amst) May 2014;17:81–97. [cited 2015 May 19] Available from: http://www.sciencedirect.com/science/article/pii/S1568786414000445.

[9] Asaithamby A, Chen DJ. Mechanism of cluster DNA damage repair in response to high-atomic number and energy particles radiation. Mutat Res 2011;711(1–2):87–99.

[10] Bee L, Fabris S, Cherubini R, Mognato M, Celotti L. The efficiency of homologous recombination and non-homologous end joining systems in repairing double-strand breaks during cell cycle progression. Cotterill S, editor. PLoS One July 11, 2013;8(7):e69061. San Francisco, USA: Public Library of Science. Available from: http://www.ncbi.nlm.nih.gov/pmc/articles/PMC3708908/.

[11] Britten RA, Peters LJ, Murray D. Biological factors influencing the RBE of neutrons: implications for their past, present and future use in radiotherapy. Radiat Res 2001;156(2):125–35.

[12] Meijer AE, Ekedahl J, Joseph B, Castro J, Harms-Ringdahl M, Zhivotovsky B, et al. High-LET radiation induces apoptosis in lymphoblastoid cell lines derived from atazia-telangiectasia patients. Int J Radiat Biol 2001;77(3):309–17.

[13] Wang H, Liu S, Zhang P, Zhang S, Naidu M, Wang Y. S-phase cells are more sensitive to high-linear energy transfer radiation. Int J Radiat Oncol Biol Phys 2009;74(4):1236–41.

[14] Rydberg B, Cooper B, Cooper PK, Holley WR, Chatterjee A. Dose-dependent misrejoining of radiation-induced DNA double-strand breaks in human fibroblasts: experimental and theoretical study for high- and low-LET radiation. Radiat Res 2005;163(5):526–34.

[15] Wang H, Wang X, Zhang P, Wang Y. The Ku-dependent non-homologous end-joining but not other repair pathway is inhibited by high linear energy transfer ionizing radiation. DNA Repair (Amst) 2008;7(5):725–33.

[16] Sutherland BM, Bennett PV, Cintron-Torres N, Hada M, Trunk J, Monteleone D, et al. Clustered DNA damages induced in human hematopoietic cells by low doses of ionizing radiation. J Radiat Res 2002;43(Suppl. 1):S149–52.

[17] Brenner DJ, Ward JF. Constraints on energy deposition and target size of multiply damaged sites associated with DNA double-strand breaks. Int J Radiat Biol 1992;61(6):737–48.

[18] Gulston M, de Lara C, Jenner T, Davis E, O'Neill P. Processing of clustered DNA damage generates additional double-strand breaks in mammalian cells post-irradiation. Nucleic Acids Res 2004;32(4):1602–9.

[19] Kempner ES. Effects of high-energy electrons and gamma rays directly on protein molecules. J Pharm Sci 2001;90(10):1637–46.

[20] Chatterjee A, Schaefer HJ. Microdosimetric structure of heavy ion tracks in tissue. Radiat Env Biophys 1976;13(3):215–27.

[21] Chatterjee A, Holley WR. Biochemical mechanisms and clusters of damage for high-LET radiation. Adv Sp Res 1992;12(2–3):33–43.

[22] Ozasa K, Shimizu Y, Suyama A, Kasagi F, Soda M, Grant EJ. Studies of the mortality of atomic bomb survivors, Report 14, 1950–2003: an overview of cancer and noncancer diseases. Radiat Res 2012;243(3):229–43. Available from: http://www.ncbi.nlm.nih.gov/pubmed/22171960.

[23] Unscear. Exposures and effects of the Chernobyl accident- ANNEX J. 2000. p. 115.

[24] Williams D. Cancer after nuclear fallout: lessons from the Chernobyl accident. Nat Rev Cancer 2002;2(7):543–9.

[25] Moore S, Stanley FKT, Goodarzi AA. The repair of environmentally relevant DNA double strand breaks caused by high linear energy transfer irradiation–no simple task. DNA Repair (Amst) May 2014;17:64–73. [cited 2015 March 27] Available from: http://www.sciencedirect.com/science/article/pii/S1568786414000251.

[26] Heslet L, Bay C, Nepper-Christensen S. Acute radiation syndrome (ARS) – treatment of the reduced host defense. Int J General Med 2012:105–15.

[27] Anno GH, Young RW, Bloom RM, Mercier JR. Dose response relationships for acute ionizing-radiation lethality. Health Phys 2003;84(5):565–75.

[28] Ichimaru M, Tomonaga M, Amenomori T, Matsuo T. Atomic bomb and leukemia. J Radiat Res 1991;32(Suppl. 2):14–9.

[29] Sawada S. Cover-up of the effects of internal exposure by residual radiation from the atomic bombing of Hiroshima and Nagasaki. Med Confl Surviv 2007;23(1):58–74.

[30] Nguyen CP, Goodman LH. Fetal risk in diagnostic radiology. Semin Ultrasound CT MR February 2012;33(1):4–10. [cited 2015 May 13] Available from: http://www.sciencedirect.com/science/article/pii/S0887217111001193.

[31] Smith G. UNSCEAR 2013 Report Volume I: Report to the General Assembly, Annex A: levels and effects of radiation exposure due to the nuclear accident after the 2011 great east-Japan earthquake and tsunami. J Radiol Prot 2014;34(3):725. Available from: http://stacks.iop.org/0952-4746/34/i=3/a=B01.

[32] Nussbaum RH. The Chernobyl nuclear catastrophe: unacknowledged health detriment. Environ Health Perspect May 2007;115(5):A238–9. National Institute of Environmental Health Sciences. Available from: http://www.ncbi.nlm.nih.gov/pmc/articles/PMC1867971/.

[33] Report of the United Nations Scientific Committee on the effects of atomic radiation to the general assembly contents Chapter I Introduction. 2013.

[34] Allen C, Ashley AK, Hromas R, Nickoloff JA. More forks on the road to replication stress recovery. J Mol Cell Biol 2011;3(1):4–12.

[35] Tsubokura M, Gilmour S, Takahashi K, Oikawa T, Yukio K. Internal radiation exposure after the Fukushima nuclear power plant disaster. JAMA 2012;308(7):669–70.

[36] Harach HR, Williams ED. Childhood thyroid cancer in England and Wales. Br J Cancer 1995;72(3):777–83.

[37] Furmanchuk AW, Averkin JI, Egloff B, Ruchti C, Abelin T, Schäppi W, et al. Pathomorphological findings in thyroid cancers of children from the Republic of Belarus: a study of 86 cases occurring between 1986 ('post-Chernobyl') and 1991. Histopathology 1992;21(5):401–8.

[38] Nikiforov YE, Rowland JM, Bove KE, Monforte-Munoz H, Fagin JA. Distinct pattern of ret oncogene rearrangements in morphological variants of radiation-induced and sporadic thyroid papillary carcinomas in children. Cancer Res 1997;57(9):1690–4.

[39] Monaco C, Visconti R, Barone MV, Pierantoni GM, Berlingieri MT, De Lorenzo C, et al. The RFG oligomerization domain mediates kinase activation and re-localization of the RET/PTC3 oncoprotein to the plasma membrane. Oncogene 2001;20(5):599–608.

[40] Powell DJ, Russell J, Nibu KI, Li G, Rhee E, Liao M, et al. The RET/PTC3 oncogene: metastatic solid-type papillary carcinomas in murine thyroids. Cancer Res 1998;58(23):5523–8.

[41] Thurston V, Williams ED. The effect of radiation on thyroid C cells. Acta Endocrinol (Copenh) 1982;99(1):72–8.

[42] Wood JW, Tamagaki H, Neriishi S, Sato T, Sheldon WF, Archer PG, et al. Thyroid carcinoma in atomic bomb survivors Hiroshima and Nagasaki. Am J Epidemiol January 1, 1969;89(1):4–14. Available from: http://aje.oxfordjournals.org/content/89/1/4.abstract.

[43] Cardis E, Gilbert E, Carpenter L, Howe G, Kato I, Armstrong B, et al. Effects of low doses and low dose rates of external radiation: cancer mortality ionizing among nuclear in three countries workers industry. Radiat Res 2009;142(2):117–32.

[44] Khait VKI, Tsyb AF, Konogorov AP, Rastopchin EM, Khait SE. Case – control analysis of leukaemia among Chernobyl accident emergency workers residing in the Russian Federation, 1986-1993. J Radiol Prot 1997;17(3):137. Available from: http://stacks.iop.org/0952-4746/17/i=3/a=003.

[45] Noshchenko AG, Zamostyan PV, Bondar OY, Drozdova VD. Radiation-induced leukemia risk among those aged 0–20 at the time of the Chernobyl accident: a case-control study in the Ukraine. Int J Cancer June 1, 2002;99(4):609–18. Wiley Subscription Services, Inc., A Wiley Company. Available from: http://dx.doi.org/10.1002/ijc.10406.

[46] Nichols KE, Levitz S, Shannon KE, Wahrer DCR, Bell DW, Chang G, et al. Heterozygous germline ATM mutations do not contribute to radiation-associated malignancies after Hodgkin's disease. J Clin Oncol 1999;17(4):1259–66.

[47] Richardson DB, Wing S, Schroeder J, Schmitz-Feuerhake I, Hoffmann W. Ionizing radiation and chronic lymphocytic leukemia. Environ Health Perspect January 21, 2005;113(1):1–5. National Institue of Environmental Health Sciences. Available from: http://www.ncbi.nlm.nih.gov/pmc/articles/PMC1253701/.

[48] Renwick A, Thompson D, Seal S, Kelly P, Chagtai T, Ahmed M, et al. ATM mutations that cause ataxia-telangiectasia are breast cancer susceptibility alleles. Nat Genet August 2006;38(8):873–5. Available from: http://dx.doi.org/10.1038/ng1837.

[49] Sakai A, Ohira T, Hosoya M, Ohtsuru A, Satoh H, Kawasaki Y, et al. Life as an evacuee after the Fukushima Daiichi nuclear power plant accident is a cause of polycythemia: the Fukushima Health Management Survey. BMC Public Health January 2014;14(1):1318. [cited 2015 April 27] Available from: http://www.biomedcentral.com/1471-2458/14/1318.

[50] Shimizu Y, Kodama K, Nishi N, Kasagi F, Suyama A, Soda M, et al. Radiation exposure and circulatory disease risk: Hiroshima and Nagasaki atomic bomb survivor data, 1950-2003. BMJ 2010;340:b5349.

[51] McFarland HI, Puig M, Grajkowska LT, Tsuji K, Lee JP, Mason KP, et al. Regulatory T cells in γ irradiation-induced immune suppression. Bansal GP, editor. PLoS One June 19, 2012;7(6):e39092. San Francisco, USA: Public Library of Science; Available from: http://www.ncbi.nlm.nih.gov/pmc/articles/PMC3378522/.

[52] McCurry J. The Fukushima Daiichi disaster: 4 years on. Lancet March 2015;385(9973):e23–4. [cited 2015 March 13] Available from: http://www.sciencedirect.com/science/article/pii/S0140673615603026.

[53] Friesecke I, Beyrer K, Wedel R, Reimers K, Fliedner TM. SEARCH: a system for evaluation and archiving of radiation accidents based on case histories. Radiat Environ Biophys 2000;39(3):213–7.

[54] Von Neubeck C, Geniza MJ, Kauer PM, Robinson RJ, Chrisler WB, Sowa MB. The effect of low dose ionizing radiation on homeostasis and functional integrity in an organotypic human skin model. Mutat Res 2015;775:10–8. Elsevier B.V. Available from: http://linkinghub.elsevier.com/retrieve/pii/S0027510715000615.

[55] Bottollier-Depois JF, Chau Q, Bouisset P, Kerlau G, Plawinski L, Lebaron-Jacobs L. Assessing exposure to cosmic radiation on board aircraft. Adv Sp Res January 2003;32(1):59–66. [cited 2015 June 3] Available from: http://www.sciencedirect.com/science/article/pii/S0273117703903711.

[56] Cucinotta FA, Durante M. Cancer risk from exposure to galactic cosmic rays: implications for space exploration by human beings. Lancet Oncol May 2006;7(5):431–5. [cited 2015 May 12] Available from: http://www.sciencedirect.com/science/article/pii/S1470204506706957.

[57] Kranz AR. Genetic and physiological damage induced by cosmic radiation on dry plant seeds during space flight. Adv Sp Res 1986;6(12):135–8.

[58] Casolino M, Bidoli V, Morselli A, Narici L, De Pascale MP, Picozza P, et al. Space travel: dual origins of light flashes seen in space. Nature April 17, 2003;422(6933):680. Available from: http://dx.doi.org/10.1038/422680a.

[59] Buonanno M, De Toledo SM, Howell RW, Azzam EI. Low-dose energetic protons induce adaptive and bystander effects that protect human cells against DNA damage caused by a subsequent exposure to energetic iron ions. J Radiat Res March 23, 2015. Available from: http://jrr.oxfordjournals.org/content/early/2015/03/23/jrr.rrv005.abstract.

[60] Bucker H, Horneck G. The biological effectiveness of HZE-particles of cosmic radiation studied in the Apollo 16 and 17 Biostack experiments. Acta Astronaut 1975;2(3–4):247–64.

[61] Li M, Holmes V, Ni H, Sanzari JK, Romero-Weaver AL, Lin L, et al. Broad-spectrum antibiotic or g-CSF as potential countermeasures for impaired control of bacterial infection associated with an SPE exposure during spaceflight. PLoS One March 20, 2015;10(3):e0120126. Public Library of Science. Available from: http://dx.doi.org/10.1371%2Fjournal.pone.0120126.

[62] Shearer WT, Zhang S, Reuben JM, Lee BN, Butel JS. Effects of radiation and latent virus on immune responses in a space flight model. J Allergy Clin Immunol 2005;115(6):1297–303.

[63] Sonnenfeld G, Shearer WT. Immune function during space flight. Nutrition October 2002;18(10):899–903. [cited 2015 May 11] Available from: http://www.sciencedirect.com/science/article/pii/S0899900702009036.

[64] Sanlorenzo M, Wehner MR, Linos E, Kornak J, Kainz W, Posch C, et al. The risk of melanoma in airline pilots and cabin crew. JAMA Dermatol 2015;151(1):51. Available from: http://archderm.jamanetwork.com/article.aspx?doi=10.1001/jamadermatol.2014.1077.

[65] Durante M, George K, Cucinotta FA. Chromosomes lacking telomeres are present in the progeny of human lymphocytes exposed to heavy ions. Radiat Res 2006;165(1):51–8.

[66] Kim R, Emi M, Tanabe K. Cancer immunoediting from immune surveillance to immune escape. Immunology May 2, 2007;121(1):1–14. Blackwell Science Inc. Available from: http://www.ncbi.nlm.nih.gov/pmc/articles/PMC2265921/.

[67] Lowe XR, Bhattacharya S, Marchetti F, Wyrobek AJ. Early brain response to low-dose radiation exposure involves molecular networks and pathways associated with cognitive functions, advanced aging and Alzheimer's disease. Radiat Res 2009;171(1):53–65.

[68] Manda K, Ueno M, Anzai K. Memory impairment, oxidative damage and apoptosis induced by space radiation: ameliorative potential of alpha-lipoic acid. Behav Brain Res March 5, 2008;187(2):387–95. [cited 2015 May 11] Available from: http://www.sciencedirect.com/science/article/pii/S0166432807005220.

[69] Shukitt-Hale B, Casadesus G, McEwen JJ, Rabin BM, Joseph JA. Spatial learning and memory deficits induced by exposure to iron-56-particle radiation. Radiat Res 2000;154(1):28–33.

[70] Cucinotta FA, Manuel FK, Jones J, Iszard G, Murrey J, Djojonegro B, et al. Space radiation and cataracts in astronauts. Radiat Res November 1, 2001;156(5):460–6. The Radiation Research Society. Available from: http://www.rrjournal.org/doi/abs/10.1667/0033-7587%282001%29156%5B0460%3ASRACIA%5D2.0.CO%3B2.

[71] Jones JA, McCarten M, Manuel K, Djojonegoro B, Murray J, Feiversen A, et al. Cataract formation mechanisms and risk in aviation and space crews. Aviat Sp Environ Med 2007;78(4 Suppl).

[72] Rastegar Z, Eckart P, Mertz M. Radiation-induced cataract in astronauts and cosmonauts. Graefe's Arch Clin Exp Ophthalmol 2002;240(7):543–7. Springer-Verlag. Available from: http://dx.doi.org/10.1007/s00417-002-0489-4.

[73] Frey MA. Radiation health: mechanisms of radiation-induced cataracts in astronauts. Aviat Space Environ Med 2009:575–6.

[74] C.P.A.Cancer. In: The 2014 cancer system performance report. 2014. Toronto.

[75] Sun Z, AbAziz A, Khairuddin Md Yusof A. Radiation-induced noncancer risks in interventional cardiology: optimisation of procedures and staff and patient dose reduction. Biomed Res Int 2013;20:976962. Hindawi Publishing Corporation; 2013 August Available from: http://www.ncbi.nlm.nih.gov/pmc/articles/PMC3762166/.

[76] Monje ML, Vogel H, Masek M, Ligon KL, Fisher PG, Palmer TD. Impaired human hippocampal neurogenesis after treatment for central nervous system malignancies. Ann Neurol 2007;62(5):515–20.

[77] Monje ML, Toda H, Palmer TD. Inflammatory blockade restores adult hippocampal neurogenesis. Science 2003;302(5651):1760–5.

[78] Ng J, Shuryak I. Minimizing second cancer risk following radiotherapy: current perspectives. Cancer Manag Res December 17, 2015;7:1–11. Dove Medical Press. Available from: http://www.ncbi.nlm.nih.gov/pmc/articles/PMC4274043/.

[79] Hanahan D, Weinberg RA. Hallmarks of cancer: the next generation. Cell 2011;144(5):646–74. Elsevier Inc. Available from: http://dx.doi.org/10.1016/j.cell.2011.02.013.

[80] Silber JH, Radcliffe J, Peckham V, Perilongo G, Kishnani P, Fridman M, et al. Whole-brain irradiation and decline in intelligence: the influence of dose and age on IQ score. J Clin Oncol 1992;10(9):1390–6.

[81] Monje ML, Mizumatsu S, Fike JR, Palmer TD. Irradiation induces neural precursor-cell dysfunction. Nat Med 2002;8(9):955–62.

[82] Merchant TE, Kiehna EN, Li C, Shukla H, Sengupta S, Xiong X, et al. Modeling radiation dosimetry to predict cognitive outcomes in pediatric patients with CNS embryonal tumors including medulloblastoma. Int J Radiat Oncol Biol Phys May 1, 2006;65(1):210–21. [cited 2015 May 29] Available from: http://www.sciencedirect.com/science/article/pii/S0360301605029536.

[83] Shetty MK. Abdominal computed tomography during pregnancy: a review of indications and fetal radiation exposure issues. Semin Ultrasound CT MR February 2010;31(1):3–7. [cited 2015 May 13] Available from: http://www.sciencedirect.com/science/article/pii/S088721710900081X.

[84] Kordolaimi SD, Efstathopoulos EP. Computed tomography radiation dosimetry: from the indicators to the indications. J Comput Assist Tomogr 2014;38(6):807–14. Available from: http://www.ncbi.nlm.nih.gov/pubmed/25055163.

[85] Bianconi E, Piovesan A, Facchin F, Beraudi A, Casadei R, Frabetti F, et al. An estimation of the number of cells in the human body. Ann Hum Biol 2013;40(6):463–71. Available from: http://www.ncbi.nlm.nih.gov/pubmed/23829164.

[86] Berrington de González A, Mahesh M, Kim K-P, Bhargavan M, Lewis R, Mettler F, et al. Projected cancer risks from computed tomographic scans performed in the United States in 2007. Arch Intern Med 2009;169(22):2071–7.

[87] Hendee WR, O'Connor MK. Radiation risks of medical imaging: separating fact from fantasy. Radiology August 1, 2012;264(2):312–21. Radiological Society of North America. Available from: http://dx.doi.org/10.1148/radiol.12112678.

[88] Linet MS, Slovis TL, Miller DL, Kleinerman R, Lee C, Rajaraman P, et al. Cancer risks associated with external radiation from diagnostic imaging procedures. CA Cancer J Clin March 1, 2012;62(2):75–100. Wiley Subscription Services, Inc., A Wiley Company. Available from: http://dx.doi.org/10.3322/caac.21132.

[89] Brenner DJ, Hall EJ. Computed tomography–an increasing source of radiation exposure. N Engl J Med 2007;357(22):2277–84.

[90] Elster J, Geitel H. Über die Radioaktivität der im Erdboden enthaltenen. Phys Z 1902;3:574.

[91] Rutherford E. Radio-activity. Cambridge Physical Press; 1904. p. 357. 370 pp.

[92] Tucker JD. Low-dose ionizing radiation and chromosome translocations: a review of the major considerations for human biological dosimetry. Mutat Res January 2008;659(3):211–20. [cited 2015 June 4] Available from: http://www.sciencedirect.com/science/article/pii/S1383574208000422.

[93] Pallis AG, Syrigos KN. Lung cancer in never smokers: disease characteristics and risk factors. Crit Rev Oncol Hematol 2013;88(3):494–503.

[94] Kreuzer M, McLaughlin J. Radon. Guidelines for indoor air quality. Sel Pollut 2010. Geneva. Available from: http://www.ncbi.nlm.nih.gov/books/NBK138712/?report=classic.

[95] Donya M, Radford M, ElGuindy A, Firmin D, Yacoub MH. Radiation in medicine: origins, risks and aspirations. Glob Cardiol Sci Pract December 31, 2014;2014(4):437–48. Qatar: Bloomsbury Qatar Foundation Journals. Available from: http://www.ncbi.nlm.nih.gov/pmc/articles/PMC4355517/.

[96] Mulloy KB, James DS, Mohs K, Kornfeld M. Lung cancer in a nonsmoking underground uranium miner. Environ Health Perspect 2001;109(3):305–9. Available from: http://www.ncbi.nlm.nih.gov/pmc/articles/PMC1240251/.

[97] Kusiak RA, Ritchie AC, Muller J, Springer J. Mortality from lung cancer in Ontario uranium miners. Br J Ind Med October 1993;50(10):920–8. Available from: http://www.ncbi.nlm.nih.gov/pmc/articles/PMC1035522/.

[98] Donoghue AM. Occupational health hazards in mining: an overview. Occup Med 2004:283–9.

[99] Radon prevention and mitigation [Internet]. Who handbook on indoor radon. Geneva: A Public Health Perspective; 2009. Available from: http://www.ncbi.nlm.nih.gov/books/NBK143225/.

[100] Chen J. Canadian lung Cancer relative risk from radon exposure for short periods in childhood compared to a lifetime. Int J Environ Res Public Health 2013;10(5):1916–26. Available from: http://www.mdpi.com/1660-4601/10/5/1916.

[101] Erickson BE. The therapeutic use of radon: a biomedical treatment in Europe; an "alternative" remedy in the United States. Dose-Response September 23, 2007;5(1):48–62. International Hormesis Society. Available from: http://www.ncbi.nlm.nih.gov/pmc/articles/PMC2477705/.

[102] Ruano-Ravina A, Pereyra MF, Castro MT, Perez-Rios M, Abal-Arca J, Barros-Dios JM. Genetic susceptibility, residential radon, and lung cancer in a radon prone area. J Thorac Oncol 2014;9(8):1073–80.

[103] Lima F, Ding D, Goetz W, Yang AJ, Baulch JE. High LET 56Fe ion irradiation induces tissue-specific changes in DNA methylation in the mouse. Environ Mol Mutagen 2014;55(3):266–77. Available from: http://dx.doi.org/10.1002/em.21832.

[104] Li JX, Chen R, Fu CL, Nie JH, Tong J. Screening of differential expressive genes in murine cells following radon exposure. J Toxicol Env Heal A 2010;73(7):499–506.

Section VIII

Bystander and Transgenerational Effects – Epigenetic Perspective

Chapter 34

Sins of Fathers Through a Scientific Lens: Transgenerational Effects

M. Merrifield, O. Kovalchuk
University of Lethbridge, Lethbridge, AB, Canada

Chapter Outline

1. INTRODUCTION

"Like mother-like daughter," "like father-like son"—these and other idioms signify a long-standing fascination of humans with similarities between parents and offspring. From ancient times, numerous theories have been proposed to explain trends and mechanisms of the inheritance of phenotypic traits. Hippocrates suggested that the parent–offspring similarity and sharing of certain phenotypic characteristics may occur due to an enigmatic blending of particles or fluids of parents. The pioneering works of Gregor Mendel established the precise rules of inheritance whereby heritable factors the nature of which was unknown at his time duplicate in parents and precisely segregate to progeny. Later on, Thomas Morgan determined that chromosomes may in turns serve as vehicles of Mendelian inheritance. Finally, upon the discovery of DNA and unraveling of its structure, DNA was unequivocally recognized as a fundamental agent of inheritance. As such, the key concept of medical genetics that seeks to link genotypes to phenotypes lies in the ability to establish links between individual genetic differences and individual phenotypic differences. For decades and until the mid-2010s, this perspective has dominated our understanding of genotypic and phenotypic variation and disease risk analysis, and thus has led to several important breakthroughs in the identification of numerous genetic underpinnings of heritable diseases. It has shaped many aspects of medicine, including those of organismal biology and evolutionary biology. Genetic components of heritable cancer syndromes and other diseases have been identified, and the presence of a genetic mutation, a polymorphism or chromosomal abnormality that promote disease form the current paradigm for disease etiology. Nevertheless, heritable components of heart disease, neuro-inflammatory and neurodegenerative disorders, diabetes, obesity, cancers, autoimmune, and other conditions remain remarkably elusive and lack the defined genetic components. This apparent "missing heritability" has sparked a lot of interest and research and has led researchers to revisit concepts of gene–gene (and more importantly gene–environment) interactions. Indeed, genome–environment interactions are equally important factors in disease etiology.

Even though the mammalian genome is rather stable because cells harbor numerous elaborate and highly efficient mechanisms that repair DNA, environmental factors have the ability to damage DNA directly by inducing genome-destabilizing mutations without altering DNA sequence, thus promoting disease via mechanisms that do not always involve direct DNA-damage or -sequence changes.

Genome Stability. http://dx.doi.org/10.1016/B978-0-12-803309-8.00034-3

Furthermore, as it has recently been established, environmental influences on disease etiology depend upon a developmental stage of an organism upon exposure to stress. Exposures during critical periods of time of organism development can alter genome activity associated with the differentiation program of cells or organ systems.

Genome stability and the ability to have healthy offspring is of an utmost importance for each individual organism and for entire populations. Understanding the rules of inheritance and the ability of the environment to influence inheritance and affect disease predisposition have been a central focus of research for many decades. Most recently, parental exposure and origins of disease have gained a lot of attention.

The analysis of genome–environment interactions has brought forth the concept of genome instability. A phenomenon of genomic or genome instability is used to describe an increased rate of acquired alterations in the genome of an organism or its offspring; the latter is referred to as transgenerational genome instability. The field of genome instability, and especially transgenerational instability, has emerged from studies that attempted to explain an unexpectedly high frequency of mutations and chromosomal damage in the progeny of irradiated somatic and germline cells.

2. RADIATION-INDUCED GENOME INSTABILITY

It has long been thought that the main factor contributing to the negative biological effects of ionizing radiation (IR) in mammalian cells, such as chromosomal aberrations, mutations, and cell death, is the result of DNA damage in directly exposed cells; that is, residual damage that has not been eradicated by DNA-repair systems in the exposed cell [1]. This paradigm has widely been challenged since 2000, mostly originating from the results of numerous in vitro studies that demonstrated the existence of delayed effects of IR exposure [2]. These delayed effects can manifest in the unexposed progeny of irradiated cells for many cell divisions (and up to 4 years) after the initial exposure [2]. The all-encompassing term given to this phenomenon is "radiation-induced genomic instability," which is used to describe the increased rate of the acquisition of alterations in the genome. Experimentally, genomic instability is observed when a cell is irradiated, then clonally expanded, and the progeny is examined genetically. As mentioned, radiation-induced genomic instability is observed generations after the initial exposure, and a number of studies have shown that this often occurs at a high frequency [3,4].

Multiple genetic end points have been utilized to evaluate radiation-induced genomic instability in a number of in vitro systems, which include, but are not limited to, chromosomal aberrations, ploidy changes, micronucleus formation, gene mutations, and amplifications, as well as increased microsatellite/ESTR (expanded simple tandem repeat) mutation rates and delayed cell death [2,5,6]. There are a number of pathways that are implicated in the initiation and perpetuation of radiation-induced genomic instability [7]. The relative contribution of the different pathways primarily depends upon the genetic background of the irradiated cell or organism [8,9], as well as the type of radiation [4].

Various in vitro systems have demonstrated a high frequency of IR-induced genomic instability by means of examining the various end points (as described earlier) that are associated with IR-induced genomic instability [2]. As of 2016, a prevailing hypothesis is that IR exposure destabilizes the genome, thus initiating a cascade of genomic events that increases the rate of point mutations, small deletions/insertions, and large rearrangements in the progeny of irradiated cells [2].

It has long been speculated that the development of genomic instability can facilitate the process of cancer initiation and/or progression [10], and indeed the loss of genomic stability is believed to be a hallmark of many cancers, as well as an important prerequisite for cancer formation [11–13]. Therefore, the general assumption is that there is a link between the induction of IR-induced genomic instability and cancer due to an increase in the accumulation of multiple genetic events within a cell that ultimately enhances radiation-induced carcinogenesis. This assumption is also supported by the findings of epidemiological studies which suggest that some types of radiation-induced cancers may follow a relative risk model in which IR exposure enhances the rate at which cancers develop, instead of inducing a specific cohort of new tumors [14]. The demonstration of IR-induced genomic instability in somatic cell–culture systems has greatly increased interest in research on the potential long-term effects of exposure. The research area that has likely benefited the most is the potential long-term effects associated with germline exposure to IR and the transmission of adverse effects (eg, genomic instability) to future generations.

2.1 Transgenerational Effects and Transgenerational Genome Instability

Initially, in vitro data have provided overwhelming evidence for the delayed effects of IR exposure manifested in the progeny of irradiated cells (ie, genomic instability) for many cell divisions, which may ultimately enhance the carcinogenic potential of these cells. Moreover, these data suggest that genomic instability can also be induced in the irradiated germline and, therefore, may be transmitted to future generations. If this is the case, the offspring of irradiated parents become genetically unstable, which results in a plethora of transgenerational effects such as the elevated mutation rates and a predisposition to cancer. Many publications have indeed characterized a wide variety of phenotypic traits observed in the offspring of irradiated parents, implicating the elevated mutation rates [15–19]. Such studies have been reinforced through various molecular techniques used to assess transgenerational genomic instability.

The first evidence for a *transgenerational effect* associated with IR exposure was demonstrated by Luning and colleagues, where the elevated rates of dominant lethal mutations (early and late embryonic death) were observed upon the intraperitonial injection of a plutonium salt solution to male mice [20]. Accordingly, an increase in dominant lethality was not only found in the germline of directly irradiated male mice, but also in the germline of their nonexposed first-generation progeny (F1). The offspring of irradiated male mice have also been shown to be reproductively challenged, exhibiting the reduced fertilization rates of both in vivo and in vitro fertilization [16,21] as well as the increased levels of prenatal mortality in the F2 generation [22]. An increase in teratogenic effects was also shown, since the number of malformed F2 fetuses was significantly higher in the paternally exposed group compared to a control one [22].

The elegant studies by Nomura have not only demonstrated that paternal irradiation leads to an increase in malformations in the progeny of irradiated parents, but they have also shown a significant increase in the incidence of cancer in the offspring [23,24]. Several additional transgenerational studies have also found a significant increase in cancer incidence among the offspring of paternally irradiated mice after the secondary exposure to known carcinogens [23,25,26]. The predisposition of the offspring of IR-exposed fathers to cancer has been investigated in human populations, where the data obtained have mainly been inconclusive [27,28]; however, two independent studies have shown a clustering of extremely high leukemia rates in children whose fathers had been exposed to radiation after working at a nuclear processing plant in the town of Sellafield [29,30].

Adding to the classical evidence of transgenerational impacts, the majority of data since 2000 have arisen from the application of an array of molecular techniques used to characterize genotypic alterations in unexposed offspring. Mainly, genotypic alterations found in the progeny of irradiated parents have included chromosomal aberrations, micronuclei formation, increased microsatellite/ESTR mutations, and altered gene-expression patterns, which are all key hallmarks of genomic instability seen in somatic cells [2,31,32]. Therefore, the manifestation of such alterations has collectively been termed *transgenerational genome instability*. Dubrova and colleagues have made a significant contribution to our current understanding of radiation-induced transgenerational genome instability by pioneering the investigation of transgenerational mutation rates within repetitive sequences of the genome [18,33]. These repetitive sequences were initially termed minisatellites, but now they are known as expanded simple tandem repeat (ESTR) loci because they are extended (500–16, 000 bp) stretches of relatively short (4–6 bp) repeats that are less stable than true minisatellites which generally consist of longer (6–100 bp) repeats [18]. Barber and colleagues studied mutation rates of two ESTR loci in the germline of F1 and F2 offspring of male mice exposed at either the premeiotic or postmeiotic stages of spermatogenesis [34]. They found an increased mutation rate in the germline of F1 offspring, which was similarly maintained in the germline of the F2 offspring in both pre/postmeiotic germ cell exposure groups. Furthermore, the elevated mutation rates were seen in all three of the mouse strains studied, and within each strain, male and female offspring (both F1 and F2) of irradiated fathers equally demonstrated the elevated mutation rates [34]. Further analysis of the unexposed F1 progeny showed that high ESTR mutation rates were observed along with the elevated levels of mutations in protein-coding genes in the germline as well as in somatic tissues such as spleen and bone marrow [35]. The observed transgenerational instability is not specific to one particular strain of mice; in fact, it has been observed in the F1 and F2 offspring of irradiated males from four different inbred strains of mice [32,34,36].

Furthermore, Barber and colleagues have also shown that spontaneous levels of SSBs and DSBs are significantly higher in the unexposed F1 offspring; however, the efficiency of DNA repair was not compromised [35]. Likewise, Koturbash and colleagues found that DNA DSBs were higher in the thymus of offspring of irradiated fathers [37]. They also analyzed Rad51 and Ku70 protein levels as indicators of homologous recombination (HR) and nonhomologous end joining (NHEJ) repair pathways, respectively. In contrast to the results of Barber and colleagues, they found evidence of a compromised HR-repair pathway indicated by downregulation of Rad51, while NHEJ was unaffected [37]. This may not necessarily result in a decrease in DNA-repair efficiency but may impact the accuracy and quality of DNA repair. Furthermore, changes in expression levels of Rad51, be it up or down, have been associated with genome instability and cancer [38–40].

In their study, Baulch and colleagues analyzed the F3 offspring of males irradiated at the B-type spermatogonial stage and found altered protein kinase activities and protein levels of p53 and p21 [36,41]. p21 is a target of p53 that arrests or slows cell-cycle progression [42]. Further investigation including the fourth-generation offspring revealed similar changes in the kinase signaling activity and protein levels of p53 and p21, although the magnitude and direction of changes in each end point differed between generations and within generations [43]. This finding alone highlights the phenotypic variability observed in the offspring of exposed males.

Filkowski and colleagues reported the existence of genome instability in the germline of male mice subjected to whole-body irradiation and their progeny, whereby parental irradiation led to the reactivation of long interspersed nuclear elements 1 (LINE1) and short interspersed nuclear elements B2 (SINE B2) [44].

Transgenerational radiation-induced effects seem to be paternal in nature, and up to now, the long-term genetic effects of maternal irradiation remain under-investigated. Dubrova and colleagues undertook an in-depth study to establish the effects of radiation exposure on mutation induction in the germline of radiation-exposed females and the potential of induction of radiation-induced transgenerational effects in their nonexposed offspring [45]. To address this question, adult female BALB/c and CBA/Ca mice were given 1 Gy of acute X-rays and mated with unexposed males, and the frequency of mutations at ESTR loci in the germline of directly exposed females and somatic tissues of the progeny was analyzed. Surprisingly, irradiation did not affect the frequency of ESTR mutations in the germline of exposed females and their progeny. Thus, in sharp contrast to the effect of paternal irradiation that resulted in an increase in the ESTR mutation frequency in the offspring of irradiated males, maternal irradiation did not impact genome stability of their F(1) offspring. Therefore, the transgenerational effects of maternal high-dose acute irradiation are likely to be negligible [45].

Interestingly, the study of the effects of in utero irradiation also revealed sex-based differences in the induction of transgenerational genome instability. In a large-scale study, Barber and colleagues studied the effects of in utero irradiation on mutation rates at the ESTR DNA loci in directly exposed mice and their first-generation (F(1)) offspring [46]. The analysis revealed that the ESTR mutation frequencies in the germline and somatic tissues of male and female mice irradiated at 12 days of gestation remained highly elevated during adulthood, especially due to the high frequency of singleton mutations, suggesting that fetal irradiation leads to genomic instability both in utero and during adulthood. Furthermore, the ESTR mutation frequency was significantly increased in the F(1) offspring of prenatally irradiated male mice as compared to controls, proving that fetal exposure leads to transgenerational genomic instability. Contrarily, female in utero exposure did not affect genome stability in the F(1) offspring [46]. Even though radiation-induced transgenerational instability is predominantly paternal in nature, some effects appear to be synergistic when both male and female parents are exposed [37].

Transgenerational radiation-induced effects were also observed in rainbow trout. In trout, unlike in mammals, maternal and paternal irradiation may be equally important in causing transgenerational effects [47]. Additionally, the IR-induced transgenerational effects were reported to occur in *Caenorhabditis elegans* [48], Daphnia [49], medaka fish [50], and other organisms.

Most importantly, the transgenerational effects were seen in human populations exposed to the environmental or medical irradiation, albeit data from human populations are much less clear and somewhat ambiguous. As such, the analysis of mutation rates in genomic repeat elements has also been applied to study the transgenerational IR effects in human populations, namely in individuals living in the vicinity of the Chernobyl reactor accident and nuclear test sites (Semipalatinsk, Kazakhstan) [51–53]. In all of these studies, they found an increase in mutation rates among the progeny of exposed parents. Taken together, these data support the hypothesis that exposure to IR can induce germline genomic instability that may predispose future generations to an increased risk of genetic diseases, infertility, and even cancer.

2.2 Bystander Effects

Adding to the complexity of radiation responses, several studies determined that radiation effects can be seen not only in the irradiated cells and their progeny but also in the distal naive "bystander cells" that received distress signals from the exposed cells as well as in the progeny of naive bystanders. Some initial evidence of a bystander effect has been obtained from studies performed at the beginning of the 20th century. Murphy and Morton, whose research interests were devoted to the study of lymphoid cells, showed altered morphological changes in lymphoid cells after culturing them with serum from radiation-exposed animals [54]. Additionally, in 1954, Parsons and colleagues reported the presence of soluble "clastogenic" factors in the circulating blood of patients who underwent radiotherapy [55]. These factors were found to be able to induce damage in the unexposed cultured cells [56–59]. Such clastogenic activity has also been demonstrated in the plasma from patients who received high-dose radiotherapy and from individuals accidentally exposed to radiation from the Chernobyl accident. Similar to genomic instability, bystander effects manifest themselves as the induction of gross chromosomal rearrangements, chromosome aberrations, sister chromatid exchanges, deletions, duplications, mutations, amplifications, and cell death [60]. Bystander effects occur in the whole organisms in vivo; and 2008 studies showed that localized cranial exposure causes an in vivo bystander response not only in somatic tissues but in the male germline as well [61]. Bystander damage to the germline caused by localized cranial radiation has transgenerational consequences causing profound effects in the unexposed progeny [61].

Therefore, environmental as well as diagnostic and therapeutic radiation exposures can lead to a wide array of effects in the unexposed progeny.

3. MECHANISMS OF TRANSGENERATIONAL EFFECTS: EPIGENETIC CHANGES

The aforementioned radiation-induced effects did not segregate in a Mendelian manner, and therefore they were proposed to be epigenetic in nature. Epigenetic alterations are meiotically heritable and mitotically stable alterations in gene expression that occur without changes in DNA sequence; they include DNA methylation, histone modifications, and the ncRNA-mediated regulation of gene expression [62].

3.1 DNA Methylation

DNA methylation was the first epigenetic alteration identified, and it is the most widely studied epigenetic mechanism. In mammals, DNA is methylated at the carbon 5 of cytosine residues to form 5-methyl-cytosines (5meC) which is established by de novo DNA methyltransferases (DNMT3a, DNMT3b, and DNMT3L), and is subsequently maintained by DNMT1 [63–65]. The de novo DNA methylation of transposons in the germline is dependent on DNMT3L, an isoform of DNMT3a and DNMT3b that lacks the methylation activity [66]. DNA methylation is known to be associated with inactive chromatin states and in most cases, with the repression of gene expression [67–69]. A proper regulation of DNA methylation is critically important for the normal development, cell proliferation, and the maintenance of genomic stability [62,70,71]. The global loss of DNA methylation has been linked to the activation of transposable elements, the elevated chromosome breakage, aneuploidy, the increased mutation rates, and therefore to the phenomenon of genomic instability [69,71,72]. In addition, the altered global DNA-methylation pattern is a well-known characteristic of cancer cells, and the global loss of cytosine methylation was the first epigenetic abnormality discovered in cancer cells [73–75]. The DNA methylation profile of cancer cells is frequently characterized by the global genome hypomethylation as well as by the concurrent hypermethylation of selected CpG islands within gene promoters (eg, tumor suppressors) [62,72,76,77].

Direct IR exposure has been reported to affect DNA-methylation patterns. Acute exposures to low-LET radiation, such as X-rays and/or γ-rays, have been noted to result in the global genomic DNA hypomethylation [78]. Since the early 2000s, IR exposure has been found to lead to the profound dose-dependent and sex- and tissue-specific global hypomethylation [79–82]. The loss of methylation was also associated with radiation-induced alterations in the expression of DNA methyltransferases, especially de novo methyltransferases DNMT3a and DNMT3b [80,83]. Most importantly, the radiation-induced global genomic DNA-hypomethylation patterns appear to be linked to genomic instability in exposed animals [79,80,82,83].

DNA methylation also plays a role in radiation-induced bystander effects. Kaup and colleagues lead the way in showing the importance of DNA methylation in the maintenance of radiation-induced bystander effects [84]. They have demonstrated that dysregulation of DNA-methylation patterns occurs in nonirradiated cells and can persist for 20 passages when they are treated with the medium from irradiated cells [84]. These bystander cells marked with aberrant methylation patterns exhibited numerous end points characteristic of genome instability [84]. The same pattern of genomic instability and a significant loss of nuclear DNA methylation was also observed in 3D human tissue models [85].

Much insight into the role of such epigenetic changes in bystander effects and transgenerational effects in vivo has come from the pioneering work of the Kovalchuk's and Engelward's laboratories. By demonstrating that radiation exposure limited to half of the body leads to the elevated levels of DNA strand breaks and the altered levels of key proteins involved in establishing and maintaining methylation marks in lead-shielded tissues at least 0.7 cm from the irradiated tissue, they produced the first data to clearly demonstrate that epigenetically regulated bystander effects occur in vivo [86]. Using localized cranial X-irradiation in a rat model, it was also shown that IR exposure can induce the profound global DNA hypomethylation in distant bystander tissues (the spleen) 24 h after exposure [87]. Importantly, these changes were still observed 7 months after exposure [87]. This is relevant to carcinogenesis due to the fact that epigenetic manifestations of bystander effects persisted over a long period of time (in humans, it was roughly equal to 10 years). Again, a profound and persistent reduction of methylation in the bystander spleen was paralleled by a decrease in the levels of key proteins involved in the establishment and maintenance of methylation patterns (eg, DNMT3a, DNMT1, and the methyl-binding protein 2 (MeCP2)). This was believed to contribute to the reactivation of the LINE1 retrotransposon observed in the bystander spleen [87]. Such hypomethylation was also manifested in the bystander germline of cranially exposed mice [61].

Consequently, the involvement of the same type of epigenetic effectors (the global DNA methylation and associated proteins) in transgenerational effects induced from the paternal whole body and localized exposure to IR has also been studied [37,44,61]. The paternal whole-body and cranial IR exposure were shown to result in a significant global loss of DNA methylation in the thymus, bone marrow, and the spleen of F1 offspring [37,44,61]. Whole-body exposure also resulted in a specific hypomethylation of LINE1 and SINE B2 in the germline of exposed males, which was further observed in the thymus of unexposed offspring [44]. The thymus of the progeny of paternal whole-body exposures to IR and bone marrow of the offspring of fathers exposed to cranial IR, in which the most pronounced decrease in DNA methylation was observed, also exhibited a significant decrease in the expression of DNMT1, DNMT3a, DNMT3b, and the methyl-binding protein MeCP2 [37,44,61]. The global loss of DNA methylation and the altered levels of methyltransferases and methyl-binding proteins can lead to the activation of transposable elements, contributing to genomic instability [88–90]. Accordingly, it can also be suggested that the global loss of DNA methylation observed in the progeny of irradiated fathers may influence retrotransposons and satellite DNA, thus underlying transgenerational genome instability. If such hypothesis is corroborated, it may help elucidate the increased mutation rates in satellite DNA and ESTR loci observed in the progeny of exposed parents [32]. Even though these epigenetic alterations are the well-characterized consequences of radiation exposure, the underlying molecular mechanism that drives these alterations, especially the site-specific changes in DNA-methylation patterns, remain elusive. Such molecular mechanisms may very likely be the main contributors to IR-induced epigenetic alterations associated with germline genomic instability, and therefore they would be strongly implicated in facilitating the epigenetic inheritance of transgenerational IR effects.

3.2 Histone Modifications

Changes in DNA methylation do not occur as isolated events because they are closely connected to other components of chromatin structure, such as histones, histone variants, and histone modifications [62,72]. The main histone modifications include acetylation, methylation, phosphorylation, and ubiquitination [91]. There is a vast complexity of epigenetic control that can be exhibited from such modifications since each of these modifications has the differing transcriptional consequences compounded by further control that depends on the type of residue to be modified and the extent of modification (eg, mono-, di-, and trimethylated) [72,92,93]. Studies in 2005 indicated that the IR-induced global loss of DNA methylation may correlate with changes in histone methylation, specifically with the loss of histone H4 lysine trimethylation [83].

One of the best studied histone modifications following IR exposure is the phosphorylation of histone H2AX at serine 139 (γH2AX). γH2AX is possibly one of the earliest cellular responses to DSB and IR exposure. The formation of γH2AX is crucial for the repair of DSBs and for the maintenance of genome stability [94–96]. The involvement of H2AX phosphorylation in bystander and transgenerational IR effects has also been suggested. The elevated levels of γH2AX have been reported in somatic and notably germline bystander tissues in vivo, and this elevation has subsequently been observed in the offspring of exposed fathers [35,37,61,86,87].

3.3 Small RNA-Mediated Events

Epigenetic mechanisms also include small noncoding RNAs [97]. Among those, two types are of a particular interest: microRNAs (miRNAs) and Piwi-interacting RNAs (piRNAs). MicroRNAs are abundant, small (~21-25 nt) single-stranded noncoding RNAs that regulate gene expression primarily at the posttranscriptional level (eg, posttranscriptional gene silencing, PTGS). Initially, miRNAs are endogenously transcribed as a part of a primary transcript (pri-miRNA) that is able to form one or more hairpin structures (miRNA stem loops) formed by complementary sequences within the transcript. miRNA genes can be transcribed independently or clustered with others and transcribed as a polycistron [98]. There is also a large number of intragenic miRNAs transcribed from within introns or exons of protein-coding and noncoding genes [99]. These primary transcripts are then processed by the RNase III enzyme Drosha in the nucleus into stem-loop-structured miRNA precursors (pre-miRNA) that are about 70 nt long. They are then exported to the cytoplasm where Dicer (the RNase III enzyme) generates a characteristic dsRNA (21–25 nt in length) that is separated into two strands, one of which is incorporated into a member of the Argonaute protein family (AGO2), a central component of the microRNA ribonucleoprotein complex (miRNP) commonly known as the RNA-induced silencing complex (RISC) [100]. To control the translation of specific mRNAs, the miRNA-guided RISC complex binds to the 3′-UTR (untranslated region) of target mRNAs with a similar sequence structure, thus serving as a translational repressor that regulates protein synthesis by targeting specific mRNAs [101]. As of 2016, it is believed that miRNAs exhibiting a high degree of complementarity to their target mRNAs are able to repress translation through mRNA cleavage. However, most miRNAs have imperfections between the complementary sequences, and therefore repress translation without mRNA cleavage [102,103]. Although the precise nature of such regulation remains unclear, it is suggested that the main mechanisms include alterations of poly(A) tail length and the binding of regulatory proteins to the UTRs of target mRNAs [97,104]. One or many miRNAs can coordinate the expression of single/multiple genes, resulting in a complex mechanism for posttranscriptional gene regulation. Consequently, miRNAs can play a key role in numerous biological contexts, including cellular differentiation, proliferation, apoptosis, and even a predisposition to cancer [105–107]. The altered levels of miRNAs have been reported in a variety of cancers [108,109].

Not unexpectedly, miRNAs are also involved in IR-induced responses in vivo [44,87,110–119] and in radiation-induced germline and transgenerational effects [44,117]. Tamminga and colleagues reported that radiation exposure significantly affected miRNA expression in testes [117]. Radiation exposure caused DNA damage and led to the ATR/Rfx1-mediated increase in miR-709 expression in exposed testes. This miRNA targeted the Brother of the Regulator of Imprinted Sites (BORIS), an important regulator of DNA methylation and imprinting.

Filkowski and colleagues showed that irradiation led to the upregulation of the miR-29 family in the exposed male germline, which caused a decrease in the expression of de novo methyltransferase, DNMT3a, and a profound hypomethylation of LINE1 and SINE B2 [44]. Epigenetic changes in the male germline led to deleterious effects in the somatic thymus tissue of the progeny of exposed animals, including hypomethylation of LINE1 and SINE B2 associated with a significant decrease in the levels of lymphoid-specific helicase (LSH) that is crucial for the maintenance of methylation and the silencing of repetitive elements. Moreover, the thymus tissue of the progeny of exposed parents exhibited a significant upregulation of miR-468 that targeted LSH and led to its decreased expression in the thymus. The study suggested that miR-468-mediated suppression of LSH led to an aberrant methylation of LINE1 and SINE B2 [44].

Recently, a novel small RNA pathway has been characterized, providing evidence for yet another small RNA-mediated epigenetic effector. Known as the Piwi/piRNA pathway, it has several unique features that make it quite suitable as a mediator of epigenetic memory in germ cells. Here, key features of the piRNA pathway is introduced, followed by a further discussion in the context of spermatogenesis in the rodent germline.

3.3.1 piRNA Biogenesis and Role in Maintaining Genome Stability

Being initially characterized in *Drosophila* [120], the central component of the pathway is a large class of short, single-stranded, noncoding RNAs (~26–31 nt) and their Piwi protein partners, a subclass of the Argonaute protein family. Both Piwi-interacting RNAs (piRNAs) and Piwi proteins have expression patterns that are largely restricted to germ cells in nearly all multicellular animals studied [121]. Piwi proteins are required for the production of their piRNA partners and are essential for various stages of spermatogenesis, the self-renewal of germ stem cells, and transposon silencing [121,122]. The best studied function of the piRNA pathway is to maintain genomic integrity by the suppression of transposable elements (TEs) via transcriptional gene silencing (TGS) [121]. TGS occurs through piRNA-mediated de novo methylation of regulatory regions of retrotransposons in embryonic germ cells; methylation is believed to be subsequently maintained in germ and somatic cells throughout the life of an organism [123,124]. While mutations in the DNMT family members affected cytosine methylation, the piRNA pathway remained largely unaffected [123]. In contrast, the loss of the piRNA pathway prevents the recognition and silencing of TE by DNMT3L, thus supporting a model in which the piRNA pathway acts upstream of DNMT3L, DNMT3a, and DNMT3b to establish patterns of DNA methylation on TEs [123]. PTGS also contributes to this process because the piRNA-guided Piwi proteins also mediate the cleavage of active transposon mRNAs from which primary piRNAs are likely to be derived through a process known as the "ping-pong" amplification cycle [125,126]. However, it is important to note that the majority of mouse and rat piRNAs are not enriched in sequences derived from transposons and repeats. In mice and rats, repeats are underrepresented because only about 17% of all piRNAs map to repetitive elements, while a random distribution should yield close to 40%, which is the proportion of repetitive sequences in the genome [127,128]. In mammals, piRNAs tend to cluster within certain regions of the genome, and a large number of piRNAs are derived from intergenic regions, but are also distributed among exonic, intronic, and intergenic repeat sequences [104]. The distinguishing feature of these clusters of uniquely mapping piRNAs is the pronounced strand bias, which leads to the suggestion that the biogenesis of piRNAs involves a long, single-stranded precursor [129]. Since piRNA sequences correspond to a variety of genomic regions, the piRNA pathway may be involved in a more complex system regulating the expression of a plethora of genes other than repetitive elements.

Indeed, several studies in the mid-2000s suggest that the piRNA pathway is not limited to the repression of transposable and repetitive elements, and it plays the diverse and complex roles in regulating gene expression at all known levels of epigenetic control. Piwi proteins and piRNAs together have been associated with mRNA and mRNA cap-binding proteins in polysomes and ribonucleoproteins (RNP) which play a central role in translational control. However, the molecular mechanisms that achieve this translational regulation and the resulting outcomes remain largely unclear [104,122,130]. The biochemically purified endogenous rat piRNA complex has been shown to exhibit the RNA cleavage activity, presumably facilitated by the rat Piwi protein, Riwi [131]. On the other hand, mouse Piwi proteins may actually be responsible for the stability of a subset of mRNAs, and the positive regulation of translation [130,132]. In addition, the Piwi protein in mice (Miwi) is required not only for piRNA but also for a particular subset of miRNAs [104]. Thus, the piRNA pathway may be involved in miRNA-mediated translational control. One common feature of *Piwi* gene mutations in mice is an increase in DNA damage marked by γH2AX foci, thus suggesting a possible link to DNA-damage repair/checkpoints [133,134].

It has been proposed that such dsDNA breaks are a result of overactive transposons; however, this relationship is not fully understood as dsDNA breaks could also be the cause of transposon activity rather than a result of it [135]. The presence of RecQ1 in the rat Piwi protein complexes is consistent with a possible role of mammalian Piwi-type proteins in DNA-repair processes [131].

RecQ is a family of helicase enzymes that have highly conserved roles in dsDNA-break repair through recombination [136]. The ability of the piRNA pathway to mediate epigenetic control of gene expression at the level of histone modifications has also been described. Human cells have been transiently transfected with a human Piwi (Piwi-like4/Hiwi2) gene containing a vector construct which induces histone H3K9 methylation at the p16Ink41 locus, resulting in a significant downregulation of p16 gene expression [137]. A 2009 study has provided some intriguing evidence for the production and function of a particular subset of abundant piRNAs which are depleted in the TE content and do not engage in the ping-pong cycle [138]. They reported a substantial population of piRNAs derived from UTR of protein-coding genes. These genic piRNAs arise preferentially from 3′-UTRs produced by a piRNA biogenesis pathway that does not require the ping-pong components and are conserved across *Drosophila*, mice, and *Xenopus* [138]. This breakthrough finding and the previously discussed studies provide overwhelming evidence for an additional and much larger breadth of piRNA-mediated gene regulation, although the role of piRNAs in TGS of TEs still remains unexplained.

3.3.2 piRNAs as Mediators of the Epigenetic Memory

The piRNA/Piwi pathway has several features that make it suitable as a mediator of the epigenetic memory in germ cells. Being mainly characterized by its ability to exert TGS by driving methylation of TE, it clearly has the ability to affect genome stability in future generations. Moreover, even though this novel small RNA pathway has been shown to play a role in many epigenetic alterations observed in response to IR, no experiments have been conducted to examine a possible role and response of this pathway to IR exposure. Because this pathway is mainly restricted to the male germline in mammals, it provides a novel mechanism to facilitate the paternal epigenetic inheritance of IR-induced genomic instability. It can also provide some insight into the observed loss of LINE1 and global DNA methylation not only in the germline of exposed males, but more importantly, in the next generation [37,44,61]. The understanding of and how the piRNA pathway responds to IR exposure can also potentially corroborate and help elucidate the increased mutation rates observed in satellite DNA and ESTR loci in the somatic and germline tissue of the progeny of exposed parents [32].

Very little is known about the role of the piRNA pathway in the production/inheritance of IR-induced genomic instability. Unpublished data from our laboratory show that radiation exposure causes profound alterations in piRNA profiles, affecting several piRNA clusters. Changes in piRNA levels are associated with the altered levels of DNA methylation of the corresponding piRNA loci in the exposed germline and in the progeny of exposed animals. Therefore, piRNAs may hold the key to understanding epigenetic mechanisms of germline and radiation-induced transgenerational genomic instability.

4. TRANSGENERATIONAL EFFECTS CAUSED BY OTHER MUTAGENS

While transgenerational effects and genome instability upon IR exposure have been mostly studied, since the mid-2000s, numerous studies appeared that showed the induction of transgenerational effects by a wide array of other chemical mutagens.

Parental exposure to urethane, 4-nitroquinoline 1-oxide and 7,12-dimethylbenz(a)anthracene led to an elevated cancer risk and mutations in the offspring [139,140]. Transgenerational effects were also reported upon exposure to anticancer drugs. The F2 offspring of males exposed to cyclophosphamide were reported to exhibit genome instability and an increase in postimplantation loss and congenital malformations [141]. Cyclophosphamide or a combination of cyclophosphamide and vinblastine caused behavioral alterations in the first- and second-generation offspring of male rats [15,45]. Paternal exposure to anticancer chemotherapy altered the quality of germ cells and profoundly affected embryo development, thus causing transgenerational effects [142]. Embryos born upon paternal cyclophosphamide exposure exhibited the elevated levels of DNA damage and dramatic alterations in the levels of DNA repair and homologous recombination genes [143]. Furthermore, our unpublished data show that paternal exposure to mitomycin C and cyclophosphamide alter gene and protein expression in the frontal cortex and the whole brain of unexposed progeny.

A large number of extensive studies have focused on transgenerational effects of environmental teratogenic agents such as endocrine disruptors. Seminal studies were conducted by Skinner and colleagues. They have shown that in utero exposure of rats to vinclozolin or methoxychlor during the period of gonadal sex determination leads to an increased infertility rate and a decreased spermatogenic capacity in the F_1 male progeny. Moreover, this altered reproductive capacity was transmitted via the male germline to a majority of male offspring up to the F_4 generation [144], and manifested as transgenerational (F1–F4) spermatogenic cell apoptosis and subfertility [145]. Moreover, Skinner and colleagues have established that exposure to endocrine disruptors leads to transgenerational reprogramming of the testis transcriptome and the development of transgenerational diseases such as spermatogenic defects, prostate disease, kidney disease, and cancer [146,147]. Transgenerational effects caused by toxicants are epigenetic in nature. Exposure to pesticides dichlorodiphenyltrichloroethane (DDT) and methoxychlor (MXC) leads to the occurrence of transgenerational sperm epimutation signatures [148] that may in turn promote genome instability [149] (also reviewed in Refs. [150–152]).

Transgenerational transmission of the effects of gestational alcohol exposure has been reported. Prenatal alcohol exposure increases the risk for alcoholism by increasing the propensity to consume alcohol and by altering a neurophysiological response to alcohol [153]. Alcohol consumption modifies the sperm epigenome and thus leads to instability and behavioral effects in the progeny [154]. In 2015, deleterious transgenerational effects were reported for alcohol, opiates, cocaine, marijuana, and nicotine [155]. Exposure to these agents causes epigenetic changes in the genome that are transmitted to the next generation [155–158]. A wide array of other life-style factors such as diet, obesity, nutritional deficiencies, and stress make their mark on the germline and cause deleterious effects in the progeny [159–163].

5. CONCLUSIONS AND OUTLOOK

In sum, parental exposure to a wide array of environmental agents affects the germline and, therefore, cause transgenerational effects in unexposed progeny. Transgenerational effects can span numerous generations and transgenerational genome instability is a key mechanism underlying transgenerational effects. Studies have reported the causes, existence, and molecular processes affected in the germline and in the progeny of exposed parents. Environmental agents were shown to cause DNA damage, altered DNA-methylation levels and deregulated gene and small RNA expression in the germline of exposed parents. In the progeny, these cause the aberrant setting of DNA-methylation marks, altered gene expression and genome instability, which result in a wide array of downstream "snowball effects," such as the accumulation of mutations, genomic rearrangements, and further genome destabilization (Fig. 34.1). While numerous studies have proposed that transgenerational effects are epigenetic in nature, the precise mechanisms of transgenerational genome instability remain to be fully elucidated. Future research in this area must rely on the use of microarray technology, next-generation sequencing, and bioinformatic approaches in order to extract the functionally relevant, causal changes influencing genetic and epigenetic reprogramming and genomic stability across generations. Such research has both practical and fundamental value, as it may offer an understating of how genotoxic factors contribute to complex diseases by altering our epigenome across generations.

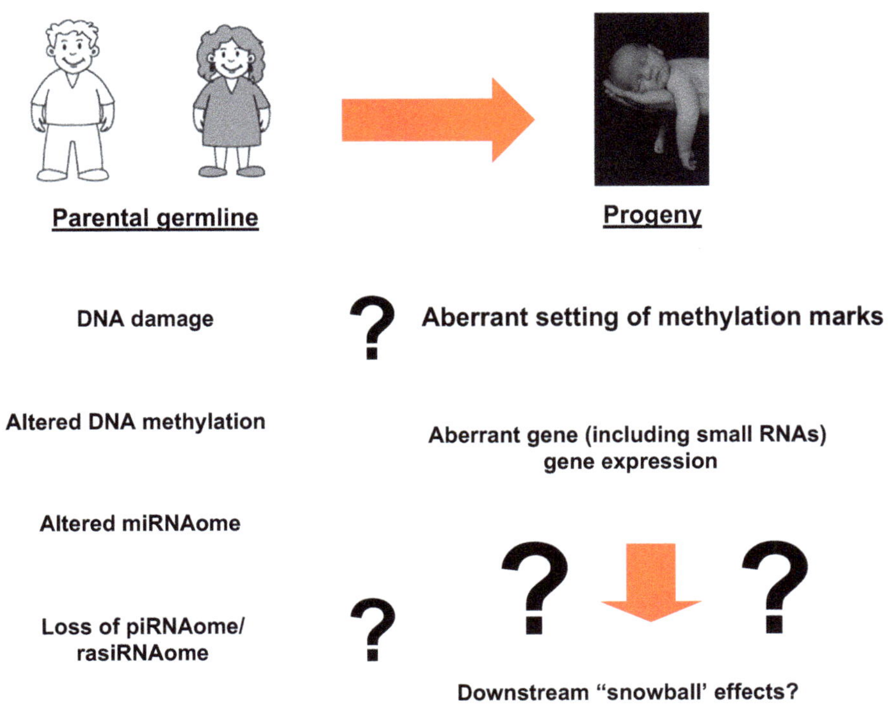

FIGURE 34.1 Transmission of stress exposure effects from the germline to progeny—potential mechanisms of transgenerational genome instability.

While this review was being written, several breakthrough articles have emerged that suggest the novel roles of small RNAs, especially TRNAs and tRNA fragments, in the germline and potentially in the transgenerational effects induced by diet deficiencies. Sperm tRNAs might mediate the transcriptional cascade effect and influence metabolic gene expression through the embryo to adulthood [164,165]. Further studies are needed to dissect the roles of small RNAs and small RNA fragments in transgenerational effects.

GLOSSARY

F1 generation The generation resulting from a cross of parental generation (the first set of parents).
Ping-pong cycle Cycle of piRNA production first identified from studies in *Drosophila*.
Transgenerational inheritance The transmittance of genetic and epigenetic information from one generation of an organism to the next ones.
Transgenerational effects A wide array of health effects that occur when environmental exposures or toxicants pass from parent to offspring.
Transgenerational genome instability Elevated frequency of mutations and genomic rearrangements transmitted from the germline of exposed parents to the progeny.

LIST OF ABBREVIATIONS

DDT Dichlorodiphenyltrichloroethane
DNMT DNA methyltransferases (3a, DNMT3b, and DNMT3L)
ESTR Expanded simple tandem repeat
LET Linear energy transfer
LINEs Long interspersed nuclear elements
miRNAs microRNAs
miRNP microRNA ribonucleoprotein complex
MIWI Mouse PIWI protein
MXC Methoxychlor
piRNA Piwi-interacting RNA
pri-miRNA Primary transcript miRNA
PTGS Posttranscriptional gene silencing
RISC RNA-induced silencing complex
RIWI Rat PIWI protein
RNP Ribonucleoproteins
SINEs Short interspersed nuclear elements
TEs Transposable elements
TGS Transcriptional gene silencing

REFERENCES

[1] Little JB. Radiation-induced genomic instability. Int J Radiat Biol 1998;74(6):663–71.
[2] Morgan WF. Non-targeted and delayed effects of exposure to ionizing radiation: I. Radiation-induced genomic instability and bystander effects in vitro. Radiat Res 2003;159(5):567–80.
[3] Limoli CL, Corcoran JJ, Milligan JR, Ward JF, Morgan WF. Critical target and dose and dose-rate responses for the induction of chromosomal instability by ionizing radiation. Radiat Res 1999;151(6):677–85.
[4] Limoli CL, Ponnaiya B, Corcoran JJ, Giedzinski E, Kaplan MI, Hartmann A, et al. Genomic instability induced by high and low LET ionizing radiation. Adv Space Res 2000;25(10):2107–17.
[5] Morgan WF. Non-targeted and delayed effects of exposure to ionizing radiation: II. Radiation-induced genomic instability and bystander effects in vivo, clastogenic factors and transgenerational effects. Radiat Res 2003;159(5):581–96.
[6] Morgan WF. Is there a common mechanism underlying genomic instability, bystander effects and other nontargeted effects of exposure to ionizing radiation? Oncogene 2003;22(45):7094–9.
[7] Kaplan MI, Limoli CL, Morgan WF. Perpetuating radiation-induced chromosomal instability. Radiat Oncol Investig 1997;5(3):124–8.
[8] Paquette B, Little JB. In vivo enhancement of genomic instability in minisatellite sequences of mouse C3H/10T1/2 cells transformed in vitro by X-rays. Cancer Res 1994;54(12):3173–8.
[9] Watson GE, Lorimore SA, Clutton SM, Kadhim MA, Wright EG. Genetic factors influencing alpha-particle-induced chromosomal instability. Int J Radiat Biol 1997;71(5):497–503.
[10] Cheng KC, Loeb LA. Genomic instability and tumor progression: mechanistic considerations. Adv Cancer Res 1993;60:121–56.
[11] Goldberg Z. Clinical implications of radiation-induced genomic instability. Oncogene 2003;22(45):7011–7.
[12] Little JB. Genomic instability and radiation. J Radiol Prot 2003;23(2):173–81.
[13] Loeb LA, Loeb KR, Anderson JP. Multiple mutations and cancer. Proc Natl Acad Sci USA 2003;100(3):776–81.

[14] Little JB. Radiation carcinogenesis. Carcinogenesis 2000;21(3):397–404.

[15] Auroux MR, Dulioust EJ, Nawar NN, Yacoub SG, Mayaux MJ, Schwartz D, et al. Antimitotic drugs in the male rat. Behavioral abnormalities in the second generation. J Androl 1988;9(3):153–9.

[16] Burruel VR, Raabe OG, Wiley LM. In vitro fertilization rate of mouse oocytes with spermatozoa from the F1 offspring of males irradiated with 1.0 Gy 137Cs gamma-rays. Mutat Res 1997;381(1):59–66.

[17] Vorobtsova IE. Irradiation of male rats increases the chromosomal sensitivity of progeny to genotoxic agents. Mutagenesis 2000;15(1):33–8.

[18] Dubrova YE. Radiation-induced transgenerational instability. Oncogene 2003;22(45):7087–93.

[19] Merrifield M, Kovalchuk O. Epigenetics in radiation biology: a new research frontier. Front Genet 2013;4:40.

[20] Luning KG, Frolen H, Nilsson A. Genetic effects of 239Pu salt injections in male mice. Mutat Res 1976;34(3):539–42.

[21] Lyon MF. The overall rates of dominant and recessive lethal and visible mutation induced by spermatogonial X-irradiation of mice. Genet Res 1964;5:448–67.

[22] Pils S, Muller WU, Streffer C. Lethal and teratogenic effects in two successive generations of the HLG mouse strain after radiation exposure of zygotes – association with genomic instability? Mutat Res 1999;429(1):85–92.

[23] Nomura T. Parental exposure to x rays and chemicals induces heritable tumours and anomalies in mice. Nature 1982;296(5857):575–7.

[24] Nomura T. Transgenerational carcinogenesis: induction and transmission of genetic alterations and mechanisms of carcinogenesis. Mutat Res 2003;544(2–3):425–32.

[25] Lord BI, Woolford LB, Wang L, McDonald D, Lorimore SA, Stones VA, et al. Induction of lympho-haemopoietic malignancy: impact of preconception paternal irradiation. Int J Radiat Biol 1998;74(6):721–8.

[26] Hoyes KP, Lord BI, McCann C, Hendry JH, Morris ID. Transgenerational effects of preconception paternal contamination with (55)Fe. Radiat Res 2001;156(5 Pt 1):488–94.

[27] Roman E, Doyle P, Maconochie N, Davies G, Smith PG, Beral V. Cancer in children of nuclear industry employees: report on children aged under 25 years from nuclear industry family study. BMJ 1999;318(7196):1443–50.

[28] McKinney PA, Fear NT, Stockton D, Investigators UKCCS. Parental occupation at periconception: findings from the United Kingdom Childhood Cancer Study. Occup Environ Med 2003;60(12):901–9.

[29] Gardner MJ, Snee MP, Hall AJ, Powell CA, Downes S, Terrell JD. Results of case-control study of leukaemia and lymphoma among young people near Sellafield nuclear plant in West Cumbria. BMJ 1990;300(6722):423–9.

[30] Dickinson HO, Parker L. Leukaemia and non-Hodgkin's lymphoma in children of Sellafield male radiation workers. Int J Cancer 2002;101(1):100.

[31] Dubrova YE. Long-term genetic effects of radiation exposure. Mutat Res 2003;544(2–3):433–9.

[32] Barber RC, Dubrova YE. The offspring of irradiated parents, are they stable? Mutat Res 2006;598(1–2):50–60.

[33] Dubrova YE. Germline mutation induction at mouse and human tandem repeat DNA loci. Adv Exp Med Biol 2003;518:115–29.

[34] Barber R, Plumb MA, Boulton E, Roux I, Dubrova YE. Elevated mutation rates in the germ line of first- and second-generation offspring of irradiated male mice. Proc Natl Acad Sci USA 2002;99(10):6877–82.

[35] Barber RC, Hickenbotham P, Hatch T, Kelly D, Topchiy N, Almeida GM, et al. Radiation-induced transgenerational alterations in genome stability and DNA damage. Oncogene 2006;25(56):7336–42.

[36] Baulch JE, Raabe OG, Wiley LM. Heritable effects of paternal irradiation in mice on signaling protein kinase activities in F3 offspring. Mutagenesis 2001;16(1):17–23.

[37] Koturbash I, Baker M, Loree J, Kutanzi K, Hudson D, Pogribny I, et al. Epigenetic dysregulation underlies radiation-induced transgenerational genome instability in vivo. Int J Radiat Oncol Biol Phys 2006;66(2):327–30.

[38] Sonoda E, Sasaki MS, Buerstedde JM, Bezzubova O, Shinohara A, Ogawa H, et al. Rad51-deficient vertebrate cells accumulate chromosomal breaks prior to cell death. EMBO J 1998;17(2):598–608.

[39] Thacker J. The RAD51 gene family, genetic instability and cancer. Cancer Lett 2005;219(2):125–35.

[40] Richardson C, Stark JM, Ommundsen M, Jasin M. Rad51 overexpression promotes alternative double-strand break repair pathways and genome instability. Oncogene 2004;23(2):546–53.

[41] Vance MM, Baulch JE, Raabe OG, Wiley LM, Overstreet JW. Cellular reprogramming in the F3 mouse with paternal F0 radiation history. Int J Radiat Biol 2002;78(6):513–26.

[42] Xiong Y, Hannon GJ, Zhang H, Casso D, Kobayashi R, Beach D. p21 is a universal inhibitor of cyclin kinases. Nature 1993;366(6456):701–4.

[43] Baulch JE, Raabe OG. Gamma irradiation of Type B spermatogonia leads to heritable genomic instability in four generations of mice. Mutagenesis 2005;20(5):337–43.

[44] Filkowski JN, Ilnytskyy Y, Tamminga J, Koturbash I, Golubov A, Bagnyukova T, et al. Hypomethylation and genome instability in the germline of exposed parents and their progeny is associated with altered miRNA expression. Carcinogenesis 2010;31(6):1110–5.

[45] Abouzeid Ali HE, Barber RC, Dubrova YE. The effects of maternal irradiation during adulthood on mutation induction and transgenerational instability in mice. Mutat Res 2012;732(1–2):21–5.

[46] Barber RC, Hardwick RJ, Shanks ME, Glen CD, Mughal SK, Voutounou M, et al. The effects of in utero irradiation on mutation induction and transgenerational instability in mice. Mutat Res 2009;664(1–2):6–12.

[47] Smith RW, Seymour CB, Moccia RD, Mothersill CE. Irradiation of rainbow trout at early life stages results in trans-generational effects including the induction of a bystander effect in non-irradiated fish. Environ Res 2016;145:26–38.

[48] Buisset-Goussen A, Goussen B, Della-Vedova C, Galas S, Adam-Guillermin C, Lecomte-Pradines C. Effects of chronic gamma irradiation: a multigenerational study using *Caenorhabditis elegans*. J Environ Radioact 2014;137:190–7.

[49] Sarapultseva EI, Gorski AI. Low-dose gamma-irradiation affects the survival of exposed *Daphnia* and their offspring. Dose Response 2013;11:460–8.

[50] Grygoryev D, Moskalenko O, Hinton TG, Zimbrick JD. DNA damage caused by chronic transgenerational exposure to low dose gamma radiation in Medaka fish (*Oryzias latipes*). Radiat Res 2013;180(3):235–46.

[51] Dubrova YE, Nesterov VN, Krouchinsky NG, Ostapenko VA, Neumann R, Neil DL, et al. Human minisatellite mutation rate after the Chernobyl accident. Nature 1996;380(6576):683–6.

[52] Dubrova YE, Grant G, Chumak AA, Stezhka VA, Karakasian AN. Elevated minisatellite mutation rate in the post-chernobyl families from ukraine. Am J Hum Genet 2002;71(4):801–9.

[53] Dubrova YE, Bersimbaev RI, Djansugurova LB, Tankimanova MK, Mamyrbaeva Z, Mustonen R, et al. Nuclear weapons tests and human germline mutation rate. Science 2002;295(5557):1037.

[54] Murphy JB, Morton J. Action of serum from x-rayed animals on lymphoid cells in vitro. J Exp Med 1915. XX11:800.

[55] Parsons Jr WB, Watkins CH, Pease GL, Childs Jr DS. Changes in sternal marrow following roentgen-ray therapy to the spleen in chronic granulocytic leukemia. Cancer 1954;7(1):179–89.

[56] Goh K, Sumner H. Breaks in normal human chromosomes: are they induced by a transferable substance in the plasma of persons exposed to total-body irradiation? Radiat Res 1968;35(1):171–81.

[57] Hollowell Jr JG, Littlefield LG. Chromosome damage induced by plasma of x-rayed patients: an indirect effect of x-ray. Proc Soc Exp Biol Med 1968;129(1):240–4.

[58] Emerit I, Oganesian N, Sarkisian T, Arutyunyan R, Pogosian A, Asrian K, et al. Clastogenic factors in the plasma of Chernobyl accident recovery workers: anticlastogenic effect of Ginkgo biloba extract. Radiat Res 1995;144(2):198–205.

[59] Emerit I, Levy A, Cernjavski L, Arutyunyan R, Oganesyan N, Pogosian A, et al. Transferable clastogenic activity in plasma from persons exposed as salvage personnel of the Chernobyl reactor. J Cancer Res Clin Oncol 1994;120(9):558–61.

[60] Kovalchuk O, Baulch JE. Epigenetic changes and nontargeted radiation effects–is there a link? Environ Mol Mutagen 2008;49(1):16–25.

[61] Tamminga J, Koturbash I, Baker M, Kutanzi K, Kathiria P, Pogribny IP, et al. Paternal cranial irradiation induces distant bystander DNA damage in the germline and leads to epigenetic alterations in the offspring. Cell Cycle 2008;7(9):1238–45.

[62] Jaenisch R, Bird A. Epigenetic regulation of gene expression: how the genome integrates intrinsic and environmental signals. Nat Genet 2003;33(Suppl.):245–54.

[63] Robertson KD. DNA methylation, methyltransferases, and cancer. Oncogene 2001;20(24):3139–55.

[64] Rountree MR, Bachman KE, Herman JG, Baylin SB. DNA methylation, chromatin inheritance, and cancer. Oncogene 2001;20(24):3156–65.

[65] Goll MG, Bestor TH. Eukaryotic cytosine methyltransferases. Annu Rev Biochem 2005;74:481–514.

[66] Kato Y, Kaneda M, Hata K, Kumaki K, Hisano M, Kohara Y, et al. Role of the Dnmt3 family in de novo methylation of imprinted and repetitive sequences during male germ cell development in the mouse. Hum Mol Genet 2007;16(19):2272–80.

[67] Hendrich B, Tweedie S. The methyl-CpG binding domain and the evolving role of DNA methylation in animals. Trends Genet 2003;19(5):269–77.

[68] Klose RJ, Bird AP. Genomic DNA methylation: the mark and its mediators. Trends Biochem Sci 2006;31(2):89–97.

[69] Weber M, Schubeler D. Genomic patterns of DNA methylation: targets and function of an epigenetic mark. Curr Opin Cell Biol 2007;19(3):273–80.

[70] Ehrlich M. DNA methylation in cancer: too much, but also too little. Oncogene 2002;21(35):5400–13.

[71] Robertson KD. DNA methylation and chromatin - unraveling the tangled web. Oncogene 2002;21(35):5361–79.

[72] Weidman JR, Dolinoy DC, Murphy SK, Jirtle RL. Cancer susceptibility: epigenetic manifestation of environmental exposures. Cancer J 2007;13(1):9–16.

[73] Feinberg AP, Vogelstein B. Hypomethylation distinguishes genes of some human cancers from their normal counterparts. Nature 1983;301(5895):89–92.

[74] Feinberg AP, Tycko B. The history of cancer epigenetics. Nat Rev Cancer 2004;4(2):143–53.

[75] Feinberg AP. The epigenetics of cancer etiology. Semin Cancer Biol 2004;14(6):427–32.

[76] Baylin SB. DNA methylation and gene silencing in cancer. Nat Clin Pract Oncol 2005;2(Suppl. 1):S4–11.

[77] Baylin SB, Ohm JE. Epigenetic gene silencing in cancer – a mechanism for early oncogenic pathway addiction? Nat Rev Cancer 2006;6(2):107–16.

[78] Kalinich JF, Catravas GN, Snyder SL. The effect of gamma radiation on DNA methylation. Radiat Res 1989;117(2):185–97.

[79] Pogribny I, Raiche J, Slovack M, Kovalchuk O. Dose-dependence, sex- and tissue-specificity, and persistence of radiation-induced genomic DNA methylation changes. Biochem Biophys Res Commun 2004;320(4):1253–61.

[80] Raiche J, Rodriguez-Juarez R, Pogribny I, Kovalchuk O. Sex- and tissue-specific expression of maintenance and de novo DNA methyltransferases upon low dose X-irradiation in mice. Biochem Biophys Res Commun 2004;325(1):39–47.

[81] Koturbash I, Pogribny I, Kovalchuk O. Stable loss of global DNA methylation in the radiation-target tissue–a possible mechanism contributing to radiation carcinogenesis? Biochem Biophys Res Commun 2005;337(2):526–33.

[82] Loree J, Koturbash I, Kutanzi K, Baker M, Pogribny I, Kovalchuk O. Radiation-induced molecular changes in rat mammary tissue: possible implications for radiation-induced carcinogenesis. Int J Radiat Biol 2006;82(11):805–15.

[83] Pogribny I, Koturbash I, Tryndyak V, Hudson D, Stevenson SM, Sedelnikova O, et al. Fractionated low-dose radiation exposure leads to accumulation of DNA damage and profound alterations in DNA and histone methylation in the murine thymus. Mol Cancer Res 2005;3(10):553–61.

[84] Kaup S, Grandjean V, Mukherjee R, Kapoor A, Keyes E, Seymour CB, et al. Radiation-induced genomic instability is associated with DNA methylation changes in cultured human keratinocytes. Mutat Res 2006;597(1–2):87–97.

[85] Sedelnikova OA, Nakamura A, Kovalchuk O, Koturbash I, Mitchell SA, Marino SA, et al. DNA double-strand breaks form in bystander cells after microbeam irradiation of three-dimensional human tissue models. Cancer Res 2007;67(9):4295–302.

[86] Koturbash I, Rugo RE, Hendricks CA, Loree J, Thibault B, Kutanzi K, et al. Irradiation induces DNA damage and modulates epigenetic effectors in distant bystander tissue in vivo. Oncogene 2006;25(31):4267–75.

[87] Koturbash I, Boyko A, Rodriguez-Juarez R, McDonald RJ, Tryndyak VP, Kovalchuk I, et al. Role of epigenetic effectors in maintenance of the long-term persistent bystander effect in spleen in vivo. Carcinogenesis 2007;28(8):1831–8.
[88] Xu GL, Bestor TH, Bourc'his D, Hsieh CL, Tommerup N, Bugge M, et al. Chromosome instability and immunodeficiency syndrome caused by mutations in a DNA methyltransferase gene. Nature 1999;402(6758):187–91.
[89] Yu F, Zingler N, Schumann G, Stratling WH. Methyl-CpG-binding protein 2 represses LINE-1 expression and retrotransposition but not Alu transcription. Nucleic Acids Res 2001;29(21):4493–501.
[90] Jirtle RL, Skinner MK. Environmental epigenomics and disease susceptibility. Nat Rev Genet 2007;8(4):253–62.
[91] Jenuwein T, Allis CD. Translating the histone code. Science 2001;293(5532):1074–80.
[92] Cheung P, Lau P. Epigenetic regulation by histone methylation and histone variants. Mol Endocrinol 2005;19(3):563–73.
[93] He S, Dunn KL, Espino PS, Drobic B, Li L, Yu J, et al. Chromatin organization and nuclear microenvironments in cancer cells. J Cell Biochem 2008;104(6):2004–15.
[94] Rogakou EP, Pilch DR, Orr AH, Ivanova VS, Bonner WM. DNA double-stranded breaks induce histone H2AX phosphorylation on serine 139. J Biol Chem 1998;273(10):5858–68.
[95] Pilch DR, Sedelnikova OA, Redon C, Celeste A, Nussenzweig A, Bonner WM. Characteristics of gamma-H2AX foci at DNA double-strand breaks sites. Biochem Cell Biol 2003;81(3):123–9.
[96] Sedelnikova OA, Pilch DR, Redon C, Bonner WM. Histone H2AX in DNA damage and repair. Cancer Biol Ther 2003;2(3):233–5.
[97] Bernstein E, Allis CD. RNA meets chromatin. Genes Dev 2005;19(14):1635–55.
[98] Chen PY, Meister G. microRNA-guided posttranscriptional gene regulation. Biol Chem 2005;386(12):1205–18.
[99] Rodriguez A, Griffiths-Jones S, Ashurst JL, Bradley A. Identification of mammalian microRNA host genes and transcription units. Genome Res 2004;14(10A):1902–10.
[100] Zeng Y. Principles of micro-RNA production and maturation. Oncogene 2006;25(46):6156–62.
[101] Hutvagner G, Zamore PD. RNAi: nature abhors a double-strand. Curr Opin Genet Dev 2002;12(2):225–32.
[102] Yekta S, Shih IH, Bartel DP. MicroRNA-directed cleavage of HOXB8 mRNA. Science 2004;304(5670):594–6.
[103] Doench JG, Sharp PA. Specificity of microRNA target selection in translational repression. Genes Dev 2004;18(5):504–11.
[104] Grivna ST, Pyhtila B, Lin H. MIWI associates with translational machinery and PIWI-interacting RNAs (piRNAs) in regulating spermatogenesis. Proc Natl Acad Sci USA 2006;103(36):13415–20.
[105] Shivdasani RA. MicroRNAs: regulators of gene expression and cell differentiation. Blood 2006;108(12):3646–53.
[106] Chang TC, Mendell JT. microRNAs in vertebrate physiology and human disease. Annu Rev Genomics Hum Genet 2007;8:215–39.
[107] Fabbri M, Ivan M, Cimmino A, Negrini M, Calin GA. Regulatory mechanisms of microRNAs involvement in cancer. Expert Opin Biol Ther 2007;7(7):1009–19.
[108] Volinia S, Calin GA, Liu CG, Ambs S, Cimmino A, Petrocca F, et al. A microRNA expression signature of human solid tumors defines cancer gene targets. Proc Natl Acad Sci USA 2006;103(7):2257–61.
[109] Wiemer EA. The role of microRNAs in cancer: no small matter. Eur J Cancer 2007;43(10):1529–44.
[110] Dickey JS, Zemp FJ, Altamirano A, Sedelnikova OA, Bonner WM, Kovalchuk O. H2AX phosphorylation in response to DNA double-strand break formation during bystander signalling: effect of microRNA knockdown. Radiat Prot Dosim 2011;143(2–4):264–9.
[111] Dickey JS, Zemp FJ, Martin OA, Kovalchuk O. The role of miRNA in the direct and indirect effects of ionizing radiation. Radiat Environ Biophys 2011;50(4):491–9.
[112] Ilnytskyy Y, Koturbash I, Kovalchuk O. Radiation-induced bystander effects in vivo are epigenetically regulated in a tissue-specific manner. Environ Mol Mutagen 2009;50(2):105–13.
[113] Ilnytskyy Y, Zemp FJ, Koturbash I, Kovalchuk O. Altered microRNA expression patterns in irradiated hematopoietic tissues suggest a sex-specific protective mechanism. Biochem Biophys Res Commun 2008;377(1):41–5.
[114] Koturbash I, Zemp F, Kolb B, Kovalchuk O. Sex-specific radiation-induced microRNAome responses in the hippocampus, cerebellum and frontal cortex in a mouse model. Mutat Res 2011;722(2):114–8.
[115] Koturbash I, Zemp FJ, Kutanzi K, Luzhna L, Loree J, Kolb B, et al. Sex-specific microRNAome deregulation in the shielded bystander spleen of cranially exposed mice. Cell Cycle 2008;7(11):1658–67.
[116] Kovalchuk O, Zemp FJ, Filkowski JN, Altamirano AM, Dickey JS, Jenkins-Baker G, et al. microRNAome changes in bystander three-dimensional human tissue models suggest priming of apoptotic pathways. Carcinogenesis 2010;31(10):1882–8.
[117] Tamminga J, Kathiria P, Koturbash I, Kovalchuk O. DNA damage-induced upregulation of miR-709 in the germline downregulates BORIS to counteract aberrant DNA hypomethylation. Cell Cycle 2008;7(23):3731–6.
[118] Wang B, Li D, Kovalchuk A, Litvinov D, Kovalchuk O. Ionizing radiation-inducible miR-27b suppresses leukemia proliferation via targeting cyclin A2. Int J Radiat Oncol Biol Phys 2014;90(1):53–62.
[119] Wang B, Li D, Kovalchuk O. p53 Ser15 phosphorylation and histone modifications contribute to IR-induced miR-34a transcription in mammary epithelial cells. Cell Cycle 2013;12(13):2073–83.
[120] Aravin AA, Lagos-Quintana M, Yalcin A, Zavolan M, Marks D, Snyder B, et al. The small RNA profile during *Drosophila melanogaster* development. Dev Cell 2003;5(2):337–50.
[121] Aravin AA, Hannon GJ. Small RNA silencing pathways in germ and stem cells. Cold Spring Harb Symp Quant Biol 2008;73:283–90.
[122] Thomson T, Lin H. The biogenesis and function of PIWI proteins and piRNAs: progress and prospect. Annu Rev Cell Dev Biol 2009;25:355–76.
[123] Aravin AA, Sachidanandam R, Bourc'his D, Schaefer C, Pezic D, Toth KF, et al. A piRNA pathway primed by individual transposons is linked to de novo DNA methylation in mice. Mol Cell 2008;31(6):785–99.

[124] Kuramochi-Miyagawa S, Watanabe T, Gotoh K, Totoki Y, Toyoda A, Ikawa M, et al. DNA methylation of retrotransposon genes is regulated by Piwi family members MILI and MIWI2 in murine fetal testes. Genes Dev 2008;22(7):908–17.
[125] Aravin AA, Hannon GJ, Brennecke J. The Piwi-piRNA pathway provides an adaptive defense in the transposon arms race. Science 2007;318(5851):761–4.
[126] Aravin AA, Sachidanandam R, Girard A, Fejes-Toth K, Hannon GJ. Developmentally regulated piRNA clusters implicate MILI in transposon control. Science 2007;316(5825):744–7.
[127] Vagin VV, Sigova A, Li C, Seitz H, Gvozdev V, Zamore PD. A distinct small RNA pathway silences selfish genetic elements in the germline. Science 2006;313(5785):320–4.
[128] Hartig JV, Tomari Y, Forstemann K. piRNAs–the ancient hunters of genome invaders. Genes Dev 2007;21(14):1707–13.
[129] Seto AG, Kingston RE, Lau NC. The coming of age for Piwi proteins. Mol Cell 2007;26(5):603–9.
[130] Unhavaithaya Y, Hao Y, Beyret E, Yin H, Kuramochi-Miyagawa S, Nakano T, et al. MILI, a PIWI-interacting RNA-binding protein, is required for germ line stem cell self-renewal and appears to positively regulate translation. J Biol Chem 2009;284(10):6507–19.
[131] Lau NC, Seto AG, Kim J, Kuramochi-Miyagawa S, Nakano T, Bartel DP, et al. Characterization of the piRNA complex from rat testes. Science 2006;313(5785):363–7.
[132] Deng W, Lin H. miwi, a murine homolog of piwi, encodes a cytoplasmic protein essential for spermatogenesis. Dev Cell 2002;2(6):819–30.
[133] Carmell MA, Girard A, van de Kant HJ, Bourc'his D, Bestor TH, de Rooij DG, et al. MIWI2 is essential for spermatogenesis and repression of transposons in the mouse male germline. Dev Cell 2007;12(4):503–14.
[134] Kuramochi-Miyagawa S, Kimura T, Ijiri TW, Isobe T, Asada N, Fujita Y, et al. Mili, a mammalian member of piwi family gene, is essential for spermatogenesis. Development 2004;131(4):839–49.
[135] Klattenhoff C, Bratu DP, McGinnis-Schultz N, Koppetsch BS, Cook HA, Theurkauf WE. *Drosophila* rasiRNA pathway mutations disrupt embryonic axis specification through activation of an ATR/Chk2 DNA damage response. Dev Cell 2007;12(1):45–55.
[136] Hunter N. The RecQ DNA helicases: Jacks-of-all-trades or master-tradesmen? Cell Res 2008;18(3):328–30.
[137] Sugimoto K, Kage H, Aki N, Sano A, Kitagawa H, Nagase T, et al. The induction of H3K9 methylation by PIWIL4 at the p16Ink4a locus. Biochem Biophys Res Commun 2007;359(3):497–502.
[138] Robine N, Lau NC, Balla S, Jin Z, Okamura K, Kuramochi-Miyagawa S, et al. A broadly conserved pathway generates 3'UTR-directed primary piRNAs. Curr Biol 2009;19(24):2066–76.
[139] Nomura T, Nakajima H, Ryo H, Li LY, Fukudome Y, Adachi S, et al. Transgenerational transmission of radiation- and chemically induced tumors and congenital anomalies in mice: studies of their possible relationship to induced chromosomal and molecular changes. Cytogenet Genome Res 2004;104(1–4):252–60.
[140] Nomura T. Transgenerational effects from exposure to environmental toxic substances. Mutat Res 2008;659(1–2):185–93.
[141] Hales BF, Crosman K, Robaire B. Increased postimplantation loss and malformations among the F2 progeny of male rats chronically treated with cyclophosphamide. Teratology 1992;45(6):671–8.
[142] Hales BF, Barton TS, Robaire B. Impact of paternal exposure to chemotherapy on offspring in the rat. J Natl Cancer Inst Monogr 2005;34:28–31.
[143] Harrouk W, Codrington A, Vinson R, Robaire B, Hales BF. Paternal exposure to cyclophosphamide induces DNA damage and alters the expression of DNA repair genes in the rat preimplantation embryo. Mutat Res 2000;461(3):229–41.
[144] Anway MD, Cupp AS, Uzumcu M, Skinner MK. Epigenetic transgenerational actions of endocrine disruptors and male fertility. Science 2005;308(5727):1466–9.
[145] Skinner MK, Anway MD. Seminiferous cord formation and germ-cell programming: epigenetic transgenerational actions of endocrine disruptors. Ann NY Acad Sci 2005;1061:18–32.
[146] Anway MD, Rekow SS, Skinner MK. Transgenerational epigenetic programming of the embryonic testis transcriptome. Genomics 2008;91(1):30–40.
[147] Anway MD, Skinner MK. Epigenetic programming of the germ line: effects of endocrine disruptors on the development of transgenerational disease. Reprod Biomed Online 2008;16(1):23–5.
[148] Haque MM, Holder LB, Skinner MK. Genome-wide locations of potential epimutations associated with environmentally induced epigenetic transgenerational inheritance of disease using a sequential machine learning prediction approach. PLoS One 2015;10(11):e0142274.
[149] Skinner MK, Guerrero-Bosagna C, Haque MM. Environmentally induced epigenetic transgenerational inheritance of sperm epimutations promote genetic mutations. Epigenetics 2015;10(8):762–71.
[150] Skinner MK. Endocrine disruptors in 2015: epigenetic transgenerational inheritance. Nat Rev Endocrinol 2016;12(2):68–70.
[151] Skinner MK. Environmental epigenetics and a unified theory of the molecular aspects of evolution: a Neo-Lamarckian concept that facilitates Neo-Darwinian evolution. Genome Biol Evol 2015;7(5):1296–302.
[152] Skinner MK. Endocrine disruptor induction of epigenetic transgenerational inheritance of disease. Mol Cell Endocrinol 2014;398(1–2):4–12.
[153] Nizhnikov ME, Popoola DO, Cameron NM. Transgenerational transmission of the effect of gestational ethanol exposure on ethanol use-related behavior. Alcohol Clin Exp Res 2016;40.
[154] Finegersh A, Rompala GR, Martin DI, Homanics GE. Drinking beyond a lifetime: new and emerging insights into paternal alcohol exposure on subsequent generations. Alcohol 2015;49(5):461–70.
[155] Yohn NL, Bartolomei MS, Blendy JA. Multigenerational and transgenerational inheritance of drug exposure: the effects of alcohol, opiates, cocaine, marijuana, and nicotine. Prog Biophys Mol Biol 2015;118(1–2):21–33.
[156] Joya X, Manzano C, Alvarez AT, Mercadal M, Torres F, Salat-Batlle J, et al. Transgenerational exposure to environmental tobacco smoke. Int J Environ Res Public Health 2014;11(7):7261–74.

[157] Marczylo EL, Amoako AA, Konje JC, Gant TW, Marczylo TH. Smoking induces differential miRNA expression in human spermatozoa: a potential transgenerational epigenetic concern? Epigenetics 2012;7(5):432–9.
[158] Ortega-Garcia JA, Martin M, Lopez-Fernandez MT, Fuster-Soler JL, Donat-Colomer J, Lopez-Ibor B, et al. Transgenerational tobacco smoke exposure and childhood cancer: an observational study. J Paediatr Child Health 2010;46(6):291–5.
[159] Hanafi MY, Saleh MM, Saad MI, Abdelkhalek TM, Kamel MA. Transgenerational effects of obesity and malnourishment on diabetes risk in F2 generation. Mol Cell Biochem 2016;412(1–2):269–80.
[160] Stuppia L, Franzago M, Ballerini P, Gatta V, Antonucci I. Epigenetics and male reproduction: the consequences of paternal lifestyle on fertility, embryo development, and children lifetime health. Clin Epigenetics 2015;7:120.
[161] Constantinof A, Moisiadis VG, Matthews SG. Programming of stress pathways: a transgenerational perspective. J Steroid Biochem Mol Biol 2015;160.
[162] Babenko O, Kovalchuk I, Metz GA. Stress-induced perinatal and transgenerational epigenetic programming of brain development and mental health. Neurosci Biobehav Rev 2015;48:70–91.
[163] Bock J, Poeschel J, Schindler J, Borner F, Shachar-Dadon A, Ferdman N, et al. Transgenerational sex-specific impact of preconception stress on the development of dendritic spines and dendritic length in the medial prefrontal cortex. Brain Struct Funct 2014;221.
[164] Chen Q, Yan M, Cao Z, Li X, Zhang Y, Shi J, et al. Sperm tsRNAs contribute to intergenerational inheritance of an acquired metabolic disorder. Science 2016;351(6271):397–400.
[165] Sharma U, Conine CC, Shea JM, Boskovic A, Derr AG, Bing XY, et al. Biogenesis and function of tRNA fragments during sperm maturation and fertilization in mammals. Science 2016;351(6271):391–6.

Chapter 35

Genomic Instability and the Spectrum of Response to Low Radiation Doses

C. Mothersill, C. Seymour
McMaster University, Hamilton, ON, Canada

Chapter Outline

1. INTRODUCTION TO LOW RADIATION–DOSE EFFECTS

1.1 Background to the Controversy

Few scientific fields are as divisive as the area of low radiation–dose effects. In 2005, two groups even analyzed the same data sets and came to opposite conclusions [1,2]. Accusations of "cherry-picking" to support a viewpoint abound, and sometimes it would appear that belief systems take the place of science. The purpose of this chapter is to question why this should be so? Why have we no definite answers about radiation health risks after low-dose/dose-rate exposure? Why are beliefs so entrenched? Most importantly, how can we remove the rhetoric from both sides and replace it with rational argument and scientific fact?

1.2 Epidemiology Is a Blunt Tool

Key to understanding the issue is to understand how we estimate risk at present and what might be wrong with this system. The perception that the system itself is wrong has led to much of the polarization, with some groups arguing that the risks are grossly overestimated and others arguing they are underestimated [3,4,5]. The first problem is that all our risk estimates use the epidemiological data from Hiroshima/Nagasaki cohorts as the "gold standard" and use cancer as the end point. The relationship is usually considered to "fit" the linear nonthreshold (LNT) relationship between dose, which is estimated and weighted using factors to correct for relative biological effectiveness (RBE) and dose rate, and effect (cancer) [6]. The A-bomb cohorts however do not give much information below 100 mGy and the exposure was acute. The dose rate and RBE correction factors are themselves highly controversial having been estimated using inbred mice or cultured cells [7,8]

Genome Stability. http://dx.doi.org/10.1016/B978-0-12-803309-8.00035-5

which may not represent human response or the induction of effects in genetically diverse populations in the environment [9]. There is also a huge disagreement about the doses actually received by the Japanese victims, the contribution of neutrons, the impact of genetics, war and malnutrition, and so on [10–13]. There is also a controversy about the reliability of the underlying radiobiological studies which purport to give the mechanistic support to LNT—for example, the work of Calabrese [4] suggests that there was a very big political component to enshrining and defending LNT as the way dose and effect are related. The usual representation of this relationship is shown in Fig. 35.1 where the low-dose region is depicted as having multiple possible dose–response relationships. The purpose of this chapter is not to try to resolve this debate, but rather to suggest a totally new way of looking at low-dose risk, which focuses on response rather than dose. Central to this concept is the acceptance of the paradigm shift which has occurred in low-dose radiobiology since 1990 [14,15]. While this shift is recognized in radiobiology, it has yet to be accepted as having any relevance to radiation protection [16,17].

1.3 Targeted and Nontargeted Effects

The key new research which is driving the paradigm shift is the recognition that in addition to targeted (direct) effects of radiation, there is a totally different set of mechanisms operating in cells which are not the result of direct damage resulting from ionizing energy deposition in DNA in targeted cells, but rather occur as a result of signaling or other mechanisms operating at the system level [18]. These so-called *nontargeted* effects (NTEs) occur in cells, organs, or organisms which have not received a direct deposition of ionizing radiation energy (see Fig. 35.2). NTEs have been extensively reviewed by these authors and many others [14,15]. They can broadly be divided into genomic instability (GI) and bystander effects with subcategories of adaptive/hormetic-type responses, generally seen as beneficial, and stress responses leading to damaging effects such as increases in mutation, related to reactive oxygen species (ROS) generally seen as adverse [18]. Fig. 35.3 represents a possible way to visualize these multiple effects.

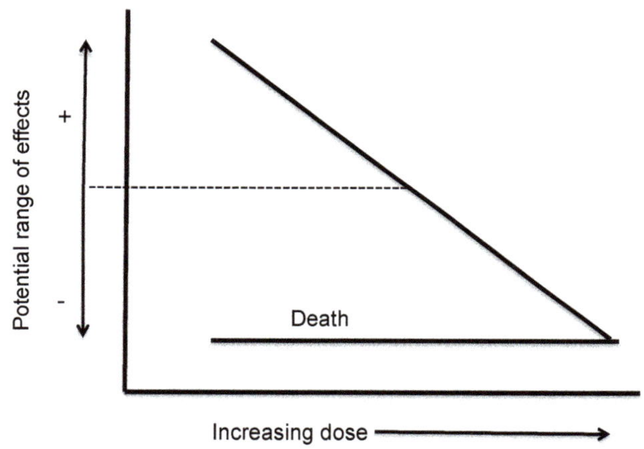

FIGURE 35.1 **The concept of a wide spectrum of possible outcomes in the low-dose region.** These may be positive or negative. The possibilities decrease as the dose increases until death is the only possible outcome.

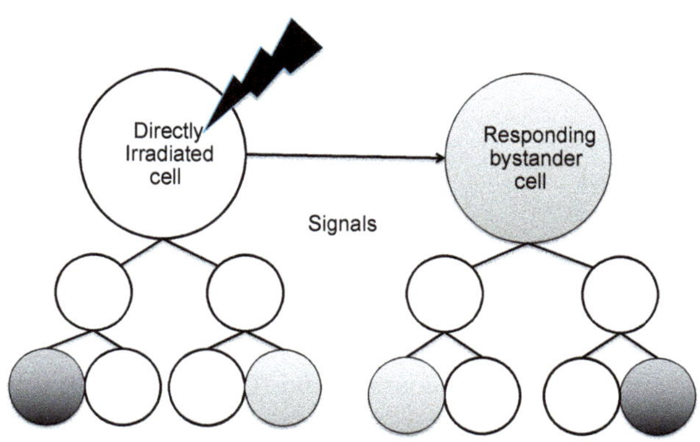

FIGURE 35.2 **Nontargeted and nonclonal effects transmitted horizontally (bystander effects) or vertically (genomic instability effects in progeny).** Light gray depicts adaptive responses and dark gray depicts stress responses.

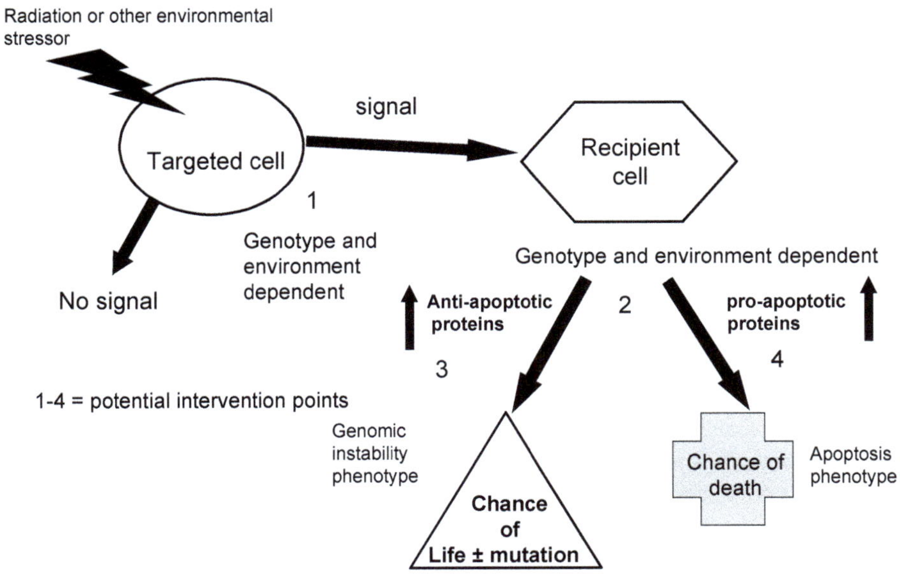

FIGURE 35.3 **Possible outcomes in the low-dose region, where nontargeted effects may or may not occur.** Numbers 1–4 indicate points where it might be possible to modulate the outcome.

1.4 Genomic Instability

GI is a concept that describes delayed genetic alterations observed in the progeny of the exposed cells many generations after the initial radiation insult [15,19,20]. The importance of the discovery of GI lies in the recognition that cell populations surviving radiation exposure and their progeny which show no evidence of mutations or altered fitness will not necessarily behave normally. Up to the 1980s, the central dogma in radiobiology was that all the damage was put into the cell by the ionizing radiation. If the cell was able to repair the damage and undergo successful mitosis—at least five times—then it (and its progeny) carried no residual damage and the population was as fit as if never irradiated [21]. The key paradigm changing finding which came from several independent studies was that the progeny of irradiated cells which have survived according to the earlier criteria show evidence of de novo, nonclonal effects, meaning that the damage was not induced by the initial energy deposition from radiation.

One of the first publications suggesting that the progeny of apparently recovered irradiated cells might also be at risk was published by Seymour et al. [19]. The team observed lethal mutations in the distant progeny of cells exposed to photons from a Cobalt-60 source. They suggested that the appearance of lethal mutations might require many successful generations to be expressed. Those investigations were reinforced by reports from Streffer and colleagues, who showed chromosomal aberrations at the second mitosis after X-irradiating a two-cell mouse embryo [22]. The Streffer group subsequently showed that 29% of the aberrations were carried from the first to the second mitosis [23]. Similar data to those presented by the Seymour group and the Streffer group were published by Mendonca and colleagues [24] where delayed cell death started after 10 successful divisions; and by Kadhim and colleagues [20], who compared the effect of X-rays against alpha particles. These studies were performed in bone marrow stem cells—which meant that clonal lineages could be followed—and revealed that alpha particles were more effective at inducing nonclonal aberrations than X-rays. Later studies revealed persistent levels of ROS in cells showing GI, suggesting that oxidative stress plays a role in perpetuating the insult [25,26].

Although GI occurs as a consequence of direct radiation exposure, it occurs in distant progeny and represents what is now referred to as the vertical transmission of radiation effects. There are several reports indicating it might also involve bystander effects (described later) because the damage in the progeny is nonclonal and therefore cannot be attributed to DNA mutations only [20,27,28]. This damage can also be induced using bystander protocols such as medium transfer, meaning no direct energy deposition is needed to trigger GI [29,30].

1.5 Bystander Effects

Another broad category of NTEs is bystander effects [31,32]. They can occur in vivo and in vitro, and refer to the horizontal transmission of radiation damage—as opposed to the vertical transmission seen in GI experiments. Bystander effects refer to effects seen in cells, tissues, or organisms which receive some type of a signal from an irradiated cell, tissue, or organism. Various protocols, possible mechanisms, and models have been extensively reviewed and readers are referred to [15,18,33,34] for both the history and mechanisms under consideration. The fundamental point of interest in this chapter is that these effects occur in the absence of a dose of radiation to the cell, tissue, or organism being examined and dominate the dose–response in the low-dose region [35,36]. What does this mean for radiation protection? How can we estimate low-dose radiation risk if unirradiated organisms merely living in a proximity to irradiated organisms [37,38,39] display the same type of effect as the directly irradiated organisms? In particular, if bystander mechanisms underlie low-dose effects and can trigger GI, does this lend support to the contention that we are underestimating low-dose risk? This position has been put forward by several authors (reviewed in [40]). However, others have found evidence for adaptive and hormetic responses [41,42,43], suggesting bystander effects could be a form of a homeostatic mechanism. Table 35.1 lists effects documented immediately after direct irradiation, in distant progeny (GI) and in bystander populations.

1.6 Adaptive/Hormetic Effects

As suggested before and in Table 35.1, the literature on NTEs documents what might be classed as desirable effects as well as adverse effects. GI is generally regarded as an undesirable effect because it involves genetic alterations [44,45]. Bystander effects were initially observed as an increase in genetic damage [46] or a decrease in cell clonogenicity [47,48]; however, several reports indicate that bystander responses may not always be harmful. Early reports of protective bystander effects in vitro include the work by Azzam and colleagues [49,50] where they showed that bystander signals induced the reduction of neoplastic transformations. There is also evidence of bystander effects inducing adaptive responses which are discussed by some authors [51]. These include the protection afforded by exposure to irradiated medium against radiation-induced cell death, GI, and micronucleus formation. Other studies looking at the bystander proteome indicated that bystander signals may confer beneficial effects by upregulating protective proteins in both rainbow trout and medaka fish exposed to bystander signals [38,52]. In mammals, similar protective effects were seen in the unirradiated left brain hemisphere of healthy Wistar rats [53]. Many authors have suggested that low doses of radiation are actually protective [49,50] and bystander mechanisms are suggested as underlying factors in radiation hormesis [54,55]. It is important at this point to distinguish between adaptive responses, hormesis, and adaptation at the individual and population levels. Adaptive responses generally refer to individuals and the responses mounted by their body systems in response to encountering a new and hazardous threat. Adaptive responses can include behavioral or physiological changes which make the individual better able to survive a future or ongoing encounter with the hazard [56]. Hormesis, on the other hand, is already there as part of the dose response [57,58]. The hormetic dose range is the range where the substance or physical agent is *not* a hazard but is present at optimal levels which if exceeded can become hazardous [59]. Examples are trace metals, which are essential for health but toxic in high amounts. Adaptation is used more as a term to describe the evolution of populations, which come to live with a hazard due to the natural selection of those genotypes or phenotypes that are most able to cope with

TABLE 35.1 List of End points Which Have Been Shown to Change in Directly Irradiated Cells, in Bystander Cells, and in Distant Progeny

Endpoint Changes	Directly Irradiated	Bystander	Distant Progeny
Reproductive death	X	X	X
Chromosomal aberration	X	X	X
Mutation	X	X	X
Mini/microsatellite	X		X
Micronucleus frequency	X	X	X
Gene expression	X	X	X
Protein expression	X	X	
Apoptosis	X	X	X
Transformation	X	X	X
Mitochondrial function	X	X	X
Calcium	X	X	
ROS	X	X	X

the altered environment. Using the example of radiation exposure, low doses may be within a natural hormesis zone and therefore be beneficial, or they may be in the adaptive zone and lead to mechanisms being induced which protect against subsequent exposures. They may also result in adaptations at the population level when radiation exposure is chronic [60,61]. Of course, a confounding factor leading to much confusion is that the dose range for different zones and transition points may be different for different species and different individuals of the same species. A key issue here is that human radiation protection seeks to protect every individual from any adverse consequence and currently does not recognize the hormesis zone, while environmental radiation protection seeks to protect populations and ecosystem structure [62] and does acknowledge at least that there are thresholds for harm if not actually beneficial doses. This leads to widely different concepts and perceptions of risk and of strategies for protection.

1.7 Generic Stress Responses

GI is often regarded as resulting from a "generic stress response" [63,64]. However, what exactly this means is not very clear. Stress is defined by Selye [65] as "the nonspecific response of the body to any demand for change." Stress is considered by Selye and many others to be necessary to trigger appropriate responses to the stressor. Calabrese et al. [66] sought to clarify the generic stress response by pointing out that there are many terms across many disciplines for what is a common occurrence in biology, that is, a small dose of a stressor can induce an adaptive response to a large dose of the same or in some cases a different stressor. The chapter went further to show that opposite effects can occur after low-dose exposures compared to high-dose exposures leading to "U"- or "J"-shaped nonlinear dose–response relationships. In radiobiology, we recognize "oxidative stress" as a generic stress resulting from excess ROS and leading to DNA, mitochondrial, and cell membrane damage [67,68]. Oxidative stress is often cited as a mechanism for deleterious low-dose effects [69]. However, others argue that the amount of ROS generated by low doses is so small in relation to that generated by oxidative metabolism, that it could not account for low-dose damage and that anyway it all gets repaired [70]. What is not considered here is again the concept of individual variation in the ability to tolerate and repair oxidative stress–induced DNA or membrane damage. Also not considered is the energy cost of repair and the dependence of repair on nutritional factors, time, age, and metabolic rate [71,72]. Other forms of stress that are thought to be associated with radiation include immune system stress and the mounting of an inflammatory response [73,74]. Both are thought to result from GI-induced changes in the bone marrow stem cells [75]. While direct doses needed to generate bone marrow stem cell damage are relatively high, bystander effects can occur at very low doses of the order of few milligray and appear to be either fully expressed or not expressed at all in the area affected by the signal, saturating at extremely low doses to the signal-generating cells [36,76,77]. This means that in theory at least, a very low dose of radiation could turn on a bystander effect in the tissue, which in certain phenotypes could lead to immune compromise and inflammatory responses. Such a mechanism has been proposed to explain the ill health seen in atomic test veterans, people who suffer from CFIDS (chronic fatigue and immune deficiency syndrome), people exposed to depleted uranium, and victims of radiation accidents, where the calculated doses are considered much too low to account for the observed level and variety of illnesses [78]. In the environment, similar mechanisms might also help to explain the reported high level of mutations in butterflies and birds from Chernobyl and Fukushima [79,80], where again, doses and dose rates are considered much too low to cause biological effects [81,82], and radiation as a cause of these phenomena is vehemently denied.

2. CONCEPT OF UNCERTAINTY

This chapter does not seek to suggest that low doses are universally "bad" or universally "good." Rather, we seek to suggest that both response extremes and everything in between *can* occur. What actually happens will depend on the context at the individual and population levels. Indeed, it is possible that an adverse outcome studied at one organization level (eg, cell death) could be beneficial if considered at a higher level of organization. This type of thinking, while obvious and accepted generally in scientific theory, does not help to resolve the debate about risk and leads to people on both sides citing papers which present sound science as "evidence" in support of their belief. Our conclusion after many years of study is that everything is true and nothing is true. In other words, the only thing to do is to accept uncertainty after low-dose exposure and move on to see where this acceptance will take us in trying to address the issues surrounding low-dose exposures to ionizing radiation. Fig. 35.4 is an attempt to convey this idea.

2.1 Spectrum of Effects

The key point is to recognize that there is a spectrum of low-dose effects and that we currently do not know the drivers that determine which ultimate outcome prevails. This is not the same as saying that the probability of random damage exists

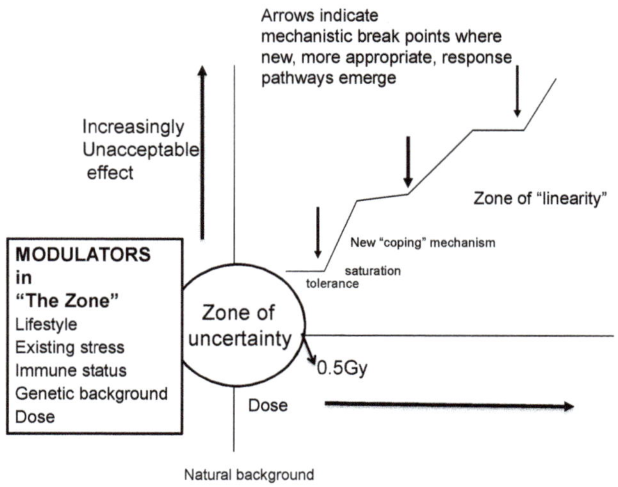

FIGURE 35.4 **The concept of an unpredictable "zone" in the dose–effect relationship where factors in addition to dose determine the response.** With an increasing dose, the unpredictable zone gives way to a series of emergent responses which optimize the outcome.

because that type of statistical analysis is concerned with estimating the chance of a mutation leading to cancer occurring and becoming fixed in DNA [83]—that is, it is a target theory-driven and dose-driven hypothesis. It is, however, what we use in radiation protection [84]. The suggestion that we are presenting is rather that a spectrum of effects ranging from truly beneficial to truly harmful can occur and could occur in the same biological system depending on factors including dose selection that impact the outcome.

2.2 Spectrum of Responses

Similarly, a spectrum of responses to the outcome(s) also can occur. These may enhance or reduce the impact, whether positive or negative, of that outcome. Examples could include a nutritional status or smoking that could impact energy delivery for repair or could compromise checkpoint proteins such as p53 [85]. Early research into factors associated with the generation of GI did identify several scenarios which favored the turning on of the phenotype in the progeny of irradiated cells. These included background genetics, time post exposure, the presence of other stressors, the availability of glucose or lactate, the point in the cell cycle of initial irradiation to progenitor cells, and the number of progenitor cells that were irradiated [86,87]. However, there were also sudden failures of hitherto reliable protocols for measuring GI that were never resolved. There was also a considerable interlaboratory variation where a protocol could not be transferred or replicated in another laboratory even by the same individual. With hindsight, these difficulties, many of which were never recorded and remain anecdotal, strongly suggest that there was an underlying randomness in the system or that we did not appreciate all the factors which determined which response prevailed. Either way, it suggests that in addition to a spectrum of effects of radiation exposure, there is also a spectrum of responses. Responses to bystander signals are also determined by genetic, epigenetic, environmental, and unknown factors [88], making it very complicated and beyond the ability of most modeling approaches to resolve [89]. In the paragraphs that follow, some of the literature which has been produced in support of what are considered to be key factors for predicting likelihood of beneficial or adverse outcomes following low-dose exposures is reviewed.

2.3 Individual Variation

The individual variation in radiosensitivity has been recognized for many years in radiotherapy and many susceptibility genes mainly associated with faulty DNA repair are known. In low-dose radiobiology, susceptibility genes are associated with extreme reactions to UV or ionizing radiation, these again mostly involve DNA repair. In the field of NTE, genetics is also known to play a role with susceptible and resistant strains of mice, cell lines, and human explants, all documented [90,91,92]. System-level variability is harder to study but bacterial populations have been shown to demonstrate cross-resistance to multiple stressors including heat and radiation [93]. Species-sensitivity distributions (SSDs) discussed later are also an ecological approach being used to determine action levels for the protection of ecosystems. However, just because the individual variation and underlying genetics have been identified as important, they do not address the problem of the *extent* to which genetics contributes to the outcome, or whether it is a determining factor always, sometimes, or ever.

2.4 The Role of Genetic Background

Proponents of the "old radiobiology" which holds that energy deposited in DNA causes strand breaks and represents the key way radiation causes damage obviously extend this theory to then suggest that damage to critical targets in DNA underlies the harm caused by ionizing radiation, and that individuals with compromised genetics are therefore most at risk because they cannot repair or detect or otherwise deal with the damage [21]. There is no doubt that at high doses of radiation, this rationale is sound and well proven but after low-dose exposure, it is likely to be far less important due to the myriad other mechanisms such as homeostasis, hormesis, adaptive mechanisms, and system-level responses. While there is probably a spectrum of gene strengths or gene dosages [94,95], it is more likely that protein-level changes induced by the radiation stress predominate meaning that enzyme kinetics, energy budgets, cofactors, and the presence or absence of activators and inhibitors are more likely to determine the outcome [96,97].

2.5 The Role of Other Stressors

One of the key issues of concern in radiation protection, particularly of non-human biota, is the problem of multiple stressors [98]. Currently, radiation is regulated as a stand-alone agent, but it is well recognized that in reality, humans and nonhumans are exposed to many stressors such as heavy metals PCBs, heat, and drought. It is one thing to realize this is an issue but quite another one to find a way of regulating in a multiple stressor environment. When DNA was the key target and double-strand breaks—the key damaging lesion, it made sense to regulate radiation separately, but now that NTEs are recognized as key low-dose effects, we have to understand how GI and bystander effects might be modulated by the presence of other stressors. Given the complexity of the issue and the few studies which address it [99,100], it is likely that a move to response- rather than dose-driven protection strategies is necessary to move forward on this issue.

2.6 The Role of Lifestyle Factors

A major gap in our knowledge concerning NTEs is the lack of information about these effects in humans. In vivo studies so far are limited to a few mouse strains, fish, and tadpoles [38,39,101]. There is research using human cell lines [102], explants [103], and bone marrow cultures [104], and some work have been done on patients with metal implants [105]. However, none of these studies were conducted with a view to looking at lifestyle factors. The explant studies of normal human bladder done by this group [106] did suggest that smoking could impact bystander effects (reducing the strength of the signal), but the numbers were very small. It appears logical though that factors leading to an increased cancer risk such as smoking might act to increase radiation-induced GI, and this is an area needing investigation. Epidemiological-type studies recording lifestyle information but monitoring GI or bystander effects as well as cancer incidence are needed.

2.7 Species-Sensitivity Distribution

This was previously referred to in the context of individual variation in radiation response. SSD is a method being developed to try to get a picture of ecosystem sensitivity to radiation. According to the US Environmental Protection Authority, "Species-sensitivity distributions (SSDs) are cumulative probability distributions of toxicity values for multiple species." For the environmental risk assessment, the chemical concentration that may be used as a hazard level can be extrapolated from SSD using a specified percentile of the distribution (http://www.epa.gov). Basically for radiation studies, SSD is calculated using the data in the literature contained in the FREDERICA database concerning radiation dose response for non-human species and builds a graph of the species present in a habitat and plots their radiosensitivity using EDR_{10} (effective dose rate causing effects in 10% of the population) or HDR_5 (hazardous dose rate affecting 5% of species at the 10% level) [107]. Most effects in radioecology concern mortality or reproductive endpoints of the effects but there is no reason that this approach could not be used to "rank" human cells/cell lines, tissues, or individuals using NTE end points. A key benefit is that this approach is effect rather than dose driven, and thus could pull out system-level effects.

3. SEARCH FOR DETERMINATORS

Much of the preceding discussion depends on being able to measure NTE reliably in a wide variety of systems at multiple levels of organization. In the next part of this chapter, the broad categories of approaches to this are discussed. Fig. 35.5 summarizes key things we know about GI and bystander effect mechanisms. From these data, attempts have been made to develop reliable determinants of response, both in the directly hit cell/tissue/organism, its progeny, and in bystanders. There are several types of determinants which is now considered.

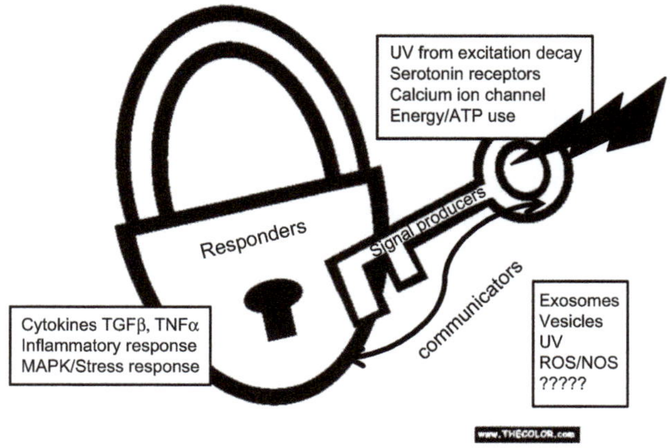

FIGURE 35.5 **Key factors involved in the mechanisms of the bystander effect grouped into signal production in the directly irradiated cell; communication of information between the targeted cell and bystanders; response transduction in bystander cells.**

3.1 Bioindicators

Bioindicator is a term taken from environmental toxicology and is defined as "an organism or biological response that reveals the presence of the pollutants by the occurrence of typical symptoms or measurable responses. These organisms (or communities of organisms) deliver information on alterations in the environment or the quantity of environmental pollutants by changing in one of the following ways: physiologically, chemically or behaviourally" [108].

A very simple bioindicator in radiobiology could be reproductive death measured using clonogenic assays [109] which can be applied to assess the level of GI in progeny or the strength of bystander signals. Population-level bioindicators could include ion fluxes through membrane channels, the integrity of gap junctions in cell membranes, or the coordinated behavior and function of mitochondria in cell populations, all of which have been documented in the literature.

3.2 Biomarkers

A biomarker is defined as "a characteristic that is objectively measured and evaluated as an indicator of normal biologic processes, pathogenic processes, or pharmacologic responses to a therapeutic intervention."

This means that the biomarker is a surrogate or marker for dose or exposure. In radiobiology, γH2AX is often referred to as a surrogate for radiation dose because it reflects the level of DNA breaks [110]. Dicentric chromosomes are also used as surrogates for dose but do not report reliably below 2 Gy [111]. Thus, biomarkers are more removed from the actual effect on a system, especially if it is complex and not linearly related to dose. This is the situation with NTE.

3.3 Biosensors

A Biosensor is defined as an analytical device, used for the detection of an analyte, that combines a biological component with a physicochemical detector. Generally, it is used to describe assays such as ELISA which use enzymes to cause reactions that make it easy to measure color change. However, the reporter assay developed to detect the strength of bystander signal could qualify as a biosensor assay [90] as could the use of fin clips, blood cells, or samples of embryos or sperm that can be incubated to allow the release of bystander signals into medium which is then assayed to determine signal strength. When used with populations of cells, sperm, or embryos, the biosensor approach can provide information about population-level responses to an insult, and thus is a useful response-driven approach for the use in nonlinear dose–response situations such as the detection of GI induction or bystander effects [112].

3.4 Signals

The ideal way to measure bystander NTE would be to know what the signal is that goes from an exposed cell or organism to another nonirradiated cell or organism. However, despite almost 20 years of looking, the actual signal or signals remain elusive. Fig. 35.5 referred to earlier shows what we know about the nature of the signal, and clearly, exosomes (possibly containing miRNA), UV or visible light, calcium, serotonin, TGFβ, p53, ROS, and NOS are all reported to be involved, but what transmits the information is unclear, and while events involved in signal production are well understood, and the response in the recipient cell is also well documented, we know little about the processes of signal transmission. It is also controversial whether there are multiple signals. The works carried out in 2007, 2012, and 2015 from our laboratory suggest that contrary to a popular belief, the signal(s) may be physical rather than chemical [113,114,115]. This work suggests that UVA is emitted by irradiated cells and that if the signal is not emitted, the bystander effect does not occur. The UVA signal is emitted from all cell types, but not all cells respond to the UVA signal. Other evidences supporting physical emissions from cells include the work by Papineni and colleagues who measure bioluminescence coming from irradiated cells [116]. However, the evidence supporting a role for exosomes or larger vesicles is equally compelling [117,118,119]. This leads us to believe that multiple mechanisms may exist.

3.5 System-Level Responses

In ecology, it has been known for many years that ecosystems work as a result of complex interactions between elements of the system. Surprisingly, in radiobiology, while this concept is well known to physiologists, radiation action was thought to involve stand-alone actions on individual cells with no communication [21]. Independent survival was a concept enshrined in target theory and its main tool of analysis—the clonogenic survival curve [21]. This all changed in the mid-1990s with several demonstrations of interdependent death and survival of irradiated cells [30]. Of course, as with most "discoveries," there was a body of research that was ignored or forgotten because it did not fit. Chief among this was the work using spheroids where cooperative repair could be demonstrated [120]. Mole [121] also suggested using modeling approaches that the proximity of at least two cells was necessary for carcinogenesis to occur. This ran contrary to the conventional wisdom that cancer originated in a single damaged cell—the clonal origin of cancer theory [122]. The application of system biology tools in radiobiology started with the realization that the microenvironment was important and that bystander effects existed [123]. Signaling is now recognized as a major factor determining the coordinated response of system elements, especially after low doses. This hierarchical theory holds that system-level signaling optimizes the system-level response to a challenge affecting lower levels of organization.

3.6 Emergent Effects

Emergence is defined as a process whereby larger entities, patterns, and regularities arise through interactions among smaller or simpler entities that themselves do not exhibit such properties. Emergence is central in theories of integrative levels and of complex systems. In system radiobiology, it refers to responses to radiation in tissues, organs, individuals, and populations that are not predictable from the behavior of individual irradiated cells [124]. Integrating complexity theory into radiobiology and radiation protection is one of the most exciting challenges in the field. Since the discovery of GI and bystander effects, it has become apparent that not only do cells not act alone but that outcomes considered at the cellular level such as death of the cell may lead to radically different consequences at higher levels of organization and if time is factored in to the experiments. New mechanisms and new responses may emerge which are not measurable at the level of the individual cell. In radioecology, the recognition of this phenomenon of emergence has led to a search for so-called "system-level biomarkers" to try to quantify impacts in complex systems. Perhaps, the act of mounting a bystander response could be considered such a biomarker?

4. CONCLUSIONS

The aim of this chapter is to consider critically what impact NTEs have on our understanding of radiation risk and what might be a way forward to reconcile the fiercely opposing views about the benefits and hazards associated with low-dose exposure. We suggest that we need to accept that a spectrum of effects occurs after low-dose exposure which cannot be predicted in relation to dose. We suggest that a response-driven approach should be considered and a search for a reliable system and individual-level determinants of response is necessary.

GLOSSARY

Adaptive response A less-damaging effect of a large dose of radiation if a small dose is administered some hours before.

Bystander effect The occurrence of radiation-type effects or responses in cells, tissues, organs, or organisms which were not irradiated but received signals from irradiated entities.

Hormesis A phenomenon where low doses of physical and chemical agents which are toxic at higher doses can be protective or "good for you."

Nontargeted effects Effects occurring in the absence of direct energy deposition in DNA.

LIST OF ACRONYMS AND ABBREVIATIONS

DNA Deoxyribonucleic acid
EDR Effective dose rate
ELISA Enzyme-linked immunosorbent assay
GI Genomic instability
γH2AX H2A histone family, member X serine phosphorylated
HDR Hazardous dose rate
LNT Linear nonthreshold
miRNA Micro-ribonucleic acid
NTE Nontargeted effects
PCB Polychlorinated biphenyl
RBE Relative biological effectiveness
ROS Reactive oxygen species
SSD Species-sensitivity distribution
TGFβ Transforming growth factor beta
UV Ultraviolet light

ACKNOWLEDGMENTS

We thank the Canadian National Science and Engineering Research Council and the Canada Research Chairs Program for financial support. We are especially grateful to Alan Cocchetto of the National CFIDS Foundation Inc. for financial support and inspiration.

REFERENCES

[1] National Research Council. Committee to assess health risks from exposure to low levels of ionizing radiation: BEIR VII- phase 2. Washington (DC): National Academies Press; 2005.
[2] Tubiana M. Dose–effect relationship and estimation of the carcinogenic effects of low doses of ionizing radiation: the joint report of the Académie des Sciences (Paris) and of the Académie Nationale de Médecine. Int J Radiat Oncol 2005;63(2):317–9.
[3] Averbeck D. Does scientific evidence support a change from the LNT model for low-dose radiation risk extrapolation? Health Phys 2009;97(5):493–504.
[4] Calabrese EJ. On the origins of the linear no-threshold (LNT) dogma by means of untruths, artful dodges and blind faith. Environ Res 2015;142:432–42.
[5] Castillo H, Schoderbek D, Dulal S, Escobar G, Wood J, Nelson R, et al. Stress induction in the bacteria *Shewanella oneidensis* and *Deinococcus radiodurans* in response to below-background ionizing radiation. Int J Radiat Biol 2015;15:1–33.
[6] Preston RJ, Boice Jr JD, Brill AB, Chakraborty R, Conolly R, Hoffman FO, et al. Uncertainties in estimating health risks associated with exposure to ionising radiation. J Radiol Prot 2013;33(3):573–88.
[7] Task Group on Radiation Quality Effects in Radiological Protection, Committee 1 on Radiation Effects, International Commission on Radiological Protection. Relative biological effectiveness (RBE), quality factor (Q) and radiation weighting factor (w(R)). A report of the International Commission on Radiological Protection. Ann ICRP 2003;33(4):1–117.
[8] Wrixon AD. New ICRP recommendations. J Radiol Prot 2008;28(2):161–8.
[9] Brechignac F. Protection of the environment: how to position radioprotection in an ecological risk assessment perspective. Sci Total Environ 2003;307(1–3):35–54.
[10] Fujikawa K, Hasegawa Y, Matsuzawa S, Fukunaga A, Itoh T, Kondo S. Dose and dose-rate effects of X rays and fission neutrons on lymphocyte apoptosis in p53(+/+) and p53(−/−) mice. J Radiat Res 2000;41(2):113–27.
[11] Dobson RL, Kwan TC. The tritium RBE at low-level exposure – variation with dose, dose rate, and exposure duration. Curr Top Radiat Res Q 1978;12(1–4):44–62.
[12] Stewart AM, Kneale GW. A-bomb survivors as a source of cancer risk estimates: confirmation of suspected bias. In: Baverstock KF, Stather JW, editors. Low-dose radiation: biological bases of risk assessment. London, New York: Taylor and Francis; 1989. p. 354–63.
[13] Kerr GD. Workshop report on atomic bomb dosimetry – residual radiation exposure: recent research and suggestions for future studies. Health Phys 2013;105(2):140–9.
[14] Brooks AL. Thirty-sixth Lauriston S. Taylor Lecture on radiation protection and measurements—from the field to the laboratory and back. Health Phys 2013;105(5):407–21.
[15] Mothersill C, Seymour C. Changing paradigms in radiobiology. Mutat Res 2012;750(2):85–95.
[16] Morgan WF, Bair WJ. Issues in low dose radiation biology: the controversy continues. A perspective. Radiat Res 2013;179(5):501–10.
[17] National Academies Press. Health risks from exposure to low levels of ionizing radiation: BEIR VII phase 2. (USA): National Academies Press; 2006.
[18] Tang FR, Loke WK. Molecular mechanisms of low dose ionizing radiation-induced hormesis, adaptive responses, radioresistance, bystander effects, and genomic instability. Int J Radiat Biol 2015;91(1):13–21.

[19] Seymour C, Mothersill C, Alper T. High yields of lethal mutations in somatic mammalian cells that survive ionizing radiation. Int J Radiat Biol 1986;50(1):167–79.

[20] Kadhim MA, MacDonald DA, Goodhead DT, Lorimore SA, Marsden SJ, Wright EG. Transmission of chromosomal instability after plutonium α-particle irradiation. Nature 1992;355(6362):738–40.

[21] Elkind MM, Whitmore GF. The radiobiology of cultured mammalian cells. New York: Gordon and Breach; 1967.

[22] Pampfer S, Streffer C. Increased chromosome aberration levels in cells from mouse fetuses after zygote X-irradiation. Int J Radiat Biol 1989;55(1):85–92.

[23] Weissenborn U, Streffer C. Analysis of structural and numerical chromosomal aberrations at the first and second mitosis after X irradiation of two-cell mouse embryos. Radiat Res 1989;117(2):214–20.

[24] Mendonca MS, Kurohara W, Antoniono R, Redpath JL. Plating efficiency as a function of time postirradiation: evidence for the delayed expression of lethal mutations. Radiat Res 1989;119(2):387–93.

[25] Clutton S, Townsend K, Walker C, Ansell J, Wright E. Radiation-induced genomic instability and persisting oxidative stress in primary bone marrow cultures. Carcinogenesis 1996;17(8):1633–9.

[26] Limoli C. Attenuation of radiation-induced genomic instability by free radical scavengers and cellular proliferation. Free Radic Biol Med 2001;31(1):10–9.

[27] Seymour C, Mothersill C. All colonies of CHO-K1 cells surviving γ-irradiation contain non-viable cells. Mutat Res 1992;267(1):19–30.

[28] Chang WP, Little JB. Delayed reproductive death as a dominant phenotype in cell clones surviving X-irradiation. Carcinogenesis 1992;13(6):923–8.

[29] Seymour CB, Mothersill C. Delayed expression of lethal mutations and genomic instability in the progeny of human epithelial cells that survived in a bystander-killing environment. Radiat Oncol Investig 1997;5(3):106–10.

[30] Mothersill C, Seymour C. Survival of human epithelial cells irradiated with cobalt 60 as microcolonies or single cells. Int J Radiat Biol 1997;72(5):597–606.

[31] Nagasawa H, Little JB. Induction of sister chromatid exchanges by extremely low doses of alpha-particles. Cancer Res 1992;52(22):6394–6.

[32] Mothersill C, Seymour CB. Medium from irradiated human epithelial cells but not human fibroblasts reduces the clonogenic survival of unirradiated cells. Int J Radiat Biol 1997;71(4):421–7.

[33] Mothersill C, Seymour C. Radiation-induced bystander effects: past history and future directions. Radiat Res 2001;155(6):759–67.

[34] Hei TK, Zhou H, Ivanov VN, Hong M, Lieberman HB, Brenner DJ, et al. Mechanism of radiation-induced bystander effects: a unifying model. J Pharm Pharmacol 2008;60(8):943–50.

[35] Seymour CB, Mothersill C. Relative contribution of bystander and targeted cell killing to the low-dose region of the radiation dose – response curve. Radiat Res 2000;153(5):508–11.

[36] Schettino G, Folkard M, Michael BD, Prise KM. Low-dose binary behavior of bystander cell killing after microbeam irradiation of a single cell with focused C(K) X rays. Radiat Res 2005;163(3):332–6.

[37] Surinov BP, Isaeva VG, Dukhova NN. Postirradiation volatile secretions of mice: syngeneic and allogeneic immune and behavioral effects. Bull Exp Biol Med 2004;138(4):384–6.

[38] Mothersill C, Bucking C, Smith RW, Agnihotri N, O'Neill A, Kilemade M, et al. Communication of radiation-induced stress or bystander signals between fish in vivo. Environ Sci Technol 2006;40(21):6859–64.

[39] Audette-Stuart M, Yankovich T. Bystander effects in bullfrog tadpoles. Radioprotection 2011;46(6):S497.

[40] Morgan WF. Non-targeted and delayed effects of exposure to ionizing radiation: I. Radiation-induced genomic instability and bystander effects in vitro. Radiat Res 2012;178(2):AV223–236.

[41] Smith RW, Mothersill C, Hinton T, Seymour CB. Exposure to low level chronic radiation leads to adaptation to a subsequent acute X-ray dose and communication of modified acute X-ray induced bystander signals in medaka (Japanese rice fish, *Oryzias latipes*). Int J Radiat Biol 2011;87(10):1011–22.

[42] Azzam EI, Little JB. The radiation-induced bystander effect: evidence and significance. Hum Exp Toxicol 2004;23(2):61–5.

[43] Calabrese EJ. Hormesis: improving predictions in the low-dose zone. EXS 2012;101:551–64.

[44] Wright EG. Manifestations and mechanisms of non-targeted effects of ionizing radiation. Mutat Res 2010;687(1–2):28–33.

[45] Kaup S, Grandjean V, Mukherjee R, Kapoor A, Keyes E, Seymour CB, et al. Radiation-induced genomic instability is associated with DNA methylation changes in cultured human keratinocytes. Mutat Res 2006;597(1–2):87–97.

[46] McAllister KA, Lorimore SA, Wright EG, Coates PJ. In vivo interactions between ionizing radiation, inflammation and chemical carcinogens identified by increased DNA damage responses. Radiat Res 2012;177(5):584–93.

[47] Mothersill C, Seymour C. Radiation-induced bystander effects: are they good, bad or both? Med Confl Surviv 2005;21(2):101–10.

[48] Sawant SG, Randers-Pehrson G, Geard CR, Brenner DJ, Hall EJ. The bystander effect in radiation oncogenesis: I. Transformation in C3H 10T1/2 cells in vitro can be initiated in the unirradiated neighbors of irradiated cells. Radiat Res 2001;155(3):397–401.

[49] Azzam EI, Toledo SMD, Raaphorst GP, Mitchel REJ. Low-dose ionizing radiation decreases the frequency of neoplastic transformation to a level below the spontaneous rate in C3H 10T1/2 cells. Radiat Res 1996;146(4):369–73.

[50] Ko M, Lao X-Y, Kapadia R, Elmore E, Redpath J. Neoplastic transformation in vitro by low doses of ionizing radiation: role of adaptive response and bystander effects. Mutat Res 2006;597(1–2):11–7.

[51] Matsumoto H, Takahashi A, Ohnishi T. Radiation-induced adaptive responses and bystander effects. Biol Sci Space 2004;18(4):247–54.

[52] Smith RW, Wang J, Mothersill CE, Hinton TG, Aizawa K, Seymour CB. Proteomic changes in the gills of wild-type and transgenic radiosensitive medaka following exposure to direct irradiation and to X-ray induced bystander signals. Biochim Biophys Acta 2011;1814(2):290–8.

[53] Smith RW, Wang J, Schültke E, Seymour CB, Bräuer-Krisch E, Laissue JA, et al. Proteomic changes in the rat brain induced by homogenous irradiation and by the bystander effect resulting from high energy synchrotron X-ray microbeams. Int J Radiat Biol 2013;89(2):118–27.
[54] Choi VWY, Cheung ALY, Cheng SH, Yu KN. Hormetic effect induced by alpha-particle-induced stress communicated in vivo between zebrafish embryos. Environ Sci Technol 2012;46(21):11678–83.
[55] Mosse IB. Genetic effects of ionizing radiation – some questions with no answers. J Environ Radioact 2012;112:70–5.
[56] Hamada N, Maeda M, Otsuka K, Tomita M. Signaling pathways underpinning the manifestations of ionizing radiation-induced bystander effects. Curr Mol Pharmacol 2011;4(2):79–95.
[57] Kudryasheva N, Rozhko T. Effect of low-dose ionizing radiation on luminous marine bacteria: radiation hormesis and toxicity. J Environ Radioact 2015;142:68–77.
[58] Szumiel I. Radiation hormesis: autophagy and other cellular mechanisms. Int J Radiat Biol 2012;88(9):619–28.
[59] Hoffmann G, Stempsey W. The hormesis concept and risk assessment: are there unique ethical and policy considerations? Hum Exp Toxicol 2008;27(8):613–20.
[60] Marquis O, Miaud C, Ficetola GF, Bocher A, Mouchet F, Guittonneau S, et al. Variation in genotoxic stress tolerance among frog populations exposed to UV and pollutant gradients. Aquat Toxicol 2009;95(2):152–61.
[61] Geras'kin S, Evseeva T, Oudalova A. Effects of long-term chronic exposure to radionuclides in plant populations. J Environ Radioact 2013;121:22–32.
[62] Bradshaw C, Kapustka L, Barnthouse L, Brown J, Ciffroy P, Forbes V, et al. Using an ecosystem approach to complement protection schemes based on organism-level endpoints. J Environ Radioact 2014;136:98–104.
[63] Burkhalter MD, Rudolph KL, Sperka T. Genome instability of ageing stem cells—induction and defence mechanisms. Ageing Res Rev 2015;23:29–36.
[64] Ermolaeva MA, Schumacher B. Systemic DNA damage responses: organismal adaptations to genome instability. Trends Genet 2014;30(3):95–102.
[65] Selye H. The stress of life. New York: McGraw-Hill; 1956.
[66] Calabrese EJ, Bachmann K, Bailer J, Bolger M, Borak J, Cai L, et al. Biological stress response terminology: integrating the concepts of adaptive response and preconditioning stress within a hormetic dose-response framework. Toxicol Appl Pharmacol 2007;222(1):122–8.
[67] Nikitaki Z, Hellweg CE, Georgakilas AG, Ravanat J-L. Stress-induced DNA damage biomarkers: applications and limitations. Front Chem 2015;3:35.
[68] Indo HP, Yen H-C, Nakanishi I, Matsumoto K, Tamura M, Nagano Y, et al. A mitochondrial superoxide theory for oxidative stress diseases and aging. J Clin Biochem Nutr 2015;56(1):1–7.
[69] Caputo F, Vegliante R, Ghibelli L. Redox modulation of the DNA damage response. Biochem Pharmacol 2012;84(10):1292–306.
[70] Ulsh BA. Checking the foundation: recent radiobiology and the linear no-threshold theory. Health Phys 2010;99(6):747–58.
[71] Kojima S. Involvement of ATP in radiation-induced bystander effect as a signaling molecule. Yakugaku Zasshi 2014;134(6):743–9.
[72] Oblong JE. The evolving role of the NAD+/nicotinamide metabolome in skin homeostasis, cellular bioenergetics, and aging. DNA Repair (Amst) 2014;23:59–63.
[73] Schaue D, Micewicz ED, Ratikan JA, Xie MW, Cheng G, McBride WH. Radiation and inflammation. Semin Radiat Oncol 2015;25(1):4–10.
[74] Rastogi S, Boylan M, Wright EG, Coates PJ. Interactions of apoptotic cells with macrophages in radiation-induced bystander signaling. Radiat Res 2013;179(2):135–45.
[75] Lorimore SA, Coates PJ, Scobie GE, Milne G, Wright EG. Inflammatory-type responses after exposure to ionizing radiation in vivo: a mechanism for radiation-induced bystander effects? Oncogene 2001;20(48):7085–95.
[76] Liu Z, Mothersill CE, McNeill FE, Lyng FM, Byun SH, Seymour CB, et al. A dose threshold for a medium transfer bystander effect for a human skin cell line. Radiat Res July 2006;166(1 Pt. 1):19–23.
[77] Rothkamm K, Löbrich M. Evidence for a lack of DNA double-strand break repair in human cells exposed to very low x-ray doses. Proc Natl Acad Sci USA 2003;100(9):5057–62.
[78] Wahab M, Nickless E, Najar-M'kacher R, Parmentier C, Podd J, Rowland R. Elevated chromosome translocation frequencies in New Zealand nuclear test veterans. Cytogenet Genome Res 2008;121(2):79–87.
[79] Hiyama A, Nohara C, Kinjo S, Taira W, Gima S, Tanahara A, et al. The biological impacts of the Fukushima nuclear accident on the pale grass blue butterfly. Sci Rep 2012;2:570–80.
[80] Mousseau TA, Moller AP. Genetic and ecological studies of animals in Chernobyl and Fukushima. J Hered 2014;105(5):704–9.
[81] Featherstone S. The swallows of Fukushima. Sci Am 2015;312(2):74–81.
[82] Garnier-Laplace J, Geras'kin S, Della-Vedova C, Beaugelin-Seiller K, Hinton T, Real A, et al. Are radiosensitivity data derived from natural field conditions consistent with data from controlled exposures? A case study of Chernobyl wildlife chronically exposed to low dose rates. J Environ Radioact 2013;121:12–21.
[83] Brenner DJ, Doll R, Goodhead DT, Hall EJ, Land CE, Little JB, et al. Cancer risks attributable to low doses of ionizing radiation: assessing what we really know. Proc Natl Acad Sci USA 2003;100(24):13761–6.
[84] Breckow J. Linear-no-threshold is a radiation-protection standard rather than a mechanistic effect model. Radiat Environ Biophys 2006;44(4):257–60.
[85] Furukawa K, Preston DL, Lönn S, Funamoto S, Yonehara S, Matsuo T, et al. Radiation and smoking effects on lung cancer incidence among atomic bomb survivors. Radiat Res 2010;174(1):72–82.
[86] Morgan WF. Is there a common mechanism underlying genomic instability, bystander effects and other nontargeted effects of exposure to ionizing radiation? Oncogene 2003;22(45):7094–9.

[87] Mothersill C, Seymour C. Mechanisms and implications of genomic instability and other delayed effects of ionizing radiation exposure. Mutagenesis 1998;13(5):421–6.

[88] Tawn EJ, Whitehouse CA, Winther JF, Curwen GB, Rees GS, Stovall M, et al. Chromosome analysis in childhood cancer survivors and their offspring—no evidence for radiotherapy-induced persistent genomic instability. Mutat Res 2005;583(2):198–206.

[89] Ballarini F, Alloni D, Facoetti A, Mairani A, Nano R, Ottolenghi A. Modelling radiation-induced bystander effect and cellular communication. Radiat Prot Dosimetry 2006;122(1–4):244–51.

[90] Mothersill C, Rea D, Wright EG, Lorimore SA, Murphy D, Seymour CB, et al. Individual variation in the production of a 'bystander signal' following irradiation of primary cultures of normal human urothelium. Carcinogenesis 2001;22(9):1465–71.

[91] Kamiya K, Ozasa K, Akiba S, Niwa O, Kodama K, Takamura N, et al. Long-term effects of radiation exposure on health. Lancet 2015;386(9992):469–78.

[92] Ponnaiya B, Cornforth MN, Ullrich RL. Radiation-induced chromosomal instability in BALB/c and C57BL/6 mice: the difference is as clear as black and white. Radiat Res 1997;147(2):121–5.

[93] Cox MM, Battista JR. *Deinococcus radiodurans*—the consummate survivor. Nat Rev Microbiol 2005;3(11):882–92.

[94] Li G, Ho VC. p53-dependent DNA repair and apoptosis respond differently to high- and low-dose ultraviolet radiation. Br J Dermatol 1998;139(1):3–10.

[95] Hall EJ, Giaccia AJ. Radiobiology for the radiologist. 6th ed. Philadelphia: Lippincott Williams & Wilkins; 2006.

[96] Lee K-F, Weng JT-Y, Hsu PW-C, Chi Y-H, Chen C-K, Liu IY, et al. Gene expression profiling of biological pathway alterations by radiation exposure. Biomed Res Int 2014;40:83–7.

[97] Choudhury A, Cuddihy A, Bristow R. Radiation and new molecular agents part I: targeting ATM-ATR checkpoints, DNA repair, and the proteasome. Semin Radiat Oncol 2006;16(1):51–8.

[98] Salbu B. Multiple stressors—a challenge for the future. J Environ Monit 2005;7(6):539.

[99] Mothersill C, Smith RW, Heier LS, Teien H-C, Land OC, Seymour CB, et al. Radiation-induced bystander effects in the Atlantic salmon (salmo salar L.) following mixed exposure to copper and aluminum combined with low-dose gamma radiation. Radiat Environ Biophys 2014;53(1):103–14.

[100] Korkaric M, Behra R, Fischer BB, Junghans M, Eggen RI. Multiple stressor effects in *Chlamydomonas reinhardtii* – toward understanding mechanisms of interaction between effects of ultraviolet radiation and chemical pollutants. Aquat Toxicol 2015;162:18–28.

[101] Morgan WF. Non-targeted and delayed effects of exposure to ionizing radiation: II. Radiation-induced genomic instability and bystander effects in vivo, clastogenic factors and transgenerational effects. Radiat Res 2003;159(5):581–96.

[102] Ryan LA, Smith RW, Seymour CB, Mothersill CE. Dilution of irradiated cell conditioned medium and the bystander effect. Radiat Res 2008;169(2):188–96.

[103] Pinho C, Timotin E, Wong R, Sur RK, Hayward JE, Farrell TJ, et al. Assessing patient characteristics and radiation-induced non-targeted effects in vivo for high dose-rate (HDR) brachytherapy. Int J Radiat Biol 2015;1:1–41.

[104] Kadhim M, Lorimore S, Townsend K, Goodhead D, Buckle V, Wright E. Radiation-induced genomic instability: delayed cytogenetic aberrations and apoptosis in primary human bone marrow cells. Int J Radiat Biol 1995;67(3):287–93.

[105] Coen N, Kadhim MA, Wright EG, Case CP, Mothersill CE. Particulate debris from a titanium metal prosthesis induces genomic instability in primary human fibroblast cells. Br J Cancer 2003;88(4):548–52.

[106] Mothersill C. p53 protein expression and increased SSCP mobility shifts in the p53 gene in normal urothelium cultured from smokers. Carcinogenesis 1997;18(6):1241–5.

[107] Garnier-Laplace J, Alonzo F, Adam-Guillermin C. Establishing relationships between environmental exposures to radionuclides and the consequences for wildlife: inferences and weight of evidence. Ann ICRP 2015;44(1 Suppl.):295–303.

[108] McCarty LS, Power M, Munkittrick KR. Bioindicators versus biomarkers in ecological risk assessment. Hum Ecol Risk Assess 2002;8(1):159–64.

[109] Puck TT. Action of x-rays on mammalian cells. J Exp Med 1956;103(5):653–66.

[110] Vandevoorde C, Franck C, Bacher K, Breysem L, Smet MH, Ernst C, et al. γ-H2AX foci as in vivo effect biomarker in children emphasize the importance to minimize x-ray doses in paediatric CT imaging. Eur Radiol 2014;25(3):800–11.

[111] Romm H, Wilkins RC, Coleman CN, Lillis-Hearne PK, Pellmar TC, Livingston GK, et al. Biological dosimetry by the triage dicentric chromosome assay: potential implications for treatment of acute radiation syndrome in radiological mass casualties. Radiat Res 2011;175(3):397–404.

[112] Kong EY, Choi VWY, Cheng SH, Yu KN. Some properties of the signals involved in unirradiated zebrafish embryos rescuing α-particle irradiated zebrafish embryos. Int J Radiat Biol 2014;90(12):1133–42.

[113] Le M, McNeill FE, Seymour C, Rainbow AJ, Mothersill CE. An observed effect of ultraviolet radiation emitted from beta-irradiated HaCaT cells upon non-beta-irradiated bystander cells. Radiat Res 2015;183(3):279–90.

[114] Mothersill C, Smith RW, Fazzari J, McNeill F, Prestwich W, Seymour CB. Evidence for a physical component to the radiation-induced bystander effect? Int J Radiat Biol 2012;88(8):583–91.

[115] Mothersill C, Moran G, McNeill F, Gow MD, Denbeigh J, Prestwich W, et al. A role for bioelectric effects in the induction of bystander signals by ionizing radiation? Dose-Response 2007;5(3):214–29.

[116] Papineni RV. X-ray irradiation induced bio luminescence: ex vivo and endoscopic imaging of radiobioluminescence. Cancer Res 2014;74(19):4921.

[117] Jella KK, Rani S, O'driscoll L, McClean B, Byrne HJ, Lyng FM. Exosomes are involved in mediating radiation induced bystander signaling in human keratinocyte cells. Radiat Res 2014;181(2):138–45.

[118] Al-Mayah AHJ, Irons SL, Pink RC, Carter DRF, Kadhim MA. Possible role of exosomes containing RNA in mediating nontargeted effect of ionizing radiation. Radiat Res 2012;177(5):539–45.

[119] Albanese J, Dainiak N. Modulation of intercellular communication mediated at the cell surface and on extracellular, plasma membrane–derived vesicles by ionizing radiation. Exp Hematol 2003;31(6):455–64.

[120] Sutherland RM, Durand RE. Radiation response of multicell spheroids–an in vitro tumour model. Curr Top Radiat Res Q 1976;11(1):87–139.

[121] Mole RH. Radiation-induced acute myeloid leukaemia: an unusually valuable experimental model for testing basic assumptions about the process of carcinogenesis. In: Iversen OH, editor. Theories of carcinogenesis. Washington (DC): Hemisphere Publishing Co.; 1988. p. 133–41.

[122] Fialkow PJ. Clonal origin of human tumors. Biochim Biophys Acta 1976;458(3):283–321.

[123] Lorimore SA, Wright EG. Radiation-induced genomic instability and bystander effects: related inflammatory-type responses to radiation-induced stress and injury? A review. Int J Radiat Biol 2003;79(1):15–25.

[124] Baverstock K. Radiation-induced genomic instability: a paradigm-breaking phenomenon and its relevance to environmentally induced cancer. Mutat Res 2000;454(1):89–109.

Chapter 36

Transgenerational Genome Instability in Plants

I. Kovalchuk
University of Lethbridge, Lethbridge, AB, Canada

Chapter Outline

1. INTRODUCTION

It is commonly accepted that cell division and organism reproduction are characterized by the faithful replication of the genetic material, DNA. The DNA sequence represents a master key that is used to reproduce cells and organisms with the identical genetic makeup. In contrast, the regulation at the epigenetic level, including epigenetic inheritance, represents a more versatile and flexible mechanism controlling gene expression and inheritance of old traits as well as the appearance of new traits. Why would it then be so important to preserve the integrity of DNA if downstream mechanisms are able to alter outcomes in terms of proteins produced, metabolites, and phenotype appearance? Epigenetic mechanisms are frequently reversible because they do not represent permanent chemical changes. In contrast, changes in nucleotide sequences such as base substitutions, deletions, and mutations are irreversible, unless a reversion mutation (mostly a single base modification) occurs. Thus, it is of an outmost importance to preserve the master key code.

Plants maintain genome integrity at all times, whether it is at the stage of active development and cell division or cell growth. An active metabolism, including photosynthesis, cellular respiration, and other physiological activities associated with the function of peroxisomes and lysosomes, poses a continuous challenge for plant genomes. These types of internal

Genome Stability. http://dx.doi.org/10.1016/B978-0-12-803309-8.00036-7

stresses result in the production of free radicals that are either damaging DNA directly or triggering changes in DNA via a variety of signaling pathways they are involved in. Radicals are also able to oxidase lipids and proteins, thus rendering them incapable of a normal cellular activity.

Being sedentary in nature, plants are at a constant state of war with the environment. Environmental stimuli represent external stresses that include, but are not limited to, changes in light intensity, temperature fluctuations, water and nutrients availability, wind and other mechanical stimuli, and an entire realm of biotic interactions that include physical and chemical influences. A good review describing different types of abiotic and biotic stresses that plants are exposed to is written by Madlung and Comai [1].

To survive these environmental pressures, organisms have to respond using the mechanisms that are already available, but they also have to develop new adaptive changes that provide advantages to them if new conditions persist. Not being able to escape external stresses, plants are limited to mechanisms of tolerance and resistance, the strategies that plants are extremely proficient at Ref. [2]. Adaptive metabolic changes in somatic cells and heritable transgenerational changes are among more sophisticated mechanisms of survival [3]. Through the process of evolution, organisms have developed efficient adaptive mechanisms of survival, and plants seem to be very efficient in doing that [4].

Plants also have the ability to maintain genome stability in the ever-changing growth environment. Many plant species seem to possess additional copies of various DNA-repair genes that often have redundant functions [5]. That is why studying DNA-repair capacity in plants using mutants is so challenging. For example, plants possess four Rad51 paralogs, AtXrcc2, AtXrcc3, AtRad51B, and AtRad51C. A mutation in any of these genes results in hypersensitivity to DNA-damaging agents such as mitomycin C [6]. Moreover, the *atrad51c* and *atxrcc3* mutants show meiotic defects and thus are difficult to propagate.

Genome integrity is maintained through a number of different mechanisms, with the direct repair of DNA damage being perhaps the most important one. There are multiple levels of control of the process of DNA-damage repair, including scanning and the identification of damage, the global or local relaxation of chromatin, the recruitment of the repairsome, actual repair steps, and the reestablishment of a similar or perhaps different status of chromatin, including changes in DNA methylation and histone modifications [7]. Since there is a possibility that chromatin compaction has a buffering ability against various factors that can damage DNA, it is plausible to think that genome stability of a given chromosomal region can be relaxed not only by choosing different DNA-repair pathways but also by introducing or removing various epigenetic modifications [8].

The control over DNA repair and genome stability is thus regulated by a variety of genetic and epigenetic factors [8]. While different DNA-repair pathways are described in great details in the other chapters of this book, the purpose of this chapter is to describe the epigenetic mechanisms that can affect and modify genome stability, with a special emphasis given to transgenerational responses.

2. GENOME STABILITY MAY DEPEND UPON THE CHOICE OF THE DSB DNA-REPAIR PATHWAY

Double-strand break (DSB) is the most dangerous DNA lesion because a single unrepaired DSB may lead to cell-cycle arrest or apoptosis. DSBs are repaired via two major repair pathways: nonhomologous end joining (NHEJ) and homologous recombination (HR) [9]. The repair via NHEJ involves a direct rejoining of break ends and does not require a significant homology between the interacting DNA molecules. In the cases when rejoining via a direct ligation process is not possible, NHEJ proteins Ku70/Ku80 search for microhomology aligning one or several complementary bases to direct repeats, and thus resulting in a removal of DNA between direct repeats. As a result, the repair via NHEJ is relatively inaccurate and is typically associated with small- and large-scale deletions, ranging from a single base pair to large DNA sequences of several thousand nucleotides [9]. Insertions and point mutations are frequent outcomes as well. In contrast, the HR mechanism requires an extensive sequence homology and the presence of a repair template. The repair via HR is quite accurate if perfectly homologous templates such as a sister chromatid or a homologous chromosome are used to prime repair synthesis. However, HR repair using a template with imperfect homology could result in gene conversion events leading to a loss of heterozygosity. In rare cases, if HR repair occurs in DNA regions containing multiple repeats, the process can result in gene translocation and duplication events as well deletions of an entire chromosome. Overall, NHEJ can be characterized as a fast-track error-prone repair mechanism, whereas HR represents a rather slow but relatively error-free process [10]. Detailed information about types of NHEJ and HR repairs in plants can be found in Chapter 12.

The balance between the occurrence of NHEJ and HR is tightly regulated and depends on such factors as the availability of repair templates, the phase of a cell cycle, the rate of cell proliferation, and even the specific function of a given cell type (reviewed in Ref. [11]). NHEJ is a predominant DNA-repair pathway, and cells use it more often while being in the

G1 phase of a cell cycle. In contrast, HR is rather a minor pathway that is more active during S and G2 phases when sister chromatids are formed [12]. Thus, HR plays an important role in the actively dividing cells and during early development of an organism. Moreover, in plants, tissues with a higher ploidy level seem to use HR less frequently. Experimental data show that older plant leaves that tend to have a larger number of genomes per single cell (due to endo-reduplication) have a lower frequency of HR when prorated to a single genome [13]. This is not surprising because an increase in ploidy results in an increase in the number of copies of potential templates available for HR repair, posing a threat to genome stability and possibly leading to large-scale deletions and chromosomal translocations. It is also possible that the genome size has also some effects on the frequency of the use of HR repair [14,15]. It may be more difficult and time-consuming to find homologous sequences in the larger genomes.

Chromatin structure in more complex genomes may also contribute to an additional difficulty for HR to occur more frequently. Since mutation frequency varies in different cell types and during different developmental stages of an organism, there definitely exist some types of chromatin-based regulation of genome stability. Further, we attempted to summarize the accumulated data on epigenetic control over the stability of highly repetitive genomes in plants.

3. EPIGENETIC REGULATION OF PLANT GENOME STABILITY

Plants dramatically differ from most other higher eukaryotes in a specific feature of their life—they are sedentary organisms. The prolonged nature of environmental conditions that has an impact on plant growth continuously poses challenges to plant defense systems, sometimes over many generations. Unlike other organisms that can leave their environment, plants cannot use escape and avoidance tactics to minimize the damaging influence of stress. It is thus logical that plants possess both short-term response systems and long-term defense strategies allowing them to cope with acute and chronic stresses. In fact, stresses typically persist for a long time, and therefore they should not be considered acute to plants. An acute stress is usually defined as a stress that affects an organism in the short term. Such definition is relative in relation to plants and can be applied only if the very same stress persists for a substantially longer period of time. When speaking about stresses that plants face, it would be more appropriate to use a term such as "high and low levels of an acute stress." A rapid alteration in homeostasis, including massive changes in the number and amount of produced metabolites, is one of the mechanisms through which plants respond to stresses. These immediate responses include flexible changes in gene expression, an increase or a decrease in the production of mRNAs, synthesis of proteins, protein and RNA degradation, synthesis and re-compartmentalization of various metabolites, balancing the salt concentration, pH, hormones, and many other events. Although these responses are critical for plant survival, their description is not the focus of this chapter; all the necessary information can be found elsewhere, including reviews by [16,17].

Many of the events described before are controlled by epigenetic mechanisms operating in somatic cells, including small RNA-mediated mRNA degradation, changes in DNA methylation and histone modifications, repositioning of histone and non-histone chromatin-binding proteins in the nuclear matrix. All these mechanisms are important for immediate plant survival. How does the epigenetic machinery protect plants and plant genomes from stress? In the following sections, we present some information on the role of each of the mechanisms mentioned in the preceding paragraphs in genome protection against stresses.

3.1 Chromatin Structure, a Response to Stress and Genome Stability

Responses to stress, including responses to genotoxic stress, involve transcriptional activation and repression of various genomic loci. Changes in chromatin structure play the most active role in this process. Chromatin decondensation involves the action of ATP-dependent remodeling complexes, covalent modifications of histones, deposition of histone variants, and/or changes in cytosine methylation. Moreover, noncoding small RNAs (smRNAs) add another level of complexity to the process as they can alter chromatin structure by directing heterochromatin formation at specific genomic sequences.

Chromatin in cells exists in different states of packaging that involve the wrapping of the DNA around the histone core. These structures, called nucleosomes, prevent the process of transcription, replication, and DNA repair from occurring. Specific histone modifications allow the unpacking of chromatin. Unpacking the damaged DNA is a double-edged sword which allows DNA-repair enzymes to fix the damage but at the same time makes DNA more vulnerable for further assaults. Specific histone modifications, primarily acetylation and methylation, make DNA more or less accessible to potential damaging agents and various rearrangements [18].

There are several experimental evidences which suggest the interdependence of DNA methylation and histone modifications. Effector proteins such as HETEROCHROMATIN PROTEIN1 (HP1) can be recruited to methylated histones: HP1 binds to methylated H3K9 and helps propagate heterochromatin to the adjacent regions of a chromosome [19]. This interaction is apparently important for a response to changes in the environmental conditions. For example, the *Arabidopsis* homolog of HP1, HETEROCHROMATIN PROTEIN1 (LHP1), is involved in regulating flowering time in response to environmental stimuli.

Methylated DNA also recruits various chromatin modifiers; methylated cytosines serve as substrates for binding of nuclear proteins named methyl-CpG-binding domain proteins (MBDs) [20]. MBDs bound to 5-methyl-cytosines recruit enzymes that modify core histone proteins and change the local chromatin structure. Similarly, HP1 protein binds methylated DNA and recruits histone modifiers [21].

3.1.1 Changes in Chromatin Structure in Response to Stress: Heterochromatin Decondensation

Nucleosome positioning, redistribution of heterochromatin and euchromatin in the nucleus, and the differential binding of chromatin-modifying proteins (excluding histones) and MBDs to DNA represent another level of complexity for an efficient response to developmental cues and environmental factors.

Chromatin condensation is critical for maintaining transcriptional gene silencing at repetitive elements. The removal of nucleosomes from specific genomic locations in response to stress could be both an active and a passive process. The fact that the original nucleosome loading and epigenetic regulation of repeats are restored fairly quickly upon recovery from stress suggests that the removal of nucleosomes can indeed be an active process. Alternatively, nucleosome loss from specific genomic positions can be associated with replication and transcription, thus representing a passive process. A study by Pecinka et al. showed that long-term exposure to heat in *Arabidopsis* resulted in the activation of some repetitive elements [22]. Surprisingly, the activation occurred without loss of DNA methylation and with only minor changes to histone modifications. Repetitive elements were primarily activated by the loss of nucleosomes and heterochromatin decondensation. The recovery from stress was characterized by nucleosome loading and transcriptional silencing. Curiously, in chromatin-assembly factor-1 (CAF-1) mutants impaired in chromatin-assembly functions, the recovery stage and nucleosome loading were considerably delayed [22]. The substantial dissociation of heterochromatin was observed beyond the recovery phase when silencing and nucleosomes had been reinstalled; the loss of heterochromatin was observed in differentiated tissues of plants exposed to heat, and it lasted in the exposed leaves until they started to show signs of senescence. Heat-induced decondensation of chromocenters and a general loss of nucleosomes presumably allowed a better accessibility of DNA to transcription complexes. A similar heterochromatin decondensation was observed in 2-day-old *Arabidopsis* plantlets in response to cell culturing, although regular chromocenters were formed in a stepwise process after a longer period in culture. The loss of heterochromatin also occurred in older plants upon floral transition in development, however, heterochromatin decondensation was not sufficient for repeat activation. Thus, local heterochromatization occurs during normal physiological and developmental processes and (un)specific responses to stress. Indeed, when plants were exposed to low-light stress, heterochromatin decondensation was more permanent and was directed toward areas with repetitive elements [23]. The reversibility of these changes was confirmed by prolonged culturing of plants exposed to low-intensity light; at a higher-light intensity, chromatin decondensation was eliminated.

Hence, is heterochromatin decondensation at genomic repeats a common response to stress? Pecinka et al. argue that it does not seem to be the case as they did not observe this phenotype after freezing (−4°C for 24h) or UV-C irradiation (3000J/m2) [22]. In fact, exposure to an abiotic stress may interfere with the plants' capacity to withstand a biotic stress. Indeed, even moderately increased temperatures can reduce biotic stress resistance by pathogens. In plants exposed to long-term heat stress, the activation of some repetitive elements is paralleled by silencing and transcriptional repression of repetitive loci carrying clusters of resistance genes [22].

Heterochromatin decondensation in response to heat stress seems not to occur equally in all tissues; the nuclei of meristematic cells do not undergo heat-induced decondensation. Actually, this does make sense. If one considers that heat stress response is transient in nature and should largely occur in somatic tissues only, the lack of changes in the meristem indicates a safeguarding mechanism for minimizing epigenetic and possibly genetic changes in the germ line. It further supports the hypothesis that decondensation is a controlled process that occurs only either during specific stages of plant development or in response to specific stresses such as heat and high light–intensity stresses. Moreover, exposure to these stresses may result in the transcriptional activation of heterochromatin-embedded genes in differentiated cells but not in dividing cells.

These results demonstrate that environmental conditions can transiently overcome epigenetic regulation and, perhaps, provide a chance for more permanent epigenetic and possibly genetic changes. The transcriptional activation of repeats occurring without DNA methylation resembles the effect of mutations in MOM1, FAS1, FAS2, BRU1, and RPA2; mutants of these plants also exhibit various degrees of activation of repetitive elements that occur without changes in methylation.

3.1.2 The Role of Chromatin-Remodeling Factors

Several chromatin-remodeling factors in plants help control gene expression and genome stability through DNA methylation and histone modifications. One of the best-known proteins, the DECREASED DNA METHYLATION1 (DDM1) protein, is a member of the SWI2/SNF2 DNA helicase family. Members of this family of proteins are involved in the control of DNA repair, recombination, gene expression, and replication [24]. It is suggested that one of the possible mechanisms of interactions of the SWI2/SNF2 family proteins with chromatin requires the disruption of DNA–histone interactions. DDM1, in particular, is involved in the regulation of DNA methylation status via changes in histone methylation as well as interactions with AtMBDs. It has been shown that the mutant of DDM1, *ddm1*, exhibits a disrupted localization of AtMBDs at chromocenters, suggesting that DDM1 may facilitate the localization of MBDs at specific nuclear domains [25].

The importance of *ddm1* for the control of DNA methylation is reflected by the fact that the *ddm1* mutant shows up to 70% reduction in global genome methylation [26]. As a consequence, this triggers the activation of transposons and retrotransposons, the transcriptional activation of a previously silent disease-resistance gene array, and the profound phenotypic instability amplified with every generation of self-propagation. The fact that *ddm1*-induced hypomethylation of various genes can be stably inherited through mitotic and meiotic cell divisions might be one of the reasons of the phenotypic instability [27]. One of the possible mechanisms of the involvement of DDM1 in the control of DNA methylation is the maintenance of CpG methylation at RNA-direct DNA methylation (RdDM)-targeted sequences after the RNA signal is removed. Although the data on genome instability in *ddm1* is scarce, one can hypothesize that the genome of *ddm1* is unstable since plants have the increased activity of transposons and retrotransposons. *ddm1* plants are more sensitive to a variety of stresses and appear to have a higher frequency of DSBs.

Another potential chromatin-remodeling factor is the nuclear MAINTENANCE OF METHYLATION 1 (MOM1) protein. It is involved in DNA methylation–independent silencing of repetitive sequences in *Arabidopsis* by preventing the transcription of 180-bp satellite repeats of transposons [28]. Curiously, in *mom1* mutants, releasing transgene silencing, the activation of transcription of 180-bp satellite repeats and *106B* dispersed repeats, and derepression of silencing of some 5S repeats occur without reducing/alternating their DNA and histone-methylation patterns. This suggests the existence of two distinct epigenetic-silencing pathways: one that is DNA-methylation dependent and the other one that is DNA-methylation independent. Although MOM1 is involved in chromatin remodeling, the mutant is not hypersensitive to the DNA-damaging agent methyl methane sulfonate (MMS). Other chromatin modifiers, such as BRU1, FAS1, FAS2, and RPA2, are also dispensable for DNA methylation, but all of them are hypersensitive to the MMS-induced DNA damage.

Other reports also indicated the link between chromatin maintenance and stress response. Mutants of a nuclear protein BRU1 involved in the maintenance of chromatin structure were highly sensitive to genotoxic stress and were characterized by an increased frequency of intrachromosomal HR [29]. Similarly, the expression of the *MIM1* gene involved in the maintenance of chromosome structure and required for the efficient HR was significantly increased by DNA-damaging agents.

Another SWI/SNF-like protein, DRD1, represents a novel plant-specific chromatin-remodeling protein that is required for RNA-directed de novo methylation of target promoters [30]. It is also necessary for the total loss of de novo DNA methylation after the RNA-silencing trigger is withdrawn. DRD1 interacts with two other factors, NRPD1b and NRPD2a, which represent subunits of a novel, plant-specific RNA polymerase, pol IVb. Together, DRD1 and the pol IVb complex act downstream of the small RNA (smRNA) biogenesis pathway (see later). Thus, they direct reversible silencing of euchromatic promoters in response to RNA signals possibly through the recruitment of DNA methyltransferases for methylation of homologous DNA sequences. It is noteworthy that among putative DRD1 targets, there are DNA glycosylases, ROS1 and DME, which are involved in active DNA demethylation. The downregulation of *ROS1* in *drd1* and *pol IVb* mutants confirms the importance of the DRD1/pol IVb pathway for the active loss of induced de novo DNA methylation [31].

3.2 The Role of DNA Methylation in the Maintenance of Plant Genome Stability and Response to Stress

DNA methylation is the most versatile mechanism involved in the regulation of gene expression, including the inheritance of specific gene expression patterns through somatic or meiotic cell divisions. Since DNA methylation is typically associated with a more restrictive chromatin state, it is highly likely that regions with higher methylation would be more stable—that is, they will have fewer mutations associated with them. Is that actually true?

There are not many reports indicating a negative correlation between DNA methylation and rearrangements. An earlier work demonstrated that DNA methylation suppresses the occurrence of HR between dispersed sequences, restricting recombination events to the gene-rich regions with a lower level of methylation [32,33]. In *Hevea brasiliensis*, the inverted correlation between DNA-methylation levels and gene rearrangements was observed [34]. In contrast, a study by Mirouze et al. could not find any significant correlation between the level of methylation and HR frequency in the *Arabidopsis* genome [35]. Also, the authors showed that the progeny of crosses between wild-type and *met1* mutant *Arabidopsis* plants impaired in the maintenance of CpG methylation showed the increased meiotic recombination frequency in the hypomethylated chromosome arms but not in the hypomethylated heterochromatic pericentromeric regions. It remains to be shown what other factors regulate the recombination frequency in the plant genome. Further studies analyzing changes in histone modifications and the binding of nonhistone chromatin proteins may allow to establish a better correlation between genome rearrangements and chromatin structure.

Stress may result in both hypo- and hypermethylation at specific genomic loci, and these changes may represent either a short-term change or a long-term strategy of response to stress [34]. Promoters of stress-responsive genes are often found to be hypomethylated [36,37], whereas methylation at other genomic loci may not be altered, and sometimes may even be increased.

Changes in DNA methylation in response to stress may occur due to many different mechanisms, including the activity of DNA methyltransferases, DNA demethylases such as ROS1, DME1, DML2, and DML3, a passive loss of methylation via the exclusion of DNA methyltransferases from the nucleus, changes in the activity of chromatin-remodeling factors and effector proteins, and many other changes in proteins regulating the chromatin structure. The regulation of methylation is a complex process, and the absence of one or several DNA methyltransferases does not necessarily result in a total loss of DNA methylation. In the *met1* plants that lack the maintenance methyltransferase, global hypomethylation is accompanied by hypermethylation at multiple transposons and repetitive element loci. The expression of DNA demethylases, *DME* and *ROS1*, is repressed in the mutant, likely as an overcompensation mechanism that prevents more extensive losses in methylation. As a result, both de novo non-CG methylation at nonrepetitive loci and RdDM-directed hypermethylation of repetitive elements are increased [38].

3.2.1 Correlation Between DNA Methylation Levels and Genome Stability

Does methylation directly influence genome stability? Unfortunately, there is no clear answer to this question. However, there exists a degree of correlation between methylation of specific cytosine nucleotides and the frequency of point mutations at these sites. Methylated cytosines are prone to frequent spontaneous deaminations as a result of which they are converted into thymines, which leads to C/G to T/A point mutations. This may explain why CG pairs occur much more rarely as compared to other nucleotide pairs. Ossowski et al. analyzed the rate of mutations in *Arabidopsis* plants self-propagated for 30 generations; it was found that a great majority of all mutations were C/G to T/A base substitutions [39]. Such bias can only be explained by a high frequency of spontaneous deamination of methylated cytosines. In contrast, the deamination of nonmethylated cytosines results in the formation of uracils that are easily recognized by the DNA-repair machinery. DNA methylation also seems to play a critical role in the control of the activity of transposable elements and in the protection of plant cells against the expression of integrated foreign DNA elements. Considering that cytosine methylation protects DNA from cleavage by an endonuclease, it can also protect it against multicopy transposable elements and aberrant gene duplications. Thus, a higher level of DNA methylation at certain loci may function as a defense mechanism against foreign invasive DNA molecules and as a protection against the cell's own transposable elements.

There also exists an inverted correlation between the level of methylation at certain genomic loci and the frequency of large chromosomal rearrangements at these loci. Although it is a common wisdom that hypomethylated loci are more prone to genomic rearrangements, there is not much data on plants that can confirm this. A higher frequency of deletions/insertions of transposable elements at long terminal repeats associated with hypomethylation was observed in the first two generations after allopolyploidization of wheat [40]. The progeny of tobacco plants exposed to tobacco mosaic virus (TMV) exhibited a higher frequency of rearrangements at *R* gene–like loci, and these changes were paralleled by hypomethylation [41].

Many stresses may directly influence the level of methylation in the genome. According to literature, salts of Cd, Ni, and Cr cause oxidative damage that induces DNA hypomethylation [42]. The mechanisms by which ROS generate hypomethylation are the activation of DNA damage–specific endonucleases, such as those associated with the formation of single-stranded breaks, makes DNA a poor acceptor of methyl groups; exposure to heavy metals results in the formation of premutagenic 8-oxo-2'-deoxyguanosine adducts which strongly inhibit methylation of adjacent cytosines; oxidative stress induces an increase in nicotinamide levels, and through its metabolite trigonelline, NIC can trigger hypomethylation in the genome [43]. Hypomethylation in response to oxidative stress triggered by heavy metal exposure can be caused by either indirect effects of heavy metals or a specific defensive mechanism by which cells regulate gene expression.

The importance of DNA methylation for the maintenance of gene-expression patterns and genome stability is reflected by the fact that DNA-repair mechanisms evolve a specific enzyme to excise methylated cytosines from DNA. ROS1 is a methylated cytosine-specific glycosylase that excises methylated cytosines through the process of base excision repair [44]. This enzyme is rather unique in plants since it combines the function of a DNA-repair enzyme with that of an active demethylating process. Curiously, in the *ros1* mutant, the expression of several transposons was found to be decreased due to an increase in methylation levels at CpNpG and CpNpN sites [45]. Active DNA demethylation is thus important in pruning methylation patterns of the genome, and even previously silent transposons need the dynamic control by methylation and demethylation. Such control is required for the plant epigenome to efficiently respond to developmental and environmental cues. For more detailed information on types of active demethylation processes, see the review by Zhu [44].

Methylation seems to be one of the most versatile epigenetic mechanisms of stress response in plants. The immediate stress response of plant somatic tissues results in changes in methylation of various areas of the genome, with genes involved in stress response being primarily hypomethylated. Exposure to cold causes demethylation and transcriptional activation of a *ZmMI1* gene in maize seedlings; the *ZmMI1* gene contains a retrotransposon-like sequence, and its activation mirrors cold-induced root-specific demethylation in the *Ac/Ds* transposon regions followed by their activation [46]. Hypomethylation in tobacco plants with *NtMET1* antisense results in the upregulation of 31 genes, with most of them being related to stress response [36]. One of the pathogen-responsive genes, *NtAlix1*, undergoes demethylation and activation in response to viral infection, thus confirming that the induction of this gene under natural stress conditions requires sequence demethylation.

The relationship between gene expression and DNA methylation was studied in hypomethylated transgenic tobacco plants expressing an anti-DNA methyltransferase sequence [47]. One of the identified genes coding for a glycerophosphodiesterase-like protein (NtGPDL) was earlier reported to be responsive to aluminum stress. Indeed, when detached leaves from wild-type tobacco plants were treated with aluminum, NtGPDL transcripts were induced within 6h, and the corresponding genomic loci were demethylated at CCGG sites within 1h. Exposure to salt and low temperature, but not to pathogen, induced similar demethylation patterns [47].

Several other reports showed changes in DNA methylation in response to stress. The nuclear genome of *Mesembryanthemum crystallinum* plants underwent a twofold increase in the level of CpNpG methylation in response to high salinity [48]. An increase in methylation was noticed in response of *M. crystallinum* plants to drought and temperature stresses upon switching from C3- to C4-type photosynthesis. A correlation between an age-dependent increase in methylation and resistance to the blight pathogen *Xanthomonas oryzae* in rice was also proposed [49]. Virus infection of tomato plants triggered changes in DNA methylation at several marker loci where the majority of polymorphisms detected were associated with genomic regions involved in defense and stress responses [50]. Exposure to heavy metal stress resulted in hypomethylation at several marker loci in hemp and clover [51]. Verhoeven et al. showed that exposure of an apomictic dandelion population to salicylic acid led to genome-wide and possibly stress-specific changes in DNA methylation in exposed plants which can be faithfully transmitted to the immediate progeny [52].

Intriguingly, secondary effects of changes in DNA methylation may include the altered frequency of genome rearrangements. Reports since 2000 have indicated that a decrease in DNA methylation at given genomic loci could attract genome rearrangements [41,53]. If these observations are accurate, then changes in DNA methylation in response to stress may have a significant impact on the rate of genetic changes in genomic loci targeted by DNA methylation. Thus, changes in DNA methylation in response to stress are not just mandatory, but in fact they could be one of the critical components of transgenerational response and ultimately of the process of directing and accelerating plant genome evolution.

3.2.2 Changes in Transposon Activity Associated With Changes in DNA Methylation and Response to Stress

A methylation-dependent activation of transposons in response to stress is a common phenomenon affecting genome stability. Exposure to cold temperatures decreases DNA methylation and as a consequence increases the rate of excision of the *Tam3* transposon [54]. The mechanism of this event is quite fascinating because the *Tam3* transposase binds the GCHCG (H = not G) sequence immediately after DNA replication and thus prevents de novo sequence methylation. A variety of abiotic and biotic stresses were shown to activate the *Tos17* (rice) [55], *Tto1* (tobacco) [56], *Tnt1* (tobacco) [57], and *BARE-1* (barley) [58] retrotransposons. Three different subfamilies of *Tnt1* retrotransposons showed different tissue-specific activation patterns and demonstrated a different inducibility by pathogen elicitors [57]. The stress-mediated activation of *Tnt1* and *Tto1* retrotransposons presumably occurs through the binding of host transcription factors to their promoter sequences that carry a similarity to sequences found in plant defense gene promoters. Thus, it can be hypothesized that adaptive processes in plants and plant genome evolution may occur through the simultaneous activation of stress-responsive genes and retrotransposons. This hypothesis was supported by the finding that in rice plants exposed to cold, the *mPing* element transposed into a rice homolog of the flowering time gene *CONSTANS* [59]. This event resulted in the alteration of flowering time in the progeny of stressed plants.

Another group of genes altered by transposable elements is the cluster of resistance genes (*R*-genes). These genes are involved in pathogen recognition and resistance due to a specific gene-for-gene interaction. As pathogens try to avoid recognition through mutations of avirulence (*Avr*) genes, plants are forced to use the same procedure with *R*-genes. Thus, there is a constant arm race between pathogens and plants. There exist many mechanisms of *R*-gene evolution, including HR and transposition. It is suggested that a number of transposable elements and their derivatives that are present at the *R*-gene loci play a significant role in a rapid diversification of this gene family [60]. It would be curious to know whether *R*-genes enjoyed a higher frequency of diversification because of the presence of transposons in their sequences, or *R*-genes could be diversified because transposons nonrandomly integrated into the *R*-gene-coding areas. Thus, the reports mentioned earlier support the long-standing hypothesis proposed by Barbara McClintock that all kinds of stresses can potentially reshape plant genomes via transposon activation [61].

3.3 The Role of Histone Modifications in the Maintenance of Genome Stability

Proteins that are associated with histone modifications can be broadly classified into writers, readers, and erasers (Fig. 36.1) [62]. Writers include enzymes involved in modifications such as acetylation, methylation, phosphorylation, and others, and their activity results in local changes in chromatin relaxation or compaction. Readers represent a set of proteins containing bromodomain, chromodomain, or Tudor domains. Most of these proteins cannot directly influence the chromatin structure, they are rather involved in recruiting other chromatin modifiers or erasers of the established chromatin marks. Finally,

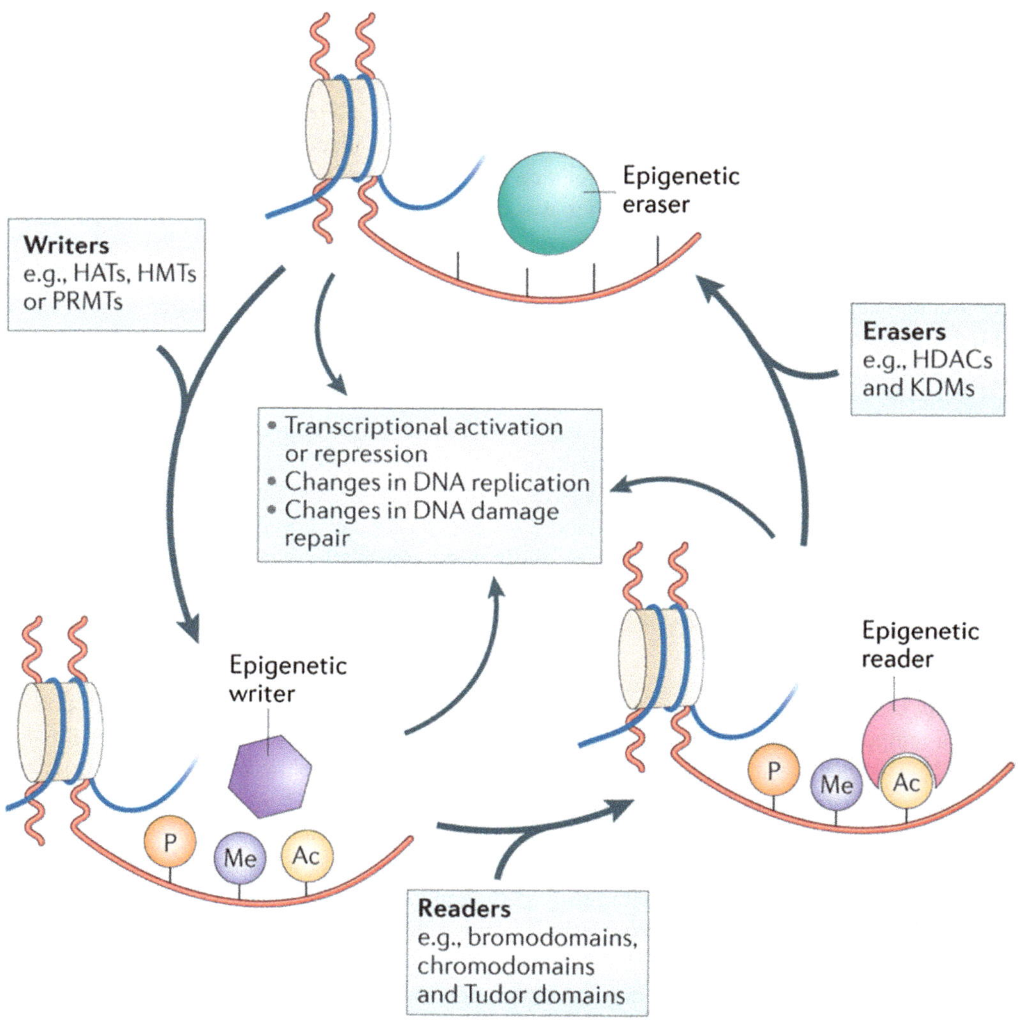

FIGURE 36.1 **Epigenetic writers, readers, and erasers.** The epigenetic regulation is a dynamic process. Epigenetic writers such as histone acetyltransferases (HATs), histone methyltransferases (HMTs), protein arginine methyltransferases (PRMTs), and kinases lay down epigenetic marks on amino acid residues on histone tails. Epigenetic readers such as proteins containing bromodomains, chromodomains, and Tudor domains bind to these epigenetic marks. Epigenetic erasers such as histone deacetylases (HDACs), lysine demethylases (KDMs), and phosphatases catalyze the removal of epigenetic marks. The addition and removal of these posttranslational modifications of histone tails lead to the addition and/or removal of other marks in a highly complicated histone code. Together, histone modifications regulate various DNA-dependent processes, including transcription, DNA replication, and DNA repair. *Reproduced from Falkenberg KJ, Johnstone RW. Histone deacetylases and their inhibitors in cancer, neurological diseases and immune disorders. Nat Rev Drug Discov 2014;13(9):673–91 with permission.*

erasers are proteins that reverse the chromatin marks and represent proteins such as histone deacetylases (HDACs) and histone demethylases (KDMs).

In plants, transcriptionally active chromatin exhibits an enhancement of H3 and H4 acetylation, trimethylation of lysine 4 from histone H3 (H3K4me3), whereas silent chromatin contains hypoacetylated H3 and H4, methylated lysine 27 (H3K27) and lysine 9 of histone H3 (H3K9) [63]. Histone acetyltransferases (HATs) and histone deacetylases modulate the expression of developmental and stress-sensitive genes.

Modifications of histones play many essential roles in the maintenance of genome stability. First, since the chromatin structure is directly correlated with the association of DNA with certain modified histones, it can be predicted that repressive histone marks, such as H3K9me and H3K27me, contribute to genome stability, whereas permissive chromatin marks, such as H3K4me, H3K36me, and H3K9ac, may contribute to genome instability. This notion may not necessarily be true because open chromatin is also associated with a higher DNA-repair capacity. A more detailed analysis of mutation rates associated with open and closed chromatin may be necessary to prove or disprove this hypothesis.

Second, high rates of exchange (removal or addition) of various histone modifications is required for a proper response to stress and DNA damage to allow either a more efficient access to damaged DNA or a more efficient transcription of loci encoding DNA-repair factors. One of the reports of 2014 demonstrated the production of DSB-induced ncRNAs (diRNAs) from genomic regions with strand breaks [64]. It demonstrated an important role of various epigenetic factors in the production of diRNAs and showed that loci with higher transcription rates have a higher frequency of diRNA production and higher repair rates.

Finally, various histone variants are playing a very critical role in DNA repair and genome stability. Phosphorylation of a specific histone variant H2AX resulting in the formation of γH2AX foci is one of the critical steps in the recognition of a strand break and the assembly of DSB-repair factors around the strand break [65].

Writers, readers, and erasers are critical for a proper response to both stress and DNA damage. HDACs, for example, have been implicated in defense against pathogens. HC toxin from *Cochliobolus carbonum* specifically targets the HDAC activity causing histone hyperacetylation in susceptible corn cultivars [66]. It should be noted, however, that among all classes of plant HDACs, proteins from the reduced potassium dependency protein 3/histone deacetylase 1 (RPD3/HDA1) and HD2 classes are the only proteins that become inhibited by the toxin [67]. In *Arabidopsis*, the *AtHDAC19* gene is induced in a similar manner by the fungus *Alternaria brassicicola* and an exogenous application of JA [68]. The overexpression of the *AtHDAC19* gene enhances fungal resistance through the apparent activation of the ethylene-responsive factor 1 (ERF1), whereas silencing of the gene increases fungal susceptibility. The expression of another HDAC, AtHDAC6, was also shown to be induced by JA application [68]. This enzyme was also shown to affect transgene silencing and DNA methylation. *AtHDAC19* is possibly involved in *Arabidopsis* resistance to a bacterial pathogen *Pseudomonas syringae*. The proposed mechanism may involve a decrease in histone acetylation through interactions of HDAC19 with WRKY38 and WRKY62, two transcription factors that repress the SA pathway [69]. The locus-specific suppression of transcription of these two WRKY genes results in the activation of the SA-dependent pathway, and thus in the resistance to bacterial pathogens.

The histone-mediated transcriptional regulation plays an important role in plant protection against stress. Besides various modifications mentioned earlier, histone variants such as H2A.Z were shown to be a part of the response to temperature stress in *Arabidopsis* [69a]. At moderately high temperatures, tight wrapping of H2A.Z and the amount of H2A.Z are reduced at the promoter of heat-responsive genes such as HSP70. A similar effect was observed in *Drosophila* where exposure to temperature stress resulted in nucleosome depletion at HSP70 loci.

3.4 ncRNAs Are Likely Involved in the Regulation of Genome Stability and DNA Repair

Many types of non-coding RNAs (ncRNAs) have been implemented in response to stress and are associated with direct or indirect regulation of genome stability. A more detailed role of ncRNAs in the regulation of DNA repair and genome stability is covered in Chapter 25.

4. TRANSGENERATIONAL RESPONSES

4.1 Types of Transgenerational Effects and Possible Mechanisms of Their Appearance

Transgenerational response is a phenomenon in which plants exhibit changes in the progeny in response to the adverse environment experienced by their parents [70]. Transgenerational changes may include alterations at many levels: DNA methylation and histone modifications, changes in transcriptome, including mRNA and ncRNA transcripts, changes in metabolome and proteome, and in stress tolerance and genome stability (reviewed in [2,71,72]. Such changes may be heritable; although it is still unclear what changes are considered to be heritable, we suggest those alterations that persist for two consecutive generations, S1 and S2, where S1 is the first progeny of plants exposed to stress, and S2 is the second generation of stressed plants. Nonheritable transgenerational changes are typically those that last for a single generation after stress exposure and disappear in the next generation if stressful conditions are not maintained. Most commonly, such changes occur due to differential seed viability/quality caused by the accumulation of metabolites/nutrients that give a certain advantage to plants grown under specific environmental conditions.

Heritability of transgenerational changes depends on the epigenetic regulation. Similarly to animals, in plants, early development reprograms epigenetic modifications such as DNA methylation and histone modifications accumulated during sporophyte development. In contrast to animals, however, reprogramming is less dramatic in plants. While in animals the majority (70–90%) of DNA methylation marks are erased, in plants, most of the marks may be faithfully retained and passed on to progeny [73]. Therefore, heritable epigenetic marks may be responsible for passing the memory of stress exposure across generations.

For a long time, heritable transgenerational changes, often referred to as "soft inheritance," were believed to be impossible or extremely rare. Hard inheritance (or Mendelian inheritance) requires mutations to occur in order to introduce a new trait. Such mutations would have to be beneficial to have a chance of becoming fixed in a population. Since mutations are extremely rare, new traits/species emerge rarely and may require many generations to become common in a certain population [2,74,75]. In contrast, soft inheritance allows an immediate response to the environment, and it is flexible (reversible) allowing the population to respond to the environment frequently and efficiently.

Moreover, it is possible that epigenetic modifications triggered by environmental stimuli are converted to genetic changes (Fig. 36.2). For example, cytosine hypomethylation or the establishment of permissive chromatin marks can lead to the increased frequency of genomic rearrangements, whereas cytosine hypermethylation may result in the increased

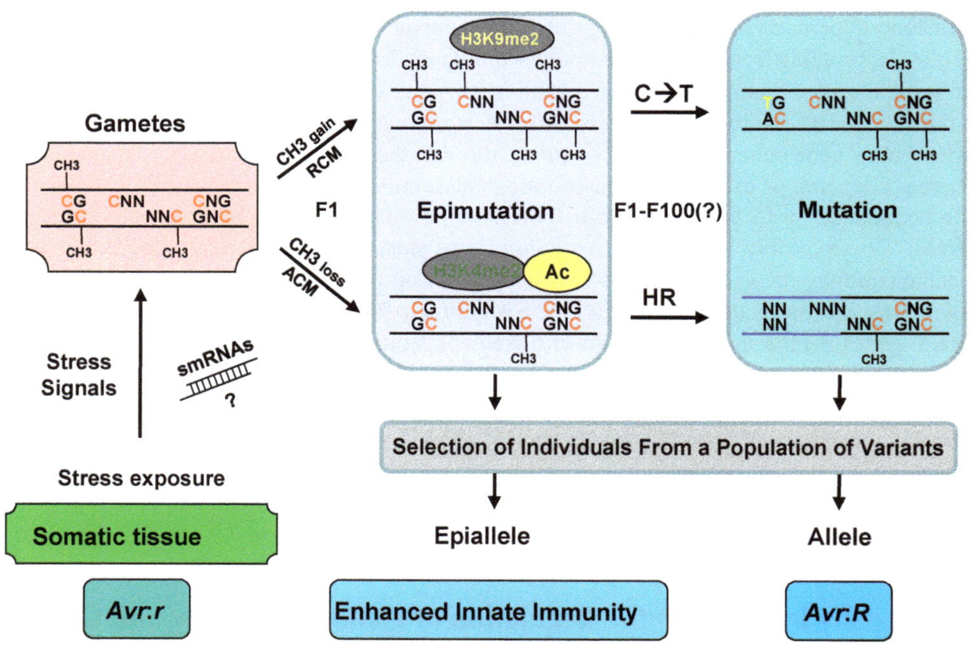

FIGURE 36.2 **Stress-induced epigenetic and genetic changes—an evolutionary perspective.** In the proposed scenario, stress generates mobile signals, for example, smRNAs, that can reach the gametes and influence DNA methylation patterns. The loss or gain of DNA methylation accompanied by repressive chromatin marks (RCMs) or active chromatin marks (ACMs) represent epimutation events. The diagram shows three types of cytosine methylation, CpG, CpNpG, and CNN. H3K9me2 exemplifies the repressive chromatin mark, whereas H3K4me2 and "Ac" (acetylation) exemplify active chromatin marks [50]. The hypermethylated regions are prone to a higher frequency of C to T mutations, whereas the hypomethylated regions have a higher frequency of homologous recombination. It is not clear how many generations are required to translate epigenetic mutations into stable genetic ones. Individuals with (epi)mutations that are beneficial for the growth in the specific environment have better chances to survive and reproduce. Thus, new epialleles and alleles are established in the population. The lower panel applies our scenario to plant–pathogen interactions. It can be hypothesized that compatible pathogen interactions in which plants do not have a functional *R*-gene (*Avr:r*) result in the cascade of the earlier described events. In the short term, epimutations/epialleles allow plants to withstand pathogen encounters through enhanced innate immunity. A long-term strategy requiring exposure to the same pathogen over multiple generations leads to the production of new resistance genes (*Avr:R*) as well as resistance to pathogens due to incompatible interactions. *Reproduced from Boyko A, Kovalchuk I. Genome instability and epigenetic modification–heritable responses to environmental stress? Curr Opin Plant Biol 2011;14(3):260–6.*

frequency of C to T point mutations due to the frequent deamination of methylated cytosine [76,77]. It is therefore prudent to suggest that in many cases, the appearance of new traits and new species in response to adverse conditions is largely driven by epigenetic mechanisms, and genetic mechanisms come second (Fig. 36.2).

4.2 Transgenerational Changes in Response to Abiotic Stress

Abiotic stress can be broadly classified as stress of nonbiological origin, such as temperature changes, water availability, exposure to toxic chemicals or radiation (UV, gamma and so on). Plants respond to stress at many levels, with the main emphasis given to mechanisms of stress survival and setting seeds. Some plants are able to tolerate abiotic stress if it is repeated in the processes known as adaptation and acclimation [78,79]. These processes operate at the somatic level, but they are also known to occur across generations. Abiotic stress is known to destabilize genomes of somatic cells. The number of stresses increases the frequency of HR in a direct or an indirect manner [37,80–82].

Work since 2000 have also demonstrated that a stress-induced increase in the frequency of somatic HR can be inherited [37,41,81,83–85]. It is important to stress out, however, that transgenerational changes in HR frequency were not always found to occur, and they likely depend on many parameters such as tests/measurements utilized, stress conditions, and plant species used. For example, one of the earlier works by Molinier et al. demonstrated that a single exposure of *Arabidopsis thaliana* plants to stress of UV radiation (UVC, specifically) results in the increased frequency of somatic HR in four consecutive nonstressed generations [86], thus representing the truly heritable epigenetic inheritance. In contrast, works by Boyko et al. [85], Kathiria et al. [87] and Rahavi et al. [88] and others provided experimental evidences that the increased frequency of somatic HR is mostly restricted to the immediate progeny of stressed plants, and if stress is not maintained, the frequency of HR drops to the endogenous level observed in unstressed plants. It is possible that the persistence of changes in the frequency of HR observed by Molinier et al. is a unique feature of a particular transgenic line used [86]. For example, the transgenic *A. thaliana* line in which an increase was observed was found to be very unstable without any stress exposure, and a simple propagation of these plants for several generations under normal conditions resulted in a dramatic increase in recombination frequency. Moreover, a work by Pecinka et al. [89] actually shows that a transgenerational increase in HR frequency occurs only in specific transgenic *Arabidopsis* lines tested and only in response to few stresses. The analysis of HR frequency in response to 10 different stresses showed that transgenerational changes occurred in response to two to three stresses, and changes were low and stochastic.

One possible explanation for such discrepancy observed in transgenerational changes in HR frequency could be the intensity of stress used for the analysis. It is possible that only a mild stress may lead to the inheritance of changes in the recombination frequency because a severe stress may have a significant negative effect on plant physiology, somatic cell death, and the negation of epigenetic factors that otherwise would lead to changes in the recombination frequency in progeny. This hypothesis was confirmed by a study that analyzed changes in the HR frequency in response to NaCl; whereas a transgenerational increase in the recombination frequency was most prominent in response to 25 mM NaCl, it was milder in response to 75 mM, and it did not exist in response to 100 mM [90]. This observation was actually consistent with the results published by Pecinka et al. [89]. The existence of response to mild rather than harsh environmental conditions is reminiscent of the long-known phenomenon of hardening in plants. Hardening in plants describes the increased tolerance to a severe stress when plants experienced a mild stress prior to exposure to a severe stress [91]. In case of a transgenerational response, this phenomenon may be referred to as transgenerational hardening.

Our work demonstrated that in most cases transgenerational changes in the recombination frequency occurred only in the immediate progeny; only two stresses tested (25 mM of salt and UVC) increased the recombination frequency in two consecutive generations, and changes in the second generation were smaller than those in the first one [85].

Would the recombination frequency increase more if plants were propagated in the presence of stresses for more than one generation? Would such changes last longer? The answers to these questions were in part obtained from the work of Rahavi et al. [88]. The authors studied changes in the recombination frequency in response to heavy metal salts such as Ni^{2+}, Cd^{2+}, and Cu^{2+}. They propagated plants on heavy metal salts for up to five generations, then stress was removed from them in each generation, starting after generation one. In most cases, an increase in the number of generations exposed to stress resulted in a higher increase in the recombination frequency, although in many cases, this increase reached the plateau already after the first or second generation of exposure. Propagating the progeny of stressed plants under normal conditions resulted in the decreasing recombination frequency, and in those cases where plants were propagated under stress conditions for more generations, a decrease in the HR frequency was less noticeable. In several cases, propagating plants in the presence of stress for three to four generations resulted in an interesting phenomenon—the removal of stress did not decrease the recombination frequency even after two generations of growth under normal conditions. These results indicate that stress memory is inherited, and the more generations are exposed to stress, the stronger and the longer lasting the memory of stress is [88].

Transgenerational responses may also depend on the timing of stress application. Since plants establish the germline relatively late during the development, exposure to stress early during development may allow to pass on the memory of stress application more efficiently (while cells are transitioning to gametes). In contrast, it is likely that stress exposure later during development when gametes are formed may not lead to the efficient generation of stress memory. This is exactly what we have observed in the experiment where we exposed *Arabidopsis* plants to heat, cold, and UVC at different time points during development: 7, 14, 21, and 28 days post germination (dpg). The analysis of HR frequency showed that the highest increase was observed in the progeny of plants exposed at 7 dpg. Similarly, the analysis of plant phenotype in the progeny showed that the progeny of plants exposed at 7 dpg had the largest seeds and the largest leaves when grown under normal conditions and when exposed to stress [92].

What type of genome instability is the most common in the progeny of stressed plants? To address this question, we have used three different transgenic lines, which allowed us to analyze the point mutation frequency, the HR frequency, and microsatellite instability [37]. Exposure to various stresses revealed that changes in the HR frequency were the most prominent among three types of genome instability that we tested. Changes in the microsatellite instability occurred in response to UVC, heat, and cold but were less prominent than changes in the recombination frequency. Finally, changes in the frequency of point mutations in the progeny were only observed in response to UVC, but not in response to any other stress. It would be interesting to analyze why changes in the HR frequency are affected the most in the progeny. HR is a mechanism of crossing over involved in a physical exchange between sister chromatids during meiosis. Such events result in gross chromosomal rearrangements and are likely the most effective in generating novel alleles [93]. If we hypothesize that transgenerational changes in genome stability in response to stress are directed at the diversification of the genome, HR should be a mechanism that is affected the most.

4.2.1 Changes in DNA Methylation in the Progeny

In the preceding paragraphs we have discussed on various epigenetic mechanisms that regulate genome stability and are responsive to stress. Transgenerational changes in DNA methylation in response to stress has been observed in many reports. One of the earliest reports by our laboratory showed that the progeny of plants exposed to ionizing radiation exhibit global genome hypermethylation [94]. Moreover, hypermethylation appeared to be dose dependent; a higher dose of radiation experienced by parental pine tree plants in Chernobyl increased the level of methylation to a higher extent in the progeny [94].

Similarly to the effect of ionizing radiation, exposure to stresses such as salt, flood, heat, cold, and UVC also resulted in hypermethylation in the progeny [85]. When plants were propagated for two generations, DNA methylation did not increase further in the second generation and stayed at the same level as in the first progeny of stressed plants. Curiously, when the progeny of salt-stressed plants were propagated under normal conditions, they maintained higher levels of methylation, but at the same time, they showed the decreased recombination frequency and the lower stress tolerance [85]. This is an interesting phenomenon. We can assume that transgenerational changes are triggered by differential expression of ncRNAs that target various genomic loci to establish differential methylation and differential gene expression, leading to changes in stress tolerance [2]. DNA methylation is maintained at a higher level no matter whether plants are or are not exposed to stress for the second time, while the recombination frequency and stress tolerance depend on the second stress exposure, which suggests that changes in DNA methylation are more robust, they are maintained in the absence of stress, and they are likely in part disconnected from the capacity to tolerate stress.

An overall increase in global genome methylation in the progeny of stressed plants does not reflect the situation at the specific loci in the entire genome because it is a mere reflection of all methylated cytosines present in the genome. A more detailed analysis on the level of individual loci in the progeny of salt-stressed plants revealed that many loci essential for stress tolerance and epigenetic regulation were either hypomethylated or hypermethylated. For example, the promoters of *SUVH2*, *SUVH5*, and *SUVH8* genes that were involved in the regulation of the chromatin structure, and the promoter of *ROS1*, a gene that helps demethylate DNA were hypermethylated, whereas the promoters of stress-responsive genes UVH3, ERF1, TUBG1, RAP2.7, and several others were hypomethylated [84]. The essential role of DNA methylation for the establishment of transgenerational stress tolerance was demonstrated by the fact that soaking seeds of the progeny of salt-stressed plants in 5-azaC, a chemical compound that modifies cytosines by preventing methylation, does not allow plants to tolerate a higher level of MMS chemical and eliminates hypermethylation [85].

4.3 Transgenerational Changes in Genome Stability, Methylation, and Stress Tolerance in Response to Biotic Stress

Biotic stress includes various plant pathogens such as bacteria, fungi, viruses, nematodes, insects, and others. Pathogen infection frequently results in changes in plant physiology, the loss of biomass, early flowering, the decreased seed set, the accumulation of protective metabolites, and many other changes. One of the first evidences that pathogens may destabilize the plant genome comes from the work of Lucht et al. [95]. *Arabidopsis* infection with *Peronospora parasitica* or treatment with chemicals such as 2,6-dichloroisonicotinic acid (INA) or benzothiadiazole (BTH) resulted in an increase in the HR frequency in stressed plants [95]. Later on, the work in our laboratory showed that infection of tobacco plants with TMV also results in an increase in the somatic recombination frequency [96]). Resistance to TMV in tobacco is conferred by the presence of the resistance gene *N* that allows cytoplasmic recognition of the virus. Lines such as Big Havana cultivars that contain the *N*-gene produce a local hypersensitive response and a systemic acquired resistance response that allow plants to localize the virus. Lines that lack the *N*-gene such as SR1 plants succumb to infection. It is important to note that resistance is temperature sensitive; at temperatures exceeding 28°C, lines that contain the *N*-gene also become sensitive. Our work showed that only infection of sensitive plants, either SR1 or Big Havana plants grown at temperatures higher than 28°C, results in the increased somatic recombination frequency. Importantly, the increase was observed in tissues that were not infected with the virus. Moreover, grafting virus-free leaves of infected plants onto naïve tobacco plants also led to an increase in the recombination frequency [96].

Next, we analyzed transgenerational changes in response to TMV. We found that the progeny of infected plants had more plants with a fully recombined luciferase transgene, which indicated that the meiotic recombination frequency also increased. In addition, we found that the somatic recombination frequency in the progeny of infected plants had also an increased rate. Similar effect was observed in response to another virus, oilseed rape mosaic virus (ORMV) [97].

We were curious why the recombination frequency increased only in plants sensitive to TMV. We hypothesized that a boost of HR may be one of the mechanisms for increasing the diversity of resistance genes that potentially leads to the generation of resistance genes that would confer the resistance to TMV. Our analysis of the SR1 genome revealed that despite the fact that SR1 plants do not have an active *N*-gene, they contain many [30–50] loci that carry a substantial (up to 65%) homology to the *N*-gene [41]. The analysis of the rearrangement frequency at these loci in the progeny of infected SR1 plants revealed an over eightfold increase as compared to the progeny of control plants. The same analysis at actin loci did not show any difference, suggesting that an increase in the rearrangement frequency is locus specific [41].

The analysis of DNA methylation in the progeny of infected plants showed global genome hypermethylation. At the same time, a decrease of methylation level was observed at resistance gene-like loci but an increased one at the actin loci [41]. It is highly likely that rearrangements at certain loci are controlled by the level of methylation; hypermethylation may prevent loci that are "irrelevant" to the response to TMV from rearrangement, whereas hypomethylation allows for the recombination and may enable a genetic diversity where it is most needed (Fig. 36.2).

Another important change observed in the progeny of infected plants was a higher tolerance to TMV infection as well as a higher tolerance to *P. syringae* and *Phytophthora nicotianae*. Thus, the progeny of TMV-infected plant have a certain degree of cross-tolerance to bacterial and fungal pathogens. The ability to delay the viral progression is likely triggered by many factors, but our research showed that these plants had a higher endogenous expression of the *PR1* gene and a higher level of callose deposition [87]. These plants were also more tolerant to chemical MMS.

Several studies confirmed our findings that infection with a pathogen leads to changes in the progeny, mainly in the form of a higher tolerance to this pathogen. Luna et al. showed that the progeny of plants that were repeatedly infected with *P. syringae* exhibited a higher tolerance to the same pathogen in the form of a reduced bacterial colonization compared to the progeny of noninfected plants [98]. These plants were also more tolerant to fungal pathogen *Hyaloperonospora arabidopsidis*. A higher pathogen tolerance was also observed in the next generation even when plants were propagated under normal conditions. Slaughter et al. analyzed the impact of a single inoculation with *P. syringae* and found that the immediate progeny had a stronger and quicker response to *Pseudomonas* infection [99]. In contrast to the study by Luna et al., Slaughter et al. found that pathogen tolerance did not persist into the next generation when the progeny of infected plants was propagated without stress [99].

A response to insects also has a transgenerational nature. Wild radish plants exposed to herbivores produce the progeny that are more resistant to herbivory [100]. Also, yellow monkeyflower plants respond to herbivory with an increased trichome density in the progeny; trichome density positively correlates with tolerance to herbivores [101,102]. The analysis of the progeny of *Arabidopsis* and tomato plants that were exposed to caterpillar herbivory showed an enhanced resistance to two out of three herbivores tested [103]. A higher tolerance to herbivores was also observed in the second generation when plants were propagated under normal conditions, but no such tolerance was observed in the third generation [103].

5. POSSIBLE MECHANISMS INVOLVED IN THE REGULATION OF TRANSGENERATIONAL INHERITANCE OF STRESS MEMORY

Which mechanisms control transgenerational changes and in particular changes in genome stability? Several mechanisms may be involved, and our experiments indicate a possible role of DNA-repair proteins and epigenetic regulators such as ncRNAs and DNA cytosine methylation.

5.1 The Potential Role of DNA-Repair Factors

As far as genome stability is concerned, a differential level of metabolites is unlikely to play any role, but it is possible that some differential transcripts accumulated in seeds developed from stressed plants may influence genome stability. For example, an increase in the level of transcript of DNA-repair genes may play a positive role. Our previous analysis indeed showed higher transcript levels of several DNA-repair genes in the progeny [84,85].

Also, it is possible that differential expression of repair genes in plants exposed to stress may also influence the recombination frequency in the progeny. It is not clear, however, whether the presence of all repair factors is essential to observe transgenerational changes in genome rearrangements. For example, an increase in the recombination frequency was observed in wild-type plants *atm* and *rad51b* but not in *ku80* plants in response to several abiotic stressors [83]. In the progeny of stressed plants, an increase in the recombination frequency was observed in wild-type plants and *ku80* mutants, whereas the *atm* mutant was partially impaired. The main changes were observed in *rad51b* mutants; the progeny of these plants completely lacked transgenerational changes in the recombination frequency. It is therefore likely that functional ATM proteins that recognize DSBs and AtRAD51B involved in the HR pathway to repair such breaks are needed for the initiation of a transgenerational signal or its transmission through gametes. Other proteins, such as KU80 and UVH3, appeared to be dispensable for transgenerational changes [83]. It is curious to draw a parallel with recent reports describing the production of diRNAs (see Chapter 25 for details). It was shown that strand breaks trigger the production of diRNA originating from the site of a strand break [64,104]. It was shown that these diRNAs are depleted in *atm* and *atr* mutants as well as in various mutants impaired in epigenetic regulation, namely DCL3, AGO2, and several others. Moreover, it was demonstrated that the NHEJ process (in which KU80 is known to participate) was not impaired when diRNAs were depleted [64,104].

5.2 The Role of Epigenetic Regulators

Several experiments in our laboratory and work of others strongly suggest the involvement of epigenetic regulators in transgenerational changes in genome instability. We have suggested earlier (see Fig. 36.2) that locus-specific changes in DNA methylation in the progeny of stressed plants are likely directed to control the expression of these loci and prevent rearrangements from occurring.

In plants, DNA methylation occurs at various sequence contexts, including symmetrical methylation at CG and CNG sites and asymmetrical methylation at CNN sites. In the latter case, there are no maintenance DNA methylation and de novo methylation events that are established through the function of ncRNAs in a sequence-specific manner. *Arabidopsis* contains four Dicer-like (DCL) proteins, among them, DCL1 that primarily functions in miRNA biogenesis and the other three, DCL2, DCL3, and DCL4, involved in biogenesis of small interfering RNAs (siRNAs) [105]. RdDM occurs through a concerted function of sequence-specific siRNAs, PolIV, DCL3, RDR6, DRM2, and several other proteins involved in two major RdDM pathways, PolIV-RdDM and RDR6-RdDM. More details on de novo RdDM silencing and self-reinforcing loops can be found in Bond and Baulcombe [106]. We hypothesized that the RdDM pathways may be responsible for transgenerational changes in methylation and genome stability.

Several experiments partially confirm our hypothesis. Boyko et al. showed that *dcl2* and *dcl3* mutants were partially impaired in a transgenerational increase in the recombination frequency in response to flood, heat, cold, and UVC as well as in a transgenerational increase in methylation [85]. Also, *dcl2* was partially impaired in transgenerational stress tolerance to MMS [85]. Similar data were reported by Rasmann et al.; the *Arabidopsis dcl2 dcl3 dcl4* triple mutant did not inherit resistance to insects in response to parental herbivory [103]. These reports support the essential role of DCLs and siRNAs in changes in DNA methylation, genome rearrangements, and the transmission of stress memory to progeny. Unlike the two aforementioned studies, report by Ito et al. showed that DCL3 may not be necessary for transgenerational transposition of *ONSEN* [107]. The heat-induced expression of *ONSEN* was higher in *dcl3* plants compared to wild-type plants. The authors suggested that DCL3 may be partially restricting the accumulation of *ONSEN* in response to heat stress in somatic tissues. In contrast, they did not find any new insertions of *ONSEN* in the progeny of heat-stressed *dcl3* plants, which was similar to the finding in wild-type plants.

The mechanism of transgenerational changes may involve several steps. First, stress response includes differential expression of mRNAs, ncRNAs, changes in DNA methylation, and histone modifications in somatic tissues. If stress occurs early during development and influences the whole plant, these changes may occur in meristem cells that will give rise to gametes. If stress occurs when gametes are established, they may also be altered in response to stress. Even if meristem cells or gametes are not altered directly, these cells may acquire the information about stress from all other somatic cells through active functions of plasmodesmata and phloem that circulate in a variety of molecules, including ncRNAs. It is possible that changes in DNA methylation and histone modifications caused by the RdDM mechanism may already occur in meristem cells or early gametes. Second, changes that occur in meristem cells or in the developing gametes have to survive reprogramming, a mechanism that erases the epigenetic marks, such as changes in DNA methylation, histone modifications, and degradation of mRNA in pollen. Epigenetic changes caused by stress also need to survive the second level of reprogramming that occurs after the fertilization event. It is possible that changes in DNA methylation occur in mature gametes or early embryos and are caused by differential expression of ncRNAs produced in gametes or embryos, or even in the endosperm. Third, it is possible that some of the differentially expressed ncRNAs may survive all reprogramming steps and trigger changes directly in the progeny. Our 2015 work in *Brassica rapa* showed that heat stress induces changes in ncRNA and mRNA expression in meristem tissues and gametes; some of these changes were propagated into the developing embryo and even into the progeny [108]. Changes observed in the somatic recombination frequency may be triggered by changes in the chromatin structure (due to either changes in DNA methylation or histone marks) or/and changes in the expression of DNA-repair genes (see earlier). Changes in stress tolerance could be the result of all factors combined, including the differential accumulation of metabolites, changes in DNA methylation, the differential expression of stress-responsive genes, the primed chromatin structure, etc. Fourth, the propagation of stress memory and the maintenance of a high frequency of HR in the next generations may require continuous stress exposure (generation after generation). This is not surprising because if changes in DNA methylation and ncRNAs that trigger it play an essential role, they need to be generated constantly to reinforce transgenerational memory and replenish the molecules depleted during reprogramming. Future research will show whether this theory has merit.

6. CONCLUDING REMARKS

Plants as any other species need to balance between genome stability and genome instability. Whereas the preservation of the genome integrity is important for passing the genetic information on to the progeny, inducing genomic variability in a random fashion or directing it at the defined genomic loci is a prerequisite for the survival of species in adverse environments.

In this chapter, we summarized various mechanisms regulating genome stability, discussed how the choice of DNA-repair pathway influences genome stability, and described the role of epigenetic factors such as DNA methylation, histone modifications, and ncRNAs in controlling genome stability. We described several experimental evidences indicating that the progeny of stressed plants exhibit a variety of changes, including changes in stress tolerance, DNA methylation, and genome stability. We further demonstrated that transgenerational changes are regulated epigenetically. We hypothesized that transgenerational changes are caused by the differential expression of ncRNAs and RdDM mechanisms causing differential changes in DNA methylation and histone modifications. Direct links between a certain type of differentially expressed siRNAs, changes in DNA methylation at specific loci targeted by these siRNAs, and genome stability remain to be established. It is unclear whether such siRNAs are passed from the progeny via gametes, or their expression is induced in the early developing embryo or in the germinated plants. It is also not clear whether the chromatin structure does have a direct effect on genome rearrangements because such links are not evidently established. It is also possible that such siRNAs are propagated in the cytoplasm without the additional transcription. Finally, it remains to be shown whether these siRNAs and rearrangements in the genome are stress specific and are indeed directed towards specific loci in the genome and promoter-specific changes at epigenetic and perhaps genetic levels.

GLOSSARY

Active chromatin marks Posttranslational histone modifications associated with high levels of gene expression and open chromatin

Epigenetic changes Heritable but reversible changes in gene expression that do not involve changes in the DNA sequence. Epigenetic changes are typically associated with reversible modifications of DNA (cytosine methylation) or histones (methylation, acetylation, etc.and so on). The differential expression of non-coding RNAs and sometimes the differential binding of non-histone chromatin modifiers may also be referred to as an epigenetic modification.

Hard or Mendelian inheritance The inheritance of traits based on the DNA sequence; according to Mendelian inheritance, new traits can only appear as a result of a mutation – —changes in the DNA sequence.

Hardening The increased tolerance to severe stress in plants after exposure to mild stress; also referred to as priming, acclimation, conditioning, etc.

Heritable transgenerational effects Transgenerational effects persisting for more than one generation, even when the stimulus causing these changes is removed; these effects are typically associated with changes in the epigenome.

Memory Genetic, epigenetic, or physiological changes that outlast stressful conditions and modify the response to subsequent stress treatments in the same or the next generation (transgenerational memory).

Non-heritable transgenerational effects (responses) Transgenerational effects that do not persist beyond the immediate generation after stress; these changes are not passed on to the subsequent generations if the stressor is removed. Such effects are typically caused by changes in seed quality due to the accumulation of metabolites, nutrients, etc.and so on, giving an advantage to the growing seedling under certain environmental conditions.

Repressive chromatin marks Posttranslational histone modifications associated with low levels of gene expression and condensed chromatin.

Soft inheritance The inheritance of traits that does not include changes in the DNA sequence but rather involves changes in gene expression, typically caused by changes in the epigenetic regulation.

Transgenerational effects (responses) Typically refers to changes in phenotype (associated with epigenetic or physiological changes) that are apparent in the progeny of an organism grown under normal conditions or in response to stress.

Transgenerational hardening A higher stress tolerance in the progeny of plants exposed to mild stress; occurs mostly due to the accumulation of nutrients and metabolites in seeds as well as due to changes in the epigenetic regulation.

LIST OF ABBREVIATIONS

CAF-1 Chromatin-assembly factor-1
DCL Dicer-like
DDM1 DECREASED DNA METHYLATION1
diRNAs DSB-induced ncRNAs
DSB Double-strand break
HATs Histone acetyltransferases
HDACs Histone deacetylases
HP1 HETEROCHROMATIN PROTEIN 1
HR Homologous recombination
KDMs Histone demethylases
MBDs Methyl-CpG-binding domain proteins
MMS Methyl methane sulfonate
MOM1 MAINTENANCE OF METHYLATION 1
ncRNAs Noncoding RNAs
NHEJ Nonhomologous end joining
ORMV Oilseed rape mosaic virus
RdDM RNA-direct DNA methylation
smRNA Small RNA
TMV Tobacco mosaic virus

REFERENCES

[1] Madlung A, Comai L. The effect of stress on genome regulation and structure. Ann Bot 2004;94(4):481–95.

[2] Boyko A, Kovalchuk I. Genome instability and epigenetic modification–heritable responses to environmental stress? Curr Opin Plant Biol 2011;14(3):260–6.

[3] Chinnusamy V, Zhu JK. Epigenetic regulation of stress responses in plants. Curr Opin Plant Biol 2009;12(2):133–9.

[4] Dassler A, Roscher C, Temperton VM, Schumacher J, Schulze ED. Adaptive survival mechanisms and growth limitations of small-stature herb species across a plant diversity gradient. Plant Biol (Stuttg) 2008;10(5):573–87.

[5] Singh SK, Roy S, Choudhury SR, Sengupta DN. DNA repair and recombination in higher plants: insights from comparative genomics of *Arabidopsis* and rice. BMC Genomics 2010;11:443.

[6] Bleuyard JY, Gallego ME, Savigny F, White CI. Differing requirements for the *Arabidopsis* Rad51 paralogs in meiosis and DNA repair. Plant J 2005;41(4):533–45.

[7] Tuteja N, Singh MB, Misra MK, Bhalla PL, Tuteja R. Molecular mechanisms of DNA damage and repair: progress in plants. Crit Rev Biochem Mol Biol 2001;36(4):337–97.

[8] Downey M, Durocher D. Chromatin and DNA repair: the benefits of relaxation. Nat Cell Biol 2006;8(1):9–10.

[9] Lieber MR. The mechanism of double-strand DNA break repair by the nonhomologous DNA end-joining pathway. Annu Rev Biochem 2010;79:181–211.

[10] Chiruvella KK, Liang Z, Wilson TE. Repair of double-strand breaks by end joining. Cold Spring Harb Perspect Biol 2013;5(5):a012757.

[11] Shrivastav M, De Haro LP, Nickoloff JA. Regulation of DNA double-strand break repair pathway choice. Cell Res 2008;18(1):134–47.

[12] Shibata A, Conrad S, Birraux J, Geuting V, Barton O, Ismail A, et al. Factors determining DNA double-strand break repair pathway choice in G2 phase. EMBO J 2011;30(6):1079–92.

[13] Boyko A, Zemp F, Filkowski J, Kovalchuk I. Double-strand break repair in plants is developmentally regulated. Plant Physiol 2006;141(2):488–97.

[14] Tiley GP, Burleigh JG. The relationship of recombination rate, genome structure, and patterns of molecular evolution across angiosperms. BMC Evol Biol 2015;15:194.

[15] Langley CH, Montgomery E, Hudson R, Kaplan N, Charlesworth B. On the role of unequal exchange in the containment of transposable element copy number. Genet Res 1988;52(3):223–35.

[16] Shinozaki K, Yamaguchi-Shinozaki K, Seki M. Regulatory network of gene expression in the drought and cold stress responses. Curr Opin Plant Biol 2003;6(5):410–7.

[17] Sung S, Amasino RM. Vernalization and epigenetics: how plants remember winter. Curr Opin Plant Biol 2004;7(1):4–10.

[18] Zhu Q, Wani AA. Histone modifications: crucial elements for damage response and chromatin restoration. J Cell Physiol 2010;223(2):283–8.

[19] Eskeland R, Eberharter A, Imhof A. HP1 binding to chromatin methylated at H3K9 is enhanced by auxiliary factors. Mol Cell Biol 2007;27(2): 453–65.

[20] Reyes JC, Hennig L, Gruissem W. Chromatin-remodeling and memory factors. New regulators of plant development. Plant Physiol 2002;130(3):1090–101.

[21] Jin B, Li Y, Robertson KD. DNA methylation: superior or subordinate in the epigenetic hierarchy? Genes Cancer 2011;2(6):607–17.

[22] Pecinka A, Dinh HQ, Baubec T, Rosa M, Lettner N, Mittelsten Scheid O. Epigenetic regulation of repetitive elements is attenuated by prolonged heat stress in *Arabidopsis*. Plant Cell 2010;22(9):3118–29.

[23] Tessadori F, van Zanten M, Pavlova P, Clifton R, Pontvianne F, Snoek LB, et al. Phytochrome B and histone deacetylase 6 control light-induced chromatin compaction in *Arabidopsis thaliana*. PLoS Genet 2009;5(9):e1000638.

[24] Havas K, Whitehouse I, Owen-Hughes T. ATP-dependent chromatin remodeling activities. Cell Mol Life Sci 2001;58(5–6):673–82.

[25] Zemach A, Li Y, Wayburn B, Ben-Meir H, Kiss V, Avivi Y, et al. DDM1 binds *Arabidopsis* methyl-CpG binding domain proteins and affects their subnuclear localization. Plant Cell 2005;17(5):1549–58.

[26] Jeddeloh JA, Stokes TL, Richards EJ. Maintenance of genomic methylation requires a SWI2/SNF2-like protein. Nat Genet 1999;22(1):94–7.

[27] Kakutani T, Munakata K, Richards EJ, Hirochika H. Meiotically and mitotically stable inheritance of DNA hypomethylation induced by ddm1 mutation of *Arabidopsis thaliana*. Genetics 1999;151(2):831–8.

[28] Vaillant I, Schubert I, Tourmente S, Mathieu O. MOM1 mediates DNA-methylation-independent silencing of repetitive sequences in *Arabidopsis*. EMBO Rep 2006;7(12):1273–8.

[29] Takeda S, Tadele Z, Hofmann I, Probst AV, Angelis KJ, Kaya H, et al. BRU1, a novel link between responses to DNA damage and epigenetic gene silencing in *Arabidopsis*. Genes Dev 2004;18(7):782–93.

[30] Kanno T, Mette MF, Kreil DP, Aufsatz W, Matzke M, Matzke AJ. Involvement of putative SNF2 chromatin remodeling protein DRD1 in RNA-directed DNA methylation. Curr Biol 2004;14(9):801–5.

[31] Penterman J, Uzawa R, Fischer RL. Genetic interactions between DNA demethylation and methylation in *Arabidopsis*. Plant Physiol 2007;145(4):1549–57.

[32] Maloisel L, Rossignol JL. Suppression of crossing-over by DNA methylation in *Ascobolus*. Genes Dev 1998;12(9):1381–9.

[33] Khrustaleva LI, de Melo PE, van Heusden AW, Kik C. The integration of recombination and physical maps in a large-genome monocot using haploid genome analysis in a trihybrid allium population. Genetics 2005;169(3):1673–85.

[34] Uthup TK, Ravindran M, Bini K, Thakurdas S. Divergent DNA methylation patterns associated with abiotic stress in *Hevea brasiliensis*. Mol Plant 2011;4(6):996–1013.

[35] Mirouze M, Lieberman-Lazarovich M, Aversano R, Bucher E, Nicolet J, Reinders J, et al. Loss of DNA methylation affects the recombination landscape in *Arabidopsis*. Proc Natl Acad Sci USA 2012;109(15):5880–5.

[36] Wada Y, Miyamoto K, Kusano T, Sano H. Association between up-regulation of stress-responsive genes and hypomethylation of genomic DNA in tobacco plants. Mol Genet Genomics 2004;271(6):658–66.

[37] Yao Y, Kovalchuk I. Abiotic stress leads to somatic and heritable changes in homologous recombination frequency, point mutation frequency and microsatellite stability in *Arabidopsis* plants. Mutat Res 2011;707(1–2):61–6.

[38] Saze H, Sasaki T, Kakutani T. Negative regulation of DNA methylation in plants. Epigenetics 2008;3(3):122–4.

[39] Ossowski S, Schneeberger K, Lucas-Lledo JI, Warthmann N, Clark RM, Shaw RG, et al. The rate and molecular spectrum of spontaneous mutations in *Arabidopsis thaliana*. Science 2010;327(5961):92–4.

[40] Kraitshtein Z, Yaakov B, Khasdan V, Kashkush K. Genetic and epigenetic dynamics of a retrotransposon after allopolyploidization of wheat. Genetics 2010;186(3):801–12.

[41] Boyko A, Kathiria P, Zemp FJ, Yao Y, Pogribny I, Kovalchuk I. Transgenerational changes in the genome stability and methylation in pathogen-infected plants: (virus-induced plant genome instability). Nucleic Acids Res 2007;35(5):1714–25.

[42] Cerda S, Weitzman SA. Influence of oxygen radical injury on DNA methylation. Mutat Res 1997;386(2):141–52.

[43] Aina R, Sgorbati S, Santagostino A, Labra M, Ghiani A, Citterio S. Specific hypomethylation of DNA is induced by heavy metals in white clover and industrial hemp. Physiol Plant 2004;121:472–80.

[44] Zhu JK. Active DNA demethylation mediated by DNA glycosylases. Annu Rev Genet 2009;43:143–66.

[45] Lei M, Zhang H, Julian R, Tang K, Xie S, Zhu JK. Regulatory link between DNA methylation and active demethylation in *Arabidopsis*. Proc Natl Acad Sci USA 2015;112(11):3553–7.

[46] Steward N, Kusano T, Sano H. Expression of ZmMET1, a gene encoding a DNA methyltransferase from maize, is associated not only with DNA replication in actively proliferating cells, but also with altered DNA methylation status in cold-stressed quiescent cells. Nucleic Acids Res 2000;28(17):3250–9.

[47] Choi CS, Sano H. Abiotic-stress induces demethylation and transcriptional activation of a gene encoding a glycerophosphodiesterase-like protein in tobacco plants. Mol Genet Genomics 2007;277(5):589–600.

[48] Dyachenko OV, Zakharchenko NS, Shevchuk TV, Bohnert HJ, Cushman JC, Buryanov YI. Effect of hypermethylation of CCWGG sequences in DNA of *Mesembryanthemum crystallinum* plants on their adaptation to salt stress. Biochem (Mosc) 2006;71(4):461–5.

[49] Sha AH, Lin XH, Huang JB, Zhang DP. Analysis of DNA methylation related to rice adult plant resistance to bacterial blight based on methylation-sensitive AFLP (MSAP) analysis. Mol Genet Genomics 2005;273(6):484–90.

[50] Mason G, Caciagli P, Accotto GP, Noris E. Real-time PCR for the quantitation of Tomato yellow leaf curl Sardinia virus in tomato plants and in *Bemisia tabaci*. J Virol Methods 2008;147(2):282–9.

[51] Panella M, Aina R, Renna M, Santagostino A, Palin L. A study of air pollutants and acute asthma exacerbations in urban areas: status report. Environ Pollut 2004;128(3):303. author reply 5.

[52] Verhoeven KJ, Jansen JJ, van Dijk PJ, Biere A. Stress-induced DNA methylation changes and their heritability in asexual dandelions. New Phytol 2010;185(4):1108–18.

[53] Bassing CH, Swat W, Alt FW. The mechanism and regulation of chromosomal V(D)J recombination. Cell 2002;109(Suppl.):S45–55.

[54] Hashida SN, Kitamura K, Mikami T, Kishima Y. Temperature shift coordinately changes the activity and the methylation state of transposon Tam3 in *Antirrhinum majus*. Plant Physiol 2003;132(3):1207–16.

[55] Hirochika H, Sugimoto K, Otsuki Y, Tsugawa H, Kanda M. Retrotransposons of rice involved in mutations induced by tissue culture. Proc Natl Acad Sci USA 1996;93(15):7783–8.

[56] Takeda S, Sugimoto K, Otsuki H, Hirochika HA. 13-bp cis-regulatory element in the LTR promoter of the tobacco retrotransposon Tto1 is involved in responsiveness to tissue culture, wounding, methyl jasmonate and fungal elicitors. Plant J 1999;18(4):383–93.

[57] Beguiristain T, Grandbastien MA, Puigdomenech P, Casacuberta JM. Three Tnt1 subfamilies show different stress-associated patterns of expression in tobacco. Consequences for retrotransposon control and evolution in plants. Plant Physiol 2001;127(1):212–21.

[58] Kalendar R, Tanskanen J, Immonen S, Nevo E, Schulman AH. Genome evolution of wild barley (*Hordeum spontaneum*) by BARE-1 retrotransposon dynamics in response to sharp microclimatic divergence. Proc Natl Acad Sci USA 2000;97(12):6603–7.

[59] Jiang N, Bao Z, Zhang X, Hirochika H, Eddy SR, McCouch SR, et al. An active DNA transposon family in rice. Nature 2003;421(6919):163–7.

[60] Ronald PC. Resistance gene evolution. Curr Opin Plant Biol 1998;1(4):294–8.

[61] McClintock B. The significance of responses of the genome to challenge. Science 1984;226(4676):792–801.

[62] Falkenberg KJ, Johnstone RW. Histone deacetylases and their inhibitors in cancer, neurological diseases and immune disorders. Nat Rev Drug Discov 2014;13(9):673–91.

[63] Kim JM, Sasaki T, Ueda M, Sako K, Seki M. Chromatin changes in response to drought, salinity, heat, and cold stresses in plants. Front Plant Sci 2015;6:114.

[64] Gao M, Wei W, Li MM, Wu YS, Ba Z, Jin KX, et al. Ago2 facilitates Rad51 recruitment and DNA double-strand break repair by homologous recombination. Cell Res 2014;24(5):532–41.

[65] Friesner JD, Liu B, Culligan K, Britt AB. Ionizing radiation-dependent gamma-H2AX focus formation requires ataxia telangiectasia mutated and ataxia telangiectasia mutated and Rad3-related. Mol Biol Cell 2005;16(5):2566–76.

[66] Brosch G, Ransom R, Lechner T, Walton JD, Loidl P. Inhibition of maize histone deacetylases by HC toxin, the host-selective toxin of *Cochliobolus carbonum*. Plant Cell 1995;7(11):1941–50.

[67] Alvarez ME, Nota F, Cambiagno DA. Epigenetic control of plant immunity. Mol Plant Pathol 2010;11(4):563–76.

[68] Zhou C, Zhang L, Duan J, Miki B, Wu K. Histone deacetylase19 is involved in jasmonic acid and ethylene signaling of pathogen response in *Arabidopsis*. Plant Cell 2005;17(4):1196–204.

[69] Kim KC, Lai Z, Fan B, Chen Z. *Arabidopsis* WRKY38 and WRKY62 transcription factors interact with histone deacetylase 19 in basal defense. Plant Cell 2008;20(9):2357–71.

[69a] Kumar SV, Wigge PA. H2A.Z-containing nucleosomes mediate the thermosensory response in *Arabidopsis*. Cell 2010;140(1):136–47.

[70] Boyko A, Kovalchuk I. Epigenetic control of plant stress response. Environ Mol Mutagen 2008;49(1):61–72.

[71] Herman JJ, Sultan SE. Adaptive transgenerational plasticity in plants: case studies, mechanisms, and implications for natural populations. Front Plant Sci 2011;2:102.

[72] Kinoshita T, Seki M. Epigenetic memory for stress response and adaptation in plants. Plant Cell Physiol 2014;55(11):1859–63.

[73] Migicovsky Z, Kovalchuk I. Epigenetic modifications during Angiosperm Gametogenesis. Front Plant Sci 2012;3:20.

[74] Youngson NA, Whitelaw E. Transgenerational epigenetic effects. Annu Rev Genomics Hum Genet 2008;9:233–57.

[75] Mirouze M, Paszkowski J. Epigenetic contribution to stress adaptation in plants. Curr Opin Plant Biol 2011;14(3):267–74.

[76] Becker C, Hagmann J, Muller J, Koenig D, Stegle O, Borgwardt K, et al. Spontaneous epigenetic variation in the *Arabidopsis thaliana* methylome. Nature 2011;480(7376):245–9.

[77] Schmitz RJ, Schultz MD, Lewsey MG, O'Malley RC, Urich MA, Libiger O, et al. Transgenerational epigenetic instability is a source of novel methylation variants. Science 2011;334(6054):369–73.

[78] Hasanuzzaman M, Nahar K, Alam MM, Roychowdhury R, Fujita M. Physiological, biochemical, and molecular mechanisms of heat stress tolerance in plants. Int J Mol Sci 2013;14(5):9643–84.

[79] Tamang BG, Fukao T. Plant adaptation to multiple stresses during submergence and following desubmergence. Int J Mol Sci 2015;16(12):30164–80.

[80] Lebel EG, Masson J, Bogucki A, Paszkowski J. Stress-induced intrachromosomal recombination in plant somatic cells. Proc Natl Acad Sci USA 1993;90(2):422–6.

[81] Boyko A, Hudson D, Bhomkar P, Kathiria P, Kovalchuk I. Increase of homologous recombination frequency in vascular tissue of *Arabidopsis* plants exposed to salt stress. Plant Cell Physiol 2006;47(6):736–42.

[82] Yao Y, Danna CH, Ausubel FM, Kovalchuk I. Perception of volatiles produced by UVC-irradiated plants alters the response to viral infection in naive neighboring plants. Plant Signal Behav 2012;7(7):741–5.

[83] Yao Y, Bilichak A, Titov V, Golubov A, Kovalchuk I. Genome stability of *Arabidopsis* atm, ku80 and rad51b mutants: somatic and transgenerational responses to stress. Plant Cell Physiol 2013;54(6):982–9.

[84] Bilichak A, Ilnystkyy Y, Hollunder J, Kovalchuk I. The progeny of *Arabidopsis thaliana* plants exposed to salt exhibit changes in DNA methylation, histone modifications and gene expression. PLoS One 2012;7(1):e30515.

[85] Boyko A, Blevins T, Yao Y, Golubov A, Bilichak A, Ilnytskyy Y, et al. Transgenerational adaptation of *Arabidopsis* to stress requires DNA methylation and the function of Dicer-like proteins. PLoS One 2010;5(3):e9514.

[86] Molinier J, Ries G, Zipfel C, Hohn B. Transgeneration memory of stress in plants. Nature 2006;442(7106):1046–9.

[87] Kathiria P, Sidler C, Golubov A, Kalischuk M, Kawchuk LM, Kovalchuk I. Tobacco mosaic virus infection results in an increase in recombination frequency and resistance to viral, bacterial, and fungal pathogens in the progeny of infected tobacco plants. Plant Physiol 2010;153(4):1859–70.

[88] Rahavi MR, Migicovsky Z, Titov V, Kovalchuk I. Transgenerational adaptation to heavy metal salts in *Arabidopsis*. Front Plant Sci 2011;2:91.

[89] Pecinka A, Rosa M, Schikora A, Berlinger M, Hirt H, Luschnig C, et al. Transgenerational stress memory is not a general response in *Arabidopsis*. PLoS One 2009;4(4):e5202.

[90] Boyko A, Kovalchuk I. Transgenerational response to stress in *Arabidopsis thaliana*. Plant Signal Behav 2010;5(8):995–8.

[91] Strimbeck GR, Schaberg PG, Fossdal CG, Schroder WP, Kjellsen TD. Extreme low temperature tolerance in woody plants. Front Plant Sci 2015;6:884.

[92] Rahavi SM, Kovalchuk I. Changes in homologous recombination frequency in *Arabidopsis thaliana* plants exposed to stress depend on time of exposure during development and on duration of stress exposure. Physiol Mol Biol Plants 2013;19(4):479–88.

[93] Lieberman-Lazarovich M, Levy AA. Homologous recombination in plants: an antireview. Methods Mol Biol 2011;701:51–65.

[94] Kovalchuk O, Burke P, Arkhipov A, Kuchma N, James SJ, Kovalchuk I, et al. Genome hypermethylation in *Pinus silvestris* of Chernobyl–a mechanism for radiation adaptation? Mutat Res 2003;529(1–2):13–20.

[95] Lucht JM, Mauch-Mani B, Steiner HY, Metraux JP, Ryals J, Hohn B. Pathogen stress increases somatic recombination frequency in *Arabidopsis*. Nat Genet 2002;30(3):311–4.

[96] Kovalchuk I, Kovalchuk O, Kalck V, Boyko V, Filkowski J, Heinlein M, et al. Pathogen-induced systemic plant signal triggers DNA rearrangements. Nature 2003;423(6941):760–2.

[97] Yao Y, Kathiria P, Kovalchuk I. A systemic increase in the recombination frequency upon local infection of *Arabidopsis thaliana* plants with oilseed rape mosaic virus depends on plant age, the initial inoculum concentration and the time for virus replication. Front Plant Sci 2013;4:61.

[98] Luna E, Bruce TJ, Roberts MR, Flors V, Ton J. Next-generation systemic acquired resistance. Plant Physiol 2012;158(2):844–53.

[99] Slaughter A, Daniel X, Flors V, Luna E, Hohn B, Mauch-Mani B. Descendants of primed *Arabidopsis* plants exhibit resistance to biotic stress. Plant Physiol 2012;158(2):835–43.

[100] Agrawal AA. Transgenerational consequences of plant responses to herbivory: an adaptive maternal effect? Am Nat 2001;157(5):555–69.

[101] Colicchio JM, Miura F, Kelly JK, Ito T, Hileman LC. DNA methylation and gene expression in *Mimulus guttatus*. BMC Genomics 2015;16:507.

[102] Holeski LM, Chase-Alone R, Kelly JK. The genetics of phenotypic plasticity in plant defense: trichome production in *Mimulus guttatus*. Am Nat 2010;175(4):391–400.

[103] Rasmann S, De Vos M, Casteel CL, Tian D, Halitschke R, Sun JY, et al. Herbivory in the previous generation primes plants for enhanced insect resistance. Plant Physiol 2012;158(2):854–63.

[104] Wei W, Ba Z, Gao M, Wu Y, Ma Y, Amiard S, et al. A role for small RNAs in DNA double-strand break repair. Cell 2012;149(1):101–12.

[105] Khraiwesh B, Zhu JK, Zhu J. Role of miRNAs and siRNAs in biotic and abiotic stress responses of plants. Biochim Biophys Acta 2012;1819(2):137–48.

[106] Bond DM, Baulcombe DC. Epigenetic transitions leading to heritable, RNA-mediated de novo silencing in *Arabidopsis thaliana*. Proc Natl Acad Sci USA 2015;112(3):917–22.

[107] Ito H, Gaubert H, Bucher E, Mirouze M, Vaillant I, Paszkowski J. An siRNA pathway prevents transgenerational retrotransposition in plants subjected to stress. Nature 2011;472(7341):115–9.

[108] Bilichak A, Ilnytskyy Y, Woycicki R, Kepeshchuk N, Fogen D, Kovalchuk I. The elucidation of stress memory inheritance in *Brassica rapa* plants. Front Plant Sci 2015;6:5.

Chapter 37

Methods for the Detection of DNA Damage

D.V. Firsanov[1,2], L.V. Solovjeva[1], V.M. Mikhailov[1], M.P. Svetlova[1]

[1]*Institute of Cytology of the Russian Academy of Sciences, Saint-Petersburg, Russia;* [2]*Nikiforov Russian Center of Emergency and Radiation Medicine of the EMERCOM of Russia, Saint-Petersburg, Russia*

Chapter Outline

1. INTRODUCTION

The sources of cellular DNA damage can be endogenous or exogenous. Among a variety of exogenous factors causing DNA damage, UV radiation from the sun, environmental chemicals, and ionizing radiation (IR) that can come from natural sources and medical procedures are well known. Endogenous sources of DNA damage are associated mainly with reactive oxygen species (ROS) production during normal cellular metabolism. The damage can be induced both in nuclear and mitochondrial DNA.

Abasic sites (AP sites) appear in DNA as a result of spontaneous hydrolysis of N-glycosidic bonds or the action of DNA glycosylases, whereas ROS leads to the induction of 7,8-dihydro-8-oxogunine (8-oxo-guanine), formamidopyrimidines, ring-saturated pyrimidines such as thymine glycols, and single-strand breaks (SSBs) or double-strand breaks (DSBs) in DNA [1,2].

IR-induced DNA damages include oxidized bases, DSBs, and SSBs: the latter are a predominant form of lesions produced by IR. DSBs are considered to be the most toxic form of DNA damage of living cells. DSBs can also arise after the action of some radiomimetic drugs like bleomycin or the inhibition of topoisomerase reaction.

Two main types of DNA lesions are generated after UV light exposure—cyclobutane pyrimidine dimers (CPD) and pyrimidine (6-4) pyrimidone photoproducts (6-4PP)—that bend the DNA molecule and interfere with DNA replication and transcription [3,4].

Genome Stability. http://dx.doi.org/10.1016/B978-0-12-803309-8.00037-9

Cellular DNA can also be damaged by chemical agents, some of which cause cancer. Chemical carcinogens are classified into two categories: direct- and indirect-acting. Direct-acting carcinogens such as nitrogen mustard, methyl nitrosourea, and dimethyl sulfate react with DNA themselves. Indirect-acting carcinogens such as benzo[a]pyrene, 2-acethylaminofluorene, and 2-naphtylamine must be metabolized before they can react with DNA. Indirect-acting carcinogens can modify DNA bases, forming DNA adducts only after conversion carried out by certain enzymes. DNA adducts lead to the incorporation of incorrect bases in daughter DNA strands during replication and the appearance of new mutations. Mutations in particular genes can induce cancer. For example, different types of cancer show a high incidence of p53 protein mutations that are associated with the loss of wild-type p53 tumor-suppression activity and contribute to malignant transformations [5].

Besides cancer, DNA lesions can lead to other deleterious biological consequences. The accumulation of DNA damage occurs during normal human aging. The development of hereditary diseases associated with the accelerated aging, such as ataxia telangiectasia, Bloom syndrome, Fanconi anemia, and others, is caused by mutations in certain genes.

A number of DNA repair protein systems are activated during cellular DNA damage response (DDR). DNA repair enzymes correct the structure of DNA, prevent the formation of mutations, and maintain DNA integrity.

Single damaged nucleotide bases that arise during DNA oxidation or the action of chemical mutagens on DNA are removed by base excision repair (BER) (see review [6]). BER includes the following steps: the recognition and removal of damaged bases by DNA glycosylase forming apurinic/apyrimidinic sites also known as abasic (AP) sites, the incision of AP site with AP endonuclease, the elimination of a sugar fragment by a phosphodiesterase, repair synthesis by DNA polymerase, and end ligation. A number of different DNA glycosylases that target deaminated, alkylated, and oxidized bases are described.

UV-induced lesions, CPDs and 6-4PPs, are repaired by nucleotide excision repair (NER). More than 30 proteins participate in this type of repair. NER eliminates a broad range of damages including not only CPDs and 6-4PPs but bulky chemical adducts and some forms of oxidative damage. Two sub-pathways of NER, global genomic repair (GGR) and transcription-coupled repair (TCR), are known (reviewed in Refs. [7,8]). Four steps of NER include: the detection of damage, excision of DNA fragment containing a lesion, gap filling with DNA polymerase, and ligation of DNA ends. The main proteins involved in NER include XPA, XPB, XPC, XPD, XPE, XPF, XPG, CSA, and CSB. Mutations in NER proteins are responsible for hereditary diseases such as xeroderma pigmentosum (XP proteins), and Cockayne syndrome (CSA and CSB proteins). TCR and GGR differ in the first step of repair. TCR is initiated when RNA polymerase stalls at the lesions, and CSA and CSB proteins enhance their recognition. XPC is a damage sensor in the GGR sub-pathway of NER.

DSBs represent the most dangerous lesions that arise after IR. In eukaryotic cells, two sub-pathways of DSB repair are known: homologous recombination (HR) and nonhomologous end joining (NHEJ). Most DSBs are eliminated by NHEJ. Classical NHEJ begins with DSB end processing by the MRE11/Rad50/NBS1 (MRN) complex and end binding by Ku70/80 and DNA-PK proteins [9]. The latter phosphorylate other proteins such as RPA, WRN, and Artemis. The break is then sealed by ligase IV and its cofactors XRCC4 and XLF. The back-up sub-pathway of NHEJ involves proteins that also participate in SSB and HR repair—MRN, PARP1, and XRCC1.

HR requires the presence of a homologous sequence as a template for repair. MRN, Exo1, and other nucleases perform the 5′- to 3′-DSB end processing. The BRCA1 protein regulates MRN complex end processing after the phosphorylation and activation of ATM and checkpoint kinases. Single-stranded DNA at the 3′-ends is bound by RPA which further allows Rad52 and Rad51 protein binding. Rad51 is a central protein of HR, and it is phosphorylated by cyclin-dependent kinases which release it to bind single-stranded DNA and form nucleofilaments for homology search in sister chromatids [10,11].

Thus, the DNA repair machinery needs to be constantly active to maintain genome integrity and avoid deleterious biological consequences of DNA damage under conditions of environmental and endogenous stresses.

The detection and measurement of DNA damage is important for experimental research and is applied in clinical assays. Methods for DNA damage detection differ in their sensitivity. This chapter mainly addresses the most sensitive immunological methods that can be used for the detection of DNA damage induced by IR, UV light, and different kinds of chemical agents. Here, we present several protocols for IR damage detection that are routinely used in our laboratory.

2. THE DETECTION OF DSBs IN CULTIVATED MAMMALIAN CELLS AND TISSUES

2.1 Phosphorylated Histone H2AX as a Marker of DSBs

H2AX is known as a variant of histone H2A in mammalian cells. Its phosphorylation on Ser139 is one of the earliest events in DDR to DSBs induced by IR, nuclease action, laser irradiation, and other agents. DNA damage activates cell cycle checkpoints that stop cell cycle progression and give time for the cells to accomplish DSB repair. The members of the phosphatidylinositol 3-kinase family (PI3)—ataxia telangiectasia-mutated (ATM) and ATM-Rad3-related (ATR)—are activated after DSB induction. ATM phosphorylates the SQ/TQ motif of its target proteins in signal transduction pathways leading to cell cycle arrest and DSB repair [12]. H2AX is phosphorylated by ATM at DSB sites in non-S-phase cells, whereas ATR phosphorylates DSBs formed at replication forks stalled at DNA lesions [13].

The phosphorylated form of H2AX (γH2AX) appears in cell nuclei within 3 min after IR and can be visualized as discrete foci using a γH2AX-specific antibody and immunofluorescence microscopy [14]. H2AX phosphorylation spreads to megabase chromatin regions surrounding DSBs. Rothkamm and Lobrich [15] have shown that the number of γH2AX foci per Gy per cell visible in the irradiated cells corresponds to the number of DSBs estimated by PFGE. Thus, the use of γH2AX as a marker of DSBs is advantageous but has some limitations. It is unknown whether γH2AX marks physical DSBs or probably registers DNA sites in which DSB ends were already sealed [16]. After completion of DSB repair, γH2AX could persist for some time at previously damaged sites, for example, due to a slow process of dephosphorylation [17].

Nevertheless, H2AX phosphorylation is successfully used for the study of spatial distribution of γH2AX foci and the estimation of kinetics of focus formation and elimination after IR exposure. The formation of γH2AX can be induced by different chemical compounds as a result of DSB induction associated with replication and repair. Monitoring γH2AX formation and cell-cycle arrest was used for the estimation of genotoxicity of different compounds used in experimental research [18]. γH2AX detection was applied for the evaluation of photogenotoxicity of chemical compounds. For that purpose, cultivated cells were treated with different environmental chemicals (benzene metabolites, photooxidized polycyclic aromatic hydrocarbons, detergents, and other substances) and exposed for 1 h to UVA at the dose correlated to the outdoor level of sunlight exposure in summer day at noon [19]. Using this approach, the risk of photocarcinogenesis after long-term exposure to low concentrations of chemicals can be predicted.

2.1.1 The Study of DSB Repair Kinetics in Cultivated Mammalian Cells

DSB repair kinetics can be studied by counting the number of γH2AX foci in individual cells after exposure to IR or other agents that induce DSBs. Bonner and colleagues [20,21] presented techniques for γH2AX detection in cultivated cells and in a variety of human and mouse tissue samples.

The maximum level of γH2AX formation was observed 1 h after IR exposure in primary human fibroblasts and Chinese hamster cells. The majority of foci were eliminated within 4–5 h after IR at the dose of 1–2 Gy [22–24]. After 7–8 h, the repair was completed, and only rare foci were seen in individual cells. These residual foci were eliminated from DNA within 1–4 days [17]. The number of spontaneous and persistent IR-induced γH2AX foci was increased in senescent cells in culture and in aging mice [25,26].

2.1.2 γH2AX Detection in Tissues of Living Organisms

The formation and elimination of DSBs after IR was studied in different mouse organs, including liver, kidney, and heart [27–30]. DSBs can be induced in noncancerous cells neighboring tumors in a so-called "bystander effect" partially associated with inflammatory cytokine production. In mice implanted with nonmetastasizing tumors, the induction of γH2AX foci was observed in distant tissues [31].

Normal and tumor tissues differ in the rate of γH2AX elimination after IR. In normal tissues, the majority of DNA damage is eliminated during 24 h post IR, while in tumors, the kinetics of γH2AX elimination is slower [32]. The analysis of DSB repair kinetics in different types of cells located in seminiferous tubules of mouse testis after total body irradiation was performed. The kinetics of DSB repair after 1 Gy irradiation was compared in undifferentiated spermatogonia, round spermatids, and somatic cells of various normal tissues. A rapid decrease of γH2AX focus number was observed in somatic tissues: only the low level of residual damage was detected 24 h post irradiation. Round spermatids and spermatogonia showed a highly increased number of γH2AX foci at 5 h post irradiation; the level of remaining foci at 24 and 48 h post irradiation was also significantly elevated indicating an impaired DSB repair capacity in these cells [33]. In mouse xenograft models of human cancer, spatiotemporal tracking of DNA damage was analyzed after intravenous injection of cell-penetrating peptide (Tat) attached covalently to ^{111}In-labeled anti-γH2AX antibody. It was found that ^{111}In-anti-γH2AX-Tat specifically targets DNA DSB and accumulates in irradiated cancer cells in vitro and in tumors in vivo following DNA damage, suggesting the potential use of radio-immunoconjugates that target γH2AX in clinical applications [34].

There are two methodological approaches for the detection of γH2AX in different mammalian tissues: immunohistochemistry and Western blotting.

There are several ways to prepare tissue sections for immunohistochemistry. Sections are prepared in the first way from paraffin-embedded formalin-fixed tissue samples and in the second way from frozen samples (eg, samples frozen in liquid nitrogen). The frozen section preparation is faster than paraffin-embedding technique and allows a good preservation of antigens. The advantage of paraffin-embedded tissue samples is that it is easier to store them. These sections are appropriate for providing details of tissue morphology. The disadvantage of paraffin blocks is that the process of tissue preservation in formalin leads to cross-linking of certain proteins in cells. The unmasking process of cross-linked antigens (antigen retrieval) is critical for binding of the antibody to its target.

The preparation of so-called tissue touch prints is an alternative to tissue section technique. The limitations of this method are the following: first, touch prints can be prepared only from soft tissues, for example, brain and liver; and second, the method of preparation does not allow visualization of tissue morphology. However, this method is easier than the preparation of frozen or paraffin-embedded tissue sections, and we recommend it for a preliminary analysis of kinetics of γH2AX elimination in tissues.

Fluorescence microscopy is used for γH2AX detection in frozen or paraffin-embedded tissue sections. γH2AX foci can be counted in individual cells in tissue sections after using primary antibodies to γH2AX and secondary fluorescently labeled antibodies. The detection of γH2AX can be also performed with the use of the peroxidase antiperoxidase (PAP) complex antibody [35] and visualization of reaction product by light microscopy. In our research, we used primary rabbit anti-γH2AX antibodies, goat antirabbit secondary antibodies followed by the PAP complex antibody produced in rabbit which were bound by secondary antibodies. Alternatively, instead of goat antirabbit IgG of the second layer which binds to the PAP complex antibody, peroxidase-conjugated antirabbit secondary antibodies or biotin-conjugated antirabbit secondary antibodies followed by peroxidase-conjugated streptavidin can be used. This method is less advantageous than fluorescence microscopy because it does not give a possibility to count foci in individual cells but only allows estimating cells as γH2AX positive or negative (the cell is considered to be positive even if it contains a minimum number of DSBs).

It should be taken into account that some tissues contain endogenous peroxidase activity which can react with substrate solution (diaminobenzidine) and produce an undesirable background. This nonspecific background can be reduced by the pretreatment of samples with hydrogen peroxide before incubation with the PAP soluble complex antibody.

Here, we present detailed protocols for frozen tissue section and touch print preparation followed by γH2AX immunostaining used in our laboratory [28,29]. Examples of γH2AX detection in a tissue section and a touch print are presented in Fig. 37.1.

2.1.2.1 Preparation of Tissue Touch Prints

1. After ether narcosis, accurately remove an organ from the animal.
2. Slice the organ into pieces and make a flat cut through a piece of tissue with tweezers and a razor blade.
3. Place the obtained surface onto a slide with polylysine and then remove it leaving a touch print.
4. After touch printing, air dry slides for 15 min to 1 h.

2.1.2.2 Preparation of Frozen Tissue Sections

1. After ether narcosis, accurately remove an organ from the animal.
2. Using tweezers and a razor blade, take an appropriate piece of the organ.

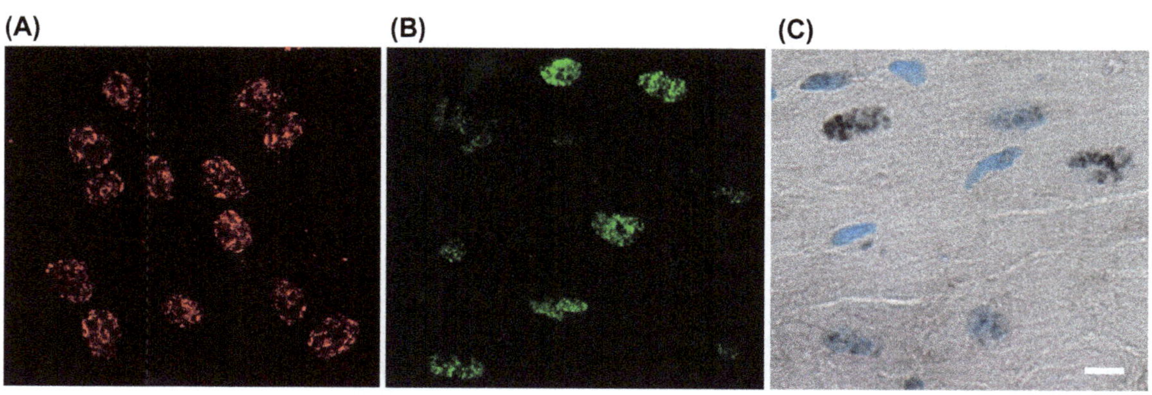

FIGURE 37.1 Immunohistochemical detection of γH2AX: (A) A touch print of Syrian hamster brain, 1 h after 5 Gy of X-ray irradiation. (B) A frozen section of Syrian hamster heart 10 μm in thickness, 1 h after 5 Gy of X-ray irradiation. Fluorescence microscopy images of cells immunostained with a mouse anti-γH2AX monoclonal antibody followed by Alexa Fluor 568 (A) and Alexa Fluor 488 (B)-conjugated secondary antibodies. (C) A frozen section of mouse heart 10 μm in thickness, 20 min after 3 Gy of X-ray irradiation. A conventional light microscopy image of cells immunostained with a peroxidase antiperoxidase soluble complex antibody (*black*). Cellular DNA was counterstained with Giemsa stain (*blue*). The majority of cells in the field of view are γH2AX-positive. The bar is 10 μm.

3. Attach the piece of the organ to a cork made from a stopper of wine bottle.
4. Immediately put the cork with the organ in liquid nitrogen.
5. Using cryostat microtome (we used Bright Co. Ltd., UK), prepare cryostat tissue sections and attach them to polylysine-treated slides.
6. Let slides to air dry for 15 min to 1 h.

2.1.2.3 Fixation and Permeabilization of Touch Prints and Tissue Sections for Fluorescence Microscopy Detection

1. Immerse slides in 2% formaldehyde (Sigma) in PBS for 20 min to fix the cells.
2. Wash slides 3 times for 5 min in PBS with gentle shaking.
3. For tissue touch prints, immerse slides in 70% ethanol chilled to −20°C for 5 min, then go to step 5.
4. For tissue sections, immerse slides in 1% Triton X-100 for 5 min; wash slides 3 times for 5 min in PBS with gentle shaking.
5. Put samples in 70% ethanol at 4°C overnight. At this step, the samples could be stored in ethanol for several days.

2.1.2.4 Fixation and Permeabilization of Tissue Sections for γH2AX Detection Using the PAP Complex Antibody

1. For fixation, immerse slides in methanol:ethanol (1:1) for 3 min.
2. Wash slides 3 times in PBS for 5 min with gentle shaking.
3. Immerse slides in 1% Triton X-100 for 5 min with gentle shaking.
4. Wash slides 2 times in PBS for 15 min with gentle shaking.

2.1.2.5 Immunohistochemical Staining of Touch Prints and Tissue Sections for Fluorescence Microscopy γH2AX Detection

1. Wash slides after ethanol storage (see Section 2.1.2.3) 2 times in PBS for 5 min with gentle shaking.
2. Block nonspecific antibody binding by incubation in PBS supplemented with 8% bovine serum albumin (BSA) at 37°C for 30 min. BSA should be prepared freshly before the experiment.
3. Wash slides in PBS for 5 min.
4. Incubate slides with primary rabbit polyclonal antibodies to γH2AX (1:200, Abcam) in 1% BSA-PBS at 37°C for 1 h. For frozen sections, sometimes it is better to incubate slides overnight at 4°C.
5. Wash slides 3 times in PBS for 5 min with gentle shaking.
6. Incubate slides with Alexa 488 (568)- or FITC-conjugated goat antirabbit antibodies (1:200, Invitrogen) at 37°C for 40 min.
7. Wash slides 3 times in PBS for 5 min with gentle shaking.
8. For nuclear staining, incubate slides with 4′,6-diamidino-2-phenylindole (DAPI) (0.05 μg/mL) in PBS in the darkness for 10 min.
9. Wash slides in PBS for 5 min with gentle shaking.
10. Mount slides in antifade solution. We used Citifluor glycerol/PBS antifade solution (Marivac, Canada).
11. Store slides in a box at 4°C.

2.1.2.6 Immunohistochemical Staining of Tissue Sections for Light Microscopy γH2AX Detection

1. Block endogenous peroxidase activity by immersing slides in 70% methanol containing 0.3% H_2O_2 at room temperature for 30 min.
2. Wash slides 3 times in PBS for 5 min with gentle shaking.
3. Block nonspecific antibody binding by incubation in PBS + 0.1% Tween 20 (PBST) supplemented with 5% bovine serum albumin (BSA-PBST) at room temperature for 30 min.
4. Wash slides 3 times in PBS for 5 min with gentle shaking.
5. Incubate slides with primary rabbit polyclonal antibodies to γH2AX (1:200, Abcam) in 1% BSA-PBST at 4°C overnight.
6. Wash slides 3 times in PBST for 10 min with gentle shaking.
7. Incubate slides with secondary goat antirabbit antibodies in 1% BSA-PBST for 1 h at room temperature.

8. Wash slides 3 times in PBST for 10 min with gentle shaking.
9. Incubate slides in PAP soluble complex antibody produced in rabbit (1:100, Sigma) in 1% BSA-PBST at 4°C overnight.
10. Wash slides 3 times in PBST for 10 min with gentle shaking.
11. Wash slides in PBS for 5 min with gentle shaking.
12. Apply diaminobenzidine (10 mg diaminobenzidine dissolve in 10 mL of PBS and mix with 0.03–0.1% solution of H_2O_2 in PBS) for 5–10 min.
13. Wash slides in running tap water. If the reaction is weak, sections could be further incubated with diaminobenzidine cobalt acetate for 5 min and then washed in running tap water.
14. Stain nuclei with Giemsa stain.
15. Dehydrate slides in the increasing ethanol concentrations (70%, 80%, 96%) for 2–5 min in each.
16. Incubate slides in xylen 3 times for 5 min at room temperature.
17. Embed slides in Canada balsam.

2.1.2.7 Western Blotting of γH2AX in Animal Tissues

Western blotting technique is often used to study the relative efficiency of γH2AX formation in different tissues. To prepare samples for electrophoresis, tissues need to be lysed in order to release the protein of interest. For the detection of nuclear proteins by conventional protocols, it is recommended to use RIPA buffer (radio-immunoprecipitation assay buffer). During lysis, proteolysis may occur which can be slowed down by keeping samples on ice and using appropriate cocktails of inhibitors. The preparation of lysate from a tissue with RIPA buffer takes about 3 h.

Here, we present an easier and faster protocol for the preparation of lysates from tissues with 4x Laemmli buffer without adding proteolysis inhibitors. The main disadvantage of this method is that protein concentration cannot be measured in the samples. However, in our hands, this method works accurate when we weigh tissue samples carefully and add the corresponding volume of buffer. The total protein concentration in the samples can be controlled using antibodies to housekeeping proteins (for example, beta-actin). The densitometry of housekeeping protein bands can be used for the comparison of protein concentrations in different samples.

The steps of the protocol are the following:

1. After ether narcosis, accurately remove an organ from the animal.
2. Using tweezers and a razor blade, take an appropriate piece of the organ.
3. Put the piece of the organ in an eppendorf tube and put it in liquid nitrogen immediately.
4. Using a ceramic mortar and liquid nitrogen, homogenize the piece of the organ to a state of powder.
5. Weigh the powder in an eppendorf tube (an empty tube should be previously weighted) and add 300 μL of 4x SDS-gel-loading buffer with b-MetOH (200 mM Tris–HCl (pH 6.8), 4% SDS, 400 mM b-MetOH, 40% Glycerol, 0.01% Bromphenol blue) to 5 mg of the tissue.
6. Mix well by vortexing and incubate 10 min at 95°C with shaking.
7. Mix well by vortexing and centrifuge at 12,000 rpm for 10 min at 4°C.
8. Take the supernatant and aliquot it into the samples.
9. Store the samples at −70°C. Alternatively, the samples could be stored at −20°C. The samples could be used for gel electrophoresis immediately; however, we recommend to freeze them first.
10. On the day of electrophoresis, take the samples from the freezer and heat them for 2 min at 95°C.
11. Load the samples on a gel along with a molecular weight marker. We recommend to use 15% SDS-PAGE for γH2AX.
12. First, run the gel at 60 V for about 15 min, and then at 120 V.
13. Transfer proteins from the gel to a Hybond-C nitrocellulose membrane (Amersham) by electroblotting for 1 h at 100 V. We recommend to use the Towbin buffer for the transfer (25 mM Tris, 192 mM glycine, 20% (v/v) methanol).
14. Block the membranes in 5% nonfat dry milk in PBS containing 0.1% Tween 20 at 4°C overnight. By our experience, overnight blocking gives the best results.
15. Wash the membranes 2 times in PBST for 10 min with shaking.
16. Incubate the membranes in the primary mouse monoclonal antibody to γH2AX (1:2000, Abcam) for 2 h at room temperature with gentle shaking.
17. Wash the membranes 3 times in PBST for 10 min with shaking.
18. Incubate the membranes with horseradish peroxidase-labeled goat antimouse IgG (1:15,000, Zymax) for 1 h at the room temperature with gentle shaking.
19. Visualize immunoblots with Immobilon Western Chemiluminescent HRP Substrate (Millipore).

2.2 Imaging of DSB Repair Proteins at Chromatin Sites Marked by γH2AX in Cultivated Mammalian Cells

H2AX located at chromatin domains surrounding DSBs is phosphorylated by ATM and DNA-PK. ATM is recruited and activated by the MRN protein complex [36]. When inactive, ATM exists in a dimeric form. In IR-damaged cells, the subunits of the dimer dissociate and become active after phosphorylation [37]. ATM is considered to be a major kinase involved in H2AX phosphorylation after DSB induction [38,39]. It plays a predominant role in γH2AX phosphorylation after IR, but in the case of ATM deficiency, it can be substituted by DNA-PK. It has been shown that H2AX phosphorylation kinetics are normal in $ATM^{-/-}$ and $DNA\text{-}PK^{-/-}$ cells [40,41]. However, the conclusion that ATM and DNA-PK can redundantly substitute each other is not supported by some research groups (reviewed in Ref. [28]).

At the sites of DSBs, γH2AX attracts a number of repair proteins which can be visualized using the corresponding antibodies. Fig. 37.2 represents an example of double immunostaining of γH2AX and 53BP1 or phospho-(Ser 2056)-DNA-PK proteins in G0 Syrian hamster fibroblasts after bleomycin treatment.

Intriguingly, all the proteins we have analyzed (53BP1, phospho-(Ser 2056)-DNA-PK, and phospho-(Ser 1981)-ATM) colocalize with γH2AX at the sites of DSBs. 53BP1 and ATM represent the proteins involved in HR, and DNA-PK is a central protein of NHEJ. It has been reported by P. Jeggo and collaborators that ATM is involved in the repair of approximately 15% of IR-induced DSBs, and 85% of DSBs are repaired in an ATM-independent manner. Moreover, these 15% of DSBs are located at the periphery of the nucleus in the area occupied by heterochromatin [42]. In contrast, we observed a uniform distribution of γH2AX colocalized with phospho-ATM within the nuclei of mammalian cells. It was proposed earlier that proteins involved in NHEJ and HR compete for DSBs [9]. Our observation that proteins of both sub-pathways accumulate at chromatin sites marked by γH2AX supports this suggestion.

In the analysis of protein colocalization in individual cells, the determination of the stage of cell cycle is sometimes necessary. For that purpose, we used staining with antibody to Ki-67, a marker of proliferation, and a nucleoside analog 5-ethynyl-2′-deoxyuridine (EdU) incorporation for the labeling of S-phase cells. Ki-67 nuclear distribution is cell-cycle dependent, and its patterns have been described for normal human fibroblasts and embryonic stem cells [43,44]. G0 cells are Ki-67-negative, and proliferative cells are Ki-67-positive. We recommend using fluorescently labeled EdU for the determination of S-phase cells. Ki-67 staining patterns in G2 cells are easily recognized; these cells contain one or two round-shaped and brightly stained nucleoli on a faint background. Thus, as a result of this approach, G0 cells are $Ki\text{-}67^-/EdU^-$, G1 cells are $Ki\text{-}67^+/EdU^-$, S-phase cells are $Ki\text{-}67^+/EdU^+$, and G2 cells $Ki\text{-}67^+/EdU^-$ (but well distinguished from G1 due to a different distribution of Ki-67). Here, we present protocols for γH2AX/DSB repair protein/Ki-67 immunostaining of cells after incorporation of EdU which is fluorescently labeled

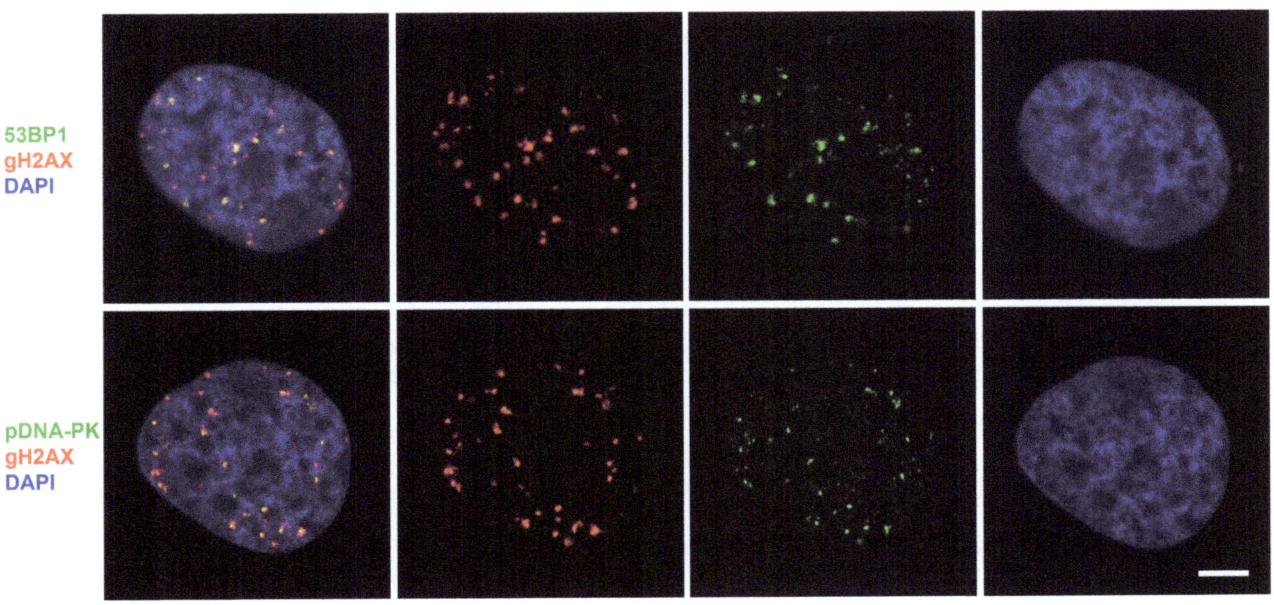

FIGURE 37.2 The visualization of DSBs by double-immunostaining of γH2AX and pDNA-PK or 53BP1 repair proteins in Syrian hamster fibroblasts. DNA in the nuclei is counterstained with DAPI. The bar is 5 μm.

by Click-iT technology with Alexa Fluor dye in a specific reaction developed by ThermoFisher Scientific (Click-iT EdU Alexa Fluor 647 Imaging Kit).

2.2.1 *Immunofluorescence Microscopy Protocol for a Simultaneous Visualization of γH2AX and pDNA-PK/pATM/53BP1 Repair Proteins in Asynchronously Growing Cells That Allows to Discriminate S-phase Cells*

2.2.1.1 Sample Preparation and Fixation

1. The day before experiment, plate 0.3–0.5 × 10^5 cells on cover-glass slides 24 mm × 24 mm in Petri dishes with a 35 mm diameter (the density of plating depends on the cell line used). For growing mammalian fibroblasts, we use Minimal essential medium supplemented with 10–12% fetal bovine serum, 100 U/mL penicillin-streptomycin mixture, and 4 mM L-glutamine. Grow the cells overnight in a CO_2 incubator under conventional conditions of cultivation at 5% CO_2.
2. Treat the cells with IR, bleomycin, or other agent of interest. Keep the cells in a CO_2 incubator for the appropriate time after the treatment. 10–30 min before fixation, add EdU to the growth medium in the optimal concentration for your cell type (10 μM final concentration is recommended by the manufacturer as the optimized concentration for A549, HeLa, and NIH/3T3 cells) (ThermoFisher Scientific, Click-iT EdU Alexa Fluor 647 Imaging Kit).
3. Rinse coverslips twice with PBS and fix with 3.7% formaldehyde in PBS for 15 min at the room temperature. Rinse the cells twice with 3% BSA solution in PBS, permeabilize with 0.5% Triton X-100 in PBS for 20 min at the room temperature, then wash the cells twice with 3% BSA in PBS.

2.2.1.2 Click-iT Technology: Fluorescent Labeling of EdU

Treat the cells on coverslips in the darkness with reagents of Click-iT reaction cocktail including Alexa Fluor 647 azide, triethylammonium salt for 30 min.

For example, for two coverslips, prepare the following mixture of components from the Kit: 86 μL 1x Click-iT reaction buffer working solution; 4 μL $CuSO_4$ solution (100 mM); 0.24 μL Alexa Fluor 647 azide working solution; 10 μL Reaction buffer additive prepared fresh by diluting 10x stock solution in deionized water.

Use 50 μL of this reaction cocktail for each coverslip within 15 min of preparation. We recommend to place the coverslips with the cells on a piece of Parafilm (Sigma) on the bottom of a Petri dish. Lay each coverslip (the surface covered with cells) on a drop of reaction cocktail. After incubation, wash the cells on coverslips twice with 3% BSA in PBS, then proceed to nuclear protein and DNA staining.

2.2.1.3 Immunostaining Procedure

1. Rinse the cells on coverslips in PBS and incubate in 1% Blocking Reagent (Roche, Cat N. 1,096,176) in PBS with 0.02% Tween 20 for 30 min.
2. Dilute all antibodies to final concentrations recommended by the manufacturer in 0.5% Blocking Reagent (Roche) solution with 0.02% Tween 20.
3. Incubate the cells on the coverslips with the mixture of the primary antibodies at 37°C for 1 h. We recommend to place coverslips in a humid chamber on a piece of Parafilm on a drop of antibodies diluted in Step 2 (40–50 μL for each coverslip 24 mm × 24 mm).

   The mixture of the primary antibodies is one of the following:

   Either: rabbit polyclonal anti-gH2AX (Abcam, 1:100) and mouse monoclonal antiphospho-ATM (S1981) (Abcam, 1:100), or mouse monoclonal anti-γH2AX (Millipore (Upstate), 1:200) and rabbit polyclonal anti-53BP1 (1:200), or mouse monoclonal anti-γH2AX (Millipore (Upstate), 1:200) and rabbit polyclonal antiphospho-DNA-PK (S2056) (Abcam, 1:100).

4. Wash the slides twice (for 30 min in total) after incubation by shaking in PBS supplemented with 0.1% Tween 20.
5. Incubate the cells on the coverslips with the mixture of secondary antibodies: Alexa Fluor 568-conjugated polyclonal goat antirabbit IgG (Invitrogen, 1:400) and Alexa Fluor 488-conjugated polyclonal goat antimouse (Invitrogen, 1:400) at 37°C for 40 min.
6. Wash the coverslips for 30 min as described in Step 4.
7. Counterstain the DNA with 0.5 μg/mL DAPI in PBS and mount in an antifade solution. We used Citifluor antifade solution (Marivac, Canada).

2.2.2 Immunofluorescent Detection of γH2AX in Different Phases of Cell Cycle as Determined Using Ki-67 Staining and EdU Incorporation

This protocol is mostly similar to the protocol described in Section 2.2.1. The difference is in the description of double-immunostaining procedure. Here, in Section 2.2.1.3 (Step 3), the content of the mixture of primary antibodies should be the following:

Mouse monoclonal anti-γH2AX (Millipore (Upstate), 1:200) and rabbit polyclonal anti-Ki-67 (Abcam, 1:200).

2.2.3 Immunofluorescence Microscopy Protocol for a Simultaneous Visualization of γH2AX and pDNA-PK/pATM/53BP1 Repair Proteins in Asynchronously Growing Cells That Allows Discriminating Cells in G0, G1, S, and G2 Phases of Cell Cycle

We suggest this protocol for double γH2AX/Ki-67 immunostaining of cells transiently transfected with a GFP-fused repair protein (pDNA-PK, pATM or 53BP1) after EdU incorporation in DNA.

In the majority of steps, this protocol is similar to the protocol described in Section 2.2.1. The difference is that the cells should be transiently transfected with a GFP-conjugated protein of your choice. Follow the protocol (Section 2.2.1) up to the description of immunostaining procedure (Section 2.2.1.3). Here, this step should be the following:

1. This step is the same as in Section 2.2.1.3.
2. This step is the same as in Section 2.2.1.3.
3. Follow Step 3 as described in Section 2.2.1.3 up to the preparation of the mixture of primary antibodies. Here, the mixture of primary antibodies should be the following:

   Mouse monoclonal anti-γH2AX (Millipore, 1:200) and rabbit polyclonal anti-Ki-67 (Abcam, 1:200).
4. This step is the same as in Section 2.2.1.3.
5. Incubate the cells on the coverslips at 37°C for 40 min with the mixture of secondary antibodies: Alexa Fluor 568-conjugated polyclonal goat antimouse IgG (Invitrogen, 1:400) and Alexa Fluor 405-conjugated polyclonal goat antirabbit (ThermoFisher Scientific, Cat No. A-31,556, dilute according to the manufacturer's recommendations).
6. Wash the coverslips for 30 min as described in Step 4.
7. Mount the cells on the slides in an antifade solution. We used Citifluor antifade solution (Marivac, Canada).

2.2.4 Microscopy and Image Acquisition

For image acquisition, the confocal Leica TCS SP5 system equipped with HCX PL APO 100x/1.4 and 40x/1.25 oil immersion objectives, 488 nm argon, 543 nm HeNe, 633 nm HeNe and 405 nm diode lasers and Leica LAS AF software were used.

Red (Alexa Fluor 568), far-red (Alexa Fluor 647), green (Alexa Fluor 488), and blue (DAPI) fluorescence were acquired sequentially to avoid fluorophore emission bleed-through artifacts.

3. γH2AX IN BIODOSIMETRY AND CLINICAL ASSAYS

Historically, the biodosimetry method was represented by cytogenetic dosimetry based on chromosome aberration analysis that offered a reliable means for estimating biological exposure to radiation. The γH2AX assay is a new and powerful tool in biodosimetry. The linear radiation dose-response relationship has been shown in experimental and clinical studies using γH2AX as a biomarker for ionizing radiation exposure. For this kind of analysis, peripheral blood lymphocytes of patients are the most easily obtainable cells. Even after low-dose radiation exposure, the measurement of γH2AX focus number is reliable [20]. Scoring of dicentric chromosomes and analysis of γH2AX focus formation in lymphocytes of healthy donors after computer tomography scans (at dose levels from 0.025 to 1 Gy) demonstrated that both methods are equally sensitive when estimating radiation-induced damage after low-dose IR [45]. γH2AX focus counting was successfully used for the estimation of radiation risks in pediatric patients with congenital heart disease who underwent cardiac catheterization procedures and received radiation doses less than 50 mSv [46]. Flow cytometry was also used to measure H2AX phosphorylation, and it was shown to be a reliable and more rapid approach than γH2AX focus scoring in individual cells [47].

γH2AX assays are helpful for monitoring a patient's progress during the treatment of cancer. The detection of γH2AX foci in lymphocytes can be used for the identification of patients with an increased radiosensitivity who have the highest risk of radiotherapy-related side effects [48]. A promising assay for monitoring the effect of anticancer therapy was based on the measurement of γH2AX in tumor cells dissociated from tumors and circulating in the blood [49]. Many other examples of using γH2AX in cancer research and clinical trials are provided in the review of Ivashkevich and colleagues [50].

4. COMET FLUORESCENCE IN SITU HYBRIDIZATION (COMET-FISH) IN THE DETECTION OF DIFFERENT TYPES OF DNA DAMAGE

The Comet assay is valuable for the elucidation of genotoxicity and DNA repair. It is widely used for the detection of IR lesions, including SSBs and DSBs. Enzyme-modified assays utilizing DNA damage-specific endonucleases such as thymine glycol DNA glycosylase–Endo III and formamidopyrimidine DNA glycosylase (FPG) were proposed for the detection of oxidized DNA lesions using the Comet technique (reviewed in Refs. [51,52]).

A simultaneous visualization of DSBs and SSBs can be performed in a two-tailed Comet assay (TT-comet) using two-dimensional electrophoresis. This modified Comet assay was used, for example, in the analysis of mammalian sperm for the evaluation of the influence of SSBs and DSBs on male infertility [53].

Comet-FISH can be applied to DNA damages induced by IR, different chemical agents, and products of cellular metabolism that can be converted to SSBs or DSBs [54]. It allows the possibility to compare the level of DNA damage and the effectiveness of its repair in certain regions of the genome. Comet-FISH is a combination of two well-known methods: the Comet assay (single-cell gel electrophoresis) and fluorescence hybridization in situ. It is a sensitive and rapid method for the detection of DNA damage. A detailed description of this technique was presented in Methods in Molecular Biology by different authors [54–56].

Briefly, cells placed in low-melting agarose on the surface of microscope slides are subjected to electrophoresis after the action of a damaging agent. In alkaline single-cell electrophoresis, comet tails are formed by DNA loops migrated from cells in the electric field. The size of a comet tail in an individual cell is proportional to the level of DNA damage. The number of fluorescent signals in comet tails gives information about the damage in the gene of interest and its repair.

Using Comet-FISH, the rate of IR-induced DSB repair was compared in TP53 and hTERT genes [57]. The TP53 gene was repaired more rapidly in normal cells than in cells of Cockayne syndrome cell line that was defective in transcription-coupled repair.

Using Comet-FISH technique with labeled single-stranded probes and a specific endonuclease, the transcription-coupled repair of CPDs in the ATM gene was documented in human fibroblasts irradiated at a low UV dose (0.1 J/m^2). 8-oxoG was also preferentially repaired in the transcribed strand of the ATM gene which was revealed using a specific glycosylase in the Comet-FISH assay [58].

The TP53 fragmentation rate was estimated by the alkaline Comet assay and Comet-FISH in lymphocytes of pharmaceutical industry workers after a prolonged exposure to phenylhydrazine, ethylene oxide, dichloromethane, and 1,2-dichloroethane. It was shown that exposure to carcinogens affected the structural integrity of TP53, and the use of personal protective equipment decreased the risk of exposure [59].

Comet-FISH has a potential to be used in understanding the impact of genotoxicity on animal physiology. The effect of hydrogen peroxide on nuclear organizer regions in Pacific oyster *Crassostea gigas* was analyzed by this technique [60].

5. METHODS FOR STUDYING DNA REPAIR AFTER UV

Two different strategies have been used for the detection of UV-induced DNA lesions. Direct methods are based on analytical chemistry requiring first the extraction and then the digestion of DNA followed by the measurement of specific DNA lesions using a specific detector coupled to chromatographic separation. Indirect biochemical methods based on the quantitation of DNA strand breaks produced by UV-lesion repair enzymes have been also developed. Detection methods based on the use of specific antibodies raised against UV lesions and UV-damage repair proteins are mostly used at the present time.

In 1991, Mori and coauthors successfully established new monoclonal antibodies (IgG class) specific to UV-induced DNA damages: CPDs and (6-4)PPs [61]. It was the first report of the simultaneous establishment of monoclonal antibodies raised for different types of UV-induced DNA damages. Antibodies specific for CPDs and (6-4)PPs have been largely used in immunological approaches (ELISA, slot blot technique, fluorescence microscopy), for in vitro and in vivo detection of UV lesions, and for studies of their repair efficiency [62,63]. These approaches offer a number of advantages, including the ability to perform analysis of a large number of samples in ELISA, and the need of a small number of cells for the quantification of immunostained cells under a fluorescent microscope. However, they are limited in sensitivity and cannot detect a relatively small number of repair events which occur within minutes after UV irradiation. A very sensitive non-radioisotopic method for the detection of oligonucleotides excised during NER was developed in 2014 [64]. The excised oligonucleotides isolated from cells were labeled with biotin, separated by gel electrophoresis, transferred to a nylon membrane, and incubated with HRP-conjugated streptavidin for chemiluminescence detection after the immobilization. Using this method, the repair of UV lesions can be detected within 6 min after UV irradiation at the dose of 10 J/m^2.

Both TCR and GGR result in the excision of about 30-base oligonucleotides containing the DNA lesion [65] followed by DNA repair synthesis (DRS) to fill the resulting gap with undamaged nucleotides and ligation [66–68]. Immunofluorescent detection of incorporated halogenated deoxyuridines, 5-iododeoxyuridine (IdU), and 5-chlorodeoxyuridine (CldU), has been used for the analysis of UV-induced DRS in mammalian cells [69,70]. It has been found that when both precursors are added simultaneously to UV-irradiated non-S-phase human fibroblasts and incubated for 2h, they label different sites in the nucleus that might be due to the compartmentalization of I-dUTP and Cl-dUTP pools. In contrast, even very short periods of IdU plus CldU labeling of S-phase cells produced IdU and CldU replication foci that were mostly overlapped [70].

The DRS-dependent incorporation of IdU is very low, but a fluorescent signal can be amplified using the tyramide signal amplification (TSA) system allowing a reliable detection of DRS foci in human cells at a very short duration (10min) of IdU labeling after UV irradiation [69]. The TSA system is an enzyme-mediated detection method that uses horseradish peroxidase (HRP) to generate a high-density labeling of a target protein or nucleic acid in situ. Taking into account that each individual repair synthesis patch is about 30-base long which is not sufficient for DRS detection using indirect immunofluorescence even with the TSA system, it is likely that the detected discrete foci of DRS represent clusters of several DRS patches [69].

In the 1980s, it was found that the proliferating cell nuclear antigen (PCNA) changes its solubility in methanol after its recruitment to DNA in undamaged S-phase cells [71] and becomes insoluble in methanol in G1/G2 cells only after UV irradiation [72], suggesting that its insolubilization may be associated with the involvement of PCNA in DNA resynthesis step of NER. The UV-induced insolubilization after Triton X-100 treatment was observed for XPA and XPB NER proteins [73]. Using a local UV-irradiation technique, a sequential immobilization of NER factors was demonstrated in irradiated spots [74,75].

H2AX phosphorylation in response to IR and chemical drugs attracts the major attention of scientists, but its role in UV-damage response is not completely characterized. It was shown by Halicka et al. that according to data obtained by flow cytometry, the highest degree of H2AX phosphorylation induced by UV occurred in S-phase cells; in G1, G2, and M cells, the degree of H2AX phosphorylation was markedly lower than that in S-phase cells, and it was strongly UV dose-dependent [76]. We demonstrate in Fig. 37.3 that CPD-positive regions colocalize almost completely with γH2AX staining after UV irradiation using polycarbonate filters with pores [77].

Immunofluorescent analysis revealed that H2AX was phosphorylated by ATR kinase at replication forks blocked by UV lesions in S-phase cells [78]. In contrast, UV-induced H2AX phosphorylation in non-S-phase cells did not occur due to DNA DSB formation, but was rather triggered by DNA repair intermediates [79,80].

It was demonstrated that ATR kinase participated in UVC induction of H2AX phosphorylation in nonreplicating cells [80]. Other researchers reported that high doses of UVA irradiation strongly induced H2AX phosphorylation in nuclei that was mediated by c-Jun N-terminal kinase (JNK), and the phosphorylation of H2AX by JNK was associated with the induction of apoptosis [81]. Thus, in non-S-phase cells, H2AX phosphorylation is NER-dependent and associated with ATR, while JNK contributes to H2AX phosphorylation after the induction of apoptosis.

Besides these approaches, a recently suggested method for monitoring the repair of UV-induced (6-4)PPs with a purified DNA damage-binding protein 2 complex (DDB2) should be mentioned. The recognition of UV-damaged DNA by

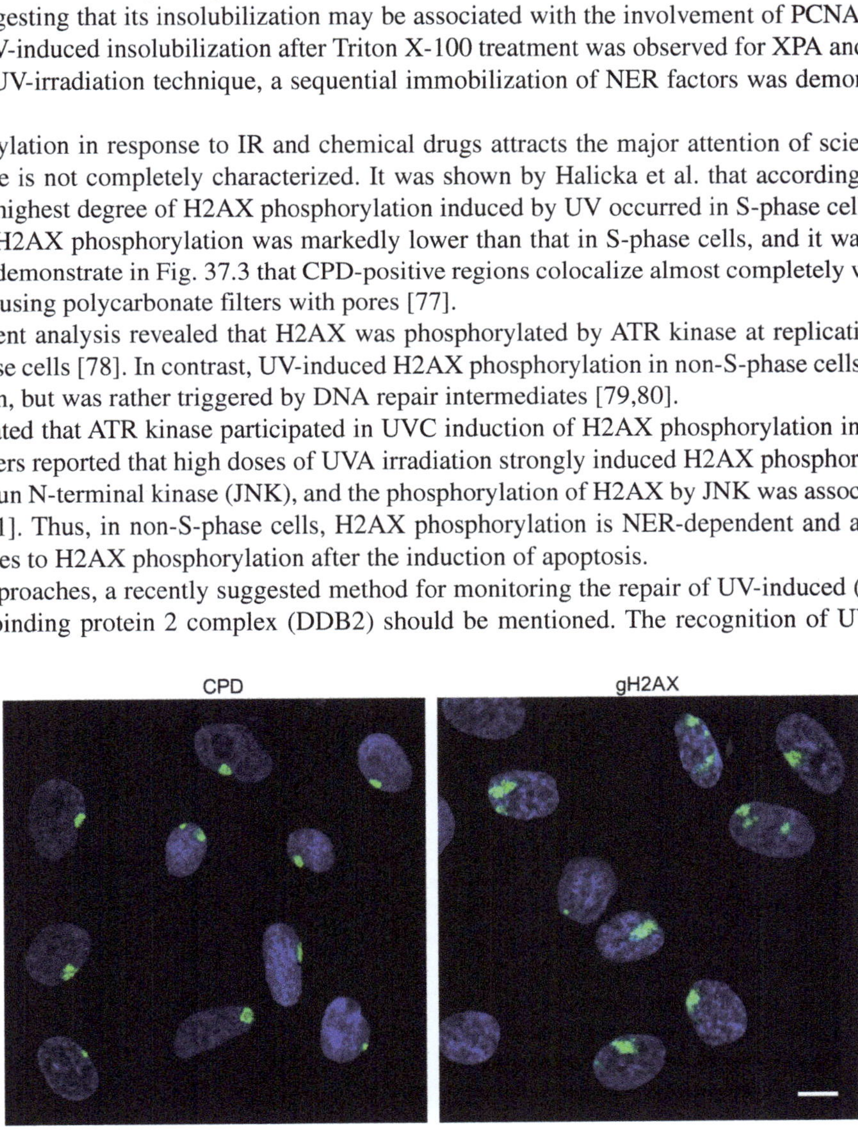

FIGURE 37.3 The visualization of UV damages after local irradiation of Syrian hamster fibroblasts. Irradiation at the dose of 100 J/m^2 was performed using isopore filters with a pore diameter 5 μm placed over the cell monolayer. Fluorescence microscopy images of fibroblasts immunostained with mouse anti-CPD antibody and mouse anti-γH2AX antibody followed by Alexa Fluor 488-conjugated secondary antibodies. The bar is 10 μm.

DDB2 is necessary for the recognition of UV lesions in GGR [82]. DDB2 binds both types of UV lesions, but it has a higher affinity for (6-4)PPs compared to CPDs [83]. FLAG-HA-tagged DDB2 protein (DDB2 proteo-probe) stably expressed in HeLa S3 cells was purified using affinity chromatography and added to cells irradiated by different sources of UV (UVA, UVB, UVC). DDB2 proteo-probe-binded (6-4)PPs that were induced preferentially by UVB and UVC but not by UVA suggest a possible use of this probe for the recognition and monitoring the repair of this certain type of UV lesions [84].

6. CONCLUSIONS

The maintenance of genome stability is important for all living organisms. In this chapter, some examples of DNA damage detection technologies are presented. A particular attention is paid to immunological methods that are commonly used for the detection and quantification of DNA damage. These methods are useful for detecting damages induced by IR, UV, and chemical carcinogens.

Here, we mainly describe γH2AX-based methods for the detection of IR-induced DNA damage and repair. The immunofluorescence microscopy technique allows the visualization and scoring of γH2AX foci and foci of different repair proteins in individual cells, and it is also useful for analysis of DSB repair. We provide several protocols for γH2AX detection in cultivated mammalian cells and tissues which might be useful for readers. The protocols for cultivated mammalian cells include the fixation and double-immunostaining of γH2AX in combination with some DSB repair proteins. The procedure of immunostaining allows the possibility to discriminate cells in different phases of cell cycle. We provide a tissue-processing technique for sectioning and γH2AX immunostaining as well as a detailed description of tissue sample preparation for electrophoresis and immunoblotting which includes some innovations in comparison with the conventionally used techniques.

The indirect approach of DSB detection utilizing γH2AX assays is important not only in basic research but also in clinical practice. Anticancer therapeutic strategies are mainly based on introducing DSBs in cancer cells. While some anticancer compounds induce DSBs directly, others produce non-DSB types of lesions that can lead to DSB formation during the process of DNA repair. The measurement of cellular γH2AX levels can be used for the estimation of chemotherapy effectiveness and radiosensitivity of patients and for the prediction of anticancer treatment toxicity. It is reasonable to expect that the importance of γH2AX assays will continue to increase in the future, and γH2AX techniques will be improved, thus giving an opportunity to broaden the scope of their applications in clinical trials.

GLOSSARY

DDB2 proteo-probe Represents the UV-induced DNA damage recognition protein DDB2 fused with FLAG-HA tag. The probe recognizes the UV-irradiated DNA in a number of assays, including cytochemistry, histochemistry, flow cytometry, slot-blotting, and DNA pull-down assays [84].

LIST OF ABBREVIATIONS

6-4PP Pyrimidine (6-4) pyrimidone photoproducts
8-oxo-guanine 7,8-Dihydro-8-oxogunine
AP sites Apurinic/apyrimidinic (abasic) sites
ATM Ataxia telangiectasia mutated
ATR ATM-Rad3 related
BER Base excision repair
BSA-PBST PBST supplemented with 5% bovine serum albumin
CldU 5-Chlorodeoxyuridine
Comet-FISH Comet fluorescence in situ hybridization
CPD Cyclobutane pyrimidine dimer
DAPI 4′,6-Diamidino-2-phenylindole
DDB2 DNA damage-binding protein 2 complex
DDR Cellular DNA damage response
DSBs Double-strand DNA breaks
EdU 5-Ethynyl-2′-deoxyuridine
GGR Global genomic repair
HR Homologous recombination
IdU 5-Iododeoxyuridine
IR Ionizing radiation

JNK c-Jun N-terminal kinase
MRN MRE11/Rad50/NBS1 complex
NER Nucleotide excision repair
NHEJ Nonhomologous end joining
PAP Peroxidase antiperoxidase
PBST PBS+0.1% Tween 20
PI3 Phosphatidylinositol 3-kinase family
ROS Reactive oxygen species
SSBs Single-strand DNA breaks
TCR Transcription-coupled repair
TSA Tyramide signal amplification
γH2AX Phosphorylated form of H2AX

ACKNOWLEDGMENTS

We are grateful to Dasha Zaytseva for careful proofreading of the manuscript. This work was supported by the grant from the Russian Foundation for Basic Research 13-04-00442a.

REFERENCES

[1] Dizdaroglu M. Oxidative damage to DNA in mammalian chromatin. Mutat Res 1992;275(3–6):331–42.
[2] Cooke MS, et al. Oxidative DNA damage: mechanisms, mutation, and disease. FASEB J 2003;17(10):1195–214.
[3] Sinha RP, et al. UV-induced DNA damage and repair: a review. Photochem Photobiol Sci 2002;1(4):225–36.
[4] Rastogi RP, et al. Molecular mechanisms of ultraviolet radiation-induced DNA damage and repair. J Nucleic Acids 2010;2010:592980.
[5] Muller PA, et al. Mutant p53 in cancer: new functions and therapeutic opportunities. Cancer Cell 2014;25(3):304–17.
[6] Kim Y-J, et al. Overview of base excision repair biochemistry. Curr Mol Pharmacol 2012;5(1):3–13.
[7] Hanawalt PC. Subpathways of nucleotide excision repair and their regulation. Oncogene 2002;21(58):8949–56.
[8] Spivak G, et al. The complex choreography of transcription-coupled repair. DNA Repair (Amst) 2014;19:64–70.
[9] Shrivastav M, et al. Regulation of DNA double-strand break repair pathway choice. Cell Res 2008;18(1):134–47.
[10] Polo SE, et al. Dynamics of DNA damage response proteins at DNA breaks: a focus on protein modifications. Genes Dev 2011;25(5):409–33.
[11] Summers KC, et al. Phosphorylation: the molecular switch of double-strand break repair. Int J Proteomics 2011;2011:373816.
[12] Osborn AJ, et al. Checking on the fork: the DNA-replication stress-response pathway. Trends Cell Biol 2002;12(11):509–16.
[13] Zou L, et al. Sensing DNA damage through ATRIP recognition of RPA-ssDNA complexes. Science 2003;300(5625):1542–8.
[14] Rogakou EP, et al. DNA double-stranded breaks induce histone H2AX phosphorylation on serine 139. J Biol Chem 1998;273(10):5858–68.
[15] Rothkamm K, et al. Evidence for a lack of DNA double-strand break repair in human cells exposed to very low x-ray doses. Proc Natl Acad Sci USA 2003;100(9):5057–62.
[16] Kinner A, et al. Gamma-H2AX in recognition and signaling of DNA double-strand breaks in the context of chromatin. Nucleic Acids Res 2008;36(17):5678–94.
[17] Löbrich M, et al. gammaH2AX foci analysis for monitoring DNA double-strand break repair: strengths, limitations and optimization. Cell Cycle 2010;9(4):662–9.
[18] Ando M, et al. Usefulness of monitoring γ-H2AX and cell cycle arrest in HepG2 cells for estimating genotoxicity using a high-content analysis system. J Biomol Screen 2014;19(9):1246–54.
[19] Ibuki Y, et al. Evaluation of chemical phototoxicity, focusing on phosphorylated histone H2AX. J Radiat Res 2015;56(2):220–8.
[20] Redon CE, et al. γ-H2AX detection in peripheral blood lymphocytes, splenocytes, bone marrow, xenografts, and skin. Methods Mol Biol 2011;682:249–70.
[21] Nakamura A, et al. Techniques for gamma-H2AX detection. Methods Enzymol 2006;409:236–50.
[22] Nazarov IB, et al. Dephosphorylation of histone gamma-H2AX during repair of DNA double-strand breaks in mammalian cells and its inhibition by calyculin A. Radiat Res 2003;160(3):309–17.
[23] Svetlova M, et al. Elimination of radiation-induced gamma-H2AX foci in mammalian nucleus can occur by histone exchange. Biochem Biophys Res Commun 2007;358(2):650–4.
[24] Solovjeva LV, et al. Forskolin decreases phosphorylation of histone H2AX in human cells induced by ionizing radiation. Radiat Res 2009;171(4):419–24.
[25] Endt H, et al. Detailed analysis of DNA repair and senescence marker kinetics over the life span of a human fibroblast cell line. J Gerontol A Biol Sci Med Sci 2011;66(4):367–75.
[26] Sedelnikova OA, et al. Delayed kinetics of DNA double-strand break processing in normal and pathological aging. Aging Cell 2008;7(1):89–100.
[27] Koike M, et al. Tissue-specific DNA-PK-dependent H2AX phosphorylation and gamma-H2AX elimination after X-irradiation in vivo. Biochem Biophys Res Commun 2008;376(1):52–5.
[28] Firsanov DV, et al. H2AX phosphorylation at the sites of DNA double-strand breaks in cultivated mammalian cells and tissues. Clin Epigenetics 2011;2(2):283–97.

[29] Firsanov D, et al. Dynamics of γH2AX formation and elimination in mammalian cells after X-irradiation. Biochimie 2012;94(11):2416–22.

[30] Gavrilov B, et al. Slow elimination of phosphorylated histone gamma-H2AX from DNA of terminally differentiated mouse heart cells in situ. Biochem Biophys Res Commun 2006;347(4):1048–52.

[31] Redon CE, et al. Tumors induce complex DNA damage in distant proliferative tissues in vivo. Proc Natl Acad Sci USA 2010;107(42):17992–7.

[32] Bañuelos CA, et al. gammaH2AX expression in tumors exposed to cisplatin and fractionated irradiation. Clin Cancer Res 2009;15(10):3344–53.

[33] Rübe CE, et al. Protecting the heritable genome: DNA damage response mechanisms in spermatogonial stem cells. DNA Repair (Amst) 2011;10(2):159–68.

[34] Cornelissen B, et al. Imaging DNA damage in vivo using gammaH2AX-targeted immunoconjugates. Cancer Res 2011;71(13):4539–49.

[35] Polak JM, et al. Introduction to immunocytochemistry. 2nd ed. Oxford (UK): BIOS Scientific Publishers Ltd.; 1997. ISBN: 1-85996-086-3.

[36] Berkovich E, et al. Roles of ATM and NBS1 in chromatin structure modulation and DNA double-strand break repair. Nat Cell Biol 2007;9(6): 683–90.

[37] Iijima K, et al. Dancing on damaged chromatin: functions of ATM and the RAD50/MRE11/NBS1 complex in cellular responses to DNA damage. J Radiat Res 2008;49(5):451–64.

[38] Burma S, et al. ATM phosphorylates histone H2AX in response to DNA double-strand breaks. J Biol Chem 2001;276(45):42462–7.

[39] Takahashi A, et al. ATM is the predominant kinase involved in the phosphorylation of histone H2AX after heating. J Radiat Res 2010;51(4):417–22.

[40] Stiff T, et al. ATM and DNA-PK function redundantly to phosphorylate H2AX after exposure to ionizing radiation. Cancer Res 2004;64(7):2390–6.

[41] Wang H, et al. Complex H2AX phosphorylation patterns by multiple kinases including ATM and DNA-PK in human cells exposed to ionizing radiation and treated with kinase inhibitors. J Cell Physiol 2005;202(2):492–502.

[42] Goodarzi AA, et al. The influence of heterochromatin on DNA double strand break repair: getting the strong, silent type to relax. DNA Repair (Amst) 2010;9(12):1273–82.

[43] Kill IR. Localisation of the Ki-67 antigen within the nucleolus. Evidence for a fibrillarin-deficient region of the dense fibrillar component. J Cell Sci 1996;109(6):1253–63.

[44] Ghule PN, et al. Cell cycle dependent phosphorylation and subnuclear organization of the histone gene regulator p220(NPAT) in human embryonic stem cells. J Cell Physiol 2007;213(1):9–17.

[45] Golfier S, et al. Dicentric chromosomes and gamma-H2AX foci formation in lymphocytes of human blood samples exposed to a CT scanner: a direct comparison of dose response relationships. Radiat Prot Dosimetry 2009;134(1):55–61.

[46] Beels L, et al. gamma-H2AX foci as a biomarker for patient X-ray exposure in pediatric cardiac catheterization: are we underestimating radiation risks? Circulation 2009;120(19):1903–9.

[47] Viau M, et al. Global quantification of γH2AX as a triage tool for the rapid estimation of received dose in the event of accidental radiation exposure. Mutat Res Genet Toxicol Environ Mutagen 2015;793:123–31.

[48] Kroeber J, et al. Distinct increased outliers among 136 rectal cancer patients assessed by γH2AX. Radiat Oncol 2015;10:36.

[49] Wang LH, et al. Monitoring drug-induced gammaH2AX as a pharmacodynamic biomarker in individual circulating tumor cells. Clin Cancer Res 2010;16(3):1073–84.

[50] Ivashkevich A, et al. Use of the γ-H2AX assay to monitor DNA damage and repair in translational cancer research. Cancer Lett 2012;327(1–2): 123–33.

[51] de Lapuente J, et al. The Comet Assay and its applications in the field of ecotoxicology: a mature tool that continues to expand its perspectives. Front Genet 2015;6:180.

[52] Shaposhnikov S, et al. Fluorescent in situ hybridization on comets: FISH comet. Methods Mol Biol 2015;1288:363–73.

[53] Cortés-Gutiérrez EI. Interpreting sperm DNA damage in a diverse range of mammalian sperm by means of the two-tailed comet assay. Front Genet 2014;5:404.

[54] Spivak G. The Comet-FISH assay for the analysis of DNA damage and repair. Methods Mol Biol 2010;659:129–45.

[55] Spivak G, et al. New applications of the Comet assay: comet-FISH and transcription-coupled DNA repair. Mutat Res 2009;681(1):44–50.

[56] Glei M, et al. Analysis of DNA damage and repair by comet fluorescence in situ hybridization (Comet-FISH). Methods Mol Biol 2014;1094:39–48.

[57] McKenna DJ. Use of the comet-FISH assay to compare DNA damage and repair in p53 and hTERT genes following ionizing radiation. PLoS One 2012;7(11):e49364.

[58] Guo J. Comet-FISH with strand-specific probes reveals transcription-coupled repair of 8-oxoGuanine in human cells. Nucleic Acids Res 2013;41(16):7700–12.

[59] Zeljezic D, et al. Evaluation of genome damage in subjects occupationally exposed to possible carcinogens. Toxicol Ind Health 2015. pii:0748233714568478, [in press].

[60] Pérez-García C, et al. Development of a comet-FISH assay for the detection of DNA damage in hemocytes of *Crassostrea gigas*. Aquat Toxicol 2015;161:189–95.

[61] Mori T, et al. Simultaneous establishment of monoclonal antibodies specific for either cyclobutane pyrimidine dimer or (6-4) photoproduct from the same mouse immunized with ultraviolet-irradiated DNA. Photochem Photobiol 1991;54(2):225–32.

[62] Perdiz D, et al. Distribution and repair of bipyrimidine photoproducts in solar UV-irradiated mammalian cells. Possible role of Dewar photoproducts in solar mutagenesis. J Biol Chem 2000;275(35):26732–42.

[63] Nikitaki Z, et al. Stress-induced DNA damage biomarkers: applications and limitations. Front Chem 2015;3:35.

[64] Choi JH, et al. Highly specific and sensitive method for measuring nucleotide excision repair kinetics of ultraviolet photoproducts in human cells. Nucleic Acids Res 2014;42(4):e29.

[65] Bowman KK, et al. Excision-repair patch lengths are similar for transcription-coupled repair and global genome repair in UV-irradiated human cells. Mutat Res 1997;385(2):95–105.

[66] Hanawalt PC. Controlling the efficiency of excision repair. Mutat Res 2001;485(1):3–13.

[67] Hoeijmakers JH. Genome maintenance mechanisms for preventing cancer. Nature 2001;411(6835):366–74.

[68] Wood RD. DNA damage recognition during nucleotide excision repair in mammalian cells. Biochimie 1999;81(1–2):39–44.

[69] Svetlova M, et al. Clustered sites of DNA repair synthesis during early nucleotide excision repair in ultraviolet light-irradiated quiescent human fibroblasts. Exp Cell Res 2002;276(2):284–95.

[70] Solovjeva L, et al. Differential incorporation of halogenated deoxyuridines during UV-induced DNA repair synthesis in human cells. DNA Repair (Amst) 2005;4(3):359–66.

[71] Madsen P, et al. S-phase patterns of cyclin (PCNA) antigen staining resemble topographical patterns of DNA synthesis. A role for cyclin in DNA replication? FEBS Lett 1985;193(1):5–11.

[72] Celis JE, et al. Increased nuclear cyclin/PCNA antigen staining of non S-phase transformed human amnion cells engaged in nucleotide excision DNA repair. FEBS Lett 1986;209(2):277–83.

[73] Svetlova M, et al. Reduced extractability of the XPA DNA repair protein in ultraviolet light-irradiated mammalian cells. FEBS Lett 1999;463(1–2):49–52.

[74] Volker M, et al. Sequential assembly of the nucleotide excision repair factors in vivo. Mol Cell 2001;8(1):213–24.

[75] Wakasugi M, et al. DDB accumulates at DNA damage sites immediately after UV irradiation and directly stimulates nucleotide excision repair. J Biol Chem 2002;277(3):1637–40.

[76] Halicka HD, et al. Histone H2AX phosphorylation after cell irradiation with UV-B: relationship to cell cycle phase and induction of apoptosis. Cell Cycle 2005;4(2):339–45.

[77] Moné MJ, et al. Local UV-induced DNA damage in cell nuclei results in local transcription inhibition. EMBO Rep 2001;2(11):1013–7.

[78] Ward IM, et al. Histone H2AX is phosphorylated in an ATR-dependent manner in response to replicational stress. J Biol Chem 2001;276(51):47759–62.

[79] Marti TM, et al. H2AX phosphorylation within the G1 phase after UV irradiation depends on nucleotide excision repair and not DNA double-strand breaks. Proc Natl Acad Sci USA 2006;103(26):9891–6.

[80] Hanasoge S, et al. H2AX phosphorylation after UV irradiation is triggered by DNA repair intermediates and is mediated by the ATR kinase. Carcinogenesis 2007;28(11):2298–304.

[81] Lu C, et al. Cell apoptosis: requirement of H2AX in DNA ladder formation, but not for the activation of caspase-3. Mol Cell 2006;23(1):121–32.

[82] Moser J, et al. The UV-damaged DNA-binding protein mediates efficient targeting of the nucleotide excision repair complex to UV-induced photo lesions. DNA Repair 2005;4(5):571–82.

[83] Treiber DK, et al. An ultraviolet light-damaged DNA recognition protein absent in xeroderma pigmentosum group E cells binds selectively to pyrimidine (6-4) pyrimidone photoproducts. Nucleic Acids Res 1992;20(21):5805–10.

[84] Dreze M, et al. Monitoring repair of UV-induced 6-4-photoproducts with a purified DDB2 protein complex. PLoS One 2014;9(1):e85896.

Chapter 38

Conserved and Divergent Features of DNA Repair: Future Perspectives in Genome Instability Research

I. Kovalchuk
University of Lethbridge, Lethbridge, AB, Canada

Chapter Outline

1. AN OVERVIEW AND COMPARISON OF DNA-REPAIR PATHWAYS IN DIFFERENT ORGANISMS

Since mid-1980s, a significant progress has been made in discovering and describing in detail various mechanisms of DNA repair in different species, including bacteria and humans. The conservation of various DNA-repair mechanisms is obvious, and the discovery of a third domain of life, the "archaea" in 1977, allowed to further cement the knowledge of DNA-repair processes.

Animals have evolved many parallel and often overlapping repair pathways to be able to deal with various types of DNA damage, including chemical modifications of the base, nucleotide misincorporation, DNA and protein cross-links, and even DNA strand breaks (Fig. 38.1) [1]. Among these pathways are mismatch repair (MMR), base excision repair (BER), and nucleotide excision repair (NER), as well as global genome repair (GGR) and transcription-coupled repair (TCR), translesion synthesis (TLS), homologous recombination (HR), and nonhomologous end joining (NHEJ) (Fig. 38.1) [1].

It would be difficult to compare all types of DNA repair in organisms, and it is beyond the scope of this chapter. We only compare major repair pathways and note major differences among them.

1.1 Direct Reversal of DNA Damage

The direct reversal of DNA damage involves repair mechanisms that restore an original molecular structure of DNA without new DNA synthesis. Three major direct reversal-repair pathways are: photolyases that repair UV-induced DNA damage,

Genome Stability. http://dx.doi.org/10.1016/B978-0-12-803309-8.00038-0

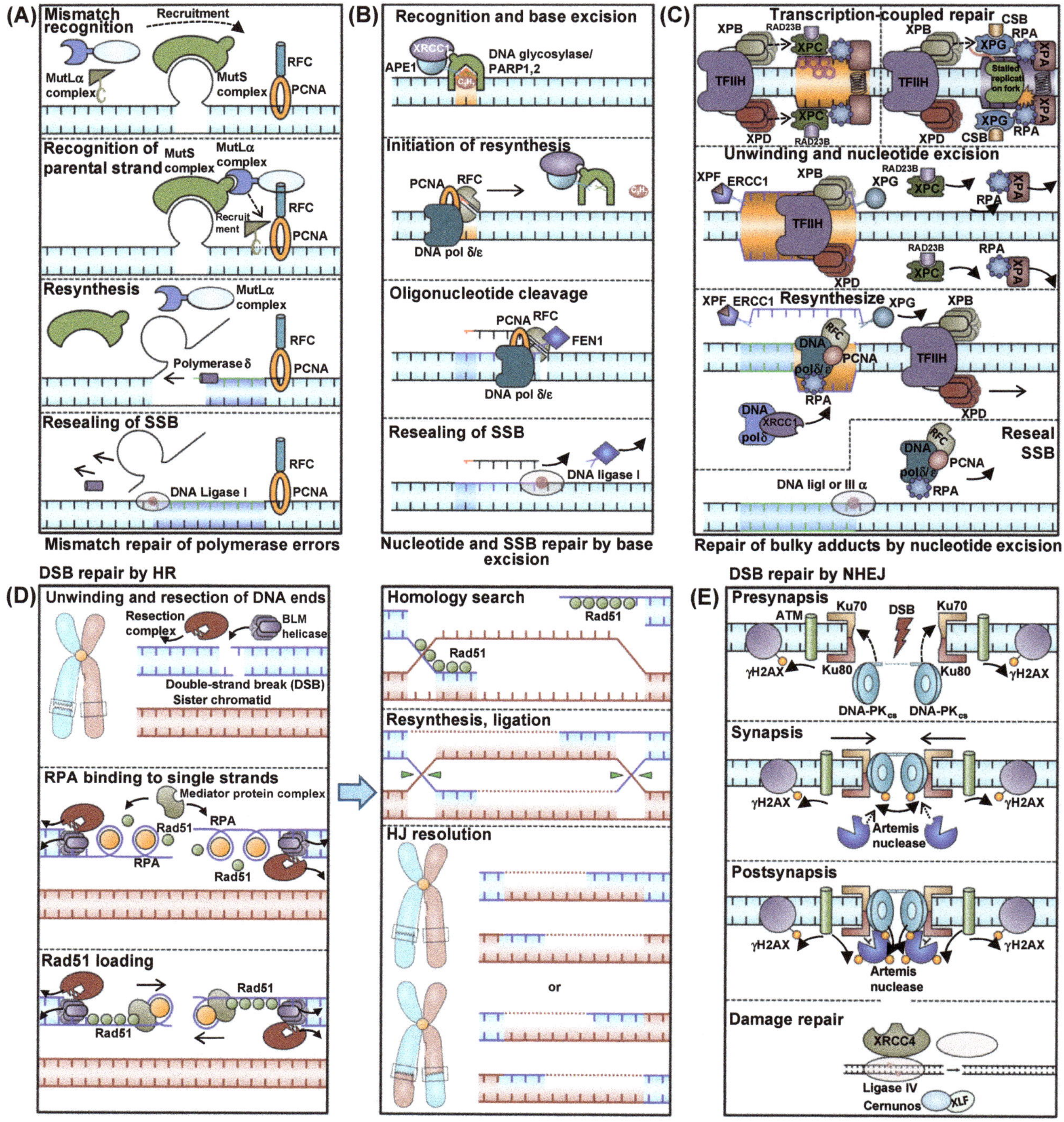

FIGURE 38.1 Mammalian DNA damage–repair pathways. Distinct DNA-repair pathways are traditionally thought to operate independently in parallel to clear different types of DNA lesions occurring in distinct cell-cycle phases. (A) Erroneous misincorporations, insertions, and deletions of nucleotides are handled by mismatch repair (MMR). (B) Single-strand breaks (SSBs) and nonhelix-distorting base modifications are repaired by base excision repair (BER). (C) Bulky helix-distorting lesions are cleared by nucleotide excision repair (NER) which can be initiated either as global-genome NER (GG-NER) or as transcription-coupled NER (TC-NER). (D) Double-strand breaks (DSBs) can be repaired through homologous recombination (HR) or (E) nonhomologous end joining (NHEJ), depending on the cell-cycle phase. In late stages of the cell cycle (S and G2) when the genomic DNA has been replicated and a homologous sister chromatid is available, DSBs are repaired by HR (D). (E) In G1 cells where no intact template is available for homology-mediated repair, the error-prone NHEJ pathway is preferred. *Reproduced from Dietlein F, Thelen L, Reinhardt HC. Cancer-specific defects in DNA repair pathways as targets for personalized therapeutic approaches. Trends Genet 2014;30(8):326–39; with permission.*

and two mechanisms for repairing alkylated bases, O^6-alkylguanine alkyl transferases (AGTs) and the AlkB-family dioxygenases [2]. These are ancient repair pathways that are highly active in bacteria (see details in Chapter 4). Most of them have retained their function in many eukaryotes.

Photolyases are proteins that reverse the UV damage and create a kind of a bridge between pyrimidine nucleotides. This type of repair process exists in prokaryotes and eukaryotes, including plants and animals, although mammals seem to lack such activity [3]. These enzymes are activated by light; they bind to a strand opposite to the damaged strand and cleave the bridge (like the dimer cyclobutane bridge formed between two thymines).

AGT enzymes function in both pro- and eukaryotes. These enzymes transfer the alkyl groups to their cysteine residues, rendering themselves inactive. A human AGT homolog is encoded by the O^6-methylguanine DNA methyltransferase (MGMT) gene. In *Escherichia coli*, the alkylation damage to phosphodiester DNA backbone, the S_p-methylphosphotriester, is repaired by the N-terminal domain of the Ada protein (N-Ada). There is no homolog of N-Ada found in eukaryotes, and it is unclear whether methylphosphotriester can be repaired in eukaryotes [4].

Oxidative dealkylation is performed by the AlkB protein. It activates dioxygen to oxidize the methyl group. The resulting oxidized product decomposes spontaneously and restores the base while releasing formaldehyde. Humans have eight homologs of AlkB, hABH1 to hABH8, with hABH2 and hABH3 being functional homologs [5]. The hABH1 protein has mitochondrial localization, with only a fraction of it being in the nucleus [6]. The authors demonstrated that hABH1 is a functional mitochondrial AlkB homolog that repairs 3-methylcytosine in single-stranded DNA and RNA.

1.2 Base Excision Repair

BER corrects DNA lesions that do not distort the DNA double helix, involving DNA damage caused by oxidation, deamination, and alkylation (Fig. 38.1). The importance of BER was recognized by the Nobel Prize Committee who in 2015 awarded Tomas Lindahl the Nobel Prize in Chemistry for his discovery of BER and detailed description of its components [7]. See Chapters 4 and 17 for more details on BER in prokaryotes and eukaryotes.

One of the initial critical steps of BER is performed by DNA glycosylases, enzymes that recognize and flip out the damaged base. DNA glycosylases are conserved among different species, although the conservation is mainly limited to the enzymatic core domain. Some exceptions also occur; for example, a prokaryote homolog of mammalian methyl purine DNA glycosylase (MPG) has been identified [8]. In addition, mammalian DNA glycosylases have extensions at the N- and C-ends of the protein that are not found in prokaryotic counterparts [9].

BER is conserved among bacteria and archaea, although the archaeal components of BER and their molecular biology are more similar to eukaryotes than to bacteria (Chapter 4). The BER pathways in some archaea have novel features, while other pathways use an additional mechanism to prevent mutations due to genomic uracil. *Ferroplasma acidarmanus* encodes a novel AGT protein (AGTendoV) that has an O^6-methyltransferase domain fused to an endonuclease V domain [10]. This bifunctional enzyme has been found in other archaeal genomes suggesting that it may be a general adaptation to their harsh environments. In addition, archaea have a unique BER ability; they use uracil scanning DNA polymerases. While bacterial polymerases replicate past uracil by inserting an adenine, archaeal replicative polymerases stall before misplaced uracils, representing a read-ahead proofreading function not found in bacteria or eukaryotes (Chapter 4).

While bacteria contain six DNA glycosylases, mammals have 11 glycosylases (Chapter 4). In mammals, BER functions in repair of both the nuclear and mitochondrial genomes, and both genomes undergo short- and long-patch repair [8]. It should be noted, however, that the ligation step of BER is performed by LIG1 upon DNA repair in the nucleus and by LIG3 in mitochondria; LIG3 seems to function only in vertebrates.

In *E. coli*, misincorporated uracil is removed mostly by the uracil-DNA-glycosylase UNG (UDG). The UDG superfamily consists of six subfamilies in prokaryotes and three members (subfamilies I, II, and III) in higher eukaryotes [11]. Only one member, the UNG protein, has been found in the human mitochondria [12].

1.3 Nucleotide Excision Repair

NER is an excision-repair pathway for repairing bulky DNA lesions (Fig. 38.1). NER mechanisms are highly conserved across all three domains of life; the major steps in GGR and TCR are conserved. At the same time, some significant differences exist.

The NER pathways in bacteria and archaea are functionally similar; however, despite the ancient nature of archaea, some aspects of their NER pathways are more similar to eukaryotic versions than to bacterial versions, and they may or may not have *uvr* homologs (Chapter 4). The presence of clear *uvr* homologs seems to coincide with their lifestyle: mesophilic archaea tend to have *uvr* genes, while hyperthermophilic archaea do not have them [13]. A universal feature, however, seems to be the presence of homologs of eukaryotic factors. Therefore, archaea such as *Methanosarcina mazei* clearly have both bacterial and eukaryotic homologs of NER.

NER is substantially more complex in eukaryotes as compared to prokaryotes. While bacteria contain three proteins involved in NER commonly known as the UvrABC system, up to 50 different proteins may be involved in NER in animals (see Chapter 17). In addition to obvious differences between the names of proteins involved and their sequence conservation, large differences exist in many steps of the process. While prokaryotes require only three proteins, eukaryotes employ many more of them with a versatile and sometimes unique function (see Chapters 4 and 17). The size of the excised DNA fragment is also substantially different: whereas in bacteria, a 12–13 nt–long fragment is removed, in eukaryotes it is more likely to be 27–29 nt. Since NER is functionally less complex in bacteria, details of the biochemistry of NER in bacteria are better understood than in eukaryotes. Please refer to Table 4.1 for a direct comparison of enzymes involved in NER in prokaryotes vs. eukaryotes.

1.4 Mismatch Repair

MMR is a complex repair process that recognizes and replaces mismatched nucleotides. It is present in all species and bacteria, excluding some *Archaea* (eg, *Actinobacteria*, *Mollicutes*, *Mycobacterium tuberculosis*, and *Helicobacter pylori*) that likely lost this repair pathway during the evolution [14]. In *E. coli*, the MMR pathway has three main functional proteins: MutS, MutL, and MutH (Chapter 4).

Mismatches in bacteria are recognized by MutS homodimer. MutS protein subunits A and B encircle mismatch-containing oligonucleotide duplexes by binding in an asymmetric manner—only one subunit contacts the mismatch [15]. In the process of evolution, eukaryotic organisms further developed this asymmetry by developing the binding through a combination of two different proteins—MutS homologs (MSHs) that form two heterodimers. Eight MSHs homologs are known to be capable of forming various dimers. MSH1, MSH6, MSH7, and MSH8 contain a special motif for the recognition of a mismatch—the GxFxE motif. While the biochemical activity of MSH8 is unknown, MSH6 forms heterodimers with MSH2, which leads to the formation of the MutSα complex, and MSH7 interacts with MSH2 to form the MutSγ complex [15]. In *Arabidopsis*, the MutSγ heterodimer binds efficiently to C/G and G/A mismatches but lacks its activity on single-nucleotide insertion-deletion loops (IDLs) [16]. In yeasts, larger IDLs are recognized by the MSH2/MSH3 heterodimer known as MutSβ [17].

In the second step, a MutL dimer binds the MutS–DNA complex, thereby stabilizing it and activating the MutH restriction endonuclease. In *E. coli*, MutL functions as a homodimer. Eukaryotes contain several MutL homologs (MLH proteins). In humans, functional MLH homologs form a heterodimer with MLH1 and PMS2 called MutLα. Two other heterodimers are known in human cells: MLH1/PMS1, referred to as MutLβ, with an unknown function [18], and MutLγ (MLH1/MLH3) which can partially substitute the activity of MutLα in vitro [19]. In *E. coli*, MutL was considered as an intermediate molecule linking MutS and MutH proteins and mediating the interaction between the mismatch recognition machinery and the strand-discrimination and excision machineries [15]. MutL in *E. coli* has no endonucleolytic activity. Such activity has likely evolved in eukaryotes. It has been discovered that MutLα possesses the endonuclease activity [20]. Kadyrov et al. [20] showed that human MutL alpha is a latent endonuclease activated in a mismatch-, MutSα-, RFC-, PCNA-, and ATP-dependent manner. See more details about the function of MMR in mammals in Chapter 18.

In the third step, MutH nicks the strand containing the incorrectly incorporated base; this protein is present in *E. coli* and *Salmonella*. MutH has no homologs in eukaryotes, and it is proposed that its activity is substituted by MutL homologs (MLH proteins).

Archaea substantially differ from bacteria in MMR. For example, in mesothermophilic archaea, the MMR pathways mirror the canonical bacterial pathways [13]. In contrast, hyperthermophilic archaea lack MutS and MutL homologs; these species also lack Uvr proteins. Nevertheless, genome replication is accurate in these organisms [13]. In fact, the error rate per genome replication is lower in the archaeon *Sulfolobus acidocaldarius* than in nearly all other microbial genomes analyzed similarly [13]. Therefore, the lack of increased mutagenesis suggests that hyperthermophiles have the same mechanism that accomplishes the same net result as MMR (see details in Chapter 4).

1.5 Double-Strand Break Repair

Double-strand break (DSB) repair is broadly divided into NHEJ and HR, and there is a great diversity and variability in these repair pathways in all three domains of life. DSB repair in mammals is presented in Fig. 38.1.

1.5.1 Nonhomologous End Joining

E. coli do not have the canonical NHEJ pathway. One hypothesis suggests that since bacteria are often in the replicative stage, they constantly possess homologous chromosomes for the efficient HR repair, thus making NHEJ nonessential [21]. Those bacteria that do have NHEJ are frequently found to be in the stationary stage when no chromosome homologs are present [22]. A 2010 work, however, demonstrated that *E. coli* strains do possess an end-joining mechanism, now called alternative end joining (A-EJ) [23]. This novel pathway does not share conserved factors with the canonical NHEJ pathways. It depends on bidirectional strand resection, the frequent use of microhomology, and nontemplated DNA synthesis. In archaea, a functional repair pathway was identified only in 2013 in a mesophilic archaeon *Methanocella paludicola*, although conserved components of NHEJ were previously identified [24,25]. The scientists described the complete NHEJ complex in archaea consisting of DNA ligase (Lig), polymerase (Pol), phosphoesterase (PE), and the Ku protein [26]. The A-EJ or A-NHEJ pathway has also been well described in mammals (see Chapter 19).

Unlike the eukaryotic NHEJ ligase IV, LigD is a large multidomain protein that contains three components within a single polypeptide: a polymerase (POL) domain, a phosphoesterase (PE) domain, and a ligase (LIG) domain [21]. Therefore, it was initially believed that bacterial Ku and LigD are sufficient to repair all DSBs generated in vivo [27]. Later on, a faithful NHEJ pathway operating specifically for 3′ overhang DSB repair and functioning independently of Ku and LigD proteins was reported [28]. It was also shown that the structure of broken ends determines the pathway and the outcome of DSB repair. In wild-type cells, this pathway joins nearly one-third of the 3′ overhangs with 100% fidelity [28].

In NHEJ occurring in vertebrates, a complex of Artemis:DNA-PKcs processes DNA ends with different configurations. The *DNA-PKcs* gene has not been found in *Arabidopsis* [29] or in other plant genomes. Instead, in yeast, plants, and vertebrates, the same job is likely performed by the MRX complex [30]. In mammals, Pol μ can add nucleotides in a template-independent manner under physiological conditions [31], whereas Pol λ can do this only when Mg^{2+} is replaced by Mn^{2+} [32]. Similarly in bacteria, the polymerase can also add nucleotides in a template-independent manner [21], likely reflecting convergent evolution.

NHEJ appears to be mechanistically different in *Saccharomyces cerevisiae* and mammals since in yeast, blunt-end ligation is inefficient, and at least one base pair of terminal microhomology may be necessary for efficient NHEJ to occur [30]. In *S. cerevisiae*, NHEJ can actually occur in the absence of Lig IV, but it is at least 10-fold less efficient than with Lig IV [33]. In contrast, Ku-deficient yeast cells repair DSBs as efficiently as wild-type yeast cells.

In mammals, there are no microhomology requirements for end joining, although occasionally single nucleotides and longer microhomologies can be found. This is a default situation for most genomic positions (excluding repetitive elements). The use of more extensive microhomology is possible in mammals when such homology is provided artificially or when some NHEJ components are lacking. In the latter case, NHEJ occurs at a slower pace with more frequent resections and alignments that use more extensive (2–3 bp) microhomology [30].

1.5.2 Homologous Recombination

The early stage of the HR mechanism is well studied and appears to be conserved in three life domains [34]. All steps of DSB repair are conserved between archaea and bacteria, although protein conservation is not as obvious as it may seem. Similarly to other repair pathways, archaea contain protein homologs from bacteria and eukarya. For example, proteins such as Mre11, Rad50, and RadA are conserved between archaea and eukarya.

In bacteria, DSBs are repaired through one of two overlapping repair pathways, one relying on the multifunctional helicase/nuclease complex RecBCD (the RecBCD pathway) and the other involving the RecF protein (the RecF pathway) [35]. In the RecBCD pathway, resection is catalyzed by the RecBCD complex, whereas in eukarya, it is done with the Mre11 and Rad50 protein complex [36]. In archaea, Rad50 and Mre11 cooperate with the HerA helicase and NurA nuclease to catalyze 3′-end resection for HR [37].

Next, the single-stranded DNA-binding protein (SSB) (known as replication protein A, RPA, in eukaryotes) binds to the 3′-ssDNA ends generated by the action of RecBCD or Mre11/Rad50, or Mre11/Rad50/HerA/NurA. SSBs and RPA function as dimers, trimers, and tetramers. In bacteria, they occur as homodimers and homotetramers, whereas in archaea, they function as monomers, dimers, and trimers [38]. In eukaryotes, RPA acts as a homodimer and heterotrimer. The DNA recombinase of the RecA family is conserved in all three domains, but it has a different name in each domain: the RecA recombinase in bacteria, RadA in archaea, and Rad51 and its paralogs in eukaryotes. RadA is more similar to Rad51 rather than to RecA. In animals, Rad51 is an essential protein that causes embryonic lethality when disrupted. Similarly, the disruption of RecA in bacteria or Rad51 in yeast is also highly deleterious to the cell, although it is not lethal [39]. Eukaryotes contain multiple Rad51 paralogs, including Rad51B, Rad51C, Rad51D, DMC1, XRCC2, and XRCC3 that cooperate with Rad51 in strand exchange. Archaea also have several RadA paralogs, including RadB, RadC2, Sso2452, and Rad55B.

Proteins involved in the late stage of HR, representing the RuvABC–Holliday junction (HJ) complex that processes recombination intermediates, have been well described in bacteria [40]. Eukaryotic cells possess two mechanisms for HJ processing: the first mechanism known as HJ dissolution relies on the activity of the BTR complex (BLM helicase–topoisomerase IIIα–RMI1–RMI2), whereas the second one requires structure-selective nucleolytic endonucleases such as MUS81–EME1 and GEN1 [41]. While the outcome of HJ dissolution is exclusively the formation of non-crossovers (NCOs), nucleolytic resolution results in the formation of NCOs as well as crossover events (COs). Mitotic cells preferentially use NCOs to prevent the loss of heterozygosity [42]; this correlates well with the fact that the HJ resolvases function late in the cell cycle to ensure a proper chromosome segregation.

Yeast cells are able to separate the recombination resulting in mitotic and meiotic DSB repair leading to CO and NCO formation [41]. In mitosis, the majority of double Holiday junctions (dHJs) are processed at early stages of the cell cycle by STR-mediated dissolution leading to NCOs. In *sgs1* mutants, however, DNA joint molecules formed between the damaged and sister chromatids persist to a later stage of the cell cycle where they are processed by Mus81–Mms4 or Yen1, resulting in a mixture of COs and NCOs as an outcome [41]. As of 2016, the temporal separation of CO and NCO formation in human cells has not been demonstrated. However, human cells deficient in BLM (Bloom's syndrome cells) have the increased frequency of CO formation [43]. CO formation in these cells occurs through the activities of MUS81–EME1, SLX1–SLX4, and GEN1 [44].

2. RECENT ADVANCES AND FUTURE DIRECTIONS IN DNA REPAIR

Despite a substantial progress in understanding the mechanisms of functioning of various DNA-repair pathways, many questions remain unanswered. We only discuss some of them.

2.1 The Remaining Questions in MMR

MMR is one of the critical repair mechanisms involved in fixing replication errors as well as regulating other repair mechanisms such as HR and DNA-damage signaling. Despite many advances in the field, several questions remain to be answered (for review, see [15]).These are:

1. Where does the MSH/MLH complex assemble? Is it assembled at the mismatch or at the strand-discrimination signal?
2. If the complex does indeed assemble at the mismatch and translocate, is the direction of its movement random, or does it depend on its initial loading on the heteroduplex?
3. What directs the process exonuclease degradation toward the mismatch rather than away from it?
4. Are all proteins involved in MMR described? Is there another exonuclease involved in MMR?

Some additional questions in MMR research were asked by Yang and Hsieh in Chapter 18. In particular, it remains to be established how the MMR machinery is recruited to newly replicated DNAs and how it is spatially positioned in relation to the advancing replisome. Other unanswered questions are: When do MutS and MutL proteins interact, and when do they function separately? How exactly is the MutL endonuclease activity targeted to the newly synthesized strand, and what is its biological scope? What is the mechanism of recruitment of replicative polymerases to single-strand gaps? In what contexts are error-prone polymerases employed instead and what are the consequences? How is MMR influenced by higher-order chromatin arrangements and nuclear architecture? Finally, how can the knowledge of MMR mechanisms improve clinical diagnostics and therapeutic outcomes?

Although the answers to the most of the abovementioned questions remain unknown, several models summarized by Li et al. [45] may have merit, and future research will show whether they hold true.

2.1.1 The Role of a Strand-Discrimination Signal in MMR

In MMR, the protein machinery has to recognize the mismatch on the strand that is opposite to the strand with the mismatch and to perform a strand-specific nick on the strand with the mismatch. One of the unanswered questions concerns the mechanism by which MMR proteins communicate between two physically separated DNA sites: the mismatch and the strand-discrimination signal [45]. The strand-discrimination signal represents a strand-specific nick that occurs in a similar manner in all three domains of life. It is not clear, however, what the source of the nicking activity in eukaryotic cells is. Several alternative models describing this process have been proposed; they can be classified into "*cis*-" or "moving" and "*trans*-" or "stationary" models (Fig. 38.2). The stationary model suggests that interactions among MMR proteins induce DNA looping that allows to bring two distant sites together; MutS or MSH heterodimers, MutSα and MutSβ remain bound

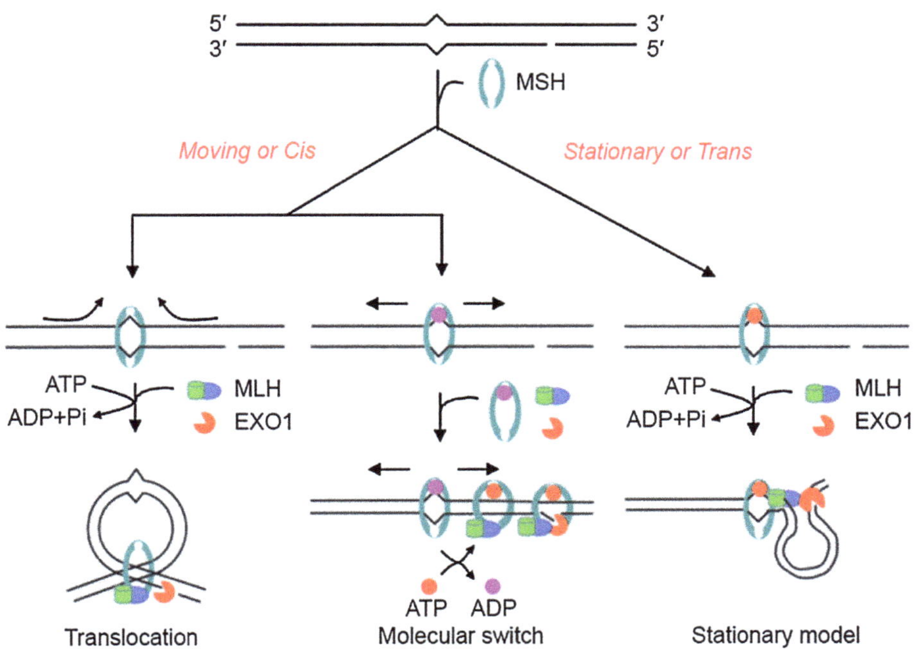

FIGURE 38.2 **Models for signaling downstream MMR events following mismatch recognition.** A schematic diagram for signaling between the mismatch and the strand discrimination signal is shown. Here, a 5′-nick is the strand discrimination signal. Similar models apply for 3′-nick-directed MMR. The "stationary" or "*trans*" model (*right*) emphasizes that MutS or its homolog (MSH) proteins remain bound at the mismatch. It is the protein–protein interactions that induce DNA bending or looping that brings the two distant sites together. The two DNA sites can cooperate in a "*trans*"-configuration. In two "*cis*" or "moving" models, one called the "translocation" model (*left*) and the other called the "molecular switch" or "sliding clamp" model (*middle*), it is hypothesized that the MSH proteins bind to the mismatch and then move away from the site to search for the strand discrimination signal. The translocation model suggests that ATP hydrolysis drives unidirectional movement of MSH proteins, resulting in the formation of an α-like loop. In the molecular switch model (*center*), binding of an MSH protein (in its ADP-bound state) to the mismatch triggers an ADP to ATP exchange that promotes bidirectional sliding of the protein away from the mismatch, thereby emptying the mismatch site for an incoming MSH protein. Mismatch excision begins when an MSH protein reaches the strand break. *Reproduced from Li GM. Mechanisms and functions of DNA mismatch repair. Cell Res 2008;18(1):85–98; with permission.*

at the mismatch [46]. The ATPase activity of MSH heterodimers is required for proofreading and verification of mismatch binding and downstream excision [47]. In support of this model, it was shown that MutS recognizes a mismatch, activating MutH to cleave the GATC site located on a separate DNA molecule that does not contain a mismatch [47]. Similarly, it was also shown that DNA excision activated by the presence of a mismatch could be initiated when a biotin–streptavidin blockade was placed between the mismatch and the preexisting nick [48].

The second model, known as a "*cis*-" or "moving" model, suggests that the MutS–MutL (or MutSα/β-MutLα) complexes bind to the mismatched site and then move away from the site to search for the strand break where exonucleases can be recruited to initiate excision [45]. There have been known two moving models: the "translocation" model and the "molecular switch" model (also known as the "sliding clamp" model) (Fig. 38.2). In the translocation model, ATP reduces the mismatch-binding affinity of MutS, and ATP hydrolysis drives unidirectional translocation of MutS proteins along the DNA helix [49]. DNA is threaded through the protein complex until the complex reaches a strand-discrimination signal in either orientation; this process forms a DNA loop (Fig. 38.2). In the molecular switch model, MutS binds to the mismatched DNA in an ADP-bound state [45]. MutS binds the mismatch, thus triggering a conformational change in these proteins and allowing an ADP to ATP exchange. ATP binding promotes a second conformational change that allows MutS to form a sliding clamp [50,51]. ATP binding also signals downstream events such as the formation of the ternary complex with MutL or MLH heterodimers in eukaryotes and sliding of the ternary complex from the mismatch to the strand break [50,51].

2.1.2 The Role of MMR in DNA-Damage Signaling

MMR plays an essential role in DNA-damage signaling. Two models have been proposed to describe details of the signaling. In the "futile DNA-repair cycle" model (Fig. 38.3), strand-specific MMR which only interacts with replicated DNA encounters DNA lesions in the template strand and engages in a futile DNA-repair cycle by activating DNA damage–signaling pathways and inducing the cell-cycle arrest and apoptosis [52]. This model was synthesized from several experiments conducted in vivo and in vitro. Exposure to MNNG was shown to induce DNA breaks/gaps, the cell-cycle arrest, and

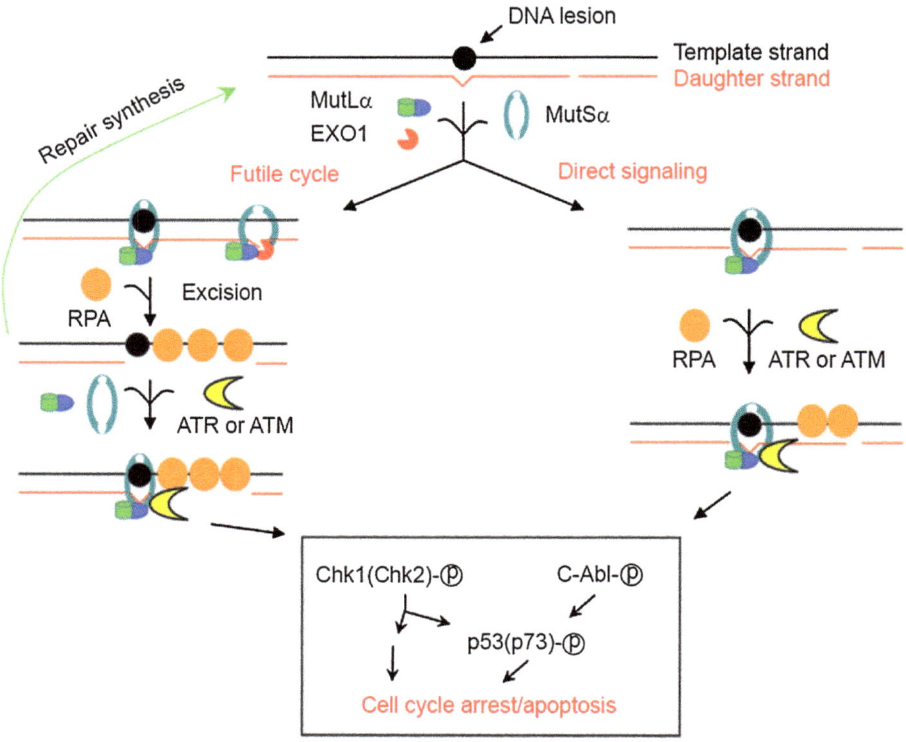

FIGURE 38.3 **Models for MMR-dependent DNA-damage signaling.** The "futile DNA-repair cycle" model (*left*) suggests that DNA adducts (*solid black circle*) induce misincorporation which triggers the strand-specific MMR reaction. Since MMR only targets the newly synthesized strand for repair, the offending adduct in the template strand cannot be removed, and it will provoke a new cycle of MMR upon repair resynthesis. Such a futile repair cycle persists and activates the ATR and/or ATM damage–signaling network to promote cell-cycle arrest and/or programmed cell death. The direct signaling model proposes that recognition of DNA adducts by MSH–MLH complexes allows the proteins to recruit ATR and/or ATM to the site, activating the downstream damage signaling. *Reproduced from Li GM. Mechanisms and functions of DNA mismatch repair. Cell Res 2008;18(1):85–98; with permission.*

persistent nuclear foci at DNA-damage sites [53]. Repair foci associated with DNA damage contain damage signaling and DNA-repair proteins such as ATR, γ-H2AX, and RPA(45). In addition, it was shown that the nicked circular heteroduplex plasmid DNA containing a single O^6-methylguanine (O^6-meG)-thymine (T) mispair can only be repaired by MMR when the lesion (O^6-meG) and the nick are on the same strand; MMR cannot repair the mismatch when the lesion and the nick are on the opposite strands, suggesting a futile repair process [54]. The direct signaling model (Fig. 38.3) proposes that hMutSα/hMutLα heterodimer triggers DNA-damage signaling directly by recruiting ATM or ATR/ARTIP to the lesion and activating a checkpoint response [45]. The functionality of this model is supported by experiments showing that ATR and ATRIP form a complex with MutSα/MutLα in the presence of O^6-meG/T which activates the ATR kinase and phosphorylates Chk1 [55] (see also Chapter 18). This model is consistent with the fact that the MutSα/MutLα complex functions as a sensor for DNA damage in mammalian cells.

2.2 The Remaining Questions in DSBs Repair

Organisms vary dramatically in the frequency of using NHEJ or HR in strand-break repair. As it has been mentioned earlier, bacteria, for example, are much less efficient in canonical NHEJ, but are very proficient in HR. Similarly, yeast also uses HR more frequently than NHEJ repair. In contrast, higher eukaryotes appear to predominantly utilize NHEJ. Such selective processing of strand breaks in yeast and higher eukaryotes can be in part explained by their genome content. The yeast genome is rather small and consists predominantly of coding sequences, and the content of noncoding and repetitive elements is relatively low; whereas higher eukaryotes have much larger genomes with a higher ratio of noncoding to coding sequences. It can be hypothesized that being the dominant repair mechanism, HR in yeasts prevents the generation of random changes in coding DNA. On the contrary, relatively small-scale deletions and insertions resulting from NHEJ repair can be tolerated by higher eukaryotes due to their large-size genomes and a low ratio of coding to noncoding sequences [56]. Another important factor to consider is the amount of time and energy required for HR to occur. In large genomes, locating a homologous template in somatic, nondividing cells requires a substantial amount of time, and instead, it can be very beneficial to use NHEJ for strand-break repair.

NHEJ is the predominant mechanism in higher eukaryotes including plants, and HR is more active during S and G2 phases as compared to other phases of the cell cycle. Now there is no doubt that genome sizes are the consequences of species evolution. The question to ask would be: how do cells decide on the preferred mechanism of DNA repair? Since mutation frequency varies among different cell types and during different developmental stages of an organism, there definitely exists some type of chromatin-based regulation of genome stability. See Chapter 28 on epigenetic regulation of genome stability in cancer for more details.

2.2.1 Nonhomologous End Joining

A good summary of the remaining questions in NHEJ research was made by Lieber [30].

2.2.1.1 Are the Two DNA Ends Held in Proximity During NHEJ or Is There Synapsis?

Depending on the size of the microhomology, there can be different requirements on Ku proteins that hold DNA ends together. In vitro experiments show that when 4 bp of terminal microhomology is used, the XRCC4:DNA ligase IV complex does not require a protein helper to bring and hold two ends of DNA together. In contrast, at 2 bp or less of terminal microhomology, the addition of Ku improves XRCC4:DNA ligase IV ligation. This is likely due to the ability of Ku to stabilize XRCC4:DNA ligase IV rather than to bring DNA ends together. Under certain salt conditions (30 mM monovalent salt or lower), DNA-PKcs appear to help bring DNA ends together. Therefore, there is an ambiguity as to the requirement of helper proteins to hold DNA ends together during NHEJ repair. Another relevant question is whether the two DNA ends generated at a single DSB are joined more efficiently than two DNA ends that arise at a distance (as in a chromosomal translocation) [30].

2.2.1.2 How Do DNA-Damage Response Proteins Interact With NHEJ Enzymes?

NHEJ is a rapid and efficient repair process, and it may repair DNA damage without the activation of DNA-damage response (DDR) proteins, ATM, the RAD50:MRE11:NBS1 complex, γ-H2AX, and 53BP1. The kinetics of the NHEJ machinery and DDR is not entirely clear and is still an area of intensive research. Many in vitro and in vivo biochemical experiments are conducted in such a way that many breaks are present simultaneously. Such conditions may artificially favor the activation of DDR pathways, therefore skewing the enzymology of NHEJ. Obviously, a competition between HR and NHEJ components may be one of the aspects that arises. In addition, systems in which multiple DSBs or SSBs are introduced trigger the activation of DDR, NHEJ, and HR pathways by altering the kinetics of the association with DNA damage, the chromatin, and nuclear scaffold, as well as with each other. Cells in which so many breaks occur simultaneously are not likely to survive in nature, and therefore, in reality, what we observe in the experimental system may not occur in nature.

2.2.1.3 Are All Components of NHEJ Accounted for?

Do we now know all proteins involved in NHEJ? It is unlikely, considering how many components in various organisms have been developed through the convergent evolution. Most obvious genetic components have been identified through experiments that show the sensitivity of cells or organisms to DNA-damaging agents such as ionizing radiation (IR) or cisplatin. Novel components may be identified by using specific experimental conditions, such as a specific mutant background, a specific cell type or/and the use of specific DNA-damaging agents.

The Werner's 3′-exonuclease/helicase enzyme is likely to be involved in DSB repair through NHEJ because WRN interacts with Ku and PARP1. However, since the mutant is not sensitive to IR and the interaction with Ku and PARP1 may simply reflect the role of WRN in replication fork repair, it is still unclear whether the WRN protein is one of the NHEJ components [57,58]. Another potential candidate can be metnase that has been proposed as a possible nuclease and helicase functioning in NHEJ [59]. Metnase appears to be present only in humans; it seems to be absent in all vertebrates tested, and it is also absent in yeast. In addition, the absence of genetic knockouts to demonstrate IR sensitivity and the absence of experiments with purified protein are also a problem. Future experiments will show whether or not Metnase is a unique NHEJ protein in humans.

2.2.2 Homologous Recombination

Many questions still remain unanswered as to the recognition of DSBs, the choice of the pathway, dynamics and kinetics of repair, and so on. How do cells choose which DNA-repair pathway to use to fix DSB? Is the choice different in different species, and what are the components responsible for the choice? What is the kinetics of HR repair in various organisms? What is the chromatin dynamics at various steps of HR repair? Specifically, what impact does chromatin have on the late stages of HR, including the formation and resolution of joint molecules? These and other questions are still to be addressed.

One of the 2015 reports clarified one step in the chromatin modification associated with DSB repair. DSBs can trigger nonproteolytic ubiquitylation of adjacent chromatin areas exposing the DNA to DNA-binding proteins. The E3 ubiquitin ligase RNF168 catalyzes ubiquitylation of H2A histones, leading to recruitment of the 53BP1 repair protein. It was demonstrated that ubiquitylation of the H1 linker histone rather than one of the core histones by RNF8 and UBC13 ubiquitin ligases is also important for DSB repair [60].

Another unresolved question is the site specificity of V(D)J cleavage. V(D)J genes are surrounded by recombination signal sequences (RSS), and the recombination is initiated by RAG2 proteins. The human genome contains a great number of RSS and RAG2-binding sites, so it is not clear how V(D)J is mediated and carried out at correct sites.

There is also an unanswered question regarding the use of HR during TLS repair. HR repair is also involved in the DNA-damage tolerance (DDT) repair pathway that promotes ssDNA-lesion bypass. This HR repair is known as the salvage HR-repair pathway. During replication, this pathway generates hyper-recombinogenic intermediates that lead to chromosome rearrangements and genome instability [61]. The salvage HR-repair pathway functions independently of the Rad5-dependent error-free tTLS-repair pathway and is repressed during the S phase [62]. It is unclear till 2016 why it is actually repressed during the S phase. It was suggested that both error-free and error-prone pathways are initiated by Rad51-mediated strand invasion, but they are different in the mechanism forming sister chromatid junctions, with Rad5 being dependent on PCNA polyubiquitination, whereas HR pathways being dependent on Rad51-mediated strand exchange that is presumably prevented by SUMOylated PCNA and Srs2 during the S phase. It only becomes derepressed when the level of sumoylation of PCNA declines in the end of the S phase [62].

Research on mammals demonstrates that meiotic HR occurs at DNA DSB hot spots and is initiated by programmed DSBs [63]. Since recombination repair occurs between homologous chromosomes, the sequence from the unbroken (undamaged) chromosome is copied. One prevailing hypothesis based on data from research on mammals suggests that over time DSB hot spots should be replaced with normal (cold) regions that should undergo the normal recombination frequency [64]. In consistence with this hypothesis, patterns of hot spots in mammals exhibit a higher divergence between species and individuals [65]. Research articles in 2015 in *Science* (vol 350, issue 6263) challenge this hypothesis, demonstrating that yeast and birds have a remarkable evolutionary stability of meiotic recombination patterns [66–68].

3. FUTURE DIRECTIONS IN RESEARCH ON DNA REPAIR, GENOME STABILITY, AND CANCER

DNA-repair capacity is often altered in cancer cells. This leads to the accumulation of many genomic alterations that allow cancer cells to prevail in the competition for resources and to survive for much longer than normal cells. Enormous variations in cancers do not allow for a complete understanding of the role that deficiencies in DNA repair and genome instability play in the development of cancers. However, advances in next-generation sequencing allowed to sequence multiple cancer cell lines and to obtain details of genomic alterations associated with these lines (Fig. 38.4). These studies showed that mutations in different cancers accumulate in distinct signaling pathways [69–72]. It was shown that inactivating mutations in the affected pathways were cancer specific (Fig. 38.4). For example, it was found that mutations in HR-repair genes were enriched in breast and ovarian cancer, as well as in chronic lymphocytic leukemia (CLL). In contrast, colorectal cancers are altered in MMR and HR-repair pathways, whereas prostate cancers are enriched in mutations inactivating HR and NER pathways (Fig. 38.4A–C,E,F) [72–74]. At the same time, cancers developing in endometrial tissues had no apparent clustering (Fig. 38.2D) [75]. Because of such heterogeneity and specificity in targeting particular DNA-repair pathways in specific cancers, it can be an arduous task to identify the subcohort of DNA repair–defective tumors for each cancer entity. Therefore, it may be more efficient to identify genetic alterations in the pathways that are predominantly altered in the respective tumor types [1].

Another area of research that might be interesting to explore in the future is the identification of proteins that play a critical role in several DNA-repair pathways. Targeting such genes in cancers may appear to be more effective. Proteins involved in distinct repair pathways may appear to be hubs of DNA-damage recognition and repair signaling [1]. The best example is the MuTSβ complex involved in MMR and HR repair. Under normal conditions, the Msh2/Msh3 heterodimer complex binds to intact dsDNA with a low efficiency, whereas when dsDNA has a 30 ssDNA overhang or contains large (~8 nt) insertion loops, the affinity increases dramatically [1]. Other enzymes, such as BRCA1, BRCA2, ERCC1, RAD50, PARP1, ATM, PTIP, XRCC1, PCNA, RFC, and RPA, share similar dual properties with Msh3.

For example, the PARP1 protein functions in both BER and A-NHEJ-repair pathways (Fig. 38.1B,E) [1]. Similarly, BRCA1 and BRCA2 proteins are involved in several DNA-repair pathways; besides being involved in HR repair, they are also important for interstrand cross-link (ICL) repair [76,77] (Fig. 38.5). Moreover, BRCA proteins might be important for the protection of replication forks, and in this role, they may be assisted by Rad51 protein [78,79]. Specifically, BRCA2-dependent fork stabilization requires the interaction between Rad51 and BRCA2 through the conserved C-terminal domain in BRCA2, and this domain is different from the BRCT domain that interacts with Rad51 for HR repair.

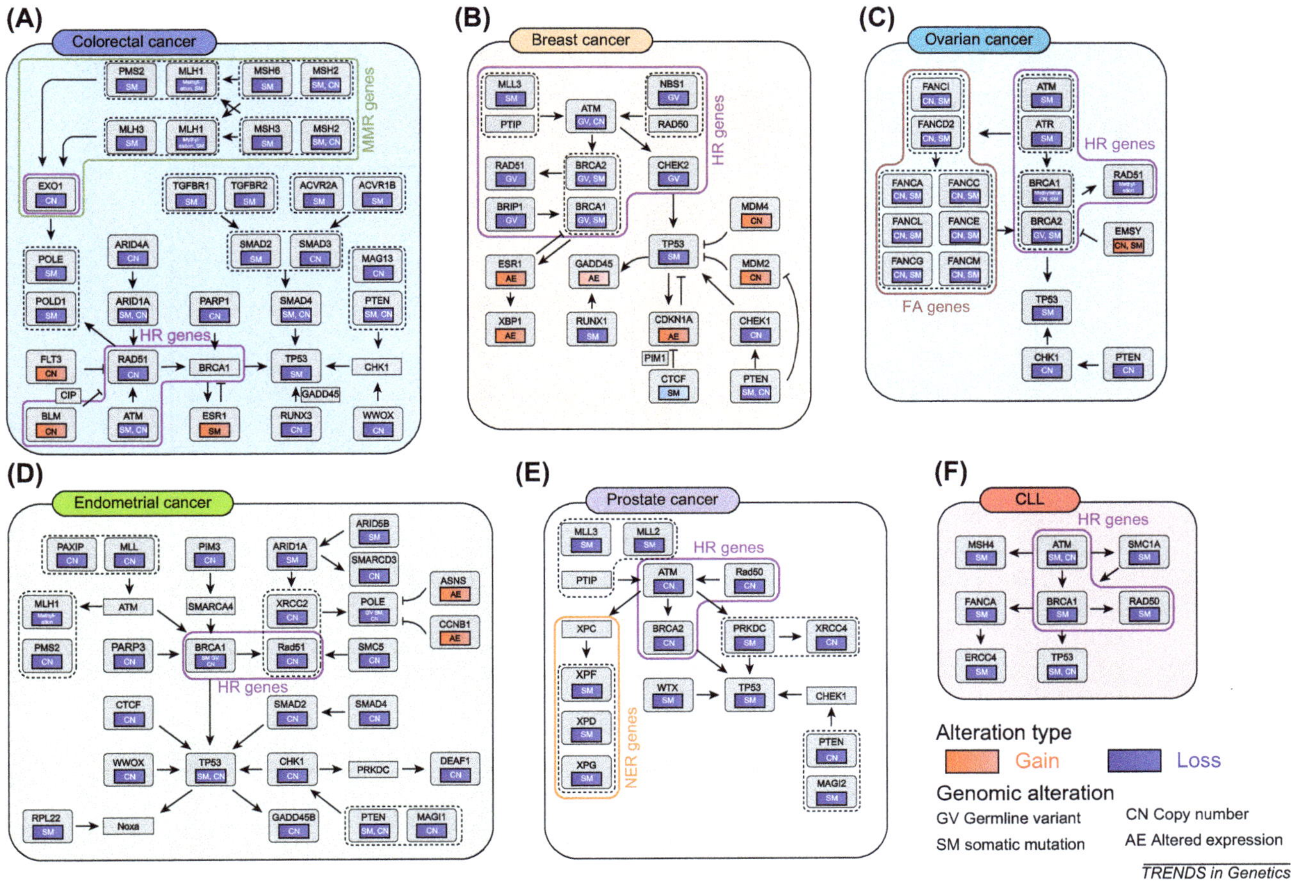

FIGURE 38.4 **Genomic aberrations in DNA-repair pathways are common in cancer.** Genes involved in different DNA damage–repair pathways are frequently altered by genomic alterations in cancer. Germline variants (GV), somatic mutations (SM), copy-number alterations (CN), and altered expression patterns (AE) in damage-repair genes are schematically represented across (A) colorectal, (B) breast, (C) ovarian, (D) endometrial, and (E) prostate cancer, as well as in (F) chronic lymphocytic leukemia (CLL). All genomic alterations were scored as significant in the respective cancer entity either by MutSig (mutations) or GISTIC (copy-number alterations) algorithms. Genomic gains are colored in red and functional losses are colored in blue. To derive entity-specific signatures of affected damage-repair pathways, genes were clustered according to their association to different pathways (*encircled*). Broken lines encircle genes encoding proteins that act together in a protein complex. *Reproduced from Dietlein F, Thelen L, Reinhardt HC. Cancer-specific defects in DNA repair pathways as targets for personalized therapeutic approaches. Trends Genet 2014;30(8):326–39; with permission.*

The knowledge about the involvement of DNA-repair proteins in multiple signaling and DNA-repair hubs is critically important because such proteins are frequently affected in cancers, including genomic and epigenomic alterations [69–72]. The prominent example is the classification of colorectal cancers (CRCs) deficient in MSH3; for therapeutic purposes, CRCs are classified as microsatellite unstable with MMR defects [70]. Such stratification, however, completely neglects the fact that MSH3 is also involved in the HR-repair pathway, and thus CRCs can be considered as HR-defective tumors that can be treated using PARP1 and/or DNA-PKcs inhibitors [80,81] (Fig. 38.5). Therefore, the identification of molecular hubs such as BRCA1, BRCA2, RPA, or MSH3 involved in multiple repair pathways will allow targeting cancers by disabling the ability of cells to deal with different types of DNA damage. For example, knowing that BRCA1 is involved in multiple DNA-repair pathways, such as HR, ICL (the Fanconi anemia–repair pathway), and mechanisms of replication fork protection would allow to treat BRCA1-deficient tumors with a combination of PARP inhibitors that target the HR pathway and platinum-based drugs that introduce cross-links. It is expected that the surrounding normal tissues will still be able to cope with these drugs by activating multiple repair pathways that function normally (such as NER and FA pathways) to cope with cisplatin damage, and the HR-repair pathway to overcome PARP inhibition [1].

It would be important to learn manipulating DNA-repair processes specifically in cancer cells. This would allow targeting them more efficiently for elimination by exposing them to several DNA-damaging agents and activating multiple repair pathways. Detailed information about DNA-repair genes that are altered in various cancers may allow a more efficient stratification of patients and designing new drugs targeting DNA repair–deficient tumors.

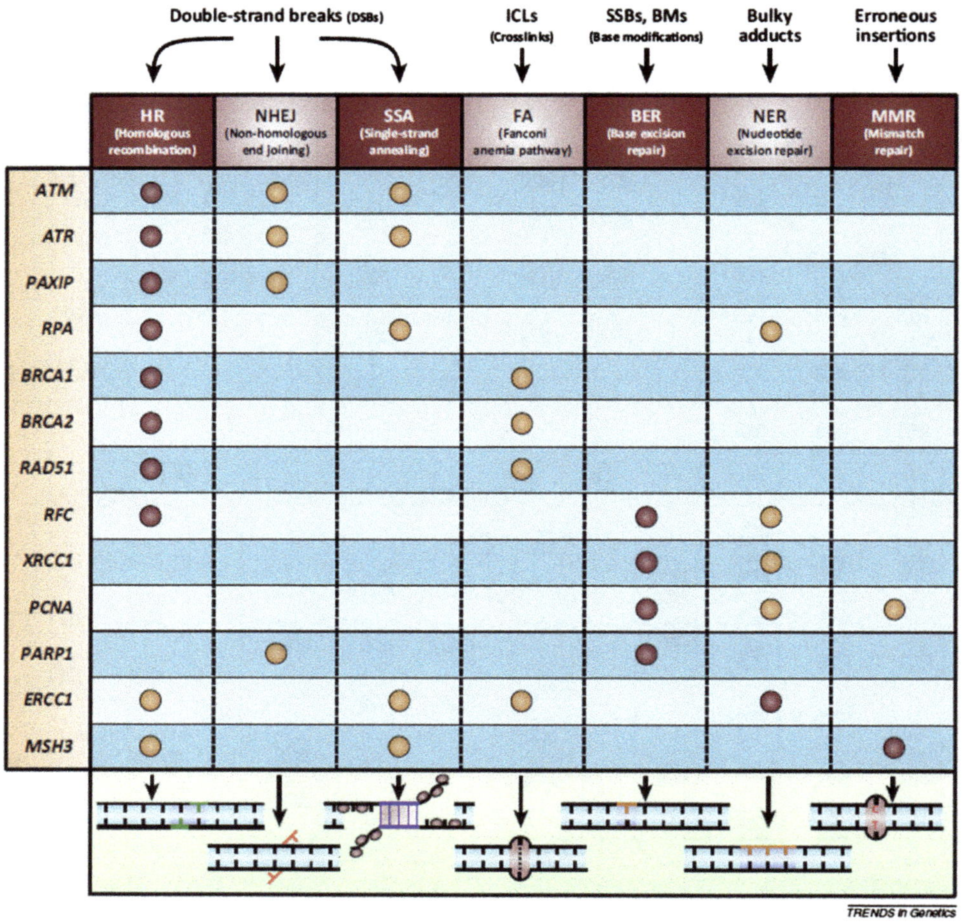

FIGURE 38.5 **Functional overlaps in DNA-repair pathways define hubs of signaling collaterals in cancer.** Multifunctional roles of proteins, which have been traditionally associated with a single DNA damage–repair pathway, are shown for six pathways. The pathway with which the gene has been primarily associated in the literature is marked by a red dot. Orange dots refer to additional roles of the protein in other pathways, which have been confirmed either by functional or biochemical experiments. These proteins are frequently affected by genomic alterations in cancer or have emerged as therapeutically amenable targets of specific kinase inhibitors. Such overlap suggests that novel druggable targets may emerge from this enhanced classification to treat DNA damage–deficient neoplastic diseases. *Reproduced from Dietlein F, Thelen L, Jokic M, Jachimowicz RD, Ivan L, Knittel G, et al. A functional cancer genomics screen identifies a druggable synthetic lethal interaction between MSH3 and PRKDC. Cancer Discov 2014;4(5):592–605; with permission.*

4. FUTURE PERSPECTIVES IN DNA-EDITING TECHNOLOGIES

The ability to delete, insert, or replace DNA sequences in the genome is critically important for both analyzing the function of various genes and reconstituting phenotypes. Gene targeting is a technique that allows sequence replacement (or deletion) by using HR. Unfortunately, for a long time, gene targeting had a low efficiency due to a low frequency of HR repair in most eukaryotes, including plants and animals. The efficiency of gene targeting can be increased dramatically when sequence-specific nicks are generated in the genome. The advent of novel engineered nucleases such as meganucleases, zinc finger nucleases (ZFNs), transcription activator–like effector-based nucleases (TALENs), and the CRISPR-Cas9 system that are able to generate such nicks in the genome allowed to develop a new field of research and technology, genome editing. See details about the CRISPRs-Cas9 system in Chapter 6 and details of using TALENs and CRISPRs for crop improvement in Chapter 13.

Some incredible uses of genome-editing technologies have been demonstrated since 2013. In fact, CRISPR was called the most important discovery of 2015 by *Science* magazine. It was used to edit the CCR5 gene encoding a cell-surface protein called C–C chemokine receptor type 5 (CCR5) in human lymphocytes; HIV is able to enter immune cells by latching onto a receptor expressed on the surface of T-helper cells and macrophages. Gene editing in a subpopulation of helper T cells followed by subculturing these cells and reintroducing them into patients allowed to achieve a significant success in HIV therapy [82].

Gantz and Bier [83] used CRISPR-Cas9 to generate a "mutagenic chain reaction" (MCR) where a mutation in one chromosome copies itself to another chromosome. The authors used *Drosophila* to show that MCR mutations efficiently spread from their chromosome of origin to the homologous chromosome, converting heterozygous mutations to homozygous mutations in most somatic and germline cells. The authors suggested that MCR technology should have broad applications; this technology can perform F1 screens for mutations in newly generated organisms, it can accelerate genetic manipulations and genome engineering due to the rapid achievement of homozygosity, it can be used to deliver transgenes to disease vectors and pest species, and it can also be used for gene therapy applications [83].

Genome editing can soon be used for the elimination of some of the carriers of dangerous diseases. For example, TEP1 gene editing by TALENs makes mosquitoes sensitive to malaria parasites, thus resulting in the elimination of infected mosquito populations in the laboratory settings [84]. It remains to be shown how these mosquitoes with edited genomes thrive in the wild.

The genome-editing technology is rather young (barely over 10 years old), with CRISPR being only a 4-year-old tool for genome editing. Hence, it is likely that many new genome-editing technologies will be discovered and developed in the nearest future. For example, a new tool based on the pentatricopeptide repeat (PPR) protein (initially suggested as a transcriptome-editing tool) [85] was proposed in 2015 to be used for genome editing [86].

The use of these editing technologies on human embryos is still debatable, so it may take years to produce the first genome-edited human. Nowadays, a great success has been achieved in genome editing, but if the CRISPR-Cas9 technology is used for in vitro or in vivo targeting in the treatment of humans, gene-targeting rates should be increased and its target specificity should be confirmed beyond doubt.

GLOSSARY

A-EJ or A-NHEJ Alternative end-joining repair of DSBs.

CO Crossover outcome of nucleolytic dissolution.

DDT The DNA damage–tolerance pathway; the translesion synthesis pathway that allows to insert the preset nucleotides base pairing with damaged nucleotides during the replication process. Usually requires switching from regular to translesion polymerases.

Genome editing Precise genetic engineering based on the ability to generate a sequence-specific nick in the genome, with the subsequent integration of desired sequences or the replacement of target sequences or the generation of random insertion/deletion (by the activity of NHEJ repair) inactivating the target gene.

HJ dissolution The mechanism of HJ processing through the BTR complex (BLM helicase–topoisomerase IIIα–RMI1–RMI2).

ICLs Interstrand cross-links; cross-links between DNA strands. Patients with Fanconi anemia are sensitive to such cross-links.

IDLs Insertion-deletion loops; DNA loops formed upon insertion/deletion of several nucleotides.

MCR Mutagenic chain reaction, the CRISPR/Cas-mediated process upon which a mutation in one chromosome copies itself to another chromosome.

NCO Non-crossover outcomes of HJ dissolution or nucleolytic dissolution.

RSS Recombination signal sequences; RSS are required for the recognition of regions involved in V(D)J recombination.

LIST OF ABBREVIATIONS

A-EJ Alternative end joining
BER Base excision repair
CLL Chronic lymphocytic leukemia
CRCs Colorectal cancers
DDT DNA-damage tolerance
DSB Double-strand break
GG-NER global-genome NER
HR Homologous recombination
ICLs Interstrand cross-links
IDLs Insertion-deletion loops
IR Ionizing radiation
MCR Mutagenic chain reaction
MMR Mismatch repair
NER Nucleotide excision repair
NHEJ Nonhomologous end joining
RSS Recombination signal sequences
SSB Single-strand break
TC-NER Transcription-coupled NER

REFERENCES

[1] Dietlein F, Thelen L, Reinhardt HC. Cancer-specific defects in DNA repair pathways as targets for personalized therapeutic approaches. Trends Genet 2014;30(8):326–39.

[2] Yi C, He C. DNA repair by reversal of DNA damage. Cold Spring Harb Perspect Biol 2013;5(1):a012575.

[3] Selby CP, Sancar A. A cryptochrome/photolyase class of enzymes with single-stranded DNA-specific photolyase activity. Proc Natl Acad Sci USA 2006;103(47):17696–700.

[4] Mishina Y, Duguid EM, He C. Direct reversal of DNA alkylation damage. Chem Rev 2006;106(2):215–32.

[5] Shankaracharya DS, Vidyarthi AS. Homology modeling and function prediction of hABH1, involving in repair of alkylation damaged DNA. Interdiscip Sci 2011;3(3):175–81.

[6] Westbye MP, Feyzi E, Aas PA, Vagbo CB, Talstad VA, Kavli B, et al. Human AlkB homolog 1 is a mitochondrial protein that demethylates 3-methylcytosine in DNA and RNA. J Biol Chem 2008;283(36):25046–56.

[7] Lindahl T. An N-glycosidase from *Escherichia coli* that releases free uracil from DNA containing deaminated cytosine residues. Proc Natl Acad Sci USA 1974;71(9):3649–53.

[8] Krokan HE, Bjoras M. Base excision repair. Cold Spring Harb Perspect Biol 2013;5(4):a012583.

[9] Hegde ML, Hazra TK, Mitra S. Functions of disordered regions in mammalian early base excision repair proteins. Cell Mol Life Sci 2010;67(21):3573–87.

[10] Kanugula S, Pauly GT, Moschel RC, Pegg AE. A bifunctional DNA repair protein from *Ferroplasma acidarmanus* exhibits O^6-alkylguanine-DNA alkyltransferase and endonuclease V activities. Proc Natl Acad Sci USA 2005;102(10):3617–22.

[11] Prakash A, Doublie S. Base excision repair in the mitochondria. J Cell Biochem 2015;116(8):1490–9.

[12] Schormann N, Ricciardi R, Chattopadhyay D. Uracil-DNA glycosylases-structural and functional perspectives on an essential family of DNA repair enzymes. Protein Sci 2014;23(12):1667–85.

[13] Grogan DW. Understanding DNA repair in hyperthermophilic archaea: persistent gaps and other reasons to focus on the fork. Archaea 2015;2015:942605.

[14] Sachadyn P. Conservation and diversity of MutS proteins. Mutat Res 2010;694(1–2):20–30.

[15] Jiricny J. Postreplicative mismatch repair. Cold Spring Harb Perspect Biol 2013;5(4):a012633.

[16] Wu SY, Culligan K, Lamers M, Hays J. Dissimilar mispair-recognition spectra of *Arabidopsis* DNA-mismatch-repair proteins MSH2*MSH6 (MutSalpha) and MSH2*MSH7 (MutSgamma). Nucleic Acids Res 2003;31(20):6027–34.

[17] Harrington JM, Kolodner RD. *Saccharomyces cerevisiae* Msh2-Msh3 acts in repair of base-base mispairs. Mol Cell Biol 2007;27(18):6546–54.

[18] Raschle M, Marra G, Nystrom-Lahti M, Schar P, Jiricny J. Identification of hMutLbeta, a heterodimer of hMLH1 and hPMS1. J Biol Chem 1999;274(45):32368–75.

[19] Cannavo E, Gerrits B, Marra G, Schlapbach R, Jiricny J. Characterization of the interactome of the human MutL homologues MLH1, PMS1, and PMS2. J Biol Chem 2007;282(5):2976–86.

[20] Kadyrov FA, Dzantiev L, Constantin N, Modrich P. Endonucleolytic function of MutLalpha in human mismatch repair. Cell 2006;126(2):297–308.

[21] Shuman S, Glickman MS. Bacterial DNA repair by non-homologous end joining. Nat Rev Microbiol 2007;5(11):852–61.

[22] Pitcher RS, Brissett NC, Doherty AJ. Nonhomologous end-joining in bacteria: a microbial perspective. Annu Rev Microbiol 2007;61:259–82.

[23] Chayot R, Montagne B, Mazel D, Ricchetti M. An end-joining repair mechanism in *Escherichia coli*. Proc Natl Acad Sci USA 2010;107(5):2141–6.

[24] Weller GR, Kysela B, Roy R, Tonkin LM, Scanlan E, Della M, et al. Identification of a DNA nonhomologous end-joining complex in bacteria. Science 2002;297(5587):1686–9.

[25] Aravind L, Koonin EV. Prokaryotic homologs of the eukaryotic DNA-end-binding protein Ku, novel domains in the Ku protein and prediction of a prokaryotic double-strand break repair system. Genome Res 2001;11(8):1365–74.

[26] Bartlett EJ, Brissett NC, Doherty AJ. Ribonucleolytic resection is required for repair of strand displaced nonhomologous end-joining intermediates. Proc Natl Acad Sci USA 2013;110(22):E1984–91.

[27] Gu J, Lieber MR. Mechanistic flexibility as a conserved theme across 3 billion years of nonhomologous DNA end-joining. Genes Dev 2008;22(4):411–5.

[28] Aniukwu J, Glickman MS, Shuman S. The pathways and outcomes of mycobacterial NHEJ depend on the structure of the broken DNA ends. Genes Dev 2008;22(4):512–27.

[29] Tamura K, Adachi Y, Chiba K, Oguchi K, Takahashi H. Identification of Ku70 and Ku80 homologues in *Arabidopsis thaliana*: evidence for a role in the repair of DNA double-strand breaks. Plant J 2002;29(6):771–81.

[30] Lieber MR. The mechanism of double-strand DNA break repair by the nonhomologous DNA end-joining pathway. Annu Rev Biochem 2010;79:181–211.

[31] Sonoda E, Hochegger H, Saberi A, Taniguchi Y, Takeda S. Differential usage of non-homologous end-joining and homologous recombination in double strand break repair. DNA Repair (Amst) 2006;5(9–10):1021–9.

[32] Ramadan K, Maga G, Shevelev IV, Villani G, Blanco L, Hubscher U. Human DNA polymerase lambda possesses terminal deoxyribonucleotidyl transferase activity and can elongate RNA primers: implications for novel functions. J Mol Biol 2003;328(1):63–72.

[33] Daley JM, Palmbos PL, Wu D, Wilson TE. Nonhomologous end joining in yeast. Annu Rev Genet 2005;39:431–51.

[34] Kowalczykowski SC, Dixon DA, Eggleston AK, Lauder SD, Rehrauer WM. Biochemistry of homologous recombination in *Escherichia coli*. Microbiol Rev 1994;58(3):401–65.

[35] Hiom KDN. A repair: common approaches to fixing double-strand breaks. Curr Biol 2009;19(13):R523–5.

[36] D'Amours D, Jackson SP. The Mre11 complex: at the crossroads of DNA repair and checkpoint signalling. Nat Rev Mol Cell Biol 2002;3(5):317–27.

[37] Hopkins BB, Paull TT. The *P. furiosus* mre11/rad50 complex promotes 5′ strand resection at a DNA double-strand break. Cell 2008;135(2):250–60.

[38] Lin Y, Lin LJ, Sriratana P, Coleman K, Ha T, Spies M, et al. Engineering of functional replication protein a homologs based on insights into the evolution of oligonucleotide/oligosaccharide-binding folds. J Bacteriol 2008;190(17):5766–80.

[39] White MF. Homologous recombination in the archaea: the means justify the ends. Biochem Soc Trans 2011;39(1):15–9.

[40] Constantinou A, Davies AA, West SC. Branch migration and Holliday junction resolution catalyzed by activities from mammalian cells. Cell 2001;104(2):259–68.

[41] Matos J, West SC. Holliday junction resolution: regulation in space and time. DNA Repair (Amst) 2014;19:176–81.

[42] Wu L, Hickson ID. The Bloom's syndrome helicase suppresses crossing over during homologous recombination. Nature 2003;426(6968):870–4.

[43] Ray JH, German J. Bloom's syndrome and EM9 cells in BrdU-containing medium exhibit similarly elevated frequencies of sister chromatid exchange but dissimilar amounts of cellular proliferation and chromosome disruption. Chromosoma 1984;90(5):383–8.

[44] Wechsler T, Newman S, West SC. Aberrant chromosome morphology in human cells defective for Holliday junction resolution. Nature 2011;471(7340):642–6.

[45] Li GM. Mechanisms and functions of DNA mismatch repair. Cell Res 2008;18(1):85–98.

[46] Guarne A, Ramon-Maiques S, Wolff EM, Ghirlando R, Hu X, Miller JH, et al. Structure of the MutL C-terminal domain: a model of intact MutL and its roles in mismatch repair. EMBO J 2004;23(21):4134–45.

[47] Junop MS, Obmolova G, Rausch K, Hsieh P, Yang W. Composite active site of an ABC ATPase: MutS uses ATP to verify mismatch recognition and authorize DNA repair. Mol Cell 2001;7(1):1–12.

[48] Wang H, Hays JB. Signaling from DNA mispairs to mismatch-repair excision sites despite intervening blockades. EMBO J 2004;23(10):2126–33.

[49] Allen DJ, Makhov A, Grilley M, Taylor J, Thresher R, Modrich P, et al. MutS mediates heteroduplex loop formation by a translocation mechanism. EMBO J 1997;16(14):4467–76.

[50] Fishel R. Mismatch repair, molecular switches, and signal transduction. Genes Dev 1998;12(14):2096–101.

[51] Jiang J, Bai L, Surtees JA, Gemici Z, Wang MD, Alani E. Detection of high-affinity and sliding clamp modes for MSH2-MSH6 by single-molecule unzipping force analysis. Mol Cell 2005;20(5):771–81.

[52] Li GM. The role of mismatch repair in DNA damage-induced apoptosis. Oncol Res 1999;11(9):393–400.

[53] Stojic L, Mojas N, Cejka P, Di Pietro M, Ferrari S, Marra G, et al. Mismatch repair-dependent G2 checkpoint induced by low doses of SN1 type methylating agents requires the ATR kinase. Genes Dev 2004;18(11):1331–44.

[54] York SJ, Modrich P. Mismatch repair-dependent iterative excision at irreparable O^6-methylguanine lesions in human nuclear extracts. J Biol Chem 2006;281(32):22674–83.

[55] Yoshioka K, Yoshioka Y, Hsieh P. ATR kinase activation mediated by MutSalpha and MutLalpha in response to cytotoxic O^6-methylguanine adducts. Mol Cell 2006;22(4):501–10.

[56] Gorbunova VV, Levy AA. How plants make ends meet: DNA double-strand break repair. Trends Plant Sci 1999;4(7):263–9.

[57] Yannone SM, Roy S, Chan DW, Murphy MB, Huang S, Campisi J, et al. Werner syndrome protein is regulated and phosphorylated by DNA-dependent protein kinase. J Biol Chem 2001;276(41):38242–8.

[58] Li B, Comai L. Displacement of DNA-PKcs from DNA ends by the Werner syndrome protein. Nucleic Acids Res 2002;30(17):3653–61.

[59] Hromas R, Wray J, Lee SH, Martinez L, Farrington J, Corwin LK, et al. The human set and transposase domain protein Metnase interacts with DNA Ligase IV and enhances the efficiency and accuracy of non-homologous end-joining. DNA Repair (Amst) 2008;7(12):1927–37.

[60] Thorslund T, Ripplinger A, Hoffmann S, Wild T, Uckelmann M, Villumsen B, et al. Histone H1 couples initiation and amplification of ubiquitin signalling after DNA damage. Nature 2015;527(7578):389–93.

[61] Bi X. Mechanism of DNA damage tolerance. World J Biol Chem 2015;6(3):48–56.

[62] Moldovan GL, Pfander B, Jentsch S. PCNA, the maestro of the replication fork. Cell 2007;129(4):665–79.

[63] de Massy B. Initiation of meiotic recombination: how and where? Conservation and specificities among eukaryotes. Annu Rev Genet 2013;47:563–99.

[64] Nicolas A, Treco D, Schultes NP, Szostak JW. An initiation site for meiotic gene conversion in the yeast *Saccharomyces cerevisiae*. Nature 1989;338(6210):35–9.

[65] Baudat F, Imai Y, de Massy B. Meiotic recombination in mammals: localization and regulation. Nat Rev Genet 2013;14(11):794–806.

[66] Lichten M. Molecular biology. Putting the breaks on meiosis. Science 2015;350(6263):913.

[67] Lam I, Keeney S. Nonparadoxical evolutionary stability of the recombination initiation landscape in yeast. Science 2015;350(6263):932–7.

[68] Singhal S, Leffler EM, Sannareddy K, Turner I, Venn O, Hooper DM, et al. Stable recombination hotspots in birds. Science 2015;350(6263):928–32.

[69] Cancer Genome Atlas Network. Comprehensive molecular portraits of human breast tumours. Nature 2012;490(7418):61–70.

[70] Cancer Genome Atlas Research Network. Comprehensive molecular characterization of clear cell renal cell carcinoma. Nature 2013;499(7456):43–9.

[71] Cancer Genome Atlas Research Network. Integrated genomic analyses of ovarian carcinoma. Nature 2011;474(7353):609–15.

[72] Grasso CS, Wu YM, Robinson DR, Cao X, Dhanasekaran SM, Khan AP, et al. The mutational landscape of lethal castration-resistant prostate cancer. Nature 2012;487(7406):239–43.

[73] Quesada V, Conde L, Villamor N, Ordonez GR, Jares P, Bassaganyas L, et al. Exome sequencing identifies recurrent mutations of the splicing factor SF3B1 gene in chronic lymphocytic leukemia. Nat Genet 2012;44(1):47–52.

[74] Puente XS, Pinyol M, Quesada V, Conde L, Ordonez GR, Villamor N, et al. Whole-genome sequencing identifies recurrent mutations in chronic lymphocytic leukaemia. Nature 2011;475(7354):101–5.

[75] Cancer Genome Atlas Research Network, Kandoth C, Schultz N, Cherniack AD, Akbani R, Liu Y, et al. Integrated genomic characterization of endometrial carcinoma. Nature 2013;497(7447):67–73.

[76] Bunting SF, Callen E, Kozak ML, Kim JM, Wong N, Lopez-Contreras AJ, et al. BRCA1 functions independently of homologous recombination in DNA interstrand crosslink repair. Mol Cell 2012;46(2):125–35.

[77] Schlacher K, Wu H, Jasin M. A distinct replication fork protection pathway connects Fanconi anemia tumor suppressors to RAD51-BRCA1/2. Cancer Cell 2012;22(1):106–16.

[78] Schlacher K, Christ N, Siaud N, Egashira A, Wu H, Jasin M. Double-strand break repair-independent role for BRCA2 in blocking stalled replication fork degradation by MRE11. Cell 2011;145(4):529–42.

[79] Hashimoto Y, Ray Chaudhuri A, Lopes M, Costanzo V. Rad51 protects nascent DNA from Mre11-dependent degradation and promotes continuous DNA synthesis. Nat Struct Mol Biol 2010;17(11):1305–11.

[80] Lord CJ, Ashworth A. Targeted therapy for cancer using PARP inhibitors. Curr Opin Pharmacol 2008;8(4):363–9.

[81] Dietlein F, Thelen L, Jokic M, Jachimowicz RD, Ivan L, Knittel G, et al. A functional cancer genomics screen identifies a druggable synthetic lethal interaction between MSH3 and PRKDC. Cancer Discov 2014;4(5):592–605.

[82] Tebas P, Stein D, Tang WW, Frank I, Wang SQ, Lee G, et al. Gene editing of CCR5 in autologous CD4 T cells of persons infected with HIV. N Engl J Med 2014;370(10):901–10.

[83] Gantz VM, Bier E. Genome editing. The mutagenic chain reaction: a method for converting heterozygous to homozygous mutations. Science 2015;348(6233):442–4.

[84] Smidler AL, Terenzi O, Soichot J, Levashina EA, Marois E. Targeted mutagenesis in the malaria mosquito using TALE nucleases. PLoS One 2013;8(8):e74511.

[85] Manna S. An overview of pentatricopeptide repeat proteins and their applications. Biochimie 2015;113:93–9.

[86] Yagi Y, Shirakawa M, Nakamura T. The challenges faced by EditForce Inc., to go beyond genome editing. Nature 2015:12. Sponsor feature.

Index

'*Note*: Page numbers followed by "f" indicate figures and "t" indicate tables.'

B

D

H

I

J

K

L

M

N

S

www.ingramcontent.com/pod-product-compliance
Ingram Content Group UK Ltd.
Pitfield, Milton Keynes, MK11 3LW, UK
UKHW051040091026
15624UKWH00011B/579

9 780128 033098